Accident Prevention Manual

for

Business & Industry

Engineering & Technology

12th Edition

NATIONAL SAFETY COUNCIL MISSION STATEMENT

The mission of the National Safety Council is to educate and influence society to adopt safety, health, and environmental policies, practices, and procedures that prevent and mitigate human suffering and economic losses arising from preventable causes.

OCCUPATIONAL SAFETY AND HEALTH SERIES

The National Safety Council Press' occupational safety and health series is composed of six volumes and two study guides developed to help readers establish, maintain, and improve safety, health, and environmental programs. These books contain the most up-to-date and reliable information on establishing priorities, collecting and analyzing data to help identify problems, and developing methods and procedures to reduce or eliminate illness and incidents, thus mitigating injury and minimizing economic loss resulting from these events.

Accident Prevention Manual for Business & Industry—4 volume set
- *Administration & Programs*
- *Engineering & Technology*
- *Environmental Management*
- *Security Management*

Study Guide: Accident Prevention Manual for Business & Industry: Administration & Programs and Engineering & Technology

Occupational Health & Safety

Fundamentals of Industrial Hygiene

Study Guide: Fundamentals of Industrial Hygiene

Some recent NSC Press additions include:

- *Injury Facts*™ (formerly *Accident Facts*®) published annually
- *Safety Culture and Effective Safety Management*
- *Safety Through Design*
- *On-Site Emergency Response Planning Guide*
- *Safety and Health Classics*
- *Lockout/Tagout: The Process of Controlling Hazardous Energy*
- *Supervisors' Safety Manual*
- *Out in Front: Effective Supervision in the Workplace*
- *Product Safety Management Guidelines*
- *OSHA Bloodborne Pathogens Exposure Control Plan* (National Safety Council/CRC/Lewis Publication)
- *Complete Confined Spaces Handbook* (National Safety Council/CRC-LEWIS Publication)

Accident Prevention Manual

FOR

Business & Industry

Engineering & Technology

12th Edition

Editors:
Philip E. Hagan, MPH, CIH, CHMM, CHCM
John F. Montgomery, Ph.D., CSP, CHCM, CHMM
James T. O'Reilly, JD

Itasca, Illinois

Project Editor: Patricia M. Dewey
Interior Design and Composition: Black Dot Group
Cover Design: Pamela J. Byers

COPYRIGHT, WAIVER OF FIRST SALE DOCTRINE

DISCLAIMER

Although the information and recommendations contained in this publication have been compiled from sources believed to be reliable, the National Safety Council makes no guarantee as to, and assumes no responsibility for, the correctness, sufficiency, or completeness of such information or recommendations. Other or additional safety measures may be required under particular circumstances.

Printed in the United States of America
07 06 05 04 11 10 9 8 7 6

Library of Congress Cataloging-in-Publication Data
Accident prevention manual for business & industry: engineering & technology/ editors, Philip E. Hagan, John F. Montgomery, James T. O'Reilly. —12th ed.
p. cm. — (Occupational safety and health series)
Includes bibliographical references and index.
ISBN 0-87912-213-7 (alk. paper)
1. Industrial safety—United States—Handbooks, manuals, etc. 2. Accidents—United States—Prevention—Handbooks, manuals, etc. I. Hagan, Philip E. II. Montgomery, John F. (John Franklin), 1944– III. O'Reilly, James T., 1947– IV. National Safety Council. V. Series.
T55 .A333 2001
363.11–dc21 00-011927

1.5M0604 Product Number: 121540000

Contents

Preface

The 12th edition of the *Accident Prevention Manual for Business & Industry: Engineering & Technology* continues a tradition begun in 1946 with the publication of the first *Accident Prevention Manual.* This Manual brings to the safety/health/environmental professional the broad spectrum of topics, specific hazards, best practices, control procedures, resources, and sources of help known in the field today. To accommodate the expansion of knowledge and topics, the Manuals are now printed in four volumes: *Administration & Programs* (12th edition), *Engineering & Technology* (12th edition), *Environmental Management* (2nd edition), and *Security Management* (1st edition).

This 12th edition builds on the excellent work of previous contributors to the National Safety Council's flagship series. Volunteer experts from many different subject areas have come together to make this book an important resource to be used in support of safety programs and related education. In addition to the expertise of National Safety Council volunteers and staff, we have received expert assistance in developing, writing, and reviewing from contributors representing various disciplines and from the editors, Philip E. Hagan, John F. Montgomery, and James O'Reilly (see Contributors section later in this Preface). If you have different ideas and want them to be considered for the 13th edition, your suggestions are welcome and can be sent to the National Safety Council, 1121 Spring Lake Drive, Itasca IL, attn. Ms. Pat Dewey.

The audience served by this text is widespread. Safety professionals with years of experience, individuals new to the field, managers tasked with safety responsibilities, and educators preparing students for careers in the field of safety will find that these volumes are a valuable source of information. Those who work in the fields of risk management and loss control, human resources, and engineering will also find programs and information that can be incorporated successfully into workplan goals and objectives that will bring added value to any organization's safety program.

NEW AND REVISED MATERIAL

The *Accident Prevention Manuals* are intended for a wide range of users: for students using them as textbooks, corporate or company managers searching for solutions to safety and health problems, new safety specialists who must plan and organize a safety and health program within a company, or for experienced safety professionals seeking to improve an operating program and to learn more about advances in the field of safety and health. To increase their usefulness, the 12th editions of the *Administration & Programs* volume and the *Engineering & Technology* volume contain new chapters and features as well as completely reviewed and revised material in all chapters.

All chapters in both volumes were reviewed, revised, and updated by safety professionals with expertise in the specific subject area. Major changes include:

Administration & Programs volume:

- Chapter 2: The Safety Professional—completely new
- Chapter 3: Safety Culture—completely new
- Chapter 4: New regulations summarized and changes in existing regulations cited
- Chapter 5: Completely new section on Protecting against Liability
- Chapter 8: Completely new section on System Safety
- Chapter 12: Expanded and revised section on shiftwork
- Chapter 15: Indoor Air Quality—completely new
- Chapter 16: Completely new sections on establishing and managing an ergonomics program and on disabilities accommodations
- Chapter 19: Workplace Violence—completely new

- Chapter 24: New sections on chemical safety, biological safety, ionizing and nonionizing radiation safety, and fiber optics safety in laboratory environments

Engineering & Technology volume:

- Chapter 1: Extensively rewritten section on safety through design
- Chapter 3: Extensively revised sections on construction safety, ladder safety, fall arrest systems, and fall protection
- Chapter 7: Extensive revision of sections on fall protection, respiratory protection, hearing protection, and eye protection
- Chapter 19: New section on power nailers and staplers

DEFINITIONS OF TERMS

As the concerns and responsibilities of safety/health/ environmental professionals expand, so must their ability to communicate and educate. Technical terms are defined in the text where they are used and also in Appendix 3, Glossary, in the *Engineering & Technology* volume. However, the terms *incident* and *accident* deserve a special note. In the years since the original publication of this manual, many theories of accident causation and definitions of the term *accident* have been advanced. The National Safety Council continues to work to increase awareness that an incident is a near-accident and that so-called accidents are not random events but rather preventable events. To that end, the term *incident* is used in its broadest sense to include incidents that may lead to property damage, work injuries, or both. The following definitions are generally used in this manual:

- **Accident:** That occurrence in a sequence of events that produces unintended injury, death, or property damage. *Accident* refers to the event, not the result of the event (see *unintentional injury*).
- **Incident:** An unintentional event that may cause personal harm or other damage. In the United States, OSHA specifies that incidents of a certain severity be recorded.
- **Near-miss incident:** For purposes of internal reporting, some employers choose to classify as "incidents" the near-miss incident; an injury requiring first aid; the newly discovered unsafe condition; fires of any size; or nontrivial incidents of damage to equipment, building, property, or product.
- **Unintentional injury:** The preferred term for accidental injury in the public health community. It refers to the result of an accident.

With proper hazard identification and evaluation, management commitment and support, preventive and corrective procedures, monitoring, evaluation, and training, unwanted events can be prevented.

ACKNOWLEDGMENTS

General Editor Phil Hagan thanks Larry Spellman, David Hibbard, Chung-an Lin, Bill Wood, and Richard Pifer for their advice and support through the years that contributed to the knowledge and experience used in the editing process for these volumes. Also greatly appreciated was the much-needed support of Gloria Lacap and Connie Lum. And, of course, he thanks Pat Dewey for providing the glue that held the whole operation together (both literally and figuratively).

General Editor John Montgomery thanks the safety and environmental professionals who have contributed to the 12th edition of the *Accident Prevention Manuals* with special thanks to Charles Foster of the Live Oak Fire Department for his excellent review of Chapter 11, Fire Protection, in the *Engineering & Technology* volume, and to John Begin of the Northwest Airlines Safety and Environmental Department for his review and contribution to Chapter 18, Emergency Preparedness, in the *Administration & Programs* volume. He also thanks his wife Karen and son Christopher for their unending support.

General Editor Jim O'Reilly thanks his colleagues, especially Rick Hackman and Jack McAneny, for their excellent suggestions and sustained effort.

Thank you for taking the time to consider and utilize some of these ideas. Of course, none of the book's comments take the place of legal advice, medical advice, or professional advice. Please be certain to discuss the contents of this text with the appropriate professional advisors. We do not offer this as a substitute for the timely, prudent expertise of your organization's regular advisors.

CONTRIBUTORS

The following safety, health, and environmental professionals have contributed to the 12th edition as editors/writers and/or reviewers of chapters or sections. The National Safety Council very much appreciates the dedication and professional expertise they have contributed to the cause of safety, health, and environmental education.

Editors/Writers

James C. Chen, attorney with Hogan & Hartson, L.L.P., Washington, DC. His practice includes regulatory compliance in the areas of environmental law, workplace safety, and transportation. He was a former enforcement attorney with the U.S. Environmental Protection Agency. His publications cover a range of subjects from food quality to pesticide safety.

J. Nigel Ellis, Ph.D., P.E., CSP, CPE 306 Country Club Dr., Wilmington, DE 19803. E-mail: www.Fall-Safety.com or NigelEllis@FallSafety.com. Author: *Introduction to Fall Protection*, 3rd Edition (ASSE, 2001).

Richard J. Hackman, CIH, QEP, Procter & Gamble, 5299 Spring Grove Avenue, Cincinnati, OH 45217.

Philip E. Hagan, MPH, CIH, CHMM, CHCM, is the Director of Safety & Environmental Management, Georgetown University, Washington, DC, and a member of the FEMA National Health and Safety Cadre. He has co-authored texts on environmental and workplace safety, training, and indoor environmental quality and presented papers on subjects ranging from hazardous waste disposal to use of computer databases in the workplace. He is a member of the American Chemical Society's task force on Laboratory Waste Management, editor of the *American Biological Safety Association Newsletter,* and peer reviewer for American National Standards for Hand Protection Selection Criteria, Classification and Performance Requirements for Chemical Protective Clothing, and Emergency Eyewash/Shower Equipment (ANSI 105-199x, ANSI 103-199x, and ANSI Z358.1-1998, respectively). E-mail: haganp@gunet.georgetown.edu.

N. Kumar Kittusamy, Sc.D., MSPH, Assistant Professor, Marshall University, Safety Technology Program, IH/Ergonomics Laboratory, 400 Hal Greer Boulevard. Huntington, WV 25755. E-mail: Kittusam@MARSHALL.EDU.

Curt L. Lewis, PE, CSP, Corporate Manager–Flight Safety Department, AMR Corporation/American Airlines, has more than 25 years experience as a professional pilot, safety engineer/director, and air safety investigator. He is a board member of the Board of Certified Safety Professionals, president of the System Safety Society, and U.S. councillor and president for the U.S. International Society of Air Safety Investigators. E-mail: Curt_Lewis@AMRCorp.COM.

John J. McAneny, CIH, The Procter & Gamble Company, Sharon Woods Technical Center, 11511 Reed Hartmen Highway, Box 215, Cincinnati, OH 45241.

Fred A. Manuele, CSP, PE, President, Hazards, Limited, retired from Marsh & McLennan as a managing director and manager of M&M Protection Consultants. He has been a member of the Board of Directors of the American Society of Safety Engineers, a member and president of the Board of Certified Safety Professionals, a member of the Board of Directors and of the Board of Delegates of the National Safety Council, and is presently Trustee of The Foundation for Safety and Health and a member of the Advisory Committee to the Institute For Safety Through Design. He is author of *On the Practice of Safety,* 2nd edition, and coeditor of *Safety Through Design* (National Safety Council, 1999). Mr. Manuele has received the Distinguished Service to Safety Award (NSC) and the honor of Fellow (ASSE) and has been inducted into the Safety and Health Hall of Fame International.

John F. Montgomery, Ph.D., CSP, CHCM, CHMM; Corporate Manager–Noise and Emissions Regulatory Programs, Environmental Department, AMR Corporation/American Airlines. He has 14 years experience with American Airlines as Manager of Ground Safety, Manager of the Environmental Department, and Manager of Noise and Emissions Regulatory Programs, and was an Assistant Professor/Lecturer at Central Missouri State University, Central Oklahoma University, Lamar University, and Texas A&M University. He is an editorial advisor (environmental) for the *Safety and Health Magazine,* is the co-editor for the 11th edition *Accident Prevention Manual,* published by the National Safety Council, 1996. He was appointed to the IATA Noise and Emissions Committee (ENTAF) as one of three representatives from North America and was appointed by ENTAF to be the IATA representative to the ICAO/CAEP/WG4 Committee. He is an author/reviewer for the United Nation's Intergovernmental Panel on Climate Change (IPCC) review of the effect of aircraft emissions on the environment.

Professor James T. O'Reilly, JD, College of Law, University of Cincinnati, Cincinnati, OH 45215, has authored 26 texts and 125 articles. His expertise was acknowledged in a March, 2000, decision of the U.S. Supreme Court. He was formerly Associate General Counsel of The Procter & Gamble Company and has served as chair of the American Bar Association's Section of Administrative Law, 1996–1997. E-mail: joreilly@fuse.net.

Lynne Zarate, MEE, is an environmental engineer at Georgetown University, Washington, DC. She has expertise in various industries, including biomechanics research, paper manufacturing, and aerospace. E-mail: zaratel@gunet.georgetown.edu.

Reviewers

John Begin, Managing Director, Safety, Health and Environment, Northwest Airlines, 5101 Northwest Drive, Department C8980 St. Paul, MN 55111-3034.

Russ Bensman, Staff Engineer, The Minster Machine Company, 240 West 5th Street, PO Box 120, Minster, OH 45865-0120. E-mail: BensmanR@MINSTER.com.

Salvatore Caccavale, CHMM, Corporate Safety & Health Manager, American National Can Company, 8770 West Bryn Maur Avene, MS 11-M, Chicago, IL 60631-3542.

Michael R. Chambers, Corporate Safety Compliance Director, Issues Manager, Eastman Kodak Corp., Colorado Division, 1100 Ridgeway Avenue, Rochester, NY 14652-6257. E-mail: mrc@kodak.com.

Dennis R. Cloutier, Product Safety Coordinator, Cincinnati Inc., Box 11111, Cincinnati, OH 45211. E-mail: dennis.cloutier@cincinnati-tools.com.

Patrick J. Conroy, president, King & Neel Consulting, Inc., 1164 Bishop Street, Suite 1710, Honolulu, HI 96853. E-mail: Patc@kingneel.com.

Bill Douglass, Douglass Consulting, 5005 Ivycrest Trail, Arlington, TX 76017.

Joseph L. Durst, Jr., 4151 East Bonanza Road, Room 205, Las Vegas, NV 89110. E-mail: Jldurst@aol.com.

T. C. Goya, Manager, Employee Safety, American Airlines, PO Box 619616 MD 5425, DFW Airport, TX 76155. E-mail: John_Goya@AMRCorp.com.

Charles M. Foster, Assistant Fire Chief, Live Oak Fire Department, 7605 Marigold Trace, San Antonio, TX 78233.

Richard Hislop, PE, CSP, is site operations manager at a research facility in Chicago and has 25 years experience in developing successful safety programs for organizations ranging from small construction contractors to international construction projects. E-mail: RDH@APS.ANL.GOV.

Alan Hoskin, Manager, Statistics, National Safety Council, 1121 Spring Lake Drive, Itasca, IL 60143-3201. E-mail: hoskina@nsc.org.

Thomas F. Joyce, Maine Department of Labor, 45 State House Station, Augusta, Maine 04333-0045. E-mail: ThomasF.Joyce @state.me.us.

Margie L. Kolbe-Mims, CSHM, Adjunct Faculty, University of Wisconsin-Whitewater; A.Mims Associates, Incorporated Safety and Health Consultants, President, has 30 years experience in the plastics manufacturing industry and in computer technology training. She is a past editorial board member for the *World Safety Organization Journal and Newsletter* and current editorial board member for the *Australia National Safety Journal-Safety Institute of Australia.* Ms Kolbe-Mims is certified by the Institute of Safety and Health Management as a Safety Manager. Mimsm@juno.com

George R. Kinsley, CSP, ERM, 704 N. Deerpath Drive, Vernon Hills, IL 60061. E-mail: Gkinsley@ermne.com.

John Kurtz, International Staple, Nail and Tool Association (ISANTA), 512 West Burlington Avenue, Suite 203, La Grange, IL 60525-2245. E-mail: isanta@ameritech.net.

John C. Langford, One Sunset Drive, Fenton, MO 63026. Johnclangford@compuserve.com.

John Leyenberger, Divisional Risk Control Director, Wal-Mart Stores, Inc., 702 Southwest 8th Street, Bentonville, AR 72716. E-mail: jfleyen@wal-mart.com.

Bradley A. McPherson, industrial hygiene, safety, and training consultant, Allegheny Energy Supply, PO Box 729, Kittanning, PA 16201.

Bob Marecek, Manager, National Safety Council Library, 1121 Spring Lake Drive, Itasca, IL 60143-3201. E-mail: marecekr @nsc.org.

Ken Mushet, MBA/TM, REM, Penn Racquet Sports, 306 South 45th Avenue, Phoenix, AZ 85043. E-mail: KMushet@ PennRacquet.com.

Terrie S. Norris, CSP, ARM, CPSI, Loss Prevention Manager, Poms & Associates, 4653 East 4th Street, Long Beach, CA 90814. Ms. Norris has more than 15 years experience in the management of safety, health, and environmental programs in the paper, rubber, and plastics, and metal fabrication industries. She holds a CSP in Comprehensive Practices and one in Management Aspects, an Associate Risk Management designation from the Insurance Institute, and is a Certified Playground Safety Inspector. E-mail: Tseagle@ptconnect.infi.net.

Bernard J. Quinn, CIH, President and CEO of AM Health and Safety, Inc., has more than 20 years experience in industrial hygiene, safety, product safety, environmental control, and radiation protection. Bjquinn@ldastorino.com

Michael S. Shanshala is currently affiliated with United Refining Company, Warren, PA, as director of safety. He has 37 years of combined experience in petroleum refining and natural gas distribution. United Refining Company, PO Box 780, Warren, PA 16365. Firesafe@penn.com

George Swartz, CSP, retired, was the director of safety and occupational health, Midas International Corp., for 23 years. He has received the Distinguished Service to Safety Award from the National Safety Council, is a Fellow of the American Society of Safety Engineers and a recipient of their Charles V. Culbertson Outstanding Volunteer Award.

George Welchel, Power Tool Institute, Inc., 1300 Sumner Avenue, Cleveland, OH 44115-2851. E-mail: powertoolinstitute. com/welcome.htm.

Jim Wilson, CSP, CHCM, CIH, REM, CPSM, president, The Provident Group, 207 West Hickory, Denton, TX 76202.

Part One

FACILITIES

To achieve continuous improvement in safety and health, companies must examine the interaction between people and the physical structures in which they work. Part One examines the design, layout, construction, and maintenance of the facilities where work is done and describes how these factors can help create safer working conditions and greater health and safety for workers. The optimum situation is to consider and design in safety features as the facilities are being planned. However, even when the design was done decades ago, the facility can be laid out, retrofitted, or rehabilitated to offer a safer working environment today.

1

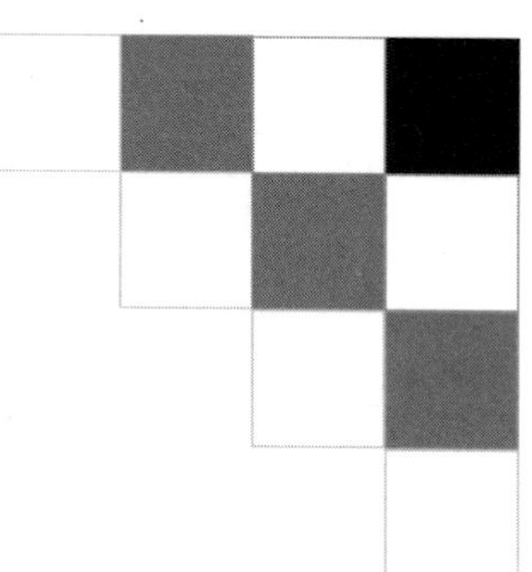

Safety Through Design

Revised by

Fred A. Manuele, CSP, PE

In the past few years, the safety through design concept has received increased attention as awareness developed that the greatest opportunity to effectively and economically eliminate or control hazards arises early in the design process. As a result, in some companies safety practitioners are called upon as advisors to design personnel.

This chapter discusses the rationale and guidelines for incorporating safety into the design process. Topics covered include:

- defining safety through design, and its model
- benefits of safety through design
- benchmarking, achieving a culture change
- relating to quality management
- integrating safety into the design process
- the role of the safety practitioner
- work procedures versus worker behavior as causal factors
- assessing risks that should be designed out of the workplace
- implementing design specifications in company operations

WHAT IS SAFETY THROUGH DESIGN?

Safety through design is not a program, with its attendant whistles, bells, slogans, and banners running to a sputtering end, to be forgotten as many programs are. What is needed is an agreed upon and well-understood concept, a way of thinking, that is translated into a process that effectively addresses hazards and risks in the design process.

Defining Safety Through Design

Safety through design is defined as the integration of hazard analysis and risk assessment methods early in the design and engineering stages and the taking of the actions necessary so that the risks of injury or damage are at an acceptable level. This concept encompasses facilities, hardware, equipment, tooling, materials, layout and configuration, energy controls, environmental concerns, and products.

The Safety Through Design Model

In Figure 1–1, emphasis is given to moving safety from an afterthought to a forethought in the design of facilities, processes, and products. Thus, considerations of hazards and risks would be moved as far "upstream" as possible in the design process. "Upstream" includes all aspects of the origination of business concepts, the relative decision making, and the design process, during which the greatest effectiveness can be achieved in hazard avoidance, elimination, or control.

As Figure 1–1 indicates, integrating consideration of hazards and risks early in the concept and design stages results in easier and less costly safety implementation, and avoids costly retrofitting in the build, operation, maintenance, and decommissioning periods.

Benefits of Safety Through Design

Data are now prevalent indicating that the following benefits will derive if decisions affecting safety, health, and the environment are integrated into the early stages of the design processes:

- significant reductions will be achieved in injuries, illnesses, damage to the environment, and their attendant costs;
- productivity will be improved;
- operating costs will be reduced; and
- expensive retrofitting to correct design shortcomings will be avoided.

A developing awareness that application of safety through design concepts impacts favorably on productivity, unit costs, and avoiding the expensive retrofitting costs has given impetus to the growth of the safety through design movement. Astute safety practitioners recognize the opportunity provided by the safety through design concepts to further incorporate safety into overall management systems. Applying the concepts brings safety practitioners into the mainstream of an entity's business.

Benchmarking for a Culture Change

To understand how far an organization has progressed in safety through design, the safety practitioner should consider proposing a benchmarking initiative. In that process, it may be determined that a culture change will be needed to effectively incorporate safety through design concepts into the design process.

Benchmarking is a standard of excellence against which other similar things are measured or judged. Benchmarking against superior performers allows learning from the experience of others without awaiting downstream experience to influence changes necessary in the design/build stages of the safety through design model.

In many organizations, a culture change will be necessary to establish effective safety through design concepts. Culture is a set of perceptions, values, beliefs, and assumptions that determines how individuals see reality and affects how they behave. (See Chapter 3, Safety Culture, in the *Administration & Programs*

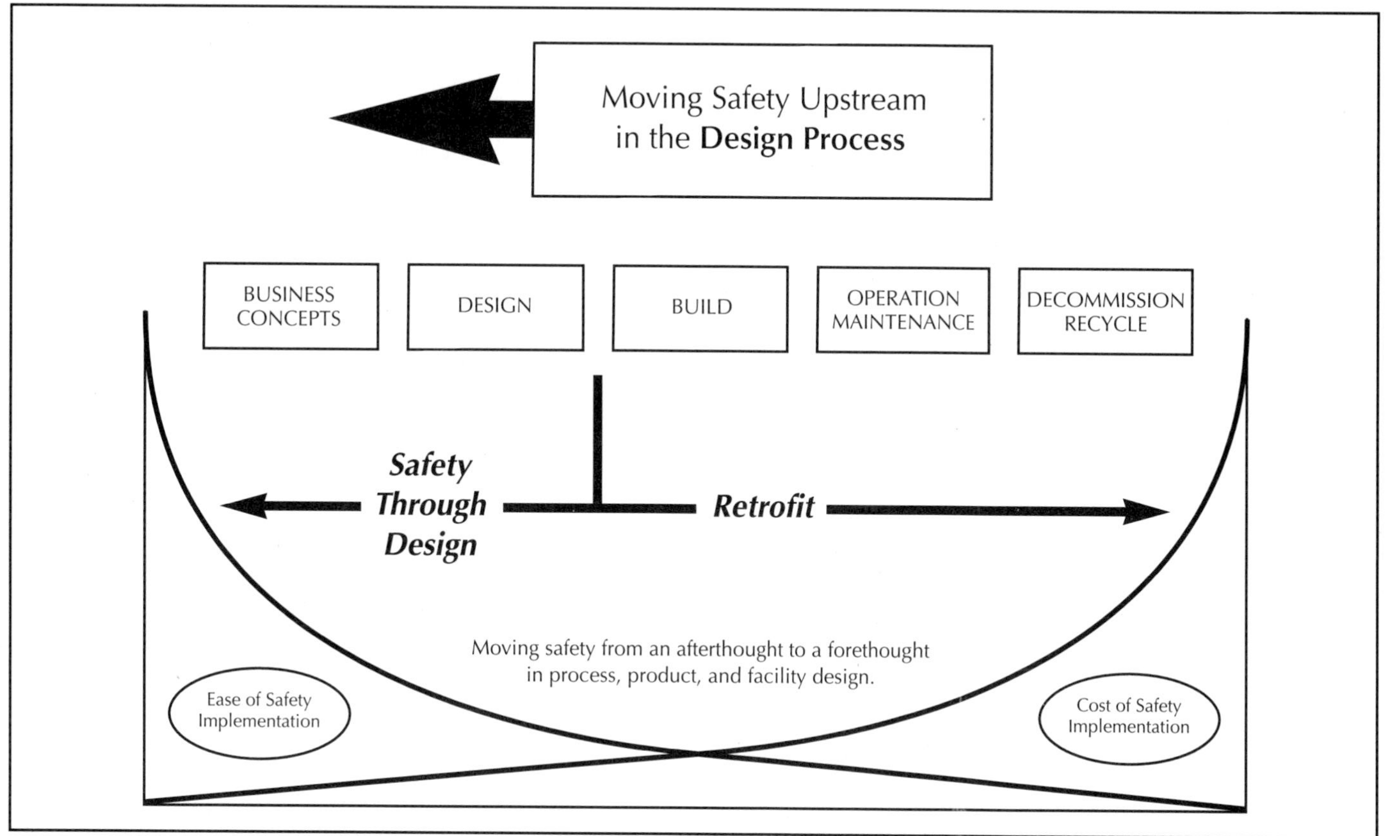

Figure 1–1. The model for safety through design. (*Source:* National Safety Council. *Safety Through Design*. Itasca, IL: 1999.)

volume and the Council's *Safety Culture and Effective Safety Management.*) Change is difficult and conformity is the norm. Since adopting safety through design concepts requires what some would consider a severe change, opposition is normal and the persuasive effort should be well-planned and supported by data that relate to the organization's goals.

Relating Safety Through Design to Quality Management

There is a remarkable correlation between quality management and safety through design principles. The same system design and continuous improvement processes that ensure that a product meets quality expectations will also ensure that safety expectations will be met. This provides a vital connection between safety through design and entity goals.

We borrow from W. Edwards Deming, who was world renowned in quality management, to support that premise. Deming stressed, again and again, that:

> Processes must be designed to achieve superior quality if that is the quality level desired, and that superior quality can not be attained otherwise.

And the same principle applies to safety. For an example of what Deming intended, we quote the fifth premise in what Deming called a "Condensation of the 14 Points of Management," as listed in his book *Out of the Crisis* (Deming 1986).

> Improve constantly and forever the system of production and service, to improve quality and productivity, and thus constantly decrease costs.

If you want superior quality, or superior safety, you must design it into new systems, and you must also maintain a continuous improvement program for the redesign of existing work places and work methods.

INTEGRATING SAFETY THROUGH DESIGN INTO THE DESIGN PROCESS

Safety through design is not and cannot be isolated from normal business planning. Safety through design is best integrated into an entity's culture by achieving recognition that the benefits to be obtained are supportive of management goals. Those benefits relate directly to business objectives of survival, competition, cost control and reduction, and return on investment.

In the design process, there is one exceptionally important and highly variable factor that must be considered, and that is the person who is to use the product, carry out the process, or staff the facility. This requires that the designer consider the mental, physical, and human factors of a very large part of the user population to achieve a design that is user friendly.

General Principles and Definitions

For the purposes of this chapter, the term design processes applies to:

- facilities, hardware, equipment, tooling, material selection, operations layout and configuration, energy control, and environmental concerns;
- work methods and procedures, personnel selection standards, training content, work scheduling, management of change procedures, maintenance requirements, and personal protective equipment needs; and
- industrial, commercial, and consumer products for human use.

In applying the safety through design concept, these definitions and principles would govern.

- Risk is defined as a measure of the probability of a hazards-related incident occurring and the severity of harm or damage that could result.
- Safety is defined as that state in which the risks are acceptable.
- In the design and redesign processes, the two distinct aspects of risk must be considered:
  - avoiding, eliminating, or reducing the *probability* of a hazards-related incident occurring; and
  - minimizing the *severity* of harm or damage, if an incident occurs.
- Minimum risk is to be sought with respect to new technology, facilities, materials and designs, in designing new production methods, and in the design of products for human use, and in all redesign endeavors.
- Minimum risk is achieved when all risks deriving from hazards are at a realistic minimum, and acceptable. Minimum risk does not mean zero risk, which is unattainable.
- All risks to which the concept of safety through design applies derive from hazards. There are no exceptions.
- Thus, hazards must be the focus of design efforts to achieve safety—a state for which the risks are judged to be acceptable.
- Hazards are most effectively and economically avoided, eliminated, or controlled if they are considered early in the design process, and where necessary as the design progresses.
- Both the technology and human activity aspects of hazards must be considered in the design decision making.
- A hazard is defined as the potential for harm: hazards include the characteristics of things and the actions or inactions of persons.
- If a hazard is not avoided, eliminated or controlled, its potential may be realized, and a hazards-related incident may occur that has the potential to, but may or may not, result in harm or damage, depending on exposures.
- A risk assessment is an analysis that addresses both the probability of a hazards-related incident occurring and the expected severity of its adverse effects.
- Hazards analyses and risk assessments must be integral parts of the design process.
- As a matter of principle, for an operation to proceed, its risks must be acceptable.

If a system—the facilities, equipment, and work methods—is not designed to minimum risk, companies cannot achieve superior results with respect to safety, even if management and personnel factors approach the ideal.

Order of Design Precedence

To achieve the greatest effectiveness in hazard avoidance, elimination, or control, companies should apply the following priorities to all design and redesign processes.

- **First priority: Design for minimum risk.** From the very beginning, the top priority should be to eliminate hazards in the design process. If an identified hazard cannot be eliminated, the associated risk is to be reduced to an acceptable level through design decisions.
- **Second priority: Incorporate safety devices.** If hazards cannot be eliminated or their risks adequately reduced through design selection, the next step is to reduce risks to an acceptable level. Companies can accomplish this step through the use of fixed, automatic, or other protective safety design features or devices. Management should establish routine functional checks of the safety devices to ensure they maintain their level of protection.

- **Third priority: Provide warning devices.** In some cases, identified hazards cannot be eliminated or their risks reduced to an acceptable level through initial design decisions or through the incorporated safety devices. Under these conditions, companies should develop systems to detect hazardous conditions and warn personnel of the hazards. Warning signals should be designed to help workers react promptly and correctly to a hazardous situation and should be standardized within all systems.
- **Fourth priority: Develop and implement operating procedures and employee training programs.** Where it is impractical to eliminate hazards or reduce their risks to an acceptable level through design selection, incorporating safety devices, or warning devices, companies should develop and implement safe operating procedures and use safety training programs.
- **Fifth priority: Use personal protective equipment.** When all other techniques cannot eliminate or control a hazard, employees should be given personal protective equipment to prevent injuries and illnesses.

However, operating procedures and training, warning signs or other written forms of advisory, and personal protective equipment should not be used as the only risk reduction method for catastrophic or critical hazards. Tasks and activities judged to be essential to safe operation may require special training and certification of personnel.

For many design situations, a combination of these five priorities will apply. However, companies should not choose a lower level of priority until they have exhausted the practical applications of higher priority levels. First and second priorities are more effective in safeguarding workers and creating safe systems because they reduce the risk by design measures that eliminate or adequately control the potential of an incident occurring and the severity of its consequences. Third and fourth priorities rely on human intervention, which means the level of safety achieved tends to depend on the knowledge and skill of the personnel involved.

Companies should develop a system to rank the risks of their operations to help them establish designing-for-safety priorities. Some risk-assessment methods include a "cost of risk reduction" factor in their formulas. Unfortunately, in those systems, companies may assign a low priority to certain low-probability, high-consequence hazards if the costs to reduce the risk are substantial. The safety professional should work to convince management not to make this sort of risk categorization. Instead, risk reduction costs should be considered separately from the hazard analysis/ risk assessment exercise.

ROLE OF THE SAFETY PRACTITIONER

The safety practitioner is often the driving force in a company for including safety decisions during design stages. Although there are numerous standards, regulations, specifications, design handbooks, and checklists that establish the minimums for specific design subjects, no specific standard clearly describes the principles to be applied in designing for safety and the goals to be achieved. The safety practitioner must work to make safety in design part of the company's philosophy and standard operating procedure.

The safety practitioner can influence the design of the workplace and work methods at three critical points:

- **In the preoperational design stage.** Before a building, system, or piece of equipment becomes operational, the safety professional has the greatest opportunity to identify and analyze hazards and to help engineers and architects design ways to avoid, control, or eliminate them. This stage can avoid costly redesigning, retrofitting, or replacing elements of the workplace.
- **In the operational stage.** After a building, system, or piece of equipment becomes operational, the safety professional can seek to make them safer through the process of continuous improvement. He or she accomplishes this task by anticipating, identifying, and evaluating current hazards and helping to control or eliminate them *before their potentials to cause injury or death are realized.*
- **In the postincident stage.** After an incident has occurred, the safety professional can still work to improve safety. By investigating the hazards related to the incident, he or she can determine the causal factors involved and can review the possible impact of design decisions on the incident. These data can then be used to improve future designs and eliminate the factors that led to the current incident.

Many companies have applied the principles of safety through design with considerable success, improving not only their safety record but also increasing their productivity and cost efficiency. Such successes eventually led the National Safety Council to create an Institute for Safety through Design. Its mission statement follows.

> To reduce the risk of injury, illness, and environmental damage by integrating decisions affecting safety, health, and the environment in all stages of the design process.

For hazards that are inherent in the design of the workplace or work methods that cannot be eliminated or controlled through redesign, companies must use the appropriate management practices to promote safety. These practices should keep the risks of employee injury or illness or environmental damage at an acceptable minimum.

Proactive versus Reactive Responses

As in other areas of safety, companies have a choice of either initiating safety through design or simply reacting to incidents and attempting to redesign equipment and procedures to prevent recurrences. Research and experience show that a company achieves the greatest effectiveness and economy when dealing with hazards in the preoperational stage of the design processes. By taking such a proactive approach, the company obtains designs that can reduce employee risk, improve productivity, and lower unit costs.

- A proactive company asks safety-related questions while a building, work system, or equipment is being designed. How well does the building manage traffic flow? Does a work system encourage employees to take dangerous shortcuts or force them to engage in risky behavior (e.g., crawl underneath a conveyor)? How can a piece of equipment be misused or malfunction, encouraging unsafe behavior? Such questions can reveal inherent flaws or hazards that can be corrected by instituting safer designs.
- This approach also shifts the emphasis away from employee behavior to the design of work practices and methods. Too often, companies and safety professionals concentrate their safety efforts on worker behavior instead of work methods and procedures.

Behavior Modification versus Workplace Redesign

Because many companies still adopt a reactive mode in designing for safety, management and safety professionals tend to focus on behavior modification or training as solutions when the problem is workplace or work methods design. Although behavior modification and training are important elements of a safety and health initiative, such measures are misdirected when applied to solve workplace or work methods design problems.

If the design of the work is overly stressful or if the work situation encourages employees to take risks, then the causal factors are principally systemic. To label the causal factors as "employee error" or "unsafe act" would be inappropriate and ineffective, as the following actual case histories illustrate.

- Bags weighing 100 lb (45 kg) were delivered to work stations on pallets. Workers slit open the bags and lifted them to shoulder height to pour the contents into hoppers. The job required a fast work pace, with workers stooping and twisting to lift the bags. Back injuries were frequent. Investigative reports always listed the causal factors as improper lifting. The corrective action was always "reinstructed the worker in proper lifting techniques."

  Obviously, this was a work methods design problem; most of the population would be overstressed in this work situation. As a result, no amount of training, reinstruction, or behavior modification would correct what was an inherently unsafe act. Yet, in a similar situation, many safety personnel would simply look to past investigation and analysis practices and recommend another employee training program on "how to lift safely."
- Because of a glitch in production scheduling, delivery of parts by a conveyor to a work station ceased. The design of the conveyor allowed parts to fall off and accumulate beneath the belt. An employee, wanting to keep up with production needs, went beneath the conveyor to retrieve the parts that had collected there. Her hair was caught in a drive belt. When the conveyor started up again, part of her scalp was torn away.

  At first, the causal factor for this incident was recorded as the unsafe act of the employee. Line workers were cautioned not to enter the space beneath the conveyor. Later, however, investigators examined the contributing factors of the production scheduling glitch and the design of the conveyor. If parts had not fallen off the conveyor, the worker would not have been tempted to retrieve them. As a result of this investigation, the design of the conveyor was modified.
- A worker failed to follow the established procedure to lock out and tag out the electrical power during a maintenance operation and was electrocuted. The incident investigation report recorded the causal factor as "employee failed to ..." However, investigators also determined that the distance to the power shutoff was 216 ft (66 m). They showed that the design of the energy system, which made the power shutoff so inconvenient, "encouraged" the employee's risky behavior. Other employees confirmed the findings by expressing their own dissatisfaction with the energy system layout, which promoted "employee error."

Had these companies studied their work processes as they were being designed, the weight of the bags, operation of the conveyor belt, and energy system layout would have raised questions of safety. Management could have foreseen the hazards represented by these work practices or procedures and could have avoided them by designing safer methods. In a reactive mode, the companies looked at design issues only after an incident had occurred.

In actual fact, companies might find that for many incidents, the casual factor labeled employee error—an unsafe act—is actually "programmed" into the prescribed work method. This is particularly the case when the design of the work is overly stressful, provokes errors, or encourages riskier actions than desired. If the work is so designed, it is reasonable to assume that the "performance deviation" is principally a systemic problem rather than a personal action problem.

Alan D. Swain, in a paper titled "Work Situation Approach to Improving Job Safety," spoke of the workers being in work situations created by management and suggested that management "forego the temptation to place the burden of accident prevention on the individual worker." As he states in his paper:

> ... [A] means of increasing occupational safety is one which recognizes that most human initiated accidents are due to the features in a work situation which define what the worker must do and how he must do it. ... [T]he situation approach, emphasizes structuring or restructuring the work situation to prevent accidents from occurring. Use of this approach requires that management recognize its responsibility (1) to provide the worker with a *safety-prone* work situation and (2) to forego the temptation to place the burden of accident prevention on the individual worker.

One of the ways to create safer workplace designs is to recognize the requirements and limitations of workers. Designers and safety professionals can use the principles of ergonomics to help them accomplish this goal.

Ergonomics and Human Factors in Safety Through Design

Alphonse Chapanis, a leading authority on designing work to fit the capabilities and limitations of people, often states that companies can benefit greatly by designing work that is not error provocative. The following excerpts are from his chapter titled "The Error-Provocative Situation" in *The Measurement of Safety Performance* (1980).

- Many work situations and equipment setups are error provocative. The evidence is clear that people make more errors with some devices than they do with others.
- Given a population of human beings with known characteristics, it is possible to design tools, appliances, and equipment that best match their capacities, limitations, and weaknesses.
- The improvement in system performance that can be realized from the redesign of equipment is usually greater than the gains that can be realized from the selection and training of personnel.
- Design characteristics that increase the probability of error include a job, situation, or system which:
  - violates operator expectations
  - requires performance beyond what an operator can deliver
  - induces fatigue
  - provides inadequate facilities or information for the operator
  - is unnecessarily difficult or unpleasant
  - is unnecessarily dangerous
- ... [A] good systems engineer can usually build a nearly infallible system out of components that individually may be no more reliable than a human being. The human factors engineer believes that with sufficient ingenuity nearly infallible systems can be built even *if* one of the components is a human being.

A central point in Dr. Chapanis' work is that "The improvement in system performance that can be realized from the redesign of equipment is usually greater than the gains that can be realized from the selection and training of personnel."

If the environment constructed by management—which includes the design of both the workplace and work methods—requires behavior that is considered unsafe, then management focus should be mainly on altering the work environment.

GUIDELINES: SAFETY THROUGH DESIGN

Because the transitions occurring in the practice of safety are leading safety professionals into a greater involvement in the design processes, introductory guidelines on designing for safety are presented here. The guidelines that follow were excerpted from Fred Manuele's "Guidelines: Designing for Safety" (see References) and were based on several sources, two of which were most significant. One is Military Standard System Safety Program Requirements, often referred to as MIL–STD–882C (1993). The second is "A Generic Thought Process for Hazard Avoidance, Elimination or Control" as presented in *On the Practice of Safety* (Figure 1–2).

Objectives of Safety Through Design

The following objectives should be considered when companies are developing a safety-through-design process.

A Generic Thought Process for Hazard Avoidance, Elimination, or Control

1. *Avoid introduction of the hazard or prevent buildup of the form of energy or hazardous materials*
   - use material handling equipment rather than manual means
   - don't elevate persons or objects
   - avoid producing or manufacturing the energy or the hazardous material
2. *Limit the hazard or the amount of energy or hazardous material*
   - consider smaller weights in material handling
   - store hazardous materials in smaller containers
   - remove unneeded objects from overhead surfaces
   - seek ways to reduce actual or potential energy input
   - use the minimum energy or material for the task (voltage, pressure, chemicals, fuel storage, heights)
3. *substitute, using the less hazardous*
   - substitute a safer substance for a more hazardous one; when hazardous materials must be used, select those with the least risk throughout the life cycle of the system
   - replace hazardous operations with less hazardous operations
   - use designs needing less maintenance
   - use designs that are easier to maintain, considering human factors
4. *Prevent unwanted energy or hazardous material buildup*
   - provide appropriate signals and controls
   - use regulators, governors, and limit controls
   - provide the required redundancy
   - control accumulation of dusts, vapors, mists, etc.
   - minimize storage to prevent excessive energy or hazardous material buildup
   - reduce operating speed (processes, equipment, vehicles)
5. *Prevent unwanted energy or hazardous material release*
   - design containment vessels, structures, elevators, and materials handling equipment to appropriate safety factors
   - consider unexpected events, including the wrong input
   - protect stored energy and hazardous material from possible shock
   - provide fail-safe interlocks on equipment, doors, valves
   - install railings on elevations
   - provide slip-resistant working surfaces
   - control traffic to avoid collisions
6. *Slow down the release of energy or hazardous material*
   - provide safety and bleed off valves
   - reduce the burning rate (using an inhibitor)
   - reduce road grade
   - provide error-forgiving road margins
7. *Separate in space or time, or both, the release of energy or hazardous materials from that which is exposed to harm*
   - isolate hazardous substances, components, and operations from other activities, areas, and incompatible materials, and from personnel
   - locate equipment so that access during operations, maintenance, repair, or adjustment minimizes personnel exposure (e.g., hazardous chemicals, high voltage, electromagnetic radiation, cutting edges)
   - arrange remote controls for hazardous operations
   - eliminate two-way traffic
   - separate vehicle from pedestrian traffic
   - provide warning systems and time delays
8. *Interpose barriers to protect the people, property, or the environment exposed to an unwanted energy or hazardous material release*
   - insulation on electrical wiring
   - guards on machines, enclosures, fences
   - shock absorbers
   - personal protective equipment
   - directed venting
   - walls and shields
   - noise controls
   - safety nets
9. *Modify the shock-concentrating surfaces*
   - padding low overheads
   - rounded corners
   - ergonomically designed tools
   - "soft" areas under playground equipment

Figure 1–2. A generic thought process for hazard avoidance, elimination, or control.

- Safety, consistent with goals, is to be designed into all processes, the workplace, work methods, and products in a proactive, cost-effective manner.
- Risk assessment is to be an integral part of the design processes.
- A fundamental design purpose is to have processes and products that are error proof or error tolerant.
- Hazards must be identified and evaluated, and then avoided, eliminated, or controlled so that the associated risks are at an acceptable level throughout the entire life cycle of processes, equipment, and products.
- Requirements for minimum risk are to be established and applied in the acquisition or acceptance of new materials, technology, or designs, and prior to the adoption of new production, test, or operating techniques.
- Actions taken to identify and eliminate hazards and to reduce their attendant risks to an acceptable level are to be documented.
- Retrofit actions required to improve safety are to be minimized through the timely inclusion of safety features during research, technology development, and in purchasing and acquisition.

- A management-of-change system is to be in place that includes identification of hazards so that an acceptable risk level is maintained when design or work methods changes are made.
- Consideration is to be given early in the design process to the risks attendant in the eventual disposal of processes and products.
- Significant safety data reflecting lessons the company has learned are to be documented and disseminated to interested personnel.

These objectives represent an ideal level toward which companies can work in their efforts to improve their safety records. The safety professional can help management decide which objectives are the most practical for the company to adopt at the beginning.

Assessing Hazard Probability and Severity

The probability that a hazard potential will be realized and a hazard-related incident will occur is described in probable occurrences per unit of time, events, population, items, or activity. Assigning a quantitative hazard probability is generally not possible early in the design process. A qualitative probability may be derived from research, analysis, evaluations of the historical safety data on similar systems, and a composite of opinions of knowledgeable people. Companies should document any supporting rationale for assigning a hazard probability.

Management, working with the safety professional, should establish appropriate definitions of hazard severity categories. This will help them establish understandable qualitative measures for incidents that might occur should a hazard potential be realized. Causal factors for those incidents would include design inadequacies, system or component failure or malfunction, work methods designs that overstress employees or are error provocative, procedural difficulties, environmental conditions, or personnel error.

A company's resources will never be sufficient to eliminate every risk, which is neither practicable nor economically feasible. In the process of designing to achieve safety, discovering a multitude of risks will be the norm. They must be categorized by probability of occurrences and severity of outcomes so that priorities can be set and actions selected and taken to achieve acceptable risk levels.

Conducting Hazard Analyses/Risk Assessments

To determine what actions are to be taken to avoid, eliminate, or control hazards, a system to determine risk levels must be applied. A good risk assessment model will enable decision makers to understand and categorize the risks and to determine the methods and costs to reduce risks to an acceptable level. For these purposes, risk is a measure of the probability and severity of adverse effects deriving from hazards.

As the top priority is to eliminate hazards in the design process, assessing only the severity of consequences will generally be sufficient during the first design stages. When hazards are not eliminated early in the design phase, management should use a risk-assessment procedure that encompasses hazard probability and severity. From this risk assessment, the company can choose which of the priority hazard control methods it will use to address the hazards that still remain.

In each of the following steps, management and the safety professional must seek the counsel and expertise of qualified, experienced personnel who are knowledgeable about the work or process. They should serve on the management team to help forge a consensus about the best methods to use to reduce or eliminate the potential harm hazards pose. This process is more effective if a hazard analysis/risk assessment scenario is written covering each of the steps. Such a scenario would include the following:

1. **Establish the analysis parameters.** The team would select a manageable task, system, or process to be analyzed and define its relationship with other tasks or systems, if appropriate.
2. **Identify the hazards.** Members of the team should concentrate on identifying hazards that could be the cause of incidents. They should then determine each hazard's potential for harm or damage, arising out of the characteristics of a job, piece of equipment, system, and the like, and the actions or inactions of employees. At this point, the team should keep an assessment of hazard potential separate from an assessment of hazard severity.
3. **Consider the failure modes.** The team should define the possible failure modes that would realize the hazard's potential and result in an incident.
4. **Describe the exposure.** The purpose of this step is to establish the number of people, the type of property, and the aspects of the environment that could be harmed or damaged, and how frequently they might be exposed to danger should the hazard be realized. This step is often based as much on the judgment and experience of team members as it is on objective data.
5. **Assess the severity of consequences.** The team makes calculated speculations regarding the number of fatalities, injuries, illnesses, value of property damaged, and extent of environmental damage that might result should hazard-related incidents occur. Historical data are of great value as a baseline. On a subjective basis, the team would need to agree on a

classification system for the severity of hazard-related consequences: e.g., catastrophic, critical, marginal, and negligible. At this point, the hazard analysis is complete.

6. **Determine the probability of the hazard being realized.** Unless empirical data are available—which is rare—the process of selecting the probability of an incident occurring is subjective. Probability has to be related to intervals of some sort, such as a unit of time or activity, events, units produced, or life cycle. Commonly used categories for assigning probability include frequent, probable, occasional, remote, and improbable.
7. **Write a concluding statement.** The team would conclude with a statement that addresses both the probability of an incident occurring and the expected severity of its adverse results.
8. **Develop proposals to remedy the hazards.** The team would then concentrate on the design and operational changes necessary to achieve an acceptable risk level.

Hazard Analysis Techniques

If hazard analyses and risk assessments are to be effectively made, those assigned the responsibility must be skilled in the use of the special analytical techniques available. At least 25 such hazard analysis techniques have been developed, each having its own advantages and disadvantages. Discussions of those techniques can be found in books such as Roland and Moriarty's *System Safety Engineering and Management;* J Stephenson's *System Safety Two Thousand;* JW Vincoli's *Basic Guide to System Safety;* and R Stephans and WW Talso's *System Safety Analysis Handbook, a Sourcebook for Safety Practitioners* (see References).

A Design Review Procedure Model

A few companies have established procedures requiring that hazards be addressed in the design processes for new or altered facilities and equipment. Figure 1–3 provides an example of an actual implementation statement that can serve as a model for those who choose to establish such design review practices. The statement can be modified to fit the circumstances of each organization.

Safety Through Design Checklist

An organization's appropriation request for new projects or major alterations may include a proposal review procedure requiring that hazards be properly addressed. If so, having a well-crafted project review checklist makes such a procedure more effective. A sample checklist is provided in Figure 1–4. It is meant to serve as a guide only and can be tailored to the needs of each organization.

SUMMARY

- Over time, the level of safety achieved will relate directly to the caliber of the initial design of facilities, hardware, equipment, tooling, operations layout, the work environment, and the work methods, and their redesign as continuous improvement is sought. The goal of continuous redesign and improvement of operations is to reduce the number of errors until operations are as error proof as human effort can make them.
- The design stage offers the greatest opportunity to anticipate, analyze, eliminate, or control hazards.
- Safety through design is defined as the integration of hazard analysis and risk assessment methods early in the design and engineering stages and the taking of the actions necessary so that the risks of injury or damage are at an acceptable level.
- In the design and redesign process, management seeks to avoid, reduce, or eliminate the probability and severity of a hazard potential being realized and causing an incident.
- Companies should apply the following priorities to design and redesign processes: design for minimum risk, incorporate safety devices, provide warning devices, develop operating procedures and employee training programs, and use personal protective equipment.
- The safety professional can influence the design of the workplace and work methods at three critical points: in the preoperational, operational, and postincident stages.
- Companies should take a proactive stance regarding safety through design. They should also examine work procedures and systems more closely as causal factors rather than assuming worker behavior is the cause of an incident.
- Companies should consider ergonomics and the strengths and limitations of workers in their design process.
- In implementing a safety through design process, companies should establish clearcut objectives, assess hazard probability/severity, conduct hazards analysis and risk assessment, establish design review procedures, and use project checklists.

REFERENCES

Accident Facts. 1996 Edition. National Safety Council, Itasca, IL, 1996.

Chapanis A. "The Error-Provocative Situation." In William E. Tarrants, ed. *The Measurement of Safety Performance.* New York: Garland Publishing, 1980.

Deming, W. Edwards. *Out of the Crisis.* Cambridge, MA: Center for Advanced Engineering Study, Massachusetts Institute of Technology, 1986.

Procedures for Design and Equipment Review

PURPOSE
To provide operations, engineering, and design personnel with guidelines and methods to foresee, evaluate, and control hazards related to occupational safety and health and the environment when considering new or redesigned equipment and process systems.

SCOPE AND DEFINITIONS
This guideline is applicable to all processes, systems, manufacturing equipment, and test fixtures regardless of size or materials used. These conditions will be necessary for an exemption from design review:

- no hazardous materials are used (as defined by 29 *CFR* 1910.1200)
- operating voltage of equipment is <15 volts and the equipment will be used in nonhazardous atmospheres and dry locations
- no hazards are present that could cause injury to personnel (e.g., overexertion, repetitive motion, error-prone situations, falls, crushing, lacerations, dismemberment, projectiles, visual injury, etc.
- pressures in vessels or equipment are <2 psi
- operating temperatures do not exceed 100 F/38 C
- no hazardous wastes as defined by 40 *CFR* 26 & 262 and/or 331 *CMR* 30 are generated
- no radioactive materials or sealed source devices are used
- if other exemptions are desired, they are to be cleared by the safety, health, and environmental professional

PHASE I—PRECAPITAL REVIEW
This review is to be completed prior to submission of a project request or a request for equipment purchase, in accord with the capital levels outlined in Bulletin XXX. Precapital reviews are crucial for planning facilities needs such as appropriateness of location, power supply, plumbing, exhaust ventilation, etc. Process and project feasibility are determined through this review. A complete "What If" hazard analysis, in accordance with Bulletin YYY, is to accompany the request. Noncapital projects should also be reviewed utilizing these procedures, but a formal "What If" hazard analysis is not required.

PHASE II—INSTALLATION REVIEW
This review requires a considerably more detailed hazards and failure analysis relative to equipment design, production systems, and operating procedures. Detailed information is documented, including equipment operating procedures and a work methods review giving emphasis to ergonomics, control systems, warning and alarm systems, etc. A "What If" system of hazard analysis may be used and documented. Other methods of hazards analysis will be applied if the hazards identified cannot be properly evaluated through the "What If" system.

The project manager shall be responsible for the establishment of a hazard review committee and for managing its functions.

HAZARD REVIEW COMMITTEE
This committee will conduct all phases of design review for equipment and processes. In addition to the project manager, members will include the safety, health, and environmental professional, the facilities engineer, the design engineer, the manufacturing engineer, and others (financial, purchasing) as needed. For particular needs, equipment design or hazard analysis talent may be recommended by the safety, health, and environmental professional.

"WHAT IF" HAZARD ANALYSIS
This method of hazard assessment utilizes a series of questions focused on equipment, processes, materials, and operator capabilities and limitations, including possible operator failures, to determine that the system is designed to a level of acceptable risk. Users of the "What If" method would be identifying the possibility of unwanted energy release or unwanted release of hazardous materials, deriving from the characteristics of facilities, equipment, and materials and from the actions or inactions of people.

Bulletin AAA contains procedures for use of a "What If" checklist. For some hazards, a "What If" checklist will be inadequate and other hazards analysis methods may be used.

RESPONSIBILITIES

Project Manager
The Project Manager will be responsible for all phases of the design review from initiation to completion. That includes initiating the design review, forming the design review committee, compiling and maintaining the required information, distributing documents, setting meeting schedules and agendas, and preparing the final design review report. Also, the project manager will be responsible for coordination and communication with

Figure 1–3. Procedures for design and equipment review.

Haddon WJ, Jr. *Preventive Medicine. The Prevention of Accidents.* Boston: 1966.

———On the escape of tigers: An ecological note. *Technology Review,* May, 1970.

Kletz T. *An Engineer's View of Human Error.* Rugby, Warwickshire, UK: Institution of Chemical Engineers, 1991.

Lowrance WW. *Of Acceptable Risk: Science and the Determination of Safety.* Los Altos, CA: William Kaufman, 1976.

Manuele F. *Guidelines: Designing for Safety.* New York: Marsh & McLennan, 1995.

———. *On the Practice of Safety.* 2nd Edition. New York: John Wiley, 1997.

all outside design, engineering, and hazard analysis consultants.

Department Manager

Department managers will see that design reviews are completed for capital expenditure or equipment purchase approvals, and previous to placing equipment or processes in operation, as required under "installation review."

Signatures of department managers shall not be placed on asset documents until they are certain that all design reviews have been properly completed, and that their findings are addressed.

Design Engineer

Whether an employee or a contractor, the design engineer shall provide to the project manager and to the review committee documentation including:

- detailed equipment design drawings
- equipment installation, operation, preventive maintenance, and test instructions
- details of and documentation for codes and design specifications
- requirements and information needed to establish regulatory permitting and/or registrations

For all of the foregoing, information shall clearly establish that the required consideration has been given to safety, health, and environmental matters.

Safety, Health, and Environmental Professional

Serving as a design review committee member, the safety, health, and environmental professional will assist in identifying and evaluating hazards in the design process and will provide counsel as to their avoidance, elimination, mitigation, or control. Special training programs for the review committee may be recommended by the safety, health, and environmental professional. Also, consultants who would complete hazards analyses other than for the "What If" system may be recommended.

ADMINISTRATIVE PROCEDURES

In this section, the administrative procedures would be set forth—such as the amount of time prior to submission of a capital expenditure or equipment purchase request to be allowed the review committee for its work, information distribution requirements, assuring that the dates for installation review meetings are planned in advance, assuring that findings of hazards analyses are addressed, and resolving differences of opinion among hazard review committee members.

Figure 1–3. *(Concluded.)*

Military Standard System Safety Program Requirements (MIL–STD–882–C). Department of Defense, Washington, DC, 1993.

National Safety Council. 1121 Spring Lake Drive, Itasca, IL 60143-3201.
Safety Through Design. Itasca, IL: 1999.
Safety Culture and Effective Safety Management. Itasca, IL: 2000.

Peters GA. *Human Error: Analysis and Control.* Goshen, NY: Aloray, 1989.

Roland HE and Moriarty B. *System Safety Engineering and Management.* 2nd ed. New York: John Wiley, 1990.

Stephans R and Talso WW, eds. *System Safety Analysis Handbook, a Sourcebook for Safety Practitioners.* Albuquerque, NM: New Mexico Chapter, System Safety Society, P.O. Box 9524, 1993.

Stephenson J. *System Safety Two Thousand: A Practical Guide for Planning, Managing, and Conducting System Safety Programs.* New York: Van Nostrand Reinhold, 1991.

Swain AD. "Work Situation Approach to Improving Job Safety." Sandia Laboratories, Albuquerque, NM.

Vincoli JW. *Basic Guide to System Safety.* New York: Van Nostrand Reinhold, 1993.

REVIEW QUESTIONS

1. The design or redesign stage of buildings, equipment, products, and work processes is the best time to ____________ hazards.
   a. Anticipate
   b. Analyze
   c. Eliminate
   d. Control
   e. All of the above
2. Define "safety through design."
3. List the order of design precedence that companies should apply to all design and redesign processes to achieve the greatest effectiveness in hazard avoidance, elimination, or control.
   a.
   b.
   c.
   d.
   e.

Review Questions continued on page 22.

SAFETY THROUGH DESIGN CHECKLIST

I. FACILITIES AND SYSTEMS

1. **PLANNING AND ADMINISTRATION OF SAFETY AND HEALTH (OSH)**

1.1 **GENERAL:** Designing facilities and systems that are inherently *safe* for use or occupancy is an integral part of every Architect/Engineer contract issued.

1.2 **OBJECTIVES:** The principal objectives are to preclude hazards, minimize hazardous exposure, and reduce systems rework and retrofit to assure the safety and health of all personnel using our facilities and systems.

1.3 **OCCUPATIONAL SAFETY AND HEALTH ACT (OSHA):** Our company has committed to a continuing program to ensure design criteria are consistent with, or exceed, OSHA requirements. The A/E shall ensure facilities are designed in compliance with Title 29, *CFR* Part 1910 OSHA Requirements, as well as company criteria. However, if conflicts between OSHA and company or other federal criteria are noted, the conflicts should be referred to the Contracting Officer Technical Representative (COTR) for resolution. The A/E should follow-up to ensure resolution is achieved though the system safety process.

1.4 **SAFETY ENGINEERING:** Customer: A list of customer special safety and health concerns (for example, confined spaces, back injury, repetitive motion, fall hazards, energy sources, hazardous materials, etc.) shall be obtained, provided to the A/E and status reports provided to the customer. Planning and design shall ensure customer safety and health needs are identified and special controls are designed into facility projects. The Facility System Safety Working Group (FSSWG) is a safety process tool to link customer safety and health needs to the designer. FSSWGs should include the safety manager, industrial hygienist, environmental engineer, planner, user, and the safety engineer, as well as any other discipline needed, based on the project. The FSSWG should develop a list of hazardous operations that are of concern and review the control methods that will be used. The safety engineer will assist in developing the list of customer safety and health concerns and coordinate with the FSSWG to determine adequacy of controls. The safety engineer will ensure that appropriate design safety reviews are conducted and recommendations/comments are provided to the project engineer or A/E.

Designer: The A/E may need to conduct their own analyses to ensure all hazards are identified, risks evaluated, and engineering controls provided. Designs should minimize hazards to construction personnel, maintenance personnel, and users. For complex or high-hazard facilities, the A/E shall prepare a Safety Assessment Report (SAR) (refer to MIL STD 882, DI-SAFT-80102A). The SAR report shall identify all customer-requested controls, residual unsafe design characteristics, quantify risk of hazards not eliminated, and identify any controls, inhibits, or safety procedures. Whenever possible the designer should try to include safety features to minimize construction safety hazards.

REFERENCES: Title 29, *CFR* Part 1910 Occupational Safety Health Standards for General Industry, *MIL STD 882* (current series) Department of Defense, System Safety Program Requirements.

II. SPECIFIC DESIGN CONSIDERATIONS

2. **AMERICANS WITH DISABILITIES**

2.1 **DESIGN ELEMENTS:** Accommodate employees with disabilities to prevent injuries and illnesses. Provide "clean offices" for sensitized individuals. Provide computer-assisted equipment and software.

2.2 **REFERENCES:** Planning and Design Policy Statement, Barrier-Free Design Accessibility Requirements; Americans with Disabilities Accessibility Guidelines (ADAG).

3. **CONFINED SPACE**

3.1 **GENERAL REQUIREMENTS:** Eliminate all possible confined spaces from the design. Remaining confined spaces shall be evaluated for accessibility, methods of isolation, maintenance, and inspection.

3.2 **REMAINING CONFINED SPACES:** For each remaining confined space, wherever possible, design the space for continuous human occupancy, prompt egress, ease of ingress, and elimination of hazardous atmosphere wherever possible.

3.3 **ACCESSIBILITY:** Design all confined spaces with multiple, large accesses. Access shall be equipped with platforms, which are large enough to support all required equipment and personnel. Access must not be restricted by objects such as pipes or ducts. Locations of ladders and scaffold mounts inside the space shall be identified. Consider fall protection issues such as anchorage points.

3.4 **ISOLATION METHODS:** Design isolation methods to isolate confined spaces from hazards. Isolation design considerations include planning for equipment removal through accesses, full isolation of confined space(s) from electrical, mechanical, hydraulic, pneumatic, chemical, thermal, radioactive, and other hazardous energy such as falling object(s); to provide for complete isolation of all flows into and out of the confined space(s) such as valve blocking, spools, double blocks and bleeds, flanges, and flushing connections.

3.5 **MAINTENANCE AND INSPECTION:** Design the confined space(s) so that maintenance and inspection can be performed from the outside or by self-cleaning systems. Position maintenance points for ease of access.

3.6 **REFERENCES:** ANSI Z 17.1 (latest version) Safety Requirements for Confined Space; Title 29 *CFR* Part 1910.146, Permit-Required Confined Spaces.

4. **ELECTRICAL SAFETY**

4.1 **GENERAL REQUIREMENTS:** Consider voltage and amperage levels, high-voltage effects, ground connections, and isolation. Design for safe operation, maintenance, and inspection.

4.2 **PROTECTIVE DEVICES AND GROUNDING:** Design ground fault circuit interrupter (GFCI) systems for all high humidity areas and areas where liquids are, or may be,

Figure 1–4. Safety through design checklist. (*Source:* National Safety Council. *Safety Through Design.* Itasca, IL: 1999.)

used in applications such as spraying pooling, or any other use by personnel. Reduce voltage and amperage levels by substituting equipment or systems; fuses, circuit breakers, and line capacities to meet anticipated future requirements. Eliminate grounding connections to gas or steam pipes, electrical conduits, and sprinkler system piping to prevent accumulation of static electricity. Engineer ground devices to ensure maximum protection for unique sites such as rooms with flammable atmospheres; to provide lightning protection on buildings and electrical structures.

4.3 **INTRINSICALLY SAFE EQUIPMENT:** Design equipment or systems with consideration of combustible gases or vapors that may be encountered and account for differences in flash points, explosive limits and ignition temperatures.

4.4 **EQUIPMENT HEAT GENERATION:** Design to accommodate for the effects caused by heat generation from equipment and to control the temperature for flammables stored near this equipment.

4.5 **ISOLATION OF HIGH VOLTAGE EQUIPMENT:** Design to isolate indoor and outdoor high voltage equipment by using enclosure vaults, security fences, lockable doors and gates. Non-insulated conductors such as bus bars on panel boards, switchboards or high voltage equipment connections in accessible areas must be enclosed or protected to eliminate or reduce electrical hazards associated with maintenance and inspection.

4.6 **ENERGY ISOLATION:** Ensure all equipment and systems can be locked out. (See lockout/tagout section.)

4.7 **REFERENCES:** NFPA 70 National Electric Code; ANSI Z244.1 (latest version) Requirements for Lockout/Tagout; Title 29 *CFR* Part 1910, Subpart S, Electrical; and 29 *CFR* Part 1910.147 Control of Hazardous Energy (lockout/tagout); ANSI C2 (National Electric Safety Code).

5. **ERGONOMICS**

5.1 **ANTHROPOMETRY:** Design for the 5th and 95th percentiles (smallest and largest) such as clearances for the 95th percentile and visual fields for the 5th and 95th percentile. Determine the range of movement expected in the job or task and provide adjustable equipment wherever possible. Eliminate gender-specific tasks and consider future workforce (older population with more females), force requirements such as grip strength and reach envelope. Eliminate spatially restricted spaces that require kneeling and crawling.

5.2 **PROCESS DESIGN:** Design layouts that do not require personnel to: twist and turn when moving an object from one conveyor belt, table or machine to another; lift an object from floor level to a conveyor, table or machine; work with elbows above waist level; use hands and elbows in a twisting motion while performing job tasks; hyper-extend or hyper-flex wrists while performing job tasks.

5.3 **WORKSTATION DESIGN:** Design for range-of-motion of the worker; for field of vision of worker; to reduce repetitive motions required per task, duration and pace of a task; to provide adequate support for back and legs (such as back supports or floor mats); adjustable work surfaces that are easily manipulated from position of the work; delivery bins to accommodate height and reach limitations; work platforms that move up and down for various operations; for powered assists to eliminate the use of extreme force; for the use of suspension devices for heavy tools; the use of diverging conveyors off main lines so that certain activities can be performed at slower rates.

5.4 **PERSONAL RISK FACTOR:** Gender—Determine and understand which tasks negatively affect which genders. Age—Determine tasks that affect workers based on age. Anthropometry—Know the range of the work force. Work method—Design proper procedures for task accomplishment. Senses—Poor vision, hearing, and smell should be considered in ergonomic design. Physical strength—The strength required to accomplish a task must be considered. Weight—The weight of the worker can affect the design of a workstation.

5.5 **UPPER EXTREMITY RISK:** Design: to eliminate the need for repetitive and/or prolonged activities; to eliminate forceful exertions with the hands such as reducing the number of hand pinch grips required in task accomplishment. Design out the need for prolonged static postures (add chairs, arm rests etc. where needed) and maintain symmetry between the worker and the task (avoid awkward postures). Eliminate the need to reach above the shoulders, reach behind the back, perform unusual twisting of wrists and other joints, maintain continued physical contact with work surfaces or use tools with excessive vibration in task accomplishment.

5.6 **ENVIRONMENTAL RISKS:** Design to maintain constant workstation temperatures and avoid excessively hot conditions or excessively cold conditions; for appropriate lighting; to eliminate excessive vibration. Consider the type of floor surface and platforms required for safety and stability at the workstation.

5.7 **REFERENCES:** ANSI A-365 (latest version) Draft Ergonomic Standard; NIOSH lifting equation.

6. **FALL PROTECTION**

6.1 **ELIMINATION OF HAZARDS:** Design stairways with standard guardrails instead of straight ladders where feasible; parapets or guardrails at roof edges; warning lines and cables to keep personnel away from fall hazards; for cranes or personnel lifts to provide safe access. Install supply and exhaust fan equipment at ground level or on roofs with sufficient space and protection to adequately perform testing and routine maintenance.

6.2 **MAINTENANCE AND INSPECTION:** Design equipment so that fall hazards are minimized during maintenance repairs, inspection or cleaning. Consider future degradation of installed equipment in maintenance and inspection activities.

6.3 **ENGINEERING CONTROLS:** Design the job or operation to eliminate work at heights. When not feasible provide prevention systems such as guardrails, catwalks, and platforms. Consider fall arrest issues such as anchorage points; compatibility of the control measures with the job tasks and work environment.

Figure 1–4. *(Continued.)*

6.4 FALL PROTECTION SYSTEMS: Install anchorage points. Design horizontal cable or I-beam trolley systems in areas where employees require continuous mobility and where platforms or guardrails are not feasible. Ensure proper test methods are used to ensure systems are capable of proper support.

6.5 REFERENCES: ANSI A92.5 (latest version) Aerial devices elevated work platforms; NFPA 101 Life Safety® Code; Title 29 *CFR* Part 1910, Subpart F, Powered Platforms, Manlifts and Vehicle-Mounted Work Platforms.

7. FIRE PROTECTION

7.1 GENERAL: Mil-Hdbk-1008 provides detailed guidance for the incorporation of fire protection engineering measures in the design and construction of DOD facilities. The handbook can be used in the purchase and preparation of facilities planning and engineering studies and design documents used for the procurement of facilities construction.

7.2 STANDARD FIRE SUPPRESSION EQUIPMENT: Design so standpipe fire suppression equipment is protected against mechanical damage and located to facilitate prompt use of hose valves, hoses, and other equipment at the time of a fire or other emergency. Locate hose outlets and connections high enough above the floor to avoid being obstructed and to be accessible to employees. Ensure that reels and cabinets are conspicuously identified.

7.3 TOTAL FLOODING SYSTEMS: Total flooding systems should provide a pre-discharge employee alarm which is capable of being perceived above ambient light or noise levels before the system discharges, giving employees time to safely exit from the discharge area.

7.4 FACILITY EVACUATION: Design doors and pathways of sufficient size to allow complete dispersal of employees from a specific department or location. Consider safe areas inside of primary facilities for personnel who cannot escape. Design internal evacuation routes so that during an emergency occupants will readily know the direction of escape from any point. Provide means of egress remote from each other to prevent the blocking of any single exit due to fire or smoke. Discharge exits directly into the street, or yard, court, or other open space that gives access away from the facility and allows for adequate space to accommodate the discharge of occupants. Mark and arrange every exit so it is clearly visible and immediately recognized as an exit in order to minimize the possibility of confusion during an evacuation. Provide adequate and reliable illumination at all exit locations with exit signs at the point of exit from the building. Mark doors, passageways, or stairways which are not (or do not lead to) an exit by a sign reading "Not an Exit" or by a sign indicating its actual character, such as "To Basement," "Storeroom," "Linen Closet," etc.

7.5 REFERENCES: Mil-Hdbk-1008, Fire Protection for Facilities Engineering, Design and Construction; Title 29 *CFR* Part 1910 Subpart L (Fire Protection); Title 29 *CFR* Part 1910, Subpart E, Means of Egress; NFPA 101, *Life Safety® Code.*

8. FIRST AID AND MEDICAL SERVICES

8.1 FACILITY DESIGN: Design medical facilities IAW MilHdbk-1191: Provide for emergency communications and ramps where needed to deliver or pick up patients by ambulance. Provide adequate ventilation in first aid facilities. Design surfaces of doors, walls, floors, and ceilings in the medical work area to be water-resistant for ease of cleaning. Ensure access doors to the medical work area are self-closing. Consider emergency access points for ambulance crews and emergency responders.

8.2 WASHING FACILITIES: Provide medical facilities with an automatic or elbow/foot operated sink for washing hands to be located in the work area. Provide adequate washing facilities using potable water, soap, and single use towels or hot air drying machines; hand-washing facilities that are readily accessible to personnel working in first-aid stations.

8.3 MEDICAL WASTE: Provide areas for the isolation of medical waste; adequate ventilation in waste storage areas. Provide adequate marking of "biohazard" storage areas.

8.4 COMPRESSED GAS: Provide safe storage of compressed gases used for medical purposes. Minimize use of equipment or fixtures having potential for an ignition source. Provide storage of compressed gases outside of the immediate medical facility whenever feasible.

8.5 PATHOGEN EXPOSURE CONTROL: Provide medical work areas that minimize splashing, spraying, spattering, and generation of droplets involving blood or other potentially infectious materials. Plan for tables, countertops, flooring, and other routinely used work surfaces to he constructed of nonporous materials to aid in exposure control.

8.6 LAUNDRY REQUIREMENTS: Design for storage and recycling of launderable materials, contaminated laundry to be bagged or contained at the location where it was used and not sorted or rinsed in the location of use. Design to accommodate wet contaminated laundry that may soak-through or leak from the bag or container.

8.7 CONTAINMENT EQUIPMENT: Plan for and/or install certified biological safety cabinets (Class I, II, or III) or other appropriate combinations of personal protection or physical containment devices for all activities involving with infectious materials that pose a threat of exposure to droplets, splashes, spills, or aerosols for pathogen control.

8.8 MAINTENANCE INSPECTION: Properly position bulky or hard-to-maintain equipment to provide ease of access. Specify that maintenance access points be positioned for safe access.

8.9 REFERENCES: NFPA 99, Health Care Facilities; NFPA 99B, Hypobaric Facilities; NFPA 101, Life Safety® Code; Mil-Hdbk-1191; Title 29 *CFR* Part 1910.151; Title 29 *CFR* Part 1910.1030; CGA Pamphlet P-2 Characteristics and Safe Handling of Medical Gas; EPA PB-86199130 Guide for Infectious Waste Management.

Figure 1–4. *(Continued.)*

9. HAZARDOUS MATERIALS

9.1 GENERAL STORAGE REQUIREMENTS: Design storage facilities to maintain and separate both hazardous and nonhazardous storage; emergency ingress and egress points from hazardous materials storage areas. Determine whether underground or above ground storage is best for the site. Locate compatible waste streams in the same region of the site to reduce potential widespread contamination. Define well installation requirements (if necessary). Consider waste compatibility, hazard classes, fire suppression guidelines, containment requirements and recovery actions needed to properly store and dispose of hazardous materials.

9.2 CHEMICAL COMPATIBILITY AND STORAGE: Design diking between incompatible materials; collection sumps under the acid and alkaline storage areas for leakage that can be drained after any incident and/or neutralized before disposal; explosion-proof rooms for Flammables/Combustibles. Design: floor drain passages; leakproofing method to prevent ground or ground water contamination; to eliminate chemical compatibility issues; to isolate materials; separate sumps for containment; eye wash and showers into appropriate locations; use a containment material to retain all emergency water.

9.3 FLAMMABLE STORAGE: Design explosion-proof rooms for flammable and combustible materials storage. Utilize spark resistant equipment, fire doors, blow-out walls, etc. Review all fire extinguishing requirements. Store fuels away and separate from oxidizers.

9.4 LOADING/UNLOADING LOCATION AND DESIGN: Place loading docks on opposite side of prevailing winds. Ensure all dock areas are designed with a sprinkler system. Provide dikes and/or containment systems for loading/off loading areas for tankers or railcars.

9.5 STORAGE ABOVE/BELOW GROUND: Provide secondary containment; spill prevention controls; overfill prevention controls; diking for tank farms. Design to protect from corrosion; leak detectors; all tanks to be grounded.

9.6 ALARM AND WARNING SYSTEMS: Design: systems to provide sufficient time for escape from the facility; alarms that are capable of being perceived above ambient noise or light levels by all personnel in the affected portions of the workplace; alarms that are distinctive and recognizable as a signal to evacuate the work area or to perform emergency actions; standardized systems to eliminate confusion in an emergency; to allow maintenance of alarm systems. Install redundant systems in areas with corrosive atmospheres or where devices are subject to unusual wear or possible destruction. Incorporate tactile devices in areas where personnel would not otherwise be able to recognize an audible or visual alarm.

9.7 REFERENCES: NFPA 30, Flammable and Combustible Liquids Code; NFPA 43D, Pesticide Storage; NFPA 49, Hazardous Chemicals Data.

10. LIGHTING

10.1 GENERAL REQUIREMENTS: Incorporate supplemental lighting around moving machinery, conveyors, steps and stairways. Consider color perception changes under yellow sodium vapor lights. Design lighting that is comfortable to the eye; provides adequate security; and adequate for the work to be performed.

10.2 EMERGENCY LIGHTING: Provide emergency lighting: where necessary for personnel to remain at machines or stations to shut down equipment in case of power failure, at stairways, passageways, or aisleways used for emergency egress.

10.3 LIGHTING IN HAZARDOUS CONDITIONS: Determine NFPA Class and Division ratings for the location where lighting is considered. Ensure all associated switches and electrical equipment are approved for the location. Portable battery powered lighting equipment, used in connection with the storage, handling, or use of flammable gases or liquids must be of the type approved for the hazardous location.

10.4 OUTSIDE LIGHTING: Locate outdoor lamps below all live power line conductors, transformers, or other electric equipment unless adequate clearances or other safeguards are provided. Eliminate high contrast light and dark areas that may contribute to accidents from people or vehicles entering or leaving lighted or non-lighted areas.

10.5 REFERENCES: NFPA 101, Life Safety Code; IESNA Lighting Handbook (latest version); IES RPO7 Practice for Industrial Lighting; IES RP-1 Practice for Office Lighting; NFPA 70, National Electric Code.

11. ENERGY CONTROL

11.1 GENERAL REQUIREMENTS: Specify or design energy isolation devices capable of being locked out. Layout machinery and equipment to ensure safe access to lockout devices and provide each machine/equipment with independent disconnects. Specify lockout devices that will hold the energy isolating devices in a "safe" or "off" position.

11.2 MACHINE AND EQUIPMENT SPECIFICATIONS: Ensure equipment and utilities have lockout capability and that any replacement, major repair, renovation, or modification of equipment will still accept lockout devices. Design emergency and non-emergency shutoff controls for easy access and usability. Integrate actuation controls with warning lights and alarms to prevent personnel exposure to hazards.

11.3 LOCKOUT DEVICES: Ensure selected devices are capable of withstanding the environment to which they are exposed and are standardized within the facility.

11.4 REFERENCES: Title 29 *CFR* Part 1910.147, Control of Hazardous Energy (lockout/tagout); ANSI Z244.1 Requirement for Lockout/Tagout; NFPA 70, National Electric Code.

Figure 1–4. *(Continued.)*

12. **MACHINE GUARDING**

12.1 **GENERAL REQUIREMENTS:** Design guards to be affixed to the machine whenever possible and ensure the guard does not itself create a hazard. Utilize structural barriers such as exterior or interior walls to isolate moving parts of machines. Specify guards that are interlocked with the source of power to the machine to prevent them from being removed by unauthorized personnel. Use vibration isolators to dampen or eliminate noise transmission and mechanical failure.

12.2 **BARRELS, CONTAINERS, AND DRUMS:** Ensure revolving barrels, containers, and drums are elevated and/or guarded by an enclosure which is interlocked with the drive mechanism.

12.3 **ANCHORING MACHINES:** Anchor all machines to prevent movement or vibration. Ensure the floor surface is appropriate for the weight and stress.

12.4 **MAINTENANCE CONSIDERATIONS:** Design maintenance access points to be located outside guards and are accessible. Provide adequate space between pieces of equipment.

12.5 **LOAD LIMITS:** Mark maximum rated limits on all machines, hoists, etc. Design out the ability for "quick-starts" to reduce mechanical stress. Install bumpers, shock absorbers, springs or other devices to lessen the effects of repeated impacts.

12.6 **REFERENCES:** OSHA 29 *CRF* 1910, Subpart O, Machinery and Machine Guarding; ANSI/ASME B15.1 (latest version) Safety Guarding and Devices.

13. **MATERIALS HANDLING**

13.1 **PLANNING FACTORS:** Consider load-handling factors for optimum storage and movement. Determine container specifications. Ensure equipment maximum load specifications are displayed prominently and that the size, weight, and center of gravity of the load are considered. Analyze types and quantities of materials to be staged indoors. Design receiving points and material handling processes to accept just-in-time deliveries.

13.2 **FACILITY-WIDE CONSIDERATIONS:** Combine compatible operations to reduce multiple handling. Minimize transport distances in the facility by analyzing material routing techniques and necessary movement. Automate process if possible.

13.3 **WORKSTATION DESIGN:** Bring workers into the design process. Plan for objects to be between shoulder and knee height. Plan for adjustable work heights. Minimize travel distances. Place items within reach distance (plan for a range). Eliminate unnecessary handling. Choose proper containers for the task. Optimize equipment/material layout. Consider relationship of worker to work station (reduce twisting, bending, reaching). Minimize negative environmental factors. Insure load stability in movement.

13.4 **MANUAL TASK:** Automate load movement by using mechanical lift tables, lifting platforms. lift trucks, cranes, hoists, etc. Have material delivered at proper height. Increase unit quantity and load per movement. Reduce weight and size for easier handling. Change load movement to pushing and pulling operation.

13.5 **INDUSTRIAL TRUCKS:** Use industrial trucks when load movement is intermittent, over long distances, or over a variable path.

13.6 **MONORAILS, HOISTS, CRANES:** Consider use of monorails and cranes when the load transfer is required from one point to another in the same general area. Consider use of hoists to facilitate the positioning, lifting, and transferring of material within a small area.

13.7 **AUTOMATED SYSTEMS:** Consider driverless, automatically guided vehicles (AGV) for small load applications. Consider automated retrieval systems (ARS) for parts and storage retrieval.

13.8 **STORED MATERIALS:** Plan for storage areas that can be kept free of tripping, fire, and other hazards. Determine and observe height limitations when stacking materials; stacking limitations for lumber, blocks, bags and bundles. Design to allow drums, barrels, and kegs to be symmetrically stacked.

13.9 **MATERIAL SHIPPING & RECEIVING POINTS:** Minimize heating and cooling losses through the use of suspended dividing flaps or other doorway enclosure techniques. Install self-leveling docks, truck levelers, forklifts, cranes, and other devices to maximize efficiency and reduce manual material movement and handling. Provide separate receiving points for hazardous and non-hazardous materials.

13.10 **RAMPS:** Provide adequate width for material transport without the removal of guardrails and toe-boards. Ensure that the slope of any gangway, ramp, or runway does not exceed specifications and rise to run ratio. Construct floor with maximum loading as a prime consideration and so that floor cannot tip, sag, or collapse under maximum anticipated loads.

13.11 **REFERENCES:** Title 29 *CFR* Part 1910, Subpart N, Materials Handling and Storage.

14. **NOISE LEVEL CONTROL**

14.1 **FACILITY EVALUATION:** Evaluate size and shape of rooms and departments, proposed layout of equipment, work stations, break areas, surface materials (e.g., ceiling/steel; walls/cinder block; floor/concrete), and noise from other sources (spillover noise).

14.2 **NOISE INTENSITY CONTROL:** Separate personnel from noise sources by the greatest reasonable distance. Install noise barriers between the source and the operators. Locate equipment and workstations so that the greatest sources of noise are not facing the operators. Take advantage of reflective walls and floor surfaces to properly channel sound. Use building construction materials that absorb sound energy. Design an enclosed room for the equipment operator. Mount equipment that may vibrate on solid foundations. Locate noisy equipment in sound enclosures to reduce ambient noise levels. Install insulating material over surfaces creating or transmitting excessive noise. Provide noise absorbent linings on air ducts and mufflers on openings.

Figure 1–4. *(Continued.)*

14.3 EQUIPMENT SELECTION: Select equipment with low vibration and noise characteristics. Stipulate permissible maximum noise levels in specifications for new equipment. Use equipment with low velocity fluids or gas flows. Where possible, lower machine or conveyor speeds or high-pressure air exhausts.

14.4 MAINTENANCE: Design for maintenance minimization of vibration and noise. Use slow-acting valves or accumulators in water systems to eliminate water hammer. Substitute high-noise processes with lower noise producing processes, such as squeezing processes instead of drop hammer processes; welding processes instead of riveting processes; chemical cleaning processes instead of high-speed grinding/polishing processes.

14.5 REFERENCES: Title 29 *CFR* Part 1910.95, Occupational Noise Exposure.

15. TOXIC MATERIALS

15.1 GENERAL REQUIREMENTS: Review Material Safety Data Sheets (MSDS) for toxic materials selected for use. Consider current Threshold Limit Values (TLVs) in the selection of materials. Design: for "just-in-time" storage and use of toxic materials on site to limit the type of substances, amount (dose) of exposure, and the number of entry routes to the body. Reduce the possibility of reactivity of chemicals in the same work area. Incorporate toxic material source reduction techniques in the design phase through closed process systems when extremely hazardous materials are used, eliminate or minimize flanges, connections, pump seals, valves, valve stems, etc., and closed process control rooms for operators. Incorporate shower and change rooms to prevent contamination outside of the facility. Ensure standardized labeling and marking of equipment, tanks, lines, etc. Ensure access ports are of adequate size to permit entry while using personal protective equipment (PPE). Provide deluge showers and eyewash fountains.

15.2 GASES: Design: ventilation systems capable of handling routine and emergency releases; safe and adequate storage and staging facilities or workstations where gases are used. Consider line systems fed to work stations to eliminate indoor gas storage and staging. Install: emergency shut-off controls for maximum accessibility during emergencies: systems to detect and warn building occupants of threatening air concentrations.

15.3 VAPORS: Install spring-loaded covers where possible on open surface tanks to reduce evaporation; fusible links on covers of all open surface tanks not having spring-loaded covers. Install systems to detect and warn building occupants of threatening air concentrations. Design ventilation systems capable of meeting indoor air quality requirements.

15.4 FUMES: Install proper ventilation in areas where hot work (welding, brazing, etc.) is performed. Design ventilation systems capable of meeting indoor air quality requirements.

15.5 AIR PARTICULATES: Install HVAC and other control systems capable of meeting indoor air quality requirements; systems to warn against concentrations reaching lower level explosive limits (LEL) or other specified action level.

15.6 ACIDS AND BASES: Design: facilities to incorporate separate staging and storage of opposing pH material; ventilation systems so that different hazard classes do not mix in the air streams.

15.7 METALS: Design to reduce, eliminate, control, or substitute noncarcinogens for carcinogens such as lead acetate and cadmium. Limit employee contact through the use of rollers, grappling devices, etc. Incorporate the use of dust filters at the point-of-operation in process to limit the spread of metal dusts through proper ventilation.

15.8 SOLVENTS: Select process solvents carefully, many different hazard types exist (flammable, acidic, alkaline, etc.). Select solvents that satisfy process requirements, and reduce engineering controls and PPE. Incorporate solvent filtration units to reduce handling and extend life if possible.

15.9 ASBESTOS: Perform an asbestos survey, including sampling and testing of suspected asbestos containing materials (ACM), for all areas of the project to be demolished, renovated or disturbed. The work shall be conducted under a safety and health plan and in full compliance with all applicable safety, health and worker protection regulations. The survey will consist of the following components.

- A review of existing building records and drawings for references to ACM used in construction, renovation, or repairs.
- An inspection of the building(s) and all associated tunnels, chases, and crawl spaces to identify those materials that may contain asbestos. The A/E will be responsible for repairing any damage to the building caused by the inspection of inaccessible spaces. The survey will include the identification of friable and nonfriable ACM. The locations of all ACM will be determined, reported, and photographed.
- Sampling and testing of the suspected materials identified in step 2. The number of samples to be taken will vary according to the condition, but sufficient samples must be taken to assure that all ACM is identified and documented. Samples will be taken of the various troweled or sprayed on surfaces, pipes, and boiler insulation, tile, siding, shingles, and other suspected materials. A lab certified by the National Voluntary Laboratory Accreditation Program (NVLAP) using polarized light microscopy will analyze samples. The results of each sample analysis shall become part of the final report.
- The A/E shall prepare comprehensive drawings and specifications as required for the removal of the ACM in a final report.
- A drawing showing all locations of ACM, the estimated quantity, and sample points. Also note locations where removal of ACM will require temporary relocation of other systems such as HVAC ducts and piping.
- The test results.
- Photographs.

Figure 1–4. *(Continued.)*

- Recommended actions with a cost estimate.
- Any other information, field notes or forms which provide pertinent data.

15.10 LEAD

(1) Perform a lead-containing paint (lead .0.5 percent by weight dry film) survey, including sampling and testing of painted surfaces, on all areas of the project to be renovated or disturbed. The number of samples to be taken will vary according to the conditions, but sufficient samples shall be taken to assure that all lead-containing paint that may be disturbed as part of this project is identified and documented. Each sample shall be removed cleanly to the substrate. The sample shall include only paint scrapings and must not be contaminated with material such as rust or mill scale, wood, concrete, or any part of the substrate. Each sample shall be sealed in a separate plastic bag or container that will not contaminate the sample, and shall be identified with the following information: project site, date, and location of sample, name and signature of person taking the sample. The sample bags or containers shall be sealed and sent to a laboratory certified by the Environmental Lead Laboratory Accreditation Program (ELLAP).

(2) For projects including total or partial demolition, the A/E will also analyze a representative core sampling for Toxicity Characteristic Leachate Procedure (TCLP) using EPA Sw-846. Sub-samples should be taken from walls, windows, floors, ceiling, doorframes, and other building components. (The sub-samples are normally taken by using a 1-inch drill bit or similar device.) The size of each sample taken will be based on the volume of that particular building component relative to the total volume of the anticipated debris. The composite sample must be thoroughly mixed before being analyzed for TCLP. If the test results are below the EPA limit of 5 ppm, the construction debris shall be identified as non-hazardous and can be disposed of at a regular construction landfill (subtitle D). For test results exceeding 5 ppm, the construction debris will be managed under Resource Conservation and Recovery Act hazardous waste regulations.

(3) Shall prepare comprehensive drawings and specifications as required for the removal and disposal of lead-containing paint. Complete documentation of the survey and all test results are required along with mapping all areas and the locations of all sample points to confirm lead-containing paint. Complete forms, field notes, photographs, and all other information including assessments, condition of paint, and anticipated physical difficulties involved with any abatement action shall be part of the survey report. A photographic record may be used to determine the validity of the proposed corrective actions if deemed necessary. The work shall be conducted under a safety and health plan and in full compliance with all applicable safety and health, and worker protection laws. The survey report shall include as a minimum the following information:

Qualifications of the laboratory.

- Type of test analysis conducted.
- Drawings showing location of samples and homogenous areas of lead-containing paint, including the estimated quantity.
- Photographs (if used).
- Results of both TCLP and total lead analysis in % concentration of lead in paint and ppm TCLP.
- Recommended actions and a detailed cost estimate.

15.11 REFERENCES: Title 29 *CFR* Part 1910, Subpart Z, Toxic and Hazardous Substances; NFPA 30, Storage of Hazardous Materials; NFPA 491, Hazardous Chemical Reactions; CGA Pamphlet: Handbook of Compressed Gases.

16. INDUSTRIAL VENTILATION

16.1 GENERAL REQUIREMENTS: Locate all emission sources. Characterize each constituent to determine specific requirements. Review: ways to reduce personnel interaction with the source reviewed, personnel location, and work practices; to determine incompatible air streams (cyanides and acids etc.). Consider effects of outside wind speed and direction: weather and seasons.

16.2 ASSESS HAZARDS: Review: Material Safety Data Sheets (MSDS) to determine specific of chemicals used in the process; exposure limits; use most conservative exposure limit determined. Mass-balances considered to determine actual loss rates. Comply with manufacturer recommendations for venting.

16.3 COMBUSTIBLE HAZARDS: Ensure: concentrations are below the levels specified; flammable or explosive limits are controlled by using intrinsically safe equipment. Incorporate explosion-venting techniques into ventilation and collection systems.

16.4 MAINTENANCE: Install gauges to measure pressure drop at exhaust ducts. Design controls that are easily accessible to operators and maintenance crews.

16.5 ALARMS AND WARNING SYSTEMS: Ensure automatic controls are installed to provide audible or visual alarms if ventilation systems fail and provide sufficient time for safe escape from the facility. Automatically shut down operation if ventilation fails. Ensure alarms are designed that are capable of being perceived above ambient interference and that tactile devices are used where personnel would not be able to recognize an audible/visual alarm.

16.6 HOODS: Design and orient to ensure ventilated materials will fall or be projected into the hoods in the direction of the airflow and that minimum airflow specifications are achieved at any point in the capture zone. Ensure airflow specifications are strong enough to overcome other forces acting on the contaminants (doors, HVAC, etc.).

16.7 DUCTWORK: Design inlet and exhaust ductwork: to be adequately supported throughout its length and is sized in accordance with good design practice(s). Provide inspection or clean-out doors in ducting as required. Provide drains as required.

16.8 MAKE-UP AIR: Provide clean "dedicated" fresh make-up air.

16.9 REFERENCES: ANSI/AIHA Z9.5 (latest version) Laboratory Ventilation; ANSI Z9.2 (latest version) Fundamentals Governing the Design and Operation of Local Exhaust Systems; Title 29 OSHA 29 *CFR* 1910.94, Ventilation; NFPA 496, Purged and Pressurized Enclosures for Electrical Equipment.

Figure 1–4. *(Concluded.)*

4. The safety practitioner can influence the design of the workplace and work methods during what three stages?
   a.
   b.
   c.

5. Explain the difference between proactive and reactive responses in designing for safety.

6. Alan D. Swain's approach to improving job safety suggests that it is management's responsibility to provide the worker with a safety-prone work situation, and avoid the temptation to place the burden of accident prevention on the ____________.

7. What is the central point in Dr. Alphonse Chapanis' work?

8. Guidelines on safety through design that companies should follow include:
   a.
   b.
   c.
   d.
   e.
   f.

9. Why is a good risk assessment model essential in conducting a hazard analysis/risk assessment?

2

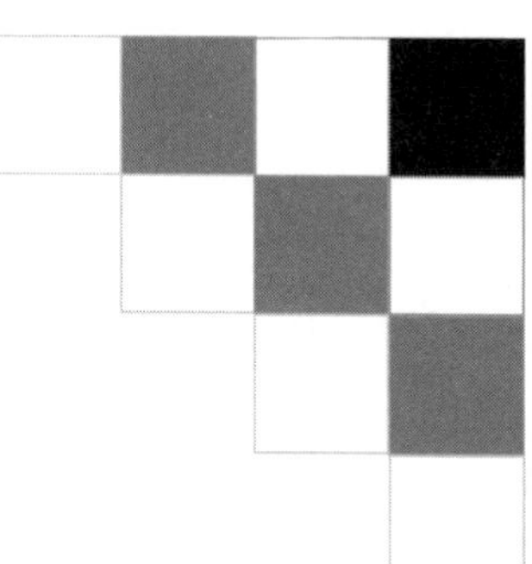

Buildings and Facility Layout

REVISED BY

Patrick J. Conroy

John C. Langford

John F. Montgomery, Ph.D., CSP, CHCM, CHMM

DESIGN FOR SAFETY

By carefully planning the design, location, and layout of a new facility or of an existing one that needs major alterations, safety and health professionals can greatly improve the safety and productivity of a facility's operations. Numerous accidents, occupational diseases, explosions, and fires can be prevented if safety measures are taken during the early planning stages. The topics covered in this chapter include:

- general considerations in designing for safety and some of the significant codes and standards involved
- safety factors to consider in selecting a facility site
- hazards and safety factors to consider in outside facilities
- safety concerns regarding facility railways
- safety design decisions when developing facility layouts
- use of lighting and color to enhance safety in the workplace
- how to make building structures safer

Ideally, safety and health professionals should conduct a safety and health study of a proposed facility while the designing and engineering are in the developmental stages. They should approach this study from the viewpoint of removing hazards rather than adding protective equipment. For example, they could suggest cutting storage of hazardous materials or substituting a less hazardous product. Also, they could suggest ways to reduce risks, such as intensifying a rate of mixing, storing a gas at a lower temperature and pressure, or simplifying the facility design. (For more detailed information on these approaches, see the References at the end of this chapter, Chapter 12, Flammable and Combustible Liquids, and Chapter 11, Fire Protection, in this volume.)

General Considerations

Effective human performance is a key factor in efficient production. Therefore, plan industrial systems with workers in mind. Safety and health professionals should ask the following questions when changes in the workplace are required:

- What will workers do?
- How should workers do it?
- Where should workers do it?
- Why should workers do it?
- What can happen to workers who do it?

Machinery is the second factor when planning facilities. A third factor is the flow of raw materials, in-process materials, component parts, and final product.

Design Considerations

Factors to consider for the general design of the workplace include the following:

- illumination
- noise and vibration control
- product flow
- ventilation (particularly of dust, vapors, and fumes)
- control of temperature and humidity
- work positions and movements of employees
- supervision and communication
- support requirements for such things as vehicles, portable ladders, material handling devices, monitoring and controlling systems, and cleaning and maintenance equipment

Factors to consider when designing machine tools and equipment include the following:

- construction and procedures
- visual displays, signs, and labels
- protective features and guards
- controls and handles
- maintenance and service needs
- safety signs

In the design stage of planning a facility, safety and health professionals should also consider the human interaction factor. They should help to provide safe and efficient ways for employees and supervisors to communicate.

Buildings, Processes, and Personnel Facilities

Major factors determining the size, shape, and type of buildings and structures include the following:

- the nature of the business and processes
- the nature of the production materials
- maintenance
- mechanical handling equipment
- climate
- working conditions
- shipping and receiving materials
- economic considerations

- personnel facilities, such as lunchrooms and medical, safety, and disaster services, which provide for employees' health, safety, and treatment of injuries

Facilities such as personnel and other employee-oriented activities should be planned and located for convenient and efficient use by workers.

Codes and Standards

Many companies follow a policy of having their safety and health specialists, as well as their insurance companies, review plans and specifications for new facilities or for those that need remodeling. Safety and health professionals should also ensure that plans include provisions for fire safety, safe work practices for electrical workers, and periodic inspections to verify the integrity and strength of the building's structure throughout the construction process. Ideally, safety and health professionals should approve and sign off on plans before they are released for bids. This policy saves the cost that would be incurred if alterations or installations were required after a facility was in operation or if the company failed to meet local and state fire, safety, and health regulations.

Most local ordinances and state or provincial laws require that governmental authorities review and approve building plans for normal and emergency exits. In some states, the proper authorities must approve plans for the installation of emergency lighting, fire alarms, and automatic sprinkler systems. Also, in certain states or provinces, exhaust and ventilating installations must also be approved.

Many national and local codes require means of controlling air-polluting industrial contaminants. In the United States, the Environmental Protection Agency has set specific emission standards for raw waste disposal and for industrial by-products. Check these codes and standards during the planning stage.

Various organizations have developed voluntary safety codes that establish standards for certain structures and equipment. Specifications for the construction of floor and wall openings and of railings, for example, are given in the American National Standards Institute's (ANSI) A1264.1, *Safety Requirements for Workplace Floor and Wall Openings, Stairs, and Railing Systems.* Proper electric wiring and electrical installations are covered in the NFPA 70, *National Electrical Code,* issued by the National Fire Protection Association (NFPA).

Requirements for fire-extinguishing equipment and fire protection standards and codes for flammable liquids and gases, combustible solids, dusts, chemicals, and explosives are provided in the *National Fire Codes,* developed by the NFPA. Be sure to check the latest edition of the standards and codes. See also Chapter 11, Fire Protection, in this volume. For the addresses of the appropriate agencies in countries other than the United States, see the Appendix, Sources of Help, in the *Administration & Programs* volume.

Remember that "design by the code" is no substitute for intelligent engineering. Codes merely establish a minimum requirement, which in many situations must be exceeded.

SITE SELECTION

By carefully planning the site of a new facility, safety and health professionals can help to ensure that the facility meets the health and safety standards recommended by local, state, and federal building codes. When planning a site for a facility, consider the following factors:

- the relationship of the new structures to climate and terrain
- space requirements
- type and size of buildings
- necessary disposal facilities
- transportation to and from facilities
- market
- labor supply
- hazards to the community

Relief models of the site, made to scale, in addition to maps can help planners design safety features into the facility's system before the construction stage begins. Safety and health professionals can also help planners spot potential safety problems in the predesign stage.

Location, Climate, and Terrain

Planners should study the climate and terrain of the site where the facility will be located. The prevailing winds, for example, may determine the best location for placing processing equipment in relation to administrative offices and to the population in the area. In areas prone to hurricanes, tornadoes, earthquakes, or floods, plans and specifications should include protective measures for personnel. Such safety factors must be designed into the facility's plans. (See the *Administration & Programs* volume, Chapter 15, Emergency Preparedness. Check also Environmental Protection Agency hazardous waste conformity issues in nonattainment areas. See Chapter 9 in the *Environmental Management* volume of the *Accident Prevention Manual.)*

Space Requirements

Fire protection codes specify minimum distances between buildings according to their size, type, and occupancy. Laws governing the storage of explosives and other highly flammable materials (15 ft. to 200 ft. depending on the materials and the amount of materials, NFPA Table 2.51.d) specify minimum distances between manufacturing areas and storage facilities for such materials. Minimum distances between both toxic and flammable materials and adjoining property are also specified.

The size of a site may be determined by both present space requirements and possible future expansion. For example, some companies, anticipating increased use of air transportation, allow space for landing fields or heliports as they acquire new property. Plans for such installations should include all necessary safety precautions, as well as ample space for outdoor storage areas in the site plan. When the area for storing materials adjacent to shops proves insufficient, provide space elsewhere. Storing materials away from the facility, however, requires added handling and transportation and increases both costs and the risk of accidents.

Parking lots are best located inside the facility's fence for the convenience, protection, and safety of employees and visitors. Since a considerable area may be necessary for employees' and visitors' vehicles, consider parking requirements during site planning.

Well-located disposal areas for solid and liquid wastes must also be provided when a site is being laid out. Plan drainage and waste disposal in relation to space, terrain, and facility needs, as well as the effect upon surrounding municipal systems.

OUTSIDE FACILITIES

When planning outside facilities, safety and health professionals should keep safety precautions in mind to reduce the chances of accidents. Several outside facilities and their safety considerations are discussed below.

Enclosures and Entrances

A fence around yards and grounds serves many purposes. A fence keeps out trespassers who may interfere with work or be injured on the property. In addition, fencing also protects employees and others from transformer stations, pits, sumps, stream banks (under certain circumstances), and similar dangerous places. A galvanized, woven-wire fence makes a good enclosure.

Enough entrances should be planned to accommodate the facility's traffic volume. Entrances should provide clearance for loaded trucks and for switching personnel riding on the sides of railroad cars. Good visibility in all directions is essential at all entrances.

Because it is unsafe for pedestrians to use the same entrances as railroad cars and motor vehicles, provide separate gates that are convenient to pedestrians' transportation and workplaces. If a pedestrian entrance must be located near railroad tracks, fence part of the right-of-way. This will prevent employees from taking shortcuts along the tracks. If pedestrian entrances must be located on busy thoroughfares, or if workers cross railroad tracks on which trains are frequently operated, install traffic signals and build subways or pedestrian bridges (Figure 2–1). Such precautions are especially important when parking lots are located at a distance from the facility.

Shipping and Receiving

Shipping and receiving facilities should mesh with the overall flow of materials within the company or facility. They should also aid the efficient flow of materials into and out of production areas. Design shipping and receiving areas to minimize building heat and cooling losses. The use of self-leveling dock boards, truck levelers, and cranes speeds up loading and unloading.

Railroad sidings—commonly used as shipping and receiving facilities—require planning, especially if it is advantageous to use bulk raw, process, and maintenance materials. Tank-car lots of hazardous materials require special consideration for pressure piping, breakaway piping, valves, pumps, derails, excess-flow valves, and vapor return lines. Each sidetrack should be protected from main line and public thoroughfares. Proper clearance between main facilities and cars should be observed. (See the Facility Railways section later in this chapter and Chapter 12, Flammable and Combustible Liquids, later in this volume.)

Roadways and Walkways

The safety and health professional should help the civil engineer design for optimum safety. Roadways in facility yards and grounds are sources of frequent accidents unless they are carefully laid out, well constructed, well surfaced and drained, and kept in good condition.

Roadways

Hauling by heavy-duty trucks requires roadways up to 50 ft (15 m) wide for two-way traffic, with ample radii at curves. Grades, in general, are limited to a maximum of 8%. A slight crown is necessary for drainage, with ditches to carry off water.

Locate roadways at least 35 ft (11 m) from buildings, especially at entrances. At loading docks, an allowance of 1½ truck lengths is desirable to make backing up easier.

Figure 2–1. A bridge provides safe crossover of a freight yard. Be sure construction and personnel protection are adequate. (Courtesy Guardian Engineering & Development Company.)

The regulation and control of traffic signs, road layout, and markings should conform to federal and state or provincial practices. U.S. Dept. of Transportation, *Manual on Uniform Traffic Control Devices for Streets and Highways,* provides guidelines on these matters.

Traffic signs and signals are essential to regulate speed and movement at hazardous locations. Stop signs are specified for railroad crossings and for entrances to main thoroughfares. SOUND YOUR HORN signs are necessary at sharp curves (blind corners)—where view is obstructed—and at entrances to buildings. Convex mirrors mounted on the sides of buildings to afford views around sharp turns or around corners of buildings help prevent accidents if roadways must be built close to buildings. Use barricades and MEN WORKING signs at construction and repair sites. Traffic signs for roadways used at night should be made of reflective or luminous materials.

Walkways

Good walkways between outside facilities help prevent injuries to employees by helping them avoid stepping on round stones or into holes and ruts in rough ground. Concrete is, therefore, preferred for sidewalks, especially in principal areas like entrances and between main buildings. To discourage shortcutting, walkways should be the shortest distance from one building to another. A walkway that must be next to railroad tracks should be separated from them by a fence or railing. Install warning signs at railroad crossings and other hazardous places. Locating walkways clear of the eaves of buildings reduces the danger of falling icicles. In the snowbelt, covered walkways increase comfort and safety. Keep walkways in good condition, especially where they cross railroad tracks, and clear them of ice and snow. If site plans call for bridges over streams, ditches, or other hazards, protect pedestrian traffic with a fence or with handrails 42 in. (1.1 m) high and intermediate rails (Figure 2–1). If the site is large enough, walkways should have provisions for employee joggers.

Trestles

If employees are required to perform duties on trestles, provide a footwalk 5 ft 1 in. (1.5 m) wide, measured from the nearest rail, on at least one side. The footwalk should have a railing with the top edge height of the top rails, or equivalent guardrail system members, 42 in. (1.1 m) plus or minus 3 in. (8 cm), above walking/working levels. Midrails shall be installed at a height midway between the top edge of the guardrail system and the walking/working surface, and, on the exposed side, toeboards should be 4 in. (10 cm) high. (See 29 *CFR* 1926.502 (b) (1) and (b) (2) (i).)

If employees travel on both sides of the track, place crosswalks at frequent and convenient locations. Metal gratings or screens installed over walkways or passages under trestles protect employees from falling materials. Openings for conveyors or hoppers require gratings, or a grizzly with bars spaced not more than 12 in. (30.5 cm) apart, to prevent employees from falling into the openings.

Parking Lots

To reduce travel within the facility's grounds, locate the parking lot between an entrance to the grounds and the employees' locker room. If possible, the location should be such that no one need cross a roadway to go from the parking lot to a building. For security, fence the entire parking area and separate it from other areas of the facility. The surface of the parking lot should be smooth and hard to prevent injuries that might occur from falls on stony or rough ground. Keep lots as level as possible with enough slope for drainage.

The use of white lines, 4 to 6 in. (10 to 15 cm) wide, to designate stalls reduces confusion, as well as the number of accidents caused when backing up. Standard stalls are 9 ft (2.7 m) wide and 20 ft (6.1 m) long. The center-to-center distance between parked vehicles depends upon the method of parking.

Angle parking has both advantages and disadvantages. The smaller the angle, the fewer the number of vehicles that can be parked in the same area. Aisle widths can be narrower, but traffic is usually restricted to one way. On the other hand, angle parking is easier for drivers and does not require a lot of space for sharp turns.

The area allowed per vehicle in parking lots varies from 200 ft^2 (19 m^2) to more than 300 ft^2 (28 m^2) if aisles are included. Large, economically laid out lots may approach the 200-ft^2 (19-m^2) figure; small or poorly configured lots may have a higher percentage of aisle space and may approach 300 ft^2 (28 m^2) per vehicle. A large, commercial parking lot, with attendant, is considered efficient if the layout keeps the space requirements to 240 ft^2(22 m^2) per vehicle.

The Americans with Disabilities Act (ADA) specifies the size, location, and number of parking stalls reserved for vehicles of disabled persons. Stalls must be 8 ft wide, next to a 5-ft wide access aisle. Two 8-ft stalls may share an access aisle. The number of accessible stalls is based upon the total number of stalls in the parking lot. For more details, see "Access to Buildings" in Chapter 13, Workers with Disabilities, and the ADA.

For orderly traffic movement, parking lots should have separate entrances for incoming and exits for outgoing vehicles. Designate such entrances with suitable signs. Control traffic at exits to heavily traveled streets either with a traffic light or an acceleration or merging lane. Be sure the parking area does not encroach on fire hydrant zones, approaches to corners, bus stops, loading zones, and clearance spaces for islands. Driveways should be a minimum of 25 ft (7.5 m) wide for two-way traffic, and the view should not be obstructed.

Install speed signs and signs limiting parking areas to employees or visitors, as needed. These signs should conform to recommended standards and should be similar to other street and highway signs. U.S. Dept. of Transportation, *Manual on Uniform Traffic Control Devices for Streets and Highways,* gives details for signs and pavement markings.

If the lot is used at night, provide adequate lighting for safety and for the prevention of theft. About 1 to 5 footcandles per ft^2 (11 to 54 lux per m^2) at a height of 36 in. (90 cm) should be adequate.

During the planning stage, include provisions for removing ice and snow from the lot.

Landscaping

Many companies landscape the grounds of both old and new facilities. Design landscaping so that trees and shrubbery do not create blind spots at roadway or walkway intersections. Proper maintenance is required to prevent bushes from creating blind spots. (See Chapter 3, Construction of Facilities, in this volume for details on grounds maintenance.)

Waste Disposal

Unsafe methods of waste disposal may cause injuries to workers and the public and damage to property. Knowing the nature of the wastes is essential for planning suitable disposal methods. These disposal methods must also conform to applicable municipal and state or provincial regulations. Since investigation, treatment, and disposal of wastes require specialized knowledge and training, safety and health professionals should consult qualified engineers.

If the use of a city or district sewage system is planned, the officials in charge of the facility shall be informed of the kind and amount of the wastes. If the wastes have properties that will interfere with operation of the sewage disposal facility, the officials may refuse to accept them.

Under no circumstances should toxic, corrosive, flammable, volatile, or radioactive wastes be drained into a public sewage system. It is especially important that facilities handling or processing radioactive materials—even in small quantities—conform exactly to local and state or provincial regulations for disposal of such wastes.

Many wastes can be disposed of on landfills, if state or provincial and local laws permit. Chemical wastes should be rendered harmless before disposal. Strong acids, for example, should be neutralized. Poisonous materials, magnesium chips, explosives, and similar substances require special procedures for safe disposal. Combustible materials, such as wood, scraps, and paper, may be burned in an incinerator. An incinerator shall conform to applicable laws, be safely located, and properly attended.

In some cases, a safety and health professional might contract a private scavenger service to dispose of waste materials. If such a service is contracted, it must be told if any of the materials is hazardous. There are special scavenger services that will take hazardous substances and dispose of them properly.

Air Pollution

Smoke and inert dust from various sources may be nuisances or even sources of danger to the public. Comply with smoke control ordinances before construction begins. Check emissions against federal and local standards to be certain that these standards will not be violated.

Toxic smoke, fumes, or dust is a serious problem in some industries. Tall stacks often are used to diffuse gases in the atmosphere. The effectiveness of this method depends on the nature and volume of the gases, the location of the facility, the prevailing wind direction, and atmospheric conditions. Rain may absorb harmful gases and cause heavy damage to crops and the environment. Conditions causing poor diffusion occasionally have temporarily shut down even those facilities that have high stacks.

Investigate the possibility of waste minimization, pollution prevention strategies, and recovering usable or marketable materials from wastes. Dust and fumes often are recovered by filters, cyclones, electronic precipitators, or similar equipment either for their intrinsic value or to keep them from polluting the atmosphere. Spray towers or other recovery equipment can often economically recover useful gases and vapors.

Confined Spaces

The hazards associated with confined spaces have been responsible for many fatal and serious injuries to workers and would-be rescuers. Each year, there are almost two fatalities for each confined space incident reported. The most common underlying cause of these accidents is insufficient preparation for undetected hazards.

A confined space is any area that is not designed for continuous human occupancy. It has limited access and ventilation. It also is susceptible to hazards such as inundation of water, gas, or solid particulate. Or, it may have sloping sides such as a bin or hopper that leads to a crusher, auger, or restriction. Other hazards include (but are not limited to) bridging of material, electrical hazards, oxygen deficiency, falling from an elevation, radiation, toxic gas or vapor, and fire or explosion.

To address the problem of confined space entry, confined spaces must first be identified. Carefully review all operations at the facility. If a company works with tanks, silos, boilers, pits, manholes, trenches, chemical storerooms, or any location that presents an access problem, it should post warnings at these locations to communicate the danger to employees.

Second, identify all the potential hazards for each location and methods to eliminate or mitigate those hazards when work must be performed at that location.

Third, develop a confined space work permit form. This form should include information that a confined space entrant and attendant must know to safely work in the confined space. Consult OSHA (29 *CFR* 1910.146) for the specific information required on this form. In some cases, it will be necessary to complete a confined space work permit each time an entry is made. Other permits may authorize entry for up to one year. Each permit is site specific.

Fourth, train personnel in the dangers and proper observance of confined spaces. Train personnel who must enter and work in these confined spaces in methods to make entry safe. This will include training in ventilation techniques and systems, respiratory protection (Figure 2–2), atmosphere testing (Figure 2–3), lockout/tagout procedures, use of protective equipment (Figure 2–4), and evacuation procedures. These personnel should be cross-trained in the duties and responsibilities of the outside attendant. The outside attendant should maintain continuous communications with the entrant and should, under no circumstances, enter the confined space to assist or rescue the entrant. The attendant is the entrant's only link to the outside world and to emergency assistance.

Finally, ensure that a trained and equipped rescue team is available to respond to an emergency. This team may consist of company personnel, or it may be a contracted service. Regardless of the source, the team must be able to respond quickly and have knowledge of the potential hazards at the client's location. Practice rescue simulations using real workers or mannequins provide necessary training for a rescue team but must be done carefully to avoid injuries (Figure 2–4).

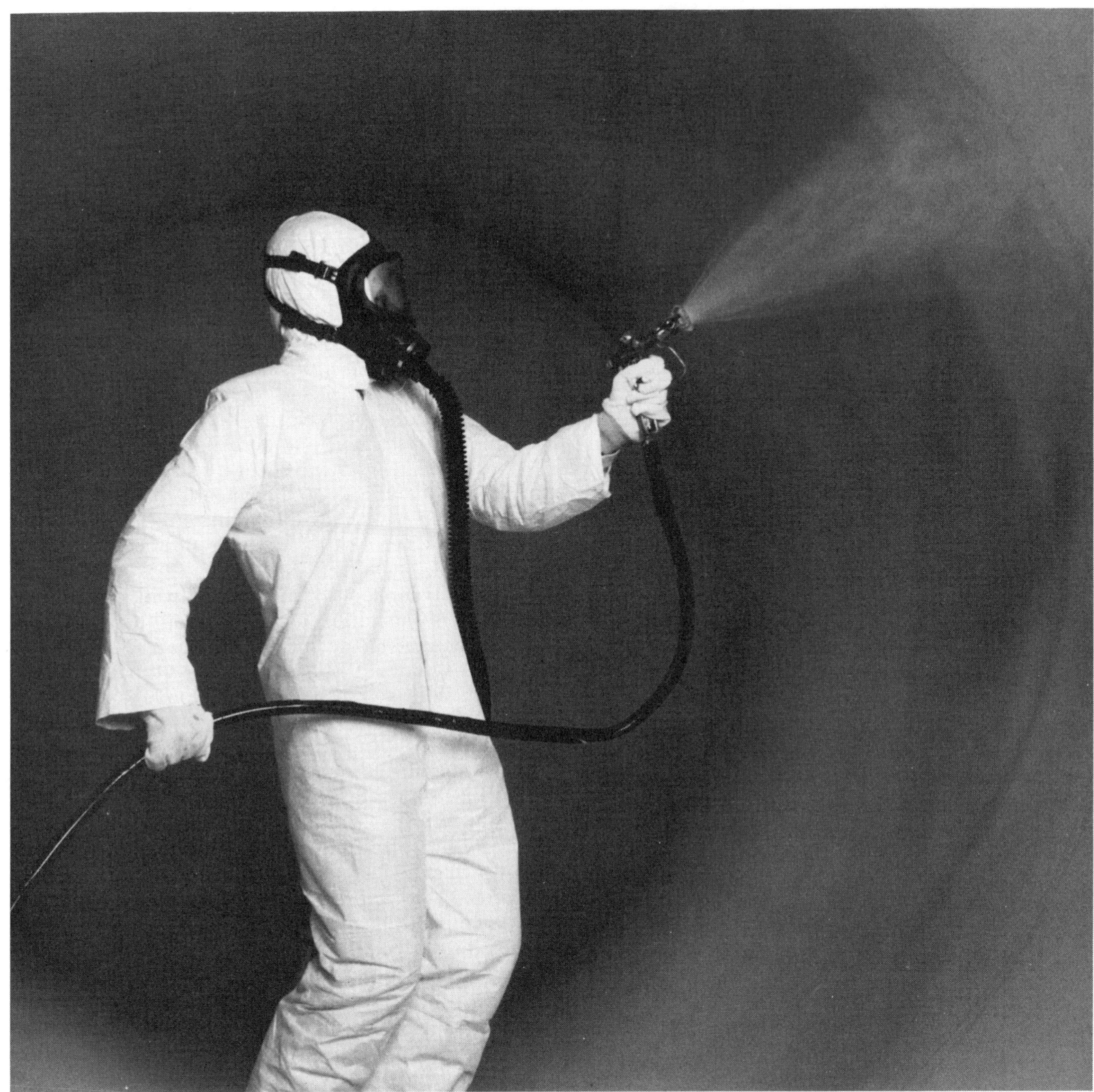

Figure 2–2. The operator is wearing a compete coverall, gloves, and a supplied air respirator while spraying inside a tank. (Courtesy Bridgeport Chemical Corporation.)

Outside Lighting

Outside lighting should serve not only to aid production but should also function as a safety measure and as part of the facility's security system. Maintenance personnel should adjust lighting as daylight hours shorten. Lighting units using different types of light sources are available for specialized applications. Consider lamp life and ease of maintenance when selecting the lighting units. Lighting equipment used in outdoor locations must withstand exposure to the elements without deteriorating. (See the section, Lighting, later in this chapter.)

Figure 2–3. Before entering a confined space, test for toxic gas or vapor and oxygen deficiency. (Courtesy Bridgeport Chemical Corporation.)

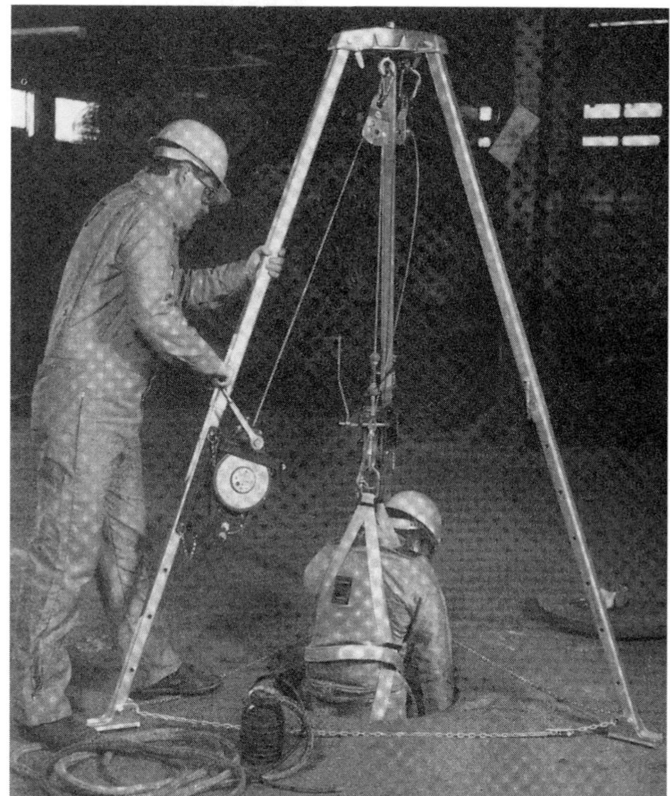

Figure 2–4. Rescue personnel should practice simulated rescues to become proficient with the equipment and procedures. This rescuer uses a hoist and winder tripod to assist the rescue of a disabled worker from a manhole. (Courtesy Miller Equipment.)

Docks and Wharves

The bottom of a sea, lake, or river is an important factor in the design and construction of docks and wharves. A soft, deep bottom limits the use of concrete and of heavy fire-resistant materials. Wood piles must be protected if marine borers are present. Flexibility and elasticity are essential in tidal waters and where waves force vessels against piers. Ice build-up from tidal action is another structural consideration.

Safety requirements on docks and wharves include good illumination for night work, a floor that will withstand heavy trucking, and traffic control equipment. When designing piers, take into account the speed and size of vehicles to be operated on them.

FACILITY RAILWAYS

Insofar as possible, eliminate facility railway hazards when a new facility is being designed. Horizontal and vertical track clearances and proper installation of track, fittings, and structures are primary considerations.

For all phases of track construction, the American Railway Engineering Association's (AREA) recommended practices are an excellent guide. However, some state or local regulations may differ from AREA recommendations for overhead clearances.

Clearances and Warning Methods

Where platforms, building entrances, or structures are located along curved track, allow additional clearance on both sides of the curve for sideways movement of railcars. Also allow additional track clearance when awnings or jalousie windows are installed in buildings adjacent to the track. Place CLOSE CLEARANCE warning signs at points where buildings or other obstructions are so close to a track that a person riding on the side of a car will not clear them. In addition, install a warning light and allow it to remain lit at all times.

Standard clearances may not be sufficient, especially if tracks pass doorways, corners of buildings, or other places where workers could walk directly onto tracks in front of moving cars. Safeguard these locations with fixed railings that force workers to detour a short distance before stepping onto the track. If a barrier railing is impractical, use hinged bars or gates that swing horizontally through an angle of not more than 90 degrees.

Another way to protect workers is to install convex mirrors at a 45-degree angle at intersections of passageways and the track. In this way, workers can see

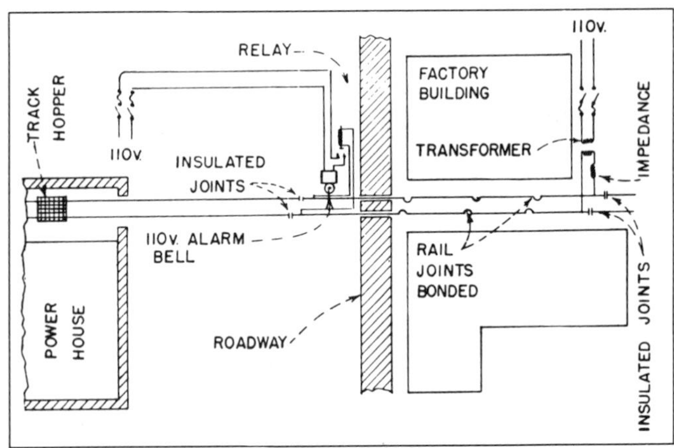

Figure 2–5. A layout of a bell warning system for blind level crossings inside the plant.

approaching railway equipment before they reach the intersection.

Various methods are used to warn workers at railway crossings *inside* facilities. Here automatic blinking lights and gongs or bells are more effective than signs. To safeguard workers on heavily traveled passageways, use both gates equipped with red lights and crossing guards provided with a whistle and a shielded red lantern. Increase the visibility of gates by painting them with alternate red and white stripes at 45-degree angles. Another means of protecting workers is the bell warning system shown in Figure 2–5.

In facility yards where facility or railroad personnel may be required to switch cars, or perform other work, at night, provide adequate lighting. Install high-intensity, high-pole lights, arranged so as to cast as few shadows as possible.

Track

Although the regulations of the Federal Railroad Administration (FRA) are not mandatory for most industrial facility railways, FRA standards and AREA recommended practices are excellent guides. However, these recommended practices usually apply only to standard gauge track. Inspect rails and fittings periodically, and repair them as required. Serious accidents could result if defects remain unrepaired.

Tracks

Where possible, arrange for tracks to be level at loading points. Workers have been killed by railcars rolling down slight slopes. If it is necessary to spot cars on grades, set brakes tightly and use car blockers, rail clamps, or track skates. Where tracks are at a dead end, install a standard bumping post or earth mound.

Derails

Install derails at the bottom of steep slopes and where sloping switch tracks connect to main lines. Also install derails on the approach to permanent shipping and receiving areas, whether located within or outside a building. Where such areas are on tracks that are open at both ends, place derails beyond both ends of the shipping and receiving area. Do not locate derails in hard-paved areas. This type of surface will defeat the purpose of the derail.

Trestles

Provide trestles with a footwalk not less than 5 ft 1 in. (1.5 m) wide. The footwalk should have a railing with the top edge height of the top rails, or equivalent guardrail system members, 42 in. (1.1 m) plus or minus 3 in. (8 cm) above walking/working levels. Midrails shall be installed at a height midway between the top edge of the guardrail system and the walking/working surface, and, on the exposed side, toeboards should be 4 in. (10 cm) high. (See 29 *CFR* 1926.502 (b) (1) and (b) (2) (i).) Railings should meet the standard side-clearance required by the railroad or the state or province. If footwalks are necessary on both sides of the track, build crosswalks to connect them. Design and build trestles to carry anticipated loads and to withstand vibration and shock.

Switches

Targets should have rounded corners to lessen the possibility of cuts, scratches, and torn clothing. Provide switch lamps or reflectorized targets if tracks are to be used at night. Install blocking in switch points and use frogs to prevent employees from getting their feet caught.

Covers in Open Areas

Cover hoppers or trackside bins into which material is dumped by spacing heavy steel bars over them. Openings at ground level for conveyors and similar equipment used to unload cars should also have covers that are kept in place when equipment is not in use. This will prevent workers from falling, or being carried, through the openings.

Cover walkways that are under trestles, in order to protect those who use them from being hit by falling materials.

Loading and Unloading

Tracks at loading and unloading areas warrant special attention because many accidents occur when loads are moved into and out of railcars. Clearance for trucks requires heavy, large bridge plates or dockboards. If the level of the dock is considerably above or below the level of railcar door openings, truck movement is more hazardous.

The dock should be wide enough to provide a temporary storage area without interfering with truck movements. A narrow dock may force trucks to turn onto dockboards at a dangerous angle. Reduce this hazard somewhat by making dockboards wider at the dock side, with flanges on the plates to turn wheels away from the edge. Portable dockboards must be securely anchored and be strong enough to carry the load imposed on them. Provide handholds or other devices that permit safe loading and unloading.

Protect cars spotted for loading or unloading from being moved by switching crews. Standard blue flags for daytime and blue lights for night use furnish warnings to train crews. Place signals between the rails at both ends of a car that is accessible from either direction. Strictly prohibit train crews from coupling engines or cars to any cars so protected. Only the employees engaged in the loading or unloading operations should remove the blue signals, and only when they are ready to release the cars. (See 49 *CFR* 218.21d.)

To warn personnel that switching operations are going on, install bells and oscillating warning lights along the tracks in work areas. Facility supervisors should turn on the warning lights and bells before switching operations are begun. Facility management is responsible for clearing employees from railroad cars before releasing the track to railroad employees for switching. Therefore, before derails and blue flags are removed, facility supervisors must make sure that

- building doors are opened and other obstructions removed to provide standard side clearance from the track area,
- all facility personnel are cleared from railroad cars and track area,
- all overhead building cranes in the area being switched have stopped operations and are clear of the tracks,
- all dockboards (bridge plates) are removed from railroad cars,
- all counterweighted, retractable, service platforms are retracted and secured,
- all equipment for moving railroad cars (cables, hooks, etc.) is removed from cars, and
- all car doors, hopper doors, etc. are closed and properly secured.

Use tracks meant for loading or unloading flammable liquids or other dangerous materials for those purposes only. For additional protection when loading or unloading dangerous materials, provide locks for switches. Specific recommendations for tank cars, grounding, rail bonding, and so on are given in Chapter 12, Flammable and Combustible Liquids, in this volume.

Figure 2–6. Plug doors should be locked before a railroad car is moved. (Courtesy Inland Steel Company.)

After unloading the car, the consignee is responsible for cleaning the car before releasing it. When pieces of crating and dunnage, nails, and strapping are left loose in the car, they become serious hazards to railroad employees and others who may have to enter it later. When railcars being loaded or unloaded are damaged or in need of repair, notify the rail carrier or switching crew so proper safety precautions may be taken.

In addition, it is the consignee's responsibility to make sure that the doors are properly closed and secured. Plug doors on box cars create a distinct hazard when the doors are not properly secured (Figure 2–6). Most rail carriers have issued instructions to switching crews not to move such cars until the doors have been secured.

Overhead-Crane Runways

A serious hazard exists at points where an overhead-crane runway crosses above a railroad track, either inside or outside a building. The cranes' loads or hook blocks may strike locomotives or cars while switching is going on.

To guard against movement of the crane near an occupied track, install a system of interlocked signal lights, with one set visible to the crane operator and the other to the switch crew. All personnel involved, especially crane operators, must be trained to respect the signals without exception. The signals can be started by hand with a key switch. The area supervisor should keep the key. The signals can also be interlocked with a derail.

Another way to control a crane's movement is to provide a zone power cutoff for the runways in the vicinity of the track. The cutoff is started by a key switch under the control of the area supervisor.

Require switch crews to get clearance from the area supervisor before moving into the overhead-crane area. Make supervisors responsible for keeping cranes clear until the switch engine and cars move out. (See also Chapter 15, Hoisting and Conveying Equipment, in this volume, for information on overhead cranes.)

Types of Motive Power

The type of motive power for a facility railway is important for the prevention of accidents and fire. Explosive gases are easily ignited by flames or sparks from fuel-fired locomotives. Where such gases may be present, use locomotives powered by electric, compressed air, or storage batteries.

Also, where ventilation is insufficient to keep the concentration of noxious and even toxic exhaust gases at a safe level, as in mines, prohibit the use of fuel-fired locomotives. Diesel engines, however, can be equipped with devices to eliminate toxic gases from the exhaust. This type of locomotive is being used in some adequately ventilated mines, but not in coal mines.

A major hazard with electric locomotives, in addition to that of sparking in explosive atmospheres, is the risk of employees touching the overhead wires or third rail. Place guards at all points where employees can contact electrified equipment, or have the wire high enough to prevent employees from coming into contact with it.

Boilers of steam locomotives are constructed in accordance with the American Society of Mechanical Engineers' (ASME) *Boiler and Pressure Vessel Code.* (See References at the end of this chapter.) They are inspected in accordance with *Laws, Rules and Instructions for Inspection and Testing in Steam Locomotives and Tenders and Their Appurtenances.* (See References at the end of this chapter.) Although the regulations of the FRA are not mandatory for most industrial facility railways, the standards are an excellent guide.

Equip diesel locomotives, which operate more quietly than steam locomotives, with bells. Some companies paint the front and rear of diesel locomotives in contrasting colors, such as yellow and black stripes. Install handrails around the outside of deck walks. Provide each locomotive with an extinguisher for oil fires.

The tanks of compressed-air locomotives should be constructed in accordance with the *ASME Boiler and Pressure Vessel Code.* Each air receiver requires an air pressure gauge, a pop safety valve, and a drainpipe with a valve at the bottom.

A battery-powered locomotive should have a deadman control. In that way, the operating lever will return automatically to the OFF position when released, or when the operator leaves the controls and the circuit is broken. Safety provisions for battery-charging rooms are covered in Chapter 17, Powered Industrial Trucks, in this volume.

Equip all facility locomotives, and cars as well, with safety appliances and with standard automatic couplers and air brakes, as required by federal law for common-carrier railroads. Maintain all equipment and appliances in sound, safe operating condition.

Tools and Appliances

Using the correct tools and appliances is important for operating a facility's railway safely. For example, automatic couplers eliminate the serious hazard of workers going between either standard- or narrow-gauge cars to insert or withdraw pins by hand.

Use only standard car wrenches to open cars with hopper bottoms. Using a special tool to close the latches on bulk cars eliminates the need of a person risking a fall from going on top of the car. Provide rerailers for narrow-gauge cars; other means of rerailing cars are generally dangerous.

Car Movers

Using hand methods to move cars often results in accidents. A safe procedure is to use a switch engine. If the ordinary hand car mover is used, however, place a shield around the bar. In that way, employees will not strike their hands or otherwise injure themselves, should the tool slip. Crowbars, push-poles, and other makeshift tools should not be used to move cars.

When a car is on a grade, test the hand brake to make sure that it takes hold and that excess slack in the brake's chain is taken up. A worker should remain on the brake platform to stop the car with the brake at the required point.

Although winch-type car pullers are used extensively, should the rope break, the operator could be killed. Because the operator is in line with the rope while operating the equipment, install a shield of steel plate or expanded metal for protection. A forged-steel hook should fasten the rope to the car.

A self-propelled, rider-operated car mover is used in many industries (Figure 2–7). This vehicle has both rubber-tired wheels and steel rail wheels. Both sets of wheels are retractable so the car mover can run either on the facility's grounds or on rails. This type of car mover is fitted with standard couplers and, depending upon the model, can develop a drawbar pull of from 8,400 to 18,000 lb (38 to 80 kN). Its use eliminates the need to use hand car movers, winches, capstans, or power equipment, none of which are designed for the purpose and all of which are often hazardous.

Figure 2–7. Self-propelled, rider-operated car mover. (Courtesy Whiting Corp.)

Safe Practices

Transportation personnel, as well as all other employees, should observe the same safety rules followed by all large railroad systems. When safety problems arise about the facility's railway, seek help and suggestions from safety professionals and from operating officers of the facility's connecting railway line. Hold safety meetings and other educational activities to inform the facility's personnel about railway hazards.

To include all the safe practices required for the safe operation of railway equipment would require a book in itself. However, the following are a few of the more important ones:

1. Stop and look both ways before crossing any track.
2. Expect trains or cars to move at any time, on any track, and in either direction.
3. Step *over* rails when crossing tracks. Never step, walk, or sit on any rail.
4. Never go between moving cars, or cars that may move, to adjust couplers or for any other purpose. (The once-common practice of kicking couplers to align them is especially dangerous.)
5. Give a hand or lamp signal to stop. Receive an acknowledgment of the signal before going between standing engines or cars.
6. To close a boxcar door, place one hand on the door handle and the other on the back end of the door.
7. Step down from cars—do not jump.

Railroads often set individual standards for operations on their own properties. Information on these standards may be available from the appropriate railroad safety department.

FACILITY LAYOUT

Size, shape, location, construction, and layout of buildings and facilities should permit the most efficient use of materials, processes, and methods. Safety and health professionals should aim for efficient production with maximum employee safety. Some principles to consider when planning internal facility layout follow:

- Employees should recognize how materials, people, and products flow through the facility.
- Employees should easily learn where things are located.
- Employees should move easily within, and to and from, the facility.
- Conveniently locate services for employees, such as the lunchroom, locker room, and personnel office.
- Locate supervision offices near the work area so that communication with employees is as easy as possible.
- Provide physical separation between areas that are extremely noisy or hazardous.

Location of Buildings and Structures

To minimize the hazards of fires and explosions, store both raw materials and finished products away from processing buildings. Storage of volatile, flammable liquids and LP gas in an area apart from processing buildings reduces the hazard of fire. In the event of fire, however, control is more easily achieved. Moreover, the cost of separate storage eventually may be less than an investment in storage in a processing building. Provide ample space between storage units and such flame sources as boilers, shops, streets, and adjoining property. Follow the codes of local and state or provincial authorities and of the NFPA in planning the location of the units of a facility.

Federal, state, and local laws govern the storage of explosives. Locate and build magazines according to the recommendations of the Institute of Makers of Explosives. The type of retardant required in relation to the distance between buildings of frame, brick, and fire-resistant construction, as well as the minimum separation distances between buildings, are given in NFPA 80A, *Exterior Fire Exposures*.

Many facilities use and store flammable liquids having flash points below 100 F (37.8 C). Plans and layouts for such facilities should conform strictly to flammable liquids handling and storage specifications developed by the NFPA 30, *Flammable and Combustible Liquids Code*, and to the requirements of local fire prevention authorities. NFPA 30 specifies the conditions under which flammable and combustible liquids of various classes can be stored in and around buildings. (See also Chapter 12, Flammable and Combustible Liquids.)

Exacting specifications must be met when storing flammable liquids outside of buildings in underground or aboveground tanks. Check current federal and state or provincial regulations regarding the amount and kinds of liquids that are allowed in underground tanks. Standards for tanks include type and thickness of material, provisions for relieving excessive internal pressure, and details about grounding, insulation, and piping. The material used to construct storage tanks depends upon the type of liquid stored, its corrosive properties, and the processing requirements.

The required distance between buildings on adjoining property and the location of an aboveground tank depends upon the content, construction, fire-extinguishing equipment, and greatest dimension (diameter or height) of the tank. A secondary containment vault, properly lined with monitors, may be required. NFPA 30 specifies four different groups of tanks and the minimum distances apart for each.

An automatic sprinkler system is considered primary protection for hazardous installations. Other safety features include spark-resistant conductive floor surfaces and grounding of equipment and structures. All tools, trucks, and similar equipment should be made from spark-resistant metal. Emergency exit doors should be provided. (See Chapter 10, Electrical Safety, and NFPA 30 for an explanation of equipment designed for use in hazardous locations.)

Layout of Equipment

A detailed flow sheet is a useful guide for laying out facilities, particularly those using dangerous and harmful materials and complicated processes. With a flow sheet, the nature of the materials and the process in each manufacturing stage can be studied and provisions made to eliminate or control hazards.

Other methods used to determine the safest and most efficient layout of machines and equipment include the two-dimensional method and the three-dimensional model. The two-dimensional method consists of templates made to scale and fitted into a plan of the site, or floor area. The most effective method is the use of three-dimensional models made to scale and set up on a scaled floor plan. The models can be rearranged until the safest and most efficient layout has been devised. Computer simulations of arrangements should also be considered.

Layout studies will show the most suitable locations for operations like spray painting, welding, and other work that generally requires separate areas. Use the three-dimensional model to anticipate and avoid congested areas. The frequent handling of materials in these areas leads to many unnecessary movements and to bad housekeeping—in itself a great source of accidents.

Insufficient headroom at aisles, platforms, pipelines, overhead conveyors, and other installations can also be discovered by studying the models. A vertical distance of at least 7 ft (2.1 m) is generally specified to provide ample clearance between passageways and stairways and overhead structures. Overhead cranes and conveyors require at least 24 in. (61 cm) of vertical and horizontal clearance.

Integrated computer systems require an area with controlled temperature, good ventilation, and electrical power. Layout studies will also reveal the best location for these systems.

Aisles

Trucks and other vehicles require ample room for movement—backing up and turning—without endangering workers and equipment. Aisles should be wide enough to permit trucks to pass without colliding. For one-way traffic, aisles should not be less than 3 ft (0.9 m) wider than the widest vehicle. Aisles for two-way traffic should not be less than 3 ft (0.9 m) wider than twice the width of the widest vehicle. These minimum widths are exceeded considerably in some new buildings where, because of anticipated heavy traffic, aisles from 12 to 20 ft (3.7 to 6.1 m) wide have been specified. A safe layout of aisles requires an absence of blind corners and an adequate radius to allow for vehicles to turn. A 6 ft (1.8 m) radius is enough for small industrial trucks (Figure 2–8).

Clearly define aisles with approved markings in either traffic paint or striping material. ANSI Z53.1, *Safety Color Code for Marking Physical Hazards,* provides guidelines on markings. (See the recommendations under Use of Color later in this chapter.) Many companies use plastic buttons that are glued or fastened to the floor with metal fasteners because of their durability.

Parking and Ramps

Designate parking areas for both hand and power trucks. Large facilities often provide garages with room for both storage and maintenance. Also provide battery charging rooms. (See Chapter 17, Powered Industrial Trucks.)

Figure 2–8. This aisle is wide enough for small industrial trucks and well marked.

If building plans include ramps for use by both pedestrians and vehicles, reserve a 3-ft (0.9-m) wide section as a walkway. Sharp turns into aisles at the top and bottom of ramps are hazardous and should be avoided. Provide an abrasive coating where floors could become slippery.

Electrical Equipment

Completely metal-enclosed and grounded unit substations have been developed for industrial facilities. However, if transformers must be installed in confined areas or near flammable materials, be sure they are noncombustible transformers. Furthermore, if these substations are in enclosed areas, provide ventilation to reduce the buildup of gases and the release of heat.

Short-circuit protective devices should have the capacity to carry the load and should be rated to open without any danger from the maximum short-circuit current. Circuit breakers, fuses, and safety switches that fail under short-circuit current may explode and cause serious damage and injuries. Use motor controls capable of lockout/tagout and other electrical safeguards.

Design safety measures for grounding systems and battery rooms. Measure every grounding system to determine whether or not the system is capable of conducting the necessary amount of return. When direct-current voltage is supplied from batteries, isolate the battery room from the work area. A battery room should be well ventilated; smoking should be prohibited.

In an up-to-date industrial electrical system, sections may be deenergized for maintenance and other work without shutting down the entire system. In addition, all electrical installations must conform to the NFPA 70, *National Electrical Code,* as well as local ordinances. (See also Chapter 10, Electrical Safety.)

Ventilating, Heating, and Air Conditioning

Ventilating, heating, and air conditioning not only provide personal comfort but also are often needed for safe processing conditions. Personal comfort affects workers' efficiency. Make every effort, therefore, not only to have general office and facility conditions comfortable, but also to eliminate—or at least reduce—poor conditions that can contribute to excessive employee fatigue and discomfort. In buildings where flammable liquids or vapors are handled, provide adequate ventilation to prevent forming explosive concentrations.

The noise and vibration caused by boilers, fans, and air conditioning equipment require that they be kept separate from general work areas. Boilers should receive adequate air and the by-products of combustion should be removed safely. Also, when locating incinerators, be sure that a negative pressure differential in a building does not cause an incinerator stack to serve as an air source.

For maintenance purposes, authorized employees should have easy access to this machinery. There must be sufficient space around the equipment for employees to replace parts. For example, there must be room to pull tubes if necessary. The National Safety Council (NSC) book, *Fundamentals of Industrial Hygiene,* 4th edition, provides more specific suggestions, as does NFPA 90A, *Installation of Air Conditioning and Ventilating Systems.*

Inside Storage

Space for raw materials and finished products may be estimated on the basis of maximum production requirements. Make allowances for shortages, seasonal shipping, and quantity purchases.

Modern mechanical handling and stacking equipment permits extensive use of vertical space through multiple decking. If this method is anticipated, be sure that flooring can support the maximum projected load. Also assure that new shelving/racking installations meet or exceed seismic code structural requirements.

Storage areas for supplies, finished products, and empty or full pallets should be easily reached by employees. Additional features should include stable piles or stacks, as well as a properly functioning fire-extinguishing system. The fire-extinguishing system's density and its area of application depend on the types of materials stored and the height of the piles or stacks.

Space near working areas for storing supplies, tools, flammable liquids, and infrequently used equipment is seldom included in layout plans. The result is that such items often are left in unsafe positions and locations. To discourage employees from leaning large materials or equipment against walls where they can fall, designate storage places for them.

Also provide space for racks, bins, and shelves. Store, above eye level, materials that project into aisles or walkways from racks, bins, and shelves. Use metal baskets or special racks provided with drip pans for storing machine parts covered with cutting oils. To store liquids, use confinement dikes, curbs, or drains. When spills occur, they will thus be prevented from affecting other areas of the facility.

Build closets for the storage of janitorial supplies such as waxes, soaps, and other cleaning supplies. Plan for these closets to have floor sinks for filling pails. Thus, personnel will not have to lift pails.

Storage of waste material may take considerable room, especially if the waste is bulky or is produced in large quantities. In buildings with basements, chutes to basement storage bins help prevent accumulations in work areas. To dispose of small quantities of sharp-edged waste, specify use of boxes with handles that are protected from contact with sharp material, or similar safety containers. (See Chapter 14, Materials Handling and Storage.)

LIGHTING

Electrical lighting properly complemented with daylight can supply a facility's lighting needs with energy savings.

Daylight

Daylight can be predicted using the procedures in *Calculation of Daylight Availability, Illuminating Engineering Society* (IES) publication RP-21. To use it to advantage, take into account the following design factors:

- variations in the amount and direction of incidental daylight
- brightness distribution of clear, cloudy, partly cloudy, and overcast skies
- variations in intensity of sunlight
- effect of local terrain, landscaping, snow reflectance, and nearby buildings on the available light
- geographical location and orientation of building

The natural light that enters and is made available for use inside a building depends on the design of the skylights and fenestration as well as the interior design and furnishings. Skylights and fenestration serve at least three useful purposes in industrial buildings: (1) provision for the admission, control, and distribution of daylight for vision; (2) provision for a distant focus to relax eye muscles; and (3) provision for reducing the claustrophobia some people experience in completely closed-in structures. Design and installation of windows should make provision for washing them easily and safely.

Skylights may also be installed to increase the natural lighting. Consideration of skylights must also include a cost-benefit analysis to evaluate the effect of the skylights on the structural, heating, ventilating, and air-conditioning systems as well as the illumination on task performance.

The amount of daylight, however, varies with the time of day, day of the year, and weather conditions. It is, therefore, necessary to install an electric lighting system that will supply the total requirements of the facility.

The primary advantage of daylighting is the reduction of electric energy required. To be effective, the daylight and electric light components must be blended to provide consistent proper illuminance for the employees' visual comfort and task performance. Consider using a lighting management control system to measure the quantity and quality of illumination.

Control systems, both manual and automatic, are available to vary the output of the electric lighting system. An electric lighting control system may make it possible to supplement the available daylight with the amount of artificial light needed to maintain the recommended levels of lighting for particular tasks. However, daylighting may be difficult to use because of the critical nature of many production tasks. An example of a good daylighting application is in warehouse and storage facilities where installation of skylights and simple photo-sensitive controls have proved effective. For a more comprehensive discussion of this subject, see the IES *Recommended Practice for Daylighting.*

Electric Lighting

The prime requirements for industrial lighting are to provide a safe working environment and to provide comfortable vision as an aid to all types of industrial operations. Under these conditions, workers are able to observe and effectively control the operations and maintenance of various types of machines and processes.

For most industrial work areas, a sufficient quantity of natural light is often not available, even under optimum daylight conditions. Therefore, electric lighting is required to maintain good conditions for vision. It is essential that the electric lighting system be designed and installed so as to provide the required level of illumination in areas adjacent to skylights, windows, or walls. Thus, good lighting may be assured over the entire work area.

Distribution of light from a luminaire (lighting unit) is important. Highly concentrated distributions make high mounting heights economically feasible. Low mounting heights, on the other hand, necessitate a widespread distribution of light.

There are four forms of electric lighting used in industrial areas:

1. General lighting produces relatively uniform illumination throughout the area involved. According to the *Guide on Interior Lighting,* publication 29.2 of the International Commission on Illumination, general lighting is designed so that the ratio of the minimum to the average illuminance should normally not be less than 0.8 in order to provide for equivalent task locations throughout the interior. The average illuminance of the general area should not be less than one-third of the average of the task areas. The average of the illuminances of adjacent interiors should not vary from each other by a ratio exceeding 5:1.
2. Care must be taken not to exceed the recommended Spacing Criterion (SC) or spacing-to-mounting-height (S/MH) ratios for the lighting equipment used. "Point-by-point" illuminance calculations are often used to check the uniformity of general lighting levels.
3. Localized general lighting is the alternative to luminaries uniformly spaced throughout the area to provide light at the working zones.
4. Supplementary lighting is used to provide higher illuminances for small or restricted areas where such illuminances cannot readily or economically be obtained by general lighting. Supplementary lighting is also used to furnish a specific brightness or a specific color, or to permit special aiming or positioning of light sources.
5. Emergency lighting must be planned carefully for every facility in accordance with ANSI/NFPA 101, *Life Safety Code.* Such lighting for means of egress, exits, and stairwells may be provided by battery-powered lighting equipment or by a complete standby generating system to power selected luminaries.

Quality of Illumination

Quality of illumination pertains to the distribution of brightness in the visual environment. Glare, diffusion, direction, uniformity, color, brightness, and brightness ratios all have a significant effect on visibility and the ability to see easily, accurately, and quickly. Installations with poor-quality lighting are uncomfortable and possibly hazardous. Moderate deficiencies are not readily detected, although the cumulative effect of even slight glare can cause significant fatigue, loss of visual efficiency, and reduced productivity.

Quantity of Illumination

The desirable quantity of light for any particular installation depends primarily upon the work that is being done. Investigations show that as the illumination of the task is increased, the ease, speed, and accuracy of accomplishing the task are also increased. Quantity of illumination is expressed in footcandles (1 footcandle equals 10.8 lux) and is measured with an illuminance meter to give a direct reading of footcandles reaching the work plane. Presently the maintained illuminances for industrial areas as recommended by the IES are given in ANSI/IES RP–7, *Practice for Industrial Lighting.* (See also the table in Appendix 1, Safety and Health Tables, in this volume.)

Glare

Glare may be defined as any brightness, within the field of vision, of such magnitude as to cause discomfort, annoyance, or loss in visual performance and visibility. Glare reduces the efficiency of the eye and often results in discomfort and fatigue. Glare may also reduce the detail of the visual task to such an extent as to seriously impair vision, thus increasing accident hazards. There are four types of glare:

1. Direct glare
2. Reflected glare
3. Discomfort glare
4. Disability glare

Direct glare is caused by a source of light within the field of vision (whether is it daylight or artificial). Direct glare may be reduced by (1) decreasing or shielding the brightness of the light source, (2) positioning the light source so that it no longer falls within the normal field of vision, or (3) increasing the brightness of the area surrounding the source of glare and against which it is seen (thus reducing the contrast between source and background brightness).

Reflected glare is caused by high-brightness images or differences in brightness reflected from shiny ceilings, walls, desk tops, materials, or from other surfaces within the field of vision. Reflected glare may be more annoying than direct glare if it occurs near or is superimposed over the task itself. For example, some tasks involve working on objects that reflect glare from the light source. Furthermore, reductions in contrast caused by "veiling" reflections often occur and can significantly reduce the contrast needed for task visibility and hence the ability to discern detail.

Reflected glare may be reduced by (1) decreasing or shielding the brightness of the light source, (2) positioning the light source or the visual tasks so that the reflected image will be directed away from the eye of the observer, (3) increasing the level of illumination by increasing the number of light sources in order to reduce the relative brightness of the glare, or (4) in special cases, changing the character of the offending surface to eliminate the specular reflection and the resultant reflected glare.

Discomfort glare is glare that produces discomfort. It does not necessarily interfere with visual performance or visibility. The publication *Discomfort Glare in the Interior Working Environment,* CIE No. 55, describes methods of glare control. (Note, CIE publications are available from Illuminating Engineering Society of North America at 345 East 47th Street, New York, New York 10017.)

Soft shadows from general illumination can accent the depth and form of various objects. Harsh shadows, however, may obscure hazardous conditions or interfere with visibility at the work area and should be avoided.

Disability glare is glare that reduces visual performance and visibility and that is accompanied by discomfort.

Safety

Safe working conditions are essential in any industrial facility and the effect of illumination on safety must be considered. Design the environment of a facility to compensate for the limitations of human capability. Any factor that aids vision increases the opportunity for a worker to detect a potential hazard and act to avert it.

In most cases where accidents are attributed to poor illumination, the cause is recorded as "very noticeable poor quality of illumination" or "practically no illumination at all." Many less tangible factors associated with poor illumination, however, are important contributing causes of industrial accidents. Some of these factors are direct glare, reflected glare from the work, and excessive shadows—all of which hamper vision, increase visual fatigue, and may lead to accidents. There are these other hazards due to bad illumination design:

1. The appearance of the regulatory safety colors under different light sources is required to be taken into account. High-pressure sodium lamps and low-pressure sodium lamps, as well as certain mercury lamps, will tend to change the appearance of the safety colors when standard paints or protective coatings are used without consideration for the color shift under these light sources. Carefully experiment with the safety colors and paints or protective coatings to obtain the proper visual response. Paint or protective coatings manufacturers should be consulted for the proper paint or protective coating formulations that will provide the safety colors in true colors under these light sources.
2. "Flicker," or the "stroboscopic effect." Flicker is the variation in light output of a lamp used on an alternating current circuit. Stroboscopic effect is most noticeable on rotating or oscillating machinery. Stroboscopic effect is particularly noticeable with high-pressure sodium lamps. All sources can be operated on three-phase power to help overcome stroboscopic effect. Avoid situations where only one or two luminaries are contributing all the illumination on a task.

Visual tasks tend to be difficult but are vital to profitable operations. Close observation of equipment and instruments and quick physical response to minute changes require high-quality illumination. Proper lighting safeguards a company's investments in both its machines and skilled personnel.

For extensive or difficult lighting installations, always consult a qualified illuminating engineer. The facility safety and health professional, however, may be generally familiar with the lamps, reflectors, and lighting requirements for industrial environments. (For a more extensive discussion of industrial lighting, see Appendix 1, Safety Tables, in this volume, and the following references: ANSI/IES RP–7, *Practice for Industrial Lighting,* and the IES *Lighting Handbook.*)

Luminaires

A wide range of luminaires permits a choice of designs for industrial applications. When selecting specific types for a proposed installation, consider

- luminance—the distribution of light from a luminaire (wide, narrow, bright, well-shielded, uplight, down-light, etc.)
- proper design of luminaire (1) to avoid objectionable glare under normal seeing conditions and (2) to produce the most efficient initial and maintained light outputs for the application
- mechanical construction that permits convenient installation and servicing
- environmental suitability for use indoors, outdoors, or in classified (hazardous) or special areas.

All interior luminaires are classified into six types: direct, semidirect, general diffuse, direct-indirect, semi-indirect, and indirect. No one system can be recommended to the exclusion of all others—each has unique characteristics to be considered for the requirements of a given application. The performance of each should be evaluated to make sure it efficiently provides the area with lighting in both the quality and quantity required for the task. Luminaire maintenance and the type of task being performed must also be carefully considered.

In many applications, a combination of lighting systems can be used. Regardless of the system or combination of systems, an increased percentage of indirect light generally results in a more glare-free and comfortable environment. Along with this benefit comes a corresponding disadvantage: For a given overall level of light, the illuminance in the work area is reduced if the light source is directed away from and reflected back to the worker's area.

This, coupled with more difficult maintenance, generally causes these indirect lighting systems to be less economical. Although all systems may find use to some degree, most production areas use direct or semi-direct lighting equipment.

Metal halide and mercury vapor lamps require observation of these safety precautions (from Philips Lighting Company, *Lamp Specification and Operating Guide,* 1/91, publication SAG 100):

- These lamps can cause serious skin burns and eye inflammation from shortwave ultraviolet radiation if the outer envelope of the lamp is broken or punctured. Do not use where people will remain for more than a few minutes unless adequate shielding (enclosed fixtures only or enclosed fixtures using UV filter glass ONLY) or other safety precautions are used. Certain lamps that will automatically extinguish when the outer envelope is broken or punctured are commercially available.
- The arc tubes of metal halide lamps are designed to operate under high pressure and at temperatures up to 900 C. If the arc tube ruptures for any reason, the outer bulb might break and pieces of extremely hot glass might be discharged into the surrounding environment, with an associated risk of property damage or personal injury. The following Good Lamp Practices must be followed to reduce the possibility of arc tube failure:
  1. Turn lamps off at least once per week for at least 15 minutes.
  2. Relamp fixtures at or before the end of rated life.
  3. Operate lamp only in its recommended position.
  4. Operate lamp with proper circuits and auxiliary equipment.
- The above Good Lamp Practices also apply to mercury vapor lamps.

Hazardous (Classified) Locations

Areas in industrial facilities may be classified as hazardous locations by the ANSI/NFPA 70, *National Electrical Code,* Chapter 5, Special Occupancies. Such areas require the use of specialized lighting equipment listed by Underwriters Laboratories (UL) publication *Electric Lighting Fixtures for Use in Hazardous (Classified) Locations,* ANSI/UL 844. Equipment listed as "UL 844" can provide the required illumination without introducing hazards to life and property. Since each type of lighting fixture is designed to meet certain requirements, these lighting fixtures are usually not interchangeable. (Consult Chapter 5 of the *National Electrical Code* to determine requirements for lighting equipment and circuits in hazardous (classified) locations.) In case of questions or doubts, consult the local electrical code inspector. (See Chapter 10, Electrical Safety, in this volume.)

Wet Locations

Lighting fixtures for use in wet or damp locations are required to be listed by Underwriters Laboratories (UL). Applicable standards are ANSI/UL 1570, 1571, and 1572.

Protective Lighting

Protective lighting is necessary for nighttime policing of outdoor areas. Such lighting discourages would-be intruders or renders them visible to facility guards. It may also reduce the risk of fire. Protective lighting, however, is not usually adequate for efficient facility operation. Therefore, protective lighting is generally treated as an auxiliary to productive lighting.

Protective lighting is achieved by supplying adequate light in border areas of buildings or, in some cases, by producing glaring light in the eyes of an intruder with no light on the guard. Arrange lighting so that concealing shadows are eliminated.

In general, there are four types of lighting units used in protective lighting systems: floodlights, street lights, Fresnel lens units, and searchlights. Infrared lamps, invisible to trespassers, are also used with infrared TV monitors. Refer to *Protective Lighting,* IES publication RP-10.

Outdoor lighting of any design (on the roofs or walls of buildings, in parking lots, on roadways, etc.) should be compatible with the surrounding neighborhood by avoiding light "spill" and directed glare onto passing motorists. Refer to *A Statement on Astronomical Light Pollution and Light Trespass,* IES publication CP-46.

Length of time for an emergency evacuation of personnel helps to influence the choice of lighting units. Correlate battery capacity of units with the number of lamps and their wattage to determine the lighting needed for the brief length of time required for complete evacuation (90 minutes) if using the recommendations of NFPA 101, *Life Safety Code* or NFPA 70, *National Electrical Code.* Where longer durations of emergency lighting are required, use engine-generator sets as the power source. Engine-generator sets for emergency lighting are required to start and provide light automatically upon failure of the normal power supply. Power is transferred from the normal to the emergency power supply by an automatic transfer switch that reverses the procedure upon reestablishment of the normal power supply.

Security of Facilities

A security system is a major consideration when planning or remodeling an industrial facility. The system should be both functional and cost-effective. Beginning with site selection, design the facility's environment for minimum loss and maximum security. Typical security considerations include the following:

- Keep the number of openings to a minimum.
- Secure all windows.
- Use protective lighting.
- Have entrances and service doors lead to a reception area.
- Install alarm systems that detect fire, fumes, vapors, and intruders.
- Limit access to docks and other receiving areas.

(See Chapter 13, Workers with Disabilities, in this volume for details about security for workers with disabilities.)

Lighting Management

The effective use of monitoring and control of the illumination may be referred to as "lighting management." The purpose for its consideration is to deliver the proper illumination when and where it is required and minimize electric energy use. Monitoring involves the use of sensors to observe illuminances and, more recently, occupancy. Controls are either ON-OFF or adjustable (and may be manual or completely automatic). In the simplest form, the occupant uses a local control to switch the lighting circuit or to adjust the illumination. The most advanced systems are microprocessor based and receive sensor inputs and respond with control outputs to enable and adjust the illumination by use of low-voltage control relays, controllable circuit breakers, and adjustable dimming systems.

Caution must be exercised in the selection and application of automatic lighting controls within industrial facilities. Where task performance is the primary objective, the worker must also have the ability to readily adjust or override the lighting controls whenever the situation arises, *at the worker's discretion.*

USE OF COLOR

The following excerpt, written by Linda Trent, is used with permission from The Sherwin-Williams Company, Cleveland, Ohio:

> At work or at play, consciously or subconsciously, people respond to the colors around them. And, in growing numbers, industrial designers and managers are paying more attention to the interactions between color, lighting, and human behavior. Managers are becoming more attuned to the industrial psychologists' message that the quality and appearance of work areas can stimulate interest or create boredom. As such, they are receptive to the idea that the proper use of color can generate a positive response to the work environment—favorably affecting workers' housekeeping efforts, safety, and overall productivity. Moreover, since it doesn't cost any more to paint work areas in scientifically chosen colors than in colors chosen entirely at random, many managers are willing to consult a professional color stylist.

Color and Light in the Workplace

The function of a paint, in addition to protecting surfaces, is to absorb or subtract some parts of the spectrum and to transmit or reflect other parts. Color is determined by which parts of the visible spectrum are reflected when absorption takes place. Consequently, color is visibly modified by different light sources. For that reason, the effects of a facility's lighting system should be a major consideration when selecting colors for the facility's interior.

Typically, a facility's light source is selected on the basis of characteristics such as the amount of light given per watt used, ease of maintenance, ease of shielding, ease of directional control, and overall cost. The light source's effect on color and color-rendering properties are usually secondary considerations. Lighting systems currently used in industrial settings include a range of fluorescent and high-intensity discharge lamps, and combinations of illuminants. The effects on color of these various light sources are detailed in Table 2–A.

Interestingly, light has an effect on color, and color also affects the quality of light. When dealing with surface colors, this effect on light is called the light-reflectance value (LRV) of color. It is an important property because the reflections from painted surfaces—ceilings, walls, machinery, and floors—act as secondary light sources. With proper color styling and recommended reflectances, work area surfaces will maximize the available light and reduce shadows.

In general, light colors reflect light while dark colors absorb light. White, followed by pastels, has the highest light reflectance value; black affords no reflectance. Often, the paint supplier refers to an LRV that is equivalent to the reflectance of a surface of material as defined in the IES *Lighting Handbook*.

In a work environment, the LRV of color can contribute significantly to seeing a task. Objects are discerned only in contrast with their surroundings. The most effective contrast can be obtained by selecting colors for their light-reflectance characteristics. In industrial settings, high-contrast conditions are provided for tasks such as inspection but are deemed unnecessary for tasks such as retrieving goods from seldom-used storage areas.

As a rule-of-thumb, surfaces in industrial facilities should be finished to provide light-reflectance values within the ranges listed in Table 2–B. Light reflectance is affected more by color than by type of materials. Keep wall reflectances within the recommended range, except when unusual conditions make alterations desirable. For example, walls in areas with high, exposed ceilings might have a high-reflectance ceiling carried down to the level of suspended light fixtures, which direct some light upward. Using this upper wall surface can increase light reflectance in the room by as much as 10% (Edmonds, 1977).

Another exception to the recommended values in Table 2–B involves peripheral areas of a room or work space. If these are not in the direct line of vision of the task and are restricted to about 10% of a worker's visual area, they will usually not affect the efficiency of the lighting system. In fact, deep tones with low-reflectance characteristics can be used as accents and focal points to make the workplace more pleasant. In addition to reflectance, several other factors affect the use of color in the workplace.

Geographical Location and Exposure

Regional color preferences can be employed in a color plan; for example, desert colors in the Southwest or colonial colors in the East. In a facility with windows, a particular exposure can be contrasted with color; examples, warm colors (reds, yellows, or oranges) for a northern exposure.

Age of the Facility

Newer buildings don't have many windows, so light colors are used to improve overall lighting and morale. In some older buildings, windows no longer access the outdoors, making the area look gloomy and the windowed-wall look cluttered. In these cases, paint the windows to increase light-reflectance values and to eliminate distractions.

Table 2–A. Color Effects of Light Sources*

Lamp Type	Appearance on Neutral Surface	Effect on "Atmosphere"	Colors Strengthened	Colors Grayed	Effect on Complexions	Remarks
FLUORESCENT						
Cool white°	white	neutral to moderately cool	orange, yellow, blue	red	pale pink	Blends with natural daylight; good color acceptance
Deluxe cool white°	white	neutral to moderately cool	all nearly equal	none appreciably	most natural	Best overall color rendition; simulates natural daylight
Warm white†	yellowish white	warm	orange, yellow	red, green	sallow	Blends with incandescent light
Deluxe warm white†	yellowish white	warm	red, orange, yellow, green	blue	ruddy	Good color rendition; simulates incandescent light
INCANDESCENT						
Filament †	yellowish white	warm	red, orange, yellow	blue	ruddiest	Good color rendering
HIGH INTENSITY DISCHARGE LAMPS						
Deluxe white mercury°	purplish white	warm, purplish	red, yellow, blue	green	ruddy	Color acceptance similar to cool-white fluorescent
Metal halide multi-vapor°	greenish to pinkish white	moderately cool, greenish	yellow, green, blue	red	grayed	Color acceptance similar to cool-white fluorescent
High-pressure sodium	golden white	warm, yellowish	yellow, orange, green	red, blue	golden	Color acceptance approaches that of warm-white fluorescent

*Table based on information from The General Electric Co.
°Greater preference at higher levels.
†Greater preference at lower levels.

Table 2–B. Reflectance Values Recommended for Facility Surfaces

	Reflectance Values	
Surface	*Manufacturing Areas (Percent)*	*Office Areas (Percent)*
Ceilings	80-90	80-90
Walls	50-65	60-70
Floors	15-30	25-40
Machinery	30-50	—
Desk tops	—	40-50

Demographics

Consider the number of female vs. male employees in the facility. Color preference studies show that men prefer blue, followed by red. Women, on the other hand, select fashion shades and subtle colors such as peach and mauve.

Noise Level

No color is going to actually diminish existing noise caused by machinery. The use of blues, greens, and neutrals, however, lessens the workers' psychological response to the noise.

Type of Equipment

Most types of equipment are painted in a medium color to reduce glare and provide good contrast for safety colors. Very large machinery, in some instances, can handle a two-tone combination that highlights operating parts and affords visual contrast.

Psychological Factors

Dark, saturated colors make surroundings appear cramped and can cause workers to feel depressed. Lighter pastels give workers a psychological lift and create the illusion of spaciousness and calmness. To reduce visually perceived clutter, paint miscellaneous structural work (pipes, I-beams, etc.) the same color as the adjacent wall or ceiling.

Safety

The Occupational Safety and Health Administration (OSHA) and ANSI have set standards for safety colors throughout industry. Table 2–C summarizes the OSHA and ANSI safety color code.

Some companies identify very closely with the colors used in their corporate logo and for packaging of products. They like to include these colors in their facilities' interiors, either in graphics or as subtle accents on doors or trim strips.

Color as Science

When scientists do research, they measure properties and dimensions. Although Sir Isaac Newton made great discoveries relating to color and light back in the 17th century, the classification of color in terms of dimensions is comparatively recent. The dimensions of color can be described as three related aspects of color—hue, value, and intensity. These terms are defined as follows:

- hue—the quality giving the color its name (red, yellow, blue, etc.)

Table 2–C. Summary of OSHA and ANSI Safety Color Code Corporate Colors°

Color	Designation
Red	Fire: Protection equipment and apparatus, including fire-alarm boxes, fire-blanket boxes, fire extinguishers, fire-exit signs, fire-hose locations, fire hydrants, and fire pumps. Danger: Safety cans or other portable containers of flammable liquids, lights at barricades and at temporary obstructions, and danger signs. Stop: Stop buttons and emergency stop bars on hazardous machines.
Orange	Dangerous Equipment: Parts of machines and equipment that may cut, crush, shock, or otherwise injure.
Yellow	Caution: Physcial hazards such as stumbling, falling, tripping, striking against, and being caught in between.
Green	Safety: First-aid equipment.
Blue	Warning: Caution limited to warning against starting, using, or moving equipment under repair.
Black on yellow	Radiation: X ray, alpha, beta, gamma, neutron, proton radiation.
Black and white	Boundaries of traffic aisles, stairways (risers, direction, and border limit lines), and directional signs.

°See full text under Section 1910.144 of Occupational Safety and Health standards. For piping colors, see ANSI Standard *Scheme for the Identification of Piping Systems,* A13.1-1981.

- value—how light or dark the color is
- intensity or saturation—how much hue the color contains in proportion to its greyness.

The first of these dimensions, hue, refers to the pure spectral hues produced when sunlight is refracted by a prism. A spectrum consists of light of different wavelengths. Each hue can be identified by its corresponding wavelength, or band of wavelengths, in the spectrum. The visible color spectrum spans between red, at about wavelength 750 nanometers (nm), and violet at wavelength 380 nm (1 nm = 10^{-9} m). Any hue can be specified according to the ratio of red, yellow, green, or blue that it perceptually contains.

Human Response to Color

When Newton defined the visual process in Opticks, he described the following three stages of color perception regarding the physics, physiology, and psychology of the process in terms that are still valid today:

1. the light entering the eye and the factors that determine the spectral composition of light (physics)
2. the response to the light in the retina and visual pathways (physiology)
3. the color actually perceived and the psychological response to the color's appearance (psychology)

Scientists who specialize in studying color are building on Newton's original observations—but using a slightly different vocabulary. Reflexive responses to color, the most verifiable of the three types of responses, result from the physical structure of the eye. The most apparent reflexive mechanism is the advancing/retreating effect of certain colors. This is caused by the human eye not focusing all wavelengths of light in the same place. Colors with longer wavelengths—red, yellow, orange—seem to move toward the observer. Cooler, darker colors—blue and green—on the other hand, have shorter wavelengths and appear to move away from the observer. This reflexive response is also used to help create the illusion of larger or smaller space. Moreover, safety engineers rely on this response by using warm, bright colors to call attention to dangerous machine parts, fire hazards, and physical hazards that might cause workers to lose their footing.

General physiological responses include all the effects of color that cannot be explained as either reflexive or conditioned responses. Of these, the photoreactive effects are the easiest to describe. For example, doctors use blue light in the treatment of hyperbilirubinemia. This is a serious condition in newborn babies, in which the blood contains potentially fatal levels of the hemoglobin by-product bilirubin that imparts to the baby a yellow hue (jaundice). Because the bilirubin is photoreactive, it breaks down when exposed to certain wavelengths of radiation. The "bililight" is a treatment used for jaundiced babies.

Some other physiological responses studied in color science are (1) the relationships between hue and perceived heat and (2) the connection between certain hues and metabolic activity. The basic contention of the former is that "warm" colors (reds, yellows, and oranges) tend to make people feel warmer, whereas "cool" colors (blues and greens) tend to make people feel cooler. Early studies suggested that the perceived temperature differences might be as great as 7 F. More carefully controlled recent studies, however, suggest the differences could be less than 1 F or completely indistinguishable. On the basis of these kinds of test results, researchers conclude that, in general, cool colors calm people while warm colors excite them.

These hue/activity correlations have practical applications in selecting colors for facility interiors. Use cool colors to ease tensions caused by highly detailed work or noisy machinery; use warm colors in a lunchroom to dispel boredom and stimulate conversation. The following list and Table 2–D give more suggestions for selecting colors to enhance productivity in the workplace.

- Use color schemes to identify and unify work areas that would otherwise be "lost" in large facilities with a variety of production activities.
- Use neutral colors of low light-reflectance values in laboratories where reflected color might prevent accurate observation of materials being tested and analyzed. However, do add interest to the area by using stronger colors on furniture and doors, where visual observation is not so critical.
- Use strong, bright colors in time-clock and locker-room areas. In nonproductive areas such as these, color can be used to provide a cheerful atmosphere for the employees at the beginning and end of the workday.
- Use colors with high reflectance values in warehouses to offset the customary low-lighting levels and to maximize the use of available light. Color code storage areas to facilitate locating materials.
- Use contrasting colors in production areas to focus attention on the task and to increase visibility.
- Use intense shades of warm colors sparingly to avoid confusing the workers or making them anxious.
- Use high-reflectance colors on stairways and sharp accent colors on rails and doors to define points of orientation.

Color-Coding

Color is used extensively for safety purposes. While never intended as a substitute for good safety measures or for use of mechanical safeguards, standard colors are used to identify specific hazards. Be sure to check the latest regulations for in-facility use, shipping, or consumer protection. In summary, they are as follows:

- Red identifies fire protection equipment, danger, and emergency stops on machines.
- Yellow is the standard color for marking (1) hazards that may result in accidents from slipping, falling, striking against something, etc., (2) flammable liquid storage cabinets, (3) a band on red safety cans, (4) materials handling equipment, such as lift trucks and gantry cranes, and (5) radiation hazard areas or containers (Safety Black on Safety Yellow). Black stripes or "checkerboard" patterns are often used with yellow.
- Green designates the location of first aid and safety equipment (other than firefighting equipment). (See also Blue, below.)
- Black and white, and combinations of them in stripes or checks, are used for housekeeping and traffic markings. They are also permitted as contrast colors.
- Orange is the standard color for highlighting dangerous parts of machines or energized equipment, such as exposed edges of cutting devices and the inside of (1) movable guards and enclosure doors and (2) transmission guards.
- Blue is used on information signs and bulletin boards that are not of a safety nature. If of a safety nature, use green except for flagging railroad cars. A blue flag is used to mark chocked cars that are unloading.
- "The radiation hazard symbol colors shall be Safety Black on Safety Yellow. All present Safety Purple on Safety Yellow or Safety Black on Safety White radiation hazard symbols may be used until replaced." ANSI Z535.1

The piping in a facility may carry harmless, valuable, or dangerous contents. Therefore, it is highly desirable to identify different piping systems. ANSI A13.1, *Scheme for the Identification of Piping Systems,* specifies standard colors for identifying pipelines and describes

Table 2–D. Characteristics and Suggested Uses for Various Colors

Color	*Impression*	*Suggested use for interiors*
Warm colors		
Red, orange, yellow	Attract attention, create excitement, promote cheerfulness, stimulate action.	Nonproductive areas, including employee entrances, corridors, lunch rooms, break areas, locker rooms, etc.
Cool colors		
Blue, turquoise, green	Cool, relaxing, refreshing, peaceful, quieting. Encourage concentration.	Production areas, maintenance shops, boiler rooms, etc.
Light colors		
Off-whites and pastel tints	Make objects seem lighter in weight; areas seem more spacious. Will usually give people a psychological lift. Reflect more light than darker tones.	Most production areas, especially small rooms, hallways, and warehouse and storage areas. Poorly illuminated rooms.
Dark colors		
Deep tones, gray, black	Make objects seem heavier; absorb light. Will make rooms appear smaller and surroundings cramped. Long exposure will create monotony and depression.	Not normally recommended for large areas because of light absorption qualities. Use should be confined to small background areas where contrast is needed.
Bright colors		
Notably yellow, as yellow-green, orange, red-orange, red	The purer these colors are, the more compellingly they attract the eye. Make objects appear larger and create excitement.	Complement to basic wall colors; small objects such as doors, columns, graphics, time clocks, time-card racks, bulletin boards, tote boxes, dollies, etc.
White		
	Pure, denotes cleanliness, reflects more light than any other color.	All ceilings and overhead structures, and rooms where maximum light reflection is needed. Can also be used on small objects for greater contrast.

methods of applying these colors to the lines. The contents of pipelines are classified in the following way:

Classification	*Color*
Fire protection	SafetyRed (7.5R; LRV 12)
Dangerous	SafetyYellow (5.0Y; LRV 69)
Safe Safety	Green (7.5G; LRV 6)
Protective materials (e.g., inert gases)	Safety Blue (2.5PB; LRV 5)

(Standard Munsell hue and light reflectance values are given for each safety color.)

Apply the proper color to the entire length of the pipe or in bands 8 to 10 in. (20 to 25 cm) wide near valves and pumps and at repeated intervals along the line. Stencil the name of the specific material in black at readily visible locations such as valves and pumps. Identify piping less than ¾ in. (1.9 cm) in diameter with enamel-on-metal tags. Other schemes may be equally effective in identifying piping networks. It should be noted that (ANSI A13.1 also recommends) highly resistant colored materials should be used where acids and other chemicals may affect paints.

There may arise situations where pipes may not be able to be color-coded or labeled. For example, some pipes may not be able to be designated to a single material. Provisions must be made to train personnel to recognize hazards of unlabeled piping.

Accident Prevention Signs

Accident prevention signs are among the most widely used safety measures in industry. Therefore, uniformity in the color and design of signs is essential. Employees may be unable to read English or may be color-blind and yet will react correctly to standard signs.

The following is a digest of requirements for signs:

DANGER—Immediate and grave danger or peril. Red oval in top panel; black or red lettering in lower panel.

CAUTION—Lesser hazards. Yellow background color; black lettering.

GENERAL SAFETY—Green background on upper panel; black or green lettering on white background on lower panel.

FIRE AND EMERGENCY—White letters on red background. Optional for lower panel: red on white background.

INFORMATION—Informational signs, bulletin boards, and railroad flags for chocked cars. Blue letters on white background.

IN-FACILITY VEHICLE TRAFFIC—Standard highway signs. (See U.S. Dept. of Transportation, *Manual on Uniform Traffic Control Devices for Streets and Highways.*)

EXIT MARKING—(See NFPA 101, *Life Safety Code,* section 5–11.)

BUILDING STRUCTURES

Structures include access areas, aisles and corridors, stairs, walkways, exits, flooring, and workstations. Stairs, runways, ramps, and other access structures are principal sources of injuries. One-fifth of all industrial injuries result from falls. Of those that take place from one level to another, the largest number of injuries occurs from falls on stairs and ladders. (See Chapter 3, Construction of Facilities.) Careful design and construction, however, can prevent many serious injuries. Also follow the standards of state and municipal governments and ANSI.

Runways, Platforms, and Ramps

Widths of runways and ramps should be adequate for anticipated traffic. Use standard railings to protect open sides. Platforms, 4 ft (1.2 m) or more above floor or ground level, should be guarded by a standard railing, The footwalk should have a railing with the top edge height of the top rails, or equivalent guardrail system members, 42 in. (1.1 m) plus or minus 3 in. (8 cm) above walking/working levels. Midrails shall be installed at a height midway between the top edge of the guardrail system and the walking/working surface, and, on the exposed side, toeboards should be 4 in. (10 cm) high. (See 29 *CFR* 1926.502 (b) (1) and (b) (2) (i).) (Figure 2–9). A convenient means of access should be provided to most locations more than 4 ft (1.2 m) above the floor.

Wire screen enclosures are recommended where materials must be stored on platforms or where persons below will be endangered by falling objects or fragments of materials. Use wire netting of No. 16 U.S. gauge wire with 1-in. mesh (1.5-mm diameter wire with 38-mm mesh) for screening. Plywood or other materials may also be used to fully enclose platforms.

Construct ramps with the least slope practicable—some states specify a slope of 1 in 10 (5°43′). Fifteen degrees (a slope of 2.68 in 10) is a recommended maximum; a slope should never exceed 20 degrees (3.64 in 10) (Figure 2–10). Check the ADA for specifications for ramp slope if there is a possibility the ramp will be used as a wheelchair access ramp.

Except where dislodged abrasives would be detrimental to equipment or process materials, use abrasive coatings or pressure-sensitive adhesive strips on ramps to help provide safe footing. Install toeboards where a ramp extends over a workplace or a passageway. On steep inclines, place cleats 16 in. (41 cm) apart. Have planks run the long way of the ramp and do not allow planks to overlap. Ramps used for wheelbarrows should have no cleats on the center plank.

Aisles and Corridors

Plan aisles and corridors to provide minimum clearances for equipment (Figure 2–11). Keep the following suggestions in mind when planning aisles and corridors:

- Keep aisles clear of structural supports, columns, etc.
- Plan traffic guides.

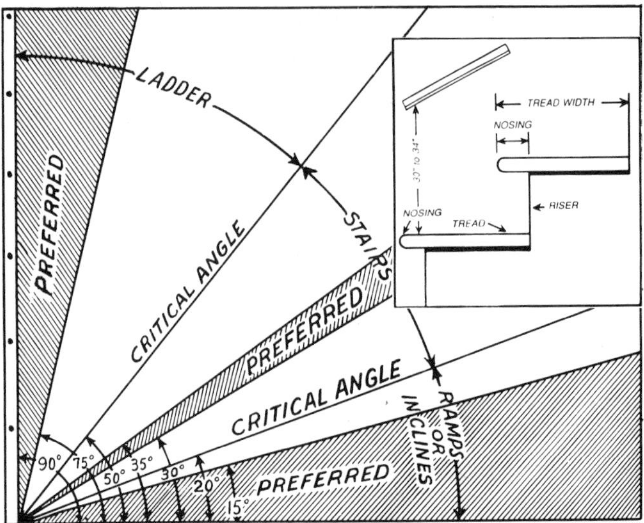

Figure 2–10. Preferred angles for fixed ladders, stairs, and ramps.

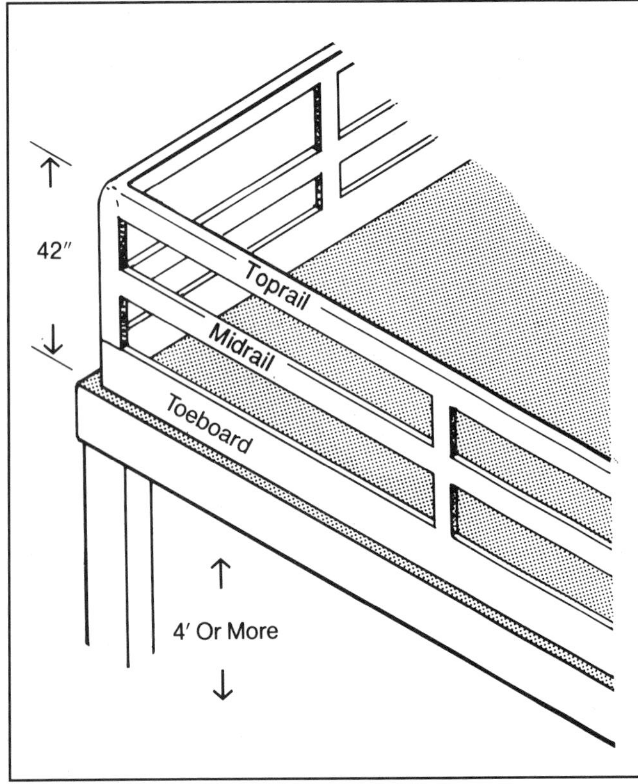

Figure 2–9. Open-sided floors or platforms more than 4 ft (1.2 m) above floor or ground level and scaffolds more than 10 ft (3.1 m) above floor or ground level should have a railing with the top edge height of the top rails, or equivalent guardrail system members, 42 in. (1.1 m) plus or minus 3 in. (8 cm) above walking/working levels. Midraills shall be installed at a height midway between the top edge of the guardrail system and the walking/working surface, and, on the exposed side, toeboards should be 4 in. (10 cm) high. (See 29 CFR 1926.502 (b) (1) and (b) (2) (i).). If persons can pass beneath or if there is moving machinery or other equipment from which falling materials could create a hazard, the guardrail should also have a 4 in. high (10 cm) toeboard. Screening can also be added.

- Locate pathways (aisles) for the shortest possible distance between work areas.
- Design aisles so users do not inadvertently hit machines and equipment.
- Avoid designing blind corners; use mirrors if necessary.
- Avoid having doors opening into corridors.
- Avoid one-way traffic.
- Avoid putting aisles against walls.

Stairways

Avoid using circular stairways. If they are absolutely necessary, however, design them with a minimum variation in tread width. Be sure to cover treads with a durable slip-resistant material.

The preferred slope for a stairway is between 30 degrees and 35 degrees from the horizontal (Figure 2–10 and Table 2–E). The most suitable slope for a fixed ladder is from 75 degrees to 90 degrees. A tread width of not less than 11 in. (28 cm), including a nosing of 1 in. (2.5 cm), is recommended. Riser height should not be more than 8 in. (20 cm) or less than 5 in. (12.5 cm) and should be constant for each flight. Stairs and landings should be able to sustain a live load of not less than 100 lb per ft^2 (48.82 kg/m^2) with a safety factor of four.

A flight of stairs having two or more risers should have a standard handrail as specified in ANSI A1264.1–1995, *Safety Requirements for Workplace Floor and Wall Openings, Stairs, and Railing Systems.* Rails should be 30 to 34 in. (76.2 to 86.4 cm) from the top surface of the stair tread, measured in line with the face of the riser. (Up to 42 in. [1.1 m] is permitted on steep angles.) An intermediate handrail is recommended

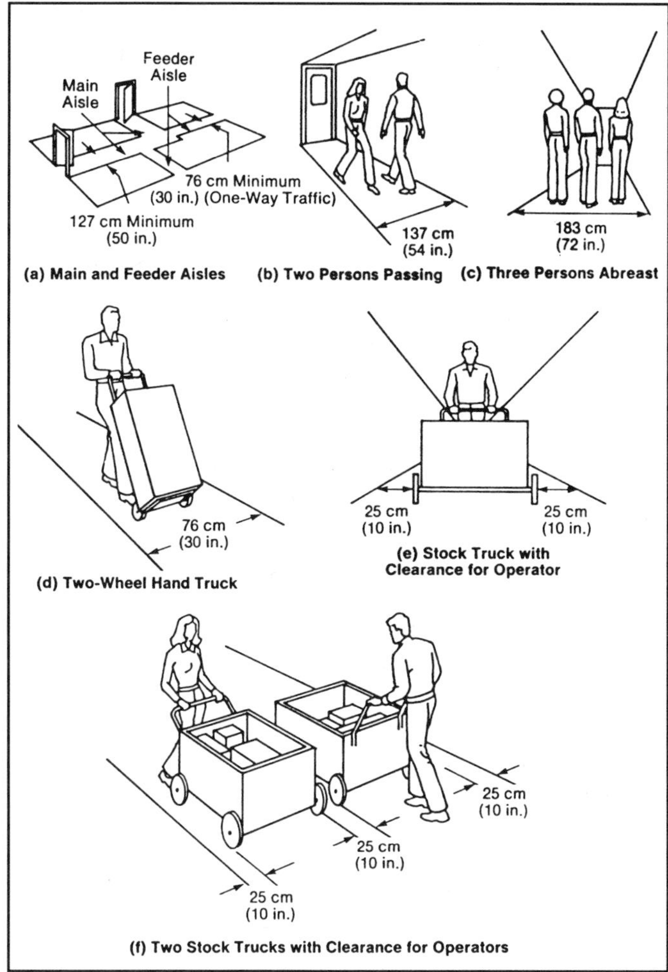

Figure 2–11. Minimum clearances for aisles and corridors. (Reprinted with permission from *Ergonomic Design for People at Work,* Eastman Kodak Company.)

Table 2–E. Slope and Dimensions of Treads and Risers

Angle of Stairway with Horizontal	*Riser (inches)*	*Tread and Nosing (inches)*
30°35′	6½	11
32°08′	6¾	10¾
33°41′	7	10½
35°16′	7¼	10¼
36°52′	7½	10
38°29′	7¾	9¾
40°08′	8	9½
41°44′	8¼	9¼
43°22′	8½	9
45°00′	8¾	8¾
46°38′	9	8½
48°16′	9¼	8¼
49°54′	9½	8

Reprinted with permission from OSHA Standards § 1910.24.
(1 in. = 2.54 cm)

for stairways more than 88 in. (2.2 m) wide. Hardwood handrails should have at least a 2 in. (5 cm) diameter (Figure 2–12). If standard black iron pipe is used, the pipe should be at least 1 in. (2.54 cm) in outside diameter. Allow clearance between the handrail and the wall of at least 1 in. (2.54 cm). To provide a smooth surface along the top and both sides of the handrail, mount rails directly on a wall or partition by means of brackets attached to the lower part of the handrail. Space brackets not more than 8 ft (2.5 m) apart. The top of the post should be capable of withstanding a load of 200 lb (91 kg) applied in any direction. Consult applicable local and state codes.

Because of space limitations, a permanent stairway sometimes has to be installed at an angle greater than 50 degrees. Such an installation (commonly called an inclined ladder or "ship's ladder") should have handrails on both sides and open risers.

Other safety precautions to keep in mind when planning stairways include the following:

- Provide adequate electrical lighting in stairways.
- Locate lights so that they do not cause glare.
- Enclose outside stairways to keep off rain, snow and ice.
- Enclose all inside stairs with partitions of fireproof or fire-resistant material.
- Install approved fire doors to prevent the spread of smoke or flames from one floor to another. Check local codes and insurance requirements about this.

Walkways

Building plans should specify elevated walkways and platforms for tanks, bins, machinery, and other places where workers must go during normal operations or for

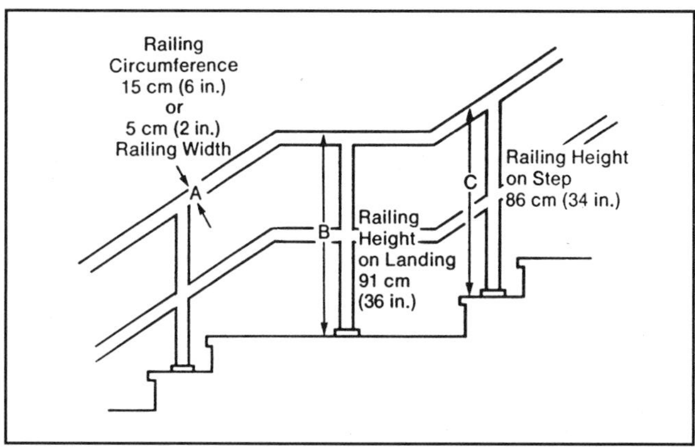

Figure 2–12. Handrail design guidelines (Reprinted with permission from *Ergonomic Design for People at Work,* Eastman Kodak Company.)

maintenance purposes. Provide conveyors with slip-resistant surfaces, crossovers, railings, and toeboards.

Equip walkways and ramps with a standard railing. A standard railing shall consist of a top rail, intermediate rail, and posts, and shall have a vertical height of 42 in. (1.1 m) nominal from the upper surface of top rail to floor, platform, runway, or ramp level. A standard toeboard shall be 4 in. (10 cm) nominal in vertical height from its top edge to the level of the floor, platform, runway, or ramp (Figure 2–9).

Exits

Locate exits so that in case of fire or other emergencies, all persons can quickly evacuate the building without loss of life. Design exits to be sufficient both in number and in size because changing or adding exits after a building is constructed is very costly. Be sure that plans conform to NFPA, federal, state or provincial, and local requirements.

NFPA 101, *Life Safety Code,* generally requires that two exits be provided on each floor, including basements. One exit from an upper floor may be an inside stairway or smokeproof tower and the other may be a moving stairway or horizontal exit. Some government regulations require two or more exits that are remote from each other to be provided for all floors of industrial buildings two or more stories high. One exit must be a stair tower made of fire-resistant material, leading directly to the outside at grade (ground) level.

To a considerable extent, the number and width of exits are determined by the building's occupancy. In high-hazard occupancy, no part of a building should be farther than 75 ft (23 m) from an exit. For medium- and low-hazard occupancy, 100 to 150 ft (30 to 45 m) is permissible. The Life Safety Code also specifies that access to exits provided by aisles, passageways, or corridors shall be convenient to every occupant and that the aggregate width of passageways and aisles shall be at least the required width of the exit. Exit doors should be clearly visible, illuminated, and equipped with signs. Exit doors should also open in the direction of exit travel. Maintain exit areas as empty, relatively safe areas that provide personnel using them a chance to get away from the facility in an emergency.

Protect with guardrails or gates, exterior door openings used for hoisting equipment or material on the outside of buildings. Because such doors are a serious hazard when open and not in use, they should be posted with NOT AN EXIT signs.

Flooring Materials

Comfort, health, and safety are closely related to the design of and specifications for floors. Because flooring requirements often vary sharply from department to department, carefully study the following factors to determine the best type of floor for a particular location:

- load
- durability
- maintenance
- noise
- dustiness
- drainage
- heat conductivity
- resilience
- electrical conductivity
- appearance
- chemical composition
- slip-resistance

Since a principal cause of floor accidents is slipperiness, determine the slip-resistant qualities of various types of floor surfaces. (See also Chapter 25, Contractor and Nonemployee Safety, in the *Administration & Programs* volume.) Inserts of various materials can reduce slipperiness in specific areas and can combat conditions that cause rapid deterioration of flooring. For example, use cast metal inserts around woodworking machines. Where acid is spilled occasionally, cover the floor surface with sheets of soft lead. Provide drainage so that the acid can be washed away. Reduce the slipping hazard at the door sills of elevators by installing steel grating filled with concrete.

Install inserts flush with the surface of the floor. An insert placed on top of the floor requires a bevel on every side from which it can be approached. The bevel should be at such an angle and low enough that a person will not trip or lose his or her balance.

Abrasive-coated fabric strips also reduce slipperiness. They are made for use on metal floor plates, the foot of stairs, stair nosings, and other high-hazard locations. An adhesive binds the strips firmly to the walking surface.

Drainage is essential where wet processes are used. In some instances, especially along passageways, floor gratings are installed to reduce slipperiness. Also consider the following safety and health factors when selecting flooring:

- Install moisture-absorbing mats or runners at entrances to reduce tracking of mud and dirt into a building.
- Install noncombustible flooring in areas where welding is performed regularly and in oven, furnace, and boiler areas.
- Protect openings in floors with railings or barriers at exposed edges. Install a top rail, intermediate rail, and toeboard, according to ANSI A1264. 1–1995,

Safety Requirements for Workplace Floor and Wall Openings, Stairs, and Railing Systems. Install sheeting or woven wire around the opening to prevent material from falling in.

There are several types of flooring materials for industrial facilities. The characteristics and uses of some flooring materials are discussed in the sections that follow.

Asphalt

Asphalt is used in various flooring materials. Dustless, plastic, odorless, and warm to the feet, asphalt provides especially good flooring in some shops. If the asphalt mixture contains silica, the flooring will be resistant to acid. Asphalt tile, often used in offices, may become slippery when washed or not properly waxed. However, it is quickly being replaced by vinyl-based sheet goods, which are not only more attractive but less trouble to keep clean. Ordinary grades of asphalt are not recommended for facility roads because they soften in hot weather and cannot accommodate heavy industrial trucks.

Paving Brick

If laid on a solid foundation like concrete, paving brick is satisfactory for heavy traffic. Cement mortar joints make a smooth surface. Foundry floors have been made from hardburned bricks laid face up on a concrete base with sand-filled joints.

Concrete

Concrete floors are used widely in warehouses and factories. A smooth concrete finish is slippery when wet. A wood float finish on a mixture of pea gravel, sand, and cement, therefore, gives concrete a roughened surface that will not crack or "dust." Concrete does not withstand acids, however, and in some types of work, employees consider concrete too hard and too cold to stand on for long periods of time. Resilient slip-resistant mats with low heat conductivity can be placed over the concrete where workers must stand in one position for considerable periods of time.

Cork Tile

Cork tile is desirable for its insulation, resiliency, quietness, high slip-resistant rating, and ability to withstand light traffic for long periods. It is not suitable, however, for wet locations or where there are heavy loadings. It is also expensive.

Asphalt-Base or Vinyl-Base Tile

Use asphalt-base or vinyl-base tile or sheet goods where cleanliness and good looks are important, such as in offices, laboratories, and workrooms. The material is easily cleaned, noiseless, and a poor conductor of heat.

Magnesite

For light traffic or where light oils are used, install magnesite flooring. It must be laid on a rigid base and should not be used where there is excessive moisture or hydrostatic pressure, as in basements. Since magnesite corrodes some metals, a coating of bituminous paint is necessary to protect metal objects, such as pipes, from contact with this.

Cast Iron Plates

Cast iron plates with checkered or otherwise roughened surfaces and laid in cement or asphalt are suitable for rough wear such as that in warehouses. However, they are noisy and highly conductive of heat and electricity. They are relatively slip-resistant unless wet or worn smooth.

Metal Grille

Floors and gratings of metal grille will not collect dust, dirt, or liquids. Because this flooring is noisy, do not use it where hand trucking is done regularly. This type of floor material is particularly useful in boiler rooms and over openings.

Parquet

Parquet is laid on an underfloor. If the material is sealed properly, little maintenance is required. Suitable for office floors, parquet wears well but may be noisy.

Rubber

Rubber flooring is resilient. It also has high dielectric strength, which is undesirable where static electricity is a problem. However, conductive types of rubber flooring are also available. Abrasive rubber flooring can be used to overcome slipperiness.

Terrazzo

Since terrazzo flooring has no joints in the surface, its use eliminates some of the difficulties encountered with some other types of flooring. Terrazzo can be made electrically conductive by grounded grilles, and it is a conductor of heat. Suitable sealers are necessary to make the floor impervious to most acids. In addition, the terrazzo mixture is slippery unless it includes abrasive aggregates.

Ceramic Glazed Tile

Facilities or laboratories requiring extreme cleanliness and sanitary conditions, such as dairies, use ceramic glazed tile. It should be laid in Trinidad asphalt or in cement with a low lime content. Two or three layers of asphalt roofing felt may be laid beneath the tile.

Wood

With the right wood, and if properly constructed, wood floors function well in various work areas. Their main drawback is that they pose a fire hazard. Plank or board floors of the softer woods are generally unsatisfactory. They are a source of numerous injuries from slipping and falling as well as from splinters. However, matched or jointed hardwood flooring, nailed to a subfloor structure or to sleepers in concrete, makes a good floor for light manufacturing. The thickness of the hardwood flooring will depend on the service to which it will be subjected. It is laid with the grain parallel to the line of truck travel.

Wood Blocks

Wood blocks meet many of the requirements for a good floor. Properly made, a floor of this type is relatively noiseless and does not become slippery or cause fatigue. If the blocks are laid on a smooth, rigid base, the floor is not likely to crack and will withstand heavy service. Wood blocks filled with creosote are necessary for floors in contact with liquids or moisture. Expansion joints are required along walls, columns, and similar places. If blocks are laid with a high-melting-point pitch, hot weather does not create problems. Oils and organic solvents, however, cause trouble because they act as a solvent on bituminous fillers and coatings.

Floor Loads

Design floors to carry anticipated loads safely. Consult a registered structural engineer, and also check ANSI A58.1, *Minimum Design Loads for Buildings and Other Structures,* to determine floor loads.

Figure 2–13 presents some floor-loading fundamentals. The ideal load is uniformly distributed over the floor area (Figure 2–13a). The same load concentrated at the center of the span (Figure 2–13b) will require twice the structural strength. Conversely stated—if a floor is designed for a given uniform load, only one-half this amount can be concentrated at the center of the span. Figure 2–13c shows the ideal location of aisles and loads.

Figure live loads and anticipated future loads. In estimating floor loads, the weight of men is figured at 160 lb (73 kg); the weight of women, at 138 lb (63 kg). The weight of equipment can be obtained from manufacturers; the weight of bulk materials, from handbooks. (When piling bulk materials, air space often results, thus reducing the overall density of the material.) Industrial trucks when loaded may weigh as much as 60,000 lb (27.2 tn). Foundations that distribute loads and vibration over larger areas, or cushion shocks with springs or vibration mounting, help reduce structural reinforcement.

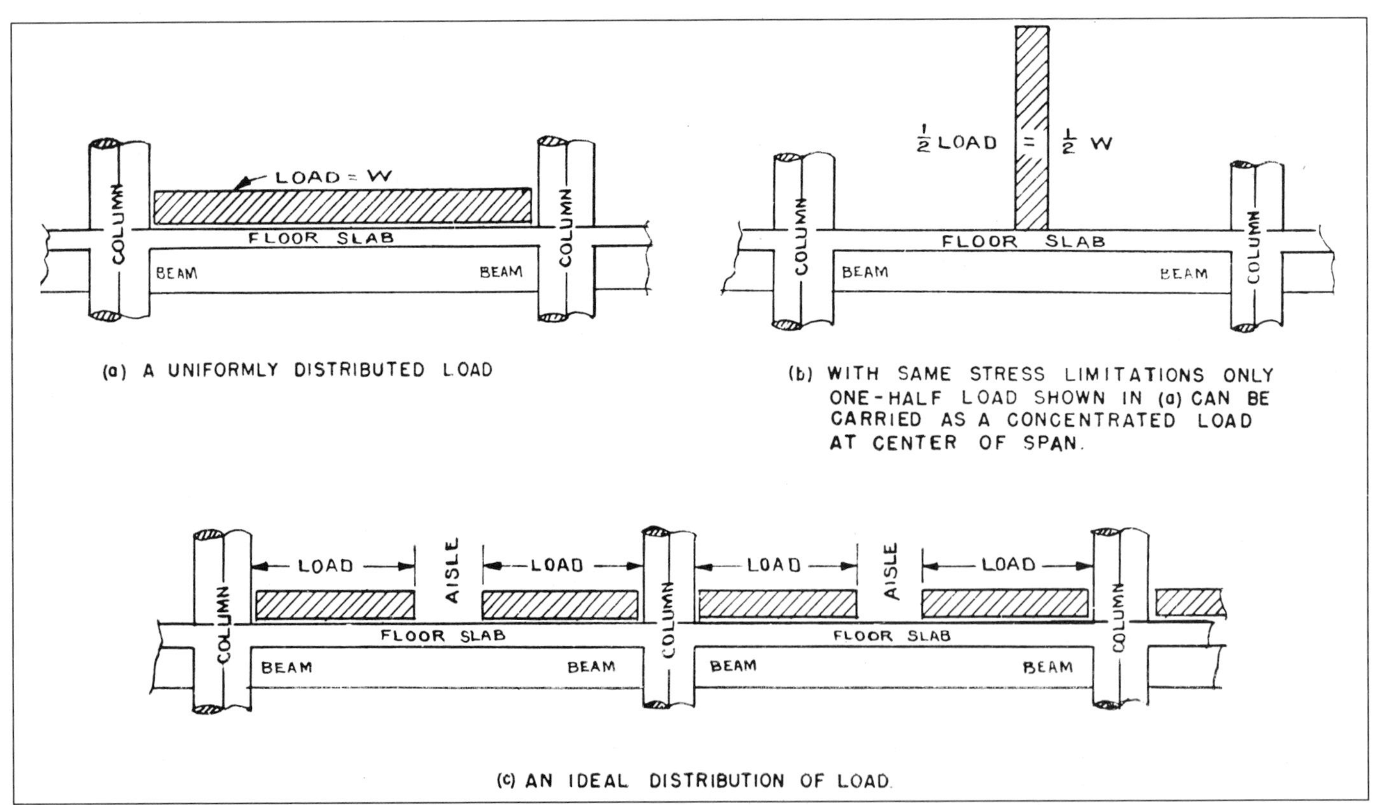

Figure 2–13. Various types of loadings. Changing the location of a load changes the total load a floor can carry.

Workstation Design

When designing employee workstations, keep the following suggestions in mind:

- Place controls where they require the least amount of movement.
- Provide lighting that is suitable to the task, as opposed to general illumination.
- Provide jigs and fixtures that relieve pressure.
- Provide a workbench so that workers may either sit or stand as needed.
- Determine work flow patterns that are normal and easy for the worker.
- Provide audio and/or visual signals from machine operators.
- Preposition materials, equipment, products, and tools.
- Place tools, controls, and materials in the employee's direct line of vision.
- If both hands are needed at the same time, motions should (1) begin at about the same time, (2) be somewhat alike for each hand, and (3) end at the same time for each hand. Both hands should not be idle at the same time.
- Provide arm motions that are smooth and continuous, not straight, irregular, and sudden.
- Design the work flow so that the work process moves smoothly.
- Facilitate the next step in the work process with antrum storage of incomplete products.
- Plan circular layouts for street line production. (See also Chapter 16, Ergonomics Program, in the *Administration & Programs* volume.)

SUMMARY

- Numerous accidents, occupational diseases, explosions, and fires can be prevented by carefully planning the design, location, and layout of a new facility or alterations to an existing one.
- Ideally, health and safety professionals should conduct a safety and health study of proposed construction in the developmental stages to remove hazards and reduce risks. They should bear in mind the role of workers, the function of machines, and the flow of materials.
- Four major factors should be considered in facility design and layout: (1) general design of the workplace; (2) compliance with appropriate codes and standards; (3) size, shape, and type of buildings, processes, and personnel facilities needed; and (4) safety procedures and fire protection standards required.
- Selection of a site involves consideration of safety issues and possible hazards to the community; studying factors of location, climate, and terrain; and understanding the space requirements of the company.
- When planning outside facilities, safety and health professionals should ensure that issues of worker safety are incorporated into designs for company grounds, shipping and receiving facilities, and all roadways, walkways, trestles, and parking lots.
- The company must develop appropriate air pollution controls and waste disposal methods that conform to municipal and state or provincial regulations.
- As far as possible, facility railway hazards should be eliminated when a new facility is being designed. Employees must know the safety regulations and practices for operating and maintaining fuel-fired, electric, diesel, compressed-air, and battery-powered locomotives.
- Facility layout of buildings and facilities should permit the most efficient use of materials, processes, and methods and minimize the hazards of fire and explosions.
- Layout of equipment can be done with the use of detailed flow sheets and must be designed to ensure maximum efficiency and safety for workers.
- Both daylight and electric lighting can supply a facility's lighting needs. Proper illumination can help to reduce accidents, minimize hazardous areas, and make buildings and grounds more secure.
- Security considerations for facilities include reducing the number of openings, securing all windows, providing protective lighting, and installing alarm systems.
- Industrial designers and managers are paying more attention to the interactions of color, lighting, and human behavior. The light-reflectance value (LRV) of color refers to its effect on light, which can contribute to workers' ability to see a task or identify color-coded materials.
- Colors also reflect regional and gender preferences, affect employees' morale, alter workers' perceptions of their surroundings, and provide an effective way to mark hazardous items and safety signs.
- Access structures (stairs, ramps, and so on) must be designed for easy use, be clearly marked, kept clear and well maintained, and safeguarded with rails, banisters, or other safety devices.
- Conduct a careful study of each department in order to choose the best flooring material. In general, floors should be slip-, scuff- and scratch-resistant, easily cleaned, and provided with adequate drainage. Floors should be designed to carry both current and anticipated future loads safely.
- To reduce strain and injuries, workstations should be well lighted, be ergonomically sound, and facilitate the work flow.

REFERENCES

American Conference of Governmental Industrial Hygienists, 6500 Glenway Avenue, Bldg. D7, Cincinnati OH, 45211.

Industrial Ventilation.

American Insurance Association, 1130 Connecticut Avenue NW, Suite 1000, Washington, DC 20036.

Recommended Good Practice Requirements of the National Board of Fire Underwriters for the Construction and Protection of Piers and Wharves.

American National Standards Institute, 11 West 42nd Street, New York, NY 10036.

Criteria for Safety Symbols, ANSI/NEMA Z535.3–1998.
Electric Lighting Fixtures for Use in Hazardous (Classified) Locations, ANSI/UL 844–1996.
Environmental and Facility Safety Signs, ANSI/NEMA Z535.2–1998.
Fluorescent Lighting Fixtures, ANSI/UL 1570–1988.
High Intensity Discharge Lighting Fixtures, ANSI/UL 1572–1990.
Incandescent Lighting Fixtures, ANSI/UL 1571–1990.
Minimum Requirements for Sanitation in Places of Employment, ANSI Z4.1–1986 (R 1995).
National Electrical Code, ANSI/NFPA 70–1999.
Practice for Industrial Lighting, ANSI/IES RP7–1990.
Safety Color Code, ANSI Z535.1–1998.
Safety Requirements for Building Construction, ANSI A10 Series.
Safety Requirements for Portable Wood Ladders, ANSI A14.1–2000.
Safety Requirements for Workplace Floor and Wall Openings, Stairs, and Railing Systems, ANSI A1264.1–1995.
Scheme for the Identification of Piping Systems, ANSI/ASME A13.1–1998.

American Society of Heating, Refrigerating, and Air Conditioning Engineers, 1791 Tullie Circle NE, Atlanta, GA 30329.

Guide and Data Book.

American Society of Mechanical Engineers, 345 East 47th Street, New York, NY 10017.

Boiler and Pressure Vessel Code.

Chowdhury J. Chemical-plant safety: an international drawing card. *Chemical Engineering* (March 16, 1987), 14–17.

DiBerardinis L, et al. *Guidelines for Laboratory Design: Health and Safety Considerations.* New York: John Wiley & Sons, 1987.

Faber Birren and Company, 500 Fifth Avenue, New York, NY 10110. *Specifications of Illumination and Color in Industry.* Reprinted from *Transactions,* American Academy of Ophthalmology and Otolaryngology.

Factory Mutual System, Engineering Division. *Handbook of Industrial Loss Prevention.* New York: McGraw-Hill Book Co.

Gausch JP and Gausch AB. Avoiding catastrophic losses: a team approach to safety assurance. *Professional Safety* (Sept. 1985), 26–32.

Illuminating Engineering Society of North America, 345 East 47th Street, New York, NY 10017.

A Statement on Astronomical Light Pollution and Light Trespass, CP–46.
Calculation of Daylight Availability, RP–21.
Glare and Lighting Design.
Lighting Handbook.
Practice for Industrial Lighting, ANSI/IES RP–7–1983.
Protective Lighting, RP–10.
Recommended Practice for Daylighting.

Institute of Transportation Engineers, 525 School Street SW, Suite 410, Washington, DC 20024.

Jones CL. *Safety in Lacquer Plants.* Hercules Powder Company, 917 Market Street, Wilmington, DE 19801.

Kletz TA. Make plants inherently safe. *Hydrocarbon Processing* (Sept. 1985), 172–80.

Krivan SP. Avoiding catastrophic loss: technical safety audit and process safety review. *Professional Safety* (Feb. 1986), 21–26.

National Fire Protection Association, 1 Batterymarch Park, Quincy, MA 02269.

Air Conditioning and Ventilating Systems, NFPA 90A, 1993.
Fire Protection Handbook, 17th edition, 1991.
Flammable and Combustible Liquids Code, NFPA 30, 1993.
General Storage, NFPA 231, 1995.
Life Safety Code, NFPA 101, 1994.
National Electrical Code, NFPA 70, 1993.
National Fire Codes, twelve volumes.
Marine Terminals, Piers, and Wharves, NFPA 307, 1995.
Protection from Exposure Fires, NFPA 80A, 1993.
Rack Storage of Materials, NFPA 231C, 1995.

National Safety Council, 1121 Spring Lake Drive, Itasca, IL 60143.

Portland Cement Association, 5420 Old Orchard Road, Skokie, IL 60077. (General.)

U.S. Environmental Protection Agency, 401 M Street SW, Washington, DC 20460.

REVIEW QUESTIONS

1. List five of the eight factors to consider for the general design of the workplace.
   a.
   b.
   c.
   d.
   e.

2. Companies should ensure that which of the following review and approve their plans and specifications for new facilities or those that need remodeling?
   a. Their safety and health specialists
   b. Their insurance companies
   c. Governmental authorities
   d. All of the above

3. Name the specific safety code for electric wiring and electrical installations, and the organization that established it.

4. List four of the six factors to consider for the design of machine tools and equipment.
   a.
   b.
   c.
   d.

5. What should be done to protect pedestrians if pedestrian entrances must be located near railroad tracks or on busy thoroughfares?

6. Knowing the nature of wastes is essential for planning appropriate disposal methods. Which of the following wastes can be disposed of by burning in an incinerator?
   a. Poisonous materials
   b. Wood and paper
   c. Magnesium chips
   d. All of the above
   e. None of the above

7. Describe the steps that should be taken to address the problem of confined spaces.
   a.
   b.
   c.
   d.
   e.

8. Which type of lighting can provide higher illumination levels for small or restricted areas?
   a. Supplementary lighting
   b. Daylight
   c. General lighting
   d. Emergency lighting

9. What are the six security factors to consider when designing a facility's environment?
   a.
   b.
   c.
   d.
   e.
   f.

10. Why do safety engineers use warm colors to call attention to dangerous machine parts, fire hazards, and physical hazards?

11. Neutral colors of low light-reflectance values should be used in what type of working environment?

12. Red is the standard color used for:
    a. Highlighting dangerous parts of machines, such as exposed cutting edges
    b. Marking flammable liquid storage cabinets
    c. The radiation hazard symbol
    d. Identifying fire protection equipment, danger, and emergency stops on machines

13. List seven factors in determining the best type of floor for a particular location.

3

Construction of Facilities

Revised by

Thomas Broderick

Patrick J. Conroy

Joseph L. Durst, Jr.

J. Nigel Ellis, Ph.D., P.E., CSP, CPE

Richard Hislop

John C. Langford

John F. Montgomery, Ph.D., CSP, CHCM, CHMM

This chapter focuses on the management of construction and demolition operations carried out in an industrial environment. It is for those who need an overview of the safety and health regulations for construction operations done on their property and sometimes in building the first facility. The following topics are covered in this chapter:

- significant on-the-job safety issues in construction projects
- how to make excavation procedures safer
- safety considerations for basic construction equiment
- safe use of hoists and cranes
- safety concerns in using formwork and falsework
- methods and equipment used in fall protection
- using process safety management to ensure safe work practices

CONTRACTED SERVICES ON PREMISES

Industrial facility owners doing construction work themselves have complete control over their employees and, therefore, can require adherence to safe practices. When a company hires an outside contractor or subcontractor, the company and the contractor must understand their legal and working relationship. The company must insist that the contractor comply with all provisions of local, state or provincial, and federal safety and health regulations that pertain to the construction work itself. (See the section on Contracts later in this section.) The construction safety and health professional must know both the risks involved in construction and those posed by the environment of the worksite.

Do not rely on local, state, or national building codes. Their main purpose is to provide minimum requirements for construction techniques to ensure a structure is built properly in accordance with recognized practices and peculiar local requirements, i.e., earthquake bracing in California, hurricane wind loads in Florida.

The basic reference manual should be the federal OSHA construction safety standard, 29 *CFR* 1926. This will provide a road map for the minimum acceptable level of safety performance for most operations on a given construction project. Also, some states have qualified OSHA plans that may impact certain construction sites. Always research particular state and local requirements before beginning a project.

Building codes and safety standards are two external forces that should be used by both contractors and safety and health professionals to create a safe project. Other equally important and often overlooked aspects are communication and training. Communication should begin with early design criteria and carry forward to each coordination meeting. Training, whether conducted by each contractor or by the safety and health professional, should become an integral part of all aspects of a project. Codes, standards, communication, and training will provide the necessary tools to complete a project in a timely and safe manner.

Architects and engineers should "think safety" when they are designing and drawing up specifications for the facility. In other words, design safety into the final product, as well as into its construction. (See Chapter 2, Buildings and Facility Layout, in this volume.) With this in mind, involve safety and health professionals at the earliest planning stages, as discussed in Chapter 2. They should discuss the proposed job with the construction department or with the outside contractors. Many safety-conscious contractors try to get a running start on a good safety program and will even insist on a safety conference before starting a job.

Analyzing the Cost of the Construction Job

Advance analysis of construction jobs is not new. In fact, some sort of analysis is made on every construction job. To bid competitively, a contractor must thoroughly analyze a job to compete with other bidders and avoid financial loss. Practical experience in running construction jobs proves that contractors cannot get very far without a set of prints and specifications that clearly define the work to be done. Accident prevention is a legal requirement and is often included in construction specifications. Unfortunately, many contractors start jobs without determining the financial advantages of accident prevention and do little advance planning to eliminate or reduce the risk of accidents.

In many cases, contractors subcontract part of the work to manufacturers and/or suppliers that install it under a purchase order. Usually the installation is done by a subsidiary of the manufacturer or by a subcontractor. Based on contract language, subcontractors and sometimes even suppliers may be required to comply. Most standard contract forms used today have a "flow-down" clause that obligates contractors to pass the requirements of the contract down to lower tiers of all levels. In those instances, be sure that suppliers comply with all the language, specifications, and general conditions of the contract that apply to the entire project.

Most contractors divide their bids into materials, labor, equipment, overhead, and profit. Insurance is an item included in overhead. The amount of insurance placed in the bid is usually a percentage of the cost of material, labor, and facility. The amount varies with the type of job and is affected to some extent by the requirements of owners, architects, and the state or province in which the job is located. This insurance item includes the expenses for the performance and payment bonds, Workers' Compensation, Social Security, unemployment insurance, employer's liability, property damage, automobile insurance, fire insurance, builders' risk insurance, and other types of coverage that the contractor may select. Insurance is considered one of the fixed charges against the job. Many people believe that, like death and taxes, little can be done about it.

In most situations the costs of Workers' Compensation insurance are not fixed but vary with the loss record of the contractor during a previous time period. Some contractors claim that there is no one item in the entire bid that can more easily produce a substantial saving. Some contracts have a "Hold Harmless" clause. A hold harmless agreement is one risk management technique that most owners insert in contracts with their contractor, where they agree to be held harmless due to any action/inaction by the contractor or the contractor's employees or their subcontractors. This may relieve the owner (contractually) from legal exposure based on the contractor's performance as included within the contract, but these clauses are not legal in all states. According to that clause, any loss is charged against the contractor and is consequently reflected in the contractor's insurance costs. Certain states may have restrictions as to how hold harmless clauses can be applied.

If such costs can be saved on insurance, the next questions are why have they been overlooked and why has there not been more enthusiastic interest by contractors generally. Perhaps the safety and health professional has inadequately analyzed the construction job or has not properly presented the job to management.

Safety and health professionals think and talk mostly in terms of hazards, usually involving personal injury. They use terms like "accident frequency," usually meaning frequency rates of disabling injuries. From this past data, especially the rates associated with the number of "lost work day" incidents and the number of "lost work days," one can see a snapshot of a particular contractor's loss experience of workers' injuries and the severity of those injuries. To get an accurate snapshot, these data should be provided for at least three years, taken from the contractor's OSHA 200 form. In addition to the contractor's OSHA 200 form, three years' Experience Modification Ratio (EMR) and actual loss data should be collected for the prime contractor and all major subcontractors.

An EMR is an insurance industry measure based on the client's loss history for three years, not including the previous year, which is used as a modifier for rating purposes. A modification of 1.00 can be termed neutral; such a company's workers' compensation premium is multiplied by one, so is used as a starting point for new clients, newly formed corporations, etc. EMRs above one indicate loss experience greater than the average for like companies in comparable industries with the same size workforce and exposures. Conversely, an EMR lower than one indicates a loss experience better than comparable companies.

Chapter 14, Environmental Management, Chapter 16, Ergonomics Program, and Chapter 17, Employee Assistance Programs, of the *Administration & Programs* volume, give a detailed discussion of how to understand accident statistics and use them to prevent accidents.

Accident prevention data has its greatest value when it influences construction plans in the following ways:

- produces a statement of the contractor's accident record in terms of cost, as one item of the bid
- defines the risk of accidents in specific and measurable terms
- sets up practical and effective safeguards to control these risks

Accident prevention is a real factor in the economic success of any construction job. It means lower costs and greater efficiency. Fortunately, we know how to prevent many accidents.

Contracts

Contracts should specify that the contractor meet certain minimum safety, health, and equipment requirements. These requirements should include provisions for protecting the company's employees and equipment, as well as the public, from construction hazards. At a minimum, note the applicable regulations and industry standards in the contract. Formally stating these minimum requirements in the contract makes subsequent enforcement easier. Depending on particular state laws and regulations, for example, in the state of California, a contractor may need to provide a site-specific safety and health program (in California this is called an Injury and Illness Prevention Plan), which focuses on a specific construction project site.

A company that hires a contractor must have a specific and agreed-upon plan for handling contractor-safety performance issues. Putting a lot of safety requirements in the contract and then not enforcing them may actually put the company at greater risk during a lawsuit. Also keep in mind that construction operations that expose host company employees may result in OSHA fines to the host company.

Note that in all construction operations the contractor must comply with applicable codes, laws, and ordinances. Some firms issue a booklet to contractors that specifies safety responsibilities and that gives details about the safety requirements specified in the general conditions of the contract. Examine contractors' previous accident records and workers' compensation records when selecting a contractor.

For example, in two general contracting companies, both with annual sales of $40 million, with 400,000 employee exposure hours, one has an EMR of 1.45 and the other has an EMR of 0.75. The EMR may indicate the depth of commitment to work safety but only within the context of all relevant data. Do not rely solely on a low EMR to select or a high EMR to reject a contracting company.

One of the safety and health professional's strongest tools may be the prequalification form. This form, sent to the contractors, requests a list of data to be received and an overview of their safety and health program. A prequalification form is a list of questions in an organized format that should offer a relevant, accurate, timely, and complete history of a company's safety and health history. Example questions or requests:
Please provide the following:

- the OSHA 200 form for the previous three years (they will usually only copy the right side, which lists cumulative totals for fatalities, lost work day incidents, number of lost work days, no lost work day incidents, number of restricted work days and the number of illnesses)
- the EMR for the previous three years, preferably on an insurance agent's or carrier's letterhead
- a copy (or synopsis) of a company's safety manual
- a copy of the written hazard communication program (the written program should be sufficient unless material safety data sheets are required by the client)
- a brief outline of safety and health training provided to supervisors. The request may be more detailed—prequalification forms can be customized to suit particular needs. With this data the client can be more confident in the selection process

Guidelines for Contractors' Safety and Health Programs

Contractors' safety and health programs provide the basis for communicating safety and health policies and administrative and technical requirements. Contractors should provide all employees with a copy of and orientation to the program. Depending on the size and complexity of the project to which the program is applied, it should cover the following requirements in clear, specific language:

- responsibilities for safety and health, including reporting hazards and accidents, obtaining and using protective equipment, conducting safety inspections, maintaining a safe and healthful work environment, and enforcing safety and health requirements
- procedures for conducting safety and health orientation and periodic (usually weekly) training sessions
- procedures for reporting accidents
- procedures for obtaining first aid and emergency treatment
- procedures for reporting work hazards
- requirements for subcontractors and suppliers
- procedures for testing and certifying equipment, if required
- physical requirements for employees
- jobsite sanitation
- the use and purpose of equipment lockout and confined space entry
- the technical requirements—personal protective equipment, hazardous materials, lighting, fire protection, welding and cutting, electrical, safe clearance, tools, material handling, rigging, machinery and equipment, pressurized systems, access, scaffolding, demolition, floor and wall openings, excavation, formwork and falsework, steel erection, blasting, etc.

A Sample Safety and Health Program

One major contractor makes an active safety and health program part of everything a worker does on a construction job. The enthusiasm generated by this program gives a real awareness of job safety as a whole and makes each worker individually safety conscious. The contractor's project manager sets up a program of safety and health measures equal to the size of the project and its conditions and hazards. This safety and health program involves the following:

- A full- or part-time safety and health professional.
- Facilities provided that have design-in protective devices to help prevent accidents, such as suitable roads, lights, barricades, signs, warning devices, and guardrails.
- Safety devices designed to lessen injury, such as safety hats, eye protection, safety belts, and similar protective equipment provided as necessary.
- A system to create and review job hazard analysis for each high-hazard operation. Such operations, in general, but customized for each project, should include confined space entry, trenching and excavations greater than five feet, demolition operations, steel erection, heavy lifts, and other types of operations that need careful planning and coordination between parties (see illustration, if necessary).
- Crew foremen who hold brief meetings (tailgate safety meetings) at least once a week, with workers under their supervision to discuss the safety operation of their crew, the safety of other workers, and specific problems related to job planning. The crew foremen should document the meetings on a "Tailgate Safety-Meeting Checklist." This list records what was discussed and the date and time. It is signed by all in attendance. Each list is to be kept on file in the construction safety office. This meeting is usually held after the project's safety committee meeting.
- A detailed list of responsibilities and accountabilities.
- A project safety committee comprising key personnel, and in which all key contractors' supervisors are also invited to participate. On some jobs, union representatives and workers are also on this committee. Membership is rotated so that the maximum number of persons will have some opportunity to participate actively in the committee's work.
- The committee meets weekly and submits a review copy of the minutes, which includes accident statistics, to the safety department in the contractor's office. The safety committee discusses accidents and incidents of the previous week. It also points out other hazards, evaluates housekeeping, and makes plans for correcting unsafe conditions and unsafe practices before they lead to an accident. Periodically, a portion of some meetings is spent on first aid instruction, CPR techniques, fire fighting, and other emergency procedures.
- Special safety meetings called by the project engineer and attended by all workers.
- Safety instructions given to all new workers as part of their first day's indoctrination and the training documented.
- A safety representative of the week, appointed on a rotating basis at each committee meeting, who acts as the project's safety inspector and submits a report to the committee.
- Adequate first aid facilities and trained personnel.
- Safety materials forwarded periodically from headquarters.
- Periodic visits to projects by management and the safety supervisor. While there, these people inspect the job and participate in the safety program. Before work starts, the project manager and staff talk with each contractor's supervisory personnel in order to explain how the contractor's work will proceed in relation to the work of others. The project's safety and health program and the part of the contractor are discussed. The contractor's key supervisor is invited to serve as a member of the project safety committee.

Preconstruction Safety and Health Conference

If the job is an addition or a renovation, acquaint the contractor's representative with (1) the facility's safety program, (2) first aid facilities, (3) emergency procedures, (4) any special safety equipment that is required because of hazards in the facility's operations, and (5) how this safety equipment can be obtained.

Before holding the conference, develop a proposed accident-prevention program. At the conferences, all involved supervisors should discuss the proposed program and develop mutual understandings about the overall aspects of the plan. Provide for periodic staff meetings of supervisors to evaluate and revise the program as required by changing conditions and new problems that may arise. Hold a safety meeting when any new contractor starts the project.

Contractors should be required to submit a job hazard analysis or safety procedures for given phases of the work before that phase is allowed to begin. These items should be a deliverable under the contract. If the job is a "hard dollar" bid, all requirements must be stated to prospective bidders in the specifications for bid. Agreements made in the preconstruction meeting may not have any legal standing (for enforcement) unless they are based on requirements in the bid specifications and the signed contract.

A suggested outline for the preconstruction safety conference follows.

A. Preconference activities
   1. Define the conference's purpose
      a. Evaluation of proposed program
      b. Discussion of job organization and operating procedures
      c. Preplanning the work and agreeing to a means for applying standard procedures
   2. Notify all parties
   3. Evaluate proposed program
   4. Decide on conference facilities

5. Determine meeting attendance
6. Determine how conference will be recorded (minutes of meetings)

B. Agenda for the conference
1. Orientation
   a. Explain why the program is necessary
   b. Advantages in terms of economy and efficiency
   c. Prescribed safety standards
   d. Review
      (1) Accident prevention agreements
      (2) General conditions of specifications on safety
      (3) Special conditions of specifications on safety
      (4) Lockout/tagout
      (5) Hazard communication
   e. Other requirements—local, state or provincial, and federal
   f. Supervision
      (1) Organization of, at project site
      (2) Functions of personnel at the site
      (3) Responsibilities and accountability
      (4) Delegated authorities
      (5) Relations regarding enforcement and discipline
2. Discussion of proposed program
   a. Plans about layout of temporary construction, site, buildings, etc.
   b. Action taken toward planning and coordinating activities between different operations and crafts
   c. Access to work areas
   d. Safety indoctrination and safety education
   e. Delegation of safety responsibilities to supervisors
   f. Integration of safety into operating methods and procedures
   g. Housekeeping program
   h. Safety factors in job-built appurtenances
   i. Traffic control and parking facilities
   j. Fire protection
   k. Lighting, ventilation, protective apparel, and medical care
   l. Safe operating conditions and maintenance of equipment
3. Discussion of follow-up procedures
   a. Methods for meeting objectives
   b. Plans for periodic readjustment of safety objectives
   c. Handling safety deficiencies
   d. Arrangements for additional meetings and periodic staff meetings
   e. Following up on agreements made in the preconstruction meeting

Three major rules to observe for a workable safety program follow:

- All agreements must be fair.
- Keep paperwork to a minimum.
- Keep the program simple and have it deal with facts.

Protection of Employees and Equipment

When construction work is being done in or around an industrial facility, the company should protect its employees and equipment from all construction hazards. These include open excavations, falling objects, welding operations, dust, dirt, temporary wiring, and temporary overhead electrical lines. Isolate the construction work from normal facility operations if at all possible. Set up barricades, fences, and guardrails and post appropriate warning signs (Figure 3–1). Through public relations, keep the company's employees and members of the public informed of the construction activities and any possible related hazards.

When a construction job is being done in an area that must be kept in operation, erect a sheeted bulkhead to keep out dust and dirt and to isolate the operation as much as possible. Flame-retardant materials may be required in some cases.

At times, construction operations must be done where flammable vapors and liquids are present or in other hazardous areas. In such instances, the contractor's employees should follow the same rigid fire prevention requirements that apply to the facility's employees. Before employees of the contractor cut, burn, or weld, the contractor should obtain a permit from the facility's engineer or safety department. In this way the necessary fire and safety regulations will be met. Contractors should also follow the usual requirements for ventilation and health protection. In an area where no flame is permitted, the contractor should (1) shut down the manufacturing process while welding and cutting are done, or (2) use screwed or bolted fittings instead of welded connections. (See Chapter 21, Welding and Cutting, and Chapter 11, Fire Protection, in this volume.)

The contractor should provide night lighting, and supplemental daytime lighting, where necessary. This is especially true in areas where open trenches or ditches create hazards in walkways and roadways. A minimum of 5 footcandles (53.8 lux) is recommended by the Illuminating Engineering Society of North America for

Figure 3–1. Typical safety stickers for plant use.

vital exterior locations or structures. (See data on illumination in the Appendix, Safety Tables, in this volume.) Lighting equipment used in a flammable or explosive atmosphere requires special hazardous location fixtures and wiring procedures.

The contractor and the facility's management should cooperate closely at all times to determine how best to prevent accidents. They should maintain especially close liaison during tie-in of new piping or equipment to existing lines and equipment, following appropriate blank-out or lockout/tagout regulations. This will ensure the safety of subsequent operations. The contractor should maintain all work in an orderly, well-kept manner.

Permits

When new building construction, modification, demolition and addition work will interface with existing operations, or when facility maintenance personnel could be included, then a decision about whose permit system will be used must be made. Permits, particularly lockout/tagout, confined space entry, and hot work, must be coordinated between the client and his or her contractor.

There are only two options: use the company's permit system or the contractor's system. Either will create additional training on the other party. If extensive interface is anticipated, then the use of the client's system(s) may be appropriate. This choice should be based on the company's accident prevention philosophy, the degree of interface anticipated, and the comfort level with the operations.

All parties must be "on the same page" to avoid incidents related to mixed systems or methods. There must be absolute agreement and understanding of what permit systems are being used. Once a system is chosen, it should be used consistently. The owner and contractor should periodically review the system's effectiveness to ensure that it is serving the intended purpose.

Machinery and Equipment

Modern construction requires a variety of machines: materials-handling equipment for site preparation, power shovels for excavating, concrete mixers or supply trucks for structural members, and many others. Do not place a machine or piece of equipment in operation until it has been inspected by a qualified person and found to be in safe operating condition. Depending on the type and use of the equipment, it may be necessary to inspect the unit daily, monthly, and/or annually. The inspection reports should be kept on file. Do not use equipment found to be defective until corrections or repairs are made.

Only designated, authorized, and properly trained personnel should operate machinery and equipment. Personnel should operate the machinery and equipment so persons and property are not endangered. Operators should observe safe speeds and load limits. Operators should not leave machinery unattended or equipment running.

Management Participation with Construction Projects

A number of studies have outlined some practices that management can employ to enact the safety and health program onsite.

Beginning with top management, they need to personally stress the importance of safety through formal and informal field contacts. They also need to talk about safety at the company level. Top managers need to be aware of the safety program and what is going on.

Contractor managers should receive specialized training with regard to the requirements and implementation of the safety program.

Middle management can add to the program by developing effective rapport with supervisors and workers, by maintaining even job and work pressure, actively supporting safety policies, and accepting some responsibility for job conditions. Also included should be new worker orientation, involvement of worker-supervisor disputes, and showing respect for supervisors.

Trucks and Other Mobile Equipment

Contractors working either inside or outside the facility should take great care to prevent their trucks and other mobile equipment from colliding with pipelines, power lines, and other equipment. In this way, they will avoid interrupting the manufacturing or processing operations. Before mobile equipment is moved, contractors should survey the area in which it is located to check for overhead wires, pipelines, excavations, invisible ground conditions, and similar hazards. Timber mats may have to be provided. Equipment with high clearances, such as cranes, should not be moved into or out of, or operated in, any area containing electric power lines until the approval of the superintendent has been obtained. No part, including the load, may reach within 10 ft (3 m) of electric lines carrying up to 50 kv and an additional 4 in. for every 10 kv above 50 kv, unless power in the lines is shut off. If local laws specify greater distances, these should prevail. (See also Chapter 15, Hoisting and Conveying Equipment.)

One method of handling construction deliveries is having a signaler serve as the eyes for the truck driver. Standard signals for ready-mix concrete trucks are shown in Figure 3–2. If there is no signaler, install reverse alarms on all heavy mobile equipment and trucks. Also be sure barricades, guardrails, and warning signs are in place to assure maximum safety.

Statistics show that a large number of construction injuries are related to material handling. Special attention should be given to delivery sequencing and material storage in order to minimize material handling exposures. To avoid extra handling and vehicle movement, contractors should notify suppliers about the exact location to make deliveries.

When trucks, bulldozers, powered wheelbarrows, and other mechanized construction equipment are to be operated within a facility, the contractor and the facility's management should agree upon the traffic flow. In that

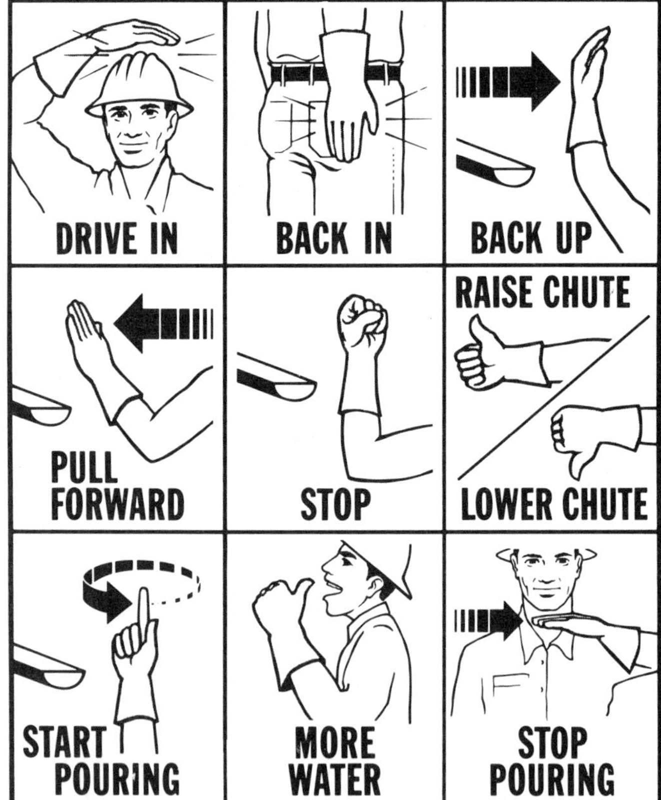

Figure 3–2. Standard signals for mixer drivers and contractor's guides. (Printed with permission from Northern California Ready Mixed Concrete & Materials Association.)

way, the areas in which the construction equipment is to be operated will be known and isolated where possible. The contractor's key personnel and the facility's affected personnel should receive drawings of these areas.

Vehicles and other equipment powered by internal combustion engines emit carbon monoxide (CO) into the air. Where engines are operated inside existing facilities, next to air inlets or any other area where CO can build up, then methods must be implemented to vent the area, provide adequate fresh air, or direct the exhaust via flexible ducting to powered vents. In any case, where engines will be operating inside, CO monitoring must be accomplished.

Sometimes it may be necessary to transport employees in trucks from one location to another within the work area. Unless this is controlled, this operation can become a major source of serious injuries. Provide seats for each person required to ride on vehicle equipment. Supply safety belts for all passengers who ride in the cabin with the driver. Do not permit workers to ride on a loaded truck or other machines not equipped for the purpose. (See National Safety Council (NSC) Industrial Data Sheet 330, *Motor Trucks for Mines, Quarries, and Construction*.)

When people are boarding or descending from a truck, the truck should be standing still. Prohibit workers from getting off or on moving equipment. Provide a boarding ladder. If necessary, provide a bus to transport employees to the worksite.

Vehicles that will be moving slower than normal traffic at night should have a yellow flashing light or four-way flashers visible from all directions. Equipment left unattended at night, adjacent to a highway in normal use, or adjacent to construction areas where work is in progress should have lights or reflectors, or barricades equipped with lights or reflectors to identify the location of the equipment.

Whenever equipment is parked, operators should set the parking brake. Equipment parked on inclines should have the wheels chocked, or track mechanism blocked, and the parking brake set. Bulldozer blades and scraper blades, end-loader buckets, dump bodies, and similar equipment should be either fully lowered or blocked when being repaired or when not in use. Personnel should not work or pass under the buckets or booms of loaders in operation. No modifications or additions that affect the capacity or safe operation of equipment should be made without the manufacturer's written approval. Tools should not be left suspended from cranes, hoists, etc.

Guarding, Safety Devices, Platforms, and Means of Access

Provide guarding for belts, pulleys, sheaves, gears, chains, shafts, clutches, drums, flywheels, and other reciprocating or rotating parts of equipment. Do not remove or make ineffective any guard, safety appliance, or device unless immediate repairs or adjustments are required, and then only after the power has been shut off. Replace guards and devices as soon as repairs and adjustments have been completed.

Properly insulate or guard current-carrying parts of electrically operated equipment. Properly ground all noncurrent-carrying metal parts. Cover high-temperature lines and equipment, located where they endanger employees or create a fire hazard, with suitable insulating materials.

Install platforms, footwalks, steps, ladders, handholds, guardrails, and toeboards on all equipment where they are needed to provide safe access. Provide suitable floors or platforms, surfaced with slip-resistant material, for all equipment operators.

Operators of equipment should be protected against the elements, falling objects, swinging loads, and similar hazards. Be sure that windows in cabs or enclosures on equipment are made of safety glass and kept in good repair at all times.

Provide safety devices to prevent unauthorized persons from starting equipment. This could simply be a key ignition system or simply blocking and locking the starter. At the end of a work shift, operators should set and lock equipment. This prevents it from being released, dropped, or activated in any way. The manufacturer's procedure for shut down should be followed.

Do not permit debris, oil, grease, oily rags, and waste to accumulate on equipment. Post safe-load capacities and operating speeds on all equipment. Be sure each piece of equipment is placed on an adequate foundation and properly secured.

Repairs

Shut down all hazardous equipment for repairs. Post suitable signs, and do not remove them until repairs have been completed. Move mobile equipment, if possible, to a safe location where operations will not interfere with the repair work. Block or crib equipment suspended in slings or supported by hoists or jacks for repairs before anyone is permitted to work underneath it. For equipment repairs made away from the source of power, such as conveyors and cable ways, use chains, blocking, or similar devices. Such precautions will prevent injury in case the equipment is accidentally started.

Before repairs on electrically powered equipment are begun, lock the main switch in the open (OFF) position. The person doing the repairs should retain the key to the switch lock. If there is more than one repair person, each should lock the main switch with a personal lock and retain the key. (See Chapter 6, Safeguarding, for complete details on the lockout/tagout procedure.)

Demolition of Structures

Facility personnel should do only minor demolition. Employ specialists in the field if structures to be removed look as though they will present a problem.

Wrecking specialists are familiar with the procedures and precautions necessary to do the work safely. They know how to protect the public and adjacent property, and they know the applicable federal, state or provincial, and municipal codes and regulations.

Following are some fundamental safety procedures and suggestions for minor demolition work:

- Make provision to keep the public and unauthorized facility employees at least 15 ft (5 m) away from the structure.
- Have a competent person make an engineering survey of the structure. Determine the condition of the framing, floors, and walls, and check for any unanticipated conditions. Check for hazardous chemicals, gases, explosives, flammable materials, asbestos, hazardous waste, electrical circuits that may be engaged, etc.
- Have a competent person or qualified engineer detail the demolition sequence.

Sometimes the conventional methods for altering or removing concrete installations are unfeasible or undesirable. In such cases, use (1) "powder cutting," a process that substitutes penetration by intense heat for concussion breakage, or (2) explosives.

EXCAVATION

To prevent injury and property damage during excavation work, make adequate protective measures part of the job. Study preexcavation conditions, such as superimposed loads, soil structure, and hydrostatic pressure, using job hazard analysis. From such a study, it is possible to evaluate changes that might occur, to prepare for situations that might develop, and to plan the job ahead. Excavation permits should be used to ensure that all the necessary precautions are taken to protect employees. The excavation must comply with the OSHA excavation standard. Protective systems for trenches or excavations twenty feet deep or deeper must be designed by a registered professional engineer.

Underground Utilities

A major hazard in urban or built-up areas is the presence of underground facilities, such as utility lines (water, electric, gas, or telephone), tanks, process piping, and sewers. If this equipment is dug into, undercut, or damaged in any way, there may be injury or death to workers, interruption of service, contamination of water, disruption of processes, and expensive delays. Many states have a one-call system for locating buried lines. In one call, all utilities can be identified and all underground utilities will be marked.

Before starting operations, consult the company or facility engineer, the utility companies' engineers, and the city's or town's engineers. The location of various facilities and their approximate depth below ground must be determined and marked by stakes in the ground or by markings on the floor.

Electronic locators can be especially helpful where an excavation would cross numerous buried obstacles. If the facilities are to be left in place, they must be protected against damage, and sometimes also against freezing.

Indicate the contents of buried tanks and piping on the location markings. If the contents are flammable or toxic, have proper protective equipment readily available in case of rupture. Also indicate the bottom depth of the tank.

Do not allow any shovel, dragline, or other digging machine to excavate close to underground facilities that must be left in place. Establish a depth limit for digging machines, and complete the excavation by hand digging. If personnel are working in a trench deeper than 5 ft (1.5 m), provide adequate bracing and shoring or have the trench slope. When hand excavation is being done, warn workers about driving picks, paving breakers, or other powered tools through buried facilities.

Specialty Contractors

Award the job to a specialty contractor whenever an excavation must be made (1) within or adjacent to a building, and (2) lower than wall or column footings and machinery or equipment foundations. The contractor's personnel should make a thorough study to determine the amount and strength of shoring required before work on an open excavation is begun. Such a study will include the nature of the soil, hydrostatic pressure, superimposed loads (both static and live), and other factors.

The depth and location of the excavation and the other characteristics will determine the need for sheet piling, shoring, and bracing, which should be designed by an engineer or other person with experience in this type of design. If underpinning, deeper support under an existing column, wall, or machine, is necessary, have it done before the open excavation is carried down to final grade.

Do not design protective systems strictly on the data from soil borings. Also consider anticipated facility-generated vibration, drying of soil, and water saturation. If cutting banks to angle of repose is used instead of shoring, that angle in many cases should be the next lower than the tables establish for the kind of soil reported.

When relying on angles of repose for protection from cave-ins of excavation or trench walls, do not rely on general soil borings. They may give a false reading not consistent with actual conditions of disturbed soil. A competent person, trained in soils analysis, must classify the soils on-site. Slopes or supports shall be based on soil type.

Barricades at Excavation Sites

Place excavated material at least 24 in. (0.6 m) from the edge of the excavation unless support system takes into account the superimposed load and the possibility of fallback. Tarpaulins, sheeted barricades, or low built-up board barricades can be used to confine the excavated material to the immediate area under construction. Do not permit excavated material to accumulate in work areas or aisles; have them trucked or otherwise removed from the building.

Barricade excavations to prevent employees and others from falling into them. When an excavation must remain open for the duration of the construction work, barricades, fences, horses, and warning signs are necessary. In some cases, watchers and flaggers may be needed. Guard the work area at night with flares, lanterns, or flashing lights.

Guard the sides of all excavations in which employees are exposed to danger from moving ground with a shoring system, sloping of the ground, or other equivalent means. The minimum slope in any soil, with the exception of solid rock, should be no less than three-fourths horizontal to one vertical.

Base shoring, sloping, and supporting systems (bracing, shoring, cribbing, etc.) on careful evaluation of factors such as (1) depth of cut; (2) soil type; (3) possible variation in water content of the material while the excavation is open; (4) anticipated changes in materials from exposure to air, sun, water, or freezing; (5) loading imposed by structures, equipment, overlaying material, or stored material; and (6) vibration from equipment, blasting, or traffic.

Have a competent person approve the excavation safeguards and inspect shoring, sloping, and supporting systems daily and after every rainstorm or other hazard-increasing occurrence. Increase protection against slides and cave-ins, if necessary. If evidence of possible cave-ins or slides is apparent, all work in the excavation should cease until the necessary precautions have been taken to safeguard the employees.

Except in hard rock, excavations below the level of the base of footing of any foundation or retaining wall should not be permitted unless the wall is underpinned and all other precautions taken to ensure the stability of the adjacent walls. If the stability of adjoining buildings or walls is endangered by excavations, provide shoring, bracing, or underpinning designed by a qualified person. Have such shoring, bracing, or underpinning inspected daily or more often, as conditions warrant. Where it is necessary to undercut the side of an excavation, safely support overhanging material.

Groundwater should be controlled. Freezing, pumping, drainage, and similar control measures should be planned and directed by a competent engineer. Consider the existing moisture balances in surrounding soils and the effects on foundations and structures if it is disturbed. When continuous operation of groundwater-control equipment is necessary, provide an emergency power source. Use diversion ditches, dikes, or other means (1) to prevent surface water from entering an excavation and (2) to provide good drainage of the area adjacent to an excavation.

Provide walkways or bridges with guardrails where people or equipment are required or permitted to cross over excavations. Where personnel are required to enter excavations over 4 ft (1.2 m) in depth, provide sufficient stairs, ladders, or ramps. Locate them so as not to require more than 25 ft (7.6 m) of lateral travel. When access to excavations more than 20 ft (6.1 m) in depth is required, provide ramps, stairs, or mechanical personnel hoists. Ladders used as accessways should extend from the bottom of the trench to not less than 3 ft (91 cm) above the surface. Provide at least two means of exit for personnel working in excavations. Where the width of the excavation exceeds 100 ft (30 m), provide two or more means of exit on each side of the excavation.

When mobile equipment is used or allowed next to excavations, install substantial stop logs or barricades. Operators of excavating or hoisting equipment should not be allowed to raise, lower, or swing loads over personnel in the excavation unless those personnel have substantial overhead protection.

In locations where oxygen deficiency or gaseous conditions are known or suspected, test the air in the excavation prior to the start of each shift, or more often if directed by the designated authority. A log of all test results should be maintained at the worksite. If air is not within specification, ventilation may have to be provided to improve the condition. Test until air is safe. A confined-space entry procedure should be mandatory in excavations with oxygen deficiencies or gaseous conditions. Where such conditions are suspected or are likely to develop in an excavation, have readily available emergency rescue equipment such as breathing apparatus, safety harness and line, and emergency medical supplies.

Trench Excavation

Trenches more than 5 ft (1.5 m) deep must be shored, laid back to a stable slope, or provided with other equivalent protection where employees may be exposed to moving ground or cave-ins per OSHA regulations. Trenches less than 5 ft (1.5 m) deep also should be protected when studies show hazardous ground movement may be expected. Bracing or shoring of trenches should progress with the excavation. See Figure 3–3 for four trench-bracing methods.

Trench boxes or shields must be rated by a professional engineer. Tabulated date showing the capability of a particular trench box must be on-site. Portable trench boxes, sliding trench boxes, or shields should be designed, constructed, and maintained to provide protection equal to or greater than the sheathing and shoring required for the situation. Cross braces or trench jacks should be in true horizontal position, secured to prevent sliding, falling, or kick-outs.

Backfilling and removal of trench supports should progress together from the bottom of the trench. Release jacks or braces slowly. In unstable soil, use ropes to pull out the jacks or braces from above, after personnel have cleared the trench.

In hand-excavated trenches, spike or bolt wooden cleats to join the ends of braces to stringers. This will prevent the braces from being knocked out of place.

In a long machine-excavated trench, a sliding trench shield may be used instead of shoring. Sliding trench shields generally are custom made to size for a specific job. They must be designed and fabricated strong enough to withstand the pressures that will be encountered. Metal, portable hydraulic shoring systems are also available.

LADDERS

Construction of all ladders should conform to the provisions of the applicable ANSI standards. Special-use climbing equipment, such as a combination stepladder-work platform, should comply with the applicable codes. Extension ladders and step ladders are rated extra heavy-duty, type IA, heavy-duty type IIA, and standard and household duty. On construction projects, unless type IA is used, the ladders must be either tested to determine that they can withstand 3.3 times maximum intended load or be manufactured and used in accordance with various applicable standards. A proactive stance, and one which is much simpler, would be to allow only type IA extra heavy-duty ladders on projects.

(For detailed information on ladders, see ANSI A14.1, *Safety Requirements for Portable Wood Ladders*; ANSI A14.2, *Safety Requirements for Portable Metal Ladders*; ANSI A14.3, *Safety Requirements for Fixed Ladders*; ANSI A14.4, *Safety Requirements for Job-Made Ladders*; and ANSI A14.5, *Safety Requirements for Portable Reinforced Plastic Ladders*.)

Fall Protection for Ladders

Fall protection for portable ladders and fixed ladder considerations begin with a stable structure; in this case a ladder that does not appreciably move when subject to forces anticipated in a fall.

Portable ladders must be tied off and stabilized at the earliest opportunity in such a way that they cannot tip to the side or slide in any way. Walk-through ladders that are tied off afford the best means of mounting and dismounting and the possibility of providing railings with no gaps for those places with multiple accesses foreseeable.

The mandatory three-point control while climbing helps ensure that a worker's center of gravity is within the siderails and capacity exists for holding the ladder if a fall occurs. When ladder rungs are provided with slip-resistance, it must be remembered that the end of the

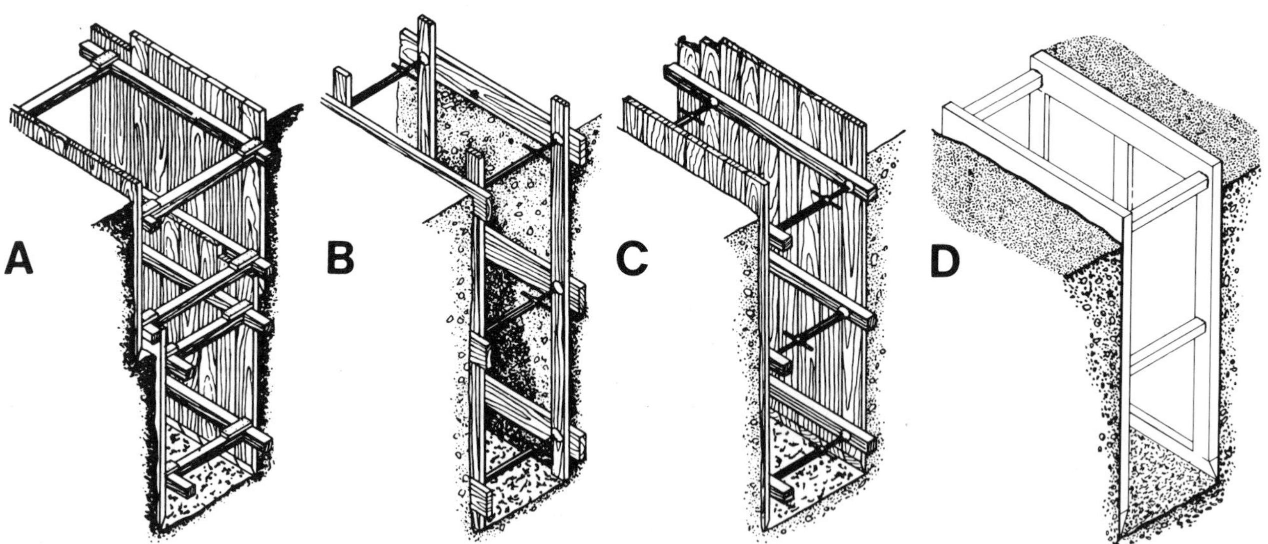

Figure 3–3. Trench bracing: *A* bracing used with two lengths of sheet piling; *B* bracing with screw jacks in hard soil; *C* screw jacks used with complete sheet piling; *D* commercially produced trench bracing.

rung is the best place to hold a ladder; if the ladder rung is rough at these points, gloves are required. Never hold the siderails of ladders because, if the foot slips off the rung, the hand will slide, resulting in a disastrous fall.

The portable ladder is preferred to be sloped at the appropriate angle for stability and this slope provides a measure of fall protection when a fall occurs. Portable usually means "can be carried by one or two persons." However, portable might also mean a several hundred pound steel ladder section carried into position by a crane or hoist. Ladders or ladder sections that are carried mechanically require fall protection during use whether they are temporarily installed by hooking on, or permanently installed by welding or bolting in recognition of the severe falling hazard.

Fixed ladders and temporary ladders that have fixed ladder characteristics, require a positive method of fall protection because free fall to the ground is a predictable consequence. The cage method of protection is not able to be part of a training program to see whether restraint by a cage is a feasible method of protection under fall conditions, hence substitute or additional protection is advisable in the form of climbing protection devices.

The use of ladders at any time when applied for working and not solely access to another level calls for serious consideration of appropriate fall protection. The higher the ladder use, the more fall protection consideration must be given.

Slip-Resistant Bases and Safety Tops

All ladders should be equipped with slip-resistant bases when there is a hazard of slipping (Figure 3–4). Slip-resistant bases are not intended as a substitute for care in placing, lashing, or holding a ladder that is being used upon metal, concrete, or slippery surfaces.

Some companies use a rope or webbing lashing to tie the top of the ladder to the pipeline or other object being worked on. Other companies replace the top rung on a portable ladder with a chain for work on cylindrical objects like poles and round columns. Such an arrangement will help prevent the ladder from slipping sideways.

Inspection and Maintenance

Inspect new ladders promptly upon receipt. Check that the ladders conform to purchase-order specifications and applicable codes. Set up an inspection program by which ladders are inspected once every three months. Keep a record of each inspection. A general inspection form is shown in Figure 3–5.

Ladders that are weak, improperly repaired, damaged, have missing rungs, or appear unsafe shall be removed from the job or site for repair or disposal. Before discarding a ladder, cut it up so no one can use it again.

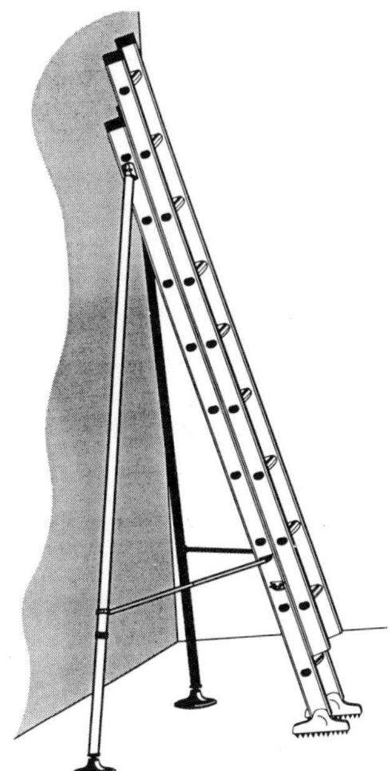

Figure 3–4. Ladder legs can be attached to a ladder for additional stability. However, the worker's reach to each side must still be limited. (Courtesy Safe-T-Legs, Inc.)

Portable ladders must be maintained in good condition at all times, and inspected frequently. Tag any ladders that have developed defects with CONDEMNED: DO NOT USE, and remove from service for repair or disposal (Figure 3–6).

Coating

ANSI A14.1, *Safety Requirements for Portable Wood Ladders*, states that "ladders may be coated with a suitable protective material. Nonconductive preservative paint may be used for identification on one side of the rails only."

One large company has the following policy:

> All wood ladders upon receipt from vendor should be delivered to the paint shop where, after approval by the purchasing department's inspector, they should be given a treatment of water-repellent preservative following the manufacturer's recommendations.

Checks, cracks, splits, and compression failures that may occur later can ordinarily be detected through a transparent coating such as clear varnish, shellac, or other clear preservative.

Markings

Mark each ladder with the name of the department to which it belongs. Some companies number their ladders consecutively so that none will be overlooked during inspection, while others stencil the date on each

LADDER INSPECTION CHECKLIST

General — Item To Be Checked	Needs Repair	Condition O.K.
Loose steps or rungs (considered loose if they can be moved at all with the hand)	❑	❑
Loose nails, screws, bolts, or other metal parts	❑	❑
Cracked, split, or broken uprights, braces, steps, or rungs	❑	❑
Slivers on uprights, rungs, or steps	❑	❑
Damaged or worn nonslip bases	❑	❑
Rusted or corroded spots	❑	❑
Stepladders		
Wobbly (from side strain)	❑	❑
Loose or bent hinge spreaders	❑	❑
Stop on hinge spreaders broken	❑	❑
Broken, split, or worn steps	❑	❑
Loose hinges	❑	❑
Extension Ladders		
Loose, broken, or missing extension locks	❑	❑
Defective locks that do not seat properly when the ladder is extended	❑	❑
Deterioration of rope, from exposure to acid or other destructive agents	❑	❑

Figure 3–5. This is an example of a general inspection checklist used for ladders.

No 600

CONDEMNED

DO NOT USE!

This item is **NOT** to be used until it has been properly repaired.

Repairs Needed:

Date ____ To Be Repaired By ____

Condemned By ____

This tag may be removed **ONLY** by the Safety Department.

No 600 Date ____

Item Condemned:

Location: ____ Dept:

Repairs Needed:

Tag Issued By:

Action Taken ____

Materials Used ____

Hours Spent ____

Date ____

Signed ____

Figure 3–6. "CONDEMNED–DO NOT USE" tag shows that a ladder or other piece of equipment should not be used until repaired.

ladder as it is put into service. Proper identification assists in inspection procedures and also in storage. Warning stickers installed by manufacturer shall not be painted, removed, or altered (Figure 3–7). (See also the section on electrical hazards, under Use of Ladders.)

Storage

Store ladders where they will not be exposed to the weather and where there is good ventilation (Figure 3–8). Do not store them near radiators, stoves, or steam pipes or in other places with excessive heat or dampness. Fiberglass ladders should be protected from direct sunlight or other ultraviolet light sources.

Ladders can be hung horizontally on brackets against a wall. To prevent warping, use more than two brackets for long ladders. Ladders can also be placed on edge on racks or on rollers, rather than stored flat. Both methods make it easy to remove ladders from storage.

Keep ladder storage space free of obstructions and accessible at all times. In that way, ladders can be obtained quickly in case of emergency.

Figure 3–7. Warning stickers added to ladders alert users to follow safe practices. (Printed with permission from Patent Scaffolding Co.)

Use of Ladders

Workers should observe certain rules when placing, ascending, and descending ladders. They should consider special hazards involved when using metal ladders.

Placement

Workers should observe the following practices when placing ladders:

- Place a ladder so that the horizontal distance from the base to the vertical plane of the support is approximately one-fourth the ladder's length between supports. For example, place a 12-ft (4-m) ladder so that the bottom (base) is 3 ft (0.9 m) away from the object against which the top is leaning (Figure 3–9).
- Do not use ladders in a horizontal position as runways or as scaffolds. Single and extension ladders are designed for use in a nearly vertical position and cannot be used in a horizontal position or with the base at a greater distance from the support than that indicated in the preceding paragraph.
- Never place a ladder in front of a door (including an overhead door) that opens toward the ladder unless the door is locked, blocked, or guarded.
- Do not place a ladder against a window pane or sash. Securely fasten a board (not with nails) across the top of the ladder to give a bearing at each side of the window. Spread attachments are available. On wide windows with a metal sash, the bearing may be across the mullions or between the window jambs.

Figure 3–8. Ladders should be stored neatly where they will not fall or cause a tripping hazard. (Courtesy International Stamping Co./Midas International Corp.)

- Place the ladder so that both side rails have secure footing. Provide solid footing on soft ground to prevent the ladder from sinking.
- Place the ladder's feet on a level base, not on movable objects.
- Never lean a ladder against unsecured backing, such as loose boxes or barrels.

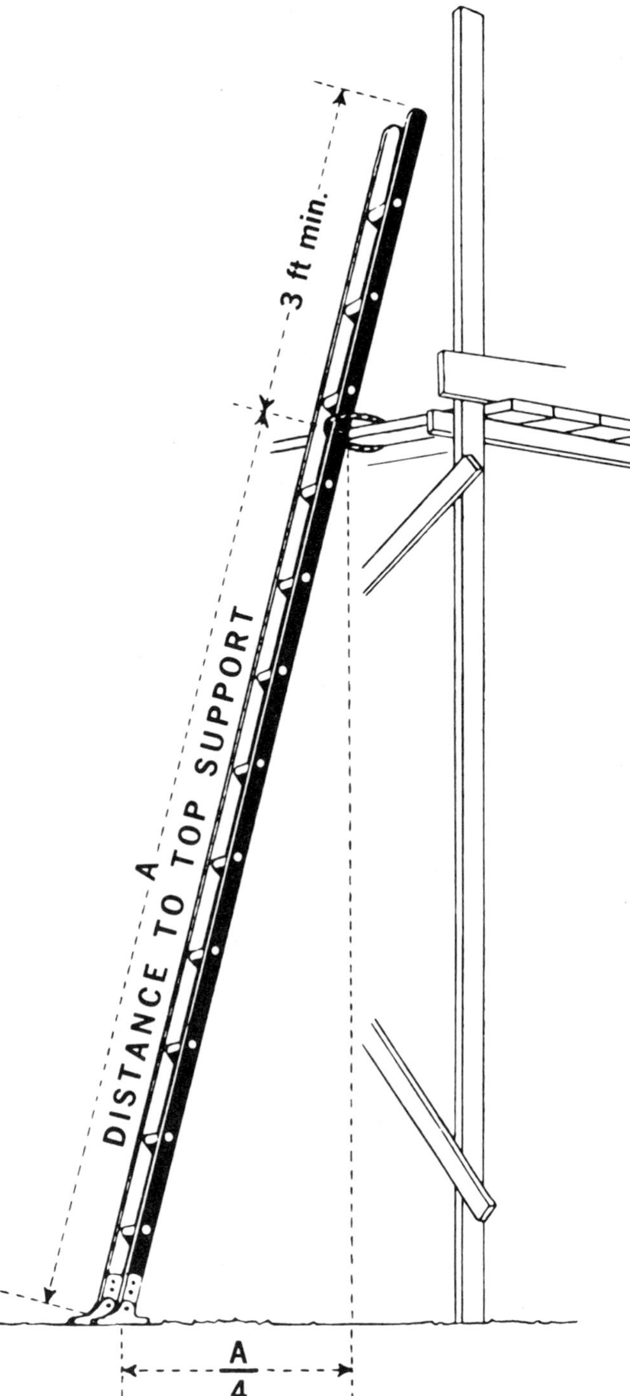

Figure 3–9. To set up a ladder safely, the base must be one-fourth the ladder length from the vertical plane of the top support. Where the ladder extends above the top landing, ladder length to the top support only is considered. Tie-off must be adequate to prevent any tipping of the ladder if the climber leans to the side.

- When using a ladder for access to high places, securely lash, bind, or otherwise secure the ladder (top and bottom) to prevent it from slipping.
- Secure both the bottom and top of a ladder to prevent displacement when using a ladder for access to a scaffold.
- Extend the ladder at least 3 ft (0.9 m) above the top landing.
- Do not place a ladder close to electrical wiring or against any operational piping (acid, chemical, sprinkler system, etc.) where damage may be done. In such cases, use nonconductive plastic ladders.
- Allow only one person at a time on a ladder.
- Do not overload a ladder or misuse it, e.g., horizontally as a platform.
- Use ladders of sufficient length so that workers do not have to stretch or reach.

Ascending or Descending Ladders

Workers must observe the following practices when ascending or descending ladders:

- Hold on to rungs (only); never hold side rails when going up or down. If material must be handled, raise or lower it with a rope either before going down or after climbing to the desired level.
- Always face the ladder when ascending or descending.
- Never slide down a ladder or run the hands behind the side rails.
- Be sure shoes are not greasy, muddy, or slippery before climbing.
- Do not climb higher than the third rung from the top on straight or extension ladders, or the second tread from the top on stepladders.
- Carry tools on a tool belt, never in the hand.

Other Recommended Practices

When using ladders, workers should observe the following general practices:

- Before using a ladder, inspect it for defects. (See Inspection and Maintenance earlier in this chapter.)
- Never use a defective ladder. Tag or mark it so that it will be repaired or destroyed (Figure 3–6).
- Do not splice or lash short ladders together. They are designed for use in their original lengths and are not strong enough for use in greater lengths.
- Do not use makeshift ladders, such as cleats fastened across a single rail.
- Be sure that a stepladder is fully open and the metal spreader locked before starting to climb.
- Keep ladders clean and free from dirt and grease, which might conceal defects.
- Do not use ladders during a strong wind except in an emergency, and then only when they are securely fastened.

- Do not leave placed ladders unattended. Remember that children may be attracted to them.
- Do not use ladders as guys, braces, or skids, or for other than their intended purposes.
- Never attempt to adjust a ladder while a user is standing on the ladder. Adjustment of extension ladders should only be made by the user when standing at the base of the ladder, so that the user may observe when the locks are properly engaged.
- The length of a straight portable ladder is to measure 30 ft (9 m) or less. Refer to ANSI 14.1 for the minimum overlap on two-section extension ladders.

Electrical Hazards and Metal Ladders

Since metal ladders conduct electricity, do not use them around energized electrical circuits or equipment, or in places where they may come in contact with electrical circuits. The importance of these electrical hazards cannot be overemphasized. Warn workers using metal ladders about these dangers. Plastic or fiberglass ladders, as well as wood ladders, should be considered for use near electrical hazards. Many construction projects forbid metal ladders.

In addition to warning workers, mark metal ladders with signs or decals reading CAUTION: DO NOT USE NEAR ELECTRICAL EQUIPMENT. These decals may be placed on the inside of the side rails at about eye level from the bottom of the ladder (Figure 3–10).

ERECTING STEEL

Erecting steel involves extensive use of cranes, derricks, hoists, ropes, and slings. For lifting heavy loads, wire rope slings are preferred to chains. With either chain or wire rope, the manufacturer's capacity rating should not be exceeded. At points where rope slings pass around sharp corners, provide padding and softeners. (See Chapter 15, Hoisting and Conveying Equipment; Chapter 16, Ropes, Chains, and Slings; and Chapter 21, Welding and Cutting, in this volume.)

Supply eye protection for workers who ream, drill, or drive wedges, shims, or pins. Provide containers for storing or carrying rivets, bolts, and driftpins. Containers must be secured to prevent falling.

Figure 3–10. Place decals (like this one) on the side rails of metal ladders.

General Safety and Health Practices

Wherever possible, observe the following precautions when erecting steel:

- Require employees to use personal protective equipment.
- Do not permit employees to ride loads, hooks, or "headache" balls.
- Do not permit employees to work near electric wires unless the wires are fully insulated.
- Remove from the job any worker who is under the influence of alcohol or drugs or who is too sick (in a doctor's opinion) to work.
- Do not allow employees to work on wet, freshly painted, or slippery steel construction.
- Have workers wear eye protection while cutting out rivets, chipping, or doing similar work. Keep adjacent areas clear of personnel, or screen such operations.
- Where it is impractical to provide temporary floors, suspend safety nets below points where employees are working, or have them use fall protection equipment. (See NSC Industrial Data Sheet 608, *Safety Nets—Fall Protection for the Construction Industry*.)
- Be sure guy cables or braces, used to hold steel while it is being erected, are guarded. This will prevent trucks or other equipment from hooking into them and pulling the steel down.

Air Tools

Operators should shut off air flow and release pressure before disconnecting pneumatic hand tools. For any adjustments or repairs, operators should never raise or lower the tool by the hose. They should always use a hand line. Tie air-hose sections together except when using automatic cutoff couplers. If an air hose must extend across a roadway, protect it from being damaged by vehicles.

Bolting

When knocking out bolts or driftpins, workers should retrieve them so they do not fall on anyone below. Workers should not throw bolts, nuts, washers, and pins. Instead, workers should place them in a bolt basket or other container and raise or lower it by a line. Provide impact wrenches with a locking device to retain the socket.

The use of high-tensile machine bolts or structural rib bolts is popular for field assembly of structural steel. However, contractors must follow the manufacturer's instructions carefully to properly install them, to apply adequate torque, and to prevent nut back-off.

Welding

Precautions to take when securing steel by welding are covered in Chapter 21, Welding and Cutting.

Drilling and Reaming

Two employees should operate drilling and reaming machines. One employee can operate these machines, however, if their handles are firmly secured to resist the torque created by the machines should the reaming or drilling bit foul.

Plumbing Up

Securely attach hooks or lashings used for plumbing up before stressing the turnbuckle. Once the turnbuckle is under stress, use a device to keep the turnbuckle from unwinding.

Place plumbing-up guys so that the bolsters, riveters, or welders can get at the connection points. Do not remove guys without first getting permission from the job superintendent. Establish a definite set of directional signals before starting to plumb a structure.

Connecting

When connectors are working together, one person should give the signals. This person should make sure everyone working on the job is in the clear. All workers should select positions that are clear of all swinging beams.

When connectors are working in pairs, they should bolt one end of the piece with two bolts before going out to connect the other end. Only one connector should go out to fasten the other end. Whenever possible, connectors should straddle the beam instead of walking along its top.

When setting columns before unhitching lifting falls, either tightly draw down the nuts on the anchor bolts or affix temporary guys. Never release a piece until the required minimum number of bolts have been installed.

Supervisors should stop connecting work during rain, high wind, or any potentially hazardous weather that might be unsafe to workers, other employees, or the public. Provide proper lighting at all times.

Erecting Steel in an Existing Facility

Erecting steel in an operating mill or facility is especially hazardous. The facility's personnel and the contractors' employees and materials cause much congestion. The operating facility's supervisor should be responsible for the various phases of the work. The steel erector and other contractors should know the facility's supervisor and others who will act as coordinators on the project.

Clearly define the work area. Identify and locate gas lines, oxygen lines, electric utilities, and electrified rails. Establish responsibility for preliminary work-closing passageways and cleaning grease or other material from crane runways, for example.

Contractors' employees and facility employees should observe all existing mill or facility safety regulations. Restrict construction areas to operating personnel. Other workers should enter the area only with permission from the superintendent or first-line supervisor.

The contractors' supervisors should obtain clearance from the mill or facility supervisor for all phases of the work. Supervisors should not delegate this responsibility to a crew member.

Where electric wires are near the work, the supervisor should determine their voltage and set the necessary clearances. The preferred procedure is to deenergize all power lines. Assume that all wires are "hot" until proven otherwise. If they cannot be deenergized, have the power company cover the lines with approved insulated rubber or plastic coverings. Note that rubber coverings on overhead power lines do not relieve the requirement for 10 ft clearance. When persons must work near hot rails, and current cannot be turned off, provide them with adequate insulation and protection. Specific electrical safety-related work practices should be prepared and all employees trained.

No one should work on an operating crane runway until the supervisor has been notified and has given permission. When work is being done on or near crane runways, place crane rail stops between the worker and the operating crane. If conditions do not permit such stops, place a safety observer in the cab of the crane to protect the worker. Use flashing lights or flags to define the work area and to warn the crane operator. Do not permit loose items to remain on cranes or crane runway girders. Fasten such items.

For additional information on erecting steel, refer to ANSI A10.13, *Safety Requirements for Steel Erection.* (See References at the end of this chapter.)

Lateral Bracing

Brace incomplete buildings, for which lateral support is not yet in place, and all free-standing walls against maximum anticipated wind pressures. Exterior masonry walls, whatever their height, whether load-bearing or nonload-bearing, are subject to wind loads beyond their designed capacity prior to the final set of the mortar or final tie into the structure. These wind loads have caused walls or sections of walls to break off and fall, causing both personal injury—sometimes fatal—and property damage (Figure 3–11).

Erect masonry walls after erecting permanently installed structural members. In that way, adequate lateral stability is provided. If this is impossible or impractical, use temporary bracing until structural members can be installed.

Usually the architect will include pilasters in the design, as well as the requirements for anchors and ties. Also, local building codes require that during the erection of walls, proper bracing and support be provided. The question is, what is proper?

During the construction of exterior masonry walls, consider two external forces—the weight (vertical force) and the wind load (horizontal force). Because the vertical load is supported by spandrels and relieving angles, the critical consideration is the horizontal load. Provide protection or bracing for this load either by screening or by simple shoring. Codes and other engineering data indicate the thickness and height of walls to withstand specific wind loads while unsupported. Check these specifications against the local wind conditions, and provide bracing when required. Use Table 3–A to check bracing that will resist pressures developed by wind at different velocities. The graph in Figure 3–12 shows wind velocities that can be withstood by concrete block walls of varying heights.

The erection of interior masonry walls presents a unique hazard both to construction workers and facility personnel. Each masonry activity should have erected a "safe zone" on the unscaffold side, which is the height of the wall plus four feet. This "safe zone" is provided to eliminate personnel from wandering into the area and being injured by falling debris. Additionally, the added four feet provide a buffer in the event the entire wall collapses.

Temporary Flooring

Where skeleton steel is being erected, maintain a tightly planked and substantial temporary floor within two stories, or 30 ft (9 m), whichever is less. Place the flooring directly under that portion of each tier of beams on which any work is being performed. The only

Figure 3–11. Wall bracing must be adequate for anticipated wind loads.

Table 3–A. Force of Wind for Given Velocities

Miles per hour (V)	*Feet per minute*	*Feet per second*	*Force in pounds per square foot ($0.004V^2$)*	*Description*
1	88	1.47	0.004	Hardly perceptible
2	176	2.93	0.014	Just perceptible
3	264	4.40	0.036	
4	352	5.87	0.064	Gentle breeze
5	440	7.33	0.1	
10	880	14.67	0.4	Pleasant breeze
15	1,320	22.0	0.9	
20	1,760	26.6	1.6	Brisk gale
25	2,200	29.3	2.5	
30	2,640	44.0	3.6	High wind
35	3,080	51.3	4.9	
40	3,520	58.6	6.4	Very high wind
45	3,960	66.0	8.1	
50	4,400	73.3	10.0	Storm
60	5,280	88.0	14.4	Great storm
70	6,160	102.7	19.6	
80	7,040	117.3	25.6	Hurricane
100	8,800	146.6	40.0	

Printed with permission from Kidder-Parker, *Architects and Builders Handbook.*

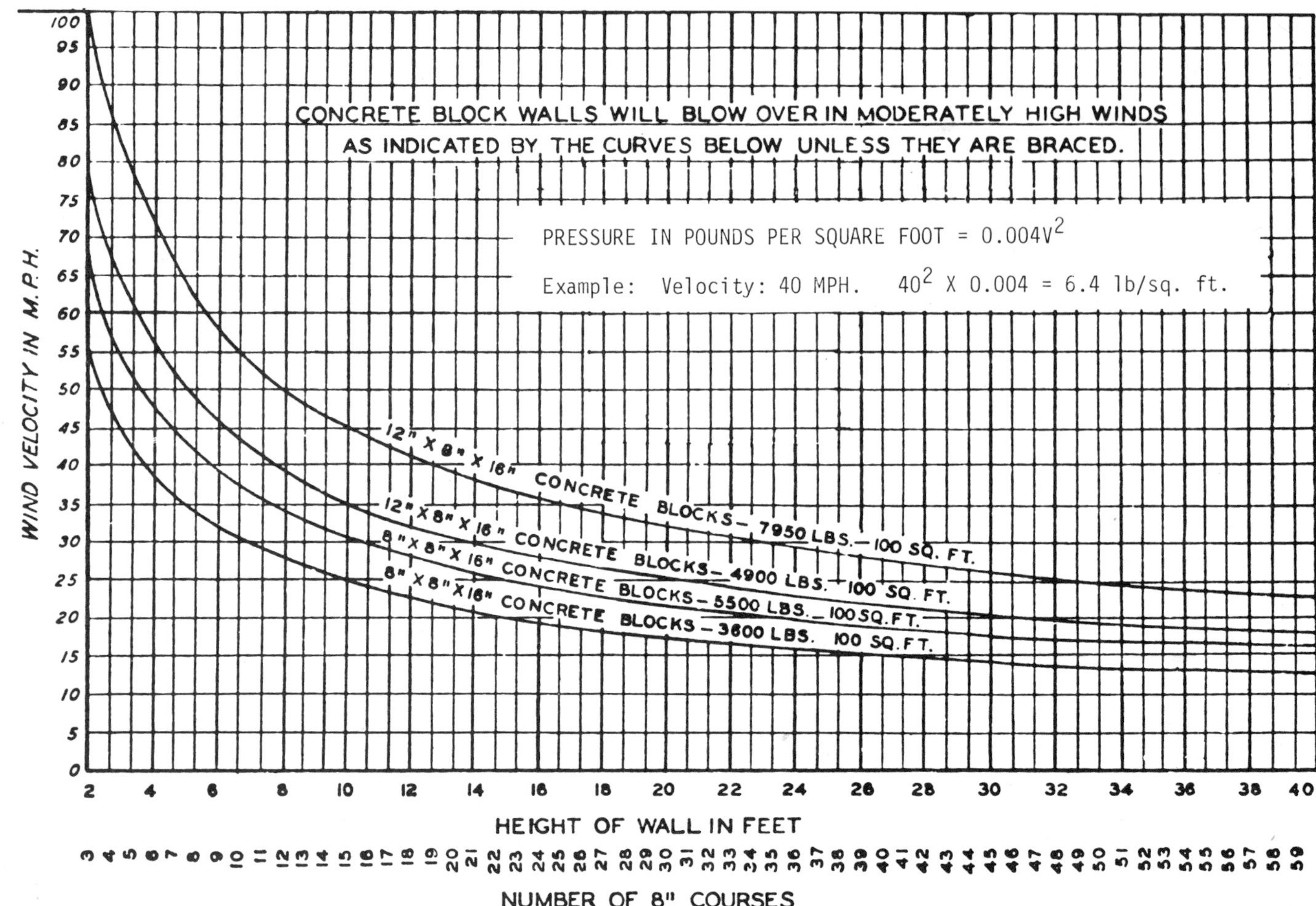

Figure 3–12. Wind velocities that can be withstood by concrete block walls of varying height. (Printed with permission from the Travelers Insurance Company.)

exception is when gathering and stacking temporary floor planks on a lower floor, in preparation for transferring such planks to an upper floor. Where such a floor is not practical, require the use of safety nets. On structures not adaptable to temporary floors, and where scaffolds are not used, install and maintain safety nets whenever the potential fall distance exceeds two stories or 25 ft (7.6 m). Workers should use safety belts during construction in a mill building or other structure (1) where no floors are contemplated, and (2) where the operation of overhead cranes will not permit temporary flooring.

The derrick, or erection, floor should be solidly planked or decked over its entire surface, except for access openings. Planking, or decking of equal strength, should be of proper thickness to carry the working load, a minimum of 50 lb/sq ft (2.4 kPa). Use planking not less than 2 in. (5 cm) thick, full-size undressed. Lay it tight and secure to prevent movement, especially displacement by the wind. Metal decking used in place of planks should be strong enough, laid tight, and tack welded to prevent movement.

When gathering and stacking temporary floor planks, employees should remove the planks successively. They should work toward the last panel of the temporary floor so that the work is always done from a planked floor. Protect employees gathering and stacking temporary floor planks from the last panel with an approved fall protection system.

Once a working floor is provided, install a safety line of 3/8-in. (9.5-mm) wire rope, or equivalent, around the perimeter of all temporarily planked or decked floors of tiered buildings or other multifloored structures. Place this line 36–42 in. (0.9–1.1 m) above the working floor, and attach it to all perimeter columns. Leave this line in place until the finished wall is installed.

Planks should overlap the bearing ends by a minimum of 12 in. (30 cm). Use wire mesh or exterior plywood around columns where planks leave an unprotected gap and do not fit tightly. All unused floor openings should be planked over or barricaded until they are needed. Replace floor planks, removed to perform work, as soon as possible, or guard the open area.

Install the permanent floors as the erection of structural members progresses. There should be not more than eight stories between the erection floor and the uppermost permanent floor, except where the structural integrity is maintained by design. At no time should there be more than two floors, or 30 ft (9 m), of unfinished bolting or welding above the foundation or uppermost permanently secured floor. The only exception to this is when the column is one continuous member and approval has been obtained from the designated authority. In no case should four floors, or 48 ft (14.6 m), be exceeded.

SCAFFOLDS AND SCAFFOLDING

A scaffold is an elevated working platform for supporting both personnel and materials. It is a temporary structure, used mainly in construction and/or maintenance work. Scaffolding is the structure—made of wood or metal—that supports the working platform. Protection against falls from stable scaffolding consists of railings, toeboards, plankings, and fall arrest systems. For more information on fall arrest systems, see Chapter 7, Personal Protective Equipment.

Fall Protection in Temporary Structures

Fall hazards are found in many built scaffold operations during erection and use. Fall protection should consist of a comprehensive fall protection plan that addresses each recognized fall hazard. A checklist should include the following:

1. complete planking of scaffold grade for a smooth surface without gaps, walkways guardrailed, no single plank access or work permitted
2. railings and toeboards
3. crossbraces supplemented to railing height
4. ladder or access climbing protection
5. fall protection for scaffold builders using adjacent structure anchorages where possible

To reasonably prevent tipover and possible falls of scaffold users, scaffolds should be tied into the structure at the first level of a tower and then periodically at higher tiers. Particular attention should be made to the covering of scaffolds outdoors with canvas or poly material that serves as a windbreak and enclosure from the elements. Additional wind forces can be gauged from Table 3–A. Fall arrest forces should be less than 900 lb using energy-absorbing equipment and self-retracting lifelines; anchorages should be near vertical to avoid bending of scaffold pieces that might extend a fall. Anchorages on adjacent structures are primary choices.

Fall protection for suspended scaffold and boatswain's chair users should take into account the likely distance of deceleration, especially at lower levels when ropes can stretch, possibly contacting a worker with the ground.

Training on fall protection systems should follow a curriculum relating to recognition and consequences of no protection, control using fall protection options and site-specific conditions and sequences. Training should be repeated or supplemented after observation for any failures in the training program to use protection or when the equipment or site conditions change.

Fall Protection Training

Training is required for fall protection systems according to U.S. OSHA 1926.503 regulations. First, workers must be trained to recognize fall hazards and the likely injurious consequences of not having protection. Workers must then be trained in the methods of fall protection that are potential or selected solutions to the fall hazards encountered. This training can be general in nature, but before work starts, it must be site-specific, related to the actual work to be done, site conditions, sequence of work, and the geometry of the workplace.

Workers must be trained in the proper use of the specific equipment that will be used onsite and also the limitations of that equipment. They must be trained in the anchorage point locations and the basis of choice of those anchorage points, which must meet regulatory requirements.

Workers must be trained in how to assemble a fall protection system, how to disassemble it, and how to store it properly. They must understand the nature of fall hazards and why eternal vigilance and following the worksite rules are the only sure ways to combat fall hazards on a broad front.

When the choice of fall protection is the use of a plan that does not involve conventional fall protection (guardrails, fall arrest equipment and nets), then each worker must know his/her role in the plan and must wear an identifying mark or color. The plan must eventually be in writing if OSHA requests the details, and therefore should always be in writing.

Readiness for work where fall hazards are encountered is best judged when workers have been tested. Often a pretest and a post-training test are advised to show both employee and employer the progress made and the proficiency gained.

The time it takes to prepare workers for encountering fall hazards should be no less than eight hours broken into modules if necessary. When the fall equipment is changed or methods are different or at a new site, the training is to be supplemented with a curriculum addressing the changes and new work plans.

Employers must certify that workers have been trained for US OSHA upon request. In Washington State, the workers must be qualified as "Competent Workers." US OSHA requires a Competent Person/ Qualified Person team to be responsible for the fall protection program.

Work must be monitored, observed and inspected for effective fall protection. This is not only the job of the employer but also the general contractor, prime contractor, and owner or owner's competent representative. The less the training time, the more observation and inspection are necessary.

Workers who do not act in accordance with their training must be returned to training promptly so that they do not endanger themselves or others. Creation of any fall hazard must be met with protection left in place for other workers who may be in the area. Housekeeping maintained on the site will help remove loose materials, which would be dangerous in suddenly windy conditions, possibly provoking a fall or other injury if a worker were hit by debris. Covering of holes in a roof suddenly when a storm approaches is something to be avoided because hurrying around fall hazards is sure to cause a fall situation. The name of the game is a comprehensive fall protection plan that begins before every phase of the job and has the support of top management.

General Safe Practices

Provide scaffolds for all work except that which can be done safely from the ground or similar footing. A competent person, trained in scaffold erection, dismantling, and alteration shall supervise all scaffold erection, dismantling, moving, or alteration. Scaffolds must also be inspected for safety at the start of every shift by a competent person.

When scaffolding is leased or purchased, safety instructions for erection and use should accompany the equipment. Never interchange the scaffolding components of different manufacturers.

All scaffolds should be designed, constructed, and maintained in accordance with the manufacturer's instructions and the applicable industry standards. Design scaffolds to support at least four times the anticipated weight of the workers and materials that will be on them.

If these structures, including such accessories as braces, brackets, trusses, screw legs, and ladders are damaged or weakened from any cause, they should be repaired or replaced immediately.

Keep scaffolds, platforms, runways, and floors free of ice, snow, grease, mud, or any other material or equipment that will make them unsafe or hazardous to persons using them. Where walkways and work surfaces are slippery, use abrasive material to assure safe footing.

Determine the width of all scaffolds, ramps, runways, and platforms by the purpose for which they are built. In no case, however, should they be less than 18 in. (46 cm) wide. They should be wide enough for passage of materials and movement of personnel.

The use of working platforms, or scaffolds, for the support of an outrigger boom, hoist, well pulley, or any other device or equipment used for hoisting materials is permissible. However, the platform, or scaffold, and the scaffolding must be reinforced and braced to withstand the additional loads. Place scaffolding on a firm, smooth foundation that will prevent sideways movement. Do not use barrels, boxes, loose bricks, or concrete blocks to support scaffolds or planks. Scaffolds should be level. The poles, legs, or uprights of scaffolds should be plumb and securely braced to prevent swaying and displacement.

Whenever work is being done above workers on a scaffold, provide overhead protection for those workers. This protection should be not more than 9 ft (3 m) above the working platform and should be made of planking or other suitable strong material.

Wood scaffold planks should not be prooftested. This may result in concealed or unrecognized damage that may cause failure later. Wood scaffold planks bear a mark, stamp, seal, or other indication of the referenced standard on usage. To check scaffold planks, the following procedure is recommended:

- Examine the plank for large knots, excessive grain slope, shakes, decay, and other defects that may render it unfit. Discard the plank upon visible or audible evidence of failure, or if it has an obvious deflection.
- Determine the safe load for a plank based on its size and species.

Means of Access

Provide a safe and convenient means for workers to gain access to the working platform. Means of access may be a portable ladder, fixed ladder, ramp or runway, or stairway.

Ladders used for access to scaffolds should conform to the requirements of the applicable ladder code. (See ANSI A14.1, ANSI A14.2, ANSI A14.3, ANSI A14.4, and ANSI A14.5. See also Chapter 2, Buildings and Facility Layout, in this volume for information concerning runways and ramps.)

Railings

Guardrails, midrails, and toeboards shall be installed on all open sides and ends of platforms more than 10 ft (3 m) above the working surface (floor). Standard railings consist of a top rail, an intermediate rail, a toeboard, and posts. They have a vertical height of approximately 42 in. (107 cm) from the upper surface of the top rail to the floor, platform, runway, or ramp level. The top rail should be smooth throughout the length of the railing. The intermediate rail should be halfway between the top rail and the floor, platform, runway, or ramp. The ends of the rails should not overhang the terminal posts except where such an overhang does not constitute a hazard.

Posts for wood railings should be made of at least 2 × 4 in. (5 × 10 cm) stock, spaced not more than 8 ft (2.4 m) apart. The top rail should be of at least 2 × 4 in. stock. The intermediate rail should be at least 1 × 6 in. (2.5 × 15 cm) stock.

Posts and top and intermediate railings for pipe railings should be at least 1½ in. (3.8 cm) in diameter, with posts spaced not more than 8 ft (2.4 m) on centers.

Posts and top and intermediate rails for structural steel railings should be of 2 × 2 × ⅜ in. (5 × 5 × 0.3 cm) angles or other metal shapes of equivalent strength. The posts should be spaced not more than 8 ft (2.4 m) on centers.

Construct the anchoring of posts and framing of members for all railings so that the completed structure shall withstand a load of at least 200 lb (91 kg) applied in any direction (except upwards) at any point on the top rail with a minimum of deflection. Provide additional strength for railings receiving heavy stress from employees trucking or handling materials. Use heavier stock, closer spacing of posts and bracing, or other means.

Other types, sizes, and arrangements of railing construction are acceptable provided they meet the following conditions:

1. a smooth-surfaced top rail at the height above floor, platform, runway, or ramp level of approximately 42 in. (107 cm)
2. a strength to withstand at least 200 lb (91 kg) top-rail pressure with a minimum deflection
3. protection between top rail and floor, platform, runway, ramp, or stair treads, equivalent at least to that afforded by a standard intermediate rail
4. elimination of overhang of rail ends unless such overhang does not constitute a hazard

A standard toeboard shall be a minimum 3.5 in. (9 cm) of vertical height from its top edge to the level of the floor, platform, runway, or ramp. It should be securely fastened in place and have not more than ¼ in. (6 mm) clearance above floor level. It may be made of any substantial material—either solid or with openings not over 1 in. (2.5 cm) in greatest dimension. Where material is piled to such a height that a standard toeboard does not provide protection, provide paneling or screening from the floor to an intermediate rail or to top rail.

Drive nails full length. Determine the quality and size of nails by values in pounds for each size and materials used as applied to the total load to be supported. Double-headed nails should not be used on decks, railings, handrails, or other walking/working surfaces.

Where persons are required to work or pass under the scaffold, screen the scaffold between the toeboard and the guardrail. The screen should extend along the entire opening and consist of No. 18 U.S. gauge ½ in. (1.25 mm) wire mesh or the equivalent.

Do not use synthetic or natural fiber ropes as guardrails. Wire rope may be used if tension is maintained to provide not more than 4 in. (10 cm) of deflection in any direction from the center line under a 200 lb (91 kg) weight. Locate support posts not more than 8 ft (2.4 m) apart.

Types of Scaffolding

Many types of scaffolding are available. The major types are wood pole scaffolding, tube and coupler scaffolding, and tubular welded frame scaffolding. Specialized scaffolding is not discussed in this Manual but is covered in governmental regulations, such as National Institute for Occupational Safety and Health's (NIOSH) *Health and Safety Guide for Masonry, Stonework, Tilesetting, Insulation, and Plastering Contractors.* (See References at the end of this chapter.) Mobile (rolling) scaffolding, swinging (suspended) scaffolding, and boatswain's chairs are discussed later in this section. (See also ANSI A10.8.1.)

Wood Pole Scaffolding

All load-carrying timber members of scaffolds shall be a minimum of 1,500 fiber construction grade lumber/stress grade. All planking shall be scaffold grades, or equivalent. The maximum permissible spans for 2 × 10 in. (5 × 25 cm) or wider planks shall be as shown in the following chart (Lumber sizes, except where otherwise stated, are nominal sizes.):

	Full Thickness Undressed Lumber			Nominal Thickness Lumber	
Working Load (lb/sq ft)	25	50	75	25	50
Permissible Span (ft)	10	8	6	8	6

Wood pole scaffolding greater than 60 ft (18 m) in height shall be designed by a qualified engineer competent in this field. It shall be constructed and erected in accordance with such design. All wood pole scaffolding 60 ft (18 m) or less in height shall be constructed and erected in accordance with appropriate regulatory standards.

Tubular Metal Scaffolding

Tubular metal scaffolding is readily available, versatile, adaptable to all scaffolding problems, and economical to use. Most manufacturers and suppliers of tubular metal scaffolding provide engineering service to help in the design of adequate scaffolding for any situation. Many suppliers also furnish erection and dismantling services. Erect such scaffolding according to the manufacturer's instructions. Do not exceed load limits, and do not use dissimilar metals together in one scaffolding.

The following are some common-sense rules for erecting, disassembling, and using metal scaffolding, recommended by the Scaffolding, Shoring, and Forming Institute. (See References at end of this chapter.)

- Post safety rules for scaffolding in a conspicuous place and make sure workers follow them.
- Abide by all regulatory codes, ordinances, and regulations.
- Inspect all equipment before use. Never use equipment that is damaged or deteriorated in any way.
- Keep equipment in good repair. Avoid using rusted equipment; its strength is not known.
- Inspect erected scaffolds regularly to be sure they are maintained in safe condition.
- Consult the scaffolding supplier when in doubt. Never take chances.
- Use extreme caution when metal scaffolds are used in the vicinity of energized electrical circuits.

Prohibit climbing of braces. Provide a ladder or stairway for entry and exit. Provide a landing platform with guardrails at intervals not to exceed 35 ft (10.6 m). Where a built-in ladder is part of a scaffold system it should conform to the requirements for ladders.

All metal scaffolds and towers shall be plumb and level. The sections of metal scaffolds shall be securely connected and all braces securely fastened.

Scaffold planks should extend over their end supports not less than 6 in. (15 cm) nor more than 12 in. (30 cm). Scaffold planks should be created at both ends to prevent movement. Securely fasten work platforms to the scaffold. To prevent movement, secure the scaffold to the building or structure at intervals not to exceed 30 ft (9 m) horizontally and 26 ft (8 m) vertically.

Tube and Clamp (Coupler) Scaffoldings

Tube and clamp (coupler) scaffolding is an assembly that consists of tubing used as posts, bearers, braces, ties, and runners. It is also the base supporting the posts and special couplers that serve to connect the uprights and to join the various parts.

Light-duty tube and coupler scaffolding should have nominal 2 in. (5 cm) O.D. steel tubing bearers. Posts should be spaced no more than 6 ft (1.8 m) apart by 10 ft (3 m) along the scaffold.

Medium-duty tube and coupler scaffolding should have posts spaced not more than 6 ft (3 m) apart by 8 ft (2 m) along the scaffold. It should have bearers of nominal 2½ in. (6 cm) O.D. steel tubing. Posts spaced not more than 5 ft (1.5 m) apart by 8 ft (3 m) along the scaffold should have bearers of nominal 2 in. (5 cm) O.D. steel tubing.

Heavy-duty tube and coupler scaffolding should have bearers of nominal 2½ in. (6 cm) O.D. steel tubing with the posts spaced not more than 6 ft (3 m) by 6½ ft (4.5 m).

The height of scaffolding will vary. It should be constructed to support four times the maximum intended load.

Install cross bracing according to the manufacturer's specifications. Spacing of bracing may vary by manufacturer.

Install lengthwise diagonal bracing on the inner and outer rows of poles at about a 45-degree angle from near the base of the first outer post upward to the extreme top of the scaffold. Where the length of the scaffold permits, duplicate such bracing beginning at every fifth post. In a similar manner, install lengthwise diagonal bracing from the last post extending back and upward toward the first post. Where conditions preclude attaching of this bracing to the posts, it may be attached to the runners.

Tubular-Welded Frame Scaffolding

Tubular welded frame scaffolding is an assembly consisting of factory-welded frames and attachable metal cross braces, leveling screws, jacks and/or baseplates, and other accessories that are available to form a complete system (Figures 3–13 and 3–14a). The requirements for tubular-welded frame scaffolding follow:

- Metal tubular frame scaffolding, including accessories such as braces, brackets, trusses, screw legs, ladders, etc., shall be designed, constructed, and erected to safely support four times the maximum rated load.

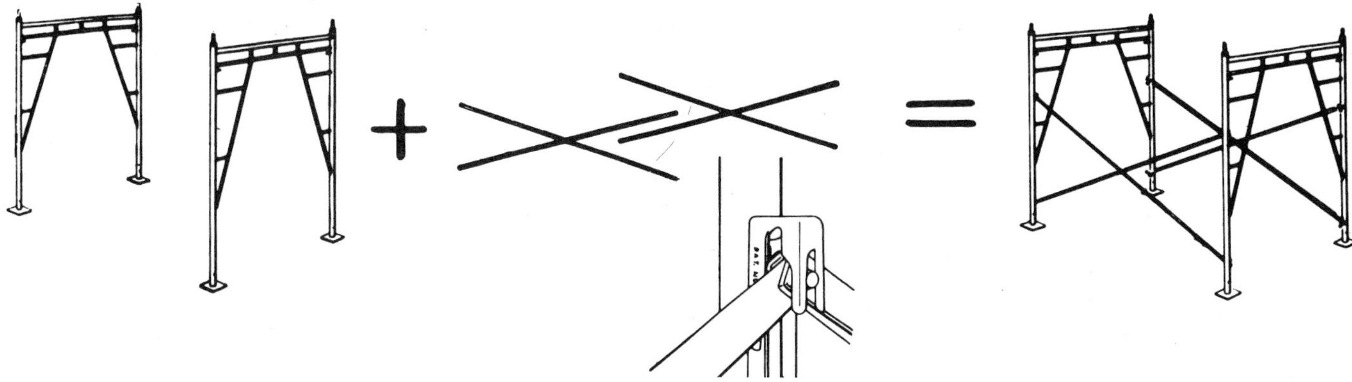

Figure 3–13. Tubular welded frame scaffold is quicker and easier to set up and is consistently safer than a wooden scaffold; it can be used over and over again for different applications. This type of scaffold can be used in conjunction with a tube and clamp scaffold to provide design flexibility. (Printed with permission from Patent Scaffolding Co.)

- Spacing of panels or frames shall be consistent with the loads imposed.
- Scaffolds shall be properly braced by cross or diagonal braces for securing vertical members together laterally. The cross braces shall be of such length to automatically square and align vertical members. In that way, the erected scaffold is always plumb, square, and rigid.
- To provide proper vertical alignment of the legs, the frames shall be placed one on top of the other with coupling or stacking pins.
- Panels shall be locked together vertically by pins or other equivalent suitable means.
- Drawings and specifications for all frame scaffolds over 125 ft (38 m) above the base plates shall be designed by a registered professional engineer.
- Guardrails are required as set forth in the wood pole scaffolding section.

Mobile (Rolling) Scaffolding

Mobile (rolling) scaffolding is caster-mounted sections of tubular metal scaffolding. Mobile scaffolding can also be made of components specifically made for the purpose (Figure 3–14b).

Observe the following requirements for mobile scaffolding:

- Mobile scaffolding should be used only on firm, level, clean surfaces.
- Free-standing mobile scaffolding platform heights should not exceed three times the smallest base dimension.
- Apply the force necessary to move the mobile scaffold as close to the base as practical.
- Stabilize the tower during movement from one location to another.
- Observe the following precautions when moving mobile scaffolding:
- Secure or remove all material and equipment from the platform before moving the scaffolding.
- Apply caster brakes at all times when scaffolding is not being moved.
- Attach casters with plain stems to the panel or adjustment screw by pins or other suitable means.
- Do not try to move rolling scaffolding without sufficient help.
- Watch out for holes in the floor and for overhead obstructions.
- Do not extend adjusting screws more than 12 in. (30 cm).
- Use horizontal bracing on scaffolding, starting with the base, at 20 ft (6 m) intervals.
- Do not use brackets on rolling scaffolds.

No person should be allowed to ride on manually propelled scaffolding unless the following conditions exist:

- The floor or surface is within 3 degrees of level and is free from pits, holes, or obstructions.
- The minimum dimension of the scaffolding base, when ready for rolling, is at least one-half of the height dimension. Outriggers, if used, are installed on both sides of the staging.
- The wheels are equipped with rubber or similar resilient tires.
- Tall metal scaffolding must be monitored to ensure it is not pushed into overhead electrical wires.

Suspended Scaffolds

Suspended scaffolds are usually factory-built. Safety factors are based on the line pull of the hoist motor(s). Each scaffold shall be installed or relocated under the supervision of a competent person.

Figure 3–14a. Scaffolding at this construction site encases the entire building so workers can reach various work areas as needed.

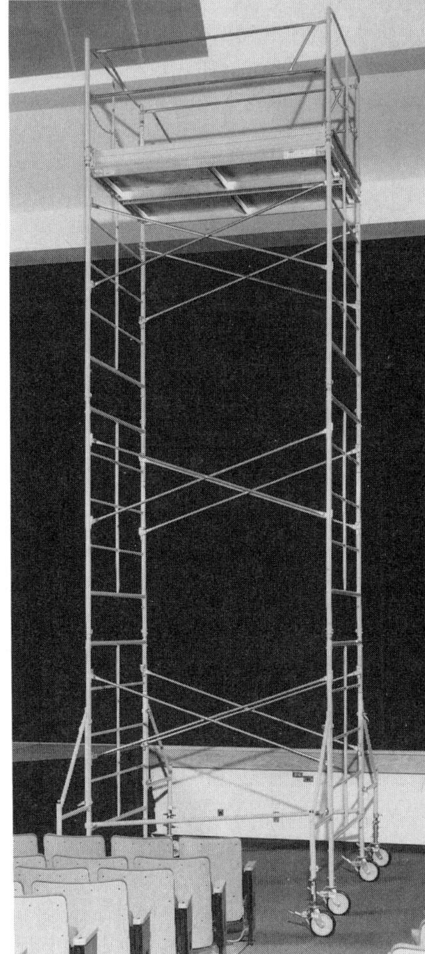

Figure 3–14b. This sectional rolling steel frame scaffold has horizontal bracing at the base. (Both photos courtesy Bil-Jax, Inc.)

Suspended scaffolds shall be securely supported from parapets or roofs, or other reliable supports, with properly placed hooks or outrigger beams strong enough to provide a minimum safety factor of four. Outrigger beams shall consist of structural metal securely fastened or anchored to the frame or floor system of the structure. Consult hoist motor manufacturer's instructions for proper sizing of outrigger beams. They shall be installed under the supervision of a competent person. All outrigger beams shall be set and maintained with their webs in a vertical position. They shall be anchored to the structure by the equivalent of U-bolts and anchor plates, washers, and nuts. A stop bolt shall be placed at each end of every outrigger beam. The outrigger beam shall rest on wood-bearing blocks. Where a single outrigger beam is used, the steel shackles of clevises with which the wire ropes are attached to the outrigger beams shall be placed directly over the hoisting drums.

Brackets should be wrought iron or mild steel. No reinforcing steel should be used as any part of a support system. Brackets should have attachments for guardrails, intermediate rails, and toeboards. Tiebacks of ½ in. (19 mm) Manila rope, or the equivalent, should serve as a secondary means of anchorage. Install the brackets at right angles to the building, whenever possible, and secure them to a structurally sound portion of the building.

Equip the fixed end of suspension ropes with a proper size thimble secured by splicing or other means and attached to the support by a closed shackle. If using wire rope clips to make the connection, they must be of the fist-grip type, properly spaced and torqued to specification. This connection also requires daily inspection by a competent person. Attach running ends to the hoisting drum. At least four turns of rope should remain on the drum at all times. Attach suspension ropes at the vertical centerline of the outrigger. The attachment should be directly over the hoisting drum.

Suspended scaffolds should be guyed, braced, or equipped with tag line to prevent swaying. Suspended scaffolds should have a standard railing.

Scaffold machines, either machine powered or hand powered, should not move when the power is stopped. Hoisting machines should be tested and listed by Underwriters Laboratories (UL) or Factory Mutual (FM). For machine powered scaffolds, use deadman controls with a nonlocking switch. A device to shut off the power should be installed ahead of the operating control. Prohibit hand-operated release mechanisms that can permit the load to descend faster than the speed rating. Also prohibit scaffold machines with cast-metal parts.

Test every suspended scaffold with twice the maximum load before it is put into operation. Inspect all anchorages of suspended scaffolds at the beginning of each shift. Inspect all wire ropes, fiber and synthetic ropes, slings, hangers, platforms, and other supporting parts before every installation. Make periodic inspections while the scaffold is in use.

Each person supported by a suspended scaffold should be protected by an approved personal fall protection system. The lifeline should be securely attached to substantial parts of the structure (not scaffold), or to securely rigged lines, which will safely suspend the employee in case of a fall. To keep the lifeline continuously attached, with a minimum of slack, to a fixed structure, change the attachment point of the lifeline as the work progresses. Note the following exception: When working on the lower tiers of a multistage suspension scaffold or where overhead protection is required, securely attach the lifeline to the scaffold. Use independent wire ropes at each end of a suspended scaffold, with approved grabbing and locking devices.

Boatswain's Chairs

Workers who use boatswain's chairs should erect them with care. The seat should not be less than 2 ft (60 cm) long by 1 ft (30 cm) wide. In wooden seats, cleats should be nailed to the underside of each end of the chair to prevent the board from splitting. The chair should be supported by a sling attached to a suspension rope.

Where blowtorches, cutting torches, or open flames are used, provide a safety harness, lanyard, and separate lifeline attached to a separate substantial portion of the structure.

HOISTS ON CONSTRUCTION SITES

Material hoists and personnel hoists are made of tubular steel. Be sure to consult with tubular steel manufacturers or suppliers for current technical data. The facility supervisor or the contractor must comply with the manufacturer's specifications and limitations.

Hoists may be erected in hoistways inside the building or in outside towers. Personnel should not be permitted to ride "material only" hoists. Many modern hoists are certified for both personnel and material. Do not permit work in or on the hoistway while the hoist is in operation.

Post the rated load capacities, recommended operating speeds, and special hazard warnings or instructions on cars and platforms. (For additional information, see ANSI A10.4, *Safety Requirements for Personnel Hoists*, and ANSI A10.5, *Safety Requirements for Material Hoists.*)

Inside Material Hoistways

If the material hoist is installed inside the building, enclose the hoistway. Solid enclosure is preferred. However, heavy wire screening, ½ in. (1.25 cm) mesh, No. 18 U.S. gauge wire, is often substituted. Partition adjacent hoistways.

Protect entrances by solid or slatted wood gates at least 5 ft (1.7 m) high and within 4 in. (10 cm) of the hoistway. Counterweight gates and use latching or locking mechanisms.

Provide protective covering of heavy planking below the cathead of all hoists. This will prevent objects from falling down the hoistway. (See Chapter 15, Hoisting and Conveying Equipment, in this volume for more details.)

Outside Material Hoistways

Hoisting towers are usually made of tubular steel and are used on construction sites. Base the design on a safety factor of at least five. Erect the tower on a level and solid foundation, and have it well guyed or fastened to the building.

Enclose the tower with heavy wire screening, and equip it with a fixed ladder extending the full height of the tower. Install standard railings and toeboards on runways connecting the tower to the building.

Material Hoist Platforms

Build material hoist platforms with a safety factor of at least five. Use timber not less than 2 in. (5 cm) for flooring. Enclose sides not used for loading with heavy wire screening, and provide 6 in. (15 cm) toeboards. Install an overhead plank covering at the crosshead. The covering can be built in hinged sections to permit handling of long material. Where wheelbarrows are handled, attach stop cleats to the floor. Equip the car with a broken rope-type of safety device. If a car's cable breaks, the car safety clamps or dogs are then thrown into position on the guide rails to stop the car.

Signal Systems

A good signal system is necessary to safely operate hoists. Electrically operated lights or bells, bells operated by pull cords, a combination of bells and lights, or telephone system can be used. Adopt standard signals, and post them at each entrance and at the engine.

Personnel Hoists

Any hoist used for carrying passengers should conform to the safety requirements of ANSI A10.4, *Safety Requirements for Personnel Hoists.* Rack and pinion hoists can be used to carry personnel or material but never both, according to strict manufacturing specifications and safety rules. Include all safety devices, including an overspeed governor, normal limit, and final limit switches, and thoroughly inspect and maintain the hoist regularly.

Temporary Use of Permanent Elevators

Permanent passenger or freight elevators in buildings under construction, modification, or demolition may be used for carrying workers or materials or both. However, they must be approved for such use and a temporary permit must be issued for the class of service. Elevators should conform to *Elevators and Escalators,* ANSI/ASME A17.1.

Towers, Masts, and Hoistway Enclosures

Design and install the tower or mast construction that forms the supports for the machinery and guide members to support the load and forces specified. For hoists located outside of a structure, the enclosures (except the one at the lowest landing) may be omitted on the side where there is no floor or scaffold adjacent to the hoistway. Enclosures on the building's side of the hoistway should be full height or a minimum of 10 ft (3 m) at each floor's landing. Enclosure at the pit should be not less than 8 ft (2 m) on all sides.

For hoists located inside a structure, enclose the hoistway throughout its height. Construct hoistway enclosures so that when they are under a horizontal pressure of 100 lb (445 N),

1. they cannot deflect more than 1 in. (2.5 cm)
2. the running clearance between the car and the hoistway enclosure is not reduced below ½ in. (2 cm), except on the sides used for loading and unloading

If on openwork, provide hoistway enclosures with an unperforated kick plate on all sides within the building or structure. The kick plate should extend not less than 12 in. (30 cm) above the level of each floor above the lowest.

Foundations of hoists should distribute the transmitted load so as not to exceed the safe load-bearing capacity of the ground upon which it is set. Anchor hoist structures to the building (or other structure) at vertical intervals not exceeding 25 ft (8 m). If tie-ins cannot be made, guy the hoist structure to adequate anchorages for stability. When wire rope is used for guys, it should be at least ½ in. (2 cm) in diameter. Tie-ins should conform to, or be equal to, the manufacturer's specifications and should remain in place until the tower or mast is dismantled.

Where multiple hoistways are used and one or more of the cars are designed solely as a material car—in accord with ANSI A10.5—personnel cars are prohibited. Each personnel hoist should be independently powered and operated. Never use Chicago booms on a hoist structure.

Doors or gates for hoistways should not be less than 6½ ft (2 m) high. If a solid door is used, it should have a vision panel, not wider than 6 in. (15 cm) and not larger than 80 in.^{2} (0.05 m^2), covered with expanded metal. Protect all hoistway entrances with substantial gates or bars.

Landing doors should lock mechanically so they cannot be operated from the landing side. At landings other than the lowest one, use locks that can be released only by a person in the car. Some doors at the lowest terminal landing automatically lock when closed with the car at the landing. In such cases, provide a means to unlock the car from the landing side to permit access to the car. Never use a hook and eye as a door-locking device.

Car Platforms

Each car should have a platform or protective covering that extends over the entire area of the car's enclosure. The covering should be nonperforated, fire retardant, and supported by the car's frame. Design the frame and the floor to handle the anticipated loads.

Car Enclosures

Enclosures and linings of cars should be made of metal or fire-retardant wood. Permanently enclose personnel hoist cars on the top and all sides, except the entrance and exit. Securely fasten this enclosure to the car's platform. Support the enclosure so that it cannot become loosened or displaced when the car's safety or buffer is engaged. Make the enclosure's walls strong enough so that their running clearance is reduced by no more than ¾ in. (2 cm) when a force of 100 lb (445 N) is applied horizontally to the walls of the enclosure.

Provide an emergency exit with an outward-opening cover in the top of all cars. The opening should be not less than 400 in.^{2} (10.16 m^2) in area, with a minimum dimension of 16 in. (0.41 m). It should provide a clear passageway unobstructed by fixed hoist equipment on or in the car. Use wire glass, or the equivalent, for vision panels. Use plain glass only for the car's operating appliances.

Do not locate a working platform on top of the hoist car, unless specifically provided in ANSI A10.4, *Safety Requirements for Personnel Hoists*. Do not place equipment that is not required for the operation of the hoist or its appliances, on the top of the hoist car. Require that the hoist be locked out per lockout/tagout procedure prior to performing repairs.

Crane Baskets

A crane-hoisted basket is a specially designed piece of equipment used when no other means are available to reach or access high work. Simply put, a crane basket is a rectangular steel cage suspended from a crane to provide workers a stable enclosure. For information on cranes used in construction, see Cranes in Chapter 15, Hoisting and Conveying Equipment, in this volume.

Minimum requirements for design, use, setup, testing and fall protection are contained within OSHA's construction standard 29 *CFR* 1926, subpart N, *Cranes, Derricks, Hoists, Elevators and Conveyers*.

Minimum requirements:

- designed by qualified person, competent in structural design
- capable of holding five times intended load
- guardrails with solid enclosure or expanded metal from toeboard to midrail
- overhead protection
- all welds by certified welder
- posted with weight of platform and maximum intended load
- a trial lift (unoccupied) with the anticipated load prior to the first lift of each day and each new lift location
- all personnel in the basket will be tied off to either the block/ball or to a structural member of the basket

Using the above and the requirements contained within subpart N will ensure each basket lift is performed in a safe manner with adequate equipment.

Wire Ropes and Sheaves

Use hoisting ropes not less than ½in. (1.25 cm) in diameter except on such equipment as small winches, like those on gin poles. In any case, they should provide a safety factor conforming to the requirements of applicable elevator codes. Inspect ropes frequently and keep them lubricated. Replace ropes when inspection discloses that wear, breakage, or corrosion has reduced their strength below the permitted safety limit.

NOTE: Where wire rope is used and fastened using clips or U-bolts, the minimum number is three clips at each termination point. Also, the clips must be installed where the "live" part of the rope contacts the "saddle" side—conversely, the "U" side, on bolt side, must contact the "dead" side. The common phrase used by millwrights for decades is "never saddle a dead horse."

Ropes shall be guarded at points where persons may come in contact with them and where objects may strike or rub against them. Keep sheaves aligned and bearings lubricated. In general, sheaves' diameter should be at least 20 times the rope's diameter. (For additional details, see Chapter 15, Hoisting and Conveying Equipment, and Chapter 16, Ropes, Chains, and Slings, in this volume.)

Hoisting Engines

Do not locate hoisting engines in public streets. If they must be located there, enclose them with barricades to protect the public. In any case, install a roof to protect the equipment and operator from the elements.

Engines should have brakes that can stop and hold 150% of the rated safe load. In addition, engines should have a pawl for holding suspended loads.

Enclose exposed gears, shafting, and couplings. Cover exposed steam pipes, and place exhaust pipes where steam cannot strike nearby persons.

Where electric hoists are used, install enclosed safety-type switches. Enclose or guard all current-carrying parts to prevent personal contact, and ground installation.

FORMWORK AND FALSEWORK

The planning and design of formwork and falsework should be in accord with provisions of the American Concrete Institute Publication ACI 347-78, *Recommended Practice for Concrete Formwork* and ANSI A10.9, *Safety Requirements for Concrete Construction and Masonry Work*.

All formwork, falsework, structural shoring, and bracing should be designed, erected, braced, and maintained to safely support all vertical and lateral loads that might be applied until such loads can be supported by the structure. Drawings and plans shall be available at the job site. All shoring equipment shall be inspected prior to erection to determine that it is as specified in the shoring layout. Any equipment found to be damaged shall not be used for shoring. Erected shoring equipment shall be inspected immediately prior to, during, and immediately after the placement of concrete. Attempts should not be made to reshore damaged, weakerned, or displaced formwork while the concrete is in a fluid state. A failure would bury workers under tons of concrete.

The sills for shoring should be sound, rigid, and capable of carrying the maximum intended load. All baseplates, shore heads, extension devices, or adjustment screws should be in firm contact with the footing sill and the form. Prohibit eccentric loads on shore heads and similar parts, unless these parts have been specifically designed for such loading.

When single post shores are used one on top of another (tiered),

- The design of the shoring shall be prepared by a qualified designer and the erected shoring inspected by an engineer qualified in structural design.
- The shores shall be vertically aligned.
- The shores shall be spliced to prevent misalignment.
- The shores shall be adequately braced in two mutually perpendicular directions at the splice level.
- Each tier shall also be diagonally braced in the same two directions.
- Adjustment of single post shores to raise formwork shall not be made after placement of concrete.

Remove and stockpile stripped forms and shoring promptly after stripping. Protruding nails, wire ties, and other form accessories not needed for later work should be pulled, cut, or by other means taken out.

Supply employees with eye and/or face protection when nailing into concrete. Require employees to wear this personal protective equipment. Provide reshoring to safely support slabs and beams after stripping. Also provide reshoring when slabs and beams are subjected to superimposed loads due to construction. When temporary storage of reinforcing rods, material, or equipment on top of formwork is needed, strengthen these areas to meet the intended loads. Do not impose construction loads on "green" concrete until engineering data are provided and allow such loads.

Adequately reinforce steel for walls, piers, columns, and similar vertical structures to prevent overturning or collapse. Do not load metal tubular frames used for shoring beyond the safe working load recommended by the manufacturer. Keep all locking devices on metal tubular frames and braces in good working order. Coupling pins should align the frame or panel legs. Pivoted cross braces should have their center pivot in place. All components should be in a condition similar to when originally manufactured.

Fasten devices for attaching the outside lateral bracing to the legs of the metal tubular shoring frames. When using tube and coupler shoring, use couplers (clamps) made of drop-forged steel, malleable iron, or structural grade aluminum. Do not use gray cast iron. Do not use couplers (clamps) if they are deformed, or broken, or have defective or missing threads on bolts, or other defects.

When checking the erected shoring frames with the shoring layout, the spacing between towels and cross brace spacing should not exceed that shown on the layout. On metal tubular frame shoring, all locking devices should be closed. On tube and coupler shoring, check all interlocking of tubular members and tightness of coupling.

FALL PROTECTION

To control recognized fall hazards, fall protection should be used. The U.S. OSHA standard 1926.500, Fall Protection, addresses the options for protection. The methods include elimination, prevention (guard rails), and fall arrest.

The residential roofing, precast, and bricklaying trades are provided the option of an OSHA fall protection plan if conventional fall protection cannot be applied. Whenever a near miss or incident occurs, the plan must be upgraded to remove or control the risk before OSHA investigates. Some trades are referred to their industry standard (e.g., excavation, electrical construction, steel erection, ladders).

Steel erection minimum standards will be those of the general contractor, owner or OSHA, whichever is stricter. OSHA has finalized 30 ft (9 m) for connectors and 15 ft (4.6 m) for all others engaged in steel erection (see the Steel Erection Negotiated Rulemaking Advisory Committee's proposed updating for *CFR* 1926.750, Subpart R) with fall protection availability for all ironworkers over 15 ft.

Fall Protection in Permanent Structures

Surround with an enclosure guard all floor and roof holes, skylights, and openings into which persons can walk. Or cover them completely with material and bracing strong enough to support any expected or potential load. Secure coverings for floor and roof openings to prevent accidental removal or displacement.

Guard every stairway and ladderway floor opening on all exposed sides except the entrance opening. Install a securely anchored standard railing with intermediate rail and toeboard.

Offset temporary stairway and ladderway entrances, or provide them with a gate to prevent anyone from walking into the opening. Guard every hatchway and chute floor opening with a hinged floor-opening cover. Equip the cover with railings attached so as to leave only one exposed side. Provide the exposed side either with a swinging gate, or offset it so that persons cannot walk into the opening.

Guard wall openings from which there is a drop of more than 4 ft (1.2 m) and the bottom of the opening is less than 3 ft (0.91 m) above the working surface, with a top rail, top rail and intermediate rail, or standard guardrail. Provide a toeboard where the bottom of the wall opening, regardless of width, is less than 4 in. (10 cm) above the working surface.

An extension platform outside a wall opening onto which materials can be hoisted for handling should have a standard railing. However, one side of an extension platform may have removable railings to enable handling of materials. If removable guardrails are provided, suitable anchor points must be installed for the attachment of personal fall arrest systems. Employees must be equipped with a proper harness and appropriate lanyard or other device, such as a retractable lifeline, and be trained to use them before the guardrails are removed.

Recent changes to fall protection on construction sites have caused confusion not only in the construction industry but with construction customers also. There are two distinct fall protection standards within OSHA's 1926. Subpart M, Fall Protection, addresses all fall applications on a construction project except during steel erection. Subpart R, Steel Erection, addresses fall protection during steel erection activities. Subpart M requires fall protection for personnel when they are exposed to a fall of 6 ft (1.8 m) or greater in height. Other specific requirements appear in Subparts L, N, and V.

There are several options available to achieve total fall protection on a construction site: personal fall arrest system; guardrails; safety net system; warning lines; or a fall protection plan (FPP) to be used in conjunction with controlled access zones (CAZ) or a safety monitoring system (SMS).

Personal Fall Arrest System

A personal fall arrest system consists of a full body harness and a deceleration lanyard. These can be used on an alternative of specialty fall devices for protection in vertical, horizontal, and sloped roof applications. Note: Never attach a lanyard to another fall arrestor, such as an SRL, to avoid excessive free fall dangers.

Vertical

Vertical lifelines with rope-grab hardware are used to provide fall protection on swing-stage scaffolds, boatswain's chairs, steel or tower work, and in elevator shafts, to name only a few applications. Matching the harness, lanyard, and rope hardware to each particular application is best accomplished with the end-user's input. There are a great many variations to vertical protection, including self-retracting lanyards, so each location requires coordination between all parties.

Horizontal

Anchorage to walklines (attaching a lanyard to a static attach point such as a structural member or to a retractable reel) have an intrinsic hazard that training and supervision can eliminate. Horizontal work attachments can create fall exposures well in excess of 6 ft (1.8 m) if the worker(s) do not keep the point of attachment at shoulder height or above, or use an attach point that would cause the lanyard to fall below the worker's shoulder height.

Sloped Roof

When a personal fall arrest system is selected, then some fall exposure is not necessarily a direct fall, but rather a slip down a slope to the edge. Using static lines or an attach point at the peak of a roof will provide fall protection as long as the employee is prevented from falling more than six feet or contacting any object during a fall. Again, there are a variety of devices available and many types of lifelines. Coordination between all parties is necessary to avoid a system that appears adequate but may create exposures greater than six feet if misused or misunderstood.

Guardrails and Safety Nets

Guardrails mean a top rail, midrail, and a toeboard. The top rail should be 42 in. (106.7 cm) high (plus or minus 3 in. or 7.6 cm). The midrail should be between top rail and toeboard, with a toeboard approximately the size of a normal 2 × 4 board. Guardrails must withstand 200 lb (90.7 kg) of force in an outward or downward force. Where screen mesh, solid panels, or vertical members are used between the top rail and the toeboard, then they must withstand a force of 150 lb (68 kg). (See also the section on Railings earlier in this chapter.)

Safety nets should be chosen when large open-web steel structures, external building protection, large span structures such as gymnasiums, mills, bridges, etc., require continuous protection for extended periods of time. Safety nets, when installed, tested, and maintained properly can provide total protection. As stated previously and repeated here, selecting nets must be a coordinated effort between parties with all hazards understood and the limitations weighed against other fall protection choices.

Warning Lines

A warning line system is restricted to low slope or flat roof work only. If personal fall arrest, rails, or nets are not used, then a warning line can be used alone or in conjunction with any of the above.

Warning lines, to protect employees from working within 6 ft (1.8 m) of a roof edge or while working within the 6-ft area between the warning line and the roof edge, creates a zone for higher awareness of the fall hazard of the roof edge. The construction of the line must be so that stanchions keep the line between 34 and 39 in. high (86 to 10 cm), resist a 16-lb force (7.3 kg), be flagged at least every 6 ft. and be of a material with a breaking strength of at least 500 lb (227 kg). These requirements are put into the standards to assure that a warning line, if chosen, will at least hold up to the daily wear and tear of construction projects.

Fall Protection Plan

A site-specific fall protection plan (FPP) is required when either controlled access zones (CAZ) or a safety monitor system (SMS) is selected. Choosing either can only be implemented when all other fall protection systems have been deemed infeasible during leading-edge work or precast concrete erection by a competent person.

The formulated FPP must be written to justify why and explain the hazards associated with each operation. Within the FPP, the reasons for choosing either CAZ or a monitor system, and why using conventional fall protection would create a greater hazard, must be explained thoroughly, including worker duties.

Controlled Access Zones

A zone demarcated with a line, rope, chain, or warning tape (minimum breaking strength of 200 lb or 91 kg) where access is restricted. The different distances for each zone during each operation are as follows:

Leading Edge:	Between 6 ft and 25 ft (1.8–7.6 m) from work area
Precast Erection:	Between 6 ft and 60 ft (1.8–18 m) from erection operation
Overhand Brick Work:	Between 10 ft and 15 ft (3–4.6 m) from working masons

All floor and roof holes, skylights, and openings into which people can walk must be guarded with enclosed guards or covered with material and bracing strong enough to support any anticipated load. Stairways, ladderway entrances, and wall openings must be guarded to prevent workers from walking into the opening.

Training

Whichever protection is chosen, training will be the single action that ties all the systems together. Fall protection training is required and must include the nature of each fall exposure, correct procedures to erect, use, maintain, and store the equipment or comply with the FPP's CAZ or safety monitoring system. Each employee's training must be documented and each retraining documented to ensure employee understanding of new hazards, new equipment, or changes to an FPP.

TEMPORARY HEATING DEVICES

When using temporary heating equipment, assign a qualified employee to operate and maintain it. Be sure to follow all the manufacturer's instructions. Each time the heater is placed in operation, check that it is functioning properly. Also, periodically check the heater when it is in operation. Equip heaters with an automatic flame-loss device to stop the flow of fuel if the flame is extinguished. Identify thermostatically controlled heaters with a warning label advising that the unit may start up at any time.

Maintain clearance around heating units to prevent ignition of combustible materials. Protect heaters against damage. The project superintendent should approve temporary heating devices prior to use. Each heater should have a safety data plate permanently affixed by the manufacturer. The plate should provide requirements or recommendations for the following:

- clearances from combustible materials
- ventilation (minimum air requirements for fuel combustion)
- fuel type and input pressure
- lighting, extinguishing, and relighting
- electrical power supply characteristics
- location, moving, and handling
- name and address of the manufacturer

Establish operating procedures to assure the following:

- proper placement and servicing
- safe clearance from combustible material
- close surveillance
- safe fuel storage and refueling
- proper maintenance
- ventilation and determination of gaseous contamination or oxygen deficiency

Heaters, when in use, should be set horizontally level, unless otherwise permitted by the manufacturer's specifications.

Heaters unsuitable for use on wood floors should be so marked. When such heaters are used, they should rest on suitable heat-insulating material, such as concrete of at least 1 in. (2.5 cm) thickness or equivalent. The insulating material should extend 2 ft (61 cm) or more in all directions. Locate heaters used near combustible tarpaulins, canvas, or similar coverings at least 10 ft (3 m) from such coverings. Securely fasten coverings to prevent them from igniting, or upsetting the heater, due to wind.

Vent gas- or liquid-fueled space-heating devices used in any enclosed building, room, or structure by a flue pipe to the exterior of the structure. Fresh air should be supplied, by natural or mechanical means, in sufficient quantities to ensure the health and safety of workers. Give particular attention to areas where heat and fumes may accumulate. When heaters are used in confined spaces, take precautions to ensure proper combustion to maintain a safe and healthful atmosphere for workers, and to limit temperature rises in the area. Locate vent pipes a safe distance from flammables and combustibles. Where vent pipes pass through combustible walls or roofs, properly insulate and securely fasten and support them to prevent accidental displacement or separation.

When heaters are used in enclosed or partially enclosed structures, have a continuous monitor for CO or test for the presence of CO within one hour after the start of each shift and at least every four hours (every two hours for solid fuel-burning heaters) thereafter. CO concentration greater than 50 ppm of air volume at workers' breathing levels requires extinguishing the heater, unless additional ventilation is provided to reduce the CO content to acceptable limits.

Provide spark arresters on all smoke stacks or burning devices having forced drafts or short stacks that permit live sparks or hot materials to escape.

Prohibit solid fuel-burning heaters in buildings and on scaffolds.

Salamanders

Salamanders and other types of portable heaters are widely used in severe weather to protect masonry, concrete, and plaster from freezing and to provide warmth for the workers. Gas- or oil-fired, electric, steam, or remotely located heaters with conducted hot air, however, are preferred to salamanders. These kinds of heaters are discussed in the sections that follow.

Solid fuel-burning salamanders are prohibited. Do not use them. Also, do not use liquid fuel-burning salamanders in confined spaces unless they are vented to the outside. Improperly installed salamanders and other open-flame heaters are particularly dangerous in tool sheds, shanties, and other small enclosed areas. They can give off large amounts of carbon monoxide (CO) and consume much oxygen. If the concentration of CO is greater than 50 parts per million (ppm) in work areas, turn the heater off or provide more ventilation. Test for CO about one hour after the start of each shift and again at least every four hours later.

The horizontal clearance between salamanders and combustibles shall be at least 2½ ft (0.76 m); overhead clearance shall be at least 6 ft (1.8 m). Keep tarpaulins and canvas or plastic coverings at least 10 ft (3 m) away. Make sure these coverings are securely fastened to keep them from blowing toward the salamander. Use flameproof materials whenever possible. To keep people away from a salamander's hot surfaces, place a noncombustible railing around it, at least 19 in. (45 cm) away from it.

Fueling Heaters

Check and follow manufacturer's instructions for fueling, as well as applicable regulations. Turn off all flames, including the pilot, if any. Use only the type of fuel specified for the unit. The unit should be cool to the touch. Before and during fueling, check all fuel lines, hoses, and connections for leaks.

Check federal and local regualtions regarding fuel storage. Indoor storage is not permitted for some fuels. If indoor storage is permitted, store only one day's supply of fuel inside a building in the vicinity of the heater. Place this fuel at least 25 ft (7.6 m) away from any source of ignition. Locate general fuel storage outside the building.

Fan-Assisted Heaters

Only use fan-assisted heaters that will not create a fire or an electrical shock during a power failure, or the failure of any electrical components. Only use power supply circuits with three- or four-wire grounding. Provide grounding continuity to all parts of the heater unit, including connection to a grounded power supply.

Natural Gas-Fueled Heaters

After assembling, test all piping, tubing, or hose of natural gas-fueled heaters for leaks, using proper soap solution. Run the heater at normal operating pressure and use soapsuds or other noncombustible means to find leaks—never use a flame. When placing a unit in operation, make sure it is working properly. Before disconnecting a heater, shut off the fuel supply at the source in order to purge the line.

A flexible gas supply line, or hose, should be no longer than 25 ft (7.6 m), and shorter if possible. Local codes, usually plumbing codes, will govern hose lengths, pressures, etc. Check all hoses and fittings to make sure they are designed for the pressure, capacity, and type of fuel being used. Hoses should have a minimum working pressure of 250 psig (1,730 kPa) and a minimum burst pressure of 1,250 psig (8,600 kPa). All hose connectors should be capable of withstanding a test pressure of 125 psig (860 kPa) without leaking, and a pull test of 400 lb (1,780 N). Securely connect hoses to the heater by mechanical means; never use "slip-end" connectors. Protect hoses and fittings from damage, and check for deterioration. Do not let hoses contact surfaces above 125 F (50 C). Locate hoses to minimize any physical damage.

Normal maintenance of natural gas-fueled heaters includes inspection of the hose's supply system for cracks, checks, abrasions, and rupture; and testing the hose, pipe, and tubing connections for leaks. Disconnect the electric power supply before repairing heaters.

Liquefied Petroleum Gas (LP-Gas) Heaters

For LP-gas heaters, follow the precautions for measuring CO concentrations, for testing operating capability, and for testing for leaks as outlined in previous sections. Use only hose labeled LP-gas or LPG for LP-gas heaters. The minimum working pressure of the hose should be 250 psig (860 kPa), the minimum burst pressure 1,250 psig (8,600 kPa). The hose should be at least 10 ft (3 m) long, but no longer than 25 ft (7.6 m), depending on local code specifications. Protect all hose from damage, deterioration, and hot surfaces.

Hose connectors should be capable of withstanding a test pressure of 500 psig (3,450 kPa) without leaking, and a pull test of 400 lb (1,780 N). Securely connect hoses to the heater by mechanical means. Never use "slip-end" connectors nor ring keepers tightened over the hose to give increased force to the metal fitting.

Equip heaters with an approved regulator in the supply line between the fuel cylinder and the heater unit. Provide cylinder connectors with an excess-flow valve to minimize the flow of gas in the event of a rupture in the fuel line.

For temporary heating, such as in concrete curing, place heaters at least 6 ft (1.8 m) away from LP-gas containers. This does not, however, prohibit the use of heaters specifically designed for attachment to the container or to the supporting standard with connecting hose less than 6 ft (1.8 m). However, the design and installation of such heaters must prevent the direct application of radiant heat onto the container. Do not direct blower-type heaters or radiant heaters toward any LP-gas container within 20 ft (6 m).

For two or more heater-container units (of either the integrated or nonintegrated type), located in an unpartitioned area on the same floor, separate the container(s) of each unit by at least 20 ft (6 m). The maximum water capacity of individual containers is 245 lb (111 kg)—nominal 100 lb (45 kg) LP-gas capacity. The total water capacity of containers manifolded together in an unpartitioned area should not be greater than 735 lb (333 kg)—nominal 300 lb (135 kg) LP-gas capacity. These containers should also be separated by at least 20 ft (6 m).

On floors on which heaters are not connected together for use, containers may be manifolded together for connection to heaters on another floor, if they meet the following two requirements:

1. The total water capacity of the containers connected to any one manifold is not greater than 2,450 lb (1,111 kg)—nominal 1,000 lb (450 kg) LP-gas capacity.
2. When more than one manifold having a total water capacity greater than 735 lb (333 kg)—nominal 300 lb (135 kg) LP-gas capacity—is located in the same unpartitioned area, separate the manifolds by at least 50 ft (15 m).

Note that these amounts of LPG may not be permitted inside buildings—check local codes.

Do not refill LP-gas cylinders inside buildings or other structures. Store cylinders outside of buildings, stand them on a firm and substantially level surface, and secure them in an upright position away from moving vehicles. Do not use container valves, connectors, regulators, manifolds, piping, and tubing as structural support for LP-gas heaters.

LP-gas heaters having inputs above 50,000 Btu per hour should be equipped with either a pilot, which must be lighted before the main burner can be turned on, or an electronic ignition.

Torches and Furnaces

Liquid-fueled blowtorches and plumbers' furnaces involve the hazards of fire and explosion. If possible, use safer equipment, such as electrically heated soldering irons, paint-remover irons, glue pots, and other devices.

Proper gas or oil space heaters listed by the American Gas Association or Underwriters Laboratories (UL) are recommended for use in areas containing combustibles (see References at the end of this chapter). Where flammable or explosive dust or vapors or hazardous products of combustion may be present, use such torches or furnaces in accord with local or federal laws and regulations.

The storage and handling of gasoline present a hazard in addition to the fire and explosion risks of the torches or furnaces themselves. Minimize this hazard by carefully observing the requirements for storage and handling of flammable liquids described in National Fire Protection Association (NFPA) 30, *Flammable and Combustible Liquids Code.*

Fire Protection

Have a qualified person make a survey of the suitability and effectiveness of fire prevention and protection measures and facilities at the project. When unusual fire hazards exist or fire emergencies develop, provide additional protection. Conspicuously post emergency telephone numbers and reporting instructions.

Never leave fires unattended and never permit open-flame devices. Prohibit all sources of ignition within 50 ft (15 m) of operations with a potential fire hazard, and conspicuously post legible NO SMOKING OR OPEN FLAME signs in the area. Prohibit smoking in all areas where flammable, combustible, or oxidizing materials are stored. Post NO SMOKING OR OPEN FLAME signs in all prohibited areas. Store insulating material with a combustible vapor barrier at least 25 ft (7.5 m) from buildings or structures. Permit only the quantity required for one day's use in buildings under construction.

Dispose of combustible waste materials in compliance with applicable fire and environmental laws and regulations. Have paint scrapings and paint-saturated debris removed from the premises on a daily basis. Shield all combustibles from the flames of torches used to cut or sweat pipe. Take precautions to protect form work and scaffolding from exposure to, and spread of, fire.

Fire Doors

During construction, give priority to building fire walls and stairway exits required for completed buildings. Fire doors, with automatic closing devices, should be hung on openings as soon as practical. Fire cut-offs should be retained in buildings undergoing alterations for demolition until operations call for their removal.

Fire Extinguishers

Provide portable fire extinguishers as follows:

1. At least one extinguisher, rated not less than 2-A, for each 3,000 ft^2 (914 m^2) of building area. Travel distance from any point to the nearest fire extinguisher should not exceed 100 ft (30 m).
2. At least one extinguisher, rated not less than 2-A, for each floor. In multistory buildings, locate at least one extinguisher adjacent to each stairway.
3. At least one extinguisher, rated not less than 10-B, provided within 50 ft (15 m) of wherever more than 5 gal (19 l) of flammable or combustible liquids, or 5 lb (2.3 kg) of flammable gas are being used. An exception is integral fuel tanks on motor vehicles.
4. At least one extinguisher, rated not less than 20-B, located outside of, but not more than 10 ft (3 m) from, the door opening into any room used for the storage of more than 60 gal (227 l) of flammable or combustible liquids.
5. At least one extinguisher, rated not less than 20-B, located not less than 25 ft (7.5 m) nor more than 75 ft (22.8 m) from any outside flammable liquid storage area.
6. At least one extinguisher, rated not less than 20-B:C, provided on all tank trucks or other vehicles used for transporting and/or dispensing flammable or combustible liquids.

Flammable Liquids

Keep flammable liquids in closed approved containers when not in use. Keep unopened containers of paints, varnishes, lacquers, thinners, and other flammable and combustible liquids in a well-ventilated location, free of excessive heat, smoke, sparks, flame, or direct rays of the sun.

All flammable and combustible liquids should be handled in safety containers with flame arresters. This requirement should not apply to those liquids that are extremely hard to pour. They may be handled in the original shipping containers. For quantities of 1 gal (3.8 l) or less, use only the original container or approved metal safety cans for storage, use, and handling. Tightly cap containers of flammable and combustible liquids.

Safety cans and other portable containers for flammable liquids having a flash point at or below 140 F (60 C) should be painted red with a yellow band around the can. Legibly write the name of the contents on the container.

Do not store flammable and combustible liquids in quantities greater than required for one day's use in buildings under construction. Storage, handling, installation, and use of LP-gases and systems should be in accord with NFPA 58, *Storage and Handling of Liquefied Petroleum Gases*.

Prohibit refueling of gasoline-operated equipment while the motor is running. Allow continuously operating equipment to be fueled only from properly protected tanks located outside the work area. Tanks should be adequately grounded and bonded to equipment to prevent static electricity buildup. Do not locate fuel-tank filler openings in such a position that spills or overflows can run down on a hot motor, exhaust pipes, or battery.

Prohibit smoking or the use of open flames on, or in the immediate vicinity of, gasoline-operated equipment while it is being refueled. Workers should not use solvents with flash points below 100 F (37.8 C) for cleaning equipment or parts. When gasoline and other highly flammable fluids are used, transfer them by approved pumps or store them in approved safety cans. Do not store gasoline, fuel oil, and other flammable or combustible liquids on equipment except in fuel tanks or approved safety cans. Locate suitable fire extinguishers on, or close to, each industrial truck. (See Chapter 12, Flammable and Combustible Liquids, and Chapter 11, Fire Protection.)

PROCESS SAFETY MANAGEMENT

If a project involves a contractor who will perform maintenance, repair, turnaround, major renovation, or specialty work with or in close proximity to a highly toxic, reactive, flammable, or explosive chemical (as listed in OSHA's general industry standard 1910.119, Appendix A), then specific process safety management actions are required. Plant owners are required to create a formal hazard assessment plan and pass the information on to contractors or create such a plan in conjunction with the contractor. These requirements are so important that it is necessary to become thoroughly familiar with the process safety management standard (29 *CFR* 1910.119, *Process Safety Management of Highly Hazardous Chemicals*). In summary, a client's responsibilities to the contractor and the contractor's responsibilities are as follows:

Client responsibilities:

- Prequalify contractors based on their safety performance and safety programs.
- Inform contract employees of particular chemical release hazards (potential release sites) of each work location.
- Familiarize all contractor personnel of emergency action plan(s) in response to accidental release of chemical(s) or other facility emergency.
- Develop and implement safe work practices to provide for the control of all known hazards dealing with contractor operations, such as lockout/tagout, confined space entry, entrance to vessels, piping, valves, etc.
- Implement procedures to control contractor personnel movement within covered process areas.
- Maintain an injury/illness log of contractor personnel in covered process areas.
- Review contractor performance per their responsibilities.

Contractor responsibilities:

- Train, document and verify understanding for each employee based upon:
  - known chemical, fire, explosive, and toxic potential of each process with which they will work
  - relevant parts of the emergency action plan
  - use of and adherence to the safe work practices for lockout/tagout, confined space entry, vessels entry, line breaking, etc.
- communicate to the owner any unusual hazard created by work or any other hazard identified during work

Using and communicating these basic understandings of the process safety management standard will lower the probability of catastrophic release of energy at the facility.

SUMMARY

- Prequalification criteria should be used to provide the necessary selection information to select a contractor who is both qualified and has the safety and health programs to meet client demands.
- Job Hazard Analysis must be included in project specifications for high hazard operations where the probability of equipment failure or personnel injury or death is significant.
- To ensure safety on facility construction sites, companies should not rely on building codes and construction safety standards alone. They must train their workers in safe working habits, good housekeeping procedures, worksite safety, and use of personal protective equipment. All construction sites must be restricted to authorized personnel only.
- Safety considerations should be calculated into the costs of construction from the beginning of the project. The safety professional can analyze the project and estimate the accident frequency potential in order to devise an appropriate accident prevention program.
- The company should ensure that the construction contract contains at least minimum safety, health, and equipment requirements on the contractor's part and that it provides for an effective safety program. The construction site itself must be made as safe as possible to protect workers and equipment from various hazards. All heavy machinery, trucks, other mobile equipment, hoisting apparatus, and conveyors must have safety devices and be operated only by trained personnel.
- Workers must be trained in the safe use of flammable liquids, power machinery and tools, and proper methods for erecting steel. Employees must wear proper protective equipment and observe general health and safety practices.
- Incomplete buildings should be supported with lateral bracing to prevent walls from breaking off and injuring workers. As construction proceeds, temporary flooring should be placed directly under the tier of beams on which work is being performed.
- In excavation work, management must do a careful survey of the site to evaluate potential hazards involving underground utilities, soil conditions, and surrounding structures. A special contractor should be hired for excavation work done within or adjacent to a building or done lower than wall or column footings and machinery or equipment foundations.
- Construction of all ladders and scaffolds should conform to the provisions of the appropriate state or provincial or local codes. Workers should have a safe access to the scaffold, and the structure should be inspected and tested regularly.

- Material and personnel hoists can be erected inside the building or in outside towers. Never permit personnel to ride on a material hoist or permit work in or on the hoistway while the hoist is in operation. All hoists should have a good signal system and be guarded to prevent material from falling on workers.
- Formwork, falsework, structural shoring, and bracing should be designed, erected, braced, and maintained in a safe manner. These items must be inspected regularly.
- All floor and roof holes, skylights, and openings into which people can walk must be guarded with enclosed guards or covered with material and bracing strong enough to support any load. Stairways, ladderway entrances, and wall openings should be guarded to prevent workers from walking into the opening.
- Heaters used on construction sites must be either gas- or oil-fired, electric, steam, or hot-air conductors. A qualified employee should be placed in charge of all temporary heating equipment to ensure safe operation and to keep the equipment in good working order.
- All construction sites must have adequate fire protection measures and fire prevention programs in force. Employees should do only minor demolition work and must observe all safety procedures for the job.
- Process safety management rules apply to projects involving highly hazardous chemicals listed in OSHA's general industry standard 1910.119, Appendix A. Formal hazard assessments and specific training requirements must be followed to create identifiable safe work practices and verifiable adherence to programs.

REFERENCES

American Concrete Institute, 22400 West Seven Mile Road, Detroit, MI 48219. *Recommended Practice for Concrete Formwork*, Publication ACI 347–78.

American Conference of Governmental Industrial Hygienists, 6500 Glenway Avenue, Bldg. D7, Cincinnati, OH 45211. *Industrial Ventilation Manual*, 17th ed.

American Gas Association, 1515 Wilson Boulevard, Arlington, VA 22209.

American Institute of Architects, 1735 New York Avenue NW, Washington, DC 20006.

American National Standards Institute, 11 West 42nd Street, New York, NY 10036.
Criteria for Safety Symbols, ANSI Z535.3–1991.
Environmental and Facility Safety Signs, ANSI Z535.2–1991.
Elevators and Escalators, ANSI/ASME A17.1–1993.
Inspector's Manual for Escalators and Elevators, ANSI/ASME A17.2–1994.
Lifting Devices, ANSI B30 Series.
Practice for Industrial Lighting, ANSI/IES RP7–1990.
Safety Color Code, ANSI Z535.1—1991.
Safety Requirements for Portable Wood Ladders, ANSI A14.1–1990.
Safety in Welding and Cutting, ANSI Z49.1–1994.
Concrete and Masonry Work, ANSI A10.9–1983.
Safety Requirements for Demolition, ANSI A10.6–1990.
Safety Requirements for Fixed Ladders, ANSI A14.3–1992.
Safety Requirements for Workplace Floor and Wall Openings, Stairs and Railing Systems, ANSI A1264.1–1989.
Safety Requirements for Job-Made Wooden Ladders, ANSI A14.4–1992.
Safety Requirements for Material Hoists, ANSI A10.5–1992.
Safety Requirements for Personnel Hoists and Employee Elevators, A10.4–1990.
Safety Requirements for Portable Reinforced Plastic Ladders, ANSI A14.5–1992.
Safety Requirements for Portable Metal Ladders, ANSI A14.2–1990.
Safety Requirements for Scaffolding, ANSI A10.8–1988.
Safety Requirements for Steel Erection, ANSI A10.13–1989.
Safety Requirements for Temporary and Portable Space Heating Devices and Equipment, ANSI A10.10–1990.
Scheme for the Identification of Piping Systems, A13.1–1981 (R1993).

Associated General Contractors of America, Inc., 1957 E Street NW, Washington, DC 20006. *Manual of Accident Prevention in Construction* Rev. ed., 1977.

Building Officials & Code Administrators International, Inc., 4051 Flossmoor Road, Country Club Hills, IL 60477. *The BOCA Basic Fire Prevention Code* (issued every three years).

Building Operations Manual. Champaign, IL: University of Illinois Press, 1976.

Dunn RL. Advanced maintenance technologies. *Plant Engineering* 41: 80–86, 1987.

Hislop RD. *Construction Site Safety: A Guide for Managing Contractors.* Boca Raton, FL: Lewis Pub., 1999.

Huntington WC. *Building Construction: Materials and Types of Construction.* 5th ed. New York: Wiley & Sons, Inc. 1981.

Illuminating Engineering Society of North America, 345 East 47th Street, New York, NY 10017.
Journal of American Insurance, 3:10–1984.

Laborers' International Union, 905 16th Street NW, Washington, DC 20006.

National Fire Protection Association, 1 Batterymarch Park, Quincy, MA 02269.
Flammable and Combustible Liquids Code, NFPA 30, 1993.
Life Safety Code, NFPA 101, 1994.
Storage and Handling of Liquefied Petroleum Gases, NFPA 58, 1995.

National Safety Council, 1121 Spring Lake Drive, Itasca, IL 60143.
Fundamentals of Industrial Hygiene, 4th ed., 1993.
Occupational Safety and Health Data Sheets.
Atmospheres in Subsurface Structures and Sewers, 12304–0550, 1987.
Belt Conveyors for Bulk Materials, 12304–0569, 1990.

Blowtorches and Plumbers' Furnaces, 12304–0470, 1990.
Flexible Insulating Protective Equipment for Electrical Workers, 12304–0598, 1991.
Load-Haul-Dump Machines in Underground Mines, 12304–0576, 1990.
Recommended Loads for Wire Rope Slings, 12304–0380, 1991.
Roller Conveyers, 12304–0528, 1991.
Safety Hats, 12304–0561, 1992.
Sidewalk Sheds, 12304–0368, 1990.

Scaffolding and Shoring Institute, 1230 Keith Bldg., Cleveland, OH 44115. Steel *Scaffolding Safety Rules*.

REVIEW QUESTIONS

1. Name the primary reference manual for most operations on a given construction project.

2. List the four general tools used by contractors and safety and health professionals to complete a project in a timely and safe manner.
a.
b.
c.
d.

3. When using temporary heating equipment, what six factors should be included in the operating procedures?
a.
b.
c.
d.
e.
f.

4. Which of the following is the preferred material to use when erecting heavy loads of steel?
a. Chains
b. Rope
c. Wire rope
d. Wire rope slings

5. Describe five of the eight general safety and health precautions to take when erecting steel.
a.
b.
c.
d.
e.

6. A ladder should be placed so that the horizontal distance between the base of the ladder and the vertical plane of the support is approximately how long?
a. One-fourth the ladder's length
b. Half the ladder's length
c. One-third the ladder's length
d. None of the above

7. Identify the regulations that require scaffold users to receive training for fall protection systems.

8. Excavation safeguards for shoring, sloping, and supporting systems should be based on careful evaluation of what five factors?
a.
b.
c.
d.
e.

9. When constructing boatswain's chairs, the worker should make sure the seats are not less than:
a. 1 ft long by 1 ft wide
b. 1.5 ft long by 1 ft wide
c. 2 ft long by 1 ft wide
d. 2.5 ft long by 1.5 ft wide

10. Which of the following is the preferred enclosure for a material hoistway installed inside a building?
a. Solid enclosure
b. ½ in. mesh, No. 18 U.S. gauge wire screening
c. 1 in. mesh, No. 18 U.S. gauge wire screening
d. Tubular steel tower

11. Describe six practices workers should follow when ascending or descending ladders.
a.
b.
c.
d.
e.
f.

12. Describe the purpose of warning lines.

4

Maintenance of Facilities

Revised by

John F. Montgomery, Ph.D., CSP, CHCM, CHMM

A sound, efficient maintenance program is essential in any industrial establishment. Such a program keeps the physical facility in good condition and avoids or removes safety and health hazards. This chapter covers the following topics:

- maintaining a building to promote worker health and safety
- preventing the problem of sick building syndrome
- dealing with safety issues in grounds maintenance
- using computer technology to help in maintenance work
- promoting safe work practices and training for maintenance crews

FACILITY MAINTENANCE

Maintenance includes (1) care to ensure long-term life of company assets, (2) routine care to maintain uninterrupted service and appearance, and (3) repair work required to restore or improve service and appearance. Maintenance is recognized as one of the keys to improved facility productivity, safety, and good public relations. Appearance—both internal and external—affects employees, customers, and the general public.

Too often, maintenance means only making repairs. Maintenance programs should place more emphasis on preventive maintenance and on the type of inspection and monitoring that discovers conditions before they result in failure and accidents. The use of computers and space-age technology has provided new tools for preventive maintenance. For example, computer programs that monitor equipment can alert personnel about maintenance needs before equipment breaks down.

The maintenance program may be supervised by the facility engineer or maintenance superintendent. However, because the safety of employees is closely tied to the condition of buildings and equipment, safety and health professionals will find that the maintenance program has an important bearing on the safety program. Therefore, do not hesitate to point out to management that equipment and structures need repairs, modification, or replacement.

Foundations

Starting from the bottom, inspect and maintain footings, column bases, foundation walls, and pits. The safety of the rest of a building, as well as that of employees and equipment, depends on a firm foundation.

Footings and Columns

Although it is hard to detect flaws in footings, it is possible to check cracks and unusual settlement of a building's columns and footings. For example, place level marks at known elevations, about 5 ft (1.5 m) above the basement floor. Check these marks periodically for signs of settlement.

Excessive settlement may threaten the stability of a building, as well as the effective working of the machines and equipment in it. Inspectors should report excessive settlement to management at once for immediate action.

Inspectors should also check the bases of columns for dry rot and rust. Dry rot around the bottom of wood columns at the basement floor level can result if the basement floor is damp, subject to water seepage, or is alternately dry and wet. Maintenance personnel should scrape away rust at the bases of steel columns and give the bases a coating of a preservative. (See also Structural Members, described in the next section.)

Foundation Walls

Inspect the inside of foundation walls for cracks, which may result from settlement of the building and shrinkage of concrete. Since these cracks are below grade, they can admit water to the basement area. If enough water comes in through large cracks, it can cause settlement of the backfilled earth around the outside of foundation walls. Sidewalks and adjacent roadways can also be damaged from such settlement.

Small cracks in foundation walls can be repaired from inside the building by applying a waterproofing material. If the cracks are relatively large, maintenance workers may have to dig down outside the building to the bottom of the wall, clean off all earth and other foreign material, and apply a waterproofing compound. They should then apply a membrane covering directly on the compound and cover the membrane with another coating of compound. When unusual settlement is noted, inspectors should take settlement readings like those for footings and columns.

Pits

Inspectors should also examine pits. Note cracks and have them repaired. Do not allow any debris or rubbish to collect in the pits. Install guardrails or covering where needed.

Structural Members

Structural members, such as joists, beams, girders, columns, and flooring, require periodic inspection and maintenance. Steel, concrete, and wood parts each need special inspections and care.

Joists, Beams, and Girders

Check joists, beams, and girders, and correct them for deflection, twisting, tipping, or other unusual conditions. In many instances, joists, beams, and girders are covered by suspended or sealed ceilings, thus making them hard to reach. In such cases, excessive deflection may be indicated only by a badly sagging floor. At least once a year, therefore, examine the entire floor system on each floor level. If major repairs are necessary, clear the floor of stored materials immediately. Before making repairs, consult a qualified engineer.

Columns

Examine building columns for unusual distortion or buckling. Avoid excessive or unusual column loadings, and check for holes cut in or through columns. If holes exist, notify a qualified engineer.

Steel Parts

Check steel I-beams, channels, columns, angles, girders, and other structural steel parts for rust once a year—more often in corrosive atmospheres. Where rust exists, scrape the steel part and paint it.

Concrete Parts

Check floor slabs, beams, girders, and columns regularly for cracks, spalling, and chipping of concrete from the reinforcing steel. Because rust may form on exposed steel, make repairs at once. Gunnite, for example, provides a protective coating for exposed steel. Any visible damage of concrete parts may evidence more serious problems. Consult a qualified engineer for further investigation.

Wood

Inspect a wood floor system for shakes, checks, and splits in joists, planks, beams, stringers, posts, and columns. Look for decay or dry rot in wood columns, joists, and other parts. Because it is important that beams, joists, and girders provide full bearing, thoroughly investigate every evidence of movement or slippage.

Walls

Masonry buildings require periodic minor repairs of walls and windows. Because mortar joints loosen and disintegrate from settlement of a building and from weathering, rake and point such joints. If these joints are not repaired, moisture will eventually seep into interior wall surfaces and cause further damage.

Exterior Walls

Inspect exterior walls made of brick, concrete, terra cotta, stone, cement or cinder block, or stucco for cracks or joint separation. Cracks result from expansion, contraction, vibration, or settlement. Fill them immediately; otherwise, water may freeze in them and cause additional damage.

If brick walls are painted on a building's exterior, and high humidity exists inside the building, excessive moisture will penetrate the brick and condense under the paint film. Should the moisture freeze, it could cause spalling and joint disintegration.

Windows

Because of settlement of a building, drying out of the wood, or improper setting of metal frames, caulked window joints may crack open, allowing moisture to enter. Before recaulking, remove all loose material, and cut out all cracks so that the new compound bonds well. Apply a suitable caulking compound with a gun that forces the compound well into the openings rather than covering just the surface joints.

Parapet and Stone Cap Repairs

Maintain and make necessary repairs to masonry, metal, and wood parapets; stone caps; and other stonework. Avoid serious repairs on masonry walls by checking them carefully once a year for cracks or spalling.

Never coat brick parapets and walls above grade, on either side, with materials such as pitch, roof paper, or asphalt roof coating. These materials do not allow walls to breathe. Such coating, which is commonly misapplied to brick parapets, causes spalling of the brick and disintegration of the mortar joints, especially in areas of the country where freezing temperatures occur.

Check stone caps and other stonework on brick walls for cracks at all mortar joints. Cracks might allow moisture to enter and eventually loosen the stone. Fill these cracks with a cement grout or mastic filler to prevent stone caps from falling off. If stone caps are on high buildings, their falling off could cause personal injuries and property damage.

Interior Walls and Ceilings

Inspect partitions, cross walls, interior sides of main walls, and ceilings as rigorously as exterior walls. Look for such defects as cracks in interior walls, holes, loose mortar in joints, broken or missing brick, and spalled or worn areas on tile or brick walls where power trucks may have frequently scraped them. To prevent damage from trucks, install standard 3 in. (8 cm) or 4 in. (10 cm) iron pipe railings near the floor level as barriers.

Ceilings require periodic painting, cleaning, and repair. Investigate unusual sags immediately and correct them. If a sag exists in a suspended ceiling, check the hangers and fastenings. However, if the sag has resulted from excessively loading the floor above, correct the situation.

Floors

Accidents owing to inadequate maintenance of floors are a major source of injuries in many facilities. Slippery conditions account for many falls by workers. Holes and other irregularities in wood and concrete floors, both inside and outside facility buildings, result in frequent injuries from stumbling and falling. In addition, they cause many truck accidents.

Maintenance and Housekeeping Procedures

Using the wrong cleaning materials, methods, and surfacing often causes even the most suitable types of flooring to deteriorate and become slippery. Do not use alkaline cleaners on terrazzo. However, mild alkaline cleaners may be used on asphalt tile. Oils are unsuitable for rubber tile and, when applied to wood floors, increase the fire hazard. To keep floors clean, safe, and sanitary, follow the recommendations of the flooring manufacturer. Standardize and spell out in detail the maintenance procedures for floors (Figure 4–1).

In general, the routine maintenance procedure for linoleum, marble, terrazzo, asphalt tile, and similar

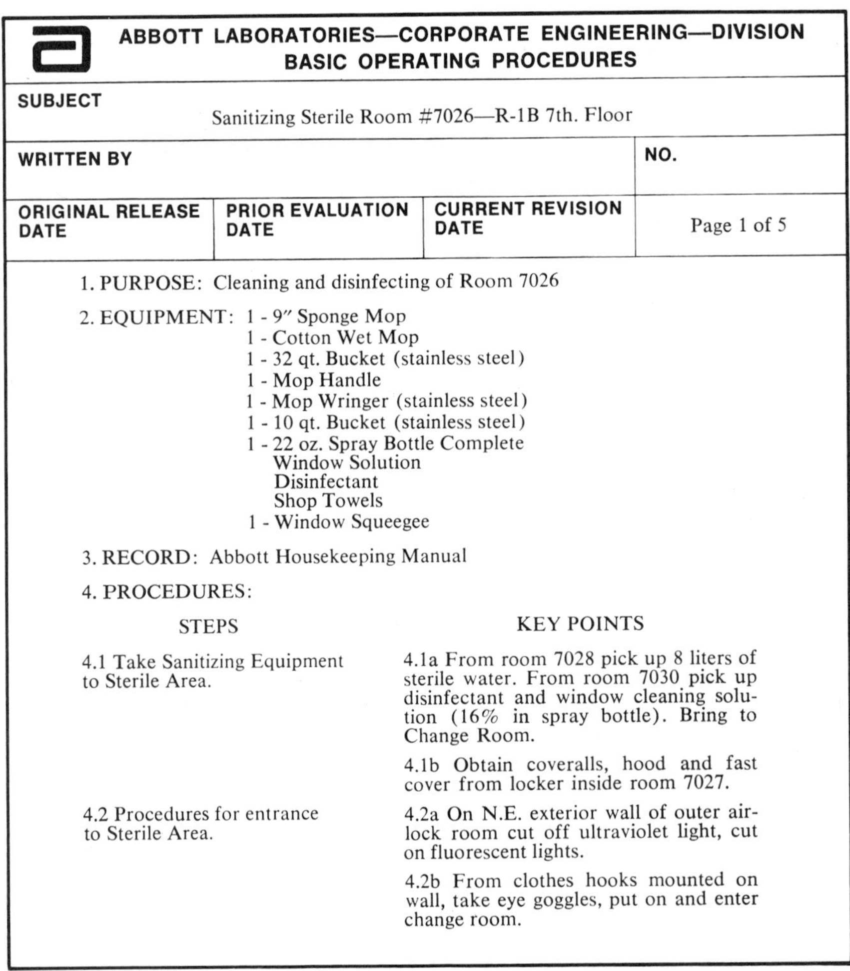

ABBOTT LABORATORIES—CORPORATE ENGINEERING—DIVISION
BASIC OPERATING PROCEDURES

SUBJECT	Sanitizing Sterile Room #7026—R-1B 7th. Floor		
WRITTEN BY			NO.
ORIGINAL RELEASE DATE	PRIOR EVALUATION DATE	CURRENT REVISION DATE	Page 1 of 5

1. PURPOSE: Cleaning and disinfecting of Room 7026

2. EQUIPMENT: 1 - 9″ Sponge Mop
1 - Cotton Wet Mop
1 - 32 qt. Bucket (stainless steel)
1 - Mop Handle
1 - Mop Wringer (stainless steel)
1 - 10 qt. Bucket (stainless steel)
1 - 22 oz. Spray Bottle Complete
Window Solution
Disinfectant
Shop Towels
1 - Window Squeegee

3. RECORD: Abbott Housekeeping Manual

4. PROCEDURES:

STEPS	KEY POINTS
4.1 Take Sanitizing Equipment to Sterile Area.	4.1a From room 7028 pick up 8 liters of sterile water. From room 7030 pick up disinfectant and window cleaning solution (16% in spray bottle). Bring to Change Room.
	4.1b Obtain coveralls, hood and fast cover from locker inside room 7027.
4.2 Procedures for entrance to Sterile Area.	4.2a On N.E. exterior wall of outer airlock room cut off ultraviolet light, cut on fluorescent lights.
	4.2b From clothes hooks mounted on wall, take eye goggles, put on and enter change room.

Figure 4–1. Standard operating procedures are spelled out in detail so that each job is completed efficiently, thoroughly, and safely. Shown here is the first page of a detailed procedure for sanitizing a sterile room. (Printed with permission from Abbott Laboratories.)

types of flooring used in offices, facilities, and the like is to (1) clean the floors with a soft brush or vacuum cleaner, and (2) when necessary, wipe them with a mop dampened with clean, cold water. Clean one section of floor at a time. If traffic in the area is heavy, rope off that section. When soap is used, remove the soapy film by thorough rinsing to avoid a slippery condition.

Ordinary wax for polishing wood, tile, and similar floor surfaces is unsuitable because of its inherently slippery nature. However, according to the manufacturers, floor oils and waxes can be used on various types of floors without adding unduly to their slipperiness, provided that users apply them as instructed. Soft floors, such as asphalt, vinyl, and linoleum, are often refinished four times a year. Hard flooring, such as concrete and terrazzo, is cleaned and sealed once a year.

Oil and grease, water, paper, sawdust, and many other foreign materials create slipping hazards on floors. Eliminate or contain leaks of oil from machines, leaks of water or other liquids from pipelines, and spillage from processing equipment by maintaining the equipment. Promptly tighten loose connections.

When leakage cannot be readily eliminated at the source, use pans and absorbent materials on floors to prevent slipping hazards. However, a study of such sources often reveals ways of keeping slippery materials from getting on the floors, such as installing splash guards on machines using cutting oils.

Promptly clean up slippery materials spilled on floors. To remove grease and oils, cover the area with slaked lime to a depth of about ¼ in. (5 mm). After two or three hours, remove the lime with a scraper or stiff brush. Sand and various commercial cleaners also can be used.

Even as innocuous a substance as coffee can cause an accident if spilled in a high-traffic area. In some offices, cups containing beverages must be either covered or placed on a tray if carried to an employee's desk.

Aisles

Keep aisles clear of machinery, equipment, and raw and manufactured materials. In many cases, the allowable floor loading has been figured on clear aisle space, with no allowance made for power trucks using aisles. For efficient and safe operations, determine whether floors and aisles are capable of sustaining the loads of power trucks. Maintain lines to indicate aisle width.

Floor Load Capacity and Load Distribution

A survey of a physical facility requires accurate data on floor load capacity (Figure 4–2).

If this data is not already known or readily obtainable from building plans, have a qualified engineer conduct a structural analysis. Rough estimates based upon experience, or conclusions reached by a casual glance at a handbook, can be dangerous. Use accurate weight data.

Load distribution is often a complex problem. Most buildings are designed to carry uniform loads. A concentrated load places greater stress on supporting members than a uniform load of equal weight. Therefore, most heavily concentrated loads, such as machines, are placed directly over beams or girders rather than on slabs or joists. As in the case of floor load capacity, a safety professional should consult a structural engineering specialist to accurately determine concentrated loads.

Overloading of Floors

Installation of heavy equipment, excessive weight and unequal distribution of stored raw and finished materials, and heavy truck transportation, may cause overloading of floors. Post signs stating allowable floor loads, and paint horizontal lines on the walls showing the maximum height to which materials may be piled (Figure 4–3).

Whether a floor is safely loaded or is overloaded depends on how closely the designed load capacity and the actual load capacity correspond.

Although evidence of overloading is not always visible, inspectors should look for it. Deflection of flooring is the most common evidence of overloading in wood and steel beams. A sag or deflection greater than 1/360th of a span's length is a warning that a floor may be overloaded.

To measure the deflection, stretch a cord between two columns of the span, 5 ft (1.5 m) down from the underside of the beams or girders at ceiling level. Measure the distance from the center of the taut cord to the bottom of the beam or girder. The difference between this measurement and 5 ft (1.5 m) is the deflection.

Floors and other structural parts show signs of overloaded floors in several ways:

- Overloaded wood beams will check and crack.
- In reinforced concrete beams and girders, concrete will spall and fall away from the tensile side.
- Wood floors will be punctured, and flat concrete slabs will crack and spall.
- Timber columns will split and crack.
- Concrete columns will spall, the concrete falling away from the reinforcing rod.
- On steel columns, flanges will be twisted.
- Bearing walls of masonry will show extensive cracking, disintegration of the bricks, bulges, and pulling away of floor joists.

Whenever any doubt exists as to the amount of deterioration of a floor's supporting parts and, therefore, of its load-carrying capacity, have a qualified structural

Figure 4–2. Communication is an important safety component in the workplace. Note the floor load capacity sign and the labels on pipes. (Courtesy International Stamping Co./Midas International Corp.)

engineer inspect it and give a determination. Post the floor loads for storage areas not originally designed for storage.

Repair Procedures

Inspect wood flooring for rot, wear, and unusual stress, indicated by sag. In many cases, a section of wood flooring may need to be replaced. Install new flooring flush with the existing flooring. Replace badly worn or loose wood-block flooring by anchored wood blocks.

To anchor loose, finished wood flooring, drill holes at an angle through the finished floor into the subfloor. Then drive flooring nails larger than the holes into the subfloor.

For concrete floors, chip out the damaged area, thoroughly clean it, and wet it down. Trowel in cement mortar (1 part Portland cement and 3 parts sand) at a minimum patch thickness of 1 in. (2.5 cm). Patches 2 in. (5 cm) thick or more may require wire mesh or reinforcing steel. For finishing, a wood float gives a less slippery surface than does a steel trowel. For proper curing, keep traffic off the patch for at least three days, unless a quick-setting cement was used.

If mixed and applied properly, epoxy resin repair materials give excellent results. A thickness of as little as $\frac{1}{8}$–$\frac{3}{16}$ in. (2–3 mm) gives an extremely tough-wearing surface. Take adequate ventilation precautions when using these materials.

Roofs

Inspect and maintain roofs and roof-mounted structures on a regular basis. Roof damage can quickly lead to structural damage of other parts of the building and equipment.

Inspection

Inspect all roofs periodically, perhaps once every six months. Check roof flashings for cracks at the parapet wall. Check roof gutters and drain connections for

Figure 4–3. Where the same type of material is stored regularly, a line can be marked on the wall to indicate the height to which material can be piled without exceeding the allowable floor load. In addition to safe floor load signs, a warning sign reading DO NOT PILE ABOVE LINE can be placed on the wall.

cracks at the roof line. Check areas around dormers, chimneys, and valleys, where metal is used, to see that the metal is tight with the roof as well as with the drain. Keep roof drains and overflow gutters through parapet walls open, because leaks will occur if water should rise above the flashings. Keep gutters clean, and check gutter areas for ice damage in early spring.

It is often difficult to find the source of a leak, since the fault may be distant from the place where the leak shows inside the building. On sloping roofs, check above and to the sides of the place where the leak shows. Flat decks laid with concrete are difficult to check because water tends to follow a crack and show up at a distant point.

Abnormal Loading

Roofs are usually designed to carry the maximum snow load expected for their locale. Roof configuration (especially multilevel) and wind frequently combine to deposit heavy drifts over portions of a roof. Ice can also cause overloading. Where practical, remove these accumulations as quickly as possible to prevent roof collapse, major leaks, or lesser damage.

Observe the following precautions to prevent roof damage:

- To avoid ice and snow buildup near drain areas, clear a path from the center of the roof to the drains.
- To allow drainage on a pitched roof with no drains, open paths leading to the roof's edge.
- Never use blowtorches or similar devices to melt ice from drains or roof surfaces.
- To avoid puncturing the roof, instruct workers to use care when removing ice and snow from it.

Roof Anchorage

Check roof anchorage during inspections. The lifting of unanchored roofs accounts for a large percentage of wind-damage loss to U.S. industry. Securely anchor the roofs of all buildings. Anchors can be readily installed, and the cost is reasonably low.

Roof-Mounted Structures

Roof inspection should include penthouses, stacks, vents, air handling units, and supports for water tanks where these structures are flashed at the main roof level. Check the roofs of penthouses because most penthouses contain elevator machinery that can be damaged by water seepage.

The windows, skylights, and monitor sash of a penthouse should have necessary reglazing, reputtying, frame caulking, or painting. When workers repair or repaint operating mechanisms, they should work on a safe scaffold and should use a safety belt or harness and a lanyard tied off to a lifeline. They should never work on or over unprotected skylights.

Repairs

Patch a leaky roof as soon as possible after the leak appears. When maintenance becomes excessive and leaks occur after each rain, replace the roof. Cold mastic applications may last about five years, whereas an application of hot pitch covered with pea gravel may last from five to ten years. Reputable roofing manufacturers and contractors will provide a guaranteed-performance bond for a number of years commensurate with the quality of the job specified.

During repair, a roof can be punctured by tools, boards with nails, stones, or other sharp objects. To prevent such puncturing, use runways and protective runboard covering. If hot pitch or roofing compound is used, workers should wear gloves, goggles, and leather or fire-resistant duck leggings.

Tanks and Towers

Maintenance of tanks and towers is important, not only for fire protection but also because structural failures may cause serious accidents. When tanks are to be cleaned or painted on the inside, strictly observe all precautions relating to entering tanks. (For a detailed discussion of the required precautions, see Cleaning Tanks in Chapter 13, Flammable and Combustible Liquids, in this volume.)

If a tank is more than 20 ft (6 m) above the ground or the building's roof, place a wood or steel balcony around the base of the tank. The balcony should support a weight of at least 100 lb/ft^2 (486 kg/m^2). For a tank not more than 15 ft (4.6 m) in diameter, the width of the balcony should be at least 18 in. (45 cm); for tank diameters greater than 15 ft, the balcony width should be 24 in. (61 cm), including railings. (See ANSI A1264.1, *Safety Requirements for Floor and Wall Openings, Stair Railing Systems, and Toeboards.)*

Stacks and Chimneys

Inspect stacks at least once every six months. They are subject to deterioration both inside and out, from weathering, high winds, lightning, settlement of the foundation, and the action of corrosive flue gases. Check a stack's ground wires for effective grounding.

A brick or concrete chimney can be protected by lightning rods of low resistance and ample current-carrying capacity. They must be installed from the top of the chimney to a good electrical ground. Lightning rods can be readily installed on chimneys when repairs call for the erection of scaffolding. Underground water pipes or buried copper plates afford good ground connections.

Fixed Ladders

The safety standard for fixed ladders (ANSI A14.3) permits use of safety systems in lieu of cage guards on tower, water tank, and chimney ladders exceeding 24 ft (7 m) in unbroken length. A landing platform shall be provided at least every 50 ft (15 m) within the length of climb. A rest platform at not over 150 ft (46 m) with a ladder safety device is used.

Safety devices for ladders allow a climber to attach a harness to a sleeve that travels along a carrier rail or cable anchored to the ladder. The sleeve is designed to lock and suspend a person who slips and starts to fall. Many safety professionals prefer such devices to cage guards.

According to ANSI A14.3, fixed ladders must meet the following general requirements. A fixed ladder must have the following characteristics:

- be designed to withstand a single concentrated load of at least 250 lb (113 kg)
- have rungs with a minimum diameter of ¾ in. (19 mm) if a metal ladder, or 1⅛ in. (28 mm) if a wood ladder
- have rungs at least 16 in. (40 cm) wide and uniformly spaced no more than 12 in. (30 cm) apart vertically
- be painted, if metal, or otherwise treated to resist deterioration
- have a preferred pitch of 75 to 90 degrees for safe use
- have a 30 in. (75 cm) clearance with at least 24 in. (61 cm) on the climbing side of the ladder, unless caged
- have at least a 7 in. (18 cm) clearance behind the ladder to provide for adequate toe space
- have side rails that extend 3½ ft (1 m) above landings
- have a clear width of 15 in. (38 cm) on every side of the center line of the ladder, unless used with cages or wells

Platforms and Loading Docks

Mechanized traffic on platforms and docks often causes damage to platform surfaces. Check wood platforms for decay or dry rot, loose or uneven planking, and weakened or broken supporting members. Make repairs immediately.

With no edge protection, concrete platforms or docks can become spalled or chipped. Ruts in the concrete's finish may cause power or hand trucks to swerve and run off a dock or into employees or material. For these reasons, provide angle-iron or channel-iron protection at the edge of the platform, and maintain it well. Resurface badly rutted platforms with concrete or epoxy cement.

Canopies

A canopy's roof should receive the same careful inspection as that given to the roof of the main building. Evidence of pulling away from the building should be noted and corrected. Because drainage of canopies is important, keep downspouts and gutters open and in good repair. Periodically scrape and paint supporting parts of canopies if they are made of wood or steel.

Sidewalks and Driveways

Repair concrete sidewalks and driveways as soon as spalling or cracking of the concrete creates a hazardous condition. In cold climate areas, make inspections in the spring after the ground has thoroughly thawed. Sections of sidewalks can be relaid, but bituminous driveways will need patching with hot tar or similar material.

A driveway's drainage is important, whether the driveway is made of concrete, asphalt, or gravel. Make repairs as soon as possible to keep damage to a minimum. To keep asphalt driveways in good condition, have them recoated periodically by someone who understands paving techniques.

Underground Utilities

Inspection and maintenance of underground utilities is especially hazardous work. Personnel should observe strict procedures and should be closely supervised.

Sewers

At least two persons should work on a sewer maintenance job. Before anyone goes into a manhole structure or sewer, obtain a confined-space entry permit and observe established confined-space entry procedures. Test for oxygen deficiency, methane, hydrogen sulfide, carbon monoxide (CO), and any other suspected atmospheric contaminant. If oxygen deficiency or any contaminant is found in concentrations approaching or exceeding the permissible exposure levels, institute other elements of the confined-space program.

Have workers use proper respiratory equipment (see Chapter 23, Respiratory Protective Equipment, in the National Safety Council's *Fundamentals of Industrial Hygiene,* 4th ed.), or provide complete ventilation with either blowers or suction fans. Blowers are preferable because their source of supply is known. Suction fans may draw poisonous gases into the area from unseen pockets or crevices. Respiratory protection may be necessary for the outside attendant as well as for the workers inside.

Utility Trenches and Tunnels

Where personnel must work in trenches over 5 ft (1.5 m) deep, shoring or sloping is required. If a trench is near machinery foundations or other superimposed loading, make a thorough design of the bracing and shoring to be installed before the work is started. When pipe or trenches must remain open overnight, place barricades, signs, and lanterns around them to protect workers, other employees, and the public.

Check atmospheric conditions in utility trenches, tunnel trenches, tunnels, and manholes before work begins. Proper ventilation is necessary for the safety of workers, who, once again, should work in pairs. Post signs to identify manholes, tunnels, and trenches known to be contaminated.

Forbid smoking in or near manholes, tunnels, or trenches. Also, forbid the use of open-flame devices, such as solder pot furnaces and welding equipment, in or near manholes, tunnels, or trenches in which tests indicate the presence of flammable gas.

Waste Disposal Facilities

When on-site waste disposal is planned in the United States, obtain a permit from the U.S. Environmental Protection Agency (EPA). The EPA specifies all requirements. Consult local regulatory bodies as well.

Other Underground Pipelines

Before starting repairs, completely drain pipelines, block off the connecting systems, and close and lock the valves. Whenever workers open a line or valve, they should watch for back pressure in the line. After a steam valve has been opened completely, it should be backed off at least one-half turn so that thermal expansion will not lock the valve in the open position. Caution workers to open valves slowly and to equalize pressure slowly. Sudden changes in pressure can wreck equipment and endanger lives.

When people are to work on an underground pipeline, they should know its approximate location before beginning to dig. Workers cutting into or working on a pipeline should know what the line is carrying and the dangers involved and should purge the line before beginning the work.

Lighting Systems

All lighting systems require regular maintenance for maximum output of light. The light output of lamps decreases as the lamps age. Consider this factor when replacing lamps.

Replacing Lamps

To reduce the number of persons who might be exposed to flying glass and dust if lamps break, replace lamp fixtures on weekends or at other times when other personnel are not near the lamps. Those who handle the lamps should wear gloves and eye protection.

To dispose of a few lamps, (1) put them in special containers for regular rubbish removal, or (2) wrap them in several thicknesses of newspaper or wrapping paper, place them outdoors in a disposal container, and crush them with a heavy stick or shovel. Break large numbers of tubes in commercially available machines, in a ventilated enclosure equipped with an exhaust and a dust collector. Wet down and remove captured dust.

Install fluorescent fixtures in which the tubes lock in place. Equip older fixtures with shields or grids beneath the tubes to prevent the tubes from falling should vibration loosen them.

To clean reflectors or glassware that cannot be taken down, shut off the current and use a cleaner that requires no rinsing. Then wipe the reflectors or glassware with a cloth.

Maintaining High Levels of Lighting

Lamp burnouts and depreciation, dirt accumulation, voltage drops, and light absorption by dirty walls and ceilings will reduce illumination levels. Because dark or dirty surfaces absorb as much as 80% of the light that strikes them, maintain a regular cleaning schedule. For dirty areas, this schedule may be every two weeks.

Use a light meter to measure illumination levels. Then check the readings with the minimum standards of illumination for industrial interiors, ANSI/IES RP7, *Practice for Industrial Lighting.* These standards, recommended by the Illuminating Engineering Society (IES) of North America, are summarized in Table 5, Levels of Illumination, in Appendix 1, Safety and Health Tables, at the end of this volume. Use these values as a guide when formulating a regular maintenance program. If cleaning lamps and replacing burnouts fail to increase the illumination to standard levels, then have a qualified illuminating engineer make a complete survey of the present system.

Using Laborsaving Devices

To simplify maintenance problems, use laborsaving devices. When possible, use disconnecting reflectors. They permit cleaning and relamping from the floor rather than at the outlet. To reach lamps mounted low, use a stepladder, which is convenient and portable. Make sure the ladder is tall enough so that workers do not stand on the top two steps during maintenance activities. Attach clips and hooks to the ladder to hold spare lamps and cleaning rags. In this way, a person can do an entire cleaning and relamping job with one trip up the ladder. Where an entire installation is cleaned frequently, use special cleaning trucks with separate compartments for cleaning solutions, warm rinse water, and clean rags.

Use maintenance platforms where a great many lamps are mounted at the same height. Some platforms are made of lightweight material and are equipped with casters so that one person can handle them easily. Others are self-propelled aerial work platforms (Figure 4–4). This type of equipment permits a person to reach several units safely without repositioning the platform.

Manufacturers have designed various devices for reaching all types and styles of lamps. One of the simplest devices is the clamp grip mounted on the end of a pole and formed to fit fluorescent tubes. Many industrial facilities use this device for fixtures with open bottoms and exposed lamps. Use these clamp-grip devices between periods of regular maintenance for replacing lamps in an emergency. They are also well suited for recessed reflector lamps.

Figure 4–4. A scissor lift with protective toeboards and railings can bring the worker to the work area. The lift is stable, but mobile, and is controlled from the platform. (Courtesy Grove Worldwide.)

Stairs and Exits

Note the following items when inspecting the condition of stairways and exits:

- appropriate exit signs
- improper or inadequate design, construction, or location
- lack of handrails
- handrails placed too low or rough handrails
- improper lighting (including emergency lighting)
- obstructions
- locked doors
- doors that open in the direction of an exit
- poor housekeeping
- wet, slippery, or damaged surfaces
- faulty treads or mats on stairs
- lack of curbing on ramps
- differentiation between
  – the way of exit access
  – the exit
  – the exit discharge

Whenever any of these defects is found during a maintenance inspection, repair or correct it immediately, if possible. Various types of slip-resistant materials can be applied directly to stairways.

Exits should not serve as storage areas. Keep exits well lit and their floors smooth. Correct hazardous conditions as soon as possible after inspection.

Check the operation of exit doors. See that they can move freely and have no obstructions and that the paths to them are clear.

Keep exit signs and lights in good repair. Test exit signs and emergency lighting that are designed to operate in the dark—in case the lighting system fails. (See National Fire Protection Association (NFPA) 101, *Life Safety Code.*)

Heating Equipment

Inspect the facility's heating equipment before winter. Be sure it is operating safely and efficiently. Breakdowns of heating equipment during cold weather are especially dangerous and costly. Observe the following precautions to reduce the possibility of breakdowns during cold weather:

- Inspect and thoroughly clean the heating systems.
- Annually inspect chimneys and vent pipes for cracks, missing mortar, and rusted holes. Repair any damage immediately.
- Keep the inside of buildings at a minimum temperature of 40 F.
- Do not leave buildings unattended for long periods.
- Daily check that the heating is operating properly.

SICK BUILDING SYNDROME

Sick building syndrome, or tight building syndrome, is a relatively new health term. See Chapter 15, Indoor Air Quality, in the *Administration & Programs* volume, and Chapter 13, Indoor Air Quality, in the *Environmental Management* volume. The causes of sick building syndrome are thought to be

- physical and chemical contaminants
- biological agents
- lack of fresh air

According to the American Conference of Governmental Industrial Hygienists (ACGIH) *Industrial Ventilation Manual,* the need for outside air to remove carbon dioxide from the body and to provide oxygen to the body is self-evident. In most situations, enough air for this purpose enters buildings by permeating walls and infiltrating around windows.

However, when 100% of the air is recirculated, gases and chemical solvents in the air are not diluted or removed from the work environment.

When the cost of losing heated and cooled air became a major concern, companies permanently sealed and weather-stripped windows, or replaced them with new, superinsulated units. Insulation was increased in older buildings. New buildings were designed and built with inoperable windows and with high-performance insulating materials. These features minimized the entry of outside air into the building. In addition, they required elaborate heating, ventilating, and air conditioning (HVAC) systems that used energy-recovery components, air washes, humidifiers, and complex air-flow mechanisms. These systems keep the environment within a sealed building livable without losing hot or cold air in the process.

Cleanup and Prevention

There are three aspects to treating sick building syndrome: contamination cleanup, preventive maintenance, and preventive design.

Remove all traces of asbestos, PCBs, and where possible, certain formaldehyde-containing insulation where they are found in amounts above government-specified limits. Removal practices must meet both OSHA and EPA standards and be conducted by a person certified for the particular materials being removed.

When bacteria have been discovered in the HVAC system, as well as in other areas of the building, take the following basic steps:

- Replace all dirty air filters.
- Empty all condensate drainage trays.
- Use hot water wherever possible to clean microbial growth from condenser coils.
- Swab down all ductwork with an antiseptic solution.
- Remove materials and treat open areas if a surface or interior area, such as one above ceiling panels or behind a wall, is infested and cannot be cleaned.
- Treat carpeting and furniture with antimicrobial solutions and thoroughly air them.
- Check all equipment in the HVAC system to make sure it operates correctly after the cleanup. Keep all drains in working order.

Preventive Maintenance

Any veteran building operator who has had to handle a stubborn case of sick building syndrome knows that a few minutes of preventive maintenance is worth hours of repairs and cleanup. The following preventive measures are recommended by experts in the field:

- Keep hot-water supply temperatures above 120 F (40 C).
- Protect all air-handling systems from environmental conditions, and provide them with a drainage system to avoid areas of standing water.
- Relative humidity should never exceed 70%.
- Avoid using air washers that use recirculating water systems.
- Humidifiers should use steam (not boiler steam) as a water source, not recirculated water.
- Avoid using spray coil systems. These are most commonly found in the southern United States.
- Find out if the building uses fan coil and heat pump units. Such a unit uses a system of fans that recirculate conditioned air through a secondary system. The secondary system uses draining pans under the coils. The coils may become a source of biological pollution.
- Air filters should have properly rated dust spot efficiencies. Check pleated glass fiber filters regularly for disintegration and possible glass fiber pollution.

Preventive design starts with the following basic elements:

- Choose an HVAC system that fits the building's size and anticipated uses.
- Allow for a generous number of intake and exhaust vents.
- Locate intake vents where they will receive the largest supply of fresh air—away from cars, buildings, and process exhausts, and as close as possible to trees, bushes, and the like.
- Fit the HVAC system with regulating generators that are flexible enough to adjust to the varying air pressures of intake and outtake vents.
- Use only steam humidifiers.
- Use prefilters to clean the air before it passes over higher-efficiency filters.
- Institute a preventive maintenance program when any system is installed. Provide for regular inspection of drain pans, filters, and any area of the HVAC that is accessible and that might fall prey to germs.

GROUNDS MAINTENANCE

To prevent accidents and injuries from tools and machines used in grounds maintenance, choose equipment for its specific purpose and use and maintain it properly. Store and use fuels and chemicals, including pesticides, rodenticides, and fertilizers, properly. Thoroughly train workers, and have them wear proper clothing and use personal protective equipment, as required.

Train maintenance workers to recognize poisonous vines, shrubs, fruits, insects, and reptiles. They should avoid contact with poison oak and ivy and should destroy all poisonous growths. After working outdoors, they should scrub their hands thoroughly. Workers should treat all cuts and scratches received outdoors with proper antiseptic covering. They should remove all foreign matter, such as glass, metal, and wire, from the grounds to be maintained.

Electric-Powered Hand Tools

Apply the same rules of safe tool use outside the facility as inside. The primary rules include the following:

- Use battery-powered tools outside, if possible.
- Use double-insulated tools connected to a circuit protected by a ground fault circuit interrupter (GFCI).
- Use the correct tool for the task at hand.
- Keep landscaping tools in good condition.
- Use tools as they were meant to be used.
- Store tools in a safe place.
- Keep cutting tools sharp and clean.
- Keep tool handles smooth and strong.
- Use shovels, spades, and other digging tools with points that are smooth and properly shaped.

Electric Landscaping Tools

Read the operator's manual carefully before using electric landscaping tools. If a tool is equipped with a three-prong plug, use a three-hole grounded receptacle and a three-wire extension cord. Use a GFCI outside. Keep visitors a safe distance from work areas, and clear the areas of debris, animals, or anything else that could inflict or suffer injury or damage if encountered while the tool is operating. Do not force the tool; it will perform best and safest at the rate for which it was designed. If the tool has a second handle, keep both hands on the tool.

Never use electric-powered tools in the rain or when grass or shrubs are wet. Don't abuse the electric cord. Never carry a tool by its cord or yank it to disconnect it from the receptacle. Always use approved cords for the location, with the proper wire size to carry the current. Be sure to avoid cutting the cord with the tool.

Always wear eye protection when using landscaping tools. Hold tools in position, ready for use, before switching on current. Always shut off current when resting one's arms, removing cuttings, or before changing the direction of a cut. If a tool becomes jammed or fails to start, always switch it off before trying to free the jam or before troubleshooting. When leaving the work scene, even for a few minutes, always shut off the current and disconnect the plug.

Electric Hedge Trimmers

The usual injuries from electric hedge trimmers are amputated fingers, serious cuts on fingers and hands, and cuts on knees and legs caused when lowering the trimmer to rest the arms. Injuries from hedge trimmers result from five types of actions: (1) changing hand position with the trimmer running, (2) holding branches away from the cutting bar, (3) removing debris from the trimmer, (4) holding the trimmer with only one hand, and (5) failing to wait for the blades to stop after turning the trimmer off.

Choose hedge trimmers with the following features:

- light enough to hold comfortably for a long time
- a large support handle for two hands to hold
- a switch that requires continuous finger pressure to maintain power
- a battery-powered or double-insulated mechanism
- the Underwriters Laboratories (UL) label

When using a trimmer, workers should get into a comfortable position, use both hands, avoid cramped spaces, take their time, and not force the tool. They should not overreach or lean off a stepladder. When they leave the work area, even for a coffee break, they should take the trimmer with them. Workers should also follow company regulations on securing tools if there is any chance that some unauthorized person might try to use them.

Lawn Trimmers and Edgers

Lawn trimmers are useful for cutting the grass around tree trunks or along fences. Edgers cut borders along the edges of sidewalks, driveways, or gardens. Each year, trimmers and edgers injure several thousand people. Workers should handle these tools cautiously because the tools have a metal cutting blade that can throw debris or cut a finger. Keep the guards on trimmers in place and in working order. Keep the blades sharp. Workers should not put their hands near a blade unless the machine is turned off.

Nylon-cord weed trimmers cannot hurt as seriously as metal-blade trimmers or edgers. However, getting hit by the cutter cord can sting. Operators should disconnect a trimmer's power cord when adjusting the cutter cord's length or changing the reel. They should take the same precautions when using a weed trimmer as they would with any electric appliance. Do not use electric edgers and trimmers in wet areas. Periodically check electric cords for cracks or breaks in the insulation.

Gasoline-Powered Equipment

Observe the following safety rules when handling gasoline:

- Never use gasoline for cleaning floors, tools, clothes, or hands. Gasoline is to be used only in engines, as a source of energy.
- Always store gasoline in an approved, closed container.
- Do not pour gasoline from one container to another. Doing so might generate a charge of static electricity, which could ignite the gasoline. To avoid generating static, maintain a metal-to-metal contact.
- Clean up gasoline spills immediately to prevent accumulation of vapors. Do not allow electric switches to be turned on until the gasoline vapors have dispersed.
- If gasoline is spilled on oneself, remove any saturated clothing immediately and keep oneself and the clothing away from sources of ignition. Wash the affected area of skin with soap and water to avoid a skin rash or irritation. If gasoline enters the eyes, flush them with water and get medical attention.
- When draining or dismantling gasoline tanks or equipment parts that likely contain gasoline, do so outdoors or in a well-ventilated area, free from sources of ignition.
- Never smoke in fueling areas, fuel-system servicing areas, bulk-fuel delivery areas, or similar areas where fuel is present.
- Never dispense gasoline into a fuel tank while the motor is running or is hot.
- Never store equipment with fuel in its tank inside a building where vapors could reach an open flame or spark. Before storing equipment in any enclosure, allow the engine to cool.
- Never run an engine indoors; doing so creates a carbon monoxide hazard.

Gasoline-powered mowers and tractors should meet the standards of ANSI/OPEI B71.1–1986, *Safety Specifications for Turf Care Equipment—Power Lawn Mowers, and Lawn and Garden Tractors.* Snow throwers should meet ANSI B71.3–1984, *Safety Specifications for Snow Throwers.*

Power Lawn Mowers

Power lawn mowers, especially the rotary type, have proven to be a mixed blessing. Although they save time and effort and leave a lawn neatly manicured, they also take lives and have caused more than 100,000 personal injuries a year. These injuries range in severity from minor cuts to amputations.

Reel mowers have several blades that shear the grass against a horizontal stationary edge. Such mowers do not need to operate as fast as rotary mowers do, so they are safer. However, they are not as good for cutting tall grass or weeds.

Safety Precautions before Mowing

Make sure the operator of a mower is well trained before using the mower. If the mower is being used for the first time in a season, have the operator review the instruction manual. Before mowing, the operator should pick up rocks, glass, tree branches and twigs, and any other objects that could become lethal missiles if thrown out by the mower blade. The operator also should observe the location of fixed objects, such as pipes, lawn sprinkler heads, and curbs, which could damage the mower or break apart and become missiles. Make any adjustments to wheel height before starting the mower. Disconnect the spark plug wire when cleaning, repairing, or inspecting the mower. Keep unauthorized persons out of the mowing area. Before starting the mower, the operator should make a quick inspection for loose nuts and bolts, check blade condition and the engine's oil level, and fill the fuel tank, using a vented can with a flex spout. The operator should wear protective footwear and safety glasses. A brimmed hat and full-length trousers and shirt will protect against sunburn.

Safety Precautions during Mowing

Instruct operators to mow in daylight or good artificial light. Operators should, as much as possible, push the mower forward and avoid pulling it backward. Pulling backward can injure the feet. When a slope or terrace must be mowed, have operators make a series of horizontal passes along the incline. If operators push the mower up the incline, the mower could drift back onto a foot. If they push the mower down the incline, they can lose their footing and fall into the mower.

Forbid operators to use the mower when the grass is wet and slippery. If the grass is damp or high, have them cut it at a slower speed, if possible, and set the cutting height higher than for dry grass. Otherwise, the discharge chute may clog up.

Rotary blades can pick up stones, pieces of wire, nails, or other objects hiding in the grass and throw them out of the discharge chute at terrific speeds. Newer models may have a guard over the discharge chute that deflects objects downward, but the guard may have to be removed if the grass catcher is used. Some mowers have guards that automatically snap back into place when the grass bag is taken off. Others require that the guard be bolted in place any time the catcher is not used.

Before removing the grass catcher to empty it, operators should shut off the engine and wait until the blade has stopped completely. Operators should also shut off the engine when attempting to free obstructions from the discharge chute, adjusting the cutting height, or performing any operation requiring a person to place hands or feet near the blade.

Riding Mowers

Riding mowers are most efficient for cutting large areas of lawn. Suggested safe practices for using a riding mower include the following:

- Fully instruct operators on using the controls and on stopping quickly. Operators should read the owner's manual at the beginning of each mowing season.
- Operators should clear from the work area objects that might be picked up and thrown and identify fixed objects that might damage the mower.
- Operators should disengage all attachment clutches and shift into neutral before attempting to start the engine.
- Operators should disengage power to attachments and stop the engine before making any repairs or adjustments. They should also disengage power to the attachments when transporting them or not using them.
- When leaving the vehicle unattended, operators should disengage the power takeoff, lower the attachments, shift into neutral, set the parking brake, stop the engine, and remove the key from the ignition.
- When mowing, operators should watch for holes in the lawn and for other hidden hazards.
- Operators should use a push mower rather than a riding mower to mow sharp corners or steep slopes. If they do mow steep slopes with a riding mower, they should mow up and down rather than across. They should avoid steep slopes altogether if the slopes are wet. To prevent tipping or loss of control on slopes or sharp turns, operators should reduce speed and avoid starting or stopping suddenly.
- Operators should not mow between large trees through which the rear wheels will not pass. Large mowers are known to turn over backward as a result of the extreme power in the rear wheels.
- When changing direction or turning around, especially on slopes, operators should use extreme caution. They should not back up without making certain it is safe to do so. They should watch for traffic when crossing or working near roadways. When using attachments, they should direct discharge of material away from anything that could be hurt or damaged by it.

- Operators should maintain the vehicle and its attachments in safe operating condition and keep safety devices in place. They should tighten nuts, bolts, and screws, especially the blade-mounting bolts. If the mower or its attachments should strike a solid object, the operator should stop and inspect the mower for damage and repair it before restarting and operating the mower. Operators should not change the engine's governor settings or overspeed the engine.

Utility Tractors

The U.S. Consumer Product Safety Commission offers the following suggestions for the purchase, safe use, and maintenance of garden tractors.

Purchase. Specify that utility tractors, including mower attachments, have safeguards for all moving parts. This will reduce the hazard of contacting belts, chains, pulleys, and gears. Buy tractors with throttle, gears, and brakes that are easy to reach and that can be operated smoothly and with minimum effort. Be sure that safety instructions are provided with the tractor. There should be warning labels on the machine itself.

Use and Maintenance. Before using the utility tractor, operators should read the owner's manual and pay attention to its recommendations. Operators should also observe the following precautions:

- Never allow children or unauthorized persons to operate the tractor, and keep unauthorized persons away from the cutting areas during operation.
- Wear sturdy, rough-soled work shoes and close-fitting slacks and shirts to avoid entanglement in the moving parts. Never operate a garden tractor in bare feet, sandals, or sneakers.
- Always turn off the machine and disconnect the spark plug wire before adjusting the machine.
- For optimum stability on slopes, drive up and down rather than across.
- Start the tractor outdoors, not in a garage where carbon monoxide can collect.
- Do not smoke near the tractor or near gasoline storage cans. Gasoline vapors can easily ignite.
- Replace or tighten all loose or broken parts, especially blades.
- Get expert servicing regularly—it may prevent serious injuries.

Accident patterns involving utility tractors include the following:

- Overturning—This can occur when driving over uneven terrain, steep slopes, or embankments. The rider can come into contact with the tractor when it overturns or sustain injuries during the fall. Utility tractors may also overturn if they are used to pull vehicles heavier than themselves out of mud or from a ditch. The front end of the tractor can rise and turn over on the operator.
- Backing up—Sometimes when a tractor backs up, it runs over bystanders. Many of the victims are young children the operator did not see.
- Igniting flammable liquids—Using gasoline around a garden tractor can be hazardous if the gasoline spills and can be ignited by a spark or heat source.

Snow Throwers

All snow throwers are potentially dangerous. Their large, exposed mechanism, which is designed to dig into the snow, is difficult to guard. With proper handling, however, snow throwers offer a safer method of removing snow than the backbreaking, heart-straining manual method. The safest snow throwers have guards on the drive chains, pulleys, and belts.

The auger at the front of the snow thrower presents the greatest hazard. Some snow throwers have an additional auger for extra throwing power. These augers, along with moving gears, drive chains, and belts, can endanger anyone tampering with a snow thrower when it is running. Injuries usually occur when the operator attempts to clear off debris while the motor is running. Even cleaning the machine with a stick can be extremely dangerous if the motor is left on. The spinning blades can pull the stick from the operator's hand and toss it back at the person with great force if the clutch lever is also held on. Some models have automatic stopping devices that take effect when the handle is released.

Although snow throwers can handle dry, powdered snow with little difficulty, they handle wet, sticky snow less effectively. Wet snow tends to clog the blades and vanes, and often jams and sticks in the chute. Snow throwers can pick up and throw ice, stones, and other hard objects.

The following safety suggestions for snow thrower operators are recommended by the Outdoor Power Equipment Institute:

- Read the operator's manual.
- Do not allow children to operate the machine or allow adults to operate it without proper instructions.
- Keep all bystanders a safe distance away.
- Disengage all clutches, and shift into neutral before starting the motor.
- Keep hands, feet, and clothing away from power-driven parts.

- Never place a hand inside the discharge chute or even near its outside edge with the engine running.
- Know the controls and how to stop the engine or how to throw the unit out of gear quickly.
- Disengage power and stop the motor before cleaning the discharge, removing obstacles, or making adjustments, or when leaving the operating position.
- Adjust height to clear gravel or crushed-rock surfaces.
- Exercise caution to avoid slipping or falling, especially when operating the machine in reverse.
- Do not operate the machine on slopes or on ground where there is a risk of slipping or falling.
- Never direct discharge at bystanders or allow anyone in front of the machine. Potentially dangerous debris may be hidden in the snow.
- Keep the machine in good working order and keep safety devices in place.

Snow Shoveling

If the area to be cleared of snow is small, or if no snow thrower is available, someone will have to shovel the snow by hand. Only someone in good physical condition and general health should do this work.

The shoveler should mentally divide the area into sections, clean one part, then rest before going on to the next section. Whenever the snow begins to feel especially heavy, the shoveler should take a break. Shovelers should keep the following information in mind:

- Wet snow is much heavier than dry snow. Govern the rate of shoveling accordingly.
- Push or sweep as much of the snow as possible.
- If an icy crust has formed on top of several inches of snow, shovel the snow in layers.
- Use small amounts of rock salt or other ice-melting materials to make the job as easy as possible.
- Dress warmly while shoveling snow, because cold itself can pose a strain on the body's circulation. Do not bundle up so heavily, however, that movement is difficult.

Pesticides

Insecticides, herbicides, fungicides, disinfectants, rodenticides, and animal repellants are all pesticides and all hazardous chemicals. Under OSHA's hazard communication standard, each worker exposed to pesticides must receive MSDSs and training. The safe use of pesticides is everyone's responsibility. The user, however, has the major responsibility, which begins the day a pesticide is selected and purchased and continues until the empty container has been disposed of properly. A U.S. Department of Agriculture county extension agent can help in choosing the proper pesticide to use for controlling certain pests.

All pesticides sold in the United States must carry an EPA registration number on the label. This label signifies that the EPA has reviewed the product and found it safe and effective when used according to directions. Every pesticide label must include a list of what the product will control, directions on how to apply the pesticide, a warning of potential hazards, and safety measures to follow.

Before using any pesticide, read the label carefully. The label describes some of the hazards involved, as well as antidotes and first-aid measures. Pesticides that bear the notice DANGER—POISON on the label are highly toxic. Breathing or ingesting them, or simply allowing them to remain on your skin, can be fatal. Pesticides marked with the word WARNING are moderately toxic and can be quite hazardous. Pesticides that have the word CAUTION on the label have low toxicity but may be harmful if eaten or grossly misused. Follow the label's instructions for mixing, handling, and applying. Be sure—do not guess when working with pesticides.

Application

Any restricted-use pesticide used around an industrial facility has to be applied by a certified handler according to the Insecticide, Fungicide, and Rodenticide Act, Public Law 92.516, 40 *CFR*. Use the least-toxic pesticide for the job in order to reduce hazards. Manufacturers have formulated different compounds to control the same pest, so use the one that is easiest on the growing plants you want to keep. For example, a severe infestation may require the use of a phosphate ester (organophosphate) insecticide, which calls for wearing protective clothing and other precautions (Figure 4–5). If necessary, consult the supplier before using phosphate ester.

Try to purchase just enough pesticide to last one season. This should cut down on storage and disposal problems. Observe the following precautions when using pesticides:

- Use a pesticide only for the purposes given on the label.
- Keep pesticides in their original, labeled containers. Check for leaks or damage to the containers.
- Mix pesticides carefully—outdoors, if possible. Keep them off skin, and avoid breathing their dust or vapors. Use protective clothing and equipment, including respirators, when using toxic chemicals.
- Set aside a special set of mixing tools—measuring spoons and a graduated measuring cup—for

Figure 4–5. When applying insecticides, be sure personnel are well trained and wear proper equipment. Phosphate ester (organophosphate) insecticides require extreme precautions to be used. Always choose the least toxic pesticide that will do the job.

use with sprays and dusts only. Keep them with your chemicals.

- Avoid spilling pesticides. Set aside a level shelf or bench in a well-ventilated area, preferably outside, for mixing chemicals. A level, uncluttered surface helps avoid spills. If chemicals do spill, wash hands at once with soap and water. Then hose down the mixing area.
- Never smoke or eat while spraying or dusting. Cover food and water containers when spraying around areas where watchdogs are kept.
- During application, stay out of the spray's drift. Do not make outside applications on a windy day.
- Avoid spraying near lakes, streams, and rivers, and make every effort to keep toxic residues from entering waterways.
- Use canister respirators that serve as proper respiratory protection.
- If a pesticide gets on skin or clothing, immediately remove the clothing and take an all-over bath or shower (be sure to shampoo). Use plenty of soap and water. Wash clothing before reusing.
- When finished using pesticides, immediately wash hands with soap and water. Do not smoke, eat, or drink without washing first.
- Never allow unauthorized personnel around treated areas, or pesticide mixing, storage, and disposal areas.

Safe Storage of Pesticides

Store all pesticides in a well-ventilated, locked area or building. Store pesticides in their original, tightly closed containers. In that way, the labels can still provide information in case of accidents. Keep soap and plenty of water close to storage areas. Seconds count when washing poisons from skin.

Do not store clothing, respirators, food, cigarettes, or drinks near pesticides. Such items might pick up poisonous fumes or dusts, or soak up spilled poisons.

Disposal of Pesticides

Dispose of all pesticides according to the instructions on the labels.

Emergency Information

If an emergency occurs when using pesticides, obtain additional advice and information on antidotes for specific pesticides from these agencies: the local poison control center, the state department of health, the county agricultural extension agent, and the regional office of the EPA. They maintain current information files on all compounds and their constituents, and on recommended treatment in case of poisoning.

COMPUTERIZED PREDICTIVE MAINTENANCE

Computerized predictive maintenance (CPM) can reduce employees' exposure to hazards, decrease equipment downtime, and optimize the effectiveness of maintenance expenditures. By reducing costs due to equipment repair and equipment downtime, an effective CPM program reduces the total controllable maintenance costs to a minimum. These savings come from the following:

- less exposure by employees to malfunctioning equipment, thus fewer accidents
- less lost production time
- fewer emergency failures of equipment
- efficient scheduling of equipment repairs and downtime

- fewer repairs and lower costs
- improved and safer use of labor
- longer life for equipment

By monitoring equipment before trouble starts, the CPM program becomes the ounce of prevention needed to avert the costs incurred from unexpected breakdowns. The benefits of a CPM program, however, typically extend beyond the maintenance operation to encompass other areas of the facility, including the safety function.

An effective CPM program not only alerts the proper personnel to potentially hazardous conditions, such as equipment failures, but also provides the record keeping required by state and federal safety regulations. The following sections describe how a typical CPM program works. (See also Chapter 9, Computers and Information Management, in the *Administration & Programs* volume.)

Computerized Assistance

The most effective and economical method of planning, scheduling, and tracking CPM tasks is to schedule them from the last service date. This process can be accomplished effectively only with computerized assistance. Use a computer software program that is consistent with the concepts and techniques on which maintenance systems consulting services are based. Principal features of the program might be as follows:

- CPM task definition and identification of the cycle on which the tasks are to be performed
- CPM specifications and procedures that define the craft, standard hours, work procedures, and measurements to be performed for each task
- automated scheduling, based on the last service date and predetermined task cycles
- CPM preparation of work orders for all crafts according to available labor
- reports, including compliance, performance, forecast, and budget planning reports
- records of equipment history for each CPM task

Maintenance Diagnostic Technology

Various studies of production costs in U.S. manufacturing and processing facilities have revealed that maintenance normally accounts for 15% to 40% of the total. Other studies show that maintenance costs represent an average of 28% of the total cost of goods sold.

More significantly, every dollar saved in maintenance drops directly to a company's bottom line. Yet although cost-effective maintenance adds to long-term profitability, top management has sought to keep costs low by avoiding investments in long-range maintenance improvements. Hence facility engineering managers have labored to improve maintenance effectiveness with minimal resources. Development of new technologies to support maintenance has lagged behind efforts to improve and automate production. But the situation is changing.

As pressure on U.S. industry to improve quality and productivity mounts, new priorities are being set. And new opportunities to improve facility maintenance through the application of technology are being discovered.

Here is a review of some of the more significant technologies available.

Laser Shaft Alignment

Laser technology has now been adapted to small, relatively inexpensive instrumentation that is useful not only in the initial alignment of machinery and shafts but also for monitoring critical shaft alignment. This new capability can help prevent many problems with rotating machinery.

Laser systems that instantly sense and display minute movements of fixed objects are now available.

Output from laser alignment systems can also be sent to computers for display, recording, printing, and analysis.

Among the advantages claimed for laser alignment systems are the following:

- They automatically send data to a microprocessor, whereas dial indications require manual input.
- A laser beam does not sag, whereas dial indicator hardware, particularly when spaces are used, may have considerable sag, which is often disregarded or incorrectly measured.
- Laser systems require no bridging hardware, so they are easy to use with a wide range of shaft diameters and space restrictions.
- Mounting distances for laser systems can range from 1 in. (2.5 cm) to more than 30 ft (9 m).
- Laser measuring devices can be adjusted during the alignment procedure. This feature gives them several times the misalignment capacity of dial indicator systems.
- Measurement resolution of 1 micron is possible.
- During alignment, only 180-degree shaft rotation is required.
- The time required to complete an alignment using lasers is normally less than half that required for other systems.
- Laser alignment normally requires less training of personnel.

- "Magna-flux" testing can be used to check for cracks in conductive materials.

Ultrasonic Testing

Instruments designed for ultrasonic testing sense ultrasound waves produced by operating machinery as well as the turbulent flow of leakage (Figure 4–6). They provide fast, accurate diagnosis of such wasteful problems as valves in a blowy mode, faulty steam traps, and vacuum and pressure leaks. Airborne ultrasonics is extremely useful in detecting mechanical problems, especially potential bearing failure.

Airborne ultrasonic testing instruments are usually battery operated for portability. Their electronic circuitry converts a narrow band of ultrasound (between 20 and 100 kHz) into the audible range so that a user can recognize the qualitative sounds of operating equipment through headphones. The intensity of the signal is usually displayed on an analog meter.

Some of these instruments include an ultrasonic transmitter. A diagnostician usually places the transmitter in a container and uses a scanner to detect areas of sonic penetration along the container's surface. This process often detects leakage in systems before they are put into service. The ultrasonic transmission method is useful in quick checks of tank seams, hatches, seals, caulking, gaskets, or building wall joints, as well as heat exchangers.

Because ultrasound is a shortwave signal, it is easily isolated in most equipment. This characteristic makes it easy to detect sounds often associated with mechanical disorders.

A shortwave signal needs greater amplitude to travel the same distance as a low-frequency sound. Therefore, the ultrasonic components of a problem sound are loudest closest to the source. By touching the surface of a section of operating equipment with a contact probe, a diagnostician can adjust the sensitivity of the instrument until the problem sound is heard. Then,

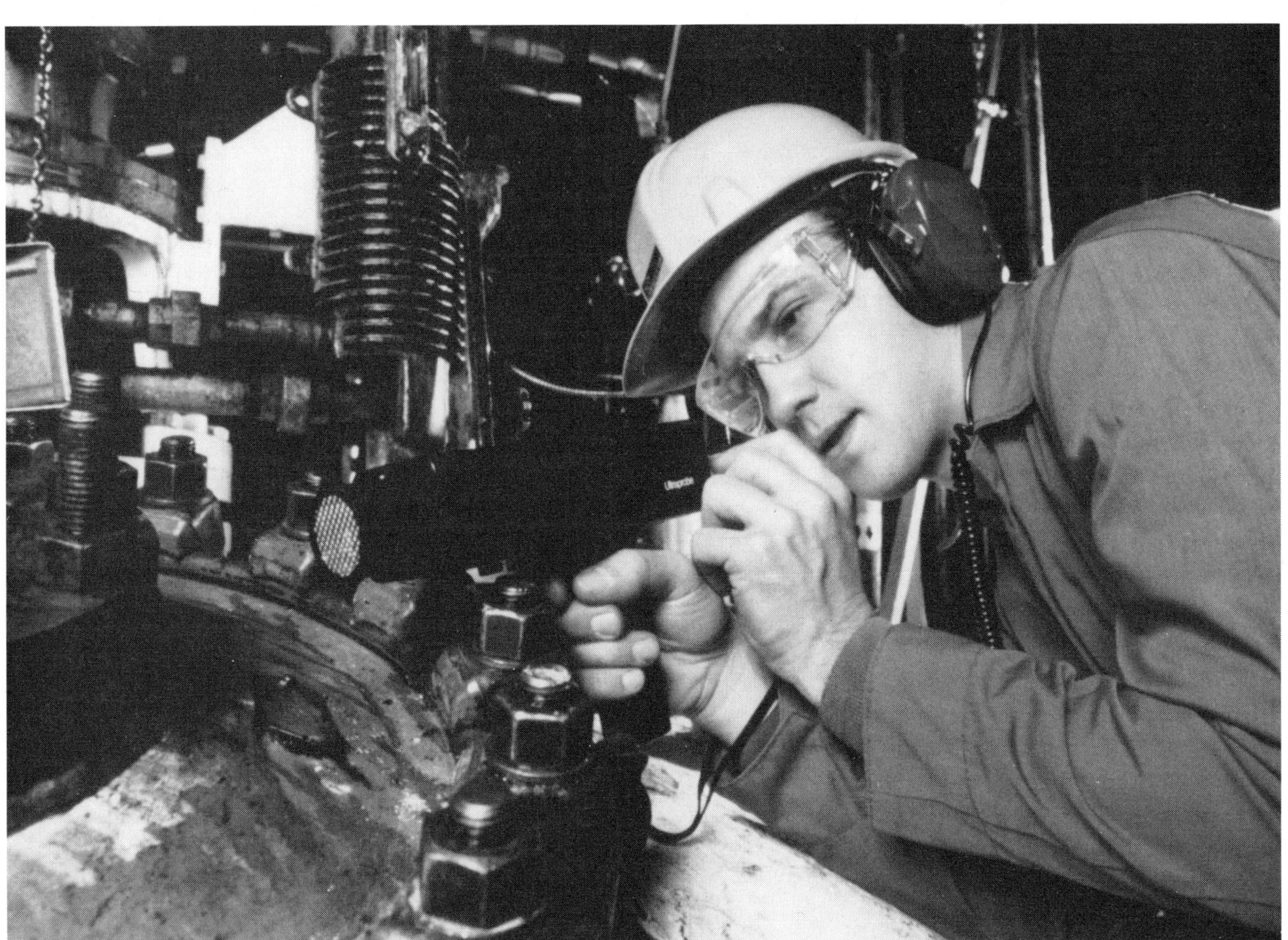

Figure 4–6. This portable inspection instrument detects leaks, line blockage, bearing failures, faulty valves, steam traps, and electrical problems through ultrasound. (Courtesy U.E. Systems, Inc.)

by listening and observing the meter, he or she can locate the loudest area.

Ball bearing problems can be evaluated by ultrasound long before they are detectable through most other methods. By the time a bearing becomes noisy enough to make a problem known audibly, it has already failed. Bearing wear is a progressive process. Even a good bearing emits a characteristic sound in the ultrasonic range. The ultrasound quality changes, and its intensity increases perceptibly, long before the bearing actually fails.

Oil Analysis

Oil analysis has become an important aid to preventive maintenance. Laboratories recommend that samples for analysis be taken at scheduled intervals. Doing so is important in order to identify trends and detect abnormalities. The length of the sampling intervals varies with different types of equipment and operating conditions.

A typical oil analysis may include tests for viscosity, water or coolant, fuel dilution, solids, fuel shoot, oxidation, nitration, total acid number, total base number, particle count, and spectrographic analysis.

Wear Particle Analysis

Wear particle analysis is related to oil analysis in that the particles to be studied are collected by drawing lubricating oil samples. But there the relationship ends. Oil analysis determines the condition of the lubricant itself. This information sometimes can then be used to draw certain conclusions about the machinery from which the sample was taken. In contrast, wear particle analysis provides direct information about wearing conditions in the machinery.

Particles in the lubricant of a machine can provide significant information about the condition of that machine. This information comes from the study of particle shapes, composition, sizes, and quantities.

Infrared Imaging

Infrared imaging provides a visual representation of the heat energy objects radiate in proportion to their temperature and emissivity. At normal temperatures, most of this energy is in the infrared spectrum and thus invisible, but it can be measured. Most imaging systems detect infrared in the range of the electromagnetic spectrum between 2 microns and 5.6 microns, or 8 microns and 14 microns.

Infrared instruments include an optical system to collect radiant energy from an object and focus it, a detector to convert the focused energy pattern to an electrical signal, and an electronic system to amplify the detector output signal and process it into a form that can be displayed.

The resulting infrared picture, or thermograph, can be interpreted directly by eye or analyzed by computer to produce additional detailed information. Advanced systems can isolate readings for separate points, calculate average readings for a defined area, produce temperature traces along a line, and make isothermal images showing thermal "contours."

The thermograph can be used directly to find and monitor thermal anomalies, or "hot spots," that indicate problems to be investigated and corrected. Or successive recorded thermal images can be used to produce quantitative surveys trending actual temperatures.

For infrared surveys to be effective as a diagnostic tool, they should be conducted by a technician thoroughly trained in operating the equipment and interpreting the imagery, which requires an understanding of the facility systems being analyzed.

Vibration Analysis

The potential maintenance benefits of vibration monitoring and analysis have been recognized for decades, but only recently has the technology become available for widespread practical use.

Machinery parameters that can be recorded include overall vibration amplitudes, vibration time waveforms, vibration amplitudes in specific frequency bands, phase, DC gap voltage, machine rotational speed or other process variables, and various qualitative observations. Many data collectors also provide graphic displays so that information can be reviewed immediately. Data can be transferred between a host computer and a data collector through direct links or modems.

For example, in an electrical group, maintenance technicians can gather baseline data on every major piece of electrical equipment at the time of installation. They can perform a vibration analysis before a motor is coupled to its load, thus providing a record on the no-load speed condition of the motor when new. Then, when they take a vibration reading during routine maintenance and compare it with the original reading, they can detect impending problems. Because maintenance costs can account for as much as 60% of total controllable facility operating costs, an effective CPM program directly affects the bottom line (Figure 4–7).

MAINTENANCE CREWS

Select maintenance employees for their experience, alertness, and mechanical ability. They should be able to learn the essential safety principles of machines or operations for which they will be responsible.

Training

Maintenance employees should have more thorough training in accident prevention than other workers.

Preventive and predictive maintenance (PPM) can reduce employee exposure to hazards, decrease equipment downtime, and optimize the effectiveness of maintenance expenditures.

As shown in Figure 1, by reducing costs due to equipment being out-of-service and repair costs, an effective PPM program reduces total controllable maintenance costs to a minimum.

These savings come from:

- Reduced employee exposure to malfunctioning equipment;
- Reduced production lost time;
- Fewer emergency failures of equipment;
- Scheduled equipment outages;
- Lower repair frequency and cost;
- Improved and safer labor utilization;
- Extended equipment life.

By allowing systematic monitoring of equipment and servicing it before trouble starts, a PPM program is the "ounce of prevention" needed to avert the costs incurred from unexpected breakdowns.

But the benefits of a PPM program typically extend beyond the maintenance operation to encompass other areas of the plant, including the safety function.

An effective PPM program, for example, not only alerts the proper authorities to potentially hazardous conditions—such as equipment failures—it also facilitates the recordkeeping required to conform to state and federal safety regulations.

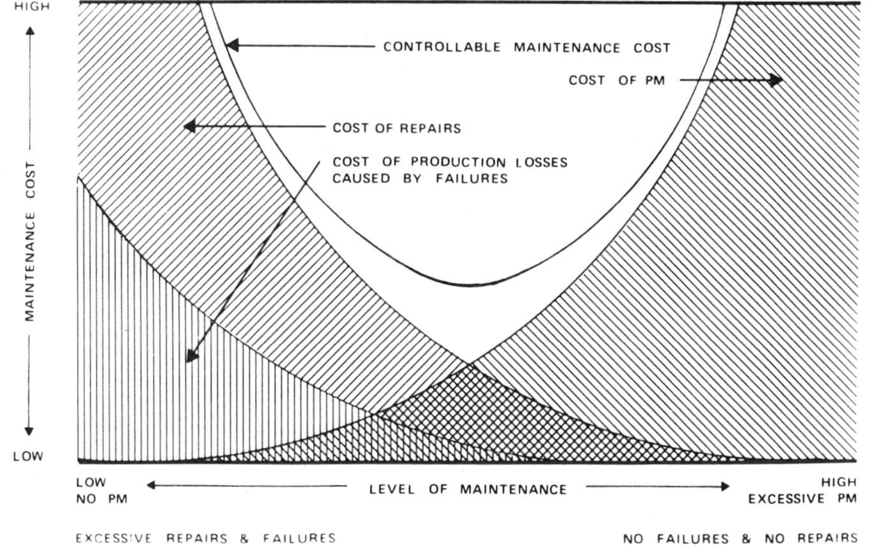

Figure 1

Figure 4–7. Controlling maintenance costs using computerized predictive maintenance.

Safety for them involves not a set pattern of activity, but a complex and constantly changing set of problems. Furthermore, they must know how to use not only ordinary tools but also ladders, protective equipment, chains, slings, ropes, and many other kinds of tools and equipment.

In order to protect themselves and others working in a maintenance area, maintenance crews must be aware of job hazards and receptive to proper training. Their training program should include first-aid and lifesaving techniques. In industries where irritating, toxic, or corrosive dusts, gases, vapors, or fluids are present, give maintenance employees special training to familiarize them with the properties of these substances and with the methods of controlling their hazards. Some companies have the purchasing department notify maintenance crews when new chemicals are purchased so the crews can plan the necessary precautions.

Before beginning nonroutine jobs, call together the maintenance crew so they can discuss the hazards involved and determine how to do the job safely. The crew should check the equipment and call the supervisor before work that looks unsafe begins. Before using tools and tackle required to do special jobs, the crew should inspect them for wear and defects. When special tools are needed to make a job safer, have the engineering department provide design and construction specifications.

For especially complicated or hazardous jobs, a safety and health professional may be called upon to help in planning. Construct scale models to determine clearances, the best methods of moving, and sequences of action. After several trials have been made and the crew has agreed on a safe procedure, record the various steps as a guide for each worker (Figure 4–1).

In the course of its daily work, the maintenance crew travels throughout the facility, becoming familiar with every machine and process. If properly selected and trained, each crew member can do much to locate and correct unsafe conditions in both the facility and its equipment.

In smaller companies, the responsibilities of maintenance personnel may include inspecting and caring for portable power tools, extension cords, and the like. If so, use special procedures and training.

Preventive Maintenance Plans

Because the function of facility maintenance is to keep equipment in top operating condition, a good maintenance system must catch breakdowns before they happen. That is, crews must use preventive maintenance.

Supervisors can set up a preventive maintenance plan at least for critical equipment and for machinery that might seriously affect the safety of workers. A good preventive maintenance program starts with a list of buildings, machinery, and equipment that require periodic inspection, adjustment, cleaning, and lubrication, as well as adjustment of guards or changes in the types of guards.

Keep detailed engineering drawings and specifications on file for each machine or structure. For machinery, specifications should include dimensions, weights, sizes, and locations of utility service connections, lubrication requirements, and details on bearings, as well as data on the power transmission or drives. For buildings, specifications should include data on general layout, services available, floor load capacities, ceiling clearances, column spacing, and many other features.

After analyzing its maintenance responsibilities, management should develop an organization chart that shows where each employee is assigned. The chart gives supervisors a comprehensive picture of their workforce and points out the status and training of key persons and reserves. The chart should include outside specialists who can be called on for unusual or hazardous jobs. This information becomes particularly valuable in emergencies, when competent people are needed in a hurry.

Inspection of Equipment

Base the schedule for inspecting equipment on the actual trouble points listed on maintenance records. The schedule can be determined by the number of inspection reports turned in by regular full-time inspectors or by maintenance crew personnel.

Inspectors must be thoroughly familiar with the equipment being inspected. This requirement is especially important for electrical equipment inspectors, because electrical equipment usually gives few indications of impending trouble. Mechanical equipment, on the other hand, often warns of deterioration by such easily recognized signs as unusual noise, unusual appearance, or substandard output.

Inspectors should use suitable instruments for making their observations. While inspectors can readily recognize burned contacts on a motor starter or hear the pounding of a worn gear, they need proper instruments to check the insulation resistance of a motor or measure the wear on a shaft. (See Chapter 20, Woodworking Machinery, and Chapter 22, Metalworking Machinery, in this volume. See also OSHA Section 1910.132, *Personal Protective Equipment, General Hazard Assessment, Training Trends.*)

Personal Protective Equipment

Maintenance workers should dress properly for their specific jobs. They should wear snug-fitting clothes with a few small pockets. Breast pockets are often sewn closed or removed to prevent items from dropping into machinery or into hard-to-reach places when the wearer leans over. Workers should not wear neckties, wristwatches, or rings or other jewelry. Workers should keep loose rags clear of moving machinery.

If workers carry so few tools that they do not need a tool bag, they should wear a special belt fitted with tool carriers. To prevent back and spine injuries from falls, workers should carry tools at the side instead of in the back portion of the belt. (See Chapter 19, Hand and Portable Power Tools, in this volume.)

Workers handling rough or sharp objects should wear gloves or hand leathers. Welding gloves, rubber gloves for electrical insulation, and chemical-resistant gloves for handling acids should be used as needed but never worn around moving machinery.

Every maintenance worker's tool kit should include an explosion-proof flashlight, since a flashlight is often needed for work in dark or gaseous places. Workers should wear side-shield safety glasses while working. Every tool kit should also contain goggles. Various kinds of goggles can provide additional protection for maintenance workers from flying objects and molten metal, injurious heat and light rays, dust and wind, and acid splashes.

When workers must climb and work in high places or must enter a manhole, bin, or tank, they should be properly trained in confined-space entry procedures and wear full personal protective equipment (PPE). (See Chapter 7, Personal Protective Equipment, for details on PPE, and Chapter 2, Buildings and Facility Layout, for more information on confined-space procedures—both chapters in this volume.)

Lockout/Tagout

The details of maintaining power presses, various other types of machinery, electrical equipment, and boilers are covered in other chapters of this volume. Safe tank entry procedures are listed in Chapter 5, Boilers and Unfired Pressure Vessels; energy isolation procedures and instructions are described in Chapter 6, Safeguarding; and electrical lockouts are discussed in Chapter 10, Electrical Safety.

Piping

Maintenance crews working on water, steam, and gas pipes must exercise extreme caution. When piping is properly identified and correctly handled, however, their hazards are greatly reduced.

Proper Identification of Piping

Accidents have occurred because of improperly identified piping. This is especially true in factories where high- and low-pressure steam lines run next to compressed-air, sprinkler-system, and sanitary lines. Protect maintenance crews from opening the wrong valves or disconnecting the wrong pipes by clearly identifying piping. Work out a system of color schemes, tags, and stencils so the contents of lines can be identified at a glance (Figure 4–8). Identification is particularly important when emergencies occur or when outside maintenance workers perform services. To prevent confusion, use a consistent color code for piping, such as the one given in ANSI A13.1, *Scheme for the Identification of Piping Systems.* (See also the section Use of Color, Chapter 2, Buildings and Facility Layout.)

Figure 4–8. This sign indicates the direction of flow and the content of the pipeline. (Courtesy W. H. Brady Company, Signmark Division.)

Proper Isolation of Piping

Before working on a pipeline, shut off the line, lock and tag the valves, relieve pressure from the section of the line, and drain it. Shutoff valves are usually of the gate type. However, if globe valves are used, the pressure side should be under the valve seat so that the packing will not have to hold the pressure when the valve is closed.

If pipelines that carry chemicals must run overhead, isolate or cover them so they will not drip on workers or materials underneath. In case chemicals do drip on workers, provide emergency showers, with plainly marked locations for them. Test the emergency showers periodically and maintain them properly. Give full instructions to maintenance crews as well as operating personnel about the location and use of emergency showers.

Take special safety precautions in maintaining pipelines, valves, and bolted flanges—especially when hazardous materials are involved. The section on Problems with Hazardous Materials in Chapter 14, Materials Handling and Storage, in this volume, contains pertinent recommendations for both operating and maintenance personnel.

When hazardous materials are encountered, the supervisor should issue special protective equipment, such as chemical-protective goggles, protective suits, rubber gloves, or respiratory-protective equipment.

To prevent their hands from slipping on piping, maintenance personnel should wipe excessive oil from pipes and fittings.

Workers should wear gloves when handling pipes and fittings, especially when the ends are threaded. They should check for burrs and file them off immediately.

Industrial Gas Lines

When industrial gases such as propane and butane are encountered, mechanics should understand their behavior, storage characteristics, and pipeline arrangement. Mechanics should have a copy of the piping layout that shows the location of safety features equipment, such as soft heads and backfire preventers. The layout should also show the location of sectional and main shutoff valves so, in case of an emergency, maintenance personnel can find these valves quickly.

Handling Pipe

When maintenance involves a considerable amount of pipe work, place lengths of pipe, valves, and fittings so the floor is not overloaded. If possible, move the material to strategic points as the job progresses. Doing this will prevent material from accumulating in one spot and will reduce the amount of handling. Avoid contacting energized electrical wires.

When lengths of pipe are transported mechanically, attach red warning flags to material that extends beyond the conveyance. If the load is allowed to stand near passageways, the flags should remain and additional warning signs or barricades should be used to prevent other workers from bumping into the load.

Long lengths of pipe being run overhead and continuously joined should be pulled up from the floor with overhead rigging to the proper location and secured with tie ropes, wires, or fixture straps. Use of overhead rigging will prevent back strains and possible falls from ladders among workers and will keep the material itself from dropping.

Crane Runways

Before mechanics work on or near an overhead crane runway, they should notify the crane operator and make sure that the operator understands what is being done and what part the operator is to take in the work. Mechanics should also provide a temporary rail stop between the crane and the point where the work is to be done. (See Chapter 15, Hoisting and Conveying Equipment, in this volume.)

When work must be done on the crane itself, the person in charge is responsible for locking out safety switches, placing warning signals to indicate that people are working above, and seeing that a careful inspection is made after the job is completed. This inspection should include a check for parts or tools left on the crane, which later might drop into the mechanism or fall on workers.

When working on a crane runway, use a hand line to eliminate the need for carrying material up a ladder. The line should be tied to a part of the crane, not attached to the worker's body. This will prevent the worker from becoming entangled in the line or from being pulled off.

Final Check: Tools and Guards

After a machine has been repaired, mechanics should, if possible, turn it over first by hand. Thus they may discover mislaid tools or materials in time to prevent them from wrecking the machinery and perhaps causing injuries. To prevent tools from being left on or near machinery after a repair job, mechanics should use a tool box with a special rack for each tool. That way, mechanics can quickly see when a tool is missing. Mechanics could also use a tool check system to help them keep track of special tools used for the job.

Mechanics should replace and securely fasten guards, pick up tools, and leave a work area in as good a condition as possible. Doing so helps promote good housekeeping.

Lubrication

As a rule, lubrication is handled by a separate maintenance schedule. Nevertheless, it is an important part of a preventive maintenance program. Supervisors should make a complete survey of the facility to determine lubrication requirements. This information should be entered in the record for each piece of machinery. Inspectors should then check the machinery for missing fittings, missing oil cups, and plugged oil holes. If possible, make repairs and upgrades at this time. For example, install automatic-feed oil cups, mechanical force-feed lubrication systems, and special fixtures. These features allow oiling personnel to reach parts in remote locations without danger from moving parts.

Provide oiling personnel with a simple diagram of each machine, with all the parts that require lubrication plainly marked. The diagram should include a table that tells the oiler the kind of lubricant to use and how often to apply it. Some companies devise a color code that gives this information. The point of lubrication, the oil cups, oil feeds, and oil holes are painted according to the color code; the oiler can then refer to a color-coded chart.

Each oiler should have a supply of necessary oils, greases, guns, fittings, and wiping cloths before starting a day's work. In large facilities, a special cart should be provided to hold all the equipment and supplies needed for the day.

Supervisors should transfer special precautions or instructions for each oiler from the master sheet to the oiler's personal schedule. The instructions would concern certain types of machines that must be stopped to assure the oiler's safety or those machines that involve exposure to electrical elements and must have power lines shut off.

Sometimes it is necessary to remove a guard from the lubrication point. Train workers to replace the guard properly in its fittings and holding devices so that it will clear all moving parts. If the guard or fittings are badly worn or damaged, workers should make a report immediately, requesting repairs or replacement.

If possible, shafting should be oiled while the machinery is at rest. Mechanics can use special pump oil cans with long spouts to reach overhead hangers or out-of-the-way bearings. Some of these oiling devices can be used without a ladder. If a ladder is used, it should not straddle machines that are running, nor should the oiler attempt to reach several countershafts from one position. Makeshift devices, such as chairs or boxes, should never be used instead of ladders.

Improper lubrication creates special problems; lack of lubrication can cause overheating of bearings, shutdowns, and fires. Mechanics must be especially careful when lubricating electrical equipment. Over lubricating motor bearings causes oil to drop or to be thrown onto the insulation of the electrical windings. The oil deteriorates the insulation, thus exposing live conductors that will arc and cause fires or cause electrically charged ungrounded surfaces. If maintenance personnel work on overlubricated equipment, they may accidentally complete the circuit to ground with their body and be fatally shocked.

In addition, if the windings become sticky from oil and if dirt accumulates on them, defects may be covered up and go unnoticed until serious trouble or a complete breakdown occurs. The oil and dirt accumulations could result in heat buildup that could in turn cause a deterioration of the electrical windings and overheating of the bearings. This could finally result in shutdown or a fire.

Shop Equipment Maintenance

Provide repair facilities for maintenance crews. The stock of spare parts and units should be large enough to meet the probable demand. Provide up-to-date machine tools, and arrange them to help the maintenance crew keep up an easy flow of work. Mechanics should use hoists to handle heavy machinery and power or hand trucks to transport material from one department to another.

The maintenance shop should have a special area that is well ventilated, especially if welding, paint spraying, or cleaning of metal machinery parts takes place in the shop. Place fire extinguishers in this area, and train the crew in using different types of extinguishers on grease, oil, or electrical fires.

Special Tools

Use spark-resistant tools of nonferrous materials in areas where flammable gases, highly volatile liquids, and other explosive substances are stored or handled. Inspectors should check nonferrous tools before each use to be sure that they have not picked up steel particles, which could produce friction when used and thus cause sparks.

Keeping Up-to-Date

The maintenance department should not overlook the potential of new products, such as cleaners, lubricants, paints, wood preservatives, insulation, floor repair materials, protective coatings, and alloys. It should also keep up with new applications for existing products. New products and applications usually lead to better, safer maintenance practices.

Increased use of mechanized equipment requires a careful review of new potential hazards, especially in high-speed equipment and processes. Use of color throughout the facility, as specified in Chapter 2, Buildings and Facility Layout, will contribute to accident prevention and help develop higher operating standards.

Mechanical devices can help offset potential hazards. Centralized lubrication, centralized spray-painting equipment, floor-cleaning machines, steam cleaners, and other devices can help make a maintenance program safer. For frequently performed maintenance jobs, provide permanent accessories, such as hoists, fixed ladders, and catwalks.

Reexamine maintenance procedures periodically for safer ways to do each job. Set up a special suggestion system for maintenance crews so they can present new ideas or corrective measures.

Engineering books and service manuals should be part of maintenance resource materials. Encourage supervisors and workers to become familiar with them and to use them when necessary.

SUMMARY

- Facility maintenance includes (1) proper long-term care of the buildings, grounds, and equipment; (2) routine care to maintain service and appearance; and (3) repair work required to restore or improve service and appearance.
- The safety and health professional should be involved in maintenance and point out hazards or faulty equipment that need attention. Maintenance inspectors should establish a regular inspection program of the facility and grounds.
- Walls and floors must be inspected for damage, defects, and wear and regularly repaired or replaced.
- Inspect and maintain roof-mounted structures regularly. Roof damage can quickly lead to structural damage of other parts of the building and equipment.
- Tanks and towers must be maintained for fire protection and to prevent serious accidents due to structural failure. Stacks and chimneys should be inspected at least once every six months.

- Platforms and loading docks, concrete sidewalks and drives, and driveways should be inspected for damage or wear and repaired on a regular basis.
- Inspection and maintenance of underground utilities should be done only by trained, closely supervised personnel using proper protective equipment and other safety devices.
- Maintenance workers should replace faulty lamps, repair broken fixtures, and dispose of all lamp-related refuse in special containers. Workers are also responsible for maintaining adequate lighting levels in the building.
- Stairways and exits should be inspected for conditions such as inadequate design or construction, improper handrails, poor lighting or housekeeping, and faulty treads or damaged surfaces. Keep exits clear and well lighted.
- To prevent accidents and injuries from grounds maintenance tools and machines, choose equipment for specific jobs and train workers in their particular responsibilities.
- Workers must know the safe-practice rules for handling and operating electric-powered hand tools and gasoline-powered equipment.
- Workers must be carefully trained in the proper operation of utility tractors and snow throwers and must observe all safety precautions and manufacturer's instructions.
- Pesticides must be carefully selected, used, stored, and disposed of to prevent accidents and injuries. Workers should protect themselves by using PPE and decontamination procedures.
- Computerized predictive maintenance (CPM) can reduce employees' exposure to hazards, decrease equipment downtime, make the most of maintenance expenditures, and create efficient schedules.
- Maintenance diagnostic technology includes a number of methods to detect facility and grounds problems before they become serious threats to a company's operations.
- Management should select maintenance crews for their experience, alertness, and mechanical abilities and train them in accident prevention.
- Supervisors should set up a preventive maintenance and inspection plan for critical equipment and machinery that might seriously affect worker safety and health.
- Maintenance workers must dress appropriately for each job, use proper PPE and tools, and know all pertinent safety procedures and practices that apply to a firm's operations. Maintenance supervisors and crews should keep up-to-date on the developments in their trade.

REFERENCES

American Conference of Governmental Industrial Hygienists, 6500 Glenway Avenue, Bldg D7, Cincinnati, OH 45211.
Industrial Ventilation Manual.

American National Standards Institute, 11 West 42nd Street, New York, NY 10036.
Criteria for Safety Symbols, ANSI/NEMA Z535.3–1998.
Environmental and Facility Safety Signs, ANSI/NEMA Z535.2–1998.
Inspectors' Manual for Escalators and Moving Walks. ANSI/ASME A17.2.3–1998.
Safety Code for Elevators and Escalators, ANSI/ASME A17.1–1996.
Safety Color Code, ANSI Z535.1–1998.
Safety Requirements for Fixed Ladders, ANSI A14.3–1992.
Safety Requirements for Material Hoists, ANSI A10.5–1992.
Safety Requirements for Portable Metal Ladders, ANSI A14.2–2000.
Safety Requirements for Portable Reinforced Plastic Ladders, ANSI A14.5–2000.
Safety Requirements for Portable Wood Ladders, ANSI A14.1–2000.
Safety Requirements for Workplace Floor and Wall Openings, Stairs, and Railing Systems, ANSI A1264.1–1995.
Safety Specifications for Consumer Turf Care Equipment—Walk-Behind Mowers and Ride-On Machines with Mowers, ANSI/OPEI B71.1–1998.
Safety Specifications for Snow Throwers, ANSI/OPEI B71.3–1995.

Claire F. Preventive and predictive maintenance—by computer, *National Safety News,* Aug., 1984.

Dunn RL. Advanced maintenance technologies. *Plant Engineering* 41: 80–86, 1987.

National Fire Protection Association, 1 Batterymarch Park, Quincy, MA 02269.
Flammable and Combustible Liquids Code, NFPA 30, 1993.
Life Safety Code, NFPA 101, 1994.

National Safety Council, 1121 Spring Lake Drive, Itasca, IL 60143.
Fundamentals of Industrial Hygiene, 4th ed., 1996.
Accident Prevention Manual for Business & Industry: Administration & Programs, 12th ed., 2001.
Accident Prevention Manual for Business & Industry: Environmental Management, 2nd ed., 2000.
Occupational Safety and Health Data Sheets:
Atmospheres in Subsurface Structures and Sewers, 12304–0550, 1987.
Blowtorches and Plumbers' Furnaces, 12304–0470, 1990.
Flexible Insulating Protective Equipment for Electrical Workers, 12304–0598, 1991.
Safety Hats, 12304–0561, 1992.

Outdoor Power Equipment Institute, 1901 L Street NW, Washington, DC 20036.

Sack TF. *Complete Guide to Building and Plant Maintenance,* 2nd ed., New York: McGraw–Hill, 1971.

Tight syndrome is a breath of stale air. *Ohio Monitor* (Oct. 1983).

Tucker G, Schneider D. *The Professional Housekeeper,* 3rd ed., New York: Van Nostrand Reinhold, 1989.

Underwriters Laboratories, Inc., 333 Pfingsten Road, Northbrook, IL 60062.

U.S. Army, Corps of Engineers, Washington, DC, *General Safety Requirements,* Manual EM385–1–1.

U.S. Department of Health and Human Services, National Institute for Occupational Safety and Health, Div. of Technical Services, 4676 Columbia Parkway, Cincinnati, OH 45226.

U.S. Department of the Interior, Environmental Protection Agency, Insecticide, Fungicide, and Rodenticide Act, Public Law No. 92-516, Oct. 21, 1972, published in *Code of Federal Regulations,* Title 40—Protection of Environment, Chapter 1, Environmental Protection Agency, Part 165.

REVIEW QUESTIONS

1. Facility maintenance can be described as:
   a. Repair work performed to restore service and appearance
   b. Preventive care to ensure long-term life of company assets
   c. Routine care to maintain uninterrupted service and appearance
   d. All of the above

2. List four items to check periodically for signs of excessive foundation settlement.
   a.
   b.
   c.
   d.

3. Which of the following is a major source of injuries when inadequately maintained?
   a. Exterior walls
   b. Windows
   c. Floors
   d. Ceilings

4. It is recommended that roofs should be inspected:
   a. Every year
   b. Every six months
   c. Every three months
   d. Every two months

5. Describe four precautions to take to prevent roof damage from ice and snow.
   a.
   b.
   c.
   d.

6. What items should be considered when inspecting the condition of exits?
   a.
   b.
   c.
   d.
   e.

7. List five precautions to take to reduce the possibility of heating equipment breakdowns during cold weather.
   a.
   b.
   c.
   d.
   e.

8. What causes sick building syndrome, also known as tight building syndrome?
   a. Physical and chemical contaminants
   b. Biological agents
   c. Lack of fresh air
   d. All of the above

9. Preventive design is an important safeguard against sick building syndrome. List five of the seven basic elements that should be included in the preventive design.
   a.
   b.
   c.
   d.
   e.

10. What are some of the hazards that can be avoided by properly training grounds maintenance workers?
   a.
   b.
   c.
   d.

11. List five actions by workers that cause injuries from electric hedge trimmers.
   a.
   b.
   c.
   d.
   e.

12. Briefly explain the purpose of a computerized predictive maintenance (CPM) program.

13. The selection of maintenance employees should be based on what three criteria?
   a.
   b.
   c.

14. The use of which of the following throughout the facility will contribute to overall accident prevention and help develop higher operating standards?
   a. Cranes
   b. Color
   c. Goggles
   d. Sprinkler system
   e. All of the above

15. What safety precautions should be taken when employees work underneath pipelines that carry chemicals?

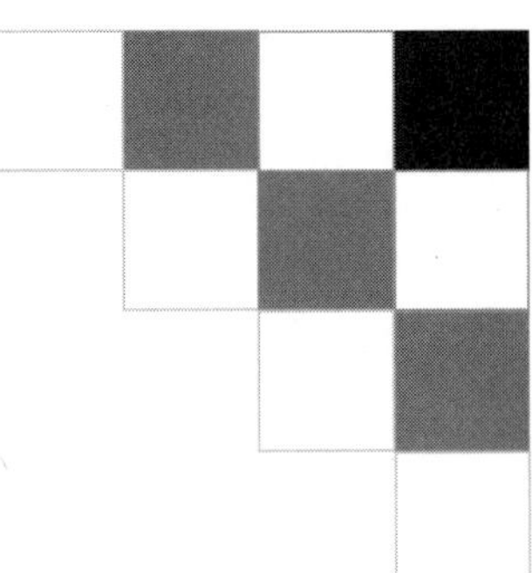

5 Boilers and Unfired Pressure Vessels

Revised by
John F. Montgomery, Ph.D., CSP, CHCM, CHMM

Boilers (fired pressure vessels) and unfired pressure vessels have many potential hazards in common, as well as hazards unique to their specific operations. These vessels hold gases, vapors, liquids, and solids at various temperatures and pressures, ranging from almost a full vacuum to pressures of thousands of pounds per square inch. In some applications, extreme pressure and temperature changes may occur in a system in rapid succession, imposing special strains. To help management create a safer environment for employees who work with such equipment, this chapter covers the following topics:

- important codes governing boilers and unfired pressure vessels
- principles of inspecting this equipment
- major safety considerations in using and maintaining boilers
- safety of high-temperature water vessels
- safety concerns for employees working with unfired pressure vessels
- hazard control in high-pressure systems
- pressure gauges as safety controls

The basic cause of furnace and related boiler explosion is the uncontrolled ignition of an excessive accumulation of fuel-air mixture within the furnace combustion chambers or ancillary areas. Many factors may contribute to the accumulation of combustibles, but among the most common is improper operation or maintenance. Another common contributing factor is human error, often due to lack of understanding, failure to follow safe operating procedures, lack of functional coordination, or other related problems.

CODES FOR BOILERS AND UNFIRED PRESSURE VESSELS

Design, fabrication, testing, and installation of boilers and unfired pressure vessels should comply with the applicable sections of the American Society of Mechanical Engineers' *Boiler and Pressure Vessel Code* (hereafter referred to as the ASME Code) and any regulations.

Synopsis of Boiler and Pressure Vessel Laws, Rules and Regulations, by States, Cities, Counties and Provinces, in the United States and Canada is available from the Uniform Boiler and Pressure Vessel Laws Society. This document details which governing bodies have made the ASME Code a legal requirement in their jurisdictions and what other compliances are required. However, if there is any question, the owners of boilers and unfired pressure vessels should check directly with the proper authority.

Compliance with the ASME Code is determined by authorized inspectors commissioned by the National Board of Boiler and Pressure Vessel Inspectors (Figure 5–1).

The ASME Code contains the following eleven sections:

I Power boilers
II Material specifications
III Nuclear power facility components
IV Heating boilers
V Nondestructive examination
VI Recommended rules for care and operation of heating boilers
VII Recommended rules for care of power boilers
VIII Pressure vessels
 Division 1
 Division 2—Alternate rules
IX Welding and brazing qualifications
X Fiberglass-reinforced plastic pressure vessels
XI Rules for in-service inspection of nuclear power facility components

These sections may be purchased from the ASME. (See References for the ASME's address.)

The minimum requirements for the design, installation, maintenance, and operation of high-pressure boilers are covered in the National Fire Protection Association's NFPA 85A, 85B, 85D, and 85E, *Boiler-Furnace Standards.* These standards also relate to associated equipment and furnace operations. When planning to install pressure vessels, secure the services of a competent pressure-vessel engineering consultant. Such a professional can survey the facility or operation to determine the requirements, design a system that will satisfy codes, and supervise installation and testing.

If the consultant arranges for the purchase of secondhand boilers or pressure vessels, have authorized inspectors check them and report whether repairs are necessary before the purchase. Make arrangements for inspectors through the facility's insurance carrier and, if necessary, make arrangements for government inspectors. (See *Synopsis of Boiler and Pressure Vessel Laws.*)

INSPECTIONS OF BOILERS AND UNFIRED PRESSURE VESSELS

Because the ASME Code covers safety of design, fabrication, and inspection only during construction of boilers and pressure vessels, the National Board of Boiler

Figure 5–1. Boilers and pressure vessels subject to ASME Code regulations must be checked during construction at all points by authorized inspectors who can certify compliance with the Code. Inspectors are commissioned by the National Board of Boiler and Pressure Vessel Inspectors. (Reprinted with permission from Lutheran General Hospital, Park Ridge, IL.)

and Pressure Vessel Inspectors has published its own *Inspection Code* (hereafter referred to as the NB Code). The NB Code provides rules and guidelines for inspection after installation, repair, alteration, derating, and rerating. Therefore, refer to the NB Code for guidance when repairing and altering boilers and pressure vessels. Also use the NB Code to supplement and expand upon the specific safety and inspection procedures discussed in the rest of this chapter.

In general, install and maintain boilers and pressure vessels in accord with manufacturers' instructions. Further, train operating personnel not only to operate equipment properly but also to make routine safety checks and to know when to call in qualified maintenance personnel.

Anticipate and avoid the following common causes of explosions in pressure vessels:

- errors in design, construction, and installation
- improper operation, human failure, and improper training of operators
- corrosion or erosion of construction materials
- mechanical breakdown, failure, or blocking of safety devices, and failure or blocking of automatic control devices
- failure to inspect thoroughly, properly, and frequently
- improper application of equipment
- lack of planned preventive maintenance

In addition to explosion hazards, boilers may also present fire hazards. They are a significant factor in fires in hotels, stores, apartment houses, and churches and synagogues. Oil-fired equipment is usually at fault, although losses from explosions of gas-fired equipment have occasionally been catastrophic.

The majority of boilers in use are automatically or semiautomatically fired, and thus they may operate unattended for long periods. Many are not maintained and checked regularly, leaving them in less than perfect condition. When fires do start in such boilers, they can gain considerable headway before being detected unless adequate precautions have been installed.

The means for controlling and containing fires from boilers include the following:

Boiler rooms should be fully enclosed and should be built with noncombustible materials, including ⅝-in. (1.6-cm) gypsum wallboard or better. Be sure to leave enough space for maintenance operations, such as pulling of tubes. Many regulations have clearance requirements that builders must meet.

Boiler rooms should have large door openings to allow easy access to, and easy installation and removal of, all boiler room equipment. Equip entryways leading to boiler rooms with 1½-hour fire-resistance-rated doors and door frames.

A noncombustible ceiling over a boiler and automatic sprinklers over the firing end of the boiler and in areas containing gas and oil pipelines are strongly recommended. If boilers are coal fired, provide automatic sprinklers over the coal augers, feeders, chutes, and indoor coal piles. (Note: Other protection features may be required for coal piles, conveyor belts, and so on.)

Proper clearance around exteriors of boiler room walls will prevent materials from being stored against the walls. Store only materials and items that pertain to the boiler room's operation in that room. If combustible items are needed in the boiler room, store minimal quantities of them in approved cabinets designed for combustible storage.

To minimize low-pressure boiler fires and explosions caused by faulty controls and safety devices, observe the following policies:

Establish a testing and servicing program in which operating controls, safety controls, and safety and relief valves are tested and maintained regularly.

To prevent damage to the valve seats, make sure that safety and relief valves are always tested under pressure (on the boiler).

Have repairs made immediately upon any indication of malfunction or leakage of operating controls, safety controls, or safety and relief valves. Never operate a boiler with a malfunctioning safety or relief valve.

Have a service organization check and service the boiler during the heating season as well as perform the normal out-of-season servicing.

Keep a boiler log. This ensures that necessary tests, maintenance, and services are performed and that records are available. Keep these records because they will provide a historical profile of the boilers.

BOILERS

In its simplest definition, a boiler is a closed vessel in which water is heated by combustion of fuel or heat from other sources. The heat forms steam, hot water, or high-temperature water (HTW) under pressure.

Design and Construction

Standard references covering the design and construction of boilers are the *Standard Handbook for Mechanical Engineers,* Theodore Baumeister, editor; and *Combustion: Fossil Power Systems,* Joseph G. Singer, editor.

Instruments

Subsection C6 of Section VII of the ASME Code states that, in general, a boiler unit should include a meter-and-control board located on the operating floor. In that way, the operator can see either the furnace door or the lighting ports of the burners and the water column of the boiler without leaving the control board. If it is not possible to see these directly, then install reliable remote-indicating equipment and back this up by having someone make a visual check when lighting is off. Such instrumentation will prevent damage of the boiler in the event of improper operation.

Economizers

Usually an integral part of the heat exchanger system, economizers are the last step in capturing as much heat as possible. The flue gas exhausted from fuel combustion is used to heat incoming cold-makeup feedwater—the hot flue gas being passed over tubes carrying the makeup water. Equip cast-iron and steel tube economizers with at least one safety valve, preferably two.

Some manufacturers recommend installing a safety relief valve under the following conditions:

- if a water bypass is piped up around the economizer, thereby potentially creating a new flow situation with water remaining in the economizer
- if a valve exists in the piping between the economizer's outlet and the drum, thus creating a possible no-flow situation

Superheaters

After the heat-transfer medium (water, steam, or other fluid) leaves the boiler unit itself, its temperature can be raised even more by being passed through a superheater. Detailed operational procedures for superheaters are given in Section VII of the ASME Code.

Air Preheaters

Fires may occur in an air preheater immediately after lighting a boiler and during periods of low-load operation. A fire can be detected by a sudden rise in the air heater's temperature.

To prevent fires, maintain proper combustion and use soot blowers properly. Do not use a soot blower when it is suspected that there is a fire in the gas passages. Such an environment could cause a serious explosion.

Chimneys

Equip chimneys—whether made of brick, concrete, or steel—with grounded lightning arresters. If chimneys are not self-supporting, fasten them to solid building structures. Any ladders added to a chimney should be of permanent construction, securely fastened to the chimney, protected with hoop enclosures, and designed to codes.

Ash Disposal Equipment

Properly guard hoistways, driving machinery, conveyors, worm gears, ash sluices, and reciprocating pumps. An alarm bell must be hooked up to the driving machinery to warn that doors are about to be opened.

Exercise special care to prevent injury to operating personnel from steam or hot water that may be present when ash gates are opened. When excess carbon is present in ash pits and it is not properly wetted down, a gas explosion can result when the gates are opened.

Never store ashes against boilers or combustible materials. Ashes contain sulfur compounds which, on contact with water, form highly corrosive acids. These acids will eat away metal surfaces.

Water Treatment

Generally, water treatment should remove dissolved O_2 and CO_2 and maintain a pH basic enough to minimize corrosion. The pH generally recommended to prevent corrosion is a maximum of 11. (NOTE: Always check and maintain the pH recommended by either the boiler manufacturer or water treatment personnel.) Because operators can sustain injuries when introducing boiler treatment compounds into feedwater, provide adequate protection against scalds and caustic burns. Using automatic feeding or softening equipment reduces the chances of injury. Consult with feedwater-treatment professionals for additional information.

Blowdown Pipes and Valves

Blowdown piping is used to remove sludge and other impurities in boiler water. If not removed, these impurities could seriously impede the efficiency and safety of the boiler. Conduct blowdown piping and boiler drains to a discharge point that will not present a hazard to operators or other personnel. All piping, operating, and discharge valves should conform to the ASME Code, Section I or Section IV.

Safety Valves and Fusible Plugs

Selection, fabrication, installation, testing, and replacement of safety valves and fusible plugs should comply with the ASME Code. Safety valves and fusible plugs are vital devices that require special attention. Boiler inspectors from insurance companies, manufacturers, and other specialists can advise on specific procedures for checking these types of equipment.

Safety valves, when properly installed, relieve excess pressure or vacuum (depending on the design) that would otherwise damage equipment or result in injury to personnel. Keep safety valves in good working order at all times, and have them checked by qualified personnel in accord with insurance company recommendations and regulations. Avoid safety valves that are set by screwing down the body, because installers usually jam them by screwing them down too tightly. Safety valves for water heaters differ from those used for boilers in that they must sense excessive temperatures as well as overpressures. (See ANSI Z21.22, *Relief Valves and Automatic Gas Shut-Off Devices for Hot Water Supply Systems.*)

If a safety valve opens, fails to reseat correctly, and cannot be freed by using the hand-lifting lever, take the boiler out of service and have the safety valve repaired or replaced. Whenever a boiler is returned to service, test the valves. To test a safety valve by hand, hold the valve wide open long enough to blow out any dirt and chips. Raise the steam pressure to at least 75% of the valve's set pressure when opening the valve with the hand-lifting lever. A more meaningful test is to raise the pressure to the safety valve's set pressure and to let the valve open.

If boilers are kept in continuous operation for several months, periodically check the safety valves. Depending on the boiler's condition, periodically repeat either the hand-lifting lever test, or test by raising the pressure at intervals during operation. Use small chains or wires that are attached to the levers of pop safety valves and that are extended over pulleys to other parts of the boiler room. A counterspring or counterweight prevents the weight of the chain or wire from pulling the valves partly open.

Do not use these testing procedures, however, on valves with a set pressure over 400 psi (2,760 kPa). Periodically check such valves according to the manufacturer's recommendations.

Support discharge pipes—individual escape pipes designed to carry discharge away from each safety valve—to prevent any stress on the safety valve. Do not connect the discharge pipe to the valve rigidly, because clearance must be allowed for the boiler's expansion.

Drain pipes from a drip pan and valve body should be located clear of the boiler setting and discharged into an open funnel that provides a clear view of the drip. Install drain pipes to avoid freezing at any point.

Fusible plugs are designed to relieve pressure and to indicate certain conditions that contribute to low water. When using these plugs, make sure they are manufactured, installed, inspected, repaired, or replaced according to the ASME Code.

See that boiler operators are trained to check the safety controls, preferably once a week, but at least once a month. Operators should use a checklist and fill out an inspection form (usually supplied by the insurance company or manufacturer).

Steam and Water Indicators

Steam gauges indicate the pressure of the steam generated. All gauges should be graduated to approximately double the pressure at which the safety valve is set, but in no case less than 1½ times that pressure. Pressure gauges installed on a multiple-boiler setup should be of the same type and graduated alike.

Good Piping Practice

To prevent accidents, install steam lines to minimize maintenance work on them. ANSI/ASME B31 Series, *Pressure Piping*, (1) prescribes minimum requirements for design, materials, fabrication, erection, testing, and inspection of various piping systems, and (2) discusses the expansion, flexibility, and supporting of lines.

As another safety consideration for good piping practice, install valves and other boiler-operating controls so they are easy to reach. Many operators and maintenance personnel have been hurt when they fell from ladders or inadequate work stands while trying to operate a rarely used and hard-to-reach valve.

If it is necessary to open lines, maintenance personnel should always assume that the lines are loaded and under pressure. Provide safe work stands. Also, establish a line-breaking permit procedure.

Placing Boilers In and Out of Service

This chapter cannot cover all details of placing boilers in and out of service. Therefore, use the ASME Code and follow all manufacturers' recommendations.

Cleaning and Maintenance

When taking a boiler out of service for a prolonged period, clean it promptly and have the National Board of Boiler and Pressure Vessel Inspectors inspect it.

Cleaning

Prompt cleaning is important. Soot gathers moisture rapidly, thus contributing to deterioration of a boiler's metal surfaces. Remove soot and fly ash as soon as the boiler has cooled. Since ashes may remain hot for days, they present a hazard to anyone entering the combustion chamber. Therefore, cautiously wet them down with a hose. When wetting down an ash pile, start at the outside and move toward the center because a jet of water driven directly into the center of a hot ash pile can cause it to explode. The operator should stay clear of any steam and dust that might come up. When removing ashes, operators should exercise care to prevent injury to other personnel from steam or hot water that may be present when the ash gates are opened. Operators should wash away the ash thoroughly and then dry the boiler's surfaces completely. Doing this prevents any remaining ash deposits from becoming cemented to the boiler.

Schedule shutdowns of boilers in continuous service in order to perform preventive maintenance. Doing this is far safer than risking an extensive shutdown caused by boiler failure. At least once a year, have each boiler, the flame-safeguard supervisory system, and other safety controls inspected during a scheduled shutdown. An authorized National Board inspector should accompany the facility inspector, who should have defective parts repaired or replaced.

Plan scheduled maintenance carefully to minimize the interruption of production. Be sure that the boilers are properly prepared before the inspector arrives. Boilers must be cool enough so that the inspector will not have to rush the work. Boilers should be clean enough so that the inspector can thoroughly examine metal parts for corrosion, pitting, cracking, and other defects. Also, have internal parts readily accessible for a close and thorough examination. Open handholes and manholes, and ventilate the boiler. Provide adequate lighting and protective equipment for work inside the boiler.

Precautions for Entering Boilers and Furnaces

General precautions for entering boilers include having proper ventilation, proper equipment, and proper protection. Observe rules for working in confined spaces. Implement a confined space permitting procedure. To ensure that no flammable or toxic gases are present, ventilate boiler systems thoroughly and then check the atmosphere with a testing instrument before permitting anyone to enter. Doing this is especially important when more than one boiler is connected to one breeching or chimney, because under certain circumstances, flue gases can come back into a boiler from other boilers.

One cannot overemphasize the need for caution when employees clean or do other maintenance work on boilers, because a great number of injuries have occurred in this work. To prevent steam, hot or high-pressure water, or hot gases from reaching employees, follow good confined space entry procedures and good lockout/tagout procedures. (See Control of All Energy Sources in Chapter 6, Safeguarding.)

Because closing and testing for leaks may not be sufficient for fuel gases, apply positive blanking or block and vent valving. Check all closed valves for leakage. Lines that are interconnected between boilers must be positively sealed off at both ends and locked out. Also, provide work stands for employees and protect them from the overhead hazard of falling ash as they enter boilers.

For ventilation, provide portable power-driven blowers that are operated outside the boiler's setting and that have canvas tubes leading in through access doors. Draft fans can be operated for short periods of time to provide ventilation.

As a precaution against electrical shock, many firms permit only 6-volt or 12-volt lights and tools inside a boiler. They connect such equipment to small, portable power transformers outside the boiler. Battery-powered lights are an even safer alternative. In all instances, properly ground all electrically operated tools and extension cords used inside a boiler. Also, make sure that this equipment is thoroughly inspected before use. (See Chapter 10, Electrical Safety, in this volume.)

When cleaning a boiler, employees should wear hard hats, safety goggles, approved dust masks, and heavy, leather-palmed gloves. They should also wear protective footwear. Personnel working in confined areas should wear a lifeline and be kept under constant observation by a spotter.

Boiler Rooms

The floors, lighting, exits, stairs, ladders, and runways of boiler rooms require special safety precautions.

Floors

Because boiler room floors can become very slippery and dirty, install a surface that can be easily cleaned. Build into the flooring ample drainage and protection against flooding.

Lighting

In addition to normal lighting, boiler rooms should have a source of emergency lighting. Keep gauges and controls especially well lit so they can be read easily. Provide well-maintained flashlights for personal use in case of power failure or other emergencies. Exits, too, should be well lit and identified.

Exits

Each boiler room should have two or more exits, remotely located from one another. If a boiler extends more than one story above ground level, the room should have an exit at the boiler runway or floor level for each story. These exits should lead to a fire escape on the outside of the building. If the boiler room is in a basement (or subbasement), exits should lead to outside stairways and runways and should have landings leading to the exit doors.

Stairs, Ladders, and Runways

Some regulatory codes require the installation of stairs, ladders, and runways around boilers that extend 10 ft (3.1 m) or more above floor level. Even if regulation does not require them, provide such access so that personnel can operate and service a boiler safely and not have to step on hot-steam lines, hot-water lines, or on valve stems or handles.

Stairs, ladders, and runways shall have standard guardrails, handrails, and toeboards. To provide a slip-resistant surface and to permit circulation of air, install runways made of steel grating. Do not locate walkways near water glasses or safety-valve discharge areas, where an operator might be accidentally scalded.

Boiler Room Emergencies

Permanently post rules for both routine and emergency boiler room operation. Be sure the rules are clear and legible (Figure 5–2). Have manufacturers supply rules applicable to their equipment. In addition, furnish all operators and substitute operators with copies of all rules. Supervisors should make sure that boiler room operators know the rules and are able to perform the necessary operations under emergency conditions.

Many facilities have only one boiler room operator. Should this employee become sick or injured, boilers may be left unattended and an accident could occur. In facilities with isolated boiler rooms operated by one person, have this person call a central location at half-hour intervals to make sure that everything is operating on schedule in the boiler room. Facilities that have a building protection patrol should have patrol officers check the boiler room. Provide an intercom system to ensure prompt response in case of an emergency. In addition, train one or more persons (a supervisor, a night-shift employee, or some other substitute) to take over the boiler room's operation in case of an emergency.

SAFETY OF HIGH-TEMPERATURE WATER

High-temperature water (HTW) is water kept in a closed system under high pressure so that it remains in liquid form rather than turning into steam. Conditions such as 400 F (200 C) and 247-psi (1700-kPa) pressure often

BOILER OPERATING INSTRUCTIONS

CAUTION

IF BOILER OVERHEATS–

- STOP FIRE. Do NOT add water.
- Call service company representative or supervisor for assistance.

IN CASE OF FLAME FAILURE–

- Do NOT restart until thoroughly vented.
- Call service company representative or supervisor for assistance.

Starting and Daily Check

1. Be sure water is at proper level.
2. Do not start fire until furnace has been thoroughly vented.
3. Use small fire during warm-up.
4. Check boiler frequently while in normal operation.
5. Test low-water fuel cutoff control on high-pressure steam boilers.

Weekly Check

Test safety valve and low-water fuel cutoff control on low-pressure steam boilers and record on Test & Maintenance Record Tag.

Monthly Check

Test safety valve and low-water fuel cutoff control on low-pressure hot water boilers and record on Test & Maintenance Record Tag.

Yearly Check

Replace or disassemble and overhaul low-water fuel cutoff control during annual boiler clean-up and repair period.

THE TRAVELERS INSURANCE COMPANIES HARTFORD, CONNECTICUT

WS-171 Rev. 6-72

Figure 5–2. An emergency procedure and checklist poster, similar to this one, should be posted permanently in the boiler room. (Reprinted with permission from The Travelers Insurance Companies.)

exist. When liquid under this pressure and temperature expands to steam at atmospheric pressure, employees can be fatally scalded.

HTW versus Steam and Cold Water

High-temperature water is different from steam and cold water when it discharges through a break in a pipe or piece of equipment. As it expands, its volume increases at a very high rate and its energy is released at a very low rate. Energy liberated in the expansion is spent in accelerating the particles of water and vapor and in pushing air out of the way so that the steam-water mixture being formed can occupy the vacated space. Therefore, practically no energy is available for rupturing equipment and imparting kinetic energy to fragments.

When steam escapes, approximately 16 times more energy is released during expansion than when HTW escapes. Therefore, considerable energy is left over to produce an explosive effect. Fragments of fracturing cast-iron valves on steam service have been known to penetrate a 10-in. (25-cm) thick brick wall. No case has been observed, however, where parts of fractured valves on HTW service have been projected any distance.

The increase in the volume of escaping HTW continues after leaving the pipe. The mixture does not form a long jet, as does escaping steam or water; rather, it spreads out almost at right angles from the centerline of the jet to form a wet fog. Nevertheless, do not let such considerations cause any feeling of false security or negligence on the part of the design engineers and operating personnel. Although HTW is safer than steam and accidents are rare, accidents still happen. Even 180-F (80-C) water can be fatal if enough of the body is exposed to it.

Causes of Failure

When equipment or piping does fail in HTW systems, it usually does so because of operating errors or mechanical forces, such as the water hammer, thermal expansion, and thermal shock, and because of faulty materials. Therefore, select and train qualified operators and allow only experienced engineers to design HTW systems. These engineers must be able to minutely analyze the entire design and to select equipment so as to avoid possible dangers.

A good design is neat and simple, but it does not overlook the essentials. Avoid overloading systems with automatic controls since, in case of a malfunction, these controls can introduce more hazards into the system than they avoid. Automatic controls also tend to turn the operator into an attendant who is incapable of operating the boiler in an emergency.

UNFIRED PRESSURE VESSELS

Unfired pressure vessels are compressed air tanks, steam-jacketed kettles, digesters, vulcanizers, and other such vessels. They can withstand internal pressure or vacuum but do not have the direct fire of burning fuel or calrod electric heaters impinging against their walls. If heat is generated in such a vessel, it is by chemical action within the vessel or by applying electric heat, steam, hot oil, or another heating medium to the vessel's contents.

Design

Unfired pressure vessels are covered in the ASME Code, Section VIII, Divisions 1 and 2. The following classes of vessels, however, are exempt from its scope:

- vessels subject to federal regulations
- vessels with a nominal capacity of 120 gal (450 l) or less of water under pressure, in which any trapped air serves only as a cushion
- vessels having an internal or external operating pressure not exceeding 15 psi (103 kPa), with no limitation on size
- vessels with an inside diameter not exceeding 6 in. (15 cm), and no limitation on pressure
- hot-water storage tanks heated by steam or other indirect means—heat input of 200,000 Btu (59,000 joules/second) or less, water temperature of 200 F (93 C) or less, and nominal capacity of 120 gal (450 l) or less

The ASME Code provides that vessels designed for pressures over 3,000 psi (20,700 kPa) may be code stamped. This is covered under Section VIII of the ASME Code, Divisions 1 and 2.

Division 1

The ASME Code, Section VIII, Division 1, normally covers vessels with ratings of 3,000 psi or less (with the exceptions just listed). Vessels may be constructed for pressures above 3,000 psi; however, design principles and construction practices, in addition to the minimum ASME Code requirements, must be considered.

Vessel designs under Division 1 rules are calculated according to the principal stress theory, and a design factor of four is provided on tensile strength. Vessels built to these specifications may be used anywhere the pressure and temperature do not exceed the ratings allowed by the ASME Code.

Before a pressure vessel is designed under Division 1 rules, consider the following questions:

- Will the material used in constructing the vessel affect, or chemically change, the material in process?
- Will the material in process affect or damage the material used in constructing the vessel?
- Will the filled vessel safely carry the weight of its contents and bear internal pressure?
- Will the vessel resist both the pressure introduced into it and any additional pressure that a chemical reaction may cause in process?
- Will the vessel withstand any intentionally or accidentally created vacuum and not collapse?

In addition to general requirements, specifications for construction of Division 1 pressure vessels should also include the following:

- working pressure range
- working temperature range
- data as to whether or not the pressure and/or temperature range is cyclic
- description of what the vessel's contents will be
- specific information that may affect fabrication and installation of the vessel, such as stress relief, radiography, welding, and other requirements

Division 2

Under Division 2 rules, vessel design is based on a detailed stress analysis. A design factor of three is provided on tensile strength. Design calculations in Division 2 are more complex than those in Division 1. However, they allow thinner wall sections and may provide for vessels used at pressures exceeding 3,000 psi.

The alternate rules of Division 2 apply only to vessels installed in a fixed location and subjected to a specific service. To obtain a vessel with an ASME stamp under these rules, a prospective purchaser must prepare a user's design specification and have it certified by a registered professional engineer who is experienced in pressure vessel design.

Other Codes

The ASME Code has been adopted by many governing bodies and therefore has the force of regulation. However, depending on the jurisdiction governing a vessel, other codes, such as the American Petroleum Institute's code and state and local codes, may be in force. These codes may impose size or service limitations more restrictive than those of the ASME Code. Always check with the local authorities before purchasing a vessel.

Secondhand Vessels

Prospective purchasers of secondhand vessels must comply with the jurisdictional requirements for secondhand vessels. Usually one of the requirements is to have the equipment inspected by an authorized National Boiler (NB) inspector. Before the vessel is purchased, obtain a written report that the equipment meets the requirements of the jurisdiction where it is to be installed. A great deal of trouble has arisen when secondhand equipment has been purchased and reinstalled before inspection. If second-hand equipment has been condemned or rejected by one state, it should not be bought, sold, or operated in another state.

Internal Inspection

Have unfired pressure vessels inspected regularly by persons who are qualified and trained for this work. Instruct inspectors to be conservative in approving borderline cases. Also be sure to check whether state or local governing bodies or insurance companies require their own people to make inspections. Often these individuals can make suggestions for refinements that not only help lower insurance rates but also increase safety.

A large company or facility may find it advantageous to employ a full-time inspection staff to administer a regular inspection program for all its pressure vessels. Such a program, coupled with good preventive maintenance, prolongs the life of vessels, piping, and so on, and prevents accidents.

The inspector or the maintenance department should keep a log of each vessel's history. Include the following in the log:

- blueprints
- manufacturer's data reports and instructions
- design data, including location of dimensional checkpoints
- installation information
- records of process changes
- vessel's historical profile, including records of all repairs and conditions found during inspections

This log will prove valuable for operating existing equipment and for designing, installing, and operating new equipment.

When the inspection is carried out by a state, province, or city inspector or by an insurance inspector, a chemical engineer from the facility, or another competent person, should accompany this inspector and describe each vessel's processes in detail. When new processes are developed, inform the inspectors and the operators in detail about these processes and how they may affect the pressure vessels.

Entry

To reduce the number of fatalities that can result from entering pressure vessels and other dangerous confined spaces, establish a safe procedure for entering tanks. At the time of this writing, OSHA published a proposed rule addressing several requirements for confined space entry procedures. Since this proposal was subject to change, we have refrained from listing the specific requirements it addresses. The hazard of working in tanks arises when workers or inspectors cannot get out of a vessel without help. Difficulty in communication compounds the problem. (See the procedures described in Chapter 12, Flammable and Combustible Liquids, Cleaning Tanks, in this volume.)

The following hazards can endanger workers in confined spaces:

- toxic materials already in the confined space or introduced later
- insufficient oxygen
- heat from fire, hot gases or liquids, or inadvertent heating
- startup of agitators or setting the confined space itself in motion

Before anyone enters a pressure vessel, make sure that it is properly drained, ventilated, and cleaned. Next, test the vessel's atmosphere for gases and oxygen content. Check for toxic atmosphere and explosivity. Disconnect and blank all connecting pipelines, or close, lock out, and tag valves on the line. All power-driven devices, such as agitators, must be positively disconnected, locked out, and tagged. (See the discussion of entry-isolation procedures in Chapter 6, Safeguarding, and Chapter 4, Maintenance of Facilities.)

After all preparations for entry are completed, the supervisor for the job should check that the vessel is safe, that all lines are closed off, that power sources are locked out, that ventilation and personal protective equipment are adequate, and that safe work procedures are planned. A "vessel entry permit" that states that all precautions have been carried out can then be issued.

When purging a tank, the vent should discharge outside into an area where no hazard will be created. In some instances, the vessels may be purged with an inert gas such as CO_2 or nitrogen. Be aware of government regulations covering the discharge of certain gases. Remember that an inert gas will not support life, so persons entering the vessel must wear air-supplied respirators or self-contained breathing equipment.

Using forced ventilation for confined spaces may be safer than requiring employees to wear respiratory protection. (See the National Safety Council's *Fundamentals of Industrial Hygiene*, and Chapter 7, Personal Protective Equipment, in this volume.) Have air blown in until tests of the exhaust and of the interior of the enclosed vessel show that the space is safe for entry, or have air continuously sucked out from the bottom by venturi action. Test all areas of the vessel for flammable and toxic gases, and for inadequate oxygen. Repeat these tests at intervals to make sure that conditions remain safe while employees are in the vessel. Introduce air to make sure that there are several changes of air per minute in the vessel.

Provide straight ladders or rope or chain ladders with rigid wood rungs. Also, do not allow employees to go into an opening that they must squeeze through. In an emergency, they may not be able to exit or be removed quickly enough to be saved.

Another necessary precaution is to have employees wear safety harnesses attached to lifelines when entering any vessel. Station an observer outside the vessel who can signal for more help, if needed. You may elect to station a person equipped to do the rescue outside, but often more than one person will be needed to help in the rescue.

Depending on the previous contents of the vessel, the person entering the vessel should be equipped with a vapor-proof flashlight or vapor-proof, low-voltage extension light. At times, this person should wear a chemical protective suit. Be sure that all inspection equipment, including tools, is made of nonsparking materials.

Cleaning

The method of cleaning depends on the use of the vessel. If it has contained petroleum or chemical products, the vessel may be filled with water, a caustic solution, or a neutralizing agent to remove sludge and other materials. Wash, steam, and/or ventilate vessels used for flammable liquids until a test with an approved explosion meter shows the level is safe. (See Chapter 12, Flammable and Combustible Liquids, in this volume.)

Hydrostatic Tests

If a pressure vessel is so constructed that an internal inspection cannot be made periodically, it should undergo a hydrostatic test if the weight of water will not in turn set up damaging stresses. In the latter case, a pneumatic test can be applied. (See the ASME Code, Section VIII, for new construction, and see the NB Code for existing vessels.)

Never use compressed gas or compressed air to test an unfired pressure vessel above its safe working pressure. Compressed gas or compressed air can be used, however, to test for leaks at pressures below the working pressure. Take great care while testing, because a vessel may fail under the test and shatter. Testing should follow procedures in the ASME and NB Codes and should be conducted under the supervision of qualified personnel.

The required pressure for a standard hydrostatic test is normally no greater than 1½ times the maximum allowable working pressure of the vessel being tested. Division 2 of the ASME Code, Section VIII, provides for the design engineer to establish upper limits, in terms of stress-intensity limits relative to the yield strength or tensile strength, or creep-rupture strength, at test temperature. Inspection of a Division 2 vessel is made at a pressure equal to the greater of the design pressures, or ¾ of the test pressure.

Isolate test areas as far as possible from other operations and provide suitable barricades to protect personnel and valuable equipment. These procedures are especially important when conducting proof tests and tests to destruction. All personnel should keep clear of a vessel under full-test pressure. Allow no one to approach it until the pressure has been reduced to, or is close to, the maximum allowable working pressure.

Detecting Cracks and Measuring Thickness

Pressure vessels used for processing gases or oily materials may have very small leaks that will not show under hydrostatic tests. To detect them, a small amount of ammonia is released inside the vessel and compressed air is then applied until a maximum pressure of 50% of the working pressure is attained. A swab soaked in hydrochloric (muriatic) acid is passed over all seams and other suspect areas. Leakage will be indicated by a white vapor (ammonium chloride), formed by the contact of escaping ammonia and the acid. Using a burning sulfur stick is also effective: a change in the flame indicates the presence of ammonia.

Radiography and ultrasonic examination are good tests for finding cracks. Inspectors also sometimes use liquid dye penetrant testing and magnetic particle testing. For example, ammonia tanks used for agriculture are subject to corrosion and to stress cracking. If such tanks are not routinely checked for cracks, catastrophic accidents could occur. The dye-penetrant tests, when made by inspectors according to manufacturers' instructions, are considered reliable.

In many other applications, it is vital to check the thickness of an unfired pressure vessel without damaging it. For this test, inspectors use instruments that measure with ultrasonic rays and electronically produced rays. Using these instruments, a qualified operator can determine the thickness of metal to within 2% or 3%. Some instruments will disclose cracks that extend to, or are slightly below, the surface. A radiograph will then allow the inspector to determine how deep the cracks are.

Another method of detecting microscopic hairline cracks is the lacquer method. When the head of a pressure vessel is suspected, it is cleaned and given a coat of clear lacquer. After the lacquer hardens, an inspector applies a hydrostatic test. Weak spots, hairline cracks, or fatigue stress cracks will show up on the head, through expansion and cracking of the lacquer.

At each inspection of pressure vessels such as vulcanizers, digesters, and autoclaves having removable cover plates, heads, or doors, inspectors should check the holding bolts, cover plate bolts, slots, and retaining rings for wear and hammer test them for soundness. (Hammer test by tapping gently with a hammer, and listen for unusual resultant sounds, indicating cracked equipment, loose bolts, and so on.) Because these parts receive considerable abuse and wear in service, periodically replace them. They are comparatively inexpensive. Gauge the width of a slot and carefully check the retaining rings and cover plates, as cracks result from the stress of improperly adjusting or closing the door plate. If cracks cannot be satisfactorily repaired, condemn the vessel.

Operator Training and Supervision

Thoroughly train employees working with pressure vessels, especially with those used in chemical processes, both in routine duty and emergency procedures. Supervisors, too, should be qualified and knowledgeable. Explain the entire process to a new employee being trained as an operator or helper. Discuss the hazards involved and just how this operation affects the entire process.

Use a checklist to make certain that no step has been overlooked in the processing cycle. The operator or helper records on a card the information obtained from recording apparatus and thermometers, as well as the time and frequency at which the valves to each pressure vessel are operated. After each complete processing cycle, the operator initials the checklist.

If the contents of a pressure vessel are being discharged to a vessel that cannot be seen by the operator, or to one run by another operator, have the operator use a whistle, bell, or light signal system. Note on the checklist the time that these signals are given and the action taken so that the wrong valve will not be opened or closed.

The supervisor of facility operations should instruct operators of vessels with cover plates or removable doors to tighten the bolts or quick-closing lugs without damaging them or the retaining rings. A torque wrench used to tighten bolts will assure uniform tightness and reduce wear and damage. Likewise, instruct operators to open cover plates or removable doors only after the vessel has been relieved of all pressure. Improperly opening vessels can be avoided more simply, however, by using a safety interlock system. (See section on Autoclaves later in this chapter.)

Operators should know where to look for wear on holding bolts, quick-opening lugs, and lug openings on cover plates or removable doors. They should also know when to notify the supervisor about a worn bolt.

So that operators will not open or close the wrong valves, tag and mark valves and pipelines as described in ANSI A13.1, *Scheme for the Identification of Piping Systems.* (See also Chapter 2, Buildings and Facility Layout, in this volume.)

Safety Devices

Because pressure vessels are used to process such a great variety of materials, equip each vessel with safety devices designed for the type of vessel and for the work it is to do. Safety valves for each vessel should be ASME/NB rated and stamped "safety valves." The vessel should be provided with safety devices that will adequately protect it against overpressure, chemical reaction, or other abnormal conditions.

Safety Valves

The old ball-and-lever safety valve has been condemned for use by all Code states because its setting is easily tampered with and it can accidentally be reset. ASME/NB-rated and -stamped safety valves of the spring-loaded type are commonly used safety devices for pressure vessels. They are used on vessels containing air, steam, gases, and liquids that will not solidify as they pass out through the safety valve's discharge.

Valves on pressure vessels containing air or steam should be large enough to discharge the contents at a rate that prevents pressure buildup as prescribed in the ASME Code. On vessels that contain liquids, the safety valve's seat should be made so that it will not collapse and the contents will not plug the discharge's opening. The safety valve's discharge line should lead to a point where it is safe and allowable to discharge. When practical, equip valves with test levers. Frequent testing prevents the valves from sticking. For vessels with dangerous contents (toxic or flammable, for example), the safety valve should not have a test lever. (See Design and Construction in the section on Boilers in this chapter.)

If liquid contents are heated, the safety valve should be designed to operate if the vessel is overpressured as the liquid expands. For pressure vessels containing hot water or in which water is heated, size the valve to relieve the contents according to the total number of Btus that can be applied to the vessel.

Rupture Disks

A frangible disk may not clog as easily as a spring-loaded safety valve and is easily and inexpensively replaced. A rupture disk may clog or become coated with material in such processes as the manufacture of varnish and other resins. At times, this coating becomes thick enough to affect the rupturing pressure of the disk, so that the disk requires replacement or cleaning with a solvent.

Check the condition of these disks at least once a year to see that they are free from any buildup of chemicals or by-products. Remember that when the disk ruptures, all pressure is relieved from the vessel. This could result in the complete loss of a product or spoiling of in-process material.

A rupture disk must function within ±5% of its specified bursting pressure at a specific temperature. Disks may be installed between a spring-loaded safety valve, or relief valve, and the pressure vessel. This prevents unnecessary corrosion of the valve and prevents it from becoming plugged by the vessel's contents. These multiple installations must be in accord with the ASME Code, Section VIII. Various designs of rupture disks are available.

Vacuum Breakers

It is just as important to protect a pressure vessel from collapsing under a vacuum as it is to protect it against bursting from overpressure. Several safety devices provide such protection. One, the mechanical vacuum breaker, similar to a spring-loaded safety valve, has a spring set at a predetermined vacuum. Another, the weight-balanced vacuum breaker, uses a weight suspended from a fulcrum attached to the gate. This vacuum breaker is generally used on pressure vessels working intermittently on pressure and vacuum. If the vacuum exceeds the setting of the weight on the fulcrum, the breaker opens. On some vessels that ordinarily work under pressure but in which a vacuum may occur because of rapid cooling (as when steam condenses), a check valve may be installed with a flap or valve disk that faces into the vessel. Whenever a vacuum occurs, the check-valve disk opens automatically.

Water Seal

A water seal is used on pressure vessels that operate on low pressure or slight vacuum, such as alcohol stills and gas holders. A water seal is a U-pipe filled with water, with one end connected to the pressure side of the vessel and the other end vented to the atmosphere.

Because the vessel operates under a pressure of only a few pounds, the degree of pressure can be regulated by the height of the water in the vent pipe. If the pressure rises above the set limit, the water is forced out of the pipe, thus relieving the pressure.

Vents

In many processes, pressure must be relieved before the pressure vessel can be opened. An easy means of relieving this pressure is to vent it to the atmosphere. Equip condensate tanks with vent pipes and safety valves. These tanks operate under very low pressure or no pressure at all, but excessive pressure can build up in them.

Vent pipes should be large enough in diameter to relieve the contents of the vessel before excess pressure can build up. Install a vent pipe, preferably with a U-bend at the atmospheric discharge, to prevent dirt from clogging the pipe. Be careful to direct the flow away from the vessel so it will not impinge on the metal in case of fire.

Also protect vent pipes in cold weather. Vapor may freeze as it leaves, thus rendering vents inoperative as safety devices. If a vent pipe is placed so that it may freeze or become clogged by dirt, install a relief valve on the pipe as added protection.

Regulating or Reducing Valves

Some vessels operate under steam pressure much lower than that obtained from the boiler or steam transmission line. A regulating or reducing valve reduces high-pressure steam to the pressure required for a specific operation.

Have a safety valve on the low-pressure side of the reducing valve. The relieving capacity of the safety valve should be enough to ensure that the pressure on the vessel being fed the steam will not exceed the vessel's safe working pressure if the reducing valve fails.

To provide protection for all pressure vessels in a battery of the same type, one reducing valve and one safety valve may be installed in the main steam line. This is the usual method in the case of steam-jacketed kettles in which ordinary pressures do not exceed 10–25 psi (70–170 kPa). Connect safety valves so there is no stop valve between them and any vessel they protect.

Autoclaves

Equip with interlocks all autoclaves, vulcanizers, retorts, digesters, and other pressure vessels that may contain large volumes of steam during operation and that must be opened for charging. An interlock will prevent the opening of the charging door until all pressure has been relieved, or prevent the pressurizing of the vessel until the door is in the fully closed position.

The most hazardous part of these vessels (autoclaves) is their closure, although they should be inspected for cracks like any other pressure vessel. Opening an autoclave with pressure in it will cause the door to be flung open with explosive violence. The contents may fire out like projectiles, and the reaction to the blowout may cause it to move back a good distance. Whether using a bolted door, a rotary-lug door, a shearing door, a clamp door, or a screwed-on door, maintain the sealing mechanism in good shape. (See Detecting Cracks and Operator Training earlier in this chapter.)

Autoclave Safety Procedures

Management must make sure personnel know the safety procedures to follow when working on pressure vessels:

- Do not weld to any part of the vessel, door, locking ring, etc.
- Do not cut, drill, or fasten to any part of vessel, door, locking ring, etc.
- Assume the vessel is pressurized any time the door is closed. That is, assume any lines or pipes connected to the autoclave are at autoclave pressure. Do not remove or adjust any autoclave fittings unless the door is open.
- At no time should a person enter the autoclave unless he or she knows there is oxygen sufficient for breathing.
- Leave no flammable materials, debris, or plastics in the autoclave. If solvents are used, use quantities in approved safety containers of less than 1 qt and ensure that no source of ignition is present (such as autoclave lights, electric heater of apparatus, cigarettes, or sparks).
- The operator must inspect the closure alignment and also inspect the vessel and closure for cracks, damage, hot spots, or any unusual conditions. If any of these conditions is observed, do not operate the autoclave. Notify maintenance or management.

Recommended Scheduled Periodic Maintenance for Autoclaves

Experts recommend the following schedule for general maintenance of most autoclaves:

- Doors—lubricate wedges quarterly; remove debris and inspect door seal weekly; check hydraulic pump door lock reservoir level every two months; check entire hydraulic system annually
- Interior—clean at least every two weeks; check thermocouple panels and screws quarterly; check all bolts, washers, and nuts for tightness annually
- Cooling System—check cooling tower fan belts biannually; inspect and clean all water filter screens biannually; drain reservoir tank biannually
- Vacuum System—check vacuum pump packing every two months; vent and drain receiver tanks annually
- Air Regulators—check and readjust air regulators every two months
- Transducers—check and recalibrate transducers every two months
- Pressure Gauges—check and recalibrate analog pressure gauges biannually

Steam-Jacketed Vessels

Steam-jacketed vessels are used to heat liquid mixtures to a moderate degree. Steam circulates between the outer and inner shells of the vessel at pressures that are usually 10–30 psi. Occasionally, the process may require that the vessels be operated at pressures up to 100 psi (690 kPa). Heat is transmitted through the inner shell to the contents.

Such vessels are used principally in commercial preparation of food, in candy manufacturing, and for cooking starch in laundries and textile mills. They are also used in the chemical industry for low-temperature "cooking." On a steam-jacketed vessel that has a tight cover, provide a separate safety valve for the inner kettle. If a steam-jacketed vessel can be completely valved off, protect the vessel with a vacuum-breaker to keep it from collapsing.

Observe the following precautions when operating steam-jacketed kettles:

- Thoroughly drain the steam space before admitting steam to the jacket. Open drain lines even though traps are installed, because water in the steam space may cause thermal shocking and water hammer leading to catastrophic failure of the trap.
- Admit steam to cold vessels slowly to allow ample time for parts to heat and expand uniformly. This precaution becomes more important as vessel size and steam temperature are increased.
- Unless automatic protection is provided, open vents when the steam supply is shut off. Opening them prevents damage to, and even collapse of, the kettle once the steam has condensed.
- Where agitators are used, paddles must not strike the kettle. Even a slight deformity in the inside may require extensive repairs. Be sure that hand stirrers are also used with care. Whether operations involve solids, semisolids, or slurry, paddles should have appropriate tolerance or clearance from vessel walls to prevent erosion or corrosion.
- See that kettles' edges are high enough, or provide guardrails, so employees will not accidentally fall in.
- Be sure that kettles are filled only to a point where undue splashing will not occur when the contents are heated or agitated. Use splash guards or loose covers.

As a general precaution to follow in maintenance work, never allow employees to climb over large open vats or kettles filled with hot, corrosive, or viscous fluids. Drain or cover the vats, provide safe work stands, and have employees wear lifelines. See that all boards in the work stands are fastened in place and not left loose. Take all other precautions to make sure that people and objects do not fall into the fluids.

Evaporating Pans

Ordinarily, evaporating pans are shallow pans containing steam coils. When the pans are in operation, the steam coils are immersed in the material being treated. If the coils become exposed, the material may overheat or ignite, and a fire or explosion may result. To prevent accidents, observe the following rules:

- Attend to the pans continuously as long as they are in operation.
- After each use, thoroughly clean the pans and coils.
- After shutting off steam in the coils, drain them to prevent the product from being drawn into lines when steam condenses and creates a vacuum. Installing a vacuum breaker would also prevent this.

HIGH-PRESSURE SYSTEMS

The hazards of high-pressure systems arise largely because of failures resulting from leaks, pulsation, vibration, and overpressure. Besides the damage that can be expected from the release of high-pressure gases if a vessel or pipe ruptures, fatal injuries can result from the blowout of high-pressure gauges or from the whiplash of broken high-pressure pipe, tubing, or hose. The potential for injury and damage from high-pressure-system accidents is very high.

Because reciprocating pumps and compressors are normally needed to generate high pressures, the inner fibers of the high-pressure system's piping and vessels are subjected to pulsating pressures. Therefore, keep the piping and vessels absolutely free of internal notches or severe scratches. These defects are stress raisers that will surely lead to cyclic stress fatigue failure. Also, prevent a pipe from bearing concentrations of stress arising from ill-planned holes or cross bores. The stress concentration factor from a radial entry to a pipe's or a cylinder's wall will reduce the pressure-endurance limit by almost one-half. Because of the pulsating pressure condition, nothing can be labeled "safe" and then be forgotten. Operators must maintain constant surveillance.

Leaks in pressurized systems can also be hazardous. Liquids expelled can easily penetrate clothing and skin. A sudden leak might instantly fill an enclosure with an explosive mixture of gases. Because most leaks occur at joints, keep the number of joints to a minimum.

Weldments are also a source of leaks. They can be joints that were welded or seams. Welds, if improperly done, are particularly prone to metallurgical weakening. For example, the area around a weldment in stainless steels may suffer chrome depletion.

Limit vibration in piping by appropriately dampening, if possible. Designers use various means to dampen in hydraulic and pressurized gas systems. Because dampening can never be fully achieved, high-pressure systems require many rigged pipe supports. Be sure that the supports are strong enough to resist deflection from any direction, and securely anchor them to the structure of the building to prevent whiplash (Figure 5–3).

Almost all Bourdon tubes in pressure gauges will eventually fail from fatigue caused by the constant pulsing of the pressure. They can fail after many cycles, or even when the gauge is new. Gauges on large vessels are not as subject to large pressure oscillations as those in lines where a compressor maintains the pressure.

PRESSURE GAUGES

Pressure gauges used at 1,800 psi (12,400 kPa) or more (except Underwriters Laboratories-listed, gas-regulator gauges) should have full-size, blowout backs; have integral sides; be front-designed to withstand internal explosion; and have either a multi-ply plastic or a double-laminated safety glass cover for the gauge faces (Figure 5–4). Tests at 3,000 psi (20,700 kPa) show that gauges not constructed in this manner will have a hole blown in the face of the gauge. Providing holes in the back of the gauge does not give enough vent area for safe clearance of gases.

Provide a substantial shield in front of high-pressure gauges. This shield should be made of acrylic plastic at least ⅝ in. (1.6 cm) thick and should meet MIL P5225B-Finish A Specification. Make sure the shield is free of scratches, gripper marks, tool chatter marks, and other stress raisers.

When mounted, a gauge should have at least 2 in. (5 cm) of clearance between itself and the item to which it attaches. If mounting the gauge flush to a backing plate, cut a hole, with a diameter at least equal to the gauge, through the plate. Of course, leave enough area to mount the face flange.

Shields mounted behind gauges should be substantial. Leave a clear area no less than 2 in. (5 cm) wide behind the gauge so vented gases can escape between the shield and the back of the gauge.

Restrict areas where high gas-pressure systems operate to all but necessary personnel. Locate reactors, pressure vessels, and heat exchangers having great hazard potential behind barricades, and install remote-control systems and other monitoring devices to operate them. Place vessels and systems undergoing tests behind barricades.

SUMMARY

- Boilers and unfired pressure vessels have many potential hazards in common that must be controlled by safety devices and safe work practices.
- Management must see to it that the design, fabrication, testing, and installation of boilers and unfired pressure vessels comply strictly with all federal, state, and local codes.
- ASME and the National Board of Boiler and Pressure Vessel Inspectors have established guidelines for inspecting this equipment. Workers should receive training to operate equipment according to safety standards and safe work practices.

Figure 5–3. Two methods of securing a high-pressure line: line was secured on either side of the fitting (lower right arrow) and at the bends (upper arrow). As a secondary protective measure, channel iron was placed over the lines.

Figure 5–4. A reference gauge has been installed for each of the four separate pressure stages of this compressor. Note the plastic shield (arrow) over the gauges.

- Boilers are closed vessels in which water is heated by fuel combustion or heat from other sources. Workers should make sure that all these elements work properly and are operated only under safe conditions.
- Management should schedule frequent cleaning and maintenance of boilers in a way that minimizes production interruptions. Workers must use extreme caution when entering boilers because of the dangers presented by high-pressure water or hot gases. Management should post rules for routine and emergency boiler room operation.
- High-temperature water (HTW) is kept in a closed system under high pressure so that it remains in liquid form instead of turning into steam. Only trained operators and engineers should work on this equipment.
- Unfired pressure vessels include air tanks, steam-jacketed kettles, digesters, and vulcanizers. Unfired pressure vessels should be inspected regularly by qualified persons. If vessels cannot be inspected internally, they should undergo a hydrostatic or pneumatic test. Cleaning methods should be selected according to the use of the vessels.
- Operators must be carefully selected and thoroughly trained to operate pressure vessels safely, to understand emergency procedures, and to use a checklist to inspect and maintain equipment. Supervisors must closely monitor all operations.
- Safety devices on all pressure vessels will help prevent accidents and injuries. Workers must be sure that these devices are properly installed and maintained. Safety precautions for steam-jacketed vessels will prevent people and objects from falling into the liquids.
- The hazards of high-pressure systems arise largely from failures caused by leaks, pulsation, vibration, and overpressure. Supervisors and workers must make sure that all elements of the systems are kept clean and in good working order and are frequently inspected.
- Pressure gauges are critical monitors of high-pressure systems and must be adequately designed for the equipment. For general safety, restrict areas where high-pressure systems operate to all but necessary personnel.

REFERENCES

American National Standards Institute, 11 West 42nd Street, New York, NY 10036.

Pressure Piping, ANSI/ASME B31 Series.
Relief Valves Hot Water Supply Systems, ANSI Z21.22–1999.
Scheme for the Identification of Piping Systems, ANSI A13.1–1981 (Erratum 1998).

American Petroleum Institute, 1220 L Street, NW, Washington, DC 20005.

American Society of Heating, Refrigerating and Air-Conditioning Engineers, 1791 Tullie Circle, NE, Atlanta, GA 30329.

Applications.
Handbook of Fundamentals.
Systems and Equipment.

American Society of Mechanical Engineers, 3 Park Avenue, New York, NY 10016–5902. Boiler and Pressure Vessel Code.

Avallone, EA, Baumeister T, eds. *Mark's Standard Handbook for Mechanical Engineers*, 10th ed. New York: McGraw-Hill, 1996.

Boiler and Pressure Vessel Laws Society, 2838 Long Beach Road, PO Box 512, Oceanside, NY 11572. *Synopsis of Boiler and Pressure Vessel Laws, Rules, and Regulations by States, Cities, Counties, and Provinces, in the United States and Canada.*

Combustion Institute, 5001 Baum Boulevard, Pittsburgh, PA 15213. (General.)

Compressed Gas Association, 1235 Jefferson Davis Highway, Arlington, VA 22202.

Cylinder Service Life: Seamless High-Pressure Cylinders, Pamphlet C-5.
High-pressure systems. Safety Guides for Laboratories, SCM–68–378, 1971.

National Board of Boiler and Pressure Vessel Inspectors, 1055 Crupper Avenue, Columbus, OH 43229.

National Board Inspection Code.

National Fire Protection Association, 1 Batterymarch Park, Quincy, MA 02269.

Fire Protection Handbook, 1997.
Life Safety Code® Handbook, 2000 Edition.

National Safety Council, 1112 Spring Lake Drive, Itasca, IL 60143. *Fundamentals of Industrial Hygiene,* 4th ed., 1996.

Occupational Safety & Health Data Sheets:*Maintenance of High-Pressure Gate and Plug Valves,* 12304–0440, 1993.

National Technical Information Service, 5285 Port Royal Road, Springfield, VA 52161.

Singer JG, ed. *Combustion: Fossil Power Systems.* 4th Edition. New York: Combustion Engineering, 1993.

Underwriters Laboratories, Inc., 333 Pfingsten Road, Northbrook, IL 60062.

U.S. Department of Labor. Occupational Safety and Health Administration, 200 Constitution Avenue, NW, Washington, DC 20210.

Code of Federal Regulations, Title 29. Section 1910.
Pressure Vessel Safety, Pub. 8–1.5, 1989.
29 CFR 1910.147, *The Control of Hazardous Energy (Lockout/Tagout).*
29 CFR 1910.146, *Permit-Required Confined Spaces.*

REVIEW QUESTIONS

1. What is the difference between the ASME Code and the NB Code for boilers and pressure vessels?
2. List four of the seven common causes of pressure vessel explosions.
   a.
   b.
   c.
   d.
3. When inspecting safety and relief valves on the boiler, the test should be performed __________ to prevent damage to the valve seats.
   a. without pressure
   b. under pressure
   c. with blowdown piping
   d. without blowdown piping
4. What is a boiler?
5. What is the maximum recommended pH for water within a boiler?
   a. 3
   b. 4
   c. 7
   d. 11
6. When should a soot blower *not* be used?
7. What is the purpose of safety valves of boilers and pressure vessels?
8. During the cleaning process, when should soot and fly ash be removed?
9. What is the proper way to wet down an ash pile?
10. Every boiler, flame-safeguard supervisory system, and other safety controls should be inspected during a scheduled shutdown period at least ________.
    a. daily
    b. weekly
    c. monthly
    d. yearly

11. What are the three general precautions for entering a boiler?

a.

b.

c.

12. What is high temperature water (HTW)?

13. What is an unfired pressure vessel?

14. List three of the six items that a vessel's history log should contain.

a.

b.

c.

15. Why would you use a hydrostatic or pneumatic test for inspection of a boiler or pressure vessel?

16. Under what condition should the safety valve of a pressure vessel *not* have a test lever?

17. What is a water seal and when is it used?

18. The most hazardous part of an autoclave is the __________.

Part Two

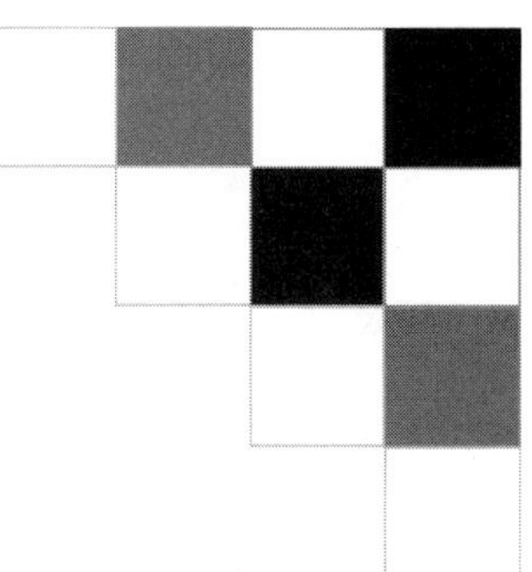

WORKPLACE EXPOSURES AND PROTECTIONS

Primary prevention is the safety strategy that focuses on reducing or eliminating injuries or illnesses by eliminating causative agents or sources. The chapters on safeguarding, industrial sanitation, electrical safety, and fire protection describe many of the primary prevention techniques appropriate to these areas. Other chapters in Part Two, such as Occupational Medical Surveillance, are based on secondary prevention techniques that are keyed to early detection and correction of problems. When potentially hazardous workplace exposures are expected and cannot be engineered out, their impact can be minimized by the appropriate use of personal protective equipment. The final chapter in Part Two deals with the safety needs and accommodations that can help keep the disabled worker safe in the workplace.

6

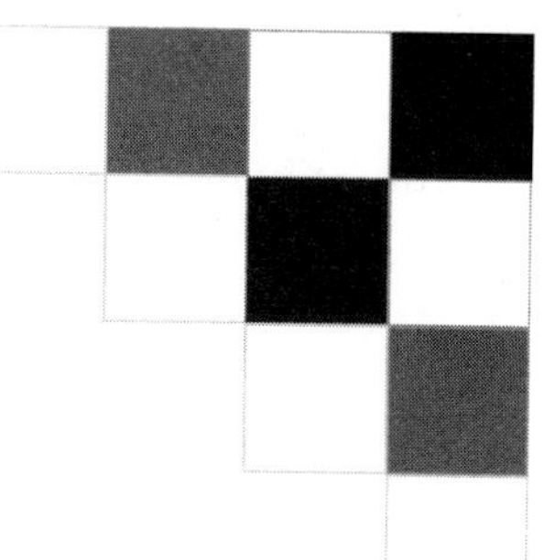

Safeguarding

Revised by

John F. Montgomery, Ph.D., CSP, CHCM, CHMM

From the early days of safety, the concept of engineering out hazards by providing safeguards has been fundamental to every occupational safety and health program and still is. Many written regulations concern the identification of mechanical hazards and then positive activity to protect the employee from the hazards.

This chapter discusses the following topics:

- creating safeguards to protect employees from hazards at the point of operation
- guarding workers against the hazards of power transmission and hazardous energy sources
- safeguarding employees during maintenance and servicing operations
- equipment and techniques used to protect workers against the hazards of robotics
- methods and equipment used to guard against excessive noise

DEFINITIONS

The terms used in this chapter have the following specific definitions:

Safeguarding

Safeguarding is any means of preventing personnel from coming in contact with the moving parts of machinery or equipment, that would potentially cause physical harm.

Device

A device is a mechanism or control designed for safeguarding at the point of operation. Devices include presence-sensing devices, movable-barrier devices, holdout or restraint devices, pull-back (out) devices, two-hand-trip devices, and two-hand control devices.

Guard

The word *guard* often refers exclusively to barriers designed for safeguarding at the point of operation. Guards include die-enclosure guards, fixed-barrier guards, interlocked-barrier guards, and adjustable-barrier guards.

Enclosure

Enclosure is safeguarding by fixed physical barriers that are mounted on or around a machine to prevent access to the moving parts. Enclosures are most effective when designed as part of the machine, but they can be bolted or welded to the frame or the floor.

Fencing

Fencing is safeguarding by means of a locked fence or rail enclosure that restricts access to a machine to authorized personnel only.

Location

Safeguarding by location results when a hazard is physically inaccessible under normal operating conditions or use. Both fencing and location are very limited as safeguarding techniques and are permitted only if caution restrictions can be met.

Nip Points or Bites

A nip point or bite is a hazardous area created by two or more mechanical parts rotating in opposite directions within the same plane and in close interaction.

Pinch Point

A pinch point is any place where a body part can be caught between two or more moving mechanical parts.

Point of Operation

A point of operation is the area on a machine where material is positioned for processing, where work is actually being performed on the material.

Power Transmission

Power transmission includes all mechanical parts, such as gears, cams, shafts, pulleys, belts, clutches, brakes, and rods, that transmit energy and motion from a source of power to equipment or a machine.

Shear Points

A shear point is a hazardous area created by the cutting movement of a mechanical part past a stationary point on a machine.

POINT-OF-OPERATION PROTECTIVE DEVICES

If all machines were alike, it would be simple to design a universal point-of-operation device/guard and install it during the manufacture of each machine. But all machines are not alike, and to further complicate the problem, purchasers of the same machine model may use it in different ways and for different production purposes. In addition, the uses of one machine may change during its lifetime. For all these reasons, effective point-of-operation devices/guards cannot always be installed by

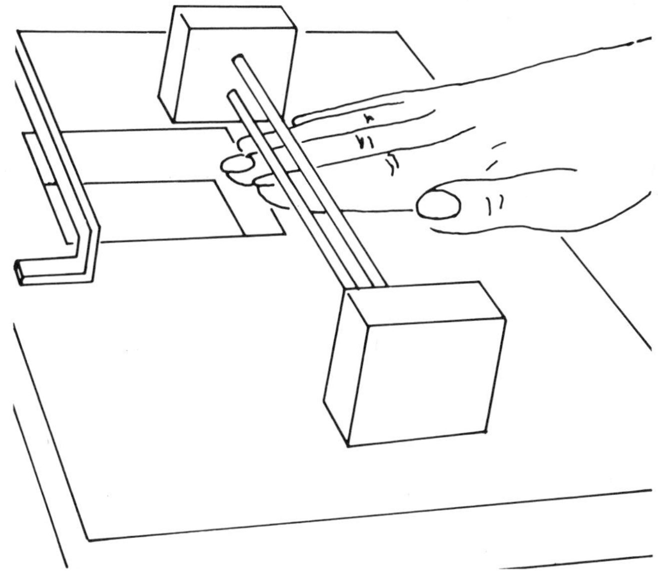

Figure 6–1. A ⅜ in. (9 mm) opening permits a small part of the fingers to slip past the guard. A smaller, ¼ in. (6 mm) opening would prohibit the fingers crossing into the danger zone.

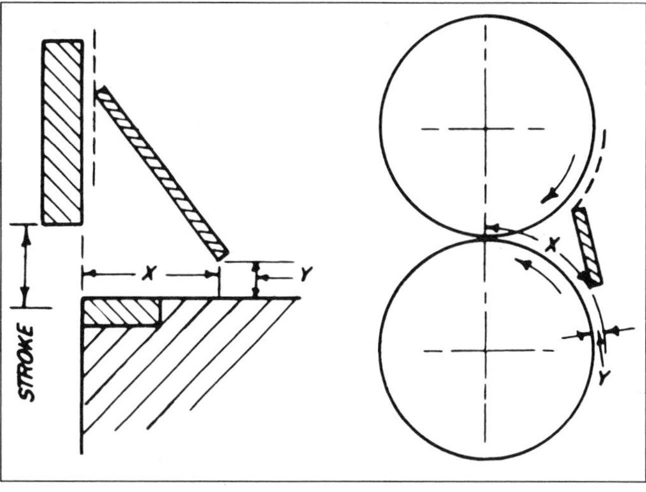

Figure 6–3. *Left:* Vertical shear hazard. *Right:* Inrunning roll hazard (no feed table used). (Printed with permission from Alliance of American Insurers; Liberty Mutual Insurance Company.)

the manufacturer. In many cases, a device/guard can be made and installed only after the user has tested the machine before release for operation.

Whenever point-of-operation protective devices/ guards are needed, a set of certain principles and conditions should be applied to the type, design, construction, and location of the safeguards. The following is a summary of these principles and other important data found in OSHA 29 *CFR* 1910.211, 212, 213, 217, and various ANSI standards (B11.1, 2, 3, etc.).

For more information on point-of-operation safeguarding related to power presses, see Chapter 23, Cold Forming of Metals, in this volume.

Openings Used for Safeguarding

It has been common practice in the design of point-of- operation safeguards to consider any opening not exceeding ⅜ in. (9 mm) as relatively safe. Such an opening would not permit entry of any significant part of a hand inside the guard (Figure 6–1). In many instances, however, a ⅜ in. (9 mm) opening is insufficient for passing material to be processed through or under the guard. Yet as the width (or height) of the opening is increased to accommodate the material, the operator is able to reach farther inside the guard. Under such conditions, it is no longer possible to prevent entry of some part of the hand within the guard. The problem is to stop movement of the hand inside the guard at a safe distance from the danger zone.

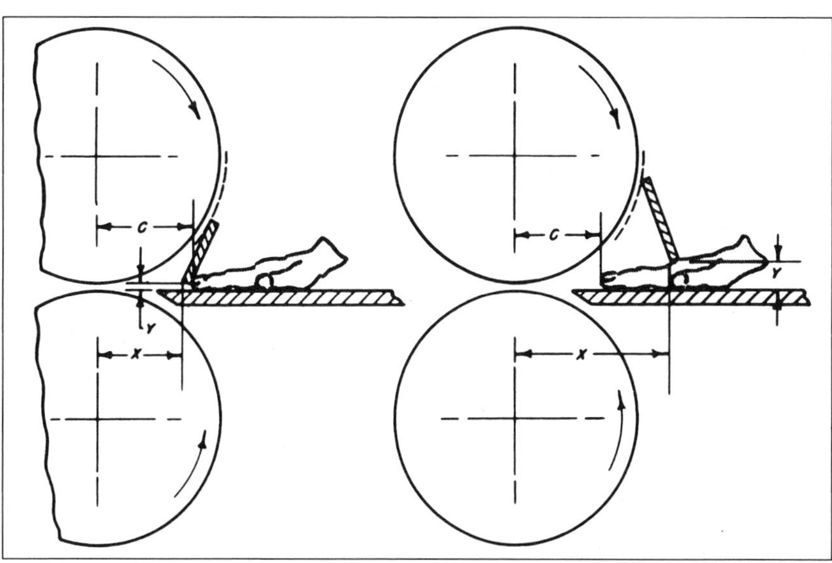

Figure 6–2. Left: A ¼ in. opening on guard stops the hand. Right: A larger opening (*Y*) placed farther back can still stop the fingers from reaching the danger zone. Both illustrations show the ends of fingers stopped approximately the same distance (*C*) from the danger zone. (Printed with permission from Alliance of American Insurers; Liberty Mutual Insurance Company.)

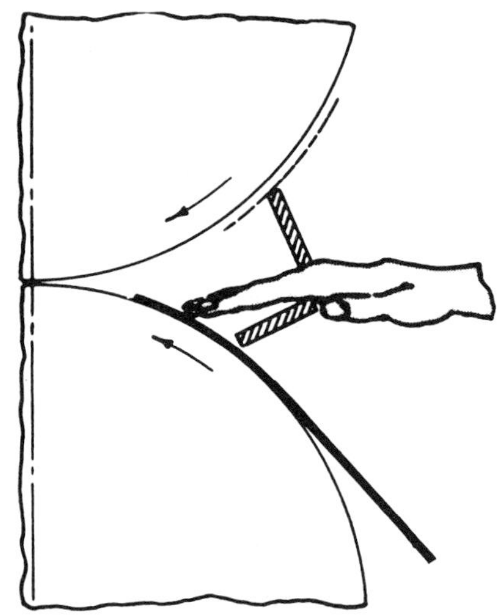

Figure 6–4. This guard protects the fingers while they position materials. (Printed with permission from Alliance of American Insurers; Liberty Mutual Insurance Company.)

Figure 6–2 shows sketches of an inrunning roll hazard where a feed table is used. There are other types of point-of-operation hazards, however, where an opening larger than that shown may be required in a guard's design.

In Figure 6–3, the proper location of the guard's distance *X* for the use of required opening *Y* must be determined. If the dimensions of the opening and its location (distance from the hazard) are properly selected, adequate safety for the operator can be established.

Some guard designers have made use of this formula:

Maximum safe opening = ¼ in. (6 mm) + ⅛ × distance to guard from danger zone

Caution: This formula is not intended for use where the distance from the guard to the danger zone exceeds 12 in. (30 cm).

Figure 6–4 illustrates a situation where it may be essential to have part of the hand and fingers extend through the guard to permit manipulation of the material inside the guard. A condition can exist where it is impossible to use hand tools or mechanical devices to carry on the operation.

Test Data

For their design analysis, the Liberty Mutual Insurance Company constructed a number of test fixtures that approximated common guards. By testing openings and hands of different sizes in these fixtures, they developed data to guide guard designers when specifying allowable openings and locations of guards. These design data are shown in Figure 6–5 with a feed table, and in Figure 6–6 without a feed table.

One set of data was established for general use. Due to variations in the size of hands and inaccuracies to be expected in maintaining a ⅜ in. (9 mm) opening, Liberty Mutual's analysts decided that for the first 1½ in. (38 mm) away from the danger line (Figure 6–7), no openings exceeding ¼ in. (6 mm) could be considered safe.

Most men and women have fingertips that will not travel far through a ⅜ in. (9 mm) opening. However, if a designer wishes to maintain a definitely safe zone beyond a ⅜ in. (9 mm) opening, the opening must be no more than ¼ in. (6 mm) wide within 1½ in. (38 mm) of the danger point (Figure 6–7).

Application of Test Data

The team from Liberty Mutual then applied the data from these tests to various situations that require guards.

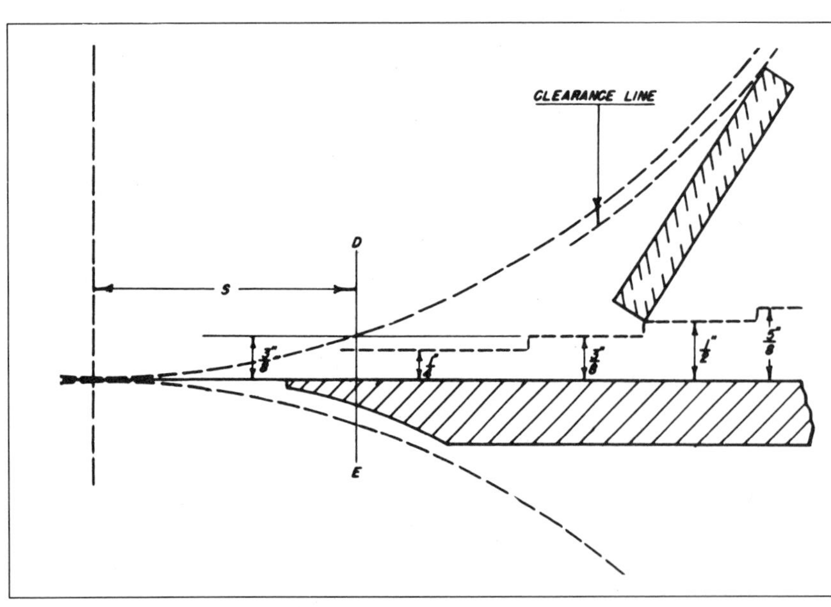

Figure 6–5. Inrunning roll nip where a feed table is used. *DE* is the stop line. *S* is the distance of ⅜ in. wide nip zone from contact point between rolls.

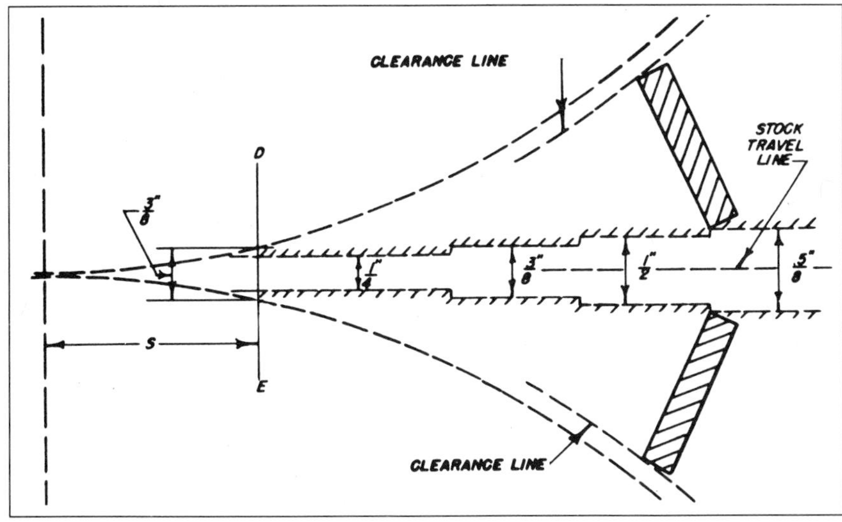

Figure 6–6. Inrunning rolls with central feed and no feed table.

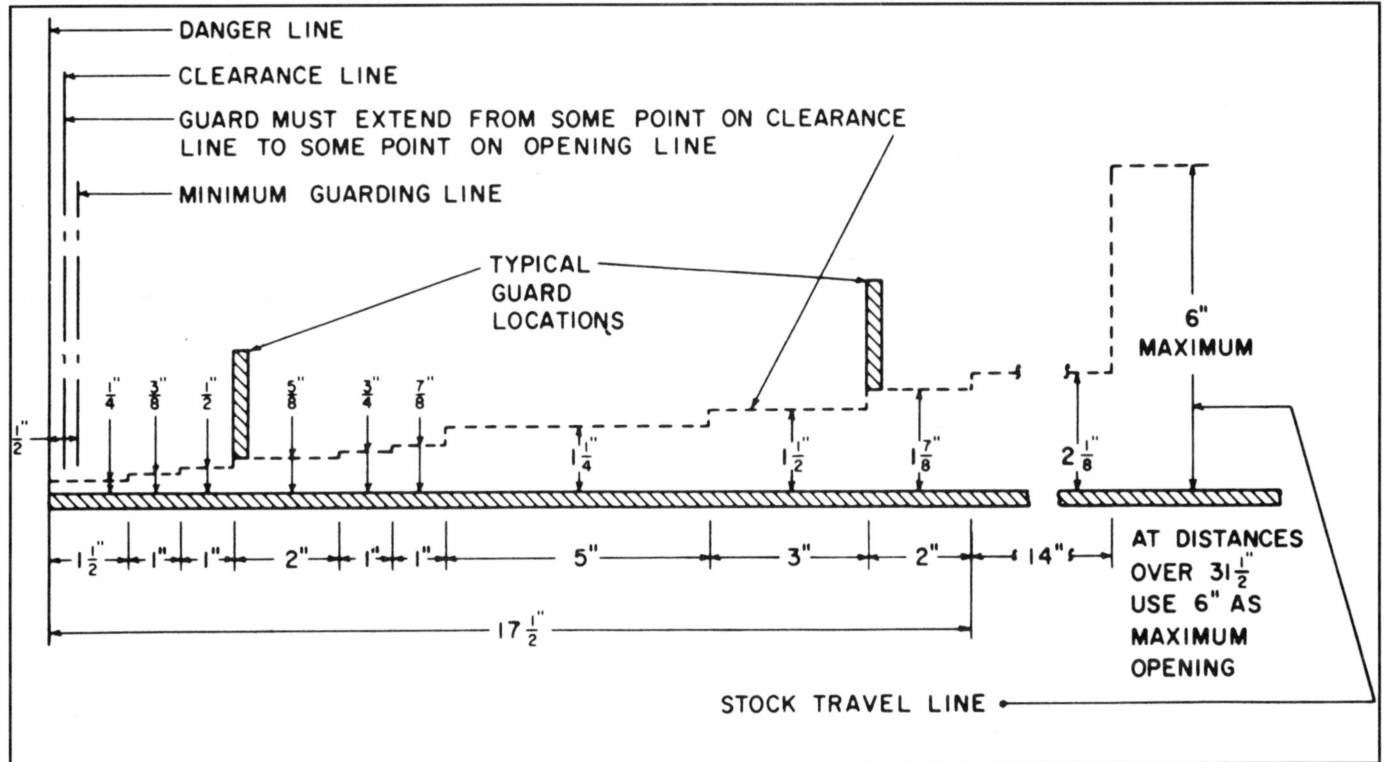

Figure 6–7. Every point-of-operation guard shall prevent entry of hands or fingers by reaching over, under, or around the guard into the point of operation. It should use fasteners not easily removable by the operator to minimize the possibility of misuse or removal of essential parts. It should offer maximum visibility of the point of operation consistent with other requirements. (Dimensions assume adult hands held flat as shown in Figure 6–2.) (Printed with permission from *Requirements for the Construction, Care, and Use of Mechanical Presses,* ANSI B11.1.)

Vertical Shear

The findings of the vertical shear portion of these tests (Figure 6–8) apply to the following items:

- all vertical openings in the guard proper, such as visibility slots, clearance slots for ejection devices, and stop gauges
- all horizontal openings in the guard proper as are necessary for feeding stock, front or sides, and for ejecting finished parts or scrap
- the installation or adjustment of a guard

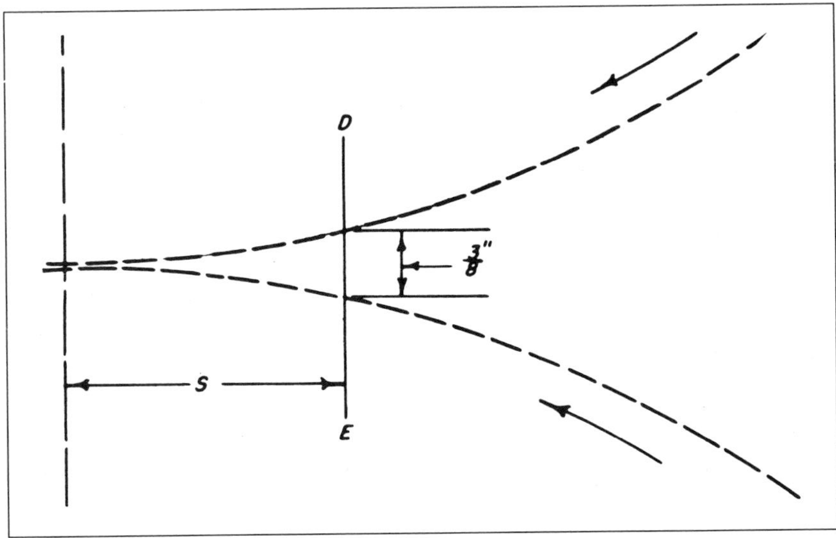

Figure 6–8. Application of test findings to guard design on inrunning rolls. See accompanying text for explanation of these figures.

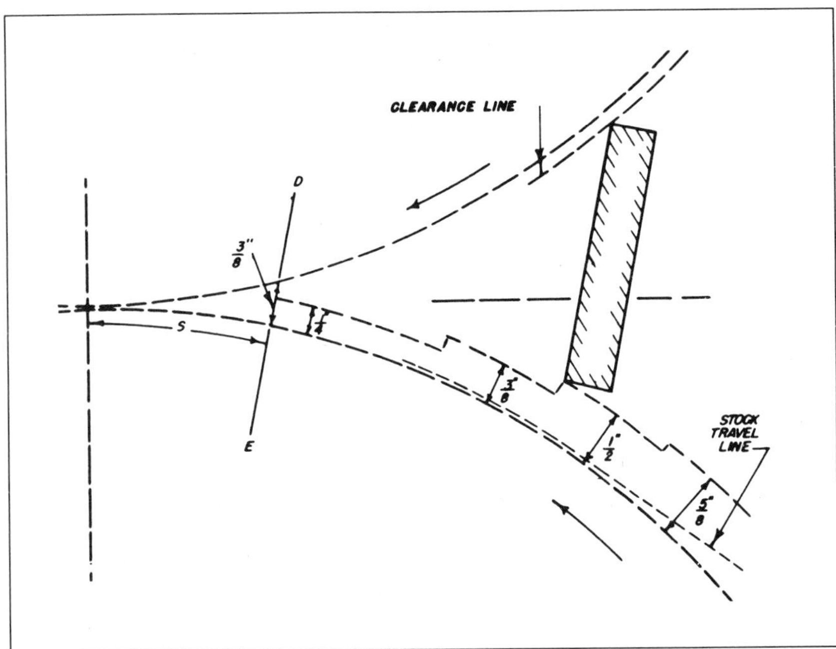

Figure 6–9. Inrunning rolls with stock traveling over one roll before entering nip zone. Caution: Where no stock is fed over the cylinder and particularly on high-speed cylinders, the dimension between the guard and the cylinder should be kept to ¼ in. to avoid creating a hazard between the moving cylinder and guard itself. Gaps or discontinuities in the cylinder surface may also create a hazard. (Printed with permission from Alliance of American Insurers; Liberty Mutual Insurance Company.)

Inrunning Rolls with Feed Table

In applying the findings of these tests to guards for inrunning rolls, the characteristics of a nip point must be considered. In Figure 6–7, the danger line represents a hazardous contact. For vertical shear exposures, it is equivalent to the shear line.

On rolls, the hazard, nip, or pinch zone is not defined by a straight line. Therefore a ⅜ in. (9 mm) width of nip zone is considered the actual nip point through which the danger line *DE* is drawn (Figure 6–8). The distance of this width of nip zone from the contact point between the rolls is designated as dimension *S*. It is recommended that rolls held less than ⅜ in. (9 mm) apart be considered as rolls in contact.

Figure 6–5 shows an inrunning-roll nip point where a feed table is used. To design a properly placed barrier guard, use the following procedure:

1. Draw a full-scale outline of the nip zone with the top surface of the feed table accurately shown. Indicate the clearance line on the top roll. If more than a ⅜ in. (9 mm) clearance is required, locate the top edge of the guard in accordance with the layout for a safe opening shown in Figure 6–9 for layout on a curved surface.

Figure 6–10. Safeguards should be made of sturdy materials. (Courtesy International Stamping Co./Midas International Corp.)

2. Determine the distance S from the center line of the rolls to a point where a ⅜ in. (9 mm) space exists between the top of the feed stock and the surface of the upper roll.
3. At this distance, begin the layout of the safe opening dimensions (Figure 6–7), up to the opening necessary for the particular guard being designed. Outline the guard section (the top edge on the clearance line on the upper roll and the bottom edge at the proper point on the layout for a safe opening), and determine the necessary dimensions for installing the guard. Also determine the width of the guard.
4. Before the guard is put in operation, check carefully for travel by hands under the guard, stability of the mounting, and rigidity of the construction.

Inrunning Rolls—Stock Traveling over One Roll before Entering Nip Zone

Figure 6–9 shows an inrunning-roll nip point where the stock travels over a portion of one roll before entering the nip zone. The stock in such an arrangement is fed either under or over a barrier.

To design a properly placed barrier guard under such conditions, use the following procedure:

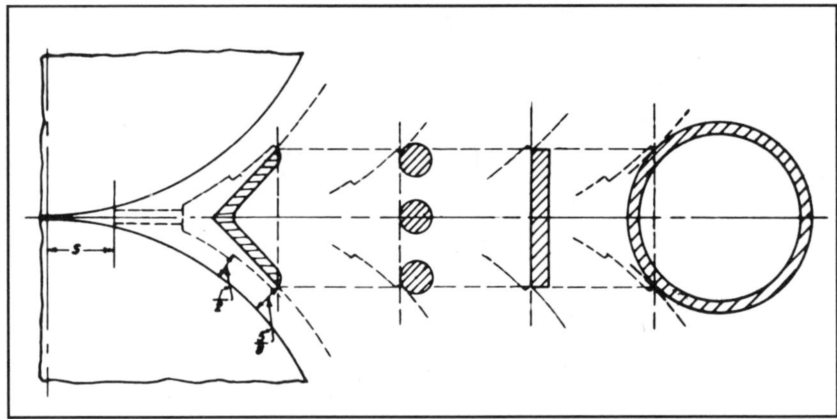

Figure 6–11. The designs for a roll nip guard can be used for feeding over or under a guard through a ⅜ in. opening.

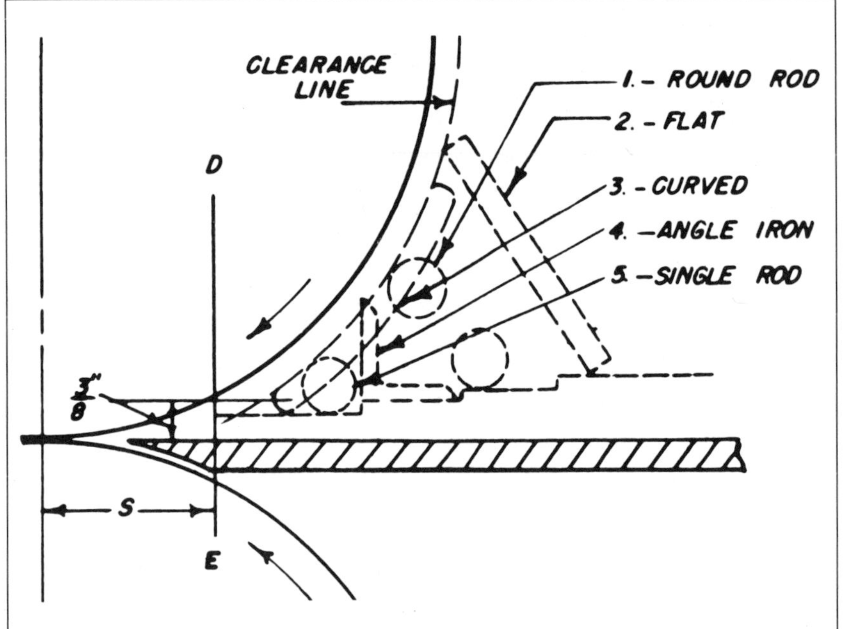

Figure 6–12. Within the outline are different designs for an effective guard on a roll nip with a feed table.

1. Draw a full-scale outline of the nip zone with the travel line of the stock indicated on the roll. Indicate the clearance line on the top roll. If more than a ⅜ in. (9 mm) clearance is required, locate the top edge of the guard in accordance with the layout for a safe opening.
2. Determine distance *S* of the center line of the roll to the point where there is a ⅜ in. (9 mm) space between the rolls.
3. At this distance, begin the layout (on the roll with the stock's travel) of the safe opening dimensions, as shown in Figure 6–7. (The layout can be made on a curved surface with ½ in. [13 mm] divider steps.) Outline the guard section (one edge touching the clearance line and the other touching the layout for a safe opening at the proper point to give the required opening). From this final layout, determine the necessary dimensions for properly locating the guard, as well as the width of the guard.
4. Before the guard is put in operation, check carefully for travel by hands under the guard, stability of mounting, and rigidity of construction.

Guard Construction

To ensure that the safe opening dimensions and effectiveness of guards are maintained, the guards should be substantially constructed and secured to minimize distortion or movement (Figure 6–10). All parts of guards should be strong enough to withstand expected stress and/or exposure. Fastenings should be secure and designed to prevent the guard from shifting and from being moved or removed. Consider any guard with openings larger than ¼ in. (6 mm) as precision construction and check it frequently for alignment and condition.

Depending on the need for visibility and rigidity, and on the method of feeding for a particular guard, the designer/engineer can use the layouts shown in Figures 6–5, 6–6, or 6–9 to select the best-suited design. Figures 6–11 and 6–12 show typical designs for guards for a roll nip zone with and without a feed table. Working from layouts similar to these, the designer/engineer can determine the location, size, and shape of the guard section. (See Chapter 20, Woodworking Machinery; Chapter 22, Metalworking Machinery; and Chapter 23, Cold Forming of Metals—all in this volume—for more specific information on guarding.)

POINT-OF-OPERATION SAFEGUARDS

Requirements for safeguarding specific machinery can be found in regulatory and related ANSI standards, such as these and the References at the end of this chapter:

- *CFR* 1910.211 through 1910.222, Subpart O—Machinery and Machine Guarding
- *Safety Requirements for the Construction, Care, and Use of Drilling, Milling, and Boring Machines*, ANSI B11.8–1983(R1994)
- *Safety Requirements for the Construction, Care, and Use of Grinding Machines*, ANSI B11.9–1975(R1997)
- *Safety Requirements for Construction, Care, and Use of Mechanical Power Presses*, ANSI B11.1–1988(R1994)
- *Safety Requirements for the Construction, Care, and Use of Metal Sawing Machines*, ANSI B11.10–1990(1997)
- *Safety Requirements for the Construction, Care, and Use of Power Press Brakes*, ANSI B11.3–1982 (R1994)

The characteristics of a proper guard include:

- integrated as a part of the machine
- well constructed, durable, and strong
- able to accommodate workpiece feeding and ejection
- protective
- easy to inspect and maintain
- relatively tamper-proof or foolproof

On the other hand, a guard should not:

- create another hazard
- interfere with production
- cause work discomfort

Types of Safeguards

A great many standardized devices as well as improvised barriers, enclosures, and tools, have been developed to protect machine operators, particularly their hands, at point-of-operation areas.

Sheet metal, perforated metal, expanded metal, heavy wire mesh, or stock may be used for various types of guards. The best practice when selecting material for new guards or barriers is to follow the requirements of the standards.

If moving parts must be visible, transparent impact plastic or safety glass can be used if the strength of metal is not required. Guards or barriers may be made of aluminum or other soft metals if resistance to rust is essential. Plastic and glass fiber barriers are usually less expensive than metal barriers, but they are lower in strength. However, these nonmetal barriers resist the effects of splashes, vapors, and fumes from corrosive substances that would react with metal.

Built-in Safeguards

Built-in machine safeguards are designed and installed by the manufacturer to be an integral part of a machine. This gives them several advantages over safeguards made by the machine user:

- Built-in safeguards conform more closely to the contours of the machine, thus making them superior in appearance and placement.
- Built-in safeguards eliminate hazards completely and permanently while withstanding daily wear and handling. Most makeshift safeguards, by contrast, offer little protection against mechanical or human failure. Yet they may give operators a false sense of security and thus may be more harmful than no safeguards at all. To make up for their inadequacy, operators must be constantly alert, which can make work unnecessarily stressful and fatiguing.
- Built-in safeguards tend to cost less than safeguards installed after a machine has been purchased, because the cost is usually spread over a large number of machines. Aside from cost savings, of course, the hazards presented by an unguarded machine make a powerful argument in favor of specifying safeguards from the manufacturer at the time a machine is ordered.

NOTE: Modifying a machine may invalidate the manufacturer's warranties. Therefore, review modification plans in cooperation with the manufacturer.

Barrier Guards

Barrier guards prevent access to the dangerous parts of the machine. They may be adjusted to accommodate different sets of tools or kinds of work. Once adjusted, barrier guards should remain fixed; they should not be moved or detached. Figure 6–7 shows the safe distance between a barrier guard and the point of operation, along with the permitted sizes of openings in a fixed guard.

Interlocking Barrier Guards

Interlocking barrier guards can be mechanical, electrical, pneumatic, or a combination of these types. The interlocking guard prevents operation of the control that sets the machine in motion until the guard is moved into a predetermined position. This prevents the operator from reaching into the point of operation.

When the guard is open and dangerous parts are exposed, the starting mechanism is locked. A locking pin or another safety device prevents the main shaft from turning or the other mechanisms from operating. Once the machine is in motion, the guard cannot be opened until the machine has come to rest or has reached a fixed position in its travel. Where neither a fixed guard nor an interlocking guard is practical, use mechanical interlocks.

An effective interlocking guard must satisfy three requirements; it must

- guard the hazardous area before the machine can be operated
- stay closed until the rotating equipment is at rest
- prevent operation of the machine if the interlocking device fails

Automatic Safeguarding Devices

An automatic barrier guard can be used, subject to the limitations outlined in Table 6–A, where neither a fixed guard nor an interlocking guard is practical. Such a guard must prevent the operator from coming in contact with the dangerous part of a machine while it is in motion, or it must be able to stop the machine in case of danger. Figure 6–13 shows the relation between safeguarding devices and fixed barrier guards. (See Chapter 23, Cold Forming of Metals, in this volume.)

All of the foregoing types of guards are suitable for protection at points of operation. Safeguarding devices can also be used in lieu of or in conjunction with barrier guards at the point of operation. Presence-sensing devices (either photoelectronic or radio frequency), pullbacks, restraints, and two-hand controls can be used under specified conditions and constraints. (See Chapter 23, Cold Forming of Metals, in this volume for further discussion of these devices.)

Substitution as a Safeguard

Like the installation of safeguards, substituting one type of machine for another can sometimes eliminate or reduce machine hazards. For example, substituting direct-drive machines or individual motors for an overhead line-shaft transmission decreases the hazards inherent in transmission equipment. Speed reducers can replace multicone pulleys. Remote-controlled automatic lubrication can eliminate the need for employees to get close to moving parts.

Matching Machine or Equipment to Operator

Safe operation of machinery involves more than eliminating or covering hazardous moving parts. The overall injury potential of the machine operation must be considered. Ask these basic questions:

- Is there a materials-handling hazard?
- Are the limitations of a person's manual effort—lifting, pushing, and pulling—recognized?
- Is the design of existing or proposed safeguards based on physiological factors and human dimensions?

Evaluate all physical or design features of a production machine and the workplace as though the machine were an extension of a person's body and could only do what that person wants it to do. To match the machine or equipment to the operator, consider the following factors:

The Workplace

Provide and arrange machines and equipment so that the operator does a minimum amount of strenuous lifting and traveling. Use conveyors, skids, jacks, or other equipment to feed raw stock. Consider using chutes or gravity feeds to remove finished stock.

The Work Height

Ensure that workstations are the optimal height for stand-up or sit-down methods of operation. For the sit-down method, determine the proper height and type of chair or stool to be used. Elbow height is the determining factor in minimizing worker fatigue. In general, an effective work level is 41 in. (1 m) from floor to work surface, with a chair height from 25 to 31 in. (0.6 to 0.8 m).

Table 6–A. Point-of-Operation Protection

Type of Safeguarding Method	*Action of Safeguard*	*Advantages*	*Limitations*	*Typical Machines on Which Used*
		ENCLOSURES OR BARRIER GUARDS		
Complete, simple fixed enclosure	Barrier or enclosure which admits the stock but which will not admit hands into danger zone because of feed opening size, remote location, or unusual shape.	Provides complete enclosure if kept in place. Both hands free. Generally permits increased production. Easy to install. Ideal for blanking on power presses. Can be used with automatic or semiautomatic feeds.	Limited to specific operations. May require special tools to remove jammed stock. May interfere with visibility.	Bread slicers Embossing presses Meat grinders Metal squaring shears Nip points of inrunning rubber, paper, textile rolls Paper corner cutters Power presses
Warning enclosures (usually adjustable to stock being fed)	Barrier or enclosure admits the operator's hand but warns him before danger zone is reached.	Makes "hard to guard" machines safer. Generally does not interfere with production. Easy to install. Admits varying sizes of stock.	Hands may enter danger zone—protection not complete at all times. Danger of operator not using guard. Often requires frequent adjustment and careful maintenance.	Band saws Circular saws Cloth cutters Dough brakes Ice crushers Jointers Leather strippers Rock crushers Wood shapers
Barrier with electric contact or mechanical stop activating mechanical or electric brake	Barrier quickly stops machine or prevents application of injurious pressure when any part of operator's body contacts it or approaches danger zone.	Makes "hard to guard" machines safer. Does not interfere with production.	Requires careful adjustment and maintenance. Possibility of minor injury before guard operates. Operator can make guard inoperative.	Calenders Dough brakes Flat roll ironers Paper box corner stayers Paper box enders Power presses Rubber mills
Enclosure with electrical or mechanical interlock	Enclosure or barrier shuts off or disengages power and prevents starting of machine when guard is open; prevents opening of the guard while machine is under power or coasting. (Interlocks should not prevent manual operation or "inching" by remote control.)	Does not interfere with production. Hands are free; operation of guard is automatic. Provides complete and positive enclosure.	Requires careful adjustment and maintenance. Operator may be able to make guard inoperative. Does not protect in event of mechanical repeat.	Dough brakes and mixers Foundry tumblers Laundry extractors, driers, and tumblers Power presses Tanning drums Textile pickers, cards
		AUTOMATIC OR SEMIAUTOMATIC FEED		
Nonmanual or partly manual loading of feed mechanism, with point of operation enclosed	Stock fed by chutes, hoppers, conveyors, movable dies, dial feed, rolls, etc. Enclosure will not admit any part of body.	Generally increases production Operator cannot place hands in danger zone.	Excessive installation cost for short run. Requires skilled maintenance. Not adaptable to variations in stock.	Baking and candy machines Circular saws Power presses Textile pickers Wood planers Wood shapers
		HAND REMOVAL/RESTRAINT DEVICES		
Hand restraints (hold-back)	A fixed bar and cord or strap with hand attachments which, when worn and adjusted, do not permit an operator to reach into the point of operation.	Operator cannot place hands in danger zone. Permits maximum hand feeding; can be used on higher-speed machines. No obstruction to feeding a variety of stock. Easy to install.	Requires frequent inspection, maintenance, and adjustment to each operator. Limits movement of operator. May obstruct space around operator. Does not permit blanking from hand-fed strip.	Embossing presses Power presses Power press brakes

Table 6–A. *(Continued.)*

Type of Safeguarding Method	*Action of Safeguard*	*Advantages*	*Limitations*	*Typical Machines on Which Used*
Hand pull-backs or pull-outs	A cable-operated attachment on slide, connected to the operator's hands to pull the hands back only if they remain in the danger zone; otherwise it does not interfere with normal operation.	Acts even in event of repeat. Permits maximum hand feeding; can be used on higher speed machines. No obstruction to feeding a variety of stock. Easy to install.	Requires unusually good maintenance and adjustment to each operator. Frequent inspection necessary. Limits movement of operator. May obstruct work space around operator. Does not permit blanking from hand-fed strip stock.	Embossing presses Power presses Power press brakes
		MISCELLANEOUS		
Limited slide travel	Slide travel limited to ¼ in. or less; fingers cannot enter between pressure points.	Provides positive protection. Requires no maintenance or adjustment.	Small opening limits size of stock.	Foot power (kick) presses Power presses
Presence-sensing device	Sensing field and brake quickly stop machine or prevent its starting if the hands are in the danger zone.	Does not interfere with normal feeding or production. No obstruction on machine or around operator.	Expensive to install. Does not protect against mechanical repeat. Limited to use on machines with means to quickly stop the machine during the operating cycle.	Embossing presses Power presses Press brakes
Type A and B gate devices	Encloses danger area before machine action starts. Stays closed until hazard ceases or stops machine if opened too soon.	Interlocked with operating cycle. Allows free access to load and unload machine. Fully encloses point of operation.	Usually limited to machines on which the part or material being processed is fully within the point-of-operation area.	Power presses Plastic injection-molding machines Compression-molding machines Die-casting machines
Special tools or handles on dies	Long-handled tongs, vacuum lifters, or hand die holders which avoid need for operator's putting hand in the danger zone.	Inexpensive and adaptable to different types of stock. Sometimes increases protection of other guards.	Operator must keep hands out of danger zone. Requires unusually good employee training and close supervision.	Dough brakes Leather die cutters Power presses Forging hammers
Special jigs or feeding devices	Hand-operated feeding devices of metal or wood which keep the operator's hands at a safe distance from the danger zone.	May speed production as well as safeguard machines. Generally economical for long jobs.	Machine itself not guarded; safe operation depends upon correct use of device. Requires good employee training, close supervision. Suitable for limited types of work.	Circular saws Dough brakes Jointers Meat grinders Paper cutters Power presses Drill presses
		TWO-HAND TRIP		
Electric	Simultaneous pressure of two hands on switch buttons in series actuates machine.			

Table 6–A. *(Concluded.)*

Type of Safeguarding Method	*Action of Safeguard*	*Advantages*	*Limitations*	*Typical Machines on Which Used*
Mechanical	Simultaneous pressure of two hands on air control valves, mechanical levers, controls interlocked with foot control, or the removal of solid blocks or stops permits normal operation of machine.	Can be adapted to multiple operation. Operator's hands away from danger zone. No obstruction to hand feeding. Does not require adjustment. Can be equipped with continuous pressure remote controls to permit "inching." Generally easy to install.	Operator may try to reach into danger zone after tripping machine. Does not protect against mechanical repeat unless blocks or stops are used. Not generally suitable for blanking operations. Must be designed to prevent tying down of one button or control which would thereby permit unsafe one-hand operation.	Dough mixers Embossing presses Paper cutters Pressing machines Power presses Power press brakes
		TWO-HAND CONTROL		
Electric	Simultaneous pressure on two-hand switches held down until dies close.	Can be adapted for multiple operators. Operators' hands away from hazard during die closing portion of stroke. No obstruction to hand feeding.	Hand buttons must be spaced far enough from point of operation to stop machine upon removal before hand can reach into the hazard. Control circuit must be designed to prevent tying down of one button. Buttons must be spaced far enough apart to prevent operation with one hand and another part of the body.	Dough mixers Embossing presses Paper cutters Pressing machines Power presses Press brakes

Controls

The position and design of machine controls, such as dials, push buttons, and levers, are important. Speed and ON-OFF controls, in particular, should be readily accessible. Controls should be standardized on similar machines so that operators can shift from one machine to another, as necessary, without having to use different controls. (See Chapter 16, Ergonomics Program, in the *Administration & Programs* volume.)

Material Handling Aids

Provide these aids to minimize manual handling of raw materials and in-process or finished parts, both to and from machines. Aids include overhead chain hoists, belt or roller conveyors, and work positioners. (See Chapter 23, Cold Forming of Metals, Figure 23–2, for work holding and positioning tools.)

Operator Fatigue

Workers can experience fatigue at a machine station, usually as the result of combined physical and mental activities, not simply from expending energy. Other contributors to fatigue include excessive speed-up, boredom from monotonous operations, and awkward work motion or operator position.

GUARDING POWER TRANSMISSIONS

All mechanical action or motion is hazardous, but to varying degrees. Rotating members, reciprocating arms, moving belts, meshing gears, cutting teeth, and parts in impact or shear are some examples of the

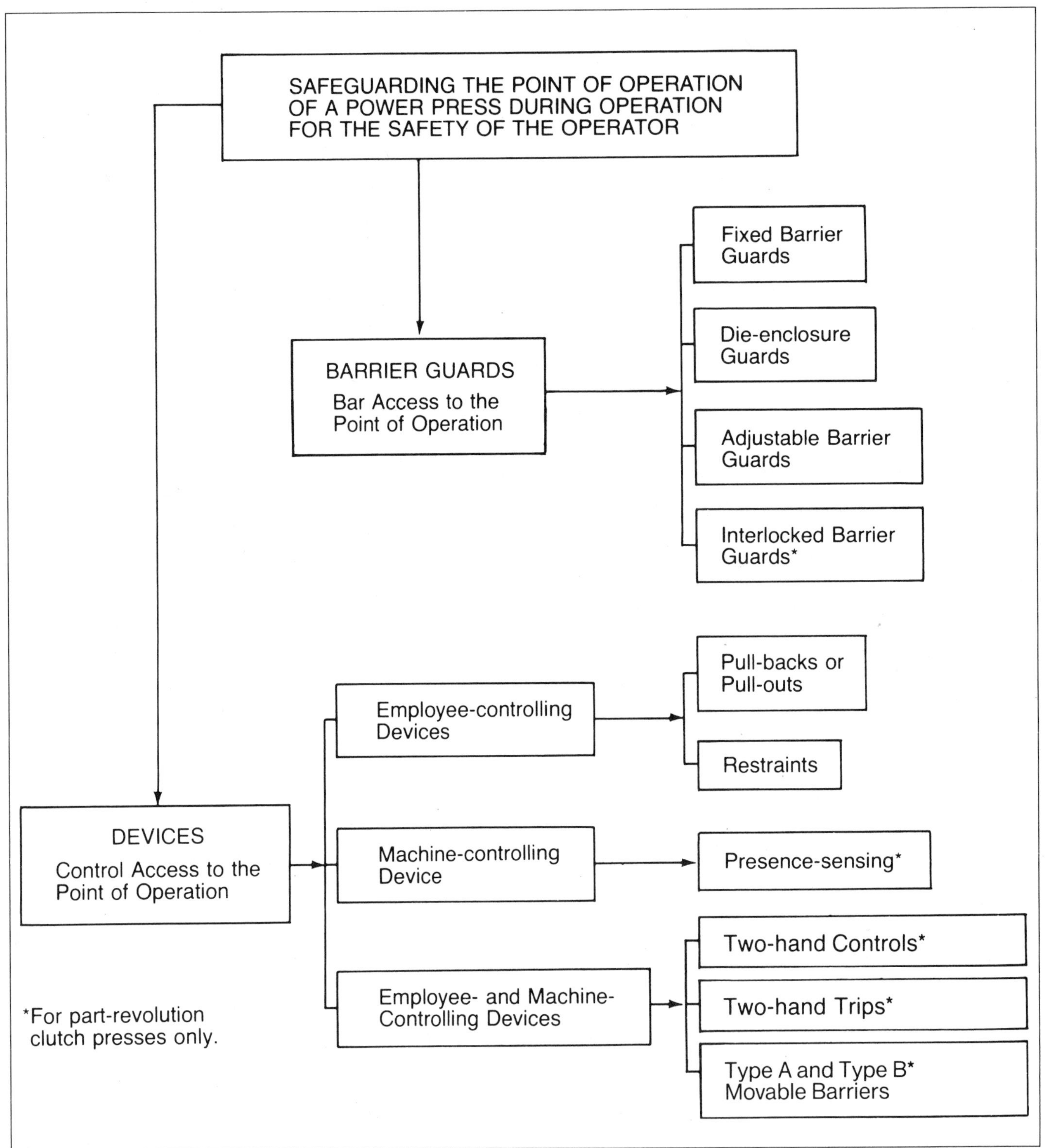

Figure 6–13. Automatic guards can protect the employee at the machine's point of operation.

types of action and motion requiring protection. They are not peculiar to any one machine or industry, but are basic to the mechanical devices used for productive purposes.

Actions or motions involving the most hazardous exposures may be classified as follows:

- rotating, reciprocating, and transverse motions

- inrunning nip points
- cutting actions
- punching, shearing, and bending actions

Methods of Guarding Actions and Motions

Whenever hazardous machine actions or motions are used, it is essential to provide protection for operators and other workers. There may be several ways to guard a situation.

Certain guarding methods may be preferable to others, but the type of operation, the size or shape of stock, the method of handling, the physical layout, the type of material, and production requirements or limitations may present important considerations. A certain flexibility in operations may also determine the feasibility of the method to be used. As a general rule, power-transmission apparatus can be protected by fixed enclosure guards.

In general, the same principles used in point-of-operation guards apply to power-transmission guards. However, openings for loading and unloading materials do not have to be considered. Power-transmission guards tend to be fixed guards that are hinged, sliding, or bolted cover plates. Do not remove them, except for maintenance, service, or adjustment and only when the equipment is properly locked out.

In general, power-transmission guards must cover all moving parts so that no part of the operator's body can come in contact with them. In many cases, a simple flat plate or box that covers the opening is all that is necessary, especially if the parts are flush with or recessed within the frame of the machine. Where parts protrude beyond the frame, it may be necessary to build a guard that conforms to the dimensions and forms of the parts being guarded (Figure 6–14). In such cases, any openings that permit shafts or other parts to pass into the machine must follow the requirements of the maximum size of permitted openings as related to the distance from the moving part (Table 6–B).

Figure 6–14. These expanded metal covers safeguard protruding moving parts and power transmissions.

Rotating, Reciprocating, and Transverse Motions

Rotating, reciprocating, and transverse motions create hazards in two general areas—at the point of operation where work is being done, and at the points where power or motion is being transmitted from one part of a mechanical linkage to another. This section will deal primarily with situations where power is being transmitted.

Any rotating object is dangerous. Even smooth, slowly rotating shafts can grip clothing or hair, and through mere skin contact force an arm or hand into a dangerous position. Incidents due to contact with rotating objects are infrequent, but the severity of injury is always great.

Common rotating mechanisms that pose hazards include collars, couplings, cams, clutches, flywheels, shaft ends, spindles, rotating bar stock, lead screws, and horizontal or vertical shafting. The danger increases when bolts, oil cups, nicks, abrasions, and projecting

Table 6–B. Standard Materials and Dimensions for Machinery Guards

	SIZE OF FILLER MATERIALS			
Material 1 in. = 2.5 cm; 1 ft = 0.305 m	*Clearance from Moving Part at All Points (inches)*	*Largest Mesh or Opening Allowable B (inches)*	*Minimum Gage (U.S. Standard) or Thickness*	*Min. Height of Guard from Floor or Platform Level (ft, in.)*
Woven Wire	Under 2	⅜	No. 16⅜ in.	8-0*
	2-4	½	No. 16½	8-0
	4-15	2	No. 12-2	8-0
Expanded Metal	Under 4	½	No. 18½ in.	8-0
	4-15	2	No. 13-2	8-0
Perforated Metal	Under 4	½	No. 20½ in.	8-0
	4-15	2	No. 14-2	8-0
Sheet Metal	Under 4		No. 22	8-0
	4-15	. . .	No. 22	8-0
Wood or Metal Strips Crossed	Under 4	⅜	¾ in. wood or No. 16 metal	8-0
	4-15	2		
Wood or Metal Strips Not Crossed	Under 4	½ the width		
	4-15	One width		
Plywood, Plastic or Equivalent	Under 4		¼ in.	
	4-15	. . .	¼ in.	8-0
Standing Railing	Min. 15			
	Max. 20	. . .	. . .	3-6

*Guards for rotating protuding objects should extend to a minimum height of 9 ft from the floor or platform.

keys or screw threads are exposed while the mechanism is rotating.

The rotating mechanism commonly occupies a stationary case or shell and consists of a revolving cylinder, a screw, agitator blades, or paddles. Examples of this type of rotating mechanism include washing machines, extractors, raw material mixers, and screw conveyors (Figure 6–15).

Reciprocating and transverse motions are hazardous because, in the back-and-forth or straight-line action, a worker may be struck or caught in a pinch or shear point between fixed or moving objects.

Inrunning Nip Points

Inrunning nip points pose a special danger existing only through the action of rotating objects. Whenever machine parts rotate toward each other, or when one rotates toward a stationary object, an inrunning nip point is formed. Objects or parts of the body may be drawn into this nip point and be bruised or crushed.

Nip point hazards include the inrunning side of rolling mills and calenders; of rolls used for bending, printing, corrugating, embossing, or feeding and conveying stock; or of a chain and sprocket, belt and pulley, gear rack, gear and pinion, or belt conveyor terminal (Figure 6–16).

Cutting Actions

Cutting action results when rotating, reciprocating, or transverse motion is imparted to a tool so that material being removed takes the form of chips. The danger of cutting action exists at the movable cutting edge of the machine as it approaches or makes contact with the material being cut. Such action occurs at the point of operation in cutting wood, metal, or other materials.

Mechanisms involving cutting action include band and circular saws, milling machines, planing or shaping machines, turning machines, boring or drilling machines, and grinding machines (Figure 6–17).

Text continues on page 164.

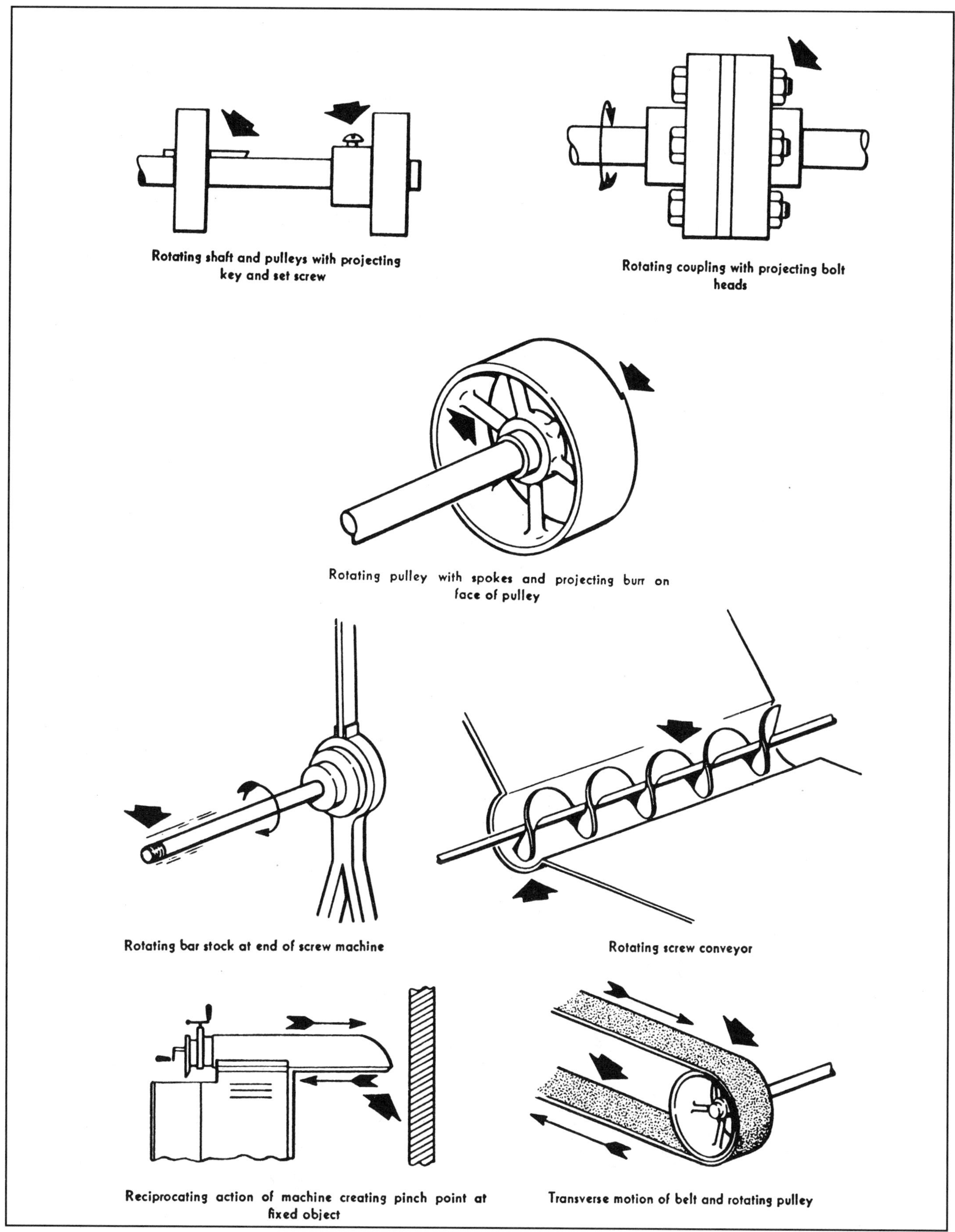

Figure 6–15. Examples of typical rotating, reciprocating, and transversing mechanisms.

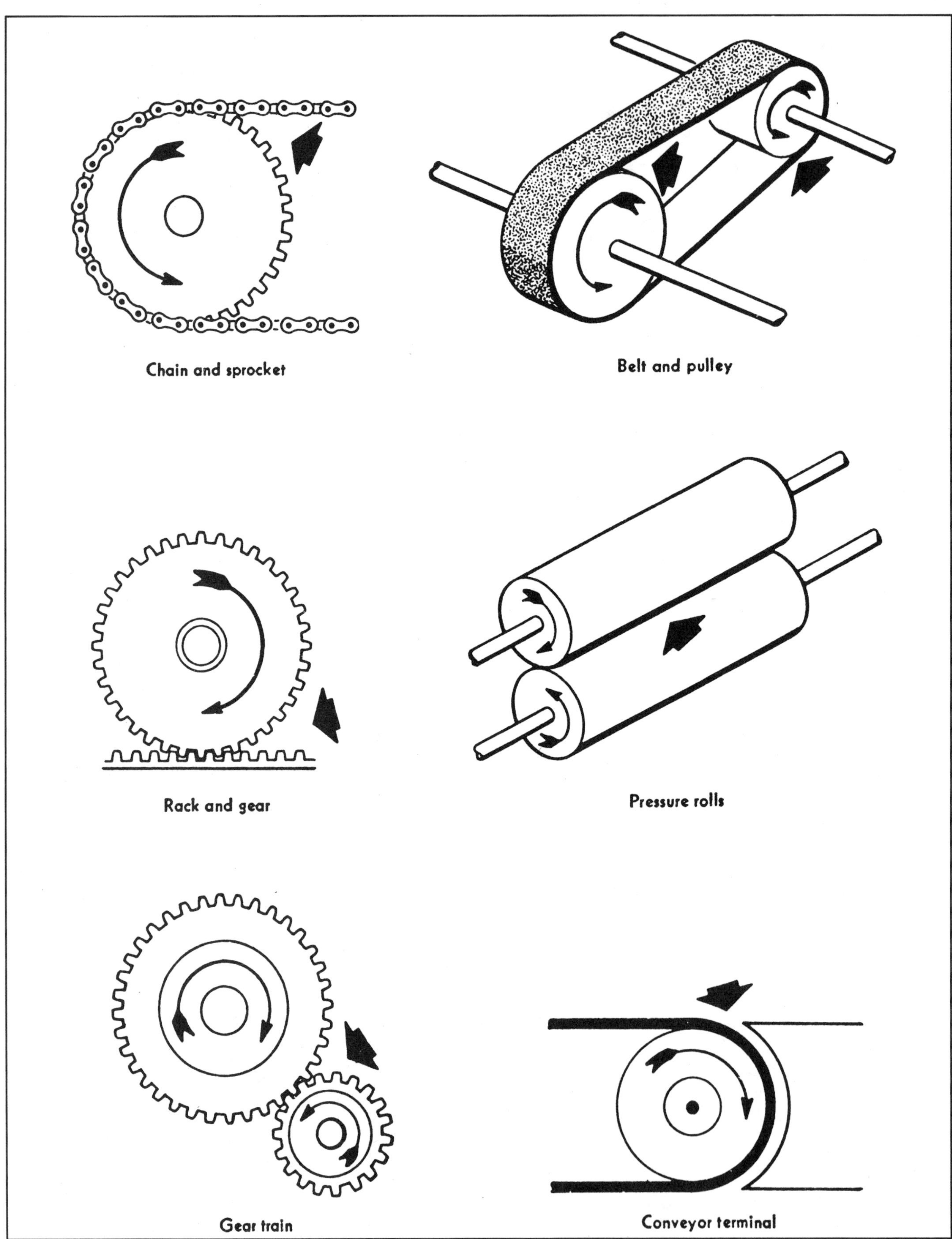

Figure 6–16. Examples of inrunning nip points.

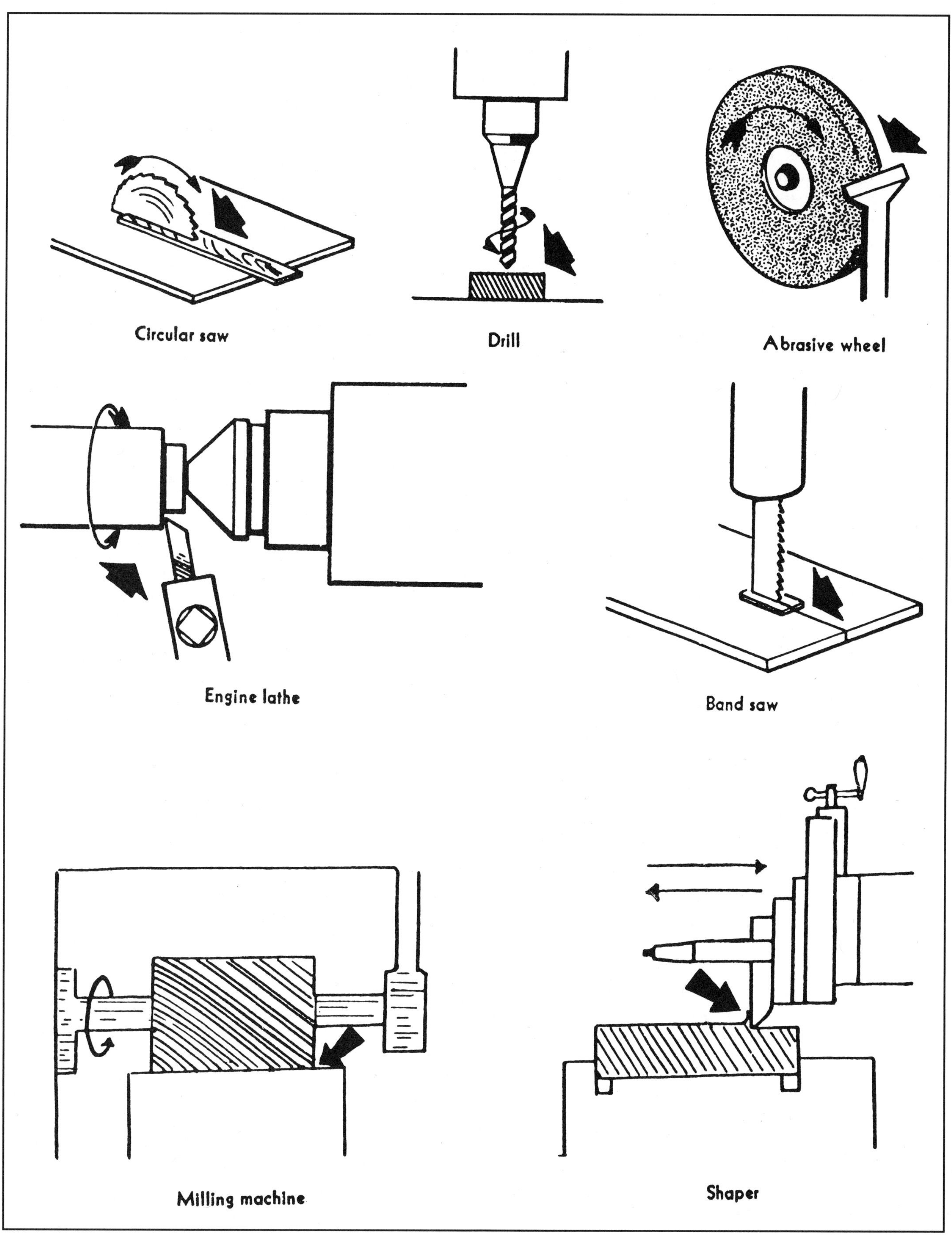

Figure 6–17. Examples of cutting actions.

Punching, Shearing, and Bending Actions

Punching, shearing, or bending action results when power is applied to a ram (plunger) or knife for the purpose of blanking, trimming, drawing, punching, shearing, or stamping metal or other materials. The danger of this type of action lies at the point of operation where stock is actually inserted, maintained, and withdrawn.

Equipment involving punching, shearing, or bending actions includes power presses, foot and hand presses, bending presses or brakes as well as squaring, guillotine, and alligator shears (Figure 6–18).

Guarding Materials

Under most circumstances, metal is the preferred material for guards. The framework of guards usually consists of structural shapes, pipe, strapping, bar, or rod stock. In general, use expanded, perforated, or solid sheet metal or wire mesh for filler material (Table 6–B). Where visibility is required, use plastic, polycarbonate or lexan.

Guards made of wood have limited application. Their lack of durability and strength, relatively high maintenance cost, and flammability are objectionable. Wood guards, especially when they become oil-soaked, can be ignited by nearby welding operations, overheated bearings, rubbing belts and defective wiring, and other sources of heat. Wood is also subject to splintering that can contaminate products or cause injury.

Where resistance to rust or damage to tools and machinery is important, use guards made of aluminum, another soft metal, or plastic. Use plastic guards where inspection of the moving parts is necessary.

When a guard cannot be made to exclude lint, provide ample ventilation. Build vents, too small to admit a hand, into the bottom of larger guards to let lint or dust drop through. Larger guards should also have self-closing access doors for cleaning by brush or vacuum hose. Consider using latches interlocked with the power source to prevent operation of the machine while the door is open.

The material selected for a guard should be substantial enough to withstand (1) internal as well as external impacts of materials, (2) parts under stress, and (3) passing pedestrians or vehicles. The parts or materials a machine is processing can become lethal projectiles. A machine located near heavy traffic aisles is vulnerable to damage. If the machine cannot be relocated, a regular enclosure guard may be insufficient to protect both people and the machine from all possible hazards. Such cases require guard railings in addition to the enclosure guard to prevent the machine from being struck by materials-handling equipment and the like.

MAINTENANCE AND SERVICING

Plan for routine maintenance when designing enclosure guards. Failure to establish, enforce, and facilitate safe maintenance and servicing procedures is probably the major cause of failure to replace a guard, especially if the maintenance has to be done frequently. It becomes easier to permanently remove the guard. This condition allows, and even promotes, an additional problem—maintenance and servicing of the machine while it is in operation. These conditions are highly hazardous for all personnel concerned. Solve these problems in one or more of the following three ways:

1. Apply engineering techniques that reduce the frequency of a job or eliminate the job. If, for example, the fittings for parts needing service were relocated

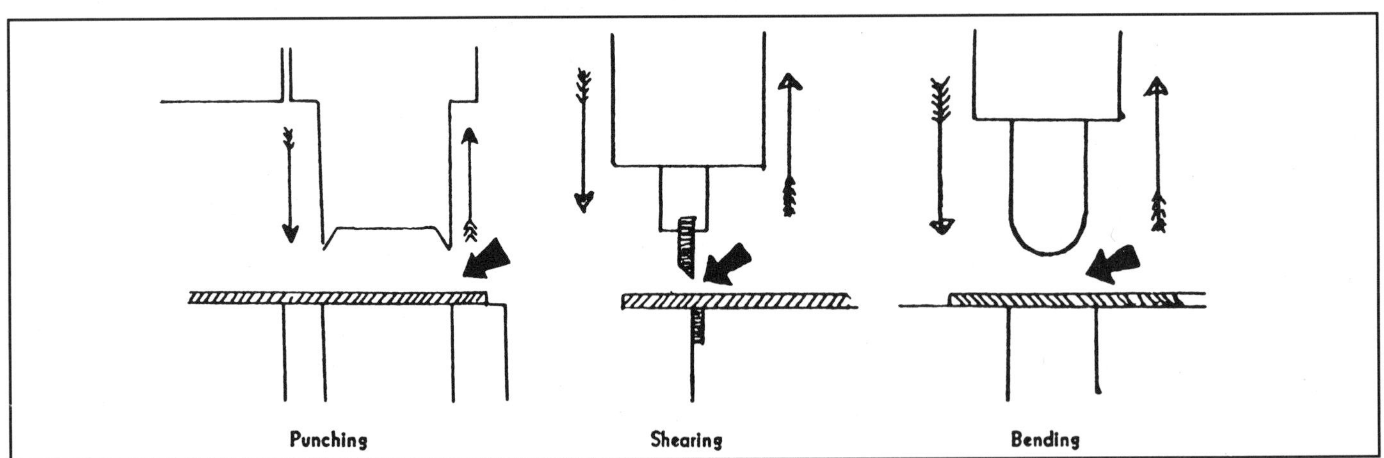

Figure 6–18. Examples of punching, shearing, and bending actions.

outside the guard, it would be unnecessary to remove the guard and would permit maintenance while the machine is operating. Alternatively, an oil or grease fitting might be lubricated by an extension through the guard. This procedure is highly recommended for operations that cannot be shut down for adjustment or maintenance.

2. Equip the machine with automatic controls for lubrication, adjustment, or service. Sophisticated equipment like this may be costly. For some machines, however, the cost is offset by savings from more reliable adjustment and maintenance procedures.

Also give thought to installing or mounting guards. The best way to mount guards securely is either by tack welding or by special bolting.

CONTROL OF HAZARDOUS ENERGY SOURCES

When maintenance and servicing are required on equipment and machines, isolate the energy sources and implement lockout/tagout procedures. The term *zero mechanical state* or *zero energy state* has often been used to describe machines with all energy sources neutralized. These terms have been incorporated into many standards. The current term indicating a machine at total rest is *energy isolation*. Types of machine energy include electrical, pneumatic, steam, hydraulic, chemical, gravity, and thermal. Energy can also take the form of the potential energy from suspended parts or springs.

NOTE: The following material is from "Control of Hazardous Energy Sources (Lockout/Tagout)," *CFR* 1910.147, and represents a minimal energy control program.

Typical Written Lockout/Tagout Procedure

Establish a procedure for controlling a hazardous energy source so that machines can be isolated from all sources of energy and potential energy.

The following minimal lockout procedure is provided to help organizations develop their own procedures. When energy-isolating devices are not lockable, tagout may be used, provided the employer complies with this procedure and provides additional training and more rigorous periodic inspections. When tagout is used and energy-isolating devices are lockable, the employer must provide full employee protection, additional training, and more rigorous periodic inspections. For more complex systems, more comprehensive procedures may need to be developed, documented, and used (Figure 6–19).

Purpose

This procedure establishes the minimum requirements for the lockout of energy-isolating devices whenever maintenance or servicing is done on machines or equipment. It shall be used to ensure that the machine or equipment is stopped, isolated from all potentially hazardous energy sources, and locked out before employees perform any servicing or maintenance where the unexpected energization or start-up of the machine, equipment, or release of stored energy could cause injury.

Compliance with This Program

All employees are required to comply with the restrictions and limitations imposed upon them during the use of lockout. Authorized employees are required to perform the lockout in accordance with this procedure. All employees, upon observing a machine or piece of equipment that is locked out to perform servicing or maintenance, shall not attempt to start, energize, or use that machine or equipment.

Sequence of Lockout

The following lockout sequence meets minimal standards for controlling a hazardous energy source.

1. Notify employees when servicing or maintenance is required on a machine or equipment. Inform them that the machine or equipment must be shut down and locked out to perform the servicing or maintenance.
2. An authorized employee shall refer to the company procedure to identify the type and magnitude of the energy that the machine or equipment utilizes, shall understand the hazards of the energy, and shall know the methods to control the energy.
3. If the machine or equipment is operating, shut it down by the normal stopping procedure (for example, depress STOP button, open switch, or close valve).
4. Deactivate the energy-isolating device(s) so that the machine or equipment is isolated from the energy source(s).
5. Lock out the energy isolating device(s) with assigned individual lock(s) (Figure 6–20).
6. Stored or residual energy (such as that in capacitors, springs, elevated machine members, rotating flywheels, hydraulic systems, and air, gas, steam, or water pressure) must be dissipated or restrained by a method such as grounding, repositioning, blocking, or bleeding down (Figure 6–21).
7. Ensure that the equipment is disconnected from the energy source(s) by first checking that no personnel

LOCKOUT : TAGOUT SYSTEM™

CAUTION: Servicing or maintenance is not permitted unless this equipment is isolated from all hazardous energy sources. This is the exclusive responsibility of designated "Authorized Employees" (**see listing at bottom) who must follow the ***complete*** "Lockout/Tagout Procedure" as published by the company. ***This sheet is limited and abbreviated; it must not be considered a substitute for the company's complete Procedure.***

LOCKOUT:TAGOUT DATA SHEET

For Equipment No. 172

Equipment Description: Blivet Sintering Machine with Taurus Automatic Feeder

HAZARDOUS ENERGY		ISOLATING DEVICES			CONTROL DEVICES (Check ✓)			
Type	Magnitude	Type	Location	I.D. No.	Lock	Tag	Both	Add'l Measures (See Below)*
Electric	440 V.	Switch	Right-Front Post	S819			✓	
Steam	170 psi	Valve	At wall 30 feet west	V47		✓		
Hydraulic	3000 psi	Valve	On pump 10 feet east	V16			✓	①

***ADDITIONAL SAFETY MEASURES (Refer To Table Above):**

① After closing and locking valve, drain hydraulic cylinder on machine 172.

****AUTHORIZED EMPLOYEES. Only the following are authorized to undertake the Lockout/Tagout Procedure on this equipment:**

Jerzi Brodofski	Andre Schoenzer		
Paul Simpson	Nancy Mahabro		
Stanley Cheng	Joe Hershko		
Walter Paley			
John Wilson			

Prepared By: George Murdock **Date:** 8/8/90 **QUESTIONS? PHONE:** 538

 Q-SIGN® V9-11

Figure 6–19. A sample of one company's documentation for a lockout/tagout procedure. Note that this is only part of a complete energy isolation program. (Courtesy Idesco Corp.)

Figure 6–20. Sample energy isolation devices with assigned individual locks. The Energy Isolation Plan tag lists all sources of hazardous energy and their isolation devices. (Courtesy Idesco Corp.)

are exposed. Then verify the isolation of the equipment by operating the push buttons or other normal operating control(s), or by testing to make certain the equipment will not operate.

CAUTION: Return operating control(s) to neutral or OFF position after verifying the isolation of the equipment.

8. The machine or equipment is now locked out.

Restoring Equipment to Service

When the servicing or maintenance is completed and the machine or equipment is ready to return to normal operating condition, the following steps shall be taken:

1. Check the machine or equipment and the immediate area around the machine or equipment to ensure that nonessential items have been removed and that the machine or equipment components are operationally intact.
2. Check the work area to ensure that all employees have been safely positioned or removed from the area.
3. Verify that the controls are in neutral.
4. Remove the lockout devices and reenergize the machine or equipment.

   NOTE: The removal of some forms of blocking may require reenergization of the machine before safe removal.
5. Notify affected employees that the servicing or maintenance is completed and the machine or equipment is ready to use.

Other Program Requirements

Setting up this program also involves meeting the following requirements:

Training

Employees taking part in this lockout/tagout procedure must be trained and retrained annually.

Lockout Equipment

Lockout and tagout devices shall be durable, marked, colored, or dyed for each facility, and supplied for the procedure. Tags must state DO NOT START, DO NOT OPERATE, or DO NOT OPEN, and must state who placed the tag, what date it was placed, and why. Tags must withstand a 50 lb pull on the attachment.

Self-Audit

Conduct an annual review of the procedure and the program.

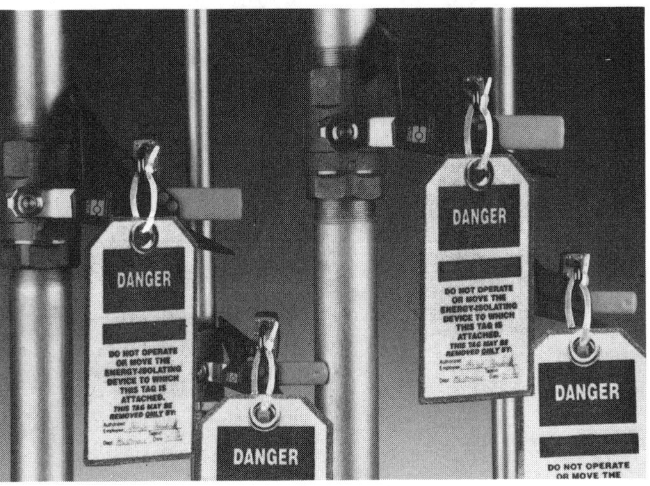

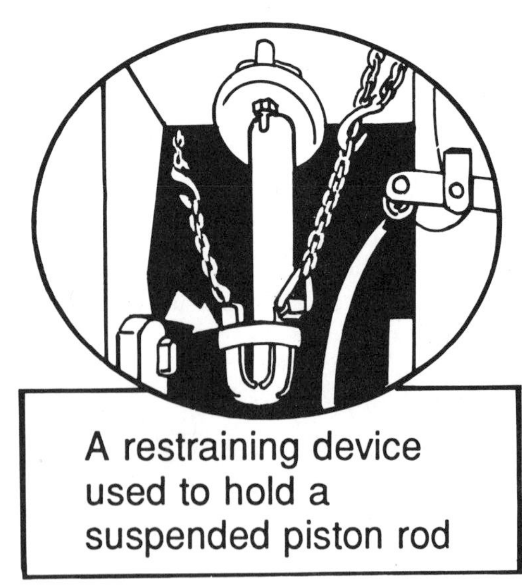

Figure 6–21. Sample devices for blocking out, blanking out, or locking out circuit breakers, pipes, or valves.

Group Lockout

If a group of employees are locking out a piece of equipment, such as an electric switch, each employee shall have an individual lock to do so.

These minimal requirements do not cover the full scope of regulatory requirements. The complete requirements are determined by a company's specific operations, machines, and equipment.

ROBOTICS SAFEGUARDING

Robots—machines specifically designed and programmed to perform certain operations—are a part of the work environment. Robots are used in a variety of applications and processes, such as spray painting, arc and spot welding, materials handling, assembly, and machine loading and unloading. They are generally defined as reprogrammable, multifunctional, mechanical manipulators and have three means of power: electromechanical, hydraulic, and pneumatic.

Hazards and Hazardous Locations

Principal hazards in using robots include the following:

- being struck by the moving parts of a robot while within the robot's operating envelope or movement zone
- entrapment between a robot's moving parts and other machinery or objects within or near the robot's movement zone
- being struck by objects or tools the robot has dropped or ejected

In order to identify where injuries can occur, the robotized workstation can be divided into two zones or volumes: the robot movement zone and the approach zone. These zones are illustrated in Figure 6–22.

If technicians stand in the robot movement zone to do a task when power is available to the robot, they are exposed to crushing, shearing, and impact injury risks. Certain regions within the movement zone, such as the region around the end effector, present increased risk.

Just outside the robot movement zone is the approach zone. The boundaries of this zone can be specified and the limits of the protected area can be known. In this zone, personnel may be exposed to thrown objects, radiation, flash, electrical hazards, or mechanical hazards of associated equipment. Furthermore, personnel in the approach zone can cross into the movement zone. Passage from the approach zone to the movement zone can be limited by the size of the openings through which personnel or working materials pass to reach the robot movement zone.

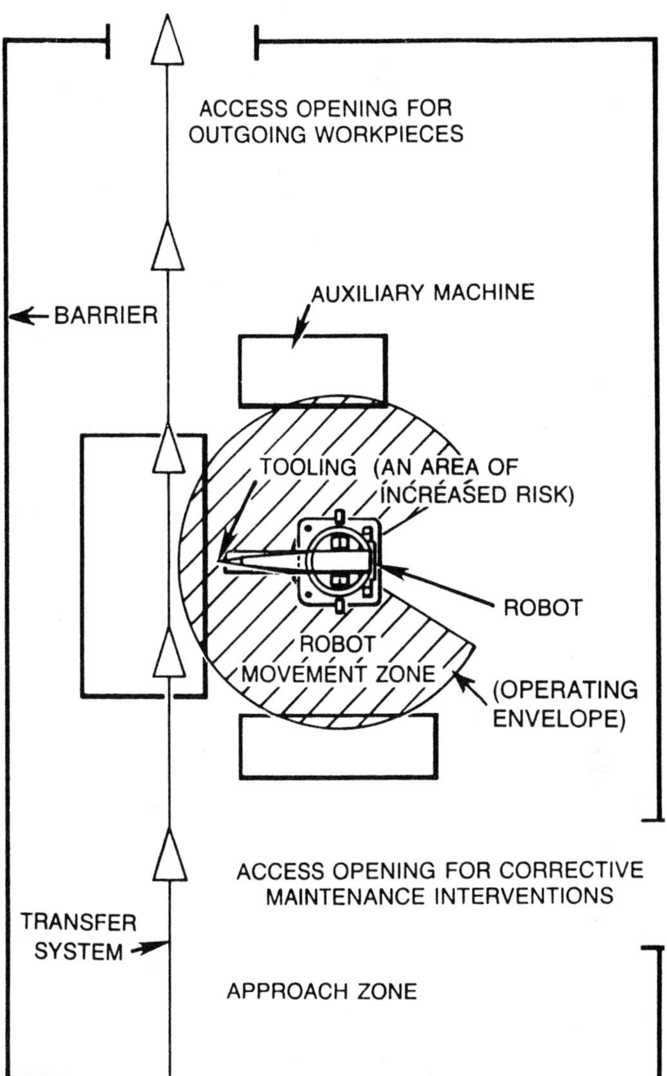

Figure 6–22. The robot movement zone (operating envelope) and the approach zone make up the danger zone of industrial robots. Note the barrier surrounding the entire operation. (Courtesy NIOSH.)

Effective design of the workstation control system minimizes the chance that the robot and associated machines could move in a way that would harm an operator who is inside the movement zone, and therefore satisfies the highest priority within the control logic. A powered robot motion is initiated by the closure of a power supply switch to an actuator, such as an electric motor or a hydraulic cylinder. This can be accomplished by any of the following:

- a planned step forward to an output condition in the control program
- a person switching the robot to automatic operation
- electromagnetic interference generating the voltage necessary for a logic switch at a microelectronic gate

- another control circuit inputting a switching signal
- a bug or an error in the control software
- a hardware failure in the switching device
- automatic restart after a power interruption

Failure to stop when commanded is also a condition that may be evaluated as a potential hazard.

If a person is present when a motion-initiating event occurs, the ensuing robot motion can lead to injury by:

- impact
- puncture
- pinch-point closure
- dragging the person over a sharp object
- pushing the person into another machine's point of operation

Personnel can be struck by:

- any part of the robot itself
- a workpiece being handled by the robot
- robot tooling

Robots manipulate many kinds of end effectors, such as grippers, welding electrodes, grinding wheels, lasers, and high pressure water jets. Problems with these end effectors are the cause of many maintenance interventions. End effectors can inflict serious cuts or burns; they may also become dangerous due to robot movement that can cause puncture wounds. A workpiece that is being handled can become a source of danger should a gripper lose its hold and allow the workpiece to fall or become a projectile.

Equipment that supplies robot power and control presents potential electrical and pressurized fluid hazards. Ruptured hydraulic lines could create dangerous high-pressure cutting streams or whipping hose hazards. A pinch point could result if control cabinets are located too close to the robot. Also, cables on the floor present tripping hazards.

Safeguarding Methods

Physically restrict a robot's movement zone to the range of motion a particular operation or installation requires. Do this by using some form of mechanical stop(s) able to withstand the force of momentum of the robot traveling at maximum speed and carrying a full load.

Install an amber light on the robot to make the robot conspicuous from all sides. This light will be on while the robot is energized, signifying that it is "live," even during periods when the robot is not moving.

"Teach" panels or pendants should contain an EMERGENCY STOP button or deadman switch allowing an operator to stop robot movement by interrupting the machine-drive power. The EMERGENCY STOP should be hard wired into the drive-power stop circuit and not be interfaced through a computer input/output register.

While a robot is in the "teach" mode of operation, limit the robot's rate of movement to 6 in. (15 cm) per second or less, as measured at the end of the robot arm when fully extended.

Program the robot so that an operator cannot place it into automatic cycle using the "teach" pendant but must instead close all interlocked gates and return to the master control panel outside the robot movement zone.

Place fixed guards, such as a 5 ft fence, around the perimeter of the robot movement zone, and design the guards to prevent inadvertent or unauthorized entry and to capture objects dropped or ejected by the robot.

Ensure that each perimeter guard allows sufficient clearance between it and the robot movement zone so that a person cannot be trapped between the two (Figure 6–22).

See that access gates in the perimeter guards are electrically interlocked to interrupt main drive power should they be opened during the automatic cycle of the robot. Program the robot so that closing the interlock gate cannot initiate the automatic cycle, which must be done only at the main control panel outside of the perimeter guards.

Place warning signs around the robot at points of access to the movement zone to make those who enter aware of the normal hazards of the robot and/or any unusual hazards that may exist. Unusual hazards can include:

- overlapping movement zones of two or more robots in close proximity
- other automation or machinery whose movement may extend into the work area

Several devices exist for sensing the presence of a person within a robot's movement zone. These devices include photoelectric cells, pressure-sensitive mats, and light or sound curtains. They are used to detect intrusion into an area where a hazard may exist.

Do not allow a person to work with a robot until he or she has received proper training in operating the robot safely as it performs a particular function. See that all persons involved in operating robots are fully aware of what happens when the control system fails, and of the safety features provided, prior to giving them hands-on experience.

Anchor a robot according to the manufacturer's recommended specifications. Locate the control panel for the robot so that an operator can easily see the robot while programming or operating the master controls.

Provide locking disconnects for all sources of energy to the robot, including electricity, air, and hydraulic power.

Provide a means to release stored energy before servicing the robot. This energy may take the form of air and hydraulic accumulators, springs, counterweights, flywheels, and the load held by the robot.

Shield all solid-state electronic devices associated with controlling the robot from possible radio-frequency interference that could cause loss of control. It is important for the robot's computer to recognize frequency interference, thus preventing unexpected movement and avoiding potential injuries. (See Chapter 25, Automated Lines, Systems, or Processes, for more information on robotics.)

GUARDS AND NOISE CONTROL

A well-designed, well-made, and well-mounted guard can be very effective in reducing unwanted noise. Noise travels primarily by conduction and vibration through the air. Sound barriers can effectively stop or lessen further activity either by absorbing, reflecting, or confining the sound waves. Since guards are usually positioned where a machine's noise originates, at either the point of operation or the point of power transmission, they can be designed as a barrier against noise as well as against personal injury. Each situation calls for inspection of the noise source and consideration of the surrounding environment. Professional assistance may be required.

SUMMARY

- Safeguarding involves identifying all hazards and potential injuries and then creating safeguards to protect employees from the hazards.
- Management should ensure that employees clearly understand safeguarding terms.
- Effective point-of-operation protection devices cannot always be installed by manufacturers. In such cases, users must design and install the devices according to the regulatory safeguarding requirements for each operation. Companies must construct point-of-operation safeguards and devices carefully and inspect them frequently.
- Guards used to protect workers from hazardous points on machines include barrier, interlocking barrier, and automatic safeguarding devices. Only fixed guards should be used for power-transmission parts. Safeguarding devices can be used instead of or in conjunction with barrier guards at points of operation.
- Power-transmission guards must cover all moving parts to prevent an operator's body from coming in contact with them. The material should withstand internal and external blows, parts under stress, or passing pedestrians and vehicles.
- During maintenance and servicing of equipment or machines, prevent unexpected, injury-causing movement by using energy isolation through lockout/tagout. The employer's program must provide full employee protection, training, and rigorous periodic inspections.
- Robotic workstations require specific barrier guards surrounding the entire operation and established safe procedures for operations in the robot movement zone and approach zone.
- Guards can help to reduce noise hazards by either absorbing or reflecting sound.

REFERENCES

Alliance of American Insurers, 3025 Highland Parkway, Suite 800, Downers Grove, IL 60515.

Safe Openings for Some Point of Operation Guards, Technical Guide 02–678.

American National Standards Institute, 11 West 42nd Street, New York, NY 10036.

Blown Film Take-Off and Auxiliary Equipment—Construction, Care, and Use, ANSI B151.4–1999.

Construction, Care, and Use of Extrusion Blow Molding Machines, ANSI B151.15-1985.

Film Casting Machines—Construction, Care, and Use, ANSI B151.2-1999.

Pneumatic Conveying Systems for Handling Combustible Materials, ANSI/NFPA 650–1998.

Safety Requirements for Bakery Equipment, ANSI Z50.1–1994.

Safety Requirements for Baling Equipment, ANSI Z245.5–1997.

Safety Requirements for the Cleaning and Finishing of Castings, ANSI Z241.3–1999.

Safety Requirements for Construction, Care, and Use of Coil Slitting Machines/Systems, ANSI B11.14–1996.

Safety Requirements for the Construction, Care, and Use of Cold Headers and Cold Formers, ANSI B11.7–1995(R2000).

Safety Requirements for the Construction, Care, and Use of Permanent Mold Casting Machines, ANSI B152.2–1982.

Safety Requirements for the Construction, Care, and Use of Horizontal Injection Molding Machines, ANSI/Society of Plastics Industry (SPI) B151.1–1997.

Safety Requirements for Construction, Care, and Use of Packaging and Packaging-Related Converting Machinery, ANSI/PMMI B155.1–2000.

Safety Requirements for the Construction, Care, and Use of Rivet Setting Equipment, ANSI B154.1–1995.

Safety Requirements for the Construction, Care, and Use of Drilling, Milling, and Boring Machines, ANSI B11.8–1983(R1994).

Safety Requirements for Workplace Floor and Wall Openings, Stairs and Railing Systems, ANSI A1264.1–1995.

Safety Requirements for the Construction, Care, and Use of Gear Cutting Machines, ANSI B11.11–1994.

Safety Requirements for the Construction, Care, and Use of Grinding Machines, ANSI B11.9–1975(R1997).

Safety Requirements for the Construction, Care, and Use of Horizontal Extrusion Presses, ANSI B11.17–1996.

Safety Requirements for Construction, Care, and Use of Hydraulic Power Presses, ANSI B11.2–1995(R2000).

Safety Requirements for Industrial Robots and Robot Systems, ANSI/Robot Industry of America (RIA) R15.06–1999.

Safety Requirements for the Construction, Care, and Use of Iron Workers, ANSI B11.5–1988(R1994).

Safety Requirements for the Construction, Care, and Use of Lathes, ANSI B11.6–1984(R 1994).

Safety Requirements for the Construction, Care, and Use of Machinery and Machine Systems for the Processing of Strip, Sheet, and Plate from Coiled Configuration, ANSI B11.18–1997.

Safety Requirements for Construction, Care, and Use of Mechanical Power Presses, ANSI B11.1–1988(R1994).

Safety Requirements for Melting and Pouring in the Metal Casting Industries, ANSI Z241.2–1999.

Safety Requirements for Construction, Care, and Use of Metal Sawing Machines, ANSI B11.10–1990(R1997).

Safety Requirements for Construction, Care, and Use of Pipe, Tube, and Shape Bending Machines, ANSI B11.15–1984(R1994).

Safety Requirements for the Construction, Care, and Use of Power Press Brakes, ANSI B11.3–1982(R1994).

Safety Requirements for the Construction, Care, and Use of Roll Forming and Roll Bending Machines, ANSI B11.12–1996.

Safety Requirements for Sand Preparation, Molding, and Coremaking in the Sand Foundry Industry, ANSI Z241.1–1999.

Safety Requirements for the Construction, Care, and Use of Shears, ANSI B11.4–1993.

Safety Requirements for the Construction, Care, and Use of Single- and Multiple-Spindle Automatic Bar and Chucking Machines, ANSI B11.13–1992(R1998).

The Use, Care, and Protection of Abrasive Wheels, ANSI B7.1–2000.

Screen Changers—Construction, Care, and Use, ANSI B151.3–1982(R1988).

Slit Tape and Monofilament Post-Extrusion Equipment—Construction, Care, and Use, ANSI B151.6–1982(R1988).

Mitchum BG. *Concepts and Techniques of Machine Safeguarding*. OSHA publication 3067. Washington, DC: U.S. Dept. of Labor, U.S. Government Printing Office.

National Fire Protection Association, 1 Batterymarch Park, Quincy, MA 02269.

Electrical Standard for Industrial Machinery, NFPA 79, 1997.

National Electrical Code®, NFPA 70, 1996.

National Safety Council, 1121 Spring Lake Drive, Itasca, IL 60143.

Power Press Safety Manual, 4th ed., 1989.

Safeguarding Concepts Illustrated, 6th ed., 1993.

Royal Society for the Prevention of Accidents, London. *Industrial Accident Prevention Bulletin*, vol. 11, No. 118.

U.S. Department of Health and Human Services, National Institute for Occupational Safety and Health, Publications Dissemination, DTS, 4676 Columbia Parkway, Cincinnati, OH 45226.

Machine Guarding—Assessment of Need, HSM 99 73 71.

Occupational Safety and Health in Vocational Education, Pub. No. 79–125.

Safe Maintenance Guide for Robotic Workstation. No. 88–108–1988.

U.S. Department of Labor, Occupational Safety and Health Administration, 200 Constitution Avenue NW, Washington, DC 20210.

Code of Federal Regulations, Title 29. Sections 1910.147, *Control of Hazardous Energy Sources (Lockout/Tagout)*, and 1910.211 through 1910.222, Subpart O–Machinery and Machine Guarding.

Principles and Techniques of Mechanical Guarding, Bulletin 2057. Washington, DC: OSHA, 1972.

REVIEW QUESTIONS

1. Define safeguarding.

2. List the six characteristics of a proper guard.
a.
b.
c.
d.
e.
f.

3. Which of the following is defined as a hazardous area created by two or more mechanical parts rotating in opposite directions within the same plane and in close interaction?
a. Pinch point
b. Nip points or bites
c. Point of operation
d. Power transmission

4. Name the four general types of safeguards.
a.
b.
c.
d.

5. Briefly explain the three advantages of built-in machine safeguards that are designed and installed by the manufacturer over safeguards made by the machine user.
a.
b.
c.

6. A point of operation is defined as:
a. All mechanical parts, such as gears, shafts, pulleys, belts, clutches, brakes, and rods, that transmit energy and motion from a source to a machine.
b. A mechanism or control designed for safeguarding.
c. The area on a machine where material is positioned for processing and where work is actually being performed on the material.
d. A hazardous area created by the cutting movement of a mechanical part past a stationary point on a machine.

7. An effective interlocking barrier guard must satisfy what three requirements?
   a.
   b.
   c.
8. Which of the following is the preferred material used for guards?
   a. Plastic
   b. Metal
   c. Shatter-resistant glass
   d. Wood
9. What are the benefits of nonmetal barriers?
10. What five factors should be considered when matching a machine or equipment to an operator?
    a.
    b.
    c.
    d.
    e.
11. List the five steps to take after a machine or equipment has been locked out for repair and is ready to return to normal operating condition.
    a.
    b.
    c.
    d.
    e.
12. Which of the following is a principal hazard when using robots?
    a. Being struck by the moving parts of a robot
    b. Entrapment between a robot's moving parts and other machinery or objects
    c. Being struck by objects or tools the robot has dropped or ejected
    d. All of the above
    e. Only a and b
13. Name three devices that exist for sensing the presence of a person within a robot's movement.
    a.
    b.
    c.
14. List three items that are required on a lockout tag.
    a.
    b.
    c.

7

Personal Protective Equipment

Revised by

Philip E. Hagan, MPH, CIH, CHMM, CHCM
Ken Mushet, MBA/TM, REM
J. Nigel Ellis, Ph.D., P.E., CSP

Methods of controlling potentially harmful exposures to hazardous substances or forms of energy found in the workplace environment typically are classified into three broad, occasionally overlapping categories: engineering controls, administrative controls, and personal protective equipment (PPE). Engineering controls are passive measures designed into the work environment to prevent contact with a harmful substance or other hazard. Common examples of engineering controls are eliminating toxic materials or using less toxic substitutes; changing process design; using barriers or guards; isolating or enclosing hazards; and using local exhaust ventilation. Administrative controls include such measures as worker rotation to minimize exposure, implementing proper housekeeping practices, and devising appropriate worker training.

This chapter covers the following topics:

- developing a program to introduce PPE in a company
- types of protective head, face and eye, and hearing protection and how to use them
- types of fall arrest systems and their use and care
- major respiratory protection equipment, including care and maintenance
- major types of footwear and hand and arm protective gear
- protective clothing for special work situations

Personal protective equipment refers to the use of respirators, special clothing, safety glasses, hard hats, or similar devices whose proper use reduces the risk of personal injury or illness resulting from occupational hazards. Generally speaking, use of PPE is the least desirable method of controlling exposure to harmful substances in the workplace environment.

Properly implemented engineering and administrative controls can greatly reduce or eliminate the hazard at the source. In contrast, when PPE is the primary control measure, the hazard is still present in the environment. The particular protective device merely provides a barrier between the hazard and the worker. Improper use or failure of the device means the worker is exposed to a direct threat to health and safety.

In some instances, PPE may be the only recourse. For example, with many physical hazards such as welding flash and sparks, dark goggles are used because no other feasible vision protection controls, short of automating the operation, are effective. On a construction site, for example, hard hats and safety shoes are necessary to protect workers exposed to hazards from falling objects or from objects that could crush a foot. However, against such risks as chemical hazards, good industrial hygiene practice dictates that PPE be used only when the more desirable engineering and administrative controls are not feasible, when PPE is an interim control method while the "higher" controls are being implemented as a supplement, or as added protection.

Clearly, management must design a safe working environment by evaluating all hazards in the work environment, assessing the need for controls, and controlling or eliminating hazards to protect workers. Such a policy means considering the worst-case analysis of conditions. For those work environment hazards that cannot be eliminated through engineering or administrative controls, PPE becomes the best protection method. It is important for management to take a strong, positive attitude toward the proper use of PPE.

A PROGRAM TO INTRODUCE PPE

Companies should conduct an assessment of hazards in the workplace, as recommended by the Occupational Safety and Health Administration (OSHA), to determine their needs for PPE to protect workers. Once management decides on the use of PPE, the following steps should be taken:

1. Write a policy on usage of the PPE and communicate it to employees and visitors as needed.
2. Select the proper type of equipment.
3. Implement a thorough training program to make certain employees know the correct use and maintenance of their equipment.
4. Enforce the use of PPE.

Policy

A written PPE program should include a policy, hazard assessment or PPE needs assessment, selection of PPE to be used, worker training and motivation in the use of PPE, and enforcement of company rules. The policy should clearly state the need for and use of PPE. It also may contain exceptions or limitations on use of PPE. Some policies or safety rules may include such details as the specific work conditions expected. Management staff must follow the same safety rules.

The following is an example of one firm's policy on wearing of PPE devices:

> For safe use of any personal protective device, it is essential the user be properly instructed in its selection, use, and maintenance. Both supervisors and workers shall be so instructed by competent persons.

Selection of Proper Equipment

After the need for PPE has been established, the next step is to select the proper type. The most important criterion is the degree of protection that a particular piece of equipment affords under various conditions.

Except for respiratory protective devices, few items of PPE available commercially are tested according to published and generally accepted performance specifications and approved by an impartial examiner. Although satisfactory performance specifications exist for certain types of PPE (notably protective helmets, devices to protect the eyes from impact and from harmful radiation, and rubber insulating gloves), there are no approving laboratories to test equipment regularly according to these specifications. (See the latest National Institute for Occupational Safety and Health (NIOSH) Certified Equipment List, which can be obtained from the U.S. Superintendent of Documents, Washington, DC.)

The Safety Equipment Institute (SEI) has formulated objective policies for third-party certification of safety equipment. SEI voluntary certification programs involve both product testing and an ongoing program of quality assurance audits. Participating manufacturers are required to submit a specific number of product models to undergo demanding performance tests in SEI-authorized independent laboratories. When the laboratory has completed the test, SEI receives a pass or fail notification. For the quality assurance program, SEI conducts an audit at a manufacturer's production facilities to ensure that products coming off the assembly line are made to the same exacting specifications as the product model actually tested for certification.

The Safety Equipment Institute's existing certification programs include (1) eye and face protection, such as goggles, face shields, spectacles, and welding helmets; (2) emergency eyewash and shower equipment; (3) firefighter's helmets; and (4) protective headwear, such as helmets. (The latest edition of the list of SEI Certified Products may be obtained by writing the Safety Equipment Institute, 1901 North Moore, Suite 501, Arlington, VA 22209.)

Proper Training

The next step is to obtain worker compliance with company requirements to wear the PPE. Several factors influence compliance, including (1) how well workers understand the need for the equipment; (2) how easy, comfortable, and convenient the equipment is to wear; (3) how effectively economic, social, and disciplinary sanctions can be used to influence the attitudes of workers; and (4) employee involvement in the decision-making process.

In organizations where workers are accustomed to wearing PPE as a condition of employment, compliance may be only a minor problem. People are issued equipment meeting the requirements of the job and are taught how and why it must be used. Thereafter, periodic checks are made until use of the issued equipment has become a matter of habit.

However, when a group of workers are issued PPE for the first time or when new devices are introduced, compliance may be a more difficult problem. The safety and health professional or management must give workers a clear and reasonable explanation as to why the equipment must be worn. If employees are required to change their traditional work procedures, they may put up resistance, whether justifiable or not. Also, management cannot ignore the fact that workers may be reluctant to use the equipment because of bravado or vanity. Having supervisors first try out new protective equipment and devices before actual adoption and asking for their feedback usually makes them more willing to persuade workers to use the equipment.

A good deal of the resistance to change can be overcome if the persons who are going to use the PPE are allowed to choose their equipment from among several styles preselected to meet job requirements. In some situations, it may be advisable to have a committee from the workforce help to select suitable devices. Management may not be able to purchase one standard type of equipment right away and may need to stock several types. In the latter case, the cost, though higher than the cost of stocking only one style, will be small compared to the potential expense of injuries resulting from failure to use the equipment. For the convenience of their employees, some companies maintain equipment stores on the facility premises.

Employees required to use PPE should be given proper training. The training program routinely should cover the following topics:

- describing what hazard and/or condition is in the work environment
- telling what has/can be/cannot be done about it
- explaining why a certain type of PPE has been selected
- discussing the capabilities and/or limitations of the PPE
- demonstrating how to use, adjust, or fit the PPE
- practicing PPE use
- explaining company policy and its enforcement
- discussing how to deal with emergencies
- discussing how PPE will be paid for, maintained, repaired, cleaned, and so on

Use and Maintenance

All equipment must be inspected before and after each use. The company should keep records of all inspections by date, with the results tabulated. Supervisors and workers should follow the recommendations of the manufacturer for inspection, maintenance, repair, and replacement of parts supplied by the manufacturer.

Enforcement

Employees need to know how the use of PPE will be enforced. Many companies have some type of progressive disciplinary action ranging from unpaid time off to termination. Management enforcement of the PPE program is critical to success.

Recognition Clubs

Several organizations sponsor recognition awards for those who avoided or minimized injury by wearing PPE. These organizations include the Wise Owl Club, sponsored by the National Society to Prevent Blindness; The Golden Shoe Club; and The Turtle Club.

Who Pays for PPE?

Employers must provide and pay for PPE required by the company for the worker to do his or her job safely and in compliance with OSHA standards. Examples would include welding gloves, respirators, hard hats, specialty glasses, specialty foot protection, and face shields. General work clothes (e.g., uniforms, pants, shirts, or blouses) not intended to function as protection against a hazard are not considered to be PPE.

For those cases where equipment is personal in nature and usable by workers away from the job, the matter of payment may be decided by labor-management negotiations. Examples would include nonspecialty safety glasses, safety shoes, and cold-weather outerwear of the type worn by construction workers. On the other hand, shoes or outerwear subject to contamination by hazardous substances which cannot be safely worn offsite must be paid for by the employer.

OSHA's 29 *CFR* 1910.132 through .138 establishes the employer's obligation to provide personal protective equipment to employees as follows:

> Protective equipment, including personal protective equipment for eyes, face, head and extremities, protective clothing, respiratory devices and protective shields and barriers, shall be provided, used and maintained in a sanitary and reliable condition wherever it is necessary by reasons of hazards of processes or environment, chemical hazards, radiological hazards or mechanical irritants encountered in a manner capable of causing injury or impairment in the function of any part of the body through absorption, inhalation, or physical contact.

Although there are some circumstances where workers in a particular trade would provide their own PPE, it is still the employer's obligation to ensure that such equipment is adequate and properly maintained.

For example, the welder's helmet is almost universally supplied by management because no one could perform the job without such protection. Although work gloves sometimes are purchased by the user, welder's or other special-purpose gloves are usually considered a necessary part of the job and are issued free. In some instances, safety shoes are offered at a partial reimbursement rate to encourage workers to buy them, while in other areas, a shoemobile service is available. The vendor's vehicle is completely equipped and stocked to fit and sell safety shoes.

The next few sections discuss seven major categories of PPE—protection for the head, eyes and face, ears, respiratory, hands, feet, and trunk. Each section provides information on the standards available or proposed, some details about the equipment available, and suggestions for selecting equipment to meet the job hazard.

HEAD PROTECTION

Employees exposed to head injury hazards must be given protective headwear. It is very important that workers wear hard hats as they are intended to be worn and not put on backwards or without the safety harness. Some operations requiring head protection include tree trimming, construction work, shipbuilding, logging, mining, electric and communication line construction or maintenance, and basic metal or chemical production.

Protective Headwear

Protective headwear is designed to absorb the shock of a blow and shield the wearer's head from the impact and penetration of falling objects. In some cases, protective headwear is required to prevent electric shock and burns. Protective headwear can also prevent the head and hair from becoming entangled in machinery or exposed to hazardous environments. Safety and health professionals should be alert to any changes in operations that may create a need for protective headwear. For example, in a slack season, a firm might transfer some employees to duties that require protective headwear. Also, construction, maintenance, and odd jobs requiring head protection often occur in the normal operations of many companies.

The American National Standards Institute (ANSI) has established standards recognized by OSHA for protective hats. For protective hats purchased prior to July 5, 1994, *ANSI Requirements for Industrial Head Protection,* Z89. 1–1969, and *ANSI Requirements for Industrial Protective Helmets for Electrical Workers,* Z89.2–1971 are applicable. The standards for protective helmets purchased after July 5, 1994, are contained in *ANSI Personnel Protection—Protective Headwear for Industrial Workers-Requirements,* Z89.1–1986. The standard is intended to be used in its entirety on a product. Identification inside the helmet shell must list the manufacturer's name, ANSI standard designation (Z89.1–1986), and either class A, B, or C. (ANSI standards are designated by a year. The helmet identification should be no more than five years old—the latest designation should be observed.)

In ANSI standard Z89.1–1986, helmet is defined as "a device that is worn to provide protection for the head, or portion thereof, against impact, flying particles, electrical shock, or any combination thereof and that includes a suitable harness" (Figure 7–1). The harness is a complete assembly that helps to maintain the helmet's position on the wearer's head. Other features may be added to adapt the helmet to specific work situations (Figures 7–2 and 7–3). Protective helmets are commonly and incorrectly called safety helmets, safety hats, and hard hats. ANSI standard Z89.1–1986 describes protective hats using the following types and classes:

- Type 1—helmets with full brim, not less than 1¼ inches wide
- Type 2—brimless helmets with a peak extending forward from the crown
- Class A—general service, limited voltage protection
- Class B—utility service, high-voltage protection
- Class C—special service, no voltage protection

Class A helmets are intended for protection against impact hazards and are typically used in heavy industrial settings.

Figure 7–1. Standard protective headgear. (Courtesy E. D. Bullard Company.)

Figure 7–2. A universal adapter with chemical goggles is added to assure that workers will wear their goggles whenever safety conditions require protective headwear. (Courtesy Mine Safety Appliances Company.)

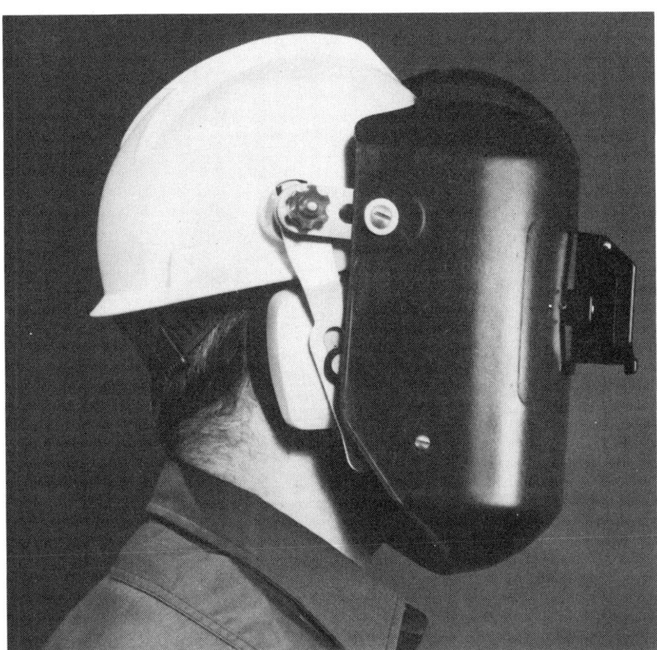

Figure 7–3. Protective welding headwear. (Courtesy Mine Safety Appliances Company.)

Class B helmets protect the wearer's head from impact, penetration by falling or flying objects and from high-voltage shock and burn. As a rule, this type of helmet is used by electrical workers.

Class C helmets are used for comfort and impact protection, usually when there is a possibility of bumping a head against a fixed object and in settings where there is no danger from electrical hazards or corrosion.

It is seldom mentioned, but the use of a chin strap adds considerably to the protection offered by a helmet. A chin strap keeps the helmet from falling off in awkward positions and keeps the helmet on during impact.

Bump Caps

A bump cap is not a helmet or hard hat. There is no standard that covers bump caps, except for each manufacturer's specification. Nonetheless, the bump cap has its place in some work environments. When the impact hazard is from bumping into stationary objects, such as low-slung pipes or catwalks, floor works, well-protected machinery or in cleaning tight spaces, and not from overhead operations, the risk of potential injury is limited by the comparatively restricted movement of the worker's head. In these cases, the bump cap is sufficient protection.

Workers who wear the bump cap and/or other protective headgear must be trained and supervised to ensure correct usage. Bump caps should never be used where ANSI–approved Class A, B, or C protective helmets are required. Because of the danger of mistaking bump caps for helmets, some organizations prohibit their use.

Hair Protection

Employees with long hair or beards who work around rotating shafts, chains, belts, or other rotating machine parts must protect their hair from contact with moving parts. Besides the danger of direct contact with the machine, which may occur when workers lean over, the hair can also be lifted into moving belts or rolls that develop heavy charges of static electricity. Because this hazard cannot be completely removed by mechanical means, workers with long hair should be required to wear protective hair coverings.

Hair nets, bandannas, and turbans are frequently unsatisfactory for hair protection because they do not cover the hair completely. Caps should cover the entire head of hair. If the wearer is exposed to sparks and hot metals, as in spot welding, the cap should be made of flame-resistant material. Some chemical facilities provide disposable flameproof caps. No standards have been accepted for caps, but they should be made of a durable fabric to withstand regular laundering and disinfecting, if they are not disposable.

To encourage its use, the cap should be as attractive as possible. It should have a simple design, be available either in a variety of head sizes or adjustable to fit all wearers, and be cool and lightweight. If dust protection is not required, the cap should be made of open weave material for better ventilation. Finally, it comes with a visor and should be worn with visor in front.

After a suitable cap has been chosen, management must enforce its use. For reasons of vanity, workers often wear the cap on the back of their heads so that part of the hair over the forehead is exposed. Sometimes this practice can be discouraged by demonstrating vividly what may happen when the hair comes in contact with a revolving spindle. Management can also use vanity in the service of safety. If workers can be shown that caps preserve hair from the effects of dust, oils, and other shop conditions, employees may be more willing to wear protective hair covering.

Maintenance

Before each use, helmets should be inspected for cracks, no matter how small, signs of impact or rough treatment, and wear that might reduce the degree of safety originally provided. Prolonged exposure to ultraviolet (UV) radiation from sunlight or other sources like welding and chemicals can shorten the life expectancy of thermoplastic helmets. Discard all helmets that show signs of chalking, cracking, or reduced surface gloss or those with broken or damaged harnesses.

Protective helmets should not be stored or carried on the rear window shelf (cradle) of a vehicle because sunlight and extreme heat may reduce its degree of protection. Also, in case of an emergency stop or collision, the helmet could become a hazardous missile inside the truck or car.

At least every 30 days, protective helmets (in particular their sweatbands and cradles) should be washed in warm, soapy water or in a suitable detergent solution recommended by the manufacturer, then rinsed thoroughly (Figure 7–4).

Before reissuing used helmets to other employees, make sure the helmets are scrubbed and disinfected. Solutions and powders are available that combine both cleaning and disinfecting. Helmets should be thoroughly rinsed with clean water and completely dried. Keep the wash solution and rinse water temperature at approximately 140 F (60 C). Do not use steam, except on aluminum helmets.

Removal of tar, paint, oil, and other materials may require the use of a solvent. Because some solvents can damage the shell, the supervisor should ask the helmet manufacturer what solvent can be used safely on the material.

Supervisors and workers should pay particular attention to the condition of the helmet's suspension webbing because it helps to absorb the shock of a blow. They should look for loose or torn cradle straps, broken sewing lines, loose rivets, defective lugs, and other defects. Sweatbands are easily replaced. Disposable helmet liners made of plastic or paper are available for hats used by many people (such as visitors). The company should stock an adequate number of crowns, sweatbands, and cradles as replacement parts. Some companies replace the complete suspension webbing at least once a year.

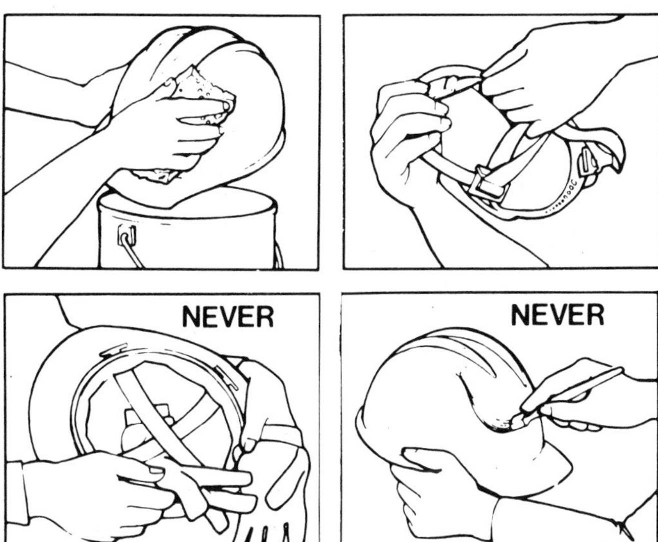

Figure 7–4. Maintenance of protective headwear. Shell should be cleaned regularly both for safety and appearance. Dirt or stains may hide hairline cracks, a reason to replace the helmet. Regular inspection of suspension system is important. Wearers should look closely for cracking, tearing, or fraying of suspension materials. Never carry anything inside the helmet. A clearance must be maintained inside the helmet for the protection system to work. Never use paint on a helmet that could affect the protective nature. Paint contains solvents which can make the shell brittle. Reflective tape is recommended for numbers or symbols.

Color-Coded Protective Helmets

Many companies use color-coded protective helmets to identify different working crews. Some colors are painted on during manufacture, and others have the color molded in. It is not recommended that paint be applied after manufacture because paint solvents may reduce the helmet's dielectric properties or affect the shell. Alterations of any sort can affect the performance of the gear. However, if painting is necessary, manufacturers should be consulted with regard to the type of paint that would be compatible with the construction of the protective helmet. Lighter colored hats are cooler to wear in the sun or under infrared energy sources.

EYE AND FACE PROTECTION

Protection of the eyes and face from injury by physical and chemical agents or by radiation is vital in any occupational health and safety program. In fact, this type of protection has the widest application and the broadest range of styles, models, and types.

The cost of acquiring and fitting eye protective devices is small when measured against the expense of eye injuries. For example, the purchase and fitting of a pair of impact-resistant spectacles may cost about $10; compensation payment for eye injury can exceed $3,600, according to the National Society to Prevent Blindness. The cost of a first-aid eye treatment may exceed $350. Some 70% of all eye injuries resulted from flying or falling objects. Contact with harmful substances, chemicals, etc., caused over 20% of injuries. Foreign bodies in the eye occurred in about 60% of the cases.

The eye and face protection standard, ANSI Z87.1-1989, *American National Standard Practice for Occupational and Educational Eye and Face Protection,* sets fairly comprehensive standards to be used for protective eye and face devices purchased after July 5, 1994.

Eye and face protective devices purchased before July 5, 1994, shall comply with the ANSI Z87.1-1968, *USA Standard for Occupational and Educational Eye and Face Protection.* In lieu of complying specifically with either ANSI standard, the employer could demonstrate that alternative protective equipment would be equally effective. These standards set performance standards, including detailed tests, for a broad range of hazards—excluding only x-ray, gamma, and high-energy particulate radiation, lasers, and masers.

Besides general requirements applying to "all occupations and educational processes," the standard provides requirements and limitations on the following:

- welding helmets
- welding hand shields
- protector selection and fitting
- flammability
- face shields
- goggles—eye cup (chipper's), dust and splash, welder's and cutter's
- spectacles—metal, plastic, and combination
- attachments and auxiliary equipment—lift fronts, chin rests, snoods, aprons, magnifiers, etc.

Selection of Eyewear

Factors that should be considered in the selection of impact-resistant eyewear include (1) level of protection afforded, (2) comfort with which they can be worn, and (3) ease of repair. Styles now available resemble more attractive, regular eyewear. Flexible glasses are preferred by many because of their light weight and convenience although they generally do not last as long as the sturdier frame and glass lens eyewear.

Proper eye protection devices should be selected and their use should be enforced to provide maximum protection for the degree of hazard involved. On certain jobs and in some locations 100% eye protection is necessary.

Face shields are not recommended by ANSI Z87.1 as basic eye protection against impact. To obtain this level of protection, face shields must be combined with basic eye protection. Face shields have their purpose, however, and are discussed later in this section. Most safety glasses should be fitted with permanent side shields.

Cup goggles should have cups large enough to protect the eye socket and to distribute any impact over a wide area of the facial bones. Cups should be flame-resistant, corrosion-resistant, and nonirritating to the skin.

If lenses are exposed to pitting from grinding wheel sparks, they should be protected by a transparent, durable coating. Welding lenses must be shielded by a cover lens of glass or plastic.

Lenses must not have appreciable distortion or prism effect. ANSI Z87.1 limits the nonparallelism between the two faces to 1/16 prism diopter (4 minutes of arc). Both the refraction in any meridian and the difference in refraction between any two meridians must be limited to 1/16 diopter. Additional information on eye protection can be found in the National Safety Council's book, *Fundamentals of Industrial Hygiene*, 4th ed.

Contact Lenses—Rumor Versus Facts

For more than 25 years a rumor has persisted that welding or other electric flashes make contact lenses stick to the eyeball. This rumor has been proved false.

Incident data and studies suggest that contact lens wearers do not appear to have problems when their eyes are properly protected in the workplace. The National Society to Prevent Blindness publishes the latest research findings as a service to both business and health and safety professionals. Their purpose is to help contact lens wearers keep their eyes in good condition. The following guidelines and recommendations for contact lens use in industry are reprinted with permission from the National Society to Prevent Blindness.

Contact lenses sometimes provide a superior means of visual rehabilitation for employees who have had a cataract removed from one or both eyes, who are highly nearsighted, or who have irregular astigmatism from corneal scars or keratoconus. Except for situations in which there exist significant risks of ocular injury, individuals may be allowed to wear contact lenses in the workplace. Generally speaking, contact lens wearers who have experienced long-term success with contacts can judge for themselves whether or not they will be able to wear contact lenses in their occupational work environment. However, contact lens wearers must conform to the prerogatives and directions of management regarding contact lens use. When the work environment entails exposures to chemicals, vapors, splashes, radiant or intense heat, molten metals, or a highly particulate atmosphere, contact lens use should be "restricted" accordingly. (Contact lens use considerations should be made on a case-by-case basis in conjunction with the guidelines of the OSHA and NIOSH.)

Recommendations

The National Society to Prevent Blindness makes the following recommendations as a service to managers who must direct contact lens use and employees who must wear them:

- A specific written management policy on contact lens use should be developed with employee consultation and involvement.
- Occupational safety eyewear meeting or exceeding ANSI Z87.1 standards should be worn at all times by individuals in designated areas.
- Employees and visitors should be advised of defined areas where contacts are allowed.
- At workstations where contacts are allowed, the type of eye protection required should be specified.
- Restrictions on contact lens wear do not apply to usual office or secretarial employees (unless they must enter hazardous areas where exposure is significant).
- A directory should be developed that lists all employees who wear contacts. This list should be maintained in the medical facility for easy access by trained first-aid personnel. Foremen or supervisors should be informed of individual employees who wear contact lenses.
- Medical and first-aid personnel should be trained in the proper procedures and equipment for removing both hard and soft contacts from conscious and unconscious workers.
- Employees should be required to keep a spare pair of contacts and/or a pair of up-to-date prescription spectacles in their possession. They will then be able to continue their job functions should they damage or lose a lens while working.
- Employees who wear contact lenses should be instructed to remove their contacts immediately if redness of the eyes, blurred vision, or pain in the eyes associated with contact lens use occurs.

Guidelines for the Use of Contact Lenses in Industrial Environments

The American Optometric Association has adopted the following policy statement concerning the use of contact lenses in industrial environments (Anthony P. Cullen, MSc, O.D., Ph.D., DSc, FCOptom, F.A.A.O.):

> Contact lenses may be worn in some hazardous environments with appropriate covering safety eyewear. Contact lenses of themselves do not provide eye protection in the industrial sense.

Most successful contact lens wearers wish to wear their contact lenses in all aspects of their lives, including the workplace. This may conflict with government- or industry-imposed restrictions on the use of contact lenses in a given industrial environment. These restrictions, in turn, may be unreasonable and discriminatory.

In risk management it is necessary to balance risk with benefits and to differentiate perceived risk from actual risk. Because both contact lens and certain environments may produce adverse ocular effects it is tempting to assume that there may be additive or synergistic effects when contact lenses are worn in that environment.

When considering the advisability of wearing contact lenses in a given industrial setting, a number of questions should be addressed:

- Is there an actual hazard?
- Does the wearing of contact lenses place the eye at greater risk than a naked eye?
- Does the removal of the contact lens increase the risk to the eye, the wearer or co-workers?
- Is the risk different for various contact lens materials and designs?
- Are there other risks to the wearer or co-workers?
- Do contact lenses decrease the efficacy of other safety strategies?

Ocular hazards are greater in some occupations than others. Those who prescribe contact lenses for industrial workers should be concerned as to the advisability of wearing the lenses in a given environment. The type of work may influence the selection of lens material and design, and wearing and replacement schedules. The following factors may be of value in making these decisions:

- toxic chemicals and/or physical agents that may be encountered
- raw material and by-products involved
- potential for ocular exposure
- protective equipment provided, available and used
- hygiene facilities available
- presence or absence of health and safety personnel

- factors that may influence compliance with cleaning and wearing schedules

An evaluation of the published material, including laboratory and human studies and well-documented case reports, indicates that contact lenses may be worn safely under a variety of environmental situations including those which, from a superficial evaluation, might appear hazardous. Indeed, some types of contact lenses may give added protection to spectacle lens and nonspectacle lens wearers in instances of certain fume exposure, chemical splash, dust, flying particles, and optical radiation. The evidence also refutes the claims that contact lenses negate the protection provided by safety equipment or make the cornea more susceptible to damage by optical radiation, in particular arc flashes. Thus, a universal ban of contact lenses in the workplace or other environments is unwarranted.

Regulations limiting the wearing of contact lenses in any given circumstance must be scientifically defensible and effectively enforceable. They should not be based on perceived hazards, random experience, isolated unverified case histories or unsubstantiated personal opinions.

Conversely, it would be imprudent for a practitioner to prescribe contact lenses in order to circumvent uncorrected visual acuity standards in those occupations where individuals may be required to function without correction on some occasions or in environments contraindicated for the type of lens prescribed.

All practitioners must stress that personal protective equipment, including safety eyewear, is not replaced by contact lenses.

Where circumstances create the necessity, eye protection must be worn.

These Guidelines were revised in May 1998 by the AOA Contact Lens Section.

Comfort and Fit

To be comfortable, eye-protective equipment must be properly fitted. Corrective spectacles should be fitted only by optometrists or ophthalmologists. An employee can be trained to adjust and maintain eye-protective equipment, however, and each employee can be taught the proper care of the device used. To give the widest possible field of vision, goggles should be fitted as close to the eyes as possible, without bringing the eyelashes in contact with the lenses.

In areas where goggles or other types of eye protection are used extensively, goggle-cleaning stations should be conveniently located. The stations should provide defogging materials and wiping tissues, along with a receptacle for discarding them. Before choosing a defogging material, test to determine the most effective type for a specific application.

Sweatbands can help to prevent eye irritation, aid visibility, and eliminate work interruptions for face mopping. Sweatbands are usually made of a soft, light, highly absorbent cellulose sponge. An elastic band holds the sweatband in place on the wearer's forehead so that it does not interfere with glasses or goggles. Evaporation from the exposed surface produces a cooling effect that increases the wearer's comfort. Wearing a sweatband should not interfere with the effectiveness of other protective equipment (respirators, protective helmets, eyewear, face shields).

Face Protection

As a general rule, face shields should be worn over suitable basic eye protection. A variety of face shields protect the face and neck from flying particles, sprays of hazardous liquids, splashes of molten metal, and hot solutions (Figure 7–5). In addition, they provide antiglare protection where required.

Three basic styles of face shields include headgear without crown protectors, with crown protectors, and with crown and chin protectors. Each of the three is available with one of these replaceable window styles:

- clear transparent
- tinted transparent
- wire screen
- combination of plastic and screen
- fiber window with a filter plate mounting

The materials used in face shields should combine mechanical strength, light weight, nonirritation to skin,

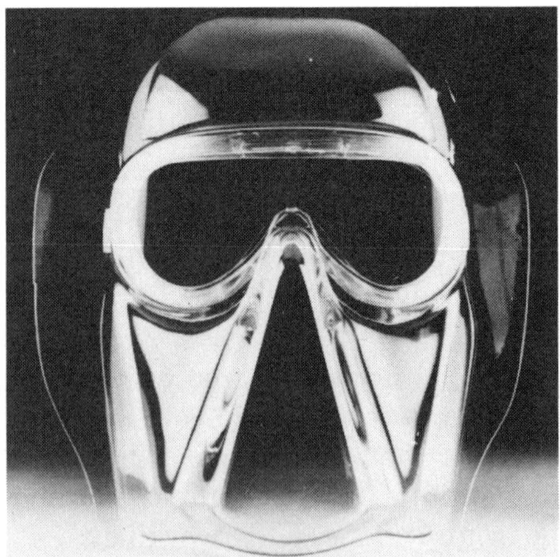

Figure 7–5. This face shield allows 160-degree peripheral vision and can be worn with prescription glasses. Its curved surfaces divert chemical splashes away from the face. (Courtesy Millennium Safety Products, Inc.)

and the ability to withstand frequent disinfecting operations. The shield should be made of noncorrosive metals and slow-burning plastics. Only optical grade (clear or tinted) plastic, which is free from flaws or distortions, should be inserted for the windows. However, plastic windows should not be used in welding operations unless they conform to the standards on transmittance of absorptive lenses, filter lenses, and plates.

On some jobs, such as pouring low-melting metals, the face shield must protect the head and face against splashes of metal. A face shield similar to an arc welder's piece but made of wire screen (which provides better ventilation than a solid shield) can be used. The plain wire will not fog under high-temperature and high-humidity conditions. A metallized plastic shield that reflects a substantial percentage of heat has been developed for jobs in which the worker is exposed to high temperatures.

Acid Hoods and Chemical Goggles

The company can provide head and face protection from splashes of acids, alkalis, or other hazardous liquids or chemicals in several ways, depending on the hazard. A hood made of chemical-resistant material with a glass or plastic window can give good protection. Some manufacturers provide a hood with replaceable inner and outer windows. In all cases, there should be a secure joint between windows and hood materials.

Although hoods are extremely hot to wear, they can be made with air lines for the wearer's comfort. If so, the wearer should have a harness or belt like that on an air-line respirator to support the hose.

If protection is necessary only from limited direct splashes, the person can wear a face shield made of a material unaffected by the liquid or a flexible-fitting chemical goggle with baffled ventilation, provided the eyes are not exposed to irritating vapor. For severe exposure potential, a face shield should be worn in connection with the flexible-fitting chemical goggles.

Face shields should be shaped to cover the whole face. They should be supported by a headband or harness, so they can be tipped back and clear the face easily. Any shield should be easily removed in case it becomes contaminated with corrosive liquid.

If goggles worn under the shield are nonventilated for protection against vapor and splashing, they should also be nonfogging. If necessary, the user can use frequent applications of antifog cleaner to avoid fogging.

Laser Beam Protection

Lasers produce monochromatic high intensity light beams, frequently capable of causing significant eye damage. A laser beam of sufficient power can theoretically produce retinal intensities at magnitudes even larger than those produced when directly viewing the sun. Exposures to this type laser beam have the potential for causing permanent blindness.

No one type of glass or plastic offers protection from all laser wavelengths. Consequently, most laser-using firms do not depend on safety glasses to protect an employee's eyes from laser burns. Some point out that laser goggles or glasses might give a false sense of security, tempting the wearer to unnecessary exposures.

Nevertheless, researchers and laser technicians frequently do need eye protection. Both spectacles and goggles are available—and glass or plastic for protection against nearly all the known laser wavelengths can be special-ordered from eyewear manufacturers. Typically, the eyewear will enjoy maximum attenuation at a specific laser wavelength, with protection falling off rather sharply at other wavelengths.

Laser protective goggles or spectacles, or an "anti-laser eyeshield," attenuate the helium-neon laser light (wavelength 6,328 angstroms [Å]) by factors of 10 (optical density [O.D.] 1), 100 (O.D. 2), 1,000 (O.D. 3), or more. An optical density of 3 or 4 still renders the beam visible in bright sunlight. Antifog-style goggles are available for use in the field, as are antifog solutions.

The American Conference of Governmental Industrial Hygienists (ACGIH) cautions that laser safety glasses or goggles should be evaluated periodically to make sure that adequate optical density is maintained at the desired laser wavelength. Laser glasses or goggles designed for protection from specific laser wavelengths must not be used with different wavelengths of laser radiation. The eyewear should clearly display the optical density values and wavelengths, which should also be marked on eyewear storage shelves.

Laser safety glasses or goggles exposed to intense energy or power density levels may lose effectiveness and should be discarded. Technical details, uses, hazards, and exposure criteria for lasers are given in *Fundamentals of Industrial Hygiene*, 4th ed., National Safety Council, Itasca, Illinois, 1996. Also see ANSI 136.1-1993, *Safe Use of Lasers*.

Eye Protection for Welding

In addition to damage from physical and chemical agents, the eyes are subject to the effects of radiant energies. Ultraviolet, visible, and infrared bands of the spectrum all can damage the eyes, and therefore require special protective measures to eliminate the hazard.

Ultraviolet radiation can produce cumulative destructive changes in the structure of the cornea and lens of the eye. Short exposures to intense UV radiation or prolonged exposures to UV radiation of low intensity will produce painful but ordinarily self-repairing corneal damage.

Radiation in the visible light band, if too intense, can cause eyestrain and headache and can destroy the tissue of the retina. Infrared radiation transmits large quantities of heat energy to the eye, causing discomfort, although the damage produced is superficial. However, extended infrared exposure has been associated with the development of cataracts.

The protective properties of filter lenses have been established by the National Bureau of Standards. The percentage transmittance of radiant energies in the three bands—UV, visible, and infrared—is established for 15 different filter lens shades (Table 7–A). Both absorptive and filter lenses are available in polycarbonate.

Welding processes (see Chapter 21, Welding and Cutting) emit radiation in three spectral bands. Depending upon the flux used and the size and temperature of the pool of melted metal, welding processes will emit more or less visible and infrared radiation; the proportion of energy emitted in the visible range increases as the temperature rises. At least one manufacturer produces an aluminized cover for the usual black welding helmet. Its purpose is to reduce infrared absorption and the resulting heat stress to the wearer.

All welding presents problems, mostly in the control of infrared and visible radiation. Heavy-gas welding and cutting operations, and arc cutting and welding exceeding 30 amperes, present additional problems in control of UV radiation. Welding helmets must be used to provide head and face protection (Figure 7–3).

Welders may choose the shade of lenses they prefer within one or two shade numbers. The most commonly used shades are No. 1.5 to No. 3.0, intended for glare from snow, ice, and reflecting surfaces and for stray flashes and reflected radiation from cutting and welding operations in the immediate vicinity (for goggles or spectacles with side shields worn under helmets in arc welding operations, particularly gas-shielded arc welding operations). Shade No. 4 is intended for the same uses as shades 1.5 to 3.0, but provides more suitable protection from greater radiation intensity.

For welding, cutting, brazing, or soldering operations, use the guide for the selection of proper shade numbers of filter lenses or windows in Chapter 21, Welding and Cutting. (Recommendations are also in ANSI Z87.1–(R1989), *Practice for Occupational and Educational Eye and Face Protection.*) To protect the filter lenses against pitting, they should be worn with a replaceable plastic or glass cover plate.

Eye protection having mild filter shade lenses or polarizing lenses and opaque side shields is adequate for protection against glare only. For conditions where hot metal may spatter and where visible glare must be reduced, management should specify a plastic face shield worn over mild filter shade spectacles with opaque side shields.

The shade of the plate in a welder's helmet can be combined with the shade of the goggle worn underneath to produce the desired total protection. This procedure has the added advantage of protecting the eyes from other welding operations or from an arc when the helmet is raised.

To protect against UV and infrared radiation and against visible glare in inspection operations, protective lenses should be installed in a hand shield or welder's

Table 7–A. Transmittances and Tolerances in Transmittance of Various Shades of Absorptive Lenses, Other Lenses, and Plates

Shade Number	Optical Density			Luminous Transmittance			Maximum Infrared Transmittance	Maximum Spectral Transmittance in the Ultraviolet and Violet			
	Maximum	Standard	Minimum	Maximum	Standard	Minimum		313 nm	334 nm	365 nm	405 nm
				Percent	Percent	Percent	Percent	Percent	Percent	Percent	Percent
1.5	0.26	0.214	0.17	67	61.5	55	25	0.2	0.8	25	65
1.7	0.36	0.300	0.26	55	50.1	43	20	0.2	0.7	20	50
2.0	0.54	0.429	0.36	43	37.3	29	15	0.2	0.5	14	35
2.5	0.75	0.643	0.54	29	22.8	18.0	12	0.2	0.3	5	15
3.0	1.07	0.857	0.75	18.0	13.9	8.50	9.0	0.2	0.2	0.5	6
4.0	1.50	1.286	1.07	8.50	5.18	3.16	5.0	0.2	0.2	0.5	1.0
5.0	1.93	1.714	1.50	3.16	1.93	1.18	2.5	0.2	0.2	0.2	0.5
6.0	2.36	2.143	1.93	1.18	0.72	0.44	1.5	0.1	0.1	0.1	0.5
7.0	2.79	2.571	2.36	0.44	0.27	0.164	1.3	0.1	0.1	0.1	0.5
8.0	3.21	3.000	2.79	0.164	0.100	0.061	1.0	0.1	0.1	0.1	0.5
9.0	3.64	3.429	3.21	0.061	0.037	0.023	0.8	0.1	0.1	0.1	0.5
10.0	4.07	3.854	3.64	0.023	0.0139	0.0085	0.6	0.1	0.1	0.1	0.5
11.0	4.50	4.286	4.07	0.0085	0.0052	0.0032	0.5	0.05	0.05	0.05	0.1
12.0	4.93	4.714	4.50	0.0032	0.0019	0.0012	0.5	0.05	0.05	0.05	0.1
13.0	5.36	5.143	4.93	0.0012	0.00072	0.00044	0.4	0.05	0.05	0.05	0.1
14.0	5.79	5.571	5.36	0.00044	0.00027	0.00016	0.3	0.05	0.05	0.05	0.1

(Reprinted with permission from ANSI standard Z87.1-1989.)

helmet. The shield should be made of a nonflammable material, which is opaque to dangerous radiation and a poor conductor of heat. A metal shield is not desirable because it becomes hot under infrared radiation.

Some tinted lenses used in special work afford no protection against infrared and UV radiation. For instance, most melters' blue glass lenses used in open-hearth furnaces and the lenses used at Bessemer converters afford no protection against either type of harmful radiation. Short exposures while using these lenses may cause no harm. However, new personnel learning these flame-reading skills should be provided with lenses that protect in these two portions of the spectrum, and all personnel should be encouraged to use them.

The chemical composition of the lens rather than its color provides the filtering effect. This factor must be considered when selecting a filtering lens.

HEARING PROTECTION

Medical professionals have long been aware of the problem of noise-induced hearing loss (NIHL) in industry. Noise, or unwanted sound, is a by-product of many industrial processes. Sound is created by pressure changes in a medium (usually air), originating from a source of vibration or turbulence. Exposure to high levels of noise can cause hearing loss. The extent of damage depends primarily on the intensity of the noise and the duration of the exposure. NIHL can be temporary or permanent. Temporary hearing loss results from short-term noise exposures while prolonged exposure to high noise levels over a period of time gradually causes permanent damage.

The American College of Occupational and Environmental Medicine's Noise and Hearing Conservation Committee has developed the following statement in response to the question "What are the distinguishing features of occupational noise-induced hearing loss?"

Occupational Noise-Induced Hearing Loss

Occupational noise-induced hearing loss, as opposed to occupational acoustic trauma, is a slowly developing hearing loss over a long period (several years) as the result of exposure to continuous or intermittent loud noise. Occupational acoustic trauma is a sudden change in hearing as a result of a single exposure to a sudden burst of sound, such as an explosive blast. The diagnosis of noise-induced hearing loss is made clinically by a physician and should include a study of the noise exposure history.

The principal characteristics of occupational noise-induced hearing loss are as follows:

- It is always sensorineural affecting hair cells in the inner ear.
- It is almost always bilateral. Audiometric patterns are usually similar bilaterally.
- It almost never produces a profound hearing loss. Usually, low-frequency limits are about 40 dB and high frequency limits about 75 dB.
- Once the exposure to noise is discontinued, there is no significant further progression of hearing loss as a result of the noise exposure.
- Previous noise-induced hearing loss does not make the ear more sensitive to future noise exposure. As the hearing threshold increases, the rate of loss decreases.
- The earliest damage to the inner ears reflects a loss at 3000, 4000, and 6000 Hz. There is always far more loss at 3000, 4000, and 6000 Hz than at 500, 1000, and 2000 Hz. The greatest loss usually occurs at 4000 Hz. The higher and lower frequencies take longer to be affected than the 3000 to 6000 Hz range.
- At stable exposure conditions, losses at 3000, 4000, and 6000 Hz will usually reach a maximal level in about 10 to 15 years.
- Continuous noise exposure over the years is more damaging than interrupted exposure to noise which permits the ear to have a rest period. (*J Occup Med* 1987,29: 981–989.)

Hearing Conservation Program

The OSHA hearing conservation standard (*CFR* 1910.95, Occupational Noise Exposure) requires a hearing conservation program for employees exposed to excessive noise. U.S. agencies and firms must develop and maintain an audiometric testing program for all employees who are exposed to noise levels in excess of 85 dB for an 8-hour time-weighted average. With the increasingly frequent use of extended hour-shifts, i.e., 10 or 12 hours, the 85 dB exposures level must be recalculated to reflect the new shift length. OSHA currently enforces a 90 dBA permissible exposure limit (Table 7–B). Exposure to 115 dBA is permitted for a maximum of 15 minutes for an 8-hour workday. No exposure above 115 dBA is permitted (29 *CFR* 1926.52).

Research demonstrates that construction workers can be exposed to noise levels of 95 to 125 dBA through daily activities (rock drilling—up to 115 dBA; chain saw—up to 125 dBA; abrasive blasting—to 112 dBA; heavy equipment operation—110 dBA; demolition—up to 117 dBA; and needle guns—up to 112 dBA.)

In fact, it is a good idea to do audiometric testing and to maintain a noise-exposure record on all employees with potential occupational exposures to noise levels of 85 dB or greater. Audiometric testing should be conducted when new employees are hired and annually thereafter. A testing program properly carried out may

Table 7–B. Permissible Noise Exposures

Duration per day, hours	Sound level dBA slow response
8	90
6	92
4	95
3	97
2	100
1½	102
1	195
½	110
¼ or less	115

Source: 29 CFR 1910.95 Table G–16.

determine whether the hearing protective devices worn by employees are actually protecting their hearing from noise damage.

The hearing conservation program may also require the use of hearing protection devices. Before requiring any employee to wear hearing protection, management should measure and evaluate the noise in the workplace. This step serves several purposes: (1) provides the physical evidence of individual exposures; (2) identifies areas where controls need to be established; (3) helps to prioritize noise-control and noise-reduction efforts, including administrative controls; (4) documents exposures in the work environment for medical-legal purposes; (5) establishes documentation for state, federal, or insurance compliance requirements; (6) provides a basis for analyzing cause-effect relationships between noise exposure and hearing status; and (7) provides insights for improving education and compliance among workers, supervisors, and managers.

When translating noise measurements into exposure estimates, remember that there is no precise safe-unsafe line of differentiation. Any unprotected encounters with steady-state or intermittent noise that exceeds about 85 dB or with impulse or impact noise that exceeds about 120 dB (peak) may overtax the auditory mechanisms of the ears. (See Gasaway, 1984, in the References.)

When noise measurement is completed, and other possible noise-control efforts are unsuccessful, then the need for hearing protection is clearly established. For explanation of noise measurement, evaluation, and control, see the National Safety Council's *Fundamentals of Industrial Hygiene*, 4th ed.

To develop an effective hearing protection program, companies need to have an accurate knowledge of the noise levels (and frequencies) that pose a hazard to workers. From the data obtained in the noise survey described above, management can select the proper hearing protection devices.

Hearing protection devices reduce (attenuate) noise levels with various degrees of success resulting in varying levels of protection at different noise frequencies. To help management choose the right devices, firms can use the U.S. Environmental Protection Act (EPA) requirement that calls for all protectors to carry a label that indicates their noise reduction rating (NRR). The number provides an estimate of a device's degree of protection and generally can be subtracted from the decibel value of noise in the workplace. This value indicates the noise level theoretically being received in the worker's ear.

However, companies should exercise some caution in applying the full NRR when using hearing protection devices to reduce occupational exposures. Because the NRR is derived under laboratory conditions, wearing conditions of the device on the job will be less than ideal and noise frequencies and sound levels will not be equal across the spectrum. When evaluating occupational noise exposure, OSHA derates the NRR by one-half for all types of hearing protection. On the other hand, NIOSH considers the performance of different types of hearing protectors and recommends subtracting from the NRR 25% for earmuffs, 50% for formable earplugs, and 70% for all other earplugs.

Types of Hearing Protection Devices

Hearing protectors in general use can be categorized as four types: enclosure (helmets), aural (ear insert), superaural (canal caps), and circumaural (earmuffs) (Figures 7–6 and 7-7).

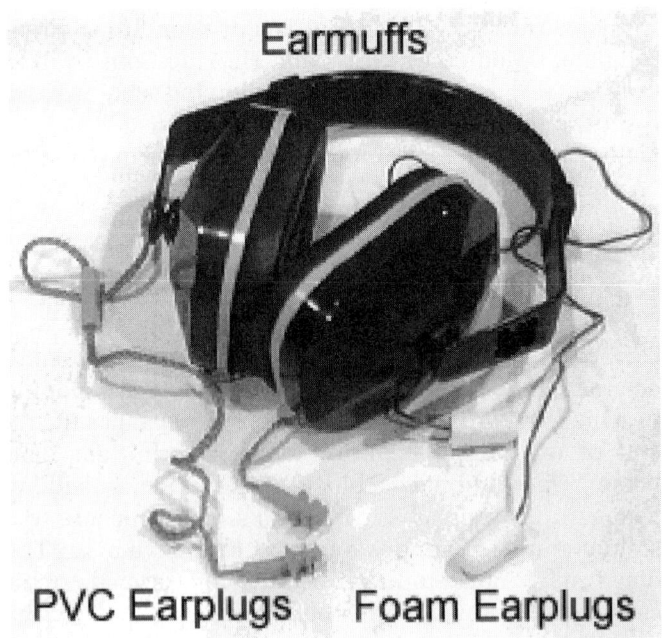

Figure 7–6. Three types of hearing protection devices.

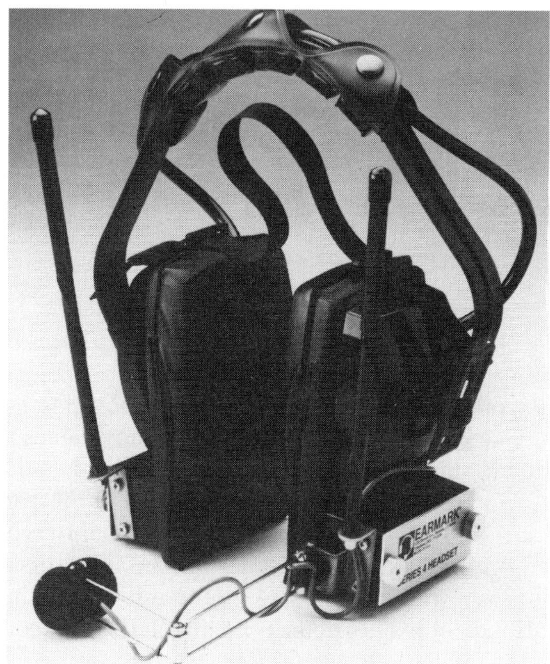

Figure 7–7. Earmuff with voice-actuated communication system. (Courtesy Earmark Inc.).

Before a company issues any ear insert, management should take certain measures: (1) each employee's ear canals should be examined for any abnormalities or irregularities; for example, certain diseases may not allow use of earplugs; (2) employees must be taught proper insertion techniques; and (3) they must be taught proper sanitation and checking techniques.

Enclosure

The enclosure hearing protector completely surrounds the head, such as an astronaut's helmet. Sound is reduced through the acoustical properties of the helmet. Additional attenuation can be achieved by wearing inserts with the enclosure helmet. Expense, temperature inside the helmet, and its bulk normally rule out general use of the enclosure hearing protector, but certain firms or industries may have specific needs for it.

Aural Insert

Commonly called inserts or earplugs, the aural insert is generally inexpensive and has a limited service life. The plug or insert falls into three broad categories: (1) formable, (2) custom molded, and (3) molded.

1. **Formable aural inserts fit all ears.** Many of the formable types are designed to be used once, then thrown away. Materials from which these disposable plugs are made include fine glass fiber, wax-impregnated cotton, and expandable plastic foam.

   Various models provide different degrees of noise reduction. Manufacturers will supply attenuation data for their products to help the safety and health professional evaluate their effectiveness for use in a given situation.
2. **Custom-molded hearing protectors.** As the name indicates, these devices are made for a specific individual. A prepared mixture is carefully placed in the person's outer ear, with a small portion extending into the ear canal; as the material sets, it conforms to the shape of the individual's ear and external ear canal. Only trained personnel should attempt the process of forming these hearing protectors.
3. **Molded (or premolded) aural inserts.** These devices are usually made from a soft silicone rubber or plastic. The most important aspect of this protector is a snug fit to provide adequate protection. Some persons may find these inserts uncomfortable because of the irregular shape of the ear.

Superaural

The superaural or canal cap hearing protector depends on sealing the external edge of the ear canal to achieve sound reduction. The caps, made of soft rubber-like material, are held in place against the edges of the ear canal by a spring band or a head suspension.

Circumaural

Cup (or earmuff) devices cover the external ear to provide an acoustical barrier between external sound and the inner ear. The attenuation provided by earmuffs varies widely due to differences in size, shape, seal material, shell mass, and type of suspension. Head size and shape also influence the attenuation characteristics of these protectors. Also, wearing other PPE such as safety helmets or safety spectacles must not compromise the efficiency of the hearing protection. Temple pieces of safety spectacles can cause noise leakage; to minimize this leakage employees can use the cable temple pieces or use aural inserts. The type of cushion used between the shell and the head also has a great influence on attenuation efficiency. Liquid or grease-filled cushions may give better noise suppression than plastic or foam rubber types but may present leakage problems.

When selecting a hearing-protection device, also consider the work area in which the employee must use it. For example, large earmuffs would not be practical for someone who works in confined areas with little headroom. For these conditions, a small or flat ear cup or insert protector would work better.

When employees must wear muff protectors in special-hazard areas (e.g., around high-voltage cables), nonconductive suspension systems may be needed in connection with muff protectors.

Another consideration when selecting a hearing-protective device is how often employees are exposed to excess noise (once a day, once a week, or infrequently). For such cases, an insert or plug device may satisfy legal requirements. If the noise exposure is relatively frequent and the employee must wear the protective device for an extended time, a muff protector might be the best choice. If noise exposures are intermittent, muff protectors are probably more desirable because it is somewhat more difficult to remove and reinsert earplugs than earmuffs.

FALL ARREST SYSTEMS

Many employees in today's workforce are tasked with duties that require work at heights above ground level. The tasks can be as simple as changing a light bulb or as difficult as painting a chimney. Both of these work situations require fall protection for the employee while the job is being done. At greater heights, as in construction or utility work, fall protection becomes mandatory under most safety regulations.

The impact from even a 4 ft (1.2 m) fall can be enough to cause serious injury. Companies can use many methods to prevent employees from falling. This section will deal only with fall protection systems and not with mobile elevated access equipment, ladders, aerial buckets, rescue equipment, and so on.

What Is Fall Arrest?

Fall arrest is defined as a means of preventing workers from experiencing disastrous falls from elevations. Fall arrest systems are usually classified aspassive or active.

Passive Fall Arrest

This system consists of components and systems, such as nets, that do not require any action on the worker's part. A properly designed passive fall arrest system, installed correctly, will protect the individual 100% of the time.

Active Fall Arrest

This system is made up of components and systems that require some manipulation by the workers to make the protection effective. These systems include harnesses, lanyards and their attachments, and component parts such as rope-grabbing devices, lifelines, self-retracting lanyards (SRLs), and so on. Active equipment will not work by itself and must be connected or employed by the individual to be protected.

When Are Fall Arrest Systems Needed?

The first factor to consider in selecting a fall arrest system is the height at which the worker will be performing the job. Some U.S. regulations provide limits, for example:

Height	General Regulation	U.S. OSHA Standard
Over 4 ft	Guardrail	1910.23
	Mid-rail	1910.23
Over 6 ft	Guardrail	1926.500
Over 25 ft	Overwater	1926.105

Second, the safety and health professional should analyze the job site and specific task to be done. If the job requires working vertically, a different or modified system will be needed than if the worker must move laterally.

Third, other factors should be addressed including rescue methods, backup systems, length of time at workstations, dry or wet conditions, number of workers needed on the job site, and environmental factors.

This complete analysis, along with a review of regulations, helps to determine the fall arrest system needed. Modern fall protection encompasses a variety of technical, medical, ergonomic, and legal issues and has become a multidisciplinary science of its own. In most cases, the introduction of a fall arrest system involves much more than simply selecting and purchasing one. Quite often, a system has to be specially designed for a particular application. In such cases the design engineer, or other competent person, has to take into account not only the individual performance of every component of the system and a proper anchor for the equipment, but also the geometry of the workplace, the environmental conditions present, and the method of post-fall rescue to be employed.

Elements of a Successful Fall Arrest Program

Some criteria need to be established for designing a fall arrest program. First, the employer must set a policy, which is clearly communicated to employees and enforced during applicable operations, that addresses these points:

- Worker qualification—Is the employee qualified to perform work at elevated conditions?
- Training—Are workers who are placed in the elevated work positions trained in the arrest system to be used?
- Selection of equipment—Is equipment being used as required to perform the job safely? Equipment purchased for the job must meet appropriate standards and if required, be certified.
- Installation of equipment—Has equipment been installed according to acceptable standards, regulations, and manufacturer's recommendations?
- Equipment maintenance and inspection—Can equipment be maintained as recommended, and

will employees inspect their personal system components daily before each use?

- Rescue procedures—Has a plan been developed to rescue any employee who has fallen while using a fall protection system?
- Job survey analysis—Has a job procedure been developed and implemented for every job in an elevated situation?

Which Fall Arrest System to Use?

Many different kinds of passive and active fall arrest systems are available. Choosing the one best suited to a particular task requires planning, forethought, and a thorough understanding of the systems on the market today. When falling hazards are identified and cannot be eliminated, management must adopt some means of control to minimize the risk of personal injury.

The analysis of elevated work tasks is intended to determine the most suitable match between required worker mobility and the capabilities of the fall arrest system. The company policy establishes what is to be done.

Next, the appropriate system and its components must be selected. A variety of equipment is available to help employers set up an effective fall arrest program. Generally, this includes nets, body support mechanisms, climbing arrest systems, vertical lifeline systems, horizontal lifeline systems, confined entry and retrieval systems, and controlled descent–emergency escape systems.

However, the proper selection and purchase of safety equipment alone does not constitute a fall arrest program. The employer also has an important responsibility for choosing and using fall equipment only for the application recommended in the literature, instructions, and on the label. Commercial equipment should never be used for applications not stated by the manufacturer.

Passive Fall Arrest Systems

Passive fall arrest systems include general all-purpose nets, personnel nets, and debris nets. These devices are easy to use and have a wide range of applications.

Nets

Properly installed, nets can be a vital part of a passive fall arrest system. Nets are designed to provide protection under and around an elevated work area where fall hazards exist. The worker is not directly involved by "wearing" fall arrest equipment; rather, the net is there to catch a falling worker before she/he hits the ground or obstructions. Two major types of nets are available, one for personnel and the other for debris. Often the two types are combined to form a dual net with twofold purpose (Figure 7–8).

Personnel Nets

Personnel nets can be used for large work crews, such as those employed on bridge construction or repair or on long-term structural projects. They also provide protection where large open areas or long leading edges expose workers to height hazards (up to 25 ft [8 m]) below the work surface by current U.S. OSHA regulations), and the use of other fall arrest equipment is deemed impractical or not feasible for the work method. The advantage of nets is that individual worker training is not required. Once installed, nets are always in place and ready for use. However, other personal fall arrest must be available during net installation and removal.

Personnel nets must be manufactured and tested in accordance with ANSI A10.11–1989, *Safety Nets Used During Construction, Repair, and Demolition Operations*, and in the U.S. *Code of Federal Regulations* (*CFR*), 1926.105 requirements. Mesh openings may not be greater than 6 in. (15 cm) by 6 in. (15 cm). Nets meeting the above requirements must bear labels displaying

Figure 7–8. These nets protect workers from falls and from falling debris. (Courtesy PearlWeave Safety Netting Corp.)

the manufacturer's name and date of manufacture, together with testing data.

Nets should be as close to the work level as possible and no lower than 25 ft (8 m) (except with bridges where U.S. OSHA considers the highest work level to be the lowest part of the bridge), and must extend outward 8 ft (2.4 m) from the structure. Nets must be tested in the field by dropping a 400-lb (181 kg) sandbag from a 25 ft (8 m) height, according to ANSI A10.11–1989, repeated at successive six-month intervals, and must not have any broken strands.

NOTE: The net may be certified to meet these requirements by a qualified person. Nets should be moved up regularly to avoid exceeding the 25 ft (8 m) limit as a building is constructed. For a personnel net to meet ANSI A10.11–1989 or OSHA regulations, the manufacturer must affix a permanent label with the following information:

- name of manufacturer
- identification of the material
- date of manufacture
- date of prototype test
- name of testing agency

Debris Nets

Debris nets are designed to catch falling debris (i.e., tools, foreign objects, falling concrete, and other construction debris) and to protect workers and pedestrians below. The strength and size of the mesh must be sufficient to catch and contain the size, weight, and impact of the objects that are likely to fall. Popular net sizes range from ¼ to ⅓ in. (6 to 8 mm) mesh. To catch large, heavy objects as well as small, light objects, the smaller mesh nets can be used in conjunction with the larger mesh and stronger personnel nets.

These net systems also can be used to catch personnel as well as debris. In these cases, personnel nets are deployed in conjunction with debris nets. The nets must be kept clear of debris to help ensure a falling worker's safety.

NOTE: A means of rescuing a fallen worker must be available.

Active Fall Arrest Systems

Active fall arrest systems include components such as fall arresters and shock absorbers, harnesses, lifelines, and SRLs. All active systems begin with an anchorage point and have some connecting components to the worker. Harnesses and components shall be used only for employee protection (as part of a personal fall arrest system or positioning device system) and not to hoist materials.

Anchor/Anchorage Points

The critical problem in all active fall protection—the anchorage point—is the position on an independent structure to which the fall arrest device or lanyard is securely attached. Supervisors and workers must also analyze all hazards below and to the side of the anchoring point to ensure that the worker does not strike or swing into any obstacles should he or she fall. The U.S. OSHA requirement for an anchorage is a 5,000-lb (2,268 kg) minimum static load strength (needed for 6 ft (2 m) of free fall). The Canadian Standard Association's (CSA) Standard Z–259.2 requires 6,000 lb (2,700 kg). The strength, location, and design must allow the worker enough mobility to perform the job.

Lanyard

A lanyard is a short flexible rope, strap, or webbing connecting the worker to the anchor. A lanyard permits limited lateral movement on the job. Its length (and placement of the anchor) determines the amount of free fall a worker experiences before the protective device stops the fall. However, retracting lifelines require 3,000 lb. Furthermore, OSHA has permitted a 2:1 safety factor for engineered systems.

Body Belts

Body belts are not acceptable as part of a personal fall arrest system. However, the use of a body belt in a positioning device system is acceptable and is regulated as follows:

- Positioning devices shall be rigged such that an employee cannot free fall more than 2 feet (0.9 m).
- Secured to an anchorage capable of supporting at least twice the potential impact load of an employee's fall or 3,000 lb (13.3 kN), whichever is greater,
- Shall be drop forged, pressed, or formed steel, or made of equivalent materials with a corrosion-resistant finish, and all surfaces and edges shall be smooth.
- Connectors, connecting assemblies, and D-rings and snaphooks shall meet requirements of 29 *CFR* 1926.501.
- Snaphooks shall be a locking type designed and used to prevent disengagement of the snaphook by the contact of the snaphook keeper by the connected member.
- Shall be inspected prior to each use for wear, damage, and other deterioration, and defective components shall be removed from service.

Harnesses

Full-body harnesses encompass the torso and are attached to other parts of the fall arrest system. A full-body harness distributes the fall arrest force over the shoulder-to-thigh body areas.

Typically made of nylon and polyester webbing straps, harnesses are the preferred worker component in fall arrest systems. By spreading the fall arrest force over the body, a harness enables the worker to avoid bodily injury. The generally accepted maximum free fall distance is 6 ft (1.8 m). A goal of 2 ft (0.6 m) is best to enable self-rescue and avoid prolonged suspension.

Retracting Lifeline Devices

Retracting lifeline devices (SRLs) are portable, self-contained devices fixed to an anchorage point above the work area. They act as an automatic taut lanyard. The lifeline rope, webbing, or cable is attached directly to the worker's body harness. The line extends out of the device as distance increases and retracts as the worker moves closer. At the moment a fall occurs, a centrifugal locking mechanism is activated to arrest the movement, thereby reducing the potential free-fall distance shock load. This device is ideal for use on sloping roofs and angular structures, because the rope is never slack and does not interfere with the surface work. The extension limit varies from 6 ft (2 m) on some devices to 300 ft (91 m) on others. Unlimited lengths are possible with a counterweight variation.

Lifelines

A horizontal lifeline is an anchoring cable rigged between two fixed anchorage points on the same level (Figure 7–9). The line may serve as a mobile fixture to attach lanyards, lifelines, or retracting lifelines. The purpose is to limit swing injuries by providing a continuously overhead fixture point as the worker moves horizontally. The important factors for nonengineered systems include (1) adequate degrees of slack; (2) 500 lb (227 kg) per worker strength (where falls up to 6 ft (2 m) are anticipated and where short horizontal lifelines [25 ft (8 m)] are used); (3) supports every 20 (6 m) to 50 ft (8 m) that preferably can be passed without detaching the line; and for engineered systems sufficient shock absorption and design strength at least twice the force calculated for the dynamic fall of an anticipated number of workers who may use the line. Lightweight, low-stretch synthetic cables serve as a practical alternative to steel cable.

Extremely careful engineering is required for all horizontal lifelines. When used with retracting lifelines, the horizontal lifeline should be arranged overhead with little sag (approximately 10 degrees to the horizontal or less, which is also usable with rope or synthetic cable). When used with lanyards, the line should be set at a maximum height of 78 in. (198 cm) at the center. The worker may not be able to travel down the horizontal lifeline slope after a fall. Slack in the line during and after a fall should be considered to determine proper clearance in relation to other obstacles and the ground.

Figure 7–9. This worker is protected by a horizontal lifeline; the lanyard extends and retracts as needed for normal operations.

Some organizations offer training programs to determine proper end-force and dynamic slack projections. Manufactured horizontal lifeline systems typically have an in-line shock absorber for reducing end forces.

Lifeline ("Dropline")

A dropline is a vertical lifeline which extends from an independent anchorage point and to which a lanyard is attached using a grabbing device. The dropline should be at least ⅝ in. (16 mm) diameter nylon, ⅝ in. (16 mm) diameter polyester, or 5/16 in. (24 mm) diameter or ⅜ in. (9 mm) diameter steel cable with a minimum breaking strength of 5,000 lb (2,449 kg). Steel cable should be used only in spark- or heat-producing work operations, although it is popular for use with work at extreme heights, such as chimney building and repair, or where a limit on the system elongation is important. Ropes always must be protected from abrasive or cutting edges, which may weaken the fibers.

Weather-protected nylon and polyester lifelines with neoprene jackets are available. Polypropylene ropes are popular with utilities because of the low moisture absorption and high dielectric constant. Ultraviolet-stabilized polypropylene ⅝ in. (16 mm) or ¾ in. (19 mm) diameter is popular for many suspended scaffold operations because of light weight and low cost; however, unstabilized ropes could be hazardous after short exposures to sunlight.

Hardware Connectors

Hardware connectors consist of bolts, shackles, D-rings, snaphooks, and metal links that connect parts of the lifeline. The U.S. OSHA regulation requires these connectors to demonstrate 4,000-lb (1,800 kg) static tensile strength of the load-bearing portion without permanent distortion.

Carabiner-type snaphooks often are an alternate oval-shaped snaphook design with an automatic twist-lock spring gate to help eliminate the possibility of roll-out. Snaphooks should be attached to compatible hardware and never to each other. These hooks must always be tested during inspection and maintenance to see if they fully close and will do so on the anchorage point without stress to the gate.

Fall Arresters and Shock Absorbers

Many types of fall arresters are available in various sizes and types of lifeline. These devices slow a worker's fall or break the fall to prevent injury. Nearly all fall arresters use friction in the rope to disperse fall energy. Often there is a gradual delay action so the body does not experience a severe jolt or shock when the fall is arrested.

If a lanyard is used as the only fall protection, a shock absorber fall arrester and a full body harness can soften the arresting force on the body. Shock absorbers may not always be required or necessary if the anchorage point height allows little or no free fall. These devices work by the unfolding and tearing of stretched woven webbing or stretching fiber.

Fall Arresting System

A fall arresting system (FAS) is engineered from eomponents and designed for a work positioning system. The purpose of a FAS is not only to stop the fall, but also to ensure that the energy gained by the body during the fall is distributed so as to prevent the wearer from being injured. A fall arresting system is composed of an independent anchorage point, a vertical lifeline (dropline), a fall arrester, a harness, and—optionally—a lanyard and a shock absorber, equipped with all the necessary hardware (snaphooks, D-rings, and so on).

It is usually easier to slowly arrest a fall and to prevent injury to the worker during and after the fall. Injuries generally occur as a result of forces acting on the body at the instant a fall is stopped or through collision with obstacles.

Work Positioning System

This is a system that permits users of harnesses and lanyards to lean to do work. Fall arrest is a separate system that does not involve the parts of a work positioning system during fall arrest.

Restraint System

This is a system that permits users of harnesses and lifelines to move up to a fall-hazard zone but restricts movements. The system must meet fall arrest requirements because of the likelihood of a free fall under some configurations of use.

Rescue System

The moments following a fall can be critical in preventing worker injuries. Organizations should develop, implement, and regularly practice rescue procedures. The following section discusses rescues for aboveground and belowground or confined spaces rescues.

Aboveground

Descent devices permanently installed or immediately available at such workstations as overhead crane cabs and grain elevator workhouses can be used effectively for lowering an injured member of the crew quickly and safely to ground level. Devices with no inherent speed control require the presence of a trained rescue team or trained co-workers to supervise the rescue operation. Automatic speed-limiting descent devices reduce or eliminate the need for trained rescue personnel; the machine itself controls the injured worker's rate of descent.

Collapsible cradles are snap-on accessories that may be used with either system. They are useful for small, cramped spaces in crane cabs and towers, but rigid stretchers should be available for victims with broken bones or internal injuries to prevent compounding the injury.

Belowground Tanks or Confined Spaces

Confined spaces are those enclosed spaces that have limited openings for entry and exit, poor ventilation, and are not designed for continuous worker occupancy. Examples of confined spaces include storage tanks, process vessels, ship compartments, pits, silos, vats, sewers, boilers, tunnels, vaults, and pipelines. (See also Chapter 2, Buildings and Facility Layout, in this volume.)

Rescue workers can be lowered into tanks or confined space by means of lifelines to locate workers or to determine if they need assistance. If manholes and tanks are deeper than 10 ft (3 m) per platform or level, fall protection is required in addition to a means of rescue under U.S. OSHA standards.

A manual or air-operated winch accessory is available for use with some steel cable retracting lifeline devices; the winch helps eliminate the need for personal attention to the victim until he or she has been raised. Alternatively, a block and tackle or ratchet winch can provide the lifting mechanism with limited human effort after the victim has been hooked up, provided a lock or overspeed mechanism is incorporated. An anchorage point, such as that provided by a 7 (2.1 m) or 10 ft (3 m) tripod, should be available before work begins at the site. (See Figure 2–6 in Chapter 2 in this volume.)

Equipment Inspection and Maintenance

The fall arrest equipment manufacturer's instructions must be incorporated into a company's inspection and preventive maintenance procedures (Table 7–C). Workers need to be trained to inspect equipment, to understand the basics of static loading of fall equipment for test purposes, and to check equipment for damage before each use.

NOTE: This section is not meant to be complete, but only to serve as a guide. Special situations such as ionized radiation, electrical conductivity, spark generation, chemicals, and so on, must also be considered. The recommendations should not be regarded as an industry consensus but rather a guideline for safety in developing company procedures.

Cleaning Fall Arrest Equipment

Fall arrest equipment must be cleaned regularly to ensure that it remains in good condition and top working order.

Synthetic Ropes and Harnesses

Washing this equipment in soapy water is the best way to remove loose debris. Rinse with fresh water and dry in a cool area away from UV light. Always make sure that labels are legible after cleaning.

NOTE: Do not use industrial solvents on synthetic materials. These chemicals can degrade the product by leaching oils used in the manufacturing process to give greater strength to the final product.

Fall Devices

Wash fall arrest devices in soapy water. If metal parts have been soiled with caked materials or paint overspray, consider using cleaning solvents such as wood alcohol or 1,1,1-trichloroethane to remove the contaminants. Labels must be legible after cleaning, or be replaced after calling the manufacturer.

Do not oil moving parts unless instructed by the manufacturer. For some parts, such as rollers or bearings subject to heavy use or dirt, such lubrication may be reasonable. On the other hand, oil could interfere with a descent device brake efficiency. Any lubrication used must have the manufacturer's approval. Many manufacturers offer reconditioning programs and retesting documentation to help owners maintain fall arrest devices in good working order. Employers should take advantage of the fall equipment manufacturer's expertise.

Storage

Keep synthetic materials away from bright light and UV light during storage, and maintain them in a cool dry place. When dyed synthetic color fades, it indicates UV exposure that may lead to equipment damage and failure. All fall arrest and body support devices must be inspected regularly and defective parts repaired or the equipment replaced.

RESPIRATORY PROTECTION

A longstanding hierarchy of controls requires employers to use engineering and work practice controls as the primary means to protect an employee's health from contaminated or oxygen-deficient air. However, if such controls are not technologically or economically feasible (or otherwise inappropriate), an employer may rely on a respiratory protection program to protect employees.

The respiratory protection program must consist of worksite-specific procedures specifying the selection, use, and care of respirators. The program must be updated as often as necessary to reflect changes in workplace conditions and respirator use.

The respiratory protection program must cover the following basic elements, as applicable:

- A written respiratory protection program containing workplace specific procedures necessary to protect the health of the employee from workplace contaminants or when the employer requires the use of respirators (29 *CFR* 1910.134).
- Procedures for selecting respirators.
- Medical evaluations of employees required to use respirators.
- Fit-testing procedures for tight-fitting respirators. Respirators fall into two general categories relating

Table 7–C. Fall Protection Equipment Maintenance and Inspection Guide

Components	Lifetime (yr) Service	Shelf	System Rating	Checking (3–6 mo)
Webbing (belts, lanyards)	2–3	7	• 5,000 lb 500–800 lb shock absorber	• Cuts, wear, burn, pull one unit
Ropes (lifelines, lanyards)	1–2	5	• 5,000 lb	• Synthetic: Pull end sample, cut in strand, worn, dirt inside • Cable: Kink, broken wire, terminations
Hardware	>5			• Cracks, distortion, wear points, corrosion
hooks,	2–3	>5		
D-rings	>5	>5	• 5,000 lb	
Locking Fall Devices				• Recertification/old models
• rope grabs	3–5		• 1,000 lb before slip	• Distortion, wear • Cleaning difficulty • Comparable parts
• retracting lifeline	3–7		• 3,000 lb line, • 350–450 lb before slip	• Operates manually as intended
• climbing device	>5		• 1,000 lb proof load	• Recertification/old models
Safety Lowering Devices				• Recertification/old models
a. retracting lifeline	3–5		• 4–9 ft/sec (3,000 lb)	• Distortion, wear
b. escape	5		• 3–6 ft/sec (1,800 lb)	• Controlled payout
c. descent-control device			• No limit (5,000 lb)	• Operates manually as intended

Metal goods should have no limit on shelf-life expectation.

(Source: Introduction to Fall Protection, J. Nigel Ellis, ASSE.)

to "fit": tight-fitting and loose-fitting. The tight-fitting respirator is designed to form an airtight seal with the face of the wearer; the loose-fitting respirator has a respiratory inlet covering that forms a partial seal (not airtight) with the face. Loose-fitting respirators need to supply enough air to maintain a slight positive pressure inside the respirator in relation to the outside environment. Each class of respirator may have tight-fitting and loose-fitting facepieces.

- Use of respirators in routine and reasonably foreseeable emergency situations.
- Procedures and schedules for cleaning, disinfecting, storing, inspecting, repairing, and otherwise maintaining respirators.
- Procedures to ensure adequate air quality, quantity and flow of breathing air for air-supplying respirators.
- Training of employees in the respiratory hazards and proper use of respirators, limitations of use and applicable maintenance procedures.
- Procedures for regularly evaluating the effectiveness of the program.

NOTE: If an employee is voluntarily using a filtering facepiece, then an employer is only required to provide a copy of Appendix D, 29 *CFR* 1910.134 to each respective user. If cases where employers allow the voluntary use of respirators other than filtering facepieces, the costs associated with ensuring use of the respirator itself does not create a hazard, such as medical evaluations and maintenance, must be provided at no cost to the employee.

The U.S. NIOSH, under authorization of the Federal Mine Safety and Health Act of 1977 and the Occupational Safety and Health Act of 1970 (OSHAct), provides a testing, approval, and certification program to ensure that safe personal protective devices and reliable industrial hazard-measuring instruments are available commercially. Use of the terms "approved" and "certified" on the product label reflects these applicable federal regulations. Employers can consult manufacturers for any necessary applications, equipment needs, repairs, and the like.

Selecting Respiratory Protection

Given the hundreds of toxic substances workers may encounter and the wide variety of respiratory protection equipment available, making the right choice of breathing equipment can be a difficult task.

The proper selection of respiratory protective equipment involves three steps: (1) identifying the hazard, (2) evaluating the hazard, and (3) finally, selecting the appropriate, approved respiratory equipment based on the first two considerations (Figure 7–10).

Identification of the Hazard

Airborne hazards that could require respiratory protection generally fall into the following basic categories (taken from the *OSHA Technical Manual,* Chapter 7):

Dusts

Particles that are formed or generated from solid organic or inorganic materials by reducing their size through mechanical processes such as crushing, grinding, drilling, abrading, or blasting.

Fumes

Particles formed when a volatilized solid, such as a metal, condenses in cool air. This physical change is often accompanied by a chemical reaction, such as oxidation. Examples are lead oxide fumes from smelting and iron oxide fumes from arc welding. A fume can also be formed when a material such as magnesium metal is burned or when welding or gas cutting is done on galvanized metal.

Mists

A mist is formed when a finely divided liquid is suspended in the air. These suspended liquid droplets can be generated by condensation from the gaseous to the liquid state or by breaking up a liquid into a dispersed state, such as by splashing, foaming, or atomizing. Examples are the oil mist produced during cutting and grinding operations, acid mists from electroplating, acid or alkali mists from pickling operations, paint spray mist from spraying operations, and the condensation of water vapor to form a fog or rain.

Gases

Gases are formless fluids that occupy the space or enclosure and that can be changed to the liquid or solid state only by the combined effect of increased pressure and decreased temperature. Examples are welding gases (such as acetylene, nitrogen, helium, and argon); carbon monoxide generated from the operation of internal combustion engines; and hydrogen sulfide, which is formed wherever there is decomposition of materials containing sulfur under reducing conditions.

Vapors

Vapors are the gaseous form of substances that are normally in the solid or liquid state at room temperature and pressure. They are formed by evaporation from a liquid or solid and can be found where parts-cleaning and painting takes place and where solvents are used.

Smoke

Smoke consists of carbon or soot particles resulting from the incomplete combustion of carbonaceous materials such as coal or oil. Smoke generally contains droplets as well as dry particles.

Oxygen Deficiency

An oxygen deficient atmosphere has an oxygen content below 0.5% by volume. Oxygen deficiency may occur in confined spaces, which include, but are not limited to, storage tanks, process vessels, towers, drums, tank cars, bins, sewers, septic tanks, underground utility tunnels, manholes, and pits.

Biological Agents

Some biological agents are pathogenic microorganisms or infectious agents that could cause disease in a susceptible population. Although respiratory protection guidelines for exposure to tuberculosis is covered by a specific standard (29 *CFR* 1910.139), it is one example of a biological agent encountered in the workplace that has caused infection in several hundred employees requiring medical treatment. TB is spread by airborne droplets when a person coughs, sneezes, or speaks and use of respiratory protection can be an effective response measure. Other biological agents, such as the bacterium *Chlamydia psittaci* and *Histoplasma capsulatum* spores, are examples of health hazards where respiratory protection could be helpful in reducing potential exposures.

Evaluation of the Hazard

The next step in a respirator selection process is a walk-through survey of the facility to identify employee groups, processes, or worker environments where the

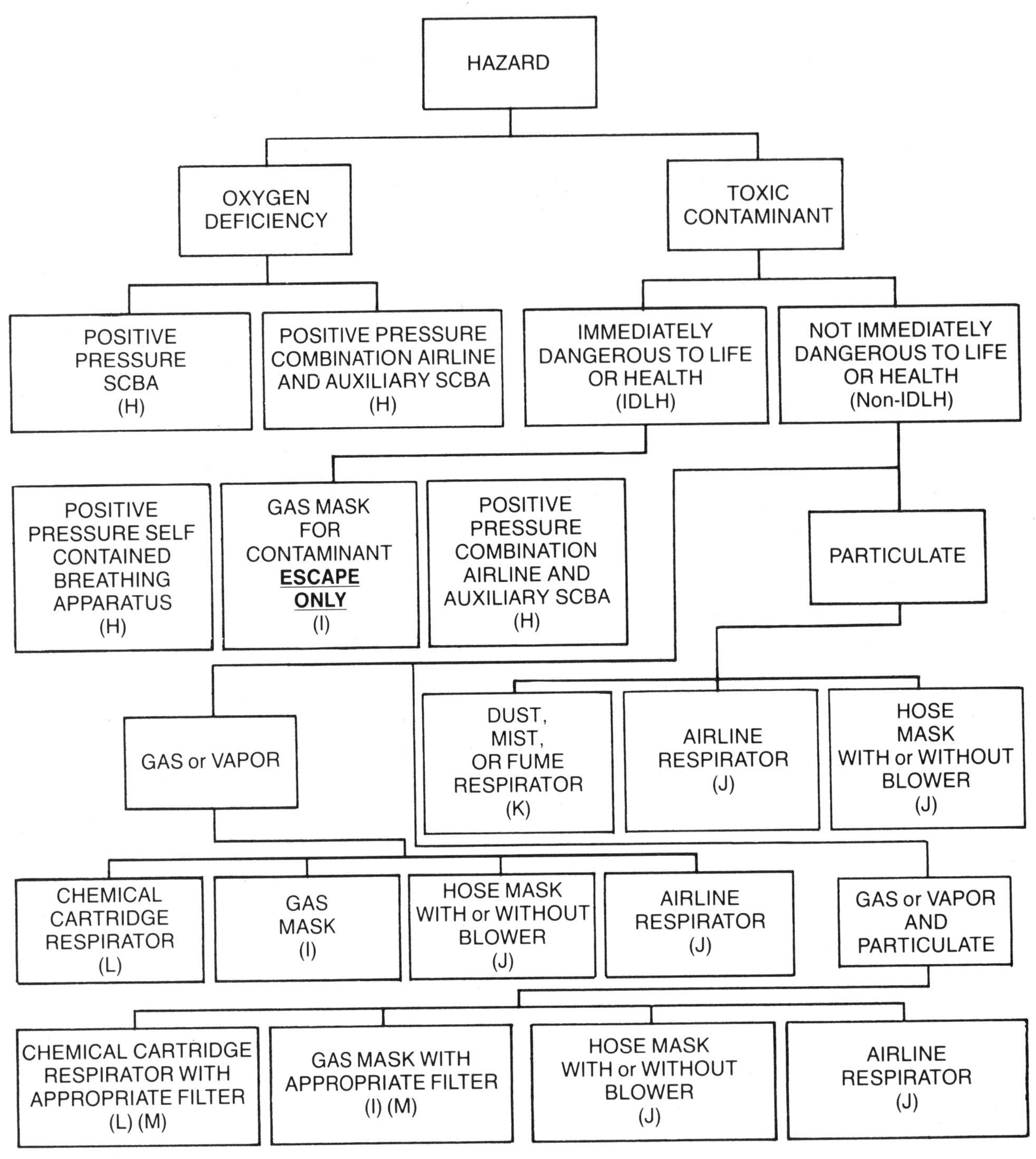

Figure 7–10. Suggested outline for selecting respiratory protective devices. Letters in parentheses refer to Subparts of Title 30 CFR, Chapter 1, Part 11.

use of respiratory protective equipment may be required. The physical and chemical nature of the identified hazard(s) must be evaluated, potential exposures quantified (concentrations, length and nature of exposure), the surrounding environment assessed, and other potential stressors identified (physical and psychological, if present). This information should be examined in the light of occupational exposure limits (both regulatory and consensus) and the availability and feasibility of appropriate respiratory protection. A hazard evaluation

Respiratory Protective Equipment Hazard Evaluation Form

Company ____________ Date ____________

Division ____________ By ____________

Department ____________ Page ____________

Employee	Job Description	Limits	Respiratory Protection Required		Respiratory Equipment Type-SCP	Remarks
			Yes	No		

Figure 7–11. This sample respiratory protective equipment hazard evaluation form specifies whether respiratory protective equipment is required for a job and what type.

form can help the safety and health professional in conducting the survey (Figure 7–11). He or she may use instruments to determine the concentration of airborne contaminants. However, only qualified individuals should be allowed to use these instruments and interpret their results. If an organization does not have inhouse qualified personnel, outside consultation will be required. Whenever processes change, a new evaluation must be conducted to quantify any new or different exposures. If hazards have worsened (or lessened), changes to the level of respiratory protection may be necessary, indicating a change in the written respiratory protection program and subsequent changes in the required respirator.

There will be one of two outcomes from the hazard evaluation. Either (1) respiratory protection is needed or (2) engineering and administrative controls are sufficient to protect the health of the workers. If the hazard evaluation indicates the need for a respiratory program, then it is important to know which respirator to select for use.

NIOSH recommends that the hazard evaluation include examining the nature and extent of the hazard, work requirements and conditions, and the characteristics and limitations of the respirators available.

When the need for a respirator is ascertained, a NIOSH-certified respirator should be selected. In some cases a specific respirator for a contaminant has not been certified. When this occurs, a NIOSH-certified respirator with no limitation prohibiting its use for that contaminant can be used only if the protective features of the respirator are appropriate for the contaminant's physical form, chemical properties, and the conditions under which it will be used. And in all cases, respirators must be chosen and used according to the limitations identified on the NIOSH certification label.

A system should be in place that provides a reliable means of protecting respirator wearers from contaminant breakthrough anytime an air-purifying respirator is used as protection against gases and vapors. The written respirator program should describe the change schedule indicating how often cartridges and/or canisters should be replaced and what information was used to make this judgement regarding service life. The service life depends upon many factors, including environmental conditions, breathing rates, cartridge filtering capacity, and the concentration of contaminants in the air. Odor thresholds and other warning properties cannot be used as the primary basis for determining the service life of gas and vapor cartridges and canisters. It would be a good idea for employers to apply a safety factor to the service life estimate to assure that the change schedule is a conservative estimate.

Effective methods for determining the service life of a respirator cartridge include:

- using cartridges with end-of-service life indicators (ESLI), or
- using an established and enforced cartridge/canister change schedule that is based on objective information or data that will ensure that canisters and cartridges are changed before the end of their service life.

For atmospheres that are immediately dangerous to life and health (IDLH), the highest level of respiratory protection and reliability is required. Only the following respirators must be provided for use in an IDLH atmosphere: either a full-facepiece pressure demand self-contained breathing apparatus (SCBA) certified for a minimum service life of 30 minutes, or a combination full-facepiece pressure demand supplied-air respirator (SAR) with an auxiliary self-contained air supply.

Protection Factors

Assigned Protection Factors (APFs) are used to indicate the level of effectiveness a respirator provides to the wearer. An employer should ensure that the APF for an assigned respirator is adequate to provide protection from the identified hazardous contaminant. To date, OSHA has not addressed APFs and employers may rely on APFs published by NIOSH and ANSI. Where there are conflicts between the NIOSH and ANSI APFs, the employer should apply the more protective APF.

Types of Respirators

The proper selection and use of a respirator will depend upon the initial determination of the concentration of the hazard or hazards present in the workplace, or the presence of an oxygen deficient atmosphere.

When using a respiratory protection program to protect the health of the employees, employers must not only provide respirators but ensure that employees use them. Respirators provide protection either by removing airborne contaminants from the air before they are inhaled or by providing an independent supply of breathable air. There are two major classifications of respirators:

- Air-supplying respirators (provide clean breathing air) and
- Air purifying respirators (remove (filter) contaminants from the air)

Air-supplying respirators provide air from a source independent of the surrounding atmosphere. They do not remove contaminants from the atmosphere. These respirators are classified by the way breathing air is supplied and regulated. Basically, these classifications are either:

- self-contained breathing apparatus (air or oxygen is carried in a tank on the worker's back, similar to SCUBA gear);
- supplied-air respirators (compressed air from a stationary source is supplied through a high-pressure hose connected to the respirator); or
- combination self-contained and supplied-air respirators.

Air-purifying respirators remove contaminants and are grouped according to the contaminant removed:

- particulate,
- vapor,
- gas, or
- combination.

Particulates are removed by filters, while vapors and gases are removed by either chemical cartridges or canisters. Filters and canisters/cartridges are typically used up during the air filtering (cleaning) process and, if replaceable, are exchanged for new ones when their effective life has ended. Filtering facepiece respirators (commonly referred to as "disposable respirators," "dust masks," or "single-use respirators") that cannot be cleaned, disinfected, or resupplied with a new filter are discarded after use.

Particulate-removing respirators are designed to filter nuisance dusts, fumes, mists, toxic dusts, radon daughters, asbestos-containing dusts, biological agents or fibers, or any combination of these substances. These respirators can be either single-use or fitted with replaceable filters that could be used until their effective life has expired. These respirators may be nonpowered or powered air-purifying.

Vapor- and gas-removing respirators use canisters or cartridges filled with a sorbent that cleans some portion of the vapors or gases from contaminated air before it can enter the breathing zone of the worker.

Combination cartridges and canisters are available to protect against particulates, vapors, and gases.

Respirators can be furthered classified according to pressure relationships between the air within the facepiece and the ambient air outside the facepiece. When the pressure is normally positive with respect to ambient air pressure throughout the breathing cycle, the respirator is a positive-pressure respirator. If the air pressure inside the facepiece is negative with respect to the ambient air pressure, the respirator is a negative-pressure respirator. Exposure evaluations should examine the concept of negative and positive pressure when considering potential contaminant leakage into the respirator.

Air-Supplying Respirators

Air-supplying respirators provide a breathing gas (usually air) to the worker. The different types are classified according to (1) the method used to supply breathing gas and (2) the method used to regulate the gas supply.

Self-Contained Breathing Apparatus

A SCBA provides a transportable supply of breathing air and affords protection against both toxic chemicals and oxygen deficiency (Figure 7–12). The wearer

Figure 7–12. Self-contained breathing apparatus. (Courtesy Mine Safety Appliances Company.)

carries enough air or oxygen for up to four hours, depending on the design. All personnel engaged in interior structural firefighting must use SCBA. SCBAs can be classified as "closed circuit" or "open circuit."

Closed Circuit

Another name for the closed-circuit SCBA is a rebreather device. After the exhaled carbon dioxide has been removed and the oxygen content has been restored by a compressed or liquid oxygen source or an oxygen-generating solid, the gas can be breathed again. The devices are designed for one- to four-hour use in oxygen-deficient or IDLH atmospheres that might exist during mine rescues or in confined spaces.

Open Circuit

An open-circuit SCBA exhausts air to the atmosphere instead of recirculating it. Compressed air is almost always the breathing gas used; compressed oxygen cannot be used in a device designed for compressed air because minute amounts of oil or other foreign matter in the device can cause an explosion. Mine safety regulations prohibit the certification of devices that permit interchangeable use of air and oxygen.

In an open-circuit SCBA, a cylinder of high-pressure compressed air (2,000–4,500 psi) supplies air to a regulator that reduces the pressure for delivery to the facepiece. The regulator is usually mounted on the facepiece or is connected by a hose to the respirator inlet. Most open-circuit SCBAs have a service life of 30 to 60 minutes. Open-circuit SCBAs are widely used in fire fighting, for industrial emergencies, and hazardous waste site work. Care must be exercised in these work environments since failure of the respirator to provide the appropriate protection may result in serious injury or death. Consequently, the employer must develop and implement specific procedures for the use of respirators in IDLH atmospheres that include the following provisions:

- At least one employee ("standby employee") should be located outside the IDLH atmosphere and maintain visual, voice, or signal line communication with the employee(s) in the IDLH atmosphere.
- Standby employee(s) must be equipped with pressure-demand or other positive-pressure SCBA, or a pressure-demand or other positive-pressure supplied-air respirator with auxiliary SCBA.
- Standby employee(s) must be equipped with appropriate retrieval equipment for lifting or removing any employee in need of such assistance from the hazardous atmosphere. When such retrieval equipment cannot be used because it would increase the overall risk resulting from entry, equivalent provisions for rescue need to be readily available.

NIOSH also certifies SCBAs with less than 30-minute service times, usually for escape use only. Escape SCBAs are also certified in combination with supplied-air air-line respirators.

Two types of open-circuit SCBAs are available, "demand" or "pressure demand." In a demand or negative pressure respirator, air at approximately 2,000 psi is supplied to the regulator through the main valve. A bypass valve passes air to the facepiece in case of regulator failure. Downstream from the main valve, a two-stage regulator reduces the pressure to approximately 50 to 100 psi.

Inhalation creates negative pressure in the facepiece, opening an admission valve and allowing air into the facepiece, but only on demand by the wearer. However, a demand SCBA offers no more protection than does an air-purifying respirator with the same facepiece. Therefore, a demand open-circuit SCBA should not be used in IDLH atmospheres.

A pressure-demand or positive-pressure regulator is designed to maintain positive pressure in the facepiece at all times. All pressure-demand devices have a special exhalation valve that maintains positive back pressure in the facepiece and opens only when the pressure exceeds that preset level (1.5–3.0 in. water pressure).

Under certain conditions of work, a momentary negative pressure may occur in the wearer's breathing zone, although the regulator still supplies additional air on demand. Because of positive pressure, any leakage should still be outward. A pressure-demand SCBA has the same service time as a demand device, if it seals well to the wearer's face. Some open-circuit SCBAs can be switched from demand to pressure-demand mode. However, the demand mode should be used only for donning and adjusting the apparatus in order to conserve air and should be switched to pressure demand for actual use.

The escape only SCBA, certified by NIOSH and the Mine Safety and Health Administration (MSHA), is usually for short duration use (3, 5, or 10 minutes), and is small in size and weight (Figure 7–13). The user wears a container of compressed air on the back hip with a readily accessible air valve. Hood or facepiece styles are available.

Supplied-Air Respirators

Supplied-air or air-line respirators are available in demand, pressure-demand, and continuous-flow models. In the past, these have been designated Type C respirators. The air-line may provide air to a facepiece, helmet, hood, or complete body suit (Figure 7–14). A demand or pressure-demand air-line respirator is very similar in operation to a demand or pressure-demand SCBA. Continuous-flow air-line respirators maintain air flow at all times. Regulations specify that a flow of

Sample MSHA/NIOSH Approval Label for Pressure-Demand SAR

PERMISSIBLE

Combination Ten Minute Self-Contained Compressed Air Breathing Apparatus for Escape Only Pressure Demand Type C Supplied Air Respirator

MINE SAFETY AND HEALTH ADMINISTRATION
NATIONAL INSTITUTE FOR OCCUPATIONAL SAFETY AND HEALTH

APPROVAL NO. TC-13F-000

ISSUED TO

ABC Company
Anywhere, U.S.A.

LIMITATIONS

Approved for respiratory protection during entry and escape from oxygen deficient atmospheres, gas, and vapors, when using air-line air supply. Approved for escape only, when using self-contained air supply. Approved for use at temperatures above -25°F.

Approved only when compressed air reservoir is fully changed with air meeting the requirements of the Compressed Air Gas Association Specifications G-7-1 for type 1, Grade D air, or equivalent specifications. The containers shall meet applicable DOT specifications.

This approval applies only when the device is supplied with respirable breathing air through 12.5 to 300 feet of hose at air pressures between 78 and 80 pounds per square inch gage or from self-contained air supply. If the supplied-air fails, open cylinder valve and proceed to fresh air immediately.

CAUTION

Use with adequate skin protection when worn in gases and vapors that poison by skin absorption (for example: hydrocyanic-acid gas). In making renewals and repairs, parts identical with those furnished by the manufacturer under the pertinent approval shall be maintained. This respirator shall be selected, fitted, used, and maintained in accordance with Mine Safety and Health Administration, and other applicable regulations.

SAMPLE

MSHA — NIOSH Approval TC-13F-000

Issued to ABC Company, February 31, 2000

The approval assembly consists of the following part numbers:

000-000
000-000
etc.

Figure 7–13. Sample MSHA/NIOSH approval label for escape only pressure-demand self-contained compressed air breathing apparatus.

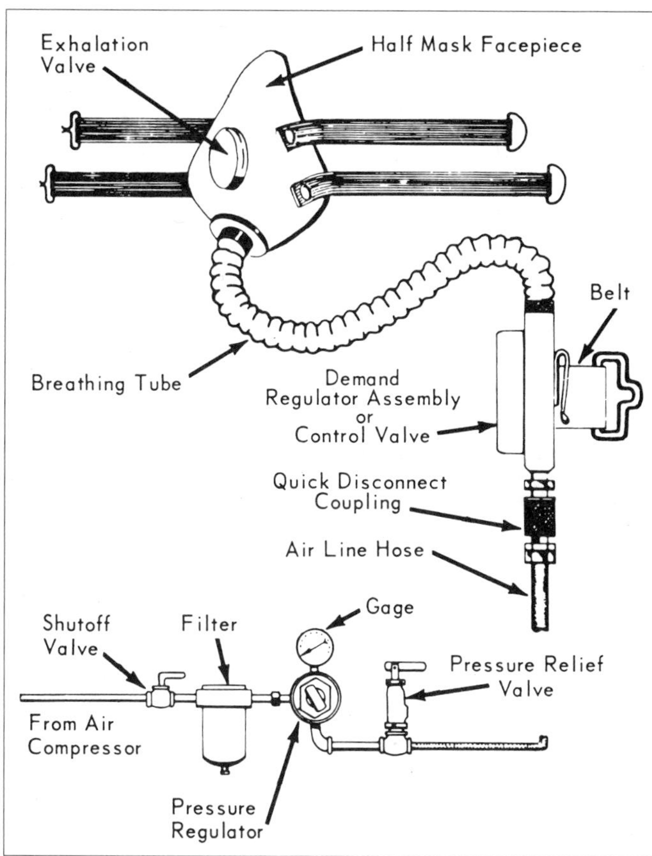

Figure 7–14. Air-line respirator parts and connections.

115 liters per minute (lpm) for a tight-fitting facepiece and 170 lpm to a loose-fitting hood or helmet must be maintained at the lowest air pressure and the longest hose length specified.

Some special valving is available with some certified air-line respirators. These incorporate vortex tubes that allow heating or cooling of the air delivered to the worker for comfort and prevention of heat or cold stress.

Air-line respirators must be used only in non-IDLH atmospheres or those from which the wearer can escape without the use of a respirator. This limitation is necessary because the air-line respirator is entirely dependent upon an air supply not carried by the wearer. If this air supply fails, the wearer might not escape immediately from a hazardous atmosphere. Another limitation of air-line respirators is that the air hose limits the wearer to a fixed distance from the air-supply source.

The air supply for air-line respirators must meet or exceed grade D or higher quality, as set forth by the *Compressed Gas Association Commodity Specification for Air*, G–7.1. To protect air quality, air from compressors must be continually monitored for carbon monoxide or the equipment must have a high temperature shut-off alarm to indicate when the compressor overheats. Overheating could introduce excess carbon monoxide into the air line. Also, where there is potential exposure to vapors or gases that poison by skin absorption, adequate skin protection must be used.

Air-Supplied Hoods

For some long-term operations that do not require a completely enclosed suit, an air-supplied hood may be used. These are particularly useful in hot, dusty environments. A vortex tube may also be used to reduce the ambient air temperature by up to 50 F (10 C). Respirable air under suitable pressure should be delivered to a hood at a volume of at least 6 ft^3/m (0.0028 m^3/s).

Abrasive Blasting Respirators

Abrasive blasting respirators are one type of air-supplied respirator. They are used to protect personnel engaged in shot, sand, or other abrasive blasting operations that involve air contaminated with high concentrations of rapidly moving abrasive particles. The requirements for abrasive blasting respirators are the same as those for an air-line respirator of the continuous-flow type, with the addition that mechanical protection from the abrasive particles is needed for the head and neck (Figure 7–15). NIOSH and MSHA staff test and certify such equipment.

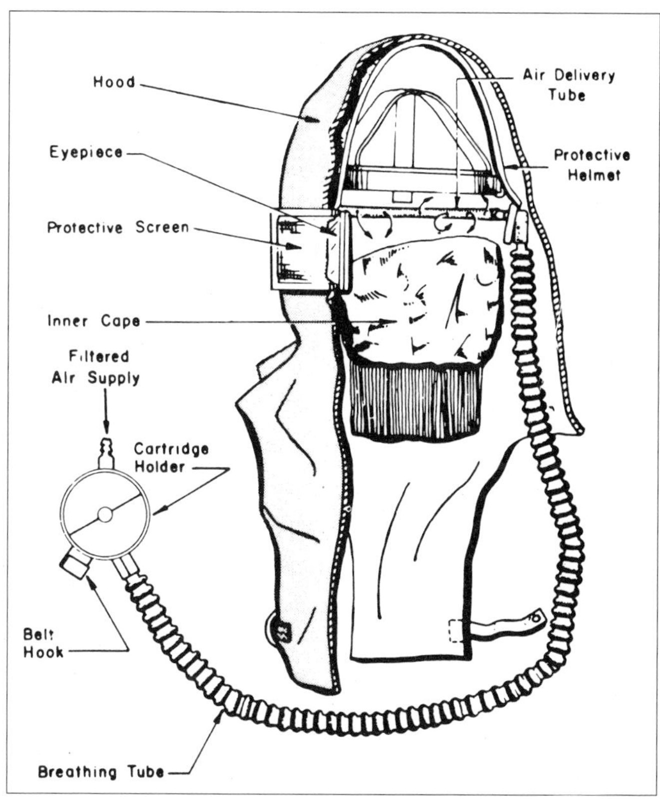

Figure 7–15. Diagram shows parts and connections for lightweight hood designs for use by persons doing abrasive blasting.

Air-Supplied Suits

The most extreme condition requiring respiratory equipment is that in which rescue or emergency repair work must be done in atmospheres that are highly corrosive to the skin and mucous membranes as well as acutely poisonous and immediately hazardous to life, such as atmospheres containing vapors of ammonia, hydrofluoric acid, or hydrochloric acid. For these conditions, a complete suit of impervious clothing, with a respirable air supply, is available (Figure 7–16).

Because no generally accepted specifications exist for such suits, companies must rely on the integrity and skill of the manufacturer. Suit material should have sufficient mechanical strength to resist rough handling and considerable abuse without tearing. The hose line supplying the air should be connected to the suit itself, as well as to the helmet. This is because wearing such a suit for a long time is not only extremely fatiguing but also dangerous unless it is well ventilated.

Personal air-conditioning devices using a vortex tube are available for air-supplied suits or hoods. These cooling and heating devices are desirable to reduce fatigue where high ambient temperatures may be encountered (as in heat-protective clothing) or where body heat may build up (as under impermeable chemical protective clothing).

The vortex device works by taking an air stream under pressure and dividing it. One portion loses heat; the other gains heat. The cold portion passes into the suit or hood; the warm portion is vented to the atmosphere, or vice versa in cold weather.

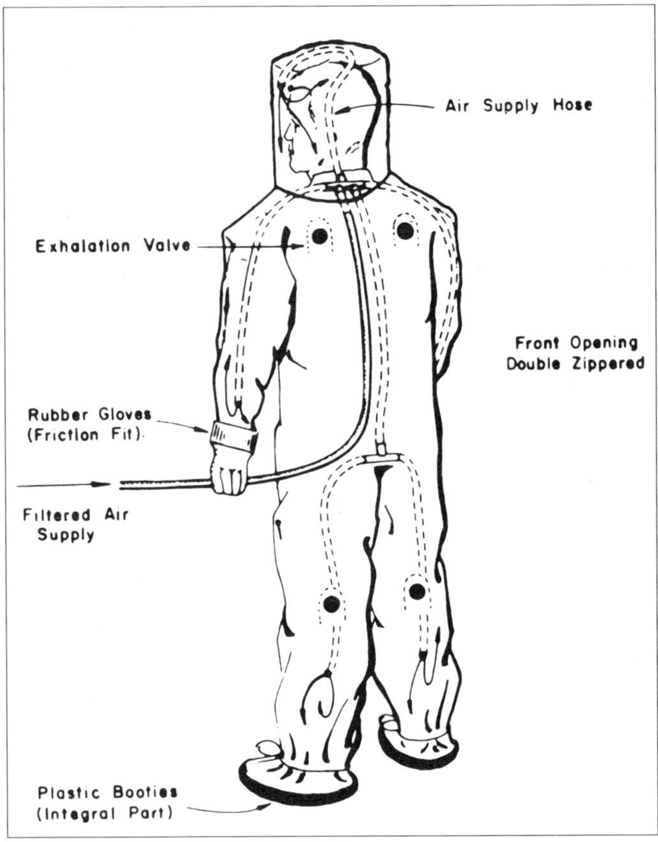

Figure 7–16. Diagram of air-supplied suit for use in corrosive chemical atmosphere.

Combination Supplied-Air SCBA Respirators

These respirators have an auxiliary air supply to protect workers against potential failure of the compressor. An air tank of 3-, 5- or 10-minute service time is typically used, mainly for emergency and escape for IDLH. The SCBA part is used only when the air line fails and the wearer must escape, or when the worker disconnects the line temporarily to change locations. A combination air line and SCBA may be used for emergency entry into a hazardous atmosphere to connect the air line only if the device is rated for 15 minutes or longer service.

Air may be supplied to a respirator from a cylinder of compressed air or directly from an air compressor. High-pressure, low-pressure, and ambient-air supply systems are available. However, the latter cannot be used to supply pressure-demand respirators as it does not develop sufficient pressure.

Air-Purifying Respirators

Air-purifying respirators can purify the air of gases, vapors, and particulates, but do not supply clean breathing air (Figure 7–17). They must never be used in oxygen-deficient atmospheres. The useful life of the air-purifying device is limited by (1) the concentration of the air contaminant, (2) breathing demand of the wearer, and (3) removal capacity of the air-purifying medium (cartridge or filter). The air-purifying respirator has a facepiece and an attached cartridge that contains specific material needed against the contaminant. This equipment is classified as either a gas and vapor respirator or a particulate respirator.

Air-purifying respirators are available in three basic configurations. The quarter-face respirator covers the mouth and nose, with the lower edge resting between the chin and mouth. The half-face respirator fits over the nose and under the chin. It seals more reliably against the face and provides better protection against toxic materials. The full-face piece covers the user from the hairline to below the chin and provides some eye protection. It is designed for use in higher concentrations of toxic materials than quarter- or half-face respirators. Mouthpiece style respirators are also available for escape use only.

Figure 7–17. Full-face airline respirator designed for use by employees working in the vicinity of hazardous emissions. (Courtesy MSA.)

Gas and vapor respirators (also known as chemical-cartridge respirators) remove gases and/or vapors by passing the contaminated air through cartridges containing charcoal or other sorbents that trap these contaminants. Cartridges must be matched to the right contaminants (Table 7–D) and are used to protect against contaminants that have adequate warning properties of smell or irritation (not exceeding certain concentrations). This characteristic allows the wearer to judge when a cartridge is no longer usable. Some cartridges are dated as well and should not be used after the expiration date.

Four major rules apply to chemical cartridge respirators:

1. They cannot be used for protection against gaseous material that is extremely toxic in very small concentrations.
2. The respirators should not be used for exposure to harmful gaseous matter that cannot clearly be detected by odor. Example: methyl chloride and hydrogen sulfide. The former is odorless; and the latter, although foul smelling, paralyzes the olfactory nerves so quickly that detection by odor is unreliable. (NOTE: Odor should not be used as a primary indicator of sorbent exhaustion.)
3. Chemical cartridge respirators should not be used against any gaseous material in concentrations highly irritating to the eyes without satisfactory eye protection, e.g., full-mask respirators.
4. Chemical cartridge respirators will not provide protection against gaseous material that is not effectively stopped by the cartridge medium used, regardless of concentrations.

Table 7–D. Color Code for Cartridges and Gas Mask Canisters

Atmospheric Contaminants to be Protected Against	*Color Assigned*
Acid gases	White
Organic vapors	Black
Ammonia gas	Green
Carbon monoxide gas	Blue
Acid gases and organic vapors	Yellow
Acid gases, ammonia, and organic vapors	Brown
Acid gases, ammonia, carbon monoxide, and organic vapors	Red
Other vapors and gases not listed above	Olive
Radioactive materials (except tritium and noble gases)	Purple
Dusts, fumes, and mists (other than radioactive materials)	Orange

Notes:

1. A purple stripe shall be used to identify radioactive materials in combination with any vapor or gas.
2. An orange stripe shall be used to identify dusts, fumes, and mists in combination with any vapor or gas.
3. Where labels only are colored to conform with this table, the canister or cartridge body shall be gray or a metal canister or cartridge body may be left in its natural metallic color.
4. The user shall refer to the wording of the label to determine the type and degree of protection the canister or cartridge will afford.

The second type of air-purifying respirator, the particulate respirator, is also known as a mechanical filter respirator. Depending upon the design, the filters can screen out dust, fog, fumes, mist, spray, or smoke by passing the contaminated air through a pad or filter. (They do not provide protection against gases, vapors, or oxygen deficiency.) They consist of a facepiece with an attached mechanical filter, papers, or similar filter substance. Many types of filters are capable of trapping a range of airborne particle classes. Filters should be changed frequently, when they become clogged, or when it becomes difficult to breathe through them. Appropriate filters and cartridges can be used together where combinations of contaminants exist.

The U.S. NIOSH has developed a new set of regulations in 42 *CFR* 84 for testing and certifying nonpowered, air-purifying, particulate-filter respirators. This new certification process ensures that applicable respirators have passed a more demanding certification test and involves only nonpowered, air-purifying, particulate-filter respirators.

This new regulation has resulted in nine classes of filters (three levels of filter efficiency, each with three categories of resistance to filter efficiency degradation). The three levels of filter efficiency are 95%, 99%, and 99.97%. The three categories of resistance to filter efficiency degradation are labeled N, R, and P (depends on the presence or absence of oil particles). The class of filter will be clearly marked on the filter, filter package, or respirator box. For example, a filter classified as N99 would mean an N-series filter that is at least 99% efficient. For chemical cartridges that include particulate filter elements (combination) a similar marking pertains only to the particulate filter element.

The new classes of nonpowered particulate respirators require new decision logic for selection of the proper respirator. A synopsis of the selection process for using the new particulate classification is outlined as follows:

- If no oil particles are present, use a filter of any series (i.e., N-, R-, or P-series).
- If oil particles (e.g., lubricants, cutting fluids, etc.) are present, use an R- or P-series filter.
- If oil particles are present and the filter is to be used for more than one work shift, use only a P-series filter.

NOTE: the following guide to these classifications: N for Not resistant to oil, R for Resistant to oil, P for oil Proof

Selection of filter efficiency (i.e., 95%, 99%, or 99.97%) depends on how much filter leakage can be accepted. Higher filter efficiency means lower filter leakage and a higher degree of protection. The choice of facepiece depends on the level of protection needed—that is, the assigned protection factor (APF) discussed earlier.

The flow chart in Figure 7–10 can be used as a guide for selection of particulate respirators.

Powered air-purifying respirators use a blower both to pass contaminated air through a filter or sorbent bed that removes the contaminant and to supply purified air to a facepiece, helmet, or hood. The purifying device may be either a filter, a cartridge, or a combination of the two. A blower, usually worn on the worker's belt, is used to force the contaminated air through the element and to the respirator facepiece. The unit supplies clean air to the worker at positive pressure, so that contaminated air does not leak into the facepiece.

Gas Masks

Gas masks have been used effectively for many years for respiratory protection against certain gases, vapors, and particulate matter that otherwise might be harmful to life or health. Gas masks are air-purifying devices designed solely to remove specific contaminants from the air; therefore, it is essential that their use be restricted to atmospheres that contain sufficient oxygen to support life. Gas masks may be used only for escape from IDLH atmospheres, never for entry into such environments. Users must assess the exposure conditions carefully before selecting a specific mask for respiratory protection. If the specific exposure concentrations are suspected of exceeding established limits, only SCBA should be used.

From a practical standpoint, gas masks are generally suitable for ventilated areas not subjected to rapid change in air-contaminant levels. They should never be used in confined spaces below or above ground where oxygen deficiency and high gas concentrations may occur. In assessing exposure conditions, workers should remember that oxygen deficiency can occur in a confined space through the displacement of air by other gases or vapors, or by means of processes (such as fire, rusting, and aerobic bacteria) that consume oxygen.

Fitting Respirators

Required fit tests must be performed before an employee uses a respirator in the workplace. Fit testing is required for all employees fulfilling any of the following criteria:

- using negative or positive pressure tight-fitting respirators
- where such respirators are required by OSHA
- where the employer requires the use of such a respirator.

Fit tests must be repeated under the following guidelines:

- at least annually,
- whenever a different respirator facepiece is used, or
- a change in the employee's physical condition could affect respirator fit.

A fit test is not required for voluntary users or for escape-only respirators.

Facial hair, jewelry, corrective glasses or goggles, or other personal protective equipment must not interfere with the seal of the facepiece of tight-fitting respirators.

1. The tight-fitting respirator is designed to form a seal with the face of the wearer. It is available in three types: quarter mask, half mask, and full facepiece.
2. The loose-fitting respirator has a respiratory inlet covering that is designed to form a partial seal with the face. These include loose-fitting facepieces, as well as hoods, helmets, blouses, or full suits, all of which cover the head completely. Because the hood is not tight-fitting, it is important that sufficient air is provided to maintain a slight positive-pressure inside the hood relative to the environment immediately outside the hood. In this way, an outward flow of air from the respirator will prevent contaminants from entering the hood.

Air-supplied masks should also be fit-tested for appropriate size. Some facilities fit-test workers who wear supplied-air respirators with tight-fitting facepieces. Determination of facepiece fit should involve both qualitative and quantitative tests. Prior to fit-testing, the employee, together with the person administering the fit-test, should check the comfort of the respirator facepiece. Make sure there is adequate time devoted to checking these items:

- proper placement on chin
- proper positioning of facepiece on nose
- comfortable strap tension
- comfortable fit across nose bridge
- ability to talk while wearing facepiece
- room for safety spectacles where required
- tendency of facepiece to slip

Qualitative Tests

In the irritant or odorous chemical agent test, the wearer is exposed to an irritant smoke, isoamyl acetate vapor, or other suitable test agent easily detectable by irritation, odor, or taste. An air-purifying respirator must be equipped with the appropriate air-purifying element. If the wearer cannot detect any penetration of the test agent, the respirator is probably tight enough.

The advantages of a qualitative test are speed, convenience, and ease of performing the test. However, these tests rely on the wearer's subjective response, so they may not be entirely reliable.

Quantitative Tests

In quantitative testing, the employee, wearing a specially designed probed respirator, stands in a test chamber and is exposed to a test atmosphere of a nontoxic, easily detectable aerosol, vapor, or gaseous test agent. Instrumentation is used to measure the leakage into the respirator.

Protection factors can be determined from quantitative fit tests by dividing the ambient airborne concentration of the challenge contaminant by the concentration inside the facepiece. For example, if the concentration outside the facepiece is 500 parts per million (ppm), and the concentration inside the respirator is 10 ppm, the protection factor would be 50. Protection factors are used in the selection process to determine the maximum use concentration (muc), which is determined by multiplying the threshold limit value (TLV) or permissible exposure level (PEL) by the protection factor.

The greatest advantage of a quantitative test is that it does not rely on a subjective response. The quantitative test is recommended when facepiece leakage must be minimized for work in highly toxic or IDLH atmospheres.

Quantitative fitting tests require expensive equipment that can be operated only by trained personnel. Because each test respirator must be equipped with a sampling probe to allow removal of a continuous air sample from the facepiece, the same facepiece cannot be worn in actual service.

Daily Fit Test

Employees who wear respirators should check the fit of their respirator each time they don it, with both negative-and positive-pressure tests.

Negative-Pressure Test

The wearer can perform this test alone in the field and should use it before entering any toxic atmosphere. The test consists of closing off the inlet of the canister, cartridges, or filters by covering with the palms or replacing the seals, or of squeezing the breathing tube so that it does not pass air; inhaling gently so that the facepiece collapses slightly; and holding the breath for 10 seconds. If the facepiece remains slightly collapsed and no inward leakage is detected, the respirator is probably tight enough. This test, of course, can be used only on respirators with tight-fitting facepieces (Figure 7–18a).

Positive-Pressure Test

This test is similar to the negative-pressure test and has the same advantages and limitations. It is conducted by closing off the exhalation valve and exhaling gently into the facepiece (Figure 7–18b). The fit is considered satisfactory if slight positive pressure can be built up inside the facepiece without any evidence of outward leakage. For some respirators, this method requires the wearer to remove the exhalation valve cover, which often disturbs the respirator fit more than the negative-pressure test does.

Therefore, this test should be used cautiously if it requires removing and replacing a valve cover. The test is easy for respirators whose valve cover has a single small port that can be closed by the palm or a finger. The wearer should perform this test before entering any hazardous environment.

Storage of Respirators

Respirators should be stored to protect them from dust, sunlight, heat, extreme cold, excessive moisture, and damaging chemicals. Unprotected respirators can sustain damaged parts or facepiece distortion that make them ineffective.

Before storing the respirator, clean or wash the device according to the manufacturer's instructions. Wiping a respirator with a cloth is not acceptable practice because fibers from the cloth can be deposited on the respirator's surface. Workers should never store the respirator with folds or creases and should never hang it by the elastic headband or place it in a position that will stretch the facepiece.

Because heat, air, light, and oil cause most rubbers to deteriorate, this equipment should be kept in a cool, dry place and protected from light and air as much as possible. Today, many new respirators are made of silicone rubber, which is more pliable and tends to become less hard with age. Many respirators come with their own plastic or metal cases. The equipment should not be stored in toolboxes or left on workbenches where it may be exposed to dust and damage by oil or other harmful materials.

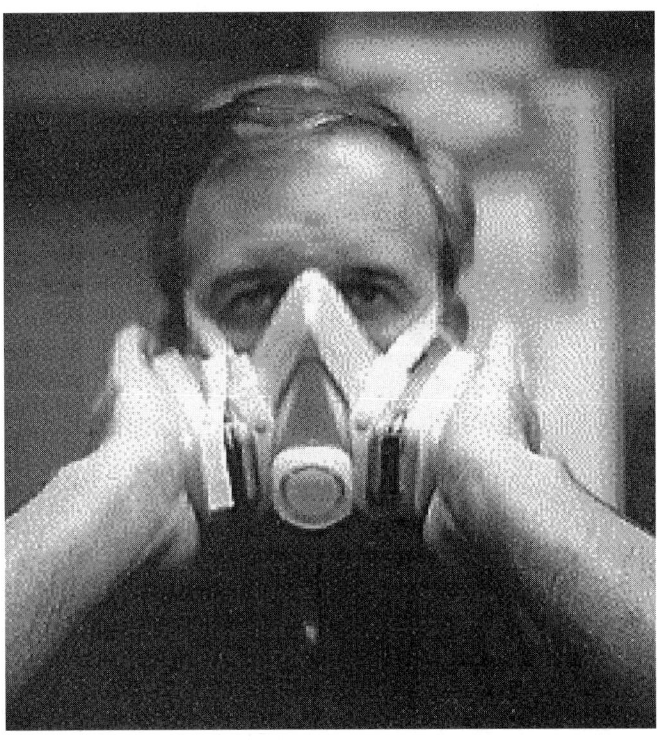

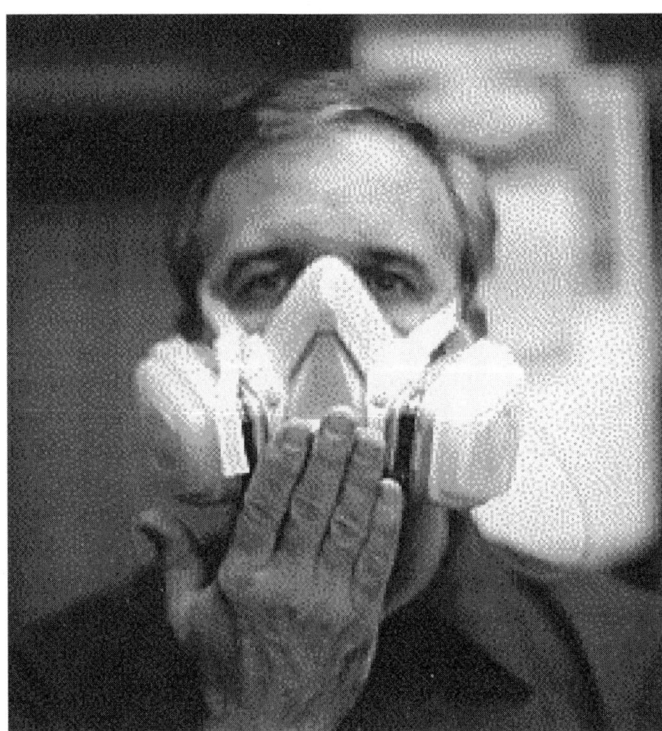

Figure 7–18. *a:* A negative-pressure test; *b:* a positive-pressure test.

After cleaning and inspection, place respirators in individual, sealable plastic bags. Store them in one layer with the facepiece and exhaust valve in normal position. Respirators should not be kept in lockers unless they can be protected from contamination, distortion, and damage.

Standard steel storage cabinets or steel wall-mounted cabinets with compartments are the best choice for storing air-purifying respirators. Special storage cabinets may be purchased from the manufacturer for SCBAs. The cabinets should be located in uncontaminated but readily accessible areas.

Maintenance of Respirators

The ongoing maintenance of the respirators themselves is an essential part of the respiratory protection program. If the equipment malfunctions because of poor maintenance, the employee may be exposed to a potentially fatal hazard.

The maintenance program should incorporate the manufacturer's instructions and should include provisions for disassembly, including the removal of the repirator's purifying elements, cleaning, sanitizing, inspecting for defects, repairing parts (if necessary), installing purifying elements, reassembling, packaging, and storing equipment.

The air-purifying elements (chemical cartridges or filters) should not be cleaned or exposed to excess moisture, including high humidity. Discard the elements if there is any question about their condition (Figure 7–19).

Supervisors should be responsible for conducting daily equipment inspections, particularly of functional parts such as exhalation valves and filter elements.

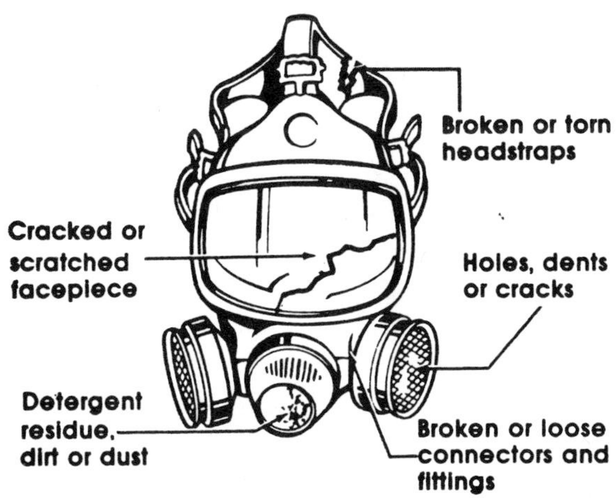

Figure 7–19. Maintenance checkpoints for air-purifying respirators.

They should see that the edges of the valves are not curled and that valve seats are smooth and clean. Inhalation and exhalation valves should be replaced periodically. In addition, users should be trained to inspect their respirators before and after each use.

Besides the daily check, weekly inspections and inspections before and after each use, conducted by trained personnel, can ensure that respirators remain in top working order. The inspector should stretch rubber parts slightly to check for fine cracks. Users need to work the rubber every so often to prevent it from setting (becoming rigid, a cause of cracking). Inspectors need to check the headband to be sure the wearer has not stretched it in securing a snug fit. Inspectors should check emergency respirators at least monthly and keep a written record of the results.

Sometimes, in an effort to reduce resistance to breathing, workers remove exhalation valves, punch holes in the filter, the rubber facepiece, or other parts. Management should investigate the reasons for this mistreatment and correct them. For instance, it may be that, in the interest of economy, filters are allowed to become completely plugged before being replaced.

Cleaning and Sanitizing

Make sure that each time an employee uses a respirator it is clean and sanitized. Actual cleaning may be done in a variety of ways.

The respiratory protection equipment should be dismantled and washed with whatever cleaner the manufacturer recommends in warm water using a brush, thoroughly rinsed in clean water, and then air dried in a clean place. Care should be taken to prevent damage from rough handling. This method is an accepted procedure for a small respirator program or where each worker cleans his or her own respirator.

A standard domestic clothes washer may be used if a rack is installed to hold the facepieces in a fixed position. (If the facepieces are placed loose in a washer, the agitator may damage them.) This method is especially valuable in large programs where respirators are used extensively.

Workers should follow the manufacturer's instructions regarding which cleaners to use. However, they should not use organic solvents, as these can deteriorate the elastomeric (rubber or silicone) facepiece. If bactericidal detergents are not available, use a disinfectant. Check with the manufacturer for disinfectants that will not damage the respirators.

Be sure that the cleaned and sanitized respirators are rinsed thoroughly in clean water no hotter than 50 C (120 F) to remove all traces of detergents and cleaners. Otherwise, skin irritation or dermatitis may result when the employee wears the respirator. Allow the respirators to air dry by themselves on a clean surface.

They may also be hung carefully on a line or in a specially designed drying cabinet. If management is unwilling or unable to run the maintenance program, the firm can contract with an outside service to maintain respirators in peak condition.

Inspection of Respirators

After cleaning and sanitizing, each respirator should be reassembled and inspected for proper working condition and repair or replacement of parts. The respirator should also be inspected routinely by the user immediately before each use to ensure that it is in proper working condition. An inspection checklist compiled by NIOSH recommends that management inspect disposable respirators for:

- holes in the filter (obtain new disposable respirator)
- straps for elasticity and deterioration (replace straps, contact manufacturer)
- metal nose clip for deterioration, if applicable (obtain new disposable respirator)

Air-purifying respirators (including quarter-mask, half-mask, full-facepiece, and gas mask) should be checked for the following items:

- facepiece
  - excessive dirt (clean all dirt from facepiece)
  - cracks, tears, or holes (obtain new facepiece)
  - distortion (allow facepiece to sit free from any constraints and see if distortion disappears—if not, obtain new facepiece)
  - cracked, scratched, or loose-fitting lenses (contact respirator manufacturer to see if replacement is possible—otherwise, obtain new facepiece)
- head straps
  - breaks or tears (replace head straps)
  - loss of elasticity (replace head straps)
  - broken or malfunctioning buckles or attachments (obtain new buckles)
  - excessively worn serrations on the head harness that might allow the facepiece to slip (replace head strap)
- inhalation and exhalation valves
  - detergent residue, dust particles or dirt on valve or valve seat (clean residue with soap and water)
  - cracks, tears, or distortion in the valve material or valve seat (contact manufacturer for instructions)
  - missing or defective valve cover (obtain valve cover from manufacturer)
- filter elements
  - proper filter for the hazard
  - missing or worn gaskets (contact manufacturer for replacement)
  - worn threads—both filter threads and facepiece threads (replace filter or facepiece, whichever is applicable)
  - cracks or dents in filter housing (replace filter)
  - deterioration of gas mask canister harness (replace harness)
  - service life indicator, expiration date, or end-of-service date
- gas mask
  - cracks or holes (replace tube)
  - missing or loose hose clamps (obtain new connectors)
  - service-life indicator on canister (or contact manufacturer to find out what indicates the end-of-service date for the canister)

Air-supplying respirators should be checked for:

- hood, helmet, blouse, or full suit (if applicable)
  - rips and torn seams (if unable to repair the tear adequately, replace)
  - headgear suspension (adjust properly for wearer)
  - cracks or breaks in face shield (replace face shield)
  - protective screen to see that it is intact and fits correctly over the face shield, abrasive blasting hoods, and blouses (obtain new screen)
- air-supply system
  - breathing air quality
  - low pressure alarm
  - breaks or kinks in air supply hoses and end fitting attachments (replace hose and/or fitting)
  - tightness of connections
  - proper setting of regulators and valves (consult manufacturer's recommendations)
  - correct operation of air-purifying elements
  - proper operation of carbon monoxide alarms or high-temperature alarms
- self-contained breathing apparatus (SCBA)
  - facepiece, head straps, valves, and breathing tube inspection checks are same as for air-purifying respirators (consult manufacturer's literature)

In some companies, maintenance service for respirators and for other kinds of PPE can be effectively provided by traveling service carts. When a number of respirators are used regularly, the organization may set up a central station for their care and maintenance, along with the care and maintenance of other items of PPE.

Each employee is given two respirators and either a locker or a hook at the central station to store the equipment.

Some facilities have found that if two respirators are assigned to each worker, equipment lasts more than twice as long. This plan works best when users cannot clean their respirators between shifts or before the next scheduled shift.

Under such a plan, users turn in marked respirators on a set schedule, depending upon use, for cleaning, disinfection, inspection, and repair. Workers wear the second until the first can be serviced and returned.

Another plan involves keeping quantities of disinfected respirators on hand for use as needed. Although this plan works well where individual needs vary, users do not readily accept the responsibility of looking after the equipment. Also, if the same respirator is worn by several persons, it must be cleaned and disinfected after each use. When not in use, the respirator must be stored in accordance with manufacturer's recommendations.

Mark respirators to indicate to whom they are assigned. Identification should be made in some form of permanent ink or paint so that workers cannot change the marking inadvertently or without effort. Medical evaluation must be conducted to ensure that employees are physically able to wear respirators before training commences.

Training

Once the right respirator has been selected, the wearer must be trained in its proper use and care. This step is important for every type of respirator. Each user should not only be trained when first acquiring the equipment but also be retrained periodically. Training sessions should include the following:

- Reasons for respiratory protection and explanation of why other controls and methods are not being used, and what efforts have been made to reduce the hazards
- Explanation of the respirator selection procedure used by the health and safety professional, including identification and evaluation of specific airborne hazards
- Proper fitting, donning, wearing, and removing of the respirator; importance of not modifying the respirator in any way that will impair or void its protective features
- Limitations, capabilities, and operation of the respirator
- Proper maintenance and storage procedures
- Wearing the respirator in a safe atmosphere to allow the user to become familiar with its characteristics
- Wearing the respirator in a test atmosphere under close supervision of the trainer to allow the wearer to simulate work activities and detect respirator leakage or malfunction
- Recognizing and coping with emergency situations
- Instructions for special use as needed
- Explanation of any regulations governing the use of respirators

The instructor should be a qualified person, such as an industrial hygienist, safety professional, nurse, or the respirator manufacturer's representative.

Medical Surveillance

Employers must provide a medical evaluation to determine each employee's fitness to wear a respirator before initial fit-testing and prior to using a respirator for the first time. Medical evaluations consist of the administration of a medical questionnaire that can be found in the mandatory Appendix C of the standard, 29 *CFR* 1910.134, or provision of a physical examination that elicits the same information as the questionnaire for the employee. An employer, who opts to provide physical examinations to his or her employees, need not also administer the medical questionnaire. These evaluations are required for all respirator users except for employees who voluntarily use dust masks and for those individuals using escape-only respirators. SCBAs are not considered escape-only respirators. Employees who refuse to be medically evaluated cannot be assigned to work in areas where they are required to wear a respirator.

HAND AND ARM PROTECTION

No one type of PPE for the extremities is suitable for the many different work situations involved in any business or industrial operation, from the laboratory to the loading dock. Thus, management and workers must select proper protection for the hands, fingers, arms, and skin based on potential exposure to identified hazards. The specific type of protection and its material depends upon the type of material being handled and the work atmosphere.

Gloves

The material to be used for gloves depends largely upon what is being handled. For most light work not involving exposure to hazardous materials or microbial contamination, a cotton glove is satisfactory and inexpensive. Rough or abrasive material requires leather gloves or leather reinforced with metal stitching for safe handling (Figure 7–20). Leather reinforced by metal stitching or

Figure 7–20. These leather or leather-reinforced work gloves protect against sharp or abrasive surfaces. (Courtesy Ansell Edmont Industrial Inc.)

metal mesh, or highly cut-resistant plastic gloves, also provide good protection from edged tools, as in butchering and similar occupations. Double gloving affords added protection. If the outer glove starts to degrade or tears, the inner glove may offer protection until the gloves are removed and replaced. It is a good idea to check the outer glove frequently, watching for signs of deterioration (color, texture change, holes, etc.) and reglove as necessary.

Many plastic and plastic-coated gloves are available in materials such as neoprene, latex, and nitrile. They are designed to protect workers from a variety of hazards. Management must give careful consideration to actual permeation tests of these gloves against hazardous chemicals. Some plastic models surpass leather in durability and effective shielding. Other types have granules or rough materials incorporated into the plastic for better gripping ability, while still others are disposable. Computer software is now available to help select the appropriate glove material. Most manufacturers can provide information regarding the rate and degree of chemical permeation through their glove materials.

Exposure to proteins from the use of natural rubber latex (NRL) products may result in adverse responses in susceptible workers. These could include irritation and several types of allergic reactions. Recommended strategies for risk reduction include reducing unnecessary exposure to NRL proteins for all workers. For example, workers in food service or landscaping industries do not need to use NRL gloves for food handling or gardening purposes.

OSHA has made the following recommendations concerning the use of NRL gloves:

- If selecting NRL gloves for worker use, designate NRL as a choice only in those situations requiring protection from infectious agents.
- When selecting NRL gloves, choose those with a lower protein content. Selecting powder-free gloves offers the additional benefit of reducing systemic allergic responses.
- Provide alternative suitable non-NRL gloves as choices for workers who are allergic to NRL gloves.

Prudent risk reduction strategies include risk surveys, mechanisms for reporting and managing cases, safe zones (nonuse areas for products containing NRL proteins), and development of policies and procedures for reducing the risk of NRL allergies in the workplace found through an initial survey.

When workers do not need complete gloves, they can wear finger stalls or cots. These are available in combinations of one or more fingers, and usually are made of rubber, duck, leather, plastic, and metal mesh. The construction of the cot depends on the degree or type of hazard to which the worker is exposed. Employees should not use gloves while working on moving machinery such as drills, saws, grinders, or other rotating and moving equipment. Machine parts might catch the glove and pull it and the worker's hand into hazardous areas.

In addition to gloves, workers also can wear mittens (including one-finger and reversible types), pads, thumb guards, finger cots, wrist and forearm protectors, elbow guards, sleeves, and capes. These protective devices are made in a wide range of materials and lengths.

Gloves or mittens having metal parts or reinforcements should never be used around electrical apparatus. Work on energized or high-voltage electric equipment requires specially made and tested rubber gloves. Workers should wear over-gloves of leather to protect the rubber gloves against wire punctures and cuts and to protect the rubber in the event of electrical flashes. Conduct frequent tests and inspections of line workers' rubber gloves and discard those that fail to meet original specifications.

Hand Leathers and Arm Protectors

For jobs requiring protection from heat or from extremely abrasive or splintery material (such as rough lumber), hand leathers or hand pads are likely to be more satisfactory than gloves. This is primarily because they can be made heavier and less flexible without discomfort.

Because hand leathers or pads are used mainly for handling heavy materials, they should not be used around moving machinery. They must always be sufficiently loose to allow workers to slip their hands and fingers out of the device if it is caught on a rough edge or nail.

For protection against heat, hand and arm protectors should be made of wool, terry, or glass fiber. Although leather can be used, it will not withstand temperatures over 150 F (65 C). Wristlets or arm protectors may be obtained in any of the materials of which gloves are made.

Impervious Clothing

For protection against dusts, vapors, and moisture of hazardous substances and corrosive liquids, the safety and health professional can choose from among many types of impervious or impermeable materials. These are fabricated into clothing of all descriptions, depending on the hazards involved. They range from aprons and bibs of sheet plastic to suits that enclose the body from head to foot and contain their own air supply.

Materials used include natural rubber, olefin, synthetic rubber, neoprene, vinyl, polypropylene, and polyethylene films and fabrics coated with these substances. Natural rubber is not suited for use with oils, greases, and many organic solvents and chemicals because it deteriorates over time. Make sure that the clothing selected will protect against the hazards involved and that it has been field tested prior to actual use. For example, some synthetic fabrics used for regular work clothing in chemical facilities, where daily contact with acids and caustic solutions would cause rapid deterioration of regular cotton clothes, are not really impervious. Such clothing cannot be used where impervious materials are indicated.

Gloves coated with synthetic rubber, synthetic elastomers, polyvinyl chloride, or other plastics offer protection against all types of petroleum products, caustic soda, tannic acid, muriatic and hydrochloric acid, and even sulfuric acid (Figure 7–21). These gloves are available in varying degrees of strength to meet individual conditions.

Gloves should be long enough to come well above the wrists, leaving no gaps between the glove and the coat or shirtsleeve. Long, flaring gauntlets should be avoided unless they are equipped with closing snaps or straps to ensure a snug fit around the wrist.

Such gauntlets offer the best protection when acids and other chemicals are being poured. If the chemicals splash, workers can receive serious acid burns unless precautions are taken. When pouring caustic substances and harmful solvents from large to small containers, workers should wear their sleeves outside the gauntlets. This prevents any spilled chemical from running down into the protective device.

In many operations, rubber gloves with extra long cuffs have been used to advantage. The cuffs of these gloves are made with a heavy ridge near the top edge,

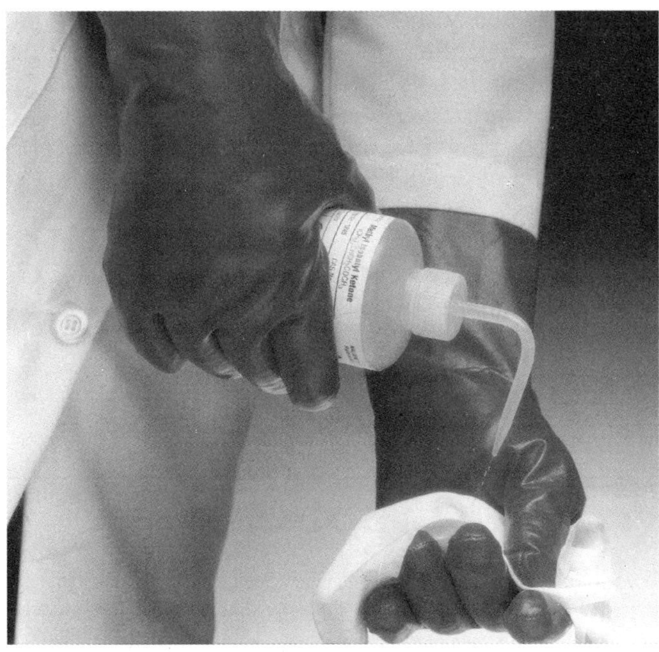

Figure 7–21. These gloves coated with synthetic rubber or plastic protect the worker from chemicals and solvents. (Courtesy Ansell Edmont Industrial.)

which, when turned back, forms a trough to catch liquids and prevent them from running down the wrist or forearm. Some are made with beads near the cuff to hold inserts which form a liquid-tight seal when inserted into a sleeve.

Where acid may splash, rubber boots or rubber shoes also should be worn. Never tuck trousers into boots when working with corrosive materials. If workers wear safety shoes inside their boots, the legs of impervious trousers should cover the tops of the shoes. These precautions keep the liquid from draining off aprons or trousers into the footwear.

Procedural Steps

After PPE is used in a corrosive atmosphere, management must establish a strict procedure for disposing of the equipment to prevent workers from coming into contact with contaminated parts. Before the equipment is removed, it should be thoroughly washed with a hose stream whether or not it has come in contact with the corrosive chemical. Boots, coats, aprons, and hats should then be taken off, followed by removal of the gloves. This is the logical order of removal if the coat has been properly put on with the sleeves outside the cuffs of the gloves.

Workers should wash their hands thoroughly before removing their face shields and goggles, then wash their hands and face again. Ideally, a complete shower and change of clothing are far more desirable.

Ensure potentially contaminated gloves are laundered thoroughly before permitting reuse or reissue. Effective laundering can be used to sanitize the fabric and prolong glove life. Gloves used in toxic chemical service must be cleaned with special care to be sure the chemicals are thoroughly removed. Gloves that have been permeated by toxic material cannot be reused.

For protection against exposure to oil and the various other compounds that rapidly attack ordinary rubber, all the equipment discussed can be obtained in plastic and synthetic rubbers.

PROTECTIVE FOOTWEAR

Specifications for protective footwear are contained in ANSI Z41–1991: *Personal Protection—Protective Footwear.* Protective footwear is classified according to its ability to meet both the requirements for compression resistance and impact resistance (see Table 7–E). All protective footwear meeting the ANSI Z41 standard contain a protective toe box. Steel, reinforced plastic, and hard rubber are materials commonly used to make protective toe boxes. For protection in wet conditions, rubber footwear is also available with toe protection. Five other types of protection are available in addition to the basic toe protection.

Regulations such as U.S. OSHA standards 29 *CFR* 1910.132—General Requirements for Personal Protective Equipment and 29 *CFR* 1910.136—Foot Protection, contain hazard assessment and foot protection requirements for employees whose work presents hazards to their feet. These hazards include objects falling onto or placed on the foot, objects rolling over the foot, sharp objects penetrating the sole of the footwear, static electricity buildup, and contact with energized electrical conductors.

Table 7–E. Minimum Requirements for Protective Footwear

Impact
I/75 = 75 ft lbf (101.7J)
I/50 = 50 ft lbf (67.8J)
I/30 = 30 ft lbf (40.7J)
Compression
C/75 = 2,500 lb (11 121 N)
C/50 = 1,750 (7 784 N)
C/30 = 1,000 (4 448 N)
Clearance (All Classifications)
Men—16.32 in. (12.7 mm)
Women—15.32 in. (11.9 mm)

Note: Any combination of impact and compression classifications will comply with standard.

Although comfort and proper fit are important factors for any footwear, it is particularly essential for protective footwear in order to encourage employee use. Management should carefully select foot protection to match the specific hazards faced by the wearer, educate employees on the need for such protection, and train workers in the proper use, care, and replacement of footwear.

Companies may arrange to have retail footwear providers supply on-site footwear sales, fitting, and training to employees through the use of portable shoemobiles.

In addition to the ANSI Z41 general requirement for impact and compression resistance protective footwear, the standard also describes additional protective options available include metatarsal footwear, conductive footwear, electrical hazard footwear, static dissipative footwear, and sole puncture resistant footwear.

Metatarsal Footwear

This footwear is designed to prevent or reduce the severity of injury to the metatarsal bones and toe areas of the foot. Metatarsal footwear meeting the ANSI Z41 standard shall have both a toe impact and compression rating and an integral metatarsal guard with an impact rating of 30, 50, or 75.

Heavy-gauge, flanged, and corrugated sheetmetal over- the-foot guards are also available for certain industrial applications. However, no standards have been developed for this equipment.

Conductive Footwear

Conductive protective toe footwear is intended to protect employees from the hazards of static electricity buildup and help equalize the electrical potential between the wearer and energized high-voltage power lines in the wearer's immediate area. This footwear is categorized into types 1 and 2.

Type 1

This conductive footwear dissipates static electricity harmlessly to the ground. It is commonly used to prevent the ignition of sensitive explosive mixtures due to static discharge. Type 1 has resistance rating of up to 500,000 ohms.

Type 2

This type of conductive footwear is intended for use by employees who work on Faraday-type shielded aerial lifts around high-voltage power transmission equipment and where induced voltage is a problem. This type of footwear has a resistance rating of not more than

10,000 ohms and helps equalize the electrical potential between the wearer and the high-voltage source.

All exposed external metal parts on types 1 and 2 conductive footwear are nonferrous. The design and materials of the soles and heels of both types are constructed to create a path through them to ground. However, personnel working near open electrical circuits should *not* use conductive footwear.

Electrical Hazard Footwear

Protective toe electrical hazard footwear is intended to protect workers against contact with exposed circuits of up to 100 volts AC/ 750 volts DC *under dry conditions.* This type of footwear is *not* intended for use in explosive or hazardous locations where conductive footwear is required. Electrical hazard footwear is intended to provide secondary electrical hazard protection on surfaces that are substantially insulated. Because the electrical insulative quality of the heel and sole provides the protection in this footwear, no metal parts should be used in these parts of the shoes.

Static Dissipative Footwear

Protective toe static dissipative footwear is designed to reduce the accumulation of excess static electricity in the body by conducting the charge to ground while maintaining a sufficiently high level of resistance to protect the wearer from electrical hazards. Type 1 static dissipative footwear has a resistance rating of between 1 million ohms (1 megohm) and 100 million ohms (100 megohms). Type 2 static dissipative footwear has the same lower limit but an upper limit of 1 billion ohms (1,000 megohms).

Sole Puncture Resistant Footwear

Protective toe footwear with sole puncture resistance reduces the risk of puncture wounds caused by sharp objects penetrating the sole of the footwear. A protective shield inserted in the sole of the footwear covers an area from the toes to overlap the crest of the heel. This shield must, of course, be flexible yet strong enough to withstand at least a 270-pound force administered by a steel test pin in accordance with ANSI Z41 test procedures.

Foundry Footwear

Specialty types of footwear are available to protect workers in the smelting and foundry industries where employees' feet can be exposed to molten metals. Such footwear often has no fasteners so they can be removed quickly in an emergency. Some NSC members in the foundry and smelting business have reported instances in which workers have suffered serious burns to their feet when unable to quickly remove ordinary footwear that came in contact with molten metal. In these occupations, the tops of the footwear should be covered by protective spats, leggings, and other devices that will prevent the entry of molten metal.

Other Features of Protective Footwear

Other protective footwear features not covered by standards are made available by various manufacturers. They include waterproofing, chemical resistance, and insulation against thermal extremes. Nearly all footwear manufacturers offer some variation on materials used in the soles and upper parts of the footwear for purposes of lengthening the life of the footwear and providing sole slip resistance under various field conditions. Manufacturers of protective footwear should be consulted about the products they offer that will best protect against the hazards that the wearer will encounter on the job.

Cleaning Rubber Boots

If rubber boots are reused by people on the next shift or job, great care should be exercised to disinfect boots after each shift or job. First, the boots are washed inside and outside with a hose containing water under pressure. Then they are dipped into a tub containing a solution of 1 part sodium hypochlorite and 19 parts water. The hose is used again for rinsing, after which the boots are ready for drying. Although other disinfecting agents can be used, this one has been satisfactory and is easily obtainable.

One method is to use a drying rack consisting of a tank with low-pressure steam coils and upright steel-pipe boot holders that permit circulation of hot air inside the boots. After the boots are washed thoroughly and dipped in the disinfecting solution, they are completely dried in about 12 minutes. When several pairs of boots need cleaning, the rack with water jets can be rearranged so that many boots can be washed and rinsed at one time.

SPECIAL WORK CLOTHING

In the modern industrial environment, exposure to fire, extreme heat, molten metal, corrosive chemicals, cold temperature, body impact, cuts from handling materials, and other specialized hazards are often part of what is known as "job exposure." Special protective clothing helps to minimize the effects of these hazards, and manufacturers usually will supply sample swatches of protective materials for testing by individual firms.

Protection against Heat and Hot Metal

Leather clothing is one of the more common forms of body protection against heat and splashes of hot metal. It also provides protection against limited impact forces and infrared and UV radiation.

Garments should be made of good quality leather, solidly constructed, and provided with fastenings to prevent gaping during body movement. Fastenings should be so designed that the wearer can remove the garment rapidly and easily. Workers should not wear turned-up cuffs or other items of clothing that can catch and hold hot metal. Garments either should have no pockets or have pockets with flaps that can be fastened shut.

For ordinary protection against hot metal, radiant heat, or flame hazards somewhat stronger than those in welding operations, wool and leather clothing is used. Specially treated clothing has been developed that is impervious to splashes of metal up to 3,000 F (1,650 C). Wool garment requirements are in general the same as those for leather, except that metal fastenings should be covered with flaps to keep them from becoming dangerously hot.

Asbestos substitutes, including fiberglass or other special high-temperature resistant materials, are available. These materials are effective when made into leggings and aprons usually worn by foundry personnel working with molten metal. Such leggings should completely encircle the leg from knee to ankle, with a flare at the bottom to cover the instep. The design of the leggings should permit rapid removal in emergencies.

If the front part of the legging is reinforced, it can provide impact protection when required. Fiberboard is the most common material used for reinforcement.

When people must work in extremely high temperatures up to 2,000 F (1,090 C)—as in furnace and oven repair, coking, slagging, firefighting, and fire rescue work—aluminized fabrics are essential. The aluminized coating reflects most of the radiant heat away, while the underlying material insulates workers against the remaining heat. Some of these suits consist of separate trousers, coats, gloves, boots, and hoods. Others are one piece from head to foot. Some suits used in industrial operations are air-fed to reduce heat and increase comfort.

Aluminized heat-resistant clothing generally falls into two classes: emergency and fire proximity suits.

Emergency suits are used when temperatures exceed 1,000 F (540 C), as in a kiln or furnace, or when workers must move through burning areas for firefighting or rescue operations. These suits are made of aluminized glass fiber with layers of quilted glass fibers and a wool lining on the inside.

Fire proximity suits are used in areas near high temperature operations, such as slagging, coking, furnace repair work with hot ingots, and firefighting where workers do not enter the flame area. These suits are seldom one-piece construction. They depend primarily on the reflective ability of an aluminized coating on a base cloth of glass fiber or synthetic fiber. Remember, never use fire proximity clothing for situations where workers are required to enter a fire.

Flame-Retardant Work Clothes

Cotton work clothing can be protected against flame or small sparks by flameproofing. One available commercial flame-retardant preparation can be applied to work clothing in ordinary laundry machinery after the garment is washed. Treating the material has two advantages: it makes the cloth highly flame resistant and adds little to the material's weight or stiffness.

Durable flame-retardant work clothes are readily available from many manufacturers. A high-temperature-resistant nylon fabric that chars rather than melts is available for the most severe work conditions. Modacrylic fabrics that resemble cotton are lightweight and have permanent fire-retardant properties. Flameproofed clothing should be marked or otherwise distinguished to reduce the chance that workers will use untreated garments by mistake.

Cleaning Work Clothing

Manufacturer's recommendations should be followed in laundering and cleaning work clothes. Excessive water temperatures or use of certain washing preparations can deteriorate the fabric or affect its properties. Spot-cleaning with organic solvents may soften or dissolve some synthetics, while chlorine bleaches will remove most flame-retardant treatment from cotton.

Workers should not use compressed air for dusting work clothes as it presents a potential for eye injury and may force contaminants into the skin. Instead, workers should use a vacuum system, which prevents dust from being spread into the air where it could be inhaled or get into workers' eyes. Many industrial laundries and industrial clothing rental agencies can advise firms on cleaning and maintaining their work clothing and body.

Protection against Impact and Cuts

A worker's body needs protection from cuts, bruises, and abrasions on most jobs where heavy, sharp, or rough material is handled. Special protectors have been developed for almost all parts of the body and are available from suppliers of safety equipment. See also the section on Hand and Arm Protection.

For example, pads made of cushioned or padded duck will protect the shoulders and back from bruises when workers carry heavy loads or objects with rough edges. Aprons of padded leather, fabric, plastic, hard fiber, or metal can protect the abdomen against blows. Similar devices of metal, hard fiber, or leather with metal reinforcements shield the worker against sharp blows with edged tools. For jobs requiring ease of movement, workers can split their aprons or equip them with fasteners so they fit snugly around the legs.

Leg protection is required on many jobs. Guards of hard fiber or metal are widely used to protect the shins against impact. Knee pads should be worn by mold loftsmen and others whose task requires continual kneeling.

Heat Stress

When selecting PPE, management must keep in mind that some types of equipment may contribute to the potential for heat stress. (See Chapter 12, Thermal Stress, in the Council's *Fundamentals of Industrial Hygiene,* 4th edition, for further information about heat stress.) Workers who wear SCBAs in combination with full body impervious suits, such as those worn on hazardous waste sites, are particularly vulnerable to heat stress. Such workers must be allowed adequate rest breaks. The employer should ensure employees are adequately hydrated and acclimatized to the heat. If necessary, cooling vortexes or vests should be supplied.

Cold-Weather Clothing

In recent years thermal insulating underwear has become popular among outdoor workers because of its lightweight protection against the cold. Thermal knit cotton patterned after regular underwear, quilted materials, or synthetic polyester fabric quilted between layers of nylon are common types of construction.

Although polyester- and nylon-quilted material does not catch fire any more easily than does cotton, once the synthetic material starts burning, it melts, forming a hot plastic mass, not unlike hot pitch, that adheres to skin and causes serious burns. Fire-retardant quilted insulating underwear is now available to combat this danger. Other special fabrics available include a nylon material that chars at a relatively high temperature and does not melt, a glass fiber material for special uses, and a breathable fabric used with a sandwich of cotton or similar material to offer excellent cold-weather protection.

When teaching workers how to dress warmly in cold weather, make sure they check not only the thermometer but wind velocity as well. The temperature may read 35 F (1.7 C), but if there is also a wind of 45 mph (72.4 km/h) it will feel like -35 F (-37 C). (See Figure 18–4, Chapter 18, Emergency Preparedness, in the *Administration & Programs* volume of this *Manual.*) A high wind chill factor, as it is known, means workers must wear more layers against the cold and protect all exposed skin surfaces from frostbite and windburn.

Special Clothing

Safety experts have developed many ingenious, highly specialized types of clothing for protection against special hazards. A partial list includes such items as:

- high visibility and night hazard clothing for construction, utility, and maintenance workers and for police and fire fighters whose work exposes them to traffic hazards
- disposable clothing made of plastic or reinforced paper for exposure to low-level nuclear radiation, for use in the drug and electronic industries, or for hazardous materials work, where contamination may be a problem
- leaded clothing made of lead glass fiber cloth, leaded rubber, or leaded plastic for laboratory workers and other personnel exposed to x-rays or gamma radiation
- electromagnetic radiation suits, which provide protection from the harmful biologic effects of electromagnetic radiation found in high-level radar fields and similar hazardous areas
- conductive clothing, made of a conductive cloth, for use by lineworkers doing barehand work on extra-high-voltage conductors; such clothing keeps the worker at the proper potential

For special applications, manufacturers have a vast number of materials they can draw upon to meet specific hazards.

SUMMARY

- Once a company decides on the use of PPE, it should develop a company policy on PPE usage for employees and visitors, select the proper equipment for the existing hazards, implement a training program, and enforce the use of PPE.
- Companies can encourage workers to use PPE by enlisting the aid of line supervisors and managers, letting employees have some choice in the type of equipment purchased, and establishing a sound training program with consistent enforcement of all rules and regulations.
- All workers exposed to head injury hazards must wear protective headwear to shield them from falling objects, blows, and electric shock and burns.
- Protective devices for the eyes and face include safety glasses, goggles, and face shields. Face shields alone generally do not provide adequate protection against eye injuries and must be combined with basic eye protective glasses or goggles.
- Management must evaluate the workplace for hearing hazards and determine the need for hearing protection

devices. Daily work in steady noise of more than 85 decibels for 8-hour shifts is considered hazardous noise exposure. Hearing protectors are categorized as enclosure, aural, superaural, and circumaural.

- Fall arrest protection, either active or passive, is defined as a means of preventing workers from experiencing disastrous falls from elevations. In selecting the right fall protection system, management should conduct a thorough job survey analysis and establish a fall protection program. Companies must also develop rescue procedures for retrieving a fallen worker from above-ground, below-ground, or confined spaces operations.
- To protect workers from airborne health hazards, management must provide respiratory protection equipment against gaseous, particulate, and combination contaminants and oxygen-deficient environments. Respirators are classified as air-supplying or air-purifying devices. All respirators must be routinely inspected, cleaned, and properly stored to ensure their protective effectiveness.
- Safety footwear includes steel, reinforced plastic, and hard rubber models, depending on the shoe design protective level required. Some jobs require conductive, nonconductive, foundry, or special-design safety shoes to protect workers' feet from injury.
- Special protective clothing is used to shield workers from such workplace hazards as heat, hot metal, chemical splashes, weather extremes, and electrical shock or burns.

REFERENCES

American College of Occupational and Environmental Medicine Noise and Hearing Conservation Committee. Guidelines for the conduct of an occupational hearing conservation program. *J Occup Med* 1987, 29: 981–989.

American Conference of Governmental Industrial Hygienists, Bldg. D–7, 6500 Glenway Avenue, Cincinnati, OH 45211.
A Guide for Control of Laser Hazards, 1976.

American National Standards Institute, 11 West 42nd Street, New York, NY 10036.
Personal Protection—Protective Footwear, ANSI Z41–1999.
Practice for Occupational and Educational Eye and Face Protection, ANSI Z87.1–1989(R 1998).
Respiratory Protection, ANSI Z88.2–1992.
Protective Headwear for Industrial Workers, ANSI Z89.1–1997.
Safe Use of Lasers, ANSI Z136.1–2000.
Safety in Welding and Cutting, ANSI Z49.1–1999.

American Society for Testing and Materials, 1916 Race Street, Philadelphia, PA 19103.
Standard Specification for Rubber Insulating Gloves, D120–70, ANSI J6.6.
Standard Specification for Rubber Insulating Sleeves, D1051–70, ANSI J6.5.

Compressed Gas Association, Inc., Crystal Gateway–1, Suite 501, 1235 Jefferson Davis Highway, Arlington, VA 22202.
Commodity Specification for Air, G–7.1.
Oxygen, C–4.
Oxygen-Deficient Atmospheres, SB–2.

DHHS. *NIOSH Guide to Industrial Respiratory Protection, NIOSH Respiratory Protection Program in Health Care Facilities.* NIOSH Publication No, 99-143, September 1999.

DHHS. *OSHA Technical Manual,* Section VIII, Chapter 2, Respiratory Protection, Appendix B-1 to 1910.134 user seal check procedures, Occupational Noise Exposure, Revised Criteria 1998. NIOSH Publication No, 98-126, June 1998.

DHHS. *TB Respiratory Protection Program In Health Care Facilities, Administrator's Guide.* NIOSH, CDC Publication No. 99-143, September 1999.

Ellis JN. *Introduction to Fall Protection.* Des Plaines, IL: American Society of Safety Engineers, 1988.

Gasaway DC. *Hearing Conservation: A Practical Manual and Guide.* Englewood Cliffs, NJ: Prentice-Hall, Inc., 1984.

Industrial Safety Equipment Association, 1901 North Moore Street, Suite 501, Alexandria, VA 22209.

Mack Publishing Company, 208 Northampton Street, Easton, PA 18042.
U.S. Pharmacopoeia.

National Fire Protection Association, 1 Batterymarch Park, Quincy, MA 02269.

National Safety Council, 1121 Spring Lake Drive, Itasca, IL 60143.
Hearing Conservation in the Workplace, 1991.
Occupational Safety and Health Data Sheets (available in the Council Library):
Flexible Insulating Protective Equipment for Electrical Workers, 12304–0598.

Plog BA, ed. *Fundamentals of Industrial Hygiene,* 4th ed. Itasca, IL: National Safety Council, 1996.

Respiratory Protection—An Employer's Manual and Guide to Industrial Respiratory Protection. National Institute for Occupational Safety and Health, Division of Technical Services. Cincinnati, OH 1978.

Respiratory Protective Devices Manual. Committee on Respirators, P.O. Box 453, Lansing, MI 48901.

SEI Certified Products. Safety Equipment Institute, 1901 North Moore Street, Suite 501, Arlington, VA 22209.

U.S. Department of the Interior, 1849 C Street, NW, Washington, DC 20240. 30 *CFR* Chapter 1, Subchapter B, Respiratory Protective Devices, Tests for Permissibility, Fees; Part 11. Note. *The Code of Federal Regulations* is available through the U.S. Government Printing Office, Washington, DC 20402.

U.S. Department of Health and Human Services, Public Health Service, Centers for Disease Control, National Institute for Occupational Safety and Health.
Criteria for a Recommended Standard: Occupational Noise Exposure. Revised Criteria, June 1998.
CFR 1910.95, Occupational Noise Exposure.
CFR 1926.52.

NIOSH Certified Equipment List, October, 1986, Pub. 87–102. Washington, DC: U.S. Superintendent of Documents, 1986.

REVIEW QUESTIONS

1. What are the three broad categories of methods used to control harmful exposures to hazardous substances?
a.
b.
c.

2. Define personal protective equipment (PPE).

3. Information on certified equipment is available through which of the following?
a. Environmental Protection Agency (EPA)
b. Safety Equipment Institute (SEI)
c. National Institute for Occupational Safety and Health (NIOSH)
d. All of the above
e. Only b and c

4. Which of the following adds considerably to the protection offered by a helmet?
a. Chin strap
b. Bump cap
c. Paint applied after manufacture of helmet
d. All of the above

5. Name the standard established by the American National Standards Institute (ANSI) for eye and face protection.

6. The aspect of protective eye lenses that provides the filtering effect against infrared and UV radiation is:
a. Color
b. Chemical composition
c. Cost
d. All of the above

7. Briefly describe the four types of hearing protection devices.
a.
b.
c.
d.

8. Name the two classifications of fall protection systems.
a.
b.

9. List three devices used in passive fall arrest systems.
a.
b.
c.

10. How often should audiometric testing be done on employees?
a. When new employees are hired
b. When new employees are hired and annually thereafter
c. Annually
d. Every two years

11. Name five components of active fall arrest systems.
a.
b.
c.
d.
e.

12. What are the three steps in selecting respiratory protective equipment?
a.
b.
c.

13. The two main categories of respirators are:
a.
b.

14. Gloves or mittens having metal parts or reinforcements should never be used around:
a. Hazardous chemicals
b. Edged tools
c. Electrical apparatus

15. Protective footwear is classified according to its ability to meet what two requirements?
a.
b.

8

Industrial Sanitation and Personnel Facilities

Revised by

Philip E. Hagan, MPH, CIH, CHMM, CHCM

This chapter will cover both the essential areas involved in employee health and the basic facilities used for employee sanitation. The topics include:

- issues of drinking water quality at the worksite
- appropriate disposal of company sewage, waste, and garbage
- keeping employee facilities safe and sanitary
- issues of nutrition and sanitation in employee food service

Companies must give adequate attention to these areas if employees are to work efficiently, with the assurance that their health and welfare are well protected.

Management should provide employees with a sanitary work environment. To make sanitation safe, efficient, orderly, and economical, management must supervise it properly and integrate it effectively with production and maintenance.

The general rules for sanitation include:

- an approved piping and storage system
- good housekeeping—as clean as the nature of the work allows
- personal cleanliness
- a good inspection system

Where wet processes are used, workers and managers must have proper drainage in the work area.

The department director or supervisor responsible for maintaining the work environment must have appropriate authority within the organization. This position should give the individual the authority to monitor the entire company or facility environment to ensure that it meets appropriate levels of cleanliness and order. Some firms appoint a director of maintenance or facility services (which may include safety and health).

DRINKING WATER

Most facilities receive water for drinking, washing, and food preparation from a municipal supply. When delivered to the facility meter, this water must meet all applicable regulatory standards and meet the criteria for classification as potable water (i.e., that it is safe and satisfactory for drinking and cooking).

In-facility Contamination

Simply because water is potable when delivered to the facility meter does not necessarily mean that it remains so. Many opportunities exist within a facility for water to become contaminated.

Adulteration of water supplies can occur due to microbiological, chemical, or physical contamination. Waterborne disease outbreaks occur from exposure to microbial agents such as *Salmonella, Campylobacter, Giardia, Legionella, Hepatitis, Listeria,* and *Cryptosporidium* (See Table 8–A for several examples). Mercury, benzene, lead, copper, and asbestos are some common chemical contaminants. Physical contamination can result in discoloration, bad taste, and high turbidities. Although a high turbidity has no direct adverse health impacts, it could interfere with the water treatment process and provide a medium for microbial growth.

One of the most common ways a water supply becomes contaminated is by mixing potable with nonpotable water. This can occur through an actual or potential physical connection between a potable water system and an unapproved water supply or other source of contamination. A cross-connection provides a pathway and backflow provides the driving force that delivers the contaminant to the potable water system. Backflow is a reverse flow condition, created by a difference in water pressures, causing water to flow back into the distribution pipes of a potable water supply from any unintended source(s). Contamination risk from cross-connection scenarios is greatest when utilities are improperly installed, poorly maintained, or old.

Other causes of contamination include improper maintenance of drinking and cooking facilities and improper installation of plumbing facilities, permitting back-siphonage of used water. Back-siphonage is a form of backflow to parts of the system that have fallen below atmospheric pressure. For example, a company pumps well water into an application tank to dilute stock chemicals, such as pesticides. If the pumping is interrupted, perhaps by a power outage, a vacuum is created by the pumping interruption. This situation could back-siphon the solution—now a toxic chemical—into surrounding wells. Such back-siphonage incidents can affect users of both company-maintained and publicly maintained water supplies. This type of contamination rarely occurs because air gaps are usually in place to prevent this from happening. An air gap is a physical separation between a potable water supply and a nonpotable water system that prevents the contamination of drinking water by back-siphonage or backflow because the nonpotable water cannot reach the potable water. An air gap is often required by code. In instances where the use of an air gap is not feasible, many other backflow prevention devices are available for use. The impact of a back-siphonage incident could be much worse if it happened at an establishment such as a food-processing facility. Just a few disease organisms can infect an entire piping network if they are back-siphoned to the community well or pump station.

Because of the serious consequences of contamination, facility management must ensure that the integrity

Table 8–A. Infectious Waterborne Diseases Caused by Drinking Contaminated Water

Diseases	*Incubation Period*	*Symptoms*	*Frequency*	*Mode*
Gastroenteritis	Variable	Lethargy, nausea, diarrhea, cramps, and other stomach ailments	Causes over ½ of waterborne disease	Sewage or chemicals in water
Bacterial shigellosis	1–7 days	Fever, vomiting, stomach cramps, diarrhea	Serious in some cases. Common	Sewage in water
Salmonellosis	6–72 hours	Abdominal pains, fever, vomiting, and nausea	Less common	Sewage in water
Typhoid fever	1–3 days	Abdominal pains, fever, chills, diarrhea or constipation, and tearing of the intestines	Rare occurrences	Sewage in surface water
Giardiasis	1–7 days	Chronic diarrhea, weight loss, intestinal and stomach gas, bloating, and anorexia	Outdoor enthusiasts commonly affected	Surface water and food
Hepatitis A	14–45 days	Jaundice, nausea, anorexia, fever, and general physical discomfort	Rare in the U.S.	Drinking and swimming

of the facility's drinking water system is maintained. If the facility has piping systems containing water that is used for nondrinking purposes, i.e., sprinklers, fire hydrants, or manufacturing processes, then these systems should be clearly identified, particularly at outlets. Facility services should see that no direct connection exists between drinking water and other water systems except through a properly installed, approved backflow prevention device. In addition, facility services should have eliminated any long dead-end runs of pipe that cannot be flushed or drained and that might serve as a reservoir for contaminated water. Finally, facility services should ensure that the location of piping approved for potable water use is easily identifiable.

Where the possibility of misusing or cross-connecting pipelines exists, all nonpotable water lines should be clearly marked as unsafe for drinking, use in food areas or personal service rooms, or washing utensils, clothes, or people. Nonpotable water may be used for cleaning work premises (other than food preparation and personal service rooms), provided it does not contain harmful concentrations of chemicals or microbial contamination.

Plumbing

Construction workers should install fixtures and faucets to prevent back-siphonage of contaminated water if the pressure drops in a supply line. Faucets and similar outlets should be at least 1 in. (2.5 cm) above the floodrim of the receptacle. To prevent backflow into the drinking water supply, it may be necessary to place surge tanks and air gaps in the fill lines to process equipment.

Another common source of water supply contamination is open joints in underground supply lines that may allow seepage from groundwater or water from leaky sewers to enter the pipes. This condition can arise where pipes are subject to vibration or corrosion and the joints between pipes open mechanically or the pipe sections crack. Codes usually prohibit installation of sewer and drinking water lines in the same trench, unless the sewer line is placed at a much greater depth and angled to prevent backflow into the water pipes.

If the supply for sprinklers and fire hydrants is the same as that for drinking water, hydrant drains or "weeps" connected directly to sewer lines may be a source of contamination. An open standpipe or reservoir may also permit contamination.

Frequently, contamination of the water supply results when a system is opened for repair or alteration. The new pipe may not be disinfected and properly flushed with clean water before being put back into service. In addition, to reduce another source of contamination, the U.S. Safe Drinking Water Act restricts the use of lead pipe, solder, or flux in the installation or repair of any public water system or in residential plumbing connected to a public water system.

Private Water Supplies

Industrial establishments in outlying districts commonly supply and treat their own water from private sources. Such water treatment installations should be built and operated under the supervision of a thoroughly trained and experienced sanitary engineer. Information in this and subsequent sections provides general background but is not a substitute for appropriate professional staff or consultants.

All underground and surface waters considered for drinking purposes should be viewed as contaminated until proved otherwise. Water supplied from private sources for the personal use of facility personnel should meet the appropriate health and environmental regulations. As a rule, groundwater collected from deep-drilled wells will be free of biological contamination but may be affected by various minerals at levels that degrade taste, odor, or other aesthetic characteristics. In

Table 8–B. MCLGs and MCLs for Inorganic Contaminants

	MCLGs	MCLs
(1) Asbestos	7 million fibers/liter (longer than 10 μm)	7 million fibers/liter (longer than 10 μm)
(2) Cadmium	0.005 mg/l	0.005 mg/l
(3) Chromium	0.1 mg/l	0.1 mg/l
(4) Lead	Zero action level = 0.015	—
(5) Mercury	0.002 mg/l	0.002 mg/l
(6) Nitrate	10 mg/l (as N)	10 mg/l (as N)
(7) Nitrite	1 mg/l (as N)	1 mg/l (as N)
(8) Total Nitrate and Nitrite	10 mg/l (as N)	10 mg/l (as N)
(9) Selenium	0.05 mg/l	0.05 mg/l

MCL: Maximum Contaminant Level (expressed as mg/l)
MCLG: Maximum Contaminant Level Goal

contrast, shallow wells are more likely to have biological and synthetic chemical contamination.

Water drawn from surface sources, such as lakes and streams, should always be treated with disinfectant and filtered for potable use. Many types of both biological and chemical contamination can come from surface water sources; therefore, companies should obtain professional advice in the design and operation of a surface water treatment system.

Because a company may have several different choices for its water supply, the supply choice will depend on daily water requirements, the amount of treatment required so that water from each source meets purity standards, and the potential each source has for additional contamination.

The daily per-person water requirements of an industrial facility can be estimated as follows: 15–20 gal (55–75 l) for drinking, lavatory, and toilet usage; 20–25 gal (75–95 l) per shower; and 5–10 gal (20–40 l) per meal if food is prepared on the premises.

Table 8–C. MCLGs and MCLs for Volatile Organic Contaminants

	MCLGs (mg/l)	MCLs (mg/l)
(1) o-Dichlorobenzene	0.6	0.6
(2) cis- 1,2-Dichloroethylene	0.07	0.07
(3) trans 1,2-Dichloroethylene	0.1	0.1
(4) 1,2-Dichloropropane	0	0.005
(5) Ethylbenzene	0.7	0.7
(6) Monochlorobenzene	0.1	0.1
(7) Styrene	0.1	0.1
(8) Tetrachloroethylene	0	0.005
(9) Toluene	1	1
(10) Xylenes (total)	10	10

Water Quality

An organization must evaluate its water supply source on the basis of the contaminants it may contain. Tables 8–B through 8–H list the limiting concentrations of two classes of contaminants. U.S. regulations have established two sets of standards relating to the quality of drinking water: primary and secondary. Primary standards are mandatory and cover all contaminants considered health hazards. These contaminants fall into the general categories of chemicals, organic chemicals, physical parameters, microbial agents, and radioactivity. Secondary standards cover the aesthetic qualities of water, such as taste, odor, and color. Although compliance with secondary standards is generally not mandatory, the EPA has given states the choice to adopt them as enforceable standards, and a prudent course of action would be to check the status in your jurisdiction. In any case, it is strongly recommended that all potable water meet these criteria as well.

Table 8–D. MCLGs and MCLs for Pesticides/PCBs

	MCLGs	MCLs (mg/l)
(1) Alachlor	Zero	0.002
(2) Atrazine	0.003 mg/l	0.003
(3) Carbofuran	0.04 mg/l	0.04
(4) Chlordane	Zero	0.002
(5) 1,2-Dibromo-3-chloropropane (DBCP)	Zero	0.0002
(6) 2,4-D	0.07 mg/l	0.071/l
(7) Ethylene dibromide (EDB)	Zero	0.00005
(8) Heptachlor	Zero	0.0004
(9) Heptachlor epoxide	Zero	0.0002
(10) Lindane	0.0002 mg/l	0.0002
(11) Methoxychlor	0.04 mg/l	0.04
(12) Polychlorinated biphenyls (PCBs) (as decachlorobiphenyl)	Zero	0.0005
(13) Toxaphene	Zero	0.003
(14) 2,4,5-TP (Silvex)	0.05 mg/l	0.05

Table 8–E. MCLGs and Treatment Technique Requirements for Other Organic Contaminants

	MCLGs	MCLs
(1) Acrylamide	Zero	Treatment technique
(2) Epichlorohydrin	Zero	Treatment technique

Table 8–F. Secondary Maximum Contaminant Levels (SMCLs)

(1) Aluminum	0.05 to 0.2 mg/l
(2) Silver	0.1 mg/l

NOTE: Because U.S. regulations are constantly being revised, managers of all potable water supplies should request updates of regulations and guidance materials from the appropriate agencies.

The equipment needed to treat water and make it potable depends on the degree of contamination and the potential for heavier contamination of the water source in the future. These factors can be evaluated only on the basis of a thorough sanitary survey of the water source. Such a survey will determine not only the type of treatment necessary but also the nature and frequency of periodic laboratory tests of the source water and the treated water.

Table 8–H. Secondary Maximum Contaminant Levels

Contaminant	Level
Aluminum	0.05–0.2 mg/l
Chloride	250 mg/l
Color	15 color units
Copper	1.0 mg/l
Corrosivity	Noncorrosive
Fluoride	2.0 mg/l
Foaming agents	0.5 mg/l
Iron	0.3 mg/l
Lead	0.015*
Manganese	0.05 mg/l
Odor	3 threshold odor number
pH	6.5–8.5
Silver	0.1 mg/l
Sulfate	250 mg/l
Total dissolved solids (TDS)	500 mg/l
Zinc	5 mg/l

*Action Level

Wells

The safest source of water is usually a drilled well with intake below the water table. Typically such wells show a reliable yield and are free from bacterial and chemical contamination. If a company uses both well and city water, facility services must be sure that no cross-connection occurs between the two systems. The local code provides specific requirements.

The wellhead should be located as far as possible from sewage lines, septic tanks, and sewage drainage

Table 8–G. Best Available Technologies to Remove Inorganic Contaminants

Inorganic contaminent	Best Available Technologies									
	Activated alumina	Coagulation/ filtration[2]	Corrosion control	Direct filtration	Diatomite filtration	Granular activated carbon	Ion exchange	Lime softening[2]	Reverse osmosis	Electro-diolysis
Asbestos		X	X	X	X					
Barium							X	X	X	X
Cadmium		X					X	X	X	
Chromium III		X					X	X	X	
Chromium VI		X					X		X	
Mercury		X[1]				X		X[1]	X[1]	
Nitrate							X		X	X
Nitrite							X		X	
Selenium IV (Selenite)	X	X						X	X	X
Selenium VI (Selenate)	X						X	X	X	

[1] BAT only if influent mercury concentrations exceed 10 µg/l. Coagulation/filtration for mercury removal includes PAC addition or post-filtration GAC column where high organic mercury is present in source water.
[2] Not 1415 BAT for small systems for variances unless treatment is currently in place.

fields or process waste-disposal systems. A 200 ft (61 m) separation is usually considered a minimum distance. Toxic chemicals that have leaked from pipelines, tanks, or lagoons or that have been spilled or disposed of on the land surface can contaminate a well at much greater distances. Such contaminants may present a serious health threat. Pesticides and herbicides spilled on or applied to land in the area of a well can also contaminate water at a considerable distance, depending upon local hydrogeology.

Government agencies can take steps to prevent contamination of underground water caused by seepage of surface waters. These agencies can require that the space between the well casing and the surrounding area be sealed with a cement grout to a minimum depth of 10 ft (3 m) below the finished ground level or floor. As a further precaution, the casing should be grout-sealed to the lowest impervious stratum it passes through.

Well installations must meet the current state well code or standards for such design considerations as siting and construction materials. Most state codes require that the top of the well casing projects at least 12 in. above the ground surface. The wellhead should not be covered over by paving or other material that would make access difficult.

Submersible, turbine, or jet pumps may be considered in any particular well. Submersibles are located in the well and do not require a pump house. Local water or sanitation agencies can add a safety factor by requiring two wells and two pumps for a water supply. In this way, if either unit fails, an automatic alternator can switch over to the other pump.

Disinfecting the Water System

The pipes, reservoirs, standpipes, pump, and well casing of a new system should be thoroughly disinfected before being placed into service. Old systems carrying treated water for the first time following extended disuse should also be disinfected, on the discharge side of the treatment facility. Likewise, a system that has been opened for repairs should be disinfected before being placed back into service.

Workers can disinfect a drinking water system more easily by filling it with water containing not less than 100 mg/l of available chlorine. The solution should remain for 24 hours in either a new system or one that has not previously carried treated water. For systems that have carried treated water and are being put back into service following minor repairs, 12 hours will be sufficient. However, if officials suspect that *Giardia lamblia* or *Cryptosporidium* are present in the water, other measures should be taken, such as filtering. Disinfection of water can be accomplished by the use of chlorine containing products, ozone, and ultraviolet radiation. Each type of disinfection has unique advantages and disadvantages. Overall effectiveness is based on such parameters as dose of disinfection product, temperature, contact time, and quality of the supply water.

A company can determine the success of a disinfecting job by measuring the residual chlorine in the solution at the end of the required time. Test kits for this purpose are commercially available and easy to use. Tests will show residual chlorine if the biological chlorine demand of the system has been met. Workers may connect the system to the drinking water supply, flush it out, and put it into service. If no residual chlorine is present, workers should drain the system and recharge it with new disinfectant solution and then repeat the procedure.

If the system contains a standpipe or reservoir, workers should add the disinfectant solution through it. Otherwise, workers can add the solution in a temporary reservoir on the supply side of the system pump and inject the solution through it. A solution containing 500 mg/l available chlorine, applied with a fog nozzle, will disinfect standpipes and covered reservoirs.

Underground water supplies can become contaminated while being developed. If so, they will have to be disinfected in a sequential fashion. After determining the 24-hour yield of the well, workers should run the test pump to clear the well of turbidity. They should then add a chlorine solution to the well that, with the 24-hour yield, will make a chlorine solution of 500 mg/l. The permanent pumping equipment should be connected to the wellhead and operated until the discharge has a distinct odor of chlorine.

Several methods are available to distribute the disinfectant solution uniformly throughout the water system. Workers may seal the well casing and inject the solution under pressure, or they may add the solution from a hose or a small pipe several levels beneath the surface of water in the well. The chlorine solution should remain in the well for 24 hours.

Water Purification

Of the several methods of water purification available, filtration and chemical disinfection are the most practical for industrial private water supplies.

Filtration

This method of water purification primarily clears turbid waters; however, it can also remove some bacterial contamination. Filtering facilities are usually two types: slow filters and rapid filters.

Slow sand filters will clarify turbid waters when operated at a rate of 25–50 gal/day/ft^2 (100–200 l/day/m^2) of filter area. Such filters should be made with 0.015–0.019 in. (0.25–0.35 mm) sand and should be at least 20 in. (50 cm) deep, but preferably 36–40 in. (90–100 cm) deep. While the filters are operating, the film that accumulates on the surface of the sand must be kept below water level because this film increases the effectiveness of the filter. Rapid sand filters, made using a uniform 0.015–0.019 in. (0.4–0.5 mm) sand with a depth of 30 in. (75 cm), will handle about 3,000 gal of turbid water a day per sq ft (12,000 l/day/m^2) of filter surface.

Both filters should be made under competent engineering supervision and operated under continuous inspection. Depending upon the water source, filters can require a presedimentation basin for preliminary water treatment. Facility services should purchase filters in pairs so that workers can remove one for cleaning and maintenance without disrupting the supply of filtered water.

Disinfection

Chlorine is generally the best available disinfecting agent for drinking water. It can be added to the water directly as a gas or as a soluble salt (calcium hypochlorite or chlorinated lime—refer to the National Safety Council's *Fundamentals of Industrial Hygiene*, 4th ed., for hazards of these chemicals).

Small-capacity chlorinators, which inject gaseous chlorine into a water system, are available and easy to operate. However, it is better to install injection pumps that supply high-concentration chlorine solutions to the system at a proper rate, because they are safer and easier to operate.

Companies must maintain standby equipment at all chlorinating stations, along with an adequate supply of spare parts. Supervisors should keep on hand gas masks that are effective against chlorine and a small bottle of ammonia to test for leaks outside areas where chlorine is stored or used. Make sure masks are inspected regularly, and train authorized employees in emergency procedures.

The chlorinator should be adjusted to leave a chlorine residue of about 0.2 mg/l in the water after 20 min of contact between chlorine and the untreated water. Test kits are available that can quickly measure residual chlorine.

It is easy to disinfect small quantities of water for emergency use in one of several ways. Commercial preparations provide good protection but must be used according to the manufacturer's instructions. Workers can use point-of-use treatment such as boiling, chlorine, iodine and filtering. Safe drinking water can be produced by vigorous boiling for one minute. As long as contamination from *Giardia lamblia* or *Cryptosporidium* is not a problem, chemical disinfection (chlorine or iodine) can be used to make water safe to drink. Safe drinking water can also be produced by boiling water for five minutes or by adding four drops of household bleach (hypochlorite solution, 4% available chlorine) or two or three drops of common tincture of iodine to a quart of water and allowing it to stand for 30 min. If the water is cloudy, the chemical disinfection dose should be doubled. Workers can lessen the water's flat medicinal taste after disinfection by pouring it from one container to another to aerate it, by adding a pinch of salt, or by letting it sit for awhile.

Water Storage

Reservoirs or standpipes for treated water should be enclosed completely and located to prevent accidental contamination. A reservoir large enough to hold a 48-hr reserve supply of treated water will provide adequate supplies in the event of unusually heavy water use or supply failure. Fit all vents with screened downspouts well above floor level. Entrance manholes should be enclosed by watertight frames at least 6 in. (15 cm) higher than the surrounding surface and fitted with watertight covers extending at least 2 in. (5 cm) down the outside of the frames. Make sure the cover is closed and locked when the entrance is not in use.

A reservoir permits full use of a small well and pump and still provides a buildup for peak demand.

WASTE DISPOSAL

Industrial facilities must ensure that a careful evaluation of all waste streams (wastewater, solid, and hazardous waste) is conducted to ensure that disposal is in accordance with regulatory requirements. Although many state and local agencies regulate the disposal of solid and hazardous wastes, the EPA has combined regulations governing solid and hazardous waste programs. The disposal of process waste requires special handling and packaging provisions if the material meets the definition of regulated hazardous waste, including documentation from "cradle to grave" (See Chapter 8, Hazardous Wastes, in NSC's *Accident Prevention Manual for Business & Industry: Environmental Management*).

Some industrial facilities must provide their own wastewater disposal systems and all must adhere to applicable regulations governing disposal. The company must also provide for proper storage and collection or disposal of solid waste.

Building Drains and Sewers

A firm's in-facility sanitary sewage collection system should conform to regulatory standards. Every fixture should be properly trapped and vented by drain(s) and stack(s) to prevent discharging sewer gases into the building and to ensure proper drainage. Facility services must see to it that traps and especially grease interceptors (such as those placed in waste pipes serving the facility cafeteria and kitchen) and interceptors designed to collect other particulate foreign materials are large enough, located for easy access, and cleaned periodically.

The building drain and sewer may be constructed of extra heavy cast-iron bell-and-spigot pipe with drainage fittings. This material is less susceptible to clogging and much easier to clean than pipes made from other materials. It also provides "insurance" when installed under floors that would be expensive to tear up. The sewer should be tight under a 10-ft (3 m) head of water. Plastic and copper plumbing materials are generally approved for drainage.

A company should provide a cleanout where the building drain passes through the building wall and at other selected locations. Local codes for trap size, permitted locations, and strainer requirements should be checked. Codes may call for installation of backwater valves to prevent backup in the sewer line. These valves should be placed where they will be accessible for inspection and cleaning.

Wastewater Disposal

The most common form of private sewage treatment involves use of a septic tank. The septic tank is comparable to the primary treatment phase of a municipal sewage treatment facility. Septic tanks are buried, watertight receptacles designed and constructed to receive wastewater; separate solids from liquids; provide limited digestion of organic matter; store solids; and allow the effluent, clarified liquid to be discharged for further treatment. To retain the solids, the septic tank uses a baffle on the discharge line. The retention of solids is essential to prevent malfunctioning of the secondary treatment component.

The effluent from the septic tank should pass through a watertight sewer that meets regulatory codes, over as short a distance as possible to the secondary treatment phase of the private sewage disposal system. The type of secondary treatment component installed may depend on the soil conditions of the proposed site. Determining the soil conditions may involve performing a soil permeability evaluation or a percolation test by a qualified individual or using some other diagnostic method approved by the state or local agencies. The secondary treatment component may consist of a subsurface seepage field, a waste oxidation lagoon, an aerated treatment system, or another type of sewage treatment system meeting state or local codes.

Sewage disposal systems should be located and constructed to prevent contaminating the groundwater or polluting the surface water (Figure 8–1). All sewage systems must be maintained to avoid causing a nuisance or health hazard to the community. For more information concerning acceptable secondary treatment and proper location, contact the local or state health department, or other regulatory agencies.

Solid Waste Disposal

Companies and facilities providing food services for their employees must arrange for proper disposal of food wastes and other refuse. Several methods of

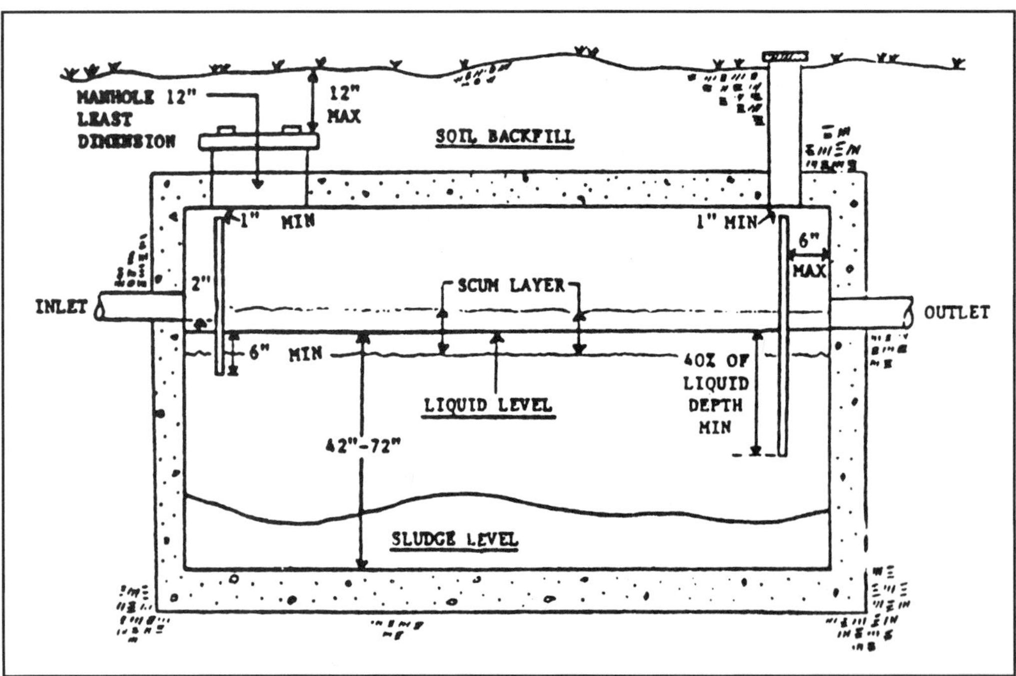

Figure 8–1. Septic tank with slip-in baffles.

disposal are available; follow local ordinances in each case.

Many facilities collect garbage and store it for pickup and disposal by municipal or private collection services. Companies should fence in any large outside garbage receptacles and compactors and lock the gates to keep children out and to prevent animals from strewing refuse about.

Metal or plastic garbage and refuse containers with tight-fitting covers should be used to keep out insects and rodents. Containers should be easy to clean and handle, and should be washed periodically with a detergent-deodorant solution. Many waste-hauling vendors require the garbage in containers to be kept in plastic bags, paper bags, or liners, because bags or liners allow the refuse to be removed easily and help keep containers clean. Place enough garbage containers in the facility so employees will not throw food waste in waste baskets or other unsuitable receptacles. Proper separation of different waste products may be required. Management should schedule frequent collections to prevent garbage from accumulating.

With the reduction of available landfill space, some states and communities are imposing mandatory waste reduction and recycling programs on businesses. The goal of these efforts is to remove recyclable materials from the waste stream and to reduce the overall amount of garbage generated. This goal can be accomplished through pollution prevention efforts such as source reduction, reuse, substitution, and composting. Management should work with their waste-hauling vendor to develop feasible programs that reduce waste and recycle materials and chemicals that can be recycled.

In areas where recycling efforts are focused, unique hazards may be present from sorting and collecting recyclable materials. Glass, metals, paper, batteries, chemicals, and even cardboard collection can pose a safety and health risk to the employees handling those materials. Use of mechanical devices is encouraged to avoid injuries from improper lifting or awkward postures. Baling or compacting machines are often used to compress and tie together cardboard or polystyrene plastic. Safeguards should be maintained to avoid eye or compression injuries. Cut hazards arise from glass and metal collection. Proper handling and storage, and proper disposal of waste batteries and chemicals are important to prevent exposure, injury, and environmental damage.

Many localities allow firms to discharge their ground food waste into the municipal sewage system. Local ordinances governing this practice will provide details. Installation of food-waste disposers in the facility kitchen can provide a convenient, efficient way to reduce the amount of garbage that must be stored and collected. Supervisors should ensure that disposers are located, grounded, and installed according to approved plumbing practices and local code requirements.

Kitchen employees should be instructed to use nonmetallic tampers, keep silverware out of the disposer, and clean the metal trap daily. Before workers clean or clear the disposer, they should make sure that it is completely stopped and disconnected from its power outlet.

Refuse Collection

Hazards in refuse collection vary with the type of equipment used and the conditions surrounding the operation. A frequent cause of accidents involves packing blades that cause injury to fingers, hands, arms, and feet. Other hazards arise from "booby traps" unwittingly laid by the companies that refuse collectors serve—loose, broken glass in a refuse container; lightweight trash cans filled with heavy objects (like chunks of concrete); heavy objects concealed by paper or other trash; or hose or other obstacles strewn along the pathway to a rubbish can. Containers that are rusted through or have unserviceable handles increase the risk of injury to refuse workers.

The National Safety Council's *Work Injury and Illness Rates* indicates that cuts, lacerations, and punctures accounted for about 14% of the lost-time injuries in the refuse collection industry; this statistic compares favorably to industry in general, where 17% of workers suffered such injuries. Wearing heavy work gloves minimizes these types of injuries.

Unique hazards may arise from sorting and collecting recyclable materials. Collecting waste glass, metals, batteries, chemicals, and even cardboard can pose a safety and health risk to the employees handling such materials. Use of mechanical devices is encouraged to avoid injuries from improper lifting or awkward postures. Baling or compacting machines are often used to compress and tie together cardboard or polystyrene plastic. Refuse collectors should maintain safeguards to avoid eye or compression injuries. Cut hazards arise from glass and metal collection. Proper handling, storage, and disposal of waste batteries and chemicals is important to prevent exposure, injury, and environmental damage. If employees are exposed to blood products, then a bloodborne pathogens program must be instituted.

Insect, Rodent, and Nuisance Bird Control

In industrial and commercial sites where insect, rodent, and nuisance bird infestations are a problem, it is always best to employ a professional pest-control operator. To obtain a list of licensed pest-control operators, companies should contact their local health agencies.

The hazards of poisonous chemicals and applications should not be risked by personnel with little or no previous experience in pest control. In some places, toxic insect-, rodent-, and bird-control chemicals may be handled only by persons with appropriate training and licensing. Because of the toxicity and resulting hazards, firms should consider or use alternative methods of pest control. These methods of pest control include removing harborages and feeding sites, pest-proofing structures, and so on. Pests contribute to the bacterial problem of waste disposal and are often the limiting factor for obtaining good sanitary practices.

Integrated Pest Management (IPM), a very popular alternative methodology used to control pests and minimize the use of pesticides, is based on a collaborative approach between housekeeping, maintenance, and pest control services. Operational and administrative intervention strategies are used to reduce the amount of pesticides needed to control pests. The implementation of an IPM program tends to be site-specific, and cost and applicability should be examined before implementation. In some settings, IPM may not be feasible.

Typical elements to be included in the implementation of a successful IPM program include sanitation practices, facility design and maintenance, record keeping, use of nonpesticide methods for pest control, minimizing the use of preventative applications of pesticides, and program monitoring and evaluation on a continuous basis. Sources of information on the implementation of IPM include state and local agricultural departments, pest management consultants, or pest management firms.

Management should notify all departments in the company of a pest-control operator's visit well in advance. When ordering the fumigation of a large facility, such as a group of buildings, railroad cars, or grain elevators, management should advise the local fire department in advance. Outside signs can be used to warn neighboring complexes or residents to keep children and pets away from treated areas.

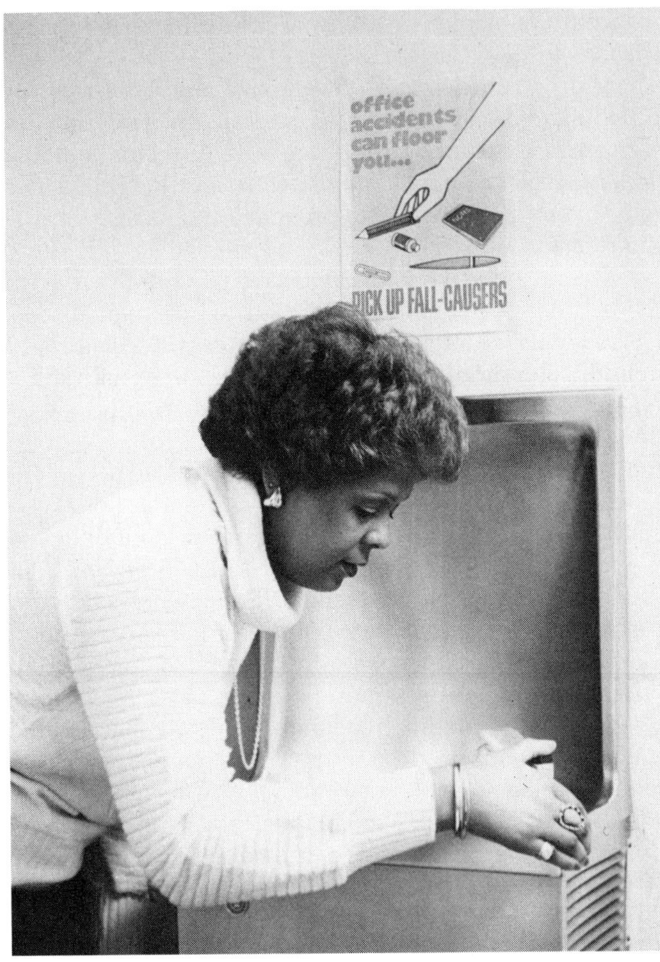

Figure 8–2. Fountains should be conveniently located so that employees can maintain their daily water intake. A safety poster can be placed near the fountain.

PERSONAL SERVICE FACILITIES

The company should conveniently locate all personal service facilities contributing to employee comfort, such as drinking fountains, washrooms, locker rooms, showers, and toilets. These facilities make up an essential part of the occupational health program in most industries.

Drinking Fountains

Sanitary drinking fountains, one to about every 50 persons, should be installed at convenient places throughout an industrial facility, in accordance with local code requirements. The fountain should have an angle jet and a lip guard (Figure 8–2), and may have a waste container for employees to discard papers or cups. It is important to be sure that the stream projector cannot be flooded or submerged should the water stream be stopped. Also, have the installer direct and project the stream so that users cannot contaminate it. In dusty areas, fountains should be covered.

The water temperature should be 50–55 F (10–13 C) for heavy manual labor, or 45 F (7 C) for less-active office work. An ice dispenser should be kept separate from water, or ice can be provided from an ice maker.

Where city water is available on construction worksites, the company can extend a water line to upper floors as the building is being erected. On each floor, a standard drinking fountain can be installed.

On worksites involved in highway, pipeline, and power-line construction, timber clearing, and the like, the drinking water source is usually so remote that it is impractical to pipe water to the job. Some companies have successfully solved this problem by using portable drinking fountains. These fountains have an insulated tank equipped with an angle jet drinking nozzle. The tank has an air pump and pressure-release valve so that it can be pumped up to the necessary operating pressure. Keep containers scrupulously clean, and sterilize them daily with steam, boiling water, or chlorine solution.

Under no circumstances should workers use a common drinking cup or ladle. If drinking cups are required, they should be single-service paper cups kept in a sanitary container at the drinking faucet, with a receptacle provided for disposal. ANSI standard A117.1–1986, *Buildings and Facilities—Providing Accessibility and Usability for Physically Handicapped People,* section 5.7, mentions modifications necessary for use by the physically challenged.

Codes generally do not permit installation of drinking fountains in any toilet room. Nor do they permit installation of bubblers as an integral part of or as an accessory to another fixture, such as a lavatory or sink.

Carafes (vacuum bottles) that are frequently used in private offices are a potential source of bacterial contamination. Employees should rinse and refill them daily and clean them periodically, using a sanitizer such as a cationic quaternary ammonium germicide.

Figure 8–3. These employees must "prep" before they can enter critical, super-clean areas. Washing hands and rubber gloves, then drying them under air dryers, prevents lint and dust contamination. (Printed with permission from Western Electric Company, Allentown, PA.)

Washrooms and Locker Rooms

ANSI Z4.1, *Minimum Requirements for Sanitation in Places of Employment*, serves as a guide to the types and sizes of washrooms, locker rooms, and accessories. A large, single washroom and locker room for each gender may be sufficient for a compact facility or an establishment employing fewer than 500 people. Washrooms in a large one-story facility generally are scattered throughout the building. If the facility consists of a series of separate buildings with only a few people working in each, the company can place all the facilities in a centralized building. This arrangement has proved successful in such establishments as chemical facilities, oil refineries, and railroad yards.

If the facility is relatively small, dressing rooms, lockers, and washrooms should be near the entrance. In a larger facility, it is better to place these facilities in a single, centrally located building or in several buildings near the work areas. In some industries, washing facilities are also used to prevent product contamination (Figure 8–3).

Washing Facilities

The company must maintain all washing facilities in a sanitary condition. Each lavatory should have hot and cold water, or at least tepid running water, and hand soap or a similar cleansing agent. Waterless skin cleansers are not substitutes for soap and water, but they are convenient for special use or where water is scarce.

For industrial occupancies of up to 100 employees, one lavatory for every 10 employees is recommended; for more than 100 employees, one lavatory for every 15 additional employees is considered adequate. In industries where workers need additional washing time, one lavatory for every five employees is recommended.

Circular washbasins (Figure 8–4) of stainless steel, stoneware, enameled iron, or other materials impervious to moisture permit a number of employees to wash at the same time by using central water sprays that are either continuous or controlled by a treadle. These basins are easy to keep clean and sterile. Their construction prevents water from splashing and spilling.

To eliminate standing water, which can transmit disease from one employee to another, make sure lavatories have no stoppers. A mixing faucet or a spray will permit employees to wash in a flowing stream with controlled temperature. Some lavatories also feature knee-actuated controls.

Wherever practicable, a thermostatic control should be installed in the hot water-supply system to keep water temperature below 140 F. Injecting live steam into tanks or lines of a cold water system (to make warm water) is dangerous, because pressure failure in such a system could release steam through the taps.

Management should establish a regular maintenance program for equipment, and train employees to report defective machinery, tools, and other equipment. For example, broken faucets and valve handles may cause serious cuts or lacerations. Handles should be

Figure 8–4. Lavatories should be supplied with running water at a controlled temperature. A sufficient number of wash-up facilities should be available. In a multiple-use lavatory, 24 in. (60 cm) of wash sink or 20 in. (50 cm) of circular basin, with water outlets for each space, is considered to be equivalent to one lavatory.

made of metal, not a breakable material, such as porcelain. If leaky faucets are repaired at once, employees will not develop the bad habit of shutting valves too tightly. Also, train employees to wash their hands after using the restroom.

Soaps

The proper type of soap is important, not only for ordinary hygiene but as a protection against dermatitis caused by cleaning agents. The soap used should have no free alkali and should have a pH less than 10.5. It should be free of mineral abrasives. Management should provide individually dispensed paste, liquid, or powder (not bar soap) for common use. Liquid or powdered soaps are preferable, because they are easy to dispense and serve as an aid to housekeeping.

Management should strongly discourage workers from removing paint, dye, and other stains with solvents or other chemicals, and especially removing grease from their hands with naphthas. Those solvents may cause a severe skin irritation.

Driers

One or more of the following means of drying hands and/or face must be convenient to the lavatories: individual paper or cloth hand towels or sections, clean individual sections of continuous cloth toweling, or warm air blowers. The firm should not allow common-use towels but should supply paper towels instead. Paper towels should be soft enough not to cause irritation and should be kept in a covered container with a disposal receptacle nearby.

Hot-air hand driers should be well-secured either to the floor or to the wall to prevent the fixtures or the electric element from coming loose. The equipment must be grounded and permanently installed without extension cords or plugs. Blowers must provide air at not less than 90 F (32 C) nor more than 140 F (60 C).

Barrier Creams

Protective or barrier creams, if properly used and reapplied frequently, provide limited protection against hand and arm irritants. There are four common types of creams, but no one cream is effective against all irritants. One principal benefit of using barrier creams is that workers must wash their hands repeatedly to remove the cream.

Lockers

Lockers should be perforated for ventilation and be large enough to permit clothing to be hung to dry. If the clothing is heavy or wet, management should provide forced circulation of hot air through the base of the lockers and out through the top, or provide hangers on elevating chains so the work clothing can be dried between shifts.

Lockers should have sloped tops to prevent workers from storing material on them (Figure 8–3). The multiple legs of lockers are serious impediments to floor cleaning; place lockers on metal frames with a minimum of floor supports. Also, anchor lockers together to prevent them from being overturned.

Persons working with toxic materials that can contaminate clothing should have separate lockers for work clothing and street clothing. These lockers preferably should be in rooms on opposite sides of the shower room so employees will have to pass through the shower room when changing from work clothing to street clothing, and vice versa.

Benches

Benches in front of the lockers should be permanently fastened to the locker base, preferably on a hinged support so they can be turned up against the faces of the lockers while the aisles are swept. Maintenance workers should check the benches regularly and keep them in good repair—free from splinters, breaks, and other imperfections.

Floors

Washrooms and locker rooms should be kept well-ventilated, warm, and at 30–60% relative humidity. Make sure heating equipment is installed to protect against burns, in compliance with state codes.

The floors of washrooms and locker rooms should be made of nonabsorbent material such as glazed brick, tile, or concrete. The floor material should continue up into the walls as a cove for at least 6 in. (15 cm) before there is a joint. The walls must connect to the floor cove with a tight joint and should be watertight to a height of at least 5 ft (1.5 m).

Select flooring material for durability and sanitation and to minimize the hazard of slipping and falling. Terrazzo, tile, marble, and polished concrete floor surfaces are particularly hazardous when wet. For safety, maintenance should establish a strict cleaning and mopping schedule so that the flooring is dry by the time workers arrive.

Concrete floors can be made much less hazardous by covering the surface with a finishing layer of abrasive grain concrete. Abrasive strips can reduce slipping hazards on old concrete floors, which have been worn smooth. Ceramic tiles are available with a skid-resistant, nonabsorbent, and watertight surface, or mats can be used.

Floors need to be inspected to be sure they are watertight. Leaky floors cause damage to joists and other structural members of the building, and if organic materials collect in them, they can attract vermin. Plastic material can cover worn wood or concrete floors to obtain a watertight surface.

Showers

Companies should install showers in work areas where employees become dirty, become wet with perspiration, or are exposed to dust or vapors on the job. Showers can also be provided to encourage employees to bicycle or walk to work. The showers should be as close to the job as possible, preferably in a separate room adjacent to the dressing rooms and locker rooms. Workers exposed to high temperatures who come off the job wet with perspiration should not have to walk through cold air when going to the shower and dressing rooms.

A company should provide one shower for every 10 employees of each sex who are required to shower during the same shift. Each shower should be supplied with hot and cold water through a mixing fixture that the user can regulate. Use an automatic regulator to maintain the maximum hot water temperature at 140 F (60 C). Deluge showers, eyewash fountains, and similar installations for emergency use are discussed in Chapter 12, Occupational Health Programs, of the *Administration & Programs* volume of this *Manual.*

Place body soap or other appropriate cleaning agents conveniently near the showers. Hot and cold water should feed a common discharge line. The company should provide employees who use showers with individual clean towels.

If a particular standard requires employees to wear protective clothing because of possible contamination with toxic material, make sure change rooms have storage facilities for street clothes. As discussed earlier under Lockers, protective clothing should be stored separately.

When the company provides clothes for workers, it should ensure that all wet clothes or those washed between shifts are dry before reuse.

The floor of the shower room and of the individual compartments should be made of nonskid material to provide good footing when it is wet. The surface can be either abrasive grain concrete or concrete with a wood-float finish.

Existing floors that were made smooth or have become smooth through long wear can be given a nonskid surface. Maintenance staff should thoroughly scrub concrete floors, using an abrasive pad and a synthetic detergent. They can apply strips of abrasive material to other types of floors to provide a slip-resistant surface. The floor throughout the shower room area should slope toward drains, preferably at the back of the shower stalls. It is not necessary to have curbs around the individual shower stalls if the floor is properly sloped. The curbs can be a tripping hazard; if used, they should be dyed or painted a contrasting color.

Do not use wood mats on shower room floors. They present a tripping hazard and may expose a worker to splinters and loose-joining members. Also, avoid placing pans of antiseptic solution at the entrance of shower stalls or shower rooms. They are useless for killing organisms and are a nuisance to keep clean.

As an item of general sanitation, shower rooms and stalls should be well-ventilated and adequately lighted to prevent the formation of mold. The floor of the shower should be mopped daily with detergent, hot water, and disinfectant to combat athlete's foot (fungus and ringworm infection). A foot-actuated spray can aid in controlling athlete's foot.

Toilets

Wall-hung, elongated-bowl flush toilets with open-front seats should be provided according to the number of employees (Table 8–I). If people other than employees are allowed to use toilet facilities, the number of toilets should be increased accordingly. There must be adequately supplied toilet paper holders in every water closet. For every three toilet facilities, the company should provide at least one lavatory in the toilet room or adjacent to it.

Table 8–I. Minimum Toilet Facilities

Number of Employees	*Minimum Number of Water Closets**
1 to 15	1
16 to 35	2
36 to 55	3
56 to 80	4
81 to 110	5
111 to 150	6
More than 150	One additional fixture for each additional 40 employees.

*Where toilet facilities will not be used by women, urinals may be provided instead of water closets, except that the number of water closets in such case shall not be reduced to less than ⅔ of the minimum specified.

Printed from OSHA Regulations, §1910.141(c)(1).

Wall-hung units are easier to keep sanitary and to clean. Codes prohibit any type that is not thoroughly washed at each discharge or that might permit siphonage of bowl contents back into the tank. Water supplied to tanks must have vacuum breakers or a positive air gap between the surface of the water in the tank and the water supply inlet.

Toilets should be no more than 200 ft (60 m) from any workstation. In multistory buildings, toilets should be only one floor above or below the work area. With toilets and lavatories placed at various points throughout the facility, the main locker room and shower room can be closed for cleaning during the work period—an advantage for the janitorial crew.

Some states require that women work no more than a certain distance from a women's rest room. Management should check for this requirement during design stages. In addition, some states require a company to install cots so women can lie down.

Ventilation is required for toilet rooms. If natural ventilation is used, windows or skylights must have a ventilation area of 6 ft^2 (0.5 m^2) for a room with one toilet, with an additional 1 ft^2 (0.1 m^2) of window ventilation space for each additional toilet. If this amount of window space is not available, the company should supply forced ventilation at a rate of three to four air changes per hour in the room.

Because windows and skylights generally do not afford sufficient light, companies should install light fixtures in all toilet rooms and washrooms. Place switches for lights, electric driers, or other equipment where they cannot be operated by anyone who is also in contact with piping or other grounded conductors. Ground fault circuit interrupters (GFCIs) should be used whenever electrical receptacles are located within 6 ft (2 m) of a sink.

Individual wall-hung urinals should be provided in the men's room. These may be substituted to the extent of one-third or less of the number of stools specified. Approved urinals must be designed so that all surfaces subject to soiling can be easily cleaned. Integral screens over the discharge openings are the major cause of chronic toilet room odors. The decomposing soil under the screen cannot be removed by any practicable method. Blow-down washout urinals are the only acceptable type. Floor-type urinals, in which the drain pipe becomes chronically offensive, and wall-hung urinals with integral screens should be replaced by the approved sanitary type (blow-down washout), thus making room deodorants unnecessary.

Management should prohibit employees from lunching in toilet rooms or in process areas where toxic or noxious materials are present. The habit of some workers to heat foods in molten lead reservoirs or other process heating equipment can be dangerous to their health and should be prohibited. Instead, companies must provide proper lunchrooms or other eating facilities outside the toilet rooms or process areas.

Covered receptacles in facility lunchrooms are the proper places to dispose of waste food and papers. Employees should never discard such refuse in the toilet rooms. If they carry cups of coffee or other drinks from the lunchroom, make sure the drinks are in covered containers or on trays to prevent spills that could create unsanitary or slippery conditions.

Portable toilets are often necessary on construction jobs. The supplier can provide waste removal and maintenance.

Janitorial Service

As a part of an overall managed facility sanitation function, the company should set up a minimum daily janitorial service for all personal service facilities. When properly designed, washrooms, shower rooms, and toilets can be thoroughly cleaned with little personal involvement in the process. Workers should mop and clean floors and fixtures with detergent and hot water at least once daily and use a sanitizing cleaner as often as necessary. The occasional use of an acid-type cleaner may be required on toilet bowls and urinals. Workers should wear rubber gloves and goggles, and thoroughly flush the fixtures following use.

When floors are being mopped, block off the area with signs reading CAUTION—WET FLOOR to prevent possible slipping accidents. This subject is also covered in Chapter 4, Maintenance of Facilities. Employee exposure to bloodborne products should be considered. If such exposure is likely, then the company should institute a bloodborne pathogens program.

FOOD SERVICE

Nutrition

Nutrition, another factor in industrial health and safety, concerns the medical and safety departments of any company or facility. Some facilities may establish food-service facilities. With care and thought, the company can deliver adequate and balanced in-facility meals. Food should be properly prepared and attractively served, with strict adherence to sanitary practices.

The nurse, working with the physician, can provide information to employees on better nutrition. The Council on Foods of the American Medical Association and the American Dietetic Association (see References) offer many excellent articles and other materials. A good breakfast, high in protein for timed energy release during the day, lessens fatigue and consequently lessens the chance of accidents.

Some organizations hire full-time dieticians to review their food-service menus and even talk to employees about nutrition, job performance, and health.

Types of Service

There are five main types of industrial food service:

1. cafeterias preparing and serving hot meals
2. canteens or lunchrooms serving sandwiches, other packaged foods, hot and cold beverages, and a few hot foods
3. mobile canteens that move through the work areas, dispensing hot and cold foods and beverages from insulated containers
4. box-lunch service
5. vending machines

Even using a mobile canteen to provide a midshift snack adds considerably to the nutrition of the average worker. If lunches are also served, the mobile canteen should carry both hot and cold foods and beverages.

The central cafeteria, with a kitchen where full meals can be prepared and served, is often the most satisfactory form of food service. In large facilities, it may be economical to supply several cafeterias from a central kitchen.

Self-service vending machines offer a wide variety of packaged ready-to-eat foods. Some machines have ovens that let the user quick-cook meals. Four important safety and health precautions should be followed in the use of microwave ovens.

1. All repairs should be made by manufacturer-authorized repair personnel.

2. Persons with cardiac pacemaker units should be warned against coming too close to microwave ovens. Most pacemakers, however, are protected against microwave radiation.
3. Interlock and leakage testing should be performed on a regular basis.
4. Do not plug vending machines or microwave ovens into extension cords.

Details are given in the National Safety Council's *Fundamentals of Industrial Hygiene,* 4th ed.

Equipment Installation and Maintenance

Proper installation and maintenance of food-heating and refrigeration equipment is essential to employee health. For example, maintenance staff must be sure that vending machines are kept sanitary and that can openers are routinely cleaned, sanitized, and kept in good working order. The company must supply sufficient utensils and adequate waste-disposal facilities in kitchens and eating areas as a necessary part of a self-service operation.

Eating Areas

The cafeteria or lunchroom should be clean and attractive to present a pleasant environment to the employee, thus encouraging its use. Employees should eat only within designated areas to help maintain proper sanitation in the workplace. This requirement will prevent insect and rodent infestations.

Minimum floor spaces for the number of people using the eating area at one time are given in Table 8–J. Where space is limited, managers can stagger lunch periods so that employees do not have to eat on the job. Employees should wash their hands before and after eating or smoking.

Kitchens

When a company sets up a cafeteria kitchen, management should pay the same attention to proper equipment and working conditions as in any other part of the facility. Although not required by some regulatory authorities, it is a good idea to use food-service equipment that complies with standards or receives approval from nationally recognized testing agencies [National Sanitation Foundation International (NSFI) and Underwriters Laboratory (UL)]. Make sure the layout conforms with local public health food service codes. To ensure that health codes are met, consult with the local public health jurisdiction early in the planning stages.

Floors made of impervious, water-resistant, easily cleaned, skid-resistant material will minimize slipping and falling hazards if water or grease is spilled. Walls and ceiling should be constructed of a durable, smooth, easily cleanable, light-colored surface. Because areas near the dishwasher and/or sink have high moisture levels, they may need special waterproof materials. The area constructed near the cooking equipment should be approved by the local fire marshal.

Artificial lighting in the food-service areas should provide at least 20 foot-candles of light on all food preparation surfaces and at equipment- or utensil-washing areas. All artificial lighting should have shielding to prevent broken glass from falling into the food.

Ranges and other heat-producing equipment require hoods and ventilation to carry away heat, grease, odors, and vapors. Because the ventilation system gets greasy, the duct work should be easily accessible for cleaning. The ventilation system must meet local fire and building codes.

Sprinkler systems and portable extinguishers should be installed for fire protection (see Chapter 11, Fire Protection). Place a fire blanket and extinguisher near any area where workers' clothing could be ignited by an open flame.

The company should install easily changeable and cleanable racks for such hand tools as knives, cleavers, and saws, and place storage racks or cabinets for utensils in convenient locations. All wall- and ceiling-mounted equipment must be easy to clean and maintain in good repair.

Table 8–J. Minimum Floor Space in Eating Areas

Number of People	*Sq Ft per Person*
25 or less	13
26 through 74	12
75 through 149	11
150 and over	10

Controlling Food Contamination

Many communities and counties have detailed food sanitary regulations governing the installation and operation of an industrial food-service facility. The company must closely adhere to regulations of the health authority who has local jurisdiction. In areas where no local health authority exists, the organization must follow the other regulatory standards or the recommendations of the U.S. Public Health Service's *Food Service Sanitation Manual* (see References).

Because of the potential for contamination during the food preparation and serving process, the Food and Drug Administration recommends the use of a proactive methodology. This methodology, known as the Hazard Analysis Critical Control Points (HACCP) system, is used to identify and eliminate food safety problems caused by biological, chemical, and physical contamination. Effective use of the HACCP system enables one to establish a list of critical control points that can be controlled to reduce or eliminate foodborne illness hazards.

Potentially hazardous foods are those that contain milk or milk products, eggs, meat, poultry, fish, shellfish, edible crustaceans, or other ingredients, including synthetic ingredients, in a form capable of supporting rapid and progressive growth of infectious or toxigenic microorganisms. Potentially hazardous foods—particularly custard-filled and cream-filled pastry, milk and milk products, egg products, fish, meat, shellfish, gravies, poultry, stuffing, sauces, dressings, and salads containing meat—should be kept at or below 40 F (4 C) unless being prepared or served. Reheating these foods after they have remained for some time at room temperature will not protect employees from foodborne illnesses.

These illnesses can be caused by a variety of bacteria, viruses, parasites, and chemicals. An example of one of these is the bacteria *Staphylococcus,* which produces a toxin in foods that is not destroyed by normal cooking temperatures and can therefore cause illness. For example, if the bacteria have been growing in food during storage at about 40 F, they will be killed by heating, but their toxins will still be capable of causing a foodborne illness. The contaminating agent of a foodborne illness usually cannot be detected through the taste, odor, texture, or appearance of the food. Therefore, proper sanitation is essential to keep food safe for consumption.

To prevent a foodborne illness, workers should observe the following guidelines:

- Wash hands frequently during food preparation and after any interruption in food preparation, after smoking, or using the restroom. Hand washing must involve using soap and warm water to be effective. Do not use common towels for hand drying, as these towels can recontaminate the hands.
- Keep hot foods hot (140 F or above) and cold foods cold (40 F or below).
- Exclude from food preparation and handling personnel with open wounds, infections, or communicable diseases transmittable by food. In addition, food service workers should be tested for exposure to hepatitis.
- Promptly place leftover food under refrigeration.
- Reheat leftovers rapidly to 165 F before serving.
- Refrigerate potentially hazardous foods until they are used.
- Eliminate flies, roaches, rodents, and other pests from the food-preparation and storage areas. Consult a professional pest-control operator for assistance.
- Never use galvanized or cadmium-plated containers for storage of moist or acid foods.
- Consult the local health authority for answers to questions or concerns on sanitation.

As a further precaution in preventing a foodborne illness, no first-aid material or personal medication should be permitted in the food-preparation area. All workers requiring immediate first aid should be seen by the company or facility nurse, or the physician; injured workers should continue on the job only at the doctor's discretion.

To prevent cross-contamination—that is, the contamination of clean items by soiled sources, such as unwashed hands or dirty shelves—it is important to handle and store containers, utensils, glassware, dishes, and silverware properly.

It is generally easier to clean utensils, dishes, and silverware by machine washing than by manual washing if the machine is kept clean and properly maintained. In either case, one of the main requirements in ensuring adequate sanitation is proper training of the food staff.

Utensils, dishes, and the like should be carefully scraped and preferably rinsed before washing. The wash cycle includes cleaning with warm water containing soap or detergent and a clear water rinse to remove any soap/detergent residue. The final stage properly sanitizes the utensils using a method such as a chemical (hypochlorite or quaternary ammonium compounds), or 170 F water. All utensils should be allowed to air dry—No towel drying is acceptable.

For effective machine washing, workers must stack the utensils in the trays loosely enough to allow the cleansing agent to reach every part. The machine should maintain the proper concentration of soap/detergent and the appropriate water temperature. The wash water must be changed periodically to prevent excessive buildup of food debris and grease, which impairs cleaning the utensils. To ensure successful operation, workers should clean the dishwashing machine, including the spray nozzles, daily to allow proper flow and distribution of the water.

Requirements for manually washing utensils are the same, except that workers must carefully scrub each utensil individually rather than simply stacking them in a tray. For manual washing of utensils, the dishwashing area must have a sink with no fewer than three compartments (Figure 8–5). The compartments should be large enough to accommodate the largest utensils, and have hot and cold potable water under pressure. The sink should be supplied with attached drain boards or movable dish tables large enough to accommodate the freshly washed utensils and dishes. All utensil washing requires the same steps: prewash/scrape, wash, clear water rinse, sanitize, and air dry.

Although several methods of sanitizing utensils are acceptable, the most common methods involve hot water or one of numerous chemicals. If using the chemical method, make sure the chemical chosen is approved for use on food-contact surfaces such as silverware, dishes, and so on.

To use the hot-water method of sanitizing, workers must maintain the temperature at 170 F or above and

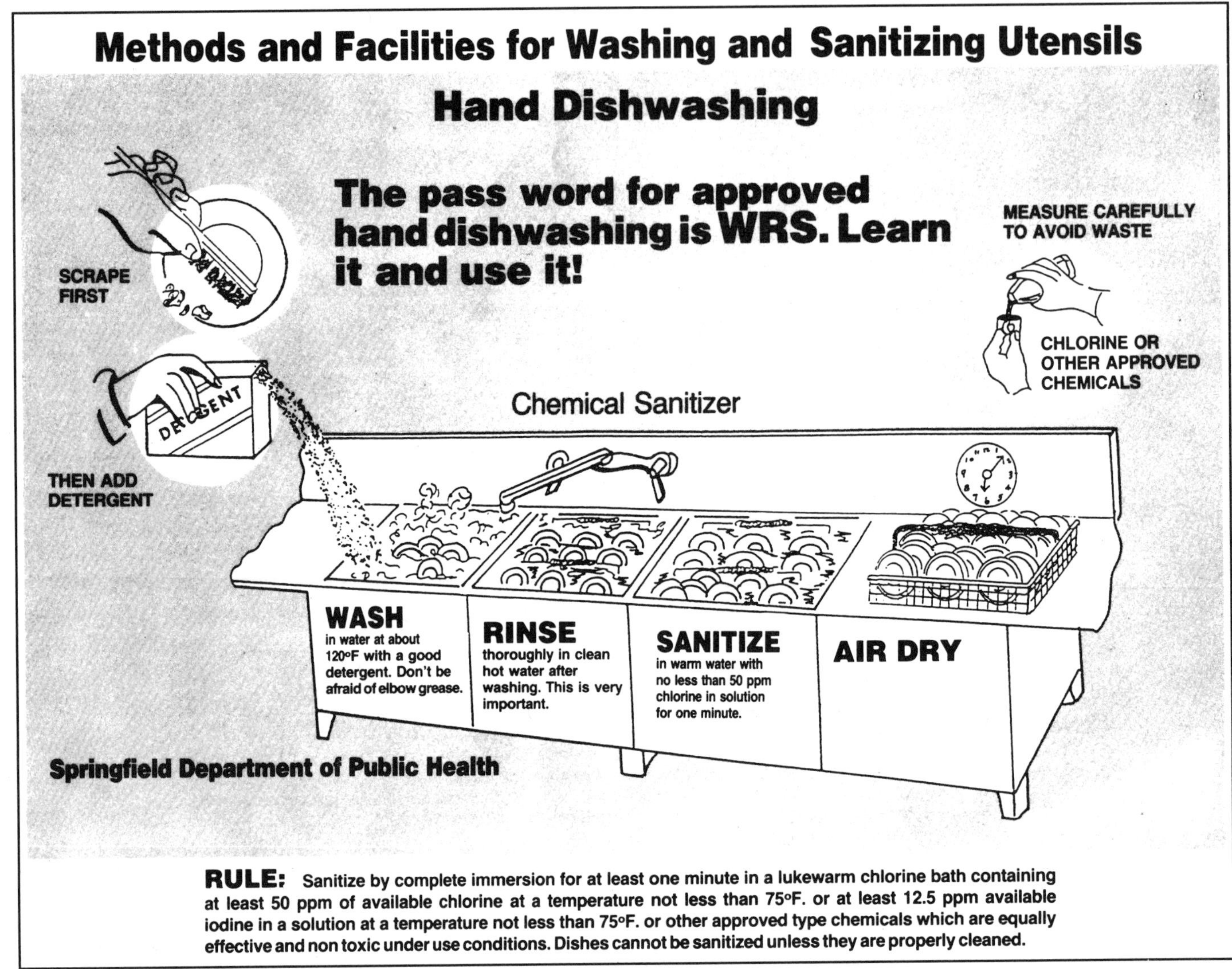

Figure 8–5. A sink with a minimum of three compartments must be provided for hand dishwashing.

keep utensils in the water for at least 30 seconds. The main problem is ensuring sufficient amounts of hot water. Long sessions of dishwashing require large amounts of hot water. Thus, the company will need a large hot-water heater with quick recovery, or it should install a booster heater on the water line or in the sanitizing tank.

Less hot water will be necessary if a chemical method is used. The most common chemical method involves immersing the utensils in a solution of hypochlorite containing at least 50 parts per million (ppm) available chlorine in water at least 75 F for a period of one (1) minute. Workers can use cationic quaternary ammonium and iodine germicides if this method achieves the same sanitary specifications as a hypochlorite solution. Any time chemicals are used, employees should have a test kit or other device that accurately measures the ppm concentration of the chemical in solution.

After utensils have been properly cleaned, they must be properly stored and handled to prevent recontamination by workers or by ordinary dust and dirt. Use of single-service containers and utensils can eliminate washing and handling. However, these items must be carefully stored to prevent accidental contamination.

SUMMARY

- Five industrial health areas must be kept sanitary and well-equipped for employee health and convenience: potable water supplies; sewage and garbage disposal; personal service facilities; food service; and heating, cooling, and ventilation.
- Management should consider establishing a separate department with its own director to ensure a sanitary work

environment. General rules for sanitation include an approved piping and storage system, good housekeeping, personal cleanliness, and a good inspection system.

- Drinking water must meet all regulatory standards of purity to avoid exposing workers to waterborne contaminants. Private water sources must be surveyed routinely to protect workers' health.
- All plumbing and fixtures should be disinfected before being put into service either for the first time or after repairs.
- Filtration and chemical disinfection are the most practical methods for purifying industrial private water supplies.
- Organizations must adhere to state and local ordinances in disposing of and recycling sewage, waste, and garbage.
- Companies providing food services for their employees must arrange for proper disposal of food refuse. All garbage-collection areas and receptacles must meet state and local codes.
- In industrial and commercial sites where insect, rodent, and nuisance bird infestations are a problem, companies should employ professional pest-control operators. Companies must ensure that refuse receptacles and liners comply with local codes.
- All personal-service facilities contributing to employee health, such as drinking fountains, washrooms, locker rooms, showers, and toilets, should be kept clean and safe for workers' use.
- Companies must maintain washrooms, showers, locker rooms, toilets, hand-washing facilities, and hand-drying mechanisms in a sanitary condition.
- Company health and safety staff can educate workers on proper nutrition and its relation to job performance and health. There are five main types of industrial food service: cafeterias, canteens or lunchrooms, mobile canteens, box lunch services, and vending machines.
- Food-service staff must receive careful training in sanitation and food-handling methods to prevent food contamination, cross-contamination, and the transmission of food-borne illnesses.

REFERENCES

American Dietetic Association, 216 West Jackson Boulevard, Chicago, IL 60606.

American Medical Association, Council on Foods, 515 North State Street, Chicago, IL 60610.

American Public Health Association, 1015 15th Street, Washington, DC 20005.

Standard Methods for the Examination of Water and Wastewater.

American National Standards Institute, 11 West 42nd Street, New York, NY 10036.

Buildings and Facilities—Providing Accessibility and Usability for Physically Handicapped People, ANSI A117.1–1986, section 5.7.

"Gas-Burning Appliances," Z21 Series.

Minimum Requirements for Sanitation in Places of Employment, ANSI Z4.1.

National Electrical Code, ANSI/NFPA 70, 1999.

"Pipe Flanges and Fittings," B16 Series.

American Water Works Association, 6666 West Quincy Avenue, Denver, CO 80235.

Bennett GW and Owens JM, eds. *Advances in Urban Pest Management.* New York: Van Nostrand Reinhold Company, 1986.

Educational Foundation of the National Restaurant Association (formerly National Institute for the Foodservice Industry), 250 South Wacker Drive, Suite 1400, Chicago, IL 60606.

Applied Foodservice Sanitation, 3rd ed., New York: John Wiley & Sons, 1987.

Environmental Management Association, 255 Detroit Avenue, Denver, CO 80206.

Freedman B. *Sanitarian's Handbook, Theory and Administrative Practice for Environmental Health,* 4th ed. New Orleans: Peerless Publishing Co., 1977, p. 525.

Illinois Department of Public Health, Office of Health Protection, Division of Food, Drugs, and Dairies, 525 West Jefferson Street, Springfield, IL 62761.

Food Service Sanitation Code, 1989, 77 Ill. Adm. Code 750.

Private Sewage Disposal Licensing Act and Code, 1986, pp. 50, 51.

National Restaurant Association, 1200 17th Street NW, Washington, DC 20036.

"A Safety Self-Inspection Program for Foodservice Operators."

National Safety Council, 1121 Spring Lake Drive, Itasca, IL 60143.

Accident Prevention Manual for Business & Industry: Environmental Management, edited by Gary R. Krieger, 2nd ed., 2000.

Fundamentals of Industrial Hygiene, 4th ed., 1996.

Work Injury and Illness Rates. Annual.

National Sanitation Foundation, P.O. Box 1468, 3475 Plymouth Road, Ann Arbor, MI 48106.

Public Health Service, U.S. Department of Health and Human Services, 200 Independence Avenue SW, Washington, DC 20201.

Food Service Sanitation Manual, Publication No. FDA 78-2081.

Robinson WH. *Urban Entomology: Insect and Mite Pests in the Human Environment.* New York: Chapman and Hall, 1996.

U.S. Environmental Protection Agency, Office of Water Program Operations, Office of Research and Development, Municipal Environmental Research Laboratory, Cincinnati, OH 45268.

Design Manual Onsite Wastewater Treatment Disposal Systems, p. 98, October 1980.

U.S. Environmental Protection Agency, Water Supply Program Division, Washington, DC 20460.

Manual for Evaluating Public Drinking Water Supply.

Water Quality Criteria, Report No. 3A. Sacramento, CA. Dept. of General Services, Office of Procurement—Stores, Documents Section, 1015 North Highland, Sacramento, CA 95662.

REVIEW QUESTIONS

1. The general rule for sanitation includes:
   a. An approved piping and storage system
   b. Good housekeeping as clean as the nature of the work allows
   c. Personal cleanliness
   d. A good inspection system
   e. All of the above
   f. Only b and c

2. How many infectious waterborne diseases are commonly caused by drinking contaminated water?
   a. Two
   b. At least five
   c. Nine
   d. At least 12

3. What is one of the most common ways a water supply becomes contaminated?

4. Faucets and similar outlets should be at least how far above the floodrim of the receptacle?
   a. 1 inch
   b. 2 inches
   c. 3 inches

5. How can a worker easily disinfect a drinking water system?

6. Name the two methods of water purification that are the most practical for industrial private-water supplies.
   a.
   b.

7. List six recyclable materials that can be hazardous to employees when sorting and collecting them.
   a.
   b.
   c.
   d.
   e.
   f.

8. What are the five personal-service facilities that contribute to employee comfort and are essential to the occupational health program in most industries?
   a.
   b.
   c.
   d.
   e.

9. Local codes require how many drinking fountains should be installed for every 50 people throughout an industrial facility?
   a. One
   b. Two
   c. Three

10. List four factors to consider when selecting a floor for washrooms and locker rooms.
    a.
    b.
    c.
    d.

11. Janitorial-service workers should mop floors and clean fixtures in all personal-service facilities with detergent and hot water at least:
    a. Once weekly
    b. Twice weekly
    c. Once daily
    d. Twice daily

12. Two important safety and health precautions that should be followed in the use of microwave ovens are:
    a.
    b.

13. List the five main types of industrial food services.
    a.
    b.
    c.
    d.
    e.

14. What is the best way for workers to combat fatigue and maintain normal blood sugar?
    a. Drink at least two cups of coffee
    b. Eat foods high in sugar
    c. Eat a breakfast or snacks high in protein

15. Food-service equipment should either receive approval from or meet established standards of which organization?

9

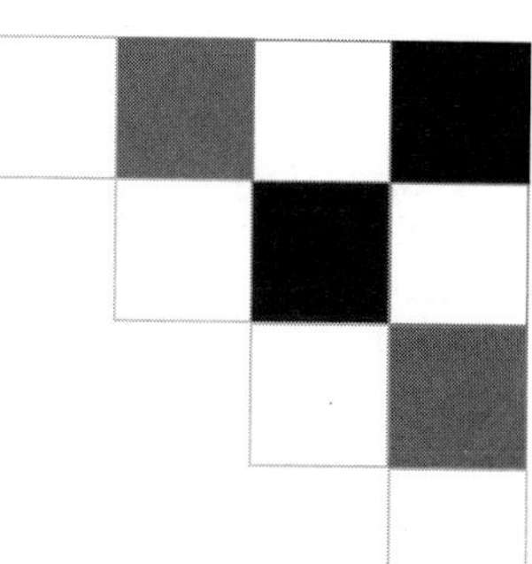

Occupational Medical Surveillance

Revised by

Philip E. Hagan, MPH, CIH, CHMM, CHCM

Medical surveillance is an organized system for collecting and using information on diseases, injuries or hazards to help prevent accidents and illnesses. Medical surveillance and screening are important components of a comprehensive worksite prevention program. This chapter discusses the benefits and limitations of occupational medical surveillance, covering the following topics:

- importance of medical surveillance in occupational health
- definitions of medical surveillance and screening
- goals of a medical surveillance program and the nature of biological monitoring
- screening for occupational cancers
- ethical and legal issues surrounding medical surveillance and screening data

BACKGROUND

Over the past three decades, the relationship between human exposures or potential exposures to environmental hazards and diseases has become the focus of considerable public concern. As a result, the discipline of medical surveillance has become an important tool in detecting and monitoring public and occupational health problems associated with such exposures.

Medical surveillance involves the identification of injuries or illnesses (e.g., asthma and skin rashes) that can be linked to exposures to toxic or hazardous substances in the workplace environment (e.g., detergent enzymes). In many cases, it is hoped that this information can then lead to a successful public health intervention (for example, a routine radioallergosorbent test (RAST) for healthy workers and removal of workers who test positive from exposure).

Thus, surveillance is an important tool used in disease detection and prevention in the worksite. Workers' exposure to hazards such as lead, animal allergens, solvents, welding fumes, bloodborne pathogens, and noise in the workplace mandates the use of a range of preventive measures, including the following:

- exposure control and monitoring of the workplace environment
- exposure assessment including biomonitoring of the worker
- periodic assessment of the health of the worker
- education of the worker regarding the hazards, health effects, and preventive measures

Communities are increasingly considering long-term medical screening and surveillance when certain individuals have an elevated risk of adverse health effects related to significant exposure(s) to toxin or toxins. For example, toxic agents such as chromium, trichloroethylene, benzene, asbestos, or plutonium in air, soil, or groundwater may spur public demand for medical monitoring for detection of latent, usually chronic disease. The potential outcomes of exposure to these toxic agents are usually, but not always, malignant diseases; however, determination of causation of disease is not a function of a medical surveillance program.

Medical surveillance is still an inexact discipline. Due to the difficulties in establishing clear cause-effect relationships between exposure and disease, in many cases, both at the workplace and in the community setting, medical screening does not always provide any clearcut benefit or improved outcome for individuals exposed to toxic agents, even for those with significant exposure. As is the case for industrial programs, the question about whether to conduct medical surveillance in a community setting requires analyses of the individuals' exposures and doses, the toxicity of the hazard, the effectiveness of available screening tests (including sensitivity, specificity, and positive predictive value), and how clinical outcomes will be affected by early detection (USPSTFG, 1996; Fleming et al., 1995 and ATSDR 1995). The issue is further complicated by the fact that most of the surveillance performed by employers and mandated by OSHA does not meet the criteria of a medical surveillance program (Silverstein, 1994).

DEFINITIONS: SURVEILLANCE AND SCREENING

Surveillance is defined in the *Dictionary of Epidemiology* as "the ongoing scrutiny [of the occurrence of disease and injury], generally using methods distinguished by their practicality, uniformity, and frequently their rapidity, rather than by complete accuracy. Its main purpose is to detect changes in trends or distributions in order to initiate investigative or control measures." According to the Centers for Disease Control and Prevention, surveillance is defined as "the ongoing systematic collection, analysis, and interpretation of health data. This information is used for planning, implementing and evaluating public health interventions and programs" (Klaucke, 1988).

Surveillance thus refers to the collection of data that are continually reviewed to evaluate and change preventive processes. OSHA medical surveillance, as is the case with most company screening programs, is implemented primarily for the early detection of disease. As a result, it is critical to select accurate tests with high predictive value. A screening program that uses tests with little or no proven value can do significant harm by failing to detect problems early.

The terms *medical surveillance, medical screening,* and *medical monitoring* have sometimes been used interchangeably. For example, OSHA medical surveillance is more accurately termed "screening" as there is no epidemiologic analysis of data, no use of data collected, and no change in preventive policies. Screening may detect actual disease (e.g., respiratory symptoms and declining lung function associated with exposure) or it may detect the risk for disease (e.g., through RAST testing or genetic testing). The term "medical monitoring" has sometimes been equated with medical surveillance, but is easily confused with the terms "environmental monitoring" and "biological monitoring." In this chapter the term medical surveillance will be understood in a broad sense, as commonly used in everyday practice, to include occupational medical screening and monitoring activities.

The ultimate goal for any program involving worker health evaluations is to benefit the population, whether it is a one-time screening of individuals or a periodic surveillance of exposure or of health-related variables in a working population. Individual health and exposure assessments are important in developing a successful screening program. In many cases, worksite programs include educational efforts related to reducing health risks for conditions, such as hypertension, for which the general population is screened. In addition, health workers might also educate workers about self-examination for testicular or breast cancer, provide smoking-cessation programs, and offer colon cancer screening. Although these conditions and behaviors are not generally caused by occupational factors, they are common among workers, and preventive actions can significantly reduce the risks of illness and death.

Occupational medical surveillance, in the strict sense, is an important public health activity. Once an exposure-related condition or a potentially exposure-related condition, such as black lung disease or carpal tunnel syndrome, is identified, health professionals must investigate the causes and frequency of the event among workers. Also, they must implement preventive measures and evaluate their impact (Rutstein, 1983).

MEDICAL SURVEILLANCE PROGRAM

Criteria for designing and implementing a medical screening/surveillance program have been developed by the World Health Organization (Spitzer and Brown, 1975) and consensus guidelines. The guidelines have been described in a number of authoritative sources, including the Canadian Task Force on the Periodic Health Examination (1994), the American Cancer Society (1993), and the United States Preventive Services Task Force (1996). The guidelines are practical strategies for providing preventive services such as screening, health care counseling, and immunizations for any age group.

The occupational health professional involved in designing a medical surveillance program should review the following exposure and health data:

- OSHA standards and surveillance requirements related to the hazardous substance
- the worker's job description
- the worker's exposure history and industrial hygiene monitoring data
- use of personal protective equipment data
- data from prior medical examinations

The occupational health professional should then carefully analyze (1) the significance and extent of exposure and the exposure dose, and (2) the toxicity of the hazard(s) and the seriousness of the target conditions. Once this analysis is completed, the health professional must consider the balance of *risks and benefits* of surveillance. To implement an effective medical surveillance program, the professional should consider three fundamental issues:

- nature of the disease ("target condition" for which monitoring is proposed), especially whether known and proven medical treatments exist for the disease and whether early disease detection will result in improved health outcomes
- accuracy, reliability, safety, and acceptability of the tests that medical science has available for early detection of the target condition
- expected prevalence of the disease in the population proposed for surveillance. If the condition is rare in the population, then most of the positive results of a screening test will be false, which would reduce the potential benefits and increase the costs and risks of surveillance

If the review and analysis show a substantial benefit for the population to be screened, then the health professional should design and develop the program, with careful adherence to accepted practice. This includes communicating results to individuals and to others, if necessary, but in strict confidence without identifying individual workers. Steps for implementing the program are listed below:

1. risk benefit analysis: analysis of exposure, toxicity, disease, testing parameters and outcome benefits (see above)

2. designation of program goals and target population
3. selection of medical tests
4. interpretation of data (individual and group)
5. selection or recommendation of treatments
6. communication of results (to individuals and worker groups, to employers following ethical guidelines)
7. program evaluation

The clinician must also ascertain that the health surveillance program is in compliance with OSHA medical screening requirements for chemicals (Table 9–A) and for certain working populations (e.g., hazardous waste workers, noise-exposed workers, those who wear respiratory protection, and fire brigade members).

Goals of Medical Surveillance

The objectives of a medical surveillance program commonly include the detection of overexposure to toxic agents and the early detection of disease among asymptomatic persons. The main goals of an occupational screening or surveillance program designed for the benefit of individual workers are presented in Table 9–B.

An occupational medical surveillance program is most commonly developed to *detect overexposure* to hazardous agents. The OSHA standard requiring blood lead levels in certain exposed workers is an example of monitoring for overexposure. Clinical exposure evaluation can be accomplished through medical history/questionnaire, examination, and biological monitoring.

Table 9–A. Substances with Medical Screening Requirements in 29 *CFR* Part 1910

Asbestos	4-Aminodiphenyl
Actinolite asbestos	Ethyleneimine
Amosite	β-Propiolactone
Anthophyllite asbestos	2-Acetylaminofluorene
Chrysotile	4-Dimethylaminoazobenzene
Crocidolite	*N*-Nitrosodimethylamine
Tremolite asbestos	Vinyl chloride
Coal tar pitch volatiles	Inorganic arsenic
Coal tar	Lead
Phenanthrene	Cadmium
Pyrene	Benzene
4-Nitrobiphenyl	Coke oven emissions
α-Naphthylamine	Cotton dust
Methyl chloromethyl ether	1,2-Dibromo-3-chloropropane
3,3′-Dichlorobenzidine/salts	Acrylonitrile
bis-Chloromethyl ether	Ethylene oxide
β-Naphthylamine	Formaldehyde
Benzidine	Methylenedianiline

Table 9–B. Purposes of an Occupational Screening Program

- Assessment of exposure and identification of overexposure (an adjunct to environmental monitoring).
- Identification of disease at an early stage. This may include recommendations for the general population (nonoccupational diseases) such as screening for hypertension, skin cancer, and high cholesterol.
- Health promotion. Education about lifestyle risk factors, safe work practices, and immunizations are included.
- Identification of exposure-related toxic effects in major body organs. Examples include OSHA mandated surveillance for lead, cadmium, and formaldehyde.
- Fitness to work. Evaluations for "baseline" medical measurements, preplacement, respirator clearance, and return to work after an injury.
- Assessment of impairment and disability.

When exposure biomonitoring is used, the expectation is that disease is entirely preventable with measures such as removing workers from hazardous areas, using engineering changes, and improving personal protective equipment.

In addition, monitoring preventive processes is particularly important. This is because in many cases, once preclinical or clinical disease can be detected, very little can be accomplished therapeutically to improve the clinical outcome. The public has been led to believe that disease detected in early stages can often be cured. We now know that a far more successful approach is risk reduction, including exposure control which prevents disease from arising in the first place. Hazard surveillance, a critical component of the risk reduction process, involves collecting and analyzing information on the type, dose, and duration of exposure.

A surveillance program can be established to *detect disease* before clinical symptoms develop or to detect symptomatic disease early. The early detection of symptomatic or asymptomatic disease is beneficial only when the treatment instituted will likely improve the disease outcome as compared to conventional treatment administered when the disease would be diagnosed without screening.

Medical surveillance of exposed populations can also yield information about the prevention of disease that can be applied to other populations with similar exposures. For example, data analysis may lead to new information on how certain chemical, physical, or biologic hazards contribute to disease or on the success of specific preventive measures.

Biological Monitoring

Biological monitoring can provide useful evidence about whether and in what way hazardous substances affect human health. Such evidence can be collected with the use of "biological markers" or "biomarkers," which are indicators of events—such as a symptom of disease or physical change—taking place in biological systems or samples. Although still primarily a research tool, biomarkers will increasingly be used as an aid in the diagnosis of specific diseases or conditions in individuals. In recent years, advances in medical science have enabled health care specialists to include damage to or mutations in genetic material and molecular constituents as biomarkers of potentially important clinical processes.

As a component of a scientifically based medical surveillance program, biological monitoring can be a valuable source of information. The American Conference of Governmental Industrial Hygienists (ACGIH) provides a list of biological exposure indices (BEIs), published annually in their booklet covering threshold limit values (TLVs). OSHA also has incorporated required biological monitoring into several standards: the Lead Standards in General Industry, 29 *CFR* 1910.1025 and in Construction, 29 *CFR* 1926.62; the Cadmium Standard, 29 *CFR* 1910.1027 and the Formaldehyde Standard, 29 *CFR* 1910.1048.

Unfortunately, the clinical use of the genetic, cytologic, and biochemical changes identified as potential markers is often poorly understood. Thus, the ability to analyze, detect, or measure these biomarkers in detail often far exceeds the existing knowledge of how these effects are related to diseases that appear in humans. A large number of biomarkers are required by OSHA or recommended in the medical literature, yet very few have been validated or have known efficacy. Thus, the design of a medical surveillance program can be challenging from this standpoint as well.

Once exposure has occurred, biomarkers, at least in theory, may assist in the evaluation and analysis of each of the following stages in exposure-related disease:

- biomarker
- internal dose
- biologically effective dose
- altered structure and/or function
- clinical disease

The precise biomarkers to be tested for must be reviewed and selected by a knowledgeable scientist/health professional. Simple detection of a chemical, adduct, or metabolite in a body fluid does not signify the presence of disease or toxicity. Similarly, in the presence of disease, biomarker detection does not imply a relationship between exposure and disease. Health professionals must exercise caution and measure the most accurate biomarker with the correct timing with respect to exposure. Not all tests for detecting substances in blood or urine are a screening priority because the substance may not be associated with serious human disease. In contrast, measurement of urine cadmium is an effort to deter the development of cadmium-induced renal disease. Yet if a worker has high urine levels and signs of kidney dysfunction, the kidney disease needs further evaluation to determine whether it is related or unrelated to cadmium exposure.

With these cautions in mind, many biological markers exist that are potentially useful in specific circumstances. Extensive literature is available on the use of biomarkers in detecting lung, immunologic, carcinogenic, hepatic, reproductive, and other disorders. Biological exposure indices (BEIs) are reference values developed as guidelines to help health professionals evaluate potential health hazards (ACGIH,1991). Biomarkers have been broadly classified into three categories: markers of exposure, markers of effect, and markers of susceptibility.

Markers of Exposure

These tests measure levels of a specific substance or its metabolites in body fluids or excreta. Examples include blood lead level, urine cadmium level, phenol level following benzene exposure, measurement of delta aminolevulinic acid following lead exposure, and DNA adducts as an indicator of carcinogen exposure.

Markers of Effect

These tests measure a biochemical, physiologic, or other alteration associated with health impairment or disease process. Examples are forced vital capacity (asthmagen), complete blood count (benzene or lead hematotoxicity), and beta-2 microglobulin in the urine (cadmium nephrotoxicity).

Markers of Susceptibility

These tests measure the body's response to exposures to specific substances. Examples include P53 mutation as a (research) indicator of lung cancer susceptibility, alpha 1-antitrypsin deficiency (susceptibility to emphysema and pulmonary irritants), and glucose 6-phosphate deficiency (susceptibility to damage by oxidant chemicals).

The concept of biological monitoring allows for increased accuracy in assessing the relationship between exposure and disease. The measurement has the advantage of offering (in theory) an assessment of all exposures (occupational and nonoccupational) and of all routes of exposure (such as cutaneous, ingestion, inhalation) contributing to internal dose.

The *limitations* of biological monitoring include:

- difficulty correlating a health risk with exposure once the exposure information is known
- short biological half-lives of some substances, which prevent accurate exposure assessment except within a limited time of exposure
- ineffective monitoring for surface active agents (for example, hazards causing skin or upper airway irritation)
- interference of tobacco, alcohol, and other agents with some test results
- measurement may reflect multiple exposure sources (air, food, water, soil, and skin contact), preventing accurate determination of occupational exposure

In addition to these drawbacks, it is difficult to assess the environmental contribution of substances that are commonly found in the body (e.g., copper or zinc). Biomarkers of effect are often not specific to a particular substance, and abnormalities may reflect causes other than the substance being investigated. For example, restrictive lung disease measured by pulmonary function testing may be due to silica or asbestos or to an underlying disease such as idiopathic pulmonary fibrosis. In the case of biomarkers that can predict future conditions or disease (e.g., alphafetoprotein for liver cancer), the test accuracy needs to be validated with longitudinal studies. Finally, the use of biomonitoring in populations with low level exposures and low incidence of disease is particularly problematic.

Combined Effects of Exposure

Another troublesome area is the fact that little is known about how the human body responds when exposed to more than one toxin or hazardous substance at the same time. The usual assumption made by health care professionals is to consider the effects of each chemical or substance independently. In fact, exposure to a single chemical rarely, if ever, occurs. As a result, researchers and occupational health care professionals are beginning to consider the synergistic effect of substances—that is, with each chemical or substance to which a worker is exposed, the total effect becomes more than the sum of the parts. As chemicals or substances combine in the body, they greatly increase the potential for an individual developing an injury or disease.

Perhaps the best-known example of synergism is that of smoking combined with asbestos. When workers are exposed to these two substances simultaneously, they greatly increase their chances of developing lung cancer. Other studies have found that carbon disulfide and environmental noise act synergistically to cause significant hearing loss in workers (Ryback, 1992).

OSHA's airborne exposure limits, as well as PELs and TLVs, were developed under a similar assumption that workers are exposed to chemicals one at a time. To determine if overexposure has occurred, health professionals can add concentrations of these substances as a fraction of their respective TLVs. If the total equals or exceeds one, then an overexposure has been detected. This is known as the mixture rule and has been published by ACGIH. The underlying assumption is that the "combined" chemicals act on the same end-organ. For example, three similar solvents could be evaluated using this principle if they act in mechanistically (toxicologically) similar ways and have the same target end-organ, e.g., central nervous system, liver.

The fact that workplace exposure involves chemical, physical, biological, and psychological hazards all interacting simultaneously complicates the health professional's efforts to determine the causes of worker complaints. As researchers develop more sophisticated models exploring the connection between synergistic effects and human health, it may be possible to devise more effective preventive measures and treatments.

SCREENING FOR OCCUPATIONAL CANCER

Medical surveillance and screening can offer some help in early detection of cancer in workers. Cancer is second only to heart disease as a cause of death in the United States. The most common cancers in men are cancer of the lung, prostate, colon, bladder, and rectum. In women, the most common cancers are breast, lung, colon, uterus, and ovary.

Factors affecting the frequency of the occurrence of cancer include age, sex, race, geography, nutrition, tobacco use, alcohol intake, infectious agents, exposure to ionizing and nonionizing radiation, specific chemical agents, and genetic factors. Table 9–C shows cancer sites for which relationships with occupational exposures are well established. The International Agency for Research in Cancer (IARC) has classified chemicals and industrial processes with respect to their potential to cause cancer in humans. OSHA and the Environmental

Table 9–C. Cancer Sites for Which Relationships with Occupational Exposures Are Well-Established

Site	Exposures
Bladder	Benzidine, B-naphthylamine, 4-aminobiphenyl, other dyes
Blood (leukemia)	Benzene (AML), radiation, x-rays
Bone	Radium
Larynx	Ethanol, isopropyl alcohol, mustard gas, asbestos
Liver	Arsenic, vinyl chloride
Lung and bronchus	Arsenic, asbestos, bis(chloromethyl)ether, chromium compounds, coal carbonization processes, coal tar pitch volatile, iron ore mining, mustard gas, nickel refining, radiation
Nasal cavity and sinuses	Isopropyl alcohol, mustard gas, nickel refining, radium, woodworking
Skin	Arsenic, coal tar products, mineral oils, radiation

Protection Agency have used other classification schemes to report and regulate potential or known carcinogens.

OSHA regulations cover 20 carcinogens and industrial processes associated with cancer. However, at this time, screening is potentially effective for only a limited number of cancers.

Lung Cancer

Lung cancer is responsible for the most deaths in the United States. A number of exposures, such as tobacco smoke, secondary smoke, and asbestos, have been linked to lung cancer. Many health professionals have screened selected populations of patients for lung cancer, commonly using chest x- rays, and sputum cytology.

Several studies have been conducted to assess the value of screening for lung cancer, including three large National Cancer Institute-sponsored clinical trials of high-risk individuals. Based on the scientific evidence at this time, the American Cancer Society, the National Cancer Institute, the United States Preventive Services Task Force, and other national and international organizations do not recommend lung cancer screening, even for individuals of high risk.

This recommendation is based primarily on the fact that early detection of lung cancer does not appear to alter the course of the disease, given current state-of-the-art treatment. As a result, a false positive test (indicating the presence of a tumor or malignancy where none exists) obtained during screening can represent a risk of its own. Workers may be subjected to extensive and invasive evaluations or treatments that can cause unpleasant and even dangerous side effects.

Bladder Cancer

Workers in the chemical and rubber industry and in dyestuff manufacturing are at risk for exposure to bladder carcinogens. In individuals who show no obvious symptoms, urinalysis for minute amounts of blood in the urine and urine cytology have been the screening tests with the most consistently accurate results. The benefits of this screening remain to be evaluated. Though still an area of active research, a common screening recommendation for workers at risk for bladder cancer is a voided urine cytology every six months. In addition, premalignant changes can be detected by biological markers, such as quantification of DNA in bladder cells. These markers are being investigated.

Genetic Events and Cancer

It is now known that multiple genetic changes occur in the development of some cancers. The early detection of cancer may soon focus on molecular markers associated with minute changes in genes or cells, which may indicate a cancer is developing. For example, small mutations of the P53 gene have been associated with lung cancer and bladder cancer in humans. A variety of new genetic markers have been found using bronchial fluid or sputum. However, the correlation of test results with actual cancer risk is poorly understood. This is an active area of research and clinical investigation.

Biomarkers of exposure (e.g., urine phenols for benzene exposure or urine arsenic levels, urinary mutagens, or DNA adducts) help measure workers' internal dose of exposure. If the dose is excessive, these results can justify using more rigorous preventive measures to reduce their health risk.

ETHICAL AND LEGAL ISSUES

The use of biological markers raises important ethical and legal concerns. Markers of susceptibility involving genetic screening in the worksite, for example, have provoked a serious debate over who has a right to the information and for what purpose. Concerns include the potential for discrimination against workers on the basis of racial or cultural characteristics and acquired or inherited genetic susceptibility.

The central legal issues address the rights of those monitored and the use of monitoring as a primary control strategy. Thus, before companies approve the use of bio-markers in health surveillance, they must carefully consider how they will handle and communicate personal health data beyond the physician-patient relationship. In addition, occupational medical surveillance raises difficult issues related to labor-management relations, labor law, and discrimination law (Ashford, 1994; Van Damme, 1995).

These ethical and legal issues are forcing companies to exercise great care when they develop their medical surveillance and monitoring programs. They must balance the benefits to be gained from these programs with the risks involved in handling confidential information.

SUMMARY

- Public concern about human exposure to environmental and occupational hazards has grown steadily over the past three decades. As a result, medical surveillance has become an important tool in detecting and monitoring injury and illness in the worksite.
- Surveillance refers to the collection of data that are continually reviewed in order to detect early signs of occupational disease. The terms medical surveillance, medical screening, and medical monitoring have sometimes been used interchangeably. In this chapter, medical surveillance is used in its broadest sense to include occupational screening and monitoring activities.
- The occupational health professional is involved in designing and implementing a medical surveillance program. To do so, the health professional should consider the nature of the disease, effectiveness of the tests used, and the expected prevalence of the disease in the population being evaluated. To implement an effective medical surveillance program, the health professional must review exposure and health data and conduct risk-benefit analyses, set program goals, select medical tests, interpret data, select or recommend treatments, communicate medical results to appropriate parties, and evaluate the program.
- The goals of medical surveillance include the detection of overexposure to a hazardous agent and the early detection of disease in order to improve clinical outcome. Environmental and medical monitoring is an important element because by the time a disease can be detected in a person, it is often very difficult to alter the course of the disease.
- Biological monitoring, which consists of testing for biomarkers of exposure, effect, and susceptibility, can assist the clinician in assessing the disease-exposure relationship. Methods for measuring or evaluating the combined effects resulting from exposure to multiple hazards are an active area of research.
- Medical surveillance programs can screen for exposure-related cancers such as lung and bladder cancer as well as investigate the potential carcinogenic effects of substances to which workers are exposed. To date, however, screening for cancers has shown only limited human health benefits.
- The ethical and legal issues raised in occupational medical surveillance programs involve confidentiality of medical data. The questions to be addressed include: Who has a right to workers' medical information and how will that information be used?

REFERENCES

American Cancer Society. Guidelines for cancer-related checkup, and update. Atlanta: American Cancer Society, 1993.

American Conference of Governmental Industrial Hygienists, Inc. *Documentation of the Threshold Limit Values and Biological Exposure Indices* vol. 3. Cincinnati: American Conference of Governmental Industrial Hygienists, 1991.

Ashford NA. Monitoring the worker and the community for chemical exposure and disease: legal and ethical considerations in the U.S. *Clin Chem,* 40 (7 Pt 2):1426–37, 1994.

Canadian Task Force on the Periodic Health Examination. Canadian guide to clinical preventive health care. Ottawa: Canada Communication Group, 1994.

Federal Register 60 (145):38840–38844, 7/28/95. ATSDR's Final Criteria for Determining the Appropriateness of a Medical Monitoring Program Under CERCLA.

Fleming L, Herzstein J, Shalat S. "Environmental Health Surveillance." Chapter 62 in Brooks S, Gochfeld M, Herzstein H, Schenker M, Jackson R, eds. *Environmental Medicine.* St. Louis: Mosby-Year Book, Inc., 1995.

Klaucke DN, et al. Guidelines for evaluating surveillance systems, *MMWR* 37(S-5):1, 1988.

Martin AA, Schenker M. Screening for lung cancer: Effective tests awaiting effective treatment. *Occup Med State of the Art Rev* 6:111, 1991.

Mullan RJ, Murthy LI. Occupational sentinel health events: An updated list for physician recognition and public health surveillance. *Am J Ind Med* 19:775, 1991.

Murthy LI, Halperin WE. Medical screening and biological monitoring: A guide to the literature for physicians. *J Occup & Environ Med* 37(2): 170–184, 1995.

Rempel D, ed. *Medical Surveillance in the Workplace. Occupational Medicine: State of the Art Reviews,* vol. 5, no.3. Philadelphia: Hanley & Belfus, Inc., July-September 1990.

Rutstein DD, et al. Sentinel health events (occupational): A basis for physician recognition and public health surveillance. *AJPH* vol. 73, no. 9: 1054–1062, September 1983.

Ryback LP. Hearing: The effects of chemicals. *Otolaryngol Head Neck Surg* 106:677–685, 1992.

Silverstein, M. Analysis of medical screening and surveillance in 21 OSHA standards: Support for a generic medical surveillance standard. *Am J Ind Med* 26:283–295, 1994.

United States Preventive Services Task Force Guide. *Guide to Clinical Preventive Services,* 2nd ed. Baltimore: Williams and Wilkins, 1996.

Van Damme K, et al. Individual susceptibility and prevention of occupational diseases: Scientific and ethical issues. *J Occup & Environ Med* 37(1):91–99, 1995.

REVIEW QUESTIONS

1. Why is medical surveillance an important tool used in disease detection and prevention in the workplace?

2. List the factors to be analyzed when deciding whether to conduct medical surveillance.
   a.
   b.
   c.
   d.

3. Which of the following are included in efforts aimed at reducing health risks for which the general population is screened?
   a. Hypertension screening
   b. Education on self-examination for testicular or breast cancer
   c. Smoking-cessation programs
   d. Colon cancer screening
   e. All of the above

4. What are the types of exposure and health data that should be reviewed by the occupational health professional when designing a medical surveillance program?
   a.
   b.
   c.
   d.
   e.

5. What are biomarkers?

6. Briefly describe five limitations of biological monitoring.
   a.
   b.
   c.
   d.
   e.

7. The occupational health professional must ensure that the health surveillance program is in compliance with the OSHA medical screening requirements for what two categories?
   a.
   b.

8. Explain the synergistic effect of substances, and give an example.

9. What are some of the ethical and legal concerns that have been raised in the use of biological markers?

10. Name the organization that classified chemicals and industrial processes with respect to their potential to cause cancer in humans.

10

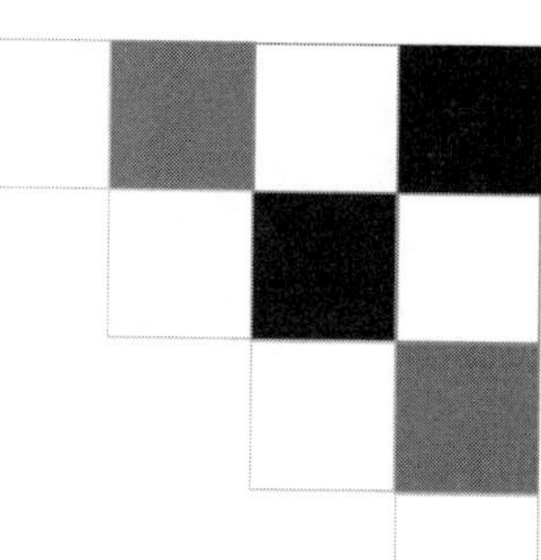

Electrical Safety

Properly used, electricity is the most versatile form of energy. Failure to establish safety precautions for the design, work practices, procedures, servicing, and maintenance operations for electrical equipment, however, often results in bodily harm (including fatalities), property damage, or both. This chapter provides an overview of basic safety considerations to minimize employees' exposure to low voltages of 600 or less. Power-distribution systems above 600 volts (v) are not addressed here. Topics covered in this chapter include:

- definitions of common electrical energy terms
- a discussion of typical electrical injuries and emergency treatment
- safety practices for using electrical equipment
- methods of grounding power equipment
- reducing hazards through prevention and inspection
- safety procedures for maintenance and repair work
- safety training for employees working with electrical equipment

In most cases, electrical and electronic equipment can be designed for both maximum safety and efficiency. However, potentially hazardous conditions, such as inadvertent contact with voltages, do exist. These hazards can occur during engineering analysis (commonly known as debugging), installation, servicing, testing, and maintenance of electrical and electronic equipment. Decreasing the exposure to the majority of these potential hazards is neither difficult nor expensive if safeguards and safe procedures are introduced in the design stage. If hazards are ignored, however, serious accidents may result.

DEFINITIONS

Before dealing with electrical equipment in any way, an understanding of a few basic electrical terms is needed. Three of these basic terms—current, voltage, and resistance—can be defined by using the following analogy: Electricity flowing through a circuit is like the flow of water through a pipe. Keep this analogy in mind, to understand these terms more easily.

Current

Think of current as the total volume of water flowing past a certain point in a given length of time. Electric current is measured in amperes (amps). The measurement used in relation to electric shock is the milliampere (mA, 0.001 ampere).

Voltage

Think of voltage as the pressure in a pipeline. Voltage is measured in volts (v). Low voltage is a rather ambiguous term depending upon whether it is being used by a safety professional, a facility electrician, or a lineman. For the purposes of this chapter, low voltage is 600 v or less. Potentially hazardous voltage is between 24 v and 600 v. Potentially lethal voltage is 50 v and above. A car battery of 12 v direct current (DC) in a dead short can be hazardous.

Resistance

Think of resistance as blockage in the water pipe. Resistance is any condition that retards current flow; it is measured in ohms.

Watt

A watt is the quantity of electricity that is consumed. It is determined by multiplying volts times amps.

Ground

A ground is an object that connects a piece of electrical equipment to earth or some conducting body that serves in place of earth. A ground serves to complete the electrical circuit and prevent the hazard of electrical shock.

Bonding

Bonding is the joining of metallic parts to form an electrically conductive path. This path assures electrical continuity.

ELECTRICAL INJURIES

Current flow and time are the factors that cause injuries in electrical shock. The severity of electrical shock is determined (1) by the amount of current that flows though the victim (Table 10–A),and (2) by the length of time that the body receives the current.

Because current flow depends on voltage and resistance, these factors are also important. In addition to current flow and time, other factors that affect the extent of injury are the parts of the body involved, and, if alternating current is used, the frequency of shock. Heat is a secondary effect on the body. Such current flow can easily be received on contact with low-voltage sources of the ordinary lighting or power circuit.

Table 10–A. Effects of Electrical Contact

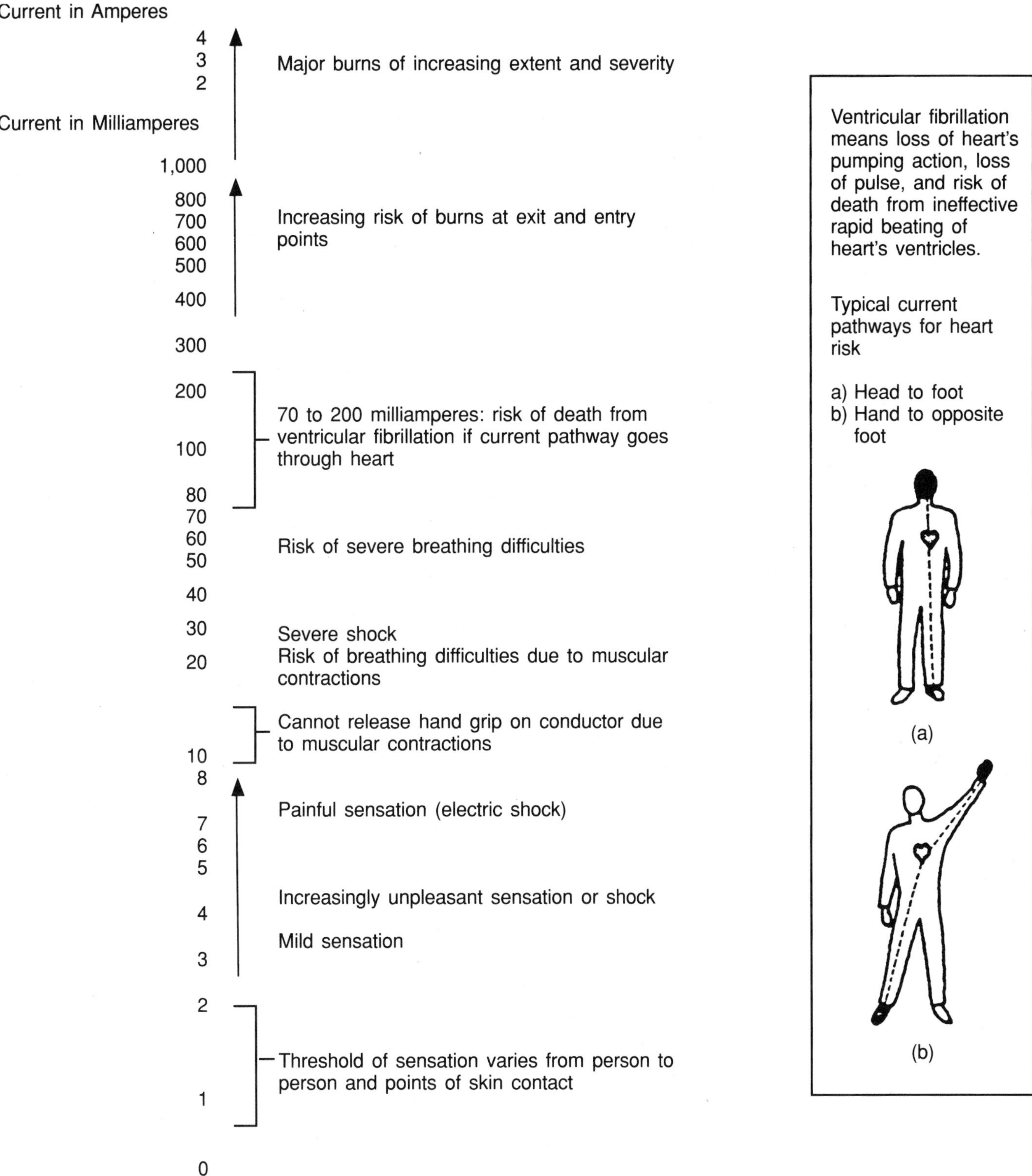

Source: *Canadian Occupational Safety*, January-February, 1988.

A person's main resistance to current flow is the skin's surface. Callous or dry skin has a fairly high resistance. A sharp decrease in resistance takes place, however, when the skin is moist (Table 10–B). Once the skin's resistance is broken down, the current flows readily through the blood and the body's tissues.

Whatever protection is offered by the skin's resistance decreases rapidly with increases in voltage. A high-voltage alternating current of 60 Hz causes violent muscular contraction, often so severe that the victim is thrown clear of the circuit. Although low voltage also results in muscular contraction, the effect is not so violent. Low voltage is dangerous, however, because it often prevents the victim from breaking the contact with the circuit.

Internal Injuries

Death or injuries from electric shock may result from the following effects of current on the body:

1. Contraction of the chest muscles, which may interfere with breathing to such an extent that death will result from asphyxiation when the contact is prolonged.
2. Temporary paralysis of the nerve center, which may result in failure to breathe, a condition that often continues long after the victim is freed from the circuit.
3. Interference with the normal rhythm of the heart, causing ventricular fibrillation. In this condition, the fibers of the heart muscle, instead of contracting in a coordinated manner—like a pump—contract separately and at different times. Blood circulation ceases and, unless proper resuscitation efforts are made, death occurs. The heart cannot spontaneously recover from this condition. It has been estimated that 50 mA is sufficient to cause ventricular fibrillation.
4. On contact with heavy current, the muscular contractions of the heart stop. In this case, the heart may resume its normal rhythm when the victim is freed from the circuit.
5. Hemorrhages and destruction of tissues, nerves, and muscles from heat due to heavy current along the electrical circuit's path through the body.
6. Severe burns may result from contact with low-voltage systems in cars, trucks, and lift trucks when metal, wrenches, or jewelry make contact with current-carrying conductors. Considerable current is likely to flow from high-voltage sources. In general, only very short exposure can be tolerated if the victim is to be revived.

Injuries from electrical shock are less severe when the current does not pass through or near nerve centers and vital organs. In the majority of electrical accidents that occur in industry, however, the current flows from the hands to the feet. Since such a path involves both the heart and the lungs, the results are usually serious.

Table 10–B. Human Resistance to Electrical Current

Body Area	*Resistance (ohms)*
Dry skin	100,000 to 600,000
Wet skin	1,000
Internal body—hand to foot	400 to 600
Ear to ear	(about) 100

Note: Data is based on limited experimental tests, and is not intended to indicate absolute values.

Skin and Eye Injuries

Another type of injury is burns from electrical flashes. Such burns are usually deep and slow to heal and may involve large areas of the body. Even persons at a reasonable distance from the arc may receive eye burns.

Where high voltages are involved, flashes of explosive violence may result. This intense arcing is caused by (1) short circuits between bus bars or cables carrying heavy current, (2) failure of knife switches, (3) opening knife switches while they are carrying a heavy load, or (4) pulling fuses in energized circuits.

Injuries from Falls

Other injuries from electrical shock include falls from one level to another. For example, a worker receives a shock from defective or malfunctioning equipment. The shock causes muscles to contract, causing the worker to lose his or her balance and fall.

Cardiopulmonary Resuscitation

Because electrical shock can stop the heart and lungs, be sure that workers involved in hazardous electrical energy levels know cardiopulmonary resuscitation (CPR) and rescue procedure. CPR training is provided by the National Safety Council First Aid Institute, the American Heart Association, the American Red Cross in the United States, and St. Johns Ambulance in Canada. Consult the telephone directory for the closest office.

Immediately apply CPR to a victim of electric shock (Figure 10–1). Continue CPR until the victim revives, or until a physician diagnoses death. The sooner CPR is started, the better the chance of revival.

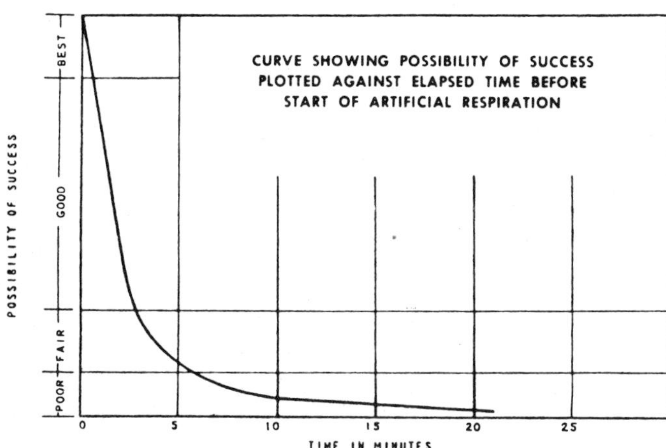

Figure 10–1. The possibility of successful revival decreases with time. (Reprinted with permission from Edison Electrical Institute Resuscitation Manual.)

ELECTRICAL EQUIPMENT

Most items of electrical equipment are designed and built for specific types of service. They operate with maximum efficiency and safety only when used for the purposes and under the conditions for which they are intended.

Selecting Electrical Equipment

When selecting electrical equipment, follow the recommendations of the various established codes and standards, such as the National Fire Protection Association (NFPA) 70, *National Electrical Code* (NEC), and the American National Standards Institute's (ANSI) C2, *National Electrical Safety Code.* In addition to these codes, check the state and local codes for industrial zoning requirements. Adherence to most of the provisions of the *National Electrical Code* is required not only by the regulators but also by many insurance companies and local governments.

Engineering consultants, manufacturers, and publications from the following groups will answer most questions concerning electrical equipment: American National Standards Institute, Canadian Standards Association, Factory Mutual System Research Organization, Illuminating Engineering Society of North America, National Fire Protection Association, and Underwriters Laboratories, Inc. (See References at the end of the chapter for addresses.) Consulting services may also be available from some of these organizations.

When ordering copies of codes or standards from publishers, give full information on the general type of equipment under consideration, the application, and the operating conditions. There is usually a charge for codes or standards.

Installing Electrical Equipment

Where space and operating requirements permit, install electrical equipment in the less congested areas of the facility or, where practical, in special rooms to which only authorized persons have access. When electrical equipment is located in the production areas of the facility, build enclosures around those parts of the equipment having exposed conductors. Also post DANGER signs.

Install floor curbing or heavy steel barriers if there is even a remote danger of industrial trucks striking critical electrical equipment. Install transformers, dead-front control boards, switches, motor starters, and other electrical equipment so that the chance of accidental contact with energized conductors is at a minimum.

Interlocks

When an interlock is used as a safety device, make sure that it is fail-safe. In other words, it must have an automatic backup in case its mechanism fails. In that way, the safety of personnel who depend upon it will not be at risk. Interlocks should meet the following criteria:

- be equipped with fail-safe features since failure of the interlock mechanism, loss of power, short circuit, or malfunction of equipment will cause the circuit to be interrupted
- have a visible disconnect, or opening, in the primary power circuit
- have a locking arrangement that makes attempts to circumvent the interlock impractical

Barriers

Barriers prevent accidental contact with electrical equipment (Figure 10–2). Make frames of wood, rolled metal shapes, angle iron, or pipe. Use wood strips, sheet metal, perforated metal, expanded metal, wire mesh, plastic, or shatterproof transparent material for filler. Dry wood and many plastics have the advantage of not conducting electricity. Unless specially treated, however, wood is combustible and may be hazardous. Ground all metal frames or guards.

Warning Signs

In addition, display warning signs near exposed current-carrying parts and in especially hazardous areas, such as high-voltage installations. These signs should be large enough to be read easily and should be visible from all approaches to the danger zone. The design of the warning sign should be in accordance with 29 *CFR* 1910.145 *Specifications for Accident Prevention Signs and Tags.*

Figure 10–2. An expanded metal screen guards the V-belt drive, belts, and sheaves of this injection pumping unit. The 5-cylinder pump is powered by a 200-hp explosion-proof electric motor. The screen guard allows easy maintenance. (Reprinted with permission from Service Pipe Line Co., Tulsa OK.)

Guarding

In many respects, standard machine-guarding practices can be applied to electrical equipment. Wiring, however, presents special hazards. Consider these hazards when planning the overall guarding program for the facility.

Install wiring in accord with the NEC unless other local requirements apply. When installing new circuits or changing existing ones, observe the requirements of national and local wiring codes.

The type of wiring depends upon the type of building construction, the size and distribution of the electrical load, exposure to dampness or corrosive vapors, location of equipment, and various other factors. For most facility conditions, grounded metal conduit is satisfactory. Other types of wiring include armored cable, nonmetallic sheathed cable, flexible metal conduit, raceways, open wiring and insulators, and concealed knob-and-tube wiring. (See NEC Article 324 for information on locations where concealed knob-and-tube wiring is not permitted.)

Use wires with insulation designed for the type of service and location. However, do not consider the insulation or weatherproofing alone as sufficient protection against shock, especially in high-voltage circuits. The NEC lists various types of insulation used on electrical conductors, as well as how to use them.

Switches

Switches, fuses, circuit breakers, ground-fault circuit interrupters, control equipment, motors, and extension cords are needed to run electrical machines safely and efficiently. Proper use of this equipment avoids hazards and prevents accidents.

Among the several types of switches are knife switches; push-button switches; snap switches; pendant switches; and enclosed, externally operated air-break switches. Many switches are designed for a specific function, such as the enclosed switch for controlling individual motors and machine tools and for lighting and power circuits. However, all switches, regardless of their function, must have approved voltage and ampere ratings that are compatible with their intended use.

Knife Switches

An open-knife switch is hazardous. Do not use it because current-carrying parts are exposed and an arc forms when the switch is opened. Enclose knife switches in grounded metal cabinets with control levers that operate outside the cabinets. A further safeguard is the safety switch; the cover must be closed before the switch can be used.

Mount a knife switch so the blades are dead when the switch is open. Install it so that gravity will not close it. Electrically interlock knife switches installed in power-switching circuits. In that way, they cannot be opened when the circuit is energized, unless switches are of the load-break type.

Where it is necessary to use disconnect switches for high-voltage, heavy-current-feeder circuits for testing floors or other service installations, locate them out of normal reach. Also, operate them only with insulated switch sticks. Where it is impractical to locate the switch out of normal reach, protect it from accidental contact by completely enclosing it, or by placing it behind a suitable fence or barricade. If it is placed behind a barrier, arrange for it to be opened or closed remotely with an insulated switch stick.

Pendant Switches

Pendant switches are used primarily where switches have not been installed on walls and where it is difficult to put them on walls. Pendant switches are also used where the switches on walls, or in cabinets, control two or more circuits. In such cases, control of individual circuits is desired.

Provide pendant switches, and in particular pendant push-button control stations, with exterior strain relief both at the point of suspension from the ceiling and at the point of entry into the enclosure. In other words, allow no pull or harmful friction on the electrical wiring or connections.

Provide a toggle switch with a button that automatically lights as soon as the circuit is open. This will help locate the switch in dark areas. However, do not rely on such a switch for illumination.

Other Switches

Push-button or snap switches are recommended because the current-carrying parts are enclosed. Install flush switches in boxes. Mount surface switches, used in open wiring and in molding work, on porcelain, composition, or other insulating subbases. These switches must have indicators to show open and closed positions.

Protective Devices

The safe current-carrying capacity of conductors is determined by their size, material, insulation, and the way in which they are installed. If conductors are forced to carry more than the maximum safe load or if heat dissipation is limited, overheating results. Overcurrent devices, such as fuses and circuit breakers, open the circuit automatically in case of excessive current flow from accidental grounds, short circuits, or overloads. Therefore, install some kind of overcurrent devices in every circuit.

Overcurrent devices should interrupt the current flow when it exceeds the conductor's capacity. Selecting properly rated equipment depends not only on the current-carrying capacity of the conductor. It also depends on the rating of the power-supply transformer, or generator, and its potential short-circuit-producing capacities. Protection of this kind, both for personnel and for equipment, is one of the important features of an electrical installation. Where higher interrupting capacity is required, use special high-capacity fuses or circuit breakers.

Fuses

Among the many types of fuses are link, plug, and cartridge fuses. Use them only in the type of circuit for which they were designed. Using the wrong type or the wrong size fuse may cause injury to personnel and damage to equipment. Overfusing is a frequent cause of overheated wiring or equipment, which can cause fires.

Before replacing fuses, lock out, tag out, and test the circuit to be sure that the voltage is turned off. Testing can save lives. Find out the cause of any short circuit or overload, and then replace a blown fuse with one of the same type and size. Never insert fuses in a live circuit.

Link Fuses

A link fuse, as its name indicates, is a strip of fusible metal that links two terminals of a fuse block. If not enclosed, it may scatter metal when it blows. Replaceable link fuses should be replaced only under the direction of qualified maintenance personnel or an electrician.

Plug Fuses

Plug fuses are used on circuits that do not exceed 30 amps at not more than 150 v to ground. In plug fuses, the fusible metal is completely enclosed. Use plug fuses that cannot be bridged inside the holder.

Cartridge Fuses

The cartridge fuse, which is widely used in industrial installations, has a fusible metal strip enclosed in a tube. Cartridge fuses that indicate when the fuse is blown, and renewable cartridge fuses in which the fusible element may be replaced, are available.

To deenergize the fuses, place a switch in any circuit that can be opened. As an additional precaution, use insulated fuse pullers.

Circuit Breakers

Circuit breakers have long been used in high-voltage circuits with large current capacities. In recent years, their use has become more common in many other kinds of circuits. They are available in a variety of types and sizes. They may operate instantly or through a delayed timing device. They may be operated manually or with power.

Circuit breakers fall into two general categories—thermal and magnetic. The thermal circuit breaker operates solely on the basis of the rise of temperature. Therefore, variations in the temperature of the room in which the circuit breaker is installed will affect the point at which it interrupts the circuit.

The magnetic circuit breaker operates only on the basis of the amount of current passing through the circuit. This type of circuit breaker has advantages where a wide fluctuation in temperature would require overrating the circuit breaker or where tripping frequencies are high.

Have experienced maintenance personnel regularly check circuit breakers. Circuit breakers should be in good operating condition at all times.

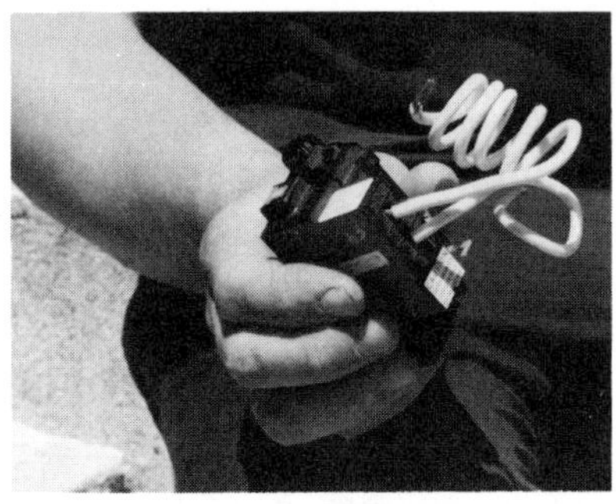

Figure 10–3. Three types of differential ground fault interrupters: *top*—circuit breaker type, *center*—outlet type, *bottom*—portable type. Current-carrying conductors pass through circular iron core of doughnut-shaped differential transformer. If the transformer senses even a portion of the current flowing to ground, it causes the circuit breaker to operate and open the circuit.

Ground-Fault Circuit Interrupters

A ground-fault circuit interrupter (GFCI) is a fast-acting, electrical circuit-interrupting device that is sensitive to very low levels of current flow to ground. The GFCI is designed to sense leaks of current large enough to cause serious personal injury. The unit operates only on line-to-ground fault currents, such as insulation leakage currents or currents likely to flow during accidental contact with a "hot" wire of a 120-v circuit and ground. It does not give protection in the event of line-to-line contact.

A receptacle GFCI that functions the same as a circuit breaker is available (Figure 10–3). However, it protects only the associated integral receptacle, plus a limited number of additional receptacles connected downstream. GFCIs are also an integral part of some extension cords.

Use GFCIs when workers are using any electrical equipment, particularly in any work environment that is, or may become, wet, and in other areas that are highly grounded. Per 29 *CFR* 1926.404(b)(1)(i), the employer shall use either a GFCI or an assured equipment ground program for branch circuits at a construction site.

Also stress off-the-job electrical safety. The NEC requires GFCIs on 15- and 20-amp, 120-v outdoor receptacles, garage circuits, and bathrooms. Consult the NEC for the specific requirements applicable to any of these occupancies.

When GFCIs are used in construction activities, locate the GFCI as close as possible to the electrical equipment it protects. Excessive length of temporary electrical wiring or long extension cords can cause ground-fault current leaks to flow by capacitive and inductive coupling. The combined leaks of current can exceed 5 mA, thus causing the GFCI to trip.

Tripping may also be caused by one or more of the following items:

- wet electrical cord-to-tool connections of extension cords
- wet power tools
- outdoor GFCIs not protected from rain or water sprays
- defective electrical equipment with case-to-hot-conductor fault
- too many power tools on one GFCI branch
- resistive heaters
- coiled extension cords (long lengths)
- poorly installed GFCIs
- defective or damaged GFCIs
- electromagnetic-induced current near high-voltage lines
- portable GFCI plugged into a GFCI-protected branch circuit
- defective power tools

Follow the manufacturer's instructions regarding the testing of the GFCI. Some recommend pushing the TEST button and resetting it monthly. Immediately replace defective units.

Remember that GFCIs do not replace a fuse or circuit breaker. Consider GFCIs as additional protection against the most common form of electrical shock and electrocution, the line-to-ground fault.

The ground fault must pass through the electronic-sensing circuit. This circuit has enough resistance to limit the current's flow to as little as 0.2 mA—well below the level where a person could feel it. Although no GFCIs protect individuals in the event of line-to-line contact, GFCIs should significantly reduce the number of electrical shock accidents, which presently account for about 1,100 deaths each year.

Control Equipment

Arrange switchboards with lockout capabilities for both alternating-current (AC) and direct-current (DC) distribution circuits so the operator can easily reach the controls. Likewise, instruments should be readable and equipment adjustable from the working area. Place switchboards so the operator will not be endangered by live or moving parts of machinery. Do not use the space behind the switchboard for storage, and keep this area clear of rubbish. Place the switchboard in a special room or use screen enclosures to keep unauthorized personnel out. Keep the doors to the enclosure locked.

Provide good lighting for the front and rear of switchboards. Be sure the lighting is ready for use at all times. Also provide an emergency source of lighting.

Switch and fuse cabinets should have close-fitting doors that are kept locked. To warn employees that electrical parts are exposed, and to remind them to close all doors to equipment, paint the inside of the doors orange. The switchboard's framework and the metal parts of guards should be grounded in accordance with NEC regulations.

Arrange connections, wiring, and equipment of switchboards and panel boards in an orderly manner. Plainly mark switches, fuses, and circuit breakers, and arrange them to afford ready identification of circuits or equipment that are supplied through them. Keep a diagram of switchboard connections and devices posted near the equipment.

Maximize protection against accidental shock from live electrical parts on switchboards, fuse panels, and control equipment by insulating the floor area within the range of the live parts. Where equipment provides exposure to 600 v or less, use special insulating mats or dry wood floors with no metal parts. The insulating mats should be moisture-resistant, nonconductive, and able to withstand mechanical abuses.

Circuits initiated by push buttons should be low voltage (600 v or less). However, push buttons may be used to control a high-voltage circuit if step-down transformers or relays are provided. They will prevent the voltage in the control part of the circuit from exceeding 250 v.

Motors

Mount motors so they do not interfere with the normal movement of personnel or materials. Motors not enclosed should be in areas free from dust, moisture, and flammable or corrosive vapors. In some instances, motors can be isolated from personnel by being mounted on overhead supports, installed below floor level, or placed in special motor rooms. If current-carrying parts of motors must be exposed and grounding is not possible, elevate them at least 8 ft (2.4 m) above the work area, or provide enclosures, barriers, or guards to prevent personnel from contacting them.

For maximum safety, each motor should be of the type and size required for the load and for the conditions under which it must operate. Avoid excessive overloading. Only use motors suitable for hazardous locations in areas that contain flammable vapors, combustibles, dusts, or fibers.

The following are common motor problems: dust, stray oil, moisture, misalignment, vibration, overload, and friction. Ways to reduce these problems are given in the sections that follow.

Dust

In facility operations, dust constantly settles on motors, housings, windings, slip rings, and commutators. Dust tries to work its way to the bearings. The problem of dust is at a minimum in the case of totally enclosed, fan-cooled motors.

Remove dust from motors before it has a chance to unite with water or oil to form a gummy mess. Therefore, wipe off motors and their housings, slip rings, and commutators at regular inspections. Occasionally, dust must be blown out of the wire-wound section with not greater than 30 psig of pressure (207 kPa) from a compressed-air unit. When this operation is performed, provide good ventilation to prevent the accumulation of dust. In some cases, use vacuum-cleaning equipment or a hand bellows. The person doing the dusting should wear eye protection and should consider using a dust mask that is appropriate for the conditions.

Oils

Oil harms commutators by deteriorating the micainsulating segments between the bars. Oil causes excessive sparking and deteriorates the insulation on the windings. This leads to an immediate danger of a burnout or breakdown. Oil also tends to hold dust, lint, and other material, thus increasing the hazard of fire.

When oil and dust have been allowed to build up, take the motor out of service and clean it according to the manufacturer's directions.

Moisture

Not all electrical insulation acts as a perfect barrier to moisture. Some types of insulation become porous with age and absorb moisture. The electrical resistance may then drop to a point where the current's leakage burns through the insulation to ground. This could cause a short circuit with the associated risk of fire or shock hazard.

Careless handling of liquids in the manufacturing process; getting motors wet during cleanup operations, especially in facilities where floors must be frequently washed with water; and the overflowing of water softener tanks, filters, dye cups, or other liquid reservoirs can cause failure and shock hazards.

To prevent damage from moisture, train personnel and relocate equipment. If the moisture hazard cannot be wholly avoided or is inherent in the work for which the motor is used, have manufacturers provide drip-proof, splash-proof, or totally enclosed designs. After a motor has been subjected to moisture and the resistance is at a dangerous point, have the motor dried before testing and returning to service.

Misalignment

Check belts and chain drives for tension, and check gearing for binding. Misalignment of the motor's shaft can cause it to spring or break, the bearings to burn out, and the motor itself to fail as a result of an overload.

Misalignment is a significant cause of motor vibration. Another cause is careless servicing that puts motors out of balance, such as loose motor-mounting bolts and worn bearings that allow the shaft to oscillate.

Overloads

A motor can be overloaded by (1) excessive friction within the motor itself, (2) using the motor for the wrong kind of job, (3) an obstruction in the driving or driven machine, or (4) pushing the machine to perform beyond the motor's capacity. If one of these conditions causes the current in a motor to exceed the current rating on its nameplate, the actual heating may increase as much as the square of the current's increase. As a result, insulation may be burned, soldered connections melted, or bearings burned out.

To safeguard against such overloads, motors have various forms of overload protection. In most cases, a thermal element is connected in the power circuit to the motor. This is accomplished (1) with integral overtemperature protective devices that will open the motor's circuit in the event of overheating, or (2) with current-sensitive protective devices that generate heat from an excessive current. These devices operate an overload relay and open the circuit to the motor. Use current-sensitive thermal elements of the proper capacity as listed in NEC. Do not use a unit of more than the recommended value. Revise motor output and production methods to control overload conditions that trip or blow these protective devices.

Friction and Wear

Follow the manufacturer's lubrication charts and instructions on types and grades of lubricants, the frequency of lubrication, and other practices. In that way, friction can be reduced and excessive wear, overheating of bearings, and possible fires from faulty lubrication can be prevented.

Extension Cords

Before using an extension cord, be sure it is listed by Underwriters Laboratories or other recognized testing laboratories. Inspect the extension cords regularly. To prevent the wire strands from breaking, avoid kinking or excessive bending of the cord. Broken strands may pierce the insulated covering and become a shock or short-circuit hazard.

Extension cords should be inspected before each use. Remove cords with cracked or worn insulation, or with damaged plugs or sockets, from service immediately. Do not permit splicing of extension cords.

Extension cords should not be connected or disconnected with an electrical load on. Store disconnected extension cords neatly coiled in a dry room at room temperature.

Extension Cords for Portable Tools

Cord for use with portable tools and equipment is made in several grades, each of which is designed for a specific type of service. Use jacketed cord with portable electric tools and with extension lamps in boilers, tanks, or other grounded enclosures. Use cords made with special types of rubber or plastic covering when the cord could come in contact with oils or solvents. Because the metal frames of portable electrical equipment should be grounded, use cord with a green-covered ground conductor and a polarized plug and receptacle.

Extension Cords for Heating Devices

Cord for heating devices, such as electric irons and water heaters, is made with an insulated covering that contains a flame-retardant or thermosetting compound, such as neoprene. Such cords are designed to resist high temperatures and, in the case of neoprene, dampness.

Flexible Cords

The various types of flexible cords and cables and their approved use and size are given in the NEC (Table 10–C). Connect flexible cords to devices and fittings so that tension will not be transmitted to joints or terminal screws. This is accomplished by special fittings, a knot in the cord, or winding with tape (Figure 10–4). All plugs that are attached to cords must have the terminal screw connections covered by suitable insulation. Avoid connecting two extension cords together.

Table 10–C. Minimum Wire Size of Extension Cord

Amps On Nameplate	Cord Length in Feet			
	25'	50'	100'	150'
0–6	18	16	16	14
10	18	16	14	12
10–12	16	16	14	12
12–16	14	12	(not recommended)	

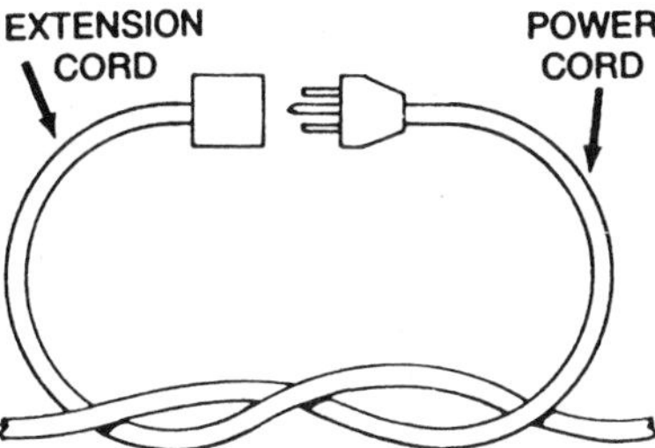

Figure 10–4. A method of securing an extension cord is to tie it as shown.

Extension Lamps

Handles of portable hand lamps must be made of nonconductive material, and there should be no metallic connection between the lamp guard and the socket's shell. For use near exposed live parts, such as the rear of switchboards, be sure that the guard itself is made of nonconductive material.

Since extension lamps are sometimes used under conditions in which a 120-v shock may prove fatal, provide safe cords and lamp holders that are kept in excellent condition. In some facilities, the shock hazard of portable lamps is eliminated by using small, portable transformers that reduce the lamp's voltage to 12 v. Special lamps for these units are rated at 75-watt capacity, which gives sufficient working light.

Most electrical equipment is designed for safe operation under limited overload conditions for varying periods of time. A continued overload, however, may introduce additional operating hazards by causing fires, short circuits, circuit failures, or mechanical failures. Be sure that designers, installers, and operators are thoroughly familiar with the limitations of their equipment. Train them to observe and report abnormal conditions.

Test Equipment

Various types of electrical- and electronic-testing equipment can detect many of these conditions before they get out of control and cause damage. Qualified technicians or specially trained maintenance personnel should make these tests. The electrical test equipment should be listed by UL or another nationally recognized laboratory.

The following types of equipment are standard and should be considered as essential testing equipment: split-core ammeter, voltmeter, ammeter, megohmmeter, ground-fault indicators and locators, wattmeter, industrial analyzer, receptacle circuit tester, receptacle tension tester, voltage detector, recording instrument, and specialized testing instruments (Figure 10–5). Specialized testing instruments, such as volt-ohm-milliammeter, oscilloscopes, and cable testers may also be used (Figures 10–6 and 10–7). Instruments such as these are generally fitted for detailed engineering work or for use where ordinary recording and indicating instruments are not accurate enough.

Specialized Processes

Among the many types of electrical equipment used in industry are the electric furnace, auxiliary heating devices, high-frequency heating equipment, electric welding equipment, x-ray laser, and ultraviolet and infrared installations. Each of these devices may introduce special operating hazards. However, protection from their electrical hazards may be secured through the same procedures recommended for use with the more common types of electrical equipment.

High-frequency heating installations range in power capacity from a few hundred watts to several hundred kilowatts. Therefore, safety considerations are of prime importance.

The resistance of the body to the flow of high-frequency current is not dependent upon the skin. At frequencies of 200 kHz to several hundred mHz, currents flow in a very thin shell on the surface of the conductor. This tendency of high-frequency current to flow on the surface is known as "skin effect" and increases as the frequency increases. Should the skin of a human being be punctured, the currents still flow on the surface and do not penetrate to the vital organs of the body.

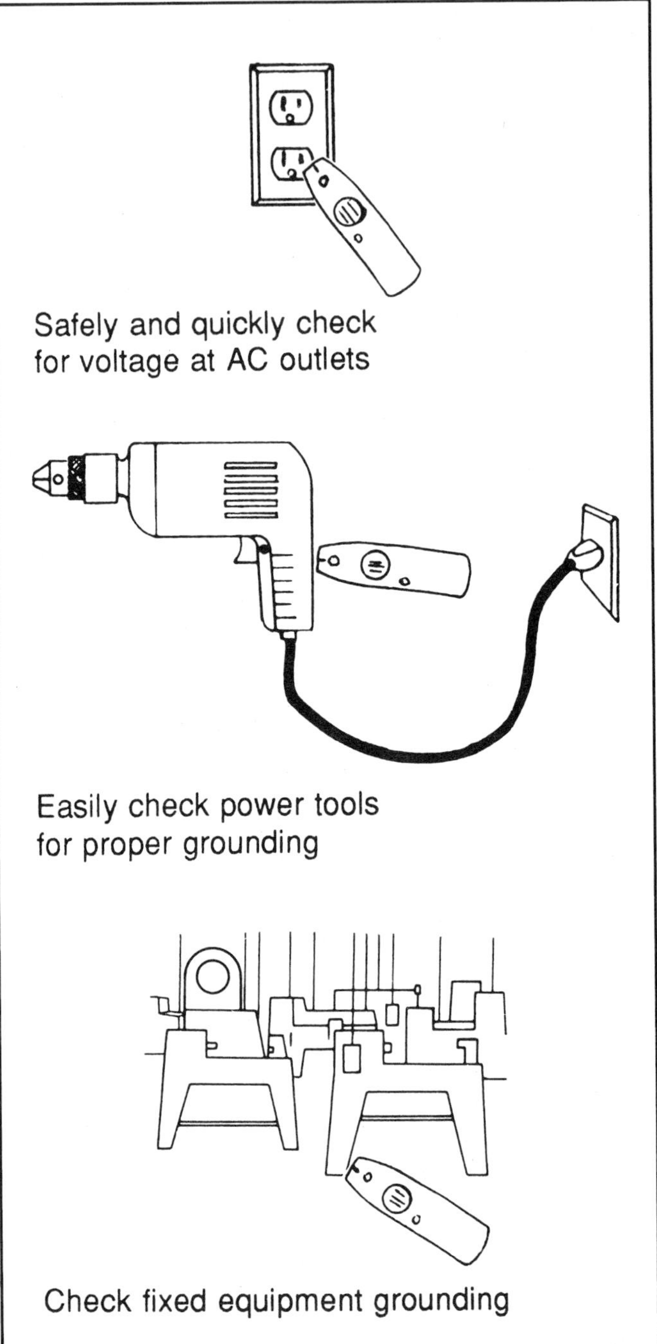

Figure 10–5. A voltage detector can be used to check outlet voltage, equipment grounding, and circuit breaker condition.

A person coming into contact with high-frequency electrical energy will in general be burned because of the natural tendency to pull away, thereby setting up an arc. High-frequency burns can also result from radio-frequency antennae or from waveguide exposure. In many cases, the burn will occur and the person will feel nothing until after the exposure. These burns are

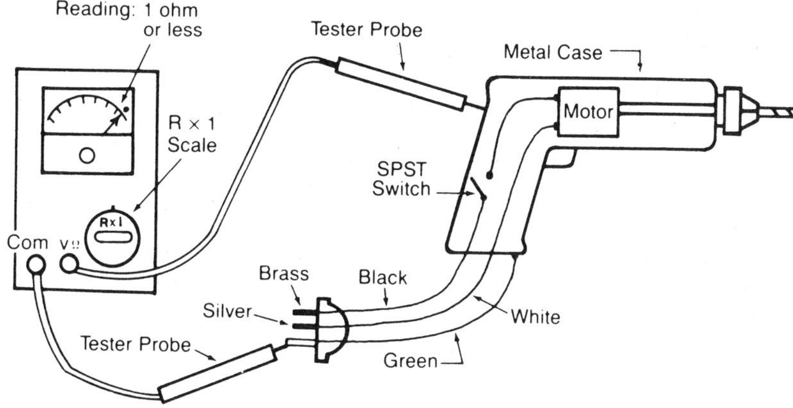

Figure 10–6. In a continuity test for checking the grounding path of an appliance, the reading on the ohmmeter should be less than 1 ohm.

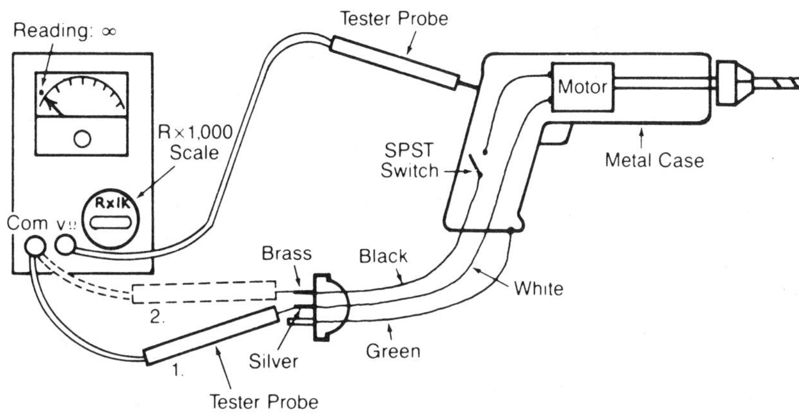

Figure 10–7. The leakage test should read near infinity on the ohmmeter when checking an appliance for current-to-case leakage.

painful and usually take longer to heal than burns from the more common thermal-heat sources.

Use the following methods to prevent high-frequency burns to operators:

- Connect interlocks, or other devices that remove power from the equipment whenever an access door is open, into the control circuit. This protects the operator from contacting any electrical or high-frequency energy.
- Have material to be heated by induction carried by revolving hopper feeds or by conveyors to the heating coil. The heating coil is enclosed with a shield and cannot be reached by the operator.
- Locate high-frequency generators some distance from the work position. The high-frequency energy is then conveyed by a waveguide or transmission line.
- Insulate conductive coils or equipment surfaces with compatible insulation materials suitable for the application. Insulation protects the user in case of accidental contact with the coils or with the equipment's surfaces.

GROUNDING

In order to understand grounding properly, note that both equipment grounding and the grounding of the electrical system itself are included under this term. (See Definitions at the beginning of this chapter.) The electrical system is grounded in order to prevent the occurrence of excessive voltages from such sources as lightning, line surges, or accidental contact with higher voltage lines. Both the electrical system and metallic enclosures are grounded to cause overcurrent devices to operate in the event of a ground fault occurring from an insulation failure. (See NFPA 70, *National Electrical Code.*)

According to the NEC, grounding is required for:

- refrigerators and similar equipment
- appliances using water, such as washers
- hand-held power tools
- motor-operated appliances, such as clippers
- any equipment used in damp areas
- portable handlamps with metallic ground guards
- some parts of nonelectrical equipment, such as frames

However, grounding is not required of the items listed above if they are:

- approved and labeled double-insulated tools
- insulated transfer tools of less than 50 v

System Grounding

AC systems operating at 50 v or more must be grounded under a variety of voltage conditions. Grounding is accomplished by bonding the identified conductor to a grounding electrode by means of an unbroken wire called a grounding electrode conductor. Identification takes the form of white or neutral grey-colored insulation.

In a home or apartment, the utility will ground the identified conductor at the transformer by attaching a plate to the bottom of the supporting pole. This identified or grounded circuit conductor is then once again grounded when it is brought to the service equipment within the home. This grounding takes the form of a grounding electrode conductor run to a metal underground water pipe, a metal building framework, a bare copper conductor encircling the building, or a concrete-encased, steel reinforcing-bar system. If none of these forms of electrodes is available, buried plates, pipes, rods, or other metal underground structures are used.

Industrial installations are grounded in a similar manner, except that the transformer may be very large and mounted on a concrete pad. In that case, the grounding electrode is a metal grid carefully placed within the ground or within the concrete pad itself. As in grounding in a home, the grounded circuit conductor brought to the service equipment within the facility or commercial building must be bonded to a grounding electrode that is specified in the NEC.

Much has been said about the resistance of the grounding electrode, and there are many erroneous ideas about its use. However, if careful attention is paid to the workmanship of the original installation, the following provide adequate resistance to ground: metal water-piping systems, building structures, buried tanks and piping systems, and concrete-encased reinforcing power systems.

Higher resistance values are encountered only when a "made" electrode such as a rod, pipe, or plate is used. In many parts of the country, low-resistance values for made electrodes are impossible to obtain. In such cases, the NEC simply requires that one additional electrode be used. Because the same term is applied both to system grounding and equipment grounding, it is often thought that the grounding electrode must have a very low resistance in order to dissipate ground faults. This is not true. Most of the current from a ground fault finds its way back to the source transformer from the service equipment via the neutral conductor, not the earth.

A grounding electrode is not intended to carry large fault currents. However, it provides a point of equalization so that large voltage differences both inside and outside the building are minimized. As stated previously, some of the primary causes of such voltage differences are lightning surges or lightning strikes on the building itself.

Once the grounded circuit conductor is taken beyond the service equipment, it must not be grounded at any point. The major exception to this rule is the case of additional transformers being used within the establishment in order to step down voltages at various locations. In such instances, the secondary conductors constitute a new or separately derived system. Then, the identified circuit conductor is to be grounded to a grounding electrode—the nearest available effectively grounded water-piping system or a building's structural metal.

The rule prohibiting the subsequent grounding of the grounded circuit conductor is frequently abused at the terminals of an ordinary parallel U-blade receptacle. Many well-meaning but uninformed maintenance personnel will connect the white, grounded circuit conductor both to the silver terminals and to the green equipment-grounding terminals on the receptacle. Such personnel think they have thereby doubled the grounding. However, in addition to violating the provisions of the NEC, they have set up a dangerous condition: If the neutral circuit is interrupted anywhere between the receptacle and its point of attachment to ground, any equipment enclosures that have been grounded by means of this receptacle now will be energized at full-line voltage.

Not all systems are required to be grounded. If trained personnel are present, some manufacturing processes can use ungrounded systems or high-impedance grounded systems. They provide a higher degree of safety by not interrupting strategic equipment when a first fault-to-ground occurs. The presence of fault-indicating equipment, together with the necessary personnel who can quickly make repairs, ensures against costly downtime of equipment and the hazardous conditions that can result.

Equipment Grounding

When the insulation on conductors fails within undergrounded metal enclosures, the enclosures are raised to line voltage and constitute a serious hazard for personnel. However, if the metal enclosure is attached to the main bonding jumper and to the service equipment with an equipment-grounding conductor, this voltage difference will not occur.

Moreover, if the fault itself has a low resistance, and the equipment-grounding conductor has been properly installed and well maintained, a large amount of current can flow. The overcurrent device that protects the circuit will then deenergize the circuit.

Fixed equipment to be grounded includes the exposed noncurrent-carrying metal parts that are likely to become energized (1) within 8 ft (2.4 m) vertically or 5 ft (1.5 m) horizontally of ground, (2) located in a damp or wet location and not isolated, (3) in electrical contact with metal, (4) in a hazardous location, (5) supplied by a metal-clad, metal-sheathed, or metal-raceway wiring method, or (6) operated with any terminal in excess of 150 v to ground. In addition, ground the exposed noncurrent-carrying metal parts of the following, regardless of voltage: certain motor frames; controller cases for motors; the electrical equipment in garages, theaters, and motion picture studios; accessible electric signs and associated equipment; and switchboard frames and structures.

Also, ground the following equipment:

- frames and tracks of electrically operated cranes
- metal frames of nonelectrically driven elevator cars that have electrical conductors
- hand-operated metal shifting ropes or cables of electric elevators
- metal enclosures around equipment carrying voltages in excess of 750 v between conductors
- mobile homes and recreational vehicles

Unlike the grounded circuit conductor, the equipment-grounding conductor may be grounded continuously along its length. The equipment-grounding conductor may be a bare conductor, the metal raceway surrounding the circuit conductors, or an insulated conductor. If the conductor is insulated, it must have a continuous green cover or a green cover with a yellow stripe in it. The equipment-grounding conductor is always attached to the green hexagonal-headed screw on receptacles, plugs, and cord connectors (Figure 10–8). Where an approved metal conduit system is used as an equipment-grounding conductor, bond the receptacle to the

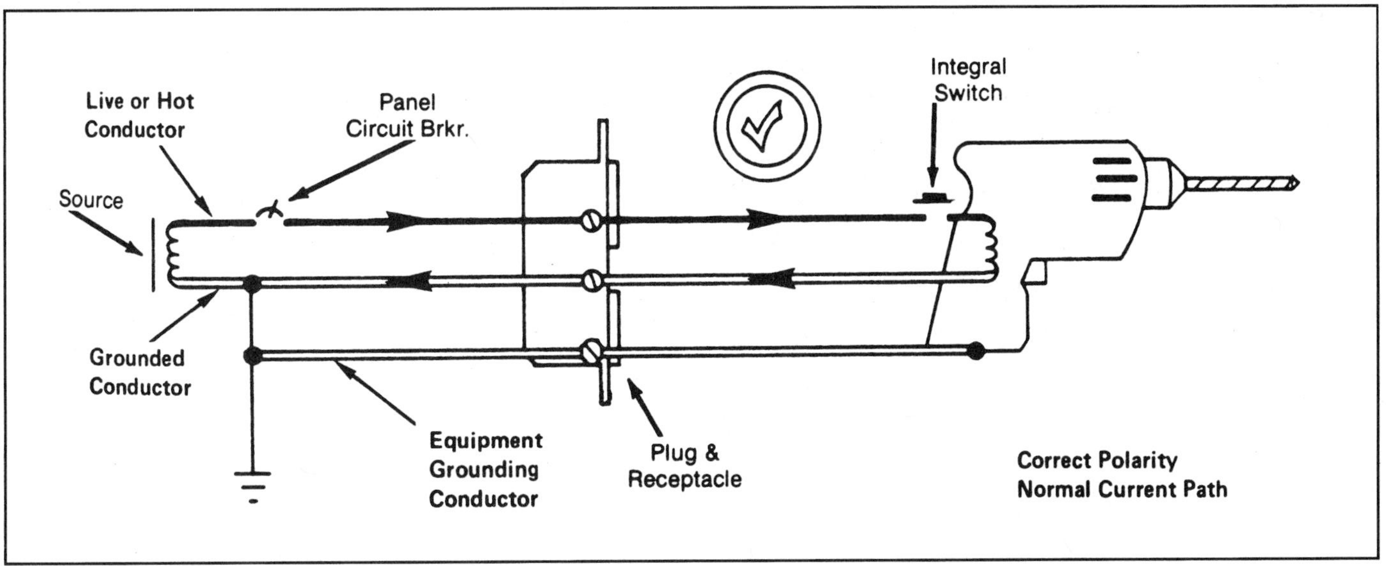

Figure 10–8. This shows the correct wiring of duplex receptacles. The green hexagonal screw is the lowest of the three screws.

box by means of a separate jumper. If this is not done, the receptacle must be listed by Underwriters Laboratories as being constructed to provide self-grounding (Figure 10–9).

Figure 10–9. Standard receptacle and plug recommended by the National Electrical Manufacturers Association. The receptacle is designed to receive a plug having three blades—a U-shaped or round grounding blade with two standard parallel polarized blades. This type of fixture also permits use of the nonpolarized standard plug.

To ground portable equipment use a separate green, insulated, equipment-grounding conductor within the portable cord. Using a separate, external-grounding conductor is much less reliable. The NEC permits this form of equipment grounding only if it is protected against physical damage and if it is part of the equipment.

Some attachment plugs are made of high-impact, transparent-plastic material. This material allows easy inspection of the plug's terminals without having to disassemble it.

Make connections to the equipment-grounding conductor and to the grounded circuit conductor so that removing receptacles, switches, lighting fixtures, and other devices will not require the interruption of either of these two conductors. The proper use of jumpers and of suitable pressure connectors will provide flexibility when maintenance is required, while still maintaining the integrity of both conductors. Solder is not permitted in the equipment-grounding circuit, nor are switches, fuses, or any other interrupting device. Always ground the equipment-grounding conductor in the same enclosure with the conductors of the circuit that it protects.

A suitable attachment plug listed for equipment-grounding circuits is available in noninterchangeable configurations for all common voltage, current, and phase combinations. These plugs each have a specific size and shape and are not interchangeable. General-duty devices may be used where they are not subjected to rough service or moisture. However, select the most rugged and watertight equipment if this kind of exposure is expected. Careful selection thus ensures against loss of grounding continuity because of damaged components or corrosion.

The size of the equipment-grounding conductor is important since this conductor carries fault current in the event that a ground fault develops. Unless sufficient current can get to the overcurrent device and cause it to operate, the circuit will not be protected. Moreover, the heating effects that occur at points of high resistance in a poor equipment-grounding system may cause fire and explosion and result in loss of life and property, as well as loss of valuable production time.

The NEC gives the proper sizes for equipment-grounding conductors used on various sizes of circuits. Increase the conductor's size if long runs are encountered and the drop of voltage exceeds an overall value of 5%.

Workmanship is one of the most overlooked aspects of good equipment grounding. The NEC states: "Connection of conductors to terminal parts shall ensure a thoroughly good connection without damaging the conductors...." Where this provision is not observed, a point of low thermal capacity will occur. Even though the resistance of the equipment-grounding circuit is measured and determined to be quite low, if the system does not have adequate thermal capacity, high-fault currents will cause these points of low thermal capacity to overheat. Arcing and fires, as well as inoperative protective equipment, could result.

Maintenance of Grounds

Only personnel with knowledge of electricity should install or repair electrical equipment. They should make certain that the green, insulated, equipment-grounding conductor is attached to the green-hexagonal binding screw, and that the white, grounded circuit conductor is attached only to the silver-colored binding screw.

Good maintenance ensures an electrically continuous equipment-grounding path from the metal enclosure of the portable equipment through the line cord, plug, receptacle, and grounding system (Figure 10–10). This path ends at the bonding jumper at the service equipment or in the enclosure of the separately derived system.

Portable testing devices, such as a three-light neon receptacle tester, provide the most convenient means for checking polarity and other circuit connections (Figure 10–11). Other metered instruments are available to measure the actual impedance of the grounding circuit. Take precautions when using testing equipment.

Use a receptacle-tension tester to inspect receptacles for deteriorating contacts (Figure 10–12). The tester employs little pointers that indicate the amount of tension that each receptacle contact will produce on the minimum-sized, Underwriters Laboratories-listed attachment plug. By using this form of inspection, the maintenance department can replace receptacles before they produce an ineffective equipment-grounding contact or before a fire occurs in the power contacts.

Check the condition of portable tools by using an insulation-resistance tester that provides for the application of 500 v DC between the motor windings and the metal enclosure. Some insulation-resistance meters provide an additional function that permits checking into continuity of the equipment-grounding conductor as well. For the industrial user who has a planned tool-maintenance program, the use of a portable appliance tester will provide automatic cycling of the portable tool through a series of timed tests. Conduct such tests in the toolroom. Check portable tools before they are issued or upon their return. To predict insulation failure, keep a record for each tool.

When a program of testing is planned, use one or more of these instruments. The entire test program can be handled by one person. To predict deteriorating trends, keep a log of the results obtained at each test point.

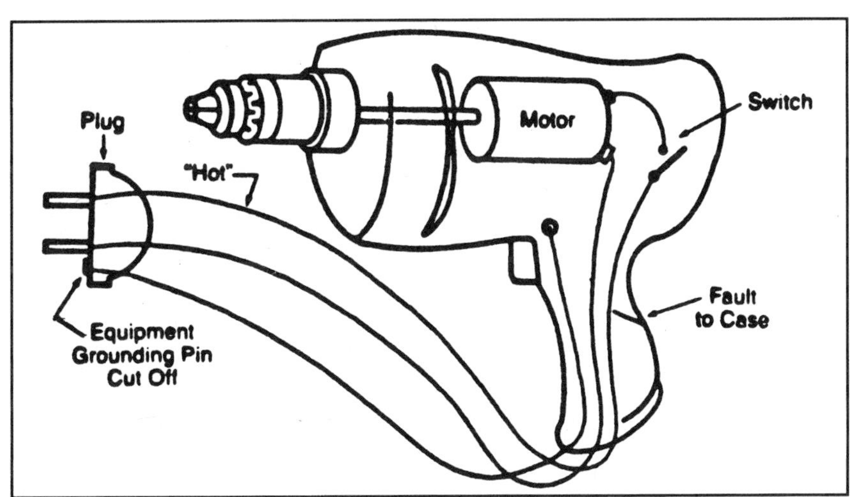

Figure 10–10. The equipment-grounding pin of this drill is removed. Cutting off the equipment-grounding conductor U-blade not only destroys the grounding path, but also may allow the plug to be inserted in the receptacle in a reverse polarity mode.

Figure 10–11. A receptacle circuit tester checks the receptacle for proper connection of ground wire, correct polarity, and faults in any of the three wires.

Figure 10–12. Use a receptacle tension tester to be sure receptacle contact tension is 8 oz or more.

Three-Wire Adapters

Maintaining a good ground and making sure that it is used properly can be a difficult task in abusive environments. In the hands of the untrained operator, the adapter is rarely connected properly. (The NEC now requires that all adapters have a wide neutral blade. It further requires that the equipment ground be made by means of a rigid tab under the cover screw on the bottom of the adapter rather than by means of a flexible pigtail.) If the grounding pin on the attachment plug should be clipped off, the operator would be holding a potentially lethal device.

Double-Insulated Tools

As an alternate to equipment grounding, the NEC recognizes the use of double-insulated appliances. Such appliances are constructed with two separate systems of insulation so that the chances of insulation failure are reduced to the lowest practical minimum.

Double-insulated appliances are of particular value in domestic situations. Many industrial users have, likewise, found them to be an effective means of reducing exposure to the hazards of electric current. In facilities that have instituted a strong safety program—where an employee looks for the grounding pin on all attachment plugs—use of double-insulated tools is now an exception to the rule. Confusion about double-insulated tools may result in the loss of some degree of safety. For example, double insulation does not protect against defects in the cord, plug, and receptacle. Continuous inspection and maintenance are required. Remember, do not use double-insulated power tools or equipment where water and a ground loop are present. Also do not use them in any situation where dampness, steam, or potential wetness can occur, unless they are protected by a GFCI. (For more detail, see Chapter 19, Hand and Portable Power Tools.)

Equipment with a dielectric housing, for example, plastic, protects the user from shock if interior wiring contacts the housing. Immersion in water, however, can allow a leakage path that may be of either high or low resistance. Handling equipment with wet hands, in high humidity, or outdoors after a rainstorm can be hazardous.

The best indication of the safety of a double-insulated tool is the Underwriters Laboratories', or other recognized testing laboratory's, label attached to the housing. The UL listing is evidence that the tool meets minimum standards.

For protection against electrical shock and to eliminate the need to ground the equipment, use self-contained battery-powered tools.

DETERMINING HAZARDOUS LOCATIONS

Hazardous (classified) locations are areas in which explosive or flammable gases or vapors, combustible dust, or ignitable fibers are present or likely to become present. Such materials can ignite as a result of electrical causes only if the following two conditions coexist:

1. The proportion of the flammable substance to oxygen in the air must permit ignition. This mixture must be present in a large enough quantity to provide an ignitable atmosphere in the vicinity of electrical equipment.
2. An electric arc, a flame escaping from an ignited substance in an enclosure, heat from an electric heater, or another source of fire, must be present at a temperature equal to or greater than the flash point of the flammable mixture.

Do not install standard electrical apparatus, considered safe for ordinary applications, in locations where flammable gases, vapors, dusts, and other easily ignitable materials are present. Sparks and electric arcs originating in such fittings have led to costly fires and explosions.

Before selecting electrical equipment and its associated wiring for a hazardous location, determine the exact nature of the flammable materials that will be present. For instance, an electrical fitting or device that is found by testing to be safe for installation in an atmosphere of combustible dust may be unsafe for operation in an atmosphere containing flammable vapors or gases.

Conduct a study of the machines or devices to be used and of the processes involving liquids, gases, or solid substances. Check their ratings. When hazards have been determined and the layout of the building inspected, decide whether one section of the facility should be classified as hazardous or whether the hazardous conditions extend to all parts.

Present the results of this study to a manufacturer of explosion-proof apparatus so that electrical equipment and wiring can be selected for safe installation. Leading manufacturers of electrical equipment often maintain staffs of engineers to guide buyers in the purchase of explosion-proof fittings.

Articles 500 through 503 of the 1990 NEC assign general requirements for electrical installations in hazardous (classified) locations, Articles 511 through 555 prescribe definitive requirements for specific types of occupancies.

Determining whether or not a hazardous situation exists in an industrial location is seldom difficult. Industries that have hazardous locations set up a formal procedure for classifying areas and for reviewing equipment and processes to determine the degree of potential hazard. Once a hazardous location has been identified and classified—and its limits defined—following NEC requirements for wiring methods and equipment compliance is relatively easy.

A major problem, however, is defining the limits of the hazardous location. How far above, below, and outward from the source does the hazardous location extend? For circumstances not covered in Articles 511 through 555, use this general rule to determine these limits: The limits of the hazardous locations are those mutually agreed upon by the owner, the owner's insurance carrier, and the authority enforcing the code.

Hazardous Locations

Hazardous locations are classified depending on the properties of the flammable vapors, liquids, or gases; combustible dusts; or fibers that may be present. Consider each room, section, or area of the facility separately.

Hazardous locations are classified as Class I, Class II, or Class III, depending on the physical properties of the combustible substance that might be present (Table 10–D). These classes are subdivided into Division 1 and Division 2, depending on the degree of likelihood that an ignitable atmosphere might be present. Combustible substances are arranged into seven groups—A through G. The grouping depends on the reaction of the substances upon contact with an ignition source—highly explosive, moderately incendiary, etc. Equipment installed in hazardous locations must be approved for the applicable class, division, and group (Table 10–E). Descriptions of the three classes and two divisions in each class follow:

- Class I locations are those in which flammable gases or vapors are present or likely to become present.
- Class II applies to locations where combustible dusts are likely to be present.
- Class III locations are those in which easily ignitable flyings such as textile fibers are present but not likely to be in suspensions in the air in sufficient concentrations to produce an easily ignitable atmosphere.
- In a Division 1 location, an ignitable atmosphere could occur at any time in the course of normal operations. Such an area represents a worst-case condition. One of the two requirements for ignition is likely to be present at any time, awaiting only a spark to ignite the flammable substance.
- In a Division 2 location, no ignitable atmosphere exists under normal operating conditions. However, an equipment malfunction, operator error, or other abnormal circumstance might create a hazardous environment.

Requirements for Division 2 electrical installations are less stringent than those for Division 1 locations because, in Division 2, two possible, but improbable, circumstances must coincide for ignition. In addition, any accidental formation of an ignitable atmosphere in a Division 2 location can usually be quickly stopped and the ignitable atmosphere dispersed. As a result, any exposure to fire and explosion is usually of short duration in a Division 2 location.

Class I Locations

Class I locations are areas in which flammable gases or vapors are present or are likely to become present in the air in quantities sufficient to produce explosive or ignitable mixtures. In general, most hazardous locations in industrial facilities fall in the Class I category.

Many common flammable substances, such as acetylene and naphtha, have been used in industrial facilities for many years. In recent years, many uses have also been found for less common ignitable gases and liquids. Hydrogen, for example, has many uses and

Table 10–D. Hazardous Location Classifications

Class I Highly flammable gases or vapors		Class II Combustible dusts		Class III Combustible fibers or flyings	
Division 1	**Division 2**	**Division 1**	**Division 2**	**Division 1**	**Division 2**
Locations where hazardous concentrations are probable, or where accidental occurrence should be simultaneous with failure of electrical equipment	Locations where flammable concentrations are possible, but only in the event of process closures, rupture, ventilation failure, etc.	Locations where hazardous concentrations are probable, where their existence would be simultaneous with electrical equipment failure, or where electrically conducting dusts are involved	Locations where hazardous concentrations are not likely, but where deposits of the dust might interfere with heat dissipation from electrical equipment, or be ignited by electrical equipment	Locations in which easily ignitable fibers or materials producing combustible flyings are handled, manufactured, or used	Locations in which such fibers of flyings are stored or handled, except in the process of manufacture

Groups:

A—Atmospheres containing acetylene
B—Atmospheres containing hydrogen or gases or vapors of equivalent hazard
C—Atmospheres containing ethyl-ether vapors, ethylene, or cyclopropane
D—Atmospheres containing gasoline, hexane, naphtha, benzene, butane, propane, alcohol, acetone, benzol, or natural gas
E—Atmospheres containing metal dust, including aluminum, magnesium, and other metals of equally hazardous characteristics
F—Atmospheres containing carbon black, coke, or coal dust
G—Atmospheres containing flour, starch, or grain dusts

must be given special consideration. Because of its wide explosive mixture range, high flame-propagation velocity, low minimum-ignition energy level, and low vapor density, a hazardous atmosphere can develop far above the hydrogen's source.

On the other hand, heavier-than-air vapors evolving from liquefied petroleum gases (LPG), such as propane and butane, also create special problems. LPG released as a liquid is highly volatile and has a low handling temperature; it readily picks up heat and creates large volumes of vapor. When released at or near ground level, the heavy vapors travel along the ground for long distances, thus extending the horizontal plane of the hazardous location. Some flammable liquids with flash points below 100 F (38 C) may produce large volumes of vapor that may spread much farther than might normally be expected.

Lighter-than-air gases usually dissipate rapidly because of their relatively low densities. Unless released in confined, poorly ventilated spaces, low-density gases seldom produce hazardous mixtures in zones that are close to grade where most electrical equipment is located.

In the case of hydrocarbons—most of which are heavier than air—the problem is not to establish the existence of a Class I location, but to define the limits of the Division 1 and Division 2 areas. Anywhere that hydrocarbons are handled, used, or stored, there is a great chance that flammable liquids, gases, and vapors will be released in large enough quantities to create a hazard. Vapor can disperse in all directions as governed by the vapor's density and the air's movement in the area. A very mild breeze can extend the limits of the hazardous locations, but the combustible mixture will not be dispersed significantly.

Table 10–E. Guidelines for Classifying Hazardous Areas

DETERMINING THE NEED FOR CLASSIFICATION

A need for classification is indicated by an affirmative answer to any of the following questions.

Class I

- Are flammable liquids, vapors, or gases likely to be present?
- Are liquids having flash points at or above 100 F likely to be handled, processed, or stored at temperatures above their flash points?

Class II

- Are combustible dusts likely to be present?
- Are combustible dusts likely to ignite as a result of storage, handling, or other causes?

Class III

- Are easily ignitable fibers or flyings present, but not likely to be in suspension in the air in sufficient quantities to produce an ignitable mixture in the atmosphere?

ASSIGNMENT OF CLASSIFICATION

Classification is determined as indicated by an affirmative answer to any question.

Class I, Division 1

- Is a flammable mixture likely to be present under normal operating conditions?
- Is a flammable mixture likely to be present frequently because of repair, maintenance, or leaks?
- Would a failure of process, storage, handling, or other equipment be likely to cause an electrical failure coinciding with the release of flammable gas or liquid?
- Is the flammable liquid, vapor, or gas piping system in an inadequately ventilated location, and does the piping system contain valves, meters, or screwed or flanged fittings that are likely to leak?
- Is the zone below the surrounding elevation or grade such that flammable liquids or vapors may accumulate?

Class II, Division 1

- Is combustible dust likely to exist in suspension in air, under normal operating conditions, in sufficient quantities to produce explosive or ignitable mixtures?
- Is combustible dust likely to exist in suspension in the air, because of maintenance or repair operations, in sufficient quantities to cause explosive or ignitable mixtures?
- Would failure of equipment be likely to cause an electrical system failure coinciding with the release of combustible dust in the air?
- Is combustible dust of an electrically conductive nature likely to be present?

Class III, Division 1

- Are easily ignitable fibers or materials producing combustible flyings handled, manufactured, or used?

Class I, Division 2

- Is the flammable liquid, vapor, or gas piping system in an inadequately ventilated location, but not likely to leak?
- Is the flammable liquid, vapor, or gas handled in an adequately ventilated location, and can the flammable substance escape only in the course of some abnormality such as failure of a gasket or packing?
- Is the location adjacent to a Division 1 location, or can the flammable substance be conducted to the location through trenches, pipes, or ducts?
- If positive mechanical ventilation is used, could failure or improper operation of ventilating equipment permit mixtures to build up to flammable concentrations?

Class II, Division 2

- Is the combustible dust likely to exist in suspension in air only under abnormal conditions, but can accumulations of dust be ignited by heat developed by electrical equipment, or by arcs, sparks, or burning materials expelled from electrical equipment?
- Are dangerous concentrations of ignitable dusts normally prevented by reliable dust-control equipment such as fans or filters?
- Is the location adjacent to a Division 1 location, and not separated by a fire wall?
- Are dust-producing materials stored or handled only in bags or containers and only stored—not used—in the area?

Class III, Division 2

- Are easily ignitable fibers or flyings only handled and stored, and not processed?
- Is the location adjacent to a Class III, Division 1 location?

DEFINING THE LIMITS OF THE CLASSIFIED LOCATION

The limits of the classified location—outward, upward, and downward from the source—must be determined by applying sound engineering judgment, experience gained on similar projects, and information from handbooks and other sources. Figure 15-7 provides an example of recommended clearances for the most common industrial-type classified locations—those involving heavier-than-air liquids, vapors, and gases.

Table is based on recommendations of the National Fire Protection Association and American Petroleum Institute.

Division 1 Locations. Class I, Division 1 locations are areas in which one or more of these following conditions exist:

- Hazardous concentrations of flammable gases or vapors exist under normal operating conditions.
- Hazardous concentrations of flammable gases or vapors may exist frequently because of leakage, repair, or maintenance operations.
- A breakdown or faulty operation of equipment might release hazardous concentrations of flammable gases or vapors and might also cause simultaneous failure of electrical equipment.

Table 10–F gives some examples of Class I, Division 1 locations.

Division 2 Locations. Class I, Division 2 locations are areas in which one or more of the following conditions prevail:

- Volatile flammable liquids or flammable gases are handled, processed, or used. However, they are confined in closed containers or systems from which they can escape only if the container accidentally ruptures or the system breaks down, or if the equipment is operated abnormally.
- Mechanical ventilation normally prevents hazardous concentrations of gases or vapors from forming. Ignitable concentrations can form only if the ventilation system fails.
- The area is adjacent to a Class I, Division 1 area from which hazardous concentrations might spread because of inadequate or unreliable ventilation supplied from a source of uncontaminated air.

Class I Groups. Maximum explosion pressures and safe operating temperatures vary widely for hazardous-location electrical installations. Therefore, the electrical equipment must be approved for the specific flammable material used in the designated hazardous location (Figure 10–13 and Table 10–G). Substances that can contribute to Class I atmosphere fall into Groups A through D. A list of these substances can be found in NFPA 497M, *Classification of Gases, Vapors and Dusts for Electrical Equipment in Hazardous (Classified) Locations.* Among the more common substances listed are:

- Group A—acetylene
- Group B—hydrogen or equivalent vapors and gases, such as manufactured gas
- Group C—ethyl-ether vapors, ethylene, cyclopropane, and similar substances
- Group D—gasoline, naphtha, benzene, hexane, butane, propane, alcohol, acetone, lacquer-solvent vapors, natural gas, and similar substances

Table 10–F. Some Examples of Class I, Division 1 Locations

Locations where volatile flammable liquids or liquefied flammable gases are transferred from one container to another.
Interiors of paint spray booths and areas adjacent to paint spray booths and other spraying operations where volatile flammable solvents are used.
Locations containing open tanks or vats of volatile flammable liquids.
Drying rooms or compartments for the evaporation of flammable solvents.
Cleaning and dyeing areas where hazardous liquids are used.
Gas generator rooms and portions of gas manufacturing plants where flammable gas may escape.
Inadequately ventilated pump and compressor rooms for flammable gas or volatile flammable liquids.
All other locations where hazardous concentrations of flammable vapors or gases are likely to form in the course of normal operations.

Class II Locations

Class II locations are areas in which combustible dusts are present or likely to become present. A potential dust-explosion hazard exists wherever combustible dusts accumulate, are handled, or are processed. Many dusts fall into the combustible category. A list of these dusts is also contained in NFPA 497M.

Division 1. Class II, Division 1 locations are ones that meet one or more of the following criteria:

- Combustible dust is in suspension in air during the course of normal operations in quantities large enough to produce explosive or ignitable mixtures.
- Mechanical failure or abnormal operation of machinery or equipment might create explosive or ignitable dust mixtures. They might also provide a source of ignition through simultaneous failure of electrical equipment, operation of protective devices, etc.
- Combustible, electrically conductive dusts may be present.

Some examples of Class II, Division 1 hazardous locations are the following:

- work areas of grain-handling and grain-storage facilities
- areas near dust-producing machinery and equipment in grain-processing facilities, starch facilities, sugar-pulverizing facilities, flour mills, melting facilities, hay-grinding facilities
- areas in which metal dusts and powders are produced, processed, handled, packed, or stored

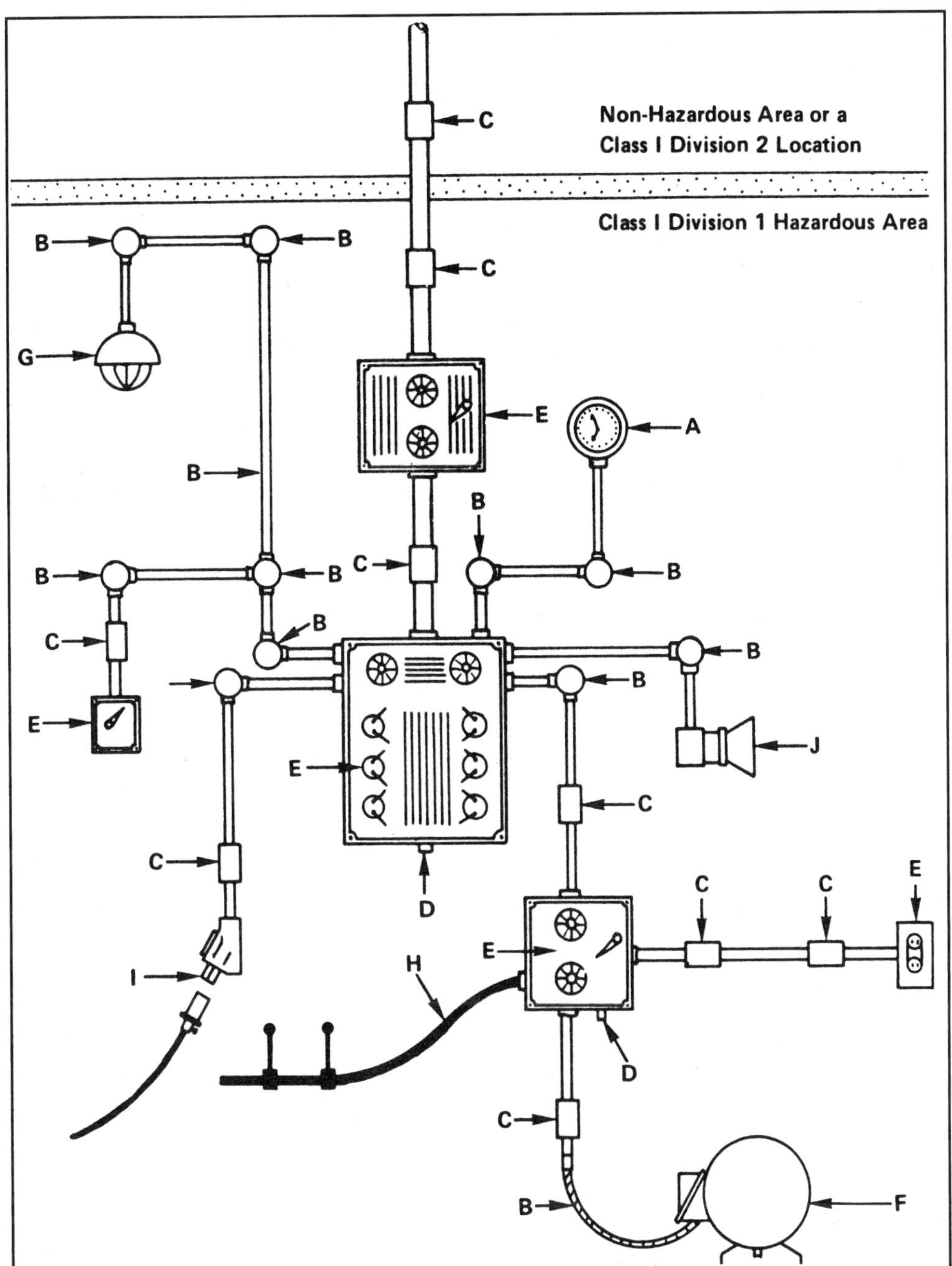

Figure 10–13. Class I, Division 1 hazardous location. (See Table 10–G.)

- coal bunkers, coal-pulverizing facilities, and areas in which coke, carbon black, and charcoal are processed, handled, or used

Dust that is carbonized or excessively dry is susceptible to spontaneous ignition. Therefore, electrical equipment installed in Class II locations should be able to operate at full load without developing surface temperatures high enough to cause excessive dehydration or carbonization of dust deposits that might form.

Division 2. In Class II, Division 2 locations, ignitable concentrations of combustible dusts are not usually found in the course of normal operations. However, abnormal conditions may allow combustible dusts to accumulate in large enough quantities and in large enough dust-to-air concentrations to permit ignition. Abnormal conditions could include failure of dust-control equipment or process equipment, or of containers, chutes, or other handling equipment. The following are usually considered as Class II, Division 2 locations:

Table 10–G. Summary of Equipment Requirements for Class I, Division 1 Hazardous Locations

A. Meters, relays, and instruments, such as voltage or current meters and pressure or temperature sensors, must be in enclosures approved for Class I, Division 1 locations. Such enclosures include explosion-proof and purged and pressurized enclosures. National Electrical Code (NEC), NFPA 70, 501-3(a).

B. Wiring methods acceptable for use in Class I, Division 1 locations include: threaded rigid metal or steel intermediate metal conduit and type MI cable. Flexible fittings, such as motor terminations, must be approved for Class I locations. All boxes and enclosures must be explosion-proof and threaded for conduit or cable terminations. All joints must be wrench tight with a minimum of five threads engaged. NEC, 501–4(a).

C. Sealing is required for conduit and cable systems to prevent the passage of gases, vapors, and flame from one part of the electrical installation to another through the conduit. Type MI cable inherently prevents this from happening by its construction; however, it must be sealed to keep moisture and other fluids from entering the cable at terminations. NEC, 501–5.

1. Seals are required where conduit passes from Division 1 to Division 2 or nonhazardous locations.
2. Seals are required within 18 in. from enclosures containing arcing devices.
3. Seals are required if conduit is 2 in. in diameter or larger entering an enclosure containing terminations, splices, or taps.

D. Drainage is required where liquid or condensed vapor may be trapped within an enclosure or raceway. An approved system of preventing accumulations or to permit periodic drainage are two methods to control condensation of vapors and liquid accumulation. NEC, 501–5(f).

E. Arcing devices, such as switches, circuit breakers, motor controllers, and fuses, must be approved for Class I locations. NEC, 501–6(a).

F. Motors shall be:

1. approved for use in Class I, Division 1 locations
2. totally enclosed with positive pressure ventilation
3. totally enclosed inert-gas-filled with a positive pressure within the enclosure, or
4. submerged in a flammable liquid or gas. NEC, 501–8(a).

G. Lighting fixtures, both fixed and portable in rigid metal conduit, must be explosion-proof and guarded against physical damage. NEC, 501–9(a).

H. Flexible cords must be designed for extra hard usage, contain a grounding conductor, be supported so that there will be no tension on the terminal connections, and be provided with seals where they enter explosion-proof enclosures. NEC, 501–11.

- areas containing only closed bins or hoppers, and enclosed spouts and conveyors
- areas containing machines and equipment from which appreciable quantities of dust would escape only under abnormal conditions
- warehouses and shipping rooms in which dust-producing materials are handled or stored only in bags or containers
- areas adjacent to Class II, Division 1 locations

Class II Groups. Electrical equipment installed in Class II locations must be approved for the class, the division, and the applicable group (Figure 10–14 and Table 10–H). For purposes of equipment approval, dusts are classified in the following three groups based on their conductivity:

- Group E—metal dusts
- Group F—carbon black, charcoal, coal, coke dust
- Group G—flour, starch, grain dust

Class III Locations

Class III locations are those in which easily ignitable fibers or flyings are present. However, these fibers are not likely to be in suspension in the air in quantities large enough to produce an ignitable atmosphere. Single fibers of organic materials such as linen, cotton tufts, and fluffy fabrics, however, are quite vulnerable to a localized heat source such as an electric spark. In pure oxygen, single fibers of cotton can be ignited by a 0.02 joule spark.

Textiles such as those used in clothing can be ignited and burned with repetitive or sustained high-energy electric sparks. Cotton and wool fabrics can be ignited in pure oxygen with a spark of 2.3 joules. In normal air, a spark of 193 joules or more is required for ignition. Silk and polyester fibers are more difficult to ignite than cotton or wool.

Fibers contaminated with oily substances can be ignited with much weaker sparks than can clean fibers. Typically, only a thousandth of the energy required to ignite a clean fabric is required for an oily sample of the same fabric. In general, the burning characteristics of fibers will be affected by the (1) specific gravity of the substance, (2) size and shape of the sample, (3) air circulation in the area, (4) oxygen concentration, and (5) relative humidity. The burning characteristics of some common fibers whose presence can cause an area to be designated as Class III are given in Table 10–I.

Division 1. Class III, Division 1 locations are ones in which easily ignitable fibers or materials, which produce combustible flyings, are manufactured, handled, or used. This classification usually includes:

- facilities that produce combustible fibers
- portions of rayon, cotton, or other textile mills
- flax-processing facilities
- clothing-manufacturing facilities
- woodworking facilities

Division 2. Class III, Division 2 locations are ones in which easily ignitable fibers are stored or handled but not manufactured or processed. An example of a Class III, Division 2 location is a textile warehouse.

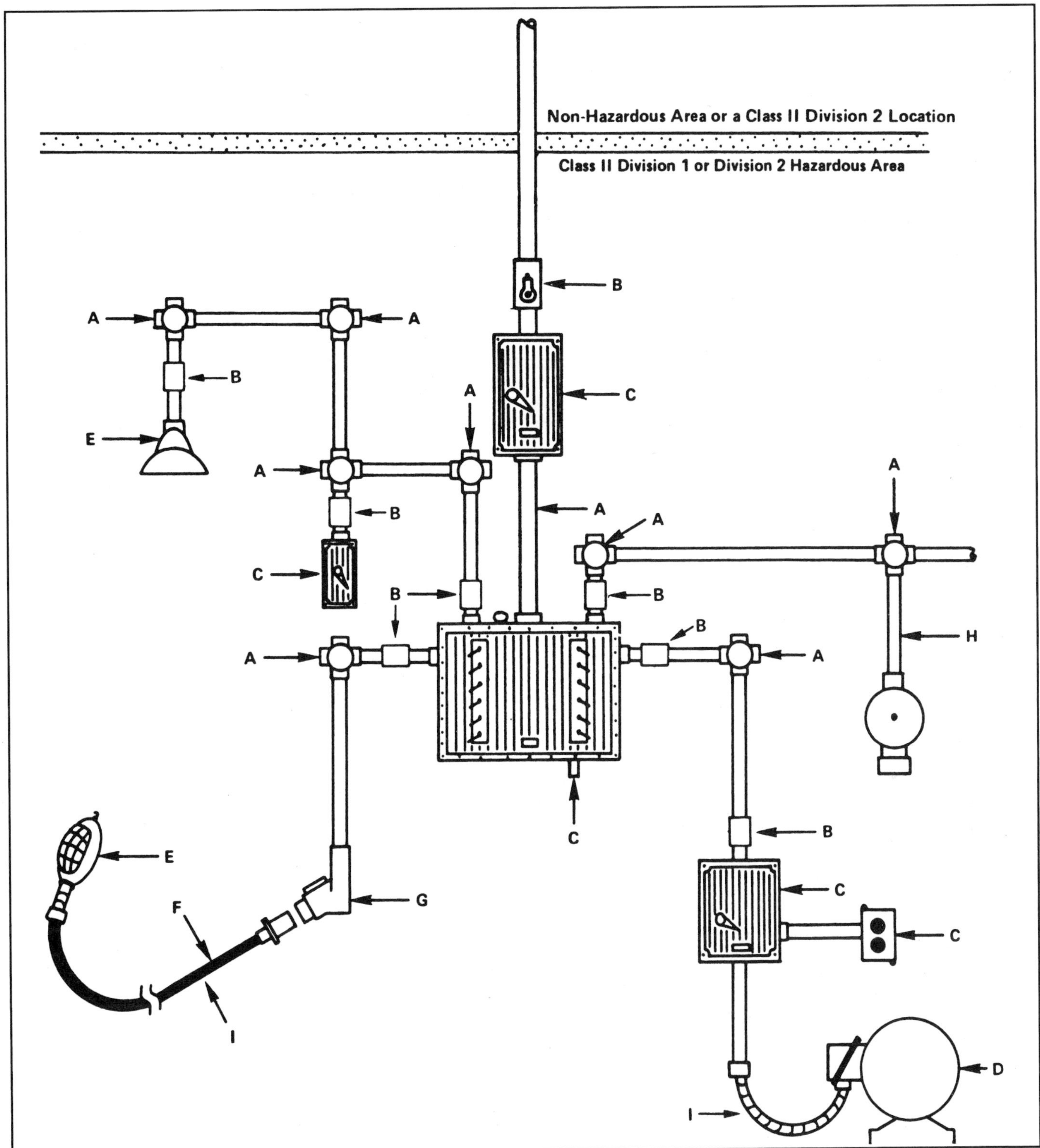

Figure 10–14. Class II hazardous locations. (See Table 10–H.)

Class III—No Groups. There are no group designations associated with Class III locations. Electrical equipment installed in Class III locations need only be approved for the applicable class and division (Figure 10–15 and Table 10–J). For equipment not subject to overloading, the maximum surface temperature of equipment under normal conditions shall not exceed 329 F (165 C); for equipment such as

Table 10–H. Summary of Class II Hazardous Locations

A. Wiring methods for Class II, Division 1 locations: boxes and fittings containing arcing and sparking parts are required to be in dust-ignition-proof enclosures. Threaded metal conduit or type MI cable with approved terminations is required for Class II, Division 1 locations. National Electrical Code (NEC), NFPA 70, 502–4(a).

In Class II, Division 2 locations, boxes and fittings are not required to be dust-ignition-proof but must be designed to minimize the entrance of dust and prevent the escape of sparks or burning material. In addition to the wiring systems suitable for Division 1 locations, the following systems are suitable for Division 2 locations: electrical metallic tubing, dust-tight wireways, and types MC and SNM cables. NEC, 502–4(b).

B. Suitable means of preventing the entrance of dust into a dust-ignition-proof enclosure must be provided where a raceway provides a path to the dust-ignition-proof enclosure from another enclosure that could allow the entrance of dust. NEC, 502–5.

C. Switches, circuit breakers, motor controllers, and fuses installed in Class II, Division 1 locations must be dust-ignition proof. NEC, 502–6. In Class II, Division 2 areas, enclosures for fuses, switches, circuit breakers, and motor controllers must be dust-tight.

D. In Class II, Division 1 locations, motors, generators, and other rotating electrical machinery must be dust-ignition proof or totally enclosed pipe ventilated. NEC, 502–8.
In Class II, Division 2 areas, rotating equipment must be one of the following types:

1. dust-ignition proof
2. totally enclosed pipe ventilated
3. totally enclosed nonventilated, or
4. totally enclosed fan cooled.

Under certain conditions, standard open-type machines and self-cleaning squirrel-cage motors may be used. NEC, 502–8(b).

E. In Class II, Division 1 locations, fixed and portable lighting must be dust-ignition proof. NEC, 502–11.
Lighting fixtures in Class II, Division 2 locations must be designed to minimize accumulation of dust and must be enclosed to prevent the release of sparks or hot metal.

In both divisions, each fixture must be clearly marked for the maximum wattage of the lamp, so that the maximum permissible surface temperature for the fixture is not exceeded. Additionally, fixtures must be protected from damage. NEC, 502–11.

F. Flexible cords in Divisions 1 and 2 are required to:

1. be suitable for extra hard usage
2. contain an equipment grounding conductor
3. be connected to terminals in an approved manner
4. be properly supported, and
5. be provided with suitable seals where necessary. NEC, 502–12.

G. Receptacles and attachment plugs used in Class II, Division 1 areas are required to be approved for Class II locations and provided with a connection for an equipment grounding conductor. NEC, 502–13.

In Division 2 areas, the receptacle must be designed so the connection to the supply circuit cannot be made or broken while the parts are exposed. This is commonly done with an interlocking arrangement between a circuit breaker and the receptacle. The plug cannot be removed until the circuit breaker is in the OFF position, and the breaker cannot be switched to the ON position unless the plug is inserted in the receptacle.

Table 10–I. Burning Characteristics of Some Common Fibers

Substance	*Specific Gravity*	*Approximate Ignition Temperature, C*	*Burning Characteristics*
Acetate	1.32	525	Melts ahead of flame
Acrilan	1.17	560	Burns readily
Arnel	1.3	525	Melts ahead of flame
Cotton	1.54 to 1.56	400	Burns rapidly
Nylon 6	1.14	530	Melts and burns
Wool	1.3	600	Melts ahead of flame

Adapted from *Wellington Sears Handbook of Industrial Textiles,* New York, N.Y., Wellington Sears Co., Inc., 1963.

transformers and motors that are subject to overloading, 248 F (120 C).

Establishing the Limits

Classifying an area as hazardous for purposes of NEC compliance is based on the possibility of flammable liquids, vapors or gases, combustible dusts, and easily ignitable fibers or flyings being present. After an area has been classified as hazardous, the next step is to determine the degree of hazard. Should the area be classified as Division 1 or Division 2? Also, the limits of the hazardous location must be defined—how far above, below, and outward from the source of the hazard does the hazardous location extend?

The safest electrical installation in a hazardous location, of course, is one that is not there at all. As much as is practical, situate electrical equipment outside the area defined as hazardous. It is, however, seldom possible or practical to locate all electrical equipment outside the hazardous area. The facility engineer is responsible for ensuring that all electrical equipment in the hazardous area, and its associated wiring, conforms to the NEC and does not significantly increase the chance of explosion.

Reducing Hazards. Two ways to reduce the chance of explosions from electrical sources are (1) remove or isolate the potential ignition source from the flammable material, and (2) control the atmosphere at the ignition source. For an explosion to occur, the following two conditions must coexist. If either of these two conditions is eliminated, the explosion hazard is reduced to zero:

1. combustible material present in a sufficient amount and the proper concentration to provide an ignitable atmosphere
2. an ignition source powerful enough to ignite the combustible materials that are present

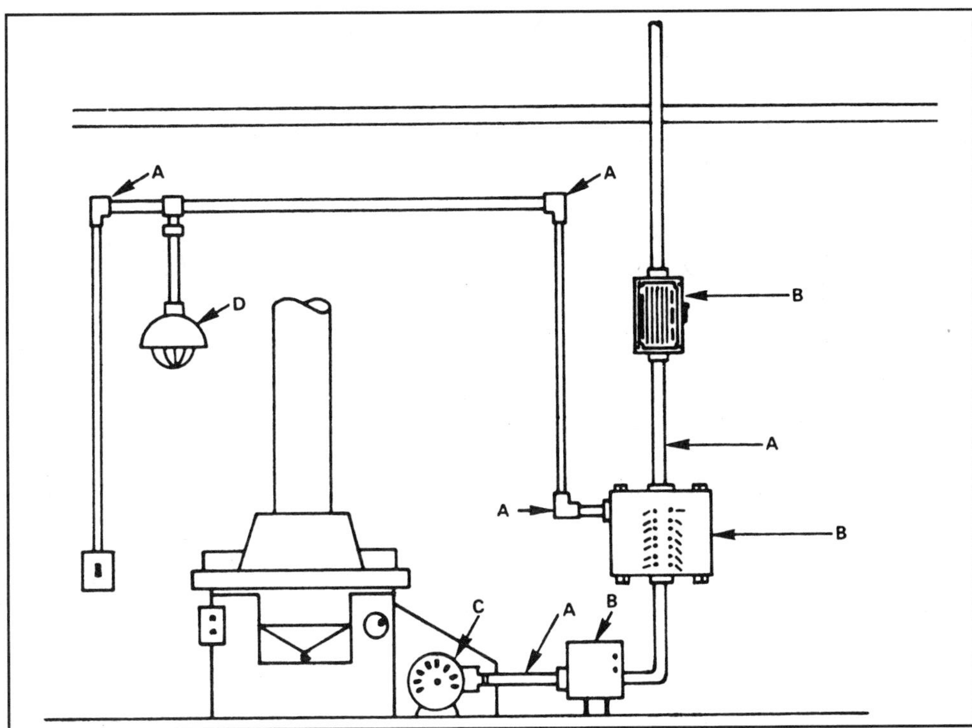

Figure 10–15. Class III hazardous locations. (See Table 10–J.)

Table 10–J. Summary of Class III Hazardous Locations

A. In Class III hazardous locations, wiring must be within a rigid metal or nonmetal conduit or be of type MI, MC, or SNM cable. Fittings and boxes shall be dust-tight. Flexible connectors are allowed. Equipment grounding shall be provided. National Electrical Code (NEC), NFPA 70, 503–3.

B. Switches, circuit breakers, motor controllers, and similar devices used in Class III hazardous locations must be within dust-tight enclosures. NEC, 503–4.

C. Motors, generators, and other rotating electric machinery must be totally enclosed nonventilated, totally enclosed pipe-ventilated, or totally enclosed fan-cooled. NEC, 503–6.

D. Lighting fixtures must have enclosures designed to minimize the entry of fibers, to prevent the escape of sparks or hot metal, and to have a maximum exposed surface temperature of less than 309 F (165 C). NEC, 503–9.

Note: Class III locations do not necessarily require equipment that is labeled as suitable for the location. Equipment in Class III locations is required only to meet certain performance requirements as outlined above.

Planning Electrical Installations. The hardest part of planning an electrical installation to conform to NEC requirements for hazardous locations is to define the limits of the hazardous area. How far should the hazardous location be considered to extend to ensure that safety is served, without taking unnecessary and expensive precautions? No hard-and-fast rules can be applied. Experience on comparable projects and an understanding of specific conditions at the job site provide a far better basis for defining limits than any theoretical study of flammable vapors, gases, dusts, or fibers.

The environmental aspects of an installation—prevailing winds, site topography, proximity to other structures and equipment, and climatic factors—can significantly affect the extent of a hazardous location. Among the factors that should be evaluated in establishing the limits are the following:

- size, shape, and construction features of the building
- existence and locations of doors and windows, as well as their manner of use
- absence or presence of walls, enclosures, and other barriers
- existence and locations of ventilation and exhaust systems
- existence and locations of drainage ditches, separators, and impounding basins
- quantity of hazardous material likely to be released
- location of possible leakages

- physical properties of the hazardous material—density, volatility, chemical stability, etc.
- frequency and type of maintenance and repair work performed on the systems containing the hazardous substance and on other equipment in the area

When establishing limits of hazardous locations, consider the area surrounding each source of hazardous material as a location and determine its individual limits. For example, consider the following rules for a Class I location:

- In the absence of walls, enclosures, or other barriers, and in the absence of air currents or other disturbing forces, a gas or vapor will be distributed in a predictable fashion.
- For heavier-than-air gases and vapors released at or near grade level, potentially hazardous concentrations are most likely to be found below grade. Heavier-than-air gases distribute themselves downward and outward. As the height above grade increases, the hazard decreases.
- For lighter-than-air gases, little or no potential exists for a hazard at or below grade. Lighter-than-air gases distribute themselves upward and outward. As the height above grade increases, the potential hazard increases. For purposes of classification, treat gases with a density greater than 75% of the density of air at standard conditions as if they were both lighter and heavier than air. Define the limits of the hazardous location accordingly.

A Sample Installation. Most hazardous locations in industrial facilities belong in Class I and involve heavier-than-air gases and vapors. Figure 10–16 provides recommendations for establishing limits for some typical situations involving such gases and vapors.

In Figure 10–16, a process pump, which handles flammable liquids at moderate pressures, is at grade elevation. Because the source of the hazard is in the open air, there are no pockets below grade level where flammable vapors can accumulate. In addition, liquid can escape only in the event of equipment failure—such as a leaking gasket. Therefore, the area surrounding the pump is classified Division 2.

A different situation prevails, though, for the trunk-line pump in Figure 10–16. Although also outdoors, the trunk-line pump operates at relatively high pressure. Thus, the chance of equipment failure and the volume of liquid that might be released increases. A below-grade trench is in the vicinity of the trunk-line pump. For these reasons, the limits of the Division 2 location surrounding the pump are extended, and the trench itself is classified as Division 1.

The outdoor tank in Figure 10–16 is installed with its base at grade level, and the tank is fitted with a floating roof. The space within the shell of the tank, above the roof, is classified as a Division 1 location. The area surrounding the tank is classified as Division 2; it extends 10 ft (3 m) above the tank. Horizontally, the Division 2 area extends 50 ft (15.2 m) from the tank or to the dike, whichever is greater.

In Figure 10–16 the indoor pump is in a building with one fully open wall but no through ventilation. The building also contains valves, meters, and other equipment and fittings that are likely to leak. The area adjacent to the pump, and near grade, is classified as Division 1. The rest of the building and the outdoor close-to-grade area next to the open wall is classified as Division 2.

The process building has numerous sources of flammable materials and is not especially well ventilated. The entire interior is classified as Division 1, and the area surrounding the building is classified as Division 2. The Division 2 area extends horizontally from the building for 50 ft (15.2 m) in every direction, and at least 10 ft (3 m) above the roof, or 25 ft (7.6 m) above the source, whichever is greater. The electrical equipment room is separate from, but immediately next to, the process building. It lies entirely within the Division 2 location that surrounds the process building for 50 ft (15.2 m) in all directions. No fire wall separates the electrical equipment building from the process building.

The interior of the electrical-equipment building, however, is classified as a safe area because its atmosphere is controlled to prevent ignition sources from connecting an ignitable mixture. A positive-pressurization system is used to keep the building purged of any flammable mixtures. Note that the air intake for the pressurization system must be at least 5 ft (1.5 m) above the classified location. This minimum distance must not be violated under any circumstances, nor may this space be invaded by any flammable mixture.

Each source of flammable material in Figure 10–16 was considered to be the focal point of a separate hazardous location. The positioning of sources of hazardous materials could cause an overlapping of Division 1 and Division 2 locations. In such cases, the more stringent classification should prevail. Thus, the area of overlap should be classified as Division 1.

Explosion-Proof Apparatus

Explosion-proof apparatus is defined in the NEC, Article 100, as

> ... apparatus enclosed in a case that is capable of withstanding an explosion of a specified gas or vapor, which may occur within it, and of

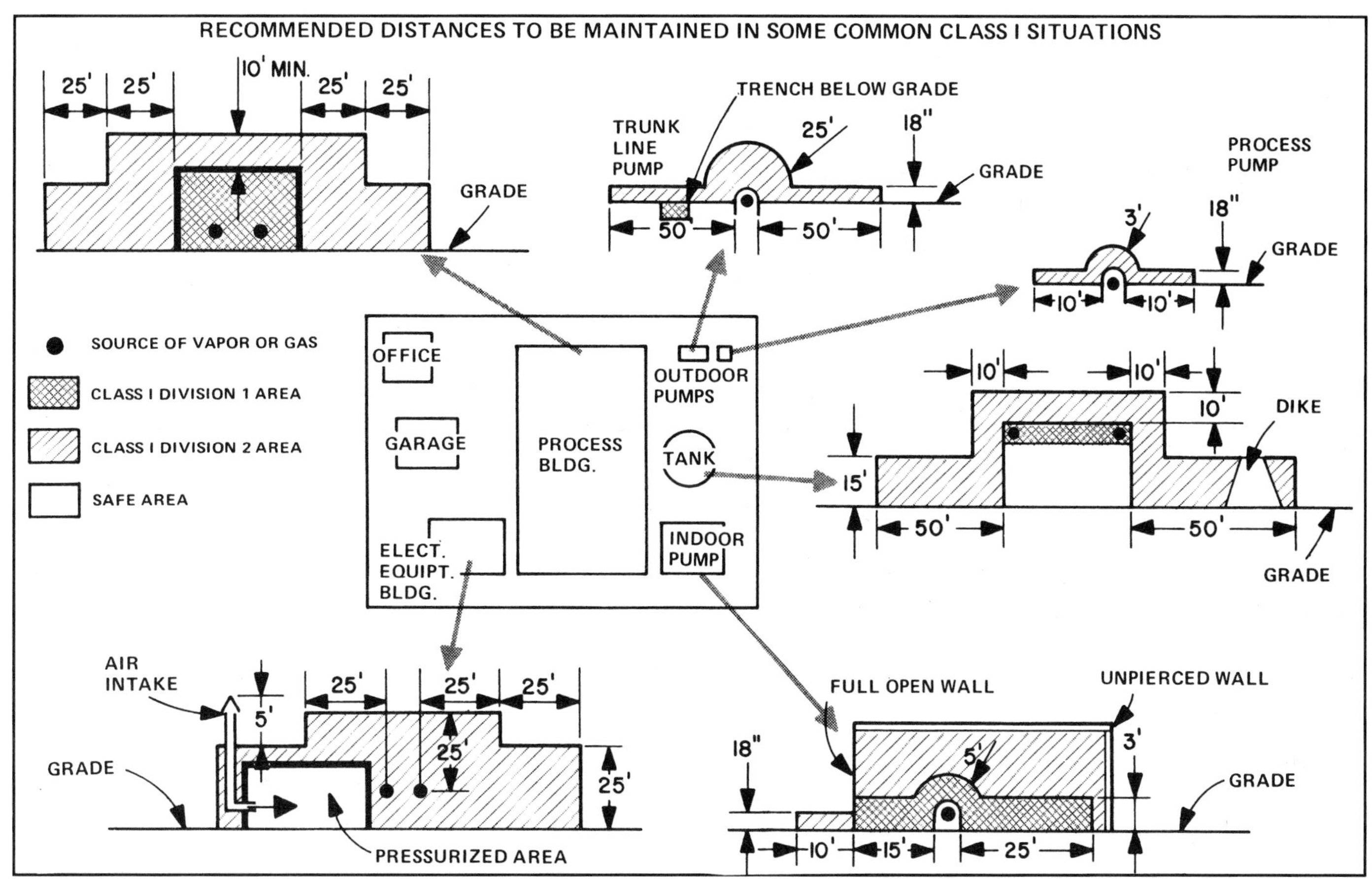

Figure 10–16. Recommended distances to be maintained in some common Class I situations, developed from standards established by NFPA and API. (Reprinted with permission from *Plant Engineering Magazine.*)

> preventing the ignition of a specified gas or vapor surrounding the enclosure by sparks, flashes, or explosion of the gas or vapor within, and which operates at such an external temperature that a surrounding flammable atmosphere will not be ignited thereby.

Only use fittings that have undergone exhaustive tests and meet the requirements of Underwriters Laboratories for use in hazardous locations (Figure 10–17). These fittings must be of durable material, provide thorough protection, and be finished for total resistance to atmospheric conditions.

Install explosion-proof fittings not only on new work but also on old wiring systems where alterations are being made or new equipment is being installed. Observing NEC requirements minimizes dangers that might result from using ordinary fittings in hazardous locations.

Explosion-proof fittings are made of durable cast material, with roomy interiors for wiring and splices. They are able to withstand high internal pressure that results from an explosion without bursting and without becoming loose. Furthermore, they are tested on the basis of a high safety factor as prescribed by Underwriters Laboratories.

It is impossible to prevent highly flammable gases from entering the interior of either an ordinary or an explosion-resisting wiring system. Gases eventually enter the entire line through the joints and through the breathing of the conduit system caused by temperature changes. Furthermore, gaseous vapors will fill every crevice whenever covers are removed. For these reasons, it is impossible (1) to provide an entirely vapor-proof switch unit, (2) to regulate temperatures, or (3) to keep the air free from flammable gases inside the fittings.

To isolate that section of the industrial facility classified as a hazardous location, positively confine the arc, heat, and explosion within the internal limits of the explosion-proof fittings. These fittings are constructed to contain the dangerous arcing, intense heat, and subsequent explosion so that the gas-laden air outside does not become ignited.

Often, control equipment can be located outside a room containing hazardous materials. In these cases,

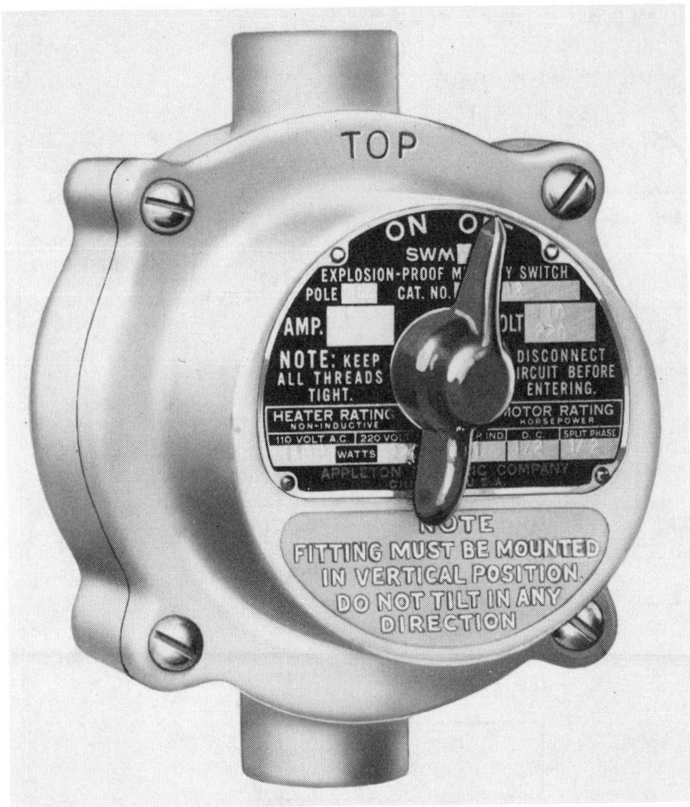

Figure 10–17. An explosion-proof and dust-tight mercury switch, listed by Underwriters Laboratories, which makes and breaks the circuit when the switch tilts the tube inside the mechanism. (Reprinted with permission from The Appleton Electric Co.)

conventional wiring equipment rather then explosion-proof fittings can be used. Thus, the cost and hazards of installation are reduced.

INSPECTION

Whenever possible before an inspection, the facilities electrician or other authorized person should deenergize equipment. Live circuits and equipment left in the operating mode are always a hazard. Always assume that a circuit is live until it is proved dead. To be sure of this, check that switches and circuit breakers carrying current to or from switchboards, buses, controls, and starting equipment are open. In the inspector's presence, the facility electrician, supervisor, or operator should conduct tests to determine that the parts to be worked on are deenergized. As an additional safeguard, the inspector should have the breakers and switches locked open, grounded, and tagged so they cannot be energized until the tests have been completed.

Upon completion of the inspection, the inspector should remove any grounds and locks/lockout tags that were used. However, before the equipment is returned to service, the chief electrician, supervisor, or operator—whichever person is in charge—should give it clearance.

If the feed-in circuit to a switchboard, bus, or other equipment must be kept live on the incoming side of a breaker or switches, the electrician should provide ample clearance, barriers, or other protection between this section and the part being worked on. Live buses, conductors, and switches should be covered with insulated blankets, specially formed insulating shields, or isolated work barriers. Inspectors should avoid working in cramped quarters, if at all possible.

Do not depend on electronically operated or remote-control circuit breakers or contractors for protection. A ground or other disturbance on the control system may permit the circuit breaker to close. If no disconnects are installed ahead of or behind the circuit breaker, block, rack out, or otherwise lock the breaker so it cannot be closed.

If the electrician must leave a circuit being tested before the test is complete, he or she should use lockout procedures. Upon returning and before continuing the tests, the electrician should check all markings, breakers, and switches to be sure that they have not been altered.

In many electrical inspections, power must be on so that the function can be determined. In many cases, instruments are designed to test functions that way. All testing equipment needs to be checked before the inspection to test correctly. Avoid wearing metal jewelry such as rings, necklaces, etc.

An electrical safety inspection program should include awareness of specific items to inspect as well as commonly violated electrical standards. Items to inspect include:

1. service entrance panel—circuit I.D., secure mounting, knockouts, connectors, clearances, live parts, rations
2. system grounding—secure connections, corrosion, access, protection, proper size
3. wiring (general)—temporary, splices, protected, J box covers, insulation, knockouts, fittings, workmanship
4. electrical equipment/machinery—grounding, wiring size, overcurrent/disconnect devices, installation, protection
5. small power tools—attachment plugs, cords, cord clamps, leakage, grounding
6. receptacles—proper polarity, adequate number, mounting, covers, connections, protection, adapters
7. lighting—grounding, connections, attachment plugs and cords, cord clamps, live parts
8. GFCI protection—wet locations, fountains, outdoor circuits, testing

9. switches—labeled or marked, covered as needed, lockable
10. extension cords—condition, plugs, receptacles, GFCI
11. protectors—fuses, circuit breakers, link fuses

Rotating and Intermittent-Start Equipment

Before working on rotating machines or on automatic and intermittent-start equipment, presumed to be out of service and stopped, inspect all the electrical-control and starting devices with the chief electrician or supervisor. For example, when inspections or repairs are to be made on motors, generators, blowers, compressors, or converters, any part of which is remote-controlled or which may automatically start, lock out circuit breakers or switches and pull the fuses.

Machinery connected to blowers, water wheels, or pumps, without check valves, may start turning even when the current to the motor has been disconnected. For this reason, block the rotor or armature before inspections are made. Follow lockout procedures on generators driven by prime movers. Lock and tag the throttle, starting valve, or other means of controlling the energy to the driving part of the unit.

When the equipment is in operation, do not remove or adjust motor brushes; do not work on contacts or other parts of the electrical equipment; and do not clean, sandpaper, or polish commutators or slip, or collector, rings.

When inspecting electrical equipment, employees should not wear loose clothing because it may become entangled in couplings, coils, or other moving parts. They should remove wristwatches, rings, and metal pens and pencils. Also, they should not use metal flashlights.

High-Voltage Equipment

In general, high-voltage equipment is more carefully guarded than low-voltage equipment because of high-voltage equipment's greater inherent hazard.

Personnel

Only authorized, trained personnel should work on high-voltage equipment. In newer installations of 2,300 v or more, attempts have been made to insulate or armor apparatus to prevent casual contact with the current-carrying parts.

Protective Equipment

Persons working on high-voltage equipment should know that rubber insulating gloves are not a good substitute for safety devices or proper procedures. Such gloves should be worn only as a supplementary measure. Before each use, check the gloves for punctures, tears, or abrasions. For an on-the-job test, roll up the cuffs and force air into the fingers and palms of the gloves. If there is an air leak, do not use the gloves. (See Chapter 7, Personal Protective Equipment, and National Safety Council (NSC) Occupational Safety and Health Data Sheet 12304–0598, *Flexible Insulating Protective Equipment for Electrical Workers,* OSHA 29 *CFR* 1910.132 and 1910.137, General Equipment PPE and Electrical Protective Equipment, for further discussion of gloves.)

Have electrical workers wear leather protectors over rubber gloves to protect the rubber from mechanical damage and from oil and grease. Provide an electrical glove-testing service. Establish regular testing intervals that are determined by the amount of use the gloves get, the type of work they are used for, and the voltages they are subject to. If there is doubt about the insulating quality of the gloves, discard them.

Private testing laboratories can be relied upon for accurate testing. In some cases, they will keep individual company records on the physical condition of gloves and the time schedule for retesting. Where no laboratory is available, small companies can usually have tests made at the local public utility. However, testing must comply with ASTM F496, *Specifications for In-Service Care of Insulating Gloves and Sleeves.*

MAINTENANCE

For safety and service, electric equipment must be well maintained. Motors, circuit breakers, moving parts of switches, and similar current-carrying devices wear out, break down, and need adjustment. Only have trained and experienced electricians make repairs on electrical circuits and electrical apparatus.

Use only high-grade electrical equipment listed as standard by Underwriters Laboratories, or by another qualified authority, in maintenance work. Inferior, unapproved equipment may become hazardous because of defective material, poor design, or poor workmanship.

If practical, deenergize conductors when maintenance work on them is necessary. Before beginning work, trained and authorized maintenance personnel should make sure that the line is dead by testing it with the voltage-testing devices provided. Do not permit workers to test a circuit with their fingers or with makeshift devices.

Instruct maintenance personnel involved with electrical work about using electrical-testing equipment and meters. They should know how to tap into the circuit; how to locate testing points from the schematic diagrams; how to use insulated meter leads, clips, and probes; and how to fuse test circuits. They should use insulated pliers, screwdrivers, testing lights, and other tools for electrical repair work.

When maintenance or repair work must be done on energized conductors, have two or more employees work together. The supervisor should instruct them on detailed procedures to be followed. The supervisor should supply maintenance crews with the proper protective equipment and see that they use it.

The kind of protective equipment needed is determined by the type of circuit, the nature of the job, and the conditions under which the work must be done. Rubber gloves, sleeves, mats, line hose, insulating platforms, safety headgear, safety glasses, safety belts, fuse tongs, and insulated switch sticks are among the more common items of protective equipment used.

Good safety practices lie not only in using the proper protective equipment, but in taking care of that equipment as well. Protective equipment should be inspected before it is used and should be checked at frequent intervals.

To prevent accidental grounding and possible severe injuries, maintenance personnel must constantly be on guard when working around electrical equipment and circuits. Grounding can easily occur if an energized loose wire contacts a water pipe, conduit, metal fixture, another wire, or anything metallic that is connected in some way to the earth.

Lockout/Tagout

The unexpected energization or start-up of machines and equipment by automatic or manual control may cause injuries from electrical shock or from, but not limited to, mechanical, chemical, or thermal radiation energies. Therefore, when electronic or electrical equipment must be serviced, maintained, repaired, or modified, open the circuit at the circuit's contact and padlock the switch in the OFF position on each energy control device. Tag the switch with a description of the work being done, the name of the person, and the department involved. Posting warning signs, or tagging alone, does not provide the positive protection that locking out equipment does (Figure 10–18).

Because of the risk of injuries or death, all authorized personnel inspecting, testing, or maintaining the electrical system and personnel affected by the aforementioned activities shall be trained in lockout/tagout procedures and follow safety-related work practices cited in 29 *CFR* 1910.331-339. (See Chapter 6, Safeguarding, for details on an energy isolation program.) The supervisor should see that the procedure is carried out and should provide the necessary keys, locks, and

Figure 10–18. Typical multiple lockout designed for six individual locks.

arrangements. No two key configurations should be the same. Check to see that each key fits only one lock.

For identification, locks may be painted various colors to indicate types of craft or to differentiate shifts. Stamp each lock with the employee's name or clock number. Have only one key for each lock. Do not keep a master key. In emergency situations, locks must be cut or torched off.

When purchasing electrical equipment, be sure that it has lockout/tagout capabilities. To make lockout systems operable, the equipment should have built-in locking devices designed for the insertion of padlocks. There are several methods of locking out equipment. In facilities where older equipment is in use, or where explosion-proof or dust-tight equipment is installed, have technicians or maintenance workers construct attachments to which locks can be applied. Electronic equipment may need plug-control lockouts. Where more than one person or group works on a piece of equipment, set up a multiple-locking device so that each person must apply his or her own padlock.

Maintain an effective program by training employees in safe lockout/tagout procedures and through constant supervision. A lockout procedure that is generally acceptable follows:

1. Alert the operator and other users of the system or equipment that is to be shut off.
2. Plan the shutdown to ensure that the system will be off.
3. Have all users place their padlocks on the control switch, lever, valve, or energy control device. Each person will not be protected unless each one puts his or her own padlock on it. Signs are no substitute for locking out the source of electrical power and all other energy sources.
4. Test the lockout/tagout procedure implemented to be sure the system is really off, and all energy sources have been deenergized, released, blocked, etc.
5. When through working, have each user remove his or her padlock equipment and process. Never permit workers to remove another's padlock. Be sure workers do not expose others to danger. Verify that the equipment is clear, and post a watch, if necessary.
6. Reenergize the system.

Removing Fuses

When it is necessary to remove a fuse, employees who are qualified, trained, and authorized shall open the operating switch to disconnect the electrical load. Extract the fuse with an insulated fuse puller. If the fuse is not protected with a switch, pull the supply end of the fuse out first. When the fuse is replaced, put the supply end in first. Replace a fuse with one of the same type and size. Do not substitute a copper wire or other conductor when replacing a fuse.

For work on lines carrying more that 600 v, open the sources of energy by tripping the circuit breaker and then opening the disconnects. Tag the conductors, and, if possible, clamp a substantial grounded conductor to each leg of the circuit before starting the work. Remove the grounding conductors just before the circuit is reenergized. As an added precaution, conduct tests with a voltmeter, or other standard, approved tester, to make sure that the circuit is clear. In emergencies, when the areas around fuse boxes are wet, maintenance workers should use wood platforms, insulated stools, or rubber boots.

Wiring

As a minimum standard, make electrical installations according to the NEC. Have a knowledgeable inspector thoroughly inspect each installation. The use of temporary wiring should be discouraged even though it may have been reasonably safe when first installed. As work progresses and repairs or installations are made, however, temporary wiring may become unsafe. It might not be properly protected from mechanical injury, and it may have been modified.

Technicians and maintenance personnel should make it a practice to inspect plugs and connections on portable electric or electronic equipment. They should check that the third wire frame ground is properly connected, that the grounding prong has not been removed, and that the cord is connected to the proper terminals. Where there is any doubt about the safe operation of a tool or instrument, remove it from service, tag it, and have qualified people inspect and repair it.

When additional equipment is being installed or operated under temporary conditions, make no taps into an existing circuit. This can only be done if an individual switch is installed in the branch line, with provision for locking the switch in the Off position and tagging it. This precaution prevents service interruptions in the main circuit when the branch's circuit power must be shut off. It also provides a means of preventing connected equipment from accidentally starting.

Frequent inspection of equipment and competent supervision of maintenance crews are extremely important. Their subsequent reports will aid the safety program considerably. Often in their routine work throughout the facility, maintenance personnel can spot hazards before they cause injuries.

EMPLOYEE TRAINING

Deviating from safe usage and installation practices with electrical and electronic equipment often results in unnecessary hazards that can cause injuries or death. Consequently, a facility's safety program must thoroughly train all employees who work with electrical and electronic equipment or who operate electrical and electronic systems. In addition to instructions on the hazards of electricity, train employees in CPR; the use of warning signs, guards, and other protective devices; and safe operational procedures. Each employee must be trained to handle emergency situations. Be sure that they are instructed never to work alone with potentially hazardous electrical equipment.

Amendments to OSHA's 29 *CFR* 1910.132, .133, and .135–.138 also address additional safety concerns in working with electrical equipment and energy sources. Three new paragraphs added to 1910.132—General Requirements discuss the employer's obligation to assess hazards in the workplace in order to select appropriate PPE for workers, to prohibit the use of defective or damaged PPE, and to use documented training sessions to teach workers how to use, care for, and repair PPE. The remaining amendments underscore the need to educate workers to use proper eye and face protection (1910.133), head protection (1910.135), foot protection (1910.136), electrical protective equipment (1910.137), and hand protection (1910.138). According to OSHA officials, the new standard is performance oriented and encourages employers to continually improve worker protection. They also believe that the new standard will improve worker acceptance of PPE by allowing better, more comfortable PPE and by ensuring that employers inform workers about selecting the proper equipment for the job.

In addition, management must develop and implement safety programs to comply with OSHA's 29 *CFR* 1910.331–.333. Both qualified and unqualified workers are covered by these safety-related work practices standards. Sections 1910.332–.333 describe the scope and content of training that employers are required to provide for workers who handle power equipment or electrical energy sources, and the selection and use of work practices to prevent injuries and fatalities to these workers. Safety programs developed by companies must meet or exceed the provisions of these standards.

In developing appropriate safety programs, management should base them on the layout of the facility's electrical system and its use of electronic equipment. The program can then be applied to the specific operations and changes that may occur within the systems.

Supervisors should be kept informed about existing and possible electrical hazards. Management should require them to maintain close supervision over all operations that involve the use of electrical or electronic equipment. Supervisors must encourage employees to report any electrical defect or problem immediately. The supervisor must then have the tool or equipment repaired or replaced at once. (See the *Administration & Programs* volume, Chapter 16, Safety Training, for more information about training programs.)

SUMMARY

- Failure to establish or use safety practices for electrical equipment can result in property damage and serious injuries or fatalities. Workers must understand clearly the need to learn and follow safe work practices.
- Severity of electrical shock is determined by (1) the amount of current that flows through the victim, (2) the length of time the body receives current, and (3) parts of the body involved.
- Electrical equipment should be installed in controlled areas and have fail-safe devices and guarding to protect workers and others nearby. All electrical equipment should be properly wired and grounded and connected to circuit breakers, ground-fault circuit interrupters, or other control equipment installed for emergency shut-off.
- In control equipment, arrange switchboards with lockout capabilities for both AC and DC circuits. Electrical motors should not interfere with the normal movement of personnel or materials. Extension cords must be inspected frequently to detect any wear, fraying, or breakage in the line.
- Make sure that workers are thoroughly trained to check their electrical equipment for safe operation, to report any abnormal conditions, to have equipment tested regularly, and to observe all safety regulations and practices to operate specialized equipment.
- Grounding includes both equipment grounding and grounding of the electrical system itself. Only trained personnel should install, test, and repair grounds.
- Standard electrical equipment should not be installed in locations where flammable gases, vapors, dusts, or other easily ignitable materials are present.
- Management should determine the specific hazards in any particular location before selecting and installing electrical equipment. Management should also reduce the hazards by removing or isolating sources of possible ignition and by controlling the atmosphere at the ignition source.
- Before inspections, electrical equipment should be deenergized whenever possible and the systems locked and tagged to prevent accidental reenergizing while tests are being conducted. Inspectors should be thoroughly trained and properly protected before inspecting high-voltage equipment.
- All components of electrical equipment must be well maintained. Only trained and experienced electricians should make repairs on electrical circuits and apparatus.
- A facility's safety program should include thorough training for all employees who work with electrical and electronic equipment or who operate electrical systems. Supervisors should know about existing and possible hazards and all employees know CPR, rescue, and other emergency procedures.

REFERENCES

American National Standards Institute, 11 West 42nd Street, New York, NY 10036.

Attachment Plugs and Receptacles, ANSI/UL 498–1993.
Rubber Insulated Wires, ANSI/UL 44-1991.
National Electrical Safety Code, ANSI C2-1990.
Relays, Breakers, Switchgear Systems Associated with Electric Power Apparatus, ANSI C37.
Specifications for Rubber Insulating Tape, D4325-83.
Safety Requirements for Lockout/Tagout of Energy Sources, ANSI Z244.1–1982. (R1993)

American Society for Testing and Material, 1916 Race Street, Philadelphia, PA 19103.

Specifications for In-Service Care of Insulating Gloves and Sleeves, F496.
Specifications for Rubber Insulating Gloves, D120.
Specifications for Rubber Insulating Tape, D4325.

Canadian Standards Association, 178 Rexdale Boulevard, Rexdale, Ontario M9W 1R3, Canada.

Factory Mutual System Research Organization, 1151 Boston-Providence Turnpike, Norwood, MA 02062.

Illuminating Engineering Society of North America, 345 East 47th Street, New York, NY 10017.

National Fire Protection Association, 1 Batterymarch Park, Quincy, MA 02269.

Classification of Gases, Vapors and Dusts for Electrical Equipment in Hazardous (Classified) Locations, NFPA 497M, 1995.
Electrical Safety Requirements for Employee Workplaces, NFPA 70E, 1993.
National Electrical Code, NFPA 70, 1993.

National Safety Council, 1121 Spring Lake Dr., Itasca, IL 60143-3210.

Electrical Inspection Illustrated, 2nd ed., 1984.

Occupational Safety and Health Data Sheets:
Electrical Switching Practices, 12304–0544, 1991.
Electromagnets Used with Crane Hoists, 12304–0359, 1985.
Electrostatic Paint Spraying and Detearing, 12304–0468, 1991.
Flexible Insulated Protective Equipment for Electrical Workers, 12304–0598, 1991.
Industrial Electric Substations, 12304–0559, 1991.
Portable Reamer-Drills, 12304–0497, 1989.

Power Tool Institute, PO Box 818, Yachats, OR 97498.

Underwriters Laboratories, Inc., 333 Pfingsten Road, Northbrook, IL 60062.

Insulated Wire, UL 44. "Electrical Appliance and Utilization Equipment Lists."
Dimensions of Attachment Plugs and Receptacles, UL 498.
Electrical Construction Materials List.
Hazardous Location Equipment List.

U.S. Department of Commerce, National Institute of Standards and Technology, Gaithersburg, MD 20899.

Safety Rules for the Installation and Maintenance of Electric Utilization Equipment, Handbook H33.

U.S. Department of Labor, Occupational Safety and Health Administration, 200 Constitution Avenue NW, Washington, DC 20210.

CFR, 1910.132–.138, Personal Protective Equipment
CFR, 1910.331–.333, Electrical Safety-Related Work Practices.

REVIEW QUESTIONS

1. What is a person's main resistance to current flow?
2. Why is low voltage more dangerous?
3. What is measured in ohms?
4. Name two factors that cause injuries in electrical shock.
5. Why is an open-knife switch hazardous?
6. How is the safe current carrying capacity of conductors determined?
7. What is the function of an overcurrent device?
8. Describe a link fuse.
9. What are the two categories of circuit breakers?
10. What is the most common safeguard against motor overload?
11. Why is an electrical system "grounded"?
12. How is grounding done?

11

Fire Protection

Revised by

John F. Montgomery, Ph.D., CSP, CHCM, CHMM

Charles M. Foster, Assistant Fire Chief

Fire protection includes procedures for preventing, detecting, and extinguishing fires. The procedures in these three areas of fire prevention aim to protect employees and property and to assure the continuity of a plant's operations. To accomplish these goals, it is necessary to develop a plantwide fire protection program.

FACILITY FIRE PROTECTION PROGRAM

The primary purpose of such a program is to prevent fires from starting. If, nonetheless, a fire does start, employees should know their role in the following procedures:

1. immediately detecting the fire and promptly transmitting an alarm
2. initiating evacuation of the building
3. confining the fire
4. extinguishing the fire.

Objectives of a Fire Protection Program

When planning a fire protection program, first make a statement of objectives for fire safety. Objectives might be stated in terms of safety for people and minimizing downtime for the plant. Fire protection systems must meet or exceed all codes and should be especially protective of areas in the plant that are vital to the continuity of its operations. When designing the plant's building and laying out its operations, incorporate greater measures of fire safety than are called for. Remember that all model code requirements and standards are minimal in nature. For example, the requirements in the Standard Building Code, the Uniform Building Code, and the National Fire Protection Association (NFPA) are minimum standards.

Architects and engineers must realize that designing for fire protection is a legitimate part of their responsibilities. They must understand (1) the special thermal load that fire puts on structural parts of buildings, and (2) the preventive measures that can be incorporated into their designs. The earlier that fire-safety objectives are identified and design decisions are made, the more effective the fire protection system can be. Also, when a fire protection item is added to a structure during the construction phase, it is usually less expensive than the same item would be if added to an already-existing structure.

Some general facts about fire protection that must be kept in mind follow:

- No facility is absolutely fire-resistive. Nearly everything can burn, given ignition, adequate fuel, and sufficient oxygen.
- Heat energy is transmitted by convection, conduction, radiation, and direct flame contact.
- Fire and flame will spread in a building both vertically and horizontally. Fire spreads vertically until the roof confines the flames; then horizontal spread begins.
- The spread of the heat, smoke, and toxic gases is possibly the greatest single danger to life and takes place in much the same manner as does the spread of fire. Smoke and toxic gases are responsible for approximately 66% of deaths from fire in buildings.
- On-site early detection of a fire is absolutely essential.
- The use to which a building is put—its occupancy—will influence the degree of fire hazards. The more hazardous the materials handled in a plant area are, the more likely is the chance for a fire to start and rapidly spread.

- The contents of a building are usually a more important factor in the start of a fire than the physical structure of the building.
- Very often only a few minutes pass between the beginning of combustion and the development of a destructive fire.
- What happens, or does not happen, in the first few minutes of a fire determines whether it can be controlled or not.
- A fire is usually (1) controlled by built-in equipment and/or (2) put out by fire fighters.
- Every fire protection device involves compromise. That is, a fire protection system always represents some trade-off involving cost, reliability, or safety. Some risk will be assumed and there will be some loss. The optimum level of fire protection is that which minimizes the cost from expected fire losses. (See the section Fire Risks, later in this chapter.)
- The cost of fire protection should have a corresponding effect in reducing the amount of loss or risk involved.
- An automatic sprinkler system is the best tool to reduce loss of life from fire because, properly installed and maintained, it is on-site, 24-hour-a-day fire protection.
- People and their actions are key elements. Probably more than half of all fire losses result from human mistakes due to inadequate training, insufficient motivation, or improper action.
- Construction alone is not adequate protection as far as life safety is concerned.

Fire Protection Engineering

Fire protection engineering is a highly developed, specialized field in which special engineering disciplines are focused. The solution of many fire protection problems requires the special training and perspective of an experienced fire protection engineer. Achieving the most efficient fire protection system requires the involvement of the architect, interior designer, urban planner, building contractor, electrical and structural engineers, fire detection-system manufacturers, building safety engineer, and local fire marshal.

The safety and health professional who is faced with special fire problems should seek specific advice from a fire protection engineering consultant. Obtain names of experts in this area from the NFPA and the Society of Fire Protection Engineers.

Authoritative fire protection literature is available from such organizations as the NFPA, the American Insurance Association, and Factory Mutual System Research Organization. (See Appendix 1, Sources of Help, in the *Administration & Programs* volume, and the NFPA *Fire Protection Handbook*. See also References, at the end of this chapter.) General information can also be obtained from trade associations, the fire insurance carrier, the local fire inspection bureau, the fire department, and government agencies.

Fire Drills

Planning for fire emergencies is not an exact science. Develop a practical emergency plan after evaluating the particular situation at hand. Prepare an emergency manual to outline procedures and drills and to assign responsibilities to each individual involved. Make prevention of personal injury and loss of life the prime objective of emergency planning.

Carefully plan and periodically carry out fire drills. Conduct them in a serious manner under rigid discipline. Training employees to leave their workplaces promptly at the proper signal and to evacuate a building speedily but without confusion is largely accomplished through fire drills. To eliminate panic in the event of an emergency and to guarantee the smooth functioning of the emergency plan, carefully develop the plan.

Post up-to-date instruction sheets, including evacuation routes, and distribute them to all employees. Maps that are posted for evacuation should also show alternate routes in case the first route is closed. In addition, when going through fire drills, from time to time, block off an exit to see how employees would react to this situation in a real fire. Also, make sure that someone is assigned to call the fire department. In many cases, all people exited a building and the fire department was never notified. Everybody thought someone else had taken care of it.

Post fire guards in each area to make sure each area has been safely evacuated. People in places such as rest rooms and high noise areas do not always hear alarms. In addition, set up lights to go on when the fire alarm system is set off. People often respond faster to lights than to alarms. Perform a roll call at an outside location to make sure that all personnel are accounted for and that no one was left behind.

Fire drills, conducted at frequent intervals, demonstrate management's concern and sincere interest in all fire prevention activities. The drills should serve as a reminder to employees and supervisors that all fire prevention practices are important. Emergency fire drills also serve as a valuable way to check the adequacy and condition of fire exits and the alarm system. Immediately and permanently correct any deficiencies.

Fire Brigades

Management cannot depend wholly on automatic fire protection equipment, municipal fire departments, or mutual-aid agreements to prevent fire losses. Fires can get out of control before a municipal fire department arrives. One method of providing additional fire protection is to form a fire brigade. If an employer chooses to develop a fire brigade, all the requirements of 29 *CFR,* 1910.156 must be met. Assign a brigade chief with the following qualifications: training and experience in fire protection and fire prevention, ability to organize, and confidence and ability to give instructions on fire prevention and fire fighting. Because fire protection demands competent and experienced leadership 24 hours a day, also assign an assistant brigade chief or captain for each shift. The assistant brigade chief or captain should have the same qualifications as the brigade chief.

Require a centrally organized plant fire brigade for each plant. The fire brigade should operate under the direction of the plant brigade chief or the brigade captain. Brigade members should be regular plant employees from all departments, thoroughly trained in using the plant's fire-fighting equipment. Electricians, engineers, mechanics, and safety and fire inspectors all have special skills, thus making them valuable members of the plant's emergency team.

The brigade chief and assistant brigade chief should conduct regularly scheduled training sessions for all members of the fire brigade (Figures 11–1 and 11–2). Also, establish continuous programs that instruct every employee on proper procedures for avoiding fire, both on and off the job.

Consult the municipal fire department when plans for a fire brigade are being considered. Find out what outside help is available and how the fire department can help the brigade in other ways, such as training. The U.S. Occupational Safety and Health Administration (OSHA) regulations contain further information on training requirements for fire brigades.

In addition, the fire brigade can receive training from the State Fire Marshal's Office or the State Fire Training Group. Most states also offer programs for industrial fire protection through land grant universities located in each state. Through such programs, one receives a standard level of training from certified instructors.

FIRE PREVENTION ACTIVITIES

The complete steps to be taken for fire prevention are extensive. Be sure to refer to the codes and standards of the NFPA, as well as local building and fire codes, to maintain the utmost in fire safety at new and existing buildings.

As a first step, all companies and plants should hold design review meetings for all new construction, as well as for changes in existing construction. Every establishment should institute procedures and regulations to assure that proper fire-extinguishing equipment is on hand and that employees are organized and trained to use the equipment correctly. In addition, regularly

Figure 11–1. A fire brigade receives training in using a hose to fight a low-level fire.

Figure 11–2. Simulations of fires in building structures similar to chemical/petroleum installations are an effective training technique.

schedule thorough inspections of all fire protection equipment. Keep a written record of these inspections.

Fire prevention includes activities directed specifically toward preventing a fire from starting. Generally, these activities include inspections, fire drills, training fire brigades, and a good communication system.

Inspections

Set up a system of periodic fire inspections for every operation. Some buildings, operations, and processes require daily inspection, while others can be inspected weekly, monthly, or at other intervals. Buildings that are well designed and provided with protective devices and construction elements intended to act as fire safety features still need a periodic, detailed inspection program.

In addition to inspections made by insurance companies, fire protection bureaus of fire departments, and the State Fire Marshal's Office, every industrial plant should include periodic self-inspection in its fire safety program. The function of these fire safety inspections is to check for proper placement and operation of fire protection equipment and to correct common causes of fire, such as poor housekeeping, improper storage of flammable materials, smoking violations, and excessive accumulations of dust or flammable material. Be sure that employees who make the inspections are trained in inspection procedures and techniques.

In some establishments or plants, the safety and health committee locates and reports fire hazards. In large plants, this job is handled by a special subcommittee of the safety committee or by a person trained to manage fire risk.

The inspector, fire chief, or other individual from the plant who is in charge of fire prevention and fire protection should establish inspection schedules, determine the routing of reports, and have a complete list of all items to be inspected. Inspection of fire equipment should cover the following items:

- control valves on piping that supplies water for fire protection
- hydrants
- fire pumps
- hose houses and associated equipment
- sprinkler system water supplies, including tanks
- automatic sprinkler systems

- special types of protection (carbon dioxide [CO_2], dry chemical, foam, or other automatic systems)
- portable fire extinguishers
- fire doors, aisles, and exits
- special hazards and operation processes
- detectors
- control room or panel checkouts
- alarm and communication systems and routines
- communication to the fire department or to other mutual aids.

In addition to inspecting fire equipment, electrical equipment, machinery, and processing equipment, check housekeeping conditions and other potential causes of fire at regular intervals. Checklists for inspections can be effective. A typical inspection form for general industrial use is shown in Figure 11–3. It can be adapted for use in various types of buildings. An inspection form helps an inspector make comments and bring recommendations to the attention of supervisors or managers. (See Chapter 5, Loss Control Programs, in the *Administration & Programs* volume, for information on inspecting for housekeeping hazards and other hazards to safety. See also the NFPA *Inspection Manual,* an excellent guide to conducting inspections.)

Conduct special inspections during and following any alterations in a plant or in a process. Make a complete seasonal inspection of equipment that will be or has been exposed to freezing temperatures. Do it early enough to replace or repair the equipment.

If the plant does not have its own fire protection expert, invite the local fire chief or fire marshal to inspect the plant's buildings, fire equipment, hazardous materials used, locations for hazardous materials, and available water supply. Ask the local fire chief how the fire department can support the plant's fire protection program. Fire departments will often use preplanning forms. Fire prevention and fire safety is a community-wide effort.

Hot-Work Permits

In an effort to establish some control over operations using flames or producing sparks, many industrial firms have instituted hot-work permit programs. These programs require that authorization be secured before equipment capable of igniting combustible materials is used outside their normal work areas.

In inaugurating a hot-work permit program, management must first develop a policy statement, such as that given in Figure 11–4. The type and extent of the hot-work permit program will depend upon the size of the plant or facility, the complexity of the operations, and the degree to which hazards are present at the work site and in the surrounding areas.

Important features of a hot-work permit program include the following:

- Inspect the area where work is to be done, and see how close combustible materials are to the work area.
- Establish fire watches if hazards warrant them. A fire watch should stay on duty for 30 minutes after all spark-producing equipment has been shut down.
- Provide fire-extinguishing equipment, usually manned by a standby employee.
- Communicate with, and coordinate the activities of, all departments concerned with fire protection.
- Isolate flammable and combustible materials from sources of ignition.
- Limit unauthorized use of flame- or spark-producing equipment.

A form or tag is generally used to administer the hot-work permit program. Although standard forms are available through insurance companies, many plants have developed special forms that relate specifically to their operation.

If hot-work is to be done by an outside contractor, establish areas of responsibility in a written contract. Ask the following questions:

- Who issues the hot-work permits and what signatures are required?
- Who provides fire watches?
- Who provides standby fire equipment?
- Who coordinates activities of the contractor that involve plant personnel?

Flame-Retardant Treatments

The burning characteristics of a number of materials can be altered. A flame-retardant treatment that penetrates or coats a material will affect its ignition and spread-of-fire characteristics. Sometimes the treatment will also change other characteristics of the material. Unfortunately, these treatments are temporary.

Many substances, such as wood, wood products, textiles, some plastics, and paper, can be successfully made flame retardant. Even a Christmas tree can be coated with a spray to change its burning characteristics.

Fabrics cannot be made noncombustible or even resistant to charring or decomposition. However, chemical treatment will reduce their flammability. (See NFPA 701, *Methods of Fire Tests for Flame-Resistant Textiles and Films.*) Some treatments merely inhibit the rapid spread of flames, whereas others prevent flames and depress the dangerous afterglow.

Where the only anticipated exposure is to small sparks, small flames, or temperatures up to 400 to 500 F (200 to 260 C), flame-retardant canvas is often preferred to chrome- tanned leather. This is especially true where flexibility, durability, strength, and resistance to abrasion are required. Chrome-tanned leather is required for

OFFICE OF PUBLIC SAFETY
SPRINGFIELD FIRE DEPARTMENT
COMMERCIAL FIRE INSPECTION

COMPANY NAME: ____________ DATE: ________ DATA CODE: ________

ADDRESS: ____________ _____SINGLE OR _____MULTIPLE OCCUPANT STRUCTURE

PHONE: ( ) ____________ FAX#: ____________

		1ST	2ND	3RD
1. OUTSIDE				
NO ADDRESS ON BUILDING	1.01			
LP GAS TANK PROTECTION	1.02			
GAS METER NOT PROTECTED	1.03			
LP-GAS TANK CLEARANCE	1.04			
2. FIRE LANE				
FIRE LANE OBSTRUCTED	2.01			
3. HOUSEKEEPING/STORAGE				
WASTE MATERIALS PROHIBITED	3.01			
WEEDS NEAR FLAM. LIQUID	3.02			
SPRINKLER CLEARANCE	3.03			
TRASH CONTAINERS	3.05			
WASTE REMOVAL	3.06			
4. EXITWAYS/STAIRWAYS/AISLE				
DISCHARGE OBSTRUCTED	4.01			
EXIT ILLUMINATION	4.02			
FLAMMABLES NEAR EXITS	4.03			
NO IDENTIFICATION	4.04			
AISLES WIDTH INADEQUATE	4.05			
IMPROPER DOOR HARDWARE	4.06			
INADEQUATE AISLES	4.07			
5. HEATING/VENTILATING				
UNIT NOT ACCESSIBLE	5.01			
CLOSED TO STORAGE	5.03			
NOT MAINTAINED	5.04			

		1ST	2ND	3RD
6. ELECTRICAL				
OBSTRUCTED PANEL	6.01			
EXTENSION CORD PROH.	6.02			
MULTIPLUG PROH.	6.03			
NO EMERGENCY LIGHTS	6.04			
NO EXIT SIGNS	6.09			
EXIT SIGN NOT ILLUMINATED	6.10			
7. FIRE WALL				
NEEDS REPAIR	7.01			
8. SPRINK/STANPIPES CONN.				
SWIVELS INOPERABLE	8.01			
CAPS NOT IN PLACE	8.02			
NOT ACCESSIBLE	8.03			
DRAINS NOT MARKED	8.04			
SPRINKLER WRENCH	8.05			
INADEQUATE HEADS	8.06			
TAMPER SWITCH	8.08			
9. COMPRESSED GASES				
CYLINDERS UNSECURED	9.01			
CYLINDER STORAGE	9.02			
WELDING/CUTTING CLEARANCE	9.03			
10. FIRE EXTINGUISHERS				
NOT OPERABLE	10.01			
IMPROPERLY LOCATED	10.02			
AT DISPENSING PUMPS	10.03			
NOT PROVIDED	10.04			

		1ST	2ND	3RD
10. FIRE EXTINGUISHERS CONT.				
NOT MARKED	10.05			
NOT MAINTAINED	10.06			
IMPROPERLY INSTALLED	10.07			
HOOD SYS. NOT MAINTAINED	10.08			
11. FIRE ALARM SYSTEM				
DETECTOR INOPERABLE	11.01			
ZONES NOT MARKED	11.02			
INSPECTION REQUIRED	11.03			
ALARM REQUIRED	11.04			
12. FLAMMABLE LIQUIDS				
IMPROPER STORAGE	12.01			
IMPROPER DISPENSING	12.02			
IGNITION SOURCE	12.03			
NO FIRE EXTINGUISHERS	12.04			
COMBUST. IN DIKE	12.05			
STORAGE CABINET REQUIRED	12.06			
LIMITED SPRAYING	12.07			
13. HAZARDOUS MATERIALS				
DEFECTIVE CONTAINERS	13.01			
SPILL NEUTRALIZING	13.02			
MSDS NOT DISPLAYED	13.03			
TOXIC STORAGE	13.04			
TANK NOT LABELED	13.05			
BUILDING LABEL	13.06			
14. MISCELLANEOUS				
OTHER (SEE REMARKS)	14.01			

BLDG. OWNER: ____________

PHONE: ( ) ____________

AFTER HOURS CONTACT PERSON

NAME: ____________

PHONE: ( ) ____________

BUS. OWNER: ____________

PHONE: ( ) ____________

TYPE OF BUSINESS OR PRODUCT: ____________

NUMBER OF EMPLOYEES: _____DAY _____NIGHT

FIRE ALARM SYSTEM: _____YES _____NO

____FULL ____PARTIAL _____MONITORED

SERVICING CO.: ____________

PHONE: ( ) ____________

LOCATION OF 1ST ANNUNCIATOR PANEL ____________

LOCATION OF 2ND FIRE CONTROL PANEL ____________

DATE OF LAST ALARM TEST: ____________

SPRINKLER: ______YES ______NO

____FULL ____PARTIAL ____MONITORED

_____WET _____DRY

SERVICING CO. ____________

DATE OF LAST FLOW TEST: ____________

BUILDING

_______LENGTH _______WIDTH

_______HEIGHT _______STORIES

BASEMENT: _______YES _______NO

UNDERGROUND TANKS _____YES _____NO

TYPES OF TANKS AND QUANTITIES:

___GAS ___DIESEL ___LPG ___WASTE OIL

PARTICIPATING PROGRAMS:		LOCATIONS:
KEY	Y OR N	____________
KNOX BOX	Y OR N	____________
TAMPERED	Y OR N	
ELEVATOR BOX	Y OR N	____________

REMARKS: ____________

	TIME IN	TIME OUT
1ST		
2ND		
3RD		

NOTIFICATIONS:

	DATE	RE-INSP	INITIALS
1ST INSP.			
2ND INSP.			
3RD INSP.			

INITIALS:

INSPECTION COMPLETED: ____________

FIRE SAFETY REVIEW: ____________

ENTERED ON COMPUTER: ____________

CHECKED THE ENTRY: ____________

I, THE UNDERSIGNED, AM IN RECEIPT OF A COPY OF THIS INSPECTION RECORD, AN AN INSPECTION BOOKLET WHICH EXPLAINS THE MARKED VIOLATIONS, AND I AM AWARE OF THE HAZARARD/S INDICATED:

____
SIGNATURE

____
PRINT NAME

COPY DISTRIBUTION: WHITE- FIRE MARSHAL'S OFFICE YELLOW- COMPANY/INSPECTOR PINK- OWNER/BUSINESS 6-86/REVISED 4-90 MMN

Figure 11–3. Typical inspection form for general industrial use. (Courtesy Springfield, Illinois, Fire Department.)

SUBJECT: HOT-WORK PERMIT PROCEDURES
TO: ALL CONCERNED

Steps will be taken immediately to put into practice a permit tag system which should provide better protection against fire from welding and other hot work in hazardous areas. This program, as outlined below, is intended to be a practical one. It must be realized by everyone that for it to be effective, the wholehearted cooperation of all concerned must be secured.

1. After an inspection of the entire plant has been made by the fire chief and after a discussion with the various department heads, areas throughout the plant will be designated as hazardous for any type of welding, burning, spark, open flame, or hot work. Such areas will be prominently marked, and before hot work is done within any such area, permit tags must be secured in order to help ensure that the area will be as free as possible from fire hazards and that proper precautions will have been taken.
2. Tags have been prepared on which pertinent information must be filled out by the parties concerned. Each employee who may do hot work will be given a supply of tags to keep with equipment. When sent to perform work in a hazardous area, the employee and the immediate foreman will check the area together to determine if necessary precautions to prevent fire have been taken.
3. It will be the maintenance foreman's responsibility to notify the foreman in the department in which the work is to be done, and together they will sign the permit for the employee to do welding or other hot work.
4. The fire chief will then check the area for firesafe working conditions, see that standby fire extinguishers are present, and assign a fire watcher when necessary. When satisfied with the precautionary measures, the fire chief will sign the permit tag and return it to the employee who is to do hot work.
5. The signed permit card will be kept by the employee doing the work until the job is completed, at which time the employee will personally check the area for fire.
6. The area will subsequently be checked by the maintenance foreman who will see that any extinguisher that may have been used is designated for recharging.
7. The completed tag will be turned in at the end of the day to the maintenance department which will collect and forward all tags to the fire chief for filing and record purposes.

Figure 11–4. Example of policy statement that can be issued by a company inaugurating a hot-work permit program in order to instruct employees.

more serious exposures, such as heavy welding, fire fighting, or foundry work.

Manufacturers of fire-retardant treatments and products can supply a fire rating to compare treated products with nontreated products. Use only fire retardants listed by Underwriters Laboratories (UL). Several commercially treated fabrics are listed by UL. After each application of flame-retardant chemicals, periodically test samples of treated materials, in a safe location, to assure satisfactory performance.

Protecting Adjacent Buildings

When a fire breaks out in a building, protect adjacent buildings by (1) closing every window facing the burning building, (2) stationing fire brigade workers with fire extinguishers or fire hoses at each window nearest the fire, and (3) stationing fire fighters on the exposed building's roof with hose lines to keep the roof wetted down and with extinguishers to put out any burning embers. Remember that radiant heat will not stop until it reaches an opaque object; therefore, water sprayed between the fire building and the exposure building must be applied directly to the exposure building.

Training Employees

Because extinguishers are effective only when fires are in their first stages, ensure that extinguishers are immediately reachable and promptly used by trained personnel. Extinguishers are only as good as the operators using them. Therefore, thoroughly train key workers on each shift.

Fire extinguisher training is intended to teach employees how to stop small fires from spreading out of control. UL considers fire extinguishers to be functioning at only 40% of their fire-fighting capacity when they are used by inexperienced operators. Such operators would include most people, for nine out of ten people have never used a fire extinguisher. This means that any fire extinguisher in the hands of a trained operator has two and a half times the fire-fighting capacity it has when used by a novice. These figures underscore the critical importance of training.

Use demonstrations to instruct employees in the use of extinguishers. A good time to do this is when extinguishers are scheduled for recharging. At the demonstrations, simulate fire conditions. Have an instructor explain the fundamentals of fire fighting and the use of the equipment. Allow employees to get the "feel" of the extinguisher. In small organizations, have everyone in the plant attend and participate in the demonstrations. In large plants, train a suitable number of employees so that personnel trained in fire fighting are distributed throughout the plant.

One of the most difficult decisions any employee faces is whether to fight a fire or to get out safely. If the following conditions are met, an employee might decide to fight a fire with an extinguisher: (1) there is a clear exit, (2) the fire brigade or department has already been called or is being called, (3) the fire is small, such as one in a wastebasket or a small tool housing, (4) the employee knows how to use the extinguisher, and (5) the extinguisher is in working order. An employee should not fight a fire if (1) the fire is clearly spreading beyond its point of origin, (2) the fire could block the exit, or (3) the employee is unsure of how to use the extinguisher.

Continue the training of employees with demonstrations, practice drills, and lectures at yearly intervals, or more often if a special fire hazard exists. Be sure that employees have printed instructions regarding the use of fire extinguishers. Photocopied or printed sheets, leaflets, or cards can give both general and detailed instructions regarding the use of the extinguishers. In addition, permanently post instructions about how to use extinguishers near or on the extinguishers themselves.

Fire extinguisher manufacturers, insurance companies, fire departments, the NFPA, and the National Safety Council (NSC) offer films, posters, and cutaway displays that are useful in explaining the construction, maintenance, and operation of portable fire extinguishers. NFPA 10, *Portable Fire Extinguishers*, gives particularly valuable information for training employees about using portable fire extinguishers (Figure 11–5). Also, local fire departments are a possible training source on the use of a portable fire extinguisher.

Remember that portable fire extinguishers are intended as a first means of defense against unwanted fires. They will not extinguish a very large fire or a very fast-moving fire. When using a portable fire extinguisher, never place yourself between the fire and an exit; always keep your back towards an exit when using it. Remember the acronym PASS: Pull, Aim, Squeeze, and Sweep.

Figure 11–5. This student is being trained in effective use of a fire extinguisher in fighting a low-level fire.

Communications

Once a fire has been detected, especially in a potentially disastrous situation, good communications are necessary (1) as a means of alerting occupants to the emergency and (2) as a way to mobilize fire protection forces, whether a plant's fire brigade, the municipal fire department, or both. A coded fire alarm system, with alarm boxes and bells, horns, or other sounding devices suitably situated, is usually needed. In very small plants, however, a steam whistle or similar device might be adequate. In any case, the alarm system is no better than employees' training in how to respond when the alarm is sounded. (See the section Fire Detection, later in this chapter.)

In addition, install a backup alarm system in case electrical power is lost. Power is often lost during fires, especially if sabotage is involved. Also install a secondary power source, especially if electric pumps augment the regular sprinkler system and standpipe system. Make sure that all persons can be alerted and that there is some way for everyone to get out of the building, regardless of the power source in use.

THE CHEMISTRY OF FIRE

Fire, or the process of combustion, is extraordinarily complex. For a fire to occur, fuel, oxygen, heat, and a chemical chain reaction must join in a symbiotic relationship (Figure 11–6). In combustion, heat energy is

Figure 11–6. "The fire tetrahedron." Oxygen, heat, fuel, and chain reactions are necessary components of a fire. Speed up the process and an explosion results.

released in a self-catalyzed reaction involving a condensed-phase fuel, a gas-phase fuel, or both. The combustion process is usually associated with rapid oxidation of a fuel by oxygen in the air. If the combustion process is confined so pressure can increase, an explosion can result. A similar process that takes place over long periods of time and at a lower temperature is called oxidation. Rusting of metal is an example of this. A fire, then, is a combustion process intense enough to emit heat and light.

In addition, a fire can be classified into two general forms or modes: flame fire and surface fire. Flame fires directly burn gaseous or vaporized fuel and include deflagrations. The rate of burning is usually high, and a high temperature is produced. The following are two types of flame fires:

- Premixed flame fires exist in a gas burner or stove and are relatively controlled.
- Diffusion flame fires refer to gases burning on mixed vapors and air. Controlling these fires is difficult.

Surface fires occur on the surfaces of a solid fuel and are often called a "glow" or "deep-embered seated" fire. Surface fires occur at the same temperature as do open flame fires.

A surface fire can be represented by a triangle composed of heat, fuel, and air. A flame fire includes an additional component: a chemical chain reaction. These two modes of fire are not mutually exclusive; they may occur alone or together.

Knowing how and why a fire burns suggests ways to control and extinguish it. The surface fire has three components that can be controlled, while the flame fire has four. Fires can be controlled in the following ways:

- Heat can be taken away by cooling.
- Oxygen can be taken away by excluding the air.
- Fuel can be removed to an area where there is not enough heat for ignition.
- The chemical reaction of the flame fire can be interrupted by inhibiting the rapid oxidation of the fuel and the concomitant production of free radicals, the lifeblood of the flame's reaction.

It is the vapor given off by the solid material (Class A fires) or the liquid (Class B fires) that actually burns, not the solid or liquid itself.

Cooling a Fire

To extinguish a fire by cooling, remove heat at a greater rate than the total heat that is being evolved by the fire. To do this, the cooling agent must reach the burning fuel directly. The cooling action may also stop the release of combustible vapors and gases. The most common and practical extinguishing agent is water applied in a solid stream or spray, or incorporated in foam. In practice, the fire is literally drowned into submission by the water. When water is the extinguishing agent in an enclosed space, the steam that is generated when the water converts from a liquid to a gas will displace oxygen to a degree.

Removing Fuel from a Fire

Often, taking the fuel away from a fire is not only difficult but dangerous. Fortunately, there are exceptions: (1) Storage tanks for flammable liquids may be arranged so that their contents can be pumped to an isolated, empty tank in case of fire. (2) When flammable gases catch fire as they are flowing from a pipe, the fire will go out if the fuel supply can be shut off. (3) In any mixture of fuel gases or vapors in air, adding an excess of air has the effect of diluting the fuel's concentration below the minimum combustible concentration point.

Limiting Oxygen in a Fire

Limit air, or oxygen, from a fire by smothering the burning area with a noncombustible material, as by covering it with a wet blanket (make sure the blanket is not made of highly combustible fibers), throwing dirt or sand on the fire, smothering it with inert gas, or covering it with a chemical or mechanical foam.

To make a wet blanket effective, keep it in place long enough for all smoldering ignition to be extinguished. Note that smothering is ineffective on substances containing their own oxygen supply, such as ammonium nitrate or nitrocellulose. Smothering is also ineffective on deep-seated materials like wood, rags, and large rolls or skids of paper.

Covering a fuel, however, can stop a fire. Many foams and some solids serve as an emulsion film, or cover, on the burning fuel, thus extinguishing the fire. If the contents of a wastebasket catch fire, drop an empty wastebasket on top of it to smother the fire.

To dilute oxygen below the concentration necessary to support combustion (15–16%), discharge CO_2 or another inert gas into the fire. The fire will then remain out (1) if the percentage of oxygen is reduced below the level of combustion long enough to allow the combustible materials to cool below their ignition temperature, and (2) if no ignition sources are present. Use an inert gas to purge operations involving flammable vapors and dusts in confined spaces where a source of ignition may exist. To ensure that the oxygen concentration remains low enough to prevent combustion, constantly monitor the flow of inert gas, the actual concentration of oxygen, or both.

Interrupting the Chain Reaction in a Fire

Analyzing the anatomy of a fire reveals that the original fuel molecules appear to combine with oxygen in a series of successive intermediate stages, called branched-chain reactions. Then the end product, combustion, occurs. The intermediate stages are responsible for the evolution of flames.

As molecules break up in these branched-chain reactions, unstable intermediate products called free radicals are formed. The concentration of free radicals determines the speed of flames. The life of the free hydroxyl radical (-OH) is very short, being on the order of 0.001 of a second, but long enough to be crucial in the combustion of fuel gases. The almost simultaneous formation and consumption of free radicals appear to be the lifeblood of the chain reaction.

Extinguishing agents, such as dry chemicals and halogenated hydrocarbons, remove the free radicals in these branched-chain reactions from their normal function as a chain carrier. The effects that various dry–chemical agents, such as sodium bicarbonate–base, potassium bicarbonate–base, and ammonium phosphate–base, have on capturing free radicals depend upon their individual molecular structure. Potassium bicarbonate dry chemical is the most effective because of the large size of the potassium ion.

Using Extinguishing Agents

Some extinguishing agents help control fire by attacking more than one of its three or four components. For example, both plain-water fog (as compared with straight-water streams) and CO_2 can react at flame temperatures with relatively slow-burning free carbon, producing carbon monoxide (CO), and as a result, decreasing black-smoke production. Because this reaction absorbs heat, it lowers the heat of the fire as well as lowering the oxygen concentration.

Matching the pace at which newer and more potent fire extinguishing agents work requires increasingly sophisticated tactics and techniques. Although a fire can be attacked from at least four different standpoints, attacking from any one of them does not necessarily result in the most rapid extinguishing time. Attacking from more than one, by using more than one agent, however, can produce a synergistic effect that hastens extinguishing. For example, fire fighters use "light water" and potassium bicarbonate–base dry chemical jointly in fighting fires from aircraft crashes.

Although extinguishing fires might sound simple, it is still very far from being an exact science. Since a fire usually has more than one source of fuel, it must be fought with several different extinguishing agents, which, by working together, complement each other.

CLASSIFICATION OF FIRES

Four general classifications of fires have been adopted by the NFPA. These classifications are based on types of combustibles and the extinguishing agent needed to combat each. (See NFPA 10, *Portable Fire Extinguishers.*)

Class A Fires

Class A fires occur in ordinary materials, such as wood, paper, excelsior, rags, and rubbish. The quenching and cooling effects of water, or of solutions containing large percentages of water, are of first importance in extinguishing these fires. Dry-chemical agents (multipurpose dry chemicals) provide both rapid knockdown of the flames and the formation of a coating that tends to retard further combustion. Where total extinguishment is mandatory, follow up with water.

Class B Fires

Class B fires occur in the vapor-air mixture over the surface of flammable liquids, such as gasoline, oil, grease, paints, and thinners. Limiting air or oxygen, i.e., inhibiting combustion, is of primary importance to stop fires of this class before they spread. Solid streams of water are likely to spread the fire. Under certain circumstances, however, water-fog nozzles may prove effective in the control but not the extinguishment, of these fires. Generally, use regular dry chemicals, multipurpose dry chemicals, CO_2, or foam for such fires.

Class C Fires

Class C fires occur in or near energized electrical equipment where nonconducting extinguishing agents must be used. Use dry chemicals or CO_2 extinguishing agents for such fires. Do not use foam or a stream of water, because both are good conductors of electricity and can expose operators to a severe shock hazard.

Class D Fires

Class D fires occur in combustible metals such as magnesium, titanium, zirconium, lithium, potassium, and sodium. Specialized techniques, extinguishing agents, and extinguishing equipment have been developed to control and extinguish fires of this type. Generally, do not use normal extinguishing agents on metal fires. In such fires, there is the danger of increasing the intensity of the fire because of a chemical reaction between some extinguishing agents and the burning metal.

Other Fires

Fires that involve certain combustible metals or reactive chemicals may require special extinguishing agents or techniques. (See NFPA 49 *Hazardous Chemicals Data*, and NFPA 325M, *Fire-Hazard Properties of Flammable Liquids, Gases and Volatile Solids*.)

FIRE RISKS

Fire protection measures can be effective only if they are based on a proper analysis and evaluation of the risk of fire. A complete evaluation is important, since a wide variety of methods and equipment exist to provide protection. The optimum level of fire protection is that which minimizes both the costs and the expected fire losses. As safety is also involved, a value must likewise be placed on saving life. Ideally, the cost of fire protection should have a concomitant effect on reducing the amount of exposure to loss or of risk involved. In time, production methods, hazards, and fire protection requirements may change. Assess these variables periodically to assure a continued optimum risk-to-protection relationship.

The process to determine the risk of fire can be covered in the following three steps:

1. Recognize hazards and potential hazards.
2. Evaluate the hazards and expected losses.
3. Evaluate the proposed preventive measures.

Fire Hazards

The information assembled while evaluating the risk of fire goes beyond the scope of the inspection for fire prevention outlined earlier in this chapter. The evaluation serves as an excellent training resource for operating and emergency personnel and as a reference when changes are made in an existing plant or when designs are drawn up for a similar plant. Therefore, it is important to develop a format and list of references for organizing a systematic fire hazard survey.

Fire Hazard Analysis

In a fire hazard analysis the following points are considered:

1. Site
   a. location and age of building
   b. accessibility for fire-fighting equipment, if not self-contained
   c. possibility of exposure
   d. amount of water supply available
   e. traffic (distance to other buildings, type and characteristics of traffic)
2. Building construction
   a. framework
   b. fire-resistance rating of materials
   c. roof materials; consider a non-combustible roof covering
   d. interior walls and windows
   e. exterior wall materials
   f. floors
   g. interior finish
   h. shape
   i. heat, ventilating, and air conditioning system
   j. use
   k. concealed spaces

  l. exterior doors and exits
  m. elevators, stairways, and chutes
  n. fuel sources for heating or for process operations
  o. storage areas
  p. raw stock
  q. finished goods
  r. electrical systems
3. Building contents
  a. materials
  b. liquids (flammability, amounts, kind)
  c. solid material (amounts and kind)
  d. location of materials
  e. basic processes
4. Management factors
  a. design for ease of housekeeping
  b. employee smoking policy
  c. overall cleanliness
5. People factor (number and characteristics)
  a. location in structure
  b. exits (number of, markings, and accessibility)
  c. activity of people (work, play, etc.)
  d. location of patrons, visitors, and employees
6. Fire protection system, if for an existing facility
  a. detection system
  b. alarm system
  c. emergency lighting communication system
  d. compartmentation
  e. extinguishing system
  f. people problems
7. After the fire
  a. follow-up or cleanup
  b. emergency plan to keep the company, plant, or operation running

Evaluation of Fire Hazards

Base the evaluation of fire hazards on the fire hazard analysis and the risk survey. Whenever the processes, operations, or uses of a plant change, evaluate the fire protection system according to the degree of change.

Determine the dollar value of potential fire losses. Also consider factors such as the loss of a key plant or operation. Such a loss to the company may be greater than the actual dollar loss.

Means exist to convert general fire hazards to some predetermined, relative-hazard scale, which then can be further matched based on some assumed levels of fire hazards or risks. However, there are many fire-risk situations that do not clearly fit predetermined assessments.

An evaluation process can take a number of different directions or follow a number of procedures. The following suggestions are intended only to help determine whether a new fire protection system or some modification of an existing one is more feasible.

Fires usually start in a facility's contents or operations rather than in its structure. Therefore, consider the following:

- What materials are flammable?
- What materials in a process or operation are most likely to ignite? burn? explode?
- What in the facility could be a source of ignition? Are any open sparks or flames present? Are high temperatures involved in any operations?
- Where are flammable and combustible materials located? Are flammable materials stored together? Do indirect connections exist? If one of the materials should burn, could the others easily ignite?
- Might any of the materials ignite because of convection or radiation?
- What toxic gases might evolve into a fire?
- How much time might it take for a fire to spread to other areas? to adjacent facilities?
- Smoke and toxic gases, and sometimes heat, are largely responsible for fire deaths. What toxic gases might evolve from the burning or smothering of contents?
- How many people are likely to be involved in the facility, in adjacent facilities, or in facilities nearby?
- Some fire hazard evaluations include determining the fire load, or total heat potential, of materials. Fire loads express the weight of combustible material per square foot (0.09 m^2) of fire area. Paper and wood have a caloric value of 7,000–8,000 Btu/lb (16–20 MJ/kg). A typical office is likely to have about 5 lb/ft^2 (2.5 kg/m^2). A flammable liquid has a heat-producing potential of about 14,000–15,000 Btu/lb (about 35 MJ/kg). Fire load does not account for the rate of heat liberation or for the distribution within a structure, and it is only a guide. The fire load can be used to develop some elements of the fire protection system.

  The exterior and interior finish materials of the facility's structure can have a fire load factor. (For further information, see the following section and the NFPA *Fire Protection Handbook*.) Many other factors, including the ventilation system, building shape, degree of openness, and amount of compartmentation affect the fire load.

FIRE-SAFE BUILDING DESIGN AND SITE PLANNING

This section is taken from *Principles of Fire Protection*, by Percy Bugbee, copyright 1990, NFPA. It is used with permission.

The object of fire safety is to protect life foremost and property secondly from the ravages of fire in a building. Building design and construction must take into account a wide range of fire safety features. Not only must the interiors and contents of buildings be protected from the dangers of fire, but the building site itself must have adequate water supplies and easy accessibility by the fire department. Architects, builders, and owners may assume that state codes provide adequate measures; however, these codes stipulate only minimal measures for fire safety. Planning and construction based upon such codes should not reduce or limit fire-safe design efforts.

Objectives of Fire Safety Design

Before a building designer can make effective decisions relating to fire safety design, the specific function of the building and the general and unique conditions that are to be incorporated into it must be clearly identified. Decisions regarding the fire safety design and construction of the building have the same objectives as do all fire protection measures, namely (1) life safety, (2) continuity of operations, and (3) property protection.

The art of probing to identify objectives is an important design function. The degree of risk that will be tolerated by the owner and the occupants is a difficult design decision; consequently, it should be identified in a clear, concise manner so that the designer can properly realize the design objectives.

Life Safety

Design considerations for life safety must address two major questions: (1) Who will use the building? (2) What will the people using the building do most of the time?

The occupied building provides a great potential for fire because of the presence of large numbers of people, any one of whom could perform a careless or malicious act resulting in fire. Appliances and mechanical or electrical equipment are a potential hazard through misuse, failure, faulty construction, or substandard installation. Accumulations of combustibles, either waiting for disposal or in storage, frequently provide a ready means by which otherwise controllable fires could spread. Store combustibles waiting for disposal in metal containers with tightfitting metal covers. Keep these containers at least 10 ft (3 m) from the building since these containers are sometimes set on fire by other persons.

The identification of specific functional patterns, constraints, and disabilities is vital in designing specific fire protection features that recognize occupant conditions and activities.

Continuity of Operations

Continuity of operations, the second major area of building design decision making, must take into consideration those specific functions conducted in a building that are vital to continued operation of the business and that cannot be transferred to another location. In this regard, the owner must identify for the designer the amount of "downtime," or the amount of time an operation can be suspended without completely suspending total operations. The degree of protection required in fire-safe building design varies with the number and scope of vital operations that are nontransferable.

Property Protection

One of the most important questions to be asked about the design of buildings with regard to protection of property is: Is there any specific high-value content that will need special design protection? The requirements with regard to protection of property within a building are often fairly easy to identify. Materials of high value that are particularly susceptible to fire and/or water damage can usually be identified in advance of building design. For example, vital records that cannot be replaced easily or quickly can be identified in advance as needing special fire protection design considerations.

Fire Hazards in Buildings

When the designer and the owner either consciously or unconsciously overlook or ignore the possibility of fire in the building to be built, the building and its occupants are endangered. The broad approach to the fire-safe design of a building requires a clear understanding of the building's function, the number and kinds of people who will be using it, and the kinds of things they will be doing. In addition, appropriate construction and protection features must be provided for the protection of the contents and, particularly for mercantile and industrial buildings, to assure the continuity of operations if a fire should occur. Too many fires disastrous to people and to property have occurred, and will continue to occur, because no one has given proper consideration to the threat of potential fire.

Smoke and Gas

Studies of fire deaths in buildings indicate that about 66% of these deaths are due to the smoke and toxic gases that evolve as products of the fire. About 34% of the deaths result from heat or contact with direct flame. The carbon monoxide developed in many fires, particularly unventilated and smoldering fires, is probably the most common cause of death. Carbon monoxide can be neither seen nor smelled. Exposure to this gas, even in small quantities, can cause impaired mental behavior. When inhaled in large quantities, the smoke given off in most fires can lead to pneumonia and other lung troubles. Smoke also obscures visibility and thus can lead to panic situations when occupants cannot see and use escape routes.

Heat and Flames

As has already been stated, heat and flames account for 34% of deaths. Although heat and flame injuries are much fewer than those caused by smoke and toxic gases, the pain and disfigurement caused by burns can also result in serious, long-term complications.

Building Elements and Contents

If the building on fire has combustible furniture, flames and toxic gases may spread so rapidly that occupants may not be able to escape. Poor construction practices, such as failure to protect shafts and other vertical openings, make the vertical spread of fire more rapid and the work of fire fighters more difficult.

Although the collapse of structural elements has not resulted in many deaths or injuries to building occupants, it is a particular hazard to fire fighters. A number of deaths and serious injuries to fire fighters occur each year because of structural failure. While some of these failures result from inherent weaknesses, many are the result of renovations to existing buildings that materially, though not obviously, affect the structural integrity of the support elements. A building should not contain surprises of this type for fire fighters.

Elements of Building Fire Safety

The fire safety of a building will depend first on what is done to prevent a fire from starting in the building, and second on what is done through design, construction, and good management to minimize the spread of fire if it happens. Good housekeeping is one of the major factors in both fire prevention and control. Keeping the fuel load down not only lessens the amount of material that can be ignited, but it also provides less material that can be consumed if a fire breaks out.

Once a fire has started, its spread will depend on the design of the building, the materials used in construction, building contents, methods of ventilation, detection and alarm facilities, and fire suppression systems, if any. Table 11–A describes the building design and construction features that influence safety. These elements are within the decision-making authority of various members of the design team, based on the assumption that their fire safety objectives are clearly defined by management, the owners, or other responsible parties, both public and private. The design and construction elements are organized in a manner that can give a quick overview of the major aspects that must be considered for fire safety. They show features that include both passive and active design and construction considerations. (A passive design element is one that requires no action to function, such as a fire wall. An active design or construction element is one that requires an action in order to function, such as a fire door that must be closed.)

The persons responsible for fire prevention are not the same ones responsible for the building design. Table 11–B describes the elements of fire prevention. Decisions concerning these elements are predominantly under the control of the building owner or occupant, or both. Table 11–B includes the elements of emergency preparedness in case of fire that are the responsibility of the owner and/or occupant.

Planning Fire Safety for Buildings

Two major categories of decisions should be made early in the design process of a building in order to provide effective fire-safe design. Early considerations should be given to both the interior building functions (discussed in this section) and exterior site planning (discussed in the next section). Building fire defenses, both active and passive, should be designed in such a way that the building itself assists in the manual suppression of fire.

Interior layout, circulation patterns, finish material, and building services are all important fire safety considerations in building design. Building design also has a significant influence on the efficiency of fire department operations. As a result, all fire suppression activities should be considered during the design phases.

Fire-Fighting Accessibility to a Building's Interior

One of the more important considerations in building design is access to the fire area. This includes access to the building itself as well as access to the building's interior.

In larger and more complex buildings, serious fires over the years have brought improvements in building design to facilitate fire department operations. The larger the building, the more important access for fire fighting becomes. In some buildings, fire fighters cannot function effectively. The spaces in which adequate fire-fighting access and operations are restricted because of architectural, engineering, or functional requirements should nonetheless be provided with effective protection. A complete automatic sprinkler system with a fire

Table 11–A. Elements of Building Fire Safety

Building Design and Construction Features Influencing Fire Safety

1. Fire Propagation
   a. Fuel load and distribution
   b. Finish materials and their location
   c. Construction details influencing fire and products of combustion movement
   d. Architectural design features (vertical/horizontal openings allowing fire spread)

2. Smoke and Fire Gas Movement
   a. Generation
   b. Movement
      —Natural air movement
      —Mechanical air movement
   c. Control
      —Barriers
      —Ventilation
      —Heating, ventilating, air conditioning
      —Pressurization
   d. Occupant protection
      —Egress
      —Temporary refuge spaces
      —Life support systems

3. Detection, Alarm, and Communication
   a. Activation
   b. Signal
   c. Communication systems
      —To and from occupants
      —To and from fire department
      —Type (automatic or manual)
      —Signal (audio or visual)

4. People Movement
   a. Occupant
      —Horizontal
      —Vertical
      —Control
      —Life support
   b. Fire fighters
      —Horizontal
      —Vertical
      —Control

5. Suppression Systems
   a. Automatic
   b. Manual (self-help; standpipes)
   c. Special

6. Firefighting Operation
   a. Access
   b. Rescue operations
   c. Venting
   d. Extinguishment
      —Equipment
      —Spatial design features
   e. Protection from structural collapse

7. Structural Integrity
   a. Building structural system (fire endurance)
   b. Compartmentation
   c. Stability

8. Site Design
   a. Exposure protection (to facility and by facility to public)
   b. Firefighting operations
   c. Personnel safety
   d. Miscellaneous (water supply, traffic, access, etc.)

Fire Emergency Considerations

1. Life Safety
   a. Toxic gases
   b. Smoke
   c. Surface flame spread

2. Continuity of Operations
   a. Structural integrity
   b. Limiting of value or separation (by passive system) of like operations

3. Structural
   a. Fire propagation
   b. Structural stability

From NFPA *Fire Protection Handbook,* 14th ed.

department connection is probably the best solution to this problem. Other methods that may be used in appropriate design situations include access panels in interior walls and floors, fixed nozzles in floors with fire department connections, and roof vents and access openings.

Ventilation

Ventilation is of vital importance in removing smoke, gases, and heat so that fire fighters can reach the seat of a blaze. It is difficult, if not impossible, to ventilate a building unless appropriate skylights, roof hatches, emergency escape exits, and similar devices are provided when the building is constructed.

Ventilation of building spaces performs the following important functions:

- Protection of life by removing or diverting toxic gases and smoke from locations where building occupants must find temporary shelter.
- Improvement of the environment in the vicinity of the fire by removal of smoke and heat. This enables fire fighters to advance close to the fire to extinguish it with a minimum of time, water, and damage.
- Control of the spread or direction of fire by setting up air currents that cause the fire to move in a desired direction. In this way occupants or valuable property can be more readily protected.

Table 11–B. Fire Prevention and Emergency Preparedness

1. Ignitors
   a. Equipment and devices
   b. Human accident
   c. Vandalism and arson
2. Ignitable Materials
   a. Fuel load
   b. Fuel distribution
   c. Housekeeping
3. Emergency Preparedness
   a. Awareness and understanding
   b. Plans for action—Evacuation or temporary refuge—Handling extinguishers
   c. Equipment
   d. Maintenance—operating manuals available

From NFPA *Fire Protection Handbook,* 14th ed.

- Provision of a release for unburned, combustible gases before they acquire a flammable mixture, thus avoiding a backdraft or smoke explosion.
- Smoke dampers should be put in air conditioning ventilation ducts in accord with NFPA 90A, *Installation of Air Conditioning and Ventilating Systems.* This standard requires smoke dampers to operate automatically upon detection of smoke, so fire will not spread through unburned areas.

Connections for Sprinklers and Standpipes

Connections for sprinklers and standpipes must be carefully located and clearly marked. The larger and taller the building becomes, the greater the volume and pressure of water that will be needed to fight a potential fire. Water damage can be very costly unless adequate measures such as floor drains and scuppers have been incorporated into the building design.

Confinement of a fire in a high-rise building can only be accomplished by careful design and planning for the whole building. As buildings increase in size and complexity, more dependence on fire detection and suppression systems is necessary. Such systems are described in detail in the NFPA references listed at the end of this chapter.

Planning Fire Safety for Sites

Proper building design for fire protection includes a number of factors outside the building itself. The site on which the building is located will influence the design, especially traffic and transportation conditions, fire department accessibility, water supply, and the exposure this facility has on the public. Inadequate water mains and poor spacing of hydrants have contributed to the loss of many buildings.

Traffic and Transportation

Fire department response time is a vital factor in building design considerations. Traffic access routes, traffic congestion at certain times of the day, traffic congestion from highway entrances and exits, and limited access highways have significant effects on fire department response distances and response time, and must be taken into account by building designers in selecting appropriate fire defenses.

Fire Department Access to the Site

Building designers must ask the question: Is the building easily accessible to fire apparatus? Ideal accessibility occurs where a building can be approached from all sides by fire department apparatus. However, such ideal accessibility is not always possible. Congested areas, topography, or buildings and structures located appreciable distances away from the street make difficult or prevent effective use of fire apparatus. When apparatus cannot come close enough to the building to be used effectively, equipment such as aerial ladders, elevating platforms, and water tower apparatus can be rendered useless.

Bridge weight loads must be taken into consideration. Many times businesses will place bridges near buildings for aesthetic purposes without taking into consideration the weight limits of fire equipment or construction equipment called in to handle problems that can arise at any time.

The matter of access to buildings has become far more complicated in recent years. The building designer must consider this important aspect during the planning stages. Inadequate attention to site details can place the building in an unnecessarily vulnerable position. If its fire defenses are compromised by preventing adequate fire department access, the building itself must make up the difference in more complete internal protection.

Water Supply to the Site

A building designer must also ask: Are the water mains adequate, are the hydrants properly located, and will gallons per minute flow be adequate? The more congested the area where the building is to be located, the more important it is to plan what the fire department may face if a fire occurs on the property. An adequate water supply delivered with the necessary pressure is required to control a fire properly and adequately. The number, location, and spacing of hydrants and the size of the water mains are vital considerations when the building designer plans fire defenses for a building.

Water mains are another concern. It is of the utmost importance not to have dead-end mains. The optimum would be that all mains be on a looped water system. In addition, the system should be valved so that if there is a breakdown or a break of any type, one section can be isolated without shutting down the whole system. Breaks can occur due to freezing in some areas, contractors cutting through mains, or almost any type of movement or impact.

Exposure Protection

Still another consideration in the design of the building is the possibility of damage from a fire in an adjoining building. The building may be exposed to heat radiated horizontally by flames from the windows of the burning neighboring building. If the exposed building is taller than the burning building, flames coming from the roof of the burning building can attack and damage the exposed building.

The damage from an exposing fire can be severe. It is dependent upon the amount of heat produced and the time of exposure, the fuel load in the exposing building, and the construction and protection of the walls and roof of the exposed building. Other factors are the distance of separation, wind direction, and accessibility of fire fighters.

Fire severity is a description of the total energy of a fire, and involves both the temperatures developed within the exposing fire and the duration of the burning. NFPA 80A, *Exterior Fire Exposures*, describes estimated minimum separation distance under light, moderate, or severe exposures. The severity of the exposure is calculated on the width and the estimated fire loadings of the buildings involved. Building designers should be aware that effective separation distances between the exposing buildings can be reduced by construction of blank fire-resistant walls (3-hr minimum), by closing wall openings with material equivalent to the wall or with ¾-hr protection and eliminating combustible projections, by using automatic deluge water curtains, and by using wired glass instead of ordinary glass.

FIRE-RESISTIVE CONSTRUCTION

It is important to remember that fire-resistive materials used in buildings provide a definite advantage. However, do not confuse them with fire-safe materials. A building constructed with fire-resistive materials can withstand a burnout of its contents without subsequent structural collapse.

Fire-resistive construction actually describes a broad range of structural systems able to withstand fires of specified intensity and duration without failure. The materials are relatively noncombustible and are given a numerical rating as to their fire resistance. Nearly every building material has a fire-resistive rating. The rating is a relative term or number that indicates the extent to which it resists the effect of a fire. Ratings are usually available for most building components, such as columns, floors, walls, doors, windows, and ceilings.

Common high-fire-resistive components include masonry load-bearing walls, reinforced-concrete or protected-steel columns, and poured-on-precast concrete floors and roofs.

Although fire-resistive structures do not, in themselves, contribute fuel to a fire, combustible trim, ceilings, and other interior finish and furnishings may produce an intense fire. However, because such combustible materials pose a serious threat to life safety, they must be considered in providing fire protection. Using approved building materials will lower the flame-spread ratings and will limit materials that can contribute to available fuel, especially in buildings without automatic sprinkler systems.

Heavy-Timber Construction

Heavy-timber construction is characterized by masonry walls, heavy-timber columns and beams, and heavy plank floors. Although not completely immune to fire the great bulk of the wooden members slows the rate of combustion. Moreover, the char that forms on wooden surfaces serves as an insulator for the wood within.

Noncombustible and Limited-Combustible Construction

Noncombustible and limited-combustible construction includes all types of structures in which the structure itself, exclusive of trim, interior finish, and contents, is noncombustible but not fire resistant. Exposed steel beams and columns, and masonry, metal, or gypsum wallboard are the most common forms of this type of construction.

Because of the tendency of steel to warp, buckle, and collapse under moderate fire exposure, noncombustible construction is relatively vulnerable to fire damage. It is, therefore, most suitable for low-hazard occupancies or ordinary hazard occupancies. To reduce susceptibility to heat collapse, load-bearing noncombusitble structural members can be protected by encasing them in concrete, covering them with gypsum wallboard, or spraying on a protective material.

Ordinary Construction

Ordinary construction consists of masonry exterior–bearing walls, or bearing portions of exterior walls, that are of noncombustible construction. Interior framing, floors, and roofs are made of wood or other combustible materials whose "bulk" is less than that required for heavy-timber construction.

If floor and roof construction and their supports have a one- hour fire-resistance rating, and all openings through floors (including stairways) are enclosed with partitions having a one-hour fire-resistance rating, then the construction is known as "protected, ordinary construction." Its occupancy should be limited to light or moderate hazards.

Even when sheathed, ordinary construction, unlike fire-resistive or noncombustible construction, still has combustible materials in concealed wall and ceiling spaces. Fire frequently originates in these concealed spaces or enters into them and then spreads rapidly throughout the entire room and building.

To prevent the free passage of flames through concealed spaces or openings, include the following safety features in the construction:

- Trim all combustible framing away from sources of heat.
- Provide effective fire barriers against the spread of fire between all subdivisions and all stories of the building.
- Provide adequate fire separation against exterior exposure.
- Fire-stop all vertical and horizontal draft openings to form effective barriers to stop or slow the spread of fire.

Wood-Frame Construction

Wood-frame construction consists primarily of wood exterior walls, partitions, floors, and roofs. Exterior walls may be stuccoed or sheathed with brick veneer or metal, or asphalt siding.

Although generally inferior to other types of construction from a fire-safety standpoint, wood-frame construction can be made reasonably safe for light-hazard, low-density occupancies. Safety can be greatly increased by suitable protection against the horizontal and vertical spread of fire, provision of safe exits, and elimination of combustible interior finishes. Install enough fire detectors to alert all occupants. Automatic sprinkler protection can greatly improve the fire safety of wood-frame buildings.

Interior Finish

The way a building fire develops and spreads and the amount of damage that ensues are largely influenced by the characteristics of the interior finish in a building. Many types of interior finish are used in buildings, and they serve many functions. Primarily they are used for aesthetic or acoustical purposes. However, insulation or protection against wear and abrasion are also considered major functions by building designers. The following statements from the National Commission on Fire Prevention and Control's report titled "America Burning" point out the need for greater concern and attention to the potential fire hazards of interior finishes.

> The modern urban environment imparts to people a false sense of security about fire. Crime may stalk the city streets, but certainly not fire, in most people's view. In part, this sense of security rests on the fact there have been no major conflagrations in American cities in more than half a century. In part, the newness of so many buildings conveys the feeling that they are invulnerable to attack by fire. Those who think only of a building's basic structure (not its contents) are satisfied, mistakenly, that the materials, such as concrete, steel, glass, aluminum, are indestructible by fire. Further, Americans tend to take for granted that those who design their products, in this case buildings, always do so with adequate attention to their safety. That assumption, too, is incorrect.

Types of Interior Finish

Interior finish is usually defined as those materials that make up the exposed interior surface of wall, ceiling, and floor constructions. The common interior finish materials are wood, plywood, plaster, wallboard, acoustical tile, insulating and decorative finishes, plastics, and various wall coverings.

While some building codes do not include floor coverings under their definitions of interior finishes, the present trend is to include them. Rugs and carpets are not subject to test and regulation under the Flammable Fabrics Act administered by the U.S. Department of Commerce. Some rugs and carpets are a factor in fire spread. (See NFPA 253, *Test for Critical Radiant Flux of Floor Coverings Systems Using a Radiant Heat Energy Source.*)

Many codes exclude trim and incidental finish from the code requirements for wall and ceiling finish. Interior finishes, however, are not necessarily limited to the walls, ceilings, and floors of rooms, corridors, stairwells, and similar building spaces. Some authorities include the linings or coverings of ducts, utility chases and shafts, or plenum spaces as interior finish as well as batt and blanket insulation, if the batt faces a stud space through which fire might spread.

Plastics

Aesthetic considerations and low cost make the use of plastic building materials desirable. All plastics are combustible, however, and no known treatment can make plastics noncombustible although some have a relatively low flame-spread rating.

Cellular plastics sprayed on walls for insulation have become popular. Fire retardants can be incorporated in many of these plastics so they can meet building code requirements. However, some plastics containing polyurethane or polystyrene have been involved in serious, rapidly spreading fires.

Wood

The physical size of wood and its moisture content are important factors that determine whether this material will provide reasonable structural integrity. Wood is the most common material used in the construction of dwellings. If a wood-frame house is subjected to a serious fire, either from burning combustibles inside the house or from an exposure fire, it will not withstand much heat and will have little structural integrity.

Heavy timber construction can resist fire very well. The timbers will char, and the resulting coating of charcoal provides an insulation for the unburned wood. Heavy timber maintains its integrity during a fire for a relatively long time, thus providing an opportunity for extinguishment. Much of the original strength of the members is retained and reconstruction is sometimes possible.

Because untreated wood-base wallboard and paneling are highly combustible, fire-retardant treatments are required by most codes. Without such fire-retardant treatments, combustible wallboard not only enables a fire to spread so fast that people may become trapped, but also contributes fuel to the fire and creates hazardous concentrations of smoke and toxic gases.

Steel

The most common building material for larger buildings is structural steel. While steel is noncombustible and contributes no fuel to a fire, it loses its strength when subjected to the high temperatures that are easily reached in a fire. The normal critical temperature of steel is 1,000 F (590 C). At this temperature, the yield stress of steel is about 60% of its value at room temperature. Buildings built of unprotected steel will collapse relatively quickly when exposed to a contents fire or an exposure fire. The lighter the steel members, the quicker will be the failure.

Another property of steel that influences its behavior in fires is expansion when the steel is heated. Walls can collapse from the movement caused by expansion of steel trusses.

Because unprotected structural steel loses its strength at high temperatures, it must be protected from exposure to the heat produced by building fires. This protection, often referred to as "fireproofing," insulates the steel from the heat. The more common methods of insulating steel are encasement of the member, application of a surface treatment, or installation of a suspended ceiling as part of a floor-ceiling assembly capable of providing fire resistance. In recent years, additional methods, such as sheet steel membrane shields around members and box columns filled with liquid, have been introduced.

In recent years, intumescent paints and coatings have been used to increase the fire endurance of structural steel. These coatings intumesce, or swell, when heated, thus forming an insulation around the steel.

Structural steel members can also be protected by sheet steel membrane shields. The sheet steel holds in place inexpensive insulation materials, thus providing a greater fire endurance. In addition, polished sheet steel has been used in recent tests to protect spandrel girders (the horizontal supports beneath the windows on many modern high-rise buildings). The shield reflects radiated heat and protects the load-carrying spandrel.

Concrete

The resistance of reinforced concrete to fire's attack will depend on the type of aggregate used to make the concrete, the moisture content, and the anticipated fire loading. Usually, reinforced concrete buildings resist fire very well; however, the heat of a fire will cause spalling (chipping and peeling away), some loss of strength of the concrete, and other deleterious effects.

Prestressed concrete is stronger than reinforced concrete and provides better fire resistance. However, prestressed concrete has a greater tendency to spall with the result that the prestressed steel may become exposed. The type of steel used for prestressing is more sensitive to elevated temperatures than the type of steel that is usually used in reinforced concrete construction. In addition, the steel used for this type of reinforced concrete construction does not regain its strength upon cooling.

Glass

Glass is a commonly used building material. Modern high-rise buildings, particularly, contain large amounts of glass. Glass is used in three primary ways in building construction: (1) for glazing, (2) for glass-fiber insulation, and (3) for glass-fiber-reinforced plastic building products.

Glass used for windows and doors has little resistance to fire. Wire-reinforced glass provides a slightly higher resistance to fire, but no glazing should be relied upon to remain intact in a fire.

Glass-fiber insulation is widely used in modern building construction. Glass fiber is popular because it is fire resistant and is an excellent insulator. However, glass fiber is often coated with a resin binder that is combustible and that can spread flames.

Glass-fiber-reinforced plastic building products such as translucent window panels are becoming more common. The glass fiber acts as reinforcement for a thermosetting resin. Usually this resin, which is combustible, comprises about 50% or more of the material. Thus, while the glass fiber itself is noncombustible, the product is highly combustible.

Gypsum

Gypsum, as reflected in products such as plaster and plasterboard, has excellent fire-resistive qualities. Gypsum is widely used because it has a high proportion of chemically combined water, which makes it an excellent, expensive, fire-resistive building material that is far superior to highly combustible fiberboard.

Masonry

Masonry (such as brick, tile, and sometimes concrete) provides good resistance to heat, and usually retains its integrity. Because of the prevalence of brick construction in European dwellings as compared with American wood-frame construction, the dwelling fire record in Europe is much more favorable than the dwelling fire record in America, according to the NFPA.

(For details about how to use masonry for fire-safe construction, see *The BOCA Basic Building Code* in References, at the end of this chapter.)

CONSTRUCTION METHODS FOR FIRE PROTECTION

Several construction methods help confine fires and control smoke. Provide for them during the design stage of building.

Confining Fire

Using stair enclosures, fire walls, and fire doors, and dividing a building into smaller units are ways to confine a fire. Plan them into the building's design.

Stair Enclosures and Fire Walls

Regardless of the type of building construction, stair enclosures are necessary to provide a safe exit path for occupants. They also retard the upward spread of fire. Under certain conditions, as where large areas or high values are involved, divide buildings horizontally with fire walls. Design fire walls to rigid specifications so that they can withstand the effects of a severe fire and of building collapse on one side. To prevent the passage of heat, protect all openings with approved closures at the same fire-exposure rating as the fire walls or greater.

Separate Units

Traditionally, dividing buildings into separate units provided functional work areas or offered occupants some degree of privacy. From the point of view of fire safety, however, dividing a building is regarded as a way to break up the total volume of a building into small cells. In such a building, fires will remain localized and can be more easily suppressed. To prevent fire from spreading from one unit to another, various building codes require (1) that units be made structurally sound enough to withstand full fire exposure without major damage, and (2) that the units' boundaries be capable of acting as nonconducting heat barriers.

Fire Doors

Fire doors are the most widely used and accepted means of protection of both vertical and horizontal openings. Fire doors are rated by testing laboratories as they are installed in a building and as assemblies (doorframe and door as a unit). The fire doors usually have a rating of ¾ to 3 hours. They may be constructed of metal or metal-clad treated wood materials and may be hinged, rolling (sliding), or curtain doors. Single or double doors may be specified.

When new construction is planned, select the proper types of fire doors. They will perform properly only in the uses for which they were designed. In addition, a fire door will perform properly only if it is installed with an approved frame, latching device, hardware, and closing device. The effectiveness of the entire assembly as a fire barrier may be destroyed if any component is omitted or if any component of lesser quality is substituted. To assure proper protection of openings, install fire doors in accord with NFPA 80, *Fire Doors and Windows*.

Fire doors are of value only if they will close or be closed at the time of a fire. Blocking or wedging fire doors open defeats their purpose. Prohibit this practice. Always check fire doors during plant or building fire prevention inspections. Be sure that door openings and the surrounding areas are clear of anything that might interfere with the free operation of fire doors. Frequently inspect the hinges, catches, latches, closers, and stay rolls of fire doors since they are especially subject to wear.

To assure proper operation, regularly test doors that have automatic closing devices. The following problems are frequently encountered in this critical test. Immediately repair any defect that would interfere with the proper operation of fire doors.

- Chains or wire ropes may have stretched.
- Hardware may be inoperative.
- Guides and bearings may need lubrication.
- Binders may be bent, thus obstructing the doorway.
- Stay rolls may have accumulations of paint.
- Fusible links may be painted.
- The hoods over rolling steel doors may be bent, thus interfering with the doors' operation.
- Where swinging doors are used in pairs, coordinators may need adjustment.

Controlling Smoke

Smoke control confines smoke, heat, and noxious gases to a limited area, dilutes them, or exhausts them, thus preventing their spread to other areas and minimizing fatalities. Smoke and hot gases generated by an uncontrolled fire, if confined within a building, can seriously impair fire-fighting operations, cause sickness and death, and can spread the fire under the roof for considerable distances from the point of origin.

The movement of smoke within a structure is determined by many factors, including building height, ceiling heights, suspended ceilings, venting, external wind force, and the direction of the wind.

Most smoke-control systems involve a combination of methods. One method of smoke control uses a physical barrier, such as a door, wall, or damper to block the smoke's movement. Smoke-control doors may be operated manually or by some automatic detection device coupled to a door closure. An alternative method is to use a pressure differential between the smoke-filled area and the protected area.

Venting is another way of removing smoke, heat, and noxious gases from a building. Plan for smoke control during the design stage of the building by specifying vents, curtain boards, and windows. For example, use smoke- and heat-venting systems consisting of curtain boards. They protect heat-banking areas under a roof. Also, use automatic or manual roof vents to release smoke and heat through the roof.

Vents are most applicable to larger areas in one-story buildings that lack sprinklers and that do not have separate units. They are also useful in windowless and underground buildings. They are not, however, a substitute for automatic sprinkler protection. Vents and draft curtains are effective for small special housings.

So many variables affect the burning of combustible material that no exact formula can be used to compute the amount of venting required. However, vent sizes and ratios have been developed from limited experiments in test buildings without sprinklers and theories about actual fire experiences. Also, a variety of prefabricated vents is available on the market. These vents can be designed to open automatically at predetermined temperatures or smoke concentrations.

Exits

Of the many factors involved in protecting life from fire, a building's exits are the most important. Nevertheless, exits are inadequate in many buildings. Consider the design of exits in a building's total fire safety system (Figure 11–7).

Management, architects, and others entrusted with the safety of employees must consider many problems when planning the emergency evacuation of buildings. In many cases, panic causes more loss of life than fire. While fire is the most common cause of panic, such dangers as boiler or air receiver explosions, fumes, or structural collapse may also threaten safe and orderly evacuation.

A building's population and degree of hazard are the major factors when designing exits. Every building or structure, and every section or area in it, shall have at least two separate means of exit. Arrange them so that the possibility of any one fire blocking all exits is minimized.

Designing exits involves more than a study of numbers, flow rate, and population densities. Safe exits require a safe path of escape from the fire with the least possible travel distance to the exit. The path should be arranged for ready use in case of an emergency. It should be large enough to permit all occupants to reach a place of safety before they are endangered by the fire or by smoke and toxic gases.

NFPA 101, *Life Safety Code*, provides a reasonable and comprehensive guide to exit requirements. If local, state or provincial, or federal codes contain more rigid recommendations, follow these codes instead.

Evacuation

Consider the following general provisions when planning for building evacuation:

- Do not design exits and other safeguards to depend solely on any single safeguard. Provide additional safeguards in case of human or mechanical failure.

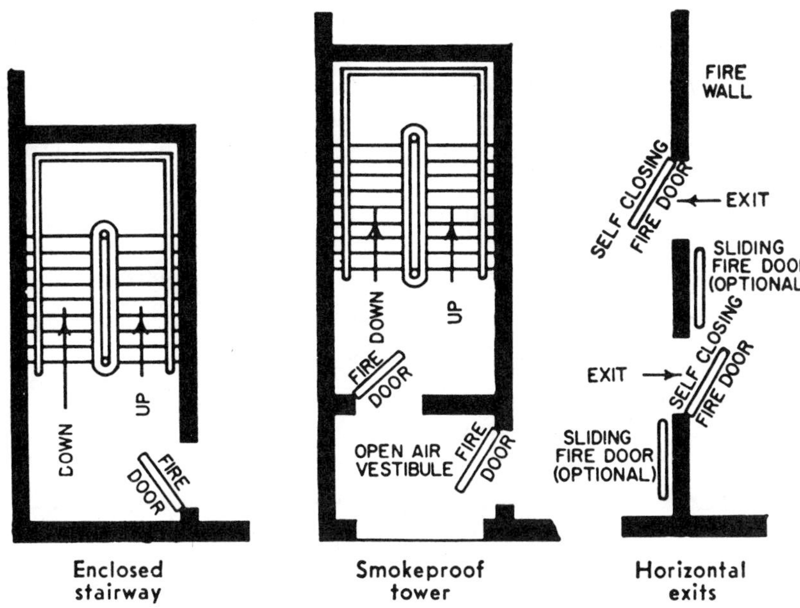

Figure 11–7. Plan views of types of exits. Stair enclosure prevents fire on any floor from trapping persons above. A smokeproof tower is better because an opening to the air at each floor largely prevents the chance of smoke on the stairway. A smokeproof tower charged with positive air pressure is more likely to prevent smoke from entering. A horizontal exit provides a quick refuge and decreases the need for a hasty flight down stairs. Horizontal sliding fire doors provided for safeguarding property values are arranged to close automatically in case of fire. Swinging doors are self-closing. Two wall openings are needed for exit in two directions. (Reprinted with permission from the National Fire Protection Association.)

- Exit doors must withstand fire and smoke during the length of time for which they are designed to be in use. Enclose or protect vertical exits and other vertical openings to afford reasonable safety to occupants while they use the exits.
- Provide alternate exits and pathways in case one exit is blocked by fire. Also, provide exits that the disabled can use.
- Provide alarm systems to alert occupants of fire or another emergency. Visual alarms are important for the hearing impaired.
- Provide exits and exit routes with adequate lighting.
- Mark exits with a readily visible sign. Mark access to exits with readily visible signs whenever an exit or exit route is not readily visible.
- To protect exiting personnel, safeguard equipment and areas of any unusual hazard that might spread fire and smoke.
- Practice an orderly exit drill procedure.
- Control psychological factors that can lead to panic.
- Select an interior finish and contents that prevent a fire from spreading fast and trapping occupants.
- Maintain adequate aisles in exit routes.
- Provide adequate space outside the building's exits. No more than 50% of occupants should be discharged to a street.
- Install exit doors a minimum of 28 in. (70 cm) wide to allow 60 persons per minute to exit. A stairway downward should allow 45 persons per minute per 22 in. (56 cm) of width to exit. Use these numbers to calculate evacuation time.

FACTORS CONTRIBUTING TO INDUSTRIAL FIRES

To eliminate the causes of fire, first determine the many ways in which fire can start. Electrical equipment, smoking, friction, open flames, and poor housekeeping are some common causes of fire.

Electrical Equipment

Install and maintain electrical equipment in accord with NFPA 70, *National Electrical Code* (NEC). Overheating of electrical equipment and arcs resulting from short circuits in improperly installed or maintained electrical equipment are two of the leading causes of fire in buildings.

Where flammable gases or vapors may be present, use only UL-listed or Factory Mutual-approved equipment. Electrical equipment that must be used in the presence of flammable liquids, gases, or dusts should be designed for the particular hazardous atmosphere.

Hazardous locations fall into three classes, which depend on the material involved. They are further divided according to the degree or severity of the hazard. Complete definitions of the classes and divisions of hazardous locations and the types of equipment designed for each are contained in the NEC, and are discussed in Chapter 10, Electrical Safety, in this volume.

Temporary or makeshift wiring, particularly if defective or overloaded, is a very common cause of electrical fires. Do not use this type of wiring. Overloaded or partially grounded wiring may also heat up enough to ignite combustibles without blowing fuses or tripping circuit breakers. Where flammable liquids are used, provide bonding and grounding with adequate and true grounds in accord with the NEC.

Portable electrical tools and extension cords should also conform to the NEC. Inspect them at frequent intervals and repair them promptly. Use waterproof cords and sockets in damp places, and use explosion-proof fixtures and lamps in the presence of highly flammable gases and vapors.

Ground or double-insulate all electrical equipment, especially portable electrical tools. Use switches, lamps, cords, fixtures, and other electrical equipment listed by a recognized testing and certifying agency. Use them only in applications for which the approval or listing was granted. (See Chapter 10, Electrical Safety.)

Protect lamp bulbs by using heavy lamp guards or adequately sealed transparent enclosures. Keep lamp bulbs away from sharp objects and secure them to prevent them from falling. Never use bare bulbs if they are exposed to flammable dusts or vapors. Always consider lamp bulbs as potential hazards in such areas. Safeguard them accordingly.

Instruct employees in the correct use of electrical equipment. Prohibit them from tampering with equipment, blocking circuit breakers, using wrong fuses, bypassing fuses, or installing equipment without authorization.

Periodically inspect and test electrical installations and all electrical equipment. This assures continued satisfactory performance and also detects deficiencies.

Smoking

Carelessly discarded cigarettes, pipe embers, and cigars are a major source of fire. Prohibit smoking, especially in woodworking shops, textile mills, flour mills, grain elevators, and places where flammable liquids or combustible products are manufactured, stored, or used.

Although it might be desirable to eliminate smoking completely in a plant, such a rule is difficult to enforce. Instead, allow smoking at specified times and in a safe place where supervision can be maintained.

Mark NO-SMOKING areas with conspicuous signs. Everyone, including supervisors and visitors, must adhere to no-smoking regulations. It may be necessary to use more than signs to draw attention to the no-smoking areas. Lines drawn on the floor, or illuminated barriers placed around areas or processes, are also effective.

In many high-hazard occupancies, smoking is permitted in special fire-safe rooms. In such cases, post instructions or warning signs to that effect. Where the exposure is severe, prohibit employees from even carrying matches, lighters, and smoking material of any kind into the danger areas.

In any case, encourage the use of safety matches, and allow smoking only in designated locations. If carrying matches into the plant is prohibited, provide special lighters in smoking rooms.

Even in plants where there is little fire hazard, provide safe butt cans and ashtrays and encourage employee use of them for matches and smoking materials. Discourage employees from throwing matches and cigarettes into places that may not be free of hazards.

Friction

Excessive heat generated by friction causes a very high percentage of industrial fires. A program of preventive maintenance on plant machinery can avert fires resulting from inadequate lubrication, misaligned bearings, or broken or bent equipment—all sources of friction.

Fires frequently result from overheated power-transmission bearings and shafting in buildings such as grain elevators, cereal, textile, and woodworking mills, and plastics and metalworking plants, where dust and lint accumulate. Make frequent inspections to see that bearings are kept well oiled and do not run hot. Keep the accumulation of flammable dust or lint on them to a minimum.

Provide drip pans beneath bearings, and clean them frequently to prevent oil from dripping to the floor or on combustible material below. Keep oil holes of bearings covered to prevent dust and gritty substances from entering the bearings, thus causing overheating.

Frictional heat sufficient to cause ignition can result from the jamming of work materials during production. Another common problem, frequently overlooked, is the tension adjustment on belt-driven machinery. If the belt is either too tight or too loose, excessive friction can cause serious overheating.

Foreign Objects or Tramp Metal

Take every precaution to keep foreign objects from entering machines or processes. They might strike sparks where there are flammable dusts, gases, or vapors, or combustible material, such as cotton lint or metal powder. For this purpose, use screens or magnetic separators such as are used in textile mills, grain elevators, and other operations where explosive mixtures of dusts are present.

Open Flames

Although open flames are probably the most obvious source of ignition for ordinary combustibles, and one would think they could be the most easily avoided, they still account for a large percentage of industrial fires. Heating equipment, torches, and welding and cutting operations are principal offenders.

Air Heaters (Gas- and Oil-Fired)

Air heaters are commonly used on construction work and often cause fires for the following reasons:

- Overheating of the air heater with resultant radiation that ignites nearby combustible materials, such as concrete form work, tarpaulins, wood structures, paper, straw, and rubbish.
- Failure to follow the One-Yard Rule: Keep anything that will burn a minimum of three feet away from a portable heater.
- Failure to insulate air heaters from floors or other combustible bases.
- Failure to provide a substantial spark shield and to use fuels which do not produce high flames or sparks. (See NFPA 211, *Chimneys, Fireplaces, Vents, and Solid Fuel Burning Appliances.*)
- Failure to provide a secure base or to anchor it properly. Vent air heaters in unventilated rooms, or enclosures to the outside, by means of an overhead hood and flue. Venting removes the toxic products of combustion and any unburned gas.

Torches

If gasoline, kerosene, liquefied petroleum gas (LP-gas), acetylene, or alcohol torches are used, place them so the flames are at least 18 in. (46 cm) from wood surfaces. Do not use them around flammable liquids, paper, excelsior, or similar material.

Portable Furnaces and Blow Torches

Provide overhead clearance of at least 4 ft (1.2 m) for portable furnaces and blow torches. Remove or protect combustible material overhead with noncombustible insulating board or sheet metal, preferably with a natural-draft hood and flue of noncombustible material.

Welding and Cutting

When possible, have welding or cutting done in special fire-safe areas or rooms with concrete or metal-plate floors. Flame impingement on concrete may cause concrete to spall. Consequently, keep work off the floor or protect the floor with a metal shield. (See NFPA 51B, *Cutting and Welding Processes.*)

In cases where welding and cutting operations are performed outside the special fire-safe areas, use hot-work permit programs to promote maximum fire-safe working conditions.

If welding must be done over wood floors, have them swept clean and wetted down, preferably covered with flame-resistant blankets, metal, or other noncombustible covering. Keep hot metal and slag from falling through floor openings, thus igniting combustible material below. Use sheet metal, flame-retardant tarpaulins, or flame-resistant curtains around welding operations to prevent sparks from reaching nearby combustible materials.

Do not permit welding or cutting in or near rooms containing any flammable liquid, vapor, or dust unless special precautions are observed and the hazards are well understood. Do not perform welding or cutting on a surface until combustible deposits have been removed. Also do not perform welding or cutting in or near closed tanks or other containers that have held flammable liquids. These containers must first be thoroughly cleansed and filled with water or purged with an inert gas. A combustible-gas indicator test must show that no trace of a flammable gas or vapor is present. Periodically repeat these tests to determine if any trace of flammable gas or vapor is released during the welding or cutting operation. If further tests show any trace, stop the work until all flammable gas or vapor is dispelled. (See NFPA 327, *Cleaning or Safeguarding Small Tanks and Containers.*)

Welding is sometimes necessary on equipment containing flammable material. However, such welding requires highly specialized procedures and should be avoided if possible. Have fire-extinguishing equipment for this type of exposure within easy reach of welding and cutting operators. If the hazard is justified, and welding must be done outside of a shop or area designated for welding, station fire watchers who will prevent sparks or molten slag from starting fires and will extinguish fires should they start. Fire watchers should remain at the work location for at least 30 minutes after the welding or cutting is completed, because some fires escape detection when they are starting. Install shields around spot welders to prevent sparks from reaching nearby combustible materials or from injuring employees.

Spontaneous Ignition

Spontaneous ignition results from a chemical reaction in which there is a slow generation of heat from oxidation of organic compounds that, under certain conditions, is accelerated until the ignition temperature of the fuel is reached. This condition is reached only where there is enough air for oxidation but not enough ventilation to carry away the heat as fast as it is generated.

Spontaneous ignition usually occurs only around quantities of bulk material packed loosely enough for a large amount of surface to be exposed to oxidation yet without adequate air circulation to dissipate heat. Exposure to high temperatures increases the tendency toward spontaneous ignition.

The presence of moisture also can advance spontaneous ignition, unless the material is wet beyond a certain point. Materials like unslaked lime promote spontaneous ignition, especially when wet. Store such chemicals in a cool, dry place away from combustible material.

At ordinary temperatures, some combustible substances oxidize slowly, and under certain conditions, they can reach their ignition point. These substances include vegetable and animal oils and fats, coal, charcoal, and some finely divided metals. Rags or wastes saturated with linseed oil or paint often cause fires too.

The best preventives against spontaneous ignition are either total exclusion of air or good ventilation. With small quantities of material, the former method is practical. With large quantities of material, such as storage piles of bituminous coal, both methods have been used with success.

Temperatures of 140 F (60 C) are considered dangerous in coal piles. If temperatures rapidly approach or exceed that figure, move the pile or rearrange it to allow better circulation of air and to avoid spontaneous ignition.

Certain agricultural products are also susceptible to spontaneous ignition. Sawdust, hay, grain, and other plant products, such as jute, hemp, and sisal fibers, may ignite spontaneously, especially if exposed to external heat or to alternate wetting and drying. Here again, the best preventives are circulation of air, removal of external sources of heat, and storage of material in smaller quantities.

Fires in iron, nickel, aluminum, magnesium, and other finely divided metals are sometimes attributed to spontaneous ignition. These fires are thought to result from the oxidation of cutting or lubricating oils, or possibly from chemical impurities.

Iron sulfide, commonly referred to as iron pyrite or pyrophoric iron, occurs as a result of lubrication of sulfur compounds in petroleum that come in contact with iron in pipes and vessels. As long as air is excluded or it is kept moist, iron sulfide does not constitute a hazard. However, when equipment containing iron sulfide is opened to the atmosphere and dried out, it will burn. Where iron sulfide may be present, make provisions to keep the inner surface wet when opening equipment. Carefully and promptly handle the disposal of accumulated deposits of iron sulfide so it will not start burning in an area where fire would create a hazard.

Housekeeping

Poor housekeeping is another factor that contributes to industrial fires. Properly collecting and storing combustibles and disposing of rubbish, as well as maintaining locker rooms, will prevent fire hazards.

Collection and Storage of Combustibles

Many industrial fires are the direct result of accumulations of oil-soaked and paint-saturated clothing, rags, waste, excelsior, and combustible refuse. Deposit such material in noncombustible receptacles with self-closing covers that are provided for this purpose and removed daily from the work areas.

Exhaust systems of effective design will remove gases, vapors, dusts, and other airborne contaminants, many of which may be fire hazards. Exhaust systems and machinery enclosures will help prevent accumulation of combustible materials on floors or machine parts. Such materials are most hazardous when airborne rather than when they have settled out.

Clean waste, although not as dangerous as oil-soaked waste, is readily combustible and should be kept in metal cans or bins with self-closing covers. Store excelsior, cotton, kapok, jute, and other highly combustible fibrous materials in covered, noncombustible containers. If large quantities of these materials are kept on hand, store them in fire-resistant rooms equipped with fire doors and automatic sprinklers. Have portable extinguishers, hose lines, or other extinguishing equipment for Class A fires available for use at such storage places.

A schedule for safe collection of all combustible waste and rubbish should be a part of the fire prevention program. Be sure that janitorial personnel or others involved in collection of waste paper in offices and service areas have fire-safe collection containers. Check collection practices to be sure that ash trays, which may contain smoldering material, are not emptied into combustible bags or cartons or into containers of combustibles.

At regular intervals, clean accumulations of all types of dust from overhead pipes, beams, and machines, especially from bearings and other heated surfaces. All organic, as well as many inorganic, materials, if ground finely enough, will burn and propagate flames. Therefore, keep roofs free from sawdust, shavings, and other combustible refuse. Cleaning of such materials should be done by vacuum removal because air blowing may create dangerous clouds of dust.

Do not store such material or allow it to accumulate in air, elevator, or stair shafts; in tunnels; in out-of-the-way corners; near electric motors or machinery; against steam pipes; or within 10 ft (3 m) of any stove, furnace, or boiler.

Rubbish Disposal

To prevent fires, federal, state or provincial, and local laws forbid certain methods of waste disposal, such as open burning, evaporation, and flushing to sewers. Fires are often caused by burning rubbish in yards near combustible buildings, sheds, lumber piles, fences, grass, or other combustible materials. If rubbish must be burned, the best and safest way is with a well-designed incinerator that meets the requirements of the environmental pollution control laws.

If flushing or dumping waste materials into sewers is prohibited, use trap or retention tanks that can be pumped out. Thus, the waste material can be disposed of properly. Also, consider chemically altering a waste material before disposing of it.

Locker Rooms

Locker rooms in which oil-soaked clothing, waste, or newspapers are kept are always a serious fire hazard. Take every precaution to prevent such combustible materials from accumulating.

Provide lockers made of metal with solid, fire-resistant sides and backs. Doors, however, should have some open spaces for ventilation. Lockers should be large enough so that air can circulate freely around clothing hung in them. Do not permit employees to leave clothing saturated with oils or paints in lockers.

Where automatic sprinklers are used, cover locker tops with screening. Also, use locker tops made of perforated metal so that water can reach burning contents if a locker fire occurs. Heavy paper pasted over the tops will keep out dust. Lockers that have sloping tops and that stand flat on the floor will prevent the accumulation of rubbish both above and below.

Explosive Atmospheres

Dusts, gases, and vapors may create explosive atmospheres. Observe the proper precautions to prevent these atmospheres from developing.

Table 11–C. Some of the More Common Potentially Explosive Dusts

Type	Example
Carbon	Coal, peat, charcoal coke, lampblack
Fertilizers	Bone meal, fish meal, blood, flour
Food products and by-products	Starches, sugars, flour, cocoa, powdered milk, grain dust
Metal powders	Aluminum, magnesium, zinc, iron
Resins, waxes, and soaps	Shellac, rosin, gum sodium resinate, soap powder, waxes
Spices, drugs, and insecticides	Cinnamon, pepper, gentian, pyrethrum, tea fluff
Wood, paper, tanning materials	Wood flour, wood dust, cellulose, cork, bark dust, wood extract
Miscellaneous	Hard rubber, sulfur, tobacco, many plastics

Source: U. S. Department of Agriculture, Technical Bulletin 490.

Dusts

A dust explosion hazard exists wherever material that will readily burn or oxidize is available in powder form. The surface, or contact, area of each dust particle is large in relation to its mass. The U.S. Department of Agriculture Technical Bulletin 490 lists 133 dusts according to their degree of explosiveness (Table 11–C). In addition, many synthetic resins and powders used in the plastics industry present a dust-explosion hazard comparable to that of coal. This group includes phenolic, urea, vinyl, and other types of resins and a number of molding compounds, primary ingredients, and fillers.

There are two ways to prevent dust explosions: (1) prevent the formation of explosive mixtures of dust and air and (2) prevent the ignition of such mixtures if their formation cannot be prevented.

Effectively prevent an explosive mixture of dusts from forming by providing an inert atmosphere. Limit oxygen to concentrations below which flame propagation does not occur. To make this method reliable, monitor the concentration of oxygen with an oxygen analyzer. This is an effective method if enclosure of the system is possible. However, do not provide inert atmospheres if operators have to enter the atmosphere to perform work.

Take extreme precautions to prevent dust from building up to explosive proportions. Extensive use of local exhaust and regular cleaning will minimize this hazard. Where possible, segregate dusty operations. Also, have dust-producing equipment totally enclosed and exhausted to prevent dust from leaking into the general work area.

Prevent ignition of an explosive mixture of dust and air by eliminating open flames, friction sparks, static electricity, welding, and excessive heat; by increasing humidity; or by using inert gas. Take every precaution to prevent overheated bearings, smoking, friction sparks, and sparks in hand tools and in grinding or welding operations. Also control sources of static electricity. (See Chapter 12, Flammable and Combustible Liquids, in this volume.)

To prevent sparks, use nonferrous material for truck wheels and bucket conveyors. Also, use pneumatic or magnetic separators to remove stones, nails, and other spark-producing foreign objects from the material being processed. Keep conveyor-belt rollers and similar moving parts properly lubricated and maintained to avoid excessive friction and heat.

As discussed earlier, use only approved dust-tight wiring, fixtures, and motors. Also, instruct employees to use only extension cords, lamps, and portable electric tools designed for protection against the hazard of dust explosions. (See NFPA, *National Fire Code*, latest edition.)

In buildings with high explosion hazards, such as grain elevators and plastics plants, provide extensive dust-collection equipment. Construct these buildings so that explosive pressures will push out hinged windows or blow out wall sections or panels designed and built to fail in a predetermined area. An explosion in such a building will not cause structural collapse. (See NFPA 8, *Deflagration Venting.*) Keep floor openings to a minimum, and seal openings for pipes and duct-work.

Have portable fire-extinguishing equipment readily available. Fog nozzles or finely divided streams of water are more effective for dust fires than are solid streams of water, which stir up dust.

Gases and Vapor

Gases and vapors that produce flammable mixtures with air or oxygen are common in industry. Such gases include hydrogen, acetylene, propane, LP-gas, CO, methane, natural gas, and manufactured gas.

Highly volatile liquids that emit flammable vapors include gasoline, benzene, naphtha, and methyl alcohol. Kerosene, turpentine, Stoddard solvent, and other liquids with flash points above 100 F (38 C) must generally be heated above normal room temperatures before they give off sufficient vapors to form ignitable concentrations.

Whenever using flammable liquids, especially when moving them from one container to another, bond them from container to container. In addition, do not transfer flammable liquids from a metal container to a plastic container, because the plastic cannot be bonded and grounded effectively. Many fires have resulted from pouring gasoline from a metal can into a plastic gasoline tank. The static electricity does not know the difference between metal and plastic.

When flammable liquids, including those with flash points above 100 F (38 C), must be handled and used, allow only minimum amounts, in safety containers, in work areas. The flash point of a liquid is the lowest temperature at which it gives off sufficient vapor to form an ignitable mixture with air and produce a flame when an ignition source is brought near the surface of the liquid. Flash-point temperatures are not always exact physical constants under work-environment conditions, because published figures are based on carefully controlled testing conditions. For instance, the amount of vapor that will accumulate in the air above a volatile liquid depends upon its relative saturation in the air above the liquid's surface. The amount of vapor given off is also directly related to the surface area of the volatile liquid. Thus, even a high-flash liquid may be dangerous if considerable surface area is exposed, as in a mist or froth. (This might be thought of as a "liquid dust.") Conversely, limiting the surface area of a liquid reduces its flash or explosion hazard.

Also consider the auto-ignition temperature of a gas or vapor in air. For example, a mixture of carbon disulfide vapor and air will ignite spontaneously at temperatures around 194 F (90 C), which is below that of an uninsulated, high-pressure steam line. However, the majority of gases and vapors will not self-ignite in air until they reach temperatures of about 500 to 900 F (260 to 482 C).

FIRE DETECTION

A look at statistical data shows that, despite good construction, cleanliness, and modern fire-fighting methods, a considerable number of losses from fire nevertheless occur. Losses would be reduced if each developing fire were detected so it could be attacked and extinguished. Thus, fire-detection devices must be a part of every fire protection system. Means of detection could be a human observer; automatic sprinklers; smoke, flame, or heat detectors; or, more likely, a combination of these. There really are no *fire* detectors, but rather heat detectors, smoke detectors, and flame detectors.

The detecting part of a fire protection system has two main tasks: (1) giving an early warning to enable building occupants to escape and (2) starting extinguishing procedures. The latter may be as simple as alerting the fire brigade or fire department, or triggering the operation of an automatic sprinkler or another suppression system.

Each automatic fire-detection system requires a sensor, which observes a physically measurable quantity of smoke, flame, or heat. For detection to occur, this quantity must undergo measurable variations when a fire begins in the vicinity of the sensor. A decision-making device coupled with the sensor then compares the measured quantity with a predetermined value. When the value is different, an alarm sounds. Thus a detector both detects and signals.

Human Observer

A human observer is a good fire-detection system for the following reason: He or she can take immediate action in a flexible way, whether calling the fire department or putting out a fire with an extinguisher. Be sure that employees report any fire that they have put out.

Early detection and notification of the responsible municipal fire department are essential. Early notification of occupants allows time to escape, and early notification of the fire department provides a better chance of stopping the fire before extensive damage occurs.

Human observers should also report malfunctioning fire alarm systems. Fire alarm systems need to be properly maintained. False alarms have a negative effect: if an actual fire occurs, the sounding alarm may be dismissed as just another false alarm.

Automatic Fire-Detection Systems

In general, there are three possible errors in any non-human fire-detection system: (1) giving a false alarm, (2) not detecting a fire, or (3) detecting it too late. The cause of false alarms may be human interference, mechanical or electrical faults, or special environmental effects.

When planning an automatic fire-detection system, use the following checklist of questions:

- What is the main purpose of the system? Typical purposes include warning the occupants of a building, protecting irreplaceable goods and highly valuable goods, protecting against interruption of production, and protecting against corrosion or radioactive contamination.
- What are the possible sources of ignition?
- What kinds of material will probably be ignited first?
- What kind of building construction is used?
- What are the environmental conditions?
- What kind of detection system has been installed and what are the reasons for choosing this system?
- How long can a fire be allowed to go undetected?

Properly engineered fire-detection systems are sound investments. Obtain the service of a fire protection consultant to engineer the system and to establish the control procedures. When planning the installation of fire detectors, use the following four steps:

1. Select the proper detector for each hazard area. For example, a computer area may require ionization or combination detectors. A warehouse may require infrared and ionization detectors. In low-risk areas, thermal detectors or a combination of detectors may be used.
2. Determine the spacing and locations of detectors in order to provide the earliest possible warning.
3. Select the best control-system arrangement to provide fast identification of the exact source of the alarm.
4. Assure that the system will notify the proper authorities, who can immediately respond to the alarm and take appropriate action. Every detection system must have its alarm signal transmitted to a point of human supervision.

From a systems analysis approach, a suitable detector and system must be fitted for a given hazard. Sensitivity, reliability, maintainability, and stability are the critical variables in the selection of a detection system.

- Sensitivity is established by the design. How well can the device detect what it is meant to detect?
- Reliability is the ability to perform an intended function when needed. In other words, the device should not malfunction.
- Maintainability means the unit is easy to service and keep at operating efficiency. A minimum of maintenance should be needed.
- Stability means the unit can sustain its sensitivity over time.

There are many types of fire detectors to handle various situations and to detect various states of the beginning of a fire. Most manufacturers and distributors offer several or all of the commonly used types. They can also engineer a combination of equipment into a coordinated system to meet the special needs of a plant.

Thermal Detectors

Thermal detectors detect the heat from a fire. There are several kinds of thermal detectors, each with a specific use. Thermal detectors are reliable for what they do. However, they can only detect the heat of a fire, which usually will not build up to significant levels until the last stage of a fire. Many fires start slowly, with little heat generated at the beginning, and will be well under way by the time a thermal detector comes into operation. They are generally used where no life hazard is involved and some fire loss can be tolerated.

Fixed-Temperature Detectors

These thermal detectors are based on a bimetallic element. They are made of two metals that have different coefficients of expansion. When heated, the element will bend to close a circuit, initiating the alarm. A thermal detector may also use a fusible, spring-loaded element, such as found in a sprinkler head, that melts at a certain temperature, releasing an arm to close a circuit. Fixed-temperature detectors are simple, inexpensive, and require a low-voltage draw.

Rate-Compensated Thermal Detectors

These detectors work by the expansion characteristics of a hollow tubular shell containing two curved expansion struts under compression, fitted with a pair of normally open, opposed contacts. When subjected to a rapid rise of heat, the shell expands and lengthens at a faster rate than the struts, thus permitting the contacts to close. When heated slowly, both the shell and the struts lengthen at about the same rate until the struts are fully extended, thus making contact at a preset temperature point.

Rate-of-Rise Thermal Detectors

These detectors use an enclosed, vented hemispherical chamber containing air at atmospheric pressure, with a small pressure-sensitive diaphragm on top. With a normal rise in temperature, the excess pressure is relieved through small vents. However, a rapid rise in temperature will deflect the diaphragm faster than the vents can operate, thus triggering an alarm. This unit responds quickly to a rapid rise in temperature.

Line Thermal Detectors

These detectors use a length of small-diameter tubing that can be as long as 1,000 ft (305 m). When exposed to the heat of a fire, air inside the tube expands, sending a pressure wave to expand a diaphragm at the end, which in turn triggers an alarm. This is an unobtrusive, inexpensive detector. For example, it can be run along the ceiling molding, where it is nearly invisible. No maintenance is needed. It can be painted over, and it will even work with the tubing broken, if the temperature rises fast enough.

Eutectic-Salt-Line Thermal Detectors

These detectors consist of pliant metal tubing containing a eutectic salt in which a wire is embedded. At a preselected temperature, the salt creates a short-circuit between the internal wire and the outside tubing, thereby triggering an alarm. This detector, along with a continuous-resistance unit, has been widely applied recently to guard against fires in jet aircraft engine nacelles. The pliant tubing can be wound around and shaped to the various components of the engines to signal any increase in temperature that might be the result of a fire caused by leaking oil, hydraulic fluid, etc.

These line detectors are also used in conveyor-belt systems where the bearings supporting a rubberized belt may ignite because of friction and lack of lubrication. If the conveyor is transporting coal, for instance, the resultant fire could be expensive and hard to extinguish.

Bulb Detection Systems

These detectors are completely mechanical. They are especially desirable in locations where the explosive nature of the fire hazard makes it wise or essential to avoid the use of electricity. These systems involve a number of bulbs containing air at atmospheric pressure. One or more of these bulbs are installed along the ceiling of the hazardous area, all connected back to a diaphragm at the control center. When a rise in temperature strikes one or more of the bulbs, it deflects the diaphragm, and a mechanical extinguishing system can be activated. This system is used as a release mechanism for CO_2 fire extinguishers in marine and industrial applications.

Smoke Detectors

Smoke detectors respond to the particles of combustion, both visible (smoke) and invisible. A conventional smoke detector operates on a light principle: It can be triggered by either a decrease or an increase in light. When smoke enters a light beam, it either absorbs light so the receiver end of the circuit registers less light, or it scatters light so that a terminal, normally bypassed by the light beam, now receives part of the light.

The more sophisticated smoke detectors are responsive both to gas or products of combustion and to smoke. These products-of-combustion detectors are capable of detecting the beginning of a fire long before there is visible smoke or flame.

Photoelectric smoke detectors are line powered and usually include lamp-supervision circuitry and an alarm in case of lamp failure. An incandescent light source may be used. Also, a high-intensity strobe lamp that generates a stronger reflection can be used so that fewer or smaller smoke particles will actuate the photocell.

Beam Photoelectric Detectors

These detectors are triggered with less light. A long beam is directed at a photocell. Rising smoke tends to obscure the beam, decreasing light transmission and sounding an alarm. These detectors are an inexpensive way to cover large spaces, such as warehouses. Beam photoelectric detectors are, however, sensitive to voltage variations, to dirt on the lamp or lens, and also to flying insects or spiders, which sometimes congregate near the lamp, seeking warmth.

Reflected-Beam Photoelectric Detectors

These detectors use a beam of light in a chamber, with the photocell normally in darkness. Should visible smoke particles enter the chamber, they scatter the light and reflect it onto the cell, causing a change in electric conductivity that results in an alarm.

Products-of-Combustion (Ionization) Detectors

These detectors sense both visible and invisible products of combustion suspended in the air. They consist of a chamber with positive and negative plates and a minute amount of radioactive material that ionizes air in the chamber. The potential between the two plates causes ions to move across the chamber, setting up a small current. When aerosols from incipient fires enter the chamber, they cling to masses of moving ions. This slows the ions' movement and increases the voltage necessary for the ions to make contact. This voltage imbalance, amplified by electrical circuitry, triggers an alarm.

Ionization detectors sense fire at the earliest practical detection stage. They are the best method for detecting slow, incipient fires in commercial buildings—a cigarette in a wastepaper basket, for example, which might be in a presmoldering condition for 30 minutes or more. The ionization detector has the additional feature of operating in the fail-safe mode. If excessive dust is present, however, the device will give a false alarm.

Single-Chamber Ionization Detectors

These detectors are most economical. The chamber of these detectors is open to the atmosphere. Current flows between two poles and gets an increased voltage from combustion aerosols. This closes the contact and sends an alarm through the relay.

Dual-Chamber Ionization Detectors

These detectors have two identical sources of radiation: one is a sealed chamber, the other is open to the atmosphere. The inner ionization chamber monitors the surrounding conditions and compensates for the effect on the ionization rate of barometric pressure, temperature, and relative humidity. This construction accepts a wider range of atmospheric variations without giving false alarms.

Low-Voltage Ionization Detectors

These detectors are relatively new. While the conventional type needs 120 v, a low-voltage detector needs only 24 v. Theoretically, an installation would cost less. Low-energy, nonarmored cable is less expensive and easier to install, with essentially no danger of short-circuits or electrical shock. Most large cities, however, require armored conduit for low-voltage detectors so the savings may be less than expected. Nevertheless, most systems specified today are low-voltage, as the low-profile detector heads are less obtrusive, while being equally sensitive and reliable.

Flame Detectors

Flame detectors respond to the optical radiant energy of combustibles. Flame detectors sense light from the flames. Sometimes they work at the ultraviolet end of the visible spectrum, but more often they work at the infrared end. To avoid false alarms from surrounding light sources, flame detectors are set to detect the typical flicker of a flame, perhaps at 5 to 30 Hz. Or there may be a few seconds delay before the alarm in order to eliminate false alarms from transient flickering light sources, such as flashlights or headlights.

Flame detectors have some very important applications, such as in large aircraft hangars and for guarding against fires in fuel and lubricant drips. In general, however, by the time flame is visible, a fire has a good foothold. Either an infrared or an ultraviolet detector can be used to sense the flame.

Infrared Detectors

These flame detectors sense a portion of the radiant infrared energy of flames. They are often used in operations requiring an extremely fast response—for example, where flammable liquids are stored or used.

Ultraviolet Detectors

These flame detectors react only to actual flame. They do not respond to glowing embers or incandescent radiation. Also, these units are insensitive to heat, infrared radiation, and ordinary illumination.

Combustion-Gas Detectors

These detectors are closest to being general-purpose detectors. Combustion-gas detectors do not rely on heat. In effect, they "smell" a potential fire by measuring the percent of gas present. Also, they do not wait for the dangerous condition of flames to occur before they sound an alarm. Most fires detected by combustion-gas detectors can be extinguished by workers on the site.

Combustion-gas detectors can usually be set to automatically sound an alarm or to set off extinguishing equipment. Some conditions may require periodic maintenance and calibration. In work areas, spacing can vary up to 40 ft (12 m) in semienclosed application.

There are areas, however, where combustion-gas detectors cannot be used. For example, in an area where a certain level of combustion gases may be tolerated at times because a particular process emits them, flash fires from nearby flammable liquids may be anticipated. Therefore, detection must be almost instantaneous if it is to be at all effective. In such an area, use a flame detector. In another location where chemicals and some plastics could generate great volumes of smoke with little combustion, use a smoke detector.

Extinguishing-System Attachments

Some automatic detection devices are often not even thought to be detectors. These devices are extinguishing-system attachments. They are, nonetheless, truly detectors. Their fire detection may be handled by water-flow indicators in a sprinkler system. These indicators may operate in response to a sudden increase or decrease in pressure or by detection of the flow of water by a vane inside the piping. They are designed to detect a flow of water that exceeds so many gallons per minute.

A second type of device indicates that the fire extinguishing system is jeopardized. The following are a few of these devices:

- Water level devices warn when a self-contained water storage supply is low, or, in the case of a low differential dry pipe valve, when the level of water is too high.
- Water temperature switches warn when the water storage supply is approaching the freezing point.
- Water-supply valve position switches signal if someone unintentionally or purposely starts to turn off the water supply. These switches should give the alarm before the handle is closed two revolutions or one-fifth the distance from the open position. However, in practice, they are often set to signal an alarm within one-half a revolution.
- Closed-circuit television is often used for a security system.

Sensor Systems

In smokestacks, storage tanks, or other areas where several variables are at work at the same time, use a linear sensor system, alone or combined with other systems, for greater accuracy in detecting fires. The common heat sensors simply provide averages. Linear sensor systems, however, give continuous point-by-point readings. They pick up and monitor trouble spots instantly over an entire area.

The concept of totally integrated fire detection brings together the entire plant's fire-fighting capacity. Thus, time and, in turn, lives and property are saved. This concept also demands the continuing close cooperation of all persons involved in building design and construction, as well as in safety engineering and fire protection.

A modern fire-detection system should not be a thermal system or a combustion-detection system or a flame- or smoke-detecting system. It should be a combination of the various types of detecting systems in one integrated system (Table 11–D).

ALARM SYSTEMS

Alarm systems can be divided into four groups: local, auxiliary, central station, and proprietary. All types of alarm systems should be equipped with a signal system that clearly communicates to all persons in the building, plant, or laboratory. Whenever an alarm is sounded in any portion of the building or area, all employees must know what the sound means. Every fire alarm system, whether it is currently in place, newly installed, or revised, should meet the following criteria:

- When alarms are audible, the alarm sound should be clearly and immediately distinguishable from other signals that might be used in a given building. Provision for alerting hearing-impaired workers is necessary.
- Strategically locate audible alarm devices so they are clearly audible to all personnel. Train personnel to recognize the signal and to respond according to that location's specific disaster control process.
- The fire alarm system should be composed of equipment that conforms to NFPA standards and is listed by UL.
- Maintain the alarm system in good working order. Test it at frequent intervals to ensure that it is working properly. The interval between tests should not exceed one month.
- All personnel should know the location of and means of contacting external fire protection sources. Also, conspicuously post this information in strategic areas. All employees should also know the proper procedures for turning in a fire alarm if they are the ones who detected the fire.

Table 11–D. Relative Sensitivity of Fire Detectors When Means of Detection Is Matched to the Class of Fire

	Class of Fire		
Type of Detector	*A*	*B*	*C*
Fixed temperature	L	H	L
Rate of rise	M	H	L
Rate compensator	M	H	L
Particulate matter	H	H	M
Visible smoke	H	L	M
Ultraviolet	L	H	H
Infrared*	L	H	L

George J. Grabowski, "Fire Detection and Activation Device for Halon Extinguishing System," National Academy of Science, 1972.

Local Alarm Systems

A local alarm consists simply of bells, horns, lights, sirens, or other warning devices right in the building. Local alarms are generally used for life protection—that is, to evacuate everyone and thereby limit injury or loss of life from the fire. A local alarm can be tied in with another system to summon the fire department.

In a large building, code local alarm signals in a recognizable pattern of sound and silence to tell where the fire is. Otherwise, the alarm will simply indicate a fire without showing where it is located, thus giving the fire a chance to grow.

A presignal alarm system is especially advantageous in preventing panic. A presignal system rings specific alarm stations before a general alarm is given. In a multibuilding operation, the manager, plant superintendent, and security chief might be alerted ahead of the general alarm. Alarms may be given for floors or building units: If a fire starts in one building, the adjacent buildings might also be alerted and evacuated, but the more remote buildings would remain undisturbed.

Local alarm systems are inexpensive, available from a wide range of suppliers, and easy to install. By themselves, however, they do not provide much protection. While they alert personnel, they do not summon the fire department.

Auxiliary Alarm Systems

Auxiliary alarm systems are even less expensive than local alarm systems. Such a system simply ties a fire detector to a nearby fire call box. In effect, it becomes a relay station triggered by fire detectors inside the building.

Central Station Systems

Central station systems are available in most major cities around the country. Operated by trained personnel, a central station continually monitors a number of establishments and, in case of an alarm, calls a nearby fire station and alerts the building's personnel. In some cases, central station personnel also go out and check on an alarm.

Central station devices are virtually always leased. The central station company installs fire detectors throughout the building, then ties them to an alarm board back at the central station, usually through leased phone lines. In rare cases, the central station may be connected directly to the fire department. Generally, however, when the attendant at the central station gets an alarm signal from a subscriber, he or she telephones the fire department.

Central station personnel are competent and expensive. Costs for leasing the equipment alone are usually set high enough to amortize the central station's entire investment in three years or less. Payment for professional operation of the station is high enough to cover the actual cost and return a profit besides.

One argument used to promote central station alarm systems is that some of the cost will be offset by a reduced insurance premium. However, a company can get as much, or more, of a reduction with its own security force.

Proprietary Alarm Systems

Proprietary alarm systems feed alarms to the building's own fire watcher or maintenance force, and, optionally, to the fire department as well. One reason for their acceptance is that insurance regulations generally require security officers.

While a proprietary system can result in substantial reductions in insurance premiums, it can mean more dramatic savings by cutting the requirements for security officers. Unless security officers are supplemented by electronic devices, insurance companies usually demand that security officers tour the building every hour. Figuring two shifts, weekends, vacations, etc., at least 13 security officers would be required. Installing an approved proprietary alarm system can cut this requirement to one patrol every four hours, freeing at least three employees on each tour of duty (for a total of nine) for more important duties elsewhere.

A proprietary alarm system's console can include much more than fire alarms. Since it will be monitored constantly, such a command console can include intrusion detection, security protection, equipment surveillance, and dozens of other monitoring functions usually ignored or haphazardly scattered throughout a building (Figure 11–8).

Some engineers recommend an electronic central processor unit (CPU), through which all remote sensors report. Located in a plant's main control room or emergency control center, the CPU receives signals not only from supervisory fire-detection equipment, but from detectors of other plant processes not necessarily associated with fire conditions. The CPU compares these parameters in ways not normally done in routine plant operations—that is, for conditions that may lead to a fire. If, for example, operating vessel pressure was to exceed a certain limit, the CPU might arm an existing suppression system.

A more complex situation might be a precipitator failure combined with low wind velocity at the chimney's output. These conditions might cause a stack fire. Signals monitored by a CPU should either sound an alert or call for a shutdown.

It is possible to assemble one's own proprietary system by buying components from a number of sources. Doing so may have a number of drawbacks, however. First, the services of a competent fire protection engineer would be required to make sure the protection is adequate. Second, components would have to carry UL of FM ratings, and then the entire system would have to be rated.

Though proprietary alarm systems are especially advantageous for large operations, they offer significant advantages to smaller establishments as well. They can be purchased outright or leased. Costs can be slashed further by installing proprietary alarm systems in a number of buildings in an area, such as factories, stores, warehouses, and office buildings. Such installation permits hiring professionals to patrol the entire complex, thus bringing costs down. Also, multiplant operations can install proprietary alarm systems in several small plants in the same area for cost efficiency. The systems can then be linked to a central control console.

System Design Considerations

Spacing, locating, and maintaining fire detectors are important design considerations. The type of building, processes, and materials used help determine where detectors are located and the type of maintenance they require.

Spacing of Detectors

Because a fire-resistive structure can be expected to withstand the burnout of its contents for a longer period of time without structural collapse than a structure constructed of combustible materials, early warning of fire is of greater importance in the latter structure. The system designer should keep this in mind. Thus, combustible buildings should have more detectors than would normally be used. Such buildings should also be tied into the local fire department or central station alarm system. Building damage will thus be greatly minimized.

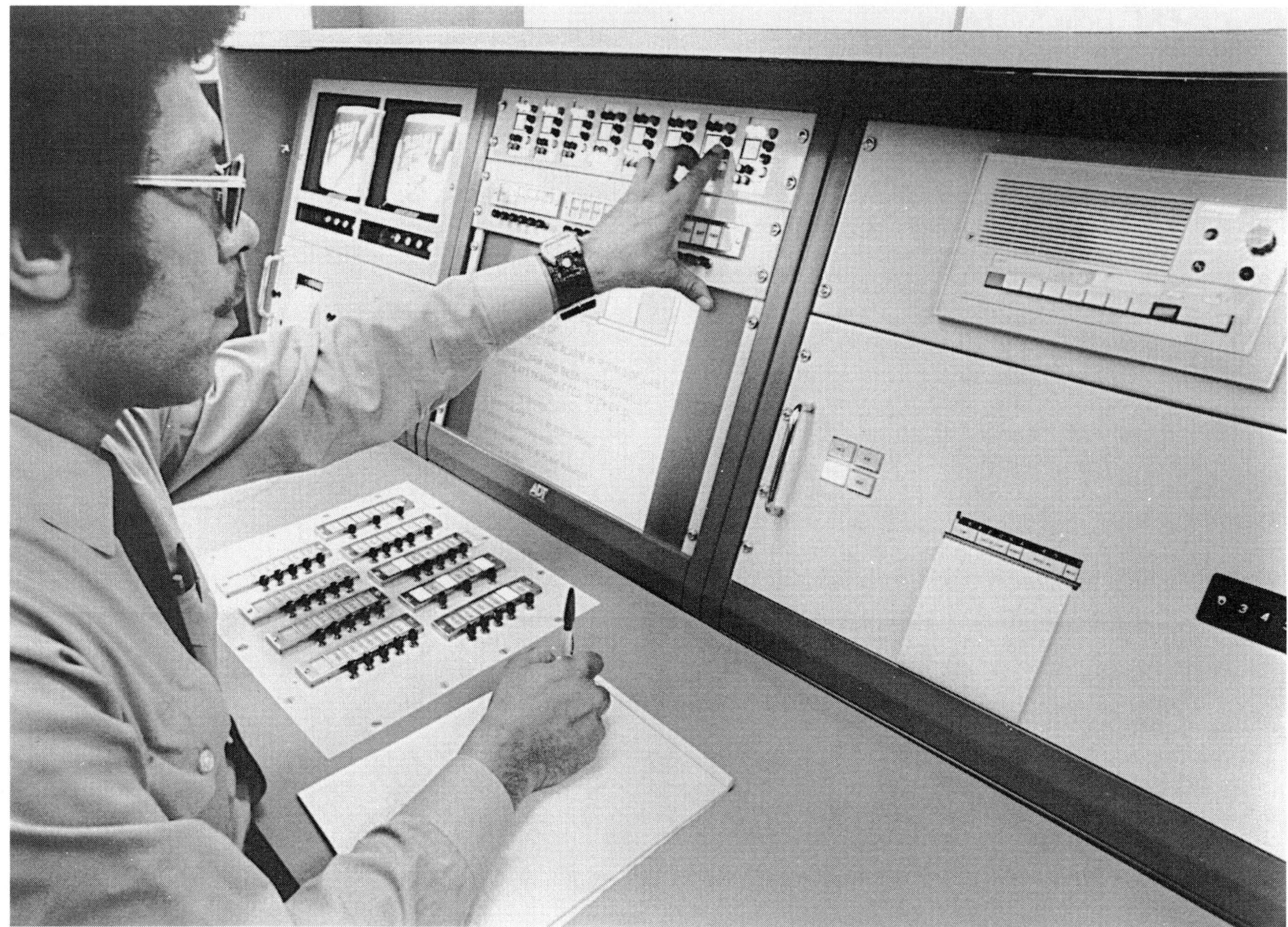

Figure 11–8. This console is part of a comprehensive proprietary system capable of detecting fire, burglary, and any unwanted fluctuations in industrial process conditions. The system incorporates closed-circuit television, portable radio communication equipment, and other instruments to handle indoor and outdoor security needs of a building or a group of buildings. (Reprinted with permission from ADT—American District Telegraph Company.)

To determine the number of detectors required for a given area, consider several factors. In general, the more detectors that are installed, the greater the coverage that is provided. If the number of detectors in a given area was doubled, the distance and the time that combustion products would have to travel from the farthest point in a room to a detector would be proportionately reduced. The exact area to be covered by an ionization detector, for example, depends upon total area, building construction, the area's contents, air movement, the value of the building and its contents, ceiling obstructions, and the cost of equipment downtime.

According to the NFPA,

> Detector spacings shall not exceed the linear maximum indicated by tests of Underwriters Laboratories Inc. and Factory Mutual Laboratories Inc. for the particular device used. Closer spacing may be required due to structural characteristics of the protected area and possible drafts or other conditions affecting detector operation. Factors such as air velocity, ceiling shape and height, ceiling material, and configuration of building structure influence the proper spacing.

Location of Smoke Detectors

The NFPA's standard also includes the location of smoke detectors:

> Smoke detectors shall be located and adjusted to operate reliably in case of smoke at any part of the protected area. The location of detectors should be based upon an engineering survey

of the application of this form of protection to the area under consideration. These features include air velocity; number of detectors to provide adequate coverage of cross-sectional area of the space with respect to travel, diffusion, or stratification of smoke; and location of detectors with respect to return, supply or circulating blowers, air conditioning facilities, temperature variations and the like. Such conditions vary with different installations and should be dealt with on the basis of experience.

Special consideration shall be given to the storage of contents of a protected space to provide unobstructed openings for the travel of smoke to the smoke detector.

Where air conditioning or ventilating equipment serves the space to be protected by a smoke detector, particular attention shall be given to the supply, return, and circulation of smoke under any condition of operation of the equipment to ensure prompt detection.

There is a temptation to try to protect many thousands of square feet of building space by spacing smoke detectors in the air-handling system. Since the products of combustion become diluted by air as they travel toward the detectors in the air-handling system, it follows that for detectors near the fan to detect smoke, the smoke must be very heavily concentrated in the occupied area of the building.

Maintenance of Detectors

Every fire protection system needs periodic maintenance and inspection by a knowledgeable, responsible person. Without periodic inspection and testing of each component, no fire protection system can be considered dependable.

This work may be done by an organization whose specialty is installing and servicing fire-detection equipment. Plant personnel also can be trained to become expert at routine inspection procedures.

Proper functioning of the system should be checked at regular intervals. Make a spot check of one or two detectors each month a regular part of the fire prevention inspection. Actuate different detectors at each inspection so that all components of the system will have been tested in the course of a year.

Every six months, inspect each detector's head screen for dust accumulation and clean it if necessary. In a dusty location, more frequent cleaning may be necessary. If so, activate a detector, and check the control unit's indications. Supervisory circuits are checked via the reset switch.

On a yearly basis, check each detector on the circuit for operation and sensitivity. Also check alarm relay contacts for proper operation.

PORTABLE FIRE EXTINGUISHERS

Equipment used to extinguish and control fires is of two types: fixed and portable. Fixed systems include water equipment, such as automatic sprinklers, hydrants, and standpipe hoses, and special pipe systems for dry chemicals, CO_2, and foam. Special pipe systems are used in areas of high fire potential where water may not be effective, such as where tanks for storage of flammable liquids and electrical equipment are located. Fixed systems, however, must be supplemented by portable fire extinguishers. These often can preclude the action of sprinkler systems. Not only can they prevent a small fire from spreading, but they can also rapidly extinguish a fire in its early stages.

Principles of Use

Even though a plant is equipped with automatic sprinklers or other means of fire protection, have portable fire extinguishers available and ready for an emergency. The term *portable* is applied to manual equipment used on small, beginning fires or used between the discovery of a fire and the functioning of automatic equipment or the arrival of professional fire fighters.

To be effective, portable extinguishers must be:

- approved by a recognized testing laboratory
- the right type for each class of fire that may occur in the area
- in sufficient quantity and size to protect against the expected exposure in the area
- located where they are easy to reach for immediate use
- maintained in operating condition, inspected frequently, checked against tampering, and recharged as required
- operable by area personnel who can find them and who are trained to use them effectively and promptly.

Classification of Fire Extinguishers

Portable extinguishers are classified to indicate their ability to handle specific classes and sizes of fires. This classification is necessary because new and improved extinguishing agents and devices are constantly being developed and because larger portable extinguishers are available. Labels on extinguishers indicate the class and relative size of fire that they can be expected to handle.

Use the following paragraphs as a guide to the selection of portable fire extinguishers for given exposures. (See NFPA 10, *Portable Fire Extinguishers.* See also the section Classification of Fires, earlier in this chapter.) Also, observe plant protection and insurance recommendations, based on fire protection requirements of the authority having jurisdiction.

- Use Class A extinguishers for ordinary combustibles, such as wood, paper, some plastics, and textiles, where a quenching-cooling effect is required.
- Use Class B extinguishers for flammable liquid and gas fires, such as oil, gasoline, paint, and grease fires, where oxygen exclusion or a flame-interrupting effect is essential.
- Use Class C extinguishers for fires involving electrical wiring and equipment where the dielectric nonconductivity of the extinguishing agent is of first importance. These units are not classified by a numeral, because Class C fires are essentially either Class A or Class B but also involve energized electrical wiring and equipment. Therefore, choose the coverage of the extinguisher for the burning fuel.
- Use Class D extinguishers for fires in combustible metals, such as magnesium, potassium, powdered aluminum, zinc, sodium, titanium, zirconium, and lithium. Persons working in areas where Class D fire hazards exist must be aware of the dangers in using Class A, B, or C extinguishers on a Class D fire. Of course they should also know the correct way to extinguish Class D fires. These units are not classified by a numerical system and are intended for special hazard protection only.

The recommendations that follow are given in NFPA 10 as a guide for marking extinguishers or extinguisher locations. They indicate which extinguisher should be used for a particular class of fire. Extinguishers suitable for more than one class of fire may be identified by multiple symbols.

Apply markings by using decals, painting, or similar methods having at least equivalent legibility and durability.

1. Extinguishers suitable for Class A fires should be identified by a triangle containing the letter *A*. If colored, the triangle should be colored green.
2. Extinguishers suitable for Class B fires should be identified by a square containing the letter *B*. If colored, the square should be colored red.
3. Extinguishers suitable for Class C fires should be identified by a circle containing the letter *C*. If colored, the circle should be colored blue.
4. Extinguishers suitable for fires involving metals should be identified by a five-pointed star containing the letter *D*. If colored, the star should be colored yellow.

Apply markings to the extinguisher on the front, of a size and form to be easily read at a distance of 3 ft (0.9 m). An easily visible picture-symbol label has been devised by the National Association of Fire Equipment Distributors (NAFED) and is now recommended in NFPA 10 (Figure 11–9). Where markings are applied to walls and panels in the vicinity of extinguishers, make them of a size and form to be easily read at a distance of 15 ft (4.6 m).

Extinguishers listed by UL are rated after physical testing. These ratings, which are indicated by a numeral and a letter, define the extinguishing potential of an extinguisher because they specify the type and size or number of extinguisher that should be installed in a specific area.

The numeral signifies the relative extinguishing potential, and the letter or letters signify the class or classes of fire on which the particular extinguisher is most

Figure 11–9. The symbols illustrated are for use on a Class A extinguisher (for extinguishing fires in trash, wood, or paper). The symbol at left is in blue. Since an extinguisher is not recommended for use on Class B or C fires, these two illustrations are in black, with a diagonal red line through them. For use on a Class A/B extinguisher, the first two illustrations would be in blue, the third would be in black with a red diagonal. For use on Class B/C, the last two illustrations would be in blue; on Class A/B/C, all three would be in blue.

effective for extinguishment. For example, a 4A:10B:C rating signifies that the extinguisher is recommended for both Class A and Class B fires and is safe to use if the fire is near or on energized electrical equipment.

Location of Extinguishers

Locate extinguishers close to likely hazards, but not so close that they would be damaged or cut off by the fire. Locate them along the normal path of exit from a building, preferably at the exits. Where highly combustible material is stored in small rooms or enclosed spaces, locate the extinguishers outside the door rather than inside. Locating them outside requires potential users to exit the room and then make a conscious decision to reenter the room and fight the fire.

Make the location of extinguishers as conspicuous as possible (Figure 11–10). For example, if an extinguisher is hung on a large column or post, paint a distinguishing red band around the post. Also, post large signs to direct attention to extinguishers. Keep the extinguishers clean. Do not paint them in any way that will camouflage them or obscure their labels and markings.

If an extinguisher is not already plainly marked to indicate the classifications of fire or types of material for which it is intended, place signs or cards indicating this information on the wall close to where it hangs. Markings indicating special uses can also be stenciled on the extinguisher or on an adjacent wall. Special labels are available from manufacturers of extinguishers and from insurance companies, NAFED, and NFPA. (See References, at the end of this chapter, for addresses.)

Figure 11–10. Fire equipment should be conspicuously located, appropriately marked, and inspected regularly.

Fire extinguishers must not be blocked or hidden by stock, finished material, or machines. Hang them as directed in NFPA 10. Then they will not be damaged by trucks, cranes, and other operations, or corroded by chemical processes; nor will they obstruct aisles or injure passersby. If the extinguishers are installed outdoors, protect them from the elements.

Make plant and warehouse aisles wide enough so that (1) mobile fire protection units can be brought close to a fire and (2) aisles can be kept free of obstructions. Mark floor spaces to allow access to fire extinguishing equipment, and protect extinguishers with bumpers or guardrails.

Extinguishers weighing over 40 lb (18 kg) should not be more than 3 ft (1 m) above the floor. Maintain a clearance of at least 4 in. (10 cm) between the bottom of the extinguisher and the floor.

Distribution of Extinguishers

The relative hazard of the occupancy, the nature of any anticipated fires, protection for special hazards, and requirements of local codes determine the minimum number and the type of portable extinguishers to be installed for each floor or area.

Class A Extinguishers

Extinguishers suitable for Class A fire hazards are installed according to the classification of occupancy: light hazard, ordinary hazard, or extra hazard (Table 11–E).

- Light-hazard occupancies include office buildings, schools (exclusive of trade schools and shops), churches, and public buildings. Because of the relatively small amount of combustibles in such buildings, incipient fires of minimal severity may be anticipated.
- Ordinary-hazard occupancies include department stores, warehouses, and manufacturing buildings where incipient fires of average severity in combustibles may be anticipated.
- Extra-hazard occupancies include some warehouses, woodworking shops, textile mills, and paper mills. Because the character or quantity of combustibles is more hazardous, extra-severe incipient fires may be anticipated.

Class B Extinguishers

Extinguisher requirements for Class B protection of a special hazard area, such as a laboratory, a kitchen, or an area where flammable liquids are stored, are in addition to the requirements of extinguishers for Class A

Table 11–E. Extinguishers Suitable for Class A Fires

	Light (Low) Hazard Occupancy	*Ordinary (Moderate) Hazard Occupancy*	*Extra (High) Hazard Occupancy*
Minimum rated single extinguisher	2-A	2-A	4-A*
Maximum floor area per unit of A	3000 sq ft	1500 sq ft	1000 sq ft
Maximum floor area for extinguisher	11,250 sq ft**	11,250 sq ft**	11,250 sq ft***
Maximum travel distance to extinguisher	75 ft	75 ft	75 ft

protection, except where the total area under consideration presents wholly Class B hazards. The requirements for fire extinguisher size and placement for Class B fires other than those in flammable liquids over ¼ in. (6 mm) deep are shown in Table 11–F.

Two or more extinguishers of lower rating, except for foam extinguishers, shall not be used to fulfill the protection requirements of Table 11–F. Up to three foam extinguishers holding 2½ gal (9.5 l) each may be used to fulfill light-hazard requirements, and up to three AFFF (aqueous film-forming foam) solution extinguishers holding about 2½ gal (9.5 l) each may be used to fulfill ordinary or extra-hazard requirements. The protection requirements may be fulfilled with extinguishers of higher ratings provided the travel distance to such larger extinguishers does not exceed 50 ft (15 m).

For flammable liquids of appreciable depth, over ¼ in. (6 mm), such as those in dip or quench tanks, the following recommendations apply:

- Provide Class B fire extinguishers on the basis of two numerical units of Class B-extinguishing potential per ft^2 (0.09 m^2) of flammable liquid surface of the largest tank to be protected within the area.
- Two or more extinguishers of lower ratings, except for foam extinguishers, shall not be used instead of the extinguisher required for the largest tank. However, up to three AFFF extinguishers may be used to fulfill these requirements.
- Portable fire extinguishers shall not be installed as the sole protection for flammable light hazard of appreciable depth (¼ in. or 6 mm) whose surface exceeds 10 ft^2 (0.9 m^2). Where personnel are trained in the extinguishment of such fires, the maximum surface area shall not exceed 20 ft^2 (1.9 m^2). Use fixed fire protection systems for protecting tanks in excess of 20 ft^2 (1.9 m^2). Portable extinguishers would then be used for putting out burning liquid spills outside the range of fire equipment or for putting out fires that originate outside the tank.
- Where approved automatic fire protection devices or systems have been installed for a flammable liquid, additional portable Class B fire extinguishers may be waived. When so waived, provide Class B extinguishers to protect areas near such protected hazards.
- Consider the travel distance between special hazards and extinguishers. Scattered or widely separated hazards shall be individually protected if the specified travel distances in Table 11–F are exceeded. Likewise, extinguishers near a hazard shall be easily reached during a fire without endangering the operator.

Table 11–F. Extinguishers Suitable for Class B Fires, for Fires in Flammable Liquids ¼ Inch and Under in Depth

Type of Hazard	*Basic Minimum Extinguisher Rating*	*Maximum Travel Distance to Extinguishers (ft)*	*(m)*
Light (low)	5-B	30	9.15
	10-B	50	15.25
Ordinary (moderate)	10-B	30	9.15
	20-B	50	15.25
Extra (high)	40-B	30	9.15
	80-B	50	15.25

Class C Extinguishers

Locate extinguishers with Class C ratings close to energized electrical equipment, which would require a nonconducting extinguishing medium if fire either directly involved or surrounded the equipment. Since such a fire is Class A or Class B, the extinguishers are sized and located on the basis of the anticipated Class A or Class B hazard. Whenever possible, deenergize electrical equipment before attacking a Class C fire.

Class D Extinguishers

Extinguishers used for Class D protection and other special fires are installed according to the size and type of the special hazard. Consider the type, quantity, and physical form of the combustible material when selecting the proper type of extinguishing agent and the method of applying it.

Selection of Extinguishers

Operating characteristics that make one type of portable fire extinguisher suitable for certain fire hazards may make the same type dangerous for others. Secure on-the-job advice from fire inspection bureaus, fire insurance carriers, and fire protection engineers when selecting extinguishers.

Do not let the extinguisher's cost be the overriding factor in the selection process. Remember, good extinguishers are worth their cost because of the protection they give. Obviously, only purchase extinguishers listed and tested according to American National Standards Institute (ANSI) standards.

Also consider the extinguisher's design and operating features, ease of maintenance, and the availability of repair service. If possible, actually operate and test the extinguisher before making the final selection. Table 11–G and Figure 11–11 give overviews of common types of extinguishers and their operating characteristics.

Although extinguishers must be installed in conformance with NFPA 10, there is some flexibility that can be used to the purchaser's advantage. For example, if a certain condition calls for Class B extinguishers, the required units could be obtained by using various sizes of dry-chemical, CO_2, or AFFF extinguishers. Consider the relative advantages and disadvantages of each of these units with respect to all the conditions in the area. Because the UL rating can vary widely for extinguishers of the same size but of different manufacturers, investigate the relative merits of any particular extinguisher available on the market.

Types of Portable Extinguishers

The following types of portable extinguishers are recommended for various types of fires: water solution, dry-chemical, CO_2, and dry-powder extinguishers. Vaporizing liquid extinguishers containing carbon tetrachloride or chlorobromomethane have been outlawed by many model codes because when these liquids are subjected to heat, they produce phosgene gas, a toxic substance.

Water Solution Extinguishers

Fire extinguishers that use water or water solutions include pump tank, stored pressure, and AFFF. These extinguishers are effective against Class A fires because of the quenching and cooling effect of water. These units cannot be used on fires in or near electrical equipment, since they can present a shock hazard to the operator.

Frequently inspect the nozzle on water extinguishers for foreign particles, such as dirt, matchsticks, and paper, that may prevent it from discharging. Also keep the pressure-relief hole in the cap of gas-cartridge and foam units free from obstruction. This hole is designed to release any residual pressure that may injure a worker when removing the cap before recharging the extinguisher.

In locations exposed to freezing temperatures, install water solution extinguishers in heated cabinets, or charge them with a nonfreezing solution. However, do not use antifreeze or salt solutions unless the equipment has been designed for such use, since these chemicals can cause rapid corrosion. Calcium chloride and other salt solutions are good conductors of electricity and therefore are especially dangerous if applied to live electrical apparatus. Check the manufacturer's recommendations in these instances.

When charging any extinguisher with an antifreeze solution, thoroughly dissolve the chemical in warm water in a separate container. Then pour it into the extinguisher through a fine strainer to remove any foreign particles that may clog the unit. More maintenance is required for extinguishers containing antifreeze solutions, because they may corrode more easily than those filled with plain water. When recharging an extinguisher with an antifreeze solution, thoroughly flush all parts, including the hose and nozzle, with plain water. To prevent freezing and clogging, be sure that all the plain water is drained off.

Dry-Chemical Extinguishers

The dry-chemical extinguisher is one of the most versatile units available. It extinguishes by interrupting the chemical flame's chain reaction. Do not confuse it with a dry-powder extinguisher.

Types of Extinguishing Agents

Four types of base extinguishing agents used in dry-chemical extinguishers are sodium bicarbonate, potassium bicarbonate (also marketed as urea potassium bicarbonate), potassium chloride, and ammonium phosphate. When recharging, use only the dry-chemical agent and cartridge recommended by the extinguisher's manufacturer.

Table 11–G. Fire Extinguisher Selection Chart

	Class A					Class A/B	Class B/C		
	Water Types		Multipurpose Dry Chemical		AFFF Foam	AFFF Foam	Carbon Dioxide	Dry Chemical Types	
	*Stored Pressure**	*Pump Tank*	*Stored Pressure*	*Cartridge Operated*	*Stored Pressure*	*Stored Pressure*	*Self Expelling*	*Stored Pressure*	*Cartridge Operated*
Sizes Available	2½ gal	2½ and 5-gal	2½-30 lb ALSO Wheeled 50-350 lb	50-30 lb ALSO Wheeled 50-350 lb	2½ gal	2½ gal	5-20 ALSO Wheeled 50-100 lb	2½-30 lb ALSO Wheeled 150-350 lb	4-30 lb ALSO Wheeled 50-350 lb
Horizontal Range (Approx.)	30 to 40 ft	30 to 40 ft	10-15 ft (Wheeled 15-45 ft)	20 to 25 ft	20 to 25 ft	3-8 ft (Wheeled 10 ft)	3-8 ft (Wheeled 10 ft)	10-15 ft (Wheeled 15-45 ft)	10-20 ft (Wheeled 15-45 ft)
Discharge Time (Approx.)	1 min.	1 to 3 min.	8-25 s (Wheeled 20-60 s)	8-25 seconds (Wheeled 20-60 s)	50 seconds	50 seconds	8-15 seconds (Wheeled 8-30 s)	8-25 seconds (Wheeled 30-60 s)	8-25 seconds (Wheeled 20-60 s)

	Class A/B/C		Class D
	Multipurpose Dry Chemical		Dry Powder
	Stored Pressure	Cartridge Operated	Cartridge Operated
Sizes Available	20½-20 lb ALSO Wheeled 150-350 lb	5-30 ALSO Wheeled 50-350 lb	30 lb ALSO Wheeled 150-350 lb
Horizontal Range (Approx.)	10-15 ft Wheeled 15-45 ft	10-20 ft (Wheeled 15-45 ft)	5 ft (Wheeled 15 ft)
Discharge Time (Approx.)	8-25 seconds (Wheeled 30-60 s)	8-25 seconds (Wheeled 20-60 s)	20 seconds (Wheeled 150 lb 70 s 350 lb 1¾ min.

*Must be protected from freezing.

Reprinted with permission from the National Association of Fire Equipment Distributors.

- Sodium bicarbonate-base dry chemical, the most common agent, is available in ordinary and foam form. The ordinary form can be used simultaneously with foam without causing the foam blanket to break down.
- Potassium bicarbonate-base dry chemical is similar in extinguishing properties to sodium bicarbonate-base dry chemical. However, it has twice the effective fire-fighting capacity on a pound-to-pound basis. This agent has good moisture repellency and is compatible with the simultaneous use of water or foam. It does not allow fuel to reflash as easily or as rapidly as sodium bicarbonate. The absence of momentary flare-up also permits the user to approach fires more closely. In addition, the potassium compound extinguishes the leading edges of contained flammable-liquid fires more easily than does the sodium compound.
- On Class B and Class C fires, ammonium phosphate-base dry chemical (multipurpose) has operating characteristics similar to sodium- and potassium-base dry chemicals. When discharged

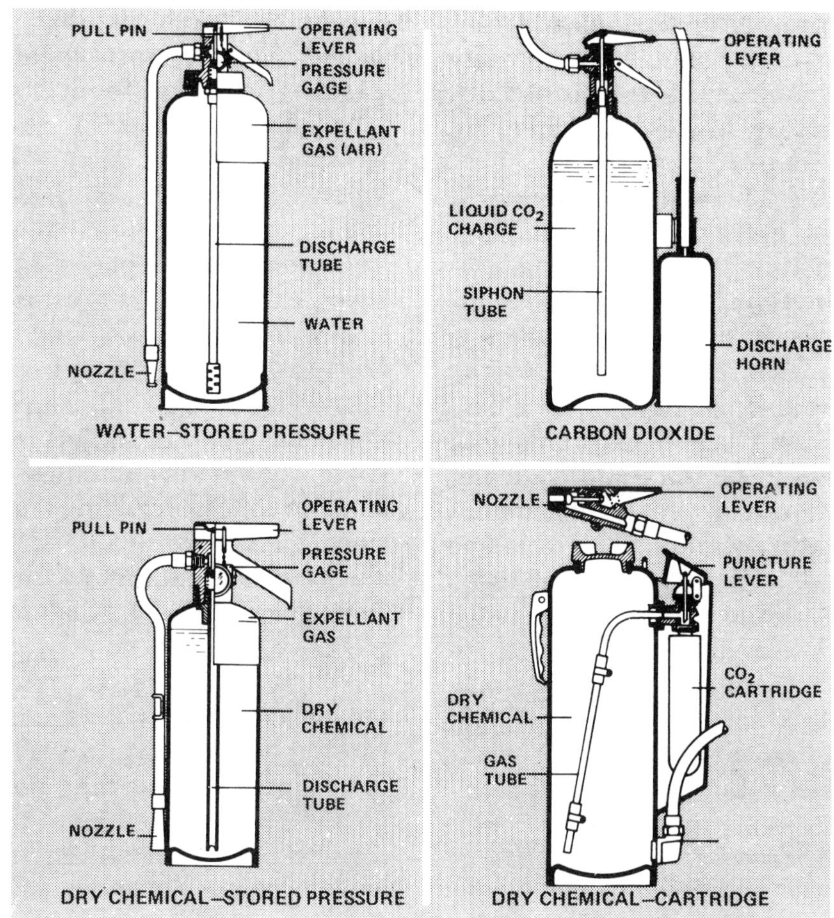

Figure 11–11. Among the major types of portable fire extinguishers are the stored-pressure water unit, the carbon dioxide unit, the stored-pressure dry-chemical unit, and the cartridge-operated dry-chemical unit. The internal parts of each type are shown and labeled. (Reprinted with permission from *Plant Engineering Magazine.*)

into a Class A fire, its chemical reaction destroys the flames. Also, a coating, formed when the extinguishing agent softens, adheres to the burning surface, thereby retarding further combustion. To obtain complete extinguishment on Class A materials, thoroughly expose all burning areas to the extinguishing agent. Because any small burning ember may be a source of reignition, properly applying multipurpose dry chemicals on Class A fires is more critical than with water-solution extinguishers. In the presence of moisture, multipurpose dry chemicals may cause corrosion when discharged on metals. Therefore, clean up the multipurpose agent immediately after the fire is extinguished. Also, never mix an ammonium phosphate-base agent with a potassium-base or sodium bicarbonate-base agent because dangerous pressure can be developed by even a trace of moisture.

Types of Extinguishers

The two basic types of dry-chemical extinguishers are defined by their propellant technique: gas cartridge or stored pressure. In the gas-cartridge extinguisher, pressure is supplied by a gas stored in a separate cylinder. In the stored-pressure extinguisher, the entire container is pressurized.

Many models of dry-chemical extinguishers have a high-velocity discharge. Therefore, take care not to aim the initial discharge directly into the burning area, since it may cause the fire to spread. For the best results when attacking a Class B fire, use a "fanning" action: Rapidly move the nozzle from side-to-side to intermix the agent thoroughly with the flames. Start well in advance of the burning edge and go beyond the burning edge on each side to avoid leaving any burning pockets behind. To minimize the possibility of reflash, continue to discharge the chemical after the fire has gone out.

Carbon Dioxide Extinguishers

CO_2 extinguishers put out fires by displacing the available oxygen. They do not leave a residue.

Dry-Powder Extinguishers

Because the use of combustible metals, such as sodium, titanium, uranium, zirconium, lithium, magnesium, sodium-potassium alloys (NaK), and other less-common metals, has increased, have dry-powder extinguishers available to fight such fires. Although dry powders are very effective on combustible metal fires, they all have certain limitations. Consider the type, quantity, and form of the metal and the existing physical conditions when selecting the proper type of dry powder and the method of application.

G-1 Powdered Agent

The oldest powdered agent is the G-l type, a graphite organic-phosphate compound. When it is applied with a scoop or shovel to a metal fire, the phosphate material generates vapors that blanket and smother the flames, and the graphite, being a good conductor of heat, cools the metal below its ignition temperature. Take care to assure that the depth of the powder's cover is adequate to provide a smothering blanket. If hot spots should occur, cover them with additional powder. Allow the burning metal to cool before attempting to dispose of the material.

Met-L-X

Another dry powder is Met-L-X. It is composed of a sodium chloride base with additives to make it free flowing, to increase water repellency, and to create the property of heat caking. This material is dispensed from a 30 lb (13.6 kg) dry-powder extinguisher similar in appearance and physical features to the cartridge-operated, dry-chemical extinguisher, or from larger wheeled or stationary units.

The technique used to extinguish a metal fire with Met-L-X is to open fully the nozzle of the extinguisher and apply a thin layer of Met-L-X over the burning mass from a safe distance, until control is established. Then, throttle the nozzle to produce a soft, heavy stream, and completely cover the burning mass with a heavy layer from close range. The heat of the fire causes the Met-L-X to cake, thus forming a crust that excludes air.

Lith-X

Lith-X is another dry-powder extinguishing agent. It is composed of a special graphite base with additives that make it free flowing so it can be discharged from an extinguisher. Lith-X was developed mainly for use on lithium fires, but is also effective on other combustible metals. Lith-X does not cake or crust when applied over a burning metal. It excludes air and conducts heat away from the burning mass, thus extinguishing the fire.

Met-L-Kyl

A problem recently developed in fire-fighting involves pyrophoric liquids, such as triethylaluminum. These liquids ignite spontaneously and the resulting fires cannot be easily extinguished by dry powder or other commonly used agents. A special material, Met-L-Kyl, has been developed, consisting of a bicarbonate-base dry chemical and an activated absorbent.

The principal of extinguishment involves the combination effect of the dry chemical, which extinguishes the flames, and the absorbent, which absorbs the remaining fuel and prevents reignition. Met-L-Kyl has been designed so that it can be discharged from an extinguisher similar to the standard cartridge-operated, dry-chemical model.

Miscellaneous Equipment

Besides the extinguishers themselves and the extinguishing agents, portable fire extinguishers sometimes require additional equipment. This is especially true if the extinguishers are large or mounted on vehicles.

Wheeled Equipment

Large portable units on wheels are commercially available. They include 50, 75, and 100 lb (23, 34, and 45 kg) CO_2 extinguishers; and 75 to 350 lb (34 to 160 kg) dry-chemical extinguishers.

Wheeled, "Twinned" Extinguishers

With the development of water-soluble, fluorocarbon surface- active agents, foaming agents are available that give water the property of floating in thin layers on liquid fuel surfaces (light water). This characteristic provides excellent protection against reflash on liquid hydrocarbon fires with only one- fourth the volume as compared to protein air foam.

A wheeled extinguisher with both Purple K dry chemical and "light water" fluorocarbon foam provides a synergistic extinguishing system. It rapidly knocks down flames and completely protects against reflash. The two extinguishing agents are simultaneously applied through duel pistol-grip nozzles.

Vehicle-Mounted Equipment

Water, foam, CO_2, and dry-chemical extinguishing agents are available in units that are mounted on vehicles. They range in size from in-plant fire vehicles able to turn in warehouse aisles to large trucks.

Fire Blankets

In some cases, fire blankets can be used to smother a small fire. Their major purpose is to extinguish burning clothing. However, they are also useful for smothering flammable-liquid fires in small, open containers.

Flame-retardant blankets are also available. The most common size of blanket is 66 by 80 in. (1.7 by 2 m). Store blankets in containers mounted on a wall or column so they can be readily pulled out.

Miscellaneous Hand Equipment

Water or antifreeze backpack tanks are available in 2½ gal (9.4 l) or 5 gal (19 l) sizes. Hand pumps are built into the hose nozzles' handles. Such units are carried on the back, and the slide-action pump is operated with both hands. The backpack unit is frequently used for combating brush fires.

Maintenance and Inspection

Give one person in the plant or establishment the responsibility for maintaining and inspecting fire-extinguishing equipment. Whether weekly or monthly, as needed, the testing and repair work should be completed in accord with NFPA 10.

If in-plant maintenance of fire extinguishing equipment is chosen, then a stock of supplies of spare parts and refills will be necessary. In some plants, specially trained personnel can test and refill extinguishers on a full-time basis. In others, this maintenance service is under contract with a service organization or the manufacturer's representative. Establish a record system and an organized plan for checking and repairing various types of extinguishers. The inspection and maintenance records should consist of at least durable tags fastened to the extinguishers. The tags should show dates of inspection (monthly) and dates of examination for recharge and other maintenance work (annually). The tags must indicate if the unit was recharged during maintenance.

Other records may be kept that list extinguishers by type, by location, and by recharge periods. The master record system should contain the history of each extinguisher, the type and quantity of refills on hand, and other pertinent information.

To discourage tampering and to make inspection easier, seal hand extinguishers or install them in cabinets. However, do not lock extinguishers in cabinets or store them in any way that would prevent immediate use in an emergency.

A mandatory requirement for keeping fire extinguishers in top condition is periodic hydrostatic pressure testing, set forth in NFPA 10. Injuries and deaths because of extinguishers rupturing during operation or recharging are the reasons for this requirement. Be sure to check NFPA 10 for details of testing and marking extinguishers.

Never allow a mass removal of extinguishers from a building. Establish the testing program so only a few scattered units are removed at a time. They should then be tested and returned to service promptly. Arrange for temporary replacement units or additional protection when extinguishers are being removed for testing. (See the section Training of Employees, earlier in this chapter, for details on training employees to use fire extinguishers.)

SPRINKLER AND WATER-SPRAY SYSTEMS

There are many types of sprinklers and water-spray systems for extinguishing fires. The type of building, operations performed in it, and materials used will help determine the type of sprinkler or water-spray system used.

Water Supply and Storage

Sprinkler systems need a reliable water supply of ample capacity and pressure for efficient fire extinguishment. Have the water supply engineered with the sprinkler protection to provide a hydraulically balanced system at the least cost. For example, to supply all sprinklers likely to open, in addition to hose streams, a volume of 500 to 3,000 gal/min (31.5 to 190 l/sec), and sometimes more, may be needed. The precise need depends on the maximum flow of water through all openings. The pressure requirement will vary. However, it should be high enough to maintain a residual pressure of 15 psig (103 kPa) while required water volume is flowing in top-story sprinklers.

Water may be supplied from the following:

- Underground supply mains from public water works.
- Automatically or manually controlled pumps drawing water from lakes, ponds, rivers, surface storage tanks, underground reservoirs, or similar adequate sources.
- Pressure tanks containing (1) water in a quantity determined by the formula in NFPA 13, *Installation of Sprinkler Systems*, and (2) compressed air for expelling the water into the piping supply system. The smallest tank for light-hazard occupancy is 2,000 gal (7,570 l).
- Elevated tanks or reservoirs that depend on gravity to force water through the system.

Consider providing at least two of these four independent water sources. Then, in case of fire, the main source furnishes water to the system immediately; it is reinforced by the second source that also can supply emergency protection if the primary source is out of service. The preferred source is a connection from a reliable public water system. Connection should be made to two different mains to provide greater volume and to give flexibility in case of failure of one water main.

A fire pump that can deliver water at high pressure over an extended time can be a second source. Locate the pump where it will not be put out of service by a fire. Be sure that this source has a reliable power supply, independent of the plant system and routed so that a fire anywhere in the protected areas will not expose the pump's power supply.

Do not use water stored for private fire protection for other purposes. Everyday use of tank-stored water necessitates constant refilling, hence the danger of accumulated sediment circulating into the hydrant and sprinkler system. Also, the varying water level may shorten the life of a wood tank or require frequent painting of a steel tank. In addition, take care that the water connections do not pollute drinking water, especially where emergency supplies are taken from a river or other nonpotable source.

Although construction and installation of water tanks are beyond the scope of this chapter, one basic consideration is worth mentioning: In areas with freezing temperatures, be sure that tanks are heated. Inspect the heating system daily during freezing weather; the control valves, weekly; and the entire system (tank, supporting tower, piping, valves, heating system, and all components and accessories), annually. (See NFPA 22, *Water Tanks for Private Fire Protection*; see also NFPA *Fire Protection Handbook.*)

To make inspection easier, wood or steel gravity tanks supported on steel towers should be ringed by a platform that is protected by a substantial steel rail. The rail should meet the requirements of providing standard guardrails and toeboards. These tanks should also have substantial steel ladders equipped with approved steel cages or basket guards.

Protect ladders of water tanks that are accessible to the public with locked gates or fences. This will discourage unauthorized persons from climbing them.

Automatic Sprinklers

Automatic sprinklers are the most extensively used and most effective installations of fixed fire-extinguishing systems. These systems are so basic and have proved so effective that most fire protection engineers consider them the most important fire-fighting tool. Nationwide figures from NFPA indicate that sprinklers have a very high efficiency rating for satisfactory extinguishment, usually over 95%.

The cost of automatic sprinkler protection is relatively small, averaging about 2% of the total plant's investment. Experience shows that a sprinkler system often can pay for itself in 10 years, and sometimes less. Insurance costs are usually 40% to 90% lower than for buildings without a sprinkler system. The cost of a sprinkler system is much less if it is installed when the building is built rather than later.

In addition to the economic factors, automatic sprinklers have an impressive lifesaving record. Loss of life by fire is rare where properly designed and maintained sprinkler systems have been installed.

However, when sprinklers do fail to operate, in perhaps 3% to 4% of the cases, the failure is caused by some readily preventable condition. Over one-third of all failures can be attributed to closed water-supply valves.

Although the primary function of the sprinkler system is to deliver water to a fire automatically, the system can also serve as a fire alarm. It can serve as an alarm when an electrical water-flow alarm switch is installed in each main riser pipe. When a fire occurs and the first sprinkler opens, the water rushing through the pipe sets off an alarm that alerts the control system or fire fighters.

Dependable sprinkler protection requires a systematic maintenance and inspection program. Such a program includes periodic inspection of water-supply valves, water-supply tests, physical inspection of the system's piping for obstructions to distribution, and similar items. (See NFPA 13A, *Inspection, Testing, and Maintenance of Sprinkler Systems*, for additional maintenance requirements.)

There are six basic types of automatic sprinkler systems: wet-pipe, dry-pipe, preaction, deluge, combined dry-pipe and preaction, and limited water-supply systems. (See NFPA 13, *Installation of Sprinkler Systems.*) The combination dry-pipe and preaction systems are used on installations that are larger than can be accommodated by one dry-pipe valve. The limited water-supply system is used for installations that do not have access to a continual or large supply of water. The other four types of sprinkler systems are discussed in detail in the following paragraphs.

Wet-Pipe System

In the wet-pipe system, which represents the greatest percentage of sprinkler installations, all parts of the system's piping are filled up to the sprinkler heads with water under pressure (Figure 11–12). Then, when heat actuates the sprinkler, water is immediately sprayed over the area below.

If a portion of the wet-pipe system is subjected to freezing temperatures, as on a loading dock, fill the exposed sections with an antifreeze solution, or connect these sprinklers to a dry-pipe system. Use a water-soluble, liquid antifreeze that is proportioned to give low-temperature protection without producing a combustible mixture, as specified in NFPA 13 and 13A. When the system is supplied from public water connections, use chemically pure glycerine (U.S. Pharmacopeia 96.5% grade) or propylene glycol antifreeze, and then add it only in accord with local health regulations.

Dry-Pipe System

The dry-pipe system generally substitutes for a wet-pipe system in areas where piping is exposed to freezing temperatures. A good rule of thumb is to use a

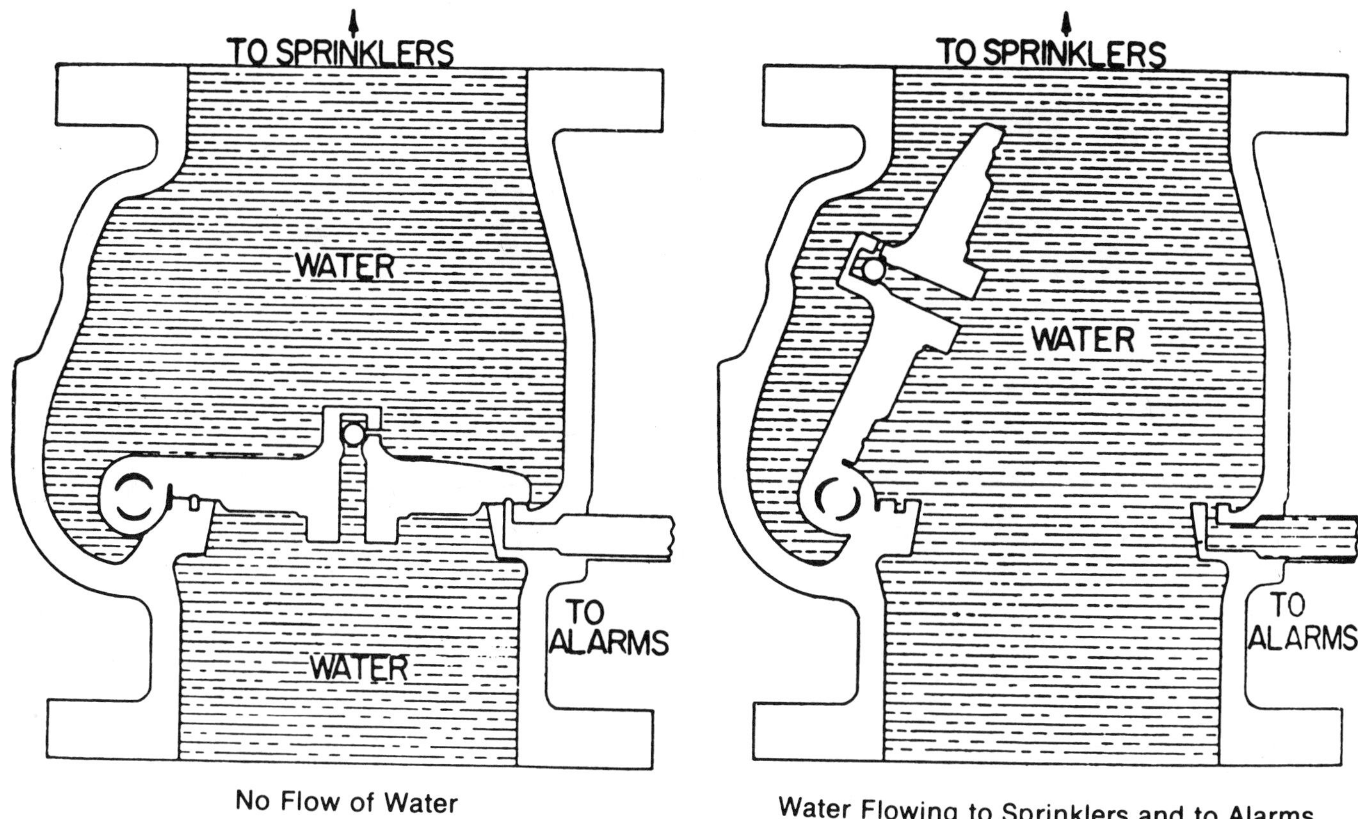

Figure 11–12. A wet-pipe sprinkler system is under water pressure at all times so that water will be discharged immediately when an automatic sprinkler operates. The automatic alarm valve triggers a warning signal when water flows through the sprinkler piping.

dry-pipe system when more than 20 sprinklers are involved. It is essential, however, to locate the dry-pipe valve and water supply line in a heated enclosure.

In the dry-pipe system, the piping contains compressed air that holds back the water by means of a dry-pipe valve (Figure 11–13). When a sprinkler opens, the air is released, the pressure drops, and the dry-pipe valve opens to admit water into the risers and branch lines. These sequential actions delay the actual wetting when compared to a wet-pipe system. Because of this delay, extra-hazard buildings are difficult to protect with a dry-pipe system. In general, more water damage may result with the dry-pipe system, because more sprinklers open than with the wet-pipe system. More sprinklers open because the fire progresses further, hence more sprinklers are tripped before the extinguishing action of the water takes effect. To reduce this delay, add quick-opening devices, such as exhausters or accelerators, to dry-pipe systems to expel the air more readily.

Keep the system's air pressure 15 to 20 psi (110 to 140 kPa) above the normal tripping pressure. High pressure delays the action of the dry-pipe valve. The system should not lose more than 10 psi (70 kPa) of air pressure per week. Doing so would require repressurizing more often than once a week. Therefore, do not use equipment that leaks. (See NFPA 13 for details of testing dry-pipe systems.)

It is essential that all parts of a dry-pipe system be installed so that they can be thoroughly drained. Therefore, where it is necessary to use inverted (pendant) sprinklers or "drop piping," use a special type of pendant sprinkler or fill the piping with an antifreeze solution.

Preaction Systems

Preaction systems are similar to dry-pipe systems. However, they react faster and hence minimize water damage in case of fire or mechanical damage to sprinklers or piping. To avoid mechanical damage to a system, connect to the piping an automatic, low-pressure air supply that compensates for minor system leakage. A rapid reduction in pressure resulting from, say, an unintentional breakage of the piping, then sends a trouble signal without tripping the water-control valve.

The preaction valve, which controls the water supply to the system's piping, is actuated by a separate fire-detection system. These fire detectors are located in the same area as the sprinkler and operate independently of

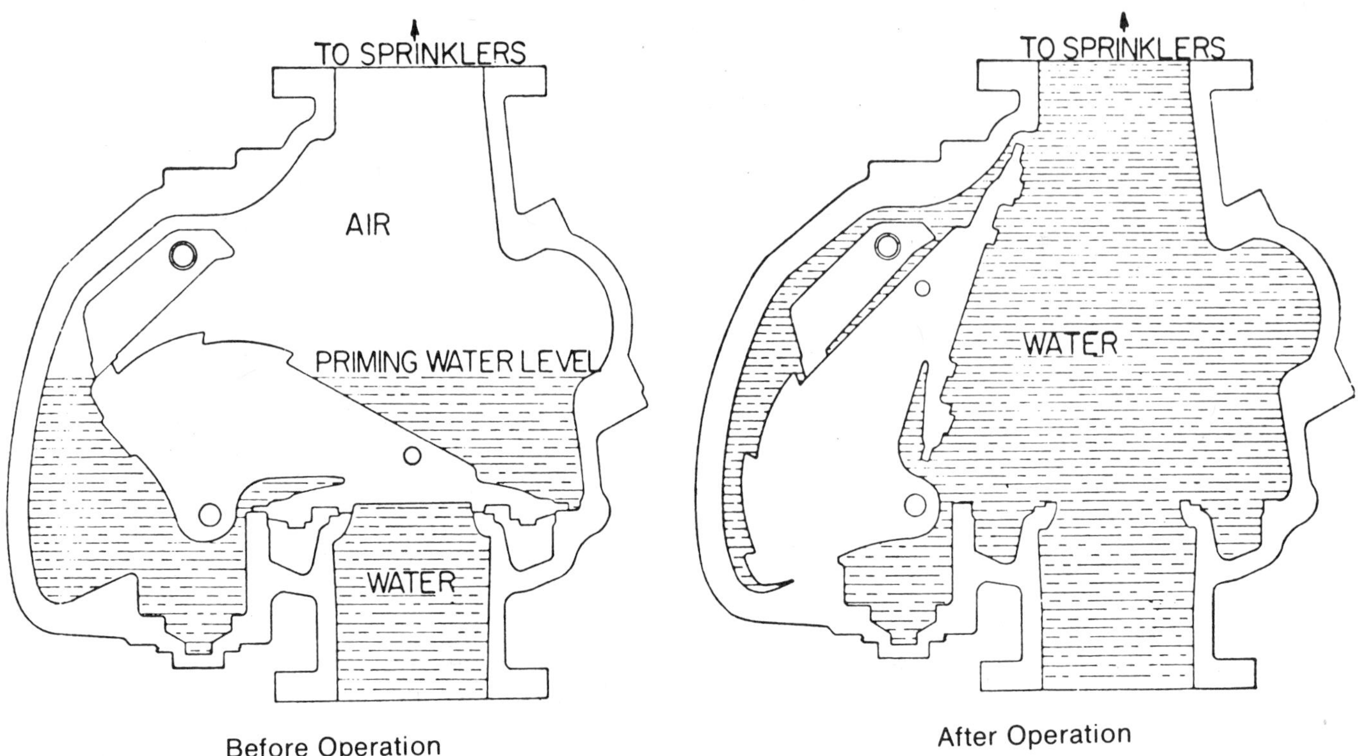

Figure 11–13. The principle of a dry-pipe system is illustrated by these simplified drawings of a dry-pipe valve. Compressed air in the sprinkler system holds the dry valve closed, preventing water from entering the sprinkler piping until the air pressure has dropped below a predetermined point. (Reprinted with permission from National Fire Protection Association.)

the sprinkler. Because the detection system is more heat sensitive than the sprinklers, the water-supply valve opens sooner than in a dry-pipe system. The water-supply valve can also be operated manually.

Usually an alarm is sounded when the valve opens and starts filling the system with water. There may then be time to put out the fire with portable equipment before the sprinklers go into action and drench the area with water. A preaction system is especially effective where valuable merchandise is handled or stored.

Deluge Systems

The deluge system wets down an entire area by admitting water to sprinklers that are open at all times. Deluge valves that control the water supply to the system are actuated by a fire-detection system located in the same area as the sprinklers. The water-supply valves can also be operated manually.

This type of system is primarily designed for extra-hazard buildings where great quantities of water may have to be applied immediately over large areas. Deluge systems are ordinarily used to best advantage where rapidly spreading fires, or flash fires, may be anticipated, such as in explosives plants, plants handling or processing nitrocellulose materials, lacquer plants, and buildings that contain large quantities of flammable materials.

Another application of a deluge system is an open system of outside sprinklers for distributing water over the roof, exterior of a building, and at windows and cornices. Such a system protects the building against fire from adjoining property. This system is usually manually operated and is used where construction is inadequately protected by design or by distance from adjacent fire hazards.

In special applications, where deluge protection is not needed over the entire area, open sprinklers and closed sprinklers may be combined in a single system. However, remember that (1) separate fire detectors are also required in the area covered by the closed sprinklers, (2) operation of a closed sprinkler will not activate the entire system, and (3) a fire in the area of the closed sprinkler will also cause water to discharge from all the open sprinklers.

Types of Sprinklers

There are many classifications of sprinklers, each of which is designed for specific applications. Upright sprinklers direct water upward against the deflector. Pendant sprinklers direct water downward against the deflector. Sidewall sprinklers discharge the major portion of water away from the nearby wall in a pattern resembling a quarter of a sphere. Sprinklers designated for quick-response are also available. However, use them only when they complement the sprinkler system. (See Factory Mutual System Research Organization and NFPA, listed in References at the end of this chapter, for additional information.)

Automatic Alarms

Make automatic alarms, operated by the flow of water through the system, a part of every standard sprinkler installation. Such an alarm may be connected to a central station fire alarm service or to the municipal fire department, or it may be a local alarm signal. Its purpose is to give prompt notice that the sprinkler system is operating. It also signals water leakage or discharge from causes other than fire.

Have the automatic alarm system tested and inspected frequently. Also have persons thoroughly familiar with the system maintain it.

Temperature Rating of Sprinklers

Select sprinklers on the basis of temperature rating and occupancy. Sprinklers are built either with heat-actuated elements of solder that melt or with special devices in which chemicals melt or expand to open them. Table 11–H shows the ratings and distinguishing colors of sprinklers.

Quick operation of sprinklers, an advantage when over a fire, may be a disadvantage elsewhere by wasting water and wetting down materials that might otherwise be unaffected.

Causes of Failure of Sprinkler Systems

Sprinklers seldom fail to control fires, but when they do, failure is usually caused by (1) not keeping all supply valves open, and (2) shutting off the supply valves prematurely during a fire. Other causes of sprinkler failure, with corresponding remedies (in italics), are as follows:

- Freezing of wet-system sprinkler pipes. *Heat the building or convert to a dry system.*
- Defective dry-pipe valve, or slow operation of dry system because of its excessive size. *Check valves at frequent intervals, "trip test" in accord with insurance company recommendations, or subdivide the system.*
- Foreign material obstructing the system. *Flush out the system on a regular schedule and provide debris-clear water at the intake through use of filter screens.*
- Improper drainage through faulty installation. *Check pitch of pipes and eliminate low spots.*
- Sprinklers obstructed by stock piled too high, sprinklers isolated by temporary partitions or shelving, and sprinklers shielded from heat. *Improve housekeeping and maintain a minimum of 18 in. (46 cm) clearance between the top of stored material and the deflector. Increase the clearance up to 36 in. (0.9 m) over large, closely packed piles of combustible cases, bales, cartons, or similar stock.*
- Corrosion of sprinklers in such locations as bleacheries, dye houses, or chemical operations. *Use sprinklers specially protected for such locations.*
- Inadequate supply of water because of faulty design or poor maintenance. *Verify that water conditions have not changed since the plant's original installation. Sturdily install and anchor the system, because explosions can jolt the piling and render the system ineffective.*

Water-Spray Systems

Water spray is effective on all types of fires where there is no hazardous chemical reaction between the water and the materials that are burning. Although these systems are independent of and supplemental to other forms of protection, they are not a replacement for automatic sprinklers.

Fixed, water-spray systems are similar to the standard deluge system except that the open sprinklers are replaced with spray nozzles. The water supply to the system can be controlled automatically or manually.

Table 11–H. Standard Temperature Ratings of Automatic Sprinklers

Rating	Operating Temperature (F)	Color	Maximum Ceiling Temperature (F)	(C)
Ordinary	135-150-160-165	Uncolored*	100	38
Intermediate	175-212	White*	150	66
High	250-280-286	Blue	225	107
Extra high	325-340-350-360	Red	300	150
Very extra high	400-415	Green	375	190
Very extra high	450	Orange	425	218
Very extra high	500	Orange	475	246

*The 135 F sprinklers of some manufacturers are half black and half uncolored. The 175 F sprinklers of the same manufacturers are yellow.

Reprinted with permission from NFPA 13, *Installation of Sprinkler Systems.*

Water-spill systems are generally used to protect flammable liquid and gas tankage, piping and equipment, cooling towers, and electrical equipment, such as transformers, oil switches, and motors. Because of its low electrical conductivity, water spray applied through fixed-piping systems on electrical equipment with voltages as high as 345,000 v has proved practical. When applied on some types of electrical equipment, however, water spray may cause short circuits by forming a continuous path of water between energized parts. In such cases, provide means for cutting off the electrical current before the water spray is applied. (See NFPA 15, *Water Spray Fixed Systems.*)

The type of water spray required depends upon the hazard and the purpose for which the protection is provided. The basic principle of water spraying is to wet the surface completely with a preselected water density, taking into consideration nozzle types, sizes, spacing, and water supply.

Water-spray systems can be designed effectively for any one, or any combination, of the following purposes:

- extinguishing fire
- controlling fire where extinguishment is not desirable, such as gas leaks
- exposure protection; that is, absorbing heat transferred from equipment by the spray
- preventing fire by having water spray dissolve, dilute, disperse, or cool flammable materials.

Because the passages in a water-spray nozzle are small in comparison with those in the ordinary sprinkler, they can easily be clogged by foreign matter in water. Therefore, strainers are ordinarily required in the supply lines of fixed-piping spray systems. The strainer baskets should have holes small enough to protect the smallest opening in the nozzles. In cases where the nozzles have extremely small water passages, they should have their own internal strainer in addition to the supply-line strainer.

Fire Hydrants

In large plants where parts of the plant are a considerable distance from public fire hydrants or where no public hydrants are available, install hydrants at convenient locations in the plant's yard. The number needed depends on the fire exposure, and the hose-laying distance to the built-up plant areas. (See NFPA 24, *Installation of Private Fire Service Mains.*)

Keep exterior fire department connections that serve sprinkler or hose systems easy to reach. The discharge ports should be at least 18 in. (46 cm) above the ground or floor level. Keep vegetation, snow, and stored materials away from hydrants or hydrant houses. In areas subjected to heavy snow, attach to each hydrant a stiff metal wire with a flag on top. In this way, the hydrant can be found even after a heavy snow. Also, protect hydrants from mechanical injury; however, this protection cannot interfere with efficient use.

Before cold weather sets in, drain or pump out hydrants if they are not the type that normally drain. Check drainage whenever hydrants are used during freezing weather. To determine whether or not hydrants are frozen, (1) partially turn the hydrant's stem (if the hydrant is frozen, the stem will not turn) or (2) lower a weight on a string into the hydrant's barrel. Approximately 4 ft of string should be attached to the weight. The weight with string should be dropped into the outlet opening. If less than 2 ft of string can be advanced into the outlet opening, the hydrant is frozen.

Frozen hydrants can best be thawed with steam introduced through the outlet by means of a steam hose that is pushed slowly down the barrel, thawing as it goes. Do not use corrosive chemicals, such as calcium chloride, caustics, or salts, to thaw frozen hydrants.

Frequently test and maintain control valves. A number of persons, including members of the plant's fire brigade, should know the location of valves and the sections of the pipe controlled by them.

Have the local fire department check connections to be sure that they are of a size and thread that will fit its equipment. If special adapters are required, supply them to fire fighters and also have them available on the premises.

Fire Hoses

Like other fire-extinguishing equipment, have hose lines available for immediate use. Be sure that they are easy to reach. Keep space around hose lines and control valves clear of obstruction. Be sure that fire brigade workers know where hose-line equipment is located and how to operate it. Keep aisles and doorways clear and wide enough for rapid use of hose-reel carts or other mobile equipment.

Woven-jacket, lined hose with an outer rubber or plastic cover is chiefly used in industries where the hose jacket must be protected against chemicals and abrasions. Lined hose with a rubber or plastic cover is available in ¾ and 1 in. (1.9 and 2.5 cm) sizes. It is generally used as a booster hose, or as a hose on chemical engines, wheeled extinguishers, and wall-mounted or vehicle-mounted pressured hose reels.

For inside use by building occupants, as opposed to standpipe systems designed for the fire department, 1½ in. (3.8 cm) unlined linen hose was often used. Only lined 1½ in. (3.8 cm) hose has been specified by NFPA. Unlined hose on an existing system may be used, provided it is maintained in good condition. However, with the introduction of lightweight, lined fire hose, which has twice the hydraulic efficiency of unlined hose, many organizations have replaced the unlined hose on their industrial standpipe installations. Unlined hose has its place, however, in such locations as office buildings, where it would be discarded and replaced if ever used.

Keep the hose on an approved swinging rack or reel, approximately 5 ft (1.4 m) above the floor or high enough so that it will not be a hazard to passersby or damaged by trucking operations. Locate hose stations intended for use by employees at the exits. (See NFPA 14, *Standpipe and Hose Systems*.)

Arrange the hose so it will not kink or tangle when pulled out. Keep one end connected to the standpipe, and equip the other end with a ⅜ or ½ in. (9.5 or 13 mm) nozzle tip or a combination spray-straight stream nozzle. To prevent kinking in use, do not place more than 150 ft (46 m) of hose at a standpipe outlet.

Except for unlined linen hose, hydrostatically test all fire hose annually. Thoroughly inspect it, dry it out, and return it to service.

Maintenance of Fire Hose for Outdoor Use

Periodically inspect and test woven-jacket, lined hose to make sure that it will be in good condition when an emergency arises. Water should be run through the hose at least twice a year. Store yard hose in standard hose houses for protection against weather. Use fire hose only for fighting fires. If hose is needed for other uses, provide separate hose.

Mildew may attack untreated hose fabric containing cotton if the hose is stored in a damp location or if it is not thoroughly dried after being wet. Fire hose is available with chemically treated fabric. Treated fabric protects against mildew and rot. Treated jackets also absorb less water, and therefore, dry more quickly. The resistance to dampness and mildew is not 100% effective, however, even when the treatment is new, and the treatment deteriorates with age.

Jackets made entirely of synthetic warp and filler are impervious to mildew and rot. It is not necessary to dry such hose. However, wash it after using it and before storing it. Hose with all-cotton jackets and with jackets made from a combination of cotton and synthetic yarns must be carefully dried.

For plant yards containing rough services that will cause heavy wear or where working pressures are above 150 psi (1,060 kPa), use double-jacket, lined hose. If hose may be subjected to acids, acid gases, and other corrosive materials, such as those found in chemical plants, use rubber-covered, woven-jacket, lined hose. For such conditions, also use hose with a neoprene-impregnated all-synthetic jacket.

Maintenance of Fire Hose for Indoor Use

Maintain and test unlined hose and woven-jacket, lined hose as follows:

- Reserve the hose for fire fighting.
- Keep hose valves tight, since leakage will rot linen hose.
- Examine hose visually each year for mildew, rot, and damage by chemicals, vermin, and abrasions. If the hose is in doubtful condition, give it a hydrostatic pressure test. Replace damaged hose.
- Give hose a pressure test after the fifth and eighth year of service. Then repeat the test every second year after the eighth year. The local fire department will often pressure-test hose. (Unlined hose cannot be pressure tested.)
- Keep hose clean. Wash woven-jacket, lined hose with laundry soap if necessary.
- Dry hose jackets thoroughly after use and keep them dry.

Hose Nozzles

Effective streams of water for fire fighting are controlled by the size and type of nozzle. The nozzle, in turn, must be supplied with the correct quantity of water at the discharge pressure for which it is designed. Nozzles are designed for solid streams, spray streams (frequently referred to as fog), or combination streams. Nozzles for special extinguishing agents, such as foam and dry chemical, are also available.

Spray Nozzles

Spray nozzles are widely used for both public and private fire protection. They make the application of water more effective under many conditions. A description of three types of nozzles follows:

1. Open nozzles of a fixed (nonadjustable) spray pattern, usually attached to shut-off valves. Some nonadjustable nozzles can be equipped with an applicator (a long pipe extension, curved at the end and fitted with a fixed-spray nozzle) for fighting fires where an extended reach is necessary.
2. Adjustable nozzles that provide variable discharges and patterns, from shutoff to straight-stream spray and from narrow- to wide-angle spray.
3. Combination nozzles in which a straight-stream, a fixed or adjustable spray, and a shutoff are selected, usually by a two- or three-way control valve.

Monitor Nozzles

Permanently mounted monitor nozzles are frequently used to protect pulpwood storage piles at paper mills, in lumber yards, in stock yards, in railway-car storage yards, and near oil storage tanks. Nozzles are often elevated to clear obstructions so that the operator can stand on a shielded platform and direct a high-pressure stream of water over a wide area. Monitor nozzles are especially useful in large, congested areas where it is impractical to lay hose lines in an emergency.

SPECIAL SYSTEMS AND AGENTS

Special hazards may require systems of extinguishment or control other than water. Each of the several systems available offers certain advantages and disadvantages to consider when making a selection. These systems are usually installed to supplement rather than replace the automatic sprinkler system. They should be engineered to fit the circumstances of the particular hazard. Install them so that their operation will shut down other processes, such as pumps and conveyors, that might intensify a fire.

The following special agents and systems are currently in use. The specific NFPA publication that covers each of these agents or systems is also listed. (See also the NFPA *Fire Prevention Handbook*.)

- Foam—NFPA 11 and NFPA 16
- Carbon dioxide—NFPA 12
- Dry chemical—NFPA 17
- Inerting—NFPA 69
- Steam smothering—NFPA 86
- Water spray and sprinkler—NFPA 15 and NFPA 13
- Wet chemical—NFPA 17
- Explosion venting—NFPA 69

Foam Systems

Foam systems are often used to protect dip tanks, oil and paint storage rooms, and asphalt coating tanks. Foam systems also have been developed to put out tank fires by subsurface injection of foam. Foam can also be used to extinguish fires in laboratories.

Fire-fighting foam is an aggregate of gas-filled bubbles formed in a water solution. It is lower in density than most flammable liquids and, as a result, forms a continuous floating blanket of material on flammable liquids. Foam extinguishes fire in these liquids by smothering and cooling the fuel, thus (1) halting the production of vapors, (2) excluding the availability of oxygen to the fire, and (3) physically separating the flames from the uninvolved fuel surface.

There are two types of foam-generation methods: (1) chemical generation and (2) mechanical, or air, generation, which includes protein and synthetic types of foam. Chemical foam is formed by a chemical reaction in which masses of bubbles of CO_2 gas and a foaming agent produce an expanded froth. Mechanical foam consists of bubbles of air produced when air and water are mechanically agitated with a foam-making agent.

Fire-fighting foams are of two major types: (1) low-expansion foam, used mainly for Class B (flammable liquids) fires, and (2) high-expansion foam, used principally for Class A (ordinary combustibles) fires.

Low-Expansion Foam

Low-expansion foam is available as chemical foam, mechanical or air-generated foam, protein foam, and synthetic or fluorinated, surface-active-agent foam.

Chemical Foam

Chemical foam is formed by the reaction in water of a mixture of aluminum sulfate with sodium bicarbonate —a foaming agent and a stabilizer. As already stated, CO_2 that evolves in the reaction produces bubbles and is trapped in the foam. Because of the high content of solids, chemical foam's chief advantage is that it is very resistant to flame and mechanical disruption. It also has an ability to "set up" and assist surface flow. Chemical foam gains much of its consistency from the hydrated aluminum hydroxide in the reacted foam mass. Very little, if any, fire-extinguishing capability is provided by the CO_2-containing bubbles.

When reacted at solution temperatures of 60 to 85 F (16 to 29 C), thick foam with volume expansions of 7:1 to 16:1 can be produced. Severe limitations of these foams are that above and below these temperatures, chemical foam-making compounds give poorer quality foams or low expansion and rapid breakdown. Dry powders, or solutions of the acid and basic components, deteriorate with storage (especially if the temperature drops below 50 F or exceeds 100 F [10–38 C]). Chemical foam cannot be transported through long pipes or hose lines or applied with devices requiring moderate pressures, because the foam breaks down.

There are four general types of equipment for producing chemical foam: self-contained units, closed generators, hopper generators, and stored-solution systems.

- In the self-contained unit, two solutions that produce foam on contact are stored independently in a single vessel and are caused to mix either manually or automatically. The amount of foam produced is determined by the quantity of foam-producing materials within the vessel.

- In the closed-generator system, chemical foam powder is stored in large hoppers, permanently fixed to foam generators, that mix the powder with water and then pump the foam or use the water pressure to force the foam to special outlets. This kind of installation may be either the one- or two-powder type. It is mainly used in storage tank farms for flammable liquids, where a single foam-producing installation can service a number of tanks.
- Hopper generators can be either permanent installations or portable. The advantage of portable generators is that they are not limited in foam production by a fixed storage of foam powder. Instead, they can be continuously refilled. Generators employing either one or two foam powders to produce foam are available. On large storage tanks for flammable liquids, permanent lines affixed to foam chambers at the top of the tank are sometimes provided. Portable foam towers are used for foam application to burning oil storage tanks. Generator and water connections can be made at a calculated safe distance from the tank at the time of a fire.
- Stored-solution systems have large, permanently installed tanks that contain two foam-producing solutions stored separately. At the time of a fire, either duplex or twin pumps force the solutions to outlets where they mix and discharge foam. Foam production is limited by the size of the system's facilities.

Because of their limitations, chemical foams, the oldest of the three types of low-expansion foams, are gradually being replaced by mechanical or air-generated foams, which include the protein and synthetic types.

Mechanical or Air-Generated Foam

Mechanical or air-generated foam is produced by the mechanical action of adding the proper amounts of a liquid concentrate into a water stream via a proportioner and then introducing and mixing air into the water-concentrate's solution. There are four basic methods of producing mechanical or air-generated foam—nozzle-aspirating systems, in-line foam-pump systems, in-line aspirating systems, and in-line compressed-air systems.

The names of these systems indicate where and how air is injected into the water-concentrate solution to produce mechanical foam. Each one of these systems uses a proportioner to introduce the foam concentrate into the water stream. There are a number of types of proportioners that can be located either at the main pump or in between the main pump and the foam maker.

Protein Concentrates

Protein concentrates consist primarily of high-molecular-weight digestion products obtained by the chemical hydrolysis of vegetable or animal protein materials. Metallic salts are included to maintain bubble strength in the presence of heat and mechanical action. Organic solvents are also added to improve the foaming characteristics and to control the freezing point. Additional additives are used to prevent corrosion, resist bacterial decomposition, and control viscosity.

Protein concentrates are available in two strengths. One is used with the proportioner at a 3%-by-volume ratio; the other, at a 6%-by-volume ratio. It is important to note that both the 3% and 6% foam liquids are satisfactory only on hydrocarbons, such as benzene, toluene, xylene, gasoline, naphtha, and kerosene.

Water-mixable solvents, such as esters, ethers, alcohols, and ketones, destroy the regular 3% and 6% foam liquids as rapidly as they are applied. Therefore, use an alcohol-resistant foam in combating fires in these materials. Where such uses are contemplated, submit a sample of the material to a supplier of alcohol-resistant foam for evaluation. Because the stabilizer in this foam is chemically different from other stabilizers, the system in which the foam is to be used must be carefully designed with regard to its limitations.

Because these foams must be used in 6% or greater concentrations, they are more costly and require greater quantities to be on hand. A 6% solution of protein foam permits vertical surfaces to be covered with an insulating blanket to aid in confining the fire. The high burnback resistance of protein foams is especially valuable when the foam must seal to the hot side of a metal tank or vessel.

Fluorinated Surfactant Foams

Fluorinated surfactant foams, sometimes referred to as "light water" or aqueous film-forming foam (AFFF), consist of a fluorinated surfactant (surface-active agent) and a foam stabilizer. AFFF is applied in 1%, 3%, or 6% solutions in either fresh water or saltwater. Unlike protein foam, which must be applied through conventional foam nozzles, AFFF may be applied with fog or water-spray nozzles. However, the resistance to flashover reignition and burnback is reduced. Special AFFF formulations also are available to overcome these limitations.

Conventional AFFF is limited to spill fires of petroleum products (nonpolar) where rapid knockdown is desired and where cost is secondary. AFFF is useful in preventing flashover of fuel vapors exposed to lingering open flames or heated surfaces. The surfactant solution floats on the surface of the fuel, forming a barely visible membrane that reduces the release of vapors and the subsequent flashover hazard.

Because air-generated foams possess low viscosity, they have fast spreading and leveling characteristics. They exclude air and halt fuel vaporization. Foam-generating devices or water-air spray devices can usually generate the easily formed foam.

To draw the concentrate from 55 gal (208 l) drums, 5 gal (19 l) pails, or smaller portable containers, install a proportioner unit and siphon hose on a standard fire hose. Some proportioner units can vary the percentage of concentrate from a plain water stream to a foam stream. A plain water stream can also be quickly obtained by removing the siphon tube from the concentrate.

Major limitations of AFFF are that it breaks down, loses burnback resistance, and fails to seal against the wall of tanks or vessels if prolonged freeburning has occurred prior to applying the foam. Therefore, the usefulness of AFFF on flammable liquids contained in metal tanks—the hot sides of which require a relatively high water retention capability—is limited. AFFF will not adhere to vertical surfaces, as protein foam will.

Foam-Water Systems

Foam-water sprinkler and spray systems are equipment for mechanically generated foam, with a deluge sprinkler system. These systems are generally used to protect hazard areas of flammable liquids. They can discharge foam or plain water selectively through aspirating devices, such as water sprinklers or foam-water-spray nozzles.

Wet-Water Foam

Wet-water foam is generated by using an aspirating nozzle or by injecting air under pressure into water containing a wetting agent. The wetting agent is a chemical compound that reduces the surface tension of the water and increases its penetrating, spreading, and emulsifying properties. UL-listed wetting agents are siphoned through a proportioner at rates not exceeding 2% by volume.

The reflective, white opaque surface formed by the air bubbles and wet-water particles makes this agent a good medium for protection from exposure fires. The cellular structure of the foam also retards heat conduction, thus affording insulation. As the foam blanket absorbs the heat away and breaks down, the wet water released from the air bubbles carries the heat away from the protected surface. Do not mix wetting agents with other wetting agents, or with mechanical or chemical foam, since it may neutralize the effect of the agents and thus destroy the fire-fighting properties.

Wet Water

The wet-water principle can also be used without generating foam. In liquid form, wet water has the same general extinguishing properties as plain water. However, as cooling ability is increased, there is a greater penetration of porous surfaces because of the wet water's reduced surface tension.

High-Expansion Foam

High-expansion foam is particularly suited for controlling and extinguishing Class A and Class B fires in confined spaces, such as basements, buildings, shafts, and sewers. High-expansion foam is a blend of surface-active agents and a synthetic detergent-foaming agent. High-expansion foam is made by introducing a small amount (usually 12%) of foam liquid into a foam generator where water and large quantities of air are mixed. Fire-fighting foams of expansions from 100:1 up to 1,000:1 can be produced. In general, one 5 gal (19 l) can of high-expansion foam can produce approximately 33 million gal (125 million l) of foam, enough to cover a football field 12 in. (0.3 m) deep. Such foams provide a unique agent for transporting water to inaccessible places, for total flooding of confined spaces, and for displacement of air, vapors, heat, and smoke.

In addition to being a superior foaming agent, high-expansion foams have an emulsifying ability to clean up petroleum product spills, as well as a wetting ability to increase the penetrating effect of water on deep-seated Class A fires. When accumulated in depth, high-expansion foams can provide an insulating barrier for the protection of exposed materials or structures not involved in a fire and thus prevent the fire's spread.

The temperature of the water and the quality of the air can affect the properties of the high-expansion foam. Water should be less than 90 F (32 C), and air for foam production must be taken where smoke and the chemical products of fire cannot be drawn into the foam generator. These products cannot be drawn into the foam generator since they can decrease the amount of foam produced, increase the drainage rate by chemical interaction with the foaming agent, and trap the gases in the bubble aggregate to create a toxic foam.

Provide adequate ventilation opposite the side of the foam's application when using high-expansion foam for fire fighting. Foam will not flow into a confined space unless provision is made for venting the displaced air and gasses.

Carbon Dioxide Extinguishing Systems

Fixed (local or flood) CO_2 systems are often installed for the protection of rooms that contain electrical equipment, flammable liquid or gas processes, dry-cleaning machinery, and other exposures. CO_2 is useful where a fire can be extinguished by displacing the oxygen content of the air or where water must not be used because of electrical hazard or the nature of the product. (See NFPA 12, *Carbon Dioxide Extinguishing Systems,* for details on installing and storing these systems.)

In the high-pressure system, CO_2 is stored in compressed-gas cylinders at normal temperatures. It is released by manual operation or by automatic devices through nozzles close to the expected source of fire. CO_2 has definite advantages, unlike water or other chemical extinguishing materials, in that it generally does not damage equipment.

In the low-pressure, fixed installation, CO_2 is stored in an insulated pressure vessel and maintained to 0 F (−18 C) by mechanical refrigeration. At this temperature the pressure is approximately 300 psi (2,070 kPa). At such low pressure, 500 lb (226 kg) to more than 125 tn (113 t) of CO_2 can be stored more economically than at a higher pressure.

Relief valves are provided in case of refrigeration failures. Liquid CO_2 is delivered through pipelines to nozzles that may have delivery capacities as high as 2,500 lb/min (15 kg/sec). As in the high-pressure CO_2 system, release can be either manual or automatic.

Fixed, local-application systems provide for extinguishing fire at its source. Total flood systems may be used in small buildings, compartments, or rooms where walls or other openings can be automatically shut when the gas is released. Provide alarms to alert persons working in areas protected by this type of system. Allow enough time to evacuate the area.

In confined locations, ventilate the area thoroughly after a fire is extinguished. There may not be enough oxygen available to sustain life.

Hand-held hose-line systems combine fixed tanks with hose reel attachments, which permit a limited range of fire fighting. Range is predicted on the length of the hose plus the effective range of CO_2.

Dry-Chemical Piped Systems

Dry-chemical piped systems have been developed for situations where quick extinguishment is needed, either in a confined area or for localized application, and where reignition is unlikely. These systems are adaptable to flammable liquid and electrical hazards. They can be operated manually or automatically, or be activated at the system or by remote control. A rate-of-rise, heat-actuated device or an electrical release controls the automatic operation.

Installations can provide for simultaneous closing of fire doors, operating valves, windows, and ventilation ducts, as well as for shutting off fans and machinery and for actuating alarms. For example, a dry chemical system installed in a vent-a-hood, above a deep vat fryer, when activated, will also shut down the natural gas or electric power to the fryer. Piped systems providing either local application or total flooding are also available.

The dry-chemical agent is neither toxic nor a conductor of electricity, nor does it freeze. In piped systems and most hand-held hose-line systems, the agent is stored in a tank that is pressurized by an inert gas cylinder when the controls are actuated. In some hand-held hose-line systems, the agent is stored in a pressurized container. Extinguishing action results mainly from the dry chemical interrupting the chemical flame's chain reaction. (See Chemistry of Fire, earlier in this chapter.)

Fixed storage tanks and pressured cylinders, similar to those used in piped systems, are available for monitored operations and for mounting on vehicles. (See NFPA 17, *Dry Chemical Extinguishing Systems.*)

Steam Systems

Automatic or manually controlled steam jet systems can be used to smother fires in closed containers or in small rooms, such as heaters, drying kilns, smoke ovens, asphalt-mixing tanks, and dry-cleaning tumbler dryers. However, such systems are practical only where a large supply of steam is continuously available. Unfortunately, most plants do not have this steam-generating capacity. In addition, steam has not been found effective on deep-seated fires that may form glowing embers, or in enclosures where the normal operating temperature is not considerably higher than air temperatures.

Consider the possibility of personal injury hazards from burns when installing steam-extinguishing systems. (See NFPA 86, *Ovens and Furnaces.*)

Inert Gas Systems

Inert gas systems can prevent fires and explosions by replacing the oxygen in the air with an inert gas, such as CO_2, nitrogen, flue gas, or other noncombustible gases, until it reaches a level (or percentage) where combustion will not take place. (See the discussion earlier in this chapter and in NFPA 69, *Explosion Prevention Systems.*)

To be effective, the inert gas must reduce the amount of oxygen in the system from the normal 21% to between 2% and 16%, depending upon the type of hazard involved and the type of inert gas. For instance, an inert gas such as CO_2 must reduce the oxygen in air to 6% to prevent fire or explosion of CO; 14% for gasoline; and 15% for common dust.

Inert gases, such as flue gas from a power plant's stacks, have been used extensively to prevent explosions. Flame producers operating on either fuel oil or gas, which yield products of combustion with a high percentage of CO_2 and nitrogen, are often used in fixed installations where large quantities of the inert gas are required, as in the purging of storage tanks, in pipelines, and in manufacturing processes with high explosion hazards. Inert gases can also be used as a means of transferring flammable liquids, for inerting the atmosphere of storage tanks of volatile flammable liquids, and for purging gas holders or pipelines. Since inert gases displace the oxygen in the air, they also have widespread use as standby emergency fire extinguishers.

Use of an inert gas in a confined space can result in an oxygen-deficient atmosphere. Therefore, before anyone is allowed to enter a confined space into which an inert gas has been introduced, thoroughly ventilate the space and test it with instruments to show if the enclosed atmosphere will support life. Otherwise, workers should wear approved breathing equipment and a harness with a lifeline for such entry. Furthermore, a watcher similarly equipped should stand by to observe workers in the confined space and to rescue them in case of an emergency.

There are three ways of applying inert gas to assure the formation of a noncombustible atmosphere within an enclosed tank or space: fixed volume, fixed rate, and variable rate.

- The fixed-volume method introduces the inert gas into the equipment chamber (1) by reducing the pressure within the chamber and allowing the inert gas to flow in until the pressure is equalized or (2) by pressuring the chamber with the inert gas and then letting off the overpressure to the atmosphere after mixing has taken place. Several pressuring cycles may be necessary to reduce the oxygen content sufficiently.
- The fixed-rate method adds a continuous supply of inert gas in amounts large enough to accommodate peak requirements. The quantity required is based on the maximum inbreathing rate that may result by sudden cooling, such as that caused by rain or a sudden drop in temperature, plus maximum product withdrawal. Although this method is relatively simple, it has the disadvantage of wasting inert gas and increasing the rate of the evaporation of the product.
- The variable-rate method supplies the inert gas to the system on a demand basis. The inert gas is continuously released to a low-volume supply line to compensate for minor pressure changes. When rapid changes take place, such as those caused by product withdrawal, a means is provided that opens a large supply line until the pressure equalizes. This type of system is extremely efficient and has the advantage of reducing the product's vapor losses by maintaining a slight positive pressure within the chamber.

Preventing Explosions

Preventing the development of explosive mixtures is the best defense against explosions. Equipment for handling and storing flammable gases should be designed, constructed, inspected, and maintained so that the danger of leakage and of explosive mixture formation is reduced to a minimum. Have qualified individuals, either from the plant's staff or from an outside source, inspect equipment at regular intervals.

Under certain conditions, ventilation will prevent excessive accumulations of gases and vapors. The best method of ventilation necessarily varies with the nature of the gas or vapor to be removed and depends upon whether the gas or vapor is heavier or lighter than air. Inasmuch as heating or cooling of the gas or vapor can change its density, design the ventilation or exhaust system or both according to operating conditions, not only according to the published density figures. (See NFPA 91, *Blower and Exhaust Systems for Dust, Stock, and Vapor Removal or Conveying*.)

Natural-draft ventilation may be through openings near the floor, near the ceiling, or both. However, the best ventilating method is a positive, local exhaust system, using explosion-proof electrical equipment and taking suction as close to the source of a vapor or gas as possible. However, use a nonexplosion-proof exhaust-fan motor if it is properly installed outside the ductwork and outside the hazardous area.

In general, flammable industrial gases, such as acetylene, CO, hydrogen, and natural gas, are lighter than air. The vapors of flammable liquids, however, are generally heavier than air. Examples are alcohol, naphtha, gasoline, benzene, amyl acetate, and carbon disulfide.

Unburned gases or flammable vapors in the combustion chambers of unit heaters, boilers, furnaces, and enameling, drying, and bakery ovens may form an explosive mixture in air. Interrupting the gas-fed pressure or extinguishing the flame or pilot light may cause an accumulation of unburned fuel.

A number of safety devices have been developed to overcome this hazard. Most of them operate automatically to provide ventilation and to control the interlocking of gas and air supplies. These actions serve as a safeguard against explosive mixtures.

On gas- and oil-fired equipment, provide ways to ventilate or purge the combustion zone thoroughly in case of flame failure. Personnel in charge of firing these devices should know the ventilating or purging time required in the event of flame failure. In the event of flame failure, the program controller should take over. Doing so forces the operator to go through the interlocked startup procedure that includes a timed pre-ignition purge cycle. Check the atmosphere inside large industrial equipment with a combustible-gas indicator before relighting.

Inspect and test gas-fired equipment, including its controls, at regular intervals, and keep it in good repair in accord with the manufacturer's recommendations. Permit only trained personnel to operate it.

Frequently inspect gas valves for leaks. If gas is present, ventilate it immediately and correct the condition before using the equipment. Follow the recommendations of the manufacturer of the equipment and of the public utility supplying the gas.

When a natural gas leak is detected, shut off the source of the natural gas and ventilate the area. In addition, refrain from turning on or off any electrical equipment in the affected area. If the electrical equipment is not intrinsically safe, the act of turning it on or off could cause an electrical arc that could ignite the accumulation of natural gas.

Suppressing Explosions

Under certain conditions, an explosion-suppression system can be used to reduce the destructive pressure of an explosion. These systems are designed to detect an explosion as it is starting and to actuate devices that suppress, vent, or take other action to prevent the full explosive force.

These systems require split-second timing. The mechanism for dispersing the suppression agent must operate at extremely high speed to fill the enclosure completely within milliseconds after detection (Figure 11–14). The suppression agent must be dispensed from the suppressors in the form of a very fine mist at a rapid speed, normally through the use of a small, secondary-explosive force. The suppression agent is normally a noncombustible liquid that is compatible with the combustion process to be suppressed.

IDENTIFICATION OF HAZARDOUS MATERIALS

Fires and other emergency situations often involve chemicals that have varying degrees of toxicity, flammability, and reactivity or stability. Make information on these relative hazards readily available to personnel confronted with such emergencies if life safety, fire prevention, and effective fire extinguishment are to be achieved. Federal hazardous materials laws require the availability of Material Safety Data Sheets (MSDS) when requested.

NFPA Identification System

A system for the quick identification of hazardous properties of chemicals has been developed by the NFPA. (See NFPA 704, *Identification of the Fire Hazards of Materials.*) For uniformity, this system recommends the use of a diamond-shaped symbol and numerals indicating the degree of hazard (Figure 11–15 and 11–16).

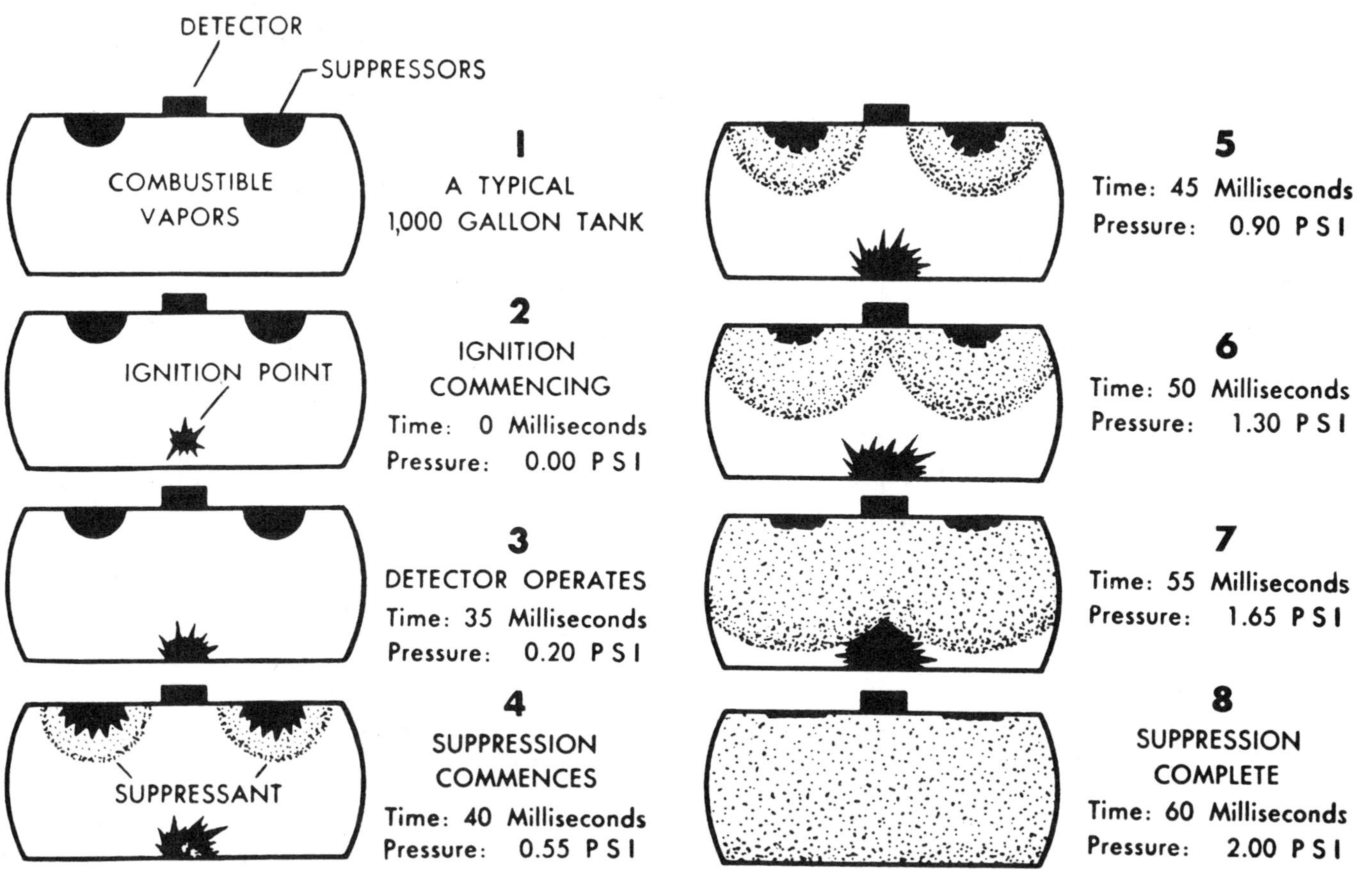

Figure 11–14. A schematic diagram of suppression of an explosion in a large cylindrical tank. (Reprinted with permission from National Fire Protection Association.)

BLUE		RED		YELLOW	
IDENTIFICATION OF HEALTH HAZARD		**IDENTIFICATION OF FLAMMABILITY**		**IDENTIFICATION OF REACTIVITY**	
Type of Possible Injury		**Susceptibility to Burning**		**Susceptibility to Release of Energy**	
Signal		**Signal**		**Signal**	
4	Materials which on very short exposure could cause death or major residual injury even though prompt medical treatment were given.	4	Materials which will rapidly or completely vaporize at atmospheric pressure and normal ambient temperature, and which will burn.	4	Materials which are readily capable of detonation or of explosive decomposition or reaction at normal temperatures and pressures.
3	Materials which on short exposure could cause serious temporary or residual injury even though prompt medical treatment were given.	3	Liquids and solids that can be ignited under almost all ambient temperature conditions.	3	Materials that are capable of detonation or explosive reaction but require a strong initiating source, or that must be heated under confinement before initiation, or react explosively with water.
2	Materials which on intense or continued exposure could cause temporary incapacitation or possible residual injury unless prompt medical treatment is given.	2	Materials that must be moderately heated or exposed to relatively high ambient temperatures before ignition can occur.	2	Materials that are normally unstable and readily undergo violent chemical changes but do not detonate; also materials that may react with water violently, or that may form potentially explosive mixtures with water.
1	Materials which on exposure would cause irritation but only minor residual injury even if no treatment is given.	1	Materials that must be preheated before ignition can occur.	1	Materials that are normally stable, but that can become unstable at elevated temperatures and pressures, or that may react with water with some release of energy, but *not* violently.
0	Materials which on exposure under fire conditions would offer no hazard beyond that of ordinary combustibles.	0	Materials that will not burn.	0	Materials that are normally stable even under fire explosive conditions, and that are not reactive with water.

Figure 11–15. The degree of hazard can be quickly identified with this system. (Adapted from National Fire Protection Association.)

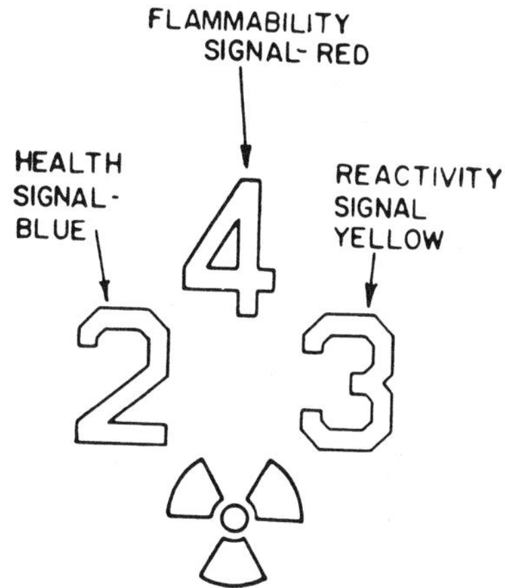

Fig. 1. For Use Where White Background is Not Necessary.

WHITE ADHESIVE-BACKED PLASTIC BACKGROUND PIECES - ONE NEEDED FOR EACH NUMERAL, THREE NEEDED FOR EACH COMPLETE SIGNAL.

Fig. 2. For Use Where White Background is Used With Numerals Made From Adhesive-Backed Plastic

WHITE PAINTED BACKGROUND, OR, WHITE PAPER OR CARD STOCK

4
2 3
W

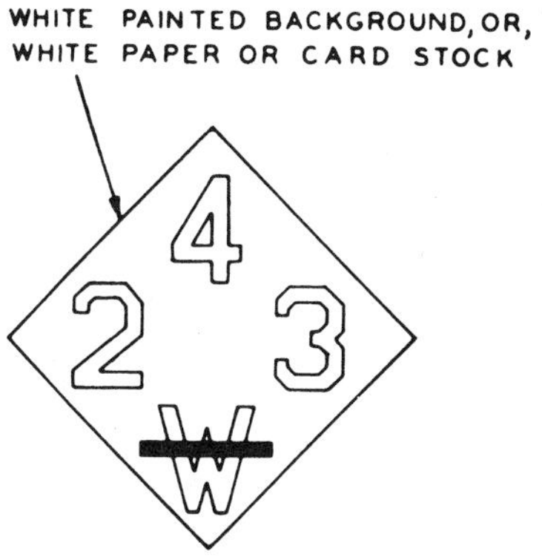

Fig. 3. For Use Where White Background is Used With Painted Numerals, or, For Use When Signal is in the Form of Sign or Placard

ARRANGEMENT AND ORDER OF SIGNALS — OPTIONAL FORM OF APPLICATION

Distance at Which Signals Must be Legible	Size of Signals Required
50 feet	1″
75 feet	2″
100 feet	3″
200 feet	4″
300 feet	6″

NOTE:
This shows the correct arrangement and order of signals used for identification of materials by hazard

Figure 11–16. NFPA hazard signal arrangement for in-plant use only. See NFPA Standard No. 704 for dimensional and other details. Meaning of number system is explained in Figure 11–15. (Reprinted with permission from National Fire Protection Association.)

The three categories of hazards are identified for each material: health, flammability, and reactivity (stability). The diamond shape also provides a place at the bottom of the diamond for special instructions, such as "W" to indicate "water reactive." The order of severity, under fire conditions, in each category is indicated numerically by five divisions ranging from four, which indicates a severe hazard, to zero, which indicates that no special hazard is involved. Colors may be used to identify each hazard category further: blue for health, red for flammability, and yellow for reactivity.

NFPA Health Hazards

A health hazard, as defined by the NFPA, is any property of a material that either directly or indirectly can cause injury or incapacitation, either temporary or permanent, from exposure by contact, inhalation, or ingestion. In general, the health hazard in fire fighting is that of a single exposure that may vary from a few seconds up to an hour. The physical exertion caused by fire fighting may intensify the effects of any exposure.

Health hazards arise from two sources: (1) the inherent properties of the material and (2) the toxic products of combustion or decomposition of the material. Common hazards from burning of ordinary combustible materials are not included.

The degrees of hazard under fire conditions are ranked according to the probable severity of the hazard to personnel, as shown in Figure 11–15. The degree of hazard should indicate (1) that people can work safely only with specialized protective equipment, (2) that they can work safely with suitable respiratory protective equipment, or (3) that they can work safely in the area with ordinary clothing.

NFPA Flammability Hazards

NFPA flammability hazards deal with the degree of susceptibility of materials to burning, even though some materials that burn under one set of conditions will not burn under others. The form or condition of a material, as well as its properties, affects the hazard. The degrees of hazard are ranked according to the susceptibility of materials to burning (Figure 11–15).

NFPA Reactivity

NFPA reactivity (instability) hazards deal with the susceptibility of materials to releasing energy. Some materials can rapidly release energy by themselves, as by self-reaction or polymerization. They can also undergo violent eruptive or explosive reaction if contacted with water or other extinguishing agents, or with certain other materials. The degrees of hazard are ranked according to ease with which energy is released and the rate and quantity of energy released (Figure 11–15).

The relative reactivity of a material is defined as follows:

- Reactive materials can enter into a chemical reaction with other stable or unstable materials. For purposes here, the other material to be considered is water and only if its reaction releases energy.
- Unstable materials are those that, in the pure state or as commercially produced, will vigorously polymerize, decompose or condense, or become self-reactive and undergo other violent chemical changes.
- Stable materials are those that can normally resist changes in their chemical composition, despite exposure to air, water, and heat as encountered in fire emergencies.

NFPA Hazard Signals

Hazard signals to show the degrees of hazard are shown in Figure 11–16. At the bottom right of the three-part diagram is an open space. This may be used to indicate additional information, such as radioactivity hazards, proper fire-extinguishing agents, skin hazards, pressurized containers, protective equipment required, or unusual reactivity with water. The recommended signal to indicate this unusual reactivity with water and to alert the fire-fighting personnel not to use water is the letter *W* with a long line through the center (Figure 11–16).

Shipping Regulations

Identification of hazardous materials during shipment is specified by regulations of the U.S. government as well as the governments of other countries, and it differs from the NFPA identification system just described. Hazardous materials carried by U.S. rail, air, water, and public highways are regulated by 49 *CFR*, Parts 170–180. "An index to the Hazardous Materials Regulations" is published by the U.S. Department of Transportation. (See References, at the end of this chapter.)

FIRE PROTECTION IN THE COMPUTER ROOM

If a fire occurs in a computer room, the numerous electrical components and outlets involved and the quantities of paper used and stored present particular hazards. Fire suppression systems must be designed with three goals in mind: (1) extinguishing fires before they can do damage or cause injury, (2) allowing workers to escape the area unharmed, and (3) protecting vulnerable electronic hardware and software.

Management should consider the following items when designing a fire protection system:

- Forms of equipment protection include hand-held fire extinguishers of the carbon dioxide type. A Class A fire extinguisher should be available for use on any paper or other ordinary combustible-type fire. Each extinguisher should be clearly labeled with its function and class.
- An automatic fire detection system with UL-listed smoke and heat detectors should be provided. Wire it to a central station supervisory service to ensure quick fire department response in an emergency.
- A dual system using both sprinklers and a total flood system is an effective method of automatic fire suppression. Using this method, primary protection is furnished by the system; if the system fails, the sprinkler system acts as secondary protection. It is important that the electric equipment is deenergized before waterflow or system "dump." This can be done manually by operators before they leave the room, or automatically by means of a single triggered activation of the system. Automatic shutdown should be provided for equipment that normally runs unattended for long periods of time. Automatic "power down" programs can be added to existing software to avoid "head crashes" that can damage computer disks.
- The sprinkler system protecting the computer room should be valved separately from other sprinkler systems in the building. Engineering needs to devise a system for drying out the equipment to reduce further damage.
- Noncombustible, waterproof covers will minimize water damage to computer equipment.

The issue of using sprinklers on computer equipment has sparked considerable controversy. According to NFPA, *Fire Protection Handbook*, 16th edition,

> Automatic sprinkler protection and water-spray fixed systems are valuable as a means of reducing fire damage, even where such electrical or

electronic equipment may be exposed. There should be little concern relative to the shock hazard, or of the water causing excessive damage to the equipment. Experience has proven that if a fire activates sprinklers, the sprinklers, if properly installed and maintained, provide for effective fire protection with virtually no hazard to personnel and with no measurable increase in damage to the equipment (as compared with the damage done by heat, flame, smoke, and manual hose streams).

The philosophy that should be followed is to use a total flood system as primary protection, then install automatic sprinklers as secondary protection.

If the computer equipment is deenergized before it gets wet, and if it is quickly dried out afterwards, little damage should result. A hand-held hair dryer is a quick, convenient way to dry out printed circuit boards. Make sure equipment is thoroughly dry before reenergizing it.

SUMMARY

- Fire protection includes procedures for preventing, detecting, and extinguishing fires to protect employees and property and to ensure continued operations. To accomplish these goals, companies must develop fire protection programs.
- The primary purpose of such programs is to prevent fires and train employees in proper procedures should a fire break out. Employees should know their roles in detecting a fire and in transmitting an alarm, evacuating a building, confining the fire, and extinguishing the fire. Fire protection programs should enable companies to reduce hazards significantly.
- Fire protection engineering is a highly developed engineering specialization. Achieving the most efficient fire protection system requires the involvement of architects, interior designers, urban planners, building contractors, electrical and structural engineers, fire detection-system manufacturers, building safety engineers, and local fire departments.
- As a first step in fire prevention, every establishment should set up a system of periodic fire inspections for every operation and ensure that proper fire-extinguishing equipment is on hand and in good working order. The person in charge of fire prevention and protection should establish the schedule, determine routing reports, maintain a complete list of inspected items, and set up regular fire drills for all personnel.
- When a fire breaks out, good communication is vital to alert workers to the emergency and to mobilize fire protection forces. Companies should have coded fire alarms with complete backup systems.
- The chemistry of fire is highly complex and can be classified into flame fire (premixed and diffusion) and surface fire. Fires can be controlled by cooling burning materials, removing oxygen from the fire, inhibiting the chemical chain reaction, and removing fuel. The objective of these methods is to interrupt the chain reaction in a fire. Fire-extinguishing agents help control fires in one or more of these ways.
- The NFPA has developed four classifications of fires. Class A fires occur in ordinary materials and can be extinguished by water or dry-chemical agents. Class B fires occur in the vapor-air mixture over the surface of flammable liquids and are extinguished by limiting air or applying dry chemicals. Class C fires occur in or near energized electrical equipment and can be put out by dry-chemical agents or CO_2. Class D fires occur in combustible metals and require special techniques and extinguishing agents and equipment to put out.
- Fire protection measures are effective only if they are based on a proper analysis and evaluation of the risk of fire. The best fire protection minimizes loss of life and property. Companies must organize a systematic fire hazard survey of all aspects of a facility and operations. The survey should aim to ensure worker safety and to estimate potential fire losses, continuity of operations, and property protection.
- In designing fire safety into a building, management must consider fire-fighting access to a building's interior, ventilation, connections for sprinklers and
- Standpipes, traffic and transportation routes, and fire department and water access to the site. Design elements must be as fire resistant as possible and help to minimize fire hazards.
- Construction methods can help to confine fires and control smoke through proper design of stairways, fire walls, fire doors, separate units, ventilation ducts, physical barriers, and fire exits.
- Companies must act to eliminate the causes of industrial fires by using only approved equipment, establishing safe work practices, and enforcing good housekeeping procedures. Workers should be trained to spot unsafe or hazardous conditions and report them immediately.
- Fire detection must be part of every fire protection system. Its two main tasks involve (1) giving an early warning to enable building occupants to escape and (2) starting extinguishing procedures. Means of fire detection can be through a human observer; automatic sprinklers; smoke, flame, or heat detectors; or, best of all, a combination of these.
- Plants or buildings should be equipped with fire alarm signal systems that clearly communicate to all personnel where the fire is located and that summon appropriate fire-fighting units. Employees must be trained to respond properly to alarm signals. Spacing, location, and maintenance of fire detectors depend on the type of building, its operations, and its materials.
- Portable fire extinguishers are listed as Class A, B, C, D, or a combination of A, B, and C, depending on the type of fire they are designed to extinguish. This equipment

must be approved by a recognized testing laboratory, located in accessible areas, clearly marked as to class and type of fire, and easily operated by workers.

- Sprinklers and water spray systems come in many varieties, depending on the type of building, the operations performed in it, and the materials used. Sprinkler systems include wet-pipe, dry-pipe, preaction, deluge, combined dry-pipe and preaction, and limited water-supply systems.
- All systems require a reliable water supply of ample capacity and pressure. Automatic sprinklers are the most common and effective of all fixed fire-extinguishing systems and can serve as fire alarms as well as fire protection. Sprinkler systems and their water supplies must be inspected and tested regularly to ensure they function properly.
- Fire hydrants, fire hoses, and hose nozzles should be available for immediate use. Hydrants are particularly effective in large plants where parts of the plant are far away from public fire hydrants or when no public hydrants are available. Fire hoses and nozzles should be inspected and maintained in good repair, and workers should be trained in their proper use during emergencies.
- Special fire hazards may require special fire-extinguishing or control agents other than water. These systems are usually installed to supplement, not replace, automatic sprinklers and other fixed or portable fire protection equipment. They must be engineered to fit the particular circumstances of a particular hazard. These special agents include foam systems, dry-chemical piped systems, steam systems, and inert gas systems.
- Foam systems are the most extensively used of these special agents. Foam can be generated by chemical or mechanical means. The principal types of foams available are low-expansion for fighting Class B fires and high-expansion for fighting Class A fires.
- In the diamond-shaped symbol of the NFPA identification system, the three categories of hazards are identified for each material: health, flammability, and reactivity, with a space for special instructions. The order of severity in each category is indicated by numbers in which four is a severe hazard and zero indicates relatively no hazard.
- Health hazards arise from inherent properties of materials and the toxic products of their combustion. Workers must use protective respiratory equipment when fighting fires involving toxic-vapor producing materials. Flammability hazards simply indicate how susceptible materials are to burning. Reactivity deals with how likely materials are to release energy in a fire.
- Hazard signals show the degree of these three categories of hazard for any material. These signals indicate such dangers as radioactivity risks, pressurized containers, or unusual reactivity with water or other chemicals.
- Identification of hazardous materials is required by federal law whenever the materials are shipped by rail, air, water, and public highways. Labels should indicate the nature of the hazard and the best means for extinguishing any fires.
- Fire protection systems for computer rooms should be designed to activate immediately after a fire is detected, to shut down the computer system automatically, to protect electrical connections and circuit boards, and to shut off automatically once the fire is extinguished.

REFERENCES

American Gas Association, 1515 Wilson Boulevard, Arlington, VA 22209. (Lists of approved gas appliances and accessories.)

American Insurance Association, 1130 Connecticut Avenue, NW, Suite 1000, Washington, DC 20036. (Publications catalog available.)
"Judging the Fire Risk," LC-PM-05-678.
"Tested Activities for Fire Prevention Committees," LC- PM-02-678.

American National Standards Institute, 11 West 42nd Street, New York, NY 10036.
Practices for Respiratory Protection for the Fire Service, ANSI Z88.5-1981.
Safety in Welding and Cutting, ANSI Z49.1-1994.

American Petroleum Institute, 1220 L Street, NW, Washington, DC 20005. (Publications and materials list available.)

American Welding Society, 550 NW Le Jeune Road, Miami, FL 33126. (Publications catalog available.)

Building Officials & Code Administrators International, Inc., 4051 West Flossmoor Road, Country Club Hills, IL 60477.
The BOCA Basic Building Code (issued every three years).
The BOCA Basic Fire Prevention Code (issued every three years).

Factory Mutual System Research Organization, 1151 Boston-Providence Turnpike, Norwood, MA 02062.
FM System Approval Guide—Equipment, Materials, Services for Conservation of Property. (Loss prevention data and books.)

Fire Equipment Manufacturers' Association, Inc., Thomas Associates, Inc., 1230 Keith Bldg., Cleveland, OH 44115.
Inspection, Recharging and Maintaining Portable Fire Extinguishers.

International Association of Fire Fighters, 1750 New York Avenue, NW, Washington, DC 20006.
"Firefighter Mortality Report," 1976.

National Association of Fire Equipment Distributors, Inc., c/o Smith, Bucklin and Associates, Mgrs., 111 East Wacker Drive, Chicago, IL 60601.
"Portable Fire Extinguisher Selection Guide," G117.

National Fire Protection Association, 1 Batterymarch Park, Quincy, MA 02269.
Fire Brigade Training Manual.
Fire Protection Handbook, latest edition.
National Fire Codes, latest edition.
NFPA Codes and Standards:
Carbon Dioxide Extinguishing Systems, NFPA 12.

Chimneys, Vents, Fireplaces, and Solid Fuel Burning Appliances, NFPA 211.
Cleaning Small Tanks and Containers, NFPA 327.
Dry Chemical Extinguishing Systems, NFPA 17.
Exhaust Systems for Air Conveying of Materials, NFPA 91.
Explosion Prevention Systems, NFPA 69.
Fire Doors and Windows, NFPA 80.
Fire Hazard Properties of Flammable Liquids, Gases, and Volatile Solids, NFPA 325.
Fire Prevention in Use of Cutting and Welding Processes, NFPA 51B.
Hazardous Chemicals Data, NFPA 49.
Identification of the Fire Hazards of Materials, NFPA 704.
Installation of Air Conditioning and Ventilating Systems, NFPA 90A.
Installation of Deluge Foam-Water Sprinkler and Foam-Water Spray Systems, NFPA 16.
Installation of Private Fire Service Mains, NFPA 24.
Installation of Sprinkler Systems, NFPA 13.
Life Safety Code, NFPA 101.
Low-Expansion Foam, NFPA 11.
Methods of Fire Tests for Flame Resistant Textiles and Films, NFPA 701.
National Electrical Code, NFPA 70.
Ovens and Furnaces, NFPA 86.
Portable Fire Extinguishers, NFPA 10.
Protection from Exposure Fires, NFPA 80A.
Standpipe and Hose Systems, NFPA 14.
Test for Critical Radiant Flux of Floor Covering Systems Using a Radiant Heat Energy Source, NFPA 253.
Venting of Deflagrations, NFPA 68.
Water Spray Fixed Systems for Fire Protection, NFPA 15.
Water Tanks for Private Fire Protection, NFPA 22.
NFPA Inspection Manual, latest edition.
Principles of Fire Protection.
"Publications Catalog."

National Safety Council, 1121 Spring Lake Drive, Itasca, IL 60143.

Society of Fire Protection Engineers, 60 Batterymarch Street, Boston, MA 02110.

Underwriters Laboratories, Inc., 333 Pfingsten Road, Northbrook, IL 60062.
Building Construction and Materials List.
Electrical Equipment List.
Fire Protection Equipment List.
Gas and Oil Equipment List.

U.S. Department of Agriculture, 14th Street and Independence Avenue, SW, Washington, DC 20250.
Technical Bulletin 49.

U.S. Department of Commerce, National Fire Prevention and Control Administration, Washington, DC 20234.

U.S. Department of Commerce, National Institute of Standards and Technology, Gaithersburg, MD 20899.

U.S. Department of Labor, Occupational Safety and Health Administration, 200 Constitution Avenue, NW, Washington, DC 20210.
Code of Federal Regulations, Title 29.
Section 1910.154.

U.S. Department of Transportation, Office of Hazardous Materials, 400 7th Street, SW, Washington, DC 20590. "An Index to the Hazardous Materials Regulations." Available from Superintendent of Documents, U.S. Government Printing Office, Washington, DC 20402.
Code of Federal Regulations, Title 49. Parts 170–180.

REVIEW QUESTIONS

1. Employees should know their role in what four general procedures in the event a fire starts in a facility?
a.
b.
c.
d.

2. What organizations can provide safety and health professional with the names of experts in the area of fire protection?

3. Which of the following are responsible for 66% of deaths from fire in buildings?
a. Contact with direct flame
b. Smoke
c. Toxic gas inhalation
d. Only a and b
e. Only b and c

4. List three techniques used to protect adjacent buildings from fire.
a.
b.
c.

5. What are the four ways fires can be controlled?
a.
b.
c.
d.

6. Under what three conditions should an employee never attempt to fight a fire with an extinguisher?
a.
b.
c.

7. Class B fires occur:
a. In ordinary materials, such as wood, paper, excelsior, rags, and rubbish
b. In or near energized electrical equipment where nonconducting extinguishing agents must be used

c. In the vapor-air mixture over the surface of flammable liquids, such as gasoline, oil, grease, paints, and thinners
d. In combustible metals such as magnesium, titanium, zirconium, lithium, potassium, and sodium

8. List the three objectives of fire safety design and construction of a building.
a.
b.
c.

9. List the four construction methods that help confine fires.
a.
b.
c.
d.

10. Fire doors should be installed in accordance with what regulation?

11. Name eight factors contributing to industrial fires.
a.
b.
c.
d.
e.
f.
g.
h.

12. Briefly describe the four critical variables in the selection of a fire-detection device.
a.
b.
c.
d.

13. How does the dry-chemical portable fire extinguisher work?

14. Fires that occur near energized electrical equipment where nonconducting extinguishing agents must be used are classified as:
a. Class A fires
b. Class B fires
c. Class C fires
d. Class D fires

15. Most fire protection engineers consider which of the following to be the most effective fire-fighting tool?
a. Fire hoses
b. Sprinkler systems
c. Automatic fire detectors
d. Alarm systems

12

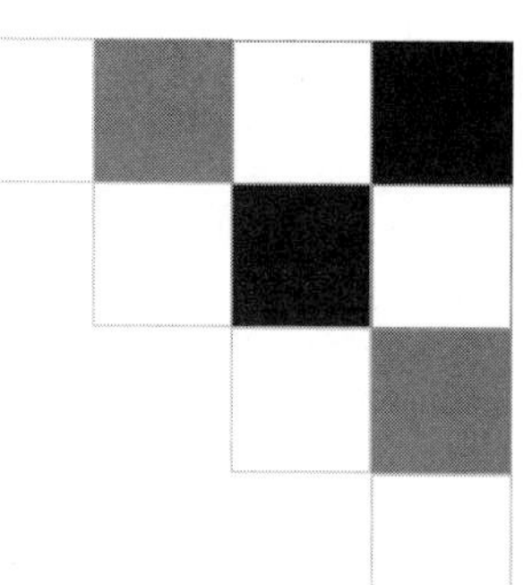

Flammable and Combustible Liquids

Revised by

John F. Montgomery, Ph.D., CSP, CHCM, CHMM

Precautions must be observed when receiving, storing, handling, and using flammable and combustible liquids. Because the specific characteristics of flammable and combustible liquids vary, and thus the handling precautions required for them, it is impossible to cover every detail of the safe handling and use of every liquid. Therefore, consult regulations, local codes, fire underwriters, the National Fire Protection Association (NFPA), trade associations, and specific handbooks for detailed information. Observe U.S. Department of Transportation (DOT) regulations and Occupational Safety and Health Administration (OSHA) standards. (See References at the end of this chapter.)

DEFINITIONS

As defined by NFPA 30, *Flammable and Combustible Liquids Code,* a flammable liquid is any liquid having a closed-cup flash point below 100 F (37.8 C) and having a vapor pressure not exceeding 40 psia (276 kPa) at 100 F. Flammable liquids are classified as shown below. (See NFPA 30 and NFPA 321, *Basic Classification of Flammable and Combustible Liquids.*) Class I shall include flammable liquids having flash points below 100 F (37.8 C) and having a vapor pressure less than or equal to 40 psia (276 kPa) at 100 F (37.8 C). These liquids may be subdivided as follows:

- **Class IA** shall include those having flash points below 73 F (22.8 C) and those having a boiling point below 100 F (37.8 C).
- **Class IB** shall include those having flash points below 73 F (22.8 C) and those having a boiling point at or above 100 F (37.8 C).
- **Class IC** shall include those having flash points at or above 73 F (22.8 C) and below 100 F (37.8 C).

Combustible liquids are those with flash points at or above 100 F (37.8 C), but below 200 F (93.3 C), closed cup. They are divided into four classes (II, III, IIIA, and IIIB). Although they do not ignite as easily as flammable liquids, they can be ignited under certain circumstances and thus must be handled with caution. Combustible liquids are classified as follows:

- **Class II** shall include those having flash points at or above 100 F (37.8 C) and below 140 F (60 C).
- **Class III** shall include those having flash points at or above 140 F (60 C) and may be subdivided as follows:
- **Class IIIA** shall include those having flash points at or above 140 F (60 C) and below 200 F (93.4 C).
- **Class IIIB** shall include those having flash points at or above 200 F (93.4 C).

The U.S. Department of Transportation, in 49 *CFR* "Hazardous Materials Regulations," Parts 170–179, defines a flammable liquid as any liquid that gives off flammable vapors at or below a temperature of 80 F (26.7 C). This definition is important because the U.S. DOT Flammable Liquid Label provides one means of identifying containers of flammable liquids for shipping, receiving, and transportation. However, there are exceptions to this definition.

Moreover, other regulatory bodies have adopted other definitions of a hazardous liquid. For example, the New York City Codes and the Federal Hazardous Substances Act define flammability by using the open-cup test. Such definitions are important if a plant must comply with regulations. Use the testing method cited by the regulation or code that applies to the liquid in question.

Some of the more common flammable and combustible liquids are gasoline, crude oils, various hydrocarbons, alcohols, and their by-products. They are all chemical combinations of hydrogen and carbon that may also contain oxygen, nitrogen, sulfur, or other elements.

Consider manufactured liquids and fluid commodities that contain flammable liquids, such as paints, floor polish, cleaning solutions, driers, and varnishes, as flammable liquids. Classify them according to the flash point of the mixture. Precautions for handling and using these liquids will differ according to their flash points, volatility, toxicity, and the percentage of flammable liquid in the mixture.

Flammable and combustible liquids vaporize and form flammable mixtures with air when left in open containers, when allowed to leak or spill, or when atomized or heated. The degree of danger depends largely on (1) the flash point of the liquid, (2) the concentration of vapors in the air (whether the vapor-air mixture is in the flammable range or not), (3) the possibility of a source of ignition at or above a temperature or energy level high enough to cause the mixture to burst into flame, and (4) the amount of vapors present.

When handling and using flammable liquids, prevent exposing large liquid surfaces to the air. The liquids themselves do not burn or explode. Instead, the vapor-air mixture, formed when liquids evaporate or are heated enough to make them emit flammable vapors, burns or explodes. Therefore, handle and store these liquids in closed containers, or systems, and avoid exposing low-flash-point liquids during use. Also, make sure that all containers are correctly labeled. (See Appendix C, Chemical Hazards, of the National Safety Council's [NSC's] *Fundamentals of Industrial Hygiene,* 4th edition, for details about many liquids.)

Terms used in this chapter will have the following meanings:

- **Autoignition temperature.** Autoignition temperature is the minimum temperature at which a flammable gas-air or vapor-air mixture will ignite from its own heat source or a contacted heated surface without the presence of an open spark or flame.

Vapors and gases will spontaneously ignite at a lower temperature in oxygen than in air. Their autoignition temperature can be lowered by the presence of catalytic substances.

- **Flammable limits.** For gases and vapors of flammable liquids, there is a minimum concentration below which propagation of flame does not occur on contact with a source of ignition. This is known as the lower flammable limit (LFL). There is also a maximum proportion of vapor or gas in air above which propagation of flame does not occur. This is known as the upper flammable limit (UFL).

  For example, a gasoline vapor-air mixture with less than approximately 1% of gasoline vapor is too lean to ignite. Similarly, if there is more than approximately 8% of gasoline vapor, the mixture will be too rich to ignite. Other gases, such as hydrogen, acetylene, and ethylene, have a much wider range of flammable limits. Treat combustible liquids according to the requirements for flammable liquids when they are heated, even though the same liquids when not heated are outside the flammable range.

  Flammable limits are determined at pressures of one atmosphere. Thus, the range will increase as temperature or pressure increases, the UFL being influenced more than the LFL.

- **Flammable range.** The flammable range lies between the LFL and UFL, expressed as the percentage of vapor or gas in air by volume. It is sometimes referred to as the explosive range.

  For example, the flammable range of gasoline vapors in air is generally taken as 1.4% to 7.6%, which is relatively narrow. Thus, a mixture of 1.4% gasoline vapor and 98.6% air is flammable, as are all the intermediate mixtures up to and including 7.6% gasoline vapor and 92.4% air. The range is therefore the difference between the limits, or 6.2% (7.6% – 1.4% = 6.2%).

- **Flash point.** The flash point is the minimum temperature at which a liquid gives off vapor concentrated enough to form an ignitable mixture with air near the surface of the liquid within a vessel specified by the appropriate testing procedure and apparatus. For liquids having viscosities less than 45 SUS (Saybolt universal seconds) and flash points below 200 F (93.4 C), the tag-closed tester is used. (See ASTM D–56–82, *Test Method for Flash Point by Tag-Closed Tester* for this procedure.) For liquids having viscosities equal to or greater than 45 SUS and flash points of 200 F (93.4 C) or higher, the Pensky-Martins closed tester is used. (See ASTM D–93–85, *Test Methods for Flash Point by Pensky-Martins Closed Tester,* for this procedure.)

  For specific liquids, such as aviation turbine fuels, paint, enamel, and varnish, alternative testing methods are available. (See NFPA 321.)

  Although other properties influence the relative hazards of flammable liquids, the flash point is the most significant factor. The relative hazard increases as the flash point decreases. The significance of this property becomes more apparent when liquids of different flash points are compared.

  At ordinary temperatures (under approximately 100 F or 37.8 C), kerosene and No. 1 fuel oil, which have flash points above 100 F (37.8 C), do not give off dangerous quantities of vapor. On the other hand, gasoline gives off vapor at a rate sufficient to form a flammable mixture with air at temperatures of about –50 F (–46 C). (Gasoline is blended to have this characteristic so car engines will start in very cold weather.)

  Any combustible liquid heated to a temperature at or above its flash point will produce ignitable vapors. Heavy fuel oil, when heated to several hundred degrees Fahrenheit, for example, may produce flammable vapors just as readily as gasoline. However, these vapors are less volatile and will condense. The characteristics of combustible liquids are also changed when they are atomized. When such liquids are heated above their flash points or atomized, treat them as flammable liquids.

  When the vapor-air mixture is below the LFL, it is too lean to ignite. The vapor concentration is too low. Conversely, a vapor-air mixture above the UFL is too rich to ignite, because not enough oxygen is present to support ignition. Be aware that vapor-air mixtures will pass in and out of the flammable range as conditions change.

- **Oxygen deficiency.** Oxygen deficiency designates an atmosphere having less than the percentage of oxygen found in normal air. Normally, air contains about 21% oxygen. When the oxygen concentration in air is reduced to approximately 16%, many individuals become dizzy, experience a buzzing in the ears, and have a rapid heart beat.

  NOTE: Testing should be done with properly calibrated instruments approved for the purpose by the National Institute for Occupational Safety and Health (NIOSH). OSHA requires that no one enter a tank or enclosed space that tests less than 19.5% oxygen, unless wearing approved breathing equipment, such as a fresh-air hose mask, self-contained breathing apparatus (SCBA), or self-generating breathing apparatus (SGBA).

- **Oxygen limits.** Usually, for an explosion to occur, enough oxygen must be present along with a vapor concentration in the flammable range. Below 12% to 14% oxygen, flammable vapors may not burn.

The actual limit depends on the inert gas used to decrease the oxygen level and on other variables. Increasing pressures and temperatures above normal ambient conditions will reduce the amount of oxygen required for combustion.

- **Propagation of flame.** Propagation of flame is the spread of flame through the entire volume of the flammable vapor-air mixture from a single source of ignition. A vapor-air mixture above or below the upper or lower flammable limit may burn at the point of ignition without propagating (spreading away) from the ignition source, or vent.
- **Rate of diffusion.** Rate of diffusion indicates the tendency of a gas or vapor to disperse into or mix with another gas or vapor, including air. This rate depends on the density of the vapor or gas as compared with that of air, which is given a value of one. Whether a vapor or gas is lighter or heavier than air determines, to a large extent, the design of a ventilation system. If the vapor or gas is heavier than air, locate the air intake duct slightly above floor level. Conversely, if the vapor or gas is lighter than air, locate the air intake duct just below ceiling level. Locate air exhaust-duct openings so they most effectively remove the vapors.
- **Vapor pressure.** Vapor pressure is the pressure exerted by a volatile liquid under any of the equilibrium conditions that may exist between the liquid and its vapors. One testing method follows ASTM D 323–72, Standard Method of Test for Vapor Pressure of Petroleum Products (Reid Method).
- **Volatility.** Volatility is the tendency or ability of a liquid to vaporize. Such liquids as alcohol and gasoline, because of their well-known tendency to evaporate rapidly, are called volatile liquids. The volatility of liquids is increased when they are heated.

GENERAL SAFETY MEASURES

Flammable and combustible liquids require careful handling. Mixing these liquids and handling them in the presence of sparks or open flames add to the hazards.

Preventing Dangerous Mixtures

Avoid unintentionally mixing flammable and combustible liquids. For example, a small amount of acetone put into a kerosene tank may lower the flash point of the tank's contents. The relatively high volatility of acetone will render the kerosene too dangerous to use. Gasoline mixed with fuel oil may change the flash point enough to make the fuel oil hazardous for home heating or similar uses. The lower-flash-point liquid can ignite, causing the higher-flash-point liquid to act as though it were a flammable liquid.

Use colors or labels or both to identify fill openings, discharge openings, and control valves on equipment containing flammable and combustible liquids. In some cases, pipelines may be painted or banded with distinctive colors and show the direction of a liquid's flow. Mark each tank with the name of the product it contains or otherwise identify the product. Keep lines from tanks of different types and classes of products separated and, preferably, provide separate pumps for different types and classes of products. (See Chapter 11, Fire Protection, for the NFPA system to identify hazardous substances.)

Use a portable container approved by Factory Mutual (FM) or Underwriters Laboratories (UL) for handling flammable liquids in quantities up to 5 gal (19 l). Clearly identify the container with the appropriate Hazcom labeling information.

Smoking

In a building or area where flammable liquids are stored, handled, or used, forbid personnel to smoke or carry strike-anywhere matches, lighters, or other spark-producing devices. The size of the restricted area will depend on the type of products handled, the design of the building, local codes, and local conditions. Conspicuously post approved NO SMOKING signs in buildings and areas where smoking is prohibited.

Static Electricity

Static electricity is generated by the contact and separation of dissimilar materials. For example, static electricity is generated when a fluid flows through a pipe or from an opening into a tank. Figure 12–1 shows several methods of generating static electricity. The principal hazards of static electricity are fire and explosion caused by spark discharges that contain enough energy to ignite flammable or explosive vapors, gases, or dust particles. Also, a worker could be injured because of an involuntary reaction caused by a static spark's shock.

A static spark poses great danger where a flammable vapor may be present in air, such as at the outlet of a flammable-liquid fill pipe, at a delivery hose nozzle, near an open flammable-liquid container, or around a tank truck's fill opening or barrel bunghole. When a difference in electrical potential is present, a spark between two bodies can occur because no good electrical-conductive path exists between them.

Bonding and Grounding

The terms *bonding* and *grounding* are sometimes used interchangeably because their meanings are not understood. The purpose of bonding is to eliminate a difference in the static-electrical-charge potential between

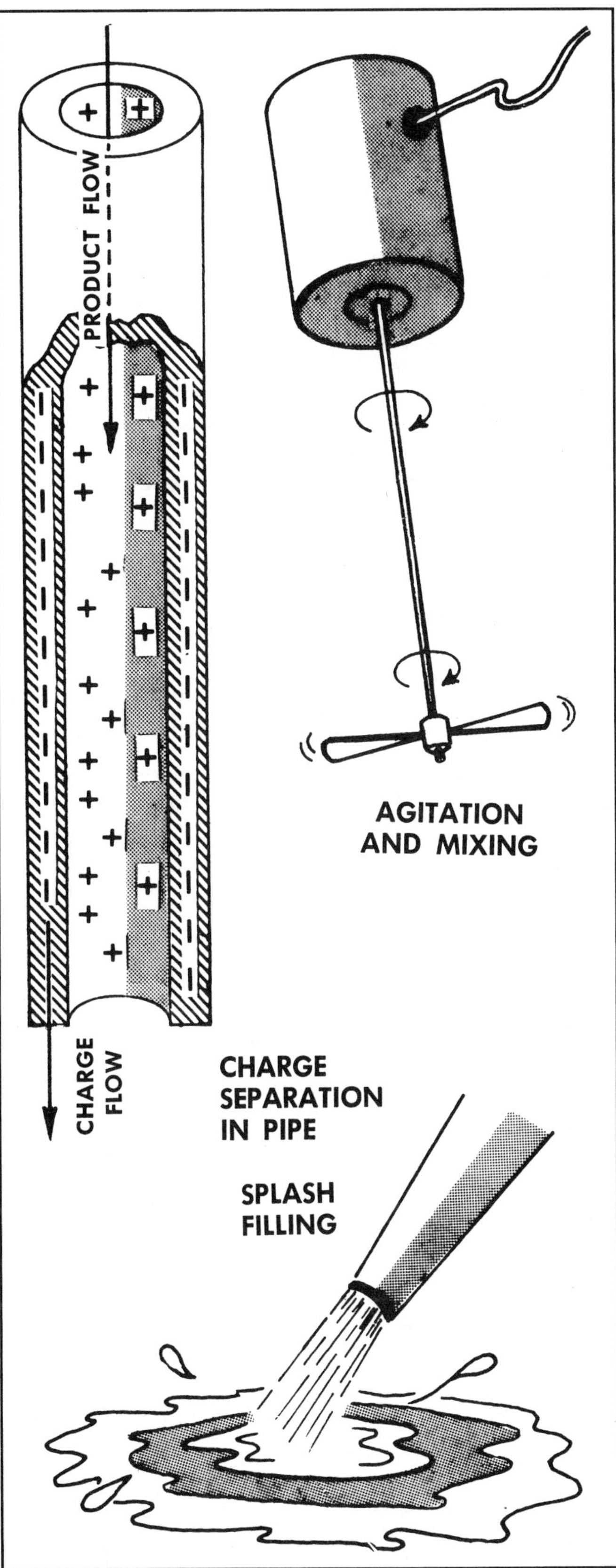

Figure 12–1. These are typical static-producing situations, including charge separation occurring in pipes.

two or more objects. The purpose of grounding is to eliminate a potential difference between an object and the ground (earth) (Figure 12–2). Hence, bonding containers for flammable liquids is necessary to prevent static electricity from causing a spark. (See NFPA 77, *Static Electricity*.) Bonding and grounding are effective only when the bonded objects are conductive materials.

Some materials, such as some plastics, can accumulate significant static electrical charges. However, such a material does not inherently allow a charge to disperse enough throughout the material to permit effective grounding or bonding. Therefore, exercise caution when using plastic construction materials to transfer flammable liquids. Use only plastics that are specifically designed and approved for this type of usage.

When two objects are bonded, charges flow freely between them and do not differ. Therefore, the likelihood of sparking between the objects is eliminated. Bonding will not eliminate the static charge, but it will equalize the potential between the objects bonded so that a spark will not occur between them.

Although bonding will eliminate a difference in potential between objects, it will not eliminate a difference in potential between these objects and the earth unless one of the objects has an adequate conductive path to the earth.

An adequate ground will continuously discharge a charged, conductive body. As a safety measure, use a ground when any doubt exists concerning a situation or when a jurisdictional authority requires it.

When flammable liquids are transferred from one container to another, provide a means of bonding between the two containers before pouring, as shown in Figures 12–3 through 12–5. Bond and ground a flammable-liquids tank truck or tank car to the loading rack (Figure 12–5).

To avoid a spark from the discharge of static electricity during flammable-liquid filling operations, provide a wire bond between the storage container and the container being filled. If a metallic path between the two containers is otherwise present, however, this procedure is unnecessary. Again, remember that some materials, such as some plastics, cannot be effectively grounded and bonded.

When loading a flammable liquid through an open dome in a tank car or tank truck, use a downspout long enough to reach the tank's bottom and thus prevent a static charge generated by the liquid's flow or splash during filling. To provide for the possibility of stray currents and to prevent a difference in static-charge potential between the fill pipe and the tank car, bond loading lines. Preferably, connect the bonding to the rails to avoid operator errors.

Fill pipes should, therefore, be in constant contact with the rim of the upper tank's opening. The loading rack should be grounded, and the tank truck or tank car should be connected to the rack with a bonding wire

CHARGED AND UNCHARGED BODIES INSULATED FROM GROUND

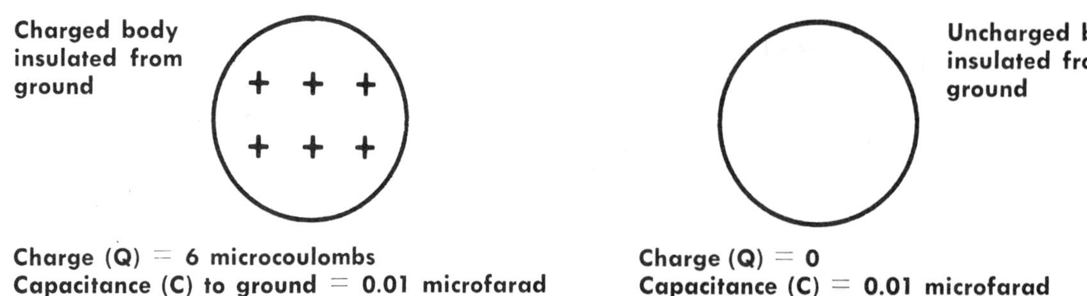

Charge (Q) = 6 microcoulombs
Capacitance (C) to ground = 0.01 microfarad
Voltage (V) to ground and uncharged body = 600 volts

Charge (Q) = 0
Capacitance (C) = 0.01 microfarad
Voltage to ground (V) = 0

Ground

BOTH INSULATED BODIES SHARE THE SAME CHARGE

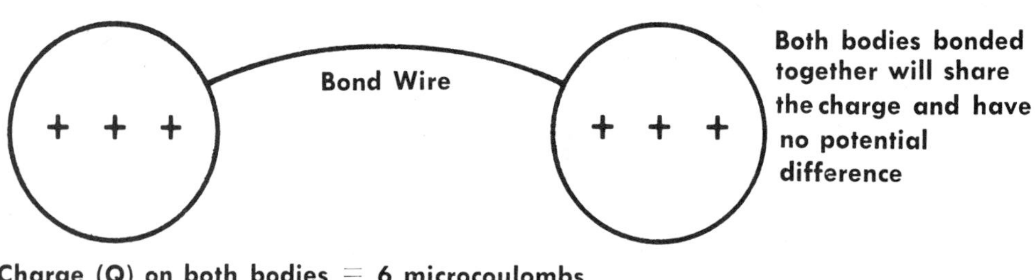

Charge (Q) on both bodies = 6 microcoulombs
Capacitance (C) to ground for both bodies = 0.02 microfarad
Voltage (V) to ground = 300 volts

Ground

BOTH BODIES ARE GROUNDED AND HAVE NO CHARGE

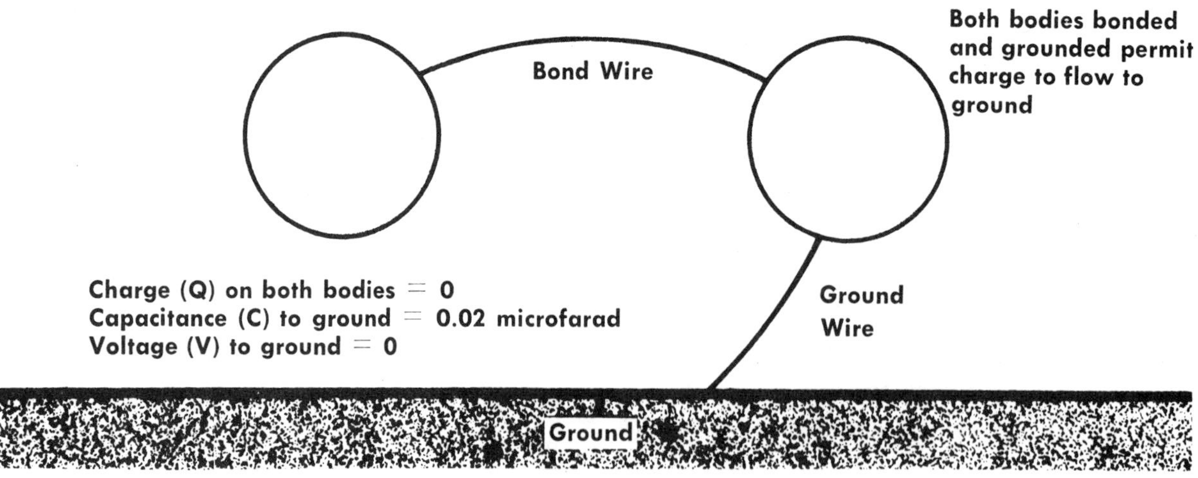

Figure 12–2. Bonding eliminates the difference in static charge potential between objects. Grounding eliminates the difference in static potential between objects and the ground. Both bonding and grounding apply only to conductive bodies and, when properly applied, can be depended on to remove the charge.

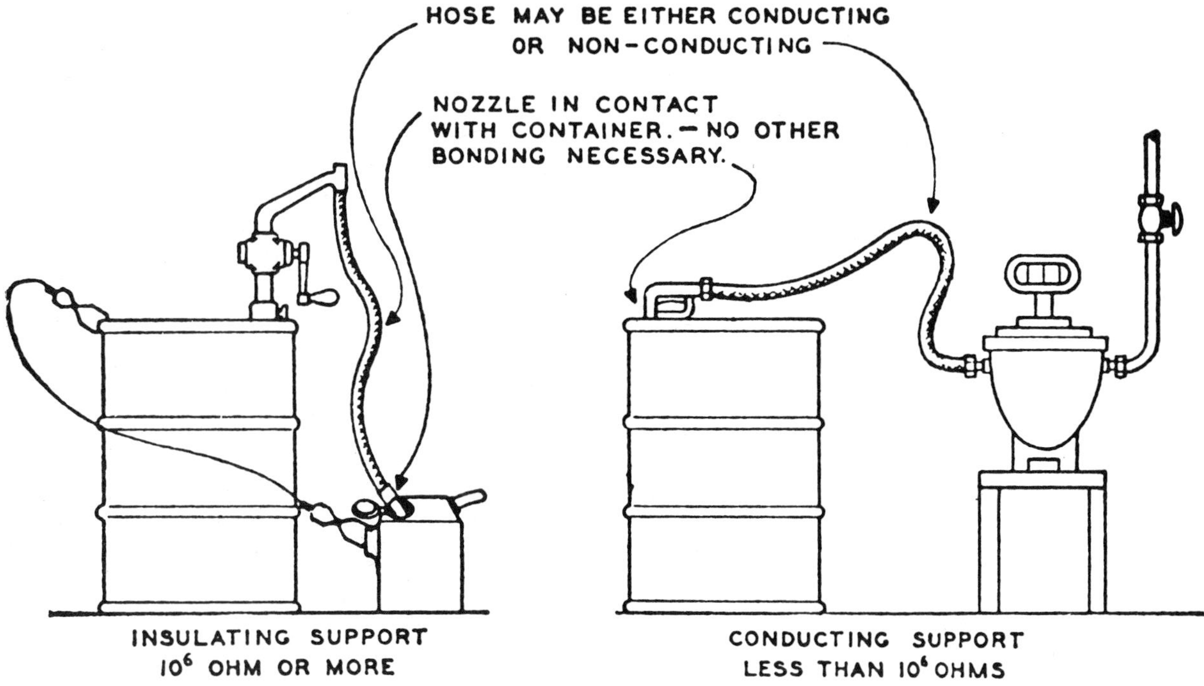

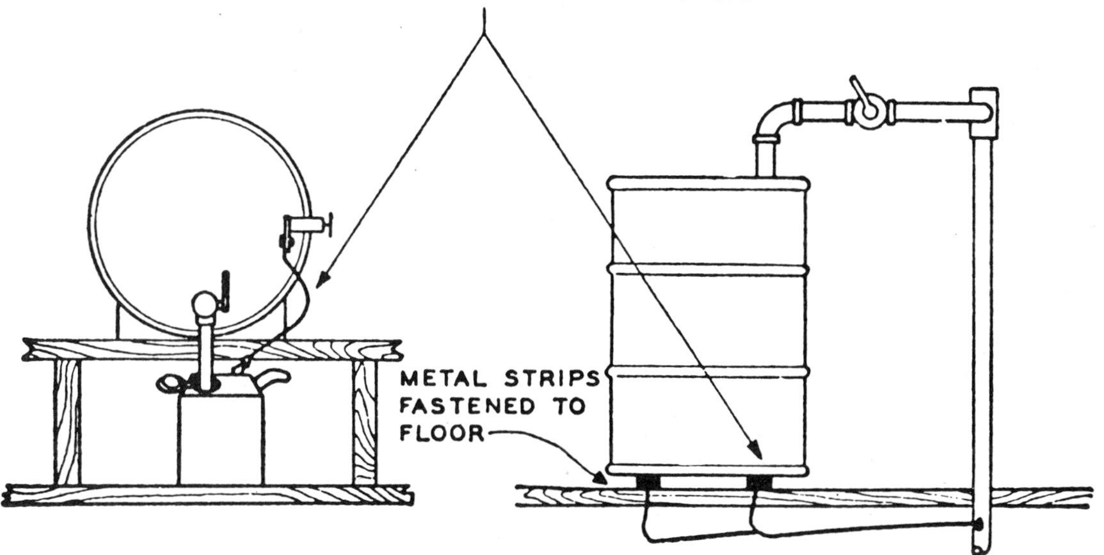

Figure 12–3. When a container is bonded during the filling process, any static electricity generated will be safely discharged. (Reprinted courtesy of National Fire Protection Association.)

(Figure 12–6). For detailed information and exceptions to this general rule, see NFPA 77, *Static Electricity,* and American Petroleum Institute (API) RP2003 82, *Protection Against Ignitions Arising out of Static, Lightning, and Stray Currents,* and NFPA 30, *Flammable and Combustible Liquids.*

Aboveground tanks used for storing flammable liquids need not be grounded unless they rest on concrete or on nonconductive supports. (See NFPA 77.) Ground wires should be uninsulated to permit easy inspection for damage (Figure 12–7).

Petroleum liquids can build up a static charge when they (1) flow through piping or filters, (2) are agitated in a tank or a container, or (3) are subjected to vigorous mechanical movement such as spraying or splashing. Proper bonding and grounding of the transfer system usually drains off the static charge to the ground as fast as it is generated. However, rapid flow

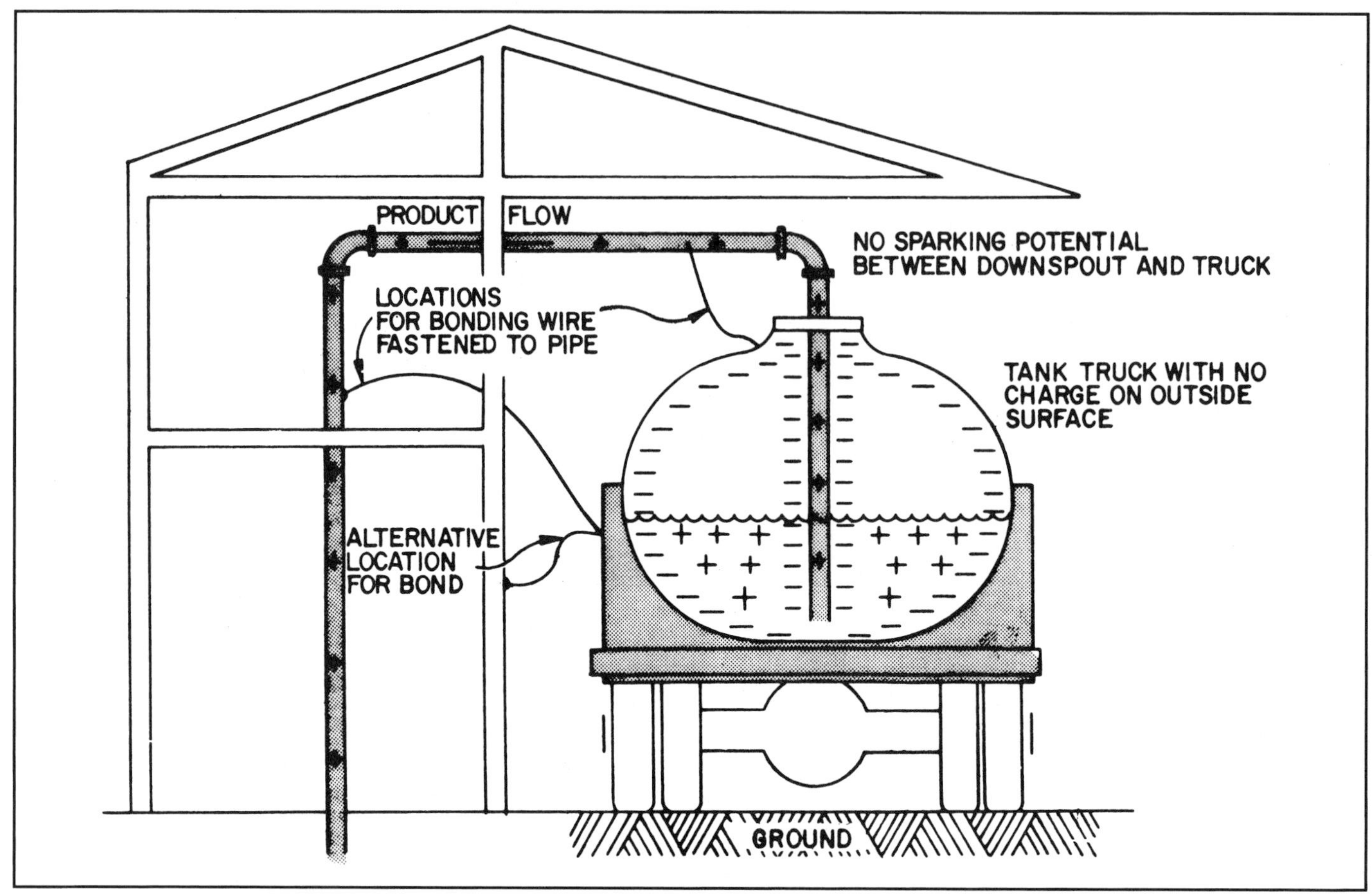

Figure 12–4. Petroleum liquids can build up electrical charges as they flow through piping and become agitated when the tank or container is filled or emptied.

rates in transfer lines can cause very high electrical potentials on the surface of liquids, regardless of a vessel's grounding. Also, some petroleum liquids—especially pure, refined products—are poor conductors of electricity. Even though the transfer system is properly grounded, a static charge may build up on the surface of the liquid in the receiving container. The charge builds up because static electricity cannot flow through the liquid to the grounded metal as fast as it is being generated. If this accumulated charge builds up high enough, a static spark with enough energy to ignite a flammable air-vapor mixture can occur. This happens when the liquid level approaches a body with different electrical potentials, as when a probe is lowered into a tank during sampling or gauging. This situation is especially dangerous when the liquid is of intermediate volatility, as is toluene or ethyl alcohol.

Control this high static charge by reducing the flow rate, using side-flow spill lines to prevent violent splashing, and using relaxation time. In such cases, it is prudent to wait for a period of time—depending on the product, filling rates, and tank size—before gauging. The lag time can last up to four hours after filling, especially in the case of large tanks.

Ground motor frames, starting or control boxes, conduits, and switches according to the requirements for installing electrical and power equipment. This grounding is for personal protection rather than for control of static charges. (See NFPA 70, *National Electrical Code* [NEC].)

Regularly check bonding and grounding systems, depending on their use, for electrical continuity. Inspect the exposed element of such a system for parts that have deteriorated because of corrosion or other damage. Many companies specify that bonds and grounds be constructed of bare-braided flexible wire. Such wire aids inspection and prevents broken wires from being concealed.

When using steam lines to clean an object in the presence of flammable liquids, bond the nozzles on the lines to the surface of the object. Also, be sure that there are no insulated conductive objects on which the steam could impinge and induce a static charge accumulation.

Flat moving belts also are sources of static electricity. If they are made of a conductive material or are coated with a conductive belt-dressing compound, however, static charges will not build up. V-belts are generally not a cause of static charge.

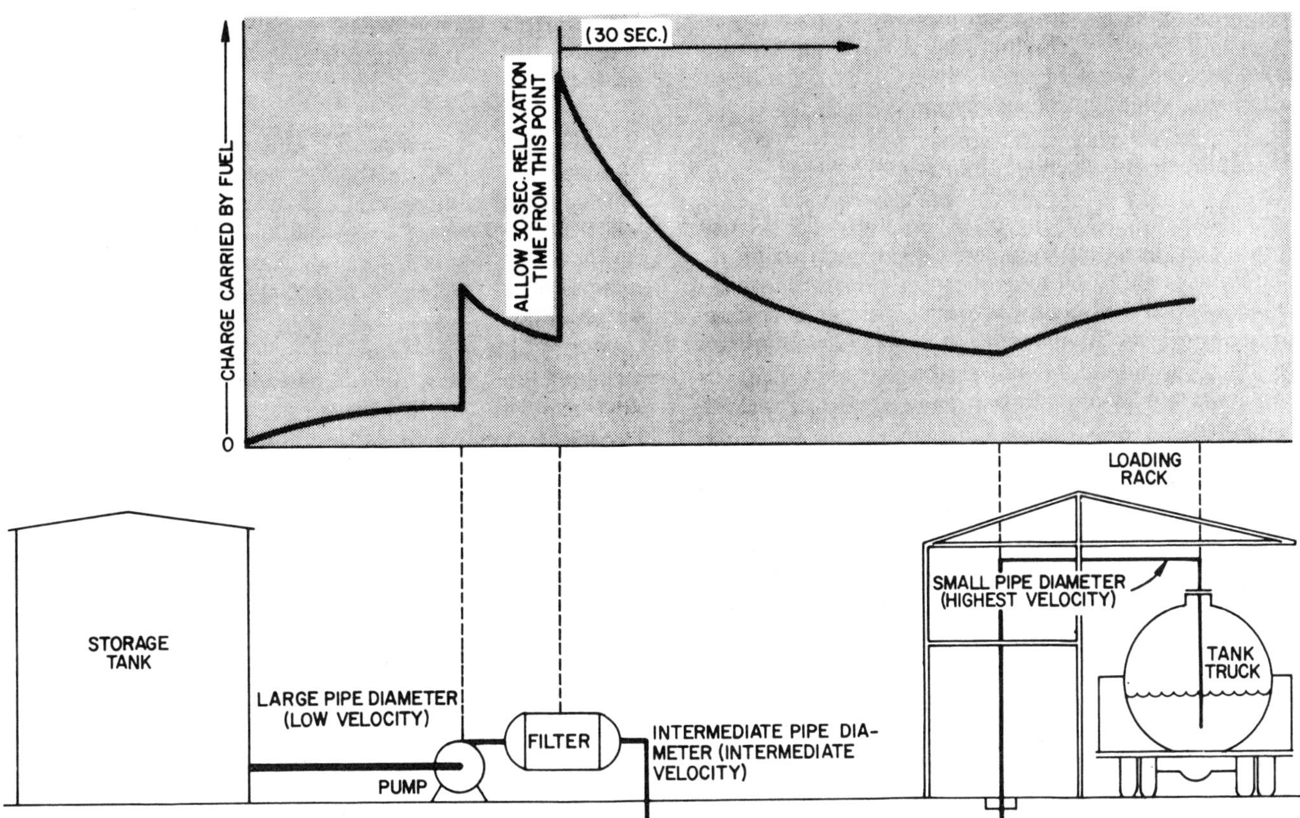

Figure 12–5. This diagram of a tank truck loading facility shows the use of a filter to dissipate the electrical static charge.

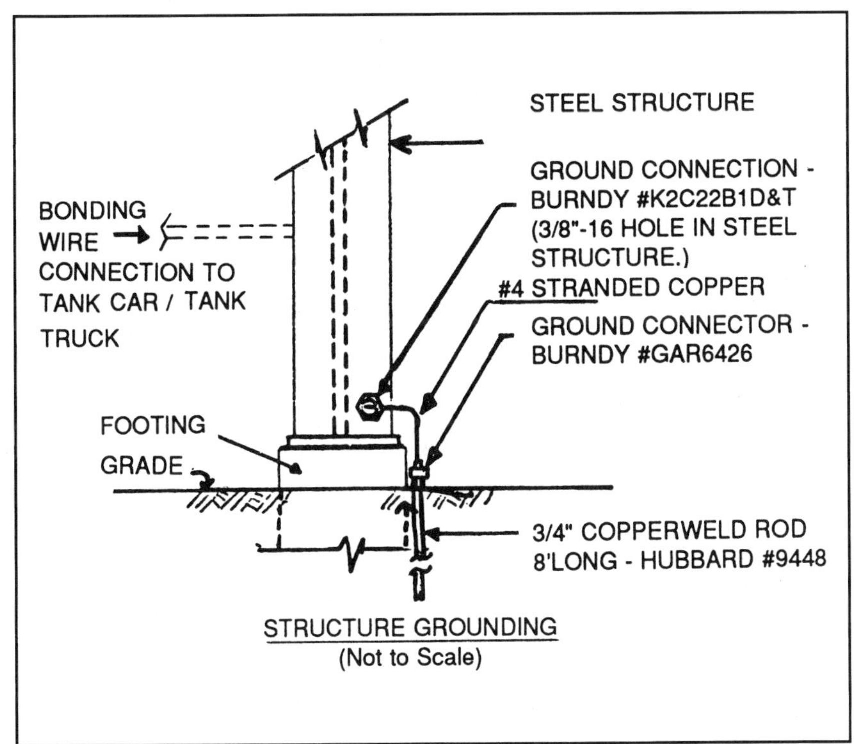

Figure 12–6. This is a detail of grounding/bonding loading rack structure. It is important that the loading rack ground cable connection to the ground rod be above grade to permit continuity and resistance testing to ensure that the rack structure is safely grounded.

Nonconductive materials, such as fabric, rubber, or plastic sheeting, passing through or over rolls also will create charges of static electricity. Static from these materials, as well as from belts, can be discharged with

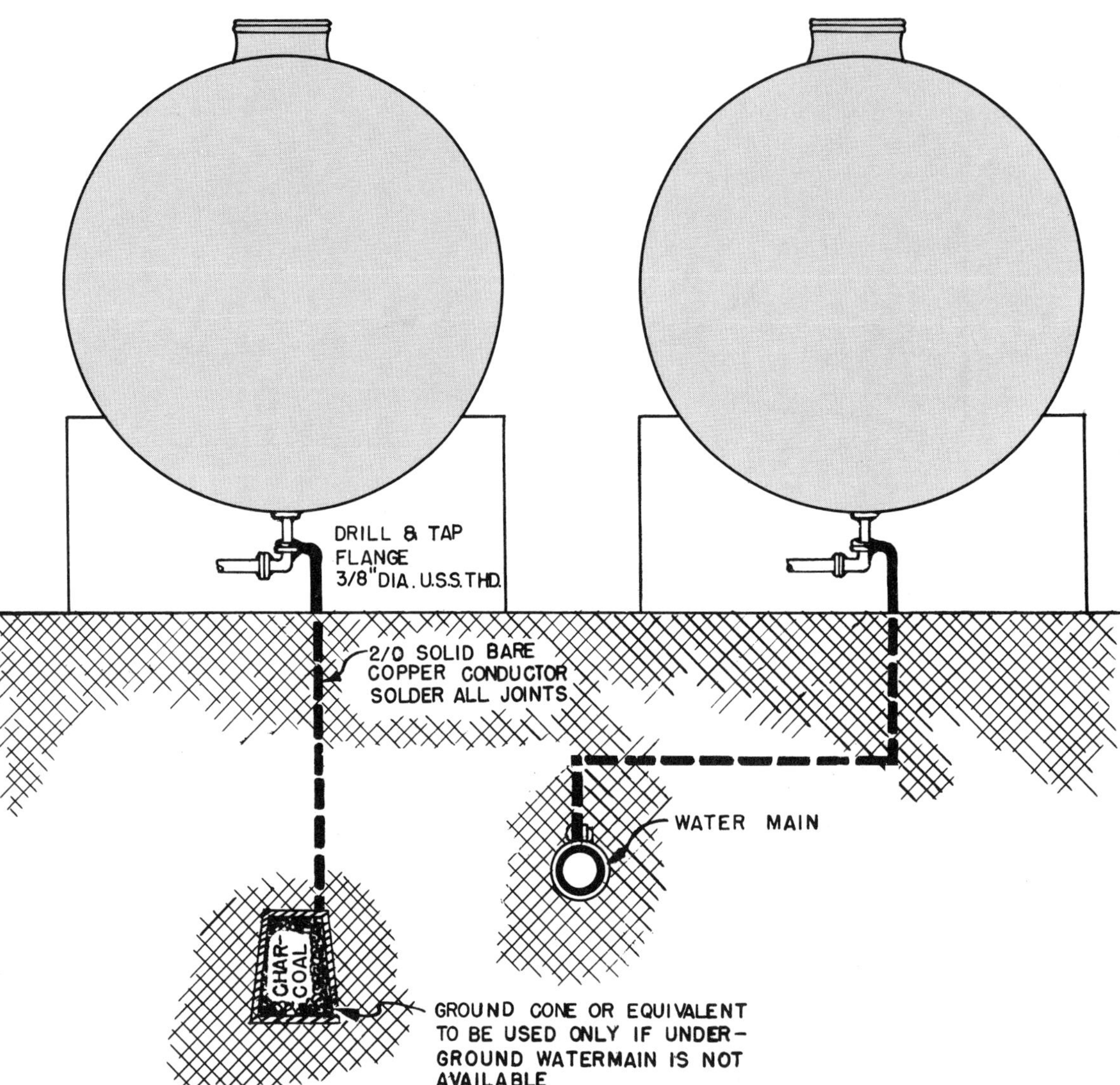

Figure 12–7. If an aboveground flammable liquid storage tank is not inherently grounded, it should be grounded to a water main, a ground cone, or its equivalent.

grounded metal combs or tinsel collectors. Radioactive substances and static neutralizers using electrical discharges are also employed for this purpose.

Electrical Equipment

Where flammable vapors exist, electricity becomes an ignition source if the proper type of electrical equipment for these atmospheres either has not been installed or has not been maintained. When using flammable liquids, consult NFPA 30, *Flammable and Combustible Liquids Code,* and the NEC, as well as local codes.

Spark-Resistant Tools

The hazard of flammable vapors or gases igniting from sparks must be recognized. However, reports of experiments as well as practical experience in the petroleum industry show no significant increase in fire safety by using spark-resistant hand tools in the presence of gasoline and similar hydrocarbon vapors.

Some materials, such as carbon disulfide, acetylene, and ethyl ether, have low ignition-energy points. As a conservative safety measure when using these and similar materials, use special tools designed to minimize the danger of sparks in hazardous locations. For example, leather-faced, plastic, and wood tools are free from the friction-spark hazard. Note, however, that metallic particles could become embedded in them.

HEALTH HAZARDS

Flammable and combustible liquids create health hazards when inhaled or when they contact skin. Irritation results from the solvent action that these liquids have on the skin's natural oils and tissue. Intoxication and other acute and chronic conditions may result from breathing vapors of flammable liquids. Atmospheres that are below flammable limits are not necessarily safe to breathe. (A chemical with an LFL of 10%, for example, would be below the flammable limit if it were present at a concentration of 1%—which is still equal to 10,000 parts per million—and may be above the standards set for one chemical.)

Vapors from flammable and combustible liquids are generally heavier than air. They will flow into pits, tank openings, confined areas, and low places, where they may displace oxygen and contaminate the normal air, thus causing a toxic as well as an explosive atmosphere. Oxygen deficiency may also occur in closed containers, such as a tank that has been closed for a long time and in which rusting has consumed the oxygen. Air out and test all confined spaces for toxic and flammable atmosphere, as well as for the level of oxygen, to determine the protective measures required before allowing personnel to enter. (See *Threshold Limit Values,* published annually by the American Conference of Governmental Industrial Hygienists, 29 *CFR* 1910.1000, and the NSC's *Fundamentals of Industrial Hygiene,* 4th edition—all of which are listed in References at the end of this chapter.)

COMBUSTIBLE-GAS INDICATORS

Unless tests prove otherwise, assume that flammable vapors and toxic mixtures are present in all tanks that have contained or have been exposed at any time to flammable and combustible liquids. Calibration tests for flammable vapor-air mixtures in tanks and other vessels can be made by using a properly calibrated, approved combustible-gas indicator. Use approved chemical-analytical methods to test for toxic substances.

Allow only experienced operators to use a combustible-gas indicator. Operators should follow the manufacturer's instructions for calibrating the unit. For an accurate measurement, correct the reading by using a calibration graph that adjusts the reading according to the atmosphere being tested. Manufacturers supply correlation graphs for many combustible gases.

NOTE: A combustible-gas indicator calibrated for the LFL will not give any reading while sampling a very highly concentrated vapor. It will give no reading because not enough oxygen may be present to cause the vapor to burn in the instrument. It will also give a negative reading while testing vessels containing high concentrations of inert gas.

The operator should check the condition and calibration of the instrument according to the manufacturer's recommended procedures. The manufacturer will generally supply the appropriate test gas to be used for the calibration check. Test gases can also be used for daily operational checks if the instrument does not have internal circuitry to perform this function. At least once a year, thoroughly examine the instrument and test its integrity. The examination can be done by the manufacturer, a recognized standards laboratory, an outside consultant, or a qualified employee on-site.

When using the instrument, do not place the sampling hose where flammable liquid, steam, or water will be drawn into the instrument. Any of these substances would put it out of service. Generally, combustible-gas indicators should not be used to test vapors from heated combustible liquids. Such vapors condense in the sampling lines or combustion chamber and give false results. If dust, particulates, moisture, or leaded gasoline is suspected to be contaminating the environment to be tested, use special filters to avoid inaccurate readings and fouling of the instrument.

A tank can be tested and found to be vapor-free before being cleaned. If it contains sludge or scale, however, flammable or combustible liquid and vapor can be released as soon as the sludge or scale is disturbed. Therefore, consider no tank free of toxic or flammable vapors as long as it contains sludge or scale.

If workers leave a tank for a period of time, such as for lunch or a shift change or overnight, test for oxygen and flammable and toxic gas before allowing workers to reenter the tank. It is essential to use a written permit form when workers must enter confined spaces—especially spaces that have contained flammable or combustible liquids or that require hot-work repairs, such as welding and cutting.

Testing devices are available for carbon monoxide (CO), benzene, hydrogen sulfide, tetraethyl lead, and other specific toxic and flammable vapors.

LOADING AND UNLOADING TANK CARS

This section relates primarily to general safety procedures for nonpressurized tank cars used to transport flammable and combustible liquids. Pressurized tank cars used to transport nonflammable and flammable gases, liquefied petroleum gas (LP gas), and other similar materials will have additional and different safety constraints.

Only trained employees should load or unload tank cars containing flammable or combustible liquids. These workers should understand the possible dangers of fire, explosion, asphyxiation, and toxic effects from exposure to flammable vapors.

Do not load or unload cars before ensuring that no exposed lights, fires, or other sources of ignition are in the area. As an added precaution, stop unloading and pumping operations during electrical storms. A rack with a gangway or bridge to the car helps workers move

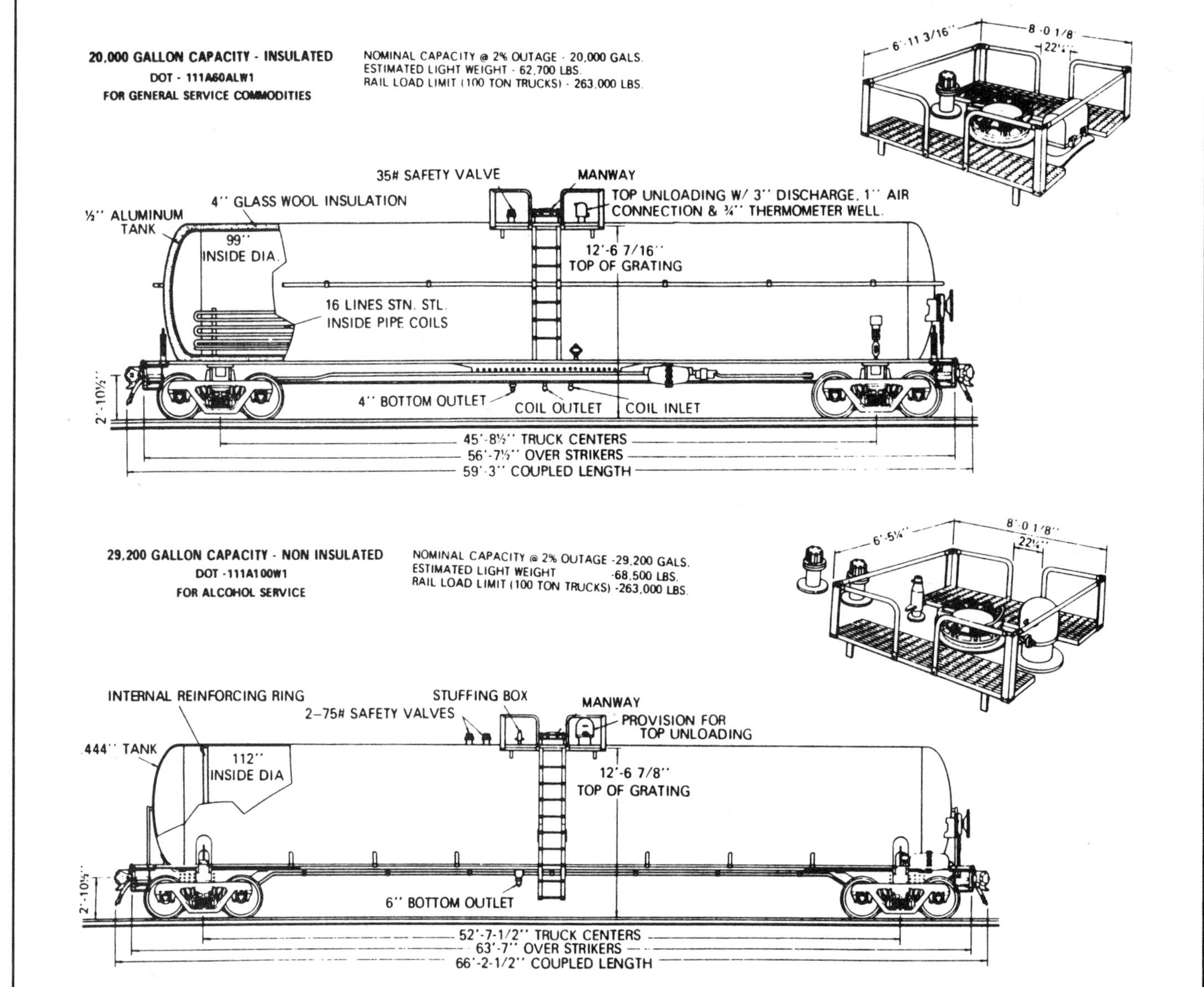

Figure 12–8. Tank cars are constructed to accommodate the various types of commodities transported.

safely when loading or unloading a car. The gangway or bridge should have guardrails (Figure 12–8).

Closely check invoices and shipping papers on incoming tank cars to make certain they match the actual numbers on the tank car. In doubtful cases, test the contents to prevent unintentional mix-ups of flammable and combustible liquids in the plant's tanks. Figure 12–8 shows the details of a typical tank car's construction.

Electrical Considerations

Provide approved electrical equipment on tank cars. Where flammable liquids are handled, be sure that this equipment is installed according to the NEC and subject to regular inspection. Bond, ground, and insulate a tank car's siding from the main rail line for protection from stray electrical currents (Figure 12–9). Usually the spur track's installation is made or supervised by the serving railroad. For applicable specifications, refer to the Association of American Railroads (AAR) publication, *Signal Manual of Recommended Practice.* (See References at the end of this chapter.)

The location of electrical power lines relative to a tank car's unloading position is an important matter discussed in the AAR's *Recommended Practice for the Prevention of Electric Sparks That May Cause Fires in Tanks or Tank Cars Containing Flammable Liquids or Flammable Compressed Gases, Due to Proximity of Wire Lines.* Observe the following provisions from this AAR circular:

1. Where any electric power line is within 20 ft (6 m) of the tank's opening, the use of a metallic gauging rod is prohibited.

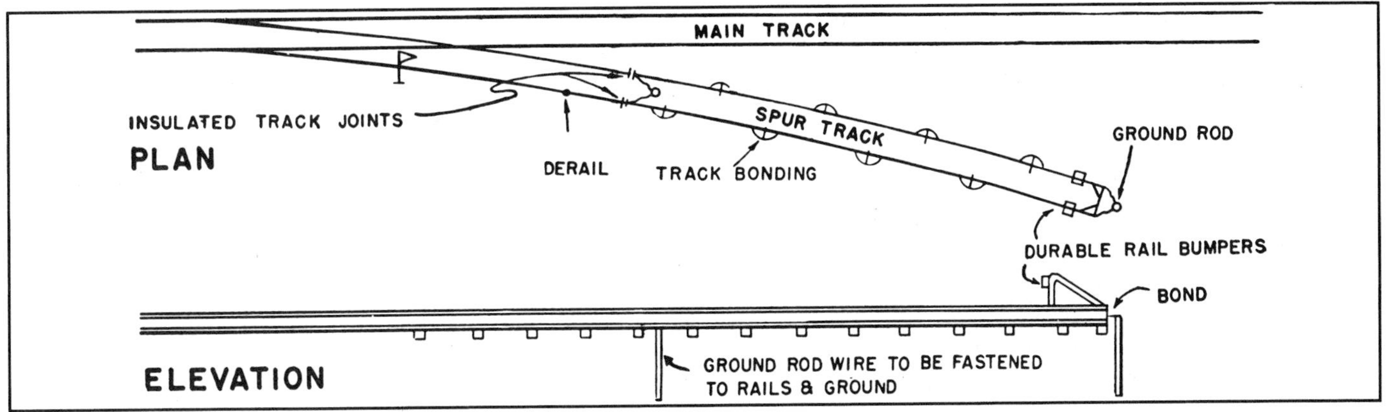

Figure 12–9. This siding is designed for unloading tank cars. It has rail joint bonding, insulated track joints, derail, and track grounding.

2. When the contents are being gauged or transferred, tank cars, wherever possible, shall not be located under or near any electric power lines.
3. Where tanks or tank cars (the contents of which are being gauged or transferred) are necessarily located under or near power lines having a span length of 150 ft (45 m) or less and operating at a voltage not exceeding 550 V between conductors, the following rules shall be observed:

   a. Where power lines pass overhead, there shall be a minimum vertical clearance of 8 ft (2.4 m) at 60 F (16 C) between the wires and the tank.
   b. Where power lines pass nearby and do not have the minimum vertical clearance specified in paragraph 3–a, there shall be a minimum horizontal clearance of 8 ft (2.4 m) between the wire lines and the tank.
   c. Openings in tanks shall be at least 6 ft (1.8 m) distant, measured horizontally from any overhead power lines.

4. Where tanks or tank cars (the contents of which are being gauged or transferred) are located under or near power lines having a span length in excess of 150 ft (45 m) or operating at a voltage in excess of 550 V between conductors, it is recommended that a special study be made by qualified persons. Based on the study, provide clearance as necessary to give adequate protection.

Spotting Cars

Use a car mover when it is necessary to spot a car by hand. The worker should stand with the handle of the bar to one side of the track's rail and should face the car with the feet well apart so as not to lose his or her balance.

After the car is spotted, with the dome outlet opposite the loading or unloading line, set the brakes and block the car's wheels. Before loading lines are connected or work is done on the car, workers should set up the "blue-flag" stop sign. (It should be at least 12 in. by 15 in. (30.5 cm by 38 cm), with words such as STOP—TANK CAR CONNECTED or STOP—MEN AT WORK.) Locate the sign about 25 ft (7.6 m) in front of the car toward the main switch so train crews cannot come onto the track without seeing it (Figure 12–9). (See Chapter 2, Buildings and Facility Layout.)

Inspection

The company receiving or shipping a tank car should make a general visual inspection of the unit and report any obvious defects to the carrier. In shipping tank cars, it is necessary that all valves and related parts be secured and that shipping papers identifying the contents should be ready. The inspector dates stenciled on the tank indicate when the unit—the safety valve or tank or both—was tested and when it is due to be tested again. Do not use tank cars that have not undergone proper testing.

Place approved containers under any leaks to prevent contamination and product loss. Frequently transfer the accumulated drippings to storage so no large quantity of flammable liquid is exposed in an open container or allowed to spread over the ground. Comply with U.S. Environmental Protection Agency (EPA) rules governing spills of hazardous substances and emergency responses. (See 29 *CFR* 1910.120.)

Relieving Pressure

Relieve the tank car of interior pressure before removing the domed manhole cover or the outlet valve cap. To relieve pressure, either raise the safety valves, open the air valve a small amount at a time, or cool the tank with water. Depress the vacuum-relief valve, if vacuum is a problem. Defer venting and unloading if a dangerous amount of vapor collects outside the car. If interior pressure is excessive, spray the car with water to cool it or allow it to stand overnight and unload it early the next morning.

Removing Covers

Loosen a screw-type dome cover by using a bar placed between the lug and the knob on the cover's top. Make two complete turns to expose the ½ in. (1.3 cm) vent holes in the threaded portion of the dome cover. If escaping vapors are heard, tighten the cover and release the pressure by raising the safety valve. Keep clear of the vapors by standing upwind. Wearing a self-contained breathing apparatus or an air-supplied respirator equipped with auxiliary SCBA may be recommended or required, depending on the chemical and the situation.

Face the dome and remove the cover. With feet well braced, use short, vigorous pushes on the bar. Transfer the cover (if not provided with a chain) and any loose tools to the walk platform or another safe place so they will not fall.

On the bolted type of dome cover, unscrew all nuts one turn and lift the cover to break any adhesion between the cover and the dome's ring. If there is a sound of escaping vapors, tighten the dome and repeat the venting operation. On the interior types of dome covers, remove all dirt, cinders, and debris before unscrewing the yoke.

When the car is unloaded through the bottom outlet valve, adjust the dome cover to allow for venting, where vapor recovery is not required. Tighten a screw-type dome cover just enough to expose the vent holes in the threaded portion of the cover.

Place a small, thin wood block under the bolted type of cover, and tighten up the interior type of dome cover in the yoke to within ½ in. (1.3 cm) of the closed position. Place a metal cover or wet burlap or canvas over the tank manhole to prevent sparks or other sources of ignition from entering and to limit the vapor escaping.

Unloading and Loading Connections

The preferred method of unloading a tank car is through the bottom, when the car is equipped with an approved bottom outlet. A tank car can also be unloaded safely through the top. When a car is to be unloaded through the bottom, close the tank's outlet valve before removing the outlet chamber cap or plug. Place an approved container under the outlet chamber and unscrew the outlet cap. (See 49 *CFR* 174.67.)

Check the condition of the outlet valve and if there is no serious leak, proceed with the unloading. If the plug does not loosen easily with a 48-in. (1.2-m) wrench, tap the bottom outlet cap, or plug, to loosen it. If the valve leaks so badly that a connection cannot be made without spilling the product, unload the car from the top.

In cold weather, if the chamber or valve becomes frozen or blocked with frozen liquid, carefully examine the outlet chamber for cracks before attempting the unloading connection. If no crack is found, make the connection for unloading, but wrap the outlet chamber with burlap or other rags. Also apply hot water or steam to it.

When heater pipes are used during unloading, apply steam slowly until it begins to exhaust at the outlet pipe. Steam pressure should not exceed that needed to bring the contents to the desired temperature and should never be high enough to overheat the contents.

Take care to avoid overheating combustible liquids, because overheating them can cause them to emit flammable vapors. Also, overheating products containing certain additives can release dangerous amounts of flammable vapors.

Before opening the car valve, carefully check the condition of the connection from the car to the storage tank. Gauge the storage tank, and watch it to prevent an overflow. Frequently check the hose and unloading line. Before loading lines are disconnected, examine the tank car to make certain it is completely empty.

A worker should be present throughout the entire loading or unloading operation and at all times while a car is connected. If it is necessary to discontinue operations before they are completed, close the outlet valve and replace the dome cover and outlet chamber cap. To prevent overflowing during transport, load cars only to the specified inches of outage. Do not allow tank cars to stand with loading or unloading connections attached after operations are completed.

If flammable liquids are spilled, stop the loading or unloading and cover spill areas with fresh dry sand or dirt or with oil absorbents, or clean them up otherwise. Never flush spills into public sewers, drainage systems, or natural waterways.

For details on unloading or loading a tank car, check with the car manufacturer, supplier, or railroad. (See Chapter 14, Materials Handling and Storage, in this volume, for information on unloading chemical tank cars.)

Placards and Shipping Papers

The U.S. DOT and its counterparts in most other countries provide specific guidelines for using placards and shipping papers. Consult the appropriate requirements for details (Figure 12–10). Do not dispose of placards, rags, waste, and blocks in the tank or in the car's body.

Figure 12–10. These placards are used on vehicles and rail cars transporting flammable or combustible liquids. The *Empty* placard can be either a separate placard, printed on the reverse side of a placard, or a composite made by covering the top triangle with a black triangle having the word *Empty* printed in white letters. *Flammable* and *Combustible* placards are red with white letters. They can be purchased through various suppliers. (Reprinted from 49 *CFR* 172. 525, .542, and .544.)

Fires

Tank car fires can be serious, especially if a spilled product burns under a tank car. Shut off all loading lines to the rack and apply a spray of water to cool the tank car, structural steel, and piping. If conditions permit, flush spilled fuel from under the tank's body to an open area or into a closed drainage system. Avoid getting too much water into the tank car and thus causing an overflow.

If a fire occurs at the dome of a tank car, and the dome is unobstructed by a fill pipe, close the dome cover to extinguish the fire. If a fill pipe is in the dome opening, do not remove the fill pipe, because removing it could splash the burning liquid and spread the fire. Use a dry chemical or carbon dioxide (CO_2) extinguisher to put out the fire. In some cases, the flames can be "blown out" with a straight stream of water directed across the opening. Take care, however, not to overflow the tank or to splash the contents and spread the fire.

To protect nearby facilities, provide drainage to limit the spread of a spill and to direct it to a safe area. The American Association of Railroads (AAR) can provide publications and training on this subject. (See References at the end of this chapter.)

LOADING AND UNLOADING TANK TRUCKS

NOTE: Tanks, trailers, and other transport vessels are the responsibility of the owner or operator of the transport vessel. Facilities for loading or unloading are the responsibility of the facility owner. Any loading or unloading procedures need to be worked out between these parties. Tank trucks, tank trailers, and tank semitrailers used for the transportation of flammable liquids should be constructed and operated according to appropriate regulatory standards. In the United States, consult 49 *CFR* 177.325 and 177.386–.399. Revision and promulgation of different regulations occurs often.

Carriers handling hazardous materials are subject to U.S. regulations under the Hazardous Materials Transportation Uniform Safety Act of 1990. Those not subject to the jurisdiction of the U.S. DOT may consult NFPA 385, *Tank Vehicles for Flammable and Combustible Liquids.*

Inspection

See that trucks are kept in good repair and inspected daily. Place special emphasis on the condition of lights, brakes, horns, rearview mirrors, bonding straps, tires, steering, and motors.

Inspect tanks and safety valves for defects in welds, piping, valves, gaskets, and interlocks. Stencil the tank of a U.S. carrier with the last current inspection date indicating that the tank has been tested according to 49 *CFR* 177. In addition, place four placards, one on each side of the tank. Keep them clean and visible at all times.

Smoking

Prohibit smoking by truck drivers or their helpers while they are driving, loading or unloading, or attending to their unit. Drivers should keep smokers and all sources of ignition away from loading or unloading operations. Provide each tank vehicle with one or more portable fire extinguishers, each having at least a 10 B, C rating.

Spotting Trucks

When a tank truck is being loaded or unloaded, see that the brakes are set, the engine stopped (unless power takeoff is required for unloading), the lights turned off, and the bonding connection made before the dome cover is opened for inspection or gauging. If trucks must be started and moved before loading, be sure that all domes are closed and latched and that the bonding connections are removed before the engine is started.

Unloading and Loading Connections

Trucks with motor-driven pumps should be shut off before loading lines are connected or disconnected. Drivers must remain at the tank controls (within 25 ft). If they must leave the site, they must stop the flammable product's discharge without endangering themselves and must remove the hose. When an unloading line must be run across a sidewalk, provide suitable warning signs.

Drivers should make sure that the tank is properly vented, that vapor recovery units are connected, and that the tank can accommodate the amount of liquid to be loaded or unloaded. Drivers should also make sure that the correct product is being handled and that the tank has previously contained the same liquid. If possible, contact personnel at the delivery site before a discharge. To assure removal of trapped liquids, especially when changing from a low-flash liquid to a higher one, completely drain and flush the tank.

Leaks

Avoid spills or overflows. If they do occur, immediately stop loading, shut off the valves, and clean up the overflow before resuming loading. Trap the liquid in containers, in an earthen or sand-diked area, or in a depression or pit, if possible. In case of a large spill, especially in urban areas, the best way to avoid endangering lives is to use portable hand pumps to discharge the product into approved drums or into another tank truck. In the United States, observe EPA spill regulations. Eliminate or control all sources of ignition in the area. Allow adequate time for all flammable vapors to dissipate before starting the truck engine.

Drivers should make every effort to park a damaged and leaking truck so that it will not endanger traffic or property. Park the truck off the highway—if possible, in a vacant lot or otherwise away from buildings and away from areas with high concentrations of people. Drivers should warn the public to keep away and should notify the police and fire departments. Drivers should not leave the truck unattended.

Drivers for carriers subject to the safety regulations of the U.S. DOT must report all broken, leaking, and damaged containers on Form F 5800.1. Immediate notification by calling 1–800–424–8802 is required if (1) a person is killed, (2) a person is injured and requires hospitalization, (3) estimated carrier or property damage exceeds $50,000, or (4) fire, breakage, spillage, or contamination involving a radioactive material occurs.

Fires

In the event of a tank-truck fire during loading or unloading, shut off the fuel supply to or from the truck, if possible. If top loading is being done, remove the spout from the tank truck. If possible, either close the dome cover or cover the opening with a wet sack or blanket.

To put out the fire, use CO_2, foam, or dry-chemical extinguishers, or use water fog. If a water supply is available, flush away the burning liquid to a safe place, but not into a public sewer, drain, or waterway. Take care to prevent water from entering the tank compartment and thus displacing the fuel. Failing to prevent this could cause more spillage and spread the fire.

Where burning vapors are escaping from leaks or vents, it may be better to let them burn until their source can be controlled. In all cases of fire, quickly inform personnel and report any emergency. Establish and rehearse fire-fighting procedures so that in an actual emergency, the procedures can be followed exactly.

STORAGE

Keep or store Class I and Class II (see Definitions, earlier in this chapter) liquids in a building used for public assembly, such as a school, church, or theater, only if they are in approved containers (Figure 12–11). Keep the containers in either a storage cabinet or a storage room that does not open to the public portion of the building. Limit the quantities stored in such locations according to NFPA 30, *Flammable and Combustible Liquids Code*. Do not store the containers where they will obstruct exits, stairways, or other areas used to safely leave the building. Also, never store them near stoves or heated pipes or expose them to sunlight or other sources of heat.

Do not store flammable or combustible liquids in open containers. Use approved containers for flammable liquids, and be sure they are closed after each use and when empty.

Store bulk Class I liquids in an underground tank or outside a building. Where permitted by local authorities, use properly protected, outside, aboveground storage tanks. No outlet from the tank should be inside a building, unless the outlet terminates in a special room (Figure 12–12).

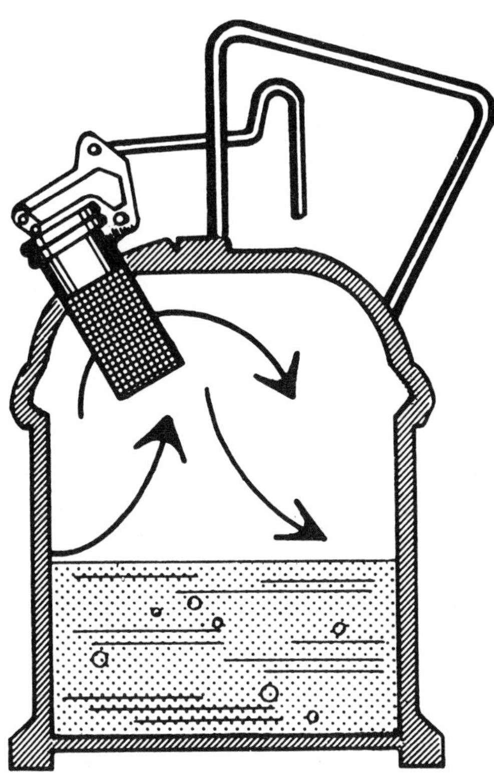

Figure 12–11. This spring-loaded cover is designed to open at 5 psi and relieve the internal vapor pressure. Only negligible losses are caused through evaporation of liquids stored in safety cans at ordinary temperature ranges.

Figure 12–12. A well-designed flammable-liquid storage room with both high- and low-level ventilation and automatic fire extinguishment. The ventilators are designed for 12 air changes per hour.

NFPA 30 specifies the quantity of each class of flammable liquids that can be stored in various locations on a plant's premises. Also, it describes the required conditions and procedures relating to such storage.

Vehicles used on a plant's property to transport flammable liquids in sealed containers should be designed to minimize damage to the containers. When employees are filling tanks and other containers, they should allow enough vapor space, or outage, above the liquid level so the liquid can safely expand when temperatures change. For example, gasoline expands at the rate of about 1% for each 14 F (8 C) rise in temperature. The recommended outage space for gasoline is 2% of a tank's or compartment's capacity. However, many jurisdictions recommend fill limits at 90% of the tank's capacity. Permanently mark high levels on containers, and place shutoffs or other control devices inside tanks.

Tank Construction

Tanks should be constructed, installed, tested, and maintained as recommended by the latest revision of NFPA 30. (See also NFPA 30A, *Automotive and Marine Station Code,* and API RP652, *Lining of Aboveground Petroleum Storage Tank Bottoms* [1991].)

Vents

Provide storage tanks with vents of a type and size recommended in NFPA 30. Vent pipes of underground tanks and vapor-recovery systems for Class I flammable liquids should terminate outside buildings. They should also be higher than the fill pipe opening and not less than 12 ft (3.7 m) above the adjacent ground level. They should discharge vertically and be located so flammable vapors cannot enter the openings of buildings or be trapped under eaves or other obstructions. Vent pipes from underground tanks storing Class II or Class III liquids should also terminate outside buildings and be higher than the fill pipe opening. Their vent outlets, however, should be above the normal level of snow. Consult U.S. EPA regulations on the discharge of regulated chemicals into the atmosphere.

Some authorities have questioned the effectiveness of brass mesh or copper screens as flame arresters in vent terminals. Underwriters Laboratories states that "under some favorable conditions, screens of fine mesh can be effective in arresting flame, but their use is not dependable. Under service conditions, they are subject to clogging and freezing, and when attempts are made to clean them there is danger of mechanical injury. The displacement of one or more wires renders the screen useless as a flame arrester." (See *Oil Tank Vents—Hazards of Screens,* Serial No. UL–31 in References, at the end of this chapter.)

A well-located vent can provide more-certain protection than screens. Safely locate vents so that escaping vapors will be properly dispersed and clear of ordinary sources of ignition. If the vents open in locations where clogging by mud wasps is possible, provide a loosely attached screen of relatively coarse mesh. (See NFPA 30 for recommended vent sizes.)

Dikes

In some locations, a flow of flammable or combustible liquid from a tank car might have serious consequences because of topography or neighboring property. In these cases, install a curb, dike, or wall around a tank or group of tanks. Design such a structure to contain at least the amount of flammable liquids that could be released by overfilling the largest tanks or by leakage from tanks, pipelines, and valves connected to these tanks. NFPA 30 gives construction details for capacity, drainage, dikes, and walls of aboveground storage tanks.

Local conditions will determine the design of dikes. For example, if a tank is close to a building and the ground's slope could cause liquid to flow toward the building, it would be prudent to construct diversion walls to direct the escaping liquid to a safe location.

A storm drain with a water-sealed catch basin is often installed for each dike enclosure. In larger plants, the drain often leads to an oil-water separator located far away from the main buildings. Here flammable or combustible liquids can be separated from water by flotation. Equip drains with control valves, and keep them closed, unless water is being drained from the area. Be sure that these systems are designed to prevent flammable liquids from entering natural water supplies, public sewers, or public drains.

Pumps

Locate flammable-liquid transfer pumps outside buildings and diked areas whenever possible. Use fire-resistant construction in buildings that house equipment for transferring flammable liquids. Provide ample ventilation, especially along the floor, where vapors might be present (Figure 12–13). Preferably, pump houses should not contain sunken pits, except small drain openings, because the danger of vapor concentrations in such areas is too great. A pump house should have a minimum of two exits, which should be kept clear of obstructions and have doors that swing outward.

Where flammable liquids are handled, partition and seal off electrical equipment from the rest of the pump house, unless the electrical equipment is approved for use in flammable atmospheres. Keep the pump room well ventilated. Also, have a well-marked master cutoff switch outside the building. All electrical equipment inside the building must conform to the National Electric Codes (NEC), as well as to local codes for indoor use.

Maintain valves and packing glands on pumps in good operating condition to prevent leaks. Excess-flow valves can also prevent large-scale spills and are sometimes required by local codes. Draining escaped liquids into a closed-pipe return system is recommended over simply catching leaking materials in drip pans. Repair leaks as soon as they are noticed.

If a centrifugal pump with a priming bleeder is used, either keep the outlet of the bleeder plugged when not in use or piped to a closed system for collection. Keep pump bearings well lubricated to prevent overheating caused by friction.

Because of the importance of good housekeeping in a pump house, provide approved containers for safe disposal of debris, rags, and other waste. See that these containers are emptied daily. Do not store tools, other than those required to operate a pump house, in the pump house. The pump house should not contain lockers or be used by employees to change clothes. Do not allow employees to loiter there. Keep all sources of ignition away from the pump house.

Locate fire extinguishers at convenient, easily identifiable points in pump houses. Have self-closing fire doors and the recommended type of automatic sprinkler or automatic fire suppression system in pump houses.

Gauging

If a tank requires manual gauging and if a walkway is not provided, arrange a way for employees to measure the contents of storage tanks so that they need not walk across the tank roof. To eliminate the need for both walking on the roof and installing a walkway, provide a platform or remote-gauging equipment. Personnel should stay off the tank roof while any pumping operation is in progress. Provide safe means of entry and exit for persons gauging tanks.

NFPA 30 states that manual-gauging openings independent of fill pipes should be equipped with vapor-tight caps or covers when they are located in a building or buried under a basement. Protect each such opening against liquid overflow and possible vapor release by means of a spring-loaded check valve or another approved device. Have only trained and qualified employees conduct manual gauging of tanks containing Class I liquids.

The float-operated gauge or the type operated by the liquid's pressure is often used on aboveground storage tanks. Newer gauging systems, based on mass or weight measurement, provide accurate indications of

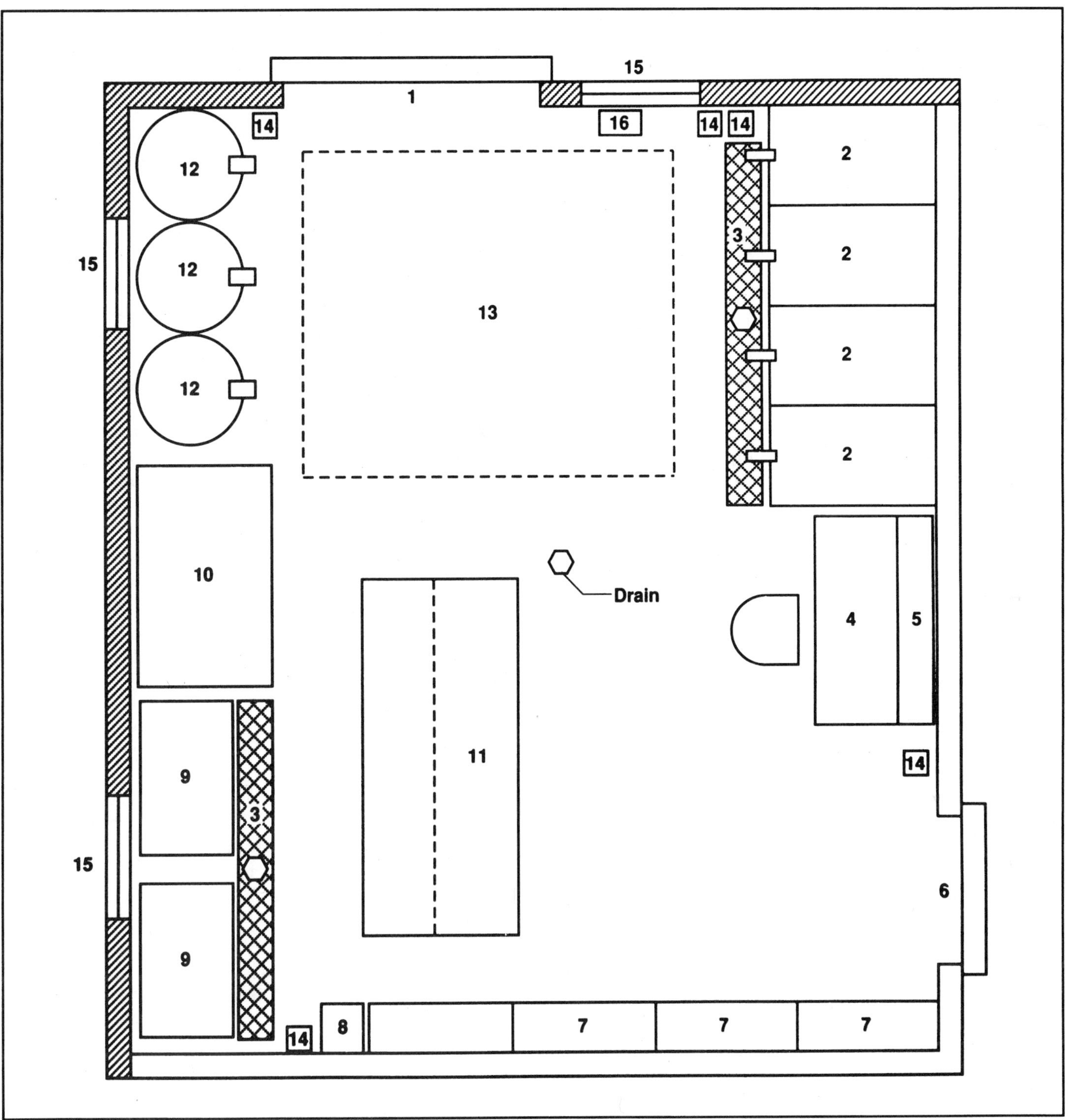

Figure 12–13. Some of the features of this oil house layout are as follows: 1 and 6: self-closing fire doors wide enough for passage of lift trucks or other materials handling equipment; 2: drum racks; 3: grating and drain; 4: desk; 5: filing and record racks; 7: individual lockers or storage cabinets; 8: waste disposal container; 9: solvent cleaning tanks; 10: purification equipment, soluble oil mixing equipment, or other special equipment; 11: cabinets and racks for equipment, supplies, and small containers; 12: grease drums with pumps; 13: parking area for oil wagons, etc.; 14: fire extinguishers; 15: ventilators; 16: container of sawdust or other absorbent.

even small leaks in aboveground storage tanks. Some companies use a sounding weight and tape or, if the tank is small, a wooden sounding rod through a gauge hatch on top of the tank.

Gauge glasses protected by metal cases are suitable when properly maintained. These are found on some flammable liquid tanks held under pressure. However, many regulations require that manual gauging be performed at regular intervals, too, in order to check the accuracy of the automatic gauging.

Tanks in Flooded Regions

Where danger from flood waters exists, install and anchor tanks according to NFPA 30, Appendix B, "Protection of Tanks Containing Flammable or Combustible Liquids in Locations That May Be Flooded."

Underground Tanks

When an underground tank is subject to heavy traffic over it, protect the tank with (1) at least 3 ft (0.9 m) of earth cover, (2) 18 in. (0.46 m) of tamped earth plus 6 in. (15 cm) of reinforced concrete, or (3) 18 in. (0.46 m) of tamped earth plus 8 in. (20 cm) of asphalt concrete.

If the underground tank is not subject to traffic over it, provide a protective cover by covering it with (1) a minimum of 2 ft (0.6 m) of earth, or (2) 1 ft (0.3 m) of earth plus 4 in. (10 cm) of reinforced concrete. The concrete cover should extend 1 ft (0.3 m) beyond the outline of the tank.

An underground tank should have a firm foundation and be surrounded with at least 6 in. (15 cm) of noncorrosive, inert materials, such as well-tamped, clean sand, earth, or gravel. Anchor the tank if there is any chance of its "floating" on rising ground water. Storage tanks inside buildings should have their fill pipes and vent pipes located outside.

NFPA 30 specifies that underground tanks or tanks under buildings should be so located with respect to existing building foundations and supports that the loads carried by the latter cannot be transmitted to the tank. The distance from any part of a tank, storing Class I liquids, to the nearest wall of any basement, pit, or cellar should be not less than 1 ft (0.3 m), and from any property line that may be built upon, not less than 3 ft (0.9 m). The distance from any part of a tank storing Class II or Class III liquids to the nearest wall of any basement, pit, cellar, or property line should be not less than 1 ft (0.3 m).

NFPA 30 also specifies that corrosion protection for an underground tank and its piping should be provided by one or more of the following methods: (1) use of protective coatings or wrapping, (2) cathodic protection, or (3) corrosion-resistant construction materials. Select the type of protection according to the area's corrosion history and a qualified engineer's judgment. Do not use cinders or other acid-forming fills around the tank. Some jurisdictions may require secondary protection and overfill safeguards for underground tanks. Some tanks may require double-walled construction with sensing devices to detect leakage between walls. Before installing the tank, give written notification to the proper authorities and get their approval.

Withdraw flammable liquids by pump. Arrange the pump and piping system so the liquid will flow back to the tank when the system is not in operation.

Aboveground Tanks

NFPA 30 sets minimum distances from property lines, public ways, and nearby buildings for aboveground tanks containing flammable or combustible liquids (Table 12–A). (See NFPA 30 for distances for aboveground tanks containing unstable liquids.) Some other recommended minimum distances are given in Table 12–B. Where end failure of a horizontal tank or vessel can expose property, place the tank with its longitudinal axis parallel to the nearest important structure's exposure.

Although rarely the case, two tank properties owned by different parties might share a common boundary. If so, and if the owners and local authorities agree, normal minimum shell-to-shell spacing can be used, such as that used within a tank farm.

Tanks storing Class I, II, or IIIA flammable or combustible liquids must be separated according to the distances shown in Table 12–C. An exception is made for tanks containing crude petroleum that stand at production facilities in isolated locations. When these tanks have individual capacities not exceeding 126,000 gal (3,000 barrels or 475,000 l) they can be as close as 3 ft (0.9 m) apart.

Space tanks used only for storing Class IIIB liquids no less than 3 ft (0.9 m) apart, unless they stand within a diked area or drainage path for a tank storing a Class I or Class II liquid. In such cases, the provisions of Table 12–C apply. For unstable liquids, the distance between such tanks shall be not less than one-half the sum of their diameters. Greater spacing may be required for tanks that stand in a diked area containing Class I or Class II liquids or stand in the drainage path of Class I or Class II liquids and are compacted in three or more rows or in an irregular pattern. The jurisdictional authority might require changes that would make tanks in the interior of the pattern easier to reach for firefighting purposes.

Many exceptions are made for petroleum tanks in refineries and producing areas, unstable flammable or combustible liquids, LP gas, and tanks arranged in a compact or irregular pattern. Consult the specific NFPA code when special situations are involved.

Position truck loading racks that dispense Class I liquids at least 25 ft (7.6 m), and those that dispense Class II or Class III liquids at least 15 ft (4.6 m), from tanks, warehouses, other plant buildings, and the nearest property line.

Table 12–A. Location of Outside Aboveground Storage Tanks from Adjoining Property or Public Way for Operating Pressures No Greater Than 2.5 psig (17 kPa)

Type of tank	Protection	Minimum distance in feet from property line that may be built on, including the opposite side of a public way and not less than 5 ft (1.5 m)	Minimum distance in feet from nearest side of any public way or from nearest important building and shall be not less than 5 ft (1.5 m)
Floating roof	Protection for exposures	½ times diameter of tank	⅙ times diameter of tank
	None	Diameter of tank but need not exceed 175 ft (54 m)	⅙ times diameter of tank
Vertical with weak roof to shell seam	Approved foam or inerting system on the tank	½ times diameter of tank	⅙ times diameter of tank
	Protection for exposures	Diameter of tank	⅓ times diameter of tank
	None	2 times diameter of tank but need not exceed 350 ft (110 m)	⅓ times diameter of tank
Horizontal and vertical, with emergency relief venting to limit pressures to 2.5 psig	Approved inerting system on the tank or approved foam system on vertical tanks	½ times Table 12-B	½ times Table 12-B
	Protection for exposures	Table 16-B	Table 12-B
	None	2 times Table 12-B	Table 12-B
	For Operating Pressures Greater Than 2.5 psig (17 kPa)		
Any type	Protection for exposures	1½ times Table 12-B but shall not be less than 50 ft (15 m)	1½ times Table 12-B but shall not be less than 25 ft (7.5 m)
	None	3 times Table 12-B but shall not be less than 50 ft (15 m)	1½ times Table 12-B but shall not be less than 25 ft (7.5 m)
Floating roof	Protection for exposures	½ times diameter of tank	⅙ times diameter of tank
	None	Diameter of tank	¼ times diameter of tank
Fixed roof	Approved foam or inerting system	Diameter of tank	⅓ times diameter of tank
	Protection for exposures	2 times diameter of tank	⅔ time diameter of tank
	None	4 times diameter of tank but need not exceed 350 ft	⅔ times diameter of tank

From National Fire Protection Association No. 30, *Flammable and Combustible Liquids Code.*

Locate tanks to avoid danger from high water levels. Tanks beside a stream without tide should, where possible, stand downstream from burnable property.

Piping materials for aboveground tanks shall be made of steel as recommended in NFPA 30. Piping can be made of materials other than steel (1) when used underground, or (2) when protected against fire exposure and located so leakage would not produce a hazard. Piping other than steel may be required by the chemical properties of the liquid handled. Cast-iron pipe is not recommended because it is brittle and can fracture under stress. When in doubt, consult the supplier, producer of the flammable or combustible liquid, or another competent authority about the suitability of the material being considered.

Piping from a storage tank should have an easy-to-reach shutoff valve at the tank. Make provisions to drain or pump the contents of the tank into another tank, or to collect the liquid within dikes or retaining walls should the tank leak or be overfilled.

Table 12–B. Reference Minimum Distances for Aboveground Outside Storage Tanks

	Floating Roof Tanks	Fixed Roof Tanks: Class I or II Liquids	Fixed Roof Tanks: Class IIIA Liquids
All tanks not over 150 ft (45 m) diameter	⅙ sum of adjacent tank diameters but not less than 3 ft (0.9 m)	⅙ sum of adjacent tank diameters but not less than 3 ft (0.9 m)	⅙ sum of adjacent tank diameter but not less than 3 ft (0.9 m)
Tanks larger than 150 ft (45 m) diameter			
Remote impounding area	⅙ sum of adjacent tank diameters	¼ sum of adjacent tank diameters	⅙ sum of adjacent tank diameters
Impounding (diking) around tanks	¼ sum of adjacent tank diameters	⅓ sum of adjacent tank diameters	¼ sum of adjacent tank diameters

From *National Fire Protection Association No. 30.*

Table 12–C. Minimum Tank Spacing (Shell-to-Shell)

Capacity Tank (gallons) (1 gal = 3.78 liters)	*Minimum Distance in Feet from Property Line Which Is or Can be Built Upon, Including the Opposite Side of a Public Way*	*Minimum Distance in Feet from Nearest Side of Any Public Way or from Nearest Important Building on the Same Property*
275 or less	5	5
276 to 750	10	5
751 to 12,000	15	5
12,001 to 30,000	20	5
30,001 to 50,000	30	10
50,001 to 100,000	50	15
100,001 to 500,000	80	25
500,001 to 1,000,000	100	35
1,000,001 to 2,000,000	135	45
2,000,001 to 3,000,000	165	55
3,000,001 or more	175	60

From *National Fire Protection Association No. 30.*

A small tank is often provided with an emergency self-closing valve, located inside the tank or at the entry point. Such a valve would automatically stop the flow of liquid in case of fire. (See Factory Mutual System, Data Sheet 7–32, *Flammable Liquid Pumping and Piping Systems.*) This valve should be closed except during loading and unloading operations. Be sure that the rope or wire and the fusible link attached to an emergency valve, which is left open, are in good condition and tested regularly for easy operation.

Besides the normal vents that take care of vacuum and pressure during pumping operations, an aboveground storage tank must have some form of emergency-relief venting. Such venting prevents buildup of excessive internal pressure in case a fire should surround the tank. In addition to a relief device, provide further protection (1) with a weak seam in the top or at the joint between the top and the shell of the tank, or (2) by some other recommended form. (See NFPA 30 for guidance on sizing normal and emergency vents.)

Except in oil refineries and oil-storage terminals, label tanks for storing Class I liquids, FLAMMABLE—KEEP FIRE AWAY, with letters at least 2 in. (5 cm) high. Also, post NO SMOKING signs. Post similar signs for Class II and, if necessary, Class III liquids. Regulations usually require additional warning labels, such as for health hazards or special precautions.

Maintain good housekeeping around storage tanks. Do not permit debris to accumulate. Also, do not store combustible materials near tanks. Eliminate the grass around tanks (or under them, if they are off the ground), or at least keep it cut.

Tanks used for storing flammable or combustible liquids should be made vapor-tight to minimize evaporation losses and prevent fires. Venting devices are normally arranged to be closed. However, they breathe automatically when liquid is pumped in or out of the tanks and as temperature fluctuates. (See API, RP 2000 82, *Venting Atmospheric and Low-Pressure Storage Tanks.*) Floating roofs with secondary seals are most effective in reducing evaporation losses and allowing expansion.

Equip large aboveground storage tanks with stairways and platforms, preferably made of steel. Having tanks with gauge tubes near the platform will reduce the need for personnel to walk on the roof of the tanks. Tanks that stand more than 1 ft (0.3 m) aboveground should have foundations of noncombustible materials. However, wood cushions can be used. For aboveground tanks, use supports made of concrete, masonry, or steel that is protected by concrete or by other approved fireproofing. Such materials will prevent supports from collapsing in case of a fire.

Overflowing of tanks presents a severe fire hazard because vapors might drift to a source of ignition, where automatic-sensing controls are absent. Operators responsible for filling tanks should maintain a constant watch on the filling rate and the level of liquid. That way, they can stop operations or divert liquid to another tank when the required level is reached. To prevent a product's release, operators must remain at the open valve while water is being drained from a tank.

It is common practice to paint flammable-liquid tanks that are exposed to the sun with aluminum, pastel, or white paint. These colors reflect the heat and help reduce the internal vapor pressure. Some regulations require the use of such paint. In some installations where highly volatile liquids are stored, water is sprayed to cool the system externally. Tanks that store viscous products, such as tar and heavy oils, may be insulated and equipped with heating coils.

Tank Fires and Their Control

Prevention is the best way to reduce the risk of fires in storage tanks. Provide tanks containing flammable liquids with a roof that floats upon the surface of the liquid—an internal or external open-top floating roof. Such a roof greatly reduces the opportunity for vapors to accumulate within the flammable range.

It only takes a few seconds for a fire inside a fixed-roof tank either to extinguish itself or to blow open the tank roof. Tank fires most commonly occur at one of the roof openings. Fire will occur only if a source of ignition, such as lightning, comes in contact with vapor being expelled from the tank because of either filling or heating. Such a fire can usually be extinguished without difficulty if the cause of the vapor's expulsion is removed by cooling or by shutting off the filling operation. Seal fires frequently occur on floating-roof tanks. These fires can usually be extinguished by portable equipment or hand foam lines from the tank platform, windgirder, or roof.

If a fire occurs near a tank, apply water immediately to the exposed tank to cool it and to reduce vaporization. This action not only saves stock but also makes ignition at the vents less likely. Because so many variables are involved in tank fires, they should be fought only by personnel trained in extinguishing them.

Inside Storage and Mixing Rooms

Flammable or combustible liquids in approved, sealed containers present a potential, rather than an active, hazard: the possibility of fire from without. Isolate inside storage rooms as much as possible. Locate them at or above grade, not immediately above a cellar or basement, and preferably along an exterior wall. The following paragraphs of this section contain a summary of NFPA 30 requirements for inside storage rooms.

Inside [storage] rooms shall be constructed to meet the selected fire resistance rating as specified in 4–4.1.4. Such construction shall comply with the test specifications given in NFPA 251, *Fire Tests of Building Construction and Materials.* Except for drains, floors shall be liquid-tight and the room shall be liquid-tight where the walls join the floor (Figure 12–14). Where an automatic fire protection system is provided, as indicated in 4–4.1.4, the system shall be designed and installed in accordance with the appropriate NFPA standard for the type of system selected.

Openings in interior walls to adjacent rooms or buildings shall be provided with normally closed, listed 1½ hr fire doors for interior walls with fire-resistance rating of 2 hr or less. Where interior walls are required to have greater than 2 hr fire-resistance rating, the listed fire doors shall be compatible with the wall

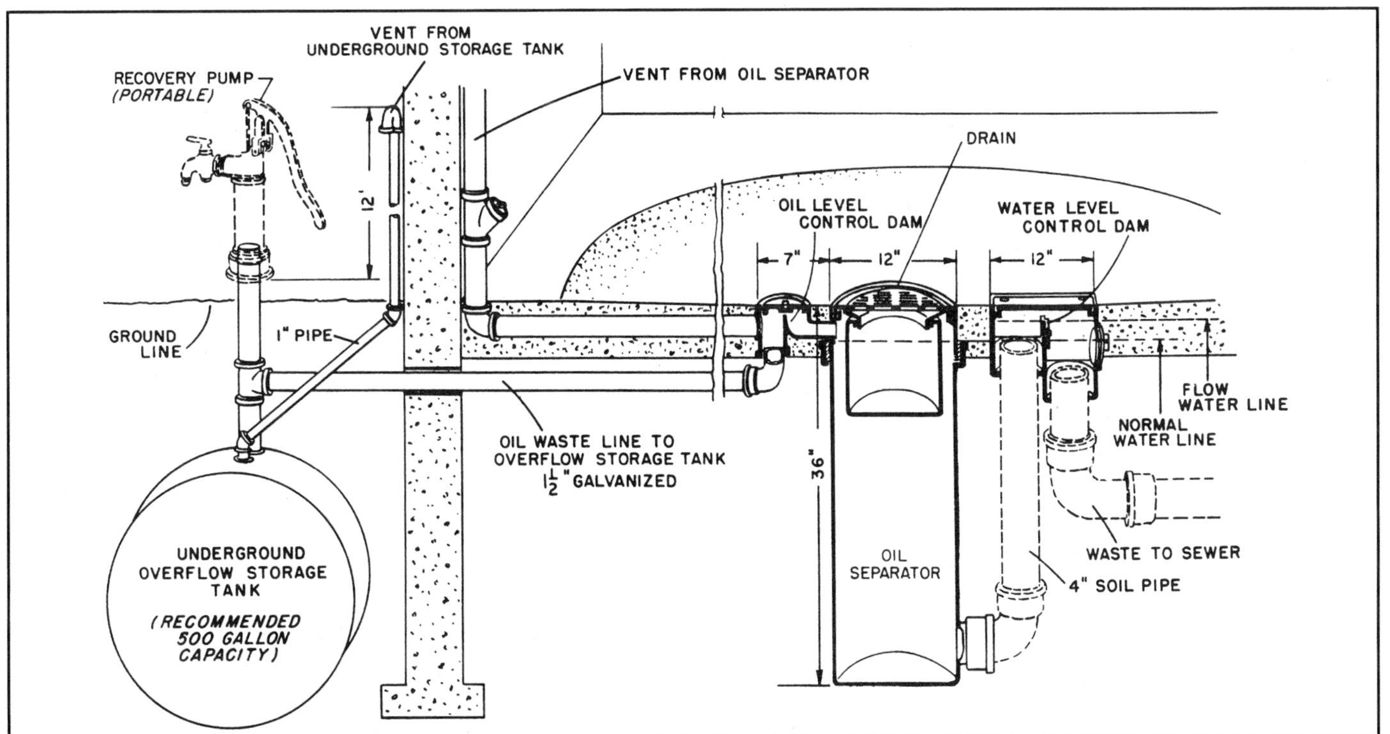

Figure 12–14. The trap and separator system shown here will prevent oil drainage from reaching the sewer and may make it possible to reclaim some of the oil house drainage.

rating. Doors may be arranged to stay open during material-handling operations if doors are designed to close automatically in a fire emergency by provision of listed closure devices. Fire doors shall be installed in accordance with NFPA 80, *Fire Doors and Windows.*

Noncombustible liquid-tight raised sills or ramps at least 4 in. (10 cm) in height or otherwise designed to prevent the flow of liquids to the adjoining areas. A permissible alternative to the sill or ramp is an open-grated trench, which drains to a safe location, across the width of the opening inside of room.

Wood at least 1 in. (2.5 cm) nominal thickness may be used for shelving, racks, dunnage, scuffboards, floor overlay, and similar installations. Storage in inside rooms shall comply with the specifications shown in Table 12–D.

Electrical wiring and equipment located in inside rooms used for Class I liquids shall be suitable for Class I, Division 2 classified locations; for Class II and Class III liquids, shall be suitable for general use. NFPA 70, *National Electrical Code,* provides information on the design and installation of electrical equipment.

Every inside room shall be provided with either a gravity or a continuous mechanical exhaust ventilation system. Mechanical ventilation shall be used if Class I liquids are dispensed within the room.

Exhaust air shall be taken from a point near a wall on one side of the room and within 12 in. (30 cm) of the floor with one or more make-up inlets located on the opposite side of the room within 12 in. (30 cm) from the floor. The location of both the exhaust and inlet air openings shall be arranged to provide, as far as practicable, air movements across all portions of the floor to prevent accumulation of flammable vapors. Exhaust from the room shall be directly to the exterior of the building without recirculation.

Exception: Recirculation is permitted where it is monitored continuously using a fail-safe system that is designed to automatically sound an alarm, stop recirculation, and provide full exhaust to the outside in the event that vapor-air mixtures in concentration over one-fourth of the lower flammable limit are detected.

If ducts are used, they shall not be used for any other purpose and shall comply with NFPA 91, *Standard for Exhaust Systems for Air Conveying of Vapors, Gases, Mists, and Noncombustible Particulate Solids.* If make-up air to a mechanical system is taken from within the building, the opening shall be equipped with a fire door or damper, as required in NFPA 91. For gravity systems, the make-up air shall be supplied from outside the building.

Mechanical ventilation systems shall provide at least one cubic foot per minute of exhaust per square foot of floor area (1 m^3 per 3 m^2), but not less than 150 CFM (4 m^3). The mechanical ventilation system for dispensing areas shall be equipped with an airflow switch or other equally reliable method which is interlocked to sound an audible alarm upon failure of the ventilation system.

In every inside room, an aisle at least 3 ft (0.90 m) wide shall be maintained so that no container is more than 12 ft (3.6 m) from the aisle. Containers over 30 gal (113.5 l) capacity storing Class I or Class II liquids shall not be stored more than one container high.

Where dispensing is being done in inside rooms, operations shall comply with the provisions of NFPA 30.

Table 12–D. Inside Storage Rooms

Automatic Fire Protection Provided*	*Fire Resistance*	*Maximum Floor Area*	*Total Allowable Quantities Gal/sq ft of floor area*
Yes	2 hours	500 sq ft	10
No	2 hours	500 sq ft	4
Yes	1 hour	150 sq ft	5
No	1 hour	150 sq ft	2

*Fire protection system shall be sprinkler, water spray, carbon dioxide, dry chemical, Halon, or other system approved by the authority having jurisdiction.

(10.7 sq ft = 1 m^2. 1 gal/sq ft = 40.5 liters/m^2).

Inside Storage Cabinets

NFPA 30 gives the following requirements for inside storage cabinets (Figure 12–15).

Not more than 120 gal (454 l) of Class I, Class II, and Class IIIA liquids may be stored in an approved storage cabinet for flammable liquids. Of this total, not more than 60 gal (227 l) may be of Class I and Class II liquids and not more than three (3) such cabinets may be located in a single fire area, except that, in an industrial occupancy, additional cabinets may be located in the same fire area if the additional cabinet, or group of not more than three (3) cabinets, is separated from other cabinets or group of cabinets by at least 100 ft (30 m).

Storage cabinets shall be designed and constructed to limit the internal temperature at the center, 1 in. (2.5 cm) from the top to not more than 325 F (162.8 C) when subjected to a 10-minute fire test with burners simulating a room fire exposure using the standard time-temperature curve as given in ASTM E152–81a. All joints and seams shall remain tight and the door shall remain securely closed during the fire test. Cabinets shall be labeled in conspicuous lettering, FLAMMABLE—KEEP FIRE AWAY. The cabinet is not required to be vented.

Metal cabinets constructed in the following manner are acceptable. The bottom, top, door, and sides of cabinet shall be at least No. 18-gauge sheet steel and double walled with 1½ in. (3.8 cm) air space. Joints shall be

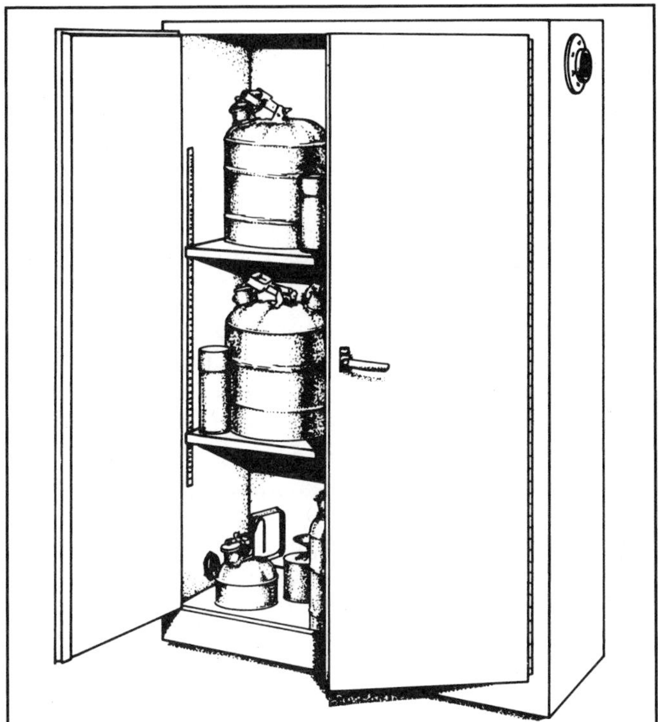

Figure 12–15. Adequate space is provided for the lubrication serviceman's equipment. The locker should be provided with top and bottom ventilation to dissipate vapors. Where flammables are to be stored in the locker, a safety locker of approved design must be used.

riveted, welded, or made tight by some equally effective means. The door shall be provided with a three-point latch arrangement and the door sill shall be raised at least 2 in. (5 cm) above the bottom of the cabinet to retain spilled liquid within the cabinet.

Wooden cabinets constructed in the following manner are acceptable. The bottom, sides and top shall be constructed of exterior grade plywood at least 1 in. (2.5 cm) in thickness, which shall not break down or delaminate under fire conditions. All joints shall be rabbeted and shall be fastened in two directions with wood screws. When more than one door is used, there shall be a rabbeted overlap of not less than 1 in. (2.5 cm). Doors shall be equipped with a means of latching and hinges shall be constructed and mounted in such a manner as to not lose their holding capacity when subjected to fire exposure. A raised sill or pan capable of containing a 2 in. (5 cm) depth of liquid shall be provided at the bottom of the cabinet to retain spilled liquid within the cabinet.

A plant manager should consult local fire prevention authorities about the type of storage and handling that is used or planned for flammable liquids.

Outside Storage Lockers

If space permits, construct storage areas for flammable liquids as separate buildings set aside from the main plant. Construction may be similar to that described for inside storage rooms. Approved prefabricated lockers may also be used. The type of product stored and the proximity to other buildings and structures will determine the best design for outdoor storage lockers. Consult local regulations for location and use of outdoor storage lockers.

CLEANING TANKS

Cleaning tanks that contained flammable or combustible liquids is extremely hazardous work. Be sure that employees involved in this work are trained, take necessary precautions, wear protective equipment as needed, and follow the proper procedures. Follow confined-space entry procedures if entry is required.

General Precautions

Clean tanks and vessels that have contained flammable and combustible liquids before inspections, repairs, entry by personnel, or changes of product. Cleaning should be supervised by a competent person who is familiar with fire and incident prevention, as well as with the requirements for tank cleaning and the hazards of the products involved. Recommendations for preventing fire, explosion, asphyxiation, and exposure to toxic materials are given in API Std 2015, *Safe Entry and Cleaning of Petroleum Storage Tanks;* and NFPA 326, *Standard for the Safeguarding of Tanks and Containers for Entry, Cleaning, or Repair.*

An industrial plant lacking proper cleaning equipment and personnel trained in tank cleaning operations should consult its supplier of flammable or combustible liquids. A confined-space program must be implemented before personnel may enter tanks. Before tank cleaning operations are started, the supervisor and crew should have the proper equipment. This may include supplied-air hose masks with blowers, SCBA, suitable clothing, safety belts, safety lines, and tools. Be sure that the equipment is appropriate, clean, and in good condition. See that all workers are trained and instructed in its use. Procedures must include a method to capture cleaning water so that it can be treated.

Before beginning repair or cleaning operations, purge the tank of all flammable vapor through ventilation or other effective means. Check the inside of the tank with an oxygen-level meter and a combustible-gas indicator before entry, during ventilation, and frequently thereafter during work. If entry into the tank is required, test the atmosphere with instruments capable of readings well below the OSHA Permissible Exposure Limit (PEL) or the Threshold Limit Value (TLV) of any toxic substances present in the contents. A chemical that tests well below the LFL can still be present in extremely toxic concentrations.

Before ventilating, remove all sources of ignition from the surrounding area. Obtain permits for appropriate safe-work-tank entry and for hot work. (See Chapter 11, Fire Protection, in this volume.) Double-block and bleed, disconnect, or blank all piping, lock out all electrical equipment, and use only lighting approved for the specific atmospheric conditions in the tank.

Consider wind and weather conditions. Do not start work (1) if wind might carry vapors into an area where they could create a hazard, or (2) if an electrical storm is threatening or in progress. Prohibit employees from smoking and carrying matches and lighters. Enforce all rules.

Bond blast-cleaning equipment to the tank to prevent static sparks. Do not use power chipping tools and rivet busters when flammable vapors from tanks may be present in the area. Bond the nozzles on water or steam lines used to free tanks of vapor to the tank shells, to prevent static accumulation. Avoid steaming large storage tanks to free gas. Keep motor trucks, gas engines, open flames, and portable electric equipment upwind at a safe distance from a tank being cleaned.

Protective Equipment

Have workers who are cleaning tanks wear impermeable rubber gloves and boots, thoroughly cleaned and in good condition. If workers are entering an atmosphere that is irritating to the skin, they may apply a special protective cream to skin that might be exposed. In an atmosphere that is dangerously irritating or corrosive to the skin, or toxic by absorption through the skin (such as tetraethyl lead), workers should wear full-body impermeable clothing and a full-face mask.

Typically, a worker entering a tank and spending an extended time there must wear a supplied-air, full-face mask or a SCBA, and a safety harness with a lifeline. The mask's facepiece should be carefully checked beforehand. Breathing equipment is unnecessary if the person in charge has determined that the vapor concentration in the tank is below the PEL or TLV and that no oxygen deficiency exists. If a test of the tank shows it to be deficient in oxygen, whether or not it is otherwise immediately dangerous, ventilate it with fresh air and check it for sufficient oxygen before a worker enters it without approved breathing apparatus.

Proper Procedures

A tank should be free of hazardous vapors and toxic materials before any work is performed inside it. (See ANSI Z117.1–1995, *Safety Requirements for Confined Spaces*.) Do not allow a worker to enter a tank with a toxic or oxygen-deficient atmosphere unless absolutely necessary, as in an emergency. When a worker must enter such a tank, he or she should be attended on the outside by another worker who is trained to monitor activities in the tank and to summon a rescue team if problems develop. Follow the confined-space entry procedures developed by plant management.

Be sure that employees engaged in tank cleaning know appropriate first aid and cardiopulmonary resuscitation (CPR). Have them periodically retrained in these skills. A tank cleaner who is overcome by vapor or toxic gas should be removed to fresh air immediately and, if necessary, given CPR until breathing resumes. Summon medical assistance, and keep the rescued worker quiet and warm until breathing and circulation return to normal.

The sense of smell cannot be relied upon for accurately estimating the amount of flammable or toxic vapor in a tank. It may, however, give warning that vapor of some kind is present. Symptoms such as intoxication, dizziness, nausea, and headache indicate the presence of a dangerous concentration of flammable or toxic vapor. Affected workers should leave the contaminated area immediately and should not return until the vapor has been cleared.

The attendant who is monitoring workers inside the tank must be trained to monitor the workers' behavior, and to know when to order evacuation. Workers should always have a clear path of exit from a tank and should be aware that they may have to leave the tank in a hurry. Always have workers use a ladder when a tank must be entered from above. Have the ladder left secured in place until the last worker is out of the tank. Workers who must enter a tank from the top should each use a body harness and a lifeline to assist with rescue work.

Do not permit blast-cleaning, burning, welding, cutting, grinding, or other spark-producing operations in a tank until (1) the area to be heated has been appropriately cleaned, (2) tests have determined that the tank's atmosphere is free from vapor or toxic hazards, and (3) the area has been cleared of any flammable or combustible liquids or solids that could be ignited by cutting or welding operations. Where any vapor is present, further ventilation or inerting by nitrogen or other appropriate oxygen-displacing inerting gases will be required before work is begun. If heavy scale is present, scrape it and probe it for flammable vapors. This must not be done while unprotected workers occupy the tank.

Do not start hot-work repairs, such as welding and cutting, until the area is inspected and a hot-work permit is issued. Even after a tank has been freed of vapor, combustible mixtures may be formed again through admission of flammable vapors or liquids from the following sources: an unblanked line or connection; a break in the bottom of the tank; sludge, sediment, or sidewall scale; or wood structures soaked with the liquid. The interior pontoons or center columns in floating-roof tanks trap quantities of vapor-releasing liquid. Also, the seals of some floating-roof tanks and some internal floating-roof tanks may contain flammable or combustible materials. For this reason, special safeguards are needed during hot work. Make periodic tests with a combustible-gas indicator during the work.

Burning or cutting can release lead fumes from leaded paint on either the inner or outer surface of the tank or from some other source. If leaded paint is present, have workers wear approved respirators for protection against lead. Also, exhaust the fumes to an area outside the tank (consult EPA regulations). Protect or remove wood supports and other combustible materials inside the tank before hot work is begun.

Ground all portable electrical equipment, or provide ground-fault circuit interrupters. Some companies use low-voltage transformers to reduce the hazard of electric shock, especially when employees are required to work in wet areas.

Cold and hot work alike can present a chemical exposure problem in a tank. Workers should wear approved respirators and appropriate eye protection if tests definitely show that wearing a full-face respirator is not necessary.

Frequently make gas tests if the presence of a hazardous gas is suspected. The concentration of gas or vapor in a tank or vessel can increase as work progresses, especially if the inside is scraped or heated during the operation.

Continual ventilation may be required in many cases where repair work is done inside a tank, such as epoxy repairs or painting. NOTE: Where fumes or toxic vapors develop from welding or from other repair work, use mechanical air movers or have workers wear air-supplied breathing apparatus. On a large tank, a door or plates may be removed for ventilation after the tank has been made gas-free. This opening can also be used to expedite removal of sediment, to increase illumination, and to allow access for workers and materials.

If a tank has been closed for some time, flammable vapors may have built up or oxygen may have become deficient because of rusting (oxidation) of the tank's metal surface. No one should enter such a tank without using an air-supplied respirator combined with SCBA, unless the tank is first tested and found to be safe for entry.

Cleaning Storage Tanks

Here is a step-by-step summary of a typical procedure for cleaning a tank that has contained a flammable or combustible liquid.

1. Remove from the vicinity of the tank all sources of ignition, such as matches, open flames, smoking materials, gas engines, welding, exposed electrical wiring, and electrical equipment.
2. Empty as much product as possible through fixed connections, without opening the tank. Continue to empty the tank by pumping, draining, and floating the tank's contents with water or another compatible solvent. Introduce the water through fixed tank connections.
3. Disconnect and blank all product, steam, foam, and similar lines. Do not rely on valves alone.
4. Undo all but four bolts on one manway. Slowly open the remaining bolts. If the product starts to leak out of the tank, close up the manway and continue to empty through fixed piping. After the product is below the level of the manway and where allowed by regulations, open the manway and either float out or pump out the remaining product to approved containers. The manway must be covered during this operation to prevent vapors from escaping into the air.
5. After all available product has been removed, the tank is ready for ventilation to remove vapors. Connect an air mover, or eductor, to the manway. Air- or steam-driven eductors are preferred. If electric eductors are used, they must be explosion-proof and inspected by a qualified person to ensure their integrity. Start the eductor on slow speed, then remove the second manway cover to provide for air intake. The eductor may then be put on full speed. Place the eductor at a low-level manway, near the vapors. Fresh air can enter from an opposite manway or from one on top of the tank. Vapors must be educted at a height of at least 12 ft (3.6 m) above ground level. This can be accomplished by connecting an air-duct tube to the eductor and by supporting the eductor along the side of the tank. (Consult EPA regulations.)
6. Various tank configurations require different methods and configurations for ventilation. Consult your supplier or refer to appropriate manuals for advice on ventilating and cleaning tanks.
7. Ventilate or steam the tank. If steam is used, cool and ventilate the tank afterward, taking care that the ventilation is adequate to prevent a vacuum from developing when the steam is turned off.
8. Bond steam lines and water-wash nozzles to the tank. Wash sludge, sediment, and scale from the tank. Let

it drain, or remove it with a pump. Thoroughly flush out and overflow the tank with water if necessary.

9. Make a gas-hazard test and, if the tank is found to have adequate oxygen and to be free of toxic or flammable vapors, check physical conditions inside the tank before allowing work to start. Otherwise, further clean and ventilate the tank as required.
10. Have the following required personal-protective equipment available: SCBA or supplied-air respirator with the escape SCBA, approved flashlights, safety belts and lifelines, flexible insulated boots and gloves, and appropriate clothing. If light is needed, use only flashlights, or electric lighting listed for combustible atmospheres by Underwriters Laboratories.
11. Test for oxygen and flammables content with a calibrated and adjusted oxygen monitor and combustible-gas indicator. Do not permit entry until the vapor concentration is 10% or less of the LFL.
12. For entering the tank, wear a fresh-air hose mask, air-supply tanks, and safety belt and lifeline if the tank's atmosphere contains more than the maximum acceptable concentration of vapor. Have ample help available outside for the number of workers inside.
13. Continue ventilation for the duration of the work in the tank. Make periodic tests for the presence of hazardous gases or flammable vapors as the work progresses.
14. If a storage tank has contained leaded gasoline and has not previously been tested and declared lead free following its last thorough cleaning, follow instructions from suppliers of tetraethyl lead.
15. If inert gas has been used for freeing gas, check for oxygen deficiency.

NOTE: Ventilation by air instead of steam is more widely used where deposits of iron sulfide are not present. Special precautions are required for servicing tanks that contain pyrophoric iron sulfide. (See API Std 2015, *Safe Entry and Cleaning of Petroleum Storage Tanks*, in References, at the end of this chapter.)

Cleaning Small Tanks and Containers

Work on a container that has held flammable or combustible liquids should be supervised by a trained supervisor who can maintain a high degree of safety during operations. If the container has held compounds such as nitrocellulose, pyroxyline solutions, nitrates, chlorates, perchlorates, or peroxides, take special precautions. The container may contain enough oxygen to support combustion. Contact the manufacturer or supplier, or refer to the manufacturer's material safety data sheets (MSDS), for specific information regarding cleaning procedures and other precautions.

Clean and steam small tanks and drums that contain flammable or combustible products only in an approved area with ample ventilation. Preferably, clean them in an outdoor area free from ignition sources. Pipes and nozzles should be electrically bonded to the containers for the flammable liquids being steamed.

First, remove the covers, plugs, and valves, and permit the tank or drum to drain into an approved container. NOTE: This drained liquid may be classified as hazardous waste. Examine the inside of the tank or drum for rags, waste, or other debris that might interfere with draining or that could retain flammable vapors. Use only lights approved for use in the NEC (Class I, Division I, Group D hazardous locations) for this inspection. (See Chapter 10, Electrical Safety, in this volume.) Mirrors may be used to reflect daylight into the tank.

Steaming

Steaming, hot chemical washes, water filling, and use of inert gas are among the common methods for cleaning and vapor-freeing small tanks and drums. If the inside of a container is clean and steam is available, the easiest cleaning method is steaming. (See American Welding Society, *Safe Practices for Welding and Cutting Containers That Have Held Combustibles.* See also NFPA 326, *Standard for the Safeguarding of Tanks and Containers for Entry, Cleaning, or Repair.*)

After draining, place the tank or drum on a steam rack or over a steam connection with the outlet holes at the lowest point. The tank or drum must rest against a steam pipe and must be bonded. Then apply an ample supply of live steam for an appropriate period of time, depending on the tank's size and contents.

The inside of the drum may also be washed with hot water if hot-work repairs are anticipated. After being cooled, test with a combustible-gas indicator. If the drum tests vapor free, hot-work repairs can then proceed. If not, clean the drum again.

If steaming will not clean the tank or drum, use a cleansing compound of sodium silicate or trisodium phosphate (washing powder), dissolved in hot water and kept at a temperature of 170 to 190 F (77 to 88 C). Slowly add hot water to overflow the container until no appreciable amount of volatile liquid, scum, or sludge appears.

Exceedingly dirty containers can require preliminary treatment with a caustic-soda solution that is agitated enough to make sure that the interior surfaces are thoroughly cleaned. Containers can then be drained, washed, and steamed. Do not use this treatment on aluminum- or zinc-coated drums. These materials could generate large amounts of hydrogen when in contact with a caustic-soda solution.

If steam or hot water is unavailable, use a cold-water solution with an increased amount of cleansing compound. The solution should be agitated to ensure thorough cleaning.

To guard against burns, especially when using steam, hot water, and caustic soda, provide workers with suitable clothing, such as boots, gloves, head coverings, face shields, and rubber aprons.

Because of their low-ignition temperature, do not steam drums that contain carbon disulfide. Make them vapor-free with a cleansing compound, and then test them for vapor.

Using Inert Gas

Small tanks can be made temporarily safe by means of an inert gas. However, this method is generally not considered as safe as steaming. Inert gas displaces oxygen and vapors but does not remove the residual source.

Portable, inert-gas generators are available for special jobs. However, they should be used only by workers who have the training and equipment to produce the proper atmosphere and to safeguard against fire and explosion.

CO_2 and nitrogen are sometimes used to make small tanks and vessels that have contained flammable liquids safe for hot-work repairs. However, use of CO_2 can lead to problems with static electricity. If entry is required, first check for oxygen deficiency and toxic vapors.

When an inert gas like CO_2 or nitrogen is used, wash the tank or vessel so it is as free as possible of flammable liquids and thoroughly flush it until the vessel overflows. When repairing, leave as much water in the tank as the work will permit. CO_2 concentrations of not less than 50%, by volume, should be used for blanketing tanks. If the tank has contained hydrogen or carbon monoxide (CO), an 80% concentration is required. Nitrogen concentrations should be 60% or higher, depending on the previous contents of the container.

Abandonment of Tanks

Thoroughly clean tanks to be permanently abandoned and make them safe from flammable vapors. Then dismantle and remove them from the premises, where allowed and according to regulations. Consult U.S. EPA underground tank regulations for details on disposal.

Cap and secure against tampering fill lines, gauge openings, and pump-suction lines of tanks taken out of service for less than 90 days. Leave the vent lines open.

DISPOSAL OF FLAMMABLE LIQUIDS

Return unused, uncontaminated flammable liquids to the vendor, salvage them for resale, or use them in some other approved way. When drummed and properly stored, most flammable liquids are stable and can be used safely for several years.

Mixtures of clean flammable liquids sometimes need to be separated before they are usable. It is best to have a recovery contractor do these separations. Used or dirty flammable liquids also can be handled by an approved recovery contractor.

If recycling or recovery of flammable liquids is infeasible, give them to a licensed disposal contractor. Observe all regulatory requirements, such as those of the Resource Conservation and Recovery Act, including saving manifests and shipping papers for future review.

COMMON USES OF FLAMMABLE AND COMBUSTIBLE LIQUIDS

Flammable and combustible liquids have many uses in industry. When workers use these liquids, be sure they know and observe the necessary precautions.

Dip Tanks

Dip tanks containing flammable liquids that are subject to ignition at ordinary temperatures and are giving off flammable vapors present a severe fire and explosion hazard. (See NFPA 34, *Dipping and Coating Processes Using Flammable or Combustible Liquids,* and 29 *CFR* 1910.108.)

Conduct dipping operations above grade in a detached one-story building of noncombustible construction or in a separate one-story section. The room should be as large as possible, adequately ventilated, away from sources of ignition, and conspicuously marked as a flammable-liquid area (Figure 12–16).

Avoid handling flammable or combustible liquids in open containers since it can be hazardous. Keep the openings of such containers as small as possible, and use a cover. The cover should be hinged-and-gravity closing or should slide on tracks and be held open by a fusible link or another heat-actuated device (Figure 12–16).

NFPA 34 states the following:

Mechanical ventilation shall be provided, and the ventilating system arranged to move air from all directions toward the vapor area's origin, past the point of operation, and then to a safe outside location. The ventilating system shall be so arranged that the failure of any ventilating fan shall automatically stop any dipping conveyor system.

Tanks with capacities of more than 500 gal (1,892 l) should have bottom drains unless the viscosity of the liquids they contain makes this requirement impractical. Install overflow pipes on tanks with capacities of more than 150 gal (568 l) to carry off any overflow liquids to an approved holding tank. Where required, protect larger tanks with automatic fire-extinguishing systems.

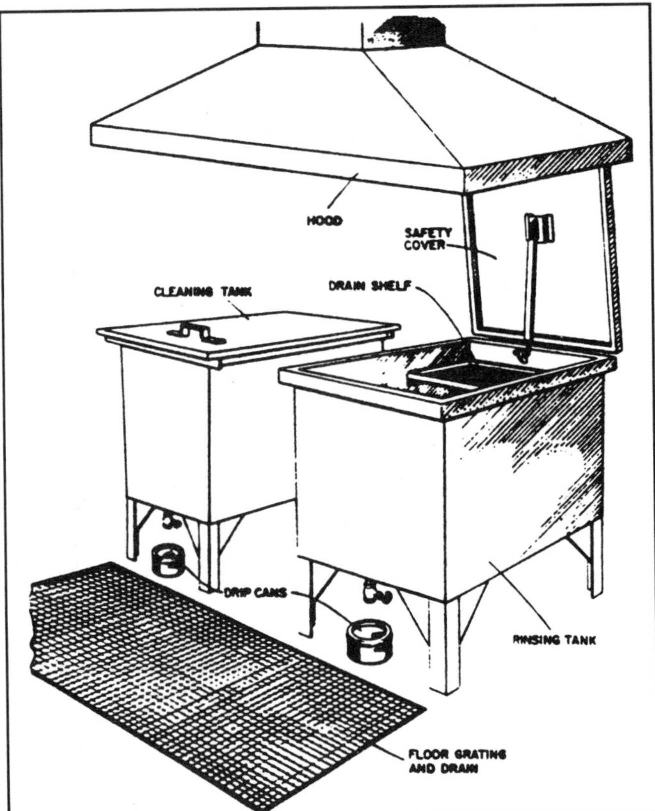

Figure 12–16. Although the overhead hood with forced ventilation is not mandatory in all cleaning tanks, it is generally good practice. Self-closing covers, with a fusible link to permit automatic closing in the event of fire, are now required by OSHA regulations.

Japanning and Drying Ovens

Ovens used for evaporating varnish, japan enamel, and other flammable and combustible liquids can present serious fire and explosion hazards. Provide ample ventilation and explosion venting for these ovens.

Drying ovens are of two types: the box oven and the continuous-conveyor oven. The box oven, which is closed while in operation, is commonly used in small-scale operations. The continuous-conveyor oven, which is open at both ends, is normally used for quantity production. Consult NFPA 86, *Ovens and Furnaces,* for details about constructing and operating drying ovens. Also consult the jurisdictional authorities.

Provide ovens that process large enough amounts of combustible materials with the proper type of fire-extinguishing equipment. Also provide safety-control devices so that the fans and conveyor in the continuous-conveyor oven will automatically stop in case of fire.

Oil Burners

Use oil burners that are approved by a recognized testing laboratory. To prevent faulty ignition or accumulation of soot, with its attendant fire hazard, use the correct type of fuel oil, as recommended by the testing laboratory or manufacturer.

Fuel oil should not have a flash point lower than 100 F (37.8 C). It should be a hydrocarbon oil, free from acid, grit, and foreign matter likely to clog or damage the burners or valves. Some plants use acid sludge or waste oil for fuel. Such fuel requires special burning equipment and procedures.

Preferably, locate an industrial fuel-oil supply tank outdoors and aboveground. Protect the tank and piping to contain leaks and spills, as required by local authorities. Provide all tanks with overfill-protection systems.

Do not use a gravity feed to burners, unless special safeguards are provided against abnormal discharges of oil at the burner. The primary hazard of oil burners is the possibility of a discharge of unburned oil into a hot fire box, where it can vaporize and form an explosive mixture. Provide approved automatic safeguards to control this hazard.

Cleaning Metal Parts

Refined petroleum solvents (over 100 F [37.8 C] flash point) are used for cleaning grease and oil from metal parts where ordinary ventilation is provided and the area is free of sources of ignition. Alkaline compounds, available under several trade names, will not cause a fire. Do not allow oil or grease to accumulate in these cleaning compounds. Even with high-flash and low-toxicity materials, provide ventilation to remove vapors. (See Factory Mutual Data Sheet 7–79, *Metal Cleaning.*)

The flammability and toxicity of gasoline and the toxicity of chlorinated solvents have been sufficient reasons for the general ban on these products for cleaning purposes.

Internal-Combustion Engines

To prevent the accumulation of rubbish, oil or fuel, and rags around industrial internal-combustion engines, practice good housekeeping. Provide proper receptacles for refuse. (See NFPA 37, *Stationary Combustion Engines and Gas Turbines.)*

Before filling a gasoline tank, shut down the engine and permit hot exhaust pipes to cool. Use approved safety cans or a hand pump with a bonded filling hose, and keep the main fuel supply in approved containers outdoors.

For engines that operate continuously and cannot be shut down for filling, locate the fuel tanks outside the engine room. In this way, vapors will not be exposed to hot engines or to exhaust. Refuel lift trucks and other mobile equipment outside buildings or in designated approved areas.

Spray Booths

Conduct paint-spraying operations in detached buildings or away from other operations, when possible. Where spraying is done in production areas, use approved spray rooms or booths with adequate ventilation. Provide an enclosed area large enough that explosive mixtures of vapor and air cannot be easily formed. Eliminate heating units, air filters, and piping that might become coated with flammable materials, or protect them against such accumulations. (See NFPA 33, *Spray Application Using Flammable or Combustible Materials.*)

Fires in spray booths and in spray-booth operations most frequently result from spontaneous ignition of spray deposits. Prevent these fires by establishing a regular cleaning schedule. The frequency of cleanings depends on how quickly deposits accumulate.

Water-wash booths have proved to be safer from fire than the dry type of booth. Water-wash booths trap the excess spray before it can enter exhaust ducts.

Electrical equipment in spraying areas should meet the requirements of the NEC for such locations. Explosion-proof lighting must be used.

Provide large spraying operations with automatic fire controls. Automatic sprinklers or CO_2 systems are most effective. Provide protection for the exhaust ducts, as well as for the spray booths. Protect the discharge heads of such equipment from overspray. Protection of automatic sprinkler heads is usually achieved by covering the heads with ordinary paper or plastic bags.

Electrostatic spraying, usually automatic, introduces a possible source of ignition in the arcing of parts to the electrodes. To overcome this hazard, parts being sprayed can be held in tight-fitting fixtures. Do not permit parts being sprayed to hang or swing and thus come close enough to induce a spark.

Liquefied Petroleum Gases

Liquefied petroleum gases (LP gases) include any material that is composed predominantly of any of the following hydrocarbons or mixtures of them: propane, propylene, butane (normal butane or isobutane), and butylenes. These gases liquefy under moderate pressure but convert to a gaseous state upon relief of the pressure. LP-gas vapor presents a hazard comparable to that of any flammable natural or manufactured gas. However, since LP-gas vapor is heavier than air, it requires adequate ventilation.

LP gases are used as fuel gases, as raw materials in chemical processes (such as the making of hydrogen), and to form special atmospheres in heat-treating furnaces. Employees should understand the properties of these gases and should be thoroughly trained in the safe practices for handling, distributing, and operating them. Develop detailed safety programs to handle any emergencies that might arise.

LP-gas storage systems should be designed and installed by experienced, reliable manufacturers who are thoroughly familiar with the hazards of these systems. Manufacturers should observe state or provincial and local codes, and the recommendations of fire prevention organizations and insurance companies. (See NFPA, the Factory Mutual System Research Organization, and the National LP Gas Association listed in References, at the end of this chapter.)

SUMMARY

- A flammable liquid is any liquid having a closed-cup flash point below 100 F (37.8 C) and a vapor pressure not exceeding 40 psia (276 kPa) at 100 F. These liquids are categorized by NFPA as Class IA through IC, depending on their flash points.
- Combustible liquids are those with flash points at or above 100 F (37.8 C) but below 200 F (93.3 C), closed cup, and are divided into Classes II, III, IIIA, and IIIB. They do not ignite as easily as flammable liquids but must be handled with the same precautions.
- The degree of danger in these liquids is determined largely by (1) flash point, (2) concentration of vapors in the air, (3) risk of ignition at or above the flash point, and (4) amount of vapors present.
- Workers should always avoid exposing large surface areas of flammable or combustible liquids to the air to avoid creating a serious fire or explosion risk. Other general safety measures include avoiding mixing flammable and combustible liquids, prohibiting smoking around or near these liquids, shielding the liquids from static electricity by bonding and grounding, and using only spark-resistant tools when working with these liquids.
- Health hazards associated with flammable and combustible liquids include skin irritation, intoxication or illness from inhaling their vapors and fumes, and oxygen deficiency in closed containers used to store these liquids. Unless tests prove otherwise, workers should assume that flammable vapors and toxic mixtures are present in all containers of these liquids.

- Loading and unloading of vessels used to transport flammable or combustible liquids should be done only by trained employees.
- Only Class I and Class II liquids can be safely stored in buildings used for public assembly. Where needed, a dike can be constructed around a tank or group of tanks for added protection. Transfer pumps should be located outside buildings and diked areas where possible to minimize hazards.
- Underground tanks should be protected against overhead traffic, built on a firm foundation, and shielded against corrosion. Aboveground tanks must be constructed an approved distance away from property lines, public ways, and nearby buildings. All tanks should be equipped with proper fire-extinguishing systems.
- Inside storage and mixing rooms should be isolated and protected as much as possible to guard against the hazard of fire from without. Outside storage lockers should be built away from the main plant whenever possible and must conform to regulations.
- Cleaning tanks that contained flammable and combustible liquids is extremely hazardous work and requires highly skilled and trained employees. Workers must wear protective equipment and understand the proper work and medical procedures to follow.
- Small tanks and containers can be cleaned by steam and made temporarily safe by means of an inert gas. Tanks to be abandoned must be thoroughly cleaned, then dismantled and removed from the premises.
- Proper disposal of flammable and combustible liquids is an important part of handling these materials safely. If recycling or recovery of these liquids is impossible, they should be burned in an EPA-approved incinerator or given to a disposal contractor.
- Because flammable and combustible liquids have many uses in industry, workers must know how to guard against fire hazards, prevent unintentional incidents, and protect their health when using flammable and combustible liquids.

REFERENCES

American Conference of Governmental Industrial Hygienists, Bldg. D–7, Glenway Avenue, Cincinnati, OH 45211.
Threshold Limit Values, published annually.

American National Standards Institute, 11 West 42nd Street, New York, NY 10036.
Fundamentals Governing the Design and Operation of Local Exhaust Systems, ANSI Z9.2–1979 (R1991).
Safety Requirements for Confined Spaces, ANSI Z117.1.

American Petroleum Institute, 1220 L Street NW, Washington, DC 20005.
Safe Entry and Cleaning of Petroleum Storage Tanks, Std 2015.
Lining of Aboveground Petroleum Storage Tank Bottoms RP 652.
Protection Against Ignitions Arising out of Static, Lightning, and Stray Currents, RP 2003 82.
Venting Atmospheric and Low-Pressure Storage Tanks, RP 2000 82.
Welded Steel Tanks for Oil Storage, SID 650.

American Society for Testing and Materials, 1916 Race Street, Philadelphia, PA 19103.
Annual Book of ASTM Standards:
Parts 23, 24, and 25, "Petroleum Products and Lubricants."
Parts 27 and 28, "Paint."
Part 30, "Soaps, Antifreezes, Polishes, Halogenated Organic Solvents, Activated Carbon, and Industrial Chemicals."
Test Methods for Flash Point by Pensky-Martens Closed Tester, D93–85.
Test Method for Flash Point by Tag-Closed Tester, D56–82.
Standard Method of Test for Vapor Pressure of Petroleum Products (Reid Method), D323–72.

American Society of Safety Engineers, 1800 East Oakton Street, Des Plaines, IL 60018.
How to Prevent Confined Space Fatalities.

American Welding Society, P.O. Box 351040, 550 LeJeune Road NW, Miami, FL 33135.
Safe Practices for Welding and Cutting Containers That Have Held Combustibles.

Association of American Railroads, Bureau for the Safe Transportation of Explosives and Other Dangerous Articles, 50 F Street NW, Washington, DC 20001.
Recommended Good Practice for Handling Collisions and Derailments Involving Hazardous Materials in Transportation, BE Pamphlet No. 1.
Recommended Good Practice for Handling Fires or Spills Involving Explosives and Other Dangerous Articles in Transportation, BE Pamphlet No. 2.
Recommended Practice for the Prevention of Electric Sparks That May Cause Fires in Tanks or Tank Cars Containing Flammable Liquids or Flammable Compressed Gases, Due to Proximity of Wire Lines, BE Circular No. 17–E.
Signal Manual of Recommended Practice.

Chemical Manufacturers Association, 2501 M Street NW, Washington, DC 20037.
"Loss Prevention Data Sheets."

Factory Mutual System Research Organization, 1151 Boston-Providence Turnpike, Norwood, MA 02062.
"Factory Mutual Loss Control Data Books" (listing available).
Factory Mutual System Approval Guide.
Flammable Liquid Pumping and Piping Systems, Data Sheet 7–32.
Metal Cleaning, Data Sheet 7–79.

Fawcett HH, Wood WS, eds. *Safety and Accident Prevention in Chemical Operations,* 2nd ed. New York: Wiley-Interscience Publishers, 1982.

Industrial Risk Insurers, 85 Woodland Street, Hartford, CT 06102.
Recommended Good Practices (Supplements to NFPA Standards).

National Fire Protection Association, 1 Batterymarch Park, Quincy, MA 02269.

Automotive and Marine Service Station Code, NFPA 30A.

Basic Classification of Flammable and Combustible Liquids, NFPA 321.

Control of Gas Hazards on Vessels, NFPA 306.

Dipping and Coating Processes Using Flammable or Combustible Liquids, NFPA 34.

Drycleaning Plants, NFPA 32.

Exhaust Systems for Air Conveying of Materials, NFPA 91.

Fire Doors and Windows, NFPA 80.

Fire Hazard Properties of Flammable Liquids, Gases, and Volatile Solids, NFPA 325.

Fire Protection Handbook.

Flammable and Combustible Liquids Code, NFPA 30.

Hazardous Chemicals Data, NFPA 49.

Identification of the Fire Hazards of Materials, NFPA 704.

Installation of Oil Burning Equipment, NFPA 31.

National Electrical Code, NFPA 70.

Ovens and Furnaces, NFPA 86.

Spray Application Using Flammable or Combustible Materials, NFPA 33.

Standard for Air Conveying of Vapors, Gases, Mists, and Noncombustible Particulate Solids, NFPA 91.

Standard for the Safeguarding of Tanks and Containers for Entry, Cleaning, or Repair, NFPA 326.

Standard Methods of Tests of Fire Endurance Building Construction and Materials, NFPA 251.

Static Electricity, NFPA 77.

Stationary Combustion Engines and Gas Turbines, NFPA 37.

Storage and Handling of Liquefied Petroleum Gases, NFPA 58.

Storage and Handling of Liquefied Petroleum Gases at Utility Gas Plants, NFPA 59.

Tank Vehicles for Flammable and Combustible Liquids, NFPA 385.

National Propane-Gas Association of America, 1600 Eisenhower Lane, Lisle, IL 60532.

National Safety Council, 1121 Spring Lake Drive, Itasca, IL 60143.

"Chemical Hazards Fact Finder."

Fundamentals of Industrial Hygiene, 4th ed., 1996.

NIOSH Pocket Guide to Chemical Hazards, 1994.

Underwriters Laboratories, Inc., 333 Pfingsten Road, Northbrook, IL 60062.

Classification of Hazards of Liquids, Research Bulletin No. 29.

Fire Protection Equipment List.

Flammable Liquids, Static Electricity Hazards, Serial No. UL–435.

Gas and Oil Equipment List.

Hazardous Location Equipment List.

The Lower Limit of Flammability and Autogenous Ignition Temperature of Certain Common Solvent Vapors Encountered in Ovens, Research Bulletin No. 43.

Oil Tank Vents—Hazards of Screens, Serial No. UL–31.

U.S. Department of Commerce, 14th Street and Constitution Avenue NW, Washington, DC 20230.

Sparking Characteristics and Safety Hazards of Metallic Materials, Technical Report No. NGF–T–1–57, PB 131131. Office of Technical Services.

Static Electricity in Nature and Industry, Bulletin No. 368.

U.S. Department of the Interior, Bureau of Mines, 2401 E Street NW, Washington, DC 20241.

Flammability Characteristics of Combustible Gases and Vapors, Bulletin 627.

Limits of Flammability of Gases and Vapors, Bulletin 503.

Mine Gases and Methods for Their Detection, Circular No. 33.

U.S. Department of Labor. Occupational Safety and Health Administration, 200 Constitution Avenue NW, Washington, DC 20210.

Code of Federal Regulations, Title 29. Subpart H, "Hazardous materials." Sections 1910.100, 1910.106, 1910.107, 1910.108, 1910.120, 1910.147, and 1910.1200.

U.S. Department of Transportation, 400 7th Street SW, Washington, DC 20590.

Code of Federal Regulations, Title 49. "Hazardous Materials Regulations," Parts 170 through 179, and "Motor Carrier Safety Regulations," Parts 393 and 397.

REVIEW QUESTIONS

1. What is a flammable liquid, as defined by NFPA 30, *Flammable and Combustible Liquids Code*?

2. When flammable and combustible liquids vaporize, forming flammable mixtures with air, the degree of danger depends on what four factors?
a.
b.
c.
d.

3. What is autoignition temperature?

4. Define flash point.

5. OSHA requires workers to wear approved breathing equipment when entering a tank or enclosed space that tests less than what percent of oxygen?
a. 19.5%
b. 20%
c. 21%
d. 21.5%

6. How is static electricity generated?

7. What is the difference between bonding and grounding?

8. Petroleum liquids can build up a static charge when they:
   a. Flow through piping or filters
   b. Are agitated in a tank or a container
   c. Are subjected to vigorous mechanical movement such as spraying or splashing
   d. All of the above

9. Which of the following would put a combustible-gas indicator out of service if drawn into the instrument?
   a. Steam
   b. Oxygen
   c. Water
   d. Flammable liquid
   e. All of the above
   f. Only a, c, and d

10. When a tank truck containing flammable liquids is being loaded or unloaded, what are the steps to take to ensure safety?
   a.
   b.
   c.
   d.

11. What four situations require drivers of tank trucks to immediately notify the U.S. Department of Transportation by phone?
   a.
   b.
   c.
   d.

12. What type of extinguishing agents should be used on a flammable liquid fire?

13. What are the three ways to protect an underground tank when it is buried under a heavily traveled roadway?
   a.
   b.
   c.

14. Why is it common practice to paint flammable-liquid tanks with aluminum, pastel, or white paint?

15. Even after a tank has been freed of vapor through proper cleaning, combustible mixtures may be formed again through admission of flammable vapors or liquids from what sources?
   a.
   b.
   c.
   d.

13

Workers with Disabilities

Revised by

James T. O'Reilly, JD

This chapter is intended to assist companies and organizations in the safe and productive placement of disabled individuals in the work force. Topics covered include the following:

- major sections of the Americans with Disabilities Act
- definitions of workers with disabilities
- discussion of reasonable accommodation
- role of the safety and health professional
- job placement issues for employers
- making jobs more accessible
- emergency procedures for workers with disabilities

Almost every worker with a physical or mental impairment can qualify for some type of job. Industry and government surveys made during the past several decades prove it is good business to hire workers with permanent disabilities. They report for work promptly, can produce as well as other workers, and their turnover rate is often lower. In the United States, the 1990 Americans with Disabilities Act provides for punitive damages against employers with 15 or more employees who discriminate against job applicants or employees with disabilities.

AMERICANS WITH DISABILITIES ACT

The 1990 U.S. Americans with Disabilities Act (ADA) encompasses the following areas: Title I—Employment Provisions, Title II—State and Local Government Provisions, Title III—Public Accommodations and Services Operated by Private Entities, Title IV—Telecommunications, and Title V—Miscellaneous Provisions.

Employment—Title I—makes it illegal to discriminate against an individual with a disability in hiring or promotion if the person is otherwise qualified for the job, and takes into account all aspects of the employment process—application procedures, the type of questions asked during an interview, the identification and delineation of essential functions of a job, accommodations, and preemployment exams.

Transportation—Title II—provides equal access for individuals with disabilities who use public transportation and other public services which have a major impact on the "employability" of people with disabilities.

Public Accommodations—Title III—makes it illegal to discriminate against individuals with disabilities in the full and equal enjoyment of goods, services, facilities, privileges, advantages, and accommodation of any place of public accommodation.

Telecommunications—Title IV—requires common carriers of telecommunications services to provide telecommunication relay services to hearing impaired and speech impaired individuals.

Miscellaneous Provisions—Title V—prohibits discrimination by state and local governments against qualified individuals with disabilities, and mandates that all government facilities, services, and communications be accessible consistent with Section 504 of the Rehabilitation Act of 1973. It also gives individuals the right to file complaints and bring private lawsuits.

Title I of the Act requires that employers make reasonable accommodations for the disabled. An employer is required to provide sufficient accommodation to allow qualified individuals with a disability to attain the same level of job performance as co-workers having similar skills and abilities. An employer is not required to employ an individual where to do so would pose a "direct threat" to the health or safety of others. The determination of direct threat must be based on the actual condition of the individual and not upon generalizations or stereotypes about the disability.

A significant aspect of Title III is that public accommodations and services operated by private entities are not required to permit individuals to participate in their services where the individual poses a direct threat to the health and safety of others which cannot be eliminated by modification of policies or practices or provision procedures or by providing auxiliary aids or services.

The law requires that companies establish affirmative action guidelines for the hiring, upgrading, promotion, award of tenure, demotion, transfer, layoff, termination, right-of-return from layoff, and rehiring of qualified disabled individuals. The law also requires employers to provide "reasonable accommodation" to modify the work environment or the job for these workers when necessary.

If an employer denies a disabled individual a specific job, the burden of proof is upon management to show that the person is unqualified because of one or more of the following reasons:

- The job would put the individual in a hazardous situation.
- Other employees would be placed in a hazardous situation if the person were on the job.
- The job requirements cannot be met by an individual with certain physical or mental limitations.
- And (for all of the above) accommodation of the job cannot reasonably be accomplished.

Developing affirmative action programs, including those for hiring workers with disabilities, is usually the responsibility of personnel other than the safety and health professional. The safety and health professional should serve as a resource person and play a critical role in evaluating the job and work environment in order to establish whether the employee can perform the essential functions of the job. Safety evaluations for the worker must include adequate access to and exit from the workplace as well as safety on the job. Safety and health professionals must be sensitive in their use of language to describe the disabled and the tasks that the disabled may be asked to perform (Figure 13–1).

Under the ADA, an employer's written job descriptions are considered evidence of the essential functions of a job if the job description existed before the job was advertised or the applicant/employee was interviewed for the job, considered for promotion, or other job-related action was taken. It is important that all essential job-related functions be defined and contained in the job description. When developing these descriptions, keep in mind intellectual as well as physical demands.

Negative Language	**Positive Language**
handicapped/handicap	disabled/person with disability
cripple/crippled by	person who has (whatever)
victim	person who uses (assistive device)
spastic	
paralytic	
afflicted/afflicted by	caused by . . .
deformed/deformed by	as a result of . . .
suffering from	
confined/restricted to a wheelchair	person who uses a wheelchair
wheelchair bound	wheelchair rider, user
deaf and dumb	pre-lingually deaf (at birth)
deaf mute	post-lingually deaf (after birth)
poor, unfortunate and similar words	

Normal—When used in the statistical sense or to express "average," the term is fine. However, this word should never be used to refer to people without apparent physical, emotional, or mental disabilities. Because most people are disabled at some time in their lives, the average person can be a disabled person. The disabilities may be inability to control one's temper, effects of past broken bones, strong prejudices, substance abuse, and other problems.

Wheelchair—People who are able to walk and run usually see a wheelchair as a confining device. In reality, it gives mobility to people who would otherwise be confined to bed. Wheelchairs come in many types, including some that have variable height controls, extra-narrow axle widths, or other features. It is often possible to build a wheelchair or other mobility device to suit a particular worker and his or her job duties. Employers should explore several options if a workplace cannot be safely or easily adapted to a standard wheelchair.

Deaf/hearing impaired—All deaf people are hearing impaired, but not all hearing-impaired people are deaf. This distinction is important as the needs of a deaf employee are very different from those of a hearing-impaired employee. Some deaf people use sign language, while others lipread. Some hearing-impaired people need relative quiet to understand verbal communication; others need a person with a deep voice to relay messages from a person with a higher-pitched voice. Ask what a particular hearing-impaired person needs.

Blind/visually impaired—Likewise, not all visually impaired people are blind or even legally blind. Some have tunnel vision, others have peripheral but no central vision. Some need strong light while others require dimmer light. Simply because a visually impaired person wears glasses does not mean their vision is 20/20. Ask what a particular visually impaired employee needs to be able to see comfortably.

(Copyright by the Right to be Proud Campaign, St. Louis, MO.)

Figure 13–1. Language Issues Regarding Disabilities

Companies should not be too quick to decide that safety and health problems are insurmountable. There are many organizations available that can help management find solutions. (See Sources of Help at the end of this chapter.)

HISTORY AND THE LAW

In the 1940s, special attention was given to employing workers with disabilities by a number of large companies that realized hiring these people was smart and sound business practice. Although some companies employed such workers before the 1940s, three events occurred in that decade to stimulate these programs and encourage other companies to institute hire-the-disabled programs.

Many individuals with disabilities were hired to help fill job vacancies left by employees who joined the military. For example, in the early 1940s, one company undertook an affirmative action program to help each returning disabled veteran to become an employable person. Other companies established similar programs.

The third event was a study published by the U.S. Department of Labor that debunked some myths about workers with disabilities being less productive, suffering more injuries, and losing more time from work than other workers. On the contrary, the Department of Labor study showed disabled workers were as productive as other workers, had lower frequency and severity of injury rates, and were absent from work only one day more per year than other workers. In researching over 11,000 workers with disabilities for almost two years, the study's authors did not find a single serious injury caused by a disabled worker, to himself/herself or to a fellow worker, that was a direct result of the disability.

One company, in a study in 1958 and another in 1981, found that of its more than 2,700 disabled workers, 96% rated average or better on safety performance; 92% rated average or better on turnover; and 85% rated average or better on attendance. After a decade or more of direct experience in hiring the disabled worker, the personnel files of many companies contain indisputable proof of the value of these employees—and of affirmative action programs—to their companies.

Rehabilitation Act of 1973

The U.S. Vocational Rehabilitation Act of 1973 (Public Law 93-112), commonly referred to as the Rehabilitation Act, was the first major civil rights law protecting the rights of persons with disabilities. This law applies to federal contractors (Section 503) and recipients of federal assistance programs (Section 504). Therefore, all employers who do work for the federal government, or receive funds from the government, are subject to this law.

Section 503 of the Act requires employers to take "affirmative action" to recruit, hire, and advance qualified individuals with disabilities. The law applies only to employers who have federal government contracts or subcontracts of $2,500 or more. Those holding contracts or subcontracts of $50,000 or more, with at least 50 employees, must prepare and maintain (review and update annually) an affirmative action program at each establishment. The program, which sets forth the employer's policies, practices, and procedures regarding disabled workers, must be available for inspection by job applicants and employees. Section 504 of the Act forbids acts of employment discrimination against qualified disabled persons by employers who receive federal funds. This Section is enforced by each department or agency that provides federal funds.

In the Rehabilitation Act Amendments of 1974, the definition of "handicapped [disabled] individual" was broadened for purposes of Section 504. With this amended definition, it became clear that Section 504 was intended to forbid discrimination against all disabled individuals, regardless of their need for or ability to benefit from vocational rehabilitation services. Thus, Section 504 reflects a national commitment to end discrimination on the basis of disability and establishes a mandate to bring persons with disabilities into the mainstream of American life.

Other U.S. federal departments and agencies also have issued regulations similar to those of the Department of Education. For example, the U.S. Department of Labor has issued regulations (29 *CFR* 32), effective November 6, 1980, which implement Section 504 of the Act for the department. All of these regulations require federal contractors and recipients of federal funds to make reasonable accommodations to the workplace when necessary for employing people with disabilities.

All records pertaining to compliance with the Act, including employment records, and any complaints and actions taken as a result, must be retained by the employer for at least one year. If the company fails to maintain complete and accurate records or fails to update the affirmative action program each year, the government can impose "appropriate sanctions" against the employer. When complaints are brought against the employer or there is some question about affirmative action programs, the employer must allow investigators to have access during normal business hours to its place of business, its books, records, rules and regulations, and accounts pertinent to compliance with the Act.

U.S. State and Local Laws

All 50 states and many local governments have now adopted building codes or legislation requiring barrier-free design or removal of barriers preventing access to the building by disabled persons. Many of these codes mandate that any public building or facility must be barrier free if the public is invited to use it for any normal purpose such as shopping, employment, recreation, lodging, or services. If accessible facilities need to be identified, the organization should use the international symbol of accessibility (Figure 13–2).

Insurance Considerations

Most companies mistakenly believe that employing disabled workers will raise their workers' compensation premiums. This is not true. Rates are based on experience by the class of industry and modified in most cases by the individual plant experience. There is no indication that losses are increased when persons with disabilities are properly placed.

Discrimination

Under the ADA, discrimination includes such actions as limiting, segregating, or classifying a job applicant or employee in a way that adversely affects the person's opportunities or status. Discrimination also includes not making reasonable accommodation for the known physical or mental limitation of an otherwise qualified person with a disability. It also includes the denial of employment because a qualified person with a disability needs reasonable accommodation.

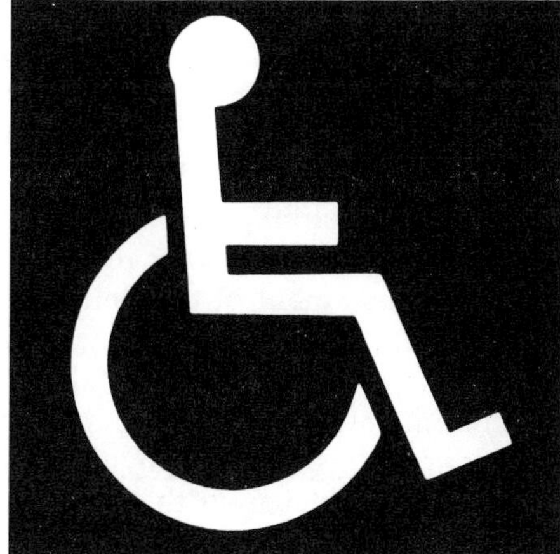

Figure 13–2. International symbol designates access for the disabled. The symbol is in white on a blue background. (Printed with permission from American National Standards Institute.)

WHO ARE DISABLED JOB-SEEKERS?

The law defines three types of disabled persons seeking employment—the disabled individual, the disabled veteran, and the qualified disabled individual.

"Disabled Individual"

ADA defines a "disabled individual" as a person who has one of the following:

1. a physical or mental impairment that substantially limits one or more of the person's major "life activities," such as:
    - ambulation
    - communication
    - education
    - employment
    - housing
    - self-care
    - socialization
    - transportation
    - vocational training
2. a record of such impairment, or
3. a perception of having such an impairment.

The term "substantially limits," as used above, has to do with the degree to which the disability affects the person's employability. A qualified disabled person who, because of the disability, finds it difficult to obtain an appropriate job or advance in a job would be considered substantially limited.

The term "physical or mental impairment" would include, but not be limited to, these conditions:

- diseases and infections
- orthopedic impairments
- visual, speech, and hearing impairments
- cerebral palsy
- epilepsy
- muscular dystrophy
- multiple sclerosis
- HIV
- cancer
- heart disease
- diabetes
- mental retardation
- emotional illness
- drug addiction
- alcoholism

It should be noted that homosexuality and bisexuality are not defined as impairments. Compulsive gambling, illegal drug use, kleptomania, and others are also not considered to be impairments.

"Disabled Veteran"

A disabled veteran is a "special handicapped individual" who:

1. is entitled to disability compensation under laws administered by the Veterans Administration for disability rated at 30% or more
2. was discharged or released from active duty due to a disability incurred or aggravated in the line of duty

A veteran with nonservice-connected disabilities is not considered a special disabled veteran but may still qualify as a disabled individual under Sections 503 and 504 of the Rehabilitation Act of 1973.

The Vietnam War had the highest proportion of disabled service personnel of any war in U.S. history. A disabled veteran of the Vietnam War is a person who was discharged or released from active duty for a service-connected disability if any part of such duty was performed between August 5, 1964, and May 7, 1975.

"Qualified Individual with a Disability"

Not every disabled person is covered by the ADA. The crucial word is qualified. A person must be capable of performing the essential functions of a job—with reasonable accommodation to the disability.

Nor is every disabled veteran and every Vietnam Era disabled veteran covered by the Rehabilitation Act of 1973 or the Vietnam Era Veterans' Readjustment Assistance Act of 1974. As with the ADA, the veteran also must be qualified, that is, capable of performing a particular job, with reasonable accommodation to the disability. Organizations should be aware of the fact that the Americans with Disabilities Act requires that a certain number of deadlines be met.

REASONABLE ACCOMMODATION

According to the ADA, an employer shall make "reasonable accommodation" to the known physical or mental limitations of an otherwise qualified disabled applicant or employee, unless the employer can demonstrate that the accommodation would impose an undue hardship. Accommodations can include modifications of equipment or facilities and alterations in processes or job descriptions. The employer may not deny any employment opportunity to a qualified disabled employee or applicant if the only basis for the denial is the need to make a reasonable accommodation.

"Undue hardship" means an action requiring significant expenses or difficulties and is determined by considering the following factors:

- overall size of the employer's operation with respect to number of employees, number and type of facilities, and size of budget
- type of operation, including the composition and structure of the work force
- nature and cost of the accommodation needed

Reasonable accommodation may include but is not limited to:

- making facilities used by employees readily accessible to and usable by disabled persons
- job restructuring, part-time or modified work schedules, acquisition or modification of equipment or devices, provision of readers or interpreters, and other similar actions

(See also Chapter 16, Ergonomics Programs, in the *Administration & Programs* volume, for a discussion of accommodations in relation to ergonomics factors.)

Three Examples

Reasonable accommodation is demonstrated in these three examples:

1. A construction equipment salesperson, whose job description required him to climb onto the equipment and demonstrate its operation during sales presentations, was given a desk job after he suffered the amputation of his arm during an off-the-job motor vehicle collision. Although his prosthetic device enabled him to operate the equipment controls, the employer had considerable concern about the man's ability to climb on and off the equipment using the prosthetic device. This resulted in the job change. Upon enactment of the Rehabilitation Act of 1973, and its amendments, his employer reinstated the man as a salesperson, making a reasonable accommodation for his handicap. The accommodation consisted of providing him with a portable climbing device so that he could get on and off the equipment safely.
2. A forklift (powered industrial truck) mechanic became blind in one eye due to a nonindustrial health problem. The man's job description required him to test drive each forklift truck after completing maintenance or repair work on it. Because the employer's

standard safety policy required that all drivers of powered industrial trucks must have binocular vision (use of both eyes), the employer at first was going to switch the man to another job, which would have lowered his earnings. Upon reviewing the requirements of the U.S. Rehabilitation Act of 1973, however, the employer provided the mechanic with a reasonable accommodation. The company altered the mechanic's specific job description to eliminate the requirement to test drive the vehicles. Instead, the job descriptions of the other mechanics were broadened to include the test driving of any vehicle repaired by the disabled mechanic.

3. A disabled individual working for an electrical appliance firm was provided with a reasonable accommodation to assemble small parts. It consisted of minor adjustments to the workbench to accommodate a wheelchair.

What might be considered as unreasonable accommodation is a company completely redesigning or altering the circuitry or operating levers of a machine to accommodate a physically disabled individual.

Accommodation Is Not New

Accommodation in employment is not a novel concept. The first applications of machine guards and ventilating fans were job accommodations. Also, the first hod carrier who lacquered and reinforced his bowler as a hard hat made a job accommodation. Job placement of employees based on medical examinations, when newly hired or returning to work after an illness or injury, is again an application of accommodation. This experience is common to every employer.

Until individuals receive adequate training in the field of rehabilitation medicine, they cannot be qualified to evaluate reasonable accommodations of the workplace, its procedures, and access for the physical or mental limitations of a disabled worker. The safety and health professional without such expertise should be one member of the team consisting of the worker's physician, the occupational physician, and a rehabilitation specialist, and, in some situations, other individuals with similar disabilities. A team approach will greatly enhance the organization's commitment to an affirmative action program and nondiscrimination. It will also contribute to an affirmative action program and nondiscrimination against disabled employees and applicants in the organization.

Job safety analysis and safe work procedures are a means of eliminating or reducing work hazards to minimize worker risk. They directly transfer to the process of accommodation. Training in safe work procedures will be important in accommodating the person with a disability to the job.

ROLE OF THE SAFETY AND HEALTH PROFESSIONAL

Affirmative action programs required by the U.S. government usually come under the responsibility of the Equal Employment Opportunity (EEO) manager or coordinator (or labor affairs personnel). As a result, the placement of qualified disabled individuals normally is under their jurisdiction.

The safety and health professional, nonetheless, should be a resource person to those responsible for job placement of qualified disabled individuals. This professional should be consulted before workers are placed and asked to evaluate any proposed reasonable accommodation. The following is an example of some of the responsibilities of the safety and health professional in relation to disabled employees.

General Responsibilities

The general responsibilities of the safety and health professional include:

- Maintaining close liaison with the EEO manager-coordinator and with medical and personnel departments when they are placing disabled employees. Rehabilitation specialists and people with similar disabilities may also be necessary members of the placement team.
- Making job safety analysis of existing work based on the job responsibilities and the abilities and limitations of the disabled employee or applicant when employing, promoting, transferring, and selecting workers with disabilities.
- Making recommendations for safety modifications of machine tools, established processes and procedures, and existing facilities and workplace environment when the company must make reasonable accommodations for disabled employees.
- As required, cooperating with the plant or building engineer or mechanical engineer and the planning, production, and maintenance departments when disabled employee accommodations are being evaluated.

Specific Responsibilities

In addition to reviewing the company's affirmative action program, the safety and health professional's specific responsibilities include the following:

Establishing specific communication channels, pertaining to disabled employees, with:

1. **EEO manager-coordinator.** Make sure this person knows the safety and health professional is part of the team and available when a job needs a safety evaluation.

2. **Medical department.** Let them know the professional will be requesting their help when evaluating a job.

3. **Personnel department.** Let them know the safety and health professional is ready when necessary to help them evaluate a job's safety and remind them of safety considerations such as:

a. Don't place a worker with a coronary condition in a job that would aggravate that condition, which is a medical evaluation process.

b. Don't place a person with a back problem in a job requiring heavy manual lifting, unless other considerations are given.

c. Make certain to place a disabled employee in a job that would be safe for that person and that will not cause a hazard to others. NOTE: Individual judgments are based on a physician's evaluation with input from the safety and health professional.

4. **Plant and mechanical engineers.** Reasonable accommodation does not necessarily mean reinstalling machines; rather, it could mean minor relocating of a machine's controls so a disabled employee could operate them properly and safely. Therefore, advise the engineers that the safety and health professional will evaluate all safety aspects of such an accommodation. Also advise them that the professional will be available for safety evaluations when they design reasonable accommodations into future facilities such as:

a. ramps for wheelchairs

b. wider door passages for wheelchairs

c. grab bars in accessible washroom facilities

d. braille numbers (within a disabled person's reach) on elevators (Figure 13–3)

e. easy access to company facilities such as lunchrooms

f. elimination of curbed crosswalks

5. **Production and maintenance departments.**

a. Because reasonable accommodations also refer to job restructuring and modifications, the safety professional should tell the production and maintenance departments that he or she will help by evaluating the safety aspects of such changes. Experienced rehabilitation experts should be consulted before the company declares it is unable to make the job suitable for a particular applicant. Often a fresh look at the situation and job from an outsider's perspective will reveal new strategies that have been missed.

b. When they delete any job specification that would arbitrarily and without justification screen out disabled individuals, the safety professional will be available if a safety evaluation is needed.

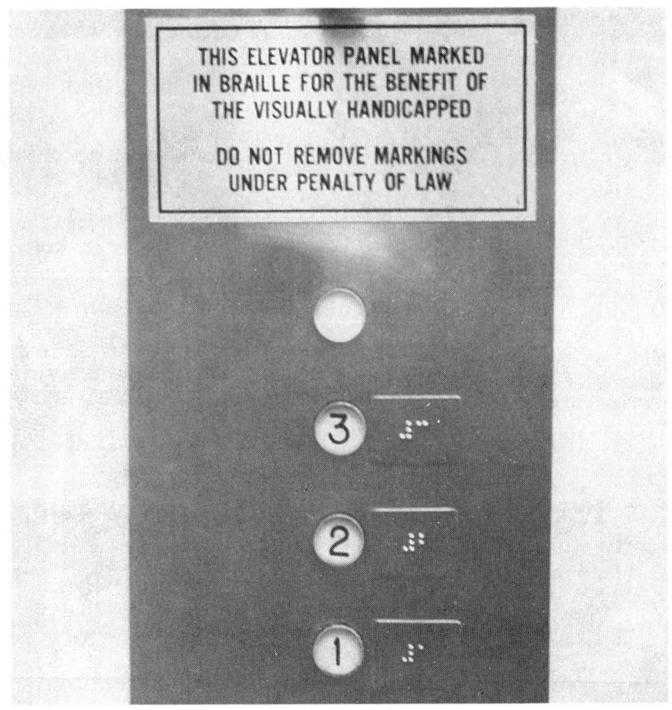

Figure 13–3. Reasonable accommodation for the disabled includes marking elevator buttons in braille. (Courtesy Governor's State University, Park Forest South, IL.)

The safety and health professional should conduct a safety evaluation of a disabled employee (in relation to the specific job or prospective job) and perform an entire job hazard analysis if needed. Rehabilitation specialists and others may be consulted as appropriate, particularly if it appears at first that the disabled employee is simply unable to perform the job. If the consensus is that the job and employee are incompatible, the company's decision, made along with the rehabilitation and other specialists, is easier to defend.

In addition, the safety and health professional should:

- Conduct a safety evaluation whenever a reasonable accommodation is being planned for a disabled employee.
- Coordinate with both the EEO manager-coordinator and the employment department to make certain any disabled employee being considered for a new position is qualified to safely and capably perform the job. This usually requires the safety and health

professional to make a safety evaluation and observe the employee during training.

- Evaluate any reported harassment of a disabled employee that affects safety. For example, name calling would not normally jeopardize the employee's safety, but pranks by other employees could result in incidents and injuries. Verbal harassment of any worker for any reason should not be tolerated. It is particularly inexcusable for a safety and health professional to stand by, claiming that such harassment does not generate a safety hazard. Angry, defensive, or depressed workers do not make sound safety judgments on the job. Further, when verbal harassment is tolerated by management, it often escalates into physical harassment.

Failure to discipline workers who are harassing others can lead to serious consequences. These can range from workplace deaths and injuries, to a decline in production, and to a loss of valued employees who may quit in disgust and resort to lawsuits that the employer will find difficult to counteract. The safety and health professional should work with the EEO manager-coordinator and other pertinent personnel to appropriately and effectively handle such situations as soon as they come to their attention.

All questions about interpretation of government requirements should be referred to the personnel responsible for implementing the affirmative action program (usually the EEO manager-coordinator) or to counsel.

The evaluation form (Figure 13–4) can help the safety and health professional perform safety evaluations for disabled employees, especially when reasonable accommodation is involved. A written report and supporting material (such as memos, blueprints, and photos) can be attached to the form to provide detailed information explaining why certain decisions were made.

Such evaluations should be kept for at least one year after the employee leaves the company. Records can be destroyed only after approval from the EEO manager coordinator or other personnel responsible for the government-required affirmative action program.

JOB PLACEMENT

Time and again, companies have found that disabled workers make for a first-class work force. When properly placed and trained so they can compete on an equal basis with others, workers with disabilities usually equal or prove to be slightly better than other employees in production and safety. Their overall attendance and job turnover records are usually superior to those without disabilities.

To place a disabled worker properly, the following requirements should be observed, where applicable, after receiving a physician's evaluation of the individual:

- The worker should meet the physical demands of the job. When necessary, the worker should receive the support of reasonable accommodation.
- The worker should not be a hazard to himself/herself. For example, a person subject to dizzy spells should not work on a ladder or scaffold or around moving machinery, where injury or death could occur.
- The worker should not be a hazard to others. For example, a person with severe vision impairment should not drive a bus or operate an overhead crane, because of the potential for personal injury or injury to others.
- The task should not aggravate the known degree of disability. A person with skin disease should not be exposed to substances that may aggravate this condition. Another example might be a worker with impaired lung function. These workers should not be exposed to substances such as smoke that will further impair lung function.
- To obtain valuable input, a conference with the individual should be held before job placement is made.

Proper placement matches the right worker to the right job on the basis of the person's ability to meet the job qualifications (Figure 13–5). As a result, the impairment virtually disappears as a factor of job performance. Moreover, employers should be aware that most disabled persons have more ability than disability, because few jobs actually require all of a worker's abilities. The job-employee match forms shown in Figure 13–5 are not only used to place disabled persons; some companies use them to place all new and transferred workers.

On the other hand, employers should also remember that each impairment can impose limitations on the type and number of activities in which the disabled person can engage. The impairment will also limit the working conditions and hazards to which this person can be exposed.

Many workers with disabilities are particularly vulnerable to tobacco smoke. Quadriplegics are endangered because they cannot cough to clear their lungs; people with heart or respiratory disease should not be exposed to environmental tobacco smoke (ETS), which contains many irritants. Workers with circulatory problems are further compromised by being exposed to nicotine, a compound that constricts the blood vessels. Other workers' disabilities may also make them hypersensitive to tobacco smoke. Clearly, prohibiting smoking in the workplace is not only reasonable but can benefit all employees, not just the disabled worker.

The safety and health professional should be aware that there are regulatory standards which, although

HANDICAPPED EMPLOYEE SAFETY EVALUATION

☐ Applicant:
☐ Employee: ______________________________ ______
(Last Name First Name M.I.) (Clock No.)

Handicap: ______________________________

Evaluation of ☐ Current job ☐ Prospective job

Job Title: ______________________________

Job Description (primary duties): ______________________________

____

____

Hazards to This Employee:
(State "none" if none) ______________________________

____

Hazards to Other Employees:
(State "none" if none) ______________________________

____

Proposed "reasonable accommodation" (if any): ______________________________

____

CONCLUSION (Based on all known factors at this time):

☐ Job is safe for this employee:
☐ as is ☐ with proposed reasonable accommodation

☐ Job is unsafe for this employee:
☐ as is ☐ with proposed reasonable accommodation

☐ No hazard to other employees:
☐ as is ☐ only with proposed reasonable accommodation

☐ Hazard to other employees:
☐ as is ☐ with proposed reasonable accommodation

Location:	Safety Supervisor (Print Name)
Date:	Safety Supervisor (Signature)

NOTE: Complete two copies of this form and give one copy to local EEO manager-coordinator. Second copy is for the safety file.

Figure 13–4. This evaluation form is an excellent administrative tool, especially when a written report and other supporting documentation are attached.

promulgated for the protection of the average employee, are not sufficient to protect employees with disabilities.

For instance, standards referring to storage of flammable and combustible liquids, 29 *CFR* 1910.106 (d)(6)(iii), include a requirement for a curb to capture spilled liquids. Disabled employees are just as vulnerable as anyone else to the hazard of chemicals spilled or escaping from a storage room. The curb can be equipped easily with a portable ramp, which not only permits a wheelchair to get in and out but allows carts to enter and exit the room. This reduces the hazard of workers transferring chemicals one by one from the cart and possibly tripping on the curb while doing so.

The "Means of Egress" standards, 29 *CFR* 1910.37, are based on the ability of an individual to move 100 ft (30 m) in 30 seconds. Perhaps some employees with disabilities cannot move that fast.

Respirators are required by the standards, for example, 29 *CFR* 1910.134, for certain jobs. Some individuals may have a physical impairment that can be

PHYSICAL CAPACITIES FORM

Leg Amputation 5″ below knee

Artificial leg—good fitting

Name ________ Sex M Age 31 Height 5′9½″ Weight 155

PHYSICAL ACTIVITIES

x	1 Walking		16 Throwing
0	2 Jumping	x	17 Pushing
0	3 Running	x	18 Pulling
x	4 Balancing		19 Handling
x	5 Climbing		20 Fingering
0	6 Crawling		21 Feeling
x	7 Standing		22 Talking
	8 Turning		23 Hearing
	9 Stooping		24 Seeing
0	10 Crouching		25 Color Vision
x	11 Kneeling		26 Depth Perception
	12 Sitting		27 Working Speed
	13 Reaching		28
x	14 Lifting		29
x	15 Carrying		30

WORKING CONDITIONS

	51 Inside	x	66 Mechanical Hazards
	52 Outside	x	67 Moving Objects
	53 Hot	0	68 Cramped Quarters
	54 Cold		69 High Places
	55 Sudden Temp. Changes		70 Exposure to Burns
x	56 Humid		71 Electrical Hazards
	57 Dry		72 Explosives
x	58 Wet		73 Radiant Energy
	59 Dusty		74 Toxic Conditions
	60 Dirty		75 Working With Others
	61 Odors		76 Working Around Others
	62 Noisy		77 Working Alone
	63 Adequate Lighting		78
	64 Adequate Ventilation		79
	65 Vibration		80

Blank Space=Full Capacity x=Partial Capacity 0=No Capacity

May work__hours per day__days per week. (IF TB, cardiac or disability requiring limited working hours.)

May lift or carry up to __**__ pounds.

DETAILS OF LIMITATIONS FOR SPECIFIC PHYSICAL ACTIVITIES:
Should not be required to walk, balance, climb, stand, kneel for prolonged periods of time.
**Should not lift heavy weights continuously.
Should not carry long distances.

PHYSICAL DEMANDS FORM

Job Title Data Processing Entry Occupational Code 4-44.110

Dictionary Title Data Processing Entry

Firm Name & Address ________

Industry ________ Industrial Code ________

Branch ________ Department ________

Company Officer ________ Analyst Wetzel Date ____

PHYSICAL ACTIVITIES

x	1 Walking		16 Throwing
	2 Jumping		17 Pushing
	3 Running		18 Pulling
	4 Balancing	x	19 Handling
	5 Climbing	x	20 Fingering
	6 Crawling		21 Feeling
	7 Standing		22 Talking
	8 Turning		23 Hearing
	9 Stooping	x	24 Seeing
	10 Crouching		25 Color Vision
	11 Kneeling		26 Depth Perception
	12 Sitting		27 Working Speed
x	13 Reaching		28
x	14 Lifting		29
x	15 Carrying		30

WORKING CONDITIONS

x	51 Inside		66 Mechanical Hazards
	52 Outside		67 Moving Objects
	53 Hot		68 Cramped Quarters
	54 Cold		69 High Places
	55 Sudden Temp. Changes		70 Exposure to Burns
	56 Humid		71 Electrical Hazards
	57 Dry		72 Explosives
	58 Wet		73 Radiant Energy
	59 Dusty		74 Toxic Conditions
	60 Dirty		75 Working With Others
	61 Odors	x	76 Working Around Others
	62 Noisy		77 Working Alone
	63 Adequate Lighting		78
	64 Adequate Ventilation		79
	65 Vibration		80

DETAILS OF PHYSICAL ACTIVITIES:
Sits at computer most of the day, reads copy, and fingers keyboard to enter data. Periodically walks short distances, reaches for, lifts, and carries small stacks of billing materials. Reaches for, handles, pushes, and pulls when organizing and filing paperwork at desk level.

Figure 13–5. Example of an employment service form used when matching workers to jobs.

affected by restricted breathing. If there is some indication of this problem, such employees must not wear a respirator until it is determined by a physician that it can be worn safely. This may preclude the individual from performing a specific job where a respirator is necessary.

The permissible exposure levels (PELs) listed in the OSHA "Air Contaminants" standards, 29 *CFR* 1910.1000, are based on the susceptibility of persons with normal breathing capacities to such contaminants. Some disabled individuals do not have normal breathing capacities and therefore are susceptible to lower levels of contaminants.

The safety and health professional must consider whether existing safety standards are sufficient to protect a particular disabled worker or candidate. More stringent standards, such as less exposure to an air contaminant than the average employee can tolerate without harm, may be necessary to protect the disabled worker or applicant. Such decisions must be made with the assistance of medical and rehabilitation consultants.

ANALYSIS OF THE JOB

Each job must be evaluated to make sure the individual being considered can do it safely. The following areas should be taken into account when making the analysis.

Physical Classification

The labor market simply does not supply only "physically perfect" workers. In fact, the percentage of workers in perfect health is relatively low—the working population now includes many persons with disabilities. Advances in medical science prolong the lives of many who would have died of war injuries or such illnesses as smallpox, tuberculosis, diabetes, and heart disease. Injuries in industry, in traffic, and in the home continue to increase the number of persons with physical disabilities.

Because the company's physician conducts physical examinations of prospective employees and makes regular plant inspections, he or she often has a better understanding of various physical and mental requirements for company jobs than do noncompany physicians. The company physician's responsibility should be to provide management with clear evaluations of each applicant's fitness for a particular job. The physician's determinations must be made only on the basis of job-worker compatibility.

Although the company physician is better able to understand job requirements, he or she may need to consult with a rehabilitation specialist.

On the other hand, safety, medical, and even many rehabilitation professionals often cannot review a work situation and ask the right questions to develop a workable compromise or accommodation for a particular disabled worker or applicant. For this reason, it is wise to consult with those who are experts in a particular disability involved.

Many support groups for disabled people exist, and the members can share experiences and insights that only those living with a particular disability can provide. Such groups can help the safety, medical, and rehabilitation specialist find creative ways to accommodate a particular job to a disabled person or suggest accommodation devices that can be made or purchased.

To locate such support groups, companies should contact their national headquarters, which can supply the names and telephone numbers of local chapters. The service directories have Rehabilitation Services and Vocational Rehabilitation listings of companies and agencies that can provide contacts for local groups. Local hospitals and physicians are also good sources, along with governmental agencies. It may be a good idea for the safety and health professional to call the groups to find out if they can provide assistance.

Many systems of employee disability classification are now in use. Generally, however, these systems use broad statements, such as "physically fit for any work"; "defect that limits applicant to certain jobs" (the defect may or may not be correctable, but may require medical supervision); and "defect that requires medical attention and is presently handicapping."

In yet another method, which approaches an ideal functional evaluation of the individual, the physician documents an employee's capacities on a form that uses identical terminology to evaluate both the physical (functional) factors and the working conditions of jobs (Figure 13–6). This effective method of presenting information from a physical examination clearly indicates the specific work capability and limitations of the individual.

Thus, the medical report is more meaningful to the placement manager because the examining physician is responsible for determining the occupational significance of physical disorders. This method, in turn, makes proper analysis of each job essential.

Job Appraisal/Job Description

Employers must know the physical requirements of jobs and the unintentional injury and health hazards involved in each one. Each job-appraisal factor can make the position unsuitable or potentially undesirable for employees with one or more types of disability. The factors to be considered in job appraisal are physical requirements, working conditions, health hazards, and injury hazards.

Physical requirements include agility, strength, exertion, vision, hearing, talking, sitting, standing, walking, running, climbing, crawling, kneeling, squatting, stooping, twisting, lifting, and handling. They should be evaluated according to quality of ability and duration of activity. For example, a job involving a considerable amount of stair climbing is unsuitable for workers with heart disease, respiratory diseases, obesity, or lower limb orthopedic disorders. Some of these people can tolerate only a small amount of stair climbing.

Both indoor and outdoor working conditions can include excessive heat or cold; excessive wetness or dryness; sudden temperature changes; and ventilation, lighting and noise problems. Also consider whether the work is performed alone, near others, with others, or as shift work or piecework. Some of these conditions could be harmful for individuals with certain disabilities. For example, work in excessive heat is generally unsuitable for persons who have had malaria, or for those with high blood pressure, heart disease, or skin disease, and for older or obese workers.

Health hazards include air pressure extremes; radiant energy (ultraviolet, infrared, radium emanations, and x-rays); silica, ETS, asbestos, dusts, and skin irritants; respiratory irritants; systemic poisons; and asphyxiants. These hazards have serious effects and can aggravate a preexisting disorder. For example, a job might involve exposure to respiratory irritants of insignificant quantities to most people; yet this condition might aggravate the disability of a person who has chronic bronchitis.

The job description needs to spell out how much lifting, how much standing, how much vision is required to do the job. Specific details will be of benefit to the employer and the applicant. (These issues were discussed in the previous section, Job Placement.)

Hazards include danger of falls from elevations; working while on moving surfaces; slipping and tripping hazards; exposure to vehicles or moving objects; objects falling from overhead; exposure to sources of foot injuries, eye injuries, cuts, abrasions, bruises, and burns; mechanical and electrical hazards; and fire and explosion hazards. These hazards could present greater dangers to workers with particular disabilities. For example, a job that may involve foot injury hazards is unsuitable for the diabetic because of his or her impaired

JOB ANALYSIS FOR PHYSICAL FITNESS REQUIREMENTS

TITLE OF POSITION	GRADE

NAME AND LOCATION OF ESTABLISHMENT	AGENCY

Does establishment have medical supervision ☐ Yes ☐ No
Is there an industrial safety branch ☐ Yes ☐ No

Refer to the manual for job analyses before using this form. Check all functional and working condition factors as well as acceptable disabilities whenever appropriate.

I. FUNCTIONAL FACTORS

L - *Little* M - *Moderate* G - *Great* O - *None*

Hands - Fingers	L	M	G	O	Arms	L	M	G	O	Legs - Feet	L	M	G	O	Body - Trunk	L	M	G	O
1. Reaching					8. Reaching					14. Walking or running					22. Sitting				
2. Pushing or pulling					9. Lifting					15. Standing					23. Bending				
3. Handling					10. Pushing or pulling					16. Sitting					24. Reaching				
4. Fingering					11. Carrying					17. Carrying					25. Lifting				
5. Climbing					12. Climbing					18. Climbing					26. Carrying				
6. Throwing					13. Throwing					19. Jumping					27. Jumping				
7. Touching					**Eyes**					20. Turning					28. Turning				
					30. Near vision					21. Lifting									
Voice					31. Far vision					**Ears**									
29. Talking					32. Color vision					33. Hearing									

II. WORKING CONDITION FACTORS

34. Inside					41. High humidity					48. Odors					55. Toxic conditions				
35. Outside					42. Low humidity					49. Body injuries					56. Infections				
36. High elevations					43. Wetness					50. Burns					57. Dust				
37. Cramped body positions					44. Air pressure					51. Electrical hazards					58. Silica dust				
38. High temperature					45. Noise					52. Explosives					59. Moving objects				
39. Low temperature					46. Vibration					53. Slippery surfaces					60. Working with others				
40. Sudden temperature changes					47. Oily					54. Radiant energy									

III. ACCEPTABLE DISABILITIES

Check appropriate square if acceptable

A - *Amputation* D - *Disability* Y - *Yes* N - *No*

Hands - Fingers	A	D	Arms	A	D	Legs - Feet	A	D	Body - Trunk	D
1 or 2 on primary hand			1 Arm			1 Leg (high)			1 Hip	
1 or 2 on secondary hand			2 Arms			2 Legs (high)			2 Hips	
More than 2 on primary hand			None ☐			1 Leg (low)			1 Shoulder	
More than 2 on secondary hand						2 Legs (low)			2 Shoulders	
1 Hand						1 Foot			Back	
2 Hands						2 Feet			None ☐	
None ☐						None ☐				

Eyes	Y	N	Ears	Y	N	Cardio - Vascular	Y	N	Tuberculosis	Y	N
Blind			Deaf			Moderate tension			Minimal (healed, stable or arrested)		
Industrially blind			Hard of hearing, 1 ear			High tension			Moderate (healed, stable or arrested)		
Blind one eye			Hard of hearing, 2 ears			Organic heart disease compensated			Far advanced (healed, stable or arrested)		
Color blind			Hearing aid acceptable						Collapse therapy		
Color blind for shades											

Figure 13–6. This form is used when analyzing jobs for fitness requirements.

circulation, which reduces sensation in the extremities, slows the healing of wounds and fractures, and increases susceptibility to gangrene of the foot.

Preemployment medical evaluation

Traditional preemployment medical examinations are prohibited under the ADA. Examinations can only be administered after an offer of employment is made and when all employees take the examination. The examinations must be restructured to comply with the ADA.

Post-offer preemployment medical examinations

These medical examinations may be allowed after an offer has been made. They also are made to determine the nature of any accommodations that may be needed.

Drug testing is not considered a medical examination and may be required before offering employment. Other regulatory tests may also be required.

ACCESS TO FACILITIES

One of a company's primary safety considerations is providing a safe and accessible parking lot for disabled employees. If people cannot find a safe place to park in the lot, they will not make it to the building. Parking spaces need to be wider than normal; details on requirements can be obtained from the Motor Vehicles departments. Both a painted symbol on the space itself and a blue-and-white sign above it should mark the parking space. The company must make sure that nondisabled employees, visitors, and others do not use these spaces. This may require rigorous enforcement.

People with mobility impairments are not the only ones who need such spaces. Employees with severe heart and lung disease, arthritis, or other chronic conditions that restrict movement will also need specially marked spaces. Hearing-impaired people, although able to walk, are at great risk walking through a parking lot and should be given special parking considerations. Finally, assistants of seriously disabled employees will need additional space in which to maneuver wheelchairs, walkers, and other equipment.

In some instances, employees may have new wheelchairs that can climb curbs and steps without assistance. However, this does not mean that curbs should not be eliminated or that some wheelchair users will not require assistance.

Safety considerations for hiring a disabled person begin with the first step in the employment process, and on the first day of work for that employee. For example, is there reasonable, safe access to the reception area, applicant-processing area, or workplace for the new employee? Curbs and stairs present barriers to those who use canes, crutches, walkers, and wheelchairs, and also increase their chances of falling. A wheelchair user cannot safely negotiate even one step without assistance, which can cause a slight risk to both the wheelchair user and the person assisting him. Wheelchair ramps are discussed under General Access later in this section.

Can disabled job applicants safely proceed to where they must go to complete an application? If already employed, can such individuals safely proceed to their workstations? Access and safety are interdependent factors that need to be reconciled when employing these people. A lack of access to the company premises has been the principal factor preventing disabled persons from seeking jobs with some organizations.

General Access

In designing access for disabled workers, do not overlook cafeteria, washroom, and restroom facilities, width of doors, height of plumbing fixtures, electrical controls, phones, and drinking fountains. Facilities designed to be truly accessible to disabled employees enhance their feelings of dignity and independence—which, in turn, can raise the morale of the entire work force.

With minimum expense, improvements and special considerations made for workers with disabilities also will benefit other employees. For instance:

Wheelchair ramps are safer than steps—for everyone. However, the slope should be correct, not too steep and without sharp turns, to make negotiating the ramp safe and easy for wheelchair users. Ramp surfaces should be slip-resistant and kept free of obstacles. All ramps should be cleaned of mud, snow, and ice, and railings must be in place where needed.

Clean and unobstructed aisles are necessary for safe wheelchair, cane, and crutch use. They also make the workplace safer for all employees and visitors. Good housekeeping improves traffic patterns and eliminates hazards.

Access to Buildings

Parking spaces 8 feet wide, next to a 5-foot-wide access aisle, should be reserved for automobiles driven by disabled personnel and visitors (Figure 13–7a). Two accessible parking spaces, however, can share a common access aisle (Figure 13–7b). Workers need room to remove and replace their wheelchairs in an auto and also to ride the chairs between aisles.

Parking spaces for disabled persons should be marked with an upright marker; symbols painted on the ground are often difficult to see or can be covered by snow or debris. At least one accessible route and entrance to the building must be provided. The entrance width should be 32 in. (80 cm); if a ramp leads to the entrance, it should be at least 36 in. (90 cm) wide. The ramp should

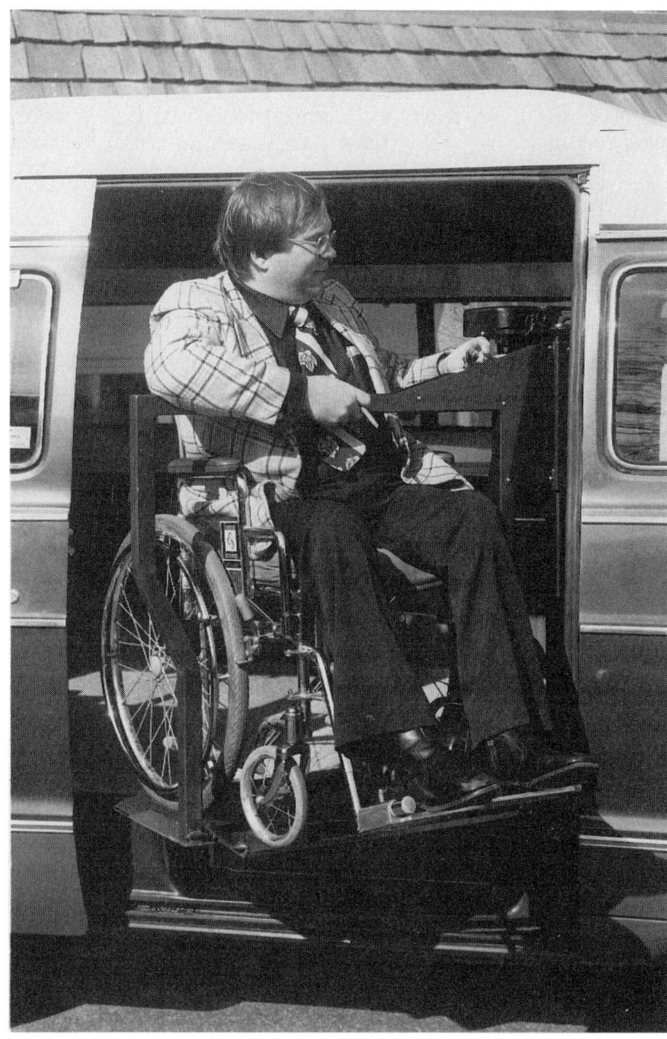

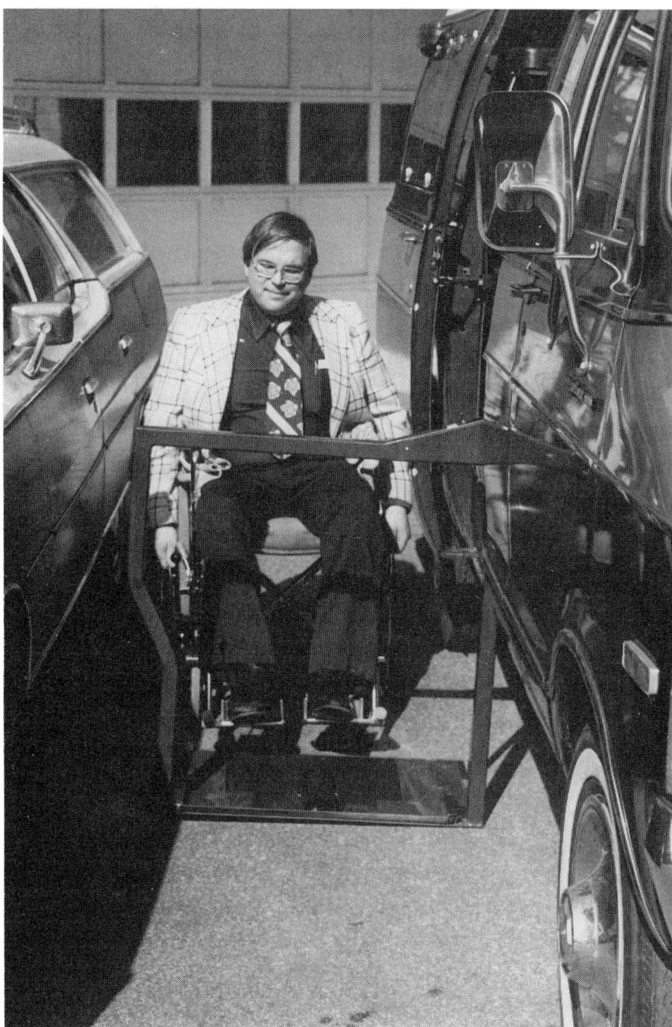

Figure 13–7a. Access aisle adjacent to a normal-width parking space is required for wheelchair clearance. Shown here is a lift that swings alongside of a van. For return, a locking, outside control box opens sliding doors and allows the user to control the lift. (Courtesy ABC Enterprises, Inc.)

be a maximum of 30 ft (9 m) long with an open, level area of at least 5 ft (1.5 m) at the bottom. Employees (and visitors) using wheelchairs, crutches, or canes can then move in and out of the building completely on their own. Protective side rails must be at least 36 in. (90 cm) apart if not adjacent to a wall to prevent persons in wheelchairs running or falling off the side.

Revolving doors are inaccessible for anyone in a wheelchair and for most people using crutches, wearing a leg cast, or even carrying bulky packages. These people need entry doors that preferably open automatically. When a doorknob is necessary, it should be 36 in. (90 cm) from the floor. It is better, however, to have a vertical grab handle on the door.

The space requirements of the average wheelchair are as follows: Most wheelchairs are 36 in. high, 26 in. wide, and 42 in. long (Figure 13–8a). They require at least 60 × 60 in. to make a 180- or 360-degree turn. However, 60 × 78 in. is preferred to make a smoother U-turn. All access aisles should be wide enough to allow a wheelchair user to make a smooth turn. "Accessibility Standards," published by the State of Illinois, contains illustrated information on maneuvering space requirements for wheelchairs (see References).

The average arm reach of people who use wheelchairs is usually 48 in. on the diagonal and 64 in. to the side. The average reach directly upward is 60 in. (Figure 13–8b). The usual maximum downward reach from the chair is 10 in. (Figure 13–8b). Shorter distances may be needed, however, to accomplish certain tasks (Figure 13–8c).

Interior Access

Both entry and interior doors should be a minimum of 32 in. (80 cm) wide. Interior doors should open by a single effort and have thresholds as level with the floor

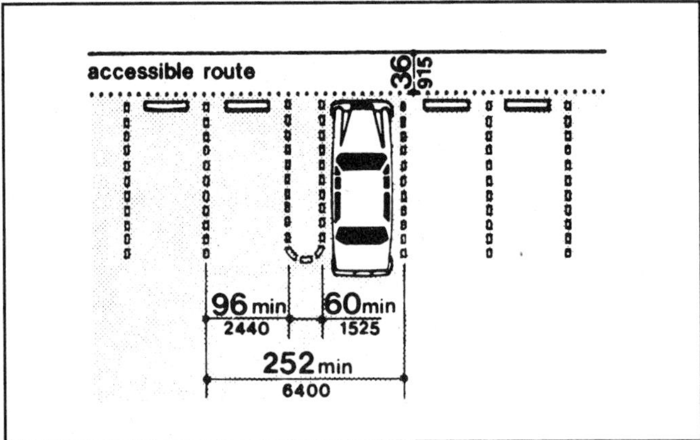

Figure 13–7b. Parking spaces for disabled persons can share a common access aisle. Aisle should be part of the accessible route to the building entrance. Overhangs from parked vehicles must not reduce the clear width of the accessible route, which must be the shortest possible distance to the entrance. (Printed with permission from American National Standards Institute.)

as possible. To make a 180- or 360-degree turn, persons in wheelchairs need an open area of 60 × 60 in. (1.5 m^2) in a typical building corridor.

Restrooms should have at least one stall wide enough for wheelchair entry. The stall should be equipped with grab bars and other fixtures no higher than 36 in. (90 cm) above floor level. The grab bars should be at 33 in. (83 cm). Stall doors should be at least 32 in. (80 cm) wide. Controls, switches, fire alarms, and other devices that might be used by a disabled individual must be within convenient reach of a wheelchair user. Restrooms should be located on each of the floors where disabled persons work.

Office Accommodations

Desk tops should be no less than 28 in. (70 cm) above the floor to accommodate wheelchairs. Metal desks usually have adjustable feet that workers can raise to the maximum. If more room is needed, the desk can be raised with additional blocks.

Chairs can be regular height, but they should be sturdy and have arms to help disabled people lift themselves up. Some individuals need a chair that will not move easily so they can stand without the chair sliding out from under them. For other disabled workers, however, casters placed on the bottom of chairs may be desirable to help them slide the chair in and out. In addition, they allow a worker to pull himself/herself from one piece of furniture to another, thereby avoiding the need to continually move in and out of the chair.

Business machines should be placed, if possible, where they will not become a barrier or obstruct traffic. This precaution is particularly important for wheelchair users.

The following accommodations may sometimes apply, depending on the individual worker and the job requirements:

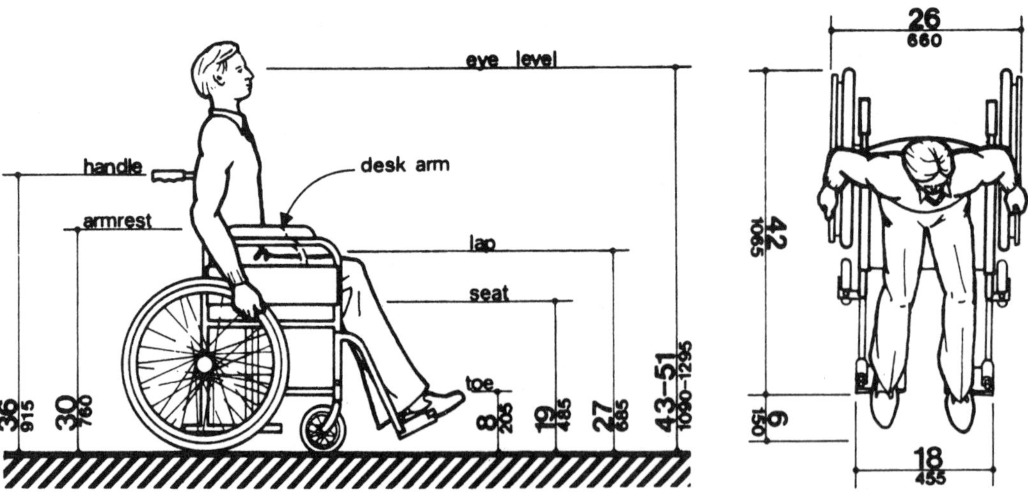

Figure 13–8a. Dimensions of adult-sized wheelchairs. Footrests can extend farther for very large people. Dimensions in this figure and in Figures 13–8b and 13–8c are in both inches and millimeters. (Printed with permission from American National Standards Institute.)

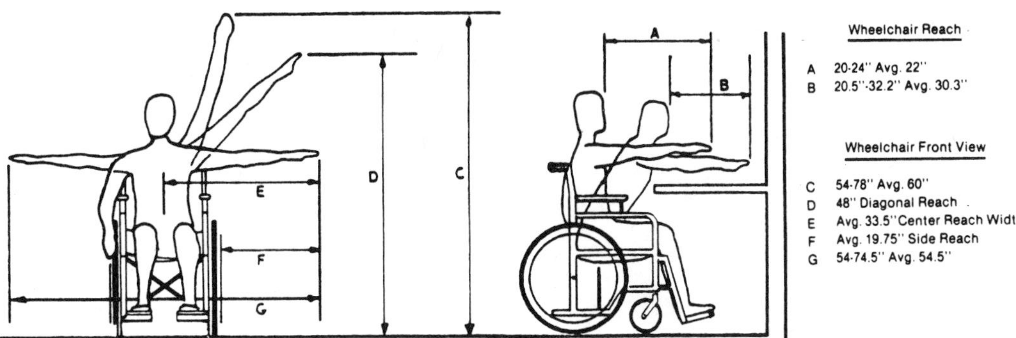

Figure 13–8b. Maximum reach from wheelchair—left: to sides; right: to front. (To convert to millimeters: 1 in.=25.4 mm.) (Printed with permission from State Board of Barrier-Free Design, Columbia, SC.)

- File cabinets should be placed so that workers can reach drawers from both the front and side. This eliminates awkward reaches by those who must use crutches, a cane, or a wheelchair.
- If books, reports, or other bulky objects must be carried from place to place, make sure a wheeled cart is kept handy so workers can load materials on the cart.
- Cords for venetian blinds, window shades, and draperies should be long enough that disabled workers can reach them easily.
- Floors must be free of extension cords, raised box receptacles, or any other raised items.
- If a disabled employee has an assistive dog, then space, exercise facilities, and water for the dog will have to be arranged. The employee will know exactly what is needed. Other employees must understand that the dog is a working animal; they should not pet or feed it without express consent from the owner. Otherwise, the dog can become confused about its role and stop being dependable as the eyes or ears of the disabled employee.

EMERGENCY PROCEDURES

One concern within the facility is adequate means of escape for all workers in an emergency. This safety requirement frequently restricts or denies disabled persons the freedom to use premises as they would wish. However, many establishments employing workers with mobility disabilities have successfully devised safe evacuation plans.

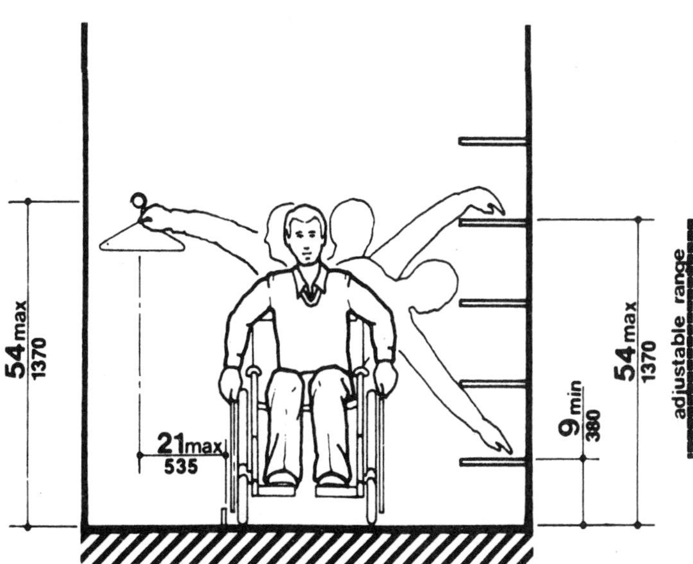

Figure 13–8c. Shorter distances are required when tasks are done. Shown here are the suggested dimensions for storage shelves and clothes racks. (Printed with permission from American National Standards Institute.)

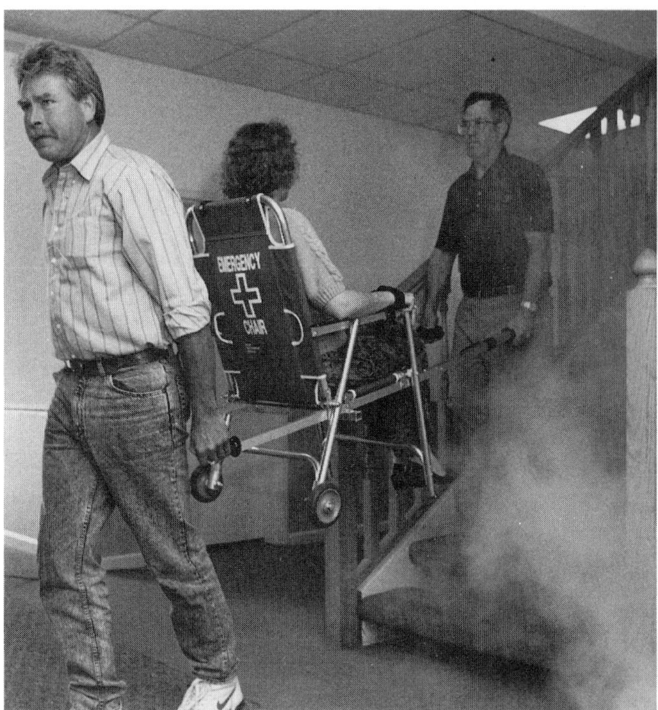

Figure 13–9. Using an evacuation chair, these workers safely transport a disabled fellow worker down fire stairs without putting themselves or the disabled worker under additional risk during an emergency. (Courtesy Safety Chair Co., Benicia, CA.)

Safety management of the disabled in many firms is well-established and provides a freedom of movement that is entirely compatible with principles of general safety. These measures include supervised use of the stairs for means of escape (this is discussed later), designated staff to assist in an emergency, strategic ramping of entrances and exits, and alarm systems suitable for vision- and hearing-impaired workers.

Employees with impaired hearing and/or vision may need additional devices to perceive an alarm. Some people with impaired hearing can hear an alarm in a lower pitch than normal; others do best with a buzzer or a device that vibrates against their skin. These employees need to be interviewed separately and a workable notification system developed. Often, all that is necessary is a "buddy" system in which a worker who can see or hear the alarm is assigned to notify the disabled employee whenever there is an emergency or drill.

One means of safely evacuating wheelchair users and permanently or temporarily disabled persons is through use of an evacuation chair (Figure 13–9). This chair is designed to ride on the ends of stair treads so one person can easily guide the disabled worker down fire stairs without putting either person in additional jeopardy during an emergency evacuation. The evacuation chair is lightweight, folds flat, and can be safely and easily stored on a wall bracket.

Without these measures, many disabled persons would be denied access to their places of employment. It is suggested that employers discuss their safety problems with fire prevention specialists and check regulatory requirements.

There are many possible ways to enable disabled people to escape safely in an emergency.

SUMMARY

- The Americans with Disabilities Act of 1990 mandates that employers provide "reasonable accommodation" for workers with disabilities when necessary. Supervisors and other management personnel must be trained to accommodate this group of workers.
- Employers who deny disabled workers a job must prove that these workers would endanger themselves or others, cannot meet job requirements, or reasonable accommodation of the workplace or job cannot be made.
- Various federal, state, and local laws also mandate affirmative action programs for the hiring and advancement of disabled persons. The law defines three types of disabled job seekers: disabled individual, disabled veteran, and qualified disabled individual.
- "Reasonable accommodation" to the known physical or mental limitations of an otherwise qualified disabled applicant or employee can include modifying equipment or changing job descriptions.
- The safety and health professional can work with the EEO coordinator to evaluate reasonable accommodation and ensure compliance with government affirmative action regulations.
- Management must evaluate each disabled job applicant to place the right person in the right job so the impairment is not a factor of job performance. More stringent standards may need to be developed to ensure safe working conditions for these employees.
- Employers must evaluate jobs in terms of physical requirements, working conditions, health hazards, and injury hazards to eliminate or control job risks that might endanger disabled workers.
- Companies can also contact self-help and support groups, governmental agencies, and private agencies to learn more about how to accommodate a job for a disabled employee and how to work with disabled persons.

SOURCES OF HELP

The overall goal of all employers should be to hire qualified disabled individuals and place them in available and safe occupations. One of the goals of the safety and health professional is to assist the employer in this worthwhile endeavor.

The following list includes government and private agencies that can help achieve this goal.

Alabama Institute for the Deaf and Blind, Talladega, AL 35160.

Arthritis Foundation, 1314 Spring Street NW, Atlanta, GA 30309. (404) 872–7100.

American Council for the Blind, 1115 Fifteenth Street NW, Suite 720, Washington, DC 20005. (202) 467–5081 or (800) 424–8666.

Architectural and Transportation Barriers Compliance Board, 1111 18th Street NW, Suite 501, Washington, DC 20036. (800) USA–ABLE (Voice) or (800) USA–ABLE (TDD).

Association for Children and Adults with Learning Disabilities, 4156 Library Road, Pittsburgh, PA 15234. (412) 341–1515.

Asthma and Allergy Foundation of America, 1717 Massachusetts Avenue NW, Suite 305, Washington, DC 20036. (202) 265–0265.

Cystic Fibrosis Foundation, 6931 Arlington Road, Suite 200, Bethesda, MD 20814. (800) 344–4823; (301) 951–4422.

Department of Justice, Office on the Americans with Disabilities Act, Civil Rights Division, P.O. Box 66228, Washington, DC 20035–6118. (202) 514–0301 (Voice); (202) 514–0381 (TDD); (202) 514–6193 (Electronic bulletin board).

Department of Transportation, 400 Seventh Street SW, Suite 501, Washington, DC 20590. (202) 366–9305 (Voice); (202) 755–7687 (TDD).

Easter Seal Society, 70 East Lake Street, Chicago, IL 60601. (312) 726–6200.

Epilepsy Foundation of America, 4351 Garden City Drive, Landover, MD 20785. (800) 332–1000; (301) 459–3700; Fax (301) 577–2684.

Equal Employment Opportunity Commission, 1801 L Street NW, Washington, DC 20507. (202) 663–4900 (Voice); (800) 800–3302 (TDD); (202) 663–4494 (TDD).

Federal Communications Commission, 1919 M Street NW, Washington, DC 20554. (202) 632–7260 (Voice); (202) 632–6999 (TDD).

George Washington University, Job Development Laboratory, Rehabilitation Research and Training Center, 2300 I Street NW, Washington, DC 20052.

Internal Revenue Service, Department of the Treasury, 1111 Constitution Avenue NW, Washington, DC 20044. (202) 566–2000.

Job Accommodation Network, West Virginia University, 809 Allen Hall, P.O. Box 6123, Morgantown, WV 26506–6123. (800) 526–7234 (U.S. other than WV—Voice and TDD); (800) 526–4698 (WV—Voice and TDD); (800) 526–2262 (Canada—Voice and TDD).

Lupus Foundation of America, 1717 Massachusetts Avenue NW, Suite 203, Washington, DC 20036. (800) 558–0121; (202) 328–4550.

Mainstream, Inc., 1200 15th Street NW, Washington, DC 20005.

Multiple Sclerosis Society, 205 East 42nd Street, New York, NY 10017. (800) 624–8236; (212) 986–3240.

Muscular Dystrophy Association, 801 Seventh Avenue, New York, NY 10019. (212) 586–0808.

Myasthenia Gravis Foundation, Inc., 53 West Jackson Boulevard, Suite 1352, Chicago, IL 60604. (800) 541–5454; (312) 427–6252.

National Association of the Deaf, 814 Thayer Avenue, Silver Spring, MD 20910–4500. (301) 587–1785 (Voice); (301) 587–1789 (TDD).

National Association for Sickle Cell Disease, 4221 Wilshire Boulevard, Suite 360, Los Angeles, CA 90010. (213) 936–7205.

National Association of Protection and Advocacy Systems, Client Assistance Program, 220 I Street NW, Suite 150, Washington, DC 20002. (202) 546–8202.

National Council on Disability, 800 Independence Avenue SW, Suite 814, Washington, DC 20591. (202) 267–3846 (Voice); (202) 267–3232 (TDD); (202) 453–4250 (Fax).

National Council of Persons with Disabilities, P.O. Box 29113, Washington, DC 20017. (202) 529–2933.

National Down Syndrome Society, 666 Broadway, New York, NY 10012. (800) 221–4602; (212) 460–9330.

National Head Injury Foundation, 333 Turnpike Road, Southborough, MA 01772. (800) 444–NHIF; (508) 485–9950.

The National Institute for Rehabilitation Engineering, 97 Decker Road, Butler, NJ 07405.

National Kidney Foundation, 30 East 33rd Street, New York, NY 10016. (800) 622–9010; (212) 889–2210.

National Organization on Disability, 910 16th Street NW, Suite 600, Washington, DC 20006. (202) 293–5960 (Voice); (202) 293–5968 (TDD); (202) 293–7999 (Fax).

National Spinal Cord Injury Association, 600 West Cummings Park, Suite 2000, Woburn, MA 01801. (800) 962–9629; (617) 935–2722.

Paralyzed Veterans of America, 801 18th Street NW, Washington, DC 20006. (202) 872–1300.

President's Committee on Employment of People with Disabilities, 1331 F Street NW, Washington, DC 20004–1107. (202) 376–6200 (Voice); (202) 376–6205 (TDD); (202) 376–6219 (Fax).

The Rehabilitation Institute of Chicago, 345 East Superior Street, Chicago, IL 60611.

Retinitis Pigmentosa Foundation Fighting Blindness, 1401 Mt. Royal Avenue, Baltimore, MD 21217–4245. (800) 638-2300.

Self Help for Hard of Hearing People, 7800 Wisconsin Avenue, Bethesda, MD 20814. (301) 657–2248.

Spina Bifida Association of America, 1700 Rockville Pike, Suite 540, Rockville, MD 20852. (800) 621-3141; (301) 770–SBAA.

Support Dogs for the Handicapped, P.O. Box 966, St. Louis, MO 63044. (314) 487–2004.

REFERENCES

ADA Compliance Guide, Thompson Publishing Group, 1725 K Street NW, Suite 200, Washington, DC 20006.

American National Standards Institute, 11 West 42nd Street, New York, NY 10036.

BNA's Americans with Disabilities Act Manual, The Bureau of National Affairs, Inc., 1231 25th Street NW, Washington, DC 20037.

Brooks WT. "Supervising Handicapped Workers for Safety."

Capital Development Board. *Accessibility Standards Illustrated.* Reprint of 1978 edition with all revisions to March 1, 1985 and Environmental Barriers Act, Public Act 84–948. Springfield: State of Illinois, 1985.

The Gallaudet Survival Guide to Signing, Gallaudet University Press, Washington, DC 20002.

Hill N, et al. "The merits of hiring disabled persons." *Business & Health.* February 1987.

———. "Hiring the handicapped: Overcoming physical & psychological barriers in the job market." *Journal of American Insurance.* Third quarter, 1986.

Mental and Physical Disability Law Reporter. American Bar Association, Commission on Mental and Physical Disability Law, 1800 M Street NW, Washington, DC 20036–5886.

President's Committee on Employment of the Handicapped, Washington, DC 20036.
"Affirmative Action to Employ Handicapped People—A Pocket Guide."
"Supervising Handicapped Employees."

The Registry of Interpreters for the Deaf, Inc., RID Publications, 8719 Colesville Road, Suite 310, Silver Spring, MD 20910–3919.

U.S. Department of Health and Human Services, Washington, DC 20202. "Nondiscrimination on the Basis of Handicap in Programs and Activities Receiving or Benefiting from Federal Financial Assistance" (45 *CFR* 84).

U.S. Department of Labor, Employment Standards Administration, Office of Federal Contract Compliance Programs, Washington, DC 20210. "Affirmative Action Obligations of Contractors and Subcontractors for Handicapped Workers" (41 *CFR* 60–741).

U.S. Department of Labor, U.S. Government Printing Office, Washington, DC 20402.

Occupational Safety and Health Act of 1970 (Public Law No. 91–596).

Occupational Safety and Health Act Regulations, Title 29, *CFR*, Chapter XVII, Part 1910.

Rehabilitation Act of 1973 (Public Law No. 93–P112).

Rehabilitation Act Amendments of 1974 (Public Law No. 93–516).

Vietnam Era Veterans' Readjustment Assistance Act of 1974 (Public Law No. 93–508), Section 38 USC 2012.

West J, ed. *The Americans with Disabilities Act—From Policy to Practice.* Milbank Memorial Fund, One East 75th Street, New York, NY 10021.

What You Absolutely Must Know about the Americans with Disabilities Act, Epstein, Becker & Green, P.C., 1227 25th Street NW, Washington, DC 20037.

Woodward RE. Industry unlocks its doors to the handicapped. *Plant Facilities.* Vol. II, No. 2 (February 1979).

———. *Comprehensive Barrier-Free Design Standard Manual.* Columbia, SC: State Board for Barrier-Free Design, P.O. Box 11954. 1979.

REVIEW QUESTIONS

1. The 1990 U.S. Americans with Disabilities Act (ADA) encompasses what general areas?
   a.
   b.
   c.
   d.
   e.
2. If an employer denies a disabled individual a specific job, management must prove that the person is unqualified because of what reasons?
   a.
   b.
   c.
   d.
3. What three occurrences in the 1940s stimulated hire-the-disabled programs?
   a.
   b.
   c.
4. List the three types of disabled persons seeking employment as defined by law.
   a.
   b.
   c.
5. Examples of "reasonable accommodation" include:
   a. Making employee facilities readily accessible to and usable by disabled persons
   b. Restructuring a job, modifying work schedules, and modifying equipment or devices
   c. Providing readers or interpreters
   d. Completely redesigning or altering the circuitry or operating levers of a machine
   e. All of the above
   f. Only a, b, and c

6. Explain four general responsibilities of the safety and health professional in relation to disabled employees.
   a.
   b.
   c.
   d.

7. The safety evaluation form for disabled employees should be kept for at least ____________ after the employee leaves the company.

8. What are the four factors employers must consider when appraising a job for employees with one or more disabilities?
   a.
   b.
   c.
   d.

9. What are six important factors to consider when designing access to facilities for disabled workers?
   a.
   b.
   c.
   d.
   e.
   f.

10. To accommodate disabled persons, the entrance width to a building should be:
    a. 28 in. (70 cm)
    b. 32 in. (80 cm)
    c. 36 in. (90 cm)
    d. None of the above

11. Desk tops should be no less than how many inches above the floor to accommodate wheelchairs?
    a. 28 in. (70 cm)
    b. 32 in. (80 cm)
    c. 36 in. (90 cm)

12. One means of safely evacuating wheelchair users and permanently or temporarily disabled persons is through the use of an ____________.

Part Three

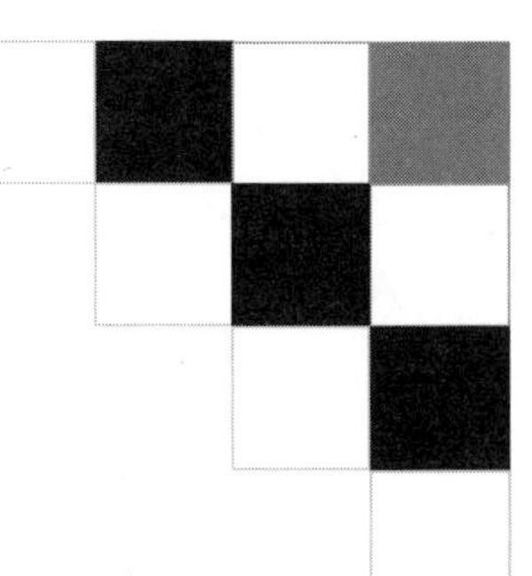

MATERIALS HANDLING

Part Three addresses incident prevention in the various forms of materials handling. Unintentional releases or falls can have serious injury consequences, and materials handling incidents and injuries are some of the most common in the workplace. The control and management of forces generated by mechanically powered equipment; hoisting and conveying equipment; or ropes, chains, and slings can have a significant impact on the overall workplace safety record. The chapters in Part Three offer constructive strategies for the safety professional who is responsible for the safe movement of products or materials in or around a workplace.

14

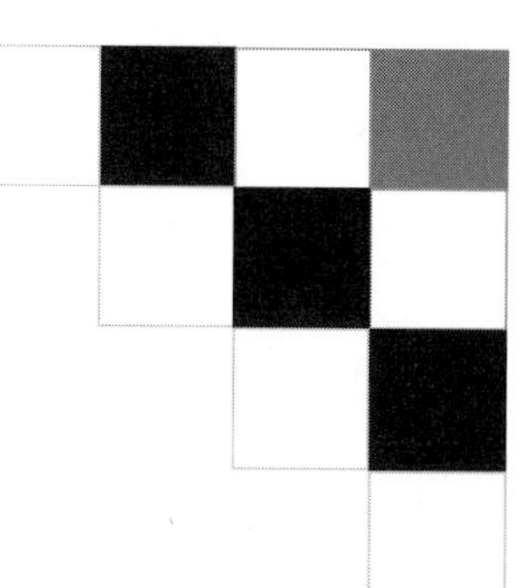

Materials Handling and Storage

Revised by

John F. Montgomery, Ph.D., CSP, CHCM, CHMM

Materials handling operations are conducted in every department, warehouse, office, or facility of a company. Materials handling is a job that almost every worker in industry performs—either as a sole duty or as part of the regular work, either by hand or with mechanical help. On an average, industry moves about 50 tn (45 t) of material for each ton of product produced.

The work environment, the need for specific training, and proper materials-handling engineering should be examined when reviewing a facility's materials-handling needs. The specific amount and extent of manual lifting involved in a particular job should also be identified before selecting and placing employees.

Mechanized materials-handling equipment is commonly used (Figure 14–1a, b) in many industries, due to the fact that material could not be processed at low cost without efficient mechanical handling. Although mechanical handling creates a new set of hazards, the net result (entirely aside from increased efficiency) is fewer injuries, lower workers' compensation expenses, and a more productive workplace. (See also Chapter 15, Hoisting and Conveying Equipment; Chapter 16, Ropes, Chains, and Slings; Chapter 17, Powered Industrial Trucks; and Chapter 25, Automated Lines, Systems, or Processes for hazards and safety techniques involved in the manual and mechanical handling of materials.) This chapter covers the following topics:

- safety measures for preventing common workplace injuries
- safety guidelines for lifting and materials handling
- types of accessories used for safe handling practices
- storage of hazardous and nonhazardous materials

Figure 14–1a. This power lift table helps prevent strains in a parts storage area.

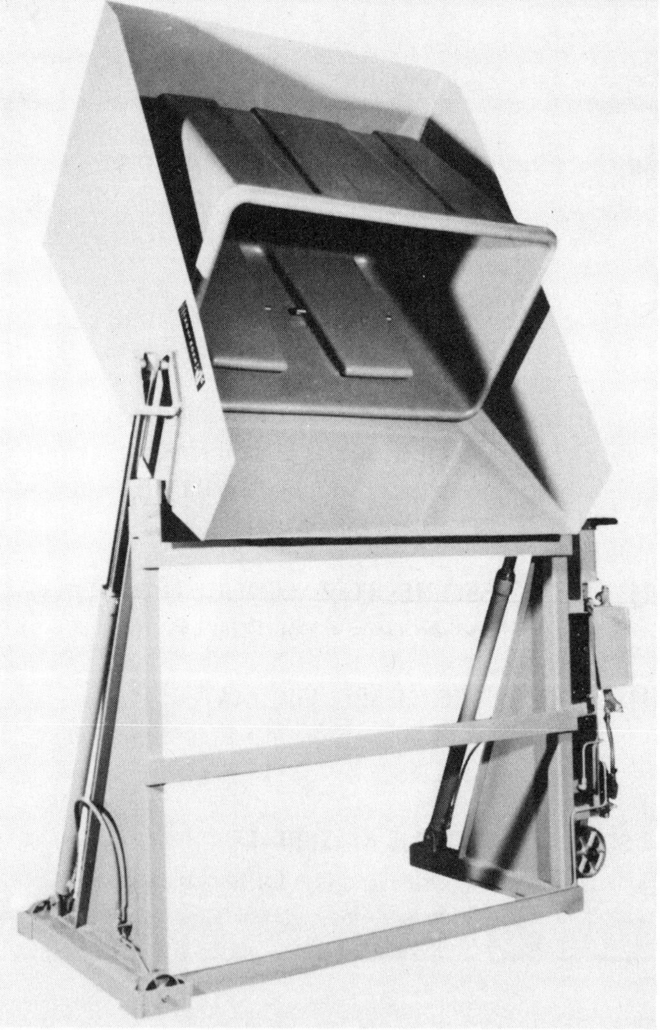

Figure 14–1b. A mobile dumper can lift materials from the floor level to table height and slide them onto the table.

- storage and handling of liquids
- safety practices for inert and flammable gas
- safety precautions in the shipping and receiving department

PREVENTING COMMON INJURIES

Handling of material accounts for 20% to 45% of all occupational injuries. These injuries could occur in every part of an operation, not just the stockroom or warehouse. Strains, sprains, fractures, and contusions are the most common injuries.

Personal Protection

Overhead hoisting equipment operators and maintenance personnel can be exposed to falls. Fall protection and safe egress should be provided where workers will foreseeably be exposed to falls. For further information, see Fall Arrest Systems in Chapter 7, Personal Protective Equipment.

Certain items of protective equipment are desirable for the prevention of various types of materials-handling injuries. Since toe and finger injuries are among the most common types, handlers should wear safety shoes and stout gloves, preferably with leather palms. Other special protective clothing, such as goggles, aprons, and leggings, for the handling of certain types of materials should also be required.

Gloves should be dry and free of grease and oil. To prevent injury from splinters, handlers should wear gloves when handling wooden crates. Clean, leather-palmed gloves give better holding power on smooth metal objects than do cotton or other types of gloves. However, it may be unsafe for workers to wear gloves near conveyors, or whenever there is a risk of catching in machinery. Take care not to bruise or squeeze the hands at doorways or other points where clearance is close. During inclement weather, special weather apparel may be necessary.

Where toxic or irritating solids are handled, workers should take daily showers to remove the material from their person before they leave the facility. Even though the exposure does not necessitate showers, encourage workers to wash thoroughly at the end of their shifts. Provide cleansing materials, shower stalls, and wash basins. Also provide washable suits of tightly woven fabric, preferably full-length coveralls, and washable caps for workers to wear. Suits, caps, socks, and underwear should be laundered daily at the facility or through a laundry service. Clothing should be laundered less frequently only at the express direction of the facility's medical department.

Materials-Handling Injuries

To gain insight into the injuries caused by materials handling, the safety professional should consider the following:

- Can the job be engineered to eliminate or reduce manual handling?
- Can the material be conveyed or moved mechanically?
- In what ways do the materials being handled (such as chemicals, dusts, rough and sharp objects) cause injury?
- Can employees be given handling aids, such as properly sized boxes, adequate trucks, or hooks, that will make their jobs safer?
- Would protective clothing, or other personal equipment, help prevent injuries?
- Would training and more effective management help reduce injuries?

These questions serve as a start for an overall appraisal of injuries caused by materials handling. Break each job into its separate tasks, and examine each task for ways to prevent injuries through job safety analysis (JSA) techniques.

Since most injuries occur to fingers and hands, give the following general pointers to employees who handle materials:

- Inspect materials for slivers, jagged or sharp edges, burrs, rough or slippery surfaces.
- Grasp objects with a firm grip.
- Keep fingers away from pinch and shear points, especially when setting down materials.
- When handling lumber, pipe, or other long objects, keep hands away from the ends to prevent them from being pinched.
- Wipe off greasy, wet, slippery, or dirty objects before trying to handle them.
- Keep hands free of oil and grease.

In most cases, have employees use gloves, hand leathers, or other hand protectors to prevent hand injuries. Urge extra caution to employees who work near moving or revolving machinery. (See Chapter 7, Personal Protective Equipment, in this volume.) In other cases, have employees use handles or holders to move objects. Provide handles for moving auto batteries, tongs for feeding material to metal-forming machinery, or baskets for carrying control-laboratory samples.

Feet and legs also sustain a large portion of materials handling injuries—the greater percentage occurring to the feet. Require that workers wear protective footwear. Safety shoes may need to be equipped with metatarsal guards for some work activities.

Workers' eyes, heads, and trunks can also be injured. When opening a wire-bound or metal-bound bale or box, workers should take special care to prevent the ends of the binding from flying loose and striking the face or body. Workers should use protective eyewear equipped with side shields and should also wear heavy gloves. The same precaution applies to handling coils, wire strapping, or cable. In many cases, special tools are available to safely cut bands, strapping, and the like. Workers should always read the labels on packages for special handling instructions. If material has the consistency of dust or is toxic, the person handling it may need to wear a respirator or other suitable personal protective equipment to prevent injury to the lungs, or use other appropriate controls provided, such as ventilation.

Manual handling of materials increases the possibility of injury and adds to the product's cost. To reduce the number of injuries caused by materials handling and to increase efficiency, minimize manual handling of material. For example, combine or eliminate operations and introduce ergonomic principles to job design. Mechanically move materials as much as possible. For those jobs that cannot be mechanized, use the safety techniques given in the next sections (see Chapter 16, Ergonomics Programs, in the *Administration & Programs* volume).

Manual Lifting

Physical differences make it impractical to establish safe lifting limits applicable to all workers. A person's height and weight, although important, do not necessarily indicate lifting capability. Some small, thin individuals can handle heavier loads than some tall, heavy persons (Kroemer, 1983). Conduct a job safety analysis and follow medical recommendations when establishing lifting standards. (See also the NIOSH *Work Practices Guide to Manual Lifting.*)

Before workers lift a heavy or bulky object and carry it to another location, they should inspect the routes over which they will move the object. In that way, workers can make sure that there are no obstructions or spills that could cause them to slip or trip. If clearance is not adequate for handling the load, workers should choose an alternate route. Next, workers should inspect the object to decide how to grasp it and, thus, avoid sharp edges, slivers, or other things that might cause injury. Workers may have to turn an object over before attempting to lift it. Also, if the object is wet or greasy, workers should wipe it dry to prevent it from slipping. If this is not practical, workers should use a rope sling or other device that will firmly grip the object.

Most lower back injuries come from tasks requiring lifting. Other activities, such as lowering, pushing, pulling, or carrying material, or twisting the body, can also cause back injuries. Back injuries are second in number after injuries to the fingers and hands.

There are several techniques for manually lifting objects with reasonable safety. However, each of these lifting techniques has limitations. Consider all three main factors in manual lifting—load location, task repetition, and load weight—when determining what is safe or unsafe to lift.

The National Institute for Occupational Health and Safety (NIOSH) has published an *Applications Manual for the Revised NIOSH Lifting Guidelines* (DHHS Publication #94–110) to assist in calculating safe lifts for most workers.

Back Belts

Back injuries account for nearly 20% of workplace injuries and are the leading cause of injuries to workers under the age of 45. Estimates put the annual cost of back injuries at $20 billion to $50 billion.

The debate over back belts focuses on two issues: (1) employees are rarely trained in proper lifting techniques or even how to use the belts correctly, and (2) wearing a back belt can give a false sense of security —people think they can lift more than they can.

Most lifting injuries occur because of excessive load or pressure on the lower back. Excessive load can occur when a load is simply too heavy to lift or places weight too far from the spine, whether in front or to the side. Lifts that involve excessive reaching or twisting can produce injury. Workers also may sustain injury when they fail to use proper lifting techniques or attempt to perform a job that is beyond their capabilities.

Back-belt manufacturers are the first to recommend that companies train employees in safe lifting techniques. They say companies should use ergonomic principles to determine whether they can reduce the lifting task by use of gravity or other lifting aids and whether changes in workstation layout can reduce lifting tasks. Job rotation and other administrative controls also can decrease the chance of injury.

Scientific studies have failed to demonstrate that back belts can prevent lifting injuries, although some studies have shown that the belts may be useful post-injury.

NIOSH summarized its review of the back-belt issue in a report entitled "Back Belts—Do They Prevent Injury?" (NIOSH publication No. 94-127) which it issued in July 1994.

According to the report, "After a review of the scientific literature, NIOSH has concluded that, because of the limitations of the studies that have analyzed workplace use of back belts, the results cannot be used to either support or refute the effectiveness of back belts in injury reduction." The report goes on to examine specific questions and issues related to back-belt use and concludes that the best way to protect workers is through ergonomic approaches designed to reduce the hazards of lifting. In NIOSH's view, belt use remains a personal decision.

A National Safety Council technical advisory report on the use of back belts, issued in September 1995 by the Ergonomic/Human Factors standing committees of the Industrial (now Business and Industry) and Labor divisions of the Council, comes to a similar conclusion. The National Safety Council stresses that several conditions and controls should be in place before companies consider the use of back belts in the workplace. Companies should reduce or mitigate risk factors through ergonomic intervention. Also, they should use engineering controls, work-method analysis, and administrative controls to reduce worker exposure to lifting injuries.

Some scientific reports on back-belt use note increased blood pressure and heart rate in some belt users due to increased intra-abdominal pressure. Therefore, the National Safety Council recommends that individuals who consider belt use be medically screened for cardiovascular and general health. To maintain strength and flexibility, companies should consider a simple exercise program for workers who use back belts. The program should be under the guidance of a healthcare professional.

For more information on the debate, see "Back Belts: The Debate Continues," *Safety & Health,* June 1996, pp. 40–41.

FALL PROTECTION

Materials-handling equipment frequently gives rise to exposure to falls by operators and maintainers of overhead hoisting systems. Provision for fall protection and egress is needed where workers will foreseeably be exposed to fall hazards.

The approach to fall hazards is as follows:

1. Recognition of each fall hazard including access to the workstation
2. Control of the fall hazard through choice from a variety of fall solutions
3. Selection and installation of fall protection including engineering if required
4. Training sufficient to fulfill fall-protection program
5. Observation of workers and enforcement of fall protection program

Recognition of fall hazards could be as follows: crane bridges, crane runways, ladder access to cranes, truss access, warehouse racks, platforms, operator cabs, access at heights, roofs, mezzanines.

Fall-protection measures include the following controls:

1. Elimination of the hazard by reorganizing the work
2. Prevention of falling by the use of guardrails, including aerial lifts
3. Fall arrest systems for horizontal and vertical travel
4. Warning lines six feet from an edge

Fall-arrest systems are often valuable when workers overreach from aerial lifts, walk catwalks, climb trusses, walk on roof edges, climb fixed ladders, or use stepladders.

In case of fire and smoke, listed controlled descent devices may be beneficial in addition to fixed ladders typically supplied according the NIOSH 76–128. These systems may also be used for the rescue of individuals from height. Crane cabs and other workstations are examples where installations may be made.

Training and inspection are vital for proper application of a fall-protection program. Workers should be given a pretest and posttest to provide feedback to worker and manager alike. Observation is vital to monitor the effects of site-specific training. The training should cover regulations, standards, the site conditions, the work to be done, the sequence of the work, the materials, the work method, and fall-arrest equipment and its proper use.

The fall protection for materials handling-related applications must be applied at the very first thought about the work and when other initial planning is undertaken.

GUIDELINES FOR LIFTING

Regardless of the approach taken to evaluate the physical stresses of lifting, a large individual variability in risk of injury and lifting performance capability exists in the population today. This means that the resulting controls must be of both an engineering and administrative nature. In other words, there are some lifting situations that are so hazardous that only a few people could be expected to be capable of safely performing them. These conditions need to be modified to reduce stresses through job redesign. On the other hand, some lifting conditions may be safely tolerated by some people, but others must be protected by an aggressive selection process, training program, and modifications of the workplace. To specifically define these conditions, two limits are provided based on epidemiological, biomechanical, physiological, and psychophysical criteria.

1. **Maximum Permissible Limit (MPL)**
   This limit is defined to best meet the four criteria:
   a. Musculoskeletal injury rates and severity rates have been shown to increase significantly in populations when work is performed above the MPL.
   b. Biomechanical compression forces on the L5/S1 disc are not tolerable over 650 kg (1,430 lb) in most workers. This would result from conditions above the MPL.

c. Metabolic rates would exceed 5.0 Kcal/minute for most individuals working above the MPL.
d. Only about 25% of men and less than 1% of women workers have the muscle strengths to be capable of performing work above the MPL.

2. **Action Limit (AL)**
   The large variability in capacities between individuals in the population indicates the need for administrative controls when conditions exceed this limit based on:
   a. Musculoskeletal injury incidence and severity rates increase moderately in populations exposed to lifting conditions described by the AL.
   b. A 350 kg (770 lb) compression force on the L5/S1 disc can be tolerated by most young, healthy workers. Such forces would be created by conditions described by the AL.
   c. Metabolic rates would exceed 3.5 for most individuals working above the AL.
   d. Over 75% of women and over 99% of men could lift loads described by the AL.

Thus, properly analyzed lifting tasks may be of three types:

1. those *above the MPL* should be viewed as unacceptable and require engineering controls
2. those *between the AL and MPL* are unacceptable without administrative or engineering controls
3. those *below the AL* are believed to represent nominal risk to most industrial workforces.

Guideline Limits

With the large number of task variables (five in this case) that modify risk during lifting, it is virtually impossible to provide a simple, yet accurate procedure for evaluating all possible jobs. This problem is further complicated by the need to satisfy four separate criteria (epidemiological, biomechanical, physiological, and psychophysical). The following guideline is the simplest form known that best satisfies the four criteria.

In algebraic form:

$$AL(kg) = 40(15/H)(1 - 0.004|V - 75|)(0.7 + 7.5/D)$$
$$(1 - F/F_{max}) - \text{metric units}$$

or

$$AL(lb) = 90(6/H)(1 - 0.01|V - 30|)(0.7 + 3/D)$$
$$(1 - F/F_{max}) - \text{U.S. customary units}$$
$$MPL = 3(AL)$$

where H = horizontal location (centimeters or inches) forward of midpoint between ankles at origin of lift
V = vertical location (centimeters or inches) at origin of lift
D = vertical travel distance (centimeters or inches) between origin and destination of lift
F = average frequency of lift (lifts/minute)
F_{max} = maximum frequency that can be sustained

For purposes of this guide, these variables are assumed to have the following limits.

1. H is between 15 cm (6 in.) and 80 cm (32 in.). Objects cannot, in general, be closer than 15 cm (6 in.) without interference with the body. Objects further than 80 cm (32 in.) cannot be reached by many people.
2. V is assumed between 0 cm and 175 cm (70 in.), representing the range of vertical reach for most people.
3. D is assumed between 25 cm (10 in.) and (200 – V) cm [(80 – V) in.]. For travel less than 25 cm, set D = 25.
4. F is assumed between 0.2 (one lift every 5 minutes) and F_{max}. For lifting less frequently than once per 5 minutes, set F = 0.

Numerous attempts have been made to train material handlers to do their work, particularly lifting, in a safe manner. Unfortunately, hopes for significant and lasting reductions of overexertion injuries through the use of training have been generally disappointing. There are several reasons for the disappointing results:

- If the job requirements are stressful, "doctoring the system" through behavioral modification will not eliminate the inherent risk. Designing a safe job is basically better than training people to behave safely in an unsafe job.
- People tend to revert to previous habits and customs if practices to replace previous ones are not reinforced and refreshed periodically.
- Emergency situations, the unusual case, the sudden quick movement, increased body weight, or impaired physical well-being may overly strain the body, since training does not include these conditions.

Thus, unfortunately, training for safe materials handling (which is not limited to lifting) should not be expected to really solve the problem. On the other hand, if properly applied and periodically reinforced, training should help to alleviate some aspects of the basic problem.

The idea of training workers in safe and proper manual materials-handling techniques has been propagated for many years. Originally it was advocated that one lift with a straight back and to unbend knees while lifting. However, the frequency and intensity of back injuries was not reduced during the last 40 years while this lifting method has been taught. Biomechanical and physiological research has shown that leg muscles used in this lifting technique do not always have the needed strength. Also, awkward and stressful postures may be assumed if this technique is applied when the object is bulky, for example. Hence, the straight back/bent knees action evolved into the "kinetic" lift, in which the back is kept mostly straight and the knees are bent, but the positions of the feet, chin, arms, hands, and torso are prescribed. Another variant is the "free-style" lift, which, however, may be better for male (but not female) workers than the straight back/bent knee technique (Garg and Saxena, 1985). It appears that no single lifting method is best for all situations (Andersson and Chaffin, 1986).

Training of proper lifting techniques is an unsettled issue. It is unclear what exactly should be taught, who should be taught, how and how often a technique should be taught. This uncertainty concerns both the objectives and methods, as well as the expected results. Claims about the effectiveness of one technique or another are frequent but are usually unsupported by convincing evidence.

A thorough review of the existing literature indicates that the issue of training for the prevention of back injuries in manual materials handling is confused at best. In fact, training may not be effective in injury prevention, or its effect may be so uncertain and inconsistent that money and effort paid for training programs might be better spent on research and implementation of techniques for worker selection and ergonomic job design. Nevertheless, according to the National Institute for Occupational Safety and Health (NIOSH),

- The importance of training in manual materials handling in reducing hazards is generally accepted. The lacking ingredient is largely a definition of what the training should be and how this early experience can be given to a new worker without harm. The value of any training program is open to question as there appear to have been no controlled studies showing a consequent drop in the manual materials handling (MMH) accident rate or the back injury rate. Yet, so long as it is a legal duty for employers to provide such training or for as long as the employer is liable to a claim of negligence for failing to train workers in safe methods of MMH, the practice is likely to continue despite the lack of evidence to support it. Meanwhile, it may be worth considering what improvements can be made to existing training techniques. (NIOSH, 1981, p. 99)
- Currently, it appears that two major training approaches are most likely to be successful. One involves training in awareness and attitude through information on the physics involved in manual materials handling and on the related biomechanical and physiological events going on in one's body. The other approach is the improvement of individual physical fitness through exercise and warm-ups (which, of course, also influences awareness and attitude, though indirectly).

Rules for Lifting

There are no comprehensive and sure-fire rules for "safe" lifting. Manual materials handling is a very complex combination of moving body segments, changing joint angles, tightening muscles, and loading the spinal column. The following DOs and DO NOTs apply, however:

- DO engineer manual lifting and lowering out of the task and workplace. If it, nevertheless, must be done by a worker, perform it between knuckle and shoulder height.
- DO be in good physical shape. If a worker is not used to lifting and vigorous exercise, he or she should not attempt to do difficult lifting or lowering tasks.
- DO think before acting. Place material conveniently within reach. Have handling aids available. Make sure sufficient space is cleared.
- DO get a good grip on the load. Test the weight before trying to move it. If it is too bulky or heavy, get a mechanical lifting aid or somebody else to help, or both.
- DO get the load close to the body. Place the feet close to the load. Stand in a stable position with the feet pointing in the direction of movement. Lift mostly by straightening the legs.
- DO NOT twist the back or bend sideways.
- DO NOT lift or lower awkwardly.
- DO NOT hesitate to get mechanical help or help from another person.
- DO NOT lift with the arms extended.
- DO NOT continue lifting when the load is too heavy.

Personnel Selection for Materials Handling

Selecting persons who are unlikely to suffer an overexertion injury is one of the three methods to reduce the risk of musculoskeletal disorders in manual materials handling. The purpose of this assessment is to place only those individuals on jobs who can do them safely. Such screening may be done before placement on a new job, or on occasion of routine examinations during employment. The basic premise is that the risk of overexertion injury for manual materials handling decreases as the handler's capability to perform such activity increases. This means that the evaluation should be designed so that it allows the administrator to match a person's capabilities for manual materials handling to the actual demands of the job. This matching process and job safety analysis requires that the administrator knows quantitatively both the job requirements and the related capabilities to be tested.

Scientists usually rely on the development and use of models. A **model** is an abstract (mathematical-physical) system that obeys specific rules and conditions and whose behavior is used to understand the real system (in this case, the worker-task) to which it is analogous in certain aspects (e.g., in physiological, biomechanical, psychophysical, or other traits). A model usually represents a theory. Without proper models, reliable and suitable methods cannot be developed. A **method** is a systematic, orderly way of arranging thoughts and executing actions. A **technique** is the specific, practical manner in which actions are done; it implements the methods that are derived from models. Evaluation techniques involve specific procedures and instruments used to obtain measurements with respect to the subject's capability to perform manual materials-handling activities.

Many models have been developed to describe the central and local limitations just discussed. In the following section, these models are simplified and categorized by major disciplines for convenience.

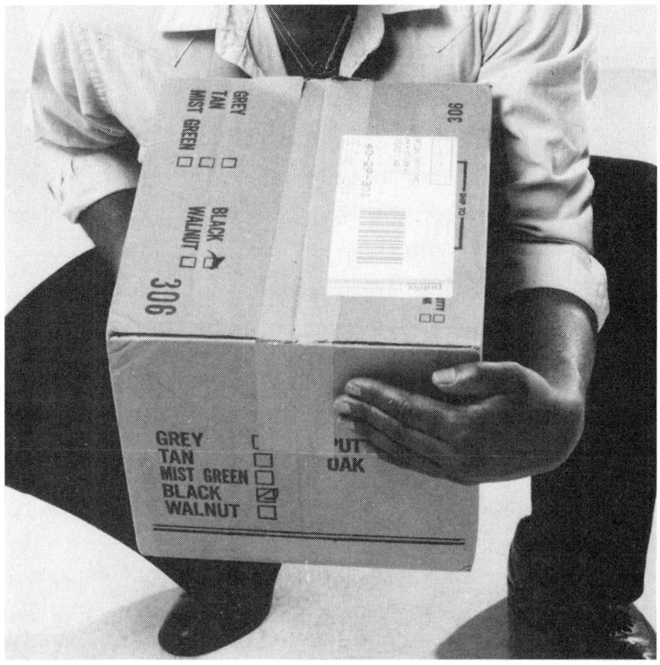

Figure 14–2. To lift, the load should be drawn close to the body, and the arms and elbows should be tucked into the side of the body. When the arms are held away from the body, they lose much of their strength and power. Keeping the arms tucked in also helps keep weight centered.

Load Held Close to the Body

While lifting and carrying objects, the worker should keep the load close to the body (Figure 14–2). The closer the load is to the body, the less it affects the lower back. To do this, the arms should be close to the body and remain straight whenever possible. Flexing the elbows and raising the shoulders imposes unnecessary strain on the upper arm and chest muscles.

When carrying an object, the arms should stay in the same position. In the case of long-distance carrying, any assistance given by the body to support the weight of an object will lessen the tension in the muscles. Carrying an object with the arms lowered assures that the weight of the object will rest against the body.

Correct Grip

An insecure grip may be caused by holding a load with the fingertips. Such a grip creates undue pressure at the ends of the digits and causes strain to certain muscles and tendons of the arm. A full-palm grip reduces local muscle stress in the arms and decreases the possibility that a load will slip (Figure 14–3). Handles or handholds are preferable. Because greasy surfaces often prevent a secure hold, wipe surfaces clean before grasping them. Use suitable, properly fitted gloves, if necessary.

Here are some lifting techniques for specific situations:

1. If the object is too bulky or too heavy to be handled by one person, get help.

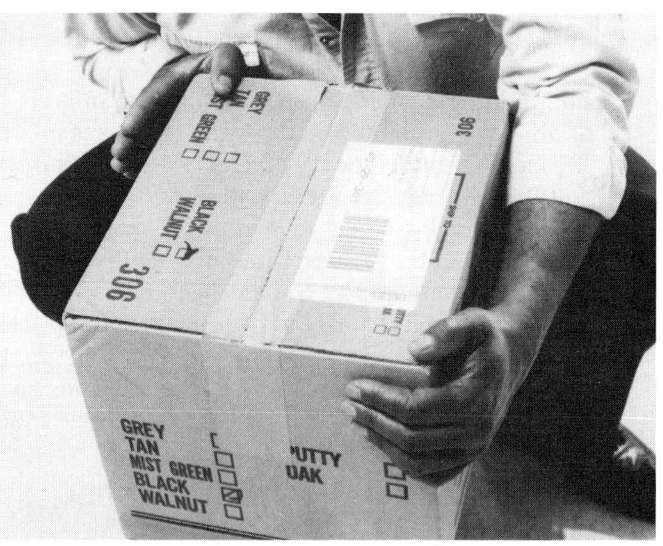

Figure 14–3. The palmar grip is one of the most important elements of correct lifting. The fingers and the hand are extended around the object to be lifted. Use the full palm; fingers alone have very little power. Pull the load in between the knees and as close to the body as possible. (The glove has been removed to show the finger positions better.)

2. Workers should not "jerk-lift" loads, as this multiplies the stress to the lower back.
3. Before lifting the load to be carried, workers should consider the distance to be traveled, any obstacles that need to be repositioned, and the length of time that the grip will have to be maintained. Workers should select a place to set the load down and rest to allow for the loss of gripping power. Pausing to rest is especially important when negotiating stairs and ramps.
4. To place an object on a bench or table, the workers should first set it on the edge and then push it far enough onto the support so that it will not fall. The object should be released gradually as it is set down. It should be moved in place by pushing with the hands and body from in front of the object. This method prevents pinched fingers.
5. Workers should securely position an object, placed on a bench or other support, so that it will not fall, tip over, or roll off. Supports should be correctly placed and strong enough to hold the load. Heavy objects should be stored at approximately waist height.
6. To raise an object above shoulder height, workers should (1) lift it to waist height, (2) rest the edge of the object on a ledge, stand, or hip, and (3) shift hand positions, so the object can be boosted after the knees are bent. Workers should straighten their knees as the object is lifted or shifted to the shoulders.
7. To change direction, the worker should lift the object to the carrying position and turn the entire body, including the feet. The worker should avoid twisting the body. In repetitive work, the worker and the object should both be positioned so that twisting the body is unnecessary when moving the object.
8. To deposit an object manually into a tight space, the worker should slide it into place with the hands in the clear—thus, avoiding pinched fingers.

Team Lifting and Carrying

When two or more people carry one object, they should adjust the load so that it rides level and so that each person carries an equal part of the load. Workers should do test lifts before proceeding. When two people carry long sections of pipe or lumber, they should walk one behind the other, carrying the material on the same shoulder and walking in step. Shoulder pads will prevent cutting into their shoulders and will also reduce fatigue.

When a team of workers carries a heavy object, such as a rail, the supervisor should make sure that proper tools are used and should provide direction for the work. Frequently, a whistle or direct command can signal "lift," "walk," and "set down." The key to safe carrying by gangs is to make every movement in unison.

Handling Specific Shapes

There are specific techniques for lifting and handling materials of various shapes.

Boxes, Cartons, and Sacks

The best way to handle boxes and cartons is to grasp the alternate top and bottom corners and to draw a corner between the legs. Banding materials should never be used to make a lift except in special cases.

Sacked materials are also grasped at opposite corners. Upon reaching an erect position, the worker should let the sack rest against the hip and belly and then swing the sack to one shoulder. As the sack reaches the shoulder, the worker should stoop slightly and put a hand on the hip so that the sack rests partly on the shoulder and partly on the arm and back. The other hand should be holding the sack at the front corner. When the sack is to be put down, the worker should swing it slowly from the shoulder until it rests against the hip and belly. While the sack is being lowered, the legs should be flexed and the back kept straight.

Barrels and Drums

Those who handle heavy barrels and drums require special training (Figures 14–4 and 14–5). A barrel is generally less hazardous to handle than a drum because the shape of the barrel aids in upending it and reducing

Figure 14–4. This spill containment caddy minimizes strains during drum transport. (Courtesy Justrite Manufacturing Company, L. L. C.)

Figure 14–5. Pneumatic devices may also reduce the need to manually handle drums. This device secures the drum by its sides and lifts, moves, and tilts the drum so it can be placed horizontally on a rack or vertically in storage.

hazards should it tip over. Workers should pay special attention to the weight and contents of barrels and drums since these factors vary greatly. Workers should wear safety shoes with metatarsal guards.

Frequently, only one person is available to handle a drum, in which case it is better to wait for help or use mechanical assistance. A commercially available drum tilter, equipped with wheels, is commonly used. An extension handle provides control and leverage during the tilting operation. The wheels allow for easy transport of the tilted drum over short distances (Figure 14–4). Another commercial device that makes tilting and transporting easier is a two-wheeled dolly equipped with large rubber tires.

Sheet Metal

Because sheet metal usually has sharp edges and corners, workers should handle it with leather gloves, hand leathers, or gloves with metal inserts. Gauntlet-type gloves or wristlets give added protection to wrists and forearms. Workers should use power equipment, such as a "grab" or a vacuum lifter for bundles of sheet metal.

Plate Glass

Workers should wear gloves or hand laps when handling flat glass. Their wrists and arms should be protected with leather cuffs and safety sleeves. The worker should wear a leather apron, leggings, safety glasses, and also safety shoes with metatarsal guards. Unless the piece of plate glass is small, the worker should carry only one piece at a time and should walk with care. The plate should be lifted carefully and carried with its bottom edge resting in the palm (turned outward) and with the other hand holding the top edge to steady it. Plate glass should never be carried under the arm because a fall might break the glass and sever an artery. Plate glass must not be carried in such a manner that it bends.

To transport larger pieces of plate glass over any distance, workers should use handling equipment specifically designed for that purpose. Equipment such as cranes equipped with vacuum frames, C-frames or spreader bars, and special wagons or dollies are normally used to transport heavy glass. If large plates must be transported by hand, two workers wearing safety hats, safety glasses, safety sleeves, cuffs, gloves, and safety shoes should do the job. For more details on safe handling of glass, see the article by Oresick listed in the References at the end of this chapter.

Long Objects

A team of workers should carry long pieces of pipe, bar stock, or lumber on their shoulders. One of the workers should guide the object when going around corners. Workers should wear shoulder pads for this operation.

Irregularly Shaped Objects

Objects with irregular shapes present special handling problems. Often the object must be turned over or up on end to secure the best possible grip. If the worker feels unable to handle the object, because of either its weight or shape, he or she should request assistance.

Miscellaneous Objects

Using slings, buckets, and bags are safe methods of securing miscellaneous material on overhead platforms. Workers should follow special safety measures when raising or lowering these materials to different elevations in order to protect workers below.

Scrap Metals

In a scrap storage area, require that the best possible housekeeping practices be observed. Irregularly shaped, jagged pieces can get tangled in such a way that strips or pieces may fly when a piece is removed from a pile. Therefore, provide workers with goggles, leather gloves or mittens, safety shoes, safety hats, and protection for their legs and body. Caution workers against stepping on objects that may roll or slide.

Heavy, Round, Flat Objects

If rolled by hand, heavy, round, flat objects, such as railcar wheels or tank covers, present considerable danger even to skilled personnel. The operation requires careful training and exacting precautions. Using a hand truck or power equipment designed for the purpose is preferred. Heavy rolls can be safely secured and handled by specially designed devices. (See Figures 17–2 and 17–3 in Chapter 17, Powered Industrial Trucks.)

Machines and Other Heavy Objects

Manual movement of heavy machinery and equipment requires special skill and knowledge. Sometimes machines or castings weighing 100 tn (90.7 t) or more must be moved from freight cars to ground level and into permanent position without the use of heavy-duty cranes or similar equipment.

Only general safety principles for such jobs can be suggested. Each task presents its own problems and requires careful study and thorough planning. Some companies build scale models of the machines and the blocking, jacks, rollers, and other equipment to be used. They then work out the procedure in miniature. In all cases, determine the safe floor load limits for areas over which the machine or part will move, as well as for the place in which it is to be installed or stored.

Select blocking and timbers with great care. Hardwoods make the best blocking and timber—preferably oak, and of the proper sizes to allow the machine to be safely blocked or cribbed as it is raised or lowered. Do not use wood that has round corners or that shows signs of dry rot.

For sufficient strength, cribbing should have a safety factor of at least four. Be cautious about the natural tendency to underestimate the load. Cribbing must be placed on a foundation in such a manner that it can be removed readily as the machine is lowered.

ACCESSORIES FOR MANUAL HANDLING

In handling materials, a variety of hand-operated accessories is available. Each tool, jig, or other device should be kept in good repair and used only for the job for which it is designed.

Hand Tools

Of all the hand tools available for the manual handling of materials, hooks, crowbars, and rollers are the most commonly used. Workers should know how to use these tools properly.

Hooks

Train the worker to use hand or packing hooks in such a way that they will not glance off hard objects and possibly injure the worker. If a hook must be carried in the worker's belt, be sure the point is covered.

Use hook handles made of hardwood, and keep them in good condition. Supervisors should check that hooks for handling logs, lumber, crates, boxes, and barrels are kept sharp and inspected daily and before each use.

Crowbars

The principal hazard in the use of a crowbar is slippage. Workers should keep their hands and gloves dry, and free of grease and oil when using a crowbar. A dull, broken crowbar is more likely to cause injury than a sharp one. The point or edge should have a good "bite." Workers should position themselves to avoid falling or pinching their hands if the bar slips or the object moves suddenly. Workers should never work astride an object. Crowbars not in use should be stored so that they will not fall or cause a tripping hazard.

Rollers

Heavy, bulky objects must often be moved by means of rollers. The principal hazard of rollers is that the fingers or toes may be pinched or crushed between a roller and the floor. Rollers should extend beyond the load to be moved and be sufficiently strong. Rollers under a load should be moved with a sledge or bar, not with a hand or foot.

Jacks

When a jack is used, workers should check the capacity plate or other marking on the jack to make sure the jack can support the load. If the identifying plate is missing, workers should determine the maximum capacity of the jack and paint it on the side. If a properly rated jack is used, it should not collapse under the load.

Workers should inspect jacks before and after each use. Any sign of hydraulic fluid leakage is sufficient reason to remove the jack from use. When a jack begins to leak, malfunction, or show any signs of wear or defects, it should be removed from service, tagged, repaired, and tested under load before being returned to service.

Workers using jacks should wear protective footwear (safety shoes) with metatarsal guards because jack handles may slip and fall or parts of machinery or equipment may become loose and drop while the load is being lifted or shifted. Furnish toweling to jack operators for removing oil from their hands and from the jack handles. In that way, they will always have a firm grip.

A heavy jack is best moved from one location to another on a dolly or special hand truck. If it has to be manually transported, it should have carrying handles. At least two workers should form a team to move it. The operating handle should never be left in the socket while a jack is being carried because it might strike other workers.

Workers should make certain that jacks are well lubricated, but only at points where lubrication is specified. They should also inspect them for broken teeth or faulty holding fixtures. Workers should never throw or drop a jack upon the floor. Such treatment may crack or distort the metal, causing the jack to break when a load is lifted.

The floor or ground surface upon which a jack is placed must be level and clean, and the safe floor load limit must not be exceeded. If the surface is earth, workers should set the jack base on substantial hardwood blocking (at least twice the size of the jack), so that it will not turn over, shift, or sink. If the surface is not perfectly level, workers should set the jack on blocking leveled by substantial shims or wedges. These devices should be placed so securely that they cannot be crushed or forced out of place.

To prevent the load from slipping, workers should avoid metal-to-metal contact between the jack head and the load. A hardwood shim, longer and wider than the face of the jack head, should be placed between the jack head and the contact surface of the load. Two-inch wood stock is suitable for this purpose. Workers should also remove oil that has collected in the bases of equipment or machines to be jacked before the operation is begun. This prevents spillage when the equipment or machines are tilted. Workers should immediately wipe up spillage of any residual oil.

Workers should never use wood or metal extenders. Instead, they should either obtain a larger jack or should place higher blocking that is correspondingly wider and longer under the jack.

All lifts should be vertical—the jack correctly centered for the lift, the base on a perfectly level surface, and the head with its shim bearing against a perfectly level meeting surface. When an emergency requires that the lifting force be applied at an angle, workers must take the following extra precautions:

1. a base of blocking, securely fastened together and to the ground, to make an immovable surface at right angles to the lift for the jack base to sit on
2. cleats on the blocking to prevent shifting of the jack base
3. a meeting surface at right angles to the direction of the lift for the jack head with its shim to bear against
4. props or guys to the load to prevent the jack from swinging sideways when lifting begins

When a jack handle is placed in the socket and before applying pressure, the worker should make sure that the area is clear and that there is ample room for an unobstructed swing of the handle. A faulty movement in the load may cause the handle to pop up and strike another worker. The person operating the handle should stand to one side so that, if the handle kicks, it will not strike any body or facial areas. When releasing a jack, the worker should keep all parts of the body clear of the movement of the handle.

After the load is raised, the worker should place metal or heavy wooden horses or blocking under it for support in case the jack should let go. The worker should never allow a raised load to remain supported only by jacks. The jack operator should immediately remove the handles of the jacks and should place them out of the way to prevent others from tripping over them.

The following special jacks require additional precautions:

- Hydraulic jacks may settle after raising a load. It is, therefore, especially important for workers to place blocking under a load that has been raised by such jacks.
- Screw jacks have a tendency to twist when a heavy load causes the floating head of the jack to bind. Jack operators, therefore, should anchor the base of a screw jack as securely as possible. In that way, the jack base will not twist and slip out from under the load when force is applied on the bars to raise the screw.

To raise a large piece of equipment with screw jacks, two or more jacks should be used. The load should be equally distributed on each jack. Operators should raise each jack a little at a time to keep the load level and the strain equal on each screw jack head. Operators should work out special signals to verify that all jacks are rising uniformly.

Hand Trucks

Many special types of hand trucks and dollies are available, such as two-wheeled trucks, flat trucks, lift trucks, and platform, refrigerator, and appliance dollies. Hand trucks can be purchased or designed for objects of various kinds and sizes. The type of hand truck most suitable for the work at hand should be used. No one hand truck is suitable for handling all types of material.

Two-Wheeled Trucks

Provide two-wheeled trucks and wheelbarrows equipped with knuckle guards to protect workers' hands from being jammed against door frames or other obstructions. Operators should also wear heavy gloves. Some two-wheeled trucks have brakes so that the worker need not hold the truck with a foot on the wheel or axle. Most commonly used trucks, however, do not have brakes.

To reduce the hazard to toes and feet, purchase trucks with wheels as far under the truck as possible. Wheel guards can be installed on many types of trucks, and operators should wear protective footwear.

Tongues of flat trucks should have counterweights, springs, or hooks to hold them vertical when not in use. If this is not possible, train workers to leave handles in such a position that they will not cause workers to trip over them.

Two-wheeled trucks look deceptively easy to handle, but workers should adhere to the following safe procedures:

- Tip the load to be lifted slightly forward so that the tongue of the truck goes under the load.
- Push the truck all the way under the load to be moved.
- Keep the center of gravity of the load as low as possible. Place heavy objects below lighter objects. When loading trucks, workers should keep their feet clear of the wheels.
- Place the load well forward so the weight will be carried by the axle not by the handles.
- Place the load so it will not slip, shift, or fall. Load only to a height that will allow a clear view ahead.
- When a two-wheeled truck or wheelbarrow is loaded in a horizontal position, raise it to traveling position by lifting with the leg muscles and keeping the back straight. Observe the same principle in setting a loaded truck or wheelbarrow down—the leg muscles should do the work.
- Let the truck carry the load. The operator should only balance and push.
- Never walk backward with a hand truck.
- For extremely bulky items or pressurized items, such as gas cylinders, strap or chain the item to the truck.
- When going down an incline, keep the truck ahead so that it can be observed at all times.
- Move trucks at a safe speed. Do not run. Keep the truck constantly under control.
- Store hand trucks with the tongue under a pallet, shelf, or table.

Four-Wheeled Trucks or Carts

Operating rules for four-wheeled trucks or carts are similar to those for two-wheeled trucks. Place extra emphasis, however, on proper loading of four-wheeled trucks. They should be evenly loaded to prevent tipping. A truck's contents should be arranged so that they will not fall or be damaged in case the truck or the load is bumped. Four-wheeled trucks should be pushed rather than pulled. Pushing an object rather than pulling on it causes less stress to the lower back and protects the worker's heel from being caught under the truck back.

Four-wheeled trucks should not be loaded so high that operators cannot see where they are going. If there are high racks on the truck, two workers should move the vehicle—one to guide the front, the other to guide the back. Handles should be placed at protected places on the racks or body of the truck. In that way, passing traffic, walls, or other objects will not crush or scrape the operator's hands.

General Precautions

Be sure that handlers of two- and four-wheeled trucks are aware of three main hazards: (1) running wheels off bridge plates or platforms, (2) colliding with other trucks or obstructions, and (3) jamming their hands between the truck and other objects.

Workers should operate trucks at a safe speed and keep them constantly under control. Special care is required at blind corners and doorways. Properly placed mirrors can aid visibility at these places.

When not in use, trucks should be stored in a designated area—not parked in aisles or other places where they could obstruct traffic or cause someone to trip. Trucks with drawbar handles should be parked with handles up and out of the way. Two-wheeled trucks should be stored on the chisel with their handles leaning against a wall or the next truck.

STORAGE OF SPECIFIC MATERIALS

Temporary and permanent storage of materials should be secure, neat, and orderly to eliminate hazards and conserve space. Materials piled haphazardly or strewn about increase the possibility of accidents to employees and of damage to materials.

The warehouse supervisor must direct the storage of raw material (and sometimes processed stock) that is kept in quantity lots for extended periods of time. The production supervisor is usually responsible for storage of limited amounts of material and stock that is kept near the processing operations for short periods of time. A good plan for storing materials reduces the amount of handling needed both to bring materials into production and to remove finished products from production to shipping.

Planning Materials Storage

When planning the storage of materials, allow adequate ceiling clearance under the sprinklers. The amount of clearance between storage and automatic sprinkler heads will vary with the material being stored and the height of storage. (See the National Fire Protection Association (NFPA) 13, *Installation of Sprinkler Systems.*) Be sure that automatic sprinkler system controls and electrical panel boxes are free and clear. Also, maintain unobstructed access to fire hoses and fire extinguishers.

Keep all the exits and aisles clear at all times. If materials are handled from the aisles, allow for the turning radius of power trucks. Employees should keep materials out of the aisles and out of the loading and unloading areas. These areas should be marked with painted or taped lines.

Use bins and racks to facilitate storage and reduce hazards (Figure 14–6). Consider adjustable storage containers to avoid placing a worker in an awkward position when he or she is returning a part. Material stored on racks, pallets, or skids is moved easily and quickly from one workstation to another with less damage to materials and fewer injuries to employees. When possible, have material that is piled on skids or pallets cross-tied.

Storage racks should be secured to the floor, the wall, and to each other. If flue spaces are provided, stock should never block the flue. If automatic sprinklers or fire-prevention devices are provided in the racks, workers should exercise special care to avoid damage to them. Racks, when damaged, should be repaired. Never allow employees to climb on racks.

In an area where the same type of material is stored continuously, it is a good idea to paint a horizontal line on the wall to indicate the maximum height to which material may be piled. This will help keep the floor load within the proper limits and the sprinkler heads in the clear. Pickup and drop stations should also be clearly marked.

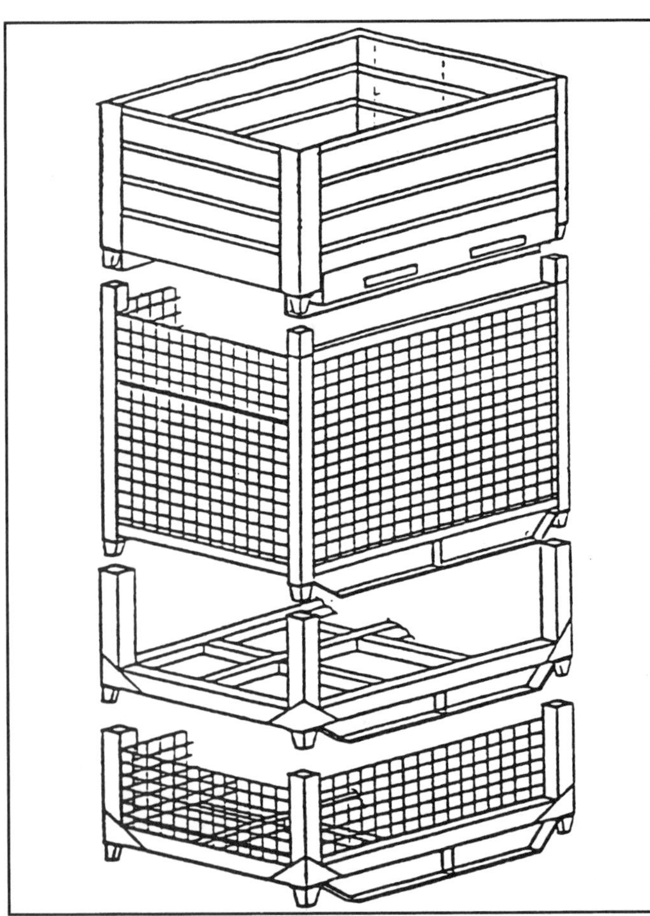

Figure 14–6. Interstacking containers like these are found in many industrial plants. To stack safely, be sure the surface is level and stable.

High bay storage is the trend for storing containers or stock of uniform size. Some European and U.S. storage facilities now have approximately 100 ft (30.5 m) high, automated storage units.

High bay facilities require unique, specially designed, high-lift materials-handling equipment. Some, up to 30 ft (9 m) high, are operated manually. Others are operated through computer control. Standards that apply to high bay storage include the American National Standards Institute (ANSI) and the American Society of Mechanical Engineers (ASME) ANSI/ASME B56.1–7 series, *Safety Standards for Powered Industrial Trucks,* and the Crawler Cranes section of ANSI/ ASME B30.5, *Mobile and Locomotive Truck Cranes.* High bay storage presents not only unique materials-handling problems but also special fire-protection problems. (See Chapter 17, Powered Industrial Trucks and NFPA 231C, *Rack Storage of Materials.*)

The protection of personnel who operate and maintain such facilities must include disaster and emergency planning, physical protection at the point of operation, lockout/tagout, and visible and audible warnings on moving equipment. Special procedures and equipment are also required for maintaining and taking physical inventories of high bay facilities.

Rigid Containers

Each type of rigid container has its own storage requirements. Methods for storing several types of these containers are given below.

Large Metal Containers and Box Pallets

There are three general types of large containers commonly used in industry: wire mesh or expanded metal containers, solid-sided metal tubs, and skids and box pallets. There are also many variations on these types.

Safe stacking of large containers requires a level and stable stacking surface. These containers must be in good condition and be able to nest or interlock with the container below (Figure 14–6). Do not intermix different types of containers (e.g., metal tubs and wood box pallets) in a stack unless they are designed for each other and full safety is assured.

A rule of thumb for stacking heights is three times the minimum base diameter of the container. For example, a container with a 2 to 3 ft (61 to 91 cm) base may be stacked 6 ft (1.8 m) high if the other stacking conditions have been met. Do not exceed weight capacities and weight-bearing capacities of the stacked containers. For visibility, do not stack containers near the corners of working (nonstorage) aisles.

Fiberboard/Cardboard Cartons

Store loaded cartons on platforms to protect them against moisture. Even low piles, when wet, will collapse. Preferably store cartons on pallets or racks. If lower cartons show any signs of being crushed, restack them.

Since the height of piles of cartons is regulated by the kind of materials in the cartons, it is not possible to establish a standard height for such piles. An important factor to consider is that the sides of cartons do not support much load. Place sheets of heavy wrapping paper between layers of cartons, therefore, to help prevent the pile from shifting. Interlocking the cartons also increases the stability of the pile.

Do not stack certain bulky materials, such as skids of paper, to maximum allowable heights in rows next to aisles, especially aisles that carry hand-truck or power-truck traffic. A good rule is to make the first row one item high unless the material can be tightly interlocked. Stretch wrapping with plastic film or banding pallets will also assist in more stable storage.

Barrels and Kegs

Piles of barrels should be symmetrical and stable, preferably in the shape of a pyramid. Block the first or bottom row to prevent the barrels from rolling. If barrels or kegs are piled on end, place planks between rows. When barrels or kegs are piled on their sides other than in pyramid form, lay them on specially constructed racks. Otherwise, lay planks between rows and block the ends of the rows. (See Portable Containers in the section Containers for Liquids later in this chapter for a discussion on drums.)

Rolled Paper and Reels

Clamp-type trucks can stack rolled paper and reels three or four high. Extreme care is required to assure even stacking. If there is any physical damage to lower rolls or reels, the material must be restacked. Band paper rolls a few inches from each end.

Compressed Gas Cylinders

For safe handling and storage of compressed gas cylinders, refer to Chapter 21, Welding and Cutting.

Uncrated Stock

Stock that does not come in its own container presents special storage problems. Examples of uncrated stock are lumber, bagged material, pipes, and sheet metal.

Lumber

Except for the amount that is needed at a given time, lumber is best stored outdoors or in a building separate from the general warehouse. Sort lumber by size and length and store it in separate piles. When piling lumber outdoors, be sure that the ground is firm. The storage area should also be well drained to remove surface water and prevent softening of the ground. Periodically check stored lumber to make sure that it has not shifted position. Cover lumber stored outdoors to prevent checking or twisting. Use a well-ventilated building for storing lumber indoors.

For long-term storage, use substantial bearings or dunnage. Concrete with spread footing extending below the frost line is a good method. For temporary storage, heavy timbers may be used to support the crosspieces. Periodically inspect the timbers for signs of deterioration, which may cause the pile to list dangerously.

If the lumber must be removed by hand, store it in low piles or in racks, and provide galleries that permit workers to reach the top of the piles. If lumber must be moved by hand to or from a higher pile, the pile should be not more than 6 ft (1.8 m) high. Provide a safe means of access to the top of the pile.

When lumber is piled and removed mechanically by forklift trucks, 20 ft (6 m) is generally considered the maximum safe height for lumber piles.

Use tie pieces not only to stabilize the lumber pile but also to provide circulation of air. Do not let tie pieces extend into walkways; cut them flush with the pile. Use tie pieces on every layer of green lumber whether stored indoors or outdoors.

Bagged Material

Cross-tie bagged material with the mouths of the bags toward the inside of the pile. Stack bags neatly. Avoid overhangs that could be ripped, thus spilling the bags' contents. This precaution also applies to stacking on pallets and stacking several tiers high by lift truck. Stretch wrapping will help contain loose commodities.

Pipe and Bar Stock

Pipe and bar stock places a heavy load on the floor. Therefore, select the floor area with load-bearing strength in mind. Because removing pipe and bar stock from racks presents a hazard to passersby, do not locate fronts of pipe and bar stock racks on main aisles. Pile pipe and other round materials in layers with strips of wood or iron between the layers. Either the strip should have blocks at one end, or the end should be turned up.

Store larger sizes of bar stock in racks that are designed to rest the bars on rollers. The center distance between rollers will be governed by the sizes of the bars and should permit their easy withdrawal. Rollers with multiple sections make withdrawal easier. Racks should incline toward the back so that bars cannot roll out. Light bar stock may be stored vertically in special racks. Special A-frame racks made of metal can hold a variety and quantity of pipes and bars safely, if they are loaded evenly and supported properly.

Materials such as lumber and pipe are particularly dangerous because of their tendency to roll or slide. Dropping, instead of placing, such objects on a pile frequently causes them to slide or bounce. Employees are likely to be injured if they attempt to stop rolling or sliding objects with their hands or feet.

Sheet Metal

Racks, similar to those used for bar stock, may be provided for plate and sheet stock, except that rollers are not always applicable. Oiled sheets require additional caution in handling.

Because sheet metal usually has sharp edges, workers should use hand leathers, leather gloves, or gloves with metal inserts, when handling it. Large quantities of sheet metal should be handled in bundles by power equipment. These bundles should be separated by strips of wood to make handling easier when the material is needed for production. Using strips of wood also lessens the chance of the material shifting or sliding in the piles.

Tin-plate strip stock is heavy and razor sharp. Should a load or partial load fall, it could badly injure anyone in the way. Observe these two measures to prevent spillage and injuries: (1) band the stock after shearing, and (2) use wooden or metal stakes around the stock tables and pallets that hold the loads. The supervisor and all who handle the bundles are responsible for making sure the load is banded properly and that the stakes are in place when the load is on the table. (Special cutting tools are required to remove bands.)

Burlap Sacking

When burlap sacking is stored in high stacks, heat is generated by the weight. This sets the stage for spontaneous combustion. One way to reduce this hazard is to cut the size of the stack by breaking it up into smaller stacks. This can be done either by making smaller stacks (which would increase the number of stacks and take more space) or by placing blocks at intervals in the stack, so that, in effect, there would be a number of small piles, one atop the other. Provide additional protection by constructing the storage room of fire-resistant materials and by having sprinklers and dust-tight lights.

Straw, excelsior, and other packing materials are usually received baled and should preferably be stored that way. Store these materials either in a separate building or in a fire-resistant room provided with sprinklers and dust-proof electric equipment. Because these materials are a fire hazard, only the amount necessary for immediate use should be taken into the packing room. Store this amount in bins made entirely of metal or made of wood lined with metal. Provide these bins with covers, and keep them closed. Large bins may have several compartments with counterweighted covers. The counterweight ropes should have fusible links to ensure automatic closing of covers in case of fire. Counterweights should be boxed in to prevent injury if the ropes break. Post NO SMOKING signs.

HAZARDOUS MATERIALS

Storage and handling of specific hazardous materials are discussed in other chapters: gases, in Chapter 21, Welding and Cutting; flammable liquids (including refrigerants) and tank car and tank truck loading and unloading, in Chapter 12, Flammable and Combustible Liquids; and NFPA hazard symbols, in Chapter 11, Fire Protection. Advise local fire departments and emergency planning committees when storing hazardous materials. Comply with EPA requirements for hazardous material and waste storage.

Containers for Liquids

This chapter deals with handling containers in which hazardous materials are stored. Drums, tanks, piping, portable containers, and tank cars are sometimes used to store hazardous liquid materials. The following sections describe safety measures for handling these materials. All storage containers need identification of the contents. The label or other marking needs to be clear and clean.

Drums

Store filled drums containing volatile liquids in a protected area out of the sun. Heat from any source causes liquids to expand. The resulting buildup of pressure could cause leaks with subsequent fire or explosion. It is recommended that an approved drum vent be installed in the bung opening as soon as a drum is received (Figure 14–7).

Reuse of drums causes problems if drums are not thoroughly cleaned beforehand. Carefully purge drums and clean them out with water, steam, and appropriate solution. Consult a chemical expert to set up safe cleaning procedures. Often the top of a drum is removed so the drum can be used as a receptacle. Never burn the top of a drum out with a torch because some liquid or vapor left behind in the drum could cause an explosion. It is much safer to follow the drum-purging methods described in Chapter 21, Welding and Cutting. Then remove the top of the drum with a mechanical opener. More details about handling and storing drums are given in the section on portable containers below. Following proper cleaning, some metal and plastic drums can be recycled by drum-recycling vendors.

Figure 14–7. This simple device automatically relieves drum pressure and can be lifted by hand to relieve vacuum. (Courtesy Justrite Manufacturing Co.)

Tanks

Design the structure of a new building to support the weight of storage tanks. However, have a structural engineer conduct an inspection before a tank is installed in an old building. Storage tanks for hazardous liquids are preferably installed outdoors. Be aware of EPA underground storage tank requirements.

There are many advantages in installing outdoor storage tanks underground. However, the danger of leaks going undetected in underground tanks containing corrosive or toxic materials probably outweighs the advantage of freedom from drips and sprays. When an outdoor tank is located in a pit, the pit should be large enough to permit easy access to all parts of the tank. Provide a permanent ladder and an access door that can be fastened shut. A confined-space entry permit may be required to enter pits and tanks.

Confined spaces would include pits and tanks (Permit-Required Confined Spaces, 29 *CFR* 1910.16). The rules for entering a confined space involve training of the employee, permits, and rescue procedures. (See also Confined Spaces in Chapter 2, Buildings and Facility Layout, and Precautions for Entering Boilers and Furnaces in Chapter 5, Boilers and Unfired Pressure Vessels.)

Install process tanks that will contain volatile or corrosive liquids only at or above grade and in areas having adequate drainage. Separate process tanks from the processing area with construction materials having a fire-resistant rating of at least two hours.

Install tanks where traffic cannot pass under them. If people must walk beneath them, install drip pans and provide drainage of the pans to a safe disposal or recovery location.

Provide tanks with permanent stairs, or ladders, and walkways that have standard guardrails and toeboards. Empty, clean, and inspect tanks for structural weaknesses at regular intervals. All storage tanks should be adequately labeled or placarded with contents and hazard warnings (NFPA 704, Identification of Fire Hazards of Materials). Keep records of each inspection. A confined-space entry permit may be required.

Tanks for holding volatile materials should be bonded and grounded, and provided with emergency venting devices. Venting should follow the provisions of NFPA 30, *Flammable and Combustible Liquids Code.* If tanks are inside buildings, vents should discharge outside the building at a location both free from any ignition source and not in contact with personnel (NFPA 704, Identification of Fire Hazards of Materials). Be sure to consider the effects of corrosion on venting devices for tanks that will contain corrosive liquids.

To minimize liquid loss and the possibility of injury from a broken fitting, connections for filling and emptying tanks are preferably made through the top. Plainly review labeled fill lines and equip them with a drain. According to NFPA 30, paragraph 5–2.4.5, use of compressed air is not permitted for the transfer of flammable and combustible liquids. To prevent priming of pumps that contain dangerous liquids and that have filling connections at the top of the tank, use self-priming pumps or pumps that generate enough suction to lift the liquid from the bottom of the tank.

Cleaning tanks can be an exceedingly dangerous operation and requires a confined-space entry permit. Establish, preferably in written form, an exact and specific cleaning procedure to be strictly followed. Specifications for tank-cleaning procedure are set forth in NFPA 327, *Cleaning and Safeguarding Small Tanks and Containers,* and the American Petroleum Institute's (API) publication Std 2015, *Safe Entry and Cleaning of Petroleum Storage Tanks.*

The procedure should be modified for toxic compounds only to the extent that more complete cleaning and personal protective equipment may be required. Many chemicals can easily permeate the material of protective clothing and be absorbed through the skin. Liquid aromatic nitro compounds and amines are solvents for rubber and are absorbed through rubber gradually, as well as through the skin. Phenolic compounds, such as carbolic acid, cresylic acid, and the cresols, are also absorbed through the skin. For such exposures, use personal protective equipment made of one of the inert synthetic rubbers or plastics. (This equipment includes gloves, aprons, boots, and respiratory- and eye-protective equipment.)

Pipelines

Install pipelines in trenches or tunnels that carry chemicals. If they must be installed overhead, isolate them so they will not drip on anyone working underneath. Pipelines for carrying flammables, however, should not be installed in tunnels. Identify all pipelines as to their content.

The following are three major sources of chemical injury in pipeline work:

- **Failure of Packing in Valve Stems or of Gaskets in Bolted Flanges.** To minimize injuries from valve packing failure, surround the valve stem with a sheet metal box or hood that will deflect spray away from the person operating the valve. So far as possible, renew packing without pressure on the valve.
- **Failure to Check That Valves Are Closed and Locked and the Lines Drained Before Tension Is Released on Flange Bolts.** The opening between the flange faces may be temporarily covered with a piece of sheet lead while the flange bolts are being loosened and the faces separated. Loosen the bolts farthest away first so that drainage will tend to go away from the worker. Insert blinds in the flanges as soon as they are opened. For lines that are opened often, use blinds permanently pivoted on a flange bolt, with one end acting as a gasket and the other as a blind.
- **Opening the Wrong Valve.** To prevent injuries and accidents from this source, identify pipelines and valves with tags, lettered markings, or distinctive colors. Distinctive colors for identifying piping have been standardized in ANSI A13.1, *Scheme for the Identification of Piping Systems,* and are described in Chapter 2, Buildings and Facility Layout. Provide specific identification of piping with a lettered legend that (1) names the material being piped, (2) summarizes the hazards involved, and (3) gives directions for safe use. Legends should be moisture-resistant and contain pigments that are colorfast. Stencils or decals may be used to apply legends. Also have valves well separated and the immediate area well lit to assure quick and easy identification.

At the conclusion of a job on a pipeline containing corrosive chemicals, wash tools and personal protective equipment thoroughly with a reagent to neutralize or remove the corrosive material. Then rinse the tools and equipment in clean water.

Portable Containers

Where raw material liquid chemicals are used in quantity, it is generally better to install pipelines and outside storage tanks than to use portable containers. Spillage is reduced and localized, thereby making it easier to handle.

Properly store portable containers, such as drums, barrels, and carboys. Keep only a minimum amount of liquid at the point of operation—only enough for use on one shift is a common rule. Store the main stock in a safe, isolated place. If the liquid is corrosive or highly toxic, isolate the storage area from the rest of the facility by impervious walls and flooring. In the storage area, post a provisional plan for safe cleanup of spillage and safe disposal of contaminated materials. Otherwise, use a separate building.

Where corrosive liquids are stored, floors should be made of cinders, or of concrete treated to decrease its solubility, or of other corrosion-resistant material. Concrete flooring is also satisfactory for storing flammable liquids. Be sure grounding is adequate if flammable liquids are involved. (See Chapter 12, Flammable and Combustible Liquids.) Allow for good floor drainage in case a container in the storage area leaks or breaks. Governmental pollution-control regulations usually prohibit draining these kinds of spills into sanitary or storm sewer systems.

The storage area must be well ventilated. Use natural ventilation whenever possible. Mechanical ventilation has no moving parts and fewer problems, but it may not always be adequate for the removal of vapors and fumes. Measurements should be made periodically to prevent exceeding maximum contaminant levels in the room.

Stack full drums in racks, preferably with a separate rack for each material. Arrange these racks to permit easy access for moving the drums in and out, as well as for ready inspection of stock (Figure 14–8).

Barrels may be stacked vertically with dunnage between the tiers. However, for more ease in handling barrels, keep them in racks similar to those used for drums. The safest way to handle drums is to use mechanically powered lift equipment with drum-lifting clamps. Transporting drums on pallets is a common practice in many companies. Workers handling these drums should use nonsparking bung nut wrenches to tighten the bung nuts and to prevent leaks during storage and transportation.

Different materials should be stored in separate designated areas, divided by wide aisles. Stack boxed carboys no higher than three. Do not use more than two tiers for carboys containing strong oxidizing agents,

Figure 14–8. Special racks facilitate safe multiple tiering of drums. The rounded corners minimize the hazard of punctures. Bungs must always be tightened to prevent leaks.

such as concentrated nitric acid or concentrated hydrogen peroxide. Before handling acid-filled carboys, inspect their nails for corrosion and the wood for weakening caused by acid. Before piling empty carboys, drain them thoroughly and replace the stoppers.

Special equipment for handling carboys should include a long-handled truck that picks up the boxed carboys under the handling cleats or between the bottom cleats provided on all standard 12- and 13-gallon (45- and 49-liter) boxed carboys. These trucks have handles long enough to keep persons who handle them away from splashes, in case carboys are dropped.

There is less danger, however, of dropping a carboy with a long-handled truck than with a standard two-wheeled truck. A long-handled truck becomes a much safer device for handling drums and barrels if it has a bed curved to fit the drum and a hook to catch the chime.

The safest way to empty a carboy is to suction the liquid with a vacuum pump or aspirator or to start a siphon with a rubber bulb or ejector. A satisfactory method of emptying a carboy is to use a carboy incliner that holds the carboy firmly by the top, as well as by the sides, and automatically returns to the neutral position on being released.

Follow prescribed safety measures when handling carboys. Do not use compressed air, even from a hand pump, on a carboy unless the carboy is enclosed by another container so that the pressures inside and outside remain the same (Figure 14–9). Never permit (1) *pouring by hand* or (2) *starting pipettes or siphons by mouth suction.* Use mechanical pumps only if they are self-priming or have sufficient suction force to start themselves.

A corrosive or poisonous liquid requires a specially identified container. The simplest is a glass or plastic jug or jar, with a good closure, placed in a metal can.

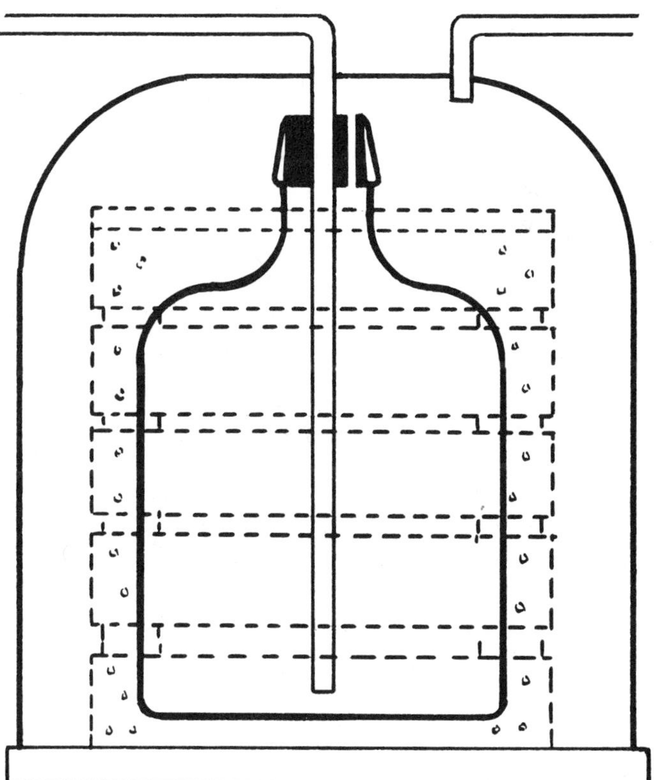

Figure 14–9. Air pressure admitted through the short pipe forces liquid from the carboy through the long pipe. Pressure is exerted on the bell cover, not the carboy.

Improvise, if necessary, a container resistant to shock by placing a jug in a metal pail and filling the space between the pail and the jug with pitch or foamed plastic. A container may be made to fit the jug with only a thin layer of padding, such as a layer of gasket rubber. Containers are also available commercially.

Keep highly toxic substances, such as cyanides and soluble oxalates, in containers of a distinctive shape if they must be handled manually. All containers must be tagged or labeled, identifying the contents and specifying the appropriate hazard warning as required by U.S. OSHA-CFR 1910.1200. Lock up toxic materials at all times, and allow them to be dispensed only by authorized personnel.

Where caustics or acids are stored, handled, or used, operable emergency showers or eyewash fountains must be available. Provide workers with chemical goggles, rubber aprons, boots, gloves, and other protective equipment necessary to handle a particular liquid. (See Chapter 7, Personal Protective Equipment, in this volume.)

Tank Cars

Isolate tank cars on sidings by derails and by blue stop flags or blue lights; chock the cars' wheels (Figure 14–10); and set hand brakes before the cars are loaded or unloaded. Before the car is opened, bond it to the loading line. Ground the track and the loading or unloading rack. Check all connections regularly. (See Chapter 12, Flammable and Combustible Liquids, in this volume for a detailed discussion of loading and unloading tank cars.)

Unload tank cars holding chemicals through the dome connection rather than through the bottom connection. If the contents are nonflammable, air pressure (not to exceed 25 psi [172 kPa]) may be used for unloading. Equip the connections with a safety valve and gauge so that the pressure on the tank can be determined at any time (Figure 14–11).

Before the car is opened, the cap of the unloading pipe should be gently backed off without being completely removed. Allow pressure in the tank car to escape gradually before the cap is entirely removed.

Equip the unloading dock with a walkway at the height of the tank cars' domes and with drawbridges that can be lowered to make a firm walkway directly to the cars' domes. Install standard handrails and slip-resistant surfaces. If corrosive materials are handled, provide emergency showers and eyewash stations along the walkway.

Some materials normally shipped in tank cars solidify at temperatures reached during shipment. Tank cars used for such materials are ordinarily equipped with steam lines for melting the contents. Thoroughly

Figure 14–10. Tank cars should have wheels chocked and be marked by an approved visible sign, flags, or blue lights. A sign stating that the tank car is connected is another good safety precaution. (Courtesy T & S Equipment Company.)

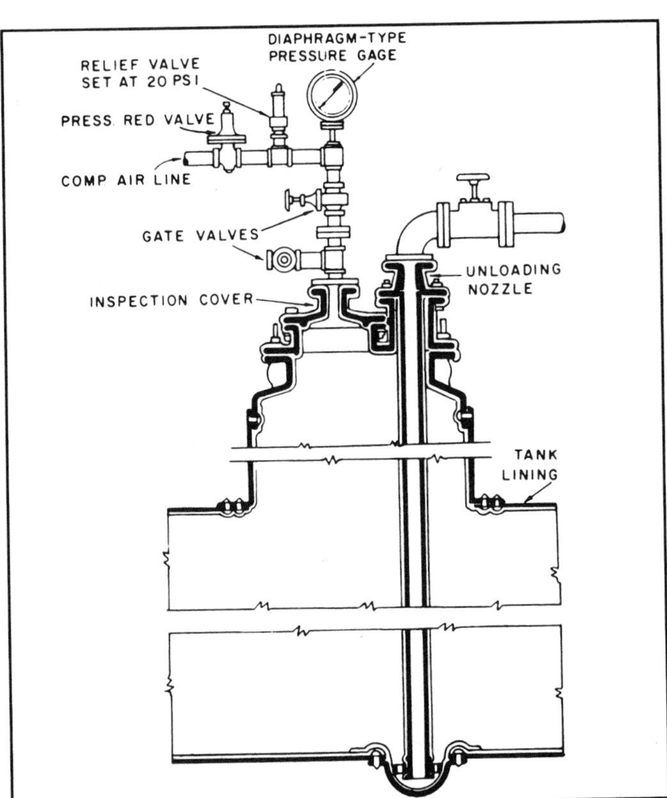

Figure 14–11. Cross section of one type of tank car designed for unloading under air pressure. Note that air pressure should not be used for discharging flammable liquids.

thaw and clear the education lines and valves before unloading is begun.

The facility's line to the unloading dock must be completely drained after unloading. To facilitate this procedure, install the line with a slope toward the storage tank so that the line drains by gravity. Should vessel entry be necessary for maintenance or cleaning, check with the supervisor about special procedures at the facility.

Containers for Solids

Bins, portable containers, and magazines are sometimes used to store hazardous solids. The following sections describe safe ways of using these containers.

Bins

When new bins are to be installed in an old structure or when new materials are to be stored in old bins, check the mechanical strength of the structure or of the bins. Solids vary in unit weight, and the more dense materials produce higher unit loads.

Sufficient slope in the cone bottom of a tank or bin is a fundamental factor to consider when designing equipment for handling bulk solids. Sufficient slope permits the solid to run freely and to prevent its arching over. Where arching takes place, devise a method to restart the flow without having a worker enter the bin from either above or below the solid material. A vibrator to shake the bottom of a small metal bin or an agitator on the bottom of the bin can start the flow. Both are standard equipment, and one or the other can be applied to bins of almost any size or shape.

Sometimes it is possible to work either from the bottom or from the top of a bin. If a person can break up the arch from the top with long tools without entering the bin, the job is reasonably safe. It is dangerous, however, when attempted from the bottom. Workers face an ever-present temptation to step inside the bin and work with a little more convenience until the material starts to flow. However, falls into open storage bins often result in injuries.

Surround bin openings at floor level or within 2 ft (61 cm) of it with standard guardrails and toeboards. If guardrails are impracticable, or if the opening is not easily accessible, like the fill opening of a high bin, cover the opening with a grating, which will not materially obstruct the opening but will prevent a worker from falling in. Many bin openings can be covered with a 2-in. (5.1-cm) mesh. Most bins can be covered by a 6-in. (15.2-cm) grating or by parallel bars on 6-in. (15.2-cm) centers.

Before entering a bin to perform any work, obtain a confined-space entry permit. This permit will outline the hazards associated with a bin, the tests required prior to entry, the names of qualified entrants and attendants, specific entry procedures, and an expiration time and date for the permit.

In most cases, tests for oxygen deficiency and the presence of flammable or toxic substances will be required. The use of approved fall protection equipment, respiratory or ventilation equipment, atmospheric monitoring equipment, and protective garments may be required. Lockout and tagout of all inundation and other associated energy sources is mandatory as well as continuous communication with a qualified attendant. Additional permits such as a hot work permit may also be required.

Empty the bin before working inside. However, if this is not possible, be alert to the possibility of bridging of the material in the bin. Work activities can cause the bridging to fail, and if this occurs, an entrant positioned on the material may be engulfed.

Where bins are filled and emptied by continuous conveyors, the control of dust is likely to be a serious problem. In filling a bin, material is generally dropped from a belt conveyor. If the material is dropped through a chute from an elevator conveyor, even more dust may be produced. Prevent the escape of dust into the rest of the facility by enclosing the bin, except for the fill opening, with a skirt of either metal or fabric and by providing an exhaust ventilation up through the filling chute. If the material is scraped from a belt conveyor, cover the conveyor at the point of discharge with an exhaust hood. Also, provide a closed chute from the discharge point to the bin.

Since the dust in these cases is released at a low speed, an inward air velocity of about 50 fpm (0.25 m/s) through all the openings is sufficient. However, there can be no seriously disturbing air currents to blow dust out of the openings, and there must be enough velocity through the rest of the system.

The same general principles apply to the discharge of bins onto conveyors. There is seldom a serious dust problem except at the loading and discharge points of the conveyors. Ventilate these points and cover them with hoods since it is seldom feasible to provide them with dust-tight enclosures or to reduce the dust by wetting it. Evaluate the air to determine if respirating protection is needed.

Combustible Solids

Where combustible materials are handled, keep the dust content of the air below the lower explosive limit. In addition to tight enclosures and dust-collection systems, good facility housekeeping will go far toward preventing disasters.

Dust explosions commonly occur as a series, not as a single shock. The first explosion uses up the dust in the air, but the shock stirs up more dust from the building. This dust is in turn set off. If the building is kept clean, this sequence cannot occur and damage will be minimal.

Exclude all sources of ignition from the area of a potentially explosive dust. Wiring, lights, and switches should comply with NFPA 61A–D, on fire and dust explosions and NFPA 70, *National Electrical Code,* for hazardous locations. Use electric motors that are totally enclosed and explosion-proof, or keep electric motors in a tight enclosure that is independently ventilated from a nonhazardous area and kept under positive pressure.

Use large and well-protected bearings. A hot bearing may ignite many types of dust. Use indirect heating systems only. Construct radiators for easy cleaning. Rigidly enforce the "NO SMOKING" rule.

Static electricity is the source of ignition in many fires. Prevent it from accumulating on most surfaces by maintaining relative humidity of 60% to 70%. If this cannot be maintained, use a ground to minimize static buildup. Also have workers wear electrostatic dispensating (ESD) protective footwear in these environments. (See ANSI Z41, *Protective Footwear.*) Remove static electricity from moving parts, such as conveyor belts and shafts, by using static collectors, grounding brushes, and conductive V-belts and leather belt dressings.

To avoid electrical shocks, ground metal bins to the conveyor frame. Check the electrical interconnection at loading and discharge points of conveyors. Use proper instrumentation to measure the effectiveness (resistance) of the grounding system. Install automatic sprinkler protection inside bins and processing equipment that contain combustible materials. Where water is undesirable, either because of reaction with or damage to the material, provide protection with inert-gas extinguishers.

If a particularly hazardous material like metal powder is being handled, completely enclose the apparatus and flood it with an inert gas like carbon dioxide, nitrogen, or helium. This process removes or reduces the oxygen content and thus prevents ignition. Explosion-vent the area, preferably with windows and skylights that swing out on friction catches.

Portable Containers

The same general rules that apply to the storage and handling of portable containers of liquids apply also to the storage and handling of portable containers of solid materials. The most popular containers for solids are 50- and 100-lb. (23- and 45-kg) paper bags. These bags are free from sifting or leaking. Handle them carefully, however, to prevent damage to machines. Have a few slipover bags available to cover the occasional broken or leaking container. This precaution both saves material and prevents skin contact with the dust.

Stack full bags on pallets or staging to prevent water damage. Interlocked stacking on pallets generally leads to better piling and fewer mechanical hazards when moving material. Protect bags from the weather—although some of the laminated bags are remarkably weather resistant.

Handle and ship large quantities of solid chemicals in bulk, cloth bags, barrels, and barrels with paper liners.

Filling bags or barrels with solids is always a potentially dusty operation. If the material being handled is finely divided or dangerous, the health and fire hazards may be serious. The simplest solution to this problem is to moisten the material so that it does not produce fine dust. You cannot use this solution, however, with some materials. Provide other methods of handling, or use hoods and exhaust ventilation.

The common way to open both cloth and paper bags is to slit one side crosswise and fold back the top and bottom. Reduce the hazard of knife cuts by keeping knives in scabbards when not in use and by using knives with hilts, guards, or returning blades.

Emptying bags and barrels involves health hazards similar to those of filling them. However, preventing skin contact with dust is somewhat harder when emptying such containers than when filling them. There is a tendency to dump them suddenly, which results in a rapid dispersion of dust that may not be trapped by the collecting systems.

Use exhaust hoods that are larger than the bags and barrels to solve this problem. In some instances, toxic materials can be handled by completely enclosing the barrel. For example, the head of the barrel is broken in after the enclosure is sealed. All dust is removed before the next barrel is put in.

Magazines

Store explosives in magazines of approved fireproof and bulletproof construction. Locate magazines at a safe distance from railroads and other buildings. Consult and follow federal, state or provincial, and local codes regarding storage of explosives. NFPA 495, *Manufacture, Transportation, Storage, and Use of Explosive Materials,* gives detailed specifications for handling and storing explosives. (See also the Institute of Makers of Explosives and E. I. du Pont de Nemours & Company in References at the end of this chapter.) Store explosives under lock and key, and maintain records of all explosives issued. Arrange storage so that the oldest explosives are used first. Advise local fire departments and emergency planning agencies when storing hazardous materials. (See also Chapter 18, Emergency Preparedness, in the *Administration & Programs* volume.)

Do not allow matches, flammable materials, metal, or metal tools in an explosives magazine. Keep floors clean and free from loose explosives. Have the floors, which are usually of wood, blind-nailed. Be sure that no nail or bolthead is exposed.

Keep magazines clean, dry, and well-ventilated. Ventilation openings should not exceed 110 in.^{2} (710 cm^2) in area and should be screened to prevent the entrance of sparks and rodents. Permit only portable lights, approved for such use, in a magazine. Do not allow fire or sparks near a magazine. Keep the surrounding ground clear of brush, leaves, grass, debris, and other flammable material. Do not expose explosives to the direct rays of the sun.

Ammonium nitrate requires special precautions including stacking limitations, air space, and ventilation. Do not permit any oils or hydrocarbons near ammonium nitrate.

Always open packages of explosives at least 50 ft (15 m) from the magazine. Use only wood wedges and wood, fiber, rawhide, zinc, babbitt metal, or rubber mallets to open cases of explosives. Never keep blasting caps or detonators of any kind in the same magazine with other explosives.

Always keep magazines of explosives and blasting supplies in a place where access to them by animals, unauthorized persons, or children is impossible. Many children have been killed or crippled because they have obtained detonators from unwatched or unguarded sources.

Containers for Gases

There are many regulations regarding the storage and handling of gas cylinders. Several government agencies and private organizations and their standards should be consulted—OSHA, National Fire Protection Association, Compressed Gas Association, and others. Care should be taken to comply with all local government codes as well.

Compressed gas cylinders should be stored in the upright position on a smooth floor and valve covers should be in place. All cylinders should be chained or otherwise fastened firmly against a wall, post, or other solid object. Different kinds of gases should either be separated by aisles or stored in separate sections of the building or storage yard. Empty cylinders must be stored apart from full cylinders.

Set up storage areas away from heavy traffic. Containers should be stored in a place where there is minimal exposure to excessive temperature, physical damage, or tampering. Never store cylinders of flammable gases near flammable or combustible substances. Before cylinders are moved, check to make certain all valves are closed. Always close the valves on empty cylinders. Never let anyone use a hammer or wrench to open valves. As stated before, cylinders should never be used as rollers to move heavy equipment. Handle all cylinders with care—a cylinder marked "empty" may not be.

To transport cylinders, use a carrier that does not allow excessive movement, sudden or violent contacts, and upsets. When a two-wheeled truck with rounded back is used, chain the cylinder upright. Never use a magnet to lift a cylinder. For short-distance moving, a cylinder may be rolled on its bottom edge, but never dragged. Cylinders should never be dropped or permitted to strike one another. Protective caps must be kept on cylinder valves when cylinders are not being used. When in doubt about how to handle a compressed gas cylinder, or how to control a particular type of gas once it is released, ask the safety professional or the safety department for advice.

STORAGE AND HANDLING OF CRYOGENIC LIQUIDS

Most gases used in facilities are also available as cryogenic liquids. Among the most common are oxygen, nitrogen, argon, helium, and hydrogen.

Liquid oxygen is frequently delivered to a facility—and even to a construction site—and then vaporized for use in flame cutting, welding, metalizing, or heating. Other uses include oxygen injection into a foundry cupola and oxygen-based processes such as paper-pulp bleaching and steelmaking.

Liquid nitrogen is also very common. A variety of processes have been developed that use the liquid primarily because of its high refrigeration values. Examples include freezing food, stripping scrap rubber from tires and cables, and removing parting lines and risers from plastic injection-molded parts. It is even used as a super-cold quencher for high-alloy steels to transform retained austenite.

The availability of large volumes of liquid helium has made possible the rapid development of superconductivity. And these examples are only a few from only some of the major industrial gases. [Most recently, cryogenic liquids are being used in conjunction with specialized ceramics in superconductor technology.]

The key to expanding use of cryogenic liquids is economics. The cost of delivering and storing the liquid is often lower than buying the gas in compressed-gas cylinders. At room temperature (70 F or 20 C) and atmospheric pressure, nitrogen occupies 700 times as much space as the same amount of nitrogen in liquid form. The reduction in cost for containers, demurrage, shipping, and storage is enormous. However, handling liquefied gases that are stored and used at very low temperatures requires some special knowledge and special precautions. To use these gases safely, the facility engineer and employees must know the specific properties of each liquified gas and its compatibility with other materials and must follow some common-sense procedures.

Characteristics of Cryogenic Liquids

A cryogenic liquid has a normal boiling point below –238 F (–150 C). (*Cryogenic* is defined in depth in the National Bureau of Standards' (NBS) Handbook 44. See References at the end of this chapter.) The most commonly used industrial gases that are transported, handled, and stored at cryogenic temperatures are oxygen, nitrogen, argon, hydrogen, and helium. Three rare atmospheric gases—neon, krypton, and xenon—are used in the liquid state. Natural gas, liquefied natural gas (LNG) or liquid methane, and carbon monoxide also are handled as cryogenic liquids, although they are not usually classified as industrial gases. Liquefied ethylene, carbon dioxide, and nitrous oxide are transported and stored as liquids, but are not classified as cryogenic.

Handling cryogenic liquids in large volumes is not new. Liquid oxygen was first shipped by tank truck in 1932, and it is common to see portable liquid containers, cryogenic trailers and trucks, and railroad tank cars hauling large quantities of liquefied gases across the country. Cryogenic tanker ships transport LNG overseas, and aircraft move other liquefied gases, especially liquid helium, from one place to another.

Many safety precautions that must be taken with compressed gases (see Chapter 21, Welding and Cutting) also apply to liquefied gases. However, some additional precautions are necessary because of the special properties exhibited by fluids at cryogenic temperatures.

Both the liquid and its boil-off vapor can rapidly freeze human tissue and can cause many common materials such as carbon steel, plastic, and rubber to become brittle or fracture under stress. Liquids in containers and piping at temperatures at or below the boiling point of liquefied air (–318 F or –194 C) can cause the surrounding air to condense to a liquid.

Extremely cold liquefied gases (helium, hydrogen, and neon) can even solidify air or other gases to which they are directly exposed. In some cases, even plugs of ice or foreign material will develop in cryogenic container vents and openings and cause the vessel to rupture. Following the supplier's operating procedures can help prevent plugging. If a plug should form, contact the supplier immediately. Do not attempt to remove the plug; move the vessel to a remote location.

All cryogenic liquids produce large volumes of gas when they vaporize. For example, 1 volume of saturated liquid nitrogen at 1 atmosphere vaporizes to 696.5 volumes of nitrogen gas at room temperature at 1 atmosphere. The volume expansion ratio of oxygen is 860.6 to 1. Liquid neon has the highest expansion ratio—1,445 to 1—of any industrial gas.

Vaporized in a sealed container, these liquids produce enormous pressures. For example, when 1 volume of liquid helium at 1 atmosphere is vaporized and warmed to room temperature in a totally enclosed container, it has the potential to generate pressure of more than 14,500 psig (100,000 kPa). Because of this high pressure, cryogenic containers usually are protected with two pressure-relief devices: a pressure-relief valve and a frangible disk.

Relief devices should function only during abnormal operation and emergencies. If they are triggered, the system should be checked for loss of insulating vacuum or for leaks. Do not tamper with the safety-valve settings. Report leaking or improperly set relief valves to the gas supplier and have them replaced or reset by qualified personnel. Similarly, all safety valves with broken seals or with any frost, ice formation, or excessive corrosion should be reported.

Most cryogenic liquids are odorless, colorless, and tasteless when vaporized to a gas. As liquids, most have no color; liquid oxygen is light blue. However, whenever the cold liquid and vapor are exposed to the atmosphere, a warning appears. As the cold boil-off gases condense moisture in the air, a fog that extends over an area larger than the vaporizing gas forms.

General Safety Practices

The properties of cryogenic liquids affect their safe handling and use (Table 14–A). The table presents data on flammability limits in air and oxygen, spontaneous ignition temperature in air at 1 atmosphere, and other information to help determine safe handling procedures. None of the gases listed is corrosive at ambient temperatures, and only carbon monoxide is toxic.

The liquids are listed by decreasing boiling point. Although xenon boils above –238 F (–150 C), it also has been included. Natural gas is not listed because it is a mixture of methane and other hydrocarbons; its boiling point depends on its composition. However, natural gas is primarily methane, and methane data are included.

General safety practices include:

- Always handle cryogenic liquids carefully. They can cause frostbite on skin and exposed eye tissue. When spilled, they tend to spread, covering a surface completely and cooling a large area. The vapors emitted by these liquids are also extremely cold and can damage delicate tissues.
- Stand clear of boiling or splashing liquid and its vapors. Boiling and splashing always occur when a warm container is charged or when warm objects

Table 14–A. Physical Properties of Cryogenic Fluids

	Xenon (Xe)	Krypton (Kr)	Methane (CH_4)	Oxygen (O_2)	Argon (Ar)	Carbon Monoxide (CO)	Nitrogen (N_2)	Neon (Ne)	Hydrogen (H_2)	Helium (He)
Boiling point, 1 atm										
F	-163	-244	-259	-297	-303	-313	-321	-411	-423	-452
C	-108	-153	-161	-183	-186	-192	-196	-246	-253	-268
Melting point, 1 atm										
F	-169	-251	-296	-362	-309	-341	-346	-416	-435	-•
C	-112	-157	-182	-219	-189	-207	-210	-249	-259	-
Density, boiling point, 1 atm lb/cu ft	191	151	26	71	87	49	50	75	4.4	7.8
Heat of vaporization, boiling point Btu/lb	41	46	219	92	70	98	85	37	193	10
Volume expansion ratio, liquid at 1 atm boiling point to gas at 60 F, 1 atm	559	693	625	861	841	-	697	1445	850	754
Flammable	No	No	Yes	No †	No	Yes	No	No	Yes	No

• Helium does not solidify at 1 atmosphere pressure.

† Oxygen does not burn, but will support combustion. However, high oxygen atmospheres substantially increase combustion rates of other materials and may form explosive mixtures with other combustibles. Flame temperatures in oxygen are higher than in air.

are inserted into a liquid. These operations should always be performed slowly to minimize boiling and splashing. If cold liquid or vapor comes in contact with the skin or eyes, first aid should be given immediately (Figure 14–12).

- Never allow any unprotected part of the body to touch uninsulated pipes or vessels that contain cryogenic fluids. The extremely cold metal will cause the flesh to stick fast to the surface and tear when withdrawn. Touching even nonmetallic materials at low temperatures is dangerous.

Tongs should be used to withdraw objects immersed in a cryogenic liquid. Objects that are soft and pliable at room temperature become hard and brittle at extremely low temperatures and will break easily.

Workers handling cryogenic liquids should use eye and hand protection to protect against splashing and cold-contact burns. Safety glasses are also recommended. If severe spraying or splashing is likely, a face shield and chemical goggles should be worn. Protective gloves should always be worn when anything that comes in contact with cold liquids and their vapors is being handled. Gloves should be loose fitting so that they can be removed quickly if liquids are spilled into them. Trousers should remain outside of boots or work shoes.

Special Precautions

Some liquefied gases require special precautions. For example, when oxygen is handled, all combustible materials, especially oil or gases, should be kept away. Smoking or open flames should never be permitted where liquid oxygen is stored or handled. NO SMOKING signs should be posted conspicuously in such areas.

Oxygen will vigorously accelerate and support combustion. Because the upper flammable limit for a flammable gas in air is higher in an oxygen-enriched air atmosphere, fire or explosion is possible over a wider range of gas mixtures.

Liquid oxygen or oxygen-rich air atmospheres should not come in contact with organic materials or flammable substances. Some organic materials—oil, grease, asphalt, kerosene, cloth, tar, or dirt containing oil or grease—react violently with oxygen and may be ignited by a hot spark. If liquid oxygen spills on asphalt or on another surface contaminated with combustibles (for example, oil-soaked concrete or gravel), no one should walk on, and no equipment should pass over, the area for at least 30 minutes after all frost or fog has disappeared.

Other special precautions against electrostatic buildup can be taken. Clothing saturated with oxygen is readily ignitable and will burn vigorously. Any clothing

Treating Cold-Contact Burns

Workers will rarely come in contact with a cryogenic liquid if proper handling procedures are used. However, in the event of contact with a liquid or cold gas, a cold-contact "burn" may occur. Actually, the skin or tissue freezes.

Medical assistance should be obtained as soon as possible. In the interim, the following emergency measures are recommended:

- Remove any clothing that may restrict circulation to the frozen area. Do not rub frozen parts, as tissue damage may result.
- As soon as is practical, immerse the affected part in warm water (not less than 105 F or more than 115 F, or 40 C to 46 C). Never use dry heat. The victim should be in a warm room, if possible.
- If the exposure has been massive and the general body temperature is depressed, the patient should be totally immersed in a warm-water bath. Supportive treatment for shock should be provided.
- Frozen tissues are painless and appear waxy and yellow. They will become swollen and painful and prone to infection when thawed. Do not rewarm rapidly. Thawing may require 15 to 60 minutes. For white people, thawing should continue until the pale blue tint of the skin turns pink or red. For black people, assess frostbite by the swelling and blistering of the skin. Reduction of swelling indicates alleviation of frostbite. Morphine or tranquilizers may be required to control the pain during thawing and should be administered under professional medical supervision.
- If the frozen part of the body thaws before the doctor arrives, cover the area with dry, sterile dressings and a large, bulky protective covering.
- Alcoholic beverages and smoking decrease blood flow to the frozen tissues and should be prohibited. Warm drinks and food may be administered.
- As with any injury or illness, monitor vital signs.

Figure 14–12. Emergency treatment for contact with a cryogenic liquid or gas.

that has been splashed or soaked with liquid oxygen, or exposed to a high gaseous-oxygen atmosphere, should be changed immediately. The contaminated systems should be aired for at least an hour until they are completely free of excess oxygen. Workers exposed to high-oxygen atmospheres should leave the area and avoid all sources of ignition until the clothing and the exposed area have been completely ventilated. Static dissipative (SD) protective footwear should be worn.

Finally, oxygen valves should be operated slowly. Abruptly starting and stopping oxygen flow may ignite contaminants in the system.

Inert Gas Precautions

The primary hazards of inert gas systems are rupture of containers, pipelines, or systems, and asphyxiation. A cryogen cannot be indefinitely maintained as a liquid even in a well-insulated container. Any liquid, or even cold vapor trapped between valves, has the potential for causing enough pressure buildup to cause violent rupture of the container or piping. The use of reliable pressure-relief devices is mandatory.

Loss of vacuum in vacuum-jacketed tanks will increase evaporation in the system, causing the relief devices to function and vent the product. Route the vented gases to a safe outdoor location. If the gases are not vented outdoors, maintain adequate ventilation; use instruments to monitor the area.

Flammable Gas Precautions

Do not permit smoking or open flames where flammable fluids are stored or handled. Clothes that minimize ignition sources should be worn in atmospheres that may contain concentrations of flammable gases.

Properly ground all major stationary equipment. Provide ground connections between stationary and mobile equipment before any flammable gas is loaded or unloaded. All electrical equipment used in or near flammable-gas loading and unloading areas, or in atmospheres that might contain explosive mixtures, should conform to NFPA 50B, *Liquefied Hydrogen Systems at Consumer Sites,* and NFPA 59A, *Production, Storage, and Handling of Liquefied Natural Gas* (LNG), or to Article 500 of the *National Electrical Code* NFPA 70. When flammable cryogenic liquids and gases are handled inside, adequate positive mechanical ventilation is necessary. Electrical equipment and wiring must conform to Article 501 of the NEC. (See Chapter 10, Electrical Safety, in this volume.)

Pipe flash-off gas from closed liquid-hydrogen containers, used or stored inside, through a laboratory hood to the outside, or vent it by other means to a safe location. If hydrogen is vented into ductwork, the ventilation system should be independent of other systems, and sources of ignition must be eliminated at the exit.

Asphyxiation

All gases except oxygen have the potential to cause asphyxiation by displacing breathable air in an enclosed workplace. The presence of these gases can only be detected by using air-monitoring instruments. Asphyxiation can be sudden or may occur slowly with workers not being aware that they are in trouble. Use and store these gases in well-ventilated areas. (See SB–2, *Oxygen-Deficient Atmospheres,* the Compressed Gas Association, in the References at the end of this chapter.)

Unless large quantities of inert gas are present, the problem is easily prevented by using proper ventilation at all times. Vent nitrogen outside to safe areas. Install analyzers with alarms to alert workers to oxygen-deficient atmospheres. Constant monitoring, sniffers, and other precautions should be used to survey the atmosphere when personnel enter enclosed areas or vessels. When it is necessary to enter an area where the oxygen content may be below 19.5%, have workers wear self-contained breathing apparatus (SCBA) or combination supplied-air respirator with emergency escape SCBA. A conventional gas mask or other air-purifying respirator will not prevent asphyxiation.

Most personnel working in or around oxygen-deficient atmospheres rely on the buddy system for protection. But, unless equipped with a portable air supply, a co-worker may also be asphyxiated when entering the area to rescue an unconscious partner. The best protection is to provide both workers with a portable supply of breathable air. Lifelines are acceptable only if the area is free of obstructions and one worker is able to lift the other rapidly and easily.

Training

The best single investment in safety is trained personnel. Some workers will need detailed training in a particular type of equipment, cryogen, or repair operation. Others will require broader training in safe handling practices for a variety of cryogenic liquids. The following subjects should be familiar to everyone involved in using, handling, storing, or transferring cryogens:

- nature and properties of the cryogen in both its liquid and gaseous states
- operation of the equipment
- approved, compatible materials
- use and care of protective equipment and clothing
- first aid and self-aid techniques to employ when medical treatment is not immediately available
- handling emergency situations such as fires, leaks, spills.

Good Housekeeping Practices

Good housekeeping is essential to safety when handling cryogens. Few cryogens are spontaneously hazardous, but each liquefied gas poses its own hazard.

Liquid oxygen may form mixtures that are shock sensitive with fuels, oils, or grease. Porous solids, such as asphalt or wood, can become saturated with oxygen and also become shock-sensitive. Ignition is more likely with weaker sparks and lower temperatures than would be required in air.

Flammable gases such as hydrogen and methane are lighter than air. At normal temperatures, they will rise. But at the first temperatures that exist just after these gases evaporate from the liquid state, the saturated vapor is heavier than air and tends to fall. Wind or forced ventilation will affect the direction of the released gases and must be considered during disposal of any leaking fluid.

The location and maintenance of safety and fire-fighting equipment are also important. Inform outside personnel also about all necessary safeguards before they enter a potentially hazardous area. In general, following good housekeeping rules and demanding that workers observe safety rules will minimize accidents.

Safe Storage and Handling Recommendations

Cryogenic liquids are stored and transported in a wide range of containers from small Dewar flasks to railroad tank cars and tank trucks (Figures 14–13 and 14–14). Only use equipment and containers designed for the intended product, service pressure, and temperature. If any questions arise about correct handling or transporting procedures, or about the compatibility of materials with a given cryogen, consult the gas supplier.

Cryogenic liquids ordinarily should not be handled in open containers unless they are specifically designed for that purpose and for the product. Cryogenic containers should be clean and made from materials suitable for cryogenic temperatures—such as austenitic stainless steels, copper, and certain aluminum alloys. Cryogens should be transferred slowly into warm lines or containers to prevent thermal shock to the piping and container, and to eliminate possible excessive pressure buildup in the system. When liquids are transferred from one container to another, the receiving container should be cooled gradually to prevent shock and reduce flashing. High concentrations of escaping gases should be vented so that they do not collect in an enclosed area. Be sure that workers observe the following rules when handling cryogens:

- Do not drop warm solids or liquids into cryogenic liquids. Violent boiling will result and liquid can splash onto personnel and equipment.
- Avoid breathing vapor from any cryogenic liquid source except for liquid-oxygen equipment designed to supply warm breathable oxygen. When cryogenic

Figure 14–13. Cryogenic liquids are stored and transported in a wide range of containers. A typical example is this flask of liquid nitrogen, an open-mouthed, unpressured, vacuum-jacketed vessel. This flask is free-standing for photographing only. Note appropriately secured flasks behind displayed flask.

liquids are being discharged from drain valves or blowdown lines, open the valves slowly to prevent splashing. Smoking should never be permitted.

Two types of portable liquid-storage vessels are generally used to hold and dispense cryogenic liquids—nonpressurized Dewar containers and pressurized liquid cylinders (Figure 14–13). Dewar containers are open-mouthed, nonpressurized, vacuum-jacketed vessels usually used to hold liquid argon, nitrogen, oxygen, or helium. Some of these containers are designed for lightweight liquids, such as helium, and for maximum holding times. Their internal support system cannot hold some of the heavier cryogens, such as argon. When using Dewar containers, be sure that no ice accumulates in the neck or on the cover. This could cause a blockage and subsequent pressure buildup. Laboratory Dewar flasks with wide-mouthed openings have no cover to protect the liquid. Most are made of metal, but some smaller units are of glass.

Liquid cylinders are pressurized containers, usually vertical vessels, designed and fabricated according to United States Department of Transportation (DOT) specifications. There are three major types of liquid cylinders: for dispensing liquid or gas, for withdrawing gas only, and for withdrawing liquid only. Each type of liquid cylinder has appropriate valves for filling and dispensing and is adequately protected with a pressure-control valve and a frangible disk.

An unusually cold outside jacket on a cryogenic vessel indicates some loss of insulating vacuum. Frost spots may appear. A vessel in this condition should be drained, removed from service, and set aside for repair. Such repairs should be handled by the manufacturer or other qualified company.

Some liquid cylinders can be handled manually, but moving them by portable handcarts is strongly recommended. Secure the cylinder to the handcart with a strap to prevent the cylinder from slipping off the cart. Vessels containing cryogens must be handled very carefully. They should not be dropped or tipped on their sides.

Transfer Lines

Many types of filling or transfer lines are used to handle the flow of cryogenic fluids from one point to another. Some of these transfer lines are small, uninsulated copper or stainless steel lines; large-diameter rigid lines; flexible hose systems; vacuum-jacketed lines; and other insulated systems.

Cryogenic liquids can be transferred by two methods. The simpler of these is gravity. In this case, the height of the stored liquid serves as the transfer medium. The other method—pressurized transfer—uses the vapor pressure of the product, or pressure from an external source, to move the liquid to the lower-pressure receiving container.

Various types of cryogenic pumps are also available. Flow rates may vary from less than one to several hundred gallons (liters) per minute. The product should be in liquid form in the transfer lines. Any vaporization of liquid within the system may cause (1) excessive pressure drop, (2) two-phase (liquid and gas)

Figure 14–14. Large volumes of cryogenic liquids can be handled easily and safely. Here, more than 50,000 gallons of liquid helium at –452 F are lifted by container carrier for loading aboard ship. Loss of helium from vaporization during a two-week voyage is so small that it is nearly undetectable. The liquid helium is surrounded by a liquid nitrogen chamber and contained in a specially insulated outer jacket.

flow, and (3) cavitation that can harm the operation of cryogenic pumps.

Short transfer lines used for intermittent service are normally not insulated. Lines used for continuous transfer of cryogens, however, are usually insulated. Be sure that all liquid transfer hoses have dust caps.

Because liquid hydrogen and liquid helium have extremely cold temperatures and low heats of vaporization, vacuum-jacketed lines are required to transfer them. To reduce costly line and flash-off losses, vacuum-jacketed lines are also occasionally used for in-facility transfer of atmospheric cryogenic fluids.

SHIPPING AND RECEIVING

In the United States the supervisor of the shipping and receiving room must be aware of DOT regulations and labels. Bills of lading or shipping forms must identify the item. Items not properly labeled should not be accepted.

Floors, Ramps, and Aisles

Floors in warehouses, storerooms, and shipping rooms must be level. Unevenness of floors may lead to the toppling of piles of stored materials. Conspicuously post safe floor-load capacities and maximum heights to which specific materials may be piled. Where bulk material, boxes, or cartons of the same weight are regularly stored, it is a good practice to paint a horizontal line on the wall indicating the maximum height to which the material may be piled.

Check the strength of floors before beginning to use power trucks. A structural expert should determine floor load capacity by studying architectural data of the facility, the age and condition of the floor members, the type of floor, and other pertinent factors.

Wherever materials are stored or transported, keep the surface of floors, platforms, and ramps in good condition. Repair damaged structures immediately; particularly watch the area around doorways and elevator entrances.

Ramps should have nonskid surfaces. When ramps are used for hand trucking, lay a slip-resistant foot strip in the center of the ramp, or in the center of each lane for two-way traffic. Ramps should have handrails and, where there is heavy trucking, substantial curbs. A separate pedestrian lane, divided from the truck lane by a handrail, is a good idea for ramps used by both pedestrians and trucks.

Aisles should be wide enough (1) to enable employees to move about freely while handling material or removing it from bins, racks, or piles and (2) to allow safe passage of loaded equipment. Use traffic-control devices, such as stop signs, to help control in-facility traffic. Mirrors, placed at blind intersections, help to prevent collisions. Warning signs and signals at such locations also serve as useful reminders, particularly to operators of power equipment. Similarly protect doorways and entrances to tunnels and elevators. Equip mobile equipment used in storage areas with backup warning devices. Clearly mark aisles and unloading areas with white paint or black and white stripes. To prevent falling and tripping, trucks not in use, material, and other objects should not stand in or extend into aisles.

Keep aisles leading to sprinkler valves and fire-extinguishing equipment clear. Do not pile materials closer than 18 in. (45.7 cm) to sprinkler heads. Closer spacing may reduce the effectiveness of the heads in the event of fire. For overly large, closely packed piles of combustible cases, bales, cartons, and similar stock, provide up to 36 in. (91.4 cm) of clearance. (See NFPA 13, *Installation of Sprinkler Systems.*) Request administrative help to prevent workers from piling materials against fire doors.

Lighting

General illumination of warehouses and storage rooms should follow ANSI/IES RP7, *Practice for Industrial Lighting,* published by the Illuminating Engineering Society (IES). (See the Appendix, Safety Tables, in this volume. See also the *IES Handbook.*)

Provide special lighting for operations requiring greater illumination. All lighting fixtures and wiring should meet the requirements of NFPA 70, *National Electrical Code.*

Stock Picking

To easily move full pallet loads, use powered industrial trucks. Shipments are typically made, however, by truck and/or railcar loads of mixed lots because many facilities produce a variety of small finished products.

In both operations, such stock is usually found in racks or bins. With the increase in high bay storage, special order-picking equipment is required. Such operations lend new efficiency to the movement of material, as well as new hazards in the areas of traffic, personal injury, and fire protection. The worker operating from a mobile order-picking truck is exposed to falls from a height, as well as to falling objects and materials-handling accidents.

Often, workers are required to climb a ladder to get small parts or stock. Workers should use only heavy-duty, materials-handling ladders. These may be on rollers, equipped with a braking mechanism with rubber feet that contact the surface as the worker's weight is imposed on the ladder. These ladders also have working platforms and standard guardrails that protect workers from falls as they reach for stock or parts.

Under no circumstances should employees be allowed to use ordinary stepladders (particularly the short 2- or 3-step stools). Heavy-duty, materials-handling equipment is required. Employees must never climb racks or shelves.

Dockboards

Dockboards are also known as bridge plates, dock plates, gangplanks, and bridge ramps. Design dockboards used in trailer and railcar loading and unloading to carry four times the heaviest expected load and to be wide enough to permit easy maneuvering of hand or power trucks. (Information given here applies to both hand and power-truck operations.)

Many modern facilities use automatic dock levelers and fixed-position hydraulic dockboards on both truck and rail docks. Dock shelters, usually found in inclement weather zones, effectively keep moisture from the dockboard and give shipping and receiving department personnel a safer working environment.

Dockboards require regularly scheduled maintenance. Most operate hydraulically and provide a solid working and walking surface for heavy industrial trucks, as well as for hand-truck operations. All shear points must be guarded. Paint the edges of movable sections yellow to denote a possible tripping hazard. When not in use, store dockboards in a safe, secure place provided for that purpose.

Trailers being loaded or unloaded by trucks must have wheel chocks at each wheel. The nose must have a special jack when the equipment operates in the forward portion of the trailer. Vertical restraints are available that will prevent trucks from slipping (Figure 14–15).

Design and maintain dockboards so they do not rock or slide when they are being used. Secure dockboards in position. Either anchor them down or equip them with devices to prevent slippage from the platform or the car threshold. The sides of dockboards should be turned up at right angles, or otherwise designed, to prevent trucks from running over the edge. Dockboards should also have a slip-resistant surface to prevent employees and trucks from slipping. They should be kept clean and free of oil, grease, water, ice, and snow.

Provide handholds, or similar devices, on dockboards to enable safe handling. Where practical, fit fork loops or lugs to the plates. In that way, they can be handled by forklift trucks. Another method for lifting steel dockboards uses a low-voltage magnet, which is hung from the forks of a forklift truck and powered by the truck's battery.

When dockboards are handled manually, workers should lower or slip them into place and not drop them. Assign enough workers to the job to permit safe and easy handling. Take extra care to prevent foot injuries.

Protective devices (wheel checks) should be used to prevent engines or car pullers from moving railroad cars while dockboards are in position and workers are on them. To warn train crews, use standard blue flags for daytime and blue lights for night work. Consult local railroad authorities about the specific warning devices.

Machines and Tools

Machines used in receiving and shipping, such as shears, saws, and nailing machines, should have protective guards. Workers should wear protective clothing—goggles, for instance—when operating a nailing or banding machine.

Figure 14–15a. Two types of vehicle restraints to secure a truck. (Courtesy Rite Hite Corp.)

Figure 14–15b.

Use high-quality tools and keep them in good condition. Keep tools sharp, and provide holsters for workers to carry them in. Provide files that have a handle on the tang.

When workers use a drawknife instead of a scraper to remove markings from cases, boxes, and barrels, they should never brace the work with their knees. Should the drawknife slip, injury would almost certainly occur.

Steel and Plastic Strapping

Train workers in both the application and removal of steel and plastic strapping—which may be either flat or round. In all cases, the workers should wear safety goggles and leather gloves. Steel-studded gloves may be required for heavy strapping.

Use equipment designed for applying and removing strapping. When operating a strapping tool, workers should face in the direction of the pull, one foot ahead of the other. Then, if the strap breaks or the tool slips, workers are in a position to protect themselves. When final tension is being attained, the worker should get out of the direct line of the strapping so that, in case of breakage, the ends of the strapping will not strike the face or body. Never use an extension "cheater bar" or a hand tool to gain more tension. Keep others away from strapping application and removal.

Workers should break or cut off excess strap beyond the tension-holding seal before a bound shipment is considered safe for further handling or shipment. Before attempting to move bound merchandise or material, the worker should examine it for broken bands or loose ends. Broken bands should be removed and safely disposed of and, if possible, replaced to keep the shipment from coming apart.

Workers should never handle a box, carton, or package by the steel bands, either manually or with a lift truck or other mechanical-handling device. Workers should check stored packages, boxes, cartons, or other material for loose or protruding banding ends that might cut or otherwise injure passersby.

To remove strapping from bound containers, workers should use a cutter designed for the work. Workers should never break the steel strap by applying leverage with a claw hammer, chisel, ax, crowbar, or other tool. Strapping should be cut square, never at an angle. Strapping cut at an angle has much sharper ends. Have containers available to dispose of plastic and steel bands separately.

Before cutting strapping, workers should make sure that no one is standing close enough to be hit by loose ends of strapping. To cut bands safely, the workers should place one gloved hand on the nearest portion of the strapping. Then, if the strapping springs, it will be held to one side and fly away from the worker's face. In addition, the face should be held out of the direct line of the strap, and workers should wear goggles.

Burlap and Sacking

Burlap and sacking are often received baled. Opening these bales is a job requiring some skill and experience. The supervisor of the department should thoroughly instruct employees in the exact procedure to be used. Burred ends of wire used to tie sacks may cause many cuts. The employee should hold down one end of the wire when making the cut and should stand clear of the free end.

Glass and Nails

Broken glass, often found in containers being unpacked, is a serious hazard in the shipping room. When unpacking glass or crockery, the worker should assume that broken material may be present. If possible, workers should wear gloves.

Companies operating large shipping rooms report many injuries from flying nails. When driving a nail, workers should start it with a few light taps, so that it will take a good hold and not fly. Workers should wear eye and face protection. Instruct workers to pull out nails that have been started at the wrong angle and drive them in properly. Nails should be driven flush so that no part of the nail projects above the surface. Not only fellow workers, but also employees of the customer or carrier, can be injured as they handle cases with poorly driven nails. Poorly driven nails may also cause loss of merchandise if the packages should come apart while being shipped.

Do not let workers leave loose nails on the floors. Loaded trucks passing over the nails may drive them into the floor with the points up. In this position, they are a serious hazard since they can easily be driven through workers' shoes and can also puncture pneumatic tires. The best practice is to have workers pick up nails from the floor throughout the day.

Instruct employees who open packing cases to bend nails over. If the packing case is to be used again or to make smaller boxes, remove the nails.

If nails are used directly from kegs, supervisors should make certain that the nails holding the keg head in place are pulled out. Have kegs used for storing nails at the workstation placed in an inclined rack so that the nails will feed out of the kegs.

Pitch and Glue

Pitch is a sealing material. To protect export shipments, parts are wrapped first with paper, plastic, or cheesecloth, and then often covered with burlap. Pitch or other material is used to seal the package. Hot pitch, however, can severely burn the skin—not only because it is hot but also because it is difficult to remove.

Only skilled personnel who have been instructed in the hazards of using pitch and know how to avoid them should work with pitch. These workers should wear goggles, face masks, gloves with sleeves rolled over the gauntlets, and aprons. Wherever possible, it is better to use cold mastic asphalt instead of hot pitch. Plastics and other sealants are often used instead of pitch, and are safer to the employee.

Labeling glue, which often contains silicate of soda, causes discomfort when it splashes into the eyes. When working with labeling glue, workers should wear goggles and use the glue brush carefully. Plastics and other sealants and adhesives are often used instead of pitch, and are safer to the employee.

Barrels, Kegs, and Drums

Projecting nails, jagged hoops and metal bands, ends of wire, splinters and slivers cause many barrel-handling injuries, some of which lead to infection. Before handling barrels and kegs, employees should inspect them and take precautions against these hazards.

One method of opening a barrel is to use a lather's hatchet or a crate-opening tool to remove the nails. Then, when the top hoop is removed and the second hoop loosened, the head can easily be removed intact. To loosen nails on a barrel with wood hoops, simply strike the hoop sharply with a hammer or hatchet near the point where the nail is located. The nail can then be easily pulled. This method of opening barrels not only preserves the barrel for future use but also prevents contaminating the barrel's contents.

Opening single-trip drums with a hammer and chisel is a frequent source of cuts and scratches. A commercial drum opener, however, will open these drums without hazard and leave a smooth rolled edge. Drum handling, storage, and opening are discussed above in the section on Hazardous Materials.

Boxes and Cartons

Employees who open boxes or cartons may incur wire punctures, nail punctures, or cuts from the device used for opening them. When wirebound boxes or nailed boxes are opened, employees should bend back wires and turn over or remove nails. Employees should wear goggles. The boxes and covers should be piled neatly out of passageways in designated storage areas.

When opening cartons, employees should use safety openers made of a protected sharpened blade. The employee slides the blade along the edge of the carton. These tools are useful only, of course, when the carton will not be reused. Cartons that are to be reused should be pried open with a flat steel pry bar, so that the flaps will not be damaged.

Construction of Boxes and Crates

Corners of boxes and crates receive more blows than other parts and, therefore, should be strongly constructed. The interlocking-corner crate is stronger than any other type and requires less lumber. Use it whenever possible.

Diagonal braces help make crate sections rigid. One diagonal brace in the section will give more stiffening than several parallel slats. The diagonal brace should extend from corner to corner and should be placed so that it does not project beyond the sides of the crate.

Skids constructed as an integral part of a box should be made of sound lumber, free from knots, and of sufficient size to support the box without breaking. The skids should be firmly attached to withstand being dragged across the floor.

The principal reason for boxes and crates falling apart is poor nailing. Use cement-coated nails, which hold better than uncoated nails, to avoid this problem. Correct nailing is essential to safe shipment. Particular care is required when using power nailers to build skids and crates. Safety glasses with side shells are a must, and, in some cases, hearing protection is required.

Broken or Damaged Containers of Consumables

Handlers should not attempt to sample or distribute food or other commodities from damaged or broken containers. These commodities could have become contaminated or tainted, or made otherwise unsafe, while en route. In one documented case that occurred in Colombia, South America, contents of insecticide and food bags became mixed during rough shipment. The consequent use of the contaminated flour caused many deaths. Careless handling of samples could also endanger health.

Loading Railcars

Heavy machinery shipped on skids should be braced inside the railcar to prevent shifting. Because workers may drive in lag screws with a hammer rather than use a wrench, do not use lag screws to fasten a skid to the car floor. Using a hammer damages the wood, thus reducing the holding power of the lag screws. Skids with large knots are hazardous when used on shipments of heavy machinery. When rollers are used to move the object, the skid is likely to break when the roller comes under a knot. (See Association of American Railroads, Pamphlet No. 21, in the References at the end of this chapter.)

Before railcars are opened, carefully inspect the doors. If damaged, repair runners and take special precautions. Use door openers, and have employees stand clear, in case an improperly loaded car is received. Do not use a forklift truck to open doors. The angle of the fork can lift the door off its track and risk injury to employees and damage to equipment.

Workers who are opening and closing railcar doors may catch their hands between the doors and the car's doorposts. Instruct workers never to grasp the leading end of the door. This might cause their fingers to catch between the railcar's door and the side of the car. Likewise, they should keep their hands and fingers away from the doorpost when they are closing the door.

To avoid leaving hazards for railroad employees or other workers, instruct employees to clean railcars after they have been unloaded. This also avoids contamination of or damage to future lading.

SUMMARY

- Handling of material accounts for 20% to 25% of all occupational injuries.
- To reduce the number of injuries caused by materials handling, companies should minimize the manual handling of material as much as possible.
- Fall hazards need to be recognized and solutions provided. Look for proper railings, platforms, stairs, aerial lifts, ladders, runways, attach or detach slings.
- Although physical differences make it impractical to establish safe lifting limits for all workers, some general principles can be applied.
- NIOSH has established an equation to calculate Action Limits (AL) and Maximum Permissible Lifts (MPL) for manual lifting based on vertical and horizontal lifts and the distance travelled.
- Guidelines for lifting must satisfy four criteria: epidemiological, biomechanical, physiological, and psychophysical.
- Accessories for manual lifting include hand tools (hooks, crowbars, rollers), jacks, and hand trucks.
- Temporary and permanent storage of materials should be neat and orderly to eliminate hazards and to conserve space.
- Rigid containers such as metal and box pallets, fiberboard/cardboard boxes, barrels and kegs, rolled paper and reels, and compressed gas cylinders must be stored to conserve space and to provide easy access when the material is needed.
- Hazardous liquid and combustible materials stored in containers require special handling and storage methods to ensure worker safety and health.
- The storage and handling of cryogenic liquids (oxygen, nitrogen, argon, helium, and hydrogen) require careful planning and worker training. These liquids can cause frostbite on contact with skin and (except for oxygen) displace breathable air in an enclosed workspace.
- Cryogenic liquids should be stored only in the containers designed for the particular gas. Containers should be transferred slowly into a warmer environment to prevent thermal shock to the containers and equipment.
- Supervisors of shipping and receiving areas must be aware of DOT regulations and labels.
- Workers in shipping and receiving must be trained in the proper use and handling of such common items as dockboards, machines and tools, steel and plastic strapping, burlap and sacking, glass and nails, pitch and glue, barrels, kegs, drums, and boxes and cartons.
- Common personal protective equipment used in materils handling includes safety shoes, gloves, goggles, aprons, and leggings to protect against the most common injuries to hands, feet, extremities, and eyes.

REFERENCES

American Petroleum Institute, 1220 L Street, NW, Washington, DC 20005.
Safe Entry and Cleaning of Petroleum Storage Tanks, Std. 2015.

American National Standards Institute, 11 West 42nd Street, New York, NY 10036.
Mobile and Locomotive Cranes, ANSI/ASME B30.5–1994.
Practice for Industrial Lighting, ANSI/IES RP7–1990.
Precautionary Labeling of Hazardous Industrial Chemicals, ANSI Z129.1–2000.
Safety Requirements for Fixed Ladders, ANSI A14.3–1992.
Safety Standards for Powered Industrial Trucks, ANSI/ASME B56.1–7–1992–1993.
Personal Protection—Protective Footwear, ANSI Z41–1999.
Scheme for the Identification of Piping Systems, ANSI A13.1–1981 (Erratum 1998).

Association of American Railroads, Operating Transportation Division, 50 F Street NW, Washington, DC 20001.
Minimum Load Standard for Machinery in Closed Cars, Pamphlet No. 21. 6/95.

Braver-Mann S. *A Comparison of the Dynamic Balanced One-Hand Lift and the Two-Hand Stoop Lift,* Unpublished Master's Thesis, University of Iowa, Iowa City, IA, 1985.

Chemical Manufacturers Association, 2501 M Street, NW, Washington, DC 20037.

Compressed Gas Association, Inc., 1235 Jefferson Davis Highway, Arlington, VA 22202.
Oxygen-Deficient Atmospheres, Bul. SB-2.

E. I. du Pont de Nemours & Company, Inc., Explosives Department, 1007 Market Street, Wilmington, DE.
Blaster's Handbook, 1989.

Illuminating Engineering Society of North America, 345 East 47th Street, New York, NY 10017.
IES Handbook.

Institute of Makers of Explosives, 1120 19th Street NW, Suite 310, Washington, DC 20036.
Safety in the Handling and Use of Explosives.

Kroemer KHE. *Material Handling: Loss Control Through Ergonomics,* 2nd edition. Schaumburg, IL: Alliance of American Insurers, 1983.

Lovested GE. Materials Handling Safety in Industry.*Materials Handling Handbook,* 2nd ed. New York: John Wiley & Sons, 1985.

National Fire Protection Association, 1 Batterymarch Park, Quincy, MA 02269.
Safeguarding of Tanks and Containers for Entry, Cleaning or Repair, NFPA 326, 1999.
Flammable and Combustible Liquids Code, NFPA 30, 1996
Identification of Fire Hazards of Materials for Emergency Response, NFPA 704, 1996.
Installation of Sprinkler Systems, NFPA 13, 1999.
Liquefied Hydrogen Systems at Consumer Sites, NFPA 50B, 1999.
Explosive Materials Code, NFPA 495, 1996.
National Electrical Code, ®NFPA 70, 1996.
National Fire Codes, Vol. 2, "Flammable and Combustible Liquids," 2000.
Powered Industrial Trucks, Including Type Designations, Areas of Use, Conversions, Maintenance, and Operations, NFPA 505, 1999.
Prevention of Fires and Dust Explosions in Agricultural and Food Products Facilities, NFPA 61, 1999.
Production, Storage, and Handling of Liquefied Natural Gas (LNG), NFPA 59A, 1996.
Safeguarding of Tanks and Containers for Entry, Cleaning, or Repair, NFPA 326, 1999.
Venting of Deflagrations, NFPA 68, 1998.

National Institute of Standards and Technology, Gaithersburg, MD 20899.
Specifications, Tolerances, and Regulations for Commercial Weighing and Measuring Devices, Handbook 44.

National Safety Council, 1121 Spring Lake Drive, Itasca, IL 60143.
Belt Conveyors (Equipment), 12304–0569, 1990.
Construction Material Hoists, 12304–0511, 1987.
Dock Plates and Gangplanks, 12304–0318, 1990.
Electromagnetics Used with Crane Hoists, 12304–0359, 1985.
Front-End Loaders, 12304–0589, 1990.
Fuses and Torpedoes Used in Railroad Operations, Handling and Storage of, 12304–0639, 1990.
Handling and Storage of Sheet Metal, 12304–0434, 1991.
Handling and Storage of Solid Sulfur, 12304–0612, 1991.
Handling Bottles and Glassware in Food Processing and Food Service, 12304–0355, 1990.
Handling Large-Diameter Oil Field Pipe, 12304–0463, 1985.
Handling Liquid Sulfur, 12304–0592, 1993.
Handling Materials in the Forging Industry, 12304–0551, 1992.
Load-Haul-Dump Machines in Underground Mines, 12304–0576, 1990.
Motor Trucks for Mines, Quarries, and Construction, 12304–0330, 1990.
Pendant-Operated and Radio-Controlled Cranes, 12304–0558. 1991.
Powered Hand Trucks, 12304–0317, 1991.
Recommended Loads for Wire Rope Slings, 12304–0380, 1991.
Roller Conveyors, 12304–0528, 1991.
Scrap Ballers, 12304–0611, 1989.
Steel Plates, Handling for Fabrication, 12304–0565, 1990.
Truck Mounted Power Winches, 12304–0441, 1990.

Oresick A. "Safety techniques in glass handling." *ASSE Journal* (May 1973), 22–29.

Szymanski E. "Safe storage and handling of cryogenic liquids." *Plant Engineering Magazine* (June 14, 1979).

U.S. Department of Health and Human Services, National Institute for Occupational Safety and Health (NIOSH), 4676 Columbia Parkway, Cincinnati, OH 45226.
Preemployment Strength Testing, DHEW (NIOSH) Publication No. 77–163.
Work Practices Guide for Manual Lifting, DHEW (NIOSH) Technical Report 81–122.

U.S. Department of Labor, Occupational Safety and Health Administration (OSHA), 200 Constitution Avenue, NW Washington, DC 20210.
Code of Federal Regulations, Title 29. Section 1910.1200, Hazard Communication Standard.

U.S. Department of Transportation, Office of Hazardous Materials, Washington, DC 20590.

U.S. Department of the Treasury, Internal Revenue Service, 1111 Constitution Avenue, Washington, DC 20224.
Published Ordinances: Explosives—State Laws and Local Ordinances Relevant to Title 18 U.S.C., Chapter 40, Publication 740.

REVIEW QUESTIONS

1. Materials handling accounts for ______________ of all occupational injuries.
   a. 10% to 25%
   b. 20% to 45%
   c. 40% to 50%
   d. 50% to 55%

2. What are the three main factors to consider when determining the safety of manual lifting for a particular load?
   a.
   b.
   c.

3. What advantage is gained by holding a load close to the body?

4. What is the key to safe carrying by teams?
   a. make every movement in unison
   b. grasp the load by opposite corners
   c. use a straight-back position
   d. twist the body toward the load

5. What are the three most common hand tools for materials handling and the principal hazard for each?
   a.
   b.
   c.

6. What should be avoided to prevent slippage when using a jack and how can this be achieved?

7. What is the proper method for using wood or metal jack extenders?

8. What special precaution should be taken when using a two-wheel hand-truck for moving pressurized items (e.g., gas cylinders)?

9. Why should four-wheel hand-trucks be pushed instead of pulled?

10. What are the three main hazards to be aware of when using two- or four-wheel hand-trucks?
    a.
    b.
    c.

11. What two purposes do tie pieces serve in the stacking of lumber?
    a.
    b.

12. To prevent injuries that are caused when the wrong valve is opened, piping should be clearly labeled with which three pieces of information?
    a.
    b.
    c.

13. What color flags and lights should be used to isolate railroad tank cars during unloading?
    a. red
    b. yellow
    c. blue
    d. orange

14. To prevent the accumulation of static electricity on most surfaces in areas with airborne dust, the relative humidity should be maintained at what level?
    a. 10% to 20%
    b. 30% to 50%
    c. 60% to 70%
    d. 90% to 100%

15. What is the simplest solution to the potentially dangerous production of fine dust during filling operations?

16. A cryogenic liquid has a normal boiling point below ______________.
    a. −298 F
    b. −238 F
    c. −150 F
    d. 0 F

17. What is the main precaution that should be taken in areas where flammable fluids are stored or handled?

18. Workers should use self-contained breathing apparatus in areas where the oxygen content may be below ______________.

19. What is a Dewar container?

20. How much weight should dockboards used in railcar loading be designed to carry?

15

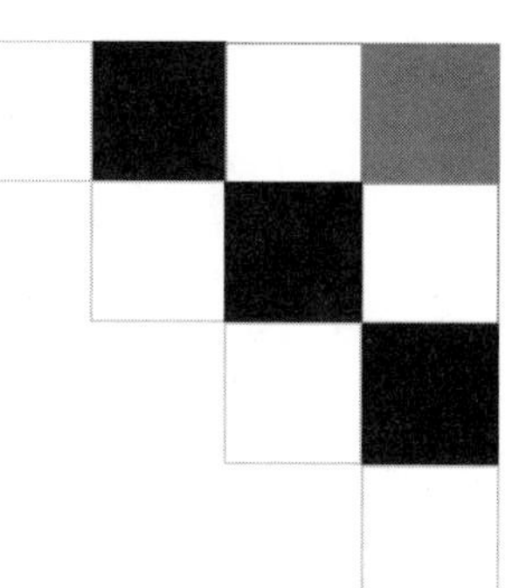

Hoisting and Conveying Equipment

Revised by

John F. Montgomery, Ph.D., CSP, CHCM, CHMM

For centuries, hoisting apparatus has been used to raise, lower, and transport heavy loads for short distances. Many thousands of hoists (electric, air, and hand-powered) are used in industry. Typically, they range from ¼ to 10 tons (226.8 kg to 9,072 kg) in capacity, but greater capacities are not unusual. Today, there are traveling cranes in steel mills, power plants, and naval shipyards able to lift hundreds of tons (907.2 kg). (See also the discussions of material and passenger hoists in Chapter 3, Construction of Facilities, and Chapter 4, Maintenance of Facilities, in this volume.) This chapter covers the following topics:

- types of hoisting apparatus and general safety rules for operating hoisting equipment
- types of cranes and general safety rules for the safe operation of each crane
- importance of proper safety training for crane operators
- types of conveyors and general safety precautions in the operation of each conveyor
- general safety codes for operating and maintaining elevators, escalators, dumbwaiters, and moving walks
- general safety precautions in the construction and operation of manlifts

HOISTING APPARATUS

The safe load capacity of each hoist shall be shown in conspicuous figures on the body of the machine. These figures should be clearly legible from the ground floor. In addition, all hoists must have a label, or labels, affixed to the hoist, hook block, or controls. These labels must explain safe operating procedures and be in a readable position.

All hoists should be securely attached to their supports (fixed member or trolley) with shackles. Hoists can be either rigid suspended or hook suspended. If the hoist is hook suspended, the support hooks should be moused or have hook latches. Latches are also recommended for load hooks. Hoist supports should be designed to bear maximum loads. Overhead hoists, operating on rails, tracks, or trolleys, should have positive stops or limiting devices on the equipment, rails, tracks, or trolleys to prevent the overrunning of safe limits. Also, the maximum load capacity of the overhead rails should be posted on the rails. Keep flanges on hoist drums with single-layer spiral grooves free of projections that could damage a rope.

Hoist operators should pick up a load only when it is directly under the hoist. Otherwise, stresses for which the hoist was not designed could be imposed upon it. If the load is not properly centered, it can swing (upon being hoisted), and injury could result. Everyone must stay out from under raised loads. When operating an overhead hoist, operators should take care to avoid injuring themselves or any other nearby workers.

CAUTION: Do not use hoists or cranes to lift, support, or otherwise transport people. The standard commercial hoist or crane does not provide a secondary means of supporting the load should the wire rope or other suspension element fail. (See the section Aerial Baskets later in this chapter.)

Examine hoists for evidence of wear, malfunction, damage, and proper operation of devices such as load hooks, ropes, brakes, clutches, and limit switches. Carefully examine deficiencies and, if determined hazardous, correct them immediately (Figure 15–1).

Electric Hoists

Rope-operated electric hoists should have nonconducting control cords unless they are grounded. Control cords should have handles of distinctly different contours. In that way, even without looking, the operator

Figure 15–1. To check the rope of this 1,000-lb (4, 450 N) electric hoist, inspect 2 to 3 ft (61 to 91 cm) increments after the rope has come to a complete stop. To check for frays, wipe a rag up and down the rope. (Courtesy Acco Hoist and Crane Division.)

will know which is the hoisting handle and which is the lowering handle.

Clearly mark each control handle "hoist" or "lower." Some companies attach an arrow to each control cord, pointing in the direction in which the load will move when the rope is pulled. Also, pass the control cords through the spreader to keep them from becoming tangled. The spreader can be a 1-in. by 3-in. (2.5-cm by 7.5-cm) board or other nonconductive material with equally spaced holes, resting upon the pull handles or lowest position. Periodically inspect control cords, usually made of fiber or light wire rope, for wear and other defects.

On the control, provide means for effecting automatic return to the OFF position. In that way, an operator must maintain a constant pull on the control rope or a push on the control button to raise or lower the load. Support the pendant station to protect the electrical conductors against strain. Also ground the station in case a ground fault occurs. Limit push-button control circuits to 150 v AC and 300 v DC.

Install a limit switch on the hoist motion. The minimum of two turns of rope shall remain on the drum when the load block is on the floor, except when a geared lower limit switch is used. One turn is then permitted. If a load block can enter a pit or hatchway in the floor, the installation of a lower travel limit switch is recommended. If that switch trips, the turns of the rope remaining on the drum can be reduced to one when the block has been stopped.

Air Hoists

After a piston air hoist has been in operation for a time, the locknut that holds the piston on its rod may become loose. When that happens, the rod will pull out of the piston, allowing the load to drop. Be sure to secure the locknut to the piston rod with a castellated nut and cotter pin. Whenever an air hoist is overhauled, check to see that the piston is well secured to the rod.

On a cylinder load balancer or hoist, do not use an ordinary hook to hang the hoist from its support. The cylinder may come unhooked if the piston rod comes in contact with an obstruction when the load is lowered. Use a clevis or other device to prevent the hook from being detached from the hoist support.

To prevent the hoist from rising or lowering too rapidly, place a choke (available from the hoist supplier) in the air-line coupling. For a rope-drum air hoist, provide a closing load-line guide.

Hand-Operated Chain Hoists

Some chain hoists are portable. Other chain hoists are either permanently hooked onto a monorail trolley or built into the trolley as an integral part. They are a good alternative for many operations that usually use a block and tackle fitted with fiber rope. Chain hoists are stronger, more dependable, and more durable than fiber-rope tackle.

There are three general types of chain hoists: spur geared, differential, and screw geared (or worm drive). The spur-geared type is the most efficient since it can pick up a load with the least effort on the part of the operator. Since the spur-geared type is free running, it tends to allow the load to run itself down. Therefore, an automatic mechanical load brake, similar to that on a crane, is provided to control the rate of descent of the load. The differential type is the least efficient. Screw-geared and differential hoists are self-locking and will automatically hold a load in position. (See Chapter 16, Ropes, Chains, and Slings, in this volume, for a description of chain hoist inspection.)

CRANES

Cranes raise, lower, and shift heavy objects through the use of a long movable arm. The hoisting apparatus of some cranes is supported on an overhead track.

Design and Construction

In the United States, the Occupational Safety and Health Administration (OSHA) requires that all new overhead and gantry cranes, constructed and installed on or after August 31, 1971, meet the design specifications of the American National Standards Institute's (ANSI) and American Society of Mechanical Engineers' (ASME) ANSI/ASME B30.2, *Overhead and Gantry Cranes (Top Running Bridge, Single or Multiple Girder, Top Running Trolley Hoist).* OSHA requirements do not cover single girder cranes (29 *CFR*, 1910.179 [6][2]). Single girder underhung cranes are covered in ANSI/ASME B30.11; single girder top-running cranes, in ANSI/ASME B30.17; overhead hoists used on single girder cranes, in ANSI/ASME B30.16. (See References at the end of this chapter for specific titles.) WARNING: Check for reverse phasing when installing a hoist or crane.

All parts of every crane (Figure 15–2), especially those subject to impact, wear, and rough usage, should be of adequate strength for its rated service. Journals and shafts should be of sufficient strength and size to bring the bearing pressure to within safe limits.

Open hooks shall not be used where there is danger of relieving the tension on the hook due to the load or to the hook catching or fouling. Each independent hoisting unit of a crane shall have brakes complying with the requirements of the ANSI B30 Series, *Safety Requirements for Cranes, Derricks, Hoists, Hooks, Jacks and Slings.*

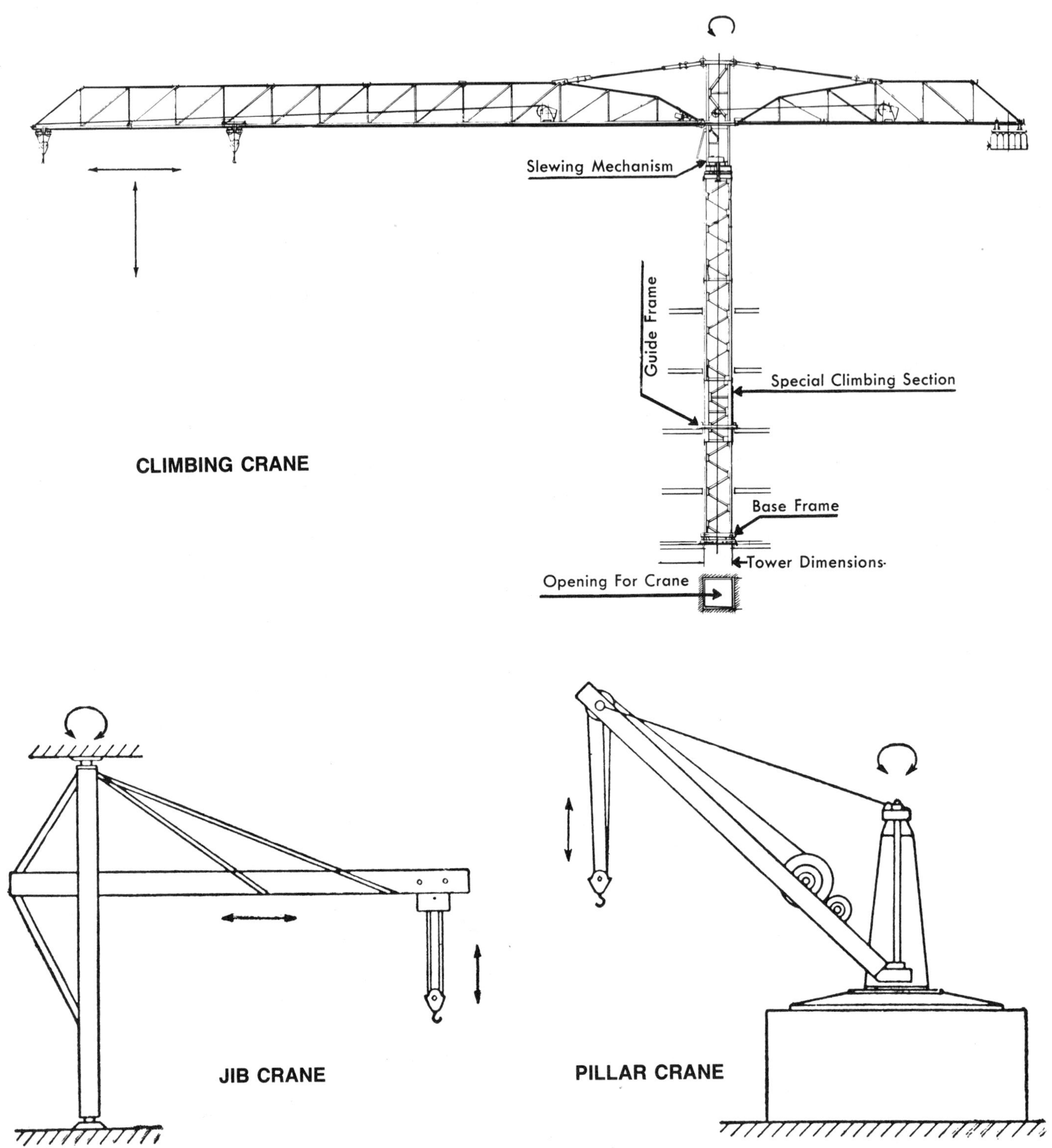

Figure 15–2a. Diagrammatic sketches of various types of cranes. (Printed with permission from ANSI Standard Series B 30, "Safety Standards for Cableways, Cranes, Derricks, Hoists, Hooks, Jacks, and Slings," except for climbing crane.)

Gantry cranes used for outdoor storage shall be provided with remotely operated rail clamps or other equivalent devices. Parking brakes are considered as minimum compliance with this rule. Apply rail clamps

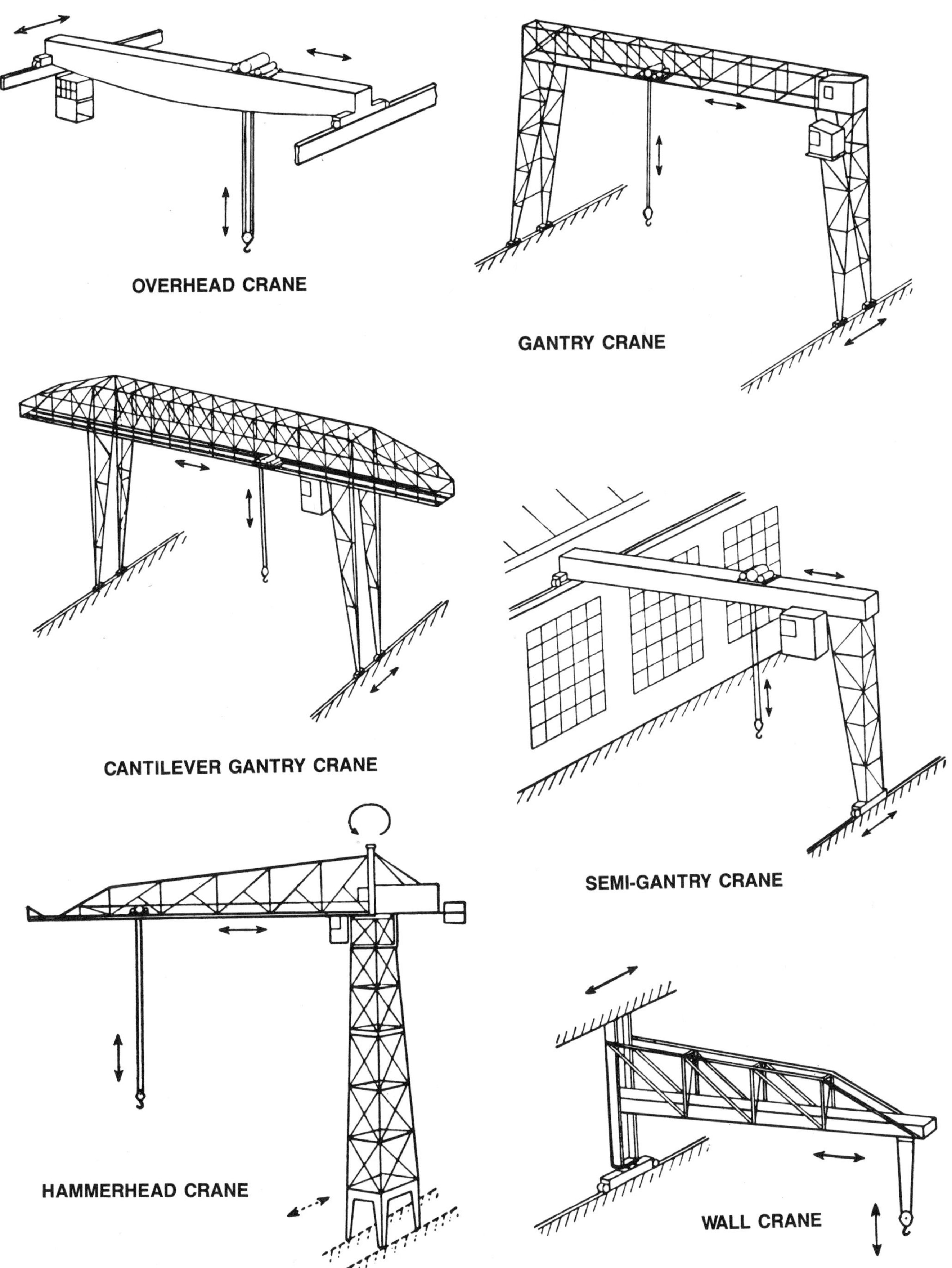

Figure 15–2b. *(Concluded.)*

only when the crane is not in motion. When rails are used as anchors, they shall be secured to withstand the resultant forces applied by the rail clamps. If the clamps act on the rail, any projection or obstruction in the clamping area must be avoided. A wind-indicating device shall be provided that will give a visible and audible alarm to the crane operator at a predetermined wind velocity.

The rated load of the crane shall be plainly marked on each side of the crane. If the crane has more than one hoisting unit, each hoist shall have its rated load marked on it or its load block. If the crane has more than one trolley unit, each trolley shall have an identification marking on it or its load block. The marking should be clearly legible from the ground or floor. The crane shall not be loaded beyond its rated capacity, except for testing.

Arrange the cab and locate control and protective equipment so that all operating handles are within convenient reach of the operator. When the operator faces either the area to be served by the load hook or the direction in which the cab is traveling, he or she should be able to reach the handles. The arrangement should allow the operator a full view of the load hook in all positions.

Mark each controller and operating lever with the action and the direction that it controls. These levers should have spring returns. In that way, if the operator releases a lever, it will automatically move into the OFF position. ANSI B30.2 requires that cranes not equipped with spring-return controllers, spring-return master switches, or momentary contact push buttons shall be provided with a device that will disconnect all motors from the line, in the event of a power failure. Such devices should not permit any motor to be restarted until the controller handle is brought to the OFF position, or a reset switch or button is operated.

Also observe the following regulations:

- All machinery, equipment, and material hoists operating on rails, tracks, or trolleys shall have positive stops or limiting devices on the equipment, rails, tracks, or trolleys to prevent overrunning the safe limits.
- All points requiring lubrication during operation shall have fittings so located or guarded as to be accessible without hazardous exposure.
- Platforms, footwalks, steps, handholds, guardrails with intermediate rails, and toeboards shall be provided on machinery and equipment to provide safe footing and accessways. Platforms and steps should be of slip-resistant material.
- Access to the crane cab and/or bridge walkway shall be provided by conveniently placed fixed ladders, stairs, or platforms whose steps leave no gap exceeding 12 in. (30.5 cm). Fixed ladders should be in conformance with ANSI A14.3, *Safety Requirements for Fixed Ladders.*
- A dry chemical or equivalent fire extinguisher shall be kept in the crane cab.

Guards and Limit Devices

If contact can be made with gears and other moving parts during normal operating conditions, they shall be totally enclosed, covered by screen guards, or placed out of reach. No overhung gears shall be used, unless means are provided to prevent their falling—should they break or work loose. The bolts in shaft couplings should be recessed so that the tops of the nuts do not project.

To prevent crushed fingers, large load hooks on cranes shall have handles. In that way, a person can hold or guide the hooks when slings are being placed on them. Also, on small cranes, the pinch points where cables pass over sheaves in the load block shall be guarded (Figure 15–3).

The hoisting motion of every crane, with the exception of boom-type cranes and derricks, must have an overtravel limit switch in the hoisting direction to stop the hoisting motion. Use lower-travel limit switches for all hoists if the load block enters pits or hatchways in the floor.

Limit devices always shall operate on a normally closed circuit and should be tested at the beginning of each shift. To make this test, the unloaded block shall be carefully run up to actuate the device. Here is a suggested testing procedure:

1. Inch block into limit switch.
2. Lower load approximately 10 ft (3 m).
3. Stop and then operate at full speed into limit switch.
4. Always conduct the test away from equipment and employees.

Hoist limit switches are operational safety devices that prevent unintended overtravel of the load block. They are not intended for constant duty. Refer questions about requirements for hoists with constant-duty limit switches to a crane or hoist manufacturer.

The hook block should be designed to lift vertically without the wire ropes of the cable twisting around each other. The hook should be of solid forged steel or built-up steel plates. Bronze and stainless steel hooks are frequently used for fire protection. Large hooks should swivel on rollers or ball bearings.

Wiring for electric equipment shall be installed according to Article 610, Cranes and Hoists, of NFPA 70, *National Electrical Code.* On electric-power-operated cranes, the power supply to the runway conductors should be controlled by a switch or circuit breaker. Locate the switch or circuit breaker on a fixed structure

Figure 15–3. This one-ton hoist is designed for light operations. Note that the push-button pendant is shaped for easy holding. (Courtesy Delmac, Inc.)

that is accessible from the floor and that can be locked in the OFF position.

Other precautions include the following:

- No guard, safety, appliance, or device shall be removed or otherwise be made ineffective in machinery or equipment, except for the purpose of making immediate repairs, lubrications, or adjustments and, then, only after the power has been turned off, except when power is necessary for making adjustments.
- All guards and devices shall be replaced immediately after completion of repairs and adjustments.
- Traveling cranes shall be equipped with a warning device that can be sounded continuously while the crane is traveling. Traveling cranes should also have a rotating or strobe light that will alert other workers in the area to the crane's movement.

Before starting maintenance or repair work on a crane, workers shall apply their personal tags and padlocks to the main power switch while it is in the OFF position. (See Glossary, *Lockout/tagout, Energy isolation,* and *Zero mechanical state [ZMS].*) When repairing cranes on a multiple-crane runway, workers should make provisions to prevent other cranes from running into the crane being repaired.

Ropes and Sheaves

Use hoisting ropes of recommended construction for crane or hoist service. The crane or rope manufacturer shall be consulted whenever a change is contemplated. The rated load divided by the number of parts of rope should not exceed 20% of the nominal breaking strength of the rope.

Inspect sheaves and drums for wear. If the grooves become enlarged from wear or corrugated from excessive rope pressure, replace them. These conditions will cause rapid wear and loss of rope strength. In cases where considerable material must be removed to regroove the drum or sheave, the strength of these parts may be impaired. In such cases, consult the hoist or crane manufacturer prior to regrooving. (See ANSI/ ASME B30.2, *Overhead and Gantry Cranes [Top Running Bridge, Single or Multiple Girder, Top Running Trolley Hoist].)*

To reduce the strain on the hoist rope where it enters the socket or anchorage on the drum, the minimum of at least two wraps shall remain on the drum when the load block is at the lowest elevation. However, one turn is permitted when a geared lower limit switch is used. Anchor the drum end of the rope to the drum by a socket arrangement approved by the crane or rope manufacturer or both. As with electric hoists, if the load block can enter a pit or hatchway in the floor, use a lower-travel limit switch. If that switch trips, reduce the turns of the rope remaining on the drum to one after the block has been stopped.

Crane and Hoist Signals

Crane movements shall always be governed by a standard of code signals that are transmitted to the crane operator by crane director (signaler). Signals may be given by any mutually understood and officially adopted method.

Hand signals, however, are preferred. A simple code of one-hand signals is appropriate for an overhead crane or bridge crane. The ANSI set of signals, adopted by many companies, is shown in Figure 15–4. (See the ANSI/ASME B30 series.) A set of one- and two-hand signals for a locomotive or crawler crane, or any other boom rig, is shown in Figure 15–5.

Signals must be discernible or audible at all times. Where visual or audible signals are inadequate, use a telephone or portable radio to communicate. A remote radio-control system for overhead cranes, which eliminates the need for hand signals, is available (Figure 15–6). The operator controls all movements of the bridge, trolley, and hoist from the plant floor. Circuits are so designed that failure of any one system component causes all crane motions to stop.

There should be only one designated person who is qualified to give crane signals to the operator. However, if signalers are changed frequently, they shall be provided with one (and only one) conspicuous armband, hat, glove, or other badge of authority. This badge of authority must be worn by the signaler currently in charge.

Operators shall not move equipment unless signals are clearly understood. The operator should move the hoisting apparatus only on signals from the proper person. A STOP signal, however, should be obeyed regardless of who gives it. Unless obedience would result in an injury, the operator should be governed absolutely by the signal. However, if an injury seems unavoidable by obeying the signal, the operator should notify the signaler at once so that corrective measures can be taken immediately.

Employees who work near cranes or assist in hooking on or arranging loads should be instructed to keep out from under loads. Supervisors shall see that this rule is strictly enforced. One manufacturing company publishes the following warning:

From a safety standpoint, one factor is paramount: Conduct all lifting operations in such a manner that, if there were an equipment failure, no personnel would be injured. This means keep out from under raised loads!

Selection and Training of Operators

Cranes, like many other pieces of equipment, present certain hazards that cannot be removed through engineering. In such instances, it is only through the exercise of intelligence, care, and good judgment that the associated risks can be reduced to an acceptable level. It is, therefore, imperative that only employees who are physically and mentally fit operate cranes. Qualified operators of hoisting equipment should meet the following minimum requirements:

- Age: The legal age for crane operations as determined by the local governing agency—generally 18 years of age.
- Language: Understand spoken and written English, as well as any other language generally used at the location.
- Physical: Pass physical examinations, including vision test for depth perception.
- Knowledge: Have basic knowledge and understanding of equipment-operating characteristics, capabilities, and limitations including: equipment-rate capacity, the effect of variables on that capacity, safety features, required operating procedures, and requirements established by local, state, and federal agencies (U.S. Department of Energy, 1988).
- Skill: Demonstrate skill in manipulating and controlling equipment through all phases of operation.

The crane operator should have a preemployment physical examination that emphasizes acuity of vision, depth and color perception, hearing, muscular coordination, and reaction time. Screening for drugs should also be done at this time. The physical examination should be required annually thereafter.

Employees selected to operate a crane must be able to learn and understand basic safety information for equipment and personnel. They must also learn the special requirements for the safe handling and use of the equipment that they will operate. Training for crane operators requires two parts: (1) an information exchange in which rules, regulations, requirements, limits, and dos and don'ts are discussed and explained, and (2) onsite, practical training in which safe operation is explained, demonstrated by the instructor, and tried by the trainee.

The initial operator-qualification process should include the following:

1. Actual hands-on training on the equipment for which the employee plans to qualify. The training should be conducted under the direction of a qualified crane operator.
2. Review of the trainee's knowledge through both written and oral examinations. The trainee should also demonstrate his or her skills for the instructor.
3. Placing a record of the training course, examination scores, and authorization permit to operate the specific hoisting equipment in the individual's training file. The record should identify the specific equipment that the individual is qualified to operate.

Some companies require both operators and local

Text continues on page 453.

HOIST. With forearm vertical, forefinger pointing up, move hand in small horizontal circle.

LOWER. With arm extended downward, forefinger pointing down, move hand in small horizontal circles.

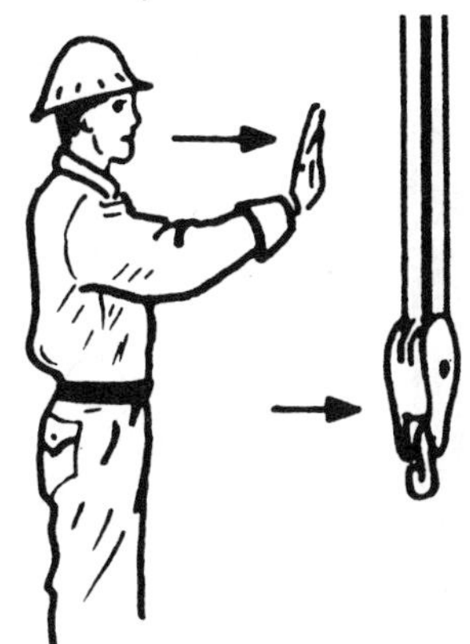

BRIDGE TRAVEL. Arm extended forward, hand open and slightly raised, make pushing motion in direction of travel.

TROLLEY TRAVEL. Palm up, fingers closed, thumb pointing in direction of motion, jerk hand horizontally.

STOP. Arm extended, palm down, move arm back and forth.

EMERGENCY STOP. Both arms extended, palms down, move arms back and forth.

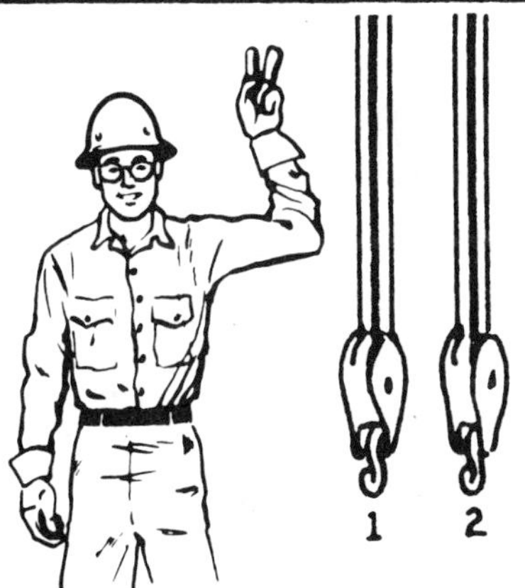

MULTIPLE TROLLEYS. Hold up one finger for block marked "1" and two fingers for block marked "2". Regular signals follow.

MOVE SLOWLY. Use one hand to give any motion signal and place other hand motionless in front of hand giving the motion signal. (Hoist slowly shown as example.)

MAGNET IS DISCONNECTED. Crane operator spreads both hands apart palms up.

Figure 15–4. Standard hand signals for controlling operation of overhead gantry cranes. (Printed with permission from ANSI Standard Series B 30, "Safety Standards for Cableways, Cranes, Derricks, Hoists, Hooks, Jacks, and Slings," The American Society of Mechanical Engineers, New York.)

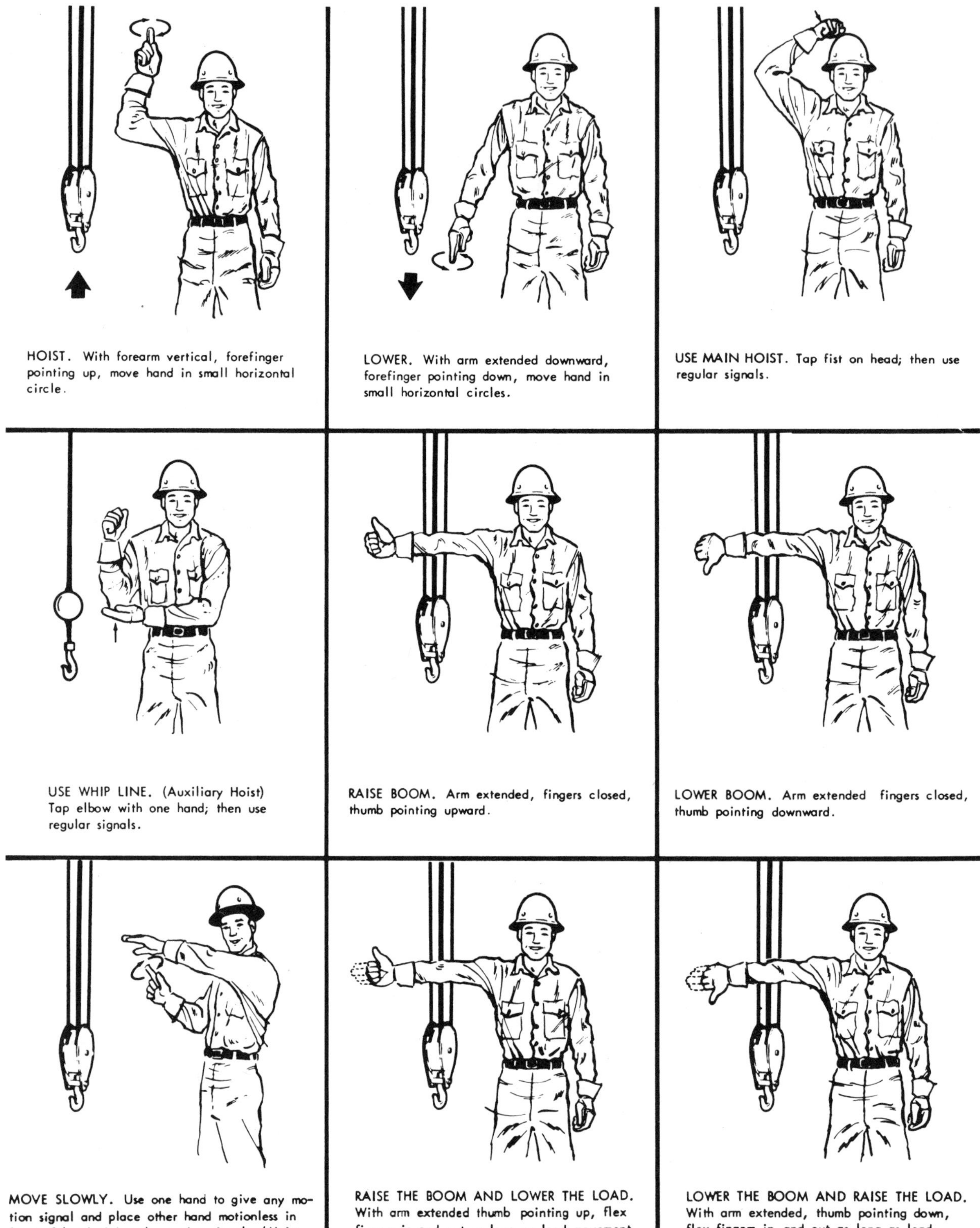

Figure 15–5. Standard hand signals suitable for crawler, locomotive, and truck boom cranes. One-hand signals for extending or retracting boom (not shown above): Extend boom—one fist in front of chest with thumb tapping chest. Retract boom—one fist in front of chest, thumb pointing outward and heel of fist tapping chest. (Printed with permission from ANSI Standard Series B30, "Safety Standards for Cableways, Cranes, Derricks, Hoists, Hooks, Jacks, and Slings," The American Society of Mechanical Engineers, New York.)

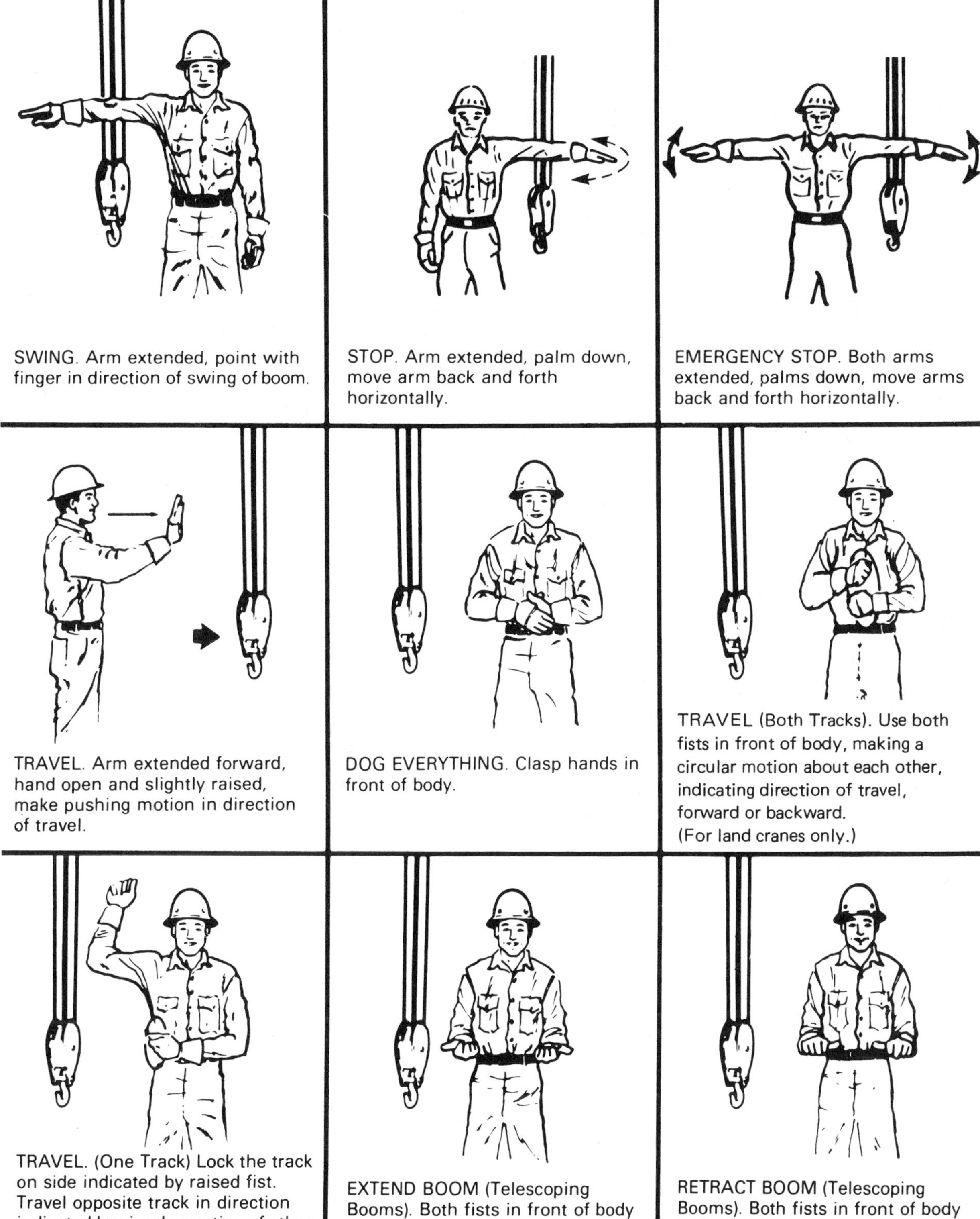

Figure 15–5. *(Concluded.)*

rigging personnel to have authorization permits. These are renewable at intervals of one or two years upon reexamination.

Finally, establish a system for documenting training and proficiency levels. Maintain this system to assure that operator's skills are up-to-date.

Figure 15–6. Operator controls overhead crane by means of hand-held remote radio (lower right.)

Inspection

Overhead and gantry cranes must be inspected according to the procedure given in ANSI/ASME B30.2, *Overhead and Gantry Cranes.* Hooks must be inspected in accordance with the procedure in ANSI/ASME B30.10.

Hooks having any of the following deficiencies shall be removed from service unless a qualified person approves their continued use and initiates corrective action. Hooks approved for continued use shall be subjected to periodic inspection.

1. Crack(s)
2. Wear exceeding 10% (or as recommended by the manufacturer) of the original sectional dimension.
3. A bend or twist exceeding 10 degrees from the plane of the unbent hook.
4. For hooks without latches, an increase in throat opening exceeding 15% (or as recommended by the manufacturer); for hooks with latches, an increase of the dimension between a fully opened latch and the tip section of the hook exceeding 8% (or as recommended by the manufacturer).
5. If a latch that is provided becomes inoperative because of wear or deformation, and is required for the service involved, it shall be replaced or repaired before the hook is put back into service. If the latch fails to fully close the throat opening, the hook shall be removed from service or moused until repairs are made.

A crane operator shall not attempt to make repairs. Any condition that might make the crane unsafe to operate should be reported to the supervisor. Certain faults may be so dangerous that the crane should be shut down at once and not operated until the faults are corrected.

A list of unsafe conditions to be checked by operators of overhead traveling cranes prior to each shift follows. Many companies have developed checklists to be completed and signed daily before work.

- bearing: loose, worn
- brakes: shoe wear
- bridge: alignment out of true (indicated by screeching or squealing wheels)
- bumpers on bridge: loose, missing, improper placement of
- collector shoes or bars: worn, pitted, loose, broken
- controllers: faulty operation because of electrical or mechanical defects
- couplings: loose, worn
- drum: rough edges on cable grooves
- end stops on trolley: loose, missing, improper placement of
- footwalk: condition
- gears: lack of lubrication or foreign material in gear teeth (indicated by grinding or squealing)
- guards: bent, broken, lost
- hoisting cable: broken wires
- hook block: chipped sheave wheels
- hooks: straightening
- lights (warning or signal): burned out, broken
- limit switch: functioning improperly
- lubrication: overflowing on rails, dirty cups
- mechanical parts (rivets, covers, etc.): loose
- overload relay: frequent tripping of power
- rails (trolley or runway): broken, chipped, cracked
- wheels: worn (indicated by bumpy riding)

Many companies believe in performance (operating) tests for all hoisting equipment. They make sure that all hoisting equipment satisfactorily completes a performance test before it is placed in service. The test should be repeated (1) at least every year; (2) prior to unusual or critical lifts; and (3) after alteration, modification, or reassembly. Record the test results and keep them available for review. In addition, regular preshift and on-shift inspections should be made and periodic load tests performed to provide added assurance that the equipment is safe to operate.

Operating Rules

The following operating rules for crane operators are from the Crane Manufacturers Association of America, Inc.:

One measure of a good crane operator is the smoothness of operation of the crane. Jumpy and jerky operation, flying starts, quick reversals, and sudden stops are the "trademarks" of the careless operator. The good operator knows and follows these tried and tested rules for safe, efficient crane handling.

1. Crane controls should be moved smoothly and gradually to avoid abrupt, jerky movements of the load. Slack must be removed from the sling and hoisting ropes before the load is lifted.
2. Center the crane over the load before starting the hoist to avoid swinging the load as the lift is started. Do not swing loads suspended from the crane to reach areas not under the crane.
3. Crane hoisting ropes should be kept vertical. Cranes shall not be used for side pulls.
4. Never lower the block below the point where less than two full wraps of rope remain on the hoisting drum. Should all the rope be unwound from the drum, be sure it is rewound in the correct direction and seated properly in the drum grooves, otherwise the rope will be damaged and the hoist limit switch will not operate to stop the hoist in the high position.
5. Be sure everyone in the immediate area is clear of the load and aware that a load is being moved. Activate the warning device (if provided) when raising, lowering, or moving loads wherever people are working to make them aware that a load is being moved.
6. Do not make lifts beyond the rated load capacity of the crane, sling chains, rope slings, etc.
7. Do not operate the crane if limit switches are out of order or if ropes show defects or wear.
8. Make certain that before moving the load, load slings, load chains, or other load-lifting devices are fully seated in the saddle of the hook.
9. When a duplex hook (double saddle hook) is used, a double sling or choker should be used to assure that the load is equally divided over both saddles of the hook.
10. On all capacity or near capacity loads, the hoist brakes should be tested by returning the master switch or push button to the OFF position after raising the load a few inches off the floor. If the hoist brakes do not hold, set the load on the floor and do not operate the crane. Report the defect immediately to the supervisor.
11. Check to be sure that the load is lifted high enough to clear all obstructions and personnel when moving bridge or trolley.
12. At no time should a load be left suspended from the crane unless the operator is at the master switches or push button with the power on, and under this condition keep the load as close as possible to the floor to minimize the possibility of an injury if the load should drop. When the crane is holding a load, the crane operator should remain at the master switch or push button.
13. When the hitcher is used, it is the joint responsibility of the crane operator and the hitcher to see that the hitches are secure and that all loose material has been removed from the load before starting a lift.
14. Do not lift loads with any sling hooks hanging loose. (If all sling hooks are not needed, they should be properly stored or a different sling should be used.)
15. All slings or cables should be removed from the crane hooks when not in use. (Dangling cables or hooks hung in sling rings can inadvertently snag other objects when the crane is moving.)
16. Crane operators should not use limit switches to stop the hoist under normal conditions. (These are emergency devices and not to be used as operating controls.)
17. Do not block, adjust, or disconnect limit switches in order to go higher than the switch will allow.
18. Upper limit switches (and lower limit switches, when provided) should be tested when stopping the hoist at the beginning of each shift, or as frequently as otherwise directed.
19. No loads should be moved or suspended over people regardless of the attachment, mechanical, magnetic, friction, or vacuum.
20. Molten metal shall never be carried overhead where it could splash onto personnel.
21. If the electric power goes off, place your cotrollers in the OFF position and keep them there until power is again available.

22. Before closing main or emergency switches, be sure that all controllers are in the OFF position so that the crane will not start unexpectedly.
23. If plugging protection is not provided, always stop the controllers momentarily in the OFF position before reversing—except to avoid injuries. (The slight pause is necessary to give the braking mechanism time to operate.)
24. Whenever the operator leaves the crane, this procedure should be followed:
    - Raise all hooks to an intermediate position.
    - Spot the crane at an approved designated location.
    - Place all controls in the OFF position.
    - Open the main switch to the OFF position.
    - Make a visual check before leaving the crane.

    NOTE: On yard cranes (cranes on outside runways), operators should set the brake and anchor it securely so the crane will not be moved by the wind.
25. When two or more cranes are used in making one lift, it is very important that the crane operators take signals from only one designated person.
26. Never attempt to close a switch that has an OUT OF ORDER or DO NOT OPERATE card on it, regardless of whether it locked out or not. Even when a crane operator has placed the card, it is necessary to make a careful check to determine that no one else is working on the crane, before removing the card.
27. In case of emergency or during inspection, repairing, cleaning, or lubricating, a warning sign or signal should be displayed and the main switch should be locked in the OFF position. This should be done whether the work is being done by the crane operator or by others. On cab-operated cranes when others are doing the work, the crane operator should remain in the cab unless otherwise instructed by the supervisor.
28. Never move or bump another crane that has a warning sign or signal displayed. Contacts with runway stops or other cranes shall be made with extreme caution. The operator shall do so with particular care for the safety of persons on or below the crane, and only after making certain that any persons on the other cranes are aware of what is being done.
29. Do not change fuse sizes. Do not attempt to repair electrical apparatus or to make other major repairs on the crane unless qualified and specific authorization has been received.
30. Never bypass any electrical limit switches or warning devices.
31. Load limit or overload devices shall not be used to measure loads being lifted. This is an emergency device and is not to be used as a production operating control.

General Maintenance and Safety Rules

The following maintenance safety rules are taken from ANSI/ ASME B30.2, *Overhead and Gantry Cranes (Top Running Bridge, Single, or Multiple Girder, Top Running Trolley Hoist):*

- To be repaired, a crane must be moved to a location where there will be minimum interference with other cranes and operations in the area.
- All controllers should be in the OFF position.
- The main power source should be disconnected, deenergized, and locked, tagged, or flagged in the deenergized position. All other sources of energy should be neutralized so that they are in a state of energy isolation.
- WARNING or OUT OF ORDER signs should be placed on the crane, on the floor beneath, or on the hook where they are visible from the floor.
- If other cranes are in operation on the same runway, rail stops or other suitable devices shall be provided to protect the idle crane.
- Where rail stops or other devices are not available or practical, a person should be located where he [or she] can warn the operator of reaching the limit of safe distance from the idle crane.
- Where there are adjacent craneways and the repair area is not protected by wire mesh or other suitable protection, or if any hazard from adjacent operations exists, the adjacent runway must also be restricted. A signaler shall be provided when cranes on the adjacent runway pass the work area. Cranes shall come to a full stop and may proceed through the area on being given a signal from the designated person.
- Trained, qualified, and authorized personnel shall be provided to work on energized equipment when adjustments and tests are required.
- After all repairs have been completed, guards shall be reinstalled, safety devices reactivated, and maintenance equipment removed before restoring crane to service.

Overhead Cranes

An overhead crane (Figure 15–7) may be operated either from a cab or from the floor. In the latter case, control devices may be either pendant push buttons or pull

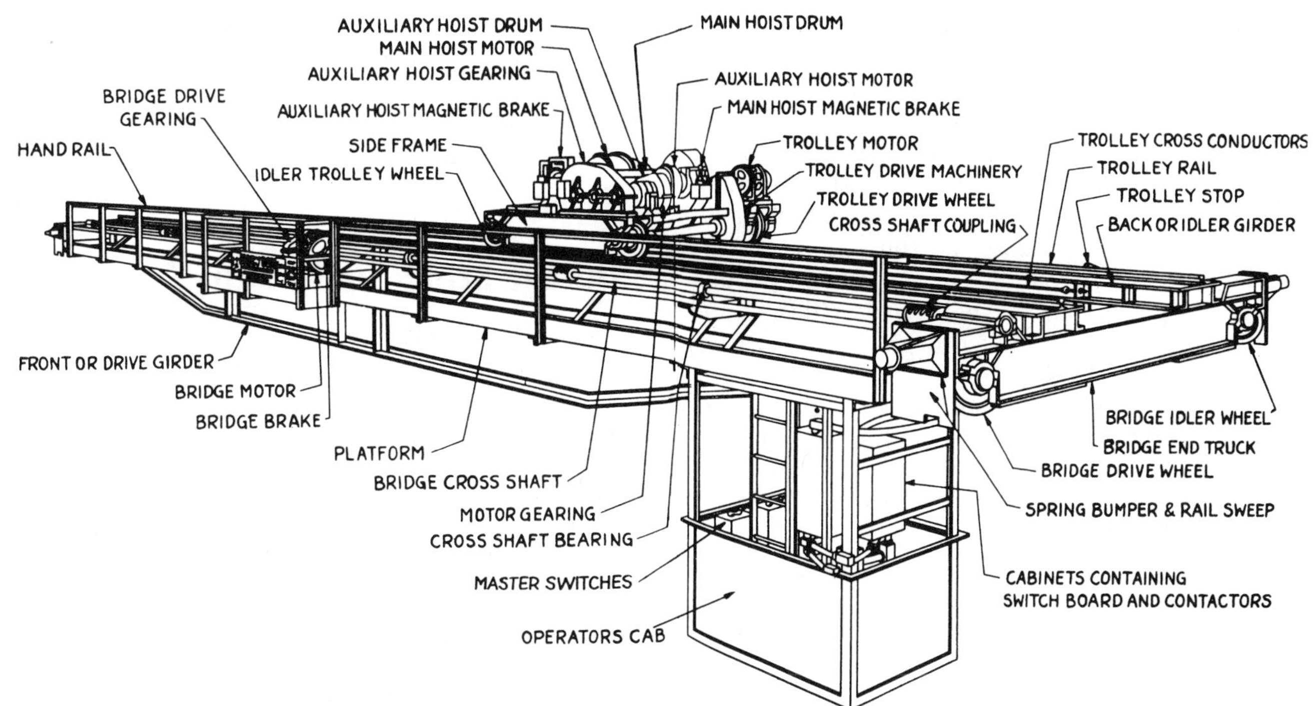

Figure 15–7. Essential parts of a typical cab-controlled overhead traveling crane. (Courtesy Shaw Box Crane and Hoist Division of Manning, Maxwell & Moore, Inc.)

ropes. (In some cases they are radio-controlled.) The control handles shall be clearly identified by signs and by shape or position so that the operator, while maintaining visual contact with the signaler, can identify each control by touch. Identify controls on all floor-operated overhead traveling cranes. Likewise, identify the controls in cab-operated cranes. If there are several cranes on the same runway or in the same building, all should have controls in identical positions so that a substitute operator will not be confused.

Safe means shall be provided for the operator to pass from the cab to the footwalk. Stairs are preferred to ladders. If ladders are used, they should meet the appropriate safety standards. Furthermore, the space that the operator must step across in going from the landing or the runway girder to the crane should not exceed 12 in. (30 cm). Safe access also shall be provided to the bridge motor and brake and to the equipment on the crane trolley. OSHA requires that a clearance of not less than 3 in. (7.6 cm) overhead and 2 in. (5.08 cm) laterally be provided and maintained between the crane and obstructions. Finally, stanchions or grab irons shall be installed to enable a person to climb safely onto the trolley.

In case of an emergency, the operator must be able to escape from the crane regardless of its location on the runway. If a fire were to occur while the crane is attached to a load, the operator would be in particular danger if he or she could not travel at once to the access landing. Install an emergency means of escape in the cab, unless a means of escape via the bridge and runway is provided. A CO_2 dry chemical or equivalent fire extinguisher should also be installed in the cab.

Strict precautions shall be taken to restrict personnel from servicing or riding the crane while it is in motion. This will prevent a person from being brushed off the crane by low beams or trusses. Service personnel must stay on the footwalk.

Footwalks and platforms should be substantial, rigidly braced, and protected on open sides with standard railings and toeboards. The footwalk should be reached by one or more fixed ladders not less than 16 in. (40 cm) wide. The outside edge of the walk should not be less than 30 in. (76 cm) from the nearest part of the trolley. The bridge walkway should have a 42-in. (1-m) high handrail, an intermediate rail, and a 4-in. (10-cm) toeboard. The space at the squaring shaft should be guarded so that a person cannot fall between the walkway and the crane girder.

A footwalk should also be provided, if headroom permits, along the entire length of the bridge of any crane that has a trolley running on the tops of the girders. This footwalk shall be on the drive side of the bridge and should have toeboards and metal handrails. Safe access to the opposite side of the trolley also shall be provided.

Flooring of walkways should be neatly fitted, leaving no openings, and shall have a slip-resistant surface. Vertical clearance between the floor of the walkway and overhead trusses, structural parts, or other permanent fixtures shall be at least 6.5 ft (2.0 m). Where such clearance is structurally impossible, built-in members should be distinctively painted, or striped and padded where necessary. Toeboards shall be provided at the edges of flooring on the trolley to prevent tools from falling to the floor below.

To guard against electric shock, a heavy rubber mat shall be provided at the control panel in the cab. The operator should have an unobstructed view of any possible position of the load hook.

The bridge truck wheels and the trolley wheels shall all have sweeps to push away a person's foot or hand. To prevent a serious impact on the crane, should a bridge wheel fail, the end frames or trucks should have safety lugs not more than 1 in. (2.5 cm) above the top of the rails.

After years of service, runway rails may become distorted. Or, the span between them may be altered from settling of the column footings. Wear on the flanges of the bridge wheels thus results. Rail alignment, therefore, should be checked every few years. If the tread of bridge truck wheels is tapered, the crane will run constantly square with the runway. Rail stops or bumpers must be so located that, when contacted, the crane bridge is square with the runway.

Electromagnets and Hook-on Devices

Electromagnets are often used with electric overhead cranes and gantry cranes, as well as with several other types of cranes. Electromagnets handle scrap iron and hot or cold ingots; they also move iron and steel products.

Do not use magnets either close to steel machines or parts or near iron materials being processed. Also keep watches and other delicate instruments out of the electromagnetic field. In that way, these objects will not become magnetized.

Label switches or switchboxes controlling power to the magnet DANGER—DO NOT OPEN SWITCH—POWER TO ELECTROMAGNET. The magnet switch must have a means of discharging the inductive load of the magnet.

The metal body of an electromagnet shall be grounded. The magnet's power supply circuit should have a battery backup system. Even with a backup system, however, never move a load, suspended from a magnet, over personnel. The switchboard, wiring, and all other electrical equipment should comply with NFPA 70, *National Electrical Code.* (See also National Safety Council Industrial Data Sheet 359, *Electromagnets Used with Crane Hoists.*)

Special hook-on or clamping devices can be designed and made for handling special shapes. A grab for positioning steel coils is shown in Figure 15–8.

Storage Bridge and Gantry Cranes

Storage bridge and gantry cranes, while similar to traveling cranes, travel on rails at ground level instead of on elevated runway girders. Gantry cranes have relatively short spans. Storage bridge cranes, however, may have a span of 300 ft (92 m) or more—sometimes with a cantilever on one or both ends. Storage bridge cranes are usually used for handling coal or ore. (See the section on Conveyors later in this chapter.) Ordinarily, a caged ladder on one of the legs of the crane provides access to the cab. Provide cranes that travel on rails with substantial rail scrapers or track clearers at each end of the trucks. Rail scrapers should be effective in both directions of travel.

There may be a serious shearing or crushing hazard in the area between cab-operated cranes and adjacent structures, or stored material. Provide a gong or other warning device that will sound intermittently from the time the travel controller handle is first moved from the OFF position until it is returned to the OFF position.

The wheel truck of gantry cranes should have adequate side clearance. On storage bridge and gantry cranes, provide bumpers made of cast steel that are at least one-half the diameter of the truck wheels in height. Fasten both wheel truck and trolley bumpers to the girder and not to the rail.

Spring bumpers usually are provided where bridge axles have antifriction bearings. If compression springs are used, they should be at least 5 in. (12.7 cm) in diameter at the point of contact. Arrange compression springs so no part can fall on a crane if a spring or guide pin breaks.

To prevent the crane from being moved down the track by a strong wind, the operator should apply rail clamps before leaving the cab, even for a short time. The holding power of the clamps should be sufficient to withstand wind pressure of 30 lb/ft^2 (1.4 k/Pa) of projected area of the crane. The electric contact rails or wires shall be so located, or so guarded, that persons normally could not come into contact with them.

So that a bridge crane can be squared to the track, the squaring shaft should have a clutch. One end of the bridge can then be moved to bring the crane into proper position, while the other end remains stationary.

All cranes used outside should have the following features:

Figure 15–8. Mounted on a 20 ton (18,144 kg) capacity crane, this grab has full access to the coil storage area. (Courtesy Harnischfeger Industries, Inc.)

- floors of the footwalk constructed to provide drainage
- an operator's cab that (1) is constructed of fire-resistant material, (2) is weatherproof, (3) has provision for heating and ventilation, (4) has ample space for control equipment, and (5) allows the operator to see signals clearly
- the floor of the cab extended to an entrance landing and equipped with a handrail and toeboard of standard construction
- a rope ladder or other means of emergency escape from the cab
- locking ratchets on wheel locks, rail clamps, and brakes so that the crane will not move in a high wind
- skew switches to prevent excessive distortion of the bridge
- a screen or other barrier (preferably nonconductive) between the contact bars and the bridge walkways to prevent unintentional contact with the current conductor.

The crane's main line switch should be so constructed that it can be locked in the OPEN position. Mount the main line switch above the cab so it can be reached conveniently from the footwalk.

Monorails

A monorail system consists of one or more independent trolleys, supported from or within an overhead track, from which hoists are suspended. All applicable safety features should conform to ANSI/ASME B30.11, *Monorails and Underhung Cranes.*

Many industries extensively use monorail hoists to raise, lower, and transport materials. There are three major groups of monorails: hand-operated monorails, semihand-operated monorails, and power-operated monorails. On the hand-operated monorail, material is raised with a hand-powered hoist, and the trolley is propelled by hand. The semihand-operated monorail has a power hoist and is moved horizontally by hand. The power-operated monorail is electrically actuated for both vertical and horizontal movements.

No attempt should be made with a monorail hoist to lift or otherwise move an object by a side pull, unless the hoist has been designed for such use. Monorail hoists, operated in swivels, should have one or more safety catches or lugs that will support the load should a suspension pin fail. Provide stops at open ends of tracks such as interlocked cranes, track openers, and track switches.

Design all trolleys for monorail systems and underhung cranes to accommodate the maximum load. Design monorail track supports and track to carry the intended loads safely. Post the maximum load capacity on the monorail system. Frequently inspect both the track and its support for signs of weakening and wear. Make necessary repairs as soon as possible.

If an electric monorail carrier or crane is operated from a cab, a fixed platform or ladder shall be provided to give the operator access to the cab. A means must also be provided for the emergency escape of the operator. The electric contact wires and the current collectors shall be so located that an operator cannot inadvertently make contact with them upon entering or leaving the cab, or while in the normal operating position inside the cab.

Jib Cranes

A jib crane is a crane capable of lifting, lowering, and rotating a load within a circular arc covered by a rotating arm or a jib. The jib, and the trolley running on it, are usually supported from a building wall, column, or pillar. A hoist (chain, air, or electric), with which the loads are lifted, is usually suspended from the trolley that travels on the jib boom.

Before a jib crane is mounted to a building wall or column of a building, the strength of the structure shall be checked by a qualified engineer to determine whether or not the column or wall to be used is strong enough to support the jib, hoist, and load. Free-standing jibs must also have good foundation support. The jib shall be braced or guyed, if necessary, to withstand the loads it is expected to carry.

A stop plate or angle iron shall be installed at the outboard end of the jib to prevent the load trolley from running off the beam. Frequently inspect this end stop to see that it is not becoming loose or rusting off.

Derricks

The principal types of derricks are the A-frame derrick, the stiff-leg derrick, and the guy derrick. Although there are other types of derricks, these are the most common. All derricks must have every part firmly anchored.

A-Frame Derricks

As the name implies, the A-frame derrick (Figure 15–9) has a frame of steel or timber shaped like the letter "A" and is erected in a vertical plane. It has a brace, or leg, that extends from the top of the A at a 45-degree angle to the ground. The sills, or lowest part of the framework, tie this brace to the bottom of the A-frame. The boom is hinged at the horizontal member of the A. The base of the A-frame and the rear brace must be firmly weighted down.

Stiff-Leg Derricks

The stiff-leg derrick (Figure 15–10) has a mast with two braces at a 90-degree angle to each other and at a 45-degree angle to the ground. Usually, steel or timber sills tie the mast and the braces together at the ground. To withstand the uplift caused by a heavy load on the boom, use sandbags, cast-iron weights, or concrete blocks to hold down the stiff legs.

The hoist engines for both stiff-leg and A-frame derricks are usually bolted to the sills. On smaller derricks, the suspended loads may be slewed by being pushed manually. The boom of a large derrick may be swung by a "bullwheel" to which cables from another drum on the hoist engine are attached. A loaded cable may whip

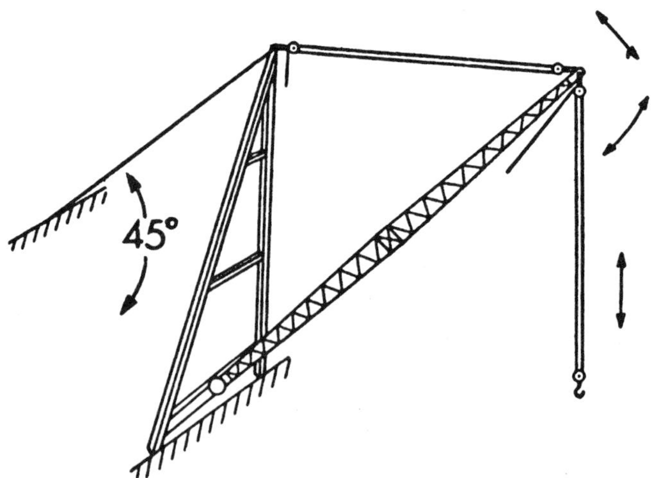

Figure 15–9. A-frame derrick. Brace is set at 45 degrees to the ground.

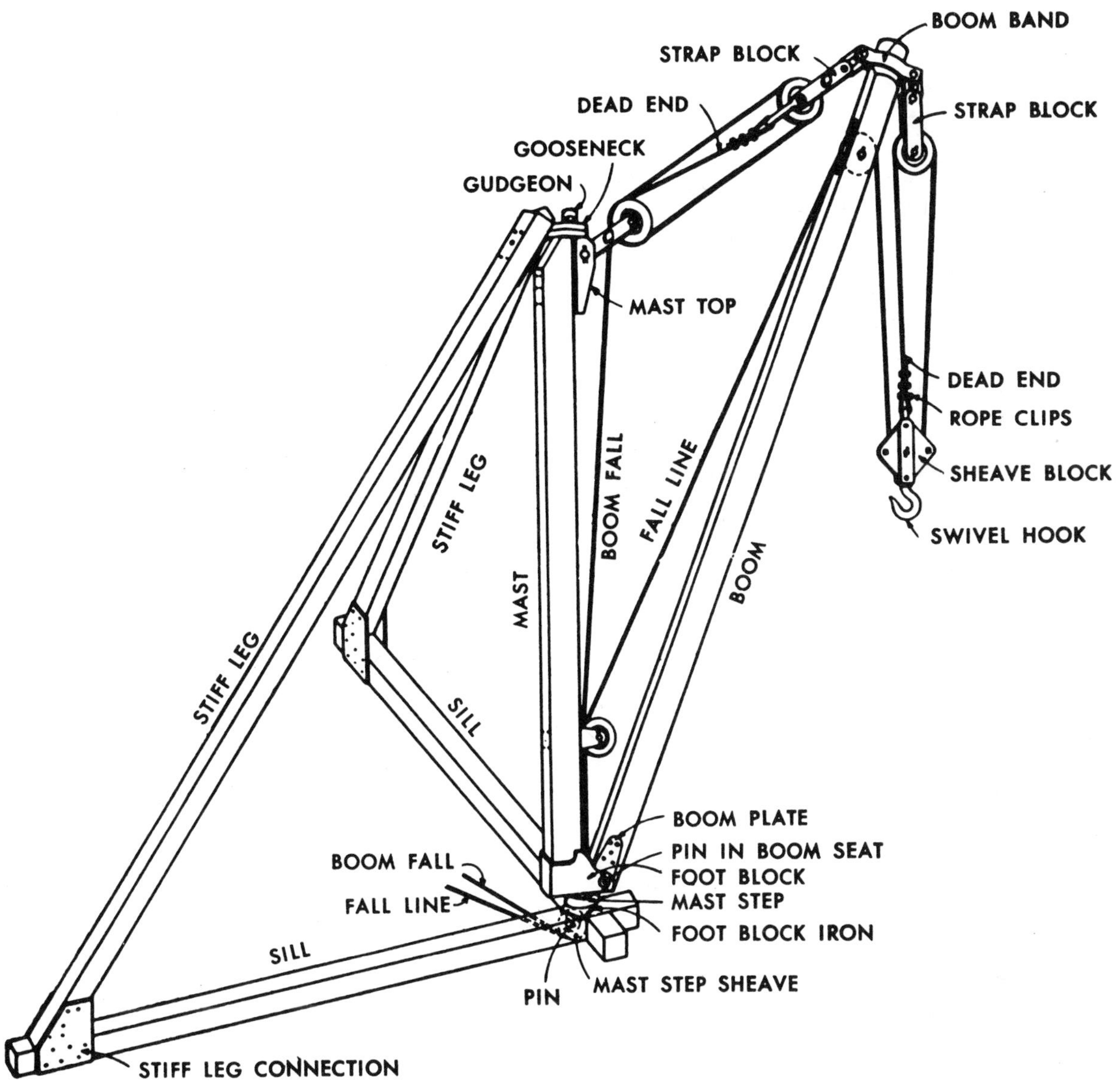

Figure 15–10. Diagram of stiff-leg derrick shows names of various parts. (Courtesy Travelers Insurance Co.)

considerably and cause severe injury. Therefore, the horizontal cables between the hoist engine and the boom hinge shall be barricaded, and workers should be prohibited from crossing over or under them.

Guy Derricks

The guy derrick (Figure 15–11) is used largely for erecting structural steel in tall buildings, especially those over 10 stories high that cannot be reached by the boom of a crawler crane operating on the ground. Such derricks usually are made of latticed steel and have an odd number of equally spaced wire rope guys, each equipped with a turnbuckle and attached to the steel beams or columns on the erection floor. If the guy derrick is erected on the ground, however, the guys should be secured to heavy steel anchors buried deep in the ground, with additional weights placed at the anchor points.

Steel beams or heavy timbers, 12 × 12 or 12 × 16 in. (30 × 30 or 30 × 40 cm), should be placed on the floor beams to support the foot of the mast. These foot blocks must be braced against the stubs of the building columns. This prevents their being "kicked" out of position when a heavy load is picked

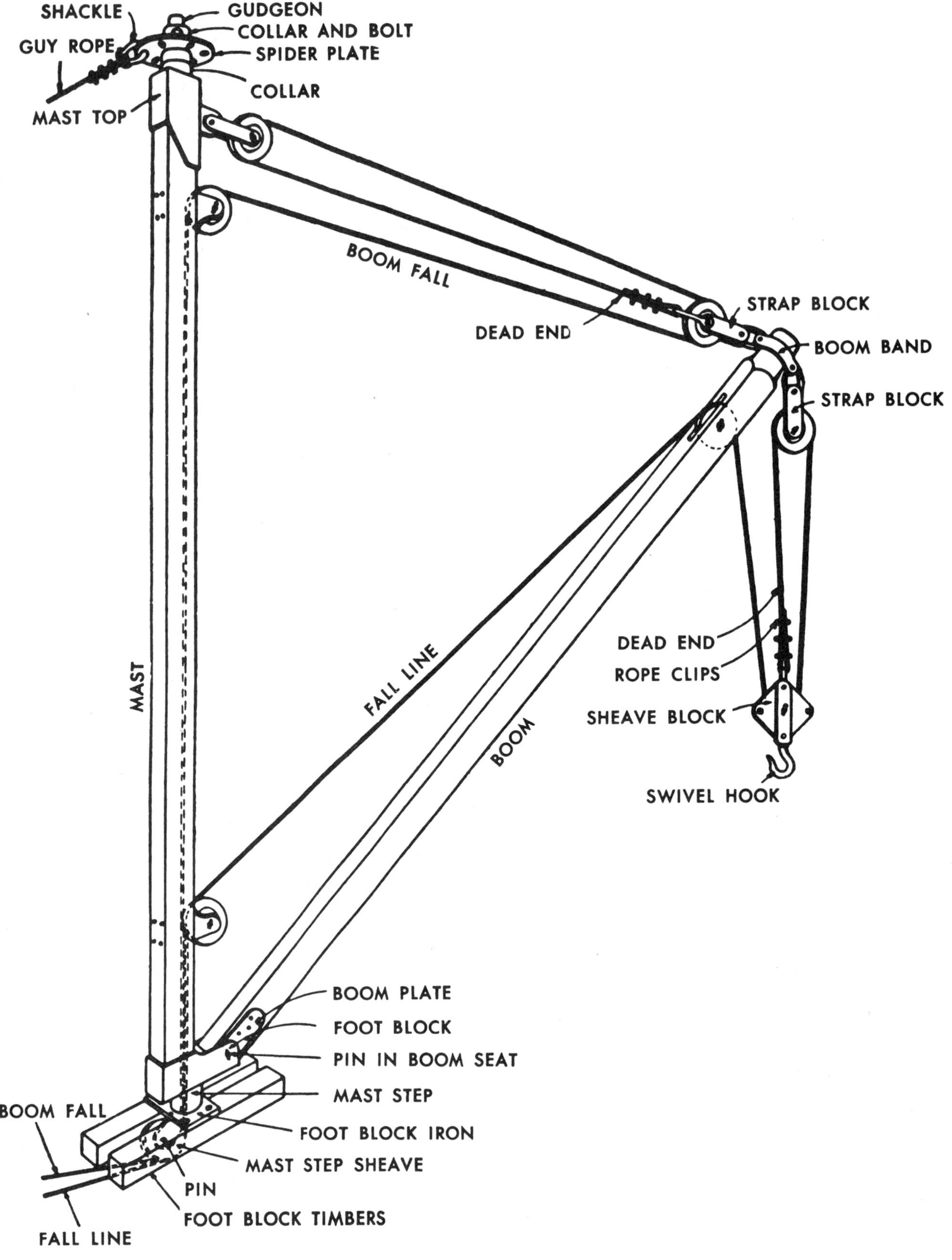

Figure 15–11. Diagram of a guy derrick. (For simplicity, only one guy is shown.) (Courtesy Travelers Insurance Co.)

up with the boom at a low angle. Wire rope and turnbuckles may be used in place of 8 × 8 in. (20 × 20 cm) timbers to secure the base of the mast.

The hoist engine, whether on the same level with the derrick base or on the ground many floors below, should be securely anchored by steel cables or shoring timbers. This prevents it from being pulled toward the base of the mast by the tension on the cables.

A unique feature of the latticed-steel guy derrick is its ability to lift itself from one erection floor to the next. First, the hinge pin is removed to disconnect the boom from the mast. The boom hoist cable (or topping lift) then lifts the boom in a vertical position and sets it on the foot blocks close to the mast. The boom is then rotated 180 degrees, the normally unused guys are secured, and the boom then stands as a guyed gin pole. (See the section on gin pole derricks below.) The load hoist of the boom then is used to pick up the mast, the mast guys being slackened off a few at a time and reattached at the upper level. When the mast is secured at the new erection floor, the boom is raised and again connected at the hinge.

To swing guy derricks, the workers push either on the suspended load or on a pipe "bull stick" attached near the base of the mast. A bullwheel can also be used as a mechanical means to swing the derrick.

Gin Pole and Breast Derricks

Other types of derricks are the gin pole and the breast derrick. The gin pole is merely a mast slightly out of plumb, with a hoisting tackle suspended from its upper end. The gin pole is supported by a number of guys, most of which are on the side away from the load. This rig is used for raising and lowering a load that needs to be moved horizontally only a few feet.

The breast derrick is a small, portable A-frame with a winch attached to it. Like the gin pole, it is erected in a nearly vertical position, with one or two guys to support it. Care must be exercised to prevent the base from slipping and causing the load to fall. Warn workers to watch their fingers when they operate the winch.

Tower Cranes

There are many design variations for the tower crane, depending upon the manufacturer and the intended use. Tower cranes can be erected on a minimum of ground area or within a building, such as, within an elevator shaft or other floor opening. To increase their range and versatility, some tower cranes are mounted on undercarriages that run on rails, rather than on a fixed base. There is also a truck-mounted tower crane.

The following procedures are causes of tower crane incidents:

- improper erection, climbing, or dismantling of the crane
- lifting loads above the rated capacity of the crane, or lifting eccentric loads
- improper bracing of the crane
- bracing against, or attaching to, material or structural members that are insecure or unable to provide the needed support
- erection within a building not designed to support the crane's weight
- operators not knowing the limitations or operating characteristics of tower cranes because of inadequate training or follow-up
- tampering with limit switches or other safety devices
- operators not receiving instructions in their native language, or in clear English
- using tower cranes during high winds

The following are general guidelines for preventing incidents with tower cranes:

- Stresses for steel used for fabrication and construction conforming to American Institute of Steel Construction specifications. (If special materials such as high-tensile steel or aluminum alloys have been used in the crane structure, the crane should bear a notice to this effect.)
- A secure attachment of counterweights; and safety ropes, rods, or chains to hold the counterweights, in addition to the basic attachment.
- The strength of the system used to anchor the rope on a winding drum with an ample safety factor exceeding the normal working load of the rope.
- Flanges of winding drums projected well above the height of the highest layer of rope wound on the drum in normal operation.
- Nonrotating hoist rope (except on receiving systems that do not require it).
- Guarding of all moving parts including pulley block and sheave guards.
- Operators who are fully trained and certified to operate the crane, and familiar with the work environment.

Operation

Only personnel of recognized ability should operate a tower crane. They should be mature in attitude, have quick responses, and be in good health. Their backgrounds should include both training and experience in the operation of this type of equipment.

Operators should possess a general knowledge of the crane's construction and the necessary knowledge of electricity, hydraulics, trade terms, parts identification, and the maintenance needs for this work. They also should have a knowledge of safety codes and standards applicable to crane operation and of any special safety recommendations of the crane manufacturer. Operators also should know the procedures involving visibility and signaling, lifting and lowering, swinging, and shutting down.

The crane operator should never stand on, or climb upon, the framework outside the cab while the crane is in operation. Prohibit climbing to the end of the jib, except when absolutely necessary. At such times, use prescribed special precautions and equipment. Operators should use fall protective equipment and other necessary personal protective equipment when necessary.

A final point needs to be made about tower crane operation and management. A tower crane is often rented, but the rental agency will not operate the crane on the job. Therefore, crane inspections prior to and during loading, and when climbing and dismounting, are extremely important.

Management needs to plan how the crane will be used—particularly when unusual loads, such as air-conditioning units, are lifted. Load weight should always be known before lifting. Management plans should also include moving the crane. Numerous unintended incidents have occurred when the crane is being lifted or climbed to another level. Finally, the plan should provide that after the job is done and the crane is being dismantled and removed, proper procedures are followed.

Communication between the crane operator and those at the loading and unloading areas must be clear, concise, and direct. While arm and hand signals are commonly used, direct voice communication can be even more effective.

Prior to initial use and each alteration or modification, cranes shall be tested by a qualified person, with 125% of rated load, unless the manufacturer recommends otherwise. Frequent and periodic inspections are to be made dependent on the service requirements of the crane. Frequent inspections consist of the following items:

- all control mechanisms for maladjustment interfering with proper operation—daily when in use
- all control mechanisms for excessive wear of components and contamination by lubricants or other foreign matter
- all crane function operating mechanisms for maladjustment interfering with proper operation and excessive wear of components
- motion-limiting devices for proper operation with the crane unloaded; each motion should be included into its limiting device or run at slow speed with care exercised
- load limiting devices for proper operation and accuracy of settings
- all hydraulic and pneumatic hoses; particularly those that flex in normal operation
- electrical apparatus for malfunctioning, signs of excessive deterioration, dirt, and moisture accumulation
- hooks and latches for deformation, chemical damage, cracks, and wear (refer to ASME B30.10)
- braces supporting crane masts (towers) and anchor bolt base connections for looseness or loss of preload
- hydraulic system for proper fluid level—daily when in use

Periodic inspections shall include the requirements of a frequent inspection plus an examination by a designated person to determine whether they constitute a hazard of the following:

- deformed, cracked, or corroded members in the crane structure and boom
- loose bolts or rivets
- cracked or worn sheaves and drums
- worn, cracked, or distorted parts such as pins, bearings, shafts, gears, rollers, locking and clamping devices, sprockets, and drive chains or belts
- excessive wear on brake and clutch system parts, linings, pawls, and ratchets
- load, wind, and other indicators for inaccuracies outside the manufacturers' recommended tolerance
- power plants for performance and compliance with safety requirements
- electrical apparatus for signs of deterioration in controllers, master switches, contacts, limiting devices, and controls
- crane hooks inspected per ASME B30.10
- travel mechanisms for malfunction, excessive wear, or damage
- hydraulic and pneumatic pumps, motors, valves, hoses, fittings, and tubing for excessive wear or damage

Light service consists of the following: irregular operation with loads generally about one-half or less of the rated load, frequent inspections monthly, and periodic inspections annually. Normal service consists of the following: operations at less than 85% rated load and not more than 10 lift cycles per hour except for isolated instances, frequent inspections weekly to monthly, and periodic inspections semiannually to annually. Heavy service consists of the following: operations at 85% to 100% of rated load or in excess of 10 lift cycles per hour as a regular specified procedure, frequent inspections daily or weekly, and periodic inspections quarterly.

Frequent inspections are most often done by the operator with no records required. Periodic inspections require records and are done by an appointed person.

Mobile Cranes

Mobile cranes are an engineering marvel. The construction industry, in particular, is able to perform tasks once nearly impossible to do. A mobile crane may have up to a 1,000-ton capacity and a length of 600 ft plus boom and jib.

Canadian (Ontario) data suggests that about one in five construction fatalities is crane-related. In general, 90% of mobile crane injuries are attributed to operator error. Other factors are:

- support failure—30%
- failure to use outrigger—20%
- crane failure—10% to 20%
- rigging—4% to 15%

In addition, of all injuries to employees, at least 25% occurred to those involved with the load, and 10% to 15% involved maintenance, refueling, etc., around the crane.

Mobile cranes include locomotive cranes, crawler cranes, wheel-mounted cranes, and industrial truck cranes. The first three types have standard designs, but industrial truck cranes have various designs, depending upon intended use.

All mobile cranes have booms with load hoists and boom hoists. In most instances, the crane swings or rotates on a turntable, which rests on a railroad car, crawler, or wheel chassis. Power is provided by electric motor, steam, gasoline, or diesel engine.

Electric-powered equipment should be grounded as specified in ANSI C2, *National Electrical Safety Code.* Repairs or adjustments shall be made only by qualified personnel. Power shall be disconnected before repairs are made. Trailing cables should be kept off the ground whenever possible and should be handled only with insulated hooks.

Gasoline-operated cranes require protection against the hazards of fire and explosion. Engines shall not be refueled while running. If refueling is done by hose connected to a tank truck or connected to drums by means of pumps, use a metallic bonding connection between the hose nozzle and the fill pipe.

If fuel is carried to the crane by hand, use safety cans. Open lights, flames, and sparks should be eliminated, and lights on the equipment should be of an approved explosion-proof type. A fire extinguisher of 5BC rating shall be kept in the cab of the rig.

To prevent the boom from being dropped unintentionally, operate the boom hoist mechanism by gearing or chain and by lowering the engine's speed. Provide a self-setting brake and locking pawl or other positive locking device. Furthermore, with the exception of some hydraulic cranes, install boom stops—preferably of the shock-absorbing type—on all cranes.

If possible, install a hoist line, two-blocking limiting device. This device could be controlled in conjunction with the load-line hoist clutches and brakes. The operator must have full control of all crane functions at all times. No attempt should be made to lift the boom by means of the hoist's load cable. Do not lift a load with the boom's hoist line unless the crane is designed for this purpose.

Load Chart

Every crane should have a capacity plate or a sign that states its safe-load capacity at various radii from the center pin of the turntable. Install this sign so it is plainly legible to the crane operator, signaler, and rigger. Mount a boom angle indicator with a freely suspended pointer actuated by gravity in front of the safe-load capacity sign on the side of the boom near the hinge. The pick must be within the limits prescribed by the load chart.

A capacity chart for the operators should indicate boom length, boom angle, and capacity. A load indicator device enables the operator to handle the load better. For locomotives using outriggers fully extended, the maximum load rating is 80% of tipping load. Where structural competence governs lifting performance, load ratings are reduced and the rating chart should indicate this. When handling loads that are limited by structural competence rather than by stability, the person responsible for the job shall make sure that the weight of the load has been determined within plus or minus 10% before it is lifted.

The operator must have safe access to and exit from the crane's cab or seat. To operate the crane safely, the operator must have an unobstructed view of the load hook and the point of operation at all times, or must rely on someone giving signals. The operator must also be able to see ahead of the crane when it is traveling on the ground, whether the chassis is moving forward or backward. On some cranes, visibility to the rear is obstructed. In such cases, the operator must use extreme caution and rely on a signaler. Other workers should always stay beyond the range of the cab's swing and out from under the boom and the load.

Other precautions include good lighting and warning devices. A crane operated after dark should have clearance lights. Floodlights should illuminate the area beneath the boom, and lights mounted on the underside of the boom are recommended. A warning bell or horn and an automatic backup alarm are necessary equipment for a wheel-mounted crane. Also, provide a warning bell or horn on crawler cranes.

When other ways are not available, it may be necessary to hoist employees to another work level or area. To do so with a crane, a number of practices need to be followed. The following information is taken from ANSI A10.28, *Safety Requirements for Work Platforms Suspended from Cranes or Derricks for Construction and Demolition Operations.*

Preparation is the key to avoiding an unintentional incident. The work platform must be designed and rigged by someone qualified to do so. A safety factor of at least five is to be built in. Identification shall include empty weight and capacity. There shall be an access gate, perimeter protection, and a grab rail. The suspension systems shall minimize tipping during use.

At each new job site, prior to hoisting employees, the platform and rigging shall be proof-tested to 125% of the platform-rated capacity for five minutes. Any deficiencies must be corrected and another proof test conducted before hoisting employees.

Prior to hoisting personnel, a trial run shall be made with the platform and rigging to the proposed work elevation to ensure (1) that boom configuration and load lines are adequate, and (2) that no interference of any kind exists and the anti-two-block protection is working, if the crane is so equipped.

The operator shall demonstrate ability to operate the crane and derrick before hoisting personnel in a suspended work platform. The operator must be comfortable in the procedure and not feel upset or unduly anxious.

Some of the crane's requirements are that live booms are not allowed and crane travel must be only on tracks unless using a portal crane. An occupied work platform must not be allowed to fall free. In addition, cranes used for hoisting personnel must have the swing brake or lock engaged and be inspected prior to the hoist, including wire rope, hook brakes, boom, and all other equipment. And finally, the total weight of the work load must not exceed 50% of the rated crane capacity.

Crawler and Wheel-Mounted Cranes

These types of cranes are frequently used in erecting structural steel for tall buildings. Extension sections may be inserted to lengthen the booms of these cranes. When the boom is assembled, check all parts of the structure for damaged or missing parts. In one documented case, a lattice bar was unintentionally omitted. Later, when a heavy load was picked up, the boom buckled at this point and several men were killed.

Locomotive Cranes

In the cab of a locomotive crane, a clear passageway should be provided from the operator's platform to an exit door near the operator's side of the cab. Doors should be hinged at the rear edge and should open outward. Sliding doors should slide to the rear to open.

The motor and all power transmission apparatus should be guarded. Enough light should be provided in the cab to permit the operator to work safely and to see the gages and indicators plainly.

Install steps and handholds for safe access to the cab. Some state laws require that footboards and handholds be provided at each end of the truck bed, or that the truck have a pilot or fender. Permit no one except necessary operating personnel on a rig while it is operating.

An on-track crane should have standard automatic couplers and uncoupling levers. It should also have air brakes as well as hand brakes. It should have rail clamps at each corner to hold it in a stored position. Provide a guard at the end of the boom to prevent the thimble on the cable from coming into contact with the sheave.

An on-track crane should be moved only on signal from an authorized signaler or switchman. This person should walk ahead of the crane to warn others and to see that switches are properly set and that the track is free of obstructions. When there is no signaler or switchman, the crane operator should move the crane only on order from the supervisor of the department in which the crane is working. The crane should not be swung across another track until the crane operator and signaler have made sure that cars are not on that track and will not be moved to it.

When moving the crane about the yard or worksite, the operator should keep the crane and boom parallel to the track, to avoid striking buildings or other structures, and should carry the boom low enough to clear overhead wires. Operators should not carry buckets and magnets on the boom when the crane is going from one location to another.

Operation of Cranes

Extended outriggers are considered a part of the counterweight on the load charts of new cranes. Separate charts state crane capacity for a traveling load and for lifting a load without using outriggers. Whether traveling or stationary, the crane's turntable shall remain level.

A boom must never be swung too rapidly. If it is, the suspended load will be swung outward by centrifugal effect. This action could cause the crane to rock, or even tip. If this occurs, the load may swing and strike a person or object.

Operating a crane on soft or sloping ground or close to the sides of trenches or excavations is dangerous. Be sure the crane is level before it is put into operation.

Outriggers can be relied upon to give stability only when used on solid ground. Use heavy timber mats whenever there is doubt as to the stability of the soil on which a crane is to be operated. Do not permit the use of makeshift methods to increase the capacity of a crane, such as using timbers with blocking or adding counterweight. If the crane tips when hoisting or lowering a load, the operator should lower the load as quickly as possible by snubbing it lightly with the brakes. Therefore, never allow workers to ride a load that is being hoisted, swung, or transported.

When operating a crane with the boom at a high angle, the operator should take care that the suspended load does not strike the boom and bend the steel lattice bars on its underside. Bending these bars will weaken the boom so that when it picks up the next heavy load, it may collapse. Likewise, if the main parts of the boom are bent even slightly, the strength of the boom may be materially reduced.

When an extended boom is used on a crane, as for erecting structural steel, the operator must use extreme care in lowering it to the ground at the end of the job. An extended boom should never be lowered to one side of the chassis or crawler. The stability of the crane is greatly reduced in that position, and the crane may even tip.

When using a boom tip extension or jib, the allowable load on the jib is limited. The operator must know its capacity. The operator must refer to the crane's capacity chart in order not to exceed load limits. The crane operator must use care when swinging with a load, especially when the jib is lowered at an angle to the main boom.

The operator must center the hook over the load to keep it from swinging while it is being lifted. When holding the hook or slings in place, while the slack is taken up, employees should keep their hands out of pinch points. A hook, or even a small piece of board, may be used for the purpose. If a person must use his or her hand, it should be placed flat against the sling to hold it. The hook-on person, the rigger, and everyone else must be in the clear before the load is lifted. Use a tag line for guiding loads. While the crane is operating, the area inside the counterweight and body frame should be roped off to prevent entry by ground personnel into the pinch point formed by these two crane parts.

Operators should never remove a heavy load from a truck by hooking a crane to the load and then having the truck pull out from under the load. If the load should prove too heavy for the crane, the crane will tip before the operator can lower the load to the ground. The load should be lifted clear of the truck body. The operator should make sure that the crane can handle it safely before the truck is moved out from under it.

A crane should never be used for jerk piling. If piling cannot be pulled by a straight, steady pull—limited to rated capacity—a pile extractor should be used. When a pile extractor is used, keep the boom angle at or less than 60 degrees above horizontal.

Consult the crane's manufacturer before modifying the equipment. Such changes shall maintain at least the same factor of safety as the originally designed equipment. Maintenance and repair work should be performed only by trained and qualified personnel. The operator, however, is responsible for keeping the unit clean. Before leaving the crane at the end of the workday, the operator should lower the load block or bucket to the ground in such a manner that it cannot be upset.

Travel of Cranes

Except for very short distances, a crane should not travel with a load suspended from the boom. When a crawler crane must travel on a public thoroughfare, the boom should point forward, and someone with a flag should walk ahead of it.

A wheel-mounted crane and a crawler crane on a semitrailer should be transported with the boom pointing toward the rear and high enough to clear an automobile. Wheel-mounted cranes with short booms may travel with the boom forward in the boom rest. One of the work crew should follow in a car or truck to keep other vehicles from traveling beneath the boom. Otherwise, if the semitrailer's wheels should roll into a low spot in the pavement, the boom might suddenly crash through a car's roof. If not disassembled, a crane being transported should have its engine running and the operator in the cab. In that way, the operator can swing the boom, when necessary, thus avoiding damage to trees, poles, or buildings when the semitrailer turns corners.

Before moving heavy, slow-moving equipment or heavy equipment on a low-slung trailer over any public or private railroad grade crossing, notify a representative of the railroad company. The railroad company can then provide flag signaling to guard against a train's striking the equipment while it is moving over the crossing, or if the equipment becomes stalled or "hung up" on the crossing. This is an important precaution for the benefit of both the equipment owner and the railroad company.

Electric Wires

Consider any overhead wire as an energized line until either the person who owns the line or the electric company indicates that it is not energized. Compliance with recommended practices, and not reliance on other devices, shall be followed in determining how close the crane and its extensions, including load, are to electric power lines. A qualified signaler shall be assigned to observe the clearances and give warning before the crane approaches the stated limits. The boom's load line and the cables of the crane shall be kept away from all electric wires, regardless of their voltage. In the United States, OSHA requires that, except where the electrical lines have been deenergized and visibly grounded at the point of work, or where insulating barriers, not a part of or an attachment to the crane, have been erected to prevent physical contact with the lines, cranes shall be operated near power lines only in accordance with the following:

- For lines rated 50 kv or below, minimum clearance between the lines and any part of the crane or load must be 10 ft (3 m).
- For lines rated over 50 kv, minimum clearance between the lines and any part of the crane or load must be either 10 ft plus 0.4 in. (3 m plus 10 mm) for each 1 kv over 50 kv, or twice the length of the line insulator but never less than 10 ft (3 m) (29 *CFR*, 1910.180 [j][1]).
- In transit and with no load and boom lowered, the clearance should be a minimum of 4 ft (1.2 m).
- At construction sites, the 4 ft (1.2 m) minimum applies only to voltages less than 50 kv. Clearance must be 10 ft (3 m) for voltages over 50 kv and up to 345 kv; 16 ft (4.9 m) for voltages up to 750 kv (29 *CFR*, 1926.550 [a][15]).

If cage-type boom guards, insulating links, or proximity warning devices are used on cranes, such devices must not be a substitute for the requirements of a specifically assigned signal person, even if such devices are required by law or regulation. In view of the complex, invisible, and lethal nature of the electrical hazard involved, and to lessen the potential of false security, devices shall be used and tested in the manner and at the intervals prescribed by the device's manufacturer.

If the boom's load line or cables unintentionally come into contact with a wire, the operator should swing the crane to get clear. If the wire has been broken and the boom cannot be cleared from it, the operator shall stay on the crane and remain calm.

If the ground is wet or damp, a crawler crane will be electrically grounded so that when the boom touches a power line, the wire will, in turn, be grounded and the power company's circuit breaker will open. Some arcing may occur. After a few seconds, however, the circuit breaker will automatically close and reenergize the wire. Again the circuit breaker may open, and again it will close. Thus, the wire may be "dead" at one instant, but live a few seconds later.

If the boom of a wheel-mounted crane on rubber tires should become tangled with a "hot" electric wire, the entire crane may be energized. In such a case, the rubber tires may or may not insulate the crane and chassis from the ground. Depending upon the voltage and the soil conditions, the tires on the crane may burn and melt, and thus lose any insulating qualities. Hence, the circuit breaker may not open, and the wire and the crane may remain energized.

Stepping from the crane to the ground is often fatal, for one hand and one foot may be in contact with the crane when the other foot touches the ground. Therefore, the operator should remain on the crane until the emergency crew from the electric company arrives and frees the crane from the live wire. However, if the gasoline tank should ignite, or if for any other reason it is impossible to remain on the crane, the operator should jump, making sure that all body parts are clear of the crane before touching the ground. (See Figure 15–12 and Occupational Safety and Health Data Sheet 743, *Mobile Cranes and Power Lines.*)

A crane that has been idle for a period of one month or more, but less than six months, shall be given an inspection by a qualified person conforming to frequent inspection criteria both of the crane and rope before being placed in service.

A crane that has been idle for a period of over six months shall be given a complete inspection by a qualified person of both the crane and rope conforming to the requirements for both frequent and periodic inspections.

Frequent inspections are done by a designated person, usually the operator, on a daily or monthly interval depending on the crane's usage. A good practice is for the operator to perform the frequent inspection routine each day before he or she starts to operate the crane. The inspection consists of the following:

- all control mechanisms for maladjustment interfering with proper operation—daily, when used
- all control mechanisms for excessive wear of components and contamination by lubricants or other foreign matter
- all safety devices for malfunctions
- all hydraulic hoses, and particularly those that flex in normal operation of crane functions, should be visually inspected once every working day, when used
- hooks and latches for deformation, chemical damage, cracks, and wear (refer to ASME B30.10)
- rope reeving for compliance with crane manufacturer's specifications
- electrical apparatus for malfunctioning, signs of excessive deterioration, dirt, and moisture accumulation
- hydraulic system for proper oil level—daily when used
- tires for recommended pressure

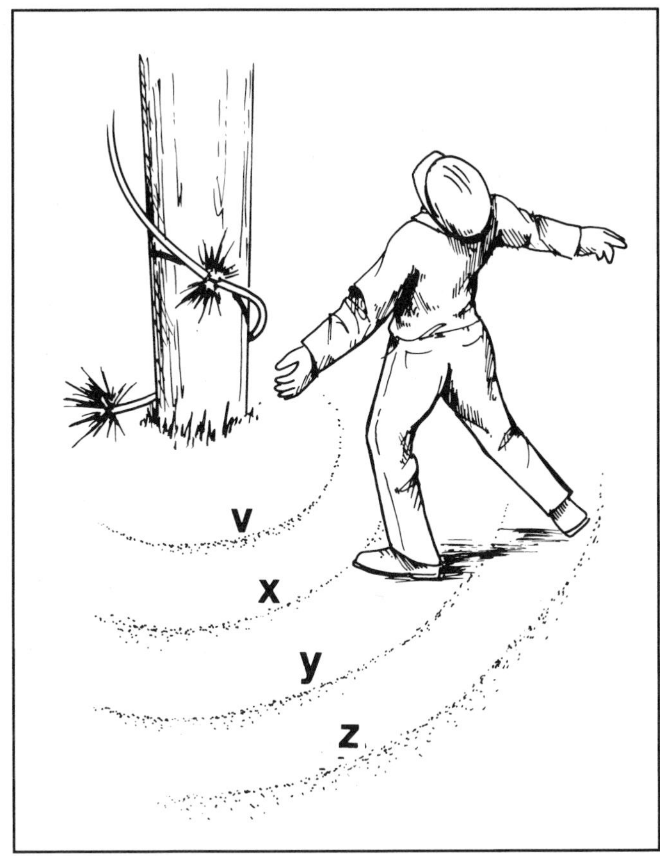

Figure 15–12. An operator on an electrified crane should jump with feet side by side and should not run (but hop with feet together) away from it, not stopping until well away. Hopping avoids the ground gradient effect: If one foot is at *y* voltage, and other at *z* voltage, the difference in voltage will cause a flow of electricity through the body.

- all running rope in service should be visually inspected once each working day to discover gross damage which may be an immediate hazard such as:
  - distortion of the rope such as kinking, crushing, unstranding, birdcaging, main strand displacement, or core protrusion; loss of rope diameter in a short rope length or unevenness of outer strands should provide evidence that the rope or ropes must be replaced
  - general corrosion
  - broken or cut strands
  - number, distribution, and type of visible broken wires (see Chapter 16, Ropes, Chains, and Slings)

Care shall be taken when inspecting sections of rapid rope deterioration such as flange points, crossover points, and repetitive pickup points on drums. Care should be taken when inspecting certain ropes such as:

- rotation resistant ropes, because the internal deterioration of rotation resistant ropes may not be readily observable
- boom hoist ropes, because of the difficulties of inspection and the important nature of these ropes

Periodic inspections are performed at intervals of one to twelve months or as specifically recommended by the manufacturer or by a qualified person. Dated records for periodic inspections shall be made on all critical items such as brakes, crane hooks, ropes, hydraulic and pneumatic cylinders, and hydraulic and pneumatic pressure valves. Records should be kept where available to appointed personnel. This inspection includes all the elements of a frequent inspection plus the following. Any deficiencies, such as those listed, shall be examined and determination made as to whether they constitute a hazard:

- deformed, cracked, or corroded members in the crane structure and entire boom
- loose bolts or rivets
- cracked or worn sheaves and drums
- worn, cracked, or distorted parts such as pins, bearing, shafts, gears, rollers, and locking devices
- excessive wear on brake and clutch system parts, linings, pawls, and ratchets
- load, boom angle, and other indicators over their full range, for any significant inaccuracies
- gasoline, diesel, electric, or other power plants for performance and compliance with safety requirements
- excessive wear of chain drive sprockets and excessive chain stretch
- crane hooks inspected for cracks
- travel steering, braking, and locking devices, for malfunction
- excessively worn or damaged tires
- hydraulic and pneumatic hose, fittings, and tubing inspected for the following:
  - evidence of leakage at the surface of the flexible hose or its junction with the metal couplings
  - blistering or abnormal deformation of the outer covering of the hydraulic or pneumatic hose
  - leakage at the threaded or clamped joints that cannot be eliminated by normal tightening or recommended procedures

  - evidence of excessive abrasion or scrubbing on the outer surface of a hose, rigid tube, or fitting

Means shall be taken to eliminate the interference of elements in contact or otherwise protect the components.

- hydraulic and pneumatic pumps and motors inspected for the following:
  - loose bolts or fasteners
  - leaks at joints between sections
  - shaft seal leaks
  - unusual noise or vibration
  - loss of operating speed
  - excessive heating of the fluid
  - loss of pressure
- hydraulic and pneumatic valves inspected for the following:
  - cracks in valve housing
  - improper return of spool to neutral position
  - leaks at spool or joints
  - sticking spools
  - failure of relief valves to attain correct pressure setting
  - relief valve pressure shall be checked as specified by manufacturer
- hydraulic and pneumatic cylinders inspected for the following:
  - drifting caused by fluid leaking across the piston
  - rod seals leakage
  - scored, nicked, or dented cylinder rods
  - dented case (barrel)
  - loose or deformed rod eyes or connecting joints
- hydraulic filters inspected for: evidence of rubber particles on the filter element may indicate hose, "O" ring, or other rubber component deterioration. Metal chips or pieces on the filter may denote failure in pumps, motors, or cylinders. Further checking will be necessary to determine the origin of the problem before corrective action can be taken.

Aerial Baskets

Aerial lift equipment is now commonly used for working above ground. These boom-mounted buckets, baskets, or platforms are used extensively in constructing and maintaining electric and telephone lines. Because of their capabilities, however, their use is increasing in harbor and port work, in the aircraft industry, in highway sign and lighting construction, and in maintenance, painting, sandblasting, and firefighting work.

Hazards of Aerial Baskets

The most frequent causes of unintentional incidents while using mobile aerial baskets include the following:

- not observing proper precautions against electrical hazards to personnel both in the basket and on the ground
- improper positioning of vehicle or outriggers, lack of sufficient blocking under outriggers, or overloading the boom, causing the apparatus to overturn or fail
- overreaching from basket or other improper work procedures
- not using proper personal protective equipment, including safety belts
- moving the truck while the boom is raised, or moving it where there is inadequate clearance for the boom
- structural or mechanical failure, or control jamming
- swinging the boom or basket against overhead obstructions or energized equipment
- moving the boom into positions that interfere with traffic
- inadequately training personnel

Operation of Aerial Baskets

The operating and maintenance instruction manuals issued by the manufacturer shall be followed. Lift controls shall be tested each day prior to use to determine that such controls are in safe working condition. Aerial baskets shall be inspected daily for defects. Mechanical equipment shall be inspected each day before using to make sure it is safe to operate. Additional safe operating procedures include the following:

- Load limits of the boom and basket shall be posted and shall not be exceeded.
- A warm-up period and test of the hydraulic system is required.
- Basket equipment approved for use on energized equipment shall be dielectrically tested periodically.
- When working near energized lines in aerial-basket trucks and aerial-ladder trucks, the trucks shall be grounded. Where grounds are not permitted by the company, barricading shall be required.
- The insulated portion of an aerial-lift boom and basket shall not be altered in any manner that might reduce its insulating value.
- Drivers of trucks with mounted aerial equipment shall be constantly aware that the vehicle has exposed equipment above the truck cab, and they shall provide necessary traveling clearance.
- The truck shall not be moved unless the boom is lowered and the basket or ladder is cradled.

- Riding in the basket while a truck is traveling shall not be permitted. (Employees may ride in the basket at the work location for short moves if the basket is returned to the cradled position for each move.)
- Available footing for the truck's wheels and outriggers shall be examined carefully to assure a stable setup. Hand brakes, chocks, and/or cribbing, when needed, shall also be used to ensure stability. The truck shall sit approximately level when viewed from the rear.
- Before lowering stabilizers, outriggers, or hydraulic jacks, the operator shall be certain there is no one in an unsafe position.
- When the boom must be maneuvered over a street or highway, necessary precautions shall be taken to avoid unintentional incidents with traffic and pedestrians.
- The operator shall always face the direction in which the basket is moving and be sure that the path of the boom or basket is clear when it is being moved.
- An employee shall not stand or sit on the top or the edge of the basket. Ladders shall not be used in the basket. While in the basket, the employee's feet shall always be on the floor of the basket.
- Employees shall not belt themselves to an adjacent pole or structure.
- Employees shall always belt themselves to the boom or ladder. Belting to the basket equipment shall be done upon entering the basket.
- When working with rubber protective equipment on energized circuits or apparatus above 300 v, the following minimum conditions shall be met in addition to all other rules governing the use of protective equipment:
  - Rubber gloves with leather protectors and rubber sleeves shall be worn.
  - An employee shall make physical contact with protective devices installed on energized primary conductors only with rubber gloves and rubber sleeves.
  - Employees shall be isolated from all grounds by using approved protective equipment or other approved devices.
  - When employees are working on the same pole or substation structure, or from a bucket truck, in no case shall they work simultaneously on energized wires or on equipment of different polarities.
- An employee shall not enter or leave the basket by walking the boom.
- Employees shall not transfer between the basket and a pole. On dual basket trucks, employees shall not transfer between the baskets.
- Employees in baskets shall not wear climbers.
- When two workers are in the basket or baskets, one of them shall be designated to operate the controls. One employee shall give all signals and make sure these signals are thoroughly understood by all people concerned.
- Baskets shall be located under or to the side of conductors or of working equipment. Raising the basket directly above energized primary conductors or equipment shall be kept to a minimum.
- Only nonconductive attachments shall be allowed on baskets.
- The operator shall be sure that hoses or lines attached to tools cannot become entangled with the levers that operate the boom.
- Air- or hydraulic-operated tools shall be disconnected from the source when not in use.
- When employees are working from the basket, take extreme care to avoid contacting poles, crossarms, or other grounded or live equipment.

Inspection of Aerial Baskets

An effective daily inspection should cover the following points. If defects are found, the inspector should report them and see that they are corrected before they develop into dangerous conditions.

- visual inspection of all attachment welds between actuating cylinders and booms or pedestals
- visual inspection of all pivot pins for security of their locking devices
- visual inspection of all exposed cables, sheaves, and leveling devices both for wear and for security of attachment
- visual inspection of hydraulic system for leaks and wear
- inspection of lubrication and of fluid levels
- visual inspection of boom and basket for cracks or abrasions
- operation of boom from ground controls through one complete cycle, listening for unusual noises and looking for deviations from normal operation.

Basket Safeguards

Equip aerial baskets with (1) safety belts and lanyards to be worn by all persons working from the baskets and (2) a means for attaching the lanyard to the equipment. In general, it is more satisfactory to anchor the lanyard to the boom. However, if this will interfere with the controls or if other considerations are involved, then anchor the lanyard to the basket. Lanyards should only be long enough to allow movement within the basket and should prevent climbing onto the rim. Such lanyards limit free fall but do not restrict work or entangle workers' feet.

Each basket operator and driver should be thoroughly trained in the use of the equipment before operating it on a job. They should know not only the particular equipment involved but also the type of work it will do in the field.

If the public is exposed to possible contact with the vehicle, set up barricades. A boom coming in contact with energized equipment might electrically charge the truck and could injure people who are walking or standing nearby.

Since many incidents with aerial baskets have been caused by inadequate footing, provide solid footing for the wheels and outriggers. Snow, ice, mud, and soft ground call for extra caution because additional firm footing under the outriggers may be necessary.

Crabs and Winches

Crabs and winches may be either hand-operated or electrically driven. Install some form of brake or safety lowering device, and anchor portable units securely against the pull of the hoisting rope or chain.

Install barricade guards to protect the operator against flying strands of wire and the recoil of broken ropes. Design such guards to protect the operator from the whipping effect of a broken rope if the point of operation cannot be totally guarded (Figure 15–13). The best defense against injury is to be well away from the direct line of pull. Also, be sure that gears are fully guarded.

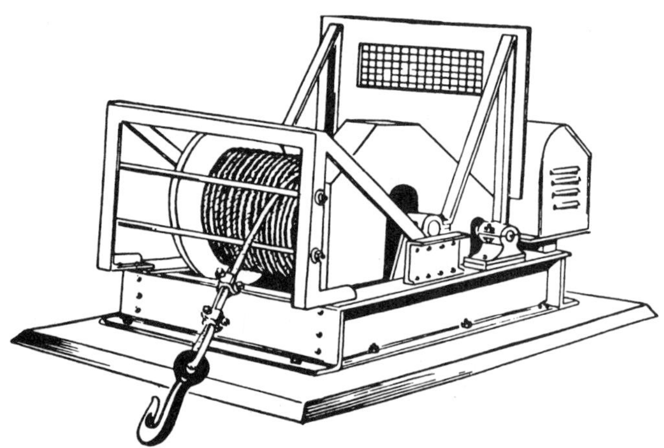

Figure 15–13. The operator of this car puller winch stands behind the shield, which protects the employee if the rope breaks.

Power-driven crabs and winches should have their moving parts encased and should be electrically grounded.

The locking pawl on the ratchet of a winch frequently presents a serious hazard to fingers, particularly when the operator attempts to disengage the pawl. To reduce this hazard, weld a small lever to the pawl so that it can be safely grasped.

Hand-operated equipment that has a crank handle instead of a handwheel poses a major danger. Operators may be struck by the revolving crank handle if they lose control while lowering a load. Provide a dog to lock the gears. A pin through the end of a crank will keep it in the socket during hoisting operations.

To lower loads rapidly, use a strap brake. Before using the brake, remove the crank or take other steps, such as replacing a spur gear and dog with a worm gear, to prevent the crank handle from flying around.

Block and Tackle

A safety factor of 10 is recommended for determining the safe working load of Manila rope (falls) in a block-and-tackle assembly. This large safety factor allows for (1) error in estimating the weight of the load, (2) vibration or shock in handling the load on the tackle, (3) loss of strength at knots and bends, and (4) deterioration of the rope due to wear or other causes.

The governing factor usually is the safe working load of the blocks, rather than of the falls (rope). By multiplying the number of sheaves and rope parts, the weight of the load that can be handled by the rope multiplies but does not correspondingly increase the strength of the blocks. Calculations show that, in most instances, using a safety factor of 10 for the rope automatically makes the load on the blocks correspond to the rope size within safe work load limits. (Mark blocks with their safe working load, as specified by their manufacturers.) The total weight on the tackle should never exceed this safe load limit. (See Chapter 16, Ropes, Chains, and Slings, in this volume.) Safe work loads for rope used in block-and-tackle assemblies are conversely 1⁄10 of the block's breaking strength, based on a safety factor of 10.

To find the required breaking strength for new rope, proceed as follows:

1. For each sheave 3 in. (7.6 cm) in diameter or larger, add 10% to the weight of the load to compensate for friction loss.
2. Divide this figure by the number of ropes or parts running from the movable block.
3. Multiply the resultant figure by a safety factor of 10.

An example for working out the procedure given above is as follows:

1. A load to be lifted weighs 2,000 lb (900 kg), and the tackle consists of two double blocks—four sheaves, four rope parts at the movable block.
2. Friction loss (10% for each sheave) = 40% or 800 lb (363 kg).
3. 2,000 + 800 = 2,800 lb (1,270 kg), which divided by 4 (the number of parts at the movable block) = 700 lb (318 kg).
4. Applying the safety factor of 10 (10 × 700) gives 7,000 lb (3,200 kg), the required breaking strength of the rope.
5. New Manila rope of ⅞ in. (22 cm) has a breaking strength of 7,700 lb (3,500 kg) and, therefore, is the proper size for the load. Synthetic fibers would have greater tensile strength. (See Table 16–A in Chapter 16, Ropes, Chains, and Slings, in this volume.)

The safe work load limit for two double blocks made for rope of ⅞ in. (22 cm) diameter is 2,000 lb (900 kg)—the equivalent of the total load in the example. This information is from a prominent manufacturer for one series of its standard blocks (regular mortise, inside iron-strapped blocks, with loose side hooks, intended for use with Manila rope).

Attach the rope to the block with a thimble and a proper eye splice. A mousing of yarn or small rope should be placed on the upper hook of a set of falls as a precaution against its unintentional detachment. Inspect blocks thoroughly and frequently, paying particular attention to parts that are subject to wear.

Figure 15–14 shows how tackle blocks should be reeved. If the sheave holes in blocks are too small to permit sufficient clearance, excessive surface wear of the rope will occur. Likewise, excessive internal friction on the fibers will occur if the diameter of the sheave is too small for the rope.

When using block and tackle in confined spaces, provide guards on the pulley block so that a person's hands cannot be caught between the pulley and the rope. When blocks and falls are used to lift heavy materials or to keep heavy loads in suspension, as on heavy-duty scaffolds, wire rope is more serviceable than fiber rope.

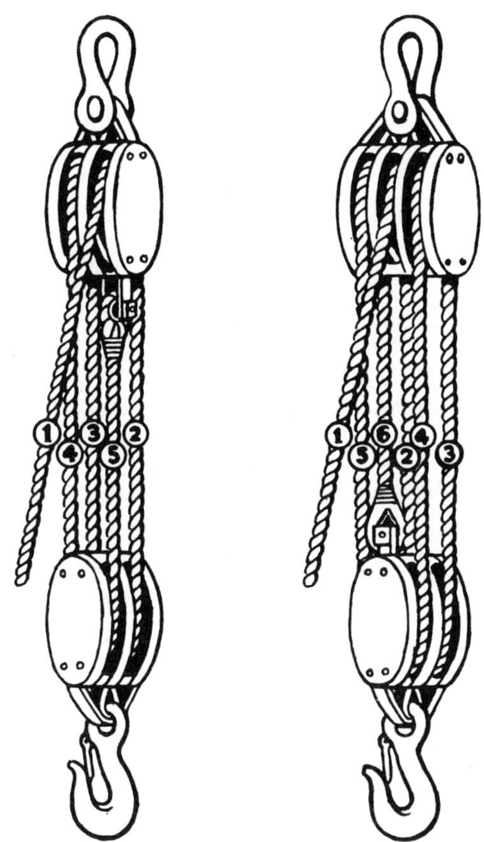

Figure 15–14. Reeving tackle blocks. Lead line and becket line should come off a middle sheave when blocks contain more than two sheaves. The upper and lower blocks will then be at right angles to each other. This eliminates tipping and the accompanying loss in efficiency.

Portable Floor Cranes or Hoists

Portable floor cranes or hoists are hoists mounted on wheels. They can be moved from place to place, either by hand or under their own power. They can raise and lower loads in a vertical line but cannot rotate around a fixed point.

Portable floor hoists are useful where overhead construction, belting, or shafting prevents the use of overhead hoists or cranes and where more expensive equipment is not justified because it would be used infrequently. These hoists are handy for placing work on machines, loading heavy material on trucks, and moving material from one location in the shop to another.

Portable floor hoists are usually operated by hand or by electric power. The lifting mechanism ordinarily is either a chain hoist or a winch with wire rope and block. Hoists operated by electric power should be effectively grounded to prevent shock in case of a short circuit. If the power cord does not have a separate ground conductor, use a special grounding wire with one end of the wire fixed permanently to the frame of the hoist. The other end of the wire is equipped with a device that can be attached to a grounded building column, water line, or other direct-to-ground connection.

To prevent foot injuries when the hoist is moved, have employees wear protective footwear and install sweep guards on the truck's wheels, where conditions permit. Truck handles on hoists should be designed to stand upright when not in use. If they project horizontally from the hoist or lie on the floor, these handles can cause workers to trip.

Tiering Hoists and Stackers

The tiering hoist—sometimes called a stacking elevator, portable elevator, tiering machine, or platform hoist—is designed to raise material in a vertical line on a moving platform. This hoist is portable and is used extensively in warehouses for piling and storing materials. It is operated either manually or electrically. Large capacity hoists usually are power driven.

Tiering hoists should have a braking device that permits safe lowering of the platform. They should also have a rachet lock (or dog) to lock the platform in position for loading and unloading operations. Do not permit workers to ride the platform because they could be crushed if the platform met an obstruction.

Tiering machines, especially the revolving type, should be operated so that they will not tip over. Basic precautions include having the machine solidly on the floor, making sure that its safe load capacity is not exceeded, and placing the load properly on the platform.

Before the machine is used, lift the casters off the floor. One type of hand-operated hoist is arranged so that the platform cannot be moved unless the machine stands solidly on the floor—on the frame and not on the casters. Material in the form of rolls and other round objects should be blocked to prevent them from rolling off the platform.

Ground power-operated tiering hoists and safeguard the wiring, preferably with armored cable. On two-section machines, provide locks to prevent the release of the upper section. To prevent shearing of fingers, install guards at gears and channel-iron guides. Some tiering machines have almost all the safeguards that a freight elevator has, including limit stops for top and bottom travel on the hoisting cable drum and for the shipper rope, if one is provided.

CONVEYORS

There has been much confusion as to what a conveyor is and what it is not, as well as confusion about the correct names for different types of conveyors. Therefore, the general definition of conveyor, given in ANSI/ASME B.20.1, *Safety Standards for Conveyors and Related Equipment,* is quoted here in full:

Conveyor. A horizontal, inclined, or vertical device for moving or transporting bulk material, packages, or objects, in a path predetermined by the design of the device, and having points of loading and discharge, fixed or selective. Included are skip hoists, and vertical reciprocating and inclined reciprocating conveyors. Typical exceptions are those devices known as industrial trucks, tractors, and trailers; tiering machines; cranes, hoists, power shovels; power scoops, bucket drag lines; platform elevators designated to carry passengers or the operator; manlifts, moving walks; moving stairways; highway or rail vehicles; cableways; or tramways, pneumatic conveyors, or integral transfer devices.

By industry agreement, nomenclature and definitions have been standardized and published in ANSI/Conveyor Equipment Manufacturers Association (CEMA) 102, *Terms and Conveyor Definitions.* Terms and definitions in this volume follow this standard.

General Precautions

The most common conveyors are of the belt, slat, apron, chain, screw, bucket, pneumatic, aerial, portable, gravity, live roll, en masse, flight, mobile, and vertical types. Design and construct these conveyors to conform with applicable codes and regulations.

Place a highly visible sign at each loading point of manually loaded conveyors traveling partially or entirely in a vertical path. The sign should show the safe load limit that can be raised or lowered. Gears, sprockets, sheaves, and other moving parts must be protected either by standard guards or must be positioned in such a way to ensure against personal injuries.

Periodically inspect the entire conveyor mechanism. Immediately replace any part showing signs of excessive wear. Pay particular attention to brakes, backstops, antirunaway devices, overload releases, and other safety devices to ensure that they are working and in good repair.

Lubricate all machine parts according to the manufacturer's instructions. Install grease nipples on long tubes or pipes. This not only permits oilers to keep a safe distance from moving parts, but also prevents shutting down the conveyor for greasing.

All conveyors within 6 ft 8 in. (2 m) of a floor or walkway surface that is a means of exit must have alternate passageways that comply with NFPA 101, *Life Safety Code,* requirements. Frequently, a work platform on a movable conveyor tripper can be used as a crossover, if properly railed.

Underpasses should have sheet metal ceilings. Where overhead conveyors dip down at workstations, install guards or handrails. Also install guards below all conveyors passing over roads, walkways, and work areas.

For conveyors that run in tunnels, pits, and similar enclosures, provide adequate drainage, lighting, ventilation, guards, and escapeways. These features are required wherever it is necessary for persons to work in or enter such areas. Provide sufficient side clearance to allow safe access and operating space for essential inspection, lubrication, repair, and maintenance operations.

Where conveyors pass through building floors, the openings should be guarded by standard handrails and toeboards. As a fire precaution, protect each opening against the spread of fire or super-heated gases from one floor to the next. To prevent the spread of fire, install doors that close automatically or install fog-type automatic sprinklers. Place the sprinklers so they provide a curtain of water fog across the opening. Where a conveyor passes through a fire wall, provide similar protection. Conveyor tunnels under stock piles of materials should be open at both ends.

If the top of a loading hopper is at or near the level of a floor or platform, protect the hopper with standard railings and toeboards or with a bar guard with openings not greater than 2 in. (5 cm) in one dimension, such as 2 in. × 12 in., 14 in. × 2 in., 2 in. × 16 in. (5.1 cm × 30.5 cm, 35.5 cm × 5.1 cm, 5.1 cm × 40.6 cm).

Elevated conveyors should have access platforms or walkways on one or both sides (Figures 15–15 and 15–16). Handrails should be 42 in. (1 m) high with an intermediate rail. Platforms should have 4 in. (10 cm) toeboards. Use checkered plate flooring or other slip-resistant surface, particularly on sloping walkways.

Sideboards along edges and at corners and turns of overhead conveyors, along with screen guards underneath high runs, will protect workers from falling material. Provide crossovers or underpasses with proper safeguards for passage over or under all conveyors. Prohibit workers from crossing over or under conveyors, except where safe passageways are provided. Forbid workers from standing or riding on conveyors.

Figure 15–16. This high-speed, slat-type conveyor system uses pushers to gently sweep items onto sort lanes. Note the access stile with handrails at the far right. (Courtesy Western Atlas, Inc.)

Operating Precautions

Locate the start button or switch for a conveyor so that the operator can see as much of the conveyor as possible. If the conveyor passes through a floor or wall, equip each side of the floor or wall with starting and stopping devices. Require that all starting buttons or switches be operated simultaneously to start the conveyor. Clearly mark these start-stop devices. Keep the area around them free of obstructions so they can be seen and reached easily. Instruct all personnel working on or near the conveyor about the location and operation of all stopping devices.

Figure 15–15. This elevated conveyor system in a quarry has a clear walkway with handrails. (Courtesy Vulcan Materials Co.)

Provide electrical or mechanical interlocking devices—or both types—that will automatically stop a conveyor when the unit it feeds (another conveyor, bin, hopper, or chute) has been stopped or is blocked. Thus, the conveyor cannot receive additional loads. If two or more conveyors operate in a series, design the controls so that if one conveyor is stopped, all conveyors feeding it are also stopped.

Locate emergency stopping devices not more than 75 ft (23 m) apart along walkways by the conveyor. For some installations, a good solution is to have a lever-operated emergency stopping device at the end of the conveyor, with a strong wire cord strung on each side of the conveyor for its entire length. A pull on the wire cord will stop the conveyor.

On conveyors where there is a possibility of reversing or running away, provide antirunaway and backstop devices. In addition, design the conveyor track so that the load (or conveyor parts) cannot slide or fall in case of mechanical or electrical failure. If such a design is not practical, install guards able to withstand the impact of shock and to hold the falling load. Design electric machines with brakes that are applied and released by the movement of operating devices, so that, if the power is interrupted with the brakes in the OFF position, the load will descend at a controlled speed.

In addition to the overload protection usually provided for electric motors, include an overload device to protect the conveyor and mechanical drive parts. Shear pins and slip or fluid couplings are examples of overload protective devices. In the event of an overload, the device must shut off the electric power quickly, disconnect the conveyor or drive parts from the power, or limit the applied torque. When a conveyor has stopped because of an overload, lock out all starting devices and remove the cause of the overload. Inspect the entire conveyor before restarting it.

With exhaust hoods, cover the loading and discharge points of a conveyor carrying material in fine or powdered form. To prevent the formation of dust clouds, provide good ventilation. Dust removal must be in accordance with government regulations. If the material is combustible, the concentration of dust must be kept below the lower explosive limit. Use only approved explosion-proof electrical fixtures. Exclude all sources of ignition from the dusty area. Ground the conveyor, and bond its parts to prevent differences in polarity. Also, ground the container into, or from which, the material is conveyed, and bond its parts.

Persons working near or on conveyors should wear close-fitting clothing that cannot become caught in moving parts. Safety shoes are recommended. If the conveyor galleries are dusty, have workers wear goggles and, if necessary, respirators.

Maintenance Precautions

Before maintenance personnel start working on a conveyor, they should lock out the main power control in the OFF position. The workers should carry the only key to their lock. If two gangs work on one conveyor, the supervisor of each gang should lock out the master switch per OSHA standard 1910.147 for group lockout/tagout.

Maintenance personnel should be able to change the position of pulleys, sprockets, or sheaves to compensate for normal working conveyor stretch and wear. Provide guards for the on-running belt at least 18 in. (55 cm) from where the belt and the head and tail pulleys touch and from where the belt and the tripper and hump pulleys touch. If the hazard is out of reach (more than 8 ft [2.4 m] above the floor or platform) or close to a wall or other obstruction, workers in the normal course of their duty should not be exposed to it.

If the skirtboards of the loading boot are close to the upper surface of the belt, a person's arm could get caught. In such a case, the belt could not be raised sufficiently to allow the arm to ride over an idler pulley under the belt. Thus, the arm undoubtedly would be badly mangled. Therefore, install guards at the sides of the conveyor at the loading boot.

Also guard the points of contact where the wheels of movable trippers roll on the rails. Where the operator travels on the tripper, provide a platform. Locate and construct the platform to protect the operator from slipping and falling and to prevent contact between the operator or his clothing and moving parts. To help prevent the operator from falling into the hopper, install handholds and railings on the platform.

At the conveyor floor above the coal bunkers in power plants, the slot—through which coal from the tripper chute is discharged into the bunker—is protected by bars placed across it about 12 in. (30 cm) apart at floor level. A piece of discarded belting of the required width and length can be placed to cover the slot, with its ends securely anchored. At the tripper platform are four pulleys that raise this belt vertically so that it passes over the access to the tripper platform. This device not only provides safety but also seals the slot and thus keeps the dust in the bunker.

If a person should fall onto a conveyor, suspend a gate or paddle as low as possible above the belt near the head pulley. In that way, the person riding on the belt could automatically pull a stop rope and quickly stop the conveyor. On belt conveyors that are at floor level or on balconies or galleries, a shield guard or housing should completely enclose each end. Guardrails and toeboards should extend the length of the conveyor.

To help remove static from belt conveyors, place tinsel or needlepoint static collectors close to the outrunning sides of the drive pulleys and idlers. Ground the pulleys and idlers, along with the shafting, through carbon or bronze brushes running on the shaft. A belt that does not move too fast can be grounded to the drive gear by a continuous strip of copper foil on the pulley side. Other belt conveyors can be grounded by being treated with conductive belt dressings.

One of the dangers of belt conveyors is that workers are tempted to clean off material that sticks to the tail drums or pulleys while they are in motion. Fixed scrapers and revolving brushes eliminate the need for cleaning by hand. To protect workers should they attempt to clean or dress the belts while they are moving, place the barrier guards directly in front of the pinch points of belts and drums. Guard the belt and drum on the side, also at a sufficient distance from the drum to prevent contact.

Belt Conveyors

A belt conveyor is an arrangement of mechanical parts that supports and propels a conveyor belt, which in turn carries bulk material. The five principal parts of a typical belt conveyor are: (1) the belt, which forms the moving and supporting surface on which the conveyed material rides; (2) the idlers, which form the supports for carrying the belt; (3) the pulleys, which support and move the belt and control its tension; (4) the drive, which imparts power to one or more pulleys to move the belt and its load; and (5) the structure, which supports and maintains the alignment of the idlers and pulleys, and supports the driving machinery.

Like all moving machinery, belt conveyors present hazards to workers, who must be safeguarded. Operators must be trained in safe work procedures around belt conveyors. Stress that workers stand clear of moving conveyors. Train them to avoid pinch points and other areas where their fingers or hands may get caught when the conveyor is moving. Place guards at all pinch points.

Other frequent causes of conveyor incidents are improper cleanup of conveyors and shoveling of spillage back on the belt. Severe injuries also arise from (1) attempting repairs or maintenance on moving conveyors, (2) attempting to cross moving belts where no crossover exists, and (3) attempting to ride a moving belt. Moreover, workers standing or working on conveyors, particularly maintenance personnel, can be injured by falls or crushed against stationary objects, particularly if a conveyor is unintentionally started.

The following injuries and hazards could be reduced by using mechanical and environmental controls around belt conveyors:

- fires from friction, overheating, static, or other electrical sources
- explosions of dust raised by combustible materials at transfer points, where belts are loaded or discharged
- respiratory and eye irritations from toxic dusts
- electrical shock from ungrounded or improperly installed controls or conductors.

Provide guards for transmission equipment and other power-driven parts in accordance with the ANSI/ASME B15.1, *Safety Standard for Mechanical Power Transmission Apparatus,* and with all state or provincial and federal regulations governing the safety, health, and welfare of employees. The ANSI/ASME B15.1 stipulates how pulleys, chains, sprockets, belts, couplings, and other parts of conveyor drives should be enclosed.

Suitable sweeps should be provided for shuttle conveyors, movable trippers, traveling plows, and hoppers and stackers. The sweeps push objects ahead of the moving pinch points between the wheels and the rails, thus guarding against nips. Again, it is imperative to comply with all state or provincial and federal regulations. Provide guards at pinch points at the head, tail, and take-up pulleys. An idler pulley becomes a hazard when skirt plates and chute skirts are so positioned as to force the belt against the idler, thus creating a pinch point.

Mechanical belt cleaners, such as fixed or tension scrapers, revolving brushes, or rubber disks, sometimes eliminate manual cleaning. They also eliminate a major reason for working around moving pulleys.

In order to lubricate a conveyor that is in continuous operation, install extension grease lines. In that way, workers cannot get caught by rollers and bearings when working around them. All grease fittings inside a guard enclosure (except those that move with the part they serve) should be fitted with extension pipes to make them accessible from outside the guard.

Slat and Apron Conveyors

The slat conveyor has one or two endless chains operating on sprockets and usually runs horizontally or at a slight incline (Figure 15–16). Attached to the chain or chains are nonoverlapping, noninterlocking slats that are closely spaced.

Apron conveyors have overlapping or interlocking plates that form a continuous moving bed. They vary greatly in size: one may be part of a bottling machine, and another may handle billets and castings in a steel mill.

Place guards at pinch points between slats or plates, and between them and the chain, sprockets, and guides. Where slats are spaced farther apart than 1 in. (2.5 cm), a serious shearing hazard exists between the slats and the conveyor substructure. When a slat conveyor is located at floor level or in working areas, the space under the top run of the slats should be filled in solid.

When designed for handling heavy material, slat and apron conveyors are usually installed flush with the floor to make loading and unloading easier. When so located, conveyors should be guarded by handrails (except at loading stations) so that workers will not step onto the moving conveyor. Openings in the floor at the loading platforms should be guarded or covered.

Chain Conveyors

Chain conveyors take many forms. However, they all carry, pull, push, haul, or tow the load either (1) directly by the chain or (2) by means of attachments, pushers, cars, or similar devices. Conveyors in which the chain itself directly moves the load are drag, rolling, and sliding chain conveyors.

Tow Conveyors

Conveyors consisting of an endless chain that is supported by trolleys from an overhead track or that runs in a track above, flush with, or under the floor, with attachments for towing trucks, dollies, or cars, are known as tow conveyors. Tow conveyors use four-wheeled carts that are held in place by a pin through the chain. These pins should be equipped with an automatic disengaging mechanism. Many pins have a horizontal bumper or kick mechanism that will disengage with 3 lb to 5 lb (1.4 kg to 2.3 kg) of pressure. Workers are consequently protected from being caught between carts.

Trolley Conveyors

An endless chain that propels a series of trolleys (Figure 15–17) supported from or within an overhead track, with the loads suspended from the trolleys, is a trolley conveyor. On trolley conveyors, guard well the return portion of the chain. Whenever possible, install guardrails along both sides of the trough to prevent a person from stepping or falling into it. Clearly designate the path of travel of such overhead chain conveyors. Provide emergency stop switches every 75 ft (23 m).

Figure 15–17. A trolley can be used to support hoists in trolley conveyors, overhead tow conveyors, overhead chain conveyors, and on jibs and monorails. This overhead monorail delivery system can carry weights up to 6,000 lb (2,722 kg). (Courtesy Litton.)

Shackle Conveyors

A shackle conveyor consists normally of a chain-type conveyor, with suspended shackles spaced evenly along the line to convey poultry or other meat products through a processing plant. In a poultry processing plant, the shackle conveyor or conveyors carry poultry (1) from the beginning of the line where live poultry is hung from each shackle, head down, (2) through all the processing phases, (3) to the end of the line where the finished product is removed for packaging and shipping. In some plants, one continuous conveyor travels through all departments and processes, while in other plants two or more similar conveyors carrying suspended shackles are used.

The speed of such conveyors varies from plant to plant, depending on the number of workers on any given conveyor line. The speed generally increases with more workers on the line. Poultry is normally processed by being hung head down and, after being cut and bled, repositioned feet down before going through scalders and beaters and then on to viscerating and cleaning.

The hazard posed by such shackle conveyors involves workers unintentionally placing a thumb or finger in one of the shackles and being dragged through a beater or scalder or to the end of a platform where, suspended by a finger, a worker could lose a thumb, a finger, or his or her life. Limit this hazard by placing EMERGENCY STOP switches at critical points, such as at the end of a platform or before entry into hazardous areas. The emergency switches are actuated by pull wires positioned below and perpendicular to the line of travel.

Screw Conveyors

Screw conveyors move bulk materials horizontally, on inclines, and vertically. They are employed in many industries and handle a variety of materials. A screw conveyor essentially consists of a continuous spiral mounted on a pipe or shaft. The screw rotates in a stationary trough, generally U-shaped. Material introduced into the trough is conveyed by the screw's rotation (Figure 15–18).

A screw conveyor that is not enclosed presents the hazard of entrapment. Feet, hands, or other portions of the body may get caught between the rotating screw and the stationary trough (Figure 15–19). When such unintended incidents occur, they can result in serious injury or death. These kinds of incidents can be avoided by setting up and enforcing an in-plant safety program and by training employees to observe the following basic safety rules:

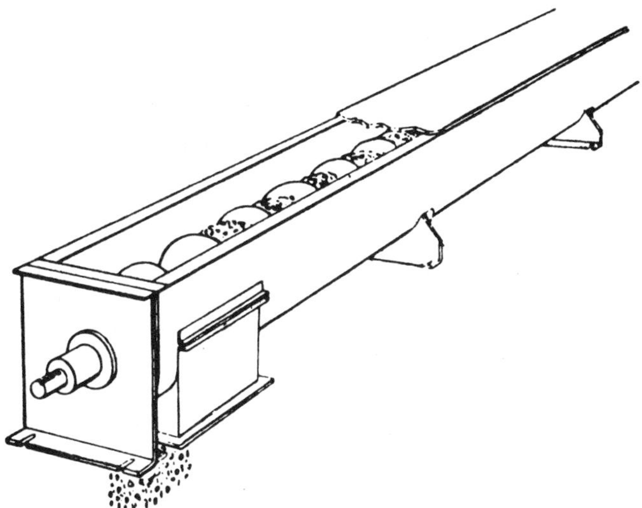

Figure 15–18. A screw conveyor with the cover cut away to show the interior. (Courtesy Conveyor Equipment Manufacturer's Association.)

Figure 15–19. This warning label is used by screw conveyor manufacturers to illustrate the nature of the hazard, the severity of an injury, and precautions to avoid it. Serious injury or death may result due to failure of personnel to follow basic safety rules.

- Covers, gratings, and guards shall be securely fastened before operating the conveyor. Conveyors should be provided with solid covers, securely fastened by bolts, spring clamps, or screw clamps. When an exposed feed opening is required, cover it with a securely fastened grating. If an open housing is functionally necessary, guard the entire conveyor with a fence or railing. Moving drive components shall be guarded.
- Never step or walk on covers, gratings, or guards.
- Lock out power before removing covers, gratings, or guards by padlocking the main disconnect in the OFF position.

More detailed information on screw conveyor operation and safety is available in Book No. 350, *Screw Conveyors,* by the Conveyor Equipment Manufacturers Association; in ANSI/ASME B20.1, *Safety Standards for Conveyors and Related Equipment,* and ANSI Z244.1, *Safety Requirements for Conveyors and Related Equipment.*

Bucket Conveyors

The three general types of bucket conveyors all have an endless belt, chain, or chains that carry either fixed or pivoted elevator buckets (Figure 15–20). Descriptions of these three types of bucket conveyors follow:

- A bucket elevator carries fixed buckets in a vertical or inclined path. The buckets discharge by the force of gravity as they pass over the head sheave or drum.
- A gravity-discharge conveyor elevator has fixed buckets and operates in vertical, inclined, and horizontal paths. The buckets act as flights while carrying material along a trough in the horizontal plane to a point of gravity-discharge.
- A pivoted-bucket conveyor also operates in horizontal, inclined, and vertical paths. The buckets remain in the carrying position until they are tipped or inverted to discharge.

To protect operating personnel, totally enclose bucket conveyors in a housing. Do not attempt to take samples when the conveyor is in motion. The pivoted-bucket conveyor has tripping devices for emptying the buckets. Frequently these tripping devices are movable so materials can be distributed evenly in the storage bins. Arrange for the remote control of levers for shifting and locking the trippers. In that way, workers will not

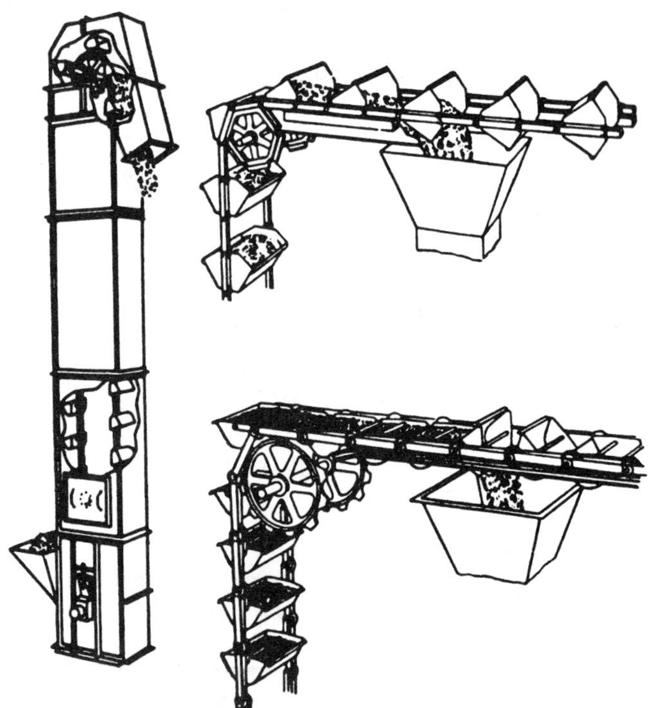

Figure 15–20. Three types of bucket conveyors. *Left:* bucket elevator. *Upper right:* gravity-discharge conveyor-elevator. *Lower right:* pivoted-bucket conveyor. (Printed with permission from Conveyor Equipment Manufacturer's Association.)

have to go on the conveyor to change the position of the trippers.

For the safety and convenience of repair personnel and operators, build a permanent footwalk alongside a conveyor that hoists material and carries it over stokers or bins. Equip the footwalk with standard handrails and toeboards, and be sure it is well lit.

Pneumatic Conveyors

A pneumatic conveyor is an arrangement of tubes or ducts through which solid objects such as cash, mail, and other small items are moved by using compressed air or a vacuum. Solid objects are placed inside a cylindrical cartridge. Compressed air, injected into the tubing system, pushes the cylinder forward at a relatively high velocity.

Bulk materials such as grain and dust are also moved by pneumatic conveyors. To convey bulk material, compressed air is injected into the piping though a nozzle below the loading chute. The air mixes with the material making it flow like a fluid through the piping.

When transporting material that could cause a dust explosion, the air velocity should exceed the critical velocity for starting a fire. In this way, an explosion from one point to another within the system will be prevented. Bond equipment electrically and ground it to prevent static electricity from being a source of fire. Additional safety precautions for working with pneumatic conveyors follow:

- A constant volume, variable pressure (positive type) blower should have a relief valve on or adjacent to it. Keep doors of blower conveyors interlocked so they cannot be opened if there is positive internal pressure. To keep conveyed material from being thrown against workers or into the working area if a gasket leaks, shield gaskets that hold the line pressures.
- Where suction lines are large enough to draw a person in, place bar guards or screening over the intake. Instruct employees to stay at a safe distance from such lines.
- Receivers and storage bins should have full-bin indicators or controls to prevent overfilling.
- A pneumatic conveyor serving an area containing contaminated air must be arranged so that no contaminated air enters the conveyor tube and is thus carried to other areas.

Aerial Conveyors

A description of the two types of aerial conveyor systems follows:

- A cableway aerial conveyor is a wire rope-supported system in which the material-handling carriers are not detached from the operating span and the travel is wholly within the span.
- A tramway aerial conveyor is a wire rope-supported system in which the travel of the material-handling carriers is continuous over the supports of one or more spans.

Aerial conveyors are used frequently in industrial plants, and particularly on large construction sites, to carry material from point to point. They also move coal and ore. Workers are generally injured by falling material when inspecting and oiling the cables and carriages, or if signals are misunderstood.

No one should work directly under the conveyor except, of course, at the loading and unloading stations. Take exceptional precautions at these points. Wherever workers must pass under the conveyor, provide a covered passageway. Provide heavy wire screens, suspended under the conveyor from pole to pole, that are wide enough to catch material that would otherwise fall on roadways or work areas.

Inspect the equipment regularly. Pay special attention to the sheave wheels and bearings, the rope fastening, the bucket latch and trunnions, and all load-sustaining parts.

For operation efficiency and protection against the weather, keep all ropes well oiled. Hauling rope should be continuously lubricated by means of a controlled drop feed from an oil reservoir at the point (one or both ends of the line) where the rope leaves the drive sheave and passes over a support sheave. In one facility, a worker had to ride a trolley carriage to oil a tramway rope used for conveying coal. An automatic lubrication system was devised to eliminate this dangerous practice.

For night work and for repairs, provide suitable lighting. Also provide a suitable signaling system, such as a telephone or electric push-button system, for every cableway aerial conveyor. On large construction jobs, a portable telephone system allows a signaler to direct the raising and lowering of loads that are out of the operator's sight.

For a tramway aerial conveyor, provide the following three control systems:

1. push-button stations and a bell signal code that indicates stop, start, slow speed, high speed, and reverse
2. an all-metallic, aerial, wire-circuit telephone with instruments at certain points along the line in addition to terminal sets
3. a second telephone circuit, which may be grounded if desired.

The U.S. Army Corps of Engineers suggests taking the following precautions when using cableway aerial conveyors:

- Keep the control console compartment locked when the cableway is in use. Permit no one in the compartment with the operator while the cableway is in operation.
- Continuously maintain at least two communication and control systems between the signaler and the cableway operator. This dual system should include voice communication by telephone and radio. Lights or bells may be included with, or substituted for, one of the voice systems.
- Only permit authorized inspection and maintenance personnel to ride cableway carriages.
- Do not permit the riding of cableway load blocks.

Portable Conveyors

Belt, flight, apron, extendable, and fixed-bucket conveyors are made as inclined portable units set on a pair of large wheels. They are used to load railroad cars and trucks with bulk materials and to raise construction material from one level to another.

Electrical equipment on portable conveyers should be weatherproof. With three-phase power, the flexible cord that is connected to the power outlet should be a four-conductor, the fourth wire being grounded in all plugs and receptacles. Arrange the cable so it cannot be run over by trucks or other machines. If it must cross a driveway, hang it on poles at a minimum height of 14 ft (4.3 m). If two or more sections of cable are required, the connectors should be kept above the ground.

Provide portable conveyors with guards as specified for the corresponding types of fixed conveyors. Skirtboards or sideboards (not less than 10 in. [25 cm] high) keep heavy material from falling over the sides and light or loose material from blowing off. This safety precaution applies to belt conveyors, as well as to other types of conveyors, because troughing of the belt gives insufficient protection against such spillage.

Fit the conveyor with a locking device to hold the conveying unit at various fixed levels, thus keeping the conveyor stable. Closely check the mechanism that raises and lowers the boom on all types of portable conveyors. This is the major hazard area of such machinery. While any positive type may be used, the self-walking worm or jack-screw type is preferred. Gear drives should be completely housed and run in oil. Thoroughly guard all chains within easy reach of workers. Devise a system for conveniently oiling chains.

In order to work in small spaces in warehouses, portable conveyors usually have booms consisting of two sections. Design these conveyors so material will not roll back from one section to the other at the transfer point.

When loading or unloading railroad cars with a portable conveyor, use a suitable safety device to prevent the cars from shifting during the operation. Do not remove the device until the crew is sure no one is in the car. A red banner or a standard blue warning sign at each end of the car is recommended as a warning to the switching crew. (See Chapter 2, Buildings and Facility Layout.)

When using portable conveyors to raise or lower construction materials, provide ample stairs or ladders in the immediate vicinity of the conveyor. Prohibit workers from walking on idle or moving conveyors.

Gravity Conveyors

Because gravity conveyors (Figure 15–21) depend wholly upon the natural pull of gravity, workers often disregard some necessary safe practices when working around them. If employees climb on a conveyor to release a blockage, they may slip on the rollers or be knocked down should the jam suddenly be released. Therefore, prohibit workers from climbing onto conveyors. In addition, the installation of steel or wood plates between rolls helps assure that neither a worker's body nor limb can fit between the rolls.

Chute Conveyors

Chute conveyors made of polished metal sheets or bars lower packing cases, cartons, and crates from one floor to another. They also move material from the sidewalk to the basement of a building through an elevator shaft in the sidewalk, or other opening.

Figure 15–21. Types of gravity conveyors. *Left:* roller conveyor. *Right:* spiral chute. (Courtesy Conveyor Equipment Manufacturer's Association.)

Inclined chute conveyors may be straight, or have a vertical curve with radius large enough to deliver packages onto the lower floor without impact or damage. Some gravity-chute conveyors are built like a spiral around a vertical pipe, with a slope at the outer edge between 18 degrees and 30 degrees.

In removing packages from the delivery end of spiral chutes, workers frequently injure their hands when they are caught by descending packages or are crushed against other packages on the delivery table. Where the chute is enclosed, place a warning sign over the delivery end. Install a simple mechanical or electrical device that signals when a package is about to be delivered from the chute. This is especially needed where the descending packages cannot be plainly seen.

Spiral chutes present a serious fire hazard because they form flues from lower floors to upper floors through which fire can quickly spread. Two methods for eliminating this hazard are (1) enclose the chute in a tower made of fire-resistant material, such as steel, concrete, or masonry, or (2) provide automatic fire doors (draft checks) where the chute passes through floors. The enclosed tower has doors at each charging station and a door at the delivery end. Keep the charging station doors closed, except when charging is being done. The door at the delivery end should close automatically in case of fire. Automatic fire doors are of two types—vertical sliding and shutter. Both types should have fusible links so that they will close automatically in case of fire.

Where an open chute is used, provide a guardrail and toeboard at each floor. Use either a movable railing or a hinged door or gate at the charging stations.

Roller or Wheel Conveyors

Roller or wheel conveyors are similar to chute conveyors, except that the angle of slope is much less (2% to 4%). These conveyors, therefore, can be used to convey packages for considerable distances on one floor. If the rollers or wheels are placed radially instead of parallel, the course of travel can be changed from a straight line to a curve.

The principal hazards in the use of roller conveyors are (1) that material may run off the edge of the rollway and fall to the floor and (2) that loads may run away. Provide a guard railing on each side of the roller conveyorway, to guide the material and prevent it from running off. Such guardrails are especially advisable at corners and turns and on elevated conveyors under which workers must work or pass. When heavy loads are conveyed, retarders, brakes, or similar devices help prevent the loads from running away. A power conveyor on which speed can be controlled could also be substituted.

Hinge a vertically swinging, hinged section of a roller or wheel conveyor to the end of a stationary section of the conveyor from which the material is flowing. This helps block the oncoming material. The open end of the conveyor line should have a stop that (1) automatically projects above the level of the rollers or wheels when the hinged section is opened and (2) automatically retracts when the hinged section is closed.

Where a horizontally swinging, hinged section occurs in a conveyor, equip the open ends of stationary sections (the two ends adjacent to the hinged section) with retractable stops. The stops prevent loads from dropping off when the hinged section is open.

Live Roll Conveyors

A live roll (or roller) conveyor consists of a series of rolls over which objects are moved by power applied through belts or chains to all or some of the rolls. Where installed at floor level or when used in work areas, design live roll conveyors to eliminate hazards from pinch points and moving parts. Such a precaution will prevent personnel from coming in contact with or crossing the conveyor.

Vertical Conveyors

Vertical conveyors handle packages or other objects in a vertical, or substantially vertical, direction. In some cases, a hinged section, interlocked to the power in the main system, will be provided for access to the workstation. The interlock should shut down the entire system until the section is restored to its position. Descriptions of three basic types of vertical conveyors follow:

- Vertical reciprocating conveyors are power- or gravity-actuated units that receive objects on a carrier or car bed, usually constructed of a power or roller conveyor, and that elevates or lowers objects to other locations.
- Suspended tray conveyors are vertical conveyors having one or more endless chains with pendant trays, cars, or carriers. These carriers receive objects at one or more levels and deliver them to another or

several levels. To prevent materials from falling on people, install guards on the underside of suspended tray conveyors.

- Vertical chain conveyors (opposed shelf types) are two or more vertical-elevating conveying units opposed to each other. Each unit consists of one or more endless chains whose adjacent facing runs operate in parallel paths. Thus, each pair of opposing shelves or brackets receives objects (usually dish trays) and delivers them to any number of levels.

Where vertical conveyors are automatically loaded and unloaded, provide guards to protect personnel from contact with moving parts. Where they are manually loaded and unloaded, install guards and safety devices, such as lintel and sill switches and deflectors.

Carriages of vertical reciprocating conveyors designed to register at a floor, balcony, gallery, or mezzanine level never should have a solid bed. This type of conveyor is not intended to carry passengers or operators, or to have its car or carriage called to a station by a manually operated push button.

POWER ELEVATORS

Before drawing up specifications for new elevators or planning major alterations for existing ones, refer to the latest edition of the American National Standards Institute/American Society of Mechanical Engineers (ANSI/ASME) A17.1, *Elevators, Escalators, and Moving Walks.* For the rest of this chapter, this standard will be referred to as the Elevator Code.

Specifications for new equipment conform to Elevator Code requirements, unless federal, state, or provincial, or local codes are more stringent. In such cases, observe the governmental regulations as the minimum requirements. Usually, the Elevator Code, however, will be stricter. Two commonly used wordings in the Elevator Code are "The elevator and associated equipment shall meet the requirements of A17.1, *Elevators, Escalators, and Moving Walks,* latest edition, except as hereinafter specifically exempted or modified." and "Except as changed or modified herein, the elevator and associated equipment shall meet the requirements of A17.1, *Elevators, Escalators, and Moving Walks,* latest edition, and also shall comply with all applicable local laws and/or ordinances."

Such paragraphs will generally take care of items not specifically spelled out in the specifications and drawings.

Types of Drives

There are two major types of power elevators: (1) electric-drive elevators and (2) hydraulic-drive elevators. However, some belt- and chain-drive machines are still in use, and repairs to existing installations are permitted by the Elevator Code. Their new installation is now prohibited by the Elevator Code.

Electric Elevators

The two general types of drives for electric elevators are traction drive and winding-drum drive. The winding-drum type, however, is now obsolete and is presently being used only in dumbwaiter elevators and in freight elevators, with restrictions as specified by the Elevator Code.

The Elevator Code requires that all drives be of the traction type. However, it permits the use of winding-drum drives on freight elevators that travel not more than 40 ft (12.2 m) at speeds not exceeding 50 ft per minute (0.25 m/s) and that are not provided with counterweights.

In the winding-drum drive, the hoisting rope is anchored in and winds on a spirally grooved drum. This is a positive drive. If the machine is not stopped at the limits of travel, the car may be pulled into the overhead structure and, should the motor be powerful enough, the ropes may be pulled from their anchorage.

In the traction-drive elevator (Figure 15–22), the hoisting rope is not attached to the drive. The elevator is moved by the traction (friction) of the ropes in grooves on the drive sheave. The grooves may be semicircular (U-groove) or undercut U-groove. Some V-groove sheaves are still in use. Generally, the V-groove or undercut U-groove is used on geared machines. The U-groove is used on gearless machines, but there are exceptions.

A simple rule to remember is that when the drive sheave has twice as many grooves as there are hoisting ropes, the drive is a double (full) wrap. The total angle of contact of each rope with the drive sheave generally is between 300 degrees and 360 degrees. The traction relation changes very little with wear.

Most companies use some form of undercut U-groove for their single-wrap drives. The friction is higher than an ordinary U-groove because of the pinching

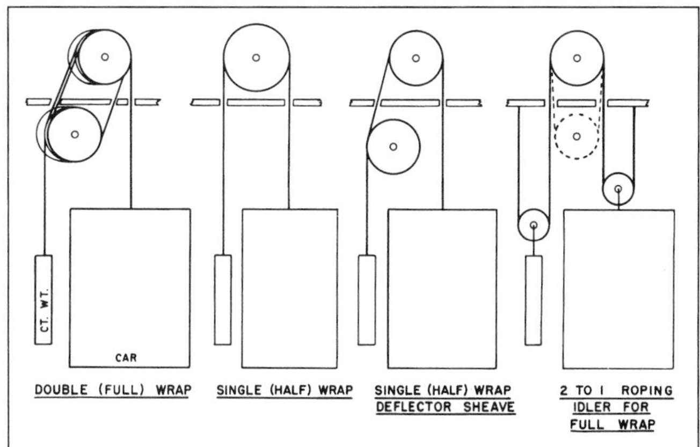

Figure 15–22. Common elevator traction drives showing double (full) and single (half) wraps.

action. The width of the undercut is varied to suit the needs of traction. The traction remains substantially constant until the groove has worn to near the bottom of the undercut. Periodically check the groove to make sure that it is not worn so much that the rope bottoms.

The V-groove likewise has higher friction when new. As the V-groove wears, however, the rope seats deeper and deeper, increasing the area of contact of rope with groove and decreasing the unit pressure. When the rope reaches the bottom of the groove, much of the driving traction probably will be lost, and a loaded (or empty) car may "break traction" and slide with the brake locked. For this reason, frequently check such grooves for wear and machine them before the rope wears the groove down to the bottom. Observe a minimum grooving bottom diameter.

Traction machines have an inherent safety feature: When the descending member (counterweight or car) bottoms, driving traction is lost and, in most cases, the ascending member cannot be pulled into the overhead. With extremely high rises, the weight of rope hanging on the down-run side may be sufficient to maintain traction after the car or counterweight has landed. Since the compensative sheave is tied down, however, there is no danger.

For safe and successful operation, the rope's tension on the car's side must bear a definite relation to the rope's tension on the counterweight's side. Normally, this ratio ranges from two-to-one to one-to-two. (See the discussion of overloads under the heading Operation, later in this section.)

Hydraulic Elevators

Hydraulic elevators are power elevators in which the energy is applied by means of a liquid under pressure in a cylinder that is equipped with a plunger or piston. Hydraulic elevators are being installed in many new buildings averaging up to six stories in height. New elevators are usually of the electrohydraulic type. In these elevators, the problems associated with older hydraulic elevators (generally lower efficiency, the difficulty of keeping the valves and stuffing boxes tight, etc.) have been eliminated by using modern technology and materials and by following proper maintenance programs.

The Elevator Code requires that new elevators be equipped with (1) anticreep-leveling devices, (2) hoistway-door locking devices, (3) electric car-door or car-gate contacts, (4) hoistway access, and (5) parking devices. This equipment is the same as that required for electric-drive elevators.

Because most electrohydraulic elevators do not have a counterweight, the motor must supply pressure to lift the entire weight of the car and the load. In addition, it must be more powerful than the motor of the traction drive of an electric-drive elevator, on which the weight of the car and part of the load is compensated for by the counterweight, to maintain the same speed. As built, the electrohydraulic elevator has all the electrical protective devices (interlocks, car-gate contacts, limit switches, and similar devices) found on an electric elevator.

Car safeties are not required on electrohydraulic elevators since they can come down no faster than the fluid can be forced out of the cylinder by the descending plunger. If counterweights and/or car safeties are on the elevator, they must comply with the Elevator Code.

Belt-Drive and Chain-Drive Elevators

The Elevator Code prohibits belt-driven and chain-driven machines from being installed. Existing elevators should be provided with electrically released brakes, terminal stopping, and safety devices as required for electric elevators. Some jurisdictions have outlawed these types of elevators while others allow their continued use.

New Elevators

The Elevator Code specifies the general requirements and load classifications for new elevators.

Requirements

When new elevators are being planned, be sure that all requirements of the latest edition of the Elevator Code are met. For example, check that there is (1) safe and convenient access to the machine room and the pit, (2) adequate lighting in the machine room and overhead spaces, and (3) convenient electric outlets on the crosshead and in the pit.

An inspection station, with slow-speed Up and Down operating buttons and an Emergency Stop switch, is required on the top of the car for maintenance personnel and inspectors to use.

The elevator must have normal- and final-limit stops, interlocks on all hoistway doors, a contact on the car door, and emergency exit or exits for the car. If a single-elevator hoistway exists, emergency access doors must be provided to the blind portions of such a hoistway. Unless the elevator is the hydraulic-plunger type, it should have a governor-actuated safety that meets the latest code requirements—city, state or provincial, federal, and/or ANSI standards.

Buffers are required to absorb the energy of descending cars and counterweights at the limits of travel. The main type of buffer in use is the spring buffer. Spring buffers are permitted only for cars whose rated speed is not in excess of 200 fpm (1.0 m/s). A few states and cities, however, permit them for higher speeds. At best, a spring buffer is a poor absorber of energy. A good spring will return approximately 95% of the stored energy so that a spring-buffer stop affords only a series of decreasing surges.

Capacity Ratings

The size, capacity, and speed of new freight elevators will depend upon the purposes for which they are to be used. When preparing specifications for new freight elevators, anticipate future load requirements as well as present ones. Therefore, find out from production personnel what size units (both freight package and freight carrier) will probably be in use 20 years hence.

The Elevator Code classifies freight elevators as follows:

- Class A—Elevators with a distributed load that is loaded by hand or by hand trucks. Here, the weight of any single piece of freight or of any single hand truck and its load is limited to a maximum of one-fourth the rated load capacity.
- Class B—Elevators used solely to carry passenger trucks and passenger automobiles up to rated capacity of the elevator.
- Class C1—Industrial truck loading in which the truck is carried by the elevator.
- Class C2—Industrial truck loading in which the truck is not carried by the elevator but used only for loading and unloading.
- Class C3—Other loading of heavy objects in which a truck is not used.

For Class C1, Class C2, and Class C3 loadings, the Elevator Code requires that the rated load of the elevator be not less than the load (including any truck) to be carried. It also requires that the elevator be provided with a two-way, automatic-leveling device.

Palletized loads increase in size year by year and lift trucks are made larger to handle them. Because of these factors, old elevators that are overloaded can start downward when the last loaded lift truck is run onto the car. In some cases, the brakes do not hold. In others, the traction relation is broken, and the hoisting ropes slip through the drive sheave. Where old elevators are used to handle heavy palletized loads, however, determine and strictly enforce safe load limits and safe operating procedures to prevent serious unintended incidents. Observe the Elevator Code to prevent such cases in new elevators.

Hoistways, Pits, and Machine Rooms

Hoistways, pits, and machine rooms are dangerous work areas. However, by following code specifications, these places can be safe for workers.

Hoistways

Most codes for building elevators require that new elevators be installed in two-hour, fire-resistant hoistways. These hoistings should have one-and-one-half-hour fire doors that fill the entire opening to prevent the rapid spread of fire from floor to floor.

At frequent intervals, clean hoistways, guide rails, and all other parts of grease and dirt to eliminate fire hazards. Projections in the hoistways should be bevelled at an angle of not less than 75 degrees with the horizontal. Windows in the walls of the hoistway enclosures are prohibited.

Do not install in or under any elevator or counterweight hoistway any pipe conveying gas or liquids, that—if discharged into the hoistway—would endanger lives. However, low-pressure steam (5 psig or less) or hot-water pipes used only for heating the hoistway and the machine room (or penthouse) are permitted, if certain conditions are met. All electrical equipment and wiring shall conform to NFPA 70/ANSI C1, *National Electrical Code.* Only such electrical wiring and equipment used directly in connection with the elevator may be installed in the hoistway.

Pits

To protect persons working in elevator pits, keep a minimum clearance of 2 ft (60 cm) between the lowest projection on the underside of the car's platform (not including guide shoes and aprons attached to the sill) and any obstruction in the pit (exclusive of compensating devices, buffers, buffer supports, and similar devices). Take these measurements when the car is resting on fully compressed buffers.

Enclose counterweight runways from a point not more than 1 ft (30 cm) above the pit floor to a point at least 7 ft (2 m) above the pit floor and adjacent pit floors, except where compensating chains or cables are used. Screen partitions, at least 7 ft (2 m) high between adjacent pits, will protect persons in one pit from cars and counterweights in adjacent pits. They will also protect employees from hazards when adjacent pits are at different levels.

Never use an elevator pit as a thoroughfare or storage space. It should be fully enclosed, and the entrances kept locked. To remove water, provide a sump pump. Drains are not to be connected directly to sewers.

Keep pits clean and free of debris. Never sweep rubbish into pits. Provide vertical ladders for easy access for cleaning pits.

Provide lighting of at least five footcandles (54 lux) at the pit's floor level. A light switch must be reachable from the pit's access door. An EMERGENCY STOP switch shall be installed in every pit and shall be reachable from the pit's access door.

Machine Rooms

Provide safe and convenient access to machine rooms. So that persons repairing or inspecting elevator hoisting machinery have sufficient room and are safe, allow at least 7 ft (2 m) of headroom between the machinery platform and the machine room's roof.

As with pits, never use elevator machine rooms as thoroughfares. The one exception would be if the elevator's equipment were in a separate locked enclosure. Rooms should be well ventilated and lighted with not less than 10 footcandles (108 lux) at floor level. Keep doors locked and affix a warning sign to prevent entry by unauthorized persons.

Keep machine rooms clean and do not use them for storage. Place small quantities of ordinary maintenance supplies in a wall cabinet. Keep a portable, Class-C fire extinguisher within reach of someone standing at the door. (See Chapter 11, Fire Protection.)

Overhead Protection

If the elevator's drive is located in the penthouse, provide a substantial grating or floor of fire-resistant construction under the drive. On all installations of overhead drives, include a cradle below the secondary sheaves if they extend below the floor or grating. For detailed information about flooring, consult the Elevator Code.

Do not hang elevator machinery, except the idler or deflecting sheaves, underneath the supporting beams at the top of the hoistway. When the governor or other devices (other than terminal-stopping switches) must be installed below the machine's floor, set them on a substantial secondary floor.

For winding-drum drives, a substantial beam or bar should be placed at the top of the counterweights' guide rails and beneath the counterweights' sheaves. This will prevent the counterweights from being drawn into the sheaves.

Hoistway Doors and Landings

The Elevator Code specifies that openings at hoistway landings of all elevators must be provided with hoistway doors that guard the full height and width of the openings.

Hoistway Doors for Passenger Elevators

Records show that most elevator unintended incidents occur at the hoistway door. Most are "tripping" incidents. Others occur when people are caught and crushed between doors or when they fall down the hoistway. These types of incidents are likely to happen on old elevators. The interlocks required on new elevators by the Elevator Code, if properly installed and maintained, practically eliminate these types of incidents.

Many serious injuries and fatalities have occurred on older elevators, with doors that needed a special key to open the doors on the corridor side, regardless of the elevator's position in the hoist. Injuries occurred when the victim assumed the elevator to be at the landing, opened the door, and stepped into the shaft. The Elevator Code permits using unlocking devices on elevators equipped with doors that are unlocked for entering the car in case of an emergency. Assign all emergency keys only to qualified personnel who will guard them and use them carefully.

To help prevent unintended incidents in hoistways of older elevators, install conduit in each hoistway. In that way, electric light can be installed opposite each opening. If the door is opened with an emergency key and the car is not at the landing, the light, which burns continuously, will help a person see that the shaft is empty. Thus, the person will not step into the shaft (Figure 15–23).

Requirements for entering hoistways can be met in either of the following ways:

Provide only two means of entry to the hoistway. One entrance should be at an upper landing to permit access to the top of the car. The other entrance should be at the lowest landing—if this landing is the normal point of entry to the shaft.

Where elevators operate in a single hoistway, provide hoistway doors that can be (1) unlocked when closed with the car at the floor, or (2) locked but possible to open from the landing, only when the car is in the landing zone. Check on this requirement with the local authorities.

In general, three types of doors are used at landings: vertical- or horizontal-sliding doors, combination sliding and swinging doors, and swinging doors. On all three types of doors use nothing less than direct-acting

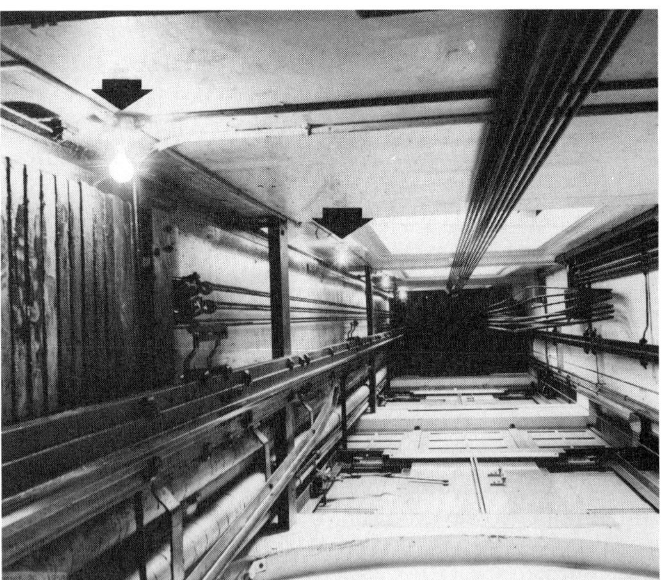

Figure 15–23. View from the pit looking up hoistways; note illumination (arrows). (Printed with permission from Architect U.S. Capitol.)

mechanical interlocks. Power-operated vertical-sliding doors and gates must operate in sequence if the elevator is used for passengers and, in all cases, if the doors close automatically.

For new elevators, according to the Elevator Code, the distance between the hoistway side of the hoistway door, opposite the car opening, and the hoistway edge of the landing threshold should be not more than ½ in. (13 mm) for swinging doors and 2¼ in. (57 mm) for sliding doors. The face of the hoistway door should not project into the hoistway beyond the edge of the landing sill. On existing elevators, if this distance exceeds 1½ in. (38 mm) for swinging doors and 2½ in. (64 mm) for sliding doors, fill in the excess space.

Do not design automatic fire doors, whose functioning depends on the action of heat, to lock (1) any landing opening in the hoistway of any elevator or (2) any exit leading to the outside of the building. Taking these precautions could save lives.

Construct and maintain the loading platform for at least 2 ft (60 cm) back from the door so persons will not readily slip. Some operators use firmly secured rubber mats, adhesive abrasive strips, or abrasive-surfaced concrete directly in front of all hoistway doors. (See NSC Industrial Data Sheet 12304–0595, *Floor Mats and Runners*.) To eliminate tripping hazards, make such surfaces flush with the surrounding floor.

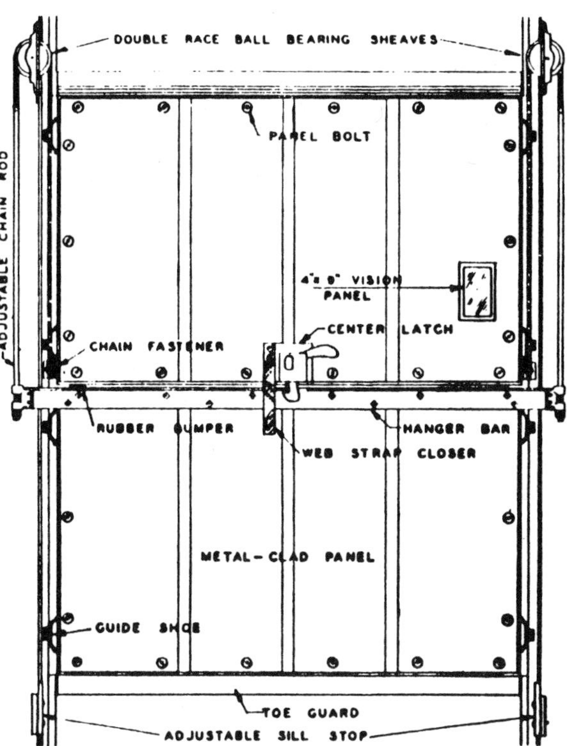

Figure 15–24. Design of a typical vertical sliding biparting steel hoistway door for freight elevators, as seen from inside the car. (Courtesy Peele Co.)

Hoistway Doors and Gates for Freight Elevators

Like passenger elevators, most freight elevators are installed in 2 hour, fire-resistant hoistways. These hoistways have 1½ hour fire doors that fill the entire opening, as required by practically all codes. However, there are some older elevators—mostly freight—that have hoistways enclosed only to a 6 or 7 ft (1.8 or 2.1 m) height. These older elevators also have hoistway gates of hardwood slats that are 5 or 6 ft (1.5 or 1.8 m) high with a clearance under them of as much as 8 to 10 in. (20 to 25 cm). Many of these elevators are shipper-rope operated, so it is necessary to reach into the hoistway to bring the car to a landing. This procedure is prohibited in new elevators.

Hoistway doors (Figure 15–24) should comply with applicable state and municipal requirements for fire resistance. Arrange doors that are closed by hand so that it is not necessary to reach in back of any panel, jamb, or sash to operate them. To ease the movement of trucks, install a door sill to fill the gap between the landing and the car.

Replace hoistway gates with doors in older elevators as soon as possible. Where gates are used, maintain them at the highest safety standard. For example, the openings in gates made of grille, lattice, or other openwork should reject a ball 2 in. (5 cm) in diameter.

The bottom of the gate should come down within 1 in. (2.5 cm) of the threshold to prevent objects from sliding under the gate into the hoistway. Where lack of headroom precludes a standard gate at the lowest landing, make the gate in two sections.

Gates should have convenient handles or straps for manual operation. However, a power attachment for closing the gate, controlled from the landing buttons, saves time and labor and decreases the possibility of leaving or propping the gate open. Power-operated doors and gates are recommended. For freight elevators, these are usually of the continuous-pressure-operation type.

The growing weight and speed of lift trucks requires that hoistway doors and gates of freight elevators be protected. A 5 tn (4,536 kg) load moving only 2 mph (2.93 fps) has an impact of 2,690 ft lb ($F = ½ MV^2$). If it is assumed that the door or gate can deflect 1 in., a force of 32,300 lb (144 kN) results. This is certainly more than any hoistway door can stand. Solutions to this problem include the following:

- Install a heavy wire rope across the opening. The truck operator must lower the rope before the truck can enter the car's platform.
- Place one person in charge of operating the elevator.

Interlocks and Electric Contacts for Both Passenger and Freight Elevators

To prevent a car from moving away from the landing, unless the hoistway door is locked in the closed position, equip hoistway doors with interlocks (Figure 15–25) that comply with the Elevator Code. Interlocks should be direct-acting mechanical/activated devices. All interlocking devices should be incapable of being plugged or made inoperative in any way. In addition, the interlock also prevents opening of the hoistway door from the landing side, except by emergency key, unless the car is within the landing zone and is either stopped or being stopped.

Locks and contacts (not interlocks) are permitted in a few cases. Contacts are required for the car door or car gate, which is considered closed if within 2 in. (5 cm) of the nearest face of the jamb. The Elevator Code defines the closed position of hoistway doors as being ⅜ in. (9.5 mm) from the jamb or between panels of center-opening doors. For vertical-slide biparting doors, however, the dimension is ¾ in. (19 mm). Under certain conditions, the elevator may be started when the doors are 4 in. (10 cm) from full closure. Design all interlocks so that the door must be locked in the closed position, as defined above, before the car can be operated.

Like any safety device, an elevator interlock or electric contact will be useless if it goes out of order easily. Therefore, use only those devices that comply with the Elevator Code and that have been either tested and listed by competent, designated testing laboratories or approved by the proper city or state authorities. Also, frequently inspect interlocking devices and see that they are maintained in proper working order.

Figure 15–25. An electromechanical interlock designed for use on biparting, manually operated steel doors on push button-controlled freight elevators.

Cars

Observe the Elevator Code safety requirements for the enclosures, top covers, doors and gates, and floors of elevator cars.

Car-Leveling Devices

Car-leveling devices automatically bring the car to a stop when the platform is level with the desired landing. A tripping hazard from uneven surfaces is thus eliminated. Moreover, in the case of freight elevators, the level surface that is automatically provided for the passage of trucks saves wear and damage to sills. It also reduces the possibility of material being jarred off trucks.

Enclosures

All elevator cars are required to be enclosed on the sides and top, except for the side or sides used to exit and enter. Openings for ventilation, an emergency exit, signals, and operating or communication equipment are allowed.

The sides and top of every passenger elevator car should be made of metal or of an approved, fire-retardant material. Although wood and openwork have been used for freight elevator cars, the Elevator Code now specifies a solid metal enclosure to a height of at least 6 ft (1.8 m) for enclosures. However, they may have perforations above the 6 ft (1.8 m) level, except in the area where the counterweight passes the car. That portion of the enclosure, for 6 in. (15 cm) on each side of the counterweight and up to the crosshead or car top, should be solid.

Fasten the enclosure securely to the floor and to the suspension sling. If the enclosure is cut away at the front to provide access to the hand rope, keep the opening low enough to prevent injury to the operator's hand.

Keep enclosures in good condition to prevent injuries to people loading and unloading the car. Broken wooden wainscoting or torn sheet metal can cause severe lacerations. One way to prevent such damage is to install bumper strips at the level of the truck's platforms and push bars. Bumper strips should be made of oak, ash, hickory, or similar tough, resilient wood. They should be at least 1 in. (25 mm) thick—2 in. (50 mm) thick where heavy trucks are used. Steel channels are also effective. Bumper strips should also be wide enough to match all heights of trucks' platforms and trucks' push bars.

Painting the inside of the car a flat black, from the floor to the top of the upper guard strip, will make bumps and scrapes less conspicuous. If the sides above the upper guard strip are painted aluminum, the high reflectance value of the paint will help overall illumination.

Top Covers

Freight elevator cars may have either solid or openwork top covers. The cover should be strong enough to sustain a load of 300 lb (135 kg) on any square area, 2 ft (61 cm) on a side, or of 100 lb (45 kg), applied at any one point. If openwork is used, the openings should be small enough to reject a ball 1½ in. (3.8 cm) in diameter. Some companies, however, recommend ¾ in. (19 mm) openings.

Wire mesh is recommended for openwork covers because it combines maximum strength with minimum weight. It also offers little interference with light and air. Do not use less than No. 10 gauge (2.5 mm dia.) steel wire.

An emergency exit of at least 400 in.^{2} (0.258 m^2) in area and not less than 16 in. (40.6 cm) on any one side should be in the top of every elevator car. Such an exit provides a clear passageway for easy escape should the car become stalled between floors or should some other emergency arise.

The exit cover should open outward and be so hinged or attached that it can be opened easily from the top of the car only. A heavy cover should be counterweighted or divided into several hinged sections, which should be kept down except when being used.

The Elevator Code now requires a clear refuge space of not less than 650 in.^{2} (0.4 m^2) on top of the car's enclosure. It shall not measure less than 16 in. (40.6 cm) on any side. The top emergency exit may open into this space provided that there is an area adjacent to the opening available for standing when the emergency exit cover is open. The minimum vertical distance in the refuge area between the top of the car's enclosure and the overhead structure or other obstruction shall not be less than 3.5 ft (1 m) when the car has reached its maximum upward movement.

Doors or Gates

To conform to requirements for new elevators, a door or gate must be provided at each entrance to the car. If the elevator is electrically operated, the door or gate should have electric contacts. The electric contacts will then prevent movement of the car unless the door or gate is within 2 in. (5 cm) of the fully closed position.

It is recommended that existing elevator cars having more than one opening be equipped with a gate at each entrance. Provide the gate with a contact to prevent its being opened when the car is in motion. When closed, the gate should guard the full width and height of the opening, except for vertical sliding gates. They should extend from a point not more than 1 in. (2.5 cm) above the car's floor to a height of at least 6 ft (1.8 m). Collapsible gates, when fully expanded, should reject a ball 3 in. (7.6 cm) in diameter. They should have convenient handles with guards that protect the operator's fingers.

Floors

Car floors should be kept in good condition. Do not use ordinary metal sheets for surfacing or repairing floors because they soon become smooth and slick. In time, even sheet metal with raised surface markings may wear smooth and then have to be replaced.

Do not let the edge of the car's platform or the edge of the landing become slippery or badly worn. At such points, cast-iron and steel sills are slipping hazards unless provided with slip-resistant surfaces. Apply abrasive strips or welded beading at those points to avoid slipperiness.

Loads

Even though the rated capacity (safe working load) of an elevator includes a safety factor, do not exceed the rated capacity. Indicate the rated capacity with a conspicuous sign inside the car. Metal signs with stamped, etched, or raised letters and figures, not less than ¼ in. (6 mm) high, are satisfactory for passenger elevators. Use a minimum of 1 in. (25 mm) high letters and figures, however, for freight elevators.

Lighting

All cars and landings shall be well lit at all times when in use. The following are minimums and can be exceeded:

At the landing edge of the car platform, when the car and loading doors are open, a car should have at least two lights that provide a minimum illumination of 5 footcandles (54 lux) for passenger elevators; 2 footcandles (27 lux) for freight elevators.

Car lights may be omitted on freight elevators with openwork car tops on cars that travel no more than 15 ft (4.57 m). However, install at least two electric lights at the top of the hoistway to furnish the minimum specified illumination.

Illumination on landing thresholds should be at least 5 footcandles (54 lux). New elevators, in addition to the regular lighting, should have emergency lighting that automatically goes on 10 seconds after the regular lighting fails.

Hoisting Ropes

Records show that it is relatively unusual for an elevator incident to be caused by the parting of the hoisting ropes. When it does happen, though, it is more likely to occur with winding-drum drives than with traction drives. Periodic resocketing intervals for winding-drum elevators are specified in the Elevator Code. Although ropes usually are installed to give the highest safety factor specified in the Elevator Code, closely inspect them to avoid the possibility of an incident. (See ANSI/ ASME A17.2, *Inspectors' Manual for Elevators and Escalators,* for wire rope inspection.)

Rope life is shortened by improper brake action. Unduly sudden stops may result from brake defects, such as heavy spring pressure or brakeshoe wear. Unnecessary starting and stopping also shorten rope life. Avoid inching for landings. In some cases, inching can be eliminated by properly adjusting the stopping devices. In most cases, however, it is caused by faulty operation—instruct the operator how to properly operate the elevator.

Ropes and rope fastenings used to hoist elevator cars must follow certain requirements. Safe operating conditions result when these requirements are observed.

Car and Counterweight Ropes

Car and counterweight ropes must be of iron (low-carbon steel) or steel having the commercial classification "Elevator Wire Rope." They can also be of rope specifically constructed for elevator use. Do not use ropes less than ⅜ in. (9.5 mm) in diameter on passenger elevators. Traction-drive elevators should have not fewer than three hoisting ropes. Winding-drum drive elevators should have not fewer than two hoisting ropes and should have two ropes for each counterweight used.

The diameter of sheaves or drums for hoisting, or counterweight ropes shall be not less than 40 times the diameter of the ropes. In practical application, however, this ratio is usually higher than 40.

On winding-drum elevators, rope should be long enough so there will be not less than one full turn of rope on the drum when the car is at the extreme limit of its overtravel. Drum ends of ropes are usually secured by babbitt-filled, tapered sockets or by clamps on the inside of the drum.

Rope Fastenings

Ends of car and counterweight-suspension ropes are usually fastened by individual babbitt-filled, tapered sockets. (See ANSI 17.1, Figure 212.9d and Table 212.9d for sketch and dimensions.) Rope sockets must develop at least 80% of the breaking strength of the rope used. Shackle rods, eyebolts, and other means used to connect sockets to the car or counterweight ropes must have a breaking strength at least equal to that of the rope.

Governor Ropes

Governor rope should be made of iron, steel, monel metal, phosphor bronze, or stainless steel. They should be of regular-lay construction not less than ⅜ in. (9.5 mm) in diameter. Rope used as a connection from the safety to the governor rope, including rope wound on the safety drum, must be corrosion-resistant.

Operating Controls

Stopping Devices

Every electric elevator must have an EMERGENCY STOP switch in the car, adjacent to the operating device, which will cut off the power. The button should be clearly identified and be red in color. Contacts of such switches should be directly opened mechanically and should not depend solely upon springs for opening contacts. Some local codes have outlawed the EMERGENCY STOP switch; consult the local authority.

Winding-drum elevators are required to have a direct-driven, adjustable, automatic, machine-limit stop mechanism. Such a mechanism will stop the car if it overruns the highest and lowest landings. In addition, limit switches are required either in the hoistways or on the car. Winding-drum elevators are also required to have a slack rope device to shut off power to the drive and to brake in case the rope becomes slack. This device must be such that it will not reset automatically when slack in the rope is removed.

Electric elevators with traction drives are required to have stopping switches in the hoistway, on the car, or in the machine room if the switches are operated by the motion of the car.

Grounding

The motor frame, and the operating cable, if insulated from the motor frame, should be grounded. All switches and wiring should conform to NFPA 70/ANSI C1, *National Electrical Code.*

Safety Devices

Every elevator car suspended by wire ropes is required to have one or more safety devices that will catch and stop the car in the event of overspeed or failure of the hoisting ropes. The safeties must be attached to the car's frame. One safety must be located within or below the lower members of the car's frame (safety plank).

Counterweight safeties should conform to the car safety requirements and are classified as follows:

- Type A, instantaneous, safeties rapidly increase their pressure on the guide rails to give a very short stopping distance. They are permitted on elevators having a rated speed of not more than 150 fpm (0.76 m/s).
- Type B safeties apply limited pressure to the guide rails, and the retarding forces are reasonably uniform after full application. They are permitted on elevators of any speed and may be used in multiples.

- Type C safeties—Type A with oil buffers between the safety plant and elevator car—are permitted on elevators with a rated speed of not more than 500 fpm (2.5 m/s).

Car Switches

Design the handle of the car switch (operating control) to return to the STOP position and lock when the operator's hand is removed.

Signal System

Every elevator, except automatic-operation and continuous-pressure-operation elevators, should have a signal system that can be operated from any landing. In this way, the car can be signaled when it is wanted at that landing.

Inspection and Maintenance

Inspections and tests must conform to requirements and regulations set forth by the particular regulatory agency. Much of the discussion in this section is based on ANSI/ASME A17.2, *Inspectors' Manual for Elevators and Escalators.* Anyone concerned with maintenance and inspection should have, and use, this manual.

Requirements for inspecting elevators, as given in ANSI/ASME A17.2, follow:

1. Acceptance inspections and tests of all new installations and alterations should be made by an inspector employed by the regulatory agency.
2. Routine inspections and tests of all installations should be made by a person qualified to perform such services, in the presence of an inspector employed or authorized by the enforcing authorities. It is recommended that periodic inspections and tests be made at intervals not longer than six months for power passenger elevators and escalators; and twelve months for power freight elevators, hand elevators, and power and hand dumbwaiters.
3. Periodic inspections and tests shall be made by a person qualified to perform such service in the presence of an inspector employed or authorized by the regulatory agency. Car and counterweight safeties, governors, and oil buffers should be given periodic inspections and tests at least every year with no load, and every five years with a full load.

Inspection Program

Careful maintenance is essential to the safe operation of elevators and all their parts. It also reduces the need for repairs. Frequent and thorough inspections by qualified outside personnel are the first requisite of an efficient maintenance program. However, minor day-to-day inspection and maintenance can be performed by qualified plant personnel.

Maintenance on a regular basis includes the following: (1) inspection of hoisting and counterweight wire ropes; (2) lubrication of oil buffers; (3) cleaning and lubrication of guide rails, controller contactors and relays, and car safety mechanisms; and (4) cleaning of hoistways, pits, machine rooms, and tops of cars.

Hoistways and Landings

Incidents occurring or originating at landings usually are due to (1) tripping or slipping at the car entrance or landing, (2) being caught inside the car, (3) falling into the hoistway, and (4) being caught by the doors.

To help prevent tripping and slipping incidents at hoistway landings, pay special attention to the following points:

- Check the leveling of the car. Improper leveling may be due to careless operation or, where cars are leveled automatically, to improper adjustment. Observe the operation of elevators and, if necessary, give instructions to the operators. With an automatic car-leveling device, the operator has no control over the final stop. Have a competent mechanic, preferably one trained or employed by the elevator manufacturer, adjust the automatic car leveler.
- Watch the condition of landing sills and floors. Landing sills should be of slip-resistant material; if they are not, they should be replaced or roughened if worn smooth. Repair broken sills, holes in flooring, worn floor coverings, and other conditions that create tripping hazards. The finished surface should be flush with the surrounding floor.
- Check the illumination of landings. Give special attention to those near building entrances where the difference between the intensity of outdoor and indoor light is noticeable. Clean globes and reflectors, and provide lamps of adequate size.
- Maintain and correctly adjust interlocks and contacts to provide the required protection at landings.

Track grooves for hoistway doors should be clean, so that the doors will move freely. There should be no excessive play. Vision panels should be clean and unbroken. Counterweights should operate freely, and sheaves should be properly aligned.

Because the entire load is transferred to the guide rails when the safety operates, the guide rails must be properly aligned at the joints and securely attached to the brackets. In addition, the bolts or welds with which the brackets are attached to the building's structure must remain tight. Alignment of rails can be checked easily by sighting along the faces; bracket bolts must be tested individually.

Check elevator cars for structural defects, such as loose bolts and other fastenings, excessive play in guide shoes, and worn or damaged flooring. Keep guide rails properly lubricated to help reduce wear between the rails and the guide shoes, except where roller guide shoes, which run on dry rails, are provided. Use the lubricant specified by the elevator manufacturer. Keep roller guide shoes clean and, if necessary, adjust them for pressure against the rails.

Test each emergency exit by opening it. Also check each exit panel to assure that it is securely fastened in the closed position. If panels are held in place by locks, keep the key in a location not reached by the public, but available to personnel for emergency use. If the panels are held by thumbscrews, they should be removable without having to use pliers.

Subject car doors or car gates to the same inspection as hoistway doors, and follow the same maintenance standards. Check contacts for adjustment in the same manner as hoistway door contacts.

Test the car-operating switch, when provided, to see that it returns to, and locks in, the neutral position, when released by the operator. On a cable-operated car, check the cable lock to make sure that, when it is locked, the cable cannot be operated. Examine all switch contacts in the car-operating device, and place the glass cover over the emergency-release switch. Emergency-release switches, however, are prohibited on automatic elevators by the Elevator Code.

Check car lights for proper operation, for loose or missing screws, and for broken or cracked lamps. Keep lamps and bulbs clean. Periodically check the emergency lighting.

Safety Devices and Limit Switches

Maintenance of safety devices is often neglected because they do not affect normal operation of the car. However, in case of an emergency, the safety of passengers depends entirely upon the proper performance of these devices. It is of the utmost importance, therefore, that they be maintained in proper working condition.

At frequent intervals, clean and lubricate all the safety devices and equipment; inspect them for worn, cracked, broken, or loose parts; and test them to determine their ability to stop and hold the car. Safety devices should be tested and adjusted only by a person qualified to perform such services. Tests should be done in the presence of an inspector employed, or authorized, by the enforcing authorities.

Limit Switches

The inspector should check limit switches for proper alignment. The mounting of switches and cams should be checked for rigidity.

Buffers

Since oil buffers lose some oil during normal usage, check the oil levels at least once every month and each time the buffer is known to have been compressed. For refilling oil buffers, use an oil of the type specified by the manufacturer. Empty, clean, and refill with fresh oil, buffers that have been submerged by floods or pit leakage. Check the alignment and the tightness of bolts in the anchorage.

Check spring buffers for alignment and for proper seating in the cups or mountings. Examine springs for deformation and permanent set.

Hoisting Machines

Examine hoisting machines carefully at each inspection. Check the machine's base for misalignment and cracks; immediately repair defects. The inspection should also include the following:

- the oil level in the motor, if used
- brake operation
- the oil level in the gear housing, if provided
- sheaves and drums for cracks and wear in the grooves

Belted Machines

At frequent intervals, examine belts for proper tension, wear, burns, condition of splices, and cuts and breaks in the surface. Check chains for excessive wear. Regularly check machine-fastening bolts, belt guards, and the fastenings of platforms under any ceiling machinery.

Inspection Routine

Persons making elevator inspections should take all necessary precautions for their own safety. ANSI/ASME A17.2, *Inspectors' Manual for Elevators and Escalators,* contains not only detailed instructions for the conduct of all tests but also comprehensive information on personal safety for inspectors working in machine rooms, in and on top of cars, and in pits.

Inspectors should wear close-fitting clothing, preferably one-piece overalls with all buttons fastened and without cuffs. Inspectors should not wear gloves, except when checking wire rope.

Inspectors must pay close attention to moving objects, such as counterweights, to hoistway projections, and to limited overhead and pit clearance. Before inspecting electrical parts, be sure that the main line disconnect switches are locked in the open position. In that way, current-carrying parts cannot be energized. When an inspector is on top of a moving car, he or she should keep one hand free to hold onto the crosshead or another part of the top of the car's frame. Inspectors should not hold hoisting ropes for support. They should attach safety belts to a fixed structure of the car or frame, not to the ropes.

If controls are not provided on top of the car for the inspector's use, the person who operates the elevator car should receive specific instructions on what to do while the inspector is inspecting the top of the car.

The order in which the various parts of the car are inspected depends upon the type of elevator and the preference of the inspector. However, inspectors can reduce the inspection time and the amount of downtime for the elevator by planning. Determine which parts of the job can be done from each part of the elevator.

A written report of each inspection should be prepared and kept on file. Such a record is especially valuable for checking the progress of defects in ropes. Each report, therefore, should give definite details concerning the ropes' condition—diameter, number of broken wires per unit length, and estimated percentage of wear.

The mechanic who corrects the trouble should make a report of each break in service. The mechanic should enter the report on the log sheet for the particular elevator. In that way, the maintenance engineer can spot defective equipment that causes repeated breaks in service, and thus correct the basic fault.

Operation

Safety and health professionals can eliminate many common causes of unintended elevator incidents by insisting on safe operating practices. Best results are obtained by assigning a properly instructed operator to full-time duty. In any case, only permit employees who have received proper instruction to act as elevator operators.

Selecting Elevator Operators

Select elevator operators with extreme care. Select operators who are mentally alert, not easily excited, and capable of carrying out instructions and of insisting upon compliance with rules.

In many cases, the lives of others depend upon the operator's efficiency. Moreover, an incompetent or poorly trained operator may damage valuable and indispensable equipment. Faulty practices, such as starting a car before the doors and gates are closed, blocking gates open, and permitting crowding on cars not provided with gates, have caused many serious injuries. Actions, such as improper loading, unnecessary starting and stopping, and reversing of the controller, have caused damage to equipment.

Some companies have found that elevator operators work more safely if they are given cards that state they have completed the training course and are authorized to operate the equipment. In some cases, company rules require that operators have such cards.

Operating Rules

Adopt definite operating rules. In the case of industrial freight elevators, post the rules in the car, and require operators to know and observe them. Many companies that employ a number of elevator operators prepare pocket-size rule books for their guidance. Plants with a number of elevators that are used under varying conditions prepare specific rules for specific elevators or groups of elevators.

Incident Investigations

Carefully investigate minor elevator incidents. Such investigations can disclose conditions or practices that, if left uncorrected, could later cause a serious injury. Therefore, determine the conditions or practices that caused the incident, and take prompt corrective action.

Overloads

Overloading elevators may result in injury to personnel, mechanical failure of the drive or of the car, or both. Many elevators still in service were built with much lower safety factors than are now required. With such elevators, the hazard of overloads is particularly great. However, even though elevators meet current safety factor requirements, the danger of overloads cannot be overemphasized. Consider the following engineering factors:

- In a traction-drive elevator, the ratio between rope tension on the driving sheave (car side) and rope tension on the counterweight side must be kept within certain limits. The counterweight normally is equal to the weight of the car plus 40% of the rated capacity. The motor torque and the brake are designed to handle this difference in weight.
- In the case of a traction-drive freight elevator, with a car weighing 8,000 lb (3,600 kg) and a rated capacity of 10,000 lb (4,500 kg), the counterweight would equal 12,000 lb (5,450 kg). That is, the weight of the car, or 8,000 lb (3,600 kg), plus 40% of the rated capacity, or 4,000 lb (1,800 kg). The motor would be designed to lift, and the brake to hold, a 6,000 lb (2,700 kg) load.
- If the elevator is overloaded by 50% of its rated capacity (for this example, 5,000 lb or 2,250 kg), the total platform load will be 15,000 lb (6,800 kg). If 4,000 lb (1,800 kg), or overbalance of the counterweight, is subtracted from the 15,000 lb (6,800 kg) on the platform, a net weight of 11,000 lb (5,000 kg) must be handled by a motor and brake designed to handle only 6,000 lb (2,725 kg). This is an 83% overload.

In such a case, the traction relation may be broken, and the motor may not even pick up the load. As the brake is lifted, the car may start to move downward, a fuse might blow, and other mechanical failures, as well as injury to personnel, may result.

In no case should an elevator be overloaded unless the manufacturer has checked the entire installation for its ability to handle the load. Failure to do so may result in serious injury or death to employees and in serious damage to valuable equipment.

For an overload of a one-piece load (a transformer, for instance) as heavy as, or heavier than, the rated capacity of the car, consult the company that installed the elevator. The company's representative should check that:

- the machine is strong enough to handle the load
- the elevator structure, including the car frame (sling), platform, and undercar safeties, is adequate
- the traction relation will not be exceeded

If the machine is otherwise strong enough for the overload, the elevator company's crew may increase the counterweight. This will maintain the traction relation while the overload is being lifted.

ELEVATOR EMERGENCY INSTRUCTIONS

1. When the alarm sounds, the elevator has stopped between floors.
2. Notify the Fire Department or Rescue Team, at once, by calling

   (Telephone Number)
   ____
3. Notify the elevator maintenance company at once by calling

   (Telephone Number)
   ____
4. Do not attempt to rescue until trained, authorized maintenance personnel arrive. Assure the stuck passengers that help is on the way and that they are safe.

Figure 15–26. This sign should be posted in each elevator car.

Emergency Procedures

Implement an emergency procedure similar to the one shown in Figure 15–26 for safely removing persons from elevators stalled between floors. As is illustrated, emergency instructions for persons involved in such incidents are spelled out on self-adhesive stickers. Mount these stickers behind protective transparent material to preserve their legibility. In addition, when deemed appropriate, post stickers in languages other than English. In gaseous or toxic environments, take tests to determine incident and injury potential. If necessary, implement other emergency procedures. (See ANSI/ASME A17.4, *Emergency Evacuation of Passengers from Elevators.*)

A telephone or other means of communication is recommended for all existing cars and on new elevators. Having this equipment may prevent occupants from panicking and may help rescuers coordinate emergency procedures. Operating elevators in an emergency and emergency signal devices are discussed in the Elevator Code.

Requirements for the Disabled

To help further increase the mobility of disabled individuals, incorporate the requirements found in the National Elevator Industry's *Minimum Passenger Elevator Requirements for the Handicapped* for all new elevators. Where practical, incorporate the requirements on existing elevators.

SIDEWALK ELEVATORS

A sidewalk elevator is a freight elevator for carrying material, exclusive of automobiles, between a landing in a sidewalk, or other area outside a building, and the floors below the sidewalk or grade level. Sidewalk elevators present hazards that are not easy to eliminate. Therefore, it is best to locate them inside the building line or in an area not open to the public. Sidewalk elevators should conform to the requirements of the Elevator Code.

Except by permission of the local authorities, the maximum dimensions of openings for sidewalk elevators should be 5 ft (1.5 m) at right angles to, and 7 ft (2.1 m) parallel with, the building line. The side of the opening nearest the building should be not more than 4 in. (10 cm) from the building wall.

Where hinged doors or covers that can be lifted vertically are provided at the sidewalk or at other areas outside the building, bow-irons or stanchions shall be provided on the car to operate such doors or covers. A loud audible signal should sound when the car is ascending.

A sidewalk elevator with winding-drum drive should have a normal terminal stopping device on the drive. A similar device should be either in the hoistway or on the operating device.

Operation

The Elevator Code requires that a sidewalk elevator be raised and lowered only from the sidewalk or other outside area. Use either a key-operated, continuous-pressure up-and-down switch, or continuous-pressure up-and-down buttons on the free end of a detachable, flexible cord, 5 ft (1.5 m) or less in length.

Hatch Covers

Automatic hatch covers, when closed, should sustain not less than 300 lb (1,136 kg) applied on any area 2 ft (610 mm) on a side, and not less than 150 lb (68 kg) applied at any point. Hatch covers must be self-closing. They are not to be fastened or held open when the car is away from the top landing.

The covers should be made of metal and should lift vertically. For hinged covers, the line of the hinges should be at right angles to the building's wall. When the covers are fully open, there should be minimum clearance of 18 in. (46 cm) between them and any obstruction.

Hatch covers should be secured to the walkway's surface. To avoid tripping hazards to passersby, no hinges, locks, or flanges should project above the closed covers.

HAND ELEVATORS

Hand-powered elevators were once used widely in storage rooms, warehouses, and private residences. Few companies manufacture hand elevators. Do not install a hand elevator where it is to be used constantly during the working day. Instead, some form of power equipment should be installed wherever an elevator is a basic part of the manufacturing process or service function.

Never apply mechanical power to hand elevators by means of rope-grips or similar attachments. If power is to be used, change the entire installation, and use all the protective devices as required for other elevators.

Hand-elevator cars in which persons can ride should not have more than one compartment nor be arranged to counterbalance another car. Hoistways, hoistway openings, pits, machinery spaces, supports, and foundations for hand elevators should conform to the requirements for power elevators.

Hoistway Doors

Conspicuously display on the landing side of hoistway doors: DANGER—ELEVATOR—KEEP CLOSED. The letters should be not less than 2 in. (5.1 cm) high. Hoistway openings may have (1) self-closing doors that extend to the floor; (2) doors made in two parts, one above the other, and so arranged that the lower part can be opened only after the upper part has been opened; or (3) doors equipped with two spring locks or latches, one located 6 ft (1.8 m) above the floor.

Safety Devices and Brakes

Hand elevators should have hand or automatic brakes that operate in either direction of motion. When the brakes are applied, they should remain locked in the ON position until released.

A hand elevator that travels more than 15 ft (4.5 m) must have a safety attached to the underside of the car's frame. The safety must be able to stop the car and sustain it and its load. If the car is to travel more than 40 ft (12 m), it must have a hand-operated brake that automatically slows the car down.

DUMBWAITERS

A dumbwaiter is defined as a hoisting and lowering mechanism equipped with a car (1) that moves in guides and has a floor area not exceeding 9 sq ft (0.8 m), (2) that has a compartment height not exceeding 4 ft (1.2 m), (3) that has a rated capacity not greater than 500 lb (225 kg), and (4) that is used exclusively for carrying materials. A dumbwaiter may be hand- or power-operated.

Hoistways and Openings

The requirements for dumbwaiter hoistways are almost the same as those for elevator hoistways. Design landing openings and doors for dumbwaiters to protect persons from falling into the hoistways. Provide the landing openings of power-driven dumbwaiters with hoistway doors that guard the full height and width of the openings.

With certain specified exceptions, power-operated dumbwaiter doors must be equipped with hoistway-unit-system hoistway-door interlocks. They will prevent the dumbwaiter from moving if any hoistway door or gate is open.

Conspicuously display on the landing side of hoistway doors of hand-operated dumbwaiters: DANGER—DUMBWAITER—KEEP CLOSED. The letters should be not less than 2 in. (5.1 cm) high.

Safety Devices and Brakes

Power-operated dumbwaiters, except hydraulic dumbwaiters, should have brakes that are automatically applied when the power is cut off or fails. The brakes should also be able to stop the car automatically within the limit of overtravel at each terminal. Hand-operated dumbwaiters should have hand-operated or automatic brakes that can sustain the weight of the car and its load.

A power-operated dumbwaiter having winding-drum drive, a travel greater than 30 ft (9 m), and a capacity in excess of 100 lb (45 kg) requires a slack-rope device. This device will cut off the power from the motor and stop the car if it is obstructed in its descent.

ESCALATORS

An escalator is a power-driven, inclined, continuous stairway for raising or lowering passengers.

Safety Devices and Brakes

EMERGENCY STOP buttons, or other types of hand-operated switches having red buttons or handles, shall be accessibly located in the right-hand newel base on new escalators, at or near the top and bottom landings. Emergency-stopping devices shall be protected from being unintentionally operated. An escalator STOP button with an unlocked cover over it, which can readily be lifted or pushed aside, shall be considered accessible. The operation of STOP buttons or switches shall interrupt the power to the drive. It shall not be possible to start the drive by pressing these buttons or switches.

Use key-operated buttons or switches to start the units. Locate them within sight of the escalator's steps. Also, provide a way to cut the escalator's power in case an ascending escalator unintentionally reverses its travel.

Each escalator should have a speed governor that will interrupt the power if the predetermined speed is exceeded—not more than 40% greater than the rated speed. The speed governor is not needed if a low-slip, alternating-current, squirrel-cage induction motor is used and if the motor is directly connected to the drive.

If a tread chain breaks, a broken chain-sensing device should cut the power. If an escalator has tightening devices that are operated by tension weights, make provisions to retain these weights in the escalator truss should they fail.

Each escalator should have an electrically released and mechanically applied brake able to stop the fully loaded escalator when it is traveling either up or down. The brake should automatically stop the escalator as soon as any safety devices begin to function.

Machinery

Every escalator machine room should have a light that can be lit without workers having to pass over or reach over any part of the machinery. The lighting control switch shall be located within easy reach of the access to such rooms. Where practicable, the light control switch shall be located on the lock, jambside of the access door. Provide reasonable access to the interior of the escalator for inspection and maintenance purposes. For the protection of maintenance personnel, install guards around all chains in escalator machinery compartments. Full lockout/tagout procedures must be followed in servicing an escalator. (See Chapter 6, Safeguarding, on lockout/tagout procedures.)

While bearings on escalators are sealed and require no oiling, the chains do require lubrication. Take care not to overlubricate. Provide an oil pan in the bottom of the truss to catch oil or grease that may drip from moving parts and to catch dust and dirt that falls between the treads.

Clean the oil pan periodically to eliminate a fire hazard from the accumulated oil-soaked dust and dirt. Attach a brush, made for this purpose, to a step axle. Draw the brush down over the drip pan to brush all the dirt to the lower end of the truss. The sweepings then can readily be removed. Reverse the unit to return the brush to the top, and then disconnect the brush.

The moving handrail will show some stretch over a period of time. Check it at intervals, and use the handrail's drive adjustment to take up the slack.

The balustrades must have handrails that move in the same direction and at about the same speed as the steps. Each moving handrail is to extend at normal handrail height, not less than 12 in. (305 mm) beyond the points of the comb plate's teeth at the upper and lower landings. Provide hand or finger guards at the point where the handrail enters the balustrade.

Inspect all parts of escalators and their drives' machinery at regular intervals to keep them well maintained. Test all safety devices for proper functioning.

Protection of Riders

Most escalators are installed in public places, and their principal hazards arise from misuse by the public.

Most escalator incidents occur when heels of shoes, fingers, or toes are caught between the surface grooves or slots on the treads and the comb plate. The width of each slot is to be not more than ¼ in. (6.3 mm) and the depth not less than ⅜ in. (9.5 mm), with a center-to-center spacing of not more than ⅜ in. (9.5 mm) between adjoining slots. In some incidents, edges of shoe soles have been caught between the step and the vertical side member (skirt guard), which should have a maximum of 3/16 in. (4.8 mm) on each side.

Because barefoot passengers have been injured on escalators, post signs warning barefoot persons not to ride the escalator. Additionally, a caution sign shall be located at the top and bottom landing of each escalator. This sign shall include the words, CAUTION, PASSENGERS ONLY, HOLD HANDRAIL, ATTEND CHILDREN, and AVOID SIDES. The sign used should be the standard depicted in Rule 805.2 of the Elevator and Escalator Code. Umbrella tips are frequently caught between the grooves and the comb plate. This type of incident sometimes results in minor injuries and damage to equipment. Other incidents have resulted from riders mishandling baggage on escalators. Riders should not place suitcases and handbags on the steps. They should always carry such items parallel to the run of the escalator.

Escalator treads and landings should be made of noncombustible material that provides a secure foothold. Some riders, through inexperience or infirmity, have trouble seeing the parting point where treads rise or descend and the point where they level off. To aid these riders, illumination should be provided for all tread surfaces. Mount additional warnings in green demarcation lights inside the truss at top and bottom to shine through the treads where they break away and come together at the trouble points.

Some manufacturers mark the edges of the steps to emphasize the lines between adjacent steps. One manufacturer, for example, adds a distinctive color strip to the edge of the step adjacent to the riser of the next step.

Post signs reading PLEASE HOLD HANDRAIL at the top and bottom of the escalator (Figure 15–27). Do not place distracting signs, such as advertising, near these critical points. At each level, use directional arrows and mark the level's number on the floor to improve the flow of traffic from the escalators.

It is extremely important that no object or construction of any kind obstruct the free flow of passengers from the area at the exit of an escalator. (This area is not a part of the escalator.) Serious injuries have occurred where the flow of traffic from the exit was restricted by a fence or barrier placed at some distance from the escalator.

Figure 15–27. Signs should be prominently displayed at the top and bottom of an escalator.

Fire Protection

Protection of escalator floor openings against the spreading of fire and smoke is required by local building codes. One of the best safeguards against the spread of fire is to divide buildings made of fire-resistive construction into limited areas in which fire can be readily controlled. Protect vertical openings from the passage of fire from story to story. This principle often is disregarded when escalators are installed.

Escalators approved as a required means of entry must be fully enclosed in accordance with local laws and ordinances. Escalators not approved as a required means of entry must have the floor openings protected by one of the following methods, in accordance with national and local standards and regulations. (See NFPA 101, *Life Safety Code.*)

MOVING WALKS

A moving passenger-carrying device, on which persons stand or walk and in which the passenger-carrying surface is uninterrupted and remains parallel to its direction of motion, is a moving walk. Criteria for the design, construction, installation, operation, inspection, and testing of moving walks are given in the Elevator Code, Part IX.

Moving walks may operate in a horizontal plane or in a slope up to a maximum of 12 degrees. However, the slope shall not exceed 3 degrees within 3 ft (91 cm) of the moving walk's entrance or exit. Operating speed and treadway width are governed by the slope.

Protection of passengers on moving walkways is the same as that on escalators. (See the section on Escalators in this chapter.)

MANLIFTS

The following are the principal hazards in the use of manlifts:

- The rider may be carried over the top.
- The rider may be unable to make an emergency stop.
- The rider may jump on or off after the step has passed the floor.
- His or her head or shoulders may strike the edges of floor openings if there is not a conical hood.
- The rider may be unable to reach the landing because of power failure and belt stoppage.
- Parts of the manlift may fail or operate unsafely.

Construction

Construct, maintain, and operate manlifts in strict compliance with the recommendations of ANSI/ASME A90.1, *Belt Manlifts.* Use a safety factor of six, based on a 200 lb (90 kg) load, on each step, on both the up and down runs. Brace all equipment securely at top, bottom, and intermediate landings. Secure the manlift's rails to prevent their spreading apart, vibrating, and becoming misaligned. Suspend the entire manlift from the top to prevent bending or buckling of the rails.

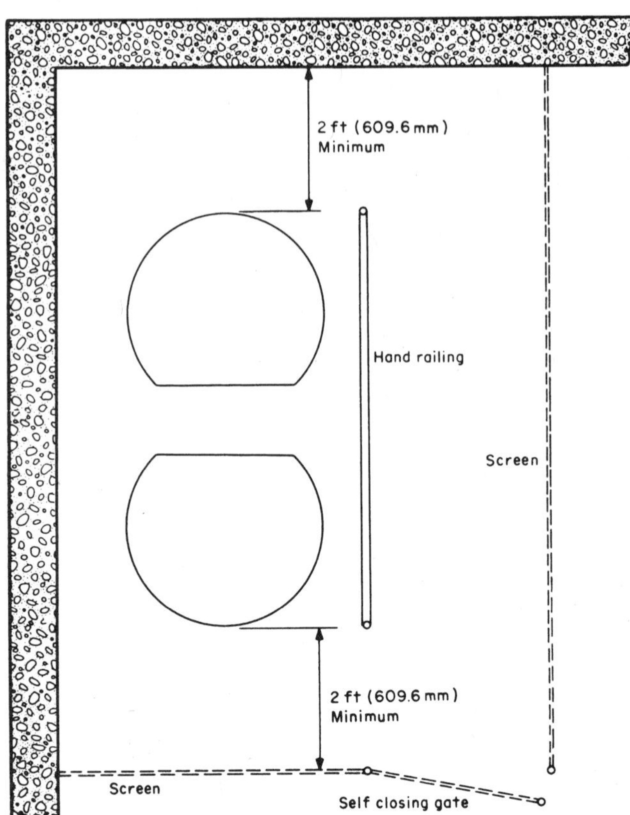

Figure 15–28. Screen enclosure for manlift floor openings (plan view). (Printed with permission from the American National Standards Institute.)

Handholds

Paint handholds, of either the open or closed type, in a bright color, such as orange or yellow. Place handholds not less than 48 in. (1.2 m), nor more than 56 in. (1.4 m), above each step. Locate steps not less than 16 ft (5 m) apart. Use slip-resistant material on the steps.

Landings

Provide floor landings or emergency landings for each 25 ft (7.6 m) of manlift travel. Between the surface of any landing and the lower edge of the conical guard suspended from the ceiling, allow at least 7 ft (2.3 m) of clearance. Allow a minimum clearance of 5 ft (1.5 m) between the center of the head pulley and the roof or other obstruction. The bottom landing on the side that is up should have steps to a platform level with the manlift's step as it rises to a horizontal position.

Floor Openings and Conical Guard Openings

At floor landings, provide standard 42 in. (1.06 m) guardrails and 4 in. (10 cm) toeboards around floor openings. Install them in such a way as to permit a landing space at least 2 ft (0.6 m) wide. Guardrails should have maze or staggered openings or self-closing gates that open away from the manlift (Figure 15–28). At each floor opening on the side that is up, install funnel-shaped (conical) guards (Figures 15–29 and 15–30).

Brakes, Safety Devices, and Ladders

Install on the motor shaft of direct-connected units and on the input shaft of belt-driven units, a brake that automatically works when the power is shut off. The brake should get its power or force from an external source. The brake must be capable of quickly stopping the manlift and holding it when the side that is down is loaded with 250 lb (110 kg) on each step. The brake should release electrically.

Provide a control rope, not less than ⅜ in. (9.5 mm) and made of Manila and/or cotton with a bronze wire center, within easy reach of both the up and down runs. When pulled in the direction of the belt's travel, the rope should cut off the power and apply the brake. Install an electromechanical device that will automatically shut off the motor's power supply and apply an electric brake should the rider fail to alight at the top landing. RESET buttons, located at the top and bottom

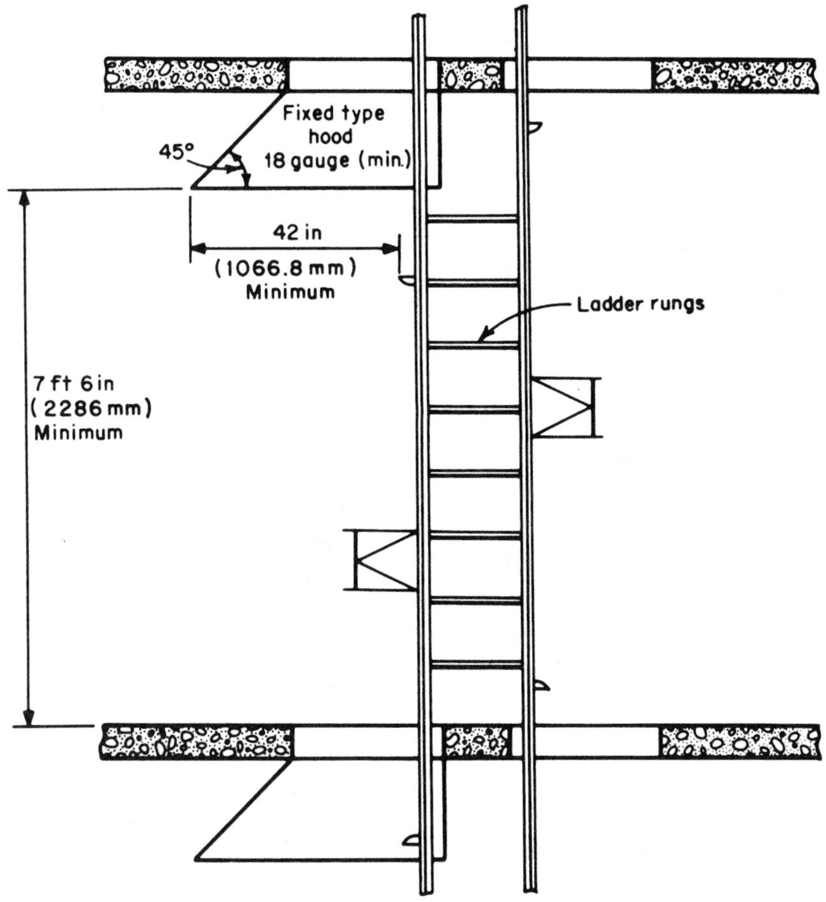

Figure 15–29. Drawing shows dimensions for a fixed flared-opening guard. (Printed with permission from the American National Standards Institute.)

Figure 15–30. In this installation, the guard is counterbalanced and will yield slightly if hit. (Courtesy Humphrey Elevator Co.)

terminals, will permit restarting the manlift after it has been shut off by the electrical safety devices.

A secondary safety control located on the top operating floor is also required. It should be set to operate when the belt has traveled 6 in. (15 cm) beyond the point of operation of the primary safety switch, in case the latter fails. The device should stop the manlift before the loaded step reaches a point 24 in. (0.6 m) above the top landing.

Provide a fixed metal ladder, accessible from both the up and down runs, for emergency exiting if the vertical distance between landings exceeds 20 ft (6 m). This ladder should meet the requirements of the local codes. However, do not provide an enclosing cage, since the ladder should be accessible from either side throughout the entire run.

Inspections

Test and inspect every manlift at least every 30 days. Use an inspection report similar to the one in Figure 15–31. Indications of a defect are such things as (1) unusual or excessive vibrations, (2) continual misalignments, and (3) "skips" when mounting steps, which indicate worn

BELT MANLIFT INSPECTION REPORT
(Weekly & Monthly)

Location ______________________ Date ______________

Manlift Make & Serial No. ______________ Code ______________

	ITEM	PERIOD Weekly	PERIOD Monthly		ITEM	PERIOD Weekly	PERIOD Monthly
1	OBSERVE MANLIFT IN OPERATION FOR POSSIBLE DEFECTS	( )	( )	19	TOP LANDING SAFETY SWITCHES	( )	( )
2	STEPS AND ROLLERS		( )	20	TOP BAR SAFETY	( )	( )
3	HAND HOLDS		( )	21	PHOTO EYE SAFETY (IF APPLICABLE)	( )	( )
4	BELT JOINT		( )	22	ON-OFF SWITCH & CONTROL ROPE (TEST UP & DOWN RUN)	( )	( )
5	BELT (LOOK FOR CUTS OR DAMAGE)	( )	( )	23	STEP CLEARANCE AT DRIVE BELT TRACKING AT DRIVE	( )	( )
6	BELT TENSION AND BELT TAKE-UP AT BOTTOM PULLEY	( )	( )	24	MOTOR		( )
7	STEP CLEARANCE AT BOOT	( )	( )	25	BRAKE		( )
8	BELT TRACK ON BOTTOM PULLEY	( )	( )	26	GEAR REDUCER & CHECK OIL LEVEL & CHANGE PER MANUAL		( )
9	BOTTOM BEARING LUBRICATION & SUPPORT		( )	27	COUPLINGS – COLLARS – KEYS		( )
10	GUIDE RAILS, PROPER ALIGNMENT FASTENINGS & SUPPORT		( )	28	HEAD SHAFT BEARINGS AND LUBRICATION		( )
11	FLOOR BRACES FOR GUIDE RAILS		( )	29	TOP PULLEY AND LAGGING	( )	( )
12	FLOOR HOODS AND OPENINGS		( )	30	OVERALL DRIVE ASS'Y AND SUPPORTS		( )
13	SAFETY SWITCHES ON MOVEABLE HOODS	( )	( )	31	SKIPPING WHEN MOUNTING STEP (CHECK DRIVE TRAIN)		( )
14	GUARD RAILS AND GATE OPERATION	( )	( )	32	VIBRATION OR MISALIGNMENT IN DRIVE		( )
15	FLOOR LANDINGS (CLEAR OF OBSTRUCTION)	( )	( )	33	GREASE BEARINGS PER MAINTENANCE MANUAL		( )
16	ILLUMINATION OF MANLIFTS AND FLOOR LANDINGS	( )	( )	34	GUIDE TRACK FREE OF FOREIGN MATERIAL AND LUBRICANTS	( )	
17	TOP & BOTTOM FLOOR WARNING SIGNS & LIGHTS	( )	( )	35			
18	BOTTOM SAFETY SWITCHES TREADLE &/ OR DROPOUTS	( )	( )	36			

IS MANLIFT BEING PROPERLY USED? ____________ BY AUTHORIZED PERSONNEL? ____________

ITEM NUMBERS ABOVE THAT WERE CORRECTED OR ARE IN NEED OF SUCH
(GIVE COMPLETE DESCRIPTION ON BACK SIDE OF THIS FORM)

INSPECTED BY ______________________ DATE ______________________

Figure 15–31. A typical checklist for inspecting a manlift. (Printed with permission from the American National Standards Institute.)

gears. Upon discovery of a defect, immediately take the manlift out of operation and do not use it until it is repaired. Each periodic inspection should cover, but not necessarily be limited to, the following items:

- steps
- lubrication
- step fastenings
- illumination
- rails
- warning signs and lights
- rail supports and fastenings
- signal equipment
- motor
- limit switches
- driving mechanism
- electric switches
- gears
- belt and belt tension
- drive pulley
- handholds and fastenings
- bottom (boot) pulley and clearance
- floor landings
- pulley supports
- guardrails
- belt splice joint
- brake
- step rollers
- key coupling keyway

Inspect the safety mechanism of a manlift daily. Check that it is operating freely and that it is free of dirt and grease. Dismantle a manlift once every year, and replace defective or excessively worn parts.

Maintain a written record of the findings at each inspection. Keep records of inspections readily available.

General Precautions

The maximum speed of a manlift belt should not exceed 80 fpm (0.4 m/s). This speed should be uniform on all manlifts throughout the plant. If a manlift carries a great deal of traffic, a maximum speed of 60 fpm (0.3 m/s) is recommended.

Number floors with large figures in full view of both ascending and descending riders. Place a constantly illuminated sign, TOP FLOOR—GET OFF, placed not more than 2 ft (0.6 m) above the top landing, in full view of ascending passengers. Use block letters at least 2 in. (5 cm) high for the sign. In addition, locate at least a 40-watt red warning light immediately below the top landing so it will shine in an ascending rider's face. The entire manlift should be illuminated at all times while it is in operation with at least 1 footcandle (11 lux) at all points and at least 5 footcandles (54 lux) at landings.

Prominently, at each landing, display signs carrying instructions for use of the manlift. Permit only authorized employees to ride manlifts, and display signs stating this at each landing. Riders should not carry on a manlift anything that cannot be placed entirely inside a pocket, a sling, or a pouch. Carefully instruct employees, particularly new ones, in the safe use of the manlift. Tell them to report immediately any defects or any irregularity in the operation of the manlift or its safety devices. Supervisors should take corrective action immediately.

SUMMARY

- Hoisting equipment must never be overloaded or used to transport people. Operators should examine hoists regularly, and repair or replace any damaged or malfunctioning parts.
- Cranes should have adequate safeguards to provide safe footing and accessways for the operator, to prevent injuries, and to limit the action of the crane arm and hoisting devices. All hoisting ropes, sheaves and drums, and other equipment should be appropriate for the crane being used. The operator's cab must protect the operator against fire and weather, be well ventilated, contain ample control equipment, and allow a clear view of signals.
- Other types of cranes, including monorails, jib cranes, derricks, tower and mobile cranes, and portable floor cranes, all have specific guidelines for safe operation and transport.
- Aerial basket lifts are commonly used for working above ground. Manufacturer's operating and maintenance instructions should be followed for safe handling of these machine parts.
- Conveyors are generally defined as a horizontal, inclined, or vertical device for moving or transporting objects. Loading points must be clearly marked, showing the safe load limit that can be raised or lowered, with safeguards used along the entire length. Maintenance personnel must lock out all power before working on a conveyor. Operators must stand clear of moving conveyors, avoid pinch points and other areas where hands or fingers can get caught, and never attempt to repair a moving belt.
- Power elevators should conform to the Elevator Code for electric-drive and hydraulic-drive elevators. The Elevator Code prohibits belt-driven and chain-driven machines from being installed. Interlocks and electric contacts for both passenger and freight elevators should be direct-acting mechanical/activated devices that cannot be inactivated.
- Companies should establish a regular program of inspection, testing, and maintenance of all elevator parts according to city, state, and federal regulations. Companies also should select elevator operators carefully and establish proper working procedures, safety rules, and emergency procedures for operators to follow.
- Dumbwaiters may be hand- or power-operated, and must be equipped with hoistway-unit-system hoistway- door interlocks. Landing openings and doors should be designed to protect people from falling into the hoistways.
- Escalators must have EMERGENCY STOP buttons or other types of hand-operated switches in easily accessible locations. Each escalator should have sensing devices that will interrupt power if the preset speed is exceeded or a tread chain breaks. The principal hazard on escalators and moving walks arise from their misuse by the public.

- Injuries on manlifts frequently occur when riders are carried over the top, jump off or on after the step has passed a floor, and are unable to reach the landing because of power failure. All manlifts must have emergency brakes, safety devices, and ladders to prevent injuries and incidents. Inspectors would test and examine manlifts every 30 days, with daily inspection of safety devices.

REFERENCES

American National Standards Institute, 11 West 42nd Street, New York, NY 10036.

Belt Manlifts, ANSI/ASME A90.1–1985.
Elevators, Escalators, and Moving Walks, ANSI/ASME A17.1–1987.
Emergency Evacuation of Passengers from Elevators, ANSI/ASME A17.4–1986.
Inspectors' Manual for Elevators and Escalators, ANSI/ASME A17.2–1988.
Monorails and Underhung Cranes, ANSI/ASME B30.11–1988.
National Electrical Safety Code, ANSI C2–1997.
Overhead and Gantry Cranes (Top Running Bridge, Single Girder, Underhung Hoist), ANSI/ASME B30.17.
Overhead and Gantry Cranes (Top Running Bridge, Single or Multiple Girder, Top Running Trolley Hoist), ANSI/ASME B30.2.
Overhead Hoists (Underhung), ANSI/ASME B30.16.
Safety Requirements for Cranes, Derricks, Hoists, Jacks, and Slings, ANSI/ASME B30 Series.
Safety Requirements for Fixed Ladders, ANSI A14.3–1992.
Safety Requirements for Workplace Floor and Wall Openings, Stairs and Railing Systems, ANSI A1264.1–1995.
Safety Requirements for Personnel Hoists, ANSI A10.4–1981.
Safety Requirements for the Lock Out/Tag Out of Energy Sources, ANSI Z244.1–1982(R1993).
Safety Requirements for Work Platforms Suspended from Cranes or Derricks, for Construction and Demolition Operations, ANSI A10.28–1998.
Safety Standards for Conveyors and Related Equipment, ANSI/ASME B20.1A–1996.
Safety Standard for Mechanical Power Transmission Apparatus, ANSI/ASME B15.1–1984.
Terms and Conveyor Definitions, ANSI/CEMA 102–1994.

Crane Manufacturers Association of America, 8720 Red Oak Boulevard, Suite 201, Charlotte, NC 28217.

Construction Safety Association of Ontario, 74 Victoria Street, Toronto, Ontario, Canada M5C 2A5.

Crane Handbook.
Mobile Crane Manual.
Rigging Manual.

Conveyor Equipment Manufacturers Association, 932 Hungerford Drive, Suite 36, Rockville, MD 20850.

Screw Conveyors, Book No. 350.

International Union of Operating Engineers, 1125 17th Street NW, Washington, DC 20036.

National Elevator Industry, Inc., 630 Third Avenue, Seventh Floor, New York, NY 10014–6705.

Minimum Passenger Elevator Requirements for the Handicapped.

National Fire Protection Association, 1 Batterymarch Park, Quincy, MA 02269.

Life Safety Code, NFPA 101, 1997.
National Electrical Code, NFPA 70, 1996.
"National Fire Codes," Vol. 4, Building Construction and Facilities.

National Safety Council, 1121 Spring Lake Drive, Itasca, IL 60143-3201.

Occupational Safety and Health Data Sheets (available in the Council Library):
Aerial Baskets, 12304–0572, 1987.
Automotive Hoisting Equipment, 12304–0437, 1979.
Belt Conveyors for Bulk Material (Part 1: *Equipment*), 12304–0569, 1990.
Belt Conveyors for Bulk Material (Part 2: *Operations*), 12304–0570, 1990.
Electromagnets Used with Crane Hoists, 12304–0359, 1985.
Flexible Insulating Protective Equipment for Electrical Workers, 12304–0598, 1985.
Floor Mats and Runners, 12304–0595, 1986.
Manlifts, 12304–0401, 1990.
Mobile Cranes and Power Lines, 12304–0743.
Pendant-Operated and Radio-Controlled Cranes, 12304– 0558, 1986.
Roller Conveyors, 12304–0528, 1985.
Tower Cranes, 12304–0630, 1983.
Truck-Mounted Power Winches, 12304–0441, 1990.
Underground Belt Conveyors, 12304–0725, 1986.

U.S. Department of Commerce, National Institute of Standards and Technology, Gaithersburg, MD 20899.

National Electrical Safety Code, NBS Handbook H30.

U.S. Department of Energy. *DOE Hoisting and Rigging Manual.* Idaho Falls ID: EG & G Idaho, Inc., October 1988.

U.S. Department of Labor, Occupational Safety and Health Administration, 200 Constitution Avenue NW, Washington, DC 20210. *Code of Federal Regulations,* Title 29. Sections 1910.179 (b)(2), 1910.180 (j)(1) and 1926.550 (a)(15).

REVIEW QUESTIONS

1. Why should a load only be lifted when it is directly under the hoist?
2. What are the three general types of chain hoists and which is more efficient?
3. What is the purpose of a spring return on an operating lever?

4. Which crane movement control hand signal should be obeyed even if it is being given by someone other than the signaler in charge?

5. What is the maximum bend or twist from the plane of the unbent hook in a crane hook that is allowable before corrective action must be taken?
   a. Any bend or twist is not permissible
   b. 5 degrees
   c. 10 degrees
   d. 20 degrees

6. What are the first three steps that should be taken before maintenance work can be performed on a crane?
   a.
   b.
   c.

7. Name and describe the three groups of monorail hoists?
   a.
   b.
   c.

8. Why should hoists or cranes not be used to lift, support, or otherwise transport people?

9. Before the first use and after modification, cranes must be tested to ______________ of the rated load unless the manufacturer recommends otherwise.
   a. 100%
   b. 125%
   c. 150%
   d. 200%

10. What is the difference between light service and heavy service?

11. What is the difference between frequent inspections and periodic inspections?

12. What are five factors that are implicated in unintended mobile crane incidents?
   a.
   b.
   c.
   d.
   e.

13. What rating should the fire extinguisher that is stored in the cab of a crane have?
   a. BC
   b. ABC
   c. 5BC
   d. UL

14. What three pieces of information should be on a crane operator's capacity chart?
   a.
   b.
   c.

15. For loads limited by structural competence, the weight should be determined to what level of precision before the load is lifted?
   a. 80%
   b. 25% of rated load
   c. ± 20 pounds
   d. ± 10% of the load weight

16. Name two purposes of a trial run before lifting personnel.

17. What three things should the authorized signaler for an on-track crane do while walking ahead of the crane?

18. For a crane operating near power lines that are rated between 50 kv and 345 kv, what is the minimum necessary clearance between the lines and any part of the crane?
   a. 4 feet
   b. 4.9 feet
   c. 10 feet
   d. 16 feet

19. What are the four benefits/purposes of using a safety factor of 10 when determining the safe working load of Manila rope?
   a.
   b.
   c.
   d.

20. What is the maximum distance between emergency stopping devices for a conveyor that operates near a walkway?
   a. 10 feet
   b. 16 feet
   c. 23 feet
   d. 75 feet

21. What are two types of chain conveyors?
   a.
   b.

22. What is the difference between chute conveyors and roller conveyors?

23. What code governs the use and design of elevators?

24. Why are car safeties not required on electrohydraulic elevators?

16

Ropes, Chains, and Slings

Revised by

Brad McPherson

Special safety precautions apply to using and storing fiber ropes, wire ropes, rope slings, chains, and chain slings. The safety and health professional should know the properties of the various types of ropes, chains, and slings used and the precautions for both use and maintenance. Figure 16–1 gives a checklist of important factors to consider when obtaining rope for a specific application. This chapter covers the following topics:

- types of fiber rope and its performance and care
- characteristics of wire rope and safety issues in its use
- safety practices in using rigging
- safety issues in using fiber and wire rope slings
- hazards and safety practices for chains and chain slings
- safety issues in using synthetic web and metal mesh slings

FIBER ROPE

Fiber rope is used extensively in handling and moving materials. Natural fiber ropes are generally made from Manila (abaca), sisal (agave), or henequen. Synthetic fiber ropes on the market include those made from nylon, polyester, and polyolefin. Manila or nylon ropes give the best uniform strength and service.

ROPE CHECKLIST	
Strength	Friction melting
Stretch with load	Combustibility
Impact load	Sunlight resistance
Permanent stretch	Latitude and altitude
Recovery from stretch	Color and type
Length	Diameter and construction of rope
Size	Frequency of use
Yardage	Storage methods
Floatability	Marine growth resistance
Flexibility	Rot resistance
Twist direction and torque	Chemical resistance
Flex life in bending	Color
Slipperiness	Aging
Texture	Contamination
Water repellency	Uniformity
Hygroscopicity	Service cost
Ruggedness in shape	Toughness against wear
Temperature resistance	

Figure 16–1. Factors that may be of significance when obtaining rope for a specified use.

Types of Fiber Ropes

Manila Fiber

The properties of Manila fiber make it the best-suited natural fiber for cordage. Manila rope is often recommended for capstan work because of its ability to pay out evenly when so used. High-grade Manila rope, when new, is firm but pliant, varies in color from ivory to light yellow, and has considerable luster. Its good reputation in fresh and salt water is well established.

Sisal Fibers

The properties of the several sisal fibers do not give sisal ropes the high general acceptance of Manila. The sisals lend themselves to use mostly in smaller ropes. For general use, they are not as satisfactory as Manila fibers, because their breaking strengths are generally lower than those of Manila. Sisal rope varies in color from white to yellowish white and lacks the gloss of high-grade Manila. The fibers are stiff and harsh and tend to splinter. This makes the ropes uncomfortable to handle.

Both sisal and Manila fibers deteriorate when in contact with acids and caustics and their mists or vapors. This deterioration is accelerated by hot, humid conditions. Both fibers lose 50% of their strength at 180 F (80 C) and burn at over 300 F (150 C).

Sisal and henequen are not as satisfactory as Manila because their strength varies from grade to grade. Sisal rope is about 80% as strong as Manila; henequen is about 50% as strong, but it resists deterioration from exposure to the air better than sisal.

Other Natural Fibers

Other natural fibers are also used in ropes but to a lesser or negligible degree and only for special reasons. These fibers include cotton, flax, coir, straw, asbestos, istle, jute, kenaf, silk, rawhide, and sansevieria.

Synthetic Fibers

Nylon, polyester, and polyolefin ropes are the major types of synthetic fiber ropes. Synthetic fiber ropes are used more often than natural fiber ropes for the following reasons:

- More is known about the properties of various synthetics. Successful use of synthetic fiber rope depends largely on selecting the synthetic with the physical properties and characteristics that most closely match the requirements of the job.
- Splices can be made readily in synthetic fiber rope and can develop nearly the full strength of the rope. Tapered splices are highly recommended for rope sizes with a diameter 1 in. (2.5 cm) or larger.

Nylon Rope

Nylon rope has more than two and a half times the breaking strength of Manila rope and about four times its working elasticity. It is, therefore, well suited to shock loading, such as is required for restraint lines. Its resistance to abrasion is remarkably high compared with that of other ropes. When nylon rope is wet or frozen, its breaking strength is reduced by 10% to 15%.

Nylon rope also is highly resistant to organisms that cause mildew and rotting and to attack by marine borers in seawater. Prolonged exposure to air results in little loss of strength. Since there is no swelling, wet nylon rope runs through blocks as easily as dry nylon rope. Although resistant to petroleum oils and most common solvents and chemicals, nylon's strength is affected by drying oils, such as linseed oil or the phenols. Nylon rope is also vulnerable to strong mineral acids, phenolic compounds, and heat.

Nylon loses some of its strength at 300 F (150 C) and all of it at 482 F (250 C), its normal melting point. Short of melting, most of nylon's strength is regained upon cooling to normal temperature. Nylon of a higher melting point is available.

Nylon, more than any other rope material, will absorb and store energy in the same manner as a spring. When nylon rope breaks, this energy makes the rope's moving ends as dangerous as a projectile. Exercise caution, therefore, when working nylon lines around corners, capstans, timber heads, and the like.

Polyester Rope

Probably the best general-purpose rope available, especially for critical uses, is made of polyester. Polyester stretches about half as much as nylon, so energy absorption is also about half as much. It is not weakened by rot, mildew, or prolonged exposure to seawater. In addition, polyester retains its full strength when wet because it does not absorb moisture. It shows little deterioration from long exposure to sunlight and resists abrasive wear well. Polyester is somewhat vulnerable to alkalis, but its resistance to ultraviolet light is good to excellent. It burns at about 480 F (250 C) and loses strength at temperatures over 390 F (200 C).

Polyolefin Rope

In general, polyolefin rope is strong and inexpensive. It floats and is unaffected by water. Polyolefin, like polyester, does not absorb moisture; therefore, it does not shrink or swell with water. It is unaffected by rot, mildew, and fungus. Polyolefin rope is also highly resistant to a wide variety of acids (except nitric acid) and alkalis, as well as to alcohol-type solvents and bleaching solutions. However, it swells and softens when exposed to hydrocarbons, particularly at temperatures above 150 F (66 C). The movement of crossed ropes, as well as other types of abrasion, must be avoided because even modest loads will cause a sawing motion that leads to a buildup of friction. Descriptions of two types of polyolefin ropes follow:

- Polypropylene rope, with a specific gravity of 0.91 and a softening point of 300 F (150 C), is made in several sizes of filaments and from film with or without longitudinal fracturing. Polypropylene rope is about 50% stronger than Manila rope, size for size. Pure polypropylene rope has rela-tively poor rendering properties. It burns at 330 F (166 C) and loses some strength at 150 F.
- Polyethylene rope, with a specific gravity of 0.95 and a softening point of 250 F (120 C), is characteristically slippery and has very little springiness. It is strong and has little stretch. Polyethylene rope also has a comparatively low softening point and low coefficient of friction.

Composite Rope

Rope made by combining several types of synthetic fibers or by combining synthetic and natural fibers is also available. Composite rope results from attempts to give the surface of the rope or strand more wear resistance, greater internal tensile strength, or more structural strength to retain its shape. Composite rope can be made to match the requirements of specific jobs.

Other Types of Rope

Rope made of paper, glass, acrylic, rayon, polyvinyl chloride, fluorocarbon, rubber, cellulose acetate, and polyurethane is also available. These types of ropes enjoy only a small percentage of the market for reasons of cost, limited use, or short supply.

Working Load

Table 16–A lists linear density, new rope tensile strength, safety factor, and working load for Manila, sisal, and synthetic ropes. Table 16–B gives the same information for double-braided nylon rope. Because the safety factor is not the same for all ropes and is based upon static loading, exercise caution when using this number.

Also be cautious when using the working load figures. Rope use, rope condition, exposure to several factors affecting rope behavior, and the degree of risk to life and property vary widely. Therefore, it is impossible to make blanket recommendations as to working loads. However, in order to provide general guidelines, working loads are tabulated for rope (1) in good condition, (2) with appropriate splices in noncritical applications, (3) under normal service conditions, and (4) under very modest dynamic loads.

Table 16–A. Specifications for Synthetic and Natural Fiber Rope

Nominal Size		Three-Strand Laid and Eight-Strand Plaited—Standard Construction											
		Manila				Polypropylene				Polyester			
Diameter	Circumference	Linear Density[1] (lbs/100 ft)	New Rope Tensile Strength[2] (lbs)	Safety Factor	Working Load[3] (lbs)	Linear Density[1] (lbs/100 ft)	New Rope Tensile Strength[2] (lbs)	Safety Factor	Working Load[3] (lbs)	Linear Density[1] (lbs/100 ft)	New Rope Tensile Strength[2] (lbs)	Safety Factor	Working Load[3] (lbs)
3/16	5/8	1.50	406	10	41	.70	720	10	72	1.20	900	10	90
1/4	3/4	2.00	540	10	54	1.20	1,130	10	113	2.00	1,490	10	149
5/16	1	2.90	900	10	90	1.80	1,710	10	171	3.10	2,300	10	230
3/8	1 1/8	4.10	1,220	10	122	2.80	2,440	10	244	4.50	3,340	10	334
7/16	1 1/4	5.25	1,580	9	176	3.80	3,160	9	352	6.20	4,500	9	500
1/2	1 1/2	7.50	2,380	9	264	4.70	3,780	9	420	8.00	5,750	9	640
9/16	1 3/4	10.4	3,100	8	388	6.10	4,600	8	575	10.2	7,200	8	900
5/8	2	13.3	3,960	8	496	7.50	5,600	8	700	13.0	9,000	8	1,130
3/4	2 1/4	16.7	4,860	7	695	10.7	7,650	7	1,090	17.5	11,300	7	1,610
13/16	2 1/2	19.5	5,850	7	835	12.7	8,900	7	1,270	21.0	14,000	7	2,000
7/8	2 3/4	22.4	6,950	7	995	15.0	10,400	7	1,490	25.0	16,200	7	2,320
1	3	27.0	8,100	7	1,160	18.0	12,600	7	1,800	30.4	19,800	7	2,820
1 1/16	3 1/4	31.2	9,450	7	1,350	20.4	14,400	7	2,060	34.4	23,000	7	3,280
1 1/8	3 1/2	36.0	10,800	7	1,540	23.8	16,500	7	2,360	40.0	26,600	7	3,800
1 1/4	3 3/4	41.6	12,200	7	1,740	27.0	18,900	7	2,700	46.2	29,800	7	4,260
1 5/16	4	47.8	13,500	7	1,930	30.4	21,200	7	3,020	52.5	33,800	7	4,820
1 1/2	4 1/2	60.0	16,700	7	2,380	38.4	26,800	7	3,820	67.0	42,200	7	6,050
1 5/8	5	74.5	20,200	7	2,880	47.6	32,400	7	4,620	82.0	51,500	7	7,350
1 3/4	5 1/2	89.5	23,800	7	3,400	59.0	38,800	7	5,550	98.0	61,000	7	8,700
2	6	108	28,000	7	4,000	69.0	46,800	7	6,700	118	72,000	7	10,300
2 1/8	6 1/2	125	32,400	7	4,620	80.0	55,000	7	7,850	135	83,000	7	11,900
2 1/4	7	146	37,000	7	5,300	92.0	62,000	7	8,850	157	96,500	7	13,800
2 1/2	7 1/2	167	41,800	7	5,950	107	72,000	7	10,300	181	110,000	7	15,700
2 5/8	8	191	46,800	7	6,700	120	81,000	7	11,600	204	123,000	7	17,600
2 7/8	8 1/2	215	52,000	7	7,450	137	91,000	7	13,000	230	139,000	7	19,900
3	9	242	57,500	7	8,200	153	103,000	7	14,700	258	157,000	7	22,400
3 1/4	10	298	69,500	7	9,950	190	123,000	7	17,600	318	189,000	7	27,000
3 1/2	11	366	82,000	7	11,700	232	146,000	7	20,800	384	228,000	7	32,600
4	12	434	94,500	7	13,500	276	171,000	7	24,400	454	270,000	7	38,600

Select a higher working load only with expert knowledge of conditions and a professional estimate of the risks involved. Factors to consider include (1) whether the rope has been subject to dynamic loading or other excessive use, (2) whether it has been inspected and found to be in good condition, (3) whether it is to be used in a recommended manner, and (4) whether the application involves high temperatures, extended periods under load, or obvious dynamic loading, such as sudden drops, snubs, or pickups. For all such applications and for applications involving more severe conditions of exposure, or for recommendations on special applications, consult the manufacturer.

Many uses of rope involve serious risk of injury to personnel or of damage to valuable property. This risk is often obvious—for example, a heavy load supported above one or more workers. An equally dangerous situation occurs if personnel are in line with a rope that is under excessive tension. Should the rope fail, it may recoil with considerable force—especially if the rope is made of nylon. Workers should be warned against standing in line with the rope. In all cases where such risks are present, or if there is any question about loads or other conditions of use, greatly reduce the working load and properly inspect the rope. Consult the manufacturer for recommendations on working loads.

Dynamic loading automatically voids the working load. Working load figures do not apply when rope is subject to significant dynamic loading. Whenever a load is picked up, stopped, moved, or swung, there is an increased force due to dynamic loading. The more rapidly or suddenly such actions occur, the greater this increase will be. In extreme cases, the force put on the rope may be two, three, or even more times the normal load involved, such as when picking up a tow on a slack line or using a rope to stop a falling object. Therefore, in applications such as towing lines, lifelines, safety lines, climbing ropes, and the like, working load as given in Tables 16–A and 16–B does not apply.

Dynamic effects are greater on a rope with little stretch, such as Manila, than on a rope with higher stretch, such as nylon. Dynamic effects are also greater on a shorter rope than on a longer one. The working load listed provides for very modest dynamic loads.

Table 16–A. *(Continued.)*

	Composite[4]				Nylon				Sisal				
Nominal Diameter (in.)	*Linear Density[1] (lbs/100 ft)*	*New Rope Tensile Strength[2] (lbs)*	*Safety Factor*	*Working Load[3] (lbs)*	*Linear Density[1] (lbs/100 ft)*	*New Rope Tensile Strength[2] (lbs)*	*Safety Factor*	*Working Load[3] (lbs)*	*Linear Density[1] (lbs/100 ft)*	*New Rope Tensile Strength[2] (lbs)*	*Safety Factor*	*Working Load[3] (lbs)*	*Nominal Diameter (in.)*
3/16	.94	720	10	72	1.00	900	12	75	1.50	360	10	36	3/16
1/4	1.61	1,130	10	113	1.50	1,490	12	124	2.00	480	10	48	1/4
5/16	2.48	1,710	10	171	2.50	2,300	12	192	2.90	800	10	80	5/16
3/8	3.60	2,440	10	244	3.50	3,340	12	278	4.10	1,080	10	108	3/8
7/16	5.00	3,160	9	352	5.00	4,500	11	410	5.26	1,400	9	156	7/16
1/2	6.50	3,960	9	440	6.50	5,750	11	525	7.52	2,120	9	236	1/2
9/16	8.00	4,860	8	610	8.15	7,200	10	720	10.4	2,760	8	345	9/16
5/8	9.50	5,760	8	720	10.5	9,350	10	935	13.3	3,520	8	440	5/8
3/4	12.5	7,560	7	1,080	14.5	12,800	9	1,420	16.7	4,320	7	617	3/4
13/16	15.2	9,180	7	1,310	17.0	15,300	9	1,700	19.5	5,200	7	743	13/16
7/8	18.0	10,800	7	1,540	20.0	18,000	9	2,000	22.5	6,160	7	880	7/8
1	21.8	13,100	7	1,870	26.4	22,600	9	2,520	27.0	7,200	7	1,030	1
1 1/16	25.6	15,200	7	2,170	29.0	26,000	9	2,880	31.3	8,400	7	1,200	1 1/16
1 1/8	29.0	17,400	7	2,490	34.0	29,800	9	3,320	36.0	9,600	7	1,370	1 1/8
1 1/4	33.4	19,800	7	2,830	40.0	33,800	9	3,760	41.7	10,800	7	1,540	1 1/4
1 5/16	35.6	21,200	7	3,020	45.0	38,800	9	4,320	47.8	12,000	7	1,710	1 5/16
1 1/2	45.0	26,800	7	3,820	55.0	47,800	9	5,320	59.9	14,800	7	2,110	1 1/2
1 5/8	55.5	32,400	7	4,620	66.5	58,500	9	6,500	74.6	18,000	7	2,570	1 5/8
1 3/4	66.5	38,800	7	5,550	83.0	70,000	9	7,800	89.3	21,200	7	3,030	1 3/4
2	78.0	46,800	7	6,700	95.0	83,000	9	9,200	108	24,800	7	3,540	2
2 1/8	92.0	55,000	7	7,850	109	95,500	9	10,600	—	—	7	—	2 1/8
2 1/4	105	62,000	8	8,850	120	113,000	9	12,600	146	32,800	7	4,690	2 1/4
2 1/2	122	72,000	7	10,300	149	126,000	9	14,000	—	—	7	—	2 1/2
2 5/8	138	81,000	7	11,600	168	146,000	9	16,200	191	41,600	7	5,940	2 5/8
2 7/8	155	91,000	7	13,000	189	162,000	9	18,000	—	—	7	—	2 7/8
3	174	103,000	7	14,700	210	180,000	9	20,000	242	51,200	7	7,300	3
3 1/4	210	123,000	7	17,600	264	226,000	9	25,200	299	61,600	7	8,800	3 1/4
3 1/2	256	146,000	7	20,800	312	270,000	9	30,000	—	—	7	—	3 1/2
4	300	171,000	7	24,400	380	324,000	9	36,000	435	84,000	7	12,000	4

[1]Linear density (lbs/100 ft) shown is "average." Maximum is 5% higher.
[2]New rope tensile strengths are based on tests of new and unused rope of standard construction in accordance with Cordage Institute Standard Test Methods.
[3]Working loads are for rope in good condition with appropriate splices, in non-critical applications, and under normal service conditions. Working loads should be exceeded only with expert knowledge of conditions and professional estimates of risk. Working loads should be reduced where life, limb, or valuable property are involved, or for exceptional service conditions such as shock, loads, sustained loads, etc.
[4]Composite rope. Materials and construction of this polyester/polypropylene composite rope conform to MIL-R-43942 and MIL-R-43952. For other composite ropes, consult the manufacturer.
(Printed from the Cordage Institute.)

This means that when a working load has been used to select a rope, the load must be handled slowly and smoothly to minimize dynamic effects and to avoid exceeding the provision for them.

Inspections

Before placing new rope in service, thoroughly inspect its entire length to determine that no part of it is damaged or defective. Any irregularity in its appearance is evidence of possible weakness. Experts disagree on what determines when a rope should be removed from service. Synthetic rope damage is not always visible.

If rope is being used under ordinary conditions, inspect it every 30 days. If it is used in critical applications, such as to support scaffolding on which employees work, inspect it more often. Inspection involves examining the entire length of the rope, inch by inch, for wear, abrasions, powdered fiber between strands, broken or cut fibers, displacement of yarns or strands, variation in size or roundness of strands, discoloration, and rotting.

To inspect the inner fibers, untwist the rope in several places to see whether the inner yarns are bright, clear, and unspotted. If exposed to acids, natural fiber ropes, such as Manila, and synthetic ropes should be scrapped or retired from critical operations. Visual inspections do not always reveal acid damage. A rope, like a chain, is only as strong as its weakest part—in the case of rope, its cross section. If there is a visible core or core damage, replace or splice out the rope.

Table 16–B. Specifications for Double-Braided Nylon Rope

Nominal Size		*Double Braided Nylon Nylon Cover-Nylon Core*			
Diameter (in.)	*Circ. (in.)*	*Linear Density[1] (lbs/100 ft)*	*New Rope Min. Strength Strength[2] (lbs)*	*Safety Factor*	*Working Load[3] (lbs)*
1/4	3/4	1.56	1,650	11	150
5/16	1	2.44	2,570	11	234
3/8	1 1/8	3.52	3,700	11	336
7/16	1 5/16	4.79	5,020	10	502
1/2	1 1/2	6.25	6,550	10	655
9/16	1 3/4	7.91	8,270	9	919
5/8	2	9.77	10,200	9	1,130
3/4	2 1/4	14.1	14,700	8	1,840
13/16	2 1/2	16.5	17,200	8	2,150
7/8	2 3/4	19.1	19,900	8	2,490
1	3	25.0	26,000	8	3,250
1 1/16	3 1/4	28.2	29,300	8	3,660
1 1/8	3 1/2	31.6	32,800	8	4,100
1 1/4	3 3/4	39.1	40,600	8	5,080
1 5/16	4	43.1	44,700	8	5,590
1 3/8	4 1/4	47.3	49,000	8	6,130
1 1/2	4 1/2	56.3	58,300	8	7,290
1 5/8	5	66.0	68,300	8	8,540
1 3/4	5 1/2	76.6	79,200	8	9,900
2	6	100	103,000	8	12,900
2 1/8	6 1/2	113	117,000	8	14,600
2 1/4	7	127	131,000	7	18,700
2 1/2	7 1/2	156	161,000	7	23,000
2 5/8	8	172	177,000	7	25,300
3	9	225	231,000	7	33,000
3 1/4	10	264	271,000	7	38,700
3 1/2	11	329	338,000	7	48,300
4	12	400	410,000	7	58,600

[1]Linear density (LD) shown is average and is determined from the equation LD=25×(Nom. diameter)^2 Tolerance is ±5%.
[2]Minimum tensile strength (MTS) is based on a large number of tests by various manufacturers and represents a value 2 standard deviations below the mean. Minimum strength is determined by the formula MTS=1057 LD$^{.995}$.
[3]Working loads are for rope in good condition with appropriate splices, in non-critical applications, and under normal service conditions. Working loads should be exceeded only with expert knowledge of conditions and professional estimates of risk. Working loads should be reduced where life, limb, or valuable property are involved, or for exceptional service conditions such as shock loads, sustained loads, etc.
(Printed from the Cordage Institute.)

Natural fiber rope loaded to more than 50% of its breaking strength will be permanently damaged; synthetics loaded to more than 65% may be damaged. Damage from overloading may be detected by examining the inside fibers. These will be broken into short lengths in proportion to the degree of overload. To make a good estimate of the strength of fibers, scratch the fibers with a fingernail—fibers of poor strength will readily part. This "fingernail test" is a quick test for chemical damage.

If the diameter of a rope is worn away more than 5%, replace the rope. In a small rope (up to ¾ in., 19 mm, in diameter), surface wear that has progressed to the center of the twisted element (yarn) may account for a loss of more than 80% of the rope's strength. In a rope with a diameter of ¾ in. (19 mm) or more, diameter surface wear may destroy the strength of the cover yarns, yet not affect the original strength of the core yarns. The remaining strength of the rope will depend on the proportion of the core yarns to the original total of yarns. If fiber samples can be secured from the rope, an estimate of the rope's strength can be made. Manually break the fiber samples, and estimate the distribution of fibers in a cross section, quartered to allow for twist configuration.

Due to slippage on a supporting surface when under high tension, synthetic ropes sometimes melt on the surface and form a skin. This skin is evidence of wear. Rope having multifilament synthetic fiber on the surface will often "fuzz." Fuzzing results from minute fiber breakage. If a rope is very fuzzy, replace it and look for the source of abrasion.

Care of Fiber Rope in Use

Safe use of rope involves recognizing the effects of such factors as chafing, cutting, elasticity, diameter-strength ratio, and general anticipated mishandling. To keep rope in good condition, observe the following precautions:

- Do not drag rope. Dragging rope wears away the outer fibers. If a rope picks up dirt and sand, abrasion within the strands of the rope will rapidly wear it out.
- Handle twisted rope so it retains the amount of twist (called balance) that the rope seeks when free and relaxed. If rotating loads and improper coiling and uncoiling change the balance, restore it by properly twisting the rope at either end. Severe imbalance can cause permanent damage; localized overtwisting causes kinking or hocking.
- Kinking strains the rope and may overstress the fibers. It may be difficult to detect a weak spot made by a kink. To prevent a new rope from kinking while it is being uncoiled, lay the rope coil on the floor with the bottom end down. Pull the bottom end up through the coil, and unwind the rope counterclockwise. If it uncoils in the other direction, turn the coil of rope over, and pull the end out on the other side.
- Avoid sharp bends over an unyielding surface since this causes extreme tension on the fibers. To make a rope fast, select an object with a smooth, round surface of sufficient diameter. If the object does have sharp corners, pad the corners. To avoid excessive bending, use sheaves or surface curvatures of suitable size for the rope's diameter (Table 16–C).
- Splice lengths of rope that must be joined. Do not knot them. A well-made splice will retain up to 100% of the strength of the rope, but a knot retains only 50% (Table 16–D).
- Thoroughly dry rope that has become wet; otherwise, it will quickly deteriorate. Do not allow wet rope to freeze. Hang up a wet rope, or lay it in a loose coil in a dry place until thoroughly dry. Rope deteriorates more rapidly if it is alternately wet and dry than if it remains wet.
- Do not use wet rope, or rope reinforced with metallic strands, near power lines or other electrical equipment. Use of such rope could inflict electric shock on workers.

Table 16–C. Sheave Sizes for Fiber Ropes of Varying Thickness

Diameter of Rope (in.)	*Diameter of Sheave (in.)*
3/4	6
7/8	7
1	8
1 1/4	10
1 3/8	11

Conversion factor: 1 in. = 2.54 cm

Table 16–D. Efficiency of Manila Fiber Rope with Splices, Hitches, and Knots

Jointure	*Percent Efficiency*
Full strength of dry rope	100
Eye splice over metal thimble	90
Short splice in rope	80
Timber hitch, round turn, half hitch	65
Bowline, slip knot, clove hitch	60
Square knot, weaver's knot, sheet bend	50
Flemish eye, overhand knot	45

Care of Fiber Rope in Storage

To maintain the strength of any rope, store it away from fumes, heat, chemicals, moisture, sunlight, and rodents. Store rope in a dry place where air can circulate freely about it. Air should not be extremely dry, however. Hang up small ropes. Lay larger ropes on gratings so air can get underneath and around them.

Do not store or use rope in an area containing acid or acidic fumes, as the rope will quickly deteriorate. Signs of deterioration from acid are dark brown or black spots on the rope.

Do not store rope unless it has been cleaned. Hang dirty rope in loops over a bar or beam, and then spray it with water to remove the dirt. The spray should not be so powerful, however, that it forces the dirt into the fibers. After washing, allow the rope to dry, and then shake it to remove the rest of the dirt.

WIRE ROPE

Wire rope is more widely used than fiber rope. This is because wire rope has greater strength and durability under severe working conditions. The physical characteristics of wire rope do not change when used in varying environments. Wire rope has controlled and predictable stretch characteristics.

Types of Wire Ropes

Wire rope consists of steel wires, strands, and core. The individual wires are cold drawn to predetermined size and breaking loads, according to required grades.

Grades include iron, tractor, mild plow steel, plow steel, improved plow steel, and extra-improved plow steel. The wires are then laid together in various geometrical arrangements, according to construction requirements for strands and classifications of wire rope (6 × 19, 6 × 37, and so on).

To make a strand, carefully selected lengths of pitch or lay are used. These lengths have a definite ratio to the length of lay or pitch used in forming the finished wire rope. After the individual strands are made, the required number are coiled around the core, which supports the load-carrying strands. The core can be made of sisal or synthetic fiber; or it can be a metallic strand core or independent wire rope core (IWRC).

The intended use of a wire rope determines the size, number, and arrangement of wires, the number of strands, the lay, and the type of core.

Classifications

The most widely used constructions of wire rope are six-strand ropes of these two classifications: 6 × 19 and 6 × 37. The 6 × 19 classification includes a variety of constructions, ranging from 15 to 26 wires per strand. Typical constructions are 6 × 19 Seale (Figure 16–2), 6 × 25 filler wire, and 6 × 19 Warrington. The 6 × 37 classification also covers a large number of designs and constructions, ranging from 27 to 49 wires per strand. Typical constructions of this classification are 6 × 41 filler wire (Figure 16–2), 6 × 37, 6 × 36 Warrington Seale, and 6 × 49 filler wire.

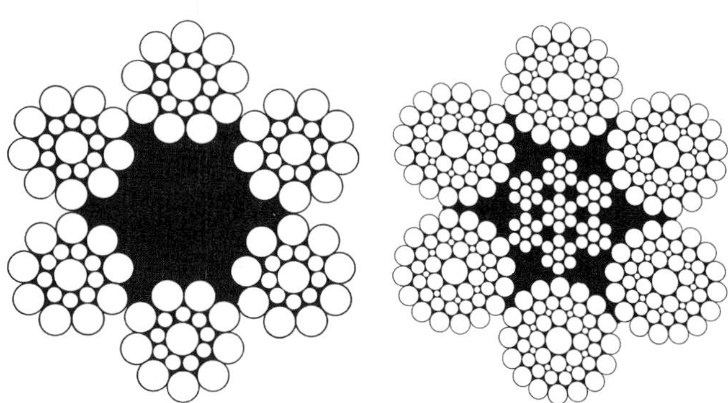

Figure 16–2. Wire rope is made from a number of individual wires grouped in strands, then laid together over a core member (fiber core, IWRC, or strand core). The number of wires per strand and the number of strands per core depend upon the expected working conditions and the amount of flexibility required. The cross section at the left above shows one construction of the 6 × 19 fiber core (FC) classification containing 114 wires. The cross section at right above shows one construction of the 6 × 37 classification containing 343 wires, including those of the independent wire core (IWRC).

Generally speaking, the more wires there are per strand, the more flexible the wire rope is. However, the fewer wires there are per strand, the more abrasion resistant and crush resistant the rope is. In ropes with large diameters, 2 in. (6.4 cm) and larger, practically all wire rope is produced in the 6 × 37 or 6 × 61 class. Therefore, because of the large number of possible rope constructions available, exercise care to select the right one (Table 16–E). (See American Iron and Steel Institute [AISI], *Wire Rope Users Manual* in References at the end of this chapter.)

Service Requirement

Depending upon service requirements and conditions, six-strand ropes may have a fiber core (FC), a wire strand core (WSC), or an IWRC. Wires in the strands may be laid in the opposite direction (regular lay) from that of the strands in the rope, or they may be laid in the same direction (lang lay) as those of the strands in the rope.

Where maximum flexibility is required, eight-strand hoisting ropes are used. They are usually of the 8 × 19 classification, with regular lay and an FC or an IWRC. Where such flexibility is not required, as in guy wires and highway guards, wire rope of 6 × 7 construction (six strands with seven wires per strand) is suitable. When selecting wire rope for a particular job, consult engineers from reliable wire rope manufacturers.

Some conditions require rope with special qualities. Fiber cores are affected by temperatures above 250 F (120 C). Under such conditions, a metallic core provides greater efficiency and safety. A zinc-coated or stainless-steel wire rope effectively resists some types of corrosion. Refer to rope manufacturers for information on specific corrosion problems.

Since preformed wire rope does not unravel, it has advantages for certain services, such as for slings to hoist heavy construction equipment. Preformed wire rope is less likely to set or kink than other types of wire rope; thus, broken wires are less likely to protrude and create a hazard to workers.

Design Factors for Rope Used in Hoisting

The operating or design factors for rope used in hoisting are calculated by dividing the nominal catalog strength of the rope by the sum of the maximum loads to be hoisted. It is normal practice to base this sum on static loads. For rope used in hoisting in mines, the

Table 16–E. Nominal Breaking Strength in Tons of Improved Plow Steel (IPS) and Extra-Improved Plow Steel (EIPS) Ropes

Diameter (in.)	*6x19 & 6x37*			*8x19*			*19x7*	
	IPS[1]		*EIPS*[2]	*IPS*		*EIPS*	*IPS*	*EIPS*
	FC[3]	*IWRC*[4]	*IWRC*	*FC*	*IWRC*	*IWRC*	*FC*	*IWRC*
3/8	6.10	6.56	7.55	5.24	5.76	6.63	5.59	6.15
7/16	8.27	8.89	10.2	7.09	7.80	8.97	7.58	8.33
1/2	10.7	11.5	13.3	9.23	10.1	11.6	9.85	10.8
9/16	13.5	14.5	16.8	11.6	12.8	14.7	12.4	13.6
5/8	16.7	17.9	20.6	14.3	15.7	18.1	15.3	16.8
3/4	23.8	25.6	29.4	20.5	22.5	25.9	21.8	24.0
7/8	32.2	34.6	39.8	27.7	30.5	35.0	29.5	32.5
1	41.8	44.9	51.7	36.0	39.6	45.5	38.3	42.2
1 1/8	52.6	56.5	65.0	45.3	49.8	57.3	48.2	53.1
1 1/4	64.6	69.4	79.9	55.7	61.3	70.5	59.2	65.1
1 3/8	77.7	83.5	96.0	67.1	73.8	84.9	71.3	78.4
1 1/2	92.0	98.9	114	79.4	87.3	100	84.4	92.8
1 5/8	107	115	132					
1 3/4	124	133	153					
1 7/8	141	152	174					
2	160	172	198					
2 1/8	179	192	221					
2 1/4	200	215	247					

[1]IPS=Improved plow steel.
[2]EIPS=Extra improved plow steel.
[3]FC=Fiber core.
[4]IWRC=Independent wire rope core.

maximum loads to be hoisted include the weight of the skip or car or cage plus the weight of the material plus the weight of the suspended rope when the skip or cage is at the lowest point in a shaft. In some cases, acceleration stresses are also considered. It is recommended that hoisting rope have at least the strength of improved plow steel. For some applications, use extra-improved plow steel, which is the strongest steel, to provide an adequate design factor and better service.

The minimum design factors for rope used in hoisting depend upon the type of service required and the federal, state or provincial, or local codes covering the particular hoisting operation. Many of these codes describe exactly how the design and operating factors should be figured. Therefore, check what codes are in force before making a final selection of wire ropes to be used in hoisting. Also, obtain the advice of a reliable wire rope manufacturer.

Inspections and Replacement

The frequency of inspections and replacement of wire rope depends on service conditions. In the United States, OSHA minimum inspection requirements for wire rope or cable include installation and yearly inspections with documentation when used as a lifting device. At regular intervals, a specially trained inspector should examine ropes on which human life depends and document such inspections. Some facilities and mines, for instance, make a daily inspection for readily observable defects, such as kinking and loose wires, and a thorough inspection weekly. For the latter inspection, the rope speed is generally less than 60 feet per minute (fpm).

The inspector checks specifically for wear of the crown wires, kinking, high strands, corrosion, loose wires, nicking, and lubrication (Figure 16–3). Rope calipers (Figure 16–4) and micrometers are used to determine changes in the cross section of rope at various locations. In most cases, sudden change in rope length and/or diameter is a warning that the wire rope is nearing the end of its useful life and that it should be removed from service. The reason for this change is general deterioration of the structure of the interior rope, such as corrosion of uninspectable wires and general deterioration of the core. A decrease in the rope's diameter is often difficult to determine.

The number of broken wires per lay is one of the principal bases for judging the condition of a rope. If most of the broken wires in a lay are concentrated in several strands, that section of the rope is weaker than it would be if the broken wires were uniformly distributed throughout all strands and along the length of the

WIRE ROPE WEAR AND DAMAGE

The evidence in these illustrations will aid the inspector in determining the actual cause of wear or damage in any wire rope.

A wire rope which has been kinked. A kink is caused by pulling down a loop in a slack line during improper handling, installation, or operation. Note the distortion of the strands and individual wires. Early rope failure will undoubtedly occur at this point.

Localized wear over an equalizing sheave. The danger of this type wear is that it is not visible during operation of the rope. This emphasizes the need of regular inspection of this portion of an operating rope.

A typical failure of a rotary drill line with a poor cut-off practice. These wires have been subjected to excessive peening causing fatigue type failures. A predetermined, regularly scheduled, cut-off practice will go far toward eliminating this type of break.

A single strand removed from a wire rope subjected to "strand nicking." This condition is the result of adjacent strands rubbing against one another and is usually caused by core failure due to continued operation of a rope under high tensile load. The ultimate result will be individual wire breaks in the valleys of the strands.

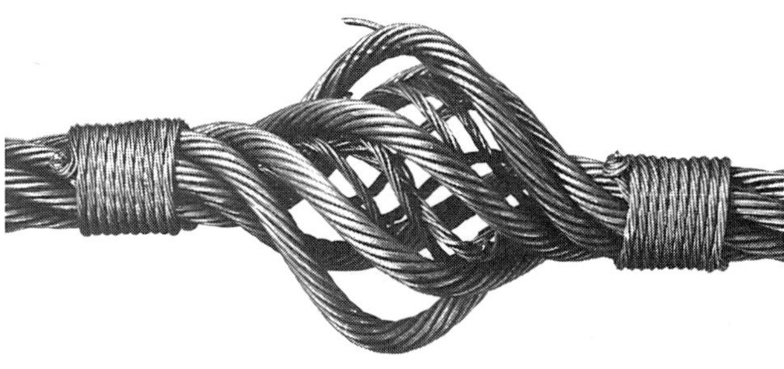

A "bird cage." Caused by sudden release of tension and resultant rebound of rope from overloaded condition. These strands and wires will not return to their original positions.

An example of a wire rope with a high strand—a condition in which one or two strands are worn before adjoining strands. This is caused by improper socketing or seizing, kinks or dog legs. Picture A is a close-up of the concentration of wear and B shows how it recurs in every sixth strand (in a six-strand rope).

Figure 16–3. Typical characteristics and causes of broken wires in wire ropes. (Courtesy Wire Rope Corp. of America, Inc.)

An illustration of a wire which has broken under tensile load in excess of its strength. It is typically recognized by the "cup and cone" appearance at the point of fracture. The necking down of the wire at the point of failure to form the cup and cone indicates that failure occurred while the wire retained its ductility.

A wire rope which has jumped a sheave. The rope itself is deformed into a "curl" as if bent around a round shaft. Close examination of the wires show two types of breaks—normal tensile "cup and cone" breaks and shear breaks which give the appearance of having been cut on an angle with a cold chisel.

A wire rope which has been subjected to repeated bending over sheaves under normal loads. This results in "fatigue" breaks in individual wires—these breaks are square and usually in the crown of the strands.

An illustration of a wire which shows a fatigue break. It is recognized by the squared off ends perpendicular to the wire. This break was produced by a torsion machine, which is used to measure the ductility. This break is similar to wire failures in the field caused by excessive bending.

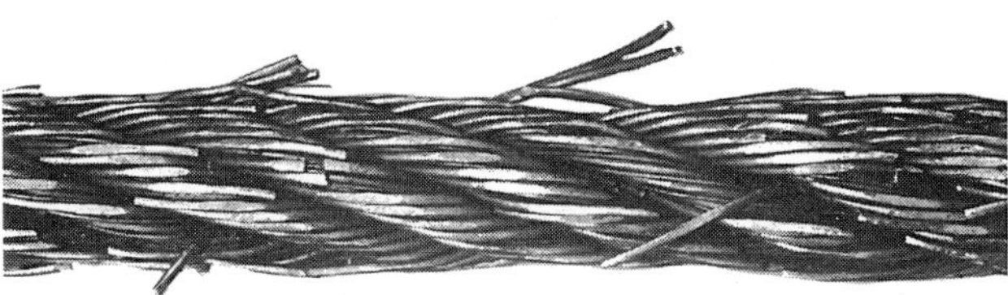

An example of "fatigue" failure of a wire rope which has been subjected to heavy loads over small sheaves. The usual crown breaks are accompanied by breaks in the valleys of the strands—these breaks are caused by "strand nicking" resulting from the heavy loads.

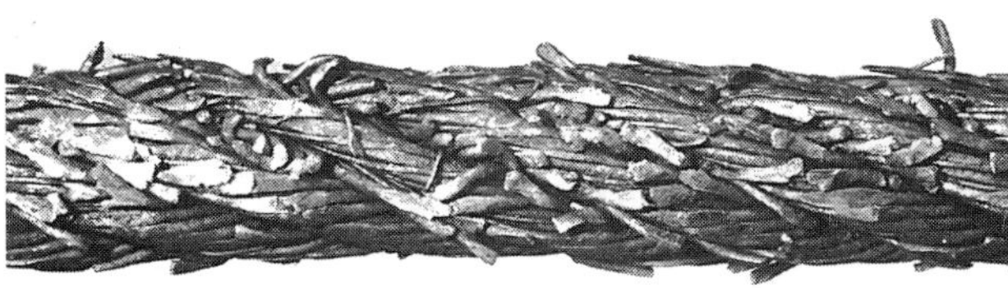

An example of a wire rope that has provided maximum service and is ready for replacement.

A close-up of a rope subjected to drum crushing. Note the distortion of the individual wires and displacement from their normal position. This is usually caused by the rope scrubbing on itself.

A fatigue break in a cable tool drill line caused by a tight kink developed in the rope during operation.

Figure 16–3. *(Concluded.)*

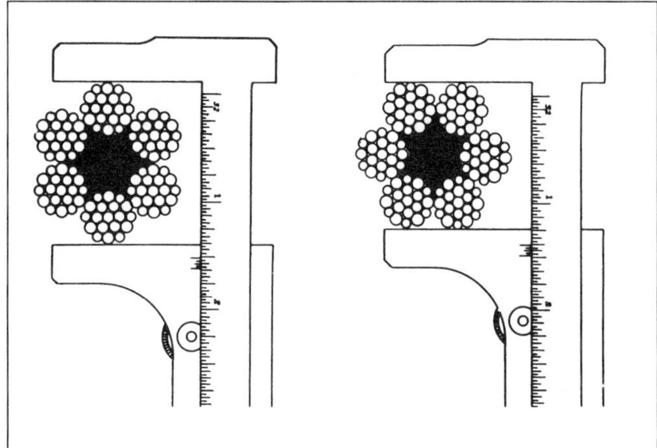

Figure 16–4. Always read the widest diameter when measuring wire rope. The correct way is shown on the left; the wrong way is shown on the right. (Courtesy Armco Steel Corporation.)

rope. If, however, the number of broken wires along the length of a rope increases rapidly between inspections, the rope is becoming fatigued and is nearing the end of its useful life.

Inspection codes may vary from state to state or from province to province with regard to rope inspection and allowable degrees of deterioration. Usually inspections are based on the number of broken wires per strand in one rope lay or on the number of broken wires per rope lay in all strands. For specific information within an operating area, consult state or provincial codes and the specific OSHA regulations covering the type of operation. Electronic inspection devices are available for determining loss of strength due to corrosion, loss of metallic area, and broken wires.

Experience and judgment of all these factors, combined with the length of time the rope has been in service and the tonnage hoisted or other work done by the rope, determine when it should be replaced. At intervals throughout the life of the rope, a short section should be cut off at the socket end. This practice has two purposes: (1) to remove wires damaged by vibration dampened at the socket, and (2) to change the positions of critical wear points throughout the system.

Care of Wire Rope in Use

Deterioration of wire ropes has a number of causes, which vary considerably in importance depending on the conditions of service (Figure 16–3). For example, corrosion often is the principal cause of deterioration of wire rope used for hoisting in wet mine shafts. Moisture and the presence of acid in the water lead to corrosion. Corrosion, particularly of the interior wires, is indicated by pitting. Corrosion accelerates wear. This highly dangerous condition is difficult to detect. Other causes of deterioration include the following:

- Wear, particularly on the crown or outside wires, can result from contact with sheaves and drums.
- Kinks are acquired from improper installation of a new rope or from hoisting with slack in the rope. A kink cannot be removed without creating a weak place.
- Fatigue, indicated by a square fracture at the end of a wire, can be caused by bending stresses from sheaves and drums with small radii; by stresses from whipping, vibration, and pounding; or by torsional stresses.
- Drying out of lubrication is often hastened by heat and operating pressure.
- Overloading, including dynamic overloading, can damage wire rope if acceleration and deceleration are factors of importance. Such damage may not become known until some time after the overload.
- Overwinding, when rope length is greater than the drum can accommodate in a single layer, can cause heavy abrasion and excessive wear at crossover points. However, successful overwinding can be achieved by using specially engineered drum grooving.
- Mechanical abuse, such as running over wire rope with equipment and permitting obstructions to remain in a rope's path of travel, can ruin the rope. It is more common for wire rope to be thrown away because of abuse than because of use.

When possible, clean a wire rope before lubricating it. Regular application of a suitable lubricant to wire rope used for hoisting prevents corrosion, wear from friction, and drying out of the core. Good lubricants are free from acids and alkali and have adhesive strength. They also have the ability to penetrate the strands. The lubricant should be insoluble under the prevailing conditions. Ropes should be dry when lubricant is applied so that the lubricant will not entrap moisture. Thin lubricants can be applied by hand. However, it is better to apply them by providing some means of dripping them on the rope or using a spray device to apply the proper quantity automatically.

Clean wire rope monthly, as is done in mine shafts, to remove dirt, abrasive particles, and corrosion-producing moisture. Do not use cleaning fluids on wire rope; they harm the core's lubricant. Light oils are sometimes used to loosen the coating of lubricant and harmful materials. Mechanical methods such as those using compressed air or a steam jet clean a rope effectively and thoroughly.

Sheaves and Drums

Fatigue of wire rope resulting from bending stresses depends upon the diameter of drums and sheaves: the larger the diameter, the more favorable a rope's service life will be. However, sometimes operators sacrifice diameter to accommodate designs and considerations for other equipment. Consider the recommendations in Table 16–F as a design base only. Many types of equipment successfully operate with smaller drum-and-sheave rope ratios, while others use much larger ratios. A case in point is the drum and sheave requirements in most mining codes, elevator codes, and shovel, hoist, and crane codes.

The safety and the service life of installations for hoisting rope can be greatly increased by using sheaves and drums of suitable size and design and by properly lubricating them. The rope and the hoisting equipment also require maintenance.

Heads, idlers, knuckles, curved sheaves, and grooved drums must have grooves that support a rope properly. Before installing a new rope, inspect the grooves, and where necessary, machine them to proper contour and groove diameter. The diameter should exceed the nominal rope diameter by the amount shown in Table 16–G. For recommended grooving for drums and sheaves, consult a wire rope manufacturer's handbook.

Sheaves

The condition and contour of sheave grooves is important for the service life of wire rope. Periodically check sheave grooves (Figure 16–5), and do not let them wear to a diameter smaller than those shown for used grooves in Table 16–G. If the grooves become more worn than this, expect a reduction in the rope's service life. Reconditioned sheave grooves should conform to

Table 16–F. Recommended Tread Diameters of Sheaves and Drums for Wire Rope

Rope Classification	Average Recommended (times rope diameter)	Minimum
6 × 7	72	42
6 × 19	45	30
6 × 37	27	18
8 × 19	31	21

Printed from *ANSI/ASME* A17.1, *Elevators, Escalators, and Moving Walks.*

Table 16–G. Groove Diameter in Relation to Wire Rope Diameter

	Amount that Groove Diameter Should Be Larger than Nominal Rope Diameter (inches)	
Rope Size (inches)	Used	New
1/4 and 5/16	1/128	1/64
3/8 to 3/4 incl.	1/64	1/32
13/16 to 1 1/8 incl.	3/128	3/64
1 3/16 to 1 1/2 incl.	1/32	1/16
1 9/16 to 2 1/4 incl.	3/64	3/32
2 5/16 and larger	1/16	1/8

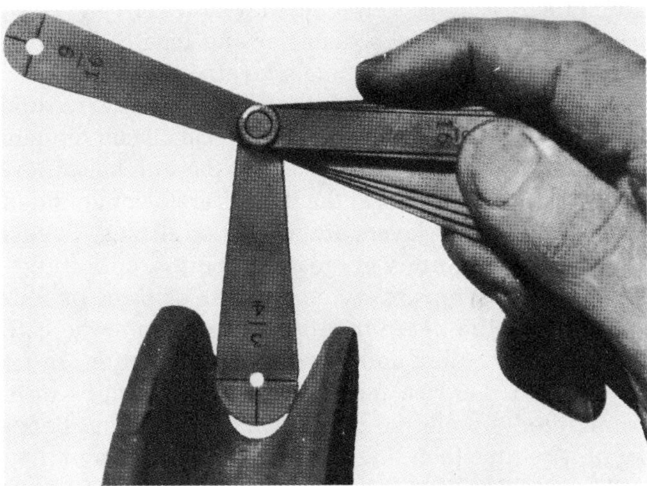

Figure 16–5. Check sheave grooves with a gauge designed with one-half the allowable oversize. If light is seen between the gauge and sheave, replace the sheave. (Courtesy Armco Steel Corporation.)

the tolerance shown in Table 16–G for new or remachined grooves.

On all new sheaves, ensure that the grooves are made for the size of rope specified. The bottom of the groove should have a 150-degree arc of support, and the sides of the groove should be tangent to the ends of the bottom arc. The depth of the groove should be one times the nominal diameter of the rope. The radius of the arc should be one-half the nominal rope diameter plus one-half the value shown in Table 16–G, for new or remachined grooves.

Check sheaves for proper alignment when they are installed. During rope changes, check the sheaves for worn bearings, broken flanges, proper groove size, smoothness, and contour. Recondition or replace heavily worn or damaged sheaves.

Sheave groove-bearing pressures can become very high, depending upon operating conditions and rope loadings. High pressures can cause excessive sheave wear and shorten the life of wire rope. It is necessary, therefore, to consider this factor and to select proper sheave materials and liners at the time of installation. For information on this subject, consult wire rope manufacturers' handbooks. (See also American Iron and Steel Institute, *Wire Rope Users Manual*, in References at the end of this chapter.)

Drums

Avoid multiple-layer winding of rope on drums, if possible. Multiple layering causes the rope to wear, thus shortening the rope's life, particularly at the point where the rope rises to the next layer. Where practical, use drums with enough diameter and length that they can take all the rope in a single layer.

Minimize crushing and excessive wear of wire rope by using spirally grooved drums that can accommodate one layer of rope. In any case, limit the number of layers to three. Rope lifters at the flanges are recommended when two or more layers are wound on drums. To distribute wear uniformly at crossover points, cut off one-and-a-quarter wraps every six months or three or four times during the life of the rope. According to the U.S. Occupational Safety and Health Administration, in no case should there be fewer than two full wraps on a drum; three is preferred. In general, avoid reverse bending of wire rope (bending first in one direction and then in the opposite) over sheaves or drums. This wears out the rope faster.

Correct fleet angle is important for even, efficient winding of wire rope. The fleet angle is the included angle between the rope winding on the drum and a line perpendicular to the drum shaft and running through the head or lead sheave (Figure 16–6).

To reduce any tendency for the rope to open-wind, do not let the fleet angle exceed 1°30′. To ensure that the rope starts back on the next layer, use a minimum angle of 0°30′ for smooth drums and 2° for grooved drums. Adhering to these specifications helps achieve uniform winding on smooth-faced drums and also increases the winding efficiency of grooved drums. For smooth-faced drums, proper direction of lay of rope for specified winding conditions helps achieve uniform winding.

Installing a wire rope on a plain-faced or smooth-faced drum requires great care. The starting position should be at the drum end so that each turn of the rope winds tightly against the preceding turn (Figure 16–7). Maintain close supervision during the entire installation process to make sure that

1. the rope is properly attached to the drum
2. appropriate tension on the rope is maintained as it is wound on the drum

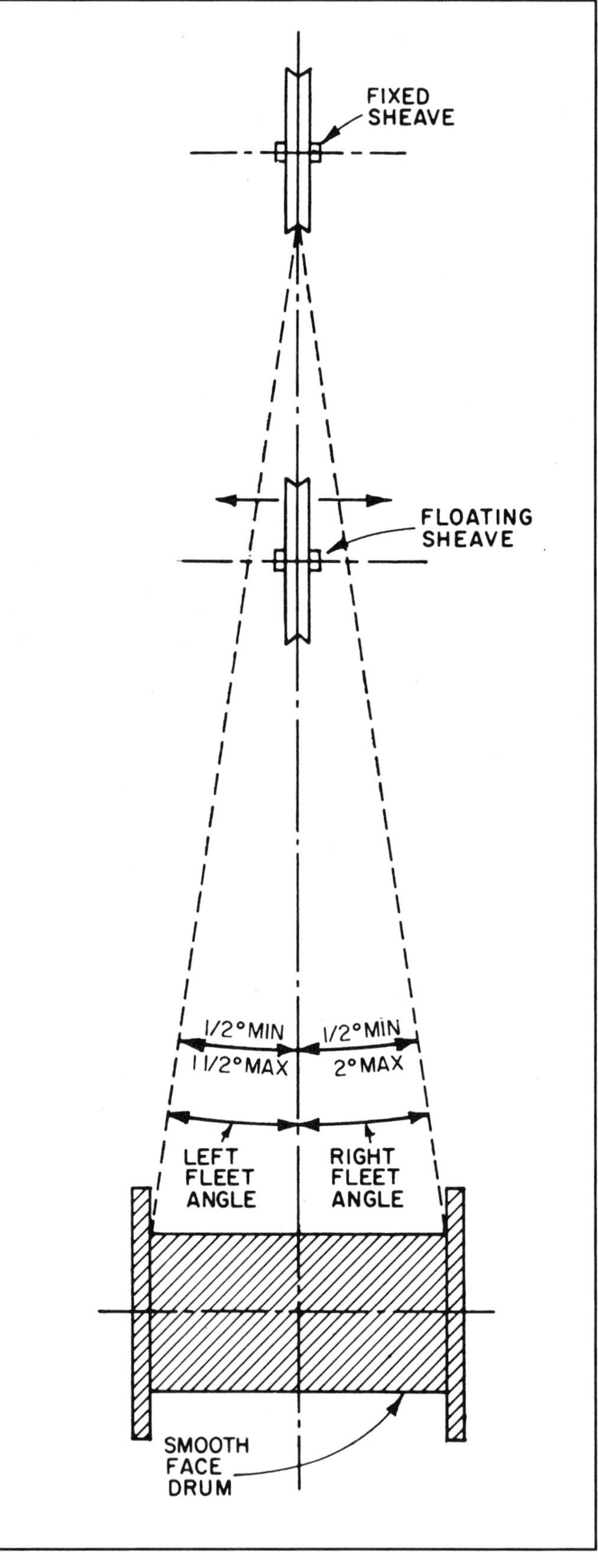

Figure 16–6. The fleet angle is graphically defined in this illustration of wire rope running from a fixed sheave, over a floating sheave, and then on to a smooth drum.

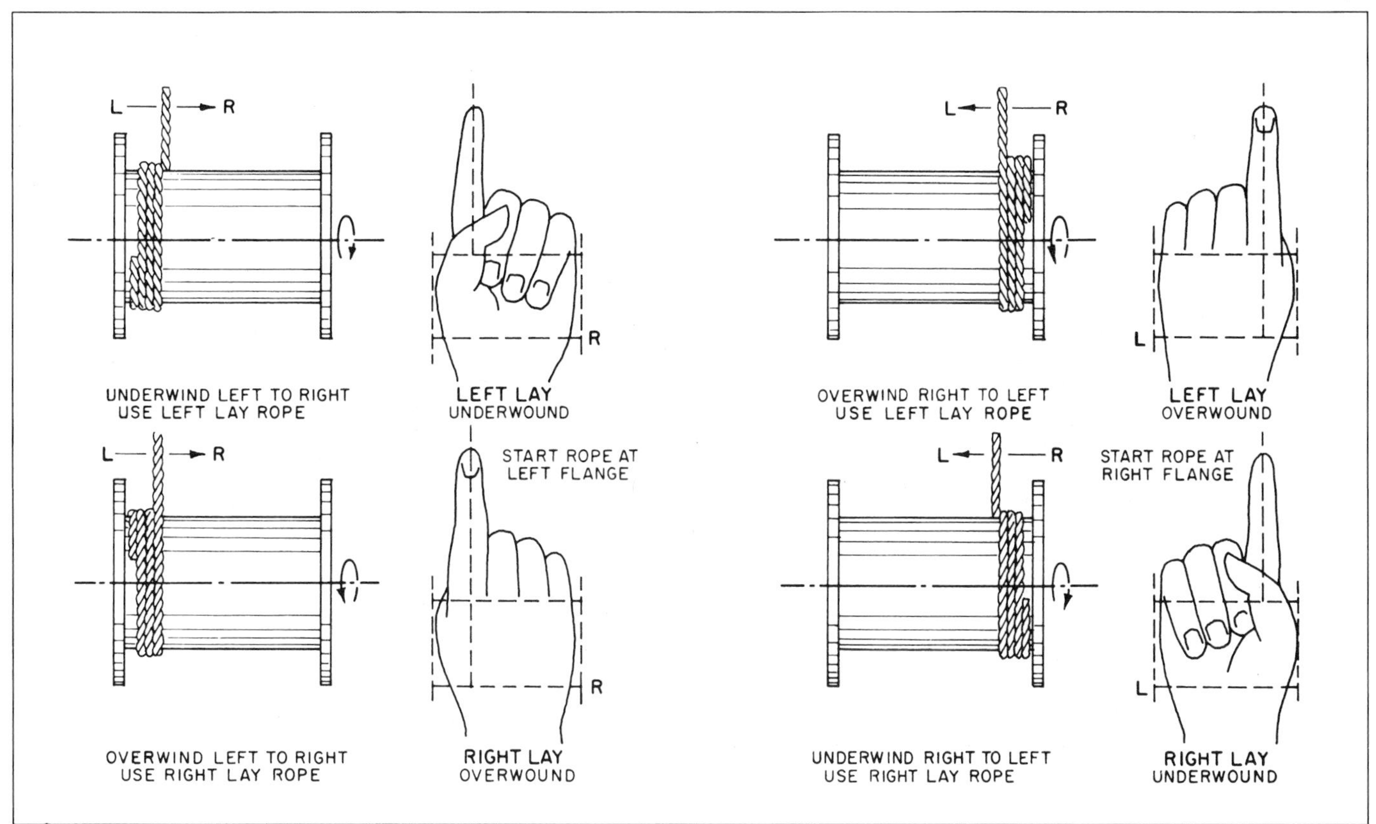

Figure 16–7. By holding the left or right hand with the index finger extended, palm up or palm down, the proper procedure for installing left- or right-lay rope on a smooth drum can be easily determined. (Courtesy American Iron and Steel Institute.)

3. each turn is guided as close to the preceding turn as possible so that there are no gaps between turns
4. there are at least two dead turns on the drum when the rope is fully unwound during normal operating cycles

Loose and uneven winding on a plain-faced or smooth-faced drum usually causes excessive wear, crushing, and distortion of the rope. Such abuse thus results in lower operating performance and a loss in the rope's effective strength. Also, on jobs that require moving and spotting a load, the operator will encounter control difficulties because the rope will pile up, pull into the pile, and fall from the pile to the drum surface. The ensuing shock can break or otherwise damage the rope.

Wire Rope Fittings

There are several ways to attach wire rope to fittings: by using pressed fittings, mechanical sleeve splices, hand-tucked splices, clips and clamps, sockets, or knots. Fittings are important for safety because they develop from 75% to 100% of the breaking strength of the rope. Manufacturers specify fittings of suitable size and design for ropes of different sizes. The strength of an attachment is attained only when the connection is made exactly according to the manufacturer's instructions (Figure 16–7). Some types of attachments, such as pressed fittings and mechanical sleeve splices that are used in making slings, must be made at either a wire rope manufacturer's facility or at a properly equipped commercial sling shop.

Efficiencies of properly made hand-tucked splices vary according to the splicer's ability and the rope's diameter but can be as high as 90% (Figure 16–8). The efficiency of mechanical sleeve splices varies from 90% to 95% when IWRC wire rope is used.

Rope often is connected to the fittings of conveyances by means of clips and clamps. Clips and clamps are rated to develop 75% to 80% of the rope's breaking strength. Figure 16–8 shows how the clips should be attached, and Table 16–H gives the number of clips and the spacing required for ropes of different sizes. It is important to retighten the nuts on all clips after the rope's first load-carrying use as well as at all subsequent regular inspection periods.

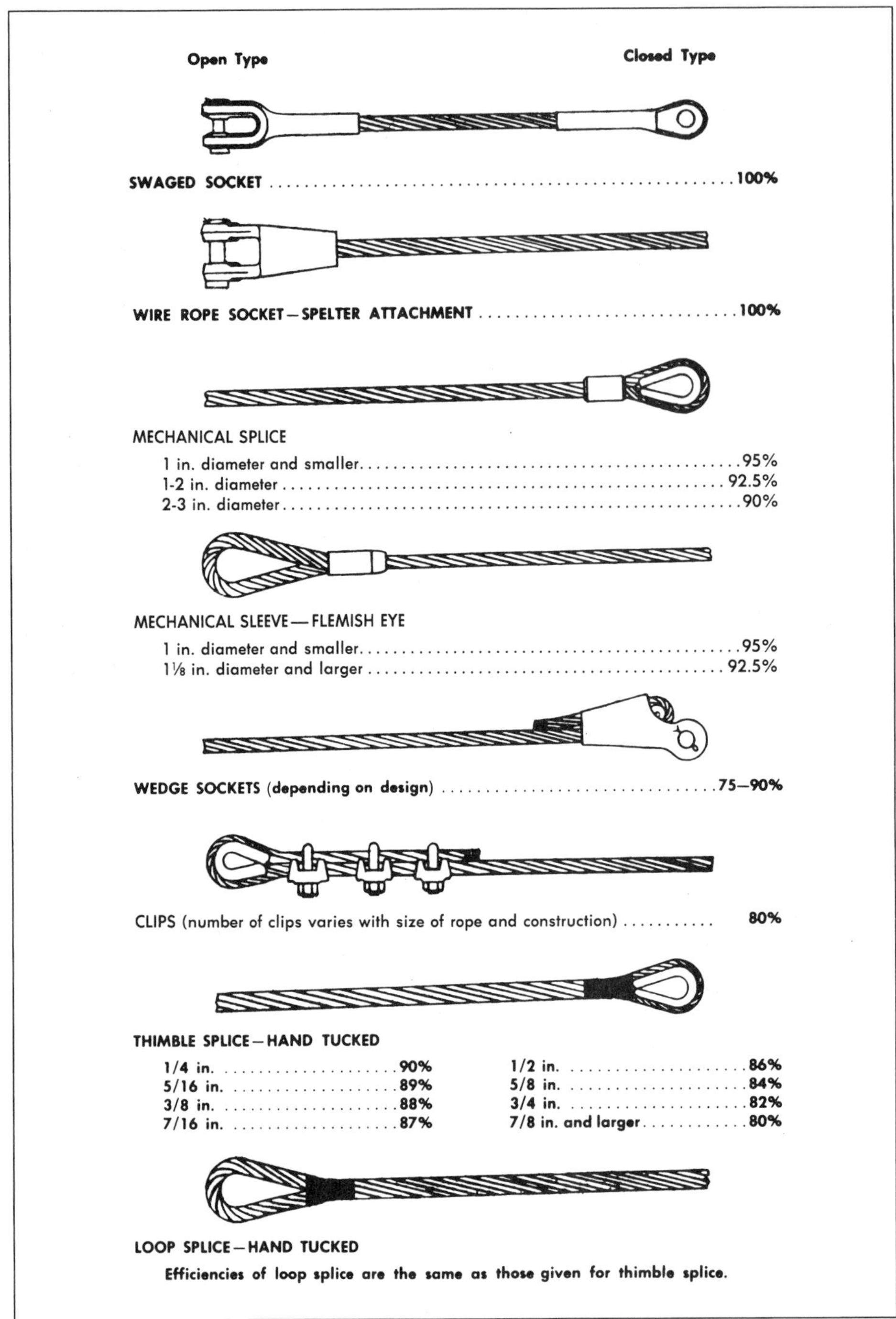

Figure 16–8. Typical efficiencies of attaching wire rope to fittings in percentages of strength of rope.

Socketing with zinc and a thermostate plastic resin will develop 100% of the rope's breaking strength. Figure 16–9 shows zinc-poured and swaged sockets. Since there is no ready way to detect flaws in the finished job, follow the recommended procedure exactly. In high-speed hoisting, fatigue is especially likely to develop with this type of attachment. Therefore, cut off and discard at frequent intervals the section adjacent to the conveyance. In some state mining laws, the required interval is every six months.

Square knots and other types of knots have low and unpredictable efficiencies—40% or less. Using them will likely result in the failure of a rope assembly, and under certain conditions, it will result in a serious

Table 16–H. Number and Spacing of Clips for Ropes of Various Sizes

Diameter of Rope (inches)	*Minimum Number of Clips*	*Length of Rope Turned Back Exclusive of Eye (inches)*	*Torque (ft-lb)*
1/8	2	3¼	—
¼	2	3¼	—
½	3	11½	65
5/8	3	12	95
¾	4	18	130
7/8	4	19	225
1	5	26	225
1 1/8	6	34	225
1¼	7	44	360
1 3/8	7	44	360
1½	8	54	360
1 5/8	8	58	430
1¾	8	61	590
2	8	71	750
2¼	8	73	750

1 in. = 2.54 cm.
1 ft-lb = 1.36 newton-meter.

The number of clips shown is based upon using right regular or Lang lay wire rope, 6×19 class or 6×37 class, fiber core or IWRC, IPS, or XIPS. If Seale construction or similar large outer wire type construction in the 6×19 class is to be used for sizes 1 in. (2.5 cm) and larger, add one additional clip.

The number of clips shown also applies to right regular lay wire rope, 8×19 class, fiber core, IPS, nominal sizes 1½ in. and smaller; and right regular lay wire rope, 18×7 class, fiber core, IPS or XIPS, nominal sizes 1¾ in. and smaller.

For other classes of wire rope not mentioned above, it may be necessary to add additional clips to the number shown.

Turn back the specified amount of rope from the thimble. Apply the first clip one base width from the dead end of the wire rope (U-bolt over dead end—live end rests in clip saddle). Tighten nuts evenly to recommended torque.

Apply the next clip as near the loop as possible. Turn on nuts firm but do not tighten.

Space additional clips if required equally between the first two. Turn on nuts—take up rope slack—tighten all nuts evenly on all clips to recommended torque.

NOTE: Apply the initial load and retighten nuts to the recommended torque. The rope will stretch and shrink in diameter when loads are applied. Inspect periodically and retighten.

The efficiency rating of a properly prepared termination for clip sizes 1/8 to 7/8 in. is approximately 80 percent and for sizes 1 to 3 in. is approximately 90 percent. This rating is based upon the catalog breaking strength of wire rope. If a pulley is used in place of a thimble for turning back the rope, add one additional clip.

accident. In the United States, OSHA regulations and other industrial and construction codes prohibit the use of knots in wire rope.

RIGGING

In lifting various materials and supplies, a number of standard chokers, slings, bridle hitches, and basket hitches can be used. Because loads vary in physical dimension, shape, and weight, a rigger needs to know what method of attachment can be used safely. It is estimated that 15% to 35% of crane accidents may involve improper rigging.

Employers must see that the personnel responsible for rigging loads receive thorough training. Riggers must (1) know the load, (2) judge distances, (3) properly select tackle and lifting gear, and (4) direct the operation.

The single most important rigging precaution is to determine the weight of the load before attempting to lift it. The weight of the load will in turn determine the lifting device, such as a crane, and the rigging gear to be used. It is also important to rig a load so that it will be stable—that is, motionless while being lifted. Properly maintaining, storing, and protecting the rigging gear will increase its life and safety.

Choker Hitch

Figure 16–10 shows this simplest of sling hitches. The sling passes entirely around the load, and the other end attaches to the hook. The choker hitch, used singly or with others, has a capacity of 75% when the choke angle is 120 degrees.

Basket Hitch

This hitch can be made with the same sling by passing the choker sling under the load, both loop eyes going to the crane hook (Figure 16–10). Slings are used in pairs when the entire load is suspended and singly when one end of the load must be raised. Rated capacities vary. A safe figure when two slings are used is to double the load rating of a choker hitch where square corners are

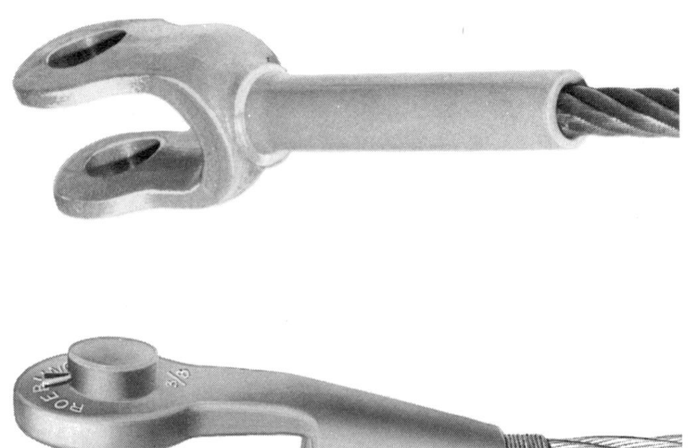

Figure 16–9. The end fittings should be of the best possible type for the specific use. Zinc-poured sockets (bottom) are efficient in straight tension but are not as fatigue-resistant as swaged sockets (above).

encountered. Reduce the load rating if sling legs deflect from a vertical position.

Another method of calculating the load rating of a basket hitch is to double the load rating of a single directly connected sling. Reduce this figure for sharp bends or other conditions that do not occur in a vertical sling. On most slings when a basket hitch is used, the rated capacities at the various angles to the vertical (Tables 16-I and 16-J) apply, provided that the radius of curvature where the sling makes contact with the load is at least 20 times the diameter of the individual component rope. Shorter radii or contact with any sharp corner will reduce the sling capacity accordingly.

FIBER AND WIRE ROPE SLINGS

The safety of a rope sling assembly depends on the following factors: the material used (fiber rope or wire rope), fittings of suitable strength for the load, the method of fastening the rope to the fittings, the type of sling (such as single-legged or three-legged), the type of hitch, and regular inspection and maintenance. Keep these factors in mind when using rope slings.

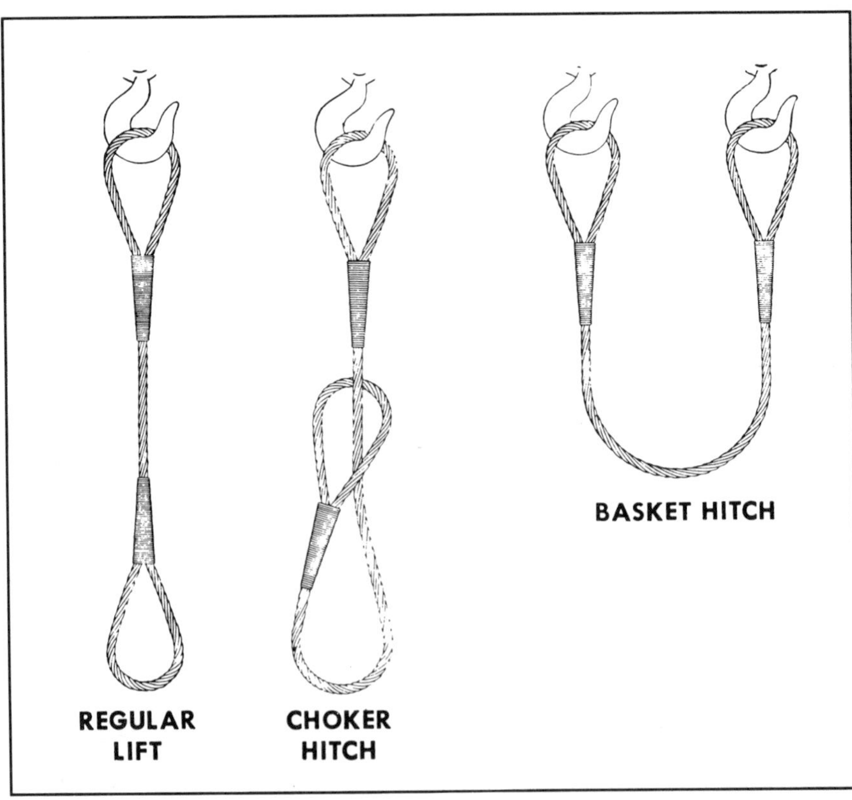

Figure 16–10. The three most common hitches for all types of slings are the regular (also called straight or vertical), the choker, and the basket. In addition, the grommet (or endless loop) type of sling can be used in a variety of configurations. The most efficient is the straight hitch. The choker hitch places a certain amount of stress at the point where the end loop encircles (or chokes) the body of the sling. Although the basket hitch results in two legs, the result is not a doubling of load capacity of the sling. (Courtesy Bethlehem Steel Co.)

Table 16–I. Rated Capacity Limits (in Tons) of Wire Rope Slings, Using Preformed Improved Plow Steel Wire Rope (depending on method of attaching the rope to the fittings)

Rope Diameter (in.)	Single Leg						Two-Leg Bridle or Basket Hitch											
	Vertical			Choker			Vertical*			30 Degrees Vertical			45 Degrees			60 Degrees Vertical		
	S	MS	HT	S	MS	HT	S	MS	HT	S	MS	HT	S	MS	HT	S	MS	HT
	6×19 Classification Construction																	
3/8	1.3	1.2	1.2	.92	.92	.92	2.6	2.4	2.4	2.3	2.1	2.0	1.8	1.7	1.7	1.3	1.2	1.2
1/2	2.3	2.2	2.0	1.6	1.6	1.6	4.6	4.4	4.0	4.0	3.8	3.5	3.3	3.1	2.8	2.3	2.2	2.0
5/8	3.6	3.4	3.0	2.5	2.5	2.5	7.2	6.8	6.0	6.2	5.9	5.2	5.1	4.8	4.2	3.6	3.4	3.0
3/4	5.1	4.9	4.2	3.6	3.6	3.6	10.0	9.8	8.4	8.7	8.5	7.3	7.1	6.9	5.9	5.0	4.9	4.2
7/8	6.9	6.6	5.5	4.8	4.8	4.8	14.0	13.0	11.0	12.0	11.0	9.5	9.9	9.3	7.8	7.0	6.6	5.5
1	9.0	8.5	7.2	6.3	6.3	6.3	18.0	17.0	14.0	16.0	15.0	12.0	13.0	12.0	10.0	9.0	8.5	7.2
1 1/8	11.0	10.0	9.0	7.9	7.9	7.9	22.0	20.0	18.0	19.0	17.0	16.0	16.0	14.0	13.0	11.0	10.0	9.0
	6×37 Classification Construction																	
1 1/4	14	13	11	9.7	9.7	9.7	28	26	22	24	23	19	20	18	16	14	13	11
1 3/8	17	15	13	12	12	12	34	30	26	29	26	23	24	21	18	17	15	13
1 1/2	20	18	16	14	14	14	40	36	32	35	31	28	28	25	23	20	18	16
1 3/4	27	25	21	19	19	19	54	50	42	47	43	36	38	35	30	27	25	21
2	34	32	28	24	24	24	68	64	56	59	55	48	48	45	40	34	32	28
2 1/4	43	40	34	30	30	30	86	80	68	74	69	59	61	57	48	43	40	34

*If slings are used to handle loads with sharp corners, pads or saddles should be used to protect the rope. The radius of bend should not be smaller than five times the diameter of the rope. If the radius of bend is smaller, a choker hitch rating should be used.

S = Socket or swaged terminal attachment.
MS = Mechanical sleeve attachment.
HT = Hand-tucked splice attachment.

Note 1. Table is based on a design factor of 5, sling angles formed by one leg and a vertical line through the crane hook, and uniform loading.
Note 2. For 3-leg bridle slings, multiply safe load limits for 2-leg bridle slings by 1.5, and for 4-leg bridle slings, multiply by 2.0.

Types of Fiber and Wire Rope Slings

Since the strength of fiber rope is affected by chemicals, freezing, high temperatures, and sharp bends, consider these factors when selecting rope for slings. OSHA 29 *CFR* 1910.184(h) stipulates, "Only fiber rope slings made from new rope shall be used. Use of repaired or reconditioned fiber rope slings is prohibited." Fiber rope is particularly suitable for handling loads that might be damaged by contact with metal slings.

Wire rope slings are usually made of extra-improved plow steel rope. If this grade of wire rope is unavailable, improved plow steel rope is used. The difference between the two grades is 15%. Normally, wire rope of IWRC construction is used in extra-improved plow steel and improved plow steel slings where mechanical loop endings or swaged or pressed-on terminations are used. In the smaller wire rope diameters up to and including 1 in. (2.54 cm), use the 6 × 19 classification wire rope. For rope diameters larger than 1 in. (2.54 cm), use the 6 × 37 classification rope. The most popular type of sling in use is the strand-laid sling made from the 6 × 19 and 6 × 37 wire rope constructions.

Another popular type of wire rope sling is the cable-laid sling. Made from multiple wire ropes laid into one rope structure, cable-laid slings offer greater flexibility than strand-laid slings. Braided slings (Figure 16–11) consist of a number of ropes braided into a single unit. They are used where flexibility, high strength, and resistance to rotation are essential. Because braided slings are braided in an open manner, they are fairly easy to inspect.

Methods of Attachment

All hooks and rings used as sling connections should develop the full rated capacity of the wire rope sling. Sockets and compression fittings, when properly attached, should develop 100% of the rated strength for wire rope. Swaged-sleeve sling endings should develop 92% to 95% of the wire rope's strength. Compression fittings and swaged-sleeve fittings are available from wire rope manufacturers or from any properly equipped sling shop.

Hand-tucked splices develop about 90% of the rope's strength in rope diameters less than ½ in. (1.27 cm), and 80% for larger diameters. Fittings used with hand-tucked slings should develop the same strength efficiencies as those used with mechanical slings. The recommended

Table 16–J. Rated Capacity Limits (in Tons) of Wire Rope Slings, Using Preformed Extra-Improved Plow Steel (depending on method of attaching the rope to the fittings)

Rope Diameter (in.)	Single Leg						Two-Leg Bridle or Basket Hitch											
	Vertical			Choker			Vertical*			30 Degrees Vertical			45 Degrees			60 Degrees Vertical		
	S	MS	HT	S	MS	HT	S	MS	HT	S	MS	HT	S	MS	HT	S	MS	HT
6x19 Classification Construction																		
3/8	1.5	1.4	1.3	1.1	1.1	1.1	3.0	2.8	2.6	2.6	2.4	2.3	2.1	1.0	1.8	1.5	1.4	1.3
1/2	2.7	2.5	2.3	1.9	1.9	1.9	5.4	5.0	4.6	4.7	4.3	4.0	3.8	3.5	3.3	2.7	2.5	2.3
5/8	4.1	3.9	3.5	2.9	2.9	2.9	8.2	7.8	7.0	7.1	6.8	6.1	5.8	5.5	4.9	4.1	3.9	3.5
3/4	5.9	5.6	4.8	4.1	4.1	4.1	12.0	11.0	9.6	10.0	9.7	8.3	8.3	7.9	6.8	5.9	5.6	4.8
7/8	8.0	7.6	6.4	5.6	5.6	5.6	16.0	15.0	13.0	14.0	13.0	11.0	11.0	11.0	9.0	8.0	7.6	6.4
1	10.0	9.8	8.3	7.2	7.2	7.2	20.0	20.0	17.0	17.0	17.0	14.0	14.0	14.0	12.0	10.0	9.8	8.3
1 1/8	13.0	12.0	10.0	9.1	9.1	9.1	26.0	24.0	20.0	23.0	21.0	17.0	18.0	17.0	14.0	13.0	12.0	10.0
6 x 37 Classification Construction																		
1 1/4	16	15	13	11	11	11	32	30	26	28	26	23	23	21	18	16	15	13
1 3/8	19	18	15	13	13	13	38	36	30	33	31	26	27	25	21	19	18	15
1 1/2	23	21	18	16	16	16	46	42	36	40	36	31	33	30	25	23	21	18
1 3/4	31	28	24	21	21	21	62	56	48	54	49	42	44	40	35	31	28	24
2	40	37	32	28	28	28	80	74	64	69	64	55	57	52	45	40	37	32
2 1/4	49	46	40	35	35	35	98	92	80	85	80	69	69	65	57	49	46	40

*If slings are used to handle loads with sharp corners, pads or saddles should be used to protect the rope. The radius of bend should not be smaller than five times the diameter of the rope. If the radius of bend is smaller, a choker hitch rating should be used.

S = Socket or swaged terminal attachment.
MS = Mechanical sleeve attachment.
HT = Hand-tucked splice attachment.

Note 1. Table is based on a design factor of 5, sling angles formed by one leg and a vertical line through the crane hook, and uniform loading.
Note 2. For 3-leg bridle slings, multiply safe load limits for 2-leg bridle slings by 1.5, and for 4-leg bridle slings, multiply by 2.0.

load rating for a sling assembly is usually based on one-fifth the calculated strength of the assembly. However, there may be cases where engineered lifts are made that do not meet this value.

Working Load

The rated load capacities as given in various sling catalogs and tables are based on newly manufactured slings. As the sling is used, factors such as abrasion, nicking, distortion, corrosion, and bending around small radii will affect the load rating. Consider these factors before lifts are made.

Here are two tips to increase the wear of wire rope slings:

- If loads that have sharp edges or sharp corners must be lifted, use pads or saddles to protect the ropes or chains.
- Thimbles spliced in the ends of slings will materially reduce wear.

Because slings can be used at various angles and since the rope stress increases rapidly with the angle of lift, it is essential to keep this in mind when ordering slings. Fortunately, most catalogs for wire rope slings have tables that give the load ratings for the most-used and most-critical angles of lift. Consult these tables for safe rigging practices.

When the rope is made into a sling and placed in position on a load, determine and carefully consider the angle formed by the ropes and the horizontal. The rated load capacity of the sling decreases sharply as the angle formed by the sling's leg and the horizontal becomes smaller. When this angle is 45 degrees, the rated load capacity has decreased to 71% of the load that can be lifted when the legs are vertical. As this angle decreases, the rated load capacity continues to decrease (Tables 16–I and 16–J).

Figure 16–12 shows how tension on a leg of a sling increases as the angle decreases from the vertical. When the angle formed by a leg and the vertical is 30 degrees, the rated load capacity is only 87% of that if both legs were vertical. For an angle of 60 degrees, the rated load capacity is only 50% of that if both legs were vertical. These losses are proportional to the cosine of the sling's angle with the vertical. The actual stress is equal to the amount of the load that a leg must support, divided by the cosine of the angle that the leg is from

Figure 16–11. Braided slings are resistant to kinking. Be sure loads are hoisted uniformly and that all slings have a minimum safety factor of five.

the vertical. To avoid excessive angles, use longer slings, if head room permits.

Throughout the shop, post tables showing rated load capacities for slings. Also, each sling should bear a tag indicating its rated load capacity (Figure 16–13).

Also consider the rated load capacity for different types of hitches. For example, a decrease of at least 25% in rated load capacity for a single-legged vertical sling occurs when a choker hitch is used. The suitable load for a basket hitch is based on the angle of the legs. For strand-laid and braided wire rope slings used in a basket hitch, the minimum diameter of curvature of the sling in contact with the load should be at least 20 times the rope's diameter; for cable-laid slings, at least 10 times.

Special clamps are used to handle steel plate, flanged castings, and similarly shaped products. Slings using horizontal and vertical clamps have the same rated load capacity as other bridle slings, provided the strength of the clamp is equal to the other components of the sling.

Inspections

Train employees to check slings daily and whenever they suspect damage after a lift. Employees should promptly report any questionable conditions in the equipment or in the assembly. In the United States, use OSHA inspection requirements for industrial slings as a guide in evaluating the sling's condition. Have a trained person make a thorough inspection at least every six months. Documentation of such inspections should be maintained.

Promptly withdraw from service slings that fail inspection requirements. Make them unusable by burning or cutting them before they are discarded.

ANGLE STRENGTH LOSS FROM RATED CAPACITY

ANGLE	FACTOR
70°	.3420
60°	.5000
50°	.6428
40°	.7660
30°	.8660
20°	.9397
10°	.9848
0°	1.0000

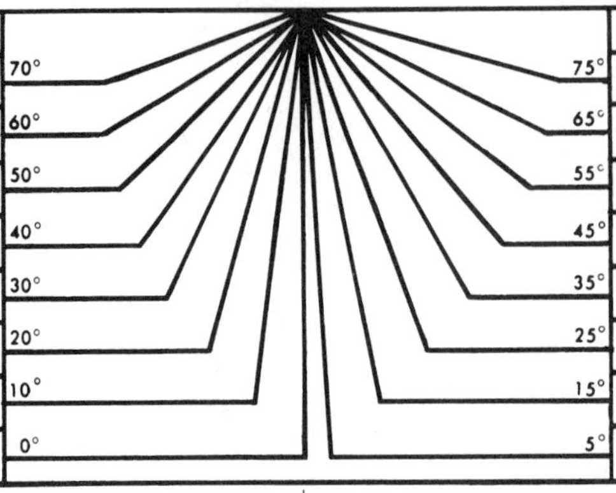

FACTOR	ANGLE
.2588	75°
.4226	65°
.5736	55°
.7071	45°
.8192	35°
.9063	25°
.9659	15°
.9962	5°

The increased angle of the sling leg reduces its capacity. See chart for loss factor. Determine the angle between the sling leg and the vertical plane. Then multiply the sling rating by the appropriate loss factor from the chart. This will determine the slings reduced rating.

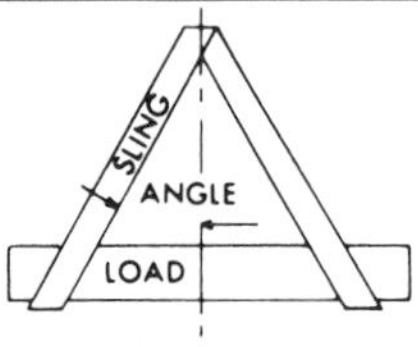

SLING ANGLE LOAD

EXAMPLE:
Assume sling capacity 2,000#
If angle = 50° then loss factor = .6428
Multiply: 2,000# x .6428
1,286# – rated capacity of sling at 50°

Figure 16–12. Increasing the angle between the sling leg and vertical increases the stress on each leg of the sling and reduces its capacity. (Courtesy Web Sling Association.)

Safe Operating Practices for Slings

The American National Standards Institute's and American Society of Mechanical Engineers' (ANSI/ASME) B30.9, *Slings*, recommends the following practices for all slings:

- Slings having suitable characteristics for the type of load, hitch, and environment shall be selected in accordance with appropriate tables. (See Tables 9–5.3 and 9–5.5 in ANSI/ASME B30.9.)
- The weight of a load shall be within the rated capacity of the sling.
- Slings shall not be shortened or lengthened by knotting or other methods not approved by the sling manufacturer.
- Slings that appear to be damaged shall not be used unless inspected and accepted as usable under Table 9–5.6. (See ANSI/ASME B30.9.)
- Slings shall be hitched in a manner providing control of the load.
- Sharp corners in contact with a sling should be padded with material of sufficient strength to minimize damage to the sling.
- Portions of the human body should be kept from between the sling and the load, and from between the sling and the crane hook or hoist hook.
- Personnel should stand clear of the suspended load.
- Personnel shall not ride the sling.
- Shock loading should be avoided.
- Slings should not be pulled from under a load when the load is resting on the sling.
- Slings should be stored in a cool, dry, and dark place to prevent environmental damage.
- Twisting and kinking the legs branches shall be avoided.
- A load applied to the hook should be centered in the bowl of the hook to prevent point loading on the hook.
- During lifting, with or without a load, personnel shall be alert for possible snagging.
- In a basket hitch, the load should be balanced to prevent slippage (Figure 16-10). When using a basket hitch, the sling's legs should contain or support the load from the sides above the center of gravity.
- Slings should be long enough so that the rated capacity is adequate when the angle of the legs is taken into consideration.
- Slings should not be dragged on the floor or over an abrasive surface.
- In a choker hitch, slings shall be long enough so the choker fitting chokes on the webbing and never on the other fitting (Figure 16–10).
- Nylon and polyester slings shall not be used at temperatures above 194 F (90 C).
- When extensive exposure to sunlight or ultraviolet light is experienced by nylon or polyester web

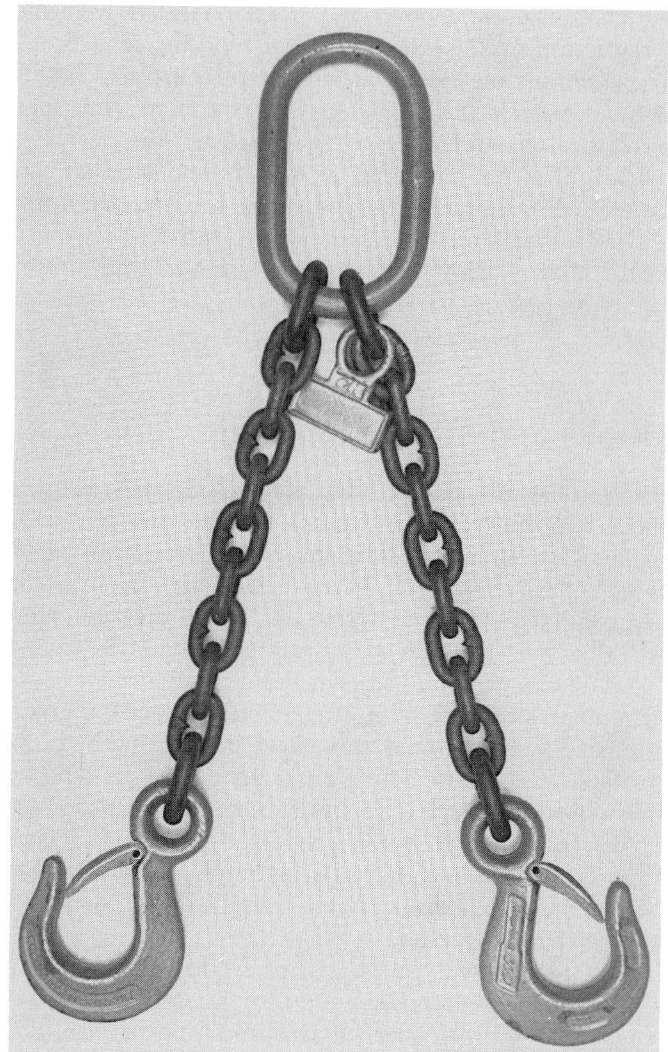

Figure 16–13. Typical double-chain sling. All components, such as the oblong master link, the body chain, and the hook, are carefully matched for compatibility. Note permanent identification tag. (Courtesy Columbus McKinnon Corporation.)

slings, the sling manufacturer should be consulted for recommended inspection procedure because of loss in strength.

CHAINS AND CHAIN SLINGS

The safety of a chain sling assembly depends on the following factors: the kind of material used, the strength of the material for the load, the method of fastening the chain to its fittings, and proper inspection and maintenance. Consider these factors when using chain slings.

Types of Chain Slings

Alloy steel has become the standard material for chain slings. Chain made from alloy steel has high resistance to abrasion and is practically immune to failure because the metal is cold worked. Special-purpose alloy chains are made from stainless steel, monel metal, bronze, and other materials. They are designed for use where resistance to corrosive substances is required, or where other special properties are desirable.

Proof coil chain, also known as common or hardware chain, is used for miscellaneous purposes where failure of the chain would not endanger human life or result in serious damage to property or equipment. Never use proof coil chain for slings.

Properties and Working Load of Alloy Steel Chain

Alloy steel chain is produced from heat-treatable alloy steel in accord with the American Society for Testing Materials' (ASTM) *Specifications for Alloy Steel Chains*, A391–1975. After heat treatment, this chain has the following mechanical properties:

Tensile Strength	115,000 psi minimum
Elongation	15% minimum

The tensile strength of alloy steel chain increases in proportion to its hardness (produced by heat treating). Resistance to abrasion also increases proportionally with hardness.

Table 16–K shows the recommended working load limits, proof test loads, and minimum breaking

Table 16–K. Working Load Limits, Proof Test Loads, and Minimum Breaking Loads for Alloy Steel Chain

Nominal Size of Chain (in.)	*Working Load Limit (lb)*	*Proof Test (lb)*	*Minimum Break (lb)*
1/4	3,250	6,500	10,000
3/8	6,600	13,200	19,000
1/2	11,250	22,500	32,500
5/8	16,500	33,000	50,000
3/4	23,000	46,000	69,500
7/8	28,750	57,500	93,500
1	38,750	77,500	122,000
1 1/8	44,500	89,000	143,000
1 1/4	57,500	115,000	180,000
1 3/8	67,000	134,000	207,000
1 1/2	80,000	160,000	244,000
1 3/4	100,000	200,000	325,000

Reprinted from *Specification for Alloy Chain*, American Society for Testing and Materials, A391-1975. *Alloy Steel Chain Specifications*, No. 3001, National Association of Chain Manufacturers.

strengths of alloy steel chain. The working load limit (safe load strength) is arrived at by dividing the breaking strength (ultimate strength) by a specified safety factor. Any number of chains can be used to lift equipment (Figure 16–14).

The values shown in Table 16–K represent the maximum loads that should ever be applied in direct tension to a length of alloy steel chain. Prior to final inspection and shipment, all alloy steel chain is tested in direct tension under the proof test loads shown in the table. (See the National Association of Chain Manufacturers in References at the end of this chapter for data on other types of chains.)

Impact conditions caused by faulty hitches, bumpy crane tracks, and slipping hookups can add materially to the stress in the chain. If severe impact loading may be encountered, use a lower working load limit, regardless of the type of chain.

Alloy steel chains are suitable for high-temperature operations. However, continuous operation at a temperature of 800 F (425 C) (the highest temperature for which continuous operation is recommended) requires a reduction of 30% in the regular working load limit. These chains may serve at temperatures up to 1,000 F (540 C) at 50% of the regular working load limit, but only for intermittent service. However, the working load limit of the chains must be permanently reduced by 15% after they have served at this high temperature. On the other hand, the general strength and working load limits of alloy steel chains are not altered appreciably by low temperatures.

Figure 16–14. Double, triple, or quad chain slings, as in this application, can be rigged to handle loads of virtually any shape or size. (Courtesy Campbell Chain Company.)

Hooks and Attachments

As a general rule, hooks, rings, oblong links, pear-shaped links, coupling links, and other attachments should be made of heat-treatable alloy steel identical or equivalent to that of the chain itself. In most cases, attachments will be installed on the chain by the chain manufacturer, who will then heat-treat and proof test the assembly.

If emergency conditions make it necessary for users to replace an attachment, they should select the grade and size with extreme care. Use high-strength, heat-treatable, alloy-connecting links of the same type as those used by the chain manufacturer. Do not use unalloyed carbon-steel hooks, repair links, rings, pear-shaped links, or other attachments. Never use homemade or makeshift bolts, rods, shackles, or hooks without safety catches.

Standard items produced from alloy steel include sling hooks, grab hooks, foundry hooks, grab links, rings, oblong links, pear-shaped links, and repair links. All such attachments, used with the recommended chain size, provide a safety factor equal to or greater than that of alloy steel chain itself. Specifications for the dimensions of these attachments will vary somewhat with the individual manufacturer.

Other useful items are handles. Many injuries have resulted from employees catching their fingers between the hook attachment and the load. To prevent such injuries, attach handles to the assembly hook or end attachment. To increase operating efficiency, use handles on large hooks, master links, and other attachments.

Inspections

Following an inspection procedure can reveal most of the causes of chain failures before failure occurs. Chain slings require three types of inspection:

1. Initial inspection. Both new and repaired slings shall be inspected before use to determine (1) that each sling meets the requirements of the purchase order, (2) that it is the correct type and has the proper rated

capacity for the application, and (3) that it has not been damaged in shipment, unpacking, or storage.

2. Frequent inspections. The sling shall be inspected by the person handling it each time it is used.
3. Periodic inspections. A semiannual or more frequent inspection by a competent person who is experienced in the inspection of chain slings. The frequency of periodic inspections shall be based on the following factors: frequency of use, severity of service conditions, and knowledge about the service life of slings used in present or similar conditions.

Documentation of such inspections should be maintained. The user of a chain should be able to detect links and hooks that have become visibly unsafe because of overloading, faulty rigging, or other unsafe practices. The competent person should have the authority to remove damaged assemblies from service so they can be reconditioned or replaced.

The best way to detect wear and stretching is by a visual, link-by-link inspection. Overall measurements of sling length, and even measurements of 1 ft to 3 ft (30 cm to 91 cm) lengths, are inadequate because not all links are affected the same. Likewise, caliper readings of only certain links can also miss wear and stretching (Table 16–L and Figure 16–15). A link-by-link inspection should be made to detect the following:

- bent links
- cracks in weld areas, in shoulders, or in any other section of links
- transverse nicks and gouges
- corrosion pits
- stretching caused by overloading

When inspecting the hook, measure between the shank and the narrowest point of the hook opening.

Figure 16–15. The links are turned to show the extreme wear at the bearing surfaces.

Whenever the throat opening exceeds 15% of the normal opening, replace the hook. Pay special attention to slings to which hooks have been added; make sure the hooks are secure.

Table 16–L. Maximum Allowable Wear at Any Point of Link

Chain Size (in.)	*Maximum Allowable Wear (in.)*
1/4	3/64
3/8	5/64
1/2	7/64
5/8	9/64
3/4	5/32
7/8	11/64
1	3/16
1 1/8	7/32
1 1/4	1/4
1 3/8	9/32
1 1/2	5/16
1 3/4	11/32

Safe Practices for Chain Slings

Follow these recognized safe practices to prevent chain failures:

- Purchase chain slings complete from the manufacturer. Whenever repairs are required, send them back to the manufacturer.
- Never anneal or normalize alloy steel chains and hooks. These processes reduce their hardness and therefore greatly reduce their strength.

- Never splice a chain by inserting a bolt between two links.
- Never put a strain on a kinked chain. Train workers to take up the slack slowly so they can see that every link in the chain seats properly.
- Do not use a hammer to force a hook over a chain link.
- Never remove the permanent identification tags that have been attached to chain slings by the manufacturer.
- Remember that decreasing the angle between the legs of a chain sling and the horizontal increases the load of the legs.
- Use chain attachments (rings, shackles, couplings, and end links) designed for the chain to which they are fastened.
- See that the load is always properly set in the bowl of the hook. Loading on or toward the point (except in the case of grab hooks or others especially designed for the purpose) overloads the hook and leads to spreading and possible failure.
- Store chains not in use in a suitable rack. Do not let them lie on the ground or floor where they can be damaged by lift trucks or other vehicles.
- Secure "out-of-balance" loads properly (Figure 16–16).

Figure 16–16. This "out-of-balance" load is secured by a double sling with chain leg adjusters (the two short chains with grab hooks attached to the master link). The chain leg passing through the bore of this casting has been protected from damage by adequate padding. (Courtesy American Chain Division of Acco.)

SYNTHETIC WEB SLINGS

Two widely used slings are the synthetic web sling and metal mesh sling. They are strong and dependable slings if properly used.

Types of Synthetic Web Slings

Nylon and polyester are the fibers most often used for synthetic web slings. Each has specific advantages and disadvantages as to stretch, strength, and chemical resistance. Synthetic web slings are useful for lifting loads that need their surfaces protected by the soft, supple web sling's surface. This usage is well suited for tubular, nonferrous, ceramic, painted, polished, and highly machined products with fine or delicate surfaces. To this end, a synthetic web sling's service life is secondary to load protection. Be warned, therefore, that synthetic web slings can be cut relatively easily and have little resistance to abrasion compared with chain or wire rope. Figure 16-17 shows several types of synthetic web slings.

Observe the following requirements when using synthetic web slings:

- According to ASTM B783–1990, *Specifications for Materials for Ferrous Powder Metallurgy Structural Parts*, the minimum breaking strength for synthetic web slings shall be five times the rated capacity (Tables 16–M and 16–N).
- Every synthetic web sling shall bear, in a legible manner, the following identification information: the name or identification of the manufacturer, the sling's code number, the rated load capacities for usable types of hitches, and the type of material (such as polyester or nylon).
- When two slings, or one sling in a basket hitch (Figure 16–10), are used to lift a load from one crane

Type I. Triangle and choker end fittings usable in vertical, choker, and basket hitches.

Type II. Triangle fittings each end usable in vertical and basket hitches only.

Type III. Flat eye ends usable in vertical, choker, and basket hitches.

Type IV. Twisted eye ends usable in vertical, choker, and basket hitches.

Type V. Endless (or grommet) usable in vertical, choker, and basket hitches.

Type VI. Reversed (or return) eye. Essentially an endless sling, butted on the sides with wear pad(s) on body. Usable in vertical, choker, and basket hitches.

Figure 16–17. Basic synthetic web sling types.

Table 16–M. Rated Capacity in Pounds for 1,600 lb/in. Web Slings Single-Ply Capacities for Various Type Slings in Vertical, Choker, and Vertical Basket Hitches

Single-Ply Capacities for Various Type Slings in Vertical, Choker, and Vertical Basket Hitches

Web Width in In.	*Types 1, 2, 3, & 4*			*Type 5 Endless*			*Type 6 Reversed Eye*		
	Hitches			*Hitches*			*Hitches*		
	Vertical	*Choker*	*Basket*	*Vertical*	*Choker*	*Basket*	*Vertical*	*Choker*	*Basket*
1	—	—	—	2,600	2,100	5,200	—	—	—
2	3,200	2,400	6,400	5,100	4,100	10,200	4,500	3,600	9,000
3	4,800	3,600	9,600	7,700	6,200	15,400	—	—	—
4	6,400	4,800	12,800	10,200	8,200	20,400	7,700	6,200	15,400
5	8,000	6,000	16,000	12,800	10,200	25,600	—	—	—
6	9,600	7,200	19,200	15,400	12,300	30,800	11,000	8,800	22,000

Rated capacities shall never be exceeded.
See manufacturer's rated capacities for multiple-ply slings.

Table 16–N. Rated Capacity in Pounds for 1,200 lb/in. Web Slings

Single-Ply Capacities for Various Type Slings in Vertical, Choker, and Vertical Basket Hitches

Web Width in In.	*Types 1, 2, 3, & 4*			*Type 5 Endless*			*Type 6 Reversed Eye*		
	Hitches			*Hitches*			*Hitches*		
	Vertical	*Choker*	*Basket*	*Vertical*	*Choker*	*Basket*	*Vertical*	*Choker*	*Basket*
1	—	—	—	1,900	1,500	3,800	—	—	—
2	2,400	1,800	4,800	3,800	3,000	7,600	3,500	2,800	7,000
3	3,600	2,700	7,200	5,800	4,600	11,600	5,000	4,000	10,000
4	4,800	3,600	9,600	7,700	6,200	15,400	6,800	5,400	13,600
5	6,000	4,500	12,000	9,600	7,700	19,200	—	—	—
6	7,200	5,400	14,400	11,500	9,200	23,000	8,000	6,400	16,000

Rated capacities shall never be exceeded.
See manufacturer's rated capacities for multiple-ply slings.

hook, the sling's capacity is reduced. The load-carrying capacity of the sling is determined by applying the appropriate factor times the hitch's capacity.

Inspections

Synthetic web slings are required by OSHA to be inspected each day before and during usage by a competent person. It is recommended that a person trained to use web slings also be trained to competently inspect them.

Wear caused by ultraviolet light can be judged only from prior experience with synthetic web slings. For this type of inspection, consult the manufacturer. However, on-site inspections should identify the following kinds of wear or damage:

- excessive abrasive wear on webbing and any fittings
- cuts, tears, snags, punctures, holes, and crushed fabric
- worn or broken stitches, particularly that of the laps
- burns, charring, melting, or weld spatter damage
- acid, caustic, or other chemical damage
- broken, distorted, or excessively worn fittings
- knots that cause doubt about the sling's safety.

Keep records of all inspections. Such records should identify the sling, the dates of inspection, and the sling's condition at the time of each inspection. Because web slings generally are not manufactured with individual identification numbers, a system of identification should be devised to facilitate record keeping. A manufacturer may be willing to stamp a company's identification number if reqested at time of purchase.

Although generally not repaired, web slings may be repaired by the manufacturer or qualified person, who shall identify the work and certify the rated load capacity. All repaired slings shall be proof tested to two times their newly rated load capacity. No temporary repairs should be made.

METAL MESH SLINGS

Metal mesh slings can safely handle sharp-edged materials (Figure 16–18), concrete in its many prestressed forms, and high-temperature materials up to 500 F (260 C). Metal mesh slings are classified as either heavy-duty, medium-duty, or light-duty. Figure 16–19 shows the sling structure and terms to identify parts of a metal mesh sling.

All metal mesh slings should be properly identified with their safe working load limit stamped in their vertical basket hitch and choker hitch (metal handle). They can be used efficiently with all weights that fall below that limit. The design factor of metal mesh slings is five to one, or five times the amount stated on the sling. All metal mesh slings are proof tested to a minimum of 200% of their rated load capacity. This removes all permanent stretching when used at the rated load capacity.

Safe Practices for Metal Mesh Slings

The safe use of metal mesh slings depends primarily on two factors: (1) use of the right sling for the right load, and (2) the construction of the sling. Any danger in the use of metal mesh slings stems mainly from improper use.

- Use elastomer-coated slings only at temperatures up to 200 F (93 C).
- Damaging the slings at the edges by faulty loading or dragging a sling out from under a load may eventually cause wear to the spirals that compose the mesh. Such abuse will reduce the sling's wire diameter and thus require the sling to be taken out of service.
- Never shorten metal mesh slings by using knots, bolts, or other unapproved methods. If shortening becomes necessary, consult the sling manufacturer on how to do it.
- Tampering with the surface of any sling can weaken it and make it highly dangerous.
- Never twist or kink the legs of a metal mesh sling or use one when the spirals are locked. A sudden jolt may break the spirals, cause the load to shift, and create havoc on the floor below. With metal mesh slings, as well as with all other slings, follow rules for certain hitches to assure the safe handling of a load.

Inspections

One of the most important precautions in the use of metal mesh slings is regular inspection by a qualified person. Base frequency of inspections on the amount of use that a sling receives and the severity of conditions. The user should inspect the sling each time it is used, thoroughly inspecting it at least once a year. Keep written inspection records that identify each sling and the items that were inspected.

Remove metal mesh slings from service if a broken weld or brazed joint is discovered along the sling edge. Also, watch for the following signs of wear. Any one of these conditions or a combination of them, if ignored, could eventually result in sling breakdown:

Figure 16–18. A four-legged, basket-hitch sling of steel mesh can take the sharp edges of lumber without failure and without damaging the wood.

- broken wires in any part of the mesh
- a loss of 25% in wire diameter due to abrasion
- a lack of sling flexibility
- cracked end fitting
- visible distortion

SUMMARY

- Workers should be trained in the special safety precautions required when using and storing various ropes, chains, and slings.
- Natural or synthetic fiber ropes are durable but tend to deteriorate when in contact with acids, caustics, and high temperatures. Nylon, polyester, and polyolefin ropes are more resistant to corrosive materials, wear and tear, and temperature extremes.
- Supervisors and workers must know the working load of the rope they are using to ensure safety. Dynamic effects are greater on ropes with little stretch and can cause them to break, thus endangering workers.
- Ropes must be regularly inspected for wear or damage and the results documented. Workers should be trained to handle ropes with care and store them properly, away from harmful substances.

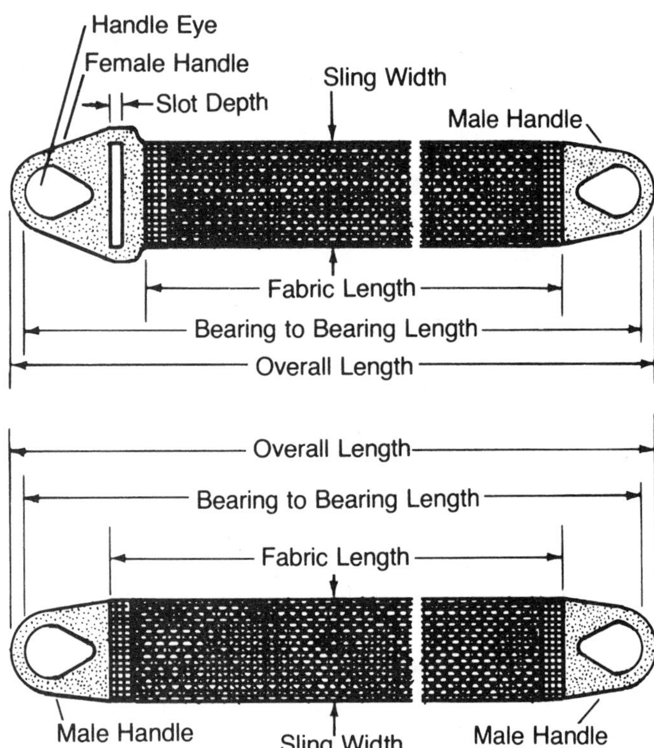

Figure 16–19. Structure and nomenclature of typical metal mesh slings.

- Wire rope has great strength and durability under severe working conditions and maintains its characteristics in widely different environments. Inspectors should check wire ropes for deterioration.
- Fiber rope slings are used for loads that might be damaged by contact with metal, while wire rope slings provide extra strength and durability.
- The safety of a chain sling assembly depends on the material used, its strength for the load handled, the method of attaching chain to fittings, and proper inspection and maintenance. Chains should be inspected daily by workers, and every six months by a trained professional.
- Synthetic web slings are useful for lifting loads that need their surfaces protected. Metal mesh slings can handle sharp-edged material, concrete, and high-temperature loads. Their safe use depends on the use of the right sling for the right load and on their construction.

REFERENCES

American Iron and Steel Institute, Committee of Wire Rope Producers, *Wire Rope Users Manual,* 1988.

American National Standards Institute, 11 West 42nd Street, New York, NY 10036.

Elevators, Escalators, and Moving Walks, ANSI/ASME A17.1–1993.
Safety Code for Cableways, Cranes, Derricks, Hoists, Hooks, Jacks, and Slings, ANSI/ASME B30 series, 1990–1993.
Slings, ANSI/ASME B30.9–1990.

American Petroleum Institute, 1220 L Street NW, Washington, DC 20005.

Recommended Practice on Application, Care, and Use of Wire Rope for Oil-Field Service, Code No. API–RP–9B.

American Society for Testing Materials, 1916 Race Street, Philadelphia, PA 19103.

Specification for Alloy Steel Chain, A391–1986.

Broderick & Bascom Rope Co., Rt. 3, Oak Grove Industrial Park, P.O. Box 844, Sedalia, MO 65301.

Rigger's Handbook, 1986. Wire Rope Handbook.

Cordage Institute, 42 North Street, Hingham, MA 02043.

Hess O. "Metal Mesh and Nylon Slings." *National Safety News,* 103: 78–79 (June 1971).

National Association of Chain Manufacturers, PO Box 3143, York, PA 17402.

Alloy Steel Chains Specifications, No. 3001.

National Fire Protection Association, 1 Batterymarch Park, Quincy, MA 02269.

Flammable and Combustible Liquids Code, NFPA 30, 1993.
National Electrical Code, NFPA 70, 1993.

National Safety Council, 1121 Spring Lake Drive, Itasca, IL 60143-3201.

Occupational Safety and Health Data Sheet: *Recommended Loads for Wire Rope Slings,* 12304–0380, 1991.

U.S. Department of the Interior, Bureau of Mines, 2401 E Street NW, Washington, DC 20241.

Recommended Procedures for Mine Hoists and Shaft Installation, Inspection, and Maintenance, Information Circular 8031.

Wire Rope Technical Board, PO Box 849, Stevensville, MD 21666.

REVIEW QUESTIONS

1. Which of the following fiber ropes give the best uniform strength and service?
   a. Manila or nylon
   b. Polyester or rayon
   c. Sisal or polyethylene
   d. Polyolefin or henequen

2. The properties of ______ fiber make it the best-suited natural fiber for cordage and is often recommended for capstan work.

3. List the three major types of synthetic fiber ropes.
   a.
   b.
   c.

4. How often should fiber rope, being used under ordinary conditions, be inspected for damage?
   a. Every two weeks
   b. Once a month
   c. Every two months
   d. None of the above

5. What is a quick way to make a good estimate of the strength of fibers in a rope, and to test for chemical damage?

6. When lengths of fiber rope must be joined, a well-made splice will retain up to 100% of the strength of the rope, but a knot retains only ___.
   a. 25%
   b. 50%
   c. 75%
   d. 85%

7. Wire rope is more widely used than fiber rope because:
   a. Wire rope has greater strength and durability under severe working conditions.
   b. The physical characteristics of wire rope do not change when used in varying environments.
   c. Wire rope has controlled and predictable stretch characteristics.
   d. All of the above
   e. Only a and c

8. Generally speaking, when there are more wires per strand, the wire rope is more ________.
   a. Flexible
   b. Crush resistant
   c. Abrasion resistant
   d. All of the above
   e. Only a and b

9. The minimum design factors for wire rope used in hoisting depend upon what two conditions?
   a.
   b.

10. List the causes of deterioration of wire ropes.
    a.
    b.
    c.
    d.
    e.
    f.
    g.
    h.

11. In the United States, OSHA regulations and other industrial and construction codes prohibit the use of ____ in wire rope.

12. In the United States, OSHA requires wire rope or cable to be inspected how often?

13. The safety of a rope sling assembly depends on what six factors?
    a.
    b.
    c.
    d.
    e.
    f.

14. Why has alloy steel become the standard material for chain slings?

15. What is the best way to detect wear and stretching of chains and chain slings?
    a. Overall measurements of sling length
    b. Even measurements of 1-ft to 3-ft (30-cm to 91-cm) lengths
    c. Caliper readings of only certain links
    d. Visual, link-by-link inspection

16. What is the difference between the usage of synthetic web slings and metal mesh slings?

17

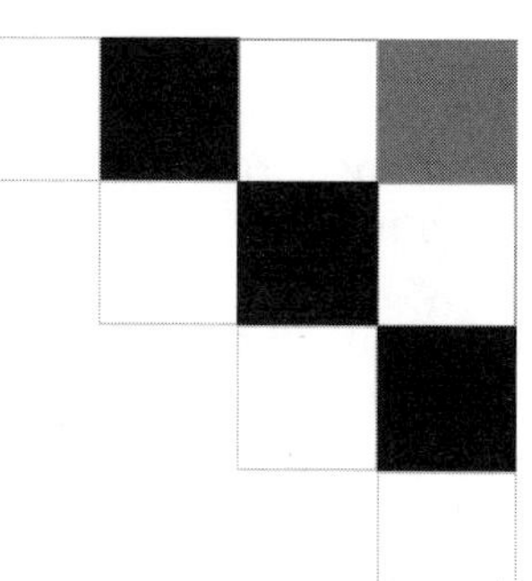

Powered Industrial Trucks

Revised by

George Swartz, CSP

John F. Montgomery, Ph.D., CSP, CHCM, CHMM

Factories, warehouses, docks, and transportation terminals use powered industrial trucks to carry, push, pull, lift, stack, and tier material. Each of these trucks requires safeguards both for the operator's protection and for the safety of other workers. The establishment of safe practices for the operation, maintenance, and inspection of powered industrial trucks also is essential. The topics covered in this chapter include:

- discussion of types of powered trucks
- methods to reduce hazards associated with these vehicles
- general safety principles in operating powered trucks
- safe practices in operating lift trucks
- important issues in inspection and maintenance
- selection and training of powered-truck operators

This chapter does not apply to compressed air or nonflammable compressed gas-operated industrial trucks, or farm vehicles, nor to vehicles intended for earth moving or over-the-road hauling. This chapter does not discuss powered industrial trucks employed in airport and air terminal areas, although many of the same procedures will apply. For information about trucks and operators in these areas, see the *Aviation Ground Operation Safety Handbook*, published by the National Safety Council. (See References at the end of this chapter.)

TYPES OF TRUCKS

Powered industrial trucks may be classified by power source, operator position, or means of engaging the load. Power sources include electric motors powered by storage batteries; engines using gasoline, liquefied petroleum gas (LP-gas), or diesel fuel; or trucks using a combination of gas or diesel and electricity. Provisions for safe operation, maintenance, and design of powered industrial trucks should meet design and construction requirements for powered industrial trucks established in the *American National Standard Powered Industrial Trucks*, Part II, ANSI B56.1—1969. All vehicles should bear a label or some other identifying mark indicating such.

Rider-Controlled Trucks

One class of powered industrial truck is designed to be controlled by an operator who rides on the truck. The widely used lift truck, with its cantilevered load engager, vertical masts, and elevating mechanism, is usually a rider-controlled truck. Some rider-controlled trucks use a platform to engage the load. Both of these trucks may be either high-lift trucks—with an elevating mechanism that permits the tiering of one load on another (Figure 17–1)—or low-lift trucks—with a mechanism that raises the load only enough to permit horizontal movement.

Figure 17–1. A high-lift fork truck for pallet storage on racks proceeds down this aisle. (Courtesy Allstate Insurance Company, Training Division.)

There are also other types of rider-controlled trucks. Straddle carriers (Figure 17–2) carry long material, such as pipe or lumber, under the truck's body, which rides on four high legs above the wheels. Powered industrial tractors draw trailers, nonpowered trucks, and other mobile loads. In warehouse operations, order-picker trucks are used to raise the operator to the desired height.

The use of attachments has increased the versatility of the lift truck. Clamps, rotators, shifters, stabilizers, pushers, pullers, up-enders, bottom dumpers, top lifters, rams, cranes, scoops, as well as other modifications, have been developed to meet specific needs. Two or more motions have been built into some attachments. For example, an attachment could clamp and rotate, or it could side-shift and push and pull (Figures 17–3). It is possible to interchange attachments so that one truck can be used for various types of loads.

Other Types of Trucks

Another category of powered industrial trucks is the motorized hand truck controlled by an operator who walks or rides behind it (Figure 17–4). It also has a platform or lifting forks to engage the load and may be either a high-lift truck for tiering or a low-lift truck to raise the load only enough for horizontal movement.

A unique powered industrial truck is the electronically controlled vehicle or automated guided vehicle (AGV), which does not need an operator (Figure 17–5). It travels over a prearranged route and is controlled by frequency sensors, a light beam, or induction tape that is outlined on or under the floor.

SAFEGUARDS

Consider the work site when purchasing or leasing industrial trucks. Working outdoors can call for long travel distances in less-than-perfect conditions. Operators are too often at the mercy of their machines. They suffer fatigue from noisy, cramped compartments, and strain from blind spots or rough terrain. Aids proven to reduce fatigue and strain include the following: backup alarm lights, headlights, turn signals, enhanced front and rear vision, noisereducing insulation, fail-safe brakes, and comfortable, wraparound seats that provide protection similar to safety belts for the operator.

Manufacturers offer operator-restraint systems on new vehicles, while other manufacturers can provide the system on request. Some older lift trucks can be retrofitted with operator-restraint systems by the manufacturers.

Safety belts can be used by lift-truck operators. Each company needs to develop its own policy of lift truck safety-belt usage. In developing a policy, the operators' safety is most important. Consider such factors as forklift truck accident data, type and design of the facility, operators' duties, entering and exiting the truck

Figure 17–2. Straddle trucks are designed to carry loads of pipes, lumber, and other long materials. (Printed with permission from Townmotor Corp.)

Figure 17–3. The special clamp on this lift truck permits handling of paper rolls without damage. (Courtesy Clark Equipment Company, Industrial Truck Division.)

Figure 17–4. The walking operator controls this electric pallet-lift by handlebar-mounted controls. (Courtesy Clark Equipment Company, Industrial Truck Division.)

easily, work area conditions, materials-handling requirements and exposure, and other factors.

When the manufacturer has provided an operator safety-restraint system on the truck, that system needs to be used as specified in the owner's manual.

Requirements

A powered industrial truck capable of lifting loads higher than the operator's head or operated in areas where there is a hazard from falling objects must be equipped with an overhead guard (Figure 17–1). This guard should not interfere with good visibility. Be sure that openings in the guard are small enough to protect the operator from being struck by material falling from an overhead load or stack. Equip trucks with overhead guards that extend beyond the operator's position. Such guards should conform with ANSI/ASME B56.1.

A load backrest extension should always be used when the type of load presents a hazard to the operator. The top of a load should not exceed the height of the backrest manufactured in accordance with U.S. Occupational Safety and Health Administration (OSHA) standard 1910.178.

To prevent particles from being thrown at the operator, install guards over exposed tires. To protect the operator when in a normal operating position, place guards over hazardous moving parts, such as chain-and-sprocket drives and exposed gears.

Although lift trucks may come with steering wheel knobs, their use is prohibited by many companies and is not recommended. If knobs are used, they should be of the mushroom type to engage the palm of the operator's hand in the horizontal position, and knobs should be mounted within the periphery of the wheel. The steering mechanism should minimize transmission of road shock to the steering wheel.

Also, confine all steering controls, except steering handles for motorized hand trucks, within the outline of the truck. Where this is not possible, provide guards to prevent injury to the operator when passing obstacles (Figure 17–6).

Figure 17–5. Some automated guided vehicles should have clearly marked aisles free of any distractions. (Courtesy Litton.)

Powered industrial trucks should have horns or other warning devices that make a distinctive sound that is loud enough to be heard clearly above other noises. This warning device should be under the operator's control. A backup alarm, however, works whenever the truck backs up. Where excessive noise could cause confusion, flashing lights mounted on the overhead guard can warn employees of approaching trucks.

Every powered industrial truck must carry a name plate showing the weight of the truck and its rated capacity as specified by ANSI/ASME B56.1 (Figure 17–7). Various parts of the vehicle may have limitations that also should be observed by the operator. Specifications of steering, braking, and other controls should also conform to ANSI B56.1.

Powered industrial trucks should also be constructed and equipped to comply with Underwriters Laboratories' Standard for Safety, No. 558 and No. 583. These standards are specified in the National Fire Protection Association's NFPA 505, *Powered Industrial Trucks, Including Type Designations, Areas of Use, Maintenance and Operation.* Users are not permitted to modify trucks without the written approval of the manufacturer.

Industrial Trucks in Hazardous Locations

The definitions of hazardous locations are given in NFPA 70, *National Electrical Code.* Class I locations are those in which flammable gases or vapors are or may be present; Class II, combustible dust; and Class III, easily ignitible fibers or flyings. ANSI/ASME B56.1 stipulates that trucks, electric- or gasoline-powered, shall not be used in certain hazardous locations unless they either comply with NFPA requirements or are specifically approved by the inspection authority for the location involved. (See NFPA 505, *Powered Industrial Trucks,* for definitions of hazardous locations.)

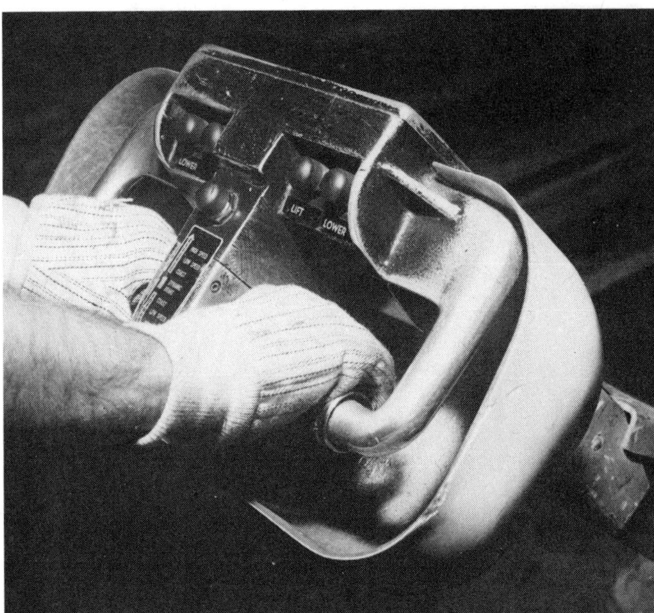

Figure 17–6. This guard protects the operator's hands from coming in contact with obstacles when the truck is being maneuvered in close quarters.

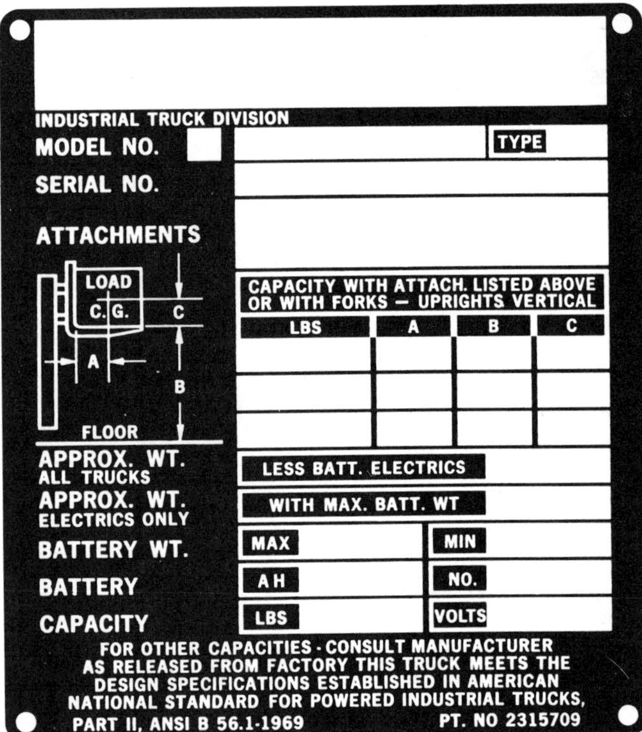

Figure 17–7. A name plate must be attached to every lift truck by the manufacturer. Pertinent identification and reference information must include the weight of the truck, its rated capacity, and model and serial numbers.

If a lift truck is operating out of doors or away from fixed fire extinguishers in the facility, an appropriate fire extinguisher must be installed on the lift truck. Train all truck operators to use it. An important advantage to operators trained to use truck-mounted fire extinguishers is that they are quickly available to combat small fires anywhere on the premises.

Each type of powered industrial truck requires specific safeguards. Select only lift trucks, straddle trucks, crane trucks, tractors and trailers, motorized hand trucks, and AGVs that are properly equipped for the safety of the operators and other workers.

Lift Trucks

If an overhead guard is attached to the rear of the truck's body, it also should be attached to the front of the body and not to the mast. An exception is made for those trucks that have the tilt cylinder as an integral part of the overhead guard construction. In such a case, the overhead guard shall be designed so as to prevent injury to the operator should the mast-tilting mechanism fail.

Forks should be locked to the carriage, and the fork extension, if used, should be designed to prevent unintentional lifting of the toe or displacement of the fork extension. Lift trucks also should be equipped, or have the means to be equipped, with mechanical hoist and tilt mechanisms to prevent overtravel of hoist and tilt motions. If the lifting systems are hydraulically driven, a relief valve should be installed in the system and suitable stops provided to prevent overtravel.

Straddle Trucks

Straddle trucks (Figure 17–2) should have horns, or other warning devices, and headlights and tail lamps for working at night. Straddle trucks also should have safe-access ladders, wheel guards, and chain-drive guards. Certain types of work may require installing a rigid overhead guard for the operator.

Operators can determine overhead clearances by observing warning devices posted in advance of overhead obstructions across railroad tracks and other passageways where straddle trucks operate. Gauge rods may be mounted on the truck at front and rear.

Straddle trucks present a special problem for operators: Because they sit so high off the ground, their angle of sight is reduced for objects immediately to the front or rear. While precautions always must be taken to avoid striking pedestrians, this especially is true when carrying long loads. Attach red flags to the ends of such loads or station signal persons in congested areas. Particular care must be taken if the truck is used after dark.

Crane Trucks

Although some crane trucks have three wheels, most usually have four wheels (Figure 17–8). One model is designed so the operator sits behind a small pillar-type jib crane mounted on a chassis. In another model, the operator stands on a platform and operates a fully or

Figure 17–8. This truck-mounted hydraulic crane is capable of swinging a full 360 degrees. Outriggers provide stability. The enclosed cab offers the operator protection plus clear visibility. (Courtesy Grove Worldwide.)

partly rotating crane. Still another type has a fixed boom, so that the entire rig must be moved from one position to another to make side motions.

An operator should drive a crane carrying a load at the lowest possible speed and should carry the load as low as possible. The operator should have a helper to hook on the load and to give signals. When a long load is being carried, the helper should walk alongside and by means of a tag line keep the load from swinging and striking against objects along the way. When the crane is traveling without a load, the operator should fasten the hook to the lower end of the boom to prevent the hook block from swinging.

Tractors and Trailers

The coupling used to make up the tractor-trailer train should incorporate all necessary safeguards. The type of coupling used depends on the construction of the trailer, the loads carried, and whether the route traveled includes sharp curves, ramps, or inclines. The coupling must be one that will not come unhitched on grades or permit the trailer to whip or cut in on curves. Also, loads on trailers should be secured to the trailer to avoid widespread scattering of material should the load shift or the trailer roll over.

Motorized Hand Trucks

In operating a motorized hand truck, the principal hazards are (1) the operator may be pinned between the truck and a fixed object, and (2) the truck may run up on the operator's heels. Operators should walk ahead of such trucks, leading them from either side of the handle and facing the direction of travel. When a truck must be driven close to a wall or other obstruction, down an incline, or onto an elevator, the operator should put the truck in reverse and walk behind it facing the direction of travel. Some motorized trucks are designed to be ridden over longer distances as well as guided while walking.

Guards are required for steering handles to prevent the operator's hand or the controls from coming into contact with obstacles when the truck must be maneuvered in close quarters (Figure 17–6). A motorized hand truck should also be equipped so that its brakes will be applied when the handle is in either the fully raised or the fully lowered position (Figure 17–9). The wheels of many motorized hand trucks can be considered guarded because they are under the frame or under the lift platform.

Hand trucks with a platform on which the operator rides in a standing position must be designed so the platform extends beyond the operator's position and is strong enough to meet the requirements of ANSI/ASME B56.1. If an enclosure for operators, similar to the one in Figure 17–3, is provided, it shall provide easy entry to and exit from the platform. To discourage operators from sitting on the truck during operation, install a prism-shaped cover over the batterybox.

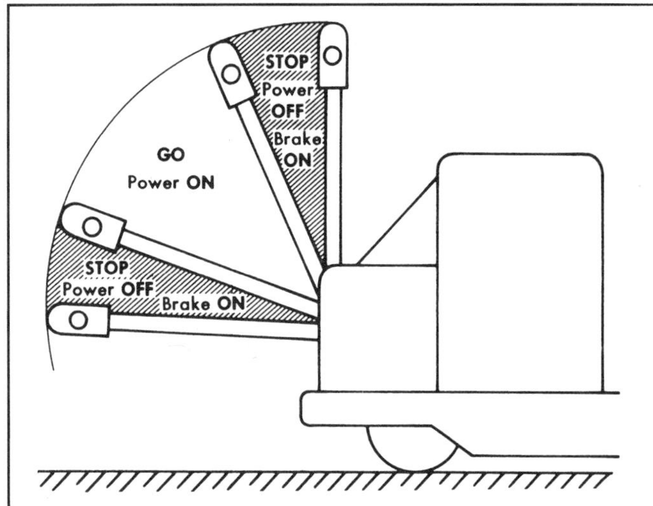

Figure 17–9. The hazard of contacting the frame or wheels of a walkie lift truck can be minimized if the "power on" area is limited to approximately the area shown.

Automated Guided Vehicles

Because trucks guided by remote control (AGVs) operate without an operator, they must be provided with some means to stop completely should someone step in front of them (Figure 17–10). Such trucks should be equipped with a lightweight, flexible bumper that, when contacted, shuts off the power and applies the brakes. Sufficient clearance between the bumper and the front of the truck is needed so the truck can come to a full stop without contacting anything in its path.

Using AGVs requires that aisles where the trucks operate be clearly marked and clear of material. Forbid employees from jumping on or off, and from riding, these trucks. Do not allow loading or unloading of AGVs that are in motion.

Design transfer conveyors so they can be moved out of the way, except when transferring loads to and from the AGVs. This avoids having a pinch point between the

Figure 17–10. Two automatically guided, driverless vehicles electronically decide which one will have the right-of-way, thus eliminating potential collisions. (Courtesy Jervis B. Webb Co.)

vehicle and the conveyor. The same logic should apply when laying out the route next to machines and columns.

GENERAL OPERATING PRINCIPLES

Operators of powered industrial trucks can prevent traffic accidents by using the same safe practices that apply to highway traffic. Operators should observe rules regarding speed, maneuvering, and loading and unloading other vehicles. Operators should consider their own safety, as well as the safety of the other workers and equipment, when using powered industrial trucks.

Speed

Excessive speed can lead to accidents both in the facility and on the road. Safe speed is the rate of travel that will permit the truck to stop well within the clear distance ahead or to make a turn without overturning. Wet or slippery floors require a slower-than-ordinary speed.

Depending on the application and operating conditions, establish specific speed limits for trucks. A speed limit for the yard will be too high for a congested warehouse. Many companies have governors installed to control vehicle speed.

Collisions between trucks and stationary objects often occur while trucks are backing up, usually when they are turning and maneuvering. In such cases, operators may be so intent on handling the load that they forget to watch where the rear of the truck is going. Because accidents caused while backing up usually result from a failure to look to the rear, operators should look in the direction of travel, maintaining a clear view of it (Figure 17–11). A back-up alarm on the counterbalanced model will help prevent collisions. Some trucks permit the operator to sit sideways to make looking backward and forward easier. Operators must be especially careful when turning because the rear wheels project beyond the truck's enclosure, thus presenting a hazard.

Operators should stop and sound the horn at blind corners, and before they pass through doorways. They should go ahead only when they can see that the way is clear. Many companies have installed large convex mirrors at blind corners so that operators and pedestrians can see each other approaching. To be effective, these mirrors have to be kept clean and properly adjusted.

Operators should avoid making quick starts, jerky stops, or quick turns. They should use extreme caution when operating on turns, railroad crossings, ramps, grades, or inclines. On descending grades, operators should keep trucks under control so that they can be brought to an emergency stop in the clear space in front of them. Operators should never use the reverse control for braking. On some models (electric), plugging is recommended.

Figure 17–11. An operator must keep alert when backing and face in the direction of travel. (Courtesy Clark Equipment Company.)

Operators should keep trucks a safe distance apart during operation; some companies specify three truck lengths. Operators must not pass other trucks traveling in the same direction at intersections, blind spots, or other dangerous locations. They must keep to the right, if aisle width permits it, without passing dangerously close to machine operators and others. Where aisles are not wide enough for continuous two-way traffic, vehicles should run in the middle of the aisle, except when in a passing situation where the vehicle then moves to the side and passes with caution.

Elevators, Bridge Plates, and Railroad Tracks

Operators should not drive trucks onto an elevator unless they have been authorized to do so. The operator should approach the elevator slowly and at a right angle to the door. The operator should enter the elevator only after the car is properly leveled and after checking to make sure the weight of the truck, load, and driver do not exceed the capacity of the elevator. After making sure that the load is lowered to the floor, the brakes are set, the power is shut off, and the controls are in neutral, the operator should get off the truck. Other personnel should stay off any elevator occupied by a lift truck.

Powered industrial trucks should be driven carefully and slowly over bridge plates that are properly secured (Figure 17–12). Trucks should cross railroad tracks diagonally whenever possible and park at least 8 ft (2.1 m) from the centerline of tracks.

Loading and Unloading

Highway trucks, trailers, and railroad cars should have their brakes set and their wheels securely blocked while they are being loaded or unloaded by powered industrial trucks. Before entering a trailer, lift-truck drivers should see that wheel chocks are squarely placed in front of the rearmost tires on dual-axle trailers. On tri-axle or quad-axle trailers, two additional chocks should be squarely placed in front of the foremost tires that are still on the ground.

When a trailer is next to a ramp or wall, place at least two wheel chocks squarely in front of the outside rear and foremost tires. When trailers are parked without the tractor, some companies also place jacks under the front of the trailer. Trailer-restraint systems also are available. (See Chapter 14, Materials Handling and Storage, Figure 14–15, for an example of two types of trailer restraints.)

Loads, whether on trucks, trailers, skids, or pallets, should be stable. Neatly pile and cross-tie objects, if their shape permits (Figure 17–13). Load irregularly shaped objects so that they cannot roll or fall off the truck. Place heavy, odd-shaped objects with the weight as low as possible. Block round objects, like pipe or shafting, and, if necessary, tie them so that they cannot roll off the truck. Loading to an excessive height not only blocks the view ahead, but makes it likely that part of the load may fall.

Figure 17–12. This hydraulic dock leveler prevents forklifts and other vehicles from rolling or being driven off the dock when the truck or trailer is not present. Note the forklift truck has a liquefied petroleum fuel storage container. (Courtesy Rite Hite Corp.)

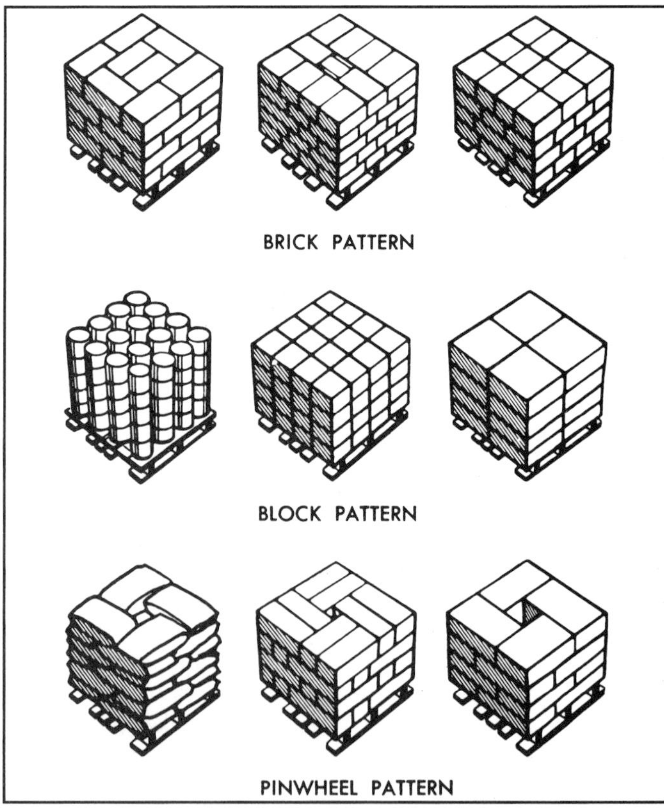

Figure 17–13. Typical pallet-loading patterns. (Printed with permission from Industrial Truck Association.)

Proper Care of Trucks

Operators should not use a powered industrial truck for any purpose other than the one for which it is designed. Common dangerous misuses include bumping skids, pushing piles of material out of the way, using makeshift connections to move heavy objects, using the forks as a hoist, and moving other trucks. Disabled trucks should not be pushed or carried by another lift truck. They should be moved by towing with a tow bar and safety chain. Never use powered industrial trucks to tow or push freight cars, and never use them to open or close freight-car doors unless the truck has a device specifically designed for this purpose.

The operator should leave a truck unattended only after the controls have been put in neutral, the power shut off, the brakes set, the key removed, or the connector plug pulled, and the load-engaging mechanism placed in a lowered and inoperative position. (OSHA defines an "unattended truck" as one where the operator is more than 25 ft [7.6 m] from it or cannot see it.) Although it is not usually good practice to leave a truck on an incline, when such action is necessary, block the wheels as an added precaution.

It is the operator's responsibility never to park a truck in an aisle or doorway, nor to obstruct material or equipment to which another worker may need access. Accidents often happen when a truck is blocking a passageway, and an unauthorized employee tries to move it.

Operators should not permit gasoline engines, such as a trailer truck, to idle for long periods in enclosed or semi-enclosed areas because exhaust vapors and combustion gases will accumulate. Concentration of carbon monoxide (CO) in areas where powered industrial trucks are operated should not exceed the levels established by regulatory authorities. Sampling for CO should be done by a qualified industrial hygienist. Direct reading instrumentation may be needed in some locations.

Catalytic exhaust purifiers, which considerably reduce the level of carbon monoxide and other noxious gases in the engine exhaust, are available. Even when exhaust purifiers are installed on lift trucks, management must provide adequate ventilation for enclosed areas and proper maintenance of the trucks in order to maintain clean air. Properly maintain exhaust purification, and do not operate batteries beyond their rated capacity.

Operator and Pedestrian Safety

Operators should keep their feet and legs inside the guard or inside the operating station of the truck. Driving with a foot or leg outside is unsafe, as is placing a hand, arm, or leg on or between the uprights of a truck. When in close quarters, operators should keep their hands where they cannot be pinched between the steering, or control, levers and projecting stationary objects. Steering handles on motorized hand trucks should have guards to protect against injuries of this kind.

Passengers must never be permitted to ride on a truck, form, coupling, or trailer. It is the operator's responsibility to keep unauthorized personnel off the truck unless a safe, designated place to ride is provided and authorized by the employer.

Looking out for pedestrians is also the truck operator's responsibility. They should sound the horn when approaching pedestrians. Discourage excessive horn-blowing, however. Having sounded the warning, the operator should proceed with caution, passing only when the pedestrians are aware of the truck's presence and are in the clear. The operator should not use the horn to "blast" a way through. They should never drive a truck directly toward anyone who is standing in front of a bench or other fixed object.

Pedestrians also have a responsibility to watch out for trucks and to get out of the way with reasonable promptness. Ill feelings and accidents can result if pedestrians refuse to move out of the way when a truck approaches—consideration by both sides is needed.

LIFT TRUCKS

Because the operation of a lift truck has some basic differences from that of an automobile or a highway truck, the prospective operator should realize that a lift truck

- is generally steered by the rear wheels
- steers more easily loaded than empty
- is driven in the reverse direction as often as in the forward
- is often steered with one hand—the other hand being used to operate the controls

Maneuvering

Because a lift truck is generally steered by the rear wheels, operators must always carefully watch the swing of the rear of the truck (Figure 17–14). Beginners usually try to turn too sharply. Some lift trucks when traveling forward, have a peculiarity known as "free turning." That is, once the turn is started, the truck tends to turn more and more sharply in smaller and smaller circles. To counteract this tendency and to slow down the sharpness of the turn, the operator must apply force on the steering wheel in the opposite direction. When such a truck is traveling in reverse, the opposite holds true—the operator must apply force in

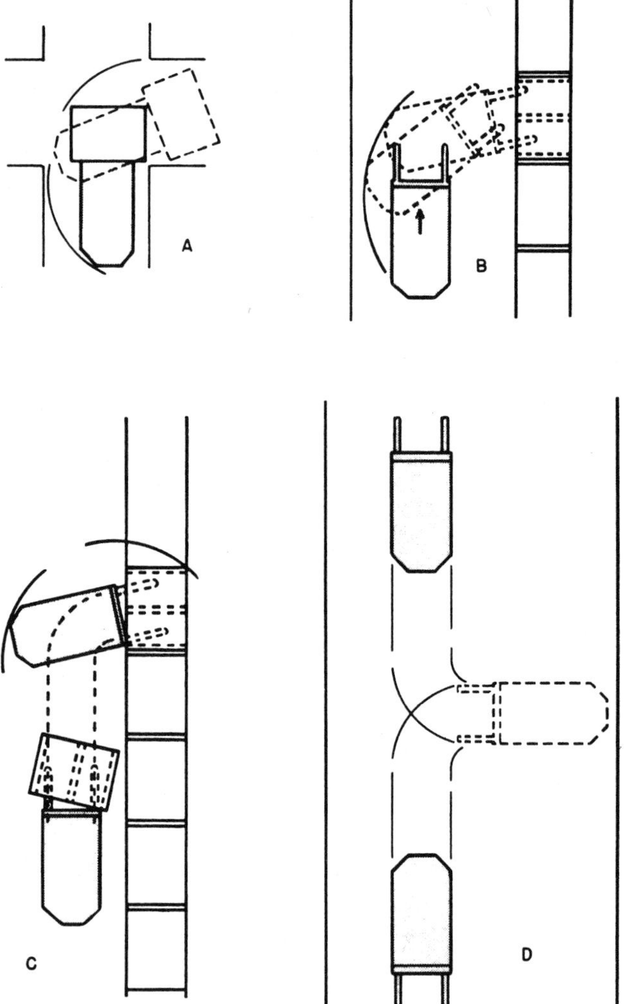

Figure 17–14. Lift-truck maneuvers. A: Turning a sharp corner. B: Turning across an aisle. C: Turning in an exceptionally narrow aisle. D: Turning around in a narrow passage. The driver should allow ample space for rear-end swing and make turns carefully.

the direction of the turn. Turns should be made smoothly and gradually at a safe speed.

Operators should learn to judge the correct aisle width for the truck size and load. They should also observe the general operating safety rules, given earlier in this chapter, as well as the specific rules for lift trucks that are discussed next.

All starts and stops should be easy and gradual to prevent the load from shifting.

The operator should be particularly careful, while either traveling or maneuvering, to avoid striking overhead structures and nearby objects, such as sprinkler piping, electrical conduit, or fixed structures. To protect critical equipment and materials, such as electrical panels, fire equipment, fire doors, and load-supporting structures, install strong barriers, posts, or curbing.

Operators should raise or lower loads only when stopped. Loaded or empty, the forks should be carried as low as possible, but high enough not to strike any raised or uneven surface. Tilting back the mast keeps the load steady and secure.

When carrying a bulky load that cannot be lowered enough to prevent its obstructing the view, the operator should drive the truck backward. Thus, the operator can see where the truck is going. Some companies have a policy that trucks are to be driven in reverse any time they are carrying a load, regardless of size. Use a spotter where necessary.

Driving on Grades

Trucks should ascend or descend grades slowly. When ascending or descending grades in excess of 10%, loaded trucks should be driven with the load toward the upgrade. Unloaded trucks should be operated on all grades with the load-engaging mechanism toward the downgrade. Keep in mind that high-lift, order-picker trucks are not designed for operation on steep grades. Consult the manufacturer's operating instructions for recommended operating procedures.

On all grades, the load and load-engaging mechanism should be tilted back, if applicable, and raised only as far as necessary to clear the road's surface. Low gear or the slowest speed should be used when the truck is descending a grade. The operator should keep clear of the edge of loading docks and ramps and never make a turn on a ramp.

Before driving over them, operators should check bridge plates to make sure they are properly secured. They should also check the floors of freight cars and trucks to be sure that they are in good condition and that they will bear the weight of the loaded truck. Operators should also check the truck or trailer to see that it is properly chocked. Failure to do this may result in the bridge plate shifting or the trailer or truck moving away from the dock (Figure 17–12).

Load Capacity

Lift trucks are rated by capacity in pounds and load center in inches. The capacity of a truck might be 5,000 lb (2,268 kg) at a 24 in. (60 cm) load center. In other words, it can pick up 5,000 lb (2,268 kg) if the center of gravity of the load is 24 in. (60 cm) from the face of the load arms. Exceeding the rated capacity is considered an overload. Every operator should be familiar with the maximum load limits of the truck being operated and should be required to observe them.

Placing extra weight on the rear of a lift truck to counterbalance an overload shall not be permitted as it may strain chains, forks, tires, axles, and the motor. It also could cause an injury. ANSI B56.1 provides more information on the stability of lift trucks. Most lift-truck manufacturers use the stability values in this standard as criteria for design factors and for determining the rated capacity of various truck models. Side stability is a critical factor in making turns at speed or on a slope or ramp. Tilting back the mast reduces side stability on high lifts and may cause tipping; allow for this factor.

Operators should never, under any circumstances, attempt to operate a truck with an overload. Such a load is dangerous because it removes weight from the wheels that steer, which affects the steering. Standing on a truck or adding counterweights to compensate for an overload should never be permitted.

Particular care should be taken not to exceed floor-load limits. The force exerted by a truck on a floor varies with the speed, load, and total weight distribution. It also is affected by the number of wheels, the wheelbase, and other factors. Refer all questions about floor capacities to a qualified architect or structural engineer.

Loading and Unloading

When standard forks are used to pick up round objects, such as rolls or drums, care must be taken to see that the forks' tips do not damage the load or push it against workers. First tilt the mast so that the tips of the forks touch the floor. Then move forward so that the forks can slide under the object. Tilting the mast backward will then cause the load to roll back against the vertical face of the forks and/or carriage and the load backrest extension—a secure carrying position. Place a block or wedge against the drum or roll.

To unload a large case or similar object without a pallet, the operator should first drive into position for stacking. The operator should place a block near the edge of a base, lower the load onto the base, and withdraw the forks so that only their tips hold up the end of the load. Next, the operator should withdraw the block, tilt the mast forward, and back away.

In attempting to pick up a palletized load, keep the forks fully and squarely seated in the pallet, an equal distance from the center stringers and well out toward the sides. Forks to be inserted in a pallet should be level, not tilted forward or back. If the forks are placed close together, the pallet tends to drop at the sides and seesaw, causing strain and instability (Figure 17–15).

When raising or lowering loads while standing still, the operator should not leave the truck in gear with the clutch depressed. The shift should be returned to neutral and the clutch disengaged.

Figure 17–15. With forks spread wide (left) load is well distributed and tends to bind itself together. The effect of placing the forks too close together is shown at right.

Tell operators of lift trucks to refuse improperly loaded skids or pallets, broken pallets, or loads too heavy for the truck. Operators should also insist on the proper identification of all chemicals before moving them and should observe the safety guidelines and regulations for handling chemicals.

When a lift truck is parked, the forks should be placed flat on the floor. Allow no one to stand or walk under elevated forks.

Using a lift truck to elevate employees (for example, to service light fixtures) should be done only if an approved safety work platform with guardrails and toeboards is secured to the forks and mast. The truck should also have an overhead guard for the operator's protection. The lift-truck operator must not leave the controls while the truck is used to lift a person. Lift trucks, without special attachments, are not generally designed to lift personnel. Special trucks are built for this purpose.

Pallets

Most companies buy ready-made pallets (Figure 17–16). No matter who builds them, however, establish procedures to inspect pallets, both before they are put into service and at regular intervals, to be sure they are safe to use. The top deckboards of pallets should be sound and securely fastened to runners. Repair or replace splintered, broken, or loose parts. Loose nails or chunks of wood can cause injury to workers and damage to trucks.

Provide a safe place, out of the way of traffic and work areas, for storing pallets. Neatly stack pallets and limit their height (no more than 6 ft per fire code) so that they are stable and secure against sliding or collapse. Also, do not leave them standing on end or in a position from which they might topple onto people or other objects. To prevent overpowering the sprinklers

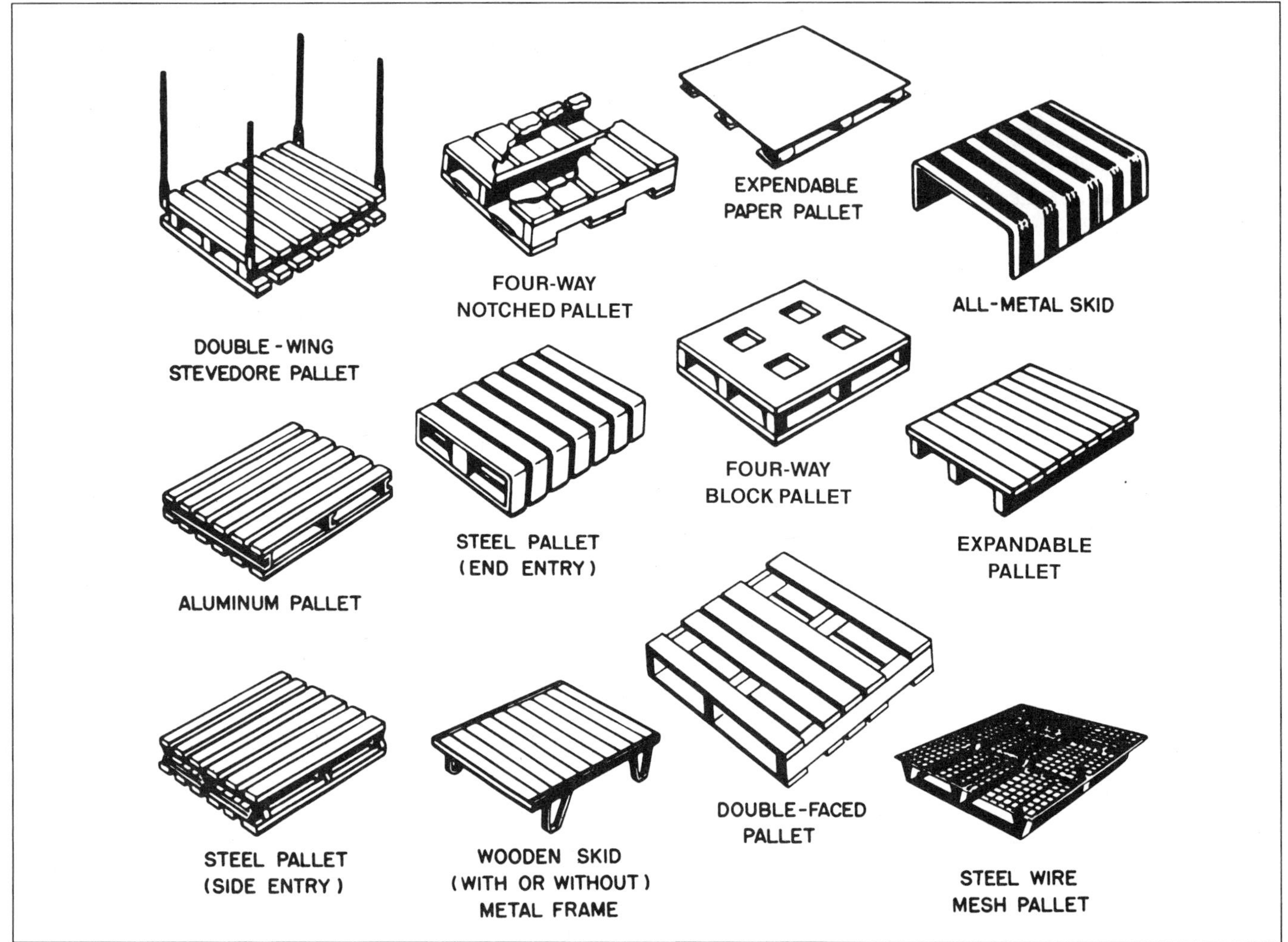

Figure 17–16. Lift-truck operators should be familiar with pallet and skid types. (Printed with permission from the Industrial Truck Association.)

in case of a fire, avoid storing large blocks of pallets within a building. Keep large stacks in outside storage, well away from buildings.

INSPECTION AND MAINTENANCE

Maintenance personnel should thoroughly inspect powered industrial trucks on a regular basis and give them a complete overhaul after regular periods of operation. Operators should make daily inspections of controls, brakes, tires, and other moving parts. They must do this at the start of each shift in multishift operations. They should use checklists (Figure 17–17 to record conditions requiring correction and keep a detailed schedule of inspections and repair records for each vehicle (Figure 17–18).

Forks should be magnafluxed on a regularly scheduled basis as determined by use. Defective brakes, controls, tires, lights, power supply, load-engaging mechanism, lift system, steering mechanism, and signal equipment should be repaired before trucks are allowed to go back into service. Prohibit operators from making repairs on trucks. Before repairs are made to any part of a powered industrial truck, the operating mechanism should be locked off.

Electric Trucks

Battery changing and charging operations for electric trucks must be performed only by trained and authorized personnel. Truck operators may or may not be so authorized, depending upon individual facility setups. Handling and charging storage batteries for electric trucks introduces several hazards.

OPERATOR'S DAILY REPORT

Battery-Powered Lift Trucks

Truck No.______________ Make ______________________ Date ________________ Shift __________

Hour Meter Reading: Start ______________ End ______________ Hours for Shift __________________

CHECK EACH ITEM	SHIFT			Explain below if not OK or any other action taken
If OK write OK	Start	During	End	
1. Battery plug connection				
2. Battery charge				
3. Battery load test				
4. Brakes – service and seat brake				
5. Lights – head, tail and warning				
6. Horn				
7. Hour meter				
8. Steering				
9. Tires				
10. Hydraulic controls				
11. Other conditions				

Remarks and additional explanation or suggestions ____

____

____

Operator's Signature ______________________________

2C97567 Printed in U.S.A. Stock No. 199.77

Figure 17–17. Operators should use a checklist to make a daily inspection of their industrial trucks.

NATIONAL SAFETY COUNCIL
FORKLIFT TRUCK OPERATORS TRAINING COURSE

INSPECTION AND MAINTENANCE LOG

Truck No.	Make	Work Done		Work Description or Remarks	Cost
		Date	Hour Meter		

2C97567 Printed in U.S.A. Stock No. 199.74

Figure 17–18. A detailed inspection and repair record should be kept for each truck.

By wearing goggles, rubber gloves, aprons, and rubber boots, operators of charging equipment can be protected against acid burns when refilling or handling batteries. Do not allow sulfuric acid used for refilling to run into ordinary cast-iron, lead, steel, or brass drains. Wood-slat mats, rubber mats, or clean floorboards will help prevent slips and falls and will protect against electric shock from the charging equipment. When racks are used to support batteries, they should be made of materials that will not generate sparks. If they are not, coat or cover them to achieve this objective.

Take precautions to prevent open flames, sparks, or electric arcs in battery-charging areas. Electrical installations should conform to the local codes and NFPA 70, *National Electrical Code.* Never lay tools or metal parts on a battery. Prohibit smoking in the charging area. (See NFPA 505, *Powered Industrial Trucks,* and NSC Data Sheet 12304–0635, *Lead-Acid Storage Batteries.*)

Charge batteries in areas designated for that purpose. Also provide facilities for (1) flushing and neutralizing spilled electrolyte, (2) fire protection, (3) protecting charging apparatus from damage by trucks, and (4) adequate ventilation to disperse flammable hydrogen gases, vapors, and fumes from batteries. Provide employees with eye protection and adequate eyewash facilities to wash their eyes should they be exposed to toxic liquids or fumes. The employer must periodically test the eyewash facilities.

Properly position trucks and apply the brake before attempting to change or charge batteries. Reinstalled batteries should be properly positioned and secured in the truck.

When charging batteries, wear eye protection and rubber gloves. Pour acid into water, never the reverse. Provide a carboy tilter or siphon for handling electrolyte. If acid or electrolyte is spilled on the worker's skin or clothing, it should be washed off immediately with plenty of water. When charging batteries, keep the vent caps in place to avoid electrolyte spray. Take care when determining whether or not vent caps are functioning. Open the battery (or compartment) cover(s) to dissipate heat. Refer to manufacturer's recommendations before performing charging and maintenance procedures.

To prevent operators from straining during manual handling of heavy or awkward batteries, provide a roller conveyor, an overhead hoist, or equivalent material handling equipment. To prevent short-circuiting, use insulated or nonconductive chains, hooks, and yokes in the hoisting mechanism. To prevent wear on insulation, which may produce arcing, watch the points where cables contact reels and suspension attachments. Battery chargers must be in the Off position while being connected and disconnected.

Gasoline-Operated Trucks

Handle and store gasoline for trucks according to the provisions of NFPA 30, *Flammable and Combustible Liquids Code.*

Safety cans used for fuel handling should be tested and approved by Factory Mutual or listed by Underwriters Laboratories. Safety cans should contain a flame-arrestant, self-closing lid. (See Chapter 12, Flammable and Combustible Liquids, in this volume.)

Fill fuel tanks on gasoline-operated trucks and tractors at designated locations, preferably in the open air, with the filling hose and equipment properly grounded and bonded. Select locations outside main buildings to lessen the chances of starting a fire.

Engines must be stopped and operators must be off trucks before they are refueled. Do not permit smoking during refueling. Gasoline tanks should be drained into grounded self-closing cans at safe locations. Workers should avoid spilling gasoline or letting the gas tank overflow during refueling. Before an attempt is made to start the engine, the gas tank's cap should be replaced and spilled fuel flushed down or allowed to vaporize.

Liquefied Petroleum Gas Trucks

The use of LP-gas as a fuel for powered industrial trucks is increasing, largely through conversion of gasolinepowered units. Before converting to LP-gas, consult the manufacturer of the truck. The manufacturer may be able to supply listed conversion units and assign a qualified representative to supervise the installation. Conversions should be attempted only by qualified mechanics who are familiar with handling LP-gas equipment and who are using listed parts.

A properly adjusted engine burning LP-gas will generally produce a substantially lower concentration of CO in the exhaust than a similar engine that uses gasoline as fuel. Only air sampling, however, can prove whether the CO concentration in an area is below the maximum allowable level. The "Threshold Limit Values" (TLV) listing of the 1989 American Conference of Governmental Industrial Hygienists (ACGIH) gives 25 ppm for an eight-hour exposure as the maximum exposure allowable. However, federal OSHA has set the allowed exposure at 50 ppm (35 ppm in Canada).

Fittings not listed by a nationally recognized agency or incorrectly installed, or connections not properly tightened before refueling is begun, may fail and release combustible gas into the air. Only use conversion units and fittings listed by an agency such as Underwriters Laboratories or Factory Mutual. The units and fittings should include all the safety features that are incorporated in LP-gas-fueled trucks listed by the testing agency. Install the units and fittings in strict conformity with requirements specified in NFPA 58, *Storage and Handling of Liquefied Petroleum Gases,* and Underwriters Laboratories' Standard for Safety No. 558, *Internal Combustion Engine-Powered Industrial Trucks.* Following these requirements will provide maximum protection against damage to the system by vibration, shock, or objects striking against it.

Only listed fuel containers, designed in accord with U.S. Department of Transportation (DOT) or ASME standards, should be used. Fill permanently mounted and removable fuel containers outdoors. However, it may be done indoors if one of the two methods specified in NFPA 58 is met. Refueling of LP-gas trucks with permanently mounted containers shall be done outdoors away from any building ventilation or air-conditioning intakes.

A special building or outside storage area is recommended for the storage of fuel containers. When the cylinders must be stored inside the building, keep them in a special room or designated safe area in accord with NFPA 58. NFPA standards permit no more than two containers for LP-gas on each industrial truck. They also require that storage inside a building not frequented by the public be limited to a total capacity for gas of 300 lb (135 kg). Enclose containers in a separate room that is well ventilated and of ample size. The walls, floor, and ceiling of the room are required to be of specified fire-resistant construction. Protect openings to other parts of the building with specified fire doors.

The proper filling of containers is of the utmost importance. The person filling the containers must be trained to handle LP-gas safely. Filling of containers from bulk storage must be done at least 10 ft (3 m) from the nearest important masonry-walled building and at least 25 ft (7.5 m) from important nonmasonry buildings and openings in masonry and nonmasonry buildings (Figure 17–19). The filling facility must conform to NFPA 58 and applicable state or provincial or local insurance regulations.

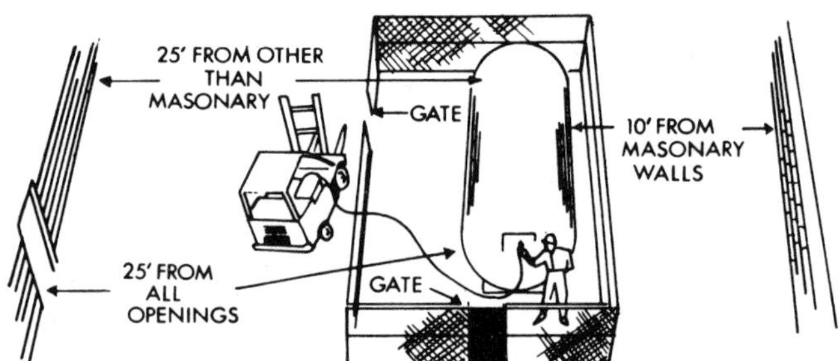

Figure 17–19. Schematic shows minimum distances allowed for industrial-truck refueling operations. (Printed with permission from the LP-Gas Association.)

Trucks themselves must comply with NFPA 505, *Powered Industrial Trucks.* Garage LP-gas-fueled trucks in a well-ventilated area. Provide ventilation at floor level because LP-gas is heavier than air. Do not garage trucks in the same room with stored cylinders.

Follow a rigid and thorough inspection and maintenance procedure for LP-gas-fueled trucks. LP-gas trucks can be stored or serviced inside garages provided that (1) the fuel system is leak-free and the container is not filled beyond the limit specified in NFPA 58, (2) the container shutoff valve is closed except when the engine is operated, and (3) the truck is not parked near inadequately vented pits or sources of heat, open flames, or other sources of ignition. Wear eye and hand protection when changing tanks. For trucks with permanently mounted fuel containers, make major repairs outdoors or in a well-ventilated, fire-resistive area provided for this purpose. Subparts of the OSHA regulations that apply to this chapter must be strictly followed.

OPERATORS

Trainees should have valid driver's licenses, good driving records and few, if any, traffic violation tickets (29 *CFR* 1910.178). They should have good attitudes toward the responsibility of operating expensive, heavy-industrial equipment in new and difficult situations. For this reason, verify the trainee's previous experience, both off and on the job, whenever possible. Driving a car and driving a forklift are not equivalent.

Selection

Trainees should meet certain physical standards and should be examined by a qualified physician familiar with the job's requirements. Minimum requirements would be 20/40 vision, corrected if necessary; good reaction time; depth perception of no less than 90% of normal; and good hearing, preferably without the benefit of a hearing aid. Give drivers a physical examination every two years, and also check their driving record off the job as well as on the job.

Trainees must appreciate the importance of the training program. Their attitude is important for good performance even if some of the finer points of the training are forgotten and if supervision is not as close as it should be.

Trainees should also show good judgment together with respect for the safety of both personnel and property. Operator trainees should understand that they carry considerable responsibility and that reckless or careless work will not be tolerated. Check operators on the job from time to time for observance of this rule. For ready identification, many companies issue badges to authorized truck operators. Badges also tend to remind operators of their responsibilities and give them pride in their job.

To orient trainees to their new responsibilities, adopt a set of rules governing the operation of powered industrial trucks. Because facility conditions and equipment vary widely, set up rules that cover specific conditions that trainees will face.

Training

Federal OSHA regulations state that "only trained and authorized operators shall be permitted to operate a powered industrial truck." Operators must receive some form of powered industrial truck training as well as pedestrian safety training. (See 29 *CFR* 1910.178(l).) A truck operated incompetently can cause severe injury or substantial property damage. No company can afford even one improperly trained truck operator.

To be effective, a training program should center around the company's policies, operating conditions, and type of trucks used. All new operators, regardless of claimed previous experience, should receive a training course. Experienced operators should be given a refresher course every two years. Information on training programs may be obtained from the National Safety Council, truck manufacturers, and trade associations.

Before starting a training program, companies should determine the problems they have experienced with industrial trucks, such as numbers and types of accidents, extent of economic losses, and operating habits of operators. This information can then be worked into the training program.

As with any program, management's support is essential if the truck operator's training is to be effective and lasting. Management must also understand that any training program will cost money and accept that fact. Materials and equipment, payment for nonproduction time and lost production, and possible damage to materials are some costs of training. Practical training can be conducted on obstacle courses (Figure 17–20).

Check standards and codes for permissible floor loads, clearances for sprinklers, travel routes of the trucks, adequacy of lighting and ventilation, and noise control. Correct unsafe or hazardous conditions before starting the training. If there are conditions that cannot be corrected, be sure instructors know about them and why the conditions cannot be changed. Trainees will not be impressed with the need for safe job performance if they feel that management permits other unsafe conditions to exist.

Supervisors also should fully understand the need for the training program. They might have to adjust their schedules to make people available for training either during or after work hours.

Figure 17–20. The driving ability of this lift-truck operator trainee is tested by maneuvering through a mockup of a crowded aisle. All training programs should contain a driving test as well as a written examination. (Courtesy Clark Equipment Co., Industrial Truck Division.)

Maintenance personnel should be involved in the program from the start. Mechanics are possible instructors because they know the trucks and how they operate. Mechanics also will understand why the new truck operators are requesting additional repairs and adjustments—they have become "safety conscious."

Other factors to consider in setting up a training program include determining

- who will be in charge
- the qualifications for both instructor(s) and trainees
- the number and length of sessions for both classroom and hands-on instruction
- the location of the sessions for both classroom and hands-on instruction
- the number of trainees in a class
- whether experienced operators who are taking a refresher course should be in the same class with trainees

- how to inform all employees about the program so they understand its importance
- how to establish and maintain a record system that stays current and can be used by regulatory officers reviewing the program

An effective training program does not end with the presentation of a certificate. Management has the responsibility to continue to maintain safe operating conditions and to insist on safe performance by all employees. To ensure continued safe work habits, keep a record of each operator's performance (Figure 17–21).

SUMMARY

- Powered trucks require safeguards for the operator's protection and for the safety of other workers. Management must establish guidelines for the operation, maintenance, and inspection of this equipment.
- Powered industrial trucks are classified by power source, operator position, or means of engaging the load.
- Factors to consider in purchasing trucks include work-site conditions, operator comfort, backup systems, and safety features such as safety belts and wraparound seats.
- NFPA standards specify certain hazardous locations, Classes I through III, in which various types of trucks should not be used unless they comply with NFPA requirements or are officially approved.
- Lift trucks should have overhead guards designed to prevent injury. Operators should realize that lift trucks are generally steered by the rear wheels, handle more easily when loaded, are used in reverse as often as forward, and are often steered with only one hand.
- Straddle trucks should be safeguarded with horns, flags, or other warning devices to protect pedestrians.
- Crane trucks should be driven at the lowest possible speed when carrying a load to maintain balance. In tractors and trailers, the coupling mechanism must be carefully safeguarded and loads secured to the trailer. Motorized hand trucks must be safeguarded for proper operation.
- Automated guided vehicles must have some means of stopping should someone step in front of them. Such trucks should be equipped with flexible bumpers that shut off power on contact.
- Operators of industrial trucks can prevent accidents by using the same safe-driving techniques they employ on the highways.
- When loading and unloading trailers, operators should make sure the brakes are on, wheels are blocked, loads are neatly stacked and stable, and loads are fastened to the trailer securely.
- Industrial trucks should not be used for any purpose other than the one for which they were designed. Operators are responsible for the care of trucks and should never leave a truck unattended, park in an aisle or doorway, idle engines for too long, or ignore mechanical problems.
- Industrial powered trucks should be inspected and overhauled regularly. Repairs, replacements, or other work should be performed only by trained mechanics wearing proper protective equipment, particularly when handling electrically powered trucks. Only authorized fuel and fuel-tank equipment should be used on these trucks.
- Training programs should center around company policies, operating conditions, and types of trucks used. Management should maintain records of each employee's driving performance.

REFERENCES

American Conference of Governmental Industrial Hygienists, 6500 Glenway Avenue, Bldg. D-7, Cincinnati, OH 45211.
"Threshold Limit Values" (latest edition).

American Insurance Association, 1130 Connecticut Avenue NW, Suite 1000, Washington, DC 20036.
Safe Handling and Use of LP-Gas.

American National Standards Institute, 11 West 42nd Street, New York, NY 10036.
Hook Type Fork and Fork Carriers for Powered Industrial Fork Lift Trucks, ANSI/ASME, B56.11.4–1992.
Low Lift and High Lift Trucks, ANSI/ASME, B56.1–1993.

National Fire Protection Association, 1 Batterymarch Park, Quincy, MA 02269.
Flammable and Combustible Liquids Code, NFPA 30, 1996.
National Electrical Code®, NFPA 70, 1996.
Powered Industrial Trucks, Including Type Designations, Areas of Use, Conversions, Maintenance and Operation, NFPA 505, 1999.
Storage and Handling of Liquefied Petroleum Gases, NFPA 58, 1998.

National Propane Gas Association of America, 1600 Eisenhower Lane, Lisle, IL 60532.

National Safety Council, 1121 Spring Lake Dr., Itasca, IL 60143-3201.
Aviation Ground Operation Safety Handbook, 5th ed.
"Fork Lift Truck Operators Training Course."
Occupational Safety and Health Data Sheets (available in the Council Library):
Powered Hand Trucks, 12304-031. 1991.

Underwriters Laboratories, Inc., 333 Pfingsten Road, Northbrook, IL 60062.
"Industrial Trucks, Electric-Battery-Powered," Standard for Safety, No. 583.
"Industrial Trucks, Internal Combustion Engine-Powered," Standard for Safety, No. 558.

OSHA. *CFR* 1910.178—Powered Industrial Trucks. Superintendent of Documents, Government Printing Office, Washington, DC 20402.

FORKLIFT TRUCK OPERATOR PERFORMANCE TEST

OPERATOR'S NAME____________________ DEPT. __________ DATE __________

1. Operator's ability to perform check-sheet inspection for safe operation of truck prior to use.

   Uses check-sheet satisfactorily ☐ Failed to check safety items ☐

2. Proper use of controls (understands proper technique and proper direction of movement of control to get desired direction of movement).

   (a) Clutch operation. Yes ☐ No ☐

   (b) Inching control (auto transmissions). Yes ☐ No ☐

   (c) Tilt control. Yes ☐ No ☐

   (d) Lift control. Yes ☐ No ☐

   (e) Attachment controls. Yes ☐ No ☐

   (f) Steering techniques for type of machine being used by operator. Yes ☐ No ☐

   (g) Proper positioning of all controls, switches, parking brakes when leaving machine unattended. Yes ☐ No ☐

   (h) Service brake. Yes ☐ No ☐

   (i) Parking brake. Yes ☐ No ☐

3. Maneuvering skills.

   (a) Smooth starting and stopping.
   Acceptable ☐ Needs practice ☐ Poor control ☐

   (b) Sharp turns forward and reverse.
   Proper speed. Yes ☐ No ☐
   Looks in direction of travel. Yes ☐ No ☐
   Carries forks low. Yes ☐ No ☐
   Clears obstacles by safe distance. Yes ☐ No ☐

4. Selecting loads.

   (a) Proper capacity for truck used. Yes ☐ No ☐

   (b) Proper size load for visibility and safety of handling. Yes ☐ No ☐

   (c) Load tilted back against back rest. Yes ☐ No ☐

   (d) Carries load low (just high enough to clear floor obstacles). Yes ☐ No ☐

Figure 17–21a. Typical form used to keep a record of an operator's performance.

5. Driving with load.

   (a) Smooth starting and stopping. Yes ☐ No ☐

   (b) Proper speed. Yes ☐ No ☐

   (c) Sounds horn at intersections and corners. Yes ☐ No ☐

   (d) Keeps to the right in aisles used for two-way traffic. Yes ☐ No ☐

   (e) Travels at least three lengths behind other vehicles. Yes ☐ No ☐

   (f) Handles load in manner to prevent product damage. Yes ☐ No ☐

6. Stacking.

   (a) Approaches loads squarely. Yes ☐ No ☐

   (b) Stacks straight and squarely. Yes ☐ No ☐

   (c) Does not tier too high. Yes ☐ No ☐

   (d) Deposits load safely; does not use excessive tilt action. Yes ☐ No ☐

   (e) When selecting top load for pickup, uses proper form spread for load. Yes ☐ No ☐

   (f) Removes load and lowers to safe level before making turn to proceed in direction of desired travel. Yes ☐ No ☐

7. Dock Safety.

   (a) Checks bridge plates (dock boards) before crossing. Yes ☐ No ☐

   (b) Checks trailers for proper wheel chocking before entering and proper jack installation of trailers where required. Yes ☐ No ☐

   (c) Checks rail freight cars for proper positioning and safe loading conditions. Yes ☐ No ☐

ADD ANYTHING THAT MAY PERTAIN TO YOUR PARTICULAR OPERATION IN MATERIAL HANDLING ON WHICH YOU MAY WANT TO TEST YOUR OPERATORS' PERFORMANCE.

2C97567 Printed in U.S.A. Stock No. 199.78

Figure 17–21b. *(Concluded.)*

———. "Proposed Rule for Training Powered Industrial Truck Operators in General, Maritime Industries." *Federal Register,* March 14, 1995:13782-13831.

Swartz G. *Forklift Safety: A Practical Guide to Preventing Powered Industrial Truck Incidents and Injuries.* Rockville, MD: Government Institutes, 1997.

U.S. Department of Health, Education, and Welfare, National Institute for Occupational Safety and Health, Division of Technical Services, 4676 Columbia Parkway, Cincinnati, OH 45226. *Outline for Training of Powered Industrial Truck Operators,* DHEW (NIOSH) Publication 78–199.

REVIEW QUESTIONS

1. Name six types of powered industrial trucks.
a.
b.
c.
d.
e.
f.

2. What aids should leased or purchased industrial trucks have to reduce driver fatigue and strain?
a.
b.
c.
d.
e.

3. Which of the following should conform with ANSI/ ASME B56.1?
a. Overhead guards on trucks
b. Name plates showing truck weight and its rated capacity
c. Specifications of steering, braking, and other controls
d. Stability of lift trucks
e. All of the above

4. Since operators in saddle trucks sit so high off the ground, their angle of sight is reduced immediately to the front and rear, posing a hazard to __________.

5. What are the two principle hazards in operating a motorized hand truck?
a.
b.

6. Operators of powered industrial trucks can prevent traffic accidents by using the same safe practices that apply to __________.

7. Describe the common, dangerous misuses of powered industrial trucks that should never be performed by operators.
a.
b.
c.
d.
e.

8. Operators should be aware of what basic differences between lift trucks and automobiles or highway trucks?
a.
b.
c.
d.
e.

9. To be effective, a training program should focus on what three aspects?
a.
b.
c.

18

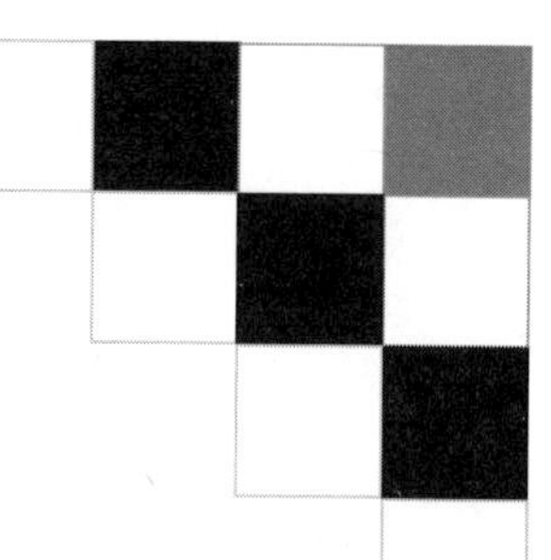

Haulage and Off-Road Equipment

Revised by

John F. Montgomery, Ph.D., CSP, CHCM, CHMM

Heavy-duty trucks are used extensively for special off-the-road operations in industries such as quarrying, mining, and construction (Figure 18–1). When on the road, they are governed by the same safe-driving rules and regulations that apply to other types of automotive equipment. This chapter discusses general safety issues and operating practices that can help companies achieve their safety goals. Topics covered include:

- general safety requirements for operating heavy-duty equipment
- operating power shovels, cranes, and similar equipment safely
- safety issues in the maintenance and operation of graders, bulldozers, and scrapers

GENERAL REQUIREMENTS

The use of heavy-duty trucks, mobile cranes, tractors, bulldozers, and other motorized equipment always presents the possibility of accidents. Workers near equipment can be injured or killed; equipment may slip over embankments, and so on (Figure 18–2). In addition, personnel who service and maintain equipment must recognize the hazard involved.

Accidents, even those that do not result in injuries, can mean serious damage to equipment, loss of production, and high replacement costs.

In general, prevention of accidents to heavy equipment requires:

1. maintenance of safety features on equipment
2. systematic equipment maintenance and repair
3. trained operators
4. trained repair and maintenance personnel

Safe and proper equipment operation instructions can be found in manufacturers' manuals. Many driving practices are the same as those necessary for the safe operation of highway vehicles. Off-the-road driving, however, involves special hazards and requires special training and safety measures.

Figure 18–1. Various types of heavy-duty equipment. (Courtesy Caterpillar, Incorporated.)

Figure 18–2a-b. Here's what a scraper looked like after it rolled down a canal berm *(left)*. The operator was belted in and was only slightly hurt; if he had tried to jump, he would have been thrown into the path of the rolling scraper. The close-up photo *(above)* shows that rollover protective structures (ROPS) also protected the windshield and operator's area from damage as well as probably saving the life of the operator. (Courtesy U. S. Dept. of Interior, Bureau of Reclamation.)

Haul Roads

Both temporary and permanent roads often are too narrow to accommodate heavy equipment and oncoming traffic, especially at curves and fills. Make sure haul roads have enough space at curves so large trucks need not cross the centerline of the road. Bank all curves toward the outer roadside.

Both temporary and permanent roadways require regular patrolling and maintenance. Too often, serious accidents, breakdowns, delays, and unnecessary maintenance expenses can be traced to neglected roadways. Provide members of road patrols with a means of summoning help and with protective equipment, such as barricades, warning signs, flags, and flares.

Seasonal conditions create road hazards that require prompt attention. Some companies provide sprinkler trucks to protect their employees against dusts and other airborne hazards during dry and windy periods.

Skidding on snow and ice is a serious hazard. Make sure snowplows or blade graders remove snow and ice as promptly and completely as possible.

When roadways are built close to high banks, inspect the slopes of the banks for loose rocks, especially after rain and freezing or thawing weather. Remove all loose rock.

Where trucks enter public highways, install signs warning both the highway traffic and other off-road vehicles. Design, color, and placement of the signs should conform to U.S. Department of Transportation, Federal Highway Administration, Washington, DC, *Manual on Uniform Traffic Control Devices for Streets and Highways,* also published as American National Standard D6.1 (see References). If operations are conducted at night, make these signs of a light-reflecting material or directly illuminate them. In situations where temporary roads cross railroad tracks, especially when high-speed trains are involved, contact the railroad representatives and place a means of warning at the crossing.

Driver Qualifications and Training

The modern heavy-duty vehicle or other off-the-road equipment is a carefully engineered and expensive piece of equipment. Only drivers who are qualified physically, mentally, and by training and experience should operate them.

No drivers should be allowed to work until management has determined their knowledge, experience, and abilities. The amount of time varies for a prospective driver or operator to become thoroughly acquainted with the equipment, safety rules, driver reports, and emergency conditions. Even experienced operators should not be permitted to operate equipment until the instructor or supervisor is satisfied with their abilities.

Because accidents caused by unsafe practices outnumber those resulting from unsafe condition of equipment and roadways, the time required for thoroughly checking and training drivers and mechanics is well warranted. After employees have been trained, they should be closely supervised to make sure they continue to work safely.

Operating Vehicles near Workers

Workers are exposed to the hazard of being struck or run over by vehicles, particularly around power shovels, concrete mixers, and other equipment, in garages, shops, dumps, and construction areas.

Backing—The "Crunch Zone"

The most dangerous movement is backing (Figure 18–3). OSHA regulations require all vehicles to have an automatic audible signaling device to warn workers when a vehicle is backing. If their vehicles are not so equipped, drivers should signal with three horn blasts.

Where a number of employees are working, the driver should ask another employee to signal whether the path is clear before the vehicle is backed or moved. The person giving the signals should always take a position within sight of the driver. Also, a standard set of signals should be devised to ensure proper communication.

Moving Forward

Serious accidents also may occur during forward movements. The hazard to workers increases with the greater height and capacity of trucks. A driver often fails to see workers crossing from the right immediately ahead of the truck. Thus, drivers should blow two blasts on the horn before starting forward. Construction vehicles with attachments such as front-end loaders and dozers should position their attachments for maximum visibility whenever moving their equipment.

Procedures for Dumping

Vehicle operations on dumps and banks involve the danger of the vehicle going over the crest while dumping a load. A person trained in proper dumping procedures for drivers is probably the best insurance against injury or damage to equipment. Drivers are required to follow this person's instructions and signals, especially in backing to dump. Signalers should always use prearranged signals.

The person responsible for dumping must know how close to the edge a vehicle can safely approach under various weather conditions. The signaler should stand on the driver's side of the vehicle clear of the backing truck and falling material, but visible to the driver. To protect the signaler further, the driver should turn and look over the left shoulder when backing to

Figure 18–3. Beware of the "crunch zone." All equipment operators must signal when backing up. All other persons must give plenty of room to vehicles, especially those with limited operator visibility—for example, this large front-end loader where the driver cannot see anything closer than 45 ft (14 m) to the rear. This photograph also illustrates the safe practice of lowering the bucket or blade on all equipment not being used. (Courtesy Inland Steel Corporation.)

have a maximum view of the area in the direction the truck is moving. To avoid hitting overhead lines or other low clearances, drivers should lower the dump box as soon as they dump their loads.

Left-hand driving also reduces the danger of going over the crest, especially in the operation of side dump trucks, since the driver is on the crest side. To help prevent the crest from caving in, stockpiles and dumps frequently are graded toward the crest so that vehicles back up the slope. Drivers also may dump loads a safe distance from the crest; the loads are then leveled by a grader or bulldozer.

Strongly built cabs, cab protectors on canopies, and safety belts are effective in preventing injuries if vehicles overturn. The operator's inclination to jump clear of a vehicle that is beginning to roll or slide over an embankment is ill advised. It is far safer to remain in the cab.

Holes, ruts, and similar rough places on dumps and roadways may cause the front wheels of a truck to cramp so the steering wheel spins, injuring fingers, arms, and ribs, particularly if the truck is not provided with power steering. Gripping the wheel on the outside and not by the spokes, driving at reduced speed, and observing the ground ahead for rough places will help the driver avoid such injuries.

Using floodlights during night operations also helps to prevent accidents.

Protective Frames for Heavy Equipment

All bulldozers, tractors, and similar equipment used in clearing operations must be equipped with substantial guards, shields, canopies, and grilles to protect the operator from falling and flying objects. Equip crawlers and rubber-tired vehicles, self-propelled pneumatic-tired earth movers, water tank trucks, and similar equipment with steel canopies and safety belts to protect operators from the hazards of rollover (Figures 18–2 and 18–4). Managers should train drivers in the use of safety equipment and require them to wear safety belts.

A canopy and its support should be able to bear at least two times the weight of the prime mover. This calculation is based on the ultimate strength of the metal and integrated loading of support members, with the resultant load applied at the point of impact. In addition, there should be a vertical clearance of 52 in. (132 cm) from the deck to the canopy where the operator enters or leaves the seat.

For more details, see National Safety Council Occupational Safety and Health Data Sheet 622, *Tractor Operation and Roll-Over Protective Structures.*

Transportation of Workers

Some jobs require workers to be transported to and from the work site. Because transporting employees can involve special risks, management should take special precautions. When such transportation occurs regularly, a bus or other vehicle designed to carry passengers should be used to avoid many of the particular risks.

Hazards are more likely to exist when transportation of workers is infrequent. Often, under these conditions, an open cargo truck of some kind is used as the transporting vehicle. If so, advise workers of the following safety procedures:

- Getting on. Workers should look before and where they step. Use every handhold available, even a helping hand from someone already on the truck. Step squarely; never at an angle. No one should ever attempt to board a moving vehicle—no matter how slow it may be traveling.
- If possible, provide benches for passengers. In no case, however, should the passengers remain standing while the vehicle is in motion. If necessary, they should sit on the truck bed.
- Avoiding horseplay. Of all the negative actions that a group of people may indulge in, horseplay is one of the most foolish and dangerous. Some companies fire anyone caught in this type of activity. Horseplay, which is inherently dangerous, may be fatal in a moving vehicle.
- Getting off. Again, workers should look before they step, then get off slowly and carefully, using every possible handhold. Under no circumstance should anyone attempt to jump off a moving or stopped vehicle. Many injuries occur as a result of jumping off a vehicle, whether it is moving or not.

Towing

Towing is a hazardous operation, especially when coupling or uncoupling equipment. Workers can be crushed when a truck or other piece of equipment moves unexpectedly while they are between the two machines.

The following safe practices are essential to prevent accidents in the coupling or uncoupling of motorized equipment:

1. No one should go between the vehicles while either one is in motion.
2. Parked vehicles must be secured against movements by having the brakes set, the wheels blocked, or both.
3. A driver should not move a vehicle while someone is between it and another vehicle, a wall, or anything else that is reasonably solid and immovable. In fact, before moving, the driver should receive an all-clear signal.
4. Tow bars are safer than tow ropes. If ropes are used, they must be in good condition and of sufficient size and length for the towing job.
5. Equipment towed on trailers should be secured to the trailer.

POWER SHOVELS, CRANES, AND SIMILAR EQUIPMENT

Safe operation of power shovels, draglines, and similar equipment begins with machine purchase. A good policy is to require in the equipment specifications that guards must cover gears and that the manufacturer must provide oiling devices, handholds, slip-resistant steps, and other safeguards. In any case, before equipment is put into operation, managers or supervisors should inspect it thoroughly and install necessary safety devices.

To keep workers from being caught between truck frames, crawler tracks, cabs, and counterweights of cranes and shovels, erect a barricade to warn nearby employees that they are close to a hazardous area. Signs, flashing lights, and other warning devices also can be used to alert people to the hazards. Barricades are easily removed when the equipment is moved.

Instruct experienced and new operators and maintenance personnel in manufacturer-recommended procedures pertaining to lubrication, adjustments, repairs, and operating practices, and make sure workers observe them (Figure 18–5). A preventive maintenance program for industrial shovels and other equipment is essential for safety and efficiency. Frequent, regular inspections and prompt repairs are the bases for effective preventive maintenance. Generally, the operator is responsible for inspecting the mechanical condition of such items as holddown bolts, brakes, clutches, clamps, hooks, and similar vital parts.

Workers should keep wire ropes, hooks, clamps, and pulleys lubricated in accordance with the manufacturers' instructions, and should inspect them daily. Equipment failures can cause serious accidents. Ropes

Figure 18–4. The scraper *(top)* and bulldozer *(bottom)* have protective bars and screens to protect operators. (Courtesy Caterpillar, Incorporated.)

Figure 18–5. Instruct both experienced and new operators and maintenance personnel in manufacturer-recommended procedures for inspection, maintenance, and repairs.

are particularly likely to develop weakness at the fastenings, at crossover points on drums, and in the sections that frequently contact the sheaves.

Grounding Systems for Powered Equipment

To prevent electrical shock, electrically powered equipment requires a good grounding system to protect workers from electrical faults in trailing cables or at the machine. Though the cable may make close physical contact with the ground's surface, the resistance to the flow of current from the equipment frame and cable to the earth usually is high because of wire insulation.

A machine-to-ground fault resistance of 100 ohms and a current of 10 amperes means an electric shock hazard of about 1,000 volts. A leakage fault current of 7–15 milliamperes is very painful and can result in the loss of muscular control. As little current as 70–200 milliamperes through the body can result in death. (See Electrical Equipment in Chapter 10, Electrical Safety, in this volume.) Because these low leakage currents can be forced through wet skin by commercial 120 volts AC, no employee should be exposed even momentarily to this electrical hazard.

A good earth-ground system may be made by driving copper-clad steel rods, or electrodes, into suitable soil for a distance of at least 8 ft (2.4 m). Rods, which are available commercially for this purpose, lower the earth resistance and provide a better ground. Since the number of rods and their spacing depend considerably on soil conditions, a company may have to increase the conductivity of the soil by applying common salt, sodium nitrate, or similar chemicals, which then are carried into the soil by rain. A grounding system having a total resistance of 1 ohm, including the cable, can be obtained in many areas through proper design and construction.

The pole line ground wire should have at least the same wire gauge size as the power wires. Wherever current is tapped from the power line, a connection is made from the pole ground wire to a ground wire in the cable. A good cable has metal shielding outside the insulating material around each conductor, and the ground conductor is in full electrical contact with the shielding. The shielding should have electrical continuity, and, if broken, should be bridged.

The ground wire in the cable is connected to the equipment frame where there should be good metal-to-metal contact. If necessary, scrape off paint or other covering to achieve good contact. A resistor between the pole line ground wire and the transformer neutral limits the amount of current to no more than 50 amperes, eliminating dangerous voltages at the powered equipment and permitting sufficient current to open the circuit breakers.

The neutral ground system permits the operation of all equipment except the machine where the fault occurs. The machine is segregated from the rest of the system when the fault immediately activates the circuit breaker. Suitable switching equipment in the neutral grounded system eliminates the hazard of several faults occurring at the same time in different phases at different locations, except for the interval required for the circuit breakers to open. Do not connect the equipment ground to the substation ground in any way; this will prevent a fault in the power supply from energizing the equipment.

Inspect circuit breakers and other devices in the grounding system monthly, and regularly check the resistance of ground rods. A megohm meter test of the cable insulation is a recommended part of cable inspection. Workers should not tape defective cable insulation; instead, replace the cable. Workers should wear rubber gloves and use insulated tongs or hooks for handling trailing power cables.

Minimum wear and damage to trailing cables is important for safety. Protect them from blasting operations as much as possible while keeping them as close to operations as practical so that a minimum length of cable is required. Workers can use tripods and wooden construction horses to keep cables off the ground. Tripods are preferable to trenches where cables cross roads.

Assign an electrician to regularly inspect and maintain the electrical parts of shovels and similar equipment, including trailing cables.

Maintenance Practices

If a power shovel is operating in a deep excavation, workers should make repairs or adjustments with the shovel in a safe position where it will not be endangered by falling or sliding rocks or earth. The operator is responsible for setting the brakes, securing the boom, lowering the dipper or bucket to the ground, taking the machine out of gear, and, before leaving the machine, exercising similar precautions to prevent accidental movement.

Before repairing any vehicle, maintenance personnel should notify the operator about its nature and location. If the work is to be done on or near moving parts, the operator should lock out the controls. Only authorized persons can remove the lock. This precaution is essential to prevent the operator from starting the equipment inadvertently. Before inspecting equipment, or checking lubricant levels, the worker should make sure the machine is safe. A caution sign, such as "inspector underneath" can be used.

If some machine parts must be in motion while employees are working on them, make sure they are turned slowly, by hand if possible, in response to guidance or on a prearranged signal if two or more persons must be involved. This precaution applies particularly to those who work around gears, sheaves, and drums. Workers who grasp ropes just ahead of the sheaves risk having their hands jerked into the sheaves. To prevent hand injuries, use a bar to guide any rope being wound on a drum.

If guards must be removed for convenience in making repairs, the job cannot be considered complete until workers replace the guards, plates, and other safety devices. Also, repair personnel should wear snug-fitting clothing, eye protection, and safety shoes.

Operating Practices

Slides of rock and material from high faces and banks cause some of the worst accidents involving power shovels. In some instances, slides have buried the shovel and operator, and in other cases, falling material has struck and injured workers while they were working around the equipment.

Some quarries that have high faces limit the height of banks to 25 ft (7.6 m) by benching and by blasting. These procedures force enough rock out from the face to reduce the pile's height. As a result, the shovel operator is able to maintain the back at a safe slope. When loading under a high face, the operator should swing the shovel to the clear side and away from the face, thereby providing a better view and reducing the injury potential.

Undercutting banks of earth, sand, gravel, and similar materials is dangerous, especially during winter and spring months. Freezing and thawing can result in a collapse of the overhanging material. To maintain a safe slope, the crew should blast the overhanging material and remove it.

The shovel operator is responsible for the safety of other employees who must work near the shovel. These workers may be struck by falling rock, squeezed between the shovel and the bank or similar pinch points, or struck by the dipper. No worker should enter a dangerous location without first notifying the operator who, in turn, should not move the equipment while the employee is passing by the machine.

Operators should fill dippers to capacity but not to overflowing, thus preventing excess material from falling on workers or spilling out of the dipper. Insofar as possible, loading should be done from the blind side. The operator should not swing a load over a vehicle nor load a truck until its driver has dismounted and is in the clear. The only exception to this rule is if the truck has a canopy designed for the protection of the driver and trucks. Rail cars should be loaded evenly so earth or rocks do not overhang the sides.

Supervisors should instruct workers to observe housekeeping practices on and around the shovel. The operator should store tools in an assigned place and keep the cab floor free of grease and oil. Ice and snow should be removed promptly, and a bulldozer should keep the area around the shovel free of rocks and ruts.

Workers should get on or off a shovel or dragline only after they have notified the operator, who, in turn, should swing the platform so workers can grasp and use steps or tread. No one should get on or off by jumping onto the tread, either while the operator is making a swing or while the equipment is stationary. Do not permit any unauthorized person on a shovel or dragline.

Mobile Cranes

The outstanding characteristic of accidents involving crawler and similar types of cranes is the severity of the injuries. Although these accidents occur with relative infrequency, the resulting injuries are about twice as serious as those caused by accidents involving other types of heavy equipment. For this reason, supervisors and managers should select crane operators on the basis of their intelligence, stability, and willingness to follow instructions.

Operators are largely responsible for the safe condition of the crane. They must regularly inspect brakes, ropes and their fastenings, and other vital parts and promptly report worn, broken, and defective parts. Like other equipment, mobile cranes should be maintained on a regular schedule.

Operators are also accountable for the safety of the oiler and bear a large measure of responsibility for preventing injuries to hookers or riggers and others working around the equipment. However, anyone working in the vicinity of a crane should learn to stay clear of the boom. In no case should anyone work or cross under the boom.

Some of the most serious accidents result from overloading cranes. Operators should never exceed the manufacturer-specified load limits for various positions of the boom. Post these load limits conspicuously in the crane cab. If operators have any doubts about the weight of a load, they should test crane capacity by first lifting the load slightly off the ground. Operating a crane on soft or sloping ground is dangerous. Always make sure the crane is level before putting it into operation. Outriggers give reliable stability only when used on solid ground. The use of makeshift methods to increase the capacity of a crane, such as timbers with blocking, is too dangerous to be permitted.

Accidents usually occur when the operator is performing more than one operation and becomes confused, distracted, or excited. Also, clutch linings may swell during wet weather, and the master clutch or the boom clutch, or both, may drag and cause the boom to be pulled over backwards. Operators should test the clutches before starting work on rainy days and adjust the clearances if necessary. Another accident cause is the sudden release of a load when the boom angle is high, for example, from the parting of a sling.

Boom stops limit the travel of the boom beyond the angle of 80 degrees above the horizontal plane and prevent the boom from being pulled backwards over the top of the machine by the boom-hoisting mechanism or by the sudden release of a heavy load suspended at a short radius. Either of these occurrences usually results in serious damage to the equipment and injuries to the operator or other workers.

Boom stops are best suited to medium-sized cranes (the 5- to 60-ton range). Boom stops should disengage the master clutch or kill the engine and stop the boom before it reaches the maximum permissible angle. One type of stop meeting these requirements has a piston and cylinder. It is spring or pneumatically actuated and is mounted on the A-frame to intercept the boom as high above the boom hinges as possible. By positive displacement of an actuator mounted on the A-frame, the boom action disengages the master clutch (or ignition breaker or compression release) by means of light rope reeved over a few small sheaves.

When a mobile crane operator must work near electric power lines, the supervisor should consult with the power company to determine whether the line can be deenergized. Most fatalities result from contact with power lines, and often the power company's service is seriously disrupted. Various states and OSHA have enacted legislation specifying the distances operators must keep their booms and wire ropes from power lines. A minimum of 10 ft (3 m) is often specified; however, the recommendations of the power company and legal requirements should be observed (Figure 18–6).

An experienced operator working with an untrained or relatively inexperienced hooker or rigger should direct the details of lifts, such as the type of sling and hitch to be used.

Figure 18–6. When mobile cranes must be operated in the vicinity of electric power lines, first consult with the utility company to determine whether the lines should be deenergized. OSHA and many state regulations require that booms and wire ropes be kept a minimum of 10 ft (3 m) away from the power lines. (Courtesy Washington State Department of Transportation.)

Although the operator usually can rely on the knowledge of an experienced rigger, the operator is responsible for the safety of a lift and can ask the supervisor for assistance.

The following safe practices are essential when handling loads.

1. Operators should center the hook over a load to keep the load from swinging when lifted.
2. Employees should keep their hands out of the pinch point when holding the hook or slings in place while the slack is taken up. A hook, or even a small piece of board, may be used for this purpose. If workers must use their hands, they should hold the sling in place with the flat of their hand.
3. The hooker, rigger, and all other personnel must be in the clear before a load is lifted.
4. Tag lines should be used for guiding loads.
5. Hookers, riggers, and others working around cranes also must keep clear of the swing of the boom and cab.
6. No load should be lifted or moved without a signal. Where the entire movement of a load cannot be seen by the operator, as in lowering a load into a pit, a trained person should be posted as a guide. To avoid confusion, use only standard hand signals.

GRADERS, BULLDOZERS, AND SCRAPERS

Many of the basic safety measures recommended for trucks also apply to graders and other types of earth-moving equipment. Operators should regularly inspect their machines and promptly report any defects and malfunctioning systems or parts. Scheduled maintenance increases the safety and efficiency of the equipment. Management should select only physically and mentally qualified individuals as operators and train them in correct operating practices, as specified in the manufacturer's manual and by company requirements. Special precautions will help to prevent injury when workers are servicing and repairing machines. In addition, employees should observe the same safety procedures that apply to other types of motorized equipment.

Maintenance

Regularly inspect brakes, controls, engine, motors, chassis, blades, blade holders, tracks, drives, hydraulic mechanisms, transmission, and other vital parts. Likewise, frequently check wheel and engine-mount bolts for tightness.

Making adjustments and repairs with the engine running is a dangerous practice, particularly when employees are working near the engine fan or adjusting a tractor clutch. Refueling should be done only with the engine stopped. Also, when inflating tires, workers must guard against the danger of a locking ring blowing off the tires.

When replacing cutting edges, operators should always block up the scraper bowl or dozer blade. After the scraper has been lifted to the desired height, place the blocks under the bottom near the ground plates. Raise apron arms to the extreme height and place a block under each arm, so the apron can drop enough to wedge each block firmly in place.

Before receiving wire rope on a drum or through sheaves, the operator should disengage the master clutch, idle the engine, and lock the brakes. The engine should be at a complete stop before working with the rope on a front-mounted drum.

If an operator is assisting a maintenance worker and working behind the scraper with the tailgate in the forward position, a block should be placed behind the tailgate so it cannot fall. This precaution is necessary in case someone should release the power control brake, permitting the tailgate to come back. When replacing ropes on scrapers, make sure the tailgate is back at the end of its travel.

General Operating Practices

Operators must look to the front, sides, and rear before moving their machine. They also should be constantly alert for employees on foot when operating near other equipment, offices, tool and supply buildings, and similar places.

Speeds are governed largely by conditions. Slow speeds are essential when driving (1) off the road and beyond the shoulder, on steep grades, and at rough places to avoid violent tilting that may throw the driver off the machine or against levers and cause serious injury; (2) in congested areas; and (3) under icy and other slippery conditions. No one other than the operator should be permitted on the vehicle at the same time.

Jumping from a standing machine can result in sprained ankles, back injuries, and other mishaps. The safe practice is to step down after looking to make sure footing is secure and no other vehicles are approaching. Ice, mud, round stones, holes, and similar conditions represent falling hazards. Likewise, deck plates and steps on equipment should be free of grease and other slipping hazards.

An operator should not drive the equipment onto a haul road without first stopping and looking both ways, whether or not the place of entry is marked with a stop sign. Generally, loaded equipment is given the right of way on job or haul roads.

Before an operator leaves the equipment, even for a short time, the bucket or blade should be lowered to the ground and the engine stopped. A safe parking location is on level ground, off a roadway, and out of the way of other equipment. The operator should never leave equipment on an inclined surface or on loose material with the engine running—the vibration can put the equipment in motion.

Procedures on Roadways

When graders, scrapers, and other earth-moving equipment are in operation along a section of a road, the precautions discussed next will help prevent accidents to the public, employees, and equipment.

Warn oncoming and outgoing traffic of danger ahead by placing barrier signs at both ends of the road section undergoing construction. Primary warning signs, such as ROAD UNDER CONSTRUCTION or BARRICADE AHEAD, should be placed 1,500 ft (460 m) from the starting point of the operation.

Orange flags or markers at the ends of blades, which may project beyond the tread of a machine, serve to warn persons and other equipment operators. An orange flag on a staff projecting at least 6 ft (1.8 m) above the rear wheel of a blade grader or mowing vehicle is recommended for operation in hilly country (Figure 18–7). Affix slow-moving vehicle emblems at the rear of

Figure 18–7. Mowers should have rollover protective structures (ROPS), slow-moving vehicle emblem, and an orange flag projecting at least 6 ft (1.8 m) high. (Courtesy Navistar International Corp.)

Figure 18–8. Close-up of a slow-moving vehicle (SMV) emblem.

these vehicles if operators must drive them even short distances on public roads (Figure 18–8).

Operators of motor graders should keep to the right side of a roadway. When blading against traffic is necessary, use flags and barricades to warn oncoming traffic. Warning signs must be placed at a considerable distance from the work area. This distance increases as the highway speed increases. Suggestions are given in the DOT Manual on *Uniform Traffic Control Devices for Streets and Highways,* ANSI D6.1. Most states use the latter standard (DOT) as minimum requirements, with additional traffic control where conditions dictate. When operations are extensive, personnel with warning devices should stand at each end of the working area so they are visible to oncoming traffic for at least 500 ft (152.5 m).

When earth-moving equipment is stopping, turning, or backing at curves, crests of hills, and similar dangerous locations, station signals or personnel with flags or other devices to warn other motorists. For safety, such movements generally require a clear view of approaching traffic for a distance of about 1,000 ft (305 m).

Use signal personnel where the working area is congested by other equipment, workers, building, excavation, and similar hazards. See Figure 18–9 for hand-signaling procedures.

Coupling and Towing Equipment

An operator should not back up to couple a tractor to a scraper, sheepsfoot roller, or other equipment without having first checked to make sure that everyone is in the clear. An operator assisted by a person on the ground should not move the equipment until signaled.

Before an employee is allowed to couple the trailing equipment, he or she should stop the tractor, put the shift lever in neutral, and set the brakes. Also, block the wheel of the equipment to be coupled.

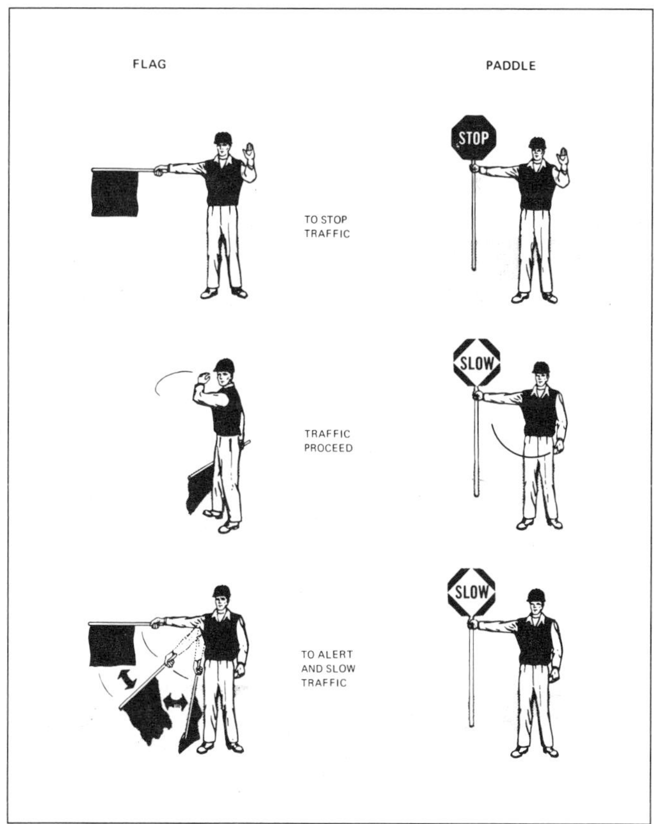

Figure 18–9. Hand-signaling devices used by flagger. Flag and background of STOP sign are bright red; diamond of SLOW sign, orange. (From ANSI D6.1–1978.)

Secure all equipment being towed not only with a regular hitch or drawbar but also with a safety chain attached to the pulling unit. A drawbar failure can result in a serious accident. When towing a scraper from one job to another, the operator should use a scraper bowl safety latch, or place a safety bolt in the beam to give maximum clearance for road projections such as at crossings. This precaution prevents the bowl from striking the ground or pavement and injuring persons or damaging equipment.

Clearing Work

Work requiring exposure to low limbs of trees or to high brush involves serious hazards. These risks can be readily eliminated by suitable protective measures and safe practices.

When using a bulldozer, equip it with a heavy, well-supported, arched steel-mesh canopy to protect the operator. (See section on Protective Frames for Heavy Equipment, earlier.) The operator also should wear goggles to shield the eyes from whipping branches.

Head protection guards against injuries from falling branches. When a bulldozer shoves hard against the butt of a large dead tree, the tree may crack in the middle or limbs may fall onto the machine. Dead branches or tops also can drop from live trees. A safe procedure to eliminate the danger is to cut the roots on three sides and then apply the power to the fourth side. Use a long rope to pull over large trees, but make sure in advance that the tractor and operator will be in the clear when the tree falls. Operators should ensure that all workers in the area are out of harm's way before pushing over any trees, bulldozing rock, and rolling logs.

Special Hazards

Fatalities can occur easily when equipment is used on dumps and fills, near excavations, and on steep slopes. The operator should keep the bulldozer blade close to the ground for balance when the machine is traveling up a steep slope.

When a worker is driving a tractor-dozer down a slope, the person should doze three or four loads of dirt to the edge of the slope, keeping the loads in front of the blade. If the dirt is lost on the way down, the operator should not lower the blade to regain the load because of the danger of overturning. Never use the blade as a brake on a steep slope except in cases of extreme emergency.

Grounds conditions will determine how close to an excavation or the crest of a dump an operator can safely work a machine. Wet weather means the operator must work at a greater distance from the edge or crest. When the ground is treacherous, assign someone to signal the driver.

Sometimes employees, the public, livestock, and property are endangered when material is pushed over the edge during side hill work. In such cases, make sure there is sufficient clearance below before the work begins.

SUMMARY

- To prevent accidental damage to heavy equipment, companies must maintain safety features and equipment, train operators, and train repair and maintenance personnel. Heavy-duty equipment requires adequate haul roads that have been constructed especially for these vehicles.
- Only qualified drivers should operate heavy-duty equipment. All heavy-duty vehicles must be equipped with substantial guards, shields, canopies, and grilles to protect the operator.
- When workers are transported, supervisors must make sure that employees know safety procedures for riding and getting on and off the vehicles. Likewise, towing requires safe practices to prevent accidents in the coupling and uncoupling of motorized equipment.
- The operation of power shovels, draglines, and similar equipment requires such safeguards as oiling devices, handholds, and slip-resistant steps. Operators are responsible for the safe working of their equipment, for inspecting all parts and reporting any problems, and for the safety of those working in areas around the machines.
- Electric-powered equipment must have a good ground system to protect workers from electrical hazards and shock. Workers also should wear protective equipment when handling power cables or other potentially dangerous equipment.
- Companies must carefully train employees in safe practices for maintenance and repair of powered equipment and in safe operating procedures. All employees should observe good housekeeping practices.
- Mobile crane accidents are infrequent but result in more severe injuries to workers. To help prevent accidents, operators must be trained to handle loads safely.
- Graders, bulldozers, and scrapers must be inspected regularly and any problems immediately corrected. As with other heavy-duty equipment, only those operators who have been carefully selected and trained should drive these vehicles.
- Operators and other employees must be trained in the proper procedures for using towing and coupling equipment and for doing clearing work.
- Operators and other workers on a construction or work site must be aware of the special hazards associated with heavy-duty equipment. Supervisors must ensure that employees observe safety practices at all times.

REFERENCES

Alliance of American Insurers, 1501 Woodfield Road, Suite 400 W, Schaumburg, IL 60173.
Code of the Road.

American Automobile Association, 1000 AAA Drive, Heathrow, FL 32746.
"Driver Training Equipment" (catalog).

American National Red Cross. *Standard First Aid and Personal Safety Manual,* 2nd ed. Washington, DC: American National Red Cross, 1979.

American National Standards Institute, 11 West 42nd Street, New York NY 10036.
Powered Industrial Trucks, ANSI/NFPA 505–1999.
Powered Industrial Trucks—Safety Standard for Low Lift and High Lift Trucks, ANSI/ASME B56.1–1993.

American Trucking Associations, Inc., 2200 Mill Road, Alexandria, VA 22314.

Association of Casualty and Surety Companies, 110 William Street, New York, NY 10038.
Guide Book, Commercial Vehicle Drivers.
Truck and Bus Drivers Rule Book.

National Safety Council, 1121 Spring Lake Dr., Itasca, IL 60143-3201.
Defensive Driving Program Materials.
Injury Facts (formerly *Accident Facts,* published annually).
National Fleet Safety Contest.
Motor Fleet Safety Manual, 4th ed., 1996.
Occupational Safety & Health Data Sheets (available in the Council Library):
Berms in Pits and Quarries, 12304–680, 1990.
Falling or Sliding Rock in Quarries, 12304–332, 1990.
Motor Trucks for Mines, Quarries, and Construction, 12304–330, 1990.
Operation of Power Shovels, Draglines, and Similar Equipment, 12304–271, 1992.
Snow Removal and Ice Control on Highways, 12304–638.
Public Employee Safety and Health Management, 1990.
Safe Driver Award Program.
"Street and Highway Maintenance."
Supervisors' Safety Manual, 9th ed., 1997.
"Vehicular Equipment Maintenance."

New York University Center for Safety Education, Washington Square, New York, NY 10003.
Publications list.

National Private Truck Council. *Driver Training Manual.* Washington, DC: National Private Truck Council, 1981.

U.S. Department of Defense, Department of the Army, The Pentagon, Washington, DC 20310.
Driver Selection and Training, TM 21–300.
Drivers' Manual. TM 21–305.

U.S. Department of Transportation, 400 Seventh Street SW, Washington, DC 20590.
Bureau of Motor Carrier Safety Regulations.
Manual on Uniform Traffic Control Devices for Streets and Highways. (Also identified as American National Standard D6.1.)
Model Curriculum for Training Tractor-Trailer Drivers.
Title 49. *Code of Federal Regulations,* Parts 390–397, *Motor Carrier Safety Regulations.*

U.S. Government Printing Office, North Capital and H Streets NW, Washington, DC 20401.
Code of Federal Regulations, Title 49—"Transportation."

REVIEW QUESTIONS

1. What signal should be made before allowing a truck to go forward?
   a. The driver should flash the truck's lights.
   b. The driver should blow two blasts on the truck's horn.
   c. The driver should turn on the truck's hazard light.
   d. The signaler should use the proper hand signals.

2. What are the three purposes of having the signaler stand on the driver's side of the vehicle during dumping procedures?
   a.
   b.
   c.

3. How much weight should the canopy and the canopy support of heavy machinery be able to bear?

4. What are five guidelines that should be followed when (un)coupling motorized equipment?
   a.
   b.
   c.
   d.
   e.

5. Ground rods or electrodes should be driven at least ______________ into the ground.
   a. 3 ft
   b. 5 ft
   c. 8 ft
   d. 12 ft

6. What is the purpose of boom stops?

7. List three of the six essential safe practices for handling loads.
   a.
   b.
   c.

8. What three conditions make driving at slow speeds extremely important?
   a.
   b.
   c.

9. What are the precautions that should be taken before stepping down from standing machinery?

10. When traveling up a slope, the operator should keep the bulldozer blade ______________ for balance.
   a. even with the top of the cab
   b. at a 75-degree angle
   c. hitched to a towbar
   d. close to the ground

Part Four

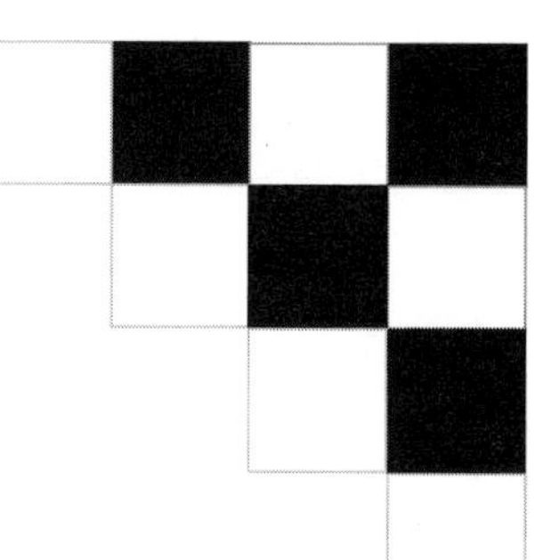

PRODUCTION OPERATIONS

The chapters in Part Four explore the hazards and injury prevention strategies associated with selected production activities. Handheld and portable power tools are significant sources of exposure to workers in a variety of production operations. Potential hazards in the manufacturing sector arise from tools and machines that weld and cut hot and cold metals. Other sources of potential exposure to hazards have developed in automated lines, systems, and processes. These chapters describe administrative controls, safe procedures, and mechanical safeguards that can reduce the hazards and incidence rates in these production operations.

19

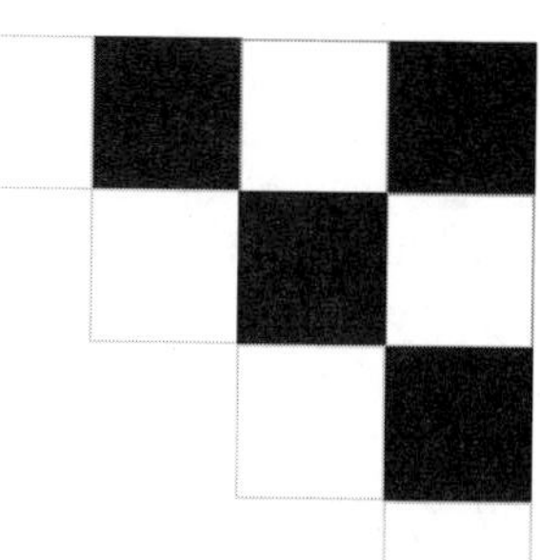

Hand and Portable Power Tools

Revised by

John Kurtz, International Staple, Nail and Tool Association

George Welchel, Power Tool Institute

Philip E. Hagan, MPH, CIH, CHMM, CHCM

Hand and power tools enable employees to apply additional force and energy to accomplish a task. These tools improve efficiency and make better products. Because of the increased force of hand and power tools, the objective of safety with these tools is to protect users from inflicting harm to themselves and others as well as to provide ergonomically designed tools (see References). Through proper selection, use, care, and supervision of hand and power tools, injuries from tools can be prevented.

This chapter covers the following topics:

- safety practices and work procedures for preventing unintentional incidents using hand and power tools
- safety issues in the repair and maintenance of these tools
- safe work methods for handling various hand tools
- hazards and safety precautions for handling portable power tools
- personal protective equipment used with hand and power tools

PREVENTING INCIDENTS

Disabilities resulting from misuse of tools or using damaged tools include loss of eyes and vision; puncture wounds from flying chips; severed fingers, tendons, and arteries; broken bones; contusions; infections from puncture wounds; ergonomic stress; as well as many other injuries.

Tool Selection

Consider all aspects of the work situation when selecting hand and portable power tools. The tools selected should (1) perform the job and (2) be usable by the employee, that is, the tool should not cause the employee any physical pain or discomfort. Ergonomics factors should be considered in tool selection. The checklist in Figure 19–1 will assist in evaluating hand tools from an ergonomics perspective. Seeking employee input in the selection of tools is also helpful.

Consider the following key points when selecting tools:

- The handle's shape and form and the material used to make it can minimize stress to employees' hands and upper body. An ergonomically well-designed tool will reduce the incidence of fatigue and injury and may improve productivity as well.
- Quality of the tool, including sharpness of the cutting edges, affects the amount of force needed to do a job. Jobs using the same tool can vary greatly in the amount of force that employees must apply.
- Power tools designed to have minimal vibration will be more comfortable to use and less likely to result in hand-arm vibration syndrome (HAVS)—also known as Reynaud's syndrome and vibration white finger. (HAVS causes numbness and blanching of the hands, and can progress to complete disability if the worker is not removed from exposure.) Provide pads and gloves for relief from the vibration; administrative controls, such as work rotation and more frequent rest breaks, can be used to reduce the time workers use vibrating tools.
- If possible, set up some foot-operated controls to avoid repeated use of hands and fingers. Excessive repetitive motion can also cause damage to the foot and leg muscles, nerves, and/or tendons.

Before placing tools with employees, also consider the workstation and the work methods. Workstations should be adjustable for each employee. In addition, workstations should allow for the full range of movements required for the job and tools. When possible, provide mechanical means of handling materials. To reduce employees' stress, allow employees to perform a variety of jobs that require use of different muscle groups.

Work methods should achieve maximum production and cause a minimum of stress to employees. Consider the following factors:

- force needed to hold and use a tool
- direction of the force
- weight of materials
- number of repetitions of an activity
- employee posture.

Changing Tools

To determine if tools should be changed, consider the following factors:

- employees' concerns about tool problems
- the facility's injury and medical records that implicate tools
- work methods
- the setup of workstations
- trends for particular jobs.

Some changes that may be necessary include providing personal protective equipment, especially gloves or pads for hand tools. Other possible changes might include rest breaks, adding employees to difficult jobs, job rotation, and body conditioning to relieve stress and fatigue. The ergonomic aspects of hand-tool design and

Hand Tool Analysis

No responses indicate potential problem areas which should receive further investigation.

	Yes	No
1. Are tools selected to limit or minimize		
• exposure to excessive vibration?	❑	❑
• use of excessive force?	❑	❑
• bending or twisting the wrist?	❑	❑
• finger pinch grip?	❑	❑
• problems associated with trigger finger?	❑	❑
2. Are tools powered where necessary and feasible?	❑	❑
3. Are tools evenly balanced?	❑	❑
4. Are heavy tools suspended or counterbalanced in ways to facilitate use?	❑	❑
5. Does the tool allow adequate visibility of the work?	❑	❑
6. Does the tool grip/handle prevent slipping during use?	❑	❑
7. Are tools equipped with handles of textured, nonconductive material?	❑	❑
8. Are different handle sizes available to fit a wide range of hand sizes?	❑	❑
9. Is the tool handle designed not to dig into the palm of the hand?	❑	❑
10. Can the tool be used safely with gloves?	❑	❑
11. Can the tool be used by either hand?	❑	❑
12. Is there a preventive maintenance program to keep tools operating as designed?	❑	❑
13. Have employees been trained		
• in the proper use of tools?	❑	❑
• when and how to report problemss with tools?	❑	❑
• in proper tool maintenace?	❑	❑

Source: NIOSH. *Elements of Ergonomics Programs.* March 1997.

Figure 19–1. A checklist such as this sample can assist in the evaluation of ergonomics factors of hand tools.

use are discussed in many of the resources listed in the chapter References. (Electrical concerns and the effects of tool vibration are addressed later in this chapter. See the sections titled Electrical Tools and Pneumatic-impact tools, respectively.)

Safety Practices

By observing the following six safety practices, most unintentional incidents with hand tools and portable power tools can be eliminated:

1. Provide proper protective equipment and have employees wear it. Eye and face protection prevents injuries from flying objects or liquids. Hand and arm protection prevents injuries from flying or sharp objects. Respiratory protective equipment can provide protection from particulates and fumes, but should be used only in conjunction with a respiratory protection program as described in 29 *CFR*, 1910.134. Power tools can also create noise levels above regulatory standards. When noise monitoring indicates that noise levels exceed OSHA standards, it will be necessary to include noise-exposed employees in a hearing conservation program with annual audiometric testing and use of hearing protection. Companies also should consider use of less noisy models of tools or other noise controls where noise levels exceed 90 dBA.
2. Select the right tool for the job. Examples of unsafe practices include (1) striking hardened striking faces of hand tools together, such as using a hammer to strike another hammer or hatchet; (2) using a claw hammer to strike a steel chisel; (3) using a file or a screwdriver to pry; (4) using a wrench instead of a hammer; and (5) using pliers instead of the proper wrench.
3. Know if a tool is in good condition and keep it in good condition. Unsafe tools include wrenches with cracked or worn jaws; screwdrivers with broken tips, or split or broken handles; hammers with chipped, mushroomed, or loose heads, and broken or split handles; mushroomed heads on chisels; dull saws; and extension cords or electrical tools with broken plugs, improper or removed grounding systems, or split insulation.
4. Properly ground power tools, using a ground-fault circuit interrupter (GFCI) protected circuit. Per 29

CFR 1926.404(b)(1)(i), the empolyer shall use either GFCI or an assured equipment ground program for branch circuits at a construction site.

5. Use tools correctly. Some common causes of injuries are (1) screwdrivers applied to objects held in the hand, (2) knives pulled toward the body, (3) failure to ground electrical equipment, (4) nail hammers striking hardened tools, and (5) using tools when work is not properly secured. (For proper uses of hand tools, see Use of Hand Tools, in this chapter.)
6. Keep tools in a safe place. Many injuries are caused by tools falling from overhead. Another source of injuries is leaving the cutting edge of knives, chisels, and other sharp tools exposed when carrying them in pockets or leaving them in tool boxes.

A safety program designed to control unintentional incidents involving hand and power tools should include the following activities:

- Train employees to select the right tools for each job, and see that the tools are available.
- Establish regular tool inspection procedures (including inspection of employee-owned tools), and provide good repair facilities to make sure that tools are kept in safe condition.
- Train and supervise employees to correctly use tools.
- Establish a procedure to control company tools. For example, set up a check-in and check-out system that evaluates tools' condition and ensures that only properly conditioned tools leave the tool-storage area.
- Provide proper storage areas in the tool room and at the workstation.
- Enforce the use of proper personal protective equipment.
- Plan each job well in advance.

Each supervisor should also check all operations to determine if special tools will do the work more safely than ordinary tools. Have special tools readily available for employees.

Central Tool Control

From the standpoint of incident prevention, central tool control assures uniform inspection and maintenance of tools by a trained employee. This employee can also distribute the correct type of personal protective equipment, such as eye and face protection, when a tool is issued.

A central control area and effective recordkeeping on tool failure and other causes of injuries help locate hazardous conditions. A central area also assures better control than does scattered storage. In that way, tools are exposed to less damage and deterioration and are not as likely to fail or create other hazards. Some companies issue each employee a set of numbered checks that are exchanged for tools from the supply room. With this system, the attendant knows the location of each tool and can recall it for inspection at regular intervals. Tools used and assigned to a workstation should be checked either between shifts by a qualified person or prior to each shift by the tool user.

The tool control attendant can help promote safety by:

- recommending or issuing the right type of tool,
- encouraging employees to turn in damaged or worn tools,
- tagging and removing from service defective tools, and
- encouraging the safe use of tools.

Set up a procedure so the tool control attendant can send tools in need of repair to a department or service firm thoroughly familiar with methods of repairing and reconditioning.

Companies that perform work at scattered locations may find that it is not always practical to maintain a central tool control area. In such cases, the supervisor should frequently inspect all tools and remove from service those found to be damaged. Many companies have each supervisor check all tools every week. Draw up a checklist for the most hazardous hand tools so inspection will be consistent (Figure 19–2).

Some workers prefer to use their own tools even though tools are furnished by the company. In this case, supervisors should examine these tools frequently to prevent the use of any that are unsafe. If worker-owned tools are found to be damaged or unsafe, supervisors should insist that they be replaced. Regulatory requirements usually state that the employer is responsible for seeing that safe tools are used, including tools and equipment that are furnished by the employee.

Tool Boxes

Tool boxes are meant to hold tools, not to be stood on, used as an anvil or a saw horse, or as a storage place for lunches. Lightweight tool boxes are made of plastic or steel, but strong, heavy-duty tool boxes are usually made of steel.

Portable Tool Boxes

Portable tool boxes may have up to five drawers, a lift-out tray, and possibly a cantilevered tray that automatically opens out when the cover is lifted. All seams should be welded and smooth with no protruding edges to catch clothing or hands. In addition to the handle on the top of the tool box's cover, look for handles at each end for those boxes designed to hold an extra-heavy

PORTABLE ELECTRIC TOOLS

Inspection Checklist

GENERAL	
Low voltage or battery powered equipment used in tanks and wet areas?	☐
Tools well maintained?	☐
Motors in good condition?	☐
Approved tools used in explosive atmospheres?	☐
Tools left where they cannot fall?	☐
CORDS	
Insulation and plugs unbroken?	☐
Cords protected against trucks and oil?	☐
Cords not in aisles?	☐
GROUNDING	
Ground wire fastener in safe condition?	☐
3-wire plug extension cord (if a 3-wire tool)?	☐
Ground wire used?	☐
Defects or minor shocks reported?	☐
Ground fault circuit interrupter used?	☐
GUARDING	
Guards used on grinders and saws?	☐
Movable guards operate freely?	☐
Eye or face protection worn?	☐

Figure 19–2. An inspection checklist card can be used for portable electric tools. Such a card encourages workers to inspect equipment before and after use. More specific cards or tags simplify the prompt recording of defects and result in better maintenance records.

load of tools. A good tool box should have a catch or a hasp at each end and should be able to be locked with either a padlock or its own built-in lock. For outdoor use look for weatherproof construction that will allow rain to drain away without entering the tool box.

Tool Chests

Tool chests are usually heavier, stronger, and, of course, have a greater capacity than tool boxes. On some models, the drawers, sometimes more than 10, can be secured with their own built-in locks. Some have a tote tray that can be removed for carrying only those tools needed for a particular job. Most tool chests are designed to be placed on top of tool cabinets.

Mobile Tool Cabinets

Mobile tool cabinets—the kind on wheels—may have 10 or more drawers. If they are designed to hold a chest, they may sometimes have 20 or more drawers. Look for a locking arrangement in which all drawers lock automatically. Also look for construction that will allow drawers, no matter how heavily loaded, to roll out freely. To prevent rolling, two wheels should lock by means of a brake. Casters should be of ball-bearing construction.

Gang Boxes

Used primarily by the construction industry, gang boxes are generally less organized than tool boxes and cabinets. For this reason, workers and supervisors should perform daily checks on all tools to ensure that any needed repairs are made.

Carrying Tools

Workers should carry their tools to and from work or the workstation in a tool box, cabinet, or other appropriate toolholder or pouch. In these ways, the worker is protected, as well as the tools themselves. Employees should never carry chisels, screwdrivers, and pointed tools with the edges or points up either in their pockets or by hand. They should carry such tools with the points and cutting edges away from their bodies.

Employees should never carry tools in any way that might interfere with the free use of both hands when climbing a ladder or other structure. Instead, have them put tools in a tool belt, pouch, or holder or hoist tools from the ground to the job in a strong bag, bucket, or similar container. Tools should be returned in the same manner—not brought down by hand, carried in pockets, or dropped to the ground.

Mislaid and loose tools cause a large number of injuries. Tools are laid down on scaffolds, on overhead piping, on top of stepladders, and in other locations from which they can fall on persons below. Leaving tools overhead is especially hazardous where there is vibration or where people are moving about.

Employees should hand tools to one another, never throw them. Workers should pass edged or pointed tools, preferably in the tool's carrying case, with the handle toward the receiver.

Workers carrying tools on their shoulders should pay close attention to clearances when turning around. They should handle the tools so they will not strike others.

MAINTENANCE AND REPAIR

When metal tools break during normal use, the causes are usually related to the tools' quality. Therefore, purchase tools of the best quality obtainable.

Inspection and Control

The toolroom attendant or tool inspector should be qualified to determine the condition of tools for further use. Never return a dull or damaged tool to stock. Keep enough tools of each kind on hand so that, when a damaged or worn tool is removed from service, it can be replaced immediately with a safe tool.

Efficient tool control requires periodic inspections of all tool operations. These inspections should cover housekeeping in the tool supply room, tool maintenance, service, number of tools in the inventory, handling routines, and condition of tools. Responsibility for such periodic inspections is usually placed with the department head and should not be delegated to others.

Hand tools receiving the heaviest wear, such as chisels, punches, wrenches, hammers, star drills, and blacksmith's tools, require frequent maintenance on a regular schedule.

Proper maintenance and repair of tools require adequate work space and equipment, such as workbenches, vises, eye and face protection, tools for repair and sharpening, and good lighting. Employees specially trained in the care of tools should be in charge of these work spaces; otherwise send tools out to a qualified shop for repairs.

Redressing Tools

Always follow the tool manufacturers' recommendations to repair, shape, and maintain tools. A properly dressed tool is a safe and efficient tool. Do not redress or reshape tools having chipped, battered, or mushroomed striking or struck surfaces. When a tool has reached this stage through normal use or abuse, discard it.

When performed, redressing should be done by a person who knows the proper dressing information for that particular tool and has the skill to use that information. If a grinder is used, take care to keep from destroying the temper of the cutting edge.

Hatchets

Redress hatchets with double bevels as illustrated in Figure 19–3a. Redress hatchets with single bevels as illustrated in Figure 19–3b.

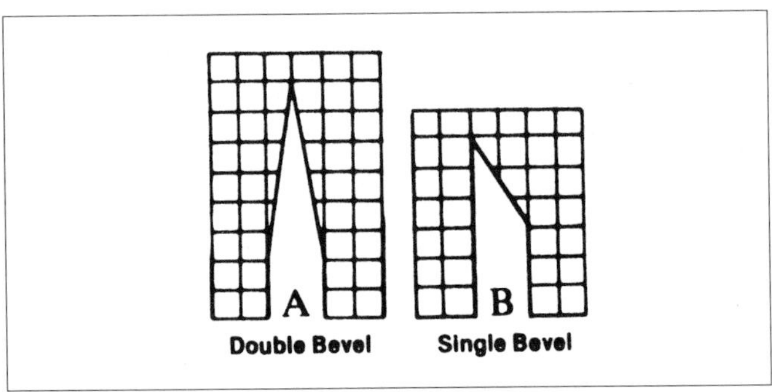

Figure 19–3. Redressing hatchets.

Flat Cold Chisels

Cold chisels are hardened on the cutting edge. Redress them with care to restore them to their original shape or to an included angle of approximately 70 degrees (Figure 19–4).

Hot Chisels

Hot chisels are tools with handles used for cutting hot metal. Redressing instructions are the same as for cold chisels.

Other Machinist's Chisels

Round nose, diamond point, and cape are other chisels commonly used in metalworking. Redressing instructions for them are the same as for flat cold chisels except that the bevel angles are different (Figure 19–4).

Punches

The working end of pin-and-rivet punches and blacksmith's punches should be redressed flat and square with the axis of the tool. The point of center punches should be redressed flat and square with the axis of the tool. The point of center punches should be redressed to an included angle of approximately 60 degrees; prick punches, to an included angle of approximately 30 degrees.

Bricklayer's Tools

Bricklayer's tools should be redressed as follows: hammer blade, 40 degrees; brick chisel, 90 degrees; and brick set, 45 degrees. Bevel slightly to remove feather edge (Figure 19–5).

Screwdrivers

Screwdrivers with a flat blade should be redressed to a flat end at a right angle to the blade. Both sides of the blade at screw contact should also be at right angles to the blade.

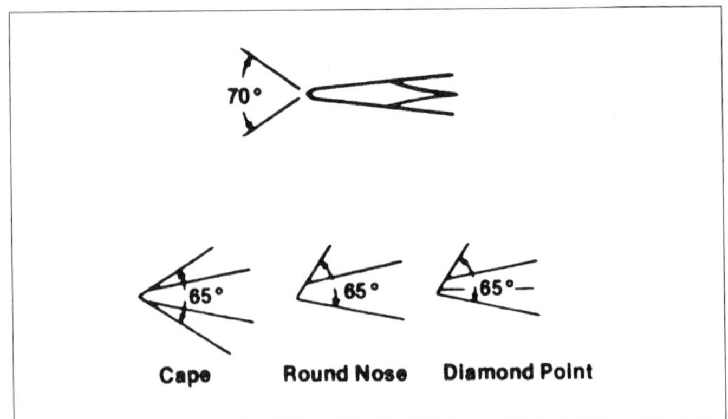

Figure 19–4. Redress cold chisels to restore the angle.

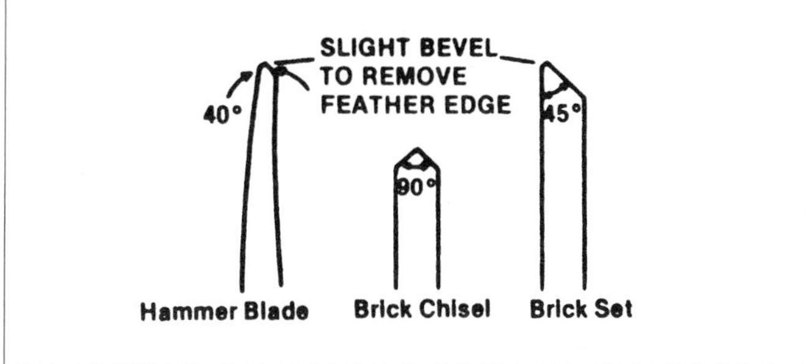

Figure 19–5. Bricklayer's tools.

Files

To keep files' cutting surfaces from clogging, use a file card often during the filing process. Never use a file without a handle. When a file becomes dull, discard it or have it resharpened.

Star Drills

Hand file all cutting edges of star drills to an included angle of approximately 70 degrees (see Figure 19–4 for included angle of 70 degrees).

Handles

Wooden handles of hand tools used for striking, such as hammers and sledges, should be made from the best straight-grained material. Hickory, ash, or maple, neatly finished and free from slivers, is preferred. Alternate materials such as fiberglass or steel with a rubber sleeve may be used.

To make sure that handles are properly attached to the tools, have them fitted to tools only by an experienced person. Poorly fitted handles make it difficult for workers to control their tools, and such handles can be dangerous as well.

Loose wooden handles in sledges, axes, hammers, cold cutters, and similar tools create hazards. No matter how tightly a handle may be wedged at the factory, both use and shrinkage can loosen it. In some cases, tapping the wedges will take up the shrinkage. In others, the head of the tool can be driven back on the handle, the wedges reset, and the protruding end of the handle cut off.

Eventually, any wooden-handled tool will need its handle replaced. Drive the old handle out. Select a new but equivalent handle. If the wood of the new handle does not bear against the head eye at all points, shave the handle until it fits snugly. Then, replace the handle in the tool, and sight along it to be sure that the head is properly centered. After the handle is firmly fitted into the head eye, wedge it according to the original pattern.

USE OF HAND TOOLS

The misuse of common hand tools is a source of many injuries to industrial workers. In many instances, injuries result because supervisors assume that everyone knows how to use common hand tools. Observations and the records of injuries show that this is not the case.

A part of every training program should, therefore, be detailed instruction in the proper use of hand tools. The following sections describe safe practices for using hand tools.

Screwdrivers

The screwdriver is probably the most commonly used and abused tool. Discourage the unsafe practice of using screwdrivers as punches, wedges, pinch bars, or pries. If used in one of these ways, they can cause injury and become unfit for the work they are intended to do. Furthermore, a broken handle, bent blade, or dull or twisted tip may cause a screwdriver to slip out of the slot and cause a hand injury.

Select a screwdriver tip that fits the screw (Figure 19–6). A sharp square-edged blade will not slip as easily as a dull, rounded one, and it requires less pressure. By redressing the tip to its original shape, it may be kept clean and sharp to permit a good grip on the head of the screw. Phillips screwdrivers and many other types of screwdrivers are safer than the flat blade screwdriver because they have less tendency to slip.

When putting in a screw, hold the work in a vise or lay it on a flat surface. This practice lessens the chance of injury to the hands, should the screwdriver slip from the workpiece.

When it is necessary to work around electrical current-bearing equipment, use an insulated screwdriver.

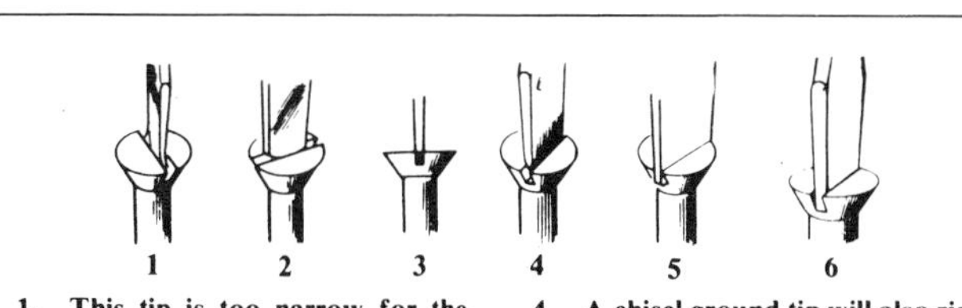

1. This tip is too narrow for the screw slot; it will bend or break under pressure.
2. A rounded or worn tip. Such a tip will ride out of the slot as pressure is applied.
3. This tip is too thick. It will only serve to chew up the slot of the screw.
4. A chisel ground tip will also ride out of the screw slot. Best to discard it.
5. This tip fits, but it is too wide and will tear the wood as the screw is driven home.
6. The right tip. This tip is a snug fit in the slot and does not project beyond the screw head.

Figure 19–6. Select the correct screwdriver to fit the screw.

However, the handle, insulated with dielectric material, is intended only as a secondary protection. Insulated blades are also intended only as a protective measure against shorting out components. Be sure that electrical current is shut off before beginning work.

Hammers

Hammers are made in different shapes and sizes, with different configurations and varying degrees of hardness. Each hammer has a specific purpose. Select hammers for their intended uses and use them only for those purposes. Proper use of most hammers involves the following basic rules:

- Always wear eye protection.
- Always strike a hammer blow squarely, with the hammer's striking face parallel with the surface being struck. Always avoid glancing blows, overstrikes, and understrikes.
- When striking another tool (chisel, punch, wedge, etc.), the striking face of the hammer should have a diameter approximately ⅜ in. (0.9 cm) larger than the struck face of the tool.
- Always use a hammer of suitable size and weight for the job. Do not use a tack hammer to drive a spike, nor a sledge hammer to drive a tack.
- Never use a hammer to strike another hammer.
- Never use a hard-surface hammer to strike another harder surface.
- Never use a hammer with a loose or damaged handle.
- Discard any hammer if it shows dents, cracks, chips, mushrooming, or excessive wear. Redressing is not recommended.

Common Nail Hammers

Designed for driving unhardened common nails, finishing nails, and nail sets, common nail hammers use the center of the hammer's face. Their shape, depth of face, and balance make them unsuitable for striking against metal, especially heavier objects, such as cold chisels.

Nail Hammers

Nail hammers are made in two patterns: curved claw and straight or ripping claw. The face is slightly crowned with the edges beveled. However, certain heavy-duty patterns may have checkered faces designed to reduce glancing blows and flying nails. Handles may be made of wood, tubular or solid steel, or fiberglass. These materials are generally furnished with rubber grips that also occasionally use wooden handles. When drawing a nail from a piece of wood with a nail hammer, place a block of wood under the head to increase the leverage, to reduce the strain on the handle, and to prevent marring of the wood.

Ball Peen Hammers

Ball peen hammers of the proper size are designed for striking chisels and punches. They are also used for riveting, shaping, and straightening unhardened metal.

Sledge Hammers

Many types and weights of sledge hammers are designed for general sledging operations such as striking wood, metal, concrete, or stone. Consult the manufacturer for specific recommendations.

Hand-Drilling Hammers

Hand-drilling hammers are designed for use with chisels, punches, star drills, and hardened nails.

Bricklayer's Hammers

The striking face of bricklayer's hammers is flat with beveled edges. The blade has a sharp, hardened cutting edge. Handles are made of wood, solid steel, or fiberglass. They also may have rubber grips.

Bricklayer's hammers are designed for setting and cutting (splitting) bricks, masonry tile, and concrete blocks. They are also used for chipping mortar from bricks. Never use a bricklayer's hammer to strike metal or to drive struck tools, including brick sets and chisels.

Riveting and Setting Hammers

Riveting hammers are designed for driving and spreading unhardened rivets on sheet-metal work. The setting hammer is designed for forming sharp corners, closing and preening seams and lock edges, and glazing points.

Other Hammers

Consult manufacturers for designs, uses, and safety programs for hammers not mentioned in this section. Other types of hammers include scaling, chipping, soft-face, nonferrous, magnetic, engineer's, blacksmith's, and spalling hammers and woodchopper's mauls.

Punches

Hand punches are made in various patterns from square, round, hexagonal, or octagonal steel stock. Punches are designed to (1) mark metal and other materials that are softer than the punch's point end, (2) drive and remove pins and rivets, and (3) align holes in different sections of materials.

Never use a punch with a mushroomed struck face or with a dull, chipped, or deformed point. Discard any punch if it is bent, cracked, or chipped. Redress the cutting edge's point to its original contour as required.

Tools for Cutting Metal

Cold chisels, hack saws, files, and snips are some hand tools used for cutting metal. Because of the material being worked with, workers must be especially careful when using metal-cutting tools.

Cold Chisels

Cold chisels have a cutting edge at one end for cutting, shaping, and removing metal that is softer than the cutting edge itself, such as cast iron, wrought iron, steel, bronze, and copper. There is a struck face on the other end of the cold chisel.

A chisel can be used to cut metal in hard-to-reach places; it will cut any metal that is softer than its own cutting edge. The cutting edge is hardened and kept sharp, at a 70-degree angle; for softer materials, a 60-degree angle is satisfactory because less pressure is required.

Types of Chisels

The following four principal kinds of chisels are used for bench metal work:

1. A flat cold chisel is most commonly used for cutting, shearing, and chipping. The width of the cutting edge determines the size. For ordinary work, a ¾-in. (1.9 cm) chisel and a 1 lb (0.45 kg) hammer are used. The flat chisel should be ground with a slightly convex cutting edge. This reduces the tendency for its corners to dig into the surfaces being chiseled and concentrates the force directly to the material being cut.
2. A diamond-point chisel cuts V-grooves and sharp interior angles.
3. A cape chisel is used for cutting keyways, slots, or square corners.
4. A round-nose chisel cuts rounded or semicircular grooves and corners that have fillets. It can draw back a drill that wandered from its intended center.

Selecting Chisels

Factors determining the selection of a cold chisel are the materials to be cut, the size and shape of the tool, and the depth of the cut to be made. The chisel should be heavy enough so that it will not buckle or spring when struck. Select a chisel just large enough for the job. In that way, the blade is used rather than only the point or corner.

As discussed earlier, use the proper hammer for the job. The striking face of the appropriate type and size hammer should have a diameter approximately ⅜ in. (0.95 cm) larger than the struck face of the chisel.

Holding Chisels

Some workers prefer to hold the chisel lightly in the hollow of the hand with the palm up, supporting the chisel with the thumb and first and second fingers. They claim that if the hammer glances from the chisel, it will strike the soft palm rather than the knuckles. Other workers think that a grip with a loose fist, keeping the fingers relaxed, holds the chisel steadier and minimizes the chances of being hit by glancing blows. Moreover, in some positions, this is the only grip that is natural or even possible. For regular use, a sponge rubber pad, forced down over the chisel, provides a protective cushion for the hand. Punch and chisel holders are commercially available. When shearing and chipping with a cold chisel, the worker should hold the tool at an angle that permits one bevel of the cutting edge to be flat against the shearing plane.

Bull chisels held by one employee and struck by another require the use of tongs or a chisel holder to guide the chisel. In that way, the person holding the chisel will not be exposed to injury. Protective holders are also commercially available.

Maintenance of Chisels

Discard any chisel that is bent, cracked, or chipped. Redress the cutting edge, or struck end, to its original contour as required.

When grinding a chisel, do not apply too much pressure to the head. The heat generated from grinding can draw the temper. Periodically immerse the chisel in cold water when grinding.

Stamping and Marking Tools

Stamping and marking tools of special alloy and design are available. If possible, workers should use holders for marking tools so they do not have to hold their fingers close to the face of the tool being struck.

Tap and Die Work

Tap and die work requires certain precautions. Firmly mount the workpiece in the vise. Use a tap wrench of the proper size. Workers should keep their hands away from broken tap ends. If a broken tap is removed by using a tap extractor or a punch and hammer, the worker should wear eye protection. When a long thread is being cut with a hand die, workers should keep their hands and arms clear of the sharp threads coming through the die.

Hack Saws

Adjust and tighten hack saw blades in the frame to prevent buckling and breaking. However, they should not be so tight that the pins that support the blade break off. Install blades with teeth pointing forward.

Use blades with 14 teeth per inch (2.5 cm) to cut soft metal; 18 teeth for tool steel, iron pipe, hard metal, and general shop use; 24 teeth for drill rods, sheet metal, copper and brass, and tubing; and 32 teeth for thin sheet metal, less than 18-gauge (0.12 cm), and for tubing. When thin metals are cut, make sure that at least three teeth are in contact with the surface being cut.

Apply pressure on the forward stroke only. Lift the saw slightly and pull back lightly in the cut to protect the teeth. Cutting speeds of 40 to 60 strokes a minute are recommended. If the blade is twisted or too much pressure is applied, the blade may break and injure the hands or arms of the user. Do not continue an old cut after changing to a new blade. The new blade may bend and break because the set of the teeth on the new blade is thicker than that of a used blade.

Files

Select the right kind of file for the job to prevent injuries, lengthen the life of the file, and increase production. Only use files with secure handles. Because the extremely hard and brittle steel of the file chips easily, never clean the file by sticking it against a vise or other metal object. Instead, use a file-cleaning card.

For the same reason, do not use a file as a hammer or as a pry. Such abuse frequently causes the file to chip or break, thus resulting in injury to the user. Do not convert a file into a center punch, chisel, or any other type of tool because the hardened steel may fracture.

Clamp the work to be filed in a vise at about waist height. Grasp the handle firmly in one hand and use the thumb and forefinger of the other to guide the point. This technique gives good control and ensures better and safer work. To file, push the file forward while bearing down on it. Release the pressure and bring the file back to its original position. If pressure is not released, the teeth will wear excessively.

Always use a file with a smooth, crack-free handle. Otherwise, if the file should slip or be struck by a revolving part of a machine, the tang may puncture the palm of the hand, the wrist, or other part of the body. Under some conditions, a clamp-on, raised-offset handle may be useful to give extra clearance for the hands.

When work to be filed is placed in a lathe, the job should be done left-handedly, with the file and hands clear of the chuck jaws or the dog. Use a fine mill file or long-angle lathe file. Take long, even strokes across the rotating work.

Hand Snips

Hand snips are divided into two groups—those for straight cuts and those for circular cuts. Snips for thicker sheets and harder materials have longer handles, alloy steel blades, and sometimes, special arrangements of levers to make cutting easier. Do not hammer on the handles or jaws of the snips. Use hand snips that are heavy enough to cut the material easily. In that way, the worker needs only one hand on the snips and can use the other to keep the edges of the cut material pulled aside. Be sure that the material is well supported before the last cut is made. In that way, the cut edges do not press against the hands. When cutting long sheet-metal pieces, push the sharp ends down next to the hand holding the snips. File off any jagged edges or slivers after cutting. (Hand snips are not designed to cut wire.) Keep jaws of snips tight and well lubricated. Select a hand snip that cuts easily and is not tiresome to use.

Workers should wear eye protection because small particles often fly with considerable force. Wearing leather or heavy canvas work gloves prevents cuts or scratches to hands caused by handling sharp edges of the sheet metal.

Cutters

Cutters used on wire, reinforcing rods, or bolts should be heavy enough for the stock. Otherwise, the jaws may be sprung or spread. Also, a chip may fly from the cutting edge and injure the user.

Cutters are designed to cut at right angles only. Do not "rock" cutters to make the cut. They are not designed to take the resulting strain. This practice can also cause the knives to chip. Cutters require frequent lubrication. To keep cutting edges from becoming nicked or chipped, do not use cutters as nail pullers or pry bars.

Cutter jaws should have the hardness specified by the manufacturer for the particular kind of material to be cut. Adjust the bumper stop behind the jaws to set cutting edges for a clearance of 0.003 in. (0.076 cm) when closed.

Tools for Cutting Wood

Workers should wear eye protection when using wood-cutting tools. When cutting wood, workers should hold tools so that, if a slip should occur, the direction of force will be away from the body. For efficient and safe work, keep wood-cutting tools sharp and ground to the proper angle. A dull tool does a poor job and may stick or bind. A sudden release may throw the user off balance or cause a hand to strike an obstruction.

Wood Chisels

Instruct inexperienced employees in the proper method of holding and using chisels. Keep wooden handles free of splinters. If the wooden handle of a chisel is designed to be struck by a wood or plastic mallet, protect the handle with a metal or leather band to prevent it from splitting. Heavy-duty or framing chisels are made with solid or molded handles and can be struck with a steel hammer.

Clamp or otherwise secure the work so it cannot move while cutting is being done. Make finish or paring cuts with hand pressure alone. Be sure that the chisel's edge is sharp and that both hands are back of the cutting edge at all times. To avoid damage to the blade of the chisel, be sure that the workpiece is free of nails. Should metal be struck by the cutting edge, a chip from the chisel might hit nearby workers.

Do not use the wood chisel as a pry or a wedge. The steel in a chisel is hard so that the cutting edge will hold. However, the steel may break if the chisel is used as a pry. When not in use, keep chisels in a rack, on a workbench, or in a slotted section of the tool box. In that way, the sharp edges are out of the way and cannot come in contact with metal surfaces.

Saws

Select saws for the work they must do. For cutting across the grain of the wood, use a crosscut saw; for cutting with the grain, use a ripping saw. The difference between them is the angle and shape of their teeth. For fast crosscut work on green wood, use a coarse saw (4 to 5 points per in. [per 2.5 cm]); for smooth, accurate cutting of dry wood, use a fine saw (8 to 10 points per in. [per 2.5 cm]). When ripping, use a coarse saw for thick stock and a fine saw for thin stock. The number of points per inch is stamped on the blade.

To prevent binding, keep blades sharp and the teeth well set. When not in use, wipe saws off with an oily rag and keep them in racks or hung by their handles to prevent their teeth from becoming dull.

Avoid sawing nails or sheetrock with a saw. Use a keyhole saw with a metal-cutting blade to cut nails; use an old saw for cutting sheetrock. Be careful not to drop a saw because the handle can break, or become loose, or the teeth can get nicked.

Axes

The double-bit axe is usually used to fell, trim, and prune trees, and to split and cut wood. It is also used for notching and shaping logs and timbers. The single-bit axe, in addition to the above uses, can be used to drive wooden stakes with its face.

The cutting edges of axes are designed for cutting wood and equally soft materials. Use a narrow-bladed axe for hard wood and a wide axe for soft wood. A sharp, well-honed axe yields better chopping speed and is much safer to use because it bites into the wood (Figure 19–7). A dull axe will often glance off the wood being cut and may strike the user in the foot or leg. Also observe the following precautions:

- Never strike an axe against metal, stone, or concrete.
- Never use an axe as a wedge or maul.
- Never strike with the side of an axe.
- Never use an axe with a loose or damaged handle.
- Use steel wedges for splitting wood, such as splitting logs for fireplace wood. Use a sledge hammer or maul for driving the wedges.

To use an axe safely, workers should lift it properly, swing it correctly, and place the stroke accurately. The proper grip for a right-handed person is to have the left hand about 3 in. (7.6 cm) from the end of the handle, and

Figure 19–7. Good honing saves labor and makes an axe safer to use. The axe should be honed after each sharpening and each use. Correct honing motion is shown here. A double bit axe can be ground to different cutting bevels for various types of work.

the right hand about three-fourths of the way up. A left-handed person should reverse the position of the hands.

Before starting to chop, a worker should make sure that there is a clear circle in which to swing the axe. Also, all vines, brush, and shrubbery within the range should be removed, especially overhead vines that may catch or deflect the axe. When using an axe, workers should wear safety shoes, eye protection, and pants of durable material.

Protect axe blades with a sheath or metal guard when possible. When the blade cannot be guarded, workers should carry axes at their sides. They should carry single-edged axes with the blade pointed down.

Hatchets

Hatchets are used for many purposes and frequently cause injuries. For example, a worker attempting to split a small piece of wood while holding the wood in his or her hand may cut his or her hand or fingers with the hatchet.

To properly start a hatchet cut, strike the wood lightly with the hatchet. Then force the blade through by striking the wood against a solid block of wood.

Miscellaneous Cutting Tools

Permit only trained employees to use planes, scrapers, bits, and drawknives. Keep these tools sharp and in good condition. When not in use, place them in a rack on the bench, or in a tool box, in such a way that will protect the user and prevent damage to the cutting edge. If knives are used on the job, consider using resistance gloves and sleeves.

Knives

Knives are more frequently the source of disabling injuries than any other hand tool. Particularly in the meat-packing industry. To prevent injuries, (1) keep knives sharp, (2) replace knives that have worn handles, and (3) use knives with retractable blades when possible.

The principal hazard in the use of knives is that workers' hands may slip from the handle onto the blade or that the knife may strike the body or the free hand. A handle guard, or a finger ring and swivel, on the handle can reduce these hazards.

When cutting, stroke away from the body. Wear a cut-resistant glove on the hand opposite the knife if that hand is used to hold the material being cut. Wear a heavy leather apron or other protective clothing when it is not possible to cut away from the body. When possible, use a rack or holder for the material to be cut. To help maintain balance, avoid jerky motions.

Be sure employees are trained and supervised. Employees who must carry knives with them on the job should keep them in sheaths or holders. They should never carry a sheathed knife on the front part of a belt. They should always carry it over the right or left hip, toward the back. This prevents severing an artery or vein in a leg in case of a fall.

Safe placing and storing of knives and other sharp hand tools is important to knife safety. When not in use, keep knives in racks and guard their edges. To protect the employee, as well as the cutting edge of the knife, keep knives separate from other tools. Supervisors should make sure that employees do not leave knives hidden under the product, under scrap paper or wiping rags, or among tools in work boxes or drawers.

Supervisors should also make certain that employees who handle knives have ample room in which to work. In that way, they are not in danger of being bumped by trucks, by the product, by overhead equipment, or by other employees. For instance, a left-handed worker should not stand close to a right-handed person. Place the left-handed person at the end of the bench or give him or her more room.

Careful job and incident analysis may suggest ways to make operating procedure, in which knives are used, safer. For instance, on some jobs special rigs, racks, or holders may be provided so it is not necessary for the operator to stand so close to the piece being cut. Also discourage the practice of wiping a dirty or oily knife on the apron or clothing. Instead, have workers wipe the blade with a towel or cloth with the sharp edge turned away from the wiping hand. Sharp knives should be washed separately from other utensils and in such a way that they will not be hidden under soapy water.

Supervisors should make sure that nothing is cut that requires excessive pressure on the knife, such as frozen meat. Food should be thawed before it is cut, or else it should be sawed. Do not use knives as substitutes for can openers, screwdrivers, or ice picks.

To cut corrugated paper, use a hooked linoleum knife. This permits good control of pressure on the cutting edge and eliminates the danger of the blade suddenly collapsing, which might happen if a pocket knife was used. Be sure that hooked knives are carried in a pouch or heavy-leather, or plastic, holder. The sharp tip must not stick out.

Prohibit horseplay around knife operations. Throwing, 'fencing,' trying to cut objects into smaller and smaller pieces, and similar practices are not only dangerous but reflect poor training and inadequate supervision.

Ring Knives

Ring knives—small, hooked knives attached to a finger—are used where string or twine must be cut frequently. Supervisors should make sure that the cutting edge is kept outside the hand, not pointed inside. A wall-mounted cutter or blunt-nose scissors would be safer.

Carton Cutters

Carton cutters are safer than hooked or pocket knives for opening cartons. They not only protect the user but also eliminate deep cuts that could damage the carton's contents. Frequently, damage to contents of soft plastic bottles may not be detected immediately. The subsequent leakage could cause chemical burns, damage to other products, or a fire.

Brad Awls

Brad awls should be started with the edges across the grain to keep the wood from splitting. Hold them at right angles to the surface to prevent them from slipping.

Tools for Materials Handling

Crowbars, hooks, shovels, and rakes are used for materials handling. Observe safety precautions when using them.

Crowbars

Whenever a crowbar is needed, use the proper size and kind of bar for the job. The crowbar should have (1) a point or toe of such shape that it will grip the object to be moved, and (2) a heel to act as a pivot or fulcrum. In some cases, a block of wood under the heel will prevent the crowbar from slipping and injuring the worker's hand.

If crowbars are stood on end when not in use, secure them so they will not fall. If they are laid on the ground when not in use, place them where they will not create a stumbling hazard.

Hooks

Keep hand hooks sharp so they will not slip when applied to a box or other object. The handle should be strong and securely attached and shaped to fit the hand. The handle and the point of long hooks should be bent on the same plane so that the bar will lie flat when not in use and, thus, not cause a tripping hazard. Shield the hook's point when it is not in use.

Shovels

Keep the edges of shovels trimmed and check the handles for splinters. Workers should wear heavy shoes with sturdy soles, preferably safety shoes. To have good balance and spring in the knees, workers should keep their feet well separated. The leg muscles should take much of the load, when shoveling.

To reduce the chance of injury, use the ball of the foot—not the arch—to press the shovel into clay or other stiff material. If the instep is used and the foot slips off the shovel, the sharp corner of the shovel may cut through the worker's shoe and into the foot.

Dipping the shovel into a pail of water occasionally keeps it free from sticky material, thus making it easier to use and less likely to cause strain. Greasing or waxing the shovel's blade also prevents some kinds of material from sticking. Chemically coated shovels prevent sticking of certain materials.

When shovels are not in use, hang them or stand them against a wall. Or, install special racks or boxes for shovels.

Rakes

Never leave a rake with the prongs turned upward where they may be stepped on, thus causing a foot injury or causing the handle to fly up and hit a worker's head. Place rakes in racks when they are not in use.

Wrenches

To ensure the safe use of all wrenches, workers must always be alert. They should be prepared for the possibility of (1) a wrench slipping off the fastener, (2) a fastener suddenly turning free, (3) a wrench breaking, or (4) a fastener breaking. Therefore, workers should brace themselves in such a way that should the wrench become free for any reason, they will not lose their balance and be injured by falling into moving machinery or falling off a platform. Workers should always try to pull the wrench toward themselves. This will provide better wrench control and less likelihood of slipping or falling.

Workers should always inspect wrenches for flaws because a previous overloading or misuse of a tool may have weakened it to the point that it will not carry a normal load. Do not grind wrenches to change their sizes or to reduce their dimensions to fit into close quarters. Instead, use a wrench of the correct size and fit. It is an unsafe practice to use shims to make the wrong wrench fit. The great variety of wrenches used for turning nuts, bolts, and fittings makes it important that workers know the purpose and limitations of each type and size.

Open-End Wrenches

Open-end wrenches have strong jaws and are satisfactory for medium-duty turning. They are susceptible to slipping if they do not fit properly or are used incorrectly. A wrench with an offset provides hand clearance and allows the worker to reach into recesses and over obstructions. A combination wrench has both jaws the same size.

Box and Socket Wrenches

Box wrenches and socket wrenches are used when a heavy pull is necessary and safety is a consideration. The greater gripping strength of the box helps to remove the

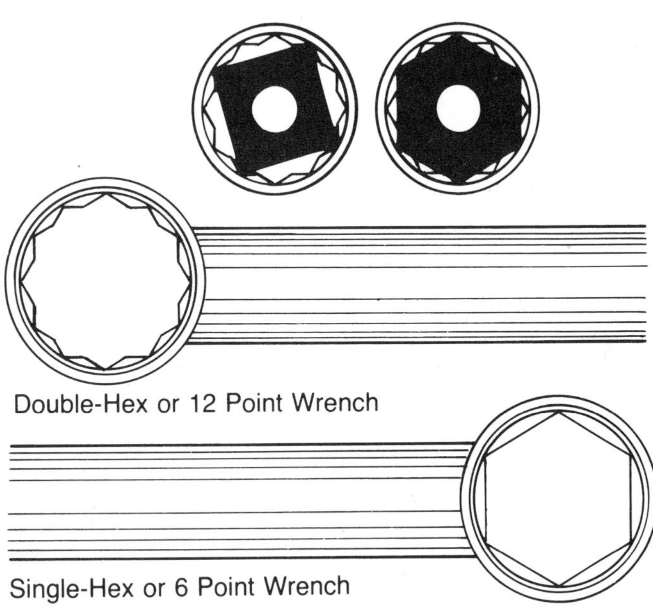

Figure 19–8. The ring of a box has 6 or 12 points. The 12-point wrench need only be turned 30 degrees before engaging a new set of flats and is therefore handy for use in confined spaces. Never use a 12-point box wrench on a square nut or bolt because it may slip.

nut quickly. Box wrenches (Figure 19–8) and socket wrenches (Figure 19–9) completely encircle the nut, bolt, or fitting. They grip the nut, bolt, or fitting at all corners, as opposed to the two corners that are gripped by an open-end wrench. Box and socket wrenches will not slip off sideways, and they eliminate the dangers of sprung jaws.

Socket wrenches and box wrenches normally come in two styles of openings—single hex or double hex (Figure 19–8). On square-headed bolts, nuts, and fittings, use either a single- or double-square design. Single- or double-hexagonal wrenches are designed for hexagonal-shaped fittings, bolts, and nuts. Ratchet handles and universal-joint fittings for socket wrenches allow them to be used where space is limited (Figure 19–9).

Never overload the capacity of the wrench by using a pipe extension on the handle or by striking the handle of a wrench with a hammer. Abuse by hammers weakens the metal of the wrench and can cause the tool to break. Special heavy-duty wrenches are available that can be used with handles as long as 3 ft (91 cm). For extra-stubborn bolts and nuts, heavy-duty, sledge-type box wrenches are available. These are of a heavy design, are properly tempered, and have a striking surface for the hammer or sledge (Figure 19–10). When possible, first use penetrating oil to loosen tight nuts.

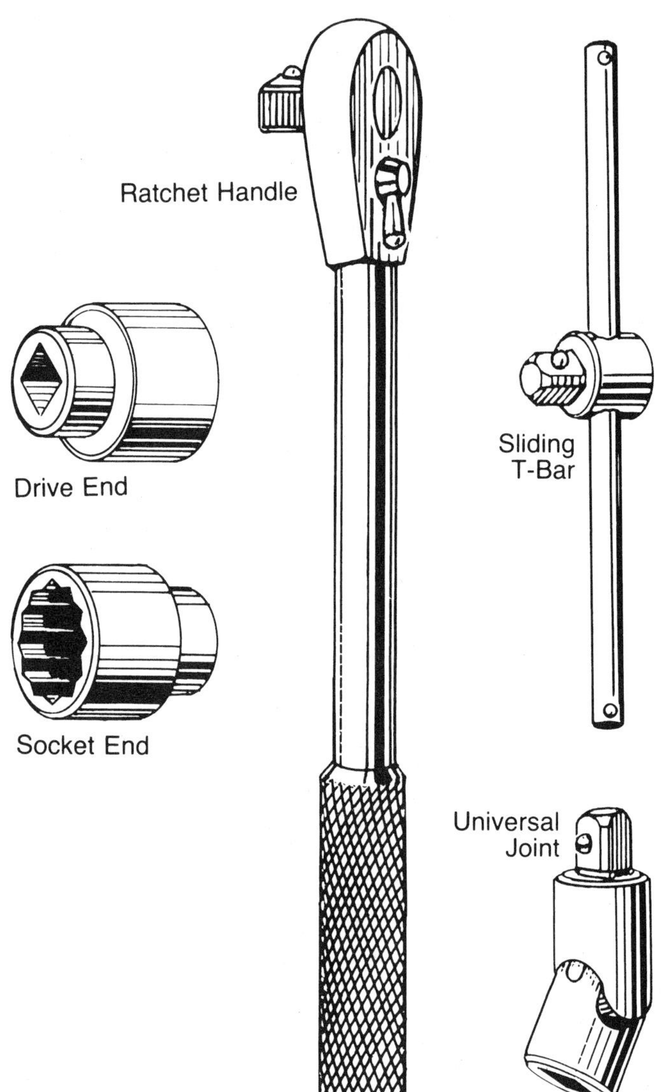

Figure 19–9. Socket wrenches (left) have a square hole in the drive end into which various handles can be fitted. The ratchet handle increases turning speed because the socket does not have to be removed from the nut between turns. The universal joint fitting provides flexibility where space is limited.

There is a correct box or socket wrench for every nut and bolt. Oversize openings will not grip the corners securely and shims should not be used to compensate for an oversize opening. The use of wrenches of the wrong size can round the corners of the bolt or cause slippage, as well as make it difficult to then apply the proper size. Be sure to use the proper tool and not to try a makeshift approach (Figure 19–11).

Figure 19–10. Using a striking-face wrench designed to be hammered is a safe approach where large nuts must be set tight or frozen nuts loosened.

Figure 19–11. The correct tool at the right time makes an otherwise hard job safe and easy.

Keep the insides of sockets clean of dirt and grime. Dirt prevents the socket from seating fully. Thus, the concentration of the pulling force at the end of the socket's opening, even with a moderate pull, can easily damage the socket or nut.

Cocking is a common cause of breakage for a socket or box wrench. Cocking is a situation in which the tool does not fit securely on the bolt or nut but instead fits at an angle. This concentrates the entire strain on a smaller area, making the tool or nut vulnerable to fracture.

Combination Wrenches

Combination wrenches have a box end and an open end. They are very handy for speeding the turning with the open end and for initial loosening or final tightening with the box end.

Torque Wrenches

Torque wrenches measure the amount of twisting force that is applied to a nut or bolt by means of a dial or calibrated arm. The torque, or twisting force, exerted on a nut or bolt is directly proportional to the length of the wrench handle and the pulling force exerted on it. Torque is usually measured in foot-pounds or newton-meters; that is, the force used times the length of the lever used to apply it. For example, a 15 lb (6.7 kg) force applied to a 2 ft (61 cm) lever gives 30 ft lb torque. The metric unit is the newton meter ($1 \text{ kgf} \times \text{m} = 9.6 \text{ N} \times \text{m}$; $1 \text{ lbf} \times \text{ft} = 1.356 \text{ N} \times \text{m}$).

Torque wrenches are used when the torque has been specified for the job or when it is important that all fasteners be fully and uniformly tightened. If a torque has been specified for a particular application, it is unsafe not to measure that torque with a torque wrench. Carefully use torque wrenches and recalibrate them frequently in accord with the manufacturer's instructions.

Adjustable Wrenches

Adjustable wrenches are generally recommended for light-duty jobs or when the proper size fixed-opening wrench is not available. Adjustable wrenches can slip, however, because it is difficult to set the correct opening size. In addition, the jaws have a tendency to work loose as the wrench is being used. These wrenches do possess one advantage—they are easily adjusted to fit metric-system nuts and bolts.

Place an adjustable wrench on the nut with the open jaws facing the user, unless the space in which the job is being done makes this method impractical. With the open jaw facing the user, the pulling force applied to the handle tends to force the movable jaw onto the nut. For that reason, and for reasons of safety, pull, do not push, wrenches (Figure 19–12). The adjusting nut is used to adjust the movable jaw. The jaws should grip the pipe about midway. According to the manufacturers, the movable jaws on adjustable wrenches are weaker than the fixed jaws.

Figure 19–12. Correct use of a wrench. The wrench is in good condition and securely gripped. The hand of the worker is braced and clear in the event that the nut should suddenly turn. To protect the worker's hand, the wrench is pulled, not pushed.

Pipe Wrenches

Workers, especially those on overhead jobs, have been seriously injured when pipe wrenches have slipped on pipes or fittings, thus causing workers to lose their balance and fall. Pipe wrenches, both straight and chain tong, should have sharp jaws. Keep them clean to prevent them from slipping.

Frequently inspect the adjusting nut of the wrench. If it is cracked, the wrench should be taken out of service. A cracked nut may break under the strain, thus causing complete failure of the wrench and possible injury to the user.

The handle of every wrench is designed to be long enough for the maximum allowable safe pressure. Do not use handle extensions, also known as cheater bars, to gain extra turning power, unless the wrench is so designed. A piece of pipe slipped over the handle to give added leverage can strain a pipe wrench or the workpiece to the breaking point. Using a makeshift extension to secure greater leverage may easily cause the wrench's head to break. Using a wrench of the wrong length is also a source of incidents. A wrench handle too small for the job does not give proper grip or leverage. An oversized wrench handle may strip the threads or suddenly break the work, thus causing a worker to slip or to fall.

Never use a pipe wrench on nuts or bolts. The corners of nuts and bolts will break the teeth of the wrench, thereby making it unsafe to use the latter on pipes and fittings. A pipe wrench also ruins the heads of nuts and bolts. Do not use a pipe wrench on valves or small brass, copper, or other soft fittings that may be crushed or bent out of shape. Do not strike a wrench with a hammer nor use it as a hammer, unless the pipe wrench is a specialized type specifically designed for such use.

Tongs

Tongs are usually bought, but some companies make their own to perform specific jobs. Often, they are designed in such a way that the hands are pinched when the tongs are closed. To prevent pinching, the end of one handle should be up-ended toward the other handle, to act as a stop, or projections can be welded on the handle ends to allow clearance for the user's hands.

Pliers

Pliers are often considered a general-purpose tool and therefore, are often used for purposes for which they were not designed. Pliers are meant for gripping and cutting operations. They are not recommended as a substitute for wrenches because their jaws are flexible and frequently slip when used for this work. Pliers also tend to round the corners of bolts' heads and nuts and to leave jaw marks on the surface. This makes it difficult to use a wrench at some future time.

Side-cutting pliers sometimes cause injuries when short ends of wire are cut. A guard over the cutting edge and the use of eye protection prevents short ends from causing injuries.

Be certain that pliers used for electrical work are insulated. In addition, employees should wear electrician's gloves, if company policy requires them to.

WARNING: The cushion grips on handles are primarily for comfort. Unless specified as insulated handles, they are not intended to give any degree of protection against electric shock and should not be used on live electric circuits.

Special Cutters

Special cutters for heavy wire, reinforcing wire, and bolts are safer than makeshift tools. The cutting edges should apply force at right angles to the wire or other work being cut. Do not use the cutter near live electrical circuits, and use it only for the rated capacity specified by the manufacturer. Workers should wear eye protection.

Special cutters include those used for cutting banding wire and strap. Do not use claw hammers and pry bars to snap metal banding material. Only cutters designed for the work provide safe and effective results.

Pullers

Pullers or knockers are the only quick, safe, and easy way to pull a gear, wheel, pulley, or bearing from a shaft. Do not use prybars and chisels because they concentrate the force at one point and tend to cock the part on the shaft. Select the correct-sized puller. The jaw capacity should be such that the jaws press tightly against the part being pulled. Use a puller with as large a pressure screw as possible.

Spark-Resistant Tools

So-called spark-resistant tools made of nonferrous materials, such as beryllium-copper alloy, are used where flammable gases, highly volatile liquids, and other explosive substances are stored or used. There is some question, however, about the ability of these materials to prevent the hazard of friction sparks igniting gasoline vapors and petroleum products. Nonferrous tools reduce the hazard from sparking but do not eliminate it.

The working edges of these tools are not as hard as steel tools. Therefore, inspect them more often before each use. Be sure that they have not picked up foreign particles that could produce friction sparks, thereby negating their value.

Soldering Irons

Soldering irons are the source of burns and of illnesses that result from inhaling the fumes. Insulated, noncombustible holders practically eliminate the fire hazard and the danger of burns from unintentional contact. Ordinary metal coverings on wooden tables are not sufficient because the metal conducts heat and may ignite the wood below.

Holders should be designed so that employees cannot unintentionally touch the hot irons if they reach for them without looking. The best holder completely encloses the heated surface and is inclined so that the weight of the iron prevents it from falling out. (See National Safety Council Data Sheet 12304–0445, *Soldering and Brazing.*)

Fumes from soldering can be toxic and/or irritating. Remove soldering fumes through local-exhaust ventilation, especially in a continuous-production operation. Lead oxides and chlorides are released when soldering with lead-tin solder and zinc-chloride flux. Lead oxides and aldehydes are released when soldering with rosin-core solder. There are different types of solder, and the hazards from each should be known before beginning work. Conduct air sampling to determine if hazardous amounts of contaminants are present.

Do not let particles of lead solder accumulate on the floor and on worktables. If the operation is such that the solder or flux may spatter, employees should wear face shields or do the work under a transparent shield. Since a primary route of exposure to lead can be ingestion, do not allow workers to eat, drink, smoke, chew gum, apply cosmetics, etc., in the work area. They should also practice good personal hygiene, for instance, wash hands before eating or leaving the workplace.

PORTABLE POWER TOOLS

Portable power tools are divided into five primary groups according to their power source: electrical, pneumatic, gasoline, hydraulic, and powder actuated. Several types of tools, such as saws, drills, and grinders, are common to the first three groups. Hydraulic tools are used mainly for compression work, and powder-actuated tools are used exclusively for penetration work, cutting, and compression.

Hazards and Safety Precautions

Portable power tools present hazards similar to stationary machines performing the same functions. In addition, portable power tools also have inherent risks. Because of the extreme mobility of power-driven tools, they can easily come in contact with the operator's body. At the same time, it is difficult to guard such equipment completely. There is also the possibility of breakage because the tool may be dropped or roughly handled. Furthermore, the source of power—electrical, mechanical, air, hydraulic, or powder cartridge—is brought close to the operator, thus creating additional potential hazards.

Typical injuries caused by portable power tools are burns, cuts, and strains. Sources of injuries include electrical shock, particles in the eyes, fires, falls, explosions of vapors or gases, and falling tools and other objects.

Observe the following precautions when using portable power tools:

- Always disconnect the tool from the source of power before changing accessories. Replace or put guards in correct adjustment before using the tool again.
- Never leave a tool in an overhead place where there is a chance that the cord or hose, if pulled, will cause the tool to fall. The cord or hose and the tool may be suspended by a tool balancer that keeps

them out of the operator's way. Cords and hoses on the floor create a stumbling or tripping hazard. Suspend them over aisles or work areas in such a way that they will not be struck by other objects or by material being handled or moved. An unexpected pull might cause the tool to jam or be dislodged from its holder and cause injury or damage to the tool. Protect them with wooden strips or special raceways if they are laid across the floor. Do not hang cords or hoses over nails, bolts, or sharp edges. Also keep them away from oil, hot surfaces, and chemicals.

- When using powder-loaded equipment for driving anchors into concrete, or when using air-driven hammers or jacks, use proper hearing protection.

All companies and manufacturers of portable power tools attach to each tool a set of operating rules or safe practices. Their use will supplement the thorough training that each power tool operator should receive. The operator manual is an excellent source of information and should be followed.

In addition, observe the following safe practices when using portable power tools:

- Store power-driven tools in secured places. Do not leave them in areas where they may be struck by passersby or be otherwise activated.
- Keep work areas heated, clean, and well lit.
- Secure, or clamp, workpieces. Normal tool use does not require a great deal of force.
- When working on a ladder, a scaffold, or in other high places, do not reach out too far. Keep the body in balance.
- Wear proper clothing for the job. Loose clothing, jewelry, and long hair may add risk to the job.
- Never use a power tool with a malfunctioning switch or part. Remove it from service, and repair or discard the tool.
- Only use accessories recommended by the manufacturer.

Selecting Tools

Replacing a hand tool with a power tool designed for the same purpose may mean substituting electrical or mechanical hazards for relatively less serious manual hazards. Therefore, anticipate the new problems and avoid as many as possible. Insist on purchasing tools designed to meet regulatory standards for safety and be sure that workers are properly trained.

Provide the tool supplier with complete information about the job on which a tool is to be used. In that way, the supplier can recommend the most appropriate tool. Factors to consider when selecting power tools include clearance in the working quarters, the type of job to be done, and the nature and thickness of materials.

Portable power tools designed to be used intermittently or on light work are generally designated as 'home-owner's grade.' Those intended for continuous operation and production service, or for heavy work, are usually identified as industrial duty.

For safe operation of portable power tools, train workers to select tools according to the tool's limitations. Workers should never tackle a job with an undersized tool. A tool that is too light may not only fail but may cause undue fatigue to the operator and create a risk of injury.

Inspection and Repair

To maintain power tools, periodically inspect them. In addition to uncovering operating defects and preventing potentially costly breakdowns, inspecting and repairing tools may prevent hazardous conditions from developing. A portable power-tool tester is an excellent means of keeping tools in good condition. Set up an inspection schedule and a system for keeping records for each tool. Tag defective tools and withdraw them from service until they are repaired.

Check electrical tools periodically (Figure 19–13). Provide a visual or external inspection at the toolroom each time a tool is returned. Also provide a thorough knock-down inspection at specified intervals. Inspect as specified by the OSHA electrical safety-related work standard.

Use colored tags to tell when the tool was last inspected. The important thing is to record the condition of the tool and to correct any unsafe conditions.

Instruct and train employees to inspect tools and to recognize and to report (and, if authorized, to correct) defects. Clearly outline the extent of this inspection and the responsibilities for correcting defects. In that way, there is neither unnecessary duplication of effort nor misunderstanding about who is responsible for maintenance. A convenient reminder of points to check is a card similar to that shown in Figure 19–2. Warn employees not to do makeshift repairs and to do no repair work unless authorized.

Clean power tools with a recommended nonflammable and nontoxic solvent. Use air drying in place of blowing with compressed air.

Electrical Tools

There are a number of rechargeable battery-powered tools available. For safety from electrical shock, they are the best possible tool. Not needing an extension

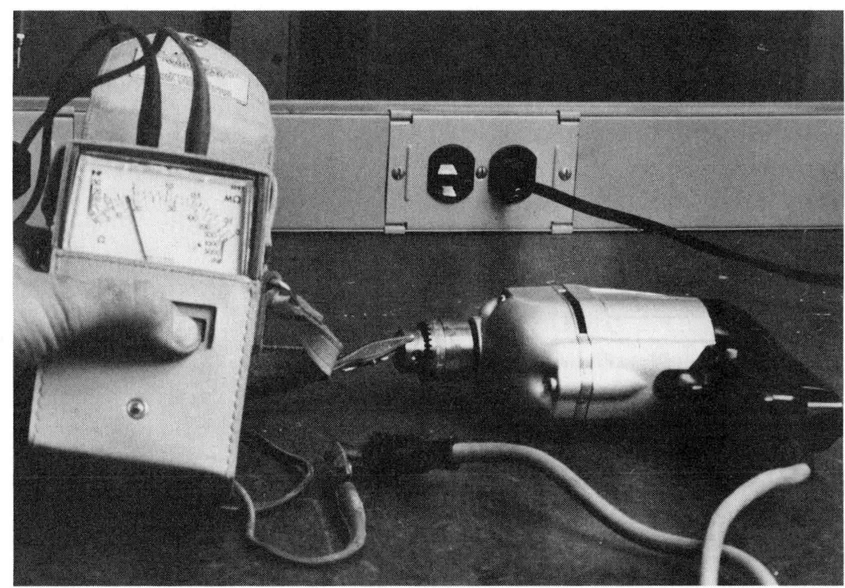

Figure 19–13. This insulation resistance tester has test leads attached to a grounding pin of the attachment plug and the chuck of the portable tool. The meter indicates the condition of the insulation of the portable tool and permits identification of impending failures. (Reprinted with permission from Daniel Woodhead Company.)

cord adds to their mobility and is a convenience feature of battery-powered tools.

Electric shock is one hazard of electrically powered tools. Types of injuries include electrical flash burns, minor shock that may cause falls, and shock that results in death. Serious electrical shock is not entirely dependent on the voltage of the power input. The ratio of the voltage to the resistance determines the amount of current that will flow and the resultant degree of hazard.

The current is regulated by the resistance to the ground of the body of the operator and by the conditions under which he or she is working. It is possible for a tool to operate with a defect or short in the wiring. The use of a ground wire protects the operator and is mandatory for all but double-insulated electrical power tools. A ground-fault circuit interrupter (GFCI) should be used with all electrical power tools, whether grounded or double insulated. (See Chapter 10, Electrical Safety, in this volume.)

Wet Locations

Low voltage of 6, 12, 24, or 42 volts through portable transformers will reduce the shock hazard in wet locations. Issue standing orders to supervisors and employees to use any available low-voltage equipment or GFCIs when working in wet locations.

Electrical tools used in wet areas or in metal tanks expose the operator to conditions favorable to the flow of current through the body, particularly if a person is wet with perspiration. Most electrical shocks from tools are caused by the failure of insulation between the current-carrying parts and the metal frames of the tools. Insulating platforms, rubber mats, and rubber gloves provide an additional factor of safety when tools are used in wet locations, such as in tanks and boilers or on wet floors.

Double-Insulated Tools

Protection from electrical shock, while using portable power tools, depends upon third-wire grounding. Double-insulated tools, however, are available that generally provide equivalent shock protection without third-wire grounding. Paragraph 250–45 of the National Fire Protection Association (NFPA) 70, *National Electrical Code*, permits double insulation for portable tools and appliances. Tools in this category are permanently marked by the words Double Insulation or Double Insulated. Units designated to this category have been tested and listed by a nationally recognized testing lab such as Underwriters Laboratories and also use a listing mark. Many U.S. manufacturers are also using the symbol shown in Figure 19–14 to denote double insulation. This symbol is also widely used in most European countries.

Conventional grounded electrical tools have a single layer of functional insulation and are metal encased. For small-capacity tools, double insulation can be provided by encasing the entire tool in a nonconductive material that is also shatterproof. The switch and gripping surface are also nonconductive. Therefore, no metal part comes in contact with the operator.

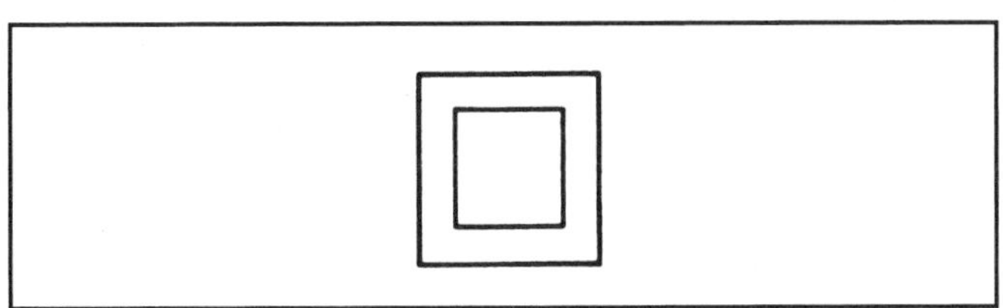

Figure 19–14. Symbol for double insulation for electrical appliances and tools.

Large-capacity electrical tools require a more rigid design to provide for greater stress requirements where more power and high-torque gearing are involved. Double-insulated tools with metal housings have an internal layer of protecting insulation that completely isolates the electrical components from the outer metal housing. This is in addition to the functional insulation found in conventional grounded tools. Thus, in addition to the functional insulation, a reinforced or protecting insulation is also incorporated into the tool. This extra, or reinforced, insulation is physically separated from the functional insulation. It is arranged so that deteriorating influences, such as temperature, contaminants, and wear, will not affect both insulations at the same time. Unless subject to immersion or extensive moisture that might nullify the double insulation, a double-insulated or all-insulated tool does not require separate ground connections. In other words, the third wire or ground wire is not needed.

Failure of insulation is harder to detect than worn or broken external wiring. Thus, frequent inspection, testing with an insulation-resistance tester, and thorough maintenance are needed. Care in handling the tool and frequent cleaning will help prevent the wear and tear that cause defects.

Grounding

Grounding portable electrical tools and using a GFCI provide a convenient way of safeguarding the operator. Thus, if there is any defect or short circuit inside the tool, the current is connected from the metal frame through a ground wire and does not pass through the operator's body. If a GFCI is used, the current is shut off before a serious shock can occur. Effectively ground all electrical power tools, except the double-insulated and cordless types. Correctly grounded tools are as safe as double-insulated or low-voltage tools, especially when used with a GFCI. Check the continuity of the ground to avoid a false sense of security.

Electrical Cords

Periodically inspect electrical cords prior to use, and keep them in good condition. Use heavy-duty plugs that clamp to the cord to prevent strain on the current-carrying parts, in case the cord is unintentionally pulled. Cover terminal screws, or connections on plugs and connectors, with proper insulation. Instruct employees not to jerk cords or wind them tightly around the tool. Also instruct workers to protect cords from sharp objects, heat, and oil, or from solvents, that might damage or soften the insulation. To ensure the continuity of grounding, use extension cords of the three-wire, grounded-connection type, with tools and equipment that require grounding. Check the wire size to ensure that the current's needs can be met. (See Cords, in this chapter, and Extension Cords, in Chapter 10, Electrical Safety.)

Cord Length (ft)	**Average Wire Size for Amp Rating of Tool**					
	0-2.0	**2.1-3.4**	**3.5-5.0**	**5.1-7.0**	**7.1-12.0**	**12.1-16.0**
25	16	16	16	16	14	14
50	16	16	16	16	14	12
100	16	16	14	12	10	—

Electric Drills

Electric drills cause injuries in several ways: (1) the torque and twisting of the tool can cause sprains and related injuries; (2) a part of the drill can be pushed into the hand, leg, or another part of the body; (3) the drill can be dropped when the operator is not actually drilling; and (4) material being drilled or parts of a broken drill can strike the eyes. Using proper eye protection will reduce the possibility of eye injuries. Although no guards are available for drill bits, some protection is afforded if drill bits are carefully chosen for the work to be done. Do not pick drill bits longer than needed to do the work.

When the operator must guide the drill with a hand, equip the drill with a sleeve that fits over the drill bit. The sleeve protects the operator's hand and also serves as a limit stop if the drill should plunge through the material.

Do not grind down oversized bits to fit small electric drills.

Observe the following precautions when using portable electrical drills:

- Be sure the trigger switch works properly. The trigger switch should turn the tool on and return it to the off position after the switch is released. If equipped with a lock-on, be sure it releases freely.
- Check carefully for loose power-cord connections and frays or damage to the cord. Replace damaged tools and extension cords immediately.
- Be sure the chuck is tightly secured to the spindle. This is especially important on reversible drills.
- Tighten the drill bit securely, as prescribed by the owner/operator's manual if there is a chuck key. Remove it from the chuck before starting the drill. A flying key can cause injuries.
- Check auxiliary handles, if they are part of the tool. Be sure they are securely installed. Always use the auxiliary drill handle when provided. It provides more control of the drill. Grasp the drill firmly by insulated gripping surfaces.
- Always wear safety goggles, or safety glasses with side shields, that comply with current national standards and a full face shield when needed. Use a dust mask for dusty working conditions. Wear hearing protection during extended periods of operation. (See 29 *CFR* 1910.133, Eye and Face Protection.)

- Always hold or brace the tool securely. Brace against stationary objects for maximum control. If drilling in a clockwise (forward) direction, brace the drill to prevent a counterclockwise reaction.
- If the drill binds in the work, release the trigger immediately, unplug the drill from the power source, and then remove the bit from the workpiece. If the drill operation could potentially bind, then do not actuate any switch lock-on.
- Never attempt to free a jammed bit by starting and stopping the drill.
- As the hole is about to be broken through, grip or brace the drill firmly, reduce pressure, and allow the bit to pass easily through the hole.
- Unplug the drill before changing bits, accessories, or attachments.
- Do not raise or lower a drill by its power cord.

Electrical Circular Saws

Electrical circular saws must have guards above and below the faceplate. The lower guard must retract automatically when cutting material. Instruct employees to use the guard as intended. Frequently check the guard to be sure that it operates freely and encloses the teeth completely when it is not cutting (Figures 19–15 and 19–16).

In the United States, OSHA standards require that frames and exposed metal parts of portable saws and other portable electrical woodworking tools that operate at over 90 volts be grounded, unless they are double-insulated or battery powered.

Figure 19–15. Portable band saw has guard that encloses the blade except for the actual work area. Hands are kept above the blade and are occupied.

Blades

Use sharp blades. Dull blades cause binding, stalling, and possible kickback. They also waste power and reduce the life of the motor and the switch. Use the correct blade for the specific application. Check that the blade has the proper size and arbor hole. Also check that the speed marked on the blade is at least as high as the no-load RPM on the saw's nameplate.

Before each use, check blades carefully for proper alignment and possible defects. Be sure that the blade's washers (flanges) are correctly assembled on the shaft and that the blade is tight.

Before each cut, be sure that the blade's guard is working. Check often to assure that guards return to their normal position quickly. If a guard seems slow to return or hangs up, repair or adjust it immediately. Never tie back or remove the guard to expose the blade.

Cords

Before starting a circular saw, be sure that the power cord and extension cord are out of the blade's path and are long enough to freely complete the cut. On the job stay constantly aware of cord location. A sudden jerk or pulling on the cord can cause loss of control of the saw and perhaps a serious injury.

Using the Saw

For maximum control, hold the saw firmly with both hands after securing the workpiece. Clamp workpieces if they are small enough. Check frequently to be sure that the clamps remain secure. Never hold a workpiece in the hand or across the leg when sawing. Avoid cutting small pieces of material that cannot be properly secured, as well as materials on which the saw's shoe cannot properly rest.

When making a blind cut (the operator cannot see behind what is being cut) be sure that hidden electrical wiring, water pipes, or any mechanical hazards are not in the blade's path. If wires are present, have a qualified person disconnect them at the power source. Contact with live wires could cause lethal shock or fire. Drain and cap

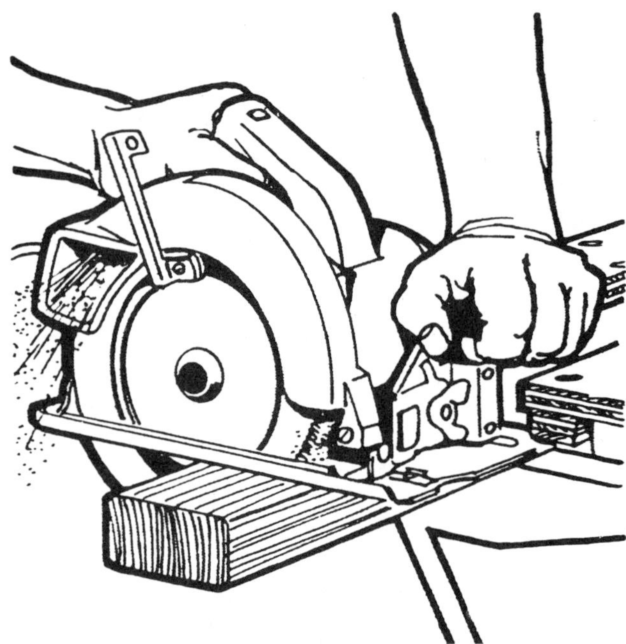

Figure 19–16. The workpiece must be securely clamped. To maintain control, use both hands to properly and safely guide the saw. (Courtesy Power Tool Institute.)

water pipes. If the tool is double insulated, hold it by the insulated grasping surfaces or the handles as provided.

Also observe the following precautions:

- Set the blade's depth to no more than ⅛ in. to ¼ in. (0.32 cm to 0.64 cm) greater than the thickness of the material being cut.
- When starting to saw, allow the blade to reach full speed before contacting the workpiece or material.
- Be alert to the chance that the blade may bind and that kickback may occur.
- If a fence or guide board is used, be certain that the blade is kept parallel with it.
- When making a partial cut, or if power is interrupted, immediately release the trigger and do not remove the saw until the blade has come to a complete stop.
- Never reach under the saw or workpiece.

Kickback

Kickback is a sudden reaction to a pinched blade that causes an uncontrolled portable tool to lift up and out of the workpiece toward the operator. Kickback is the result of misusing a tool and/or of incorrect operating procedures or conditions.

Take the following specific precautions to help prevent kickback when using any type of circular saw:

- Keep blades sharp. A sharp blade cuts more easily and tends to cut its way out of a pinching condition.
- Make sure the blade has an adequate set in the teeth. Tooth set provides clearance between the sides of the blade and the workpiece, thus minimizing the chance of binding. Some saw blades have hollow-ground sides instead of tooth set to provide clearance.
- Keep blades clean. A buildup of pitch or sap on the surface of the blades increases the blades' thickness and also increases the friction on their surface. These conditions cause an increase in the likelihood of a kickback.
- Do not cut wet wood. It produces higher friction against the blade. The blade also tends to load up with wet sawdust and has a greater chance of kickback.
- Be cautious of stock that is pitchy, knotty, or warped. Such stock is most likely to create pinching conditions and possible kickback.
- Release the switch immediately if the blade binds or the saw stalls.
- Never remove the saw from a cut while the blade is rotating.
- Never use a bent, broken, or warped blade. The chance of binding, and the resulting kickback, is greatly increased by these conditions.
- Overheating a blade can cause it to warp and, thus, result in a kickback. Buildup of sap on the blades, insufficient set, dullness, and unguided cuts, can all cause an overheated blade.
- Never use more blade protrusion than is required to cut the workpiece—⅛ in. to ¼ in. (0.32 cm to 0.64 cm) greater than the thickness of the stock is usually sufficient. On some blades, the blade gullet needs to be clear of the material surface to kick off sawdust. This minimizes the amount of the blade's surface that is exposed and reduces the chance of kickback, and severity, if any kickback does occur.
- Minimize blade pinching by placing the saw's shoe on the clamped, supported portion of the workpiece and by allowing the piece that is cut off to fall away freely.

Reciprocating Saws

The versatility of the reciprocating saw in cutting metal, pipe, wood, and other materials has made it a widely used tool. By design, it is a simple tool to handle. Its few demands for safe use, however, are very important. Observe the following precautions:

- On some blades, the blade gullet needs to be clear of the material surface to kick off sawdust.
- Without exception, use the blade specifically recommended for the job being done. Follow the owner/operator's manual recommendations.

- Operators should position themselves to maintain full control of the tools. They should avoid cutting above shoulder height.
- Use sharp blades. Dull blades can produce excessive heat, make sawing difficult, result in forcing the tool, and possibly cause an injury.
- To minimize blade flexing and to provide a smooth cut, use the shortest blades that will do the job.
- When plunge (pocket) cutting, use a blade designed for that purpose. Maintain firm contact between the saw's shoe and the material being cut (Figure 19–17).
- When making a blind cut, be sure that hidden electrical wiring or water pipes are not in the path of the cut. If wires are present, have a qualified person disconnect them at their power source to prevent the chance of lethal shock or fire. Drain and cap water pipes.
- Always hold the tool by the specific gripping surfaces.
- When making anything other than a through cut, allow the saw to come to a complete stop before removing the blade from the workpiece. This prevents both breakage of the blade and possibly losing control of the saw.
- Remember that the blade and blade clamp may be hot immediately after cutting. Avoid contact until they have cooled.

Miter Saws

These saws are used for crosscutting, mitering, and beveling wood, nonferrous metals, and plastics. They cut through the workpiece at a predetermined angle or miter.

Figure 19–17. When plunge cutting, maintain firm contact between the shoe and the workpiece. Plunge cutting requires blades designed for the purpose. (Courtesy Power Tool Institute.)

When using miter saws, operators should always wear safety goggles or safety glasses with side shields that comply with standards, and a full face shield when needed. Provide dust masks in dusty working conditions and hearing protection during extended periods of operation. Operators should not wear gloves, loose clothing, jewelry, or any dangling objects, including long hair, that may catch in rotating parts or accessories of the saw.

Be sure that all guards are in place and working. If a guard seems slow to return to its normal position or hangs up, adjust or repair it immediately. Because of the downward cutting motion, operators' safety requires that they keep their hands and fingers away from the path in which the blade travels—especially during repetitive, monotonous operations. Do not be lulled into carelessness because of a false sense of security. Blades are extremely unforgiving.

Observe the following safety precautions:

- Clean the lower guard frequently to help visibility and movement. Unplug before adjusting or cleaning.
- Use only recommended sizes of blades and RPM-rated blades.
- Do not use abrasive cut-off wheels on miter-saws. Miter-saw guards are not appropriate for abrasive cut-off wheels.
- Remember that loose blades can fly off. Regularly check and tighten the blade and blade-attachment mechanism.
- When installing or changing a blade, be sure that the blade and its related washers and fasteners are correctly positioned and secured on the saw's arbor.
- To avoid losing control or placing hands in the blades' path, hold or clamp all material securely against the fence when cutting. Do not perform operations freehand.
- Never recut small pieces, and support long material at the same height as the saw table.
- Never place hands or fingers in the path of the blade, or reach in back of the fence.
- Use the brake if one is provided, since blades coast after being turned off. To avoid contact with a coasting blade, do not reach into cutting areas until the blade comes to a full stop.
- After completing a cut, release the trigger switch and allow the blade to come to a complete stop, then raise the blade from the workpiece. If the blade stays in the cutting area after a cut is complete, injury can occur from unintentional contact.

Miter-saws have spring-loaded saw heads that return the saw's head to its up position. Adjust, repair, or replace the spring mechanism if the saw's head does not automatically return to its up position when released (Figure 19–18).

Figure 19–18. Lock the miter head in the down position when not in use (back view shown). (Courtesy Power Tool Institute.)

Jig/Saber Saws

Observe the following precautions when using jig/saber saws:

- Check carefully that the blades are adequately secured in position before plugging the saw in.
- Make sure the cord is out of the way and not in the line of the cut.
- Firmly position the saw's base plate/shoe on the workpiece before turning on the tool.
- Keep hands and fingers clear of moving parts.
- After making partial cuts, turn the tool off. Remove the blade from the workpiece only after the blade has fully stopped.
- Know what is behind a cut before making it. Be sure that hidden electrical wiring, water pipes, and hazardous objects of any kind are not in the path of the cut. If wires are present, have a qualified person disconnect them at their power source to prevent the chance of lethal shock. Drain and cap water pipes.
- Always hold the tool by the specific grasping surfaces.
- When plunge cutting, use a blade designed for the purpose and follow the manufacturer's recommended procedures.
- Throughout cutting procedures, maintain firm contact between the base and the material being cut.
- Remember that the blade and blade clamp may be hot immediately after cutting. Therefore, keep hands away from them until they have cooled down.
- Do not leave saws unattended. Unplug and secure the tool immediately after use.

Rotary Die Grinders

Rotary die grinders perform a wide variety of jobs. Be sure that operators have a thorough understanding of the procedure for which they are using these tools.

Since grinders operate at high speeds, be alert and cautious to avoid injuries from contacting the working end or from thrown objects. Always wear safety goggles or safety glasses with side shields that comply with standards and a full face shield, when needed. Use a dust respirator for dusty working conditions. Wear hearing protection during extended periods of operation. Air sampling may have to be performed to determine if exposure to air contaminants caused by grinding is within acceptable limits.

Be sure that the switch is in the OFF position before plugging the grinder in. Caution operators to hold the wheel or cutter away from themselves and co-workers when starting a grinder. Also warn operators not to use a rotary die grinder with the cutter pointing towards them. If the grinder should slip, the cutter could cause injury. Have operators remove from the area all materials and debris that might be ignited by sparks.

Use grinding wheels when working with hard materials; and use rotary files for soft materials, such as aluminum, brass, copper and wood. Using grinding wheels on soft materials will excessively load the wheel and could cause the wheel to shatter or disintegrate. Parts could fly off and cause injuries.

Before each use, check the cutter or wheel for tightness. A loose cutter or wheel can be thrown from the rotary grinder and cause serious injuries. However, do not overtighten the collet. It can damage the collet cutter or wheel.

If the grinder is dropped, inspect it for damage, such as a cracked wheel, broken collet, or bent mandrel. Repair or replace damaged parts to prevent further breakage or other objects from being thrown.

Excessive pressure during use can bend or break the collet, mandrel, or wheel/cutter. If the grinder runs smoothly when not under load, but does not run smoothly under load, then excessive pressure is being used.

When placing a mounted grinding wheel, burr, or cutter in the collet, keep the distance between the back of the wheel and the front of the collet at a minimum. This prevents both bending the shank and wheel damage that could cause injuries. Make sure the shaft is engaged in the collet at least ½ in. (1.25 cm).

Grinding Wheels, Buffers, and Wire Brushes

These tools have special functions and thus require extra care when used. Grinding wheels present especially unusual hazards. Consequently, the storage, mounting, and use of grinding wheels require thorough training and extensive knowledge. Persons using or maintaining grinding wheels should observe safety recommendations conforming to ANSI B7.1, *Safety Requirements for the Use, Care, and Protection of Abrasive Wheels,* and ANSI B74.2, *Specifications for Shapes and Sizes of Grinding Wheels, and Shapes, Sizes, and Identification of Mounted Wheels.* Improper mounting, storage, or use of abrasive wheels could turn an important useful tool into a lethal device. (See also Chapter 22, Metalworking Machinery, in this volume.)

Sanders

Belt and disk sanders cause serious skin burns when the rapidly moving abrasive touches the body. Because it is impossible to guard sanders completely, thoroughly train employees in their use. For example, the motion of the sander should be away from the body. Keep all clothing clear of moving parts. Wear dust-type safety goggles or plastic face shields (Figure 19–19). If harmful dusts are created, use a respirator approved for this type of exposure. Work in well-ventilated areas. Refer to the OSHA standard Flammable Exposure Limit for wood dust and ensure that employees receive hazard communications training.

Cleaning

Sanders require especially careful cleaning because of the dusty nature of the work. If a sander is steadily used, periodically dismantle it. Thoroughly clean it every day by blowing it out with low-pressure air.

CAUTION: Compressed air used for cleaning must be 30 psig (200 kPa) or less. The operator should wear safety goggles or work with a transparent shield between his or her body and the air blast.

Fire and Explosion Hazards

Because of the dust created in wood sanding, the fire and explosion hazard is considerable. Keep the dust to a minimum through adequate ventilation or, if the sander is so designed, with a dust collector or vacuum bag. Because of the extreme combustibility of wood dust and wood finishes, do not dispose of such dust in incinerators. There it would burn with almost explosive force.

To minimize the explosion hazard, if much wood sanding must be done, use electrical equipment designed for this exposure. Provide fire extinguishers approved for electrical or Class C fires, and tell employees what to do in case of fire. (See Chapter 20, Woodworking Machinery, in this volume.)

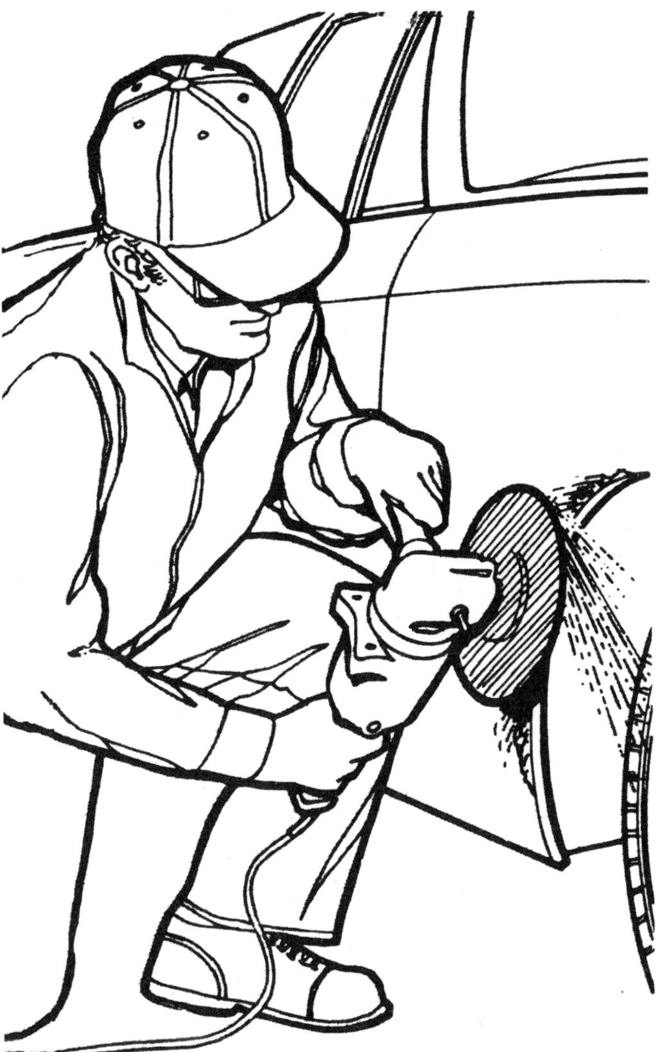

Figure 19–19. Be sure you have proper ventilation, eye protection, and dust mask, if necessary. Check the owner/operator's manual. (Courtesy Power Tool Institute.)

Operating

Observe the following precautions when using sanders:

- Before connecting a portable sander to the power supply, be sure that the switch and switch lock, if provided, are in the OFF position. If not, the sander will start immediately and loss of control could result in an injury.
- Stay constantly aware of the cord's location. Keep power-supply cords and extension cords from getting entangled with the moving parts of the sander. Damaged cords can result in an electrical shock. A cord that is contacted by a moving belt can cause loss of tool control and, thus, possible injury.
- Use abrasive belts that are the width recommended by the manufacturer.

- Always keep face and hands clear of moving parts such as belts and pulleys.
- Never lock a portable sander in the ON position when the nature of a job may require stopping the sander quickly, such as using a disk sander on an automobile's fender. The rotating disk could become jammed and result in an injury.
- Never force a portable sander. The sander's weight applies adequate pressure. Forcing the sander and providing too much pressure can cause stalling, overheating of the tool, burning of the workpiece, and possible kickback of the tool or workpiece.
- Be careful not to expose the sander to liquids. Do not use it in damp or wet locations.
- When adjusting the tracking of the belt on a portable belt sander, be sure that the sander is supported and positioned to avoid unintentional contact with the operator or adjacent objects.
- Do not work with a faulty tracking sander. Discontinue work until the problem is corrected.
- The work area should be at least 3 ft to 4 ft (0.9 m to 1.2 m) larger than the length of stock being sanded.
- Use jigs or fixtures to hold the workpiece, whenever possible.
- Always unplug sanders, and store them after use.
- Remove from the area material or debris that might be ignited by sparks from sanded metal. (See Chapter 22, Metalworking Machinery, in this volume.)

Routers

The widespread use of routers is based on their ability to perform an extensive range of smooth finishing and decorative cuts. Safety in operating a router starts by understanding that it runs at a very high speed—in the range of 20,000 RPM. This is 15 to 25 times faster than a drill.

Observe the following precautions when using a router:

- Install router bits securely and according to the owner/ operator's manual.
- Always use the wrenches provided with the tool.
- Keep a firm grip with both hands on the router at all times (Figure 19–20). Failure to do so could result in loss of control and, thus, lead to possible serious injuries.
- Hold only those gripping surfaces designated by the manufacturer.
- Always face the cutter blade's opening away from the body.
- When starting a router equipped with carbide-tipped bits, start the router beneath a workbench. This protects the operator from a possible flying cutter, should the carbide be cracked.

Figure 19–20. Grip routers firmly with two hands before turning on the switch. (Courtesy Power Tool Institute.)

- If the router is equipped with a chip shield, keep it properly installed.
- Keep hands away from bits and the cutter area when the router is plugged in.
- Do not reach underneath the work while bits are rotating. Never attempt to remove debris while the router is operating.
- Always disconnect the plug from the electrical outlet before changing bits or making any adjustments. When changing a bit immediately after use, be careful not to touch the bit or the collet with hands or fingers. They could be burned because of the heat built up from cutting.
- Make cutting-depth adjustments only according to the tool manufacturer's recommended procedures for these adjustments. Tighten adjustment locks. Make certain that the cutter's shaft is engaged in the collet at least ½ in. (1.25 cm). Check the owner/operator's manual.
- Be certain to secure clamping devices on the workpiece before operating the router.
- Be sure that the switch is in the OFF position before plugging the router into the power outlet.
- For greater control, always allow the motor to reach full speed before feeding the router into the work.
- Never force a router into the material.
- When removing a router from the workpiece, always be careful not to turn the base and bit toward the body.

- Unplug and store the router immediately after use.
- Always wear eye protection with side shields or full face protection.

Air-Powered Tools

Air hoses, air-powered grinders, and pneumatic-impact tools each have specific hazards. By observing certain precautions, however, using these tools will be much safer.

Air Hose

An air hose presents the same tripping or stumbling hazards that cords on electric tools do. A number of manufacturers offer self-storing, recoiling air hoses that work well when suspended above workstations. Persons or material unintentionally hitting the hose may cause the operator to lose his or her balance, or cause the tool to fall from an overhead place. Protect air hoses on the floor from trucks and pedestrians by laying two planks on either side of them or by building a runway over them. It is preferable, however, to suspend hoses over aisles and work areas.

Warn workers against disconnecting the air hose from the tool and using it for cleaning machines or removing dust from clothing. In the United States, regulations mandate that air pressure in excess of 30 psig (200 kPa) must not be used to clean machines and that low pressure must be used only with effective chip guarding or personal protective equipment, such as safety goggles. To remove dust from clothing, brush or use vacuum equipment.

Incidents sometimes occur when the air hose becomes disconnected and whips about. A short chain attached to the hose and to the tool's housing will keep the hose from whipping about, should the coupling break. In some cases, couplings should also have such chains between the sections of hose.

Before attempting to disconnect the air hose from the air line, shut off the air. Also release any air pressure inside the line before disconnecting it.

A safety-check valve installed in the air line at the manifold will shut off the air supply automatically if a fracture occurs anywhere in the line. However, the safety-check valve must be compatible with the air-flow rate.

If kinking or excessive wear of the hose is a problem, protect the hose with a wrapping of strip metal or wire. One objection to armored hose is that it may become dented and thus restrict the flow of air. This applies only to heavy-duty hose used in construction work, however.

Air-Powered Grinders

Air-powered grinders require the same type of guarding as electrical grinders. Maintenance of the speed regulator or governor on these machines is of particular importance to avoid overspeeding the wheel. Have a qualified person inspect the grinder at each wheel change.

Pneumatic-Impact Tools

In pneumatic-impact tools, such as riveting guns and jackhammers, the tool proper is fitted into a gun and receives its impact from a rapidly moving reciprocating piston. The piston is driven by compressed air at about 90 psig (600 kPa) pressure.

Determine noise levels caused by pneumatic-impact tools to see if protective-noise devices should be provided for workers. Check the time limits and sound-level requirements of the regulatory standards. Consider isolating such operations or substituting quieter methods.

Pneumatic-impact tools require two safety devices. The first device is an automatic-closing valve that is actuated by a trigger located inside the handle where it is reasonably safe from being unintentionally operated. The machine can operate only when the trigger is depressed. The second device is a retaining device that holds the tool in place so that it cannot be unintentionally fired from the barrel (Figure 19–21). Impress on all operators of small air hammers not to squeeze the trigger until the tool is on the workpiece.

When using pneumatic-impact tools, there is the hazard of flying chips. Have operators wear the correct eye protection for this hazard. If other employees must be in the vicinity, they should be similarly protected. Where possible, set up screens to shield other workers when chippers, riveting guns, or air drills are being used.

When two chippers are working, they should stand back to back, to prevent face cuts from flying chips. Workers should not point a pneumatic hammer at anyone, nor should they stand in front of operators who are handling pneumatic hammers.

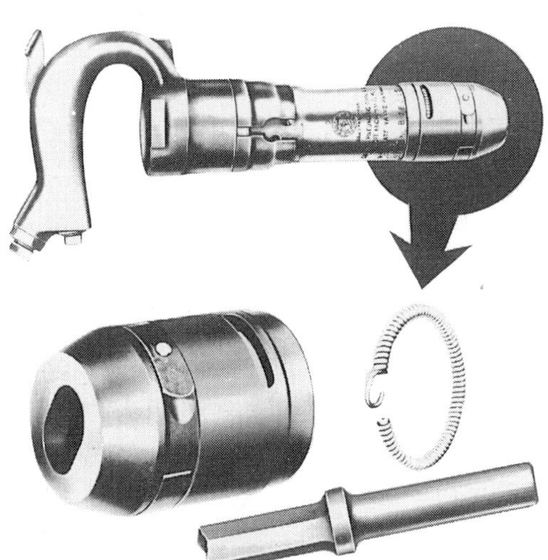

Figure 19–21. Chipping hammer safety retainer prevents the discharge of the tool. (Reprinted with permission from Chicago Pneumatic Tool Co.)

Handling heavy jackhammers causes fatigue and may cause strains. Provide jackhammer handles with heavy rubber grips to reduce vibration and fatigue. Operators should wear protective footwear with metatarsal guards to reduce the possibility of injury should the jackhammer fall. Since hand-arm vibration syndrome (HAVS) is the potential result of working with heavy vibrating equipment, supervisors should become familiar with the methods used to reduce exposure to vibration, and to symptoms of vibration-induced disease. The NSC's publication, *Occupational Vibration: Preventing Injuries and Illness*, discusses vibration-related problems.

Many incidents are caused by the steel drill breaking because the operator loses balance and falls. Also, if the steel is too hard, a particle of metal may break off and strike the operator. Follow the manufacturer's instructions for sharpening and tempering steel.

Impact Wrenches

Impact wrenches are widely used in garages, repair shops, field work, and manufacturing industries during disassembly and assembly operations. Electrical and pneumatic power are commonly used. Pneumatic power is favored in heavy-duty operations. Pneumatic-powered tools frequently generate high-impact noise levels. If noise cannot be controlled to meet noise regulatory requirements, then include workers in a hearing conservation program with annual audiometric testing and use of hearing protection. Also, enforce use of eye protection near the workstation.

Electrical power is used most often with single-drive socket operations. On single-socket operations, extension cords and air hoses left lying on the floor can cause tripping hazards. To reduce operator fatigue on repetitive operations such as assembly lines, a common practice is to suspend the unit above the point of operation or with a spring-loaded overhead on trolleys, or similar balancing device. This practice of suspending devices on balancing units could cause injury by striking the upper body during a pendulum-motion swing.

Use sockets that are specifically designated as impact- wrench sockets (Figure 19–22). Sockets and accessories that are made only for hand use will not stand up to impact-wrench use. They are subject to premature failure and breaking and, thus, could cause injuries. Impact-wrench sockets usually are identified by a black finish on the outside and have heavier section thickness.

Figure 19–22. For impact wrenches, use only sockets designated as impact wrench sockets. (Courtesy Power Tool Institute.)

Observe the following precautions when using impact wrenches:

- Never use a wire, soft pin, or nail to hold the socket onto the square spindle of the impact wrench. If the proper retaining device on the tool is broken, repair the tool.
- Avoid excessive impacting, particularly on small bolt sizes. Small bolts could easily be broken or the threads stripped. Overtorquing can cause premature failure of fasteners or other damage and, thus, lead to incidents.
- On applications where a low or critical level of torque is required, impact each fastener lightly. Then perform the final tightening with a hand torque wrench.
- If the owner/operator's manual recommends using wood-boring bits with an impact wrench, be sure to unplug the tool before changing the bits.
- Do not use an impact wrench in wet or damp environments.
- Do not use nails, bolts, or other makeshift items as substitutes for safety pins. Safety pins are usually designed to shear at definite preset pressure levels. Secure pins with retainers. Substituting poor-quality, inadequately designed shear pins or improperly using sockets could cause sudden failure and result in parts flying off.

Power Nailers and Staplers

Power nailers and staplers are fastening tools used primarily for securing wood products and other materials to wood by rapidly driving a nail or staple fastener into the materials being assembled. These tools are most commonly used in the building construction industry, but are also found in mass production applications located in manufacturing and industrial facilities. Tool capabilities and features vary with the wide variety of fastening applications.

Although established safety practices apply to all power nailers and staplers, additional precautions may be required depending upon the tool's source of power. Most of these tools are pneumatic, i.e., they are powered by compressed air from an air compressor. Some are electric. Others use a gas fuel cartridge, sometimes with a battery. Tools with a gas fuel cartridge should only be used in well-ventilated areas. Manufacturers of tools using gas fuel have guidelines for tool and gas cartridge storage. Power driven nailers and staplers are not intended for use in explosive atmospheres.

Pneumatically powered tools must only be connected to a source of clean, dry, air regulated to the manufacturer's safe operating pressure. Bottled gases must never be used with pneumatic tools since they can result in the explosion of the tool. This is due to the fact that gas cylinders, bottles, and tanks are filled to very high pressures for efficiently transporting large quantities of the gas. The tank pressures far exceed the safe operating pressures of power nailers and staplers. Failure of pressure regulators used to reduce tank pressure could result in the delivery of high-pressure gas directly to the tool, possibly resulting in an explosion. Furthermore, the use of many bottled gasses, such as oxygen, can result in combustion—also presenting the possibility of explosion if used with these pneumatic tools.

Power nailers and staplers vary in the types of operating controls and how the controls are activated before driving a fastener. The tool may have one or more trigger controls, a 'workpiece contact,' or both. If so equipped, the tool's workpiece contact must be brought into contact with the materials being assembled before the fastener can be driven. To minimize the possibility of unintentionally activating a power nailer or stapler, tool operators must always keep their fingers away from the trigger or triggers until they are ready to drive a fastener.

When selecting a power driven nailer or stapler for a given fastening application, consideration should be given to the type of operating control activation sequence desired. For example, tools may require a specific sequence of the operating controls, including activation of the trigger or triggers, and workpiece contact before the tool will drive a fastener. The operating control activation sequence can vary depending on the tool or application and may be changeable by the operator via the tool's built-in features or by a 'modification kit' available from the manufacturer.

It is mandatory that all power driven nailer and stapler operators, and other people in the immediate work area, wear eye protection. Airborne dust and debris are commonly generated during the operation of these tools. Therefore, it is imperative that both front and side protection be provided to the eye. Tool manufacturers, and the applicable ANSI safety standard for these tools, require use of eye protection meeting ANSI Z87.1, 'American National Standard for Occupational and Educational Eye and Face Protection.'

Since power-driven nailer and stapler designs vary among manufacturers, and sometimes within the manufacturers' product line, tool operators must have a thorough understanding of the operating and safety instructions for each tool they use. Besides the subjects discussed above, tool instructions typically address other topics such as installation, maintenance, inspection before use, loading, use, clearing of jams, handling between fastenings, and additional personal protective equipment. Tool users must read and understand all instructions—they will get the best results and performance.

Percussion Tools

Percussion tools, such as hammers, rotary hammers, and hammer drills, are primarily associated with masonry applications as varied as chipping, drilling, anchor setting, and breaking of pavement. They range from pistol-grip tools to large demolition hammers. Normal operating modes include hammering, hammering with rotary motion and rotation, or drilling only. Many models incorporate a combination of these modes.

The capacity of these tools is normally rated in maximum diameter that is displayed on the nameplate. Do not attempt to use a bit larger than that specified unless otherwise recommended in the owner/operator's manual. Before operating one of these tools, compare the data on the nameplate with the voltage source. Be sure that the voltage and frequency are compatible.

Observe the following precautions when using percussion tools:

- For maximum control, use the auxiliary handles provided with the tool.
- Do not tamper with clutches on those models that provide them. Have the clutch settings checked at the manufacturer's service facility, at the intervals recommended in the owner/operator's manual.
- Check for subsurface hazards, such as electrical conductors or water lines, before drilling or breaking blindly into a surface. If wires are present, have a qualified person disconnect them at the power source, or be sure to avoid them. Otherwise, there is the risk of lethal shock or fire. Drain and cap water pipes.
- Always hold the tool by the insulated grasping surfaces.
- Do not force the tool. Percussion tools are designed to hit with a predetermined force. Added pressure by the operator only causes operator fatigue, excessive bit wear, and reduced control.

- Keep the work area clear of debris.
- Always have firm footing.
- Remember to unplug the tool before changing bits or servicing.
- For percussion tools with rotating features, comply with all of the operating considerations referred to under Electric Drills, in this chapter, and Drills, in Chapter 22, Metalworking Machinery.

Special Power Tools

Tools with a flexible shaft require the same type of personal protective equipment as do direct-powered tools of the same type. Install and operate abrasive wheels to conform with ANSI B7.1, *Safety Requirements for the Use, Care, and Protection of Abrasive Wheels,* and federal regulations. Protect the flexible shaft against denting and kinking, which may damage the inner core and shaft. Whenever the tool is not in use, shut off the power. When the motor is being started, hold the tool end with a firm grip to prevent injury from sudden whipping. The abrasive wheel or buffer of the tool is difficult to guard. Because it is more exposed than the wheel or buffer on a stationary grinder, use extra care to avoid damage. Place grinder wheels on the machine or put them on a rack—do not leave them on the floor.

Hydraulic Power Tools

Hydraulic power tools are used in some industries, notably the electric utility industry, where employees work aloft from a hydraulically powered aerial-lift device. The power is obtained from the source used to operate devices such as hydraulic chain saws and compression devices. Some compression devices have a small hydraulic press that is pumped by the operator. Small leaks in the hydraulic hose or around fittings are hazards in the use of such equipment. There have been instances where employees have put a hand over a pinhole leak and had oil forced into their finger by the high pressure. Also take care to always use a hose built for the pressure involved because a rupture can cause serious consequences.

Gasoline-Powered Tools

Gasoline-powered tools are widely used in logging, construction, and other heavy industries. The best known and most commonly used tool is the chain saw.

Selecting a Chain Saw

Chain saws can be purchased in a variety of horsepowers and sizes. Some points to consider before purchasing include the size of the job, the balance of the saw, hand guards, kickback features, vibration reduction systems, and the convenience or ease of refueling.

Before beginning an operation, read the manufacturer's manual. Be sure that persons using chain saws are trained to operate the equipment according to the manufacturer's specific instructions.

Chain-Saw Hazards

Operators of chain saws are exposed to hazards similar to those encountered by workers using hand saws. Kickback is the single biggest cause of chain-saw injuries. A kickback is the sudden and potentially violent rearward and/or upward movement of the chain saw. It is often caused by the chain striking wood or other objects on the top quadrant on the tip of the chain guide bar. It can also be caused by binding or pinching in the cut.

Three types of anti-kickback devices are found on chain saws: a safety nose or guard, a safety chain, or a chain brake. The safety nose prevents contact with the chain at the end of the chain. The safety chain is designed to reduce the tendency for the chain to catch or 'hang up' in the wood. A chain brake stops the chain as the chain bar raises upward and the hand (at the handle) pivots against the brake switch.

In addition, the following hazards are specific to chain saws:

- falling while carrying a saw or when sawing
- sprains and strains from carrying, and working with, a heavy saw
- hand-arm vibration syndrome
- being cut by contact with the chain while it is in motion
- being cut by the chain when it is not in motion, either on or off the saw
- injuries from starting the gasoline engine
- inhaling exhaust fumes
- being struck by wood from overhead because a tree is vibrating
- sawdust in the eyes, especially when holding the tail stock or "stinger" end of a saw above the head
- burns from contact with a hot muffler or cylinder head
- injuries due to saws binding and kicking back at an operator
- injuries from falling trees and snags or rolling logs because the operator could not hear them above the noise of the saw's engine.

Carefully select saw operators. During their training period, inexperienced employees should work with an experienced faller under constant supervision.

Prevention of Fires

Take gasoline to the job in a sturdy, capped, UL-listed container that is painted red and labeled GASOLINE. The container should have a suitable spout for pouring gasoline into the tank, or provide a funnel for this purpose. Under no circumstances should the gasoline tank be replenished while the engine is running. Carefully wipe off any gasoline spilled on the tank or engine before starting the engine.

Additional precautions to prevent fires include the following:

- Turn off engine before refueling.
- Fill tanks only in an area with bare ground.
- Do not fill tanks where another tank was previously filled.
- Keep saws clean of gasoline, oil, and sawdust.
- Keep mufflers in good condition.
- Keep spark plugs and wire connections tight.
- Have fire extinguishers near power saws at all times.
- Keep flammable materials away from the point of the saw's cut.

Powder-Actuated Tools

Powder-actuated tools are used to make forced-entry fastenings in various construction materials (Figure 19–23). The systems are simple to use. However, there are precautions and safeguards that must be observed. Allow only trained and qualified operators to use this equipment, and then under close supervision. To become a qualified operator, a person must be thoroughly trained under supervision of a manufacturer's authorized instructor. Upon completion of the background training, the person is required to demonstrate competence through use of the system in varied applications and to pass a written examination. Upon successful completion of the examination, the instructor will issue a Qualified Operator's Card (Figure 19–24), records of which may be kept at or required by certain regulatory agencies.

Also, the possessor of a Qualified Operator's Card must be familiar with any regulations that apply to the use, maintenance, and storage of the system. Obtain more information about the use of these systems by writing to the Powder-Actuated Tool Manufacturer's Institute (see References), and from the NSC Occupational Safety & Health Data Sheet 12304–0236, *Powder-Activated Hand Tools.* (See also 29 *CFR* 1910.243(d)(1)(i), Explosive Actuated Fastening Tools.)

PERSONAL PROTECTIVE EQUIPMENT

Workers using revolving tools, such as drills, saws, and grinders, should not wear ties, gloves, loose clothing, and jewelry. Clothing should be free of oil, solvents, or frayed edges to minimize the fire hazard from sparks. The weight of most power tools makes it advisable for users to wear safety shoes to reduce the chances of injuries should the tools or workpiece fall or be dropped.

When power tools are used in overhead places, the operator should wear fall-protection devices to minimize the danger of falling, should the tool break suddenly or shock the operator, or should the operator slip. Also, attach a safety line to the tool to keep it from falling on persons below should it be dropped. (See 29 *CFR* 1926, Subpart M, Fall Protection.)

Workers operating chain saws and carrying them through the woods must be surefooted. Because falls are among the most common incidents, operators

Figure 19–23. This powder-actuated tool drives studs into concrete slab. The operator should wear eye, ear, and head protection. Note the holster (left) for carrying the tool. (Reprinted with permission from Hilti Inc.)

QUALIFIED OPERATOR OF POWDER-ACTUATED TOOLS

Make(s)________________ Model(s)________________

This certifies that ________________________
(NAME OF OPERATOR)

Card No.__________ Soc. Sec. No.________________

Has received the prescribed training in the operation of powder actuated tools manufactured by

____
(NAME OF MANUFACTURER)

Trained and issued by ________________________
(SIGNATURE OF AUTHORIZED INSTRUCTOR)

I have received instruction in the safe operation and maintenance of powder actuated fastening tools of the makes and models specified and agree to conform to all rules and regulations governing their use. Failure to comply shall be cause for immediate revocation of this card.

________________________ ____________
(SIGNATURE) (DATE)

Figure 19–24. A wallet card for qualified powder-actuated tool operators is available from tool manufacturers. A list of instructors should be maintained by each manufacturer. Based on ANSI A103—1984.

should wear proper footwear to minimize this hazard. Operators should wear sharp caulked boots. In some parts of the country, hobnailed shoes are preferred. In the winter, operators should wear rubber-soled shoes. Protective footwear is a good investment for members of a cutting crew.

Fallers and buckers should always wear protective helmets because many people have been killed by falling trees. Many other lives have been saved, however, because workers wore safety hats. Make their use mandatory.

On buffing, grinding, and sanding jobs that produce harmful dusts, provide workers with approved dust-type respirators. For operators of powder-actuated tools or jack hammers, provide hearing protection if more positive noise controls are not possible.

Employees should wear eye and face protection where flying particles present a hazard. Some companies require eye protection for all power-saw operators. In all operations where striking and struck tools are used, or where the cutting action of a tool causes particles to fly, provide eye protection that conforms to ANSI Z87.1, *Practice for Occupational and Educational Eye and Face Protection.* Minimize the hazard of flying particles by using nonferrous, soft striking tools and by shielding the job site with metal, wood, or canvas. However, eye protection is still required.

Wear eye protection or face shields when using woodworking or cutting tools, such as chisels, brace and bits, planes, scrapers, and saws. There is always the chance of particles falling or flying into the eyes. Also wear eye protection or face shields when working with grinders, buffing wheels, and scratch brushes. The unusual positions in which the wheel operates may cause particles to be thrown off in all directions. For this reason, eye protection is even more important than it is when working with stationary grinders.

Do not overlook eye protection on the following jobs:

- cutting wire and cable
- striking wrenches
- using hand drills
- chipping concrete
- removing nails from lumber
- shoveling material
- working on the leeward side of a job
- using wrenches and hammers overhead
- working on other jobs where particles of materials or debris may fall

A ballistic nylon patch that covers part of the leg has reduced injuries to that part of the body. The patches increase the time that an operator has to shut off this saw, in the event a saw's chain should come against the leg.

Provide first aid kits at the job site. Be sure that at least one individual in each work crew is trained in first aid. In areas where poisonous snakes are known the exist, provide a snake bite kit.

Take noise-level readings of the chain saws and related equipment. If necessary, provide earmuffs or earplugs for workers. Protective clothes such as vibration damping gloves may need to be used to prevent HAVS from developing. Keeping warm and dry is also an important precaution, so adequate rainwear and cold weather gear is necessary. Smoking also exacerbates vibration-induced illness and should be discouraged.

SUMMARY

- Proper selection, use, care, and supervision of hand and power tools can prevent abuse of these tools and eliminate or reduce employee injuries.
- Management should select proper hand and power tools and change tools only after careful consideration. These changes may include adding personal protective equipment, job rotation, or other adjustments.
- Five safety practices include (1) provide proper protective equipment, (2) select the right tool for the right job, (3) keep tools in good condition, (4) properly ground power tools, (5) train workers to use tools correctly, and (6) store tools in a safe place.
- Tool boxes should be used only for storing tools and not as stools, anvils, saw horses, or lunch boxes. Make sure tool boxes and cabinets are locked after each work day and that all tools are accounted for.
- A good maintenance and repair program includes tool control through periodic inspection of all tool operations. Make sure employees have adequate work space and equipment for repairs.
- Misuse of common hand tools is a source of many injuries. Workers should be trained in safe work habits and proper use of tools.
- Soldering irons can be the source of burns and illnesses that result from inhaling fumes. Soldering irons must have adequate holders to prevent burns, and workers have proper protective gear and ventilation to eliminate hazards.
- Hazards of portable power tools are associated with their mobility and energy sources. Workers should be trained in proper tool use, to select the right tool for the job, and to inspect and repair their equipment.
- The risk of electrical shock from electrically powered tools can be reduced by using battery-operated tools, properly grounding equipment, and using only approved wiring and current. All parts of electrical equipment should be inspected regularly.
- To prevent injuries associated with air hoses, workers should make sure hoses do not present tripping hazards, avoid using hoses as cleaners, and prevent unintentional disconnection of hoses from the tools.
- Pneumatic-impact tools require an automatic-closing valve and a retaining device to hold the tool in place. Workers must protect their hearing and eyesight when using these tools.
- Special power tools include hydraulic, gasoline-powered, and powder-actuated equipment. Each type has its own hazards and safety precautions for safe operation.
- Workers operating power tools must wear the correct clothing, personal protective equipment, and fall protection equipment where appropriate.

REFERENCES

American National Standards Institute, 11 West 42nd Street, New York, NY 10036.

Budzik, R.S. *Precision Sheet Metal Shop Theory,* 2nd ed. Chicago: Practical Publications, 1988.

Compressed Air and Gas Institute, 1230 Keith Building, Cleveland, OH 44115.

Grandjean, E. *Fitting the Task to the Man: An Ergonomic Approach,* 4th ed. New York: Taylor & Francis, 1988.

Grinding Wheel Institute, 30200 Detroit Road, Cleveland, OH 44145.

Jackson, A. *Tools and How to Use Them—An Illustrated Encyclopedia.* Avenal, NJ: Outlet Book Company, Inc., 1992.

McDonnell, L.P, Kaumeheiwa, A. *The Use of Hand Woodworking Tools,* 2nd edition. Albany, NY: Delmar Publications, Inc., 1978.

National Fire Protection Association, 1 Batterymarch Park, Quincy, MA 02269.
National Electrical Code, NFPA 70, 1993.

National Safety Council, 1121 Spring Lake Drive, Itasca, IL 60143-3201.
Fundamentals of Industrial Hygiene, 4th ed., 1988, Chapter 13, Ergonomics.
Occupational Vibration: Preventing Injury and Illness (booklet), 1991.
Occupational Safety and Health Data Sheets: Powder-Activated Hand Tools.

Powder-Activated Tool Manufacturer's Institute, 1000 Fairgrounds Boulevard, St. Charles, MO 63301.
Basic Training Manual.

Power Tool Institute, Inc., PO Box 818, Yachats, OR 97498.
Power Tool Safety Is Specific.

Putz-Anderson, V (ed) *Cumulative Trauma Disorders: A Manual for Musculoskeletal Diseases of the Upper Limbs.* New York: Taylor & Francis, 1988.

TPC Training Systems, 750 Lake Cook Road, Buffalo Grove, IL 60089.

Maintenance Fundamentals: Hand Tools and Portable Power Tools, 1980.

Underwriters Laboratories, Inc., 333 Pfingsten Road, Northbrook, IL 60062. *Electric Tools,* UL 45.

U.S. Bureau of Naval Personnel. *Tools and Their Uses,* Rate Training Manual, NAVPERS 10085-B. New York: Dover, 1973.

U.S. Consumer Product Safety Commission, Washington, DC. *Product Safety Fact Sheet No. 51, Chain Saws.*

U.S. Department of Commerce, Office of Technical Service, Washington, DC 20234. *Sparking Characteristics and Safety Hazards of Metallic Materials,* Technical Report No. NGF-T–1–57, PB 131131.

U.S. Department of Health and Human Services, Public Health Service, Centers for Disease Control, NIOSH,

Criteria for a Recommended Standard, Occupational Exposure to Hand-Arm Vibration, Sept. 1989.

Elements of Ergonomics Programs: A Primer Based on Workplace Evaluations of Musculoskeletal Disorders. March 1997.

U.S. Department of Labor. Occupational Safety and Health Administration, 200 Constitution Avenue NW, Washington, DC 20210.

Code of Federal Regulations, Title 29. Section 1910.242. Chapter XVII.

29 CFR 1910.243(d)(1)(i), Explosive Actuated Fastening Tools.

29 CFR 1910.133, Eye and Face Protection

29 CFR 1926.404(b)(1)(i)

29 CFR 1926, Subpart M, Fall Protection.

Walker, J.R. *Exploring Metalworking: Basic Fundamentals.* South Holland, IL. The Goodheart-Wilcox Co., Inc., 1987.

Zinngrabe, C.J, Schumacher, F.W. *Sheet Metal Hand Processes.* Albany, NY: Delmar Publications, Inc., 1974.

REVIEW QUESTIONS

1. Each year hand tools are the source of what percent of all compensable disabling injuries?
   a. 2%
   b. 6%
   c. 10%
   d. 14%

2. Most incidents involving hand tools and portable power tools can be eliminated by observing what six safety practices?
   a.
   b.
   c.
   d.
   e.
   f.

3. What is the advantage of having centralized tool control in an industrial setting?

4. When metal tools break during normal use, the causes are usually related to the tools' ________.
   a. Size
   b. Quality
   c. Handle

5. Which of the following is probably the most commonly used and abused tool?
   a. Screwdriver
   b. Hammer
   c. Wrench
   d. Pliers

6. When striking another tool, the striking face of the hammer should have a diameter approximately ____________ larger than the struck face of the tool.

7. Identify the tool that is more frequently the source of disabling injuries than any other hand tool.

8. Based on their power source, portable power tools are divided into what five primary groups?
   a.
   b.
   c.
   d.
   e.

9. What are the inherent risks of portable power tools?
   a.
   b.
   c.
   d.

10. What are three precautions that should be observed when using portable power tools?
    a.
    b.
    c.

11. Name seven ways to properly maintain power tools.
    a.
    b.
    c.
    d.
    e.
    f.
    g.

12. What is the most convenient way of safeguarding the operator of portable electrical tools?

13. Name seven of the twelve precautions that should be observed when using portable electric drills.
    a.
    b.
    c.
    d.
    e.
    f.
    g.

20

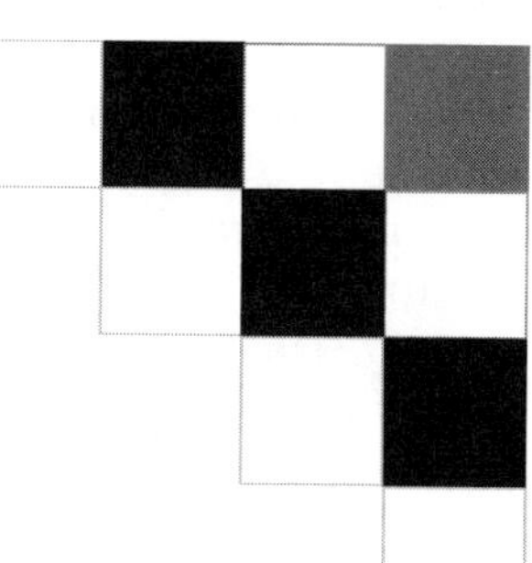

Woodworking Machinery

Joseph L. Durst, Jr.

Patrick J. Conroy

Philip E. Hagan, MPH, CIH, CHMM, CHCM

Each piece of woodworking equipment poses its own hazard potential. To reduce the possibility of serious injury, management should provide workers with correctly guarded equipment, adequate jigs and fixtures, appropriate training, and proper enforcement of established safety rules. The topics covered in this chapter include:

- general principles to be included in a safety program for woodworking machinery
- hazards and safe work practices for various types of power saws
- safety precautions and worker protection for the use of woodworking equipment

Because woodworking equipment is used in many industries, this chapter covers only the equipment and not specific woodworking industry operations.

The woodworker should know proper and safe procedures: how to choose the right machinery for the job and how to use the machinery correctly. The well-trained operator recognizes the potential for incidents and knows what to do when warning signs arise. For example, a change in noise, pitch, or any other operating characteristic of mechanical equipment should alert the trained worker to follow approved procedures for reporting or correcting a potentially hazardous situation. Supervisors and line managers should observe novice operators frequently to ensure that they are following established procedures.

GENERAL SAFETY PRINCIPLES

Companies should provide employees with equipment that meets the existing standards and regulations of the U.S. Occupational Safety and Health Administration (OSHA), the American National Standards Institute (ANSI), and the National Fire Protection Association's (NFPA) *National Electrical Code.* Purchasers must specify on their purchase orders whatever optional or accessory parts are necessary to meet the requirements for mechanical and electrical safeguarding.

Management must observe the following general safety principles. A summary of safety rules for specific machines is given in Figure 20-1.

- Maintain all machines so that while they are running at full or idle speed and with the largest cutting tool attached, they are free of excessive noise and vibration.
- Level all machines, including portable or mobile ones, and, where necessary, dampen their vibration. Secure machines to the floor or other suitable foundations to eliminate all movement or "walking."
- Secure small units to benches or stands of adequate strength and design (Table 20–A).
- Make sure the machine is constructed so that tools which are too large for the machine's design cannot be mounted on it.
- Ensure that all arbors and mandrels have firm and secure bearings and are free from slip or play.
- Regularly check the adjustment of all safety devices. Those involving electrical circuits should be actuated to make sure they operate properly. Operators should always stop and securely lock out (energy isolation) machines before cleaning, adjusting, or maintaining them. (See Control of All Energy Sources in Chapter 6, Safeguarding, in this volume.)
- Keep loose clothing, long hair, jewelry, and gloves away from rotating parts of machinery, especially from nip points and the point of operation.
- After the equipment has been completely stopped, clean work surfaces with a brush, not with the hand or a compressed air nozzle.
- If possible, make adjustments only while the machine is not running. (See Chapter 6, Safeguarding.)

Electrical Equipment

All of the metal framework on electrically driven machines should be grounded, including the motor. The framework should comply with the National Fire Prevention Association (NFPA) 70, *National Electrical Code,* and other applicable standards. There may be other local, state or provincial, and federal codes that apply to these machines. The NFPA code includes the following provisions:

- The machine shall have a cutoff device (an easily identified EMERGENCY STOP switch, panic bar, or dead-man switch) within reach of the operator in the normal operating position.
- Electrically driven equipment shall be controlled with a magnetic switch or other device that will prevent automatic restarting of the machine after a power failure. This is needed in cases where the machine, should it start automatically, would create a hazard (Figure 20–2).
- Clearly marked power controls and operating controls shall be located within easy reach of the operator and away from a hazardous area. They shall be positioned so the operator can remain at the regular work location while operating the machine.
- Each operating control shall be protected against unexpected or accidental activation (Figure 20–2).
- Each machine operated by an electric motor shall be provided with a positive means (lockout) for

SUMMARY OF SAFETY RULES FOR VARIOUS WOODWORKING TOOLS

Every operator should be trained in the safety rules covered in this chapter. As a summary, safety rules that demand close attention are listed below. Be sure the operator checks the manufacturer's manual, understands the requirements, and follows the recommended procedures.

TABLE SAW

- Feed with the body to side of stock
- Blade height
- Splitter and antikickback fingers for ripping
- Stock firm to fence
- Remove ripfence for crosscuts
- Blade guards

CIRCULAR SAW

- Blade guards
- Binding
- Blade—correct type
- Blade—tight on the arbor
- Firm support for work
- No obstructions
- Begin cut with motor at manufacturer's recommended speed for materials being cut
- Hand and finger position

RADIAL ARM SAW

- Ripsawing—direction of feed (cut) and antikickback fingers
- Blade guards
- Pull for cross cuts
- End plates on track—arm tight
- Clamp handles tight
- Material tight to fence
- Return cutter to rear of track
- Hand and finger position

BAND SAW

- Feed with body to side of stock
- Guard height 1/8-in. clearance of material
- Tension and type of blade
- Release cuts before long curves
- Stop machine to remove scrap or pull out incomplete cut
- Flat stock
- Push stick for small parts

JOINTER/PLANER

- Depth of cut
- Length of stock
- Sharp cutters
- No hands over cutters
- Push stick for small stock
- Guard

WOOD SHAPER

- Clamping workpiece
- Use correct guard
- Feed into knives—don't back off
- No feeding between fence and cutters
- Collar and starting pin work for irregular work—stock of sufficient weight
- Fence opening only enough to clear cutters
- Use stock as guard by shaping the underside of stock
- Spindle nut tight
- Shape only pieces 10 in. or longer
- Proper types of cutters

SANDER

- Keep hand from abrasive surface
- Ventilation
- Belt or disk condition
- Sand on downward side of disk

LATHE

- Stock without defects, glued joints dry
- Power off when changing speeds on V-belt lathes
- Toolrest close to stock
- Hold tools firmly in both hands
- Remove toolrest when sanding or polishing

Figure 20–1. Summary of rules for safe operation of woodworking tools. (Printed with permission from Power Tool Institute, Inc.)

Table 20–A. Typical Heights and Work Space

Machine	Table Heights (in.)		(cm)	Work Area
Band saws	46		115	On three sides—a radius equal to twice the band saw diameter (as measured from the point cut).
Circular saws	36	(Hand feed)	90	Clearance on the working side should be 3 ft (90 cm) plus the length of the stock.
	32	(Power feed)	80	
Jointers	33		85	3 ft plus the length of stock.
Lathes	41		100	Clearance of at least 30 in. (75 cm) from stand, with smaller distances on ends and backside allowable.
Radial saws	39		85	Ripping—saw table equal to twice the length of the stock. Crosscutting—saw table equal to length of the stock plus 3 ft.
Sanders	36		90	3 ft plus the length of stock.
Shapers	36		90	3 ft plus the length of stock.

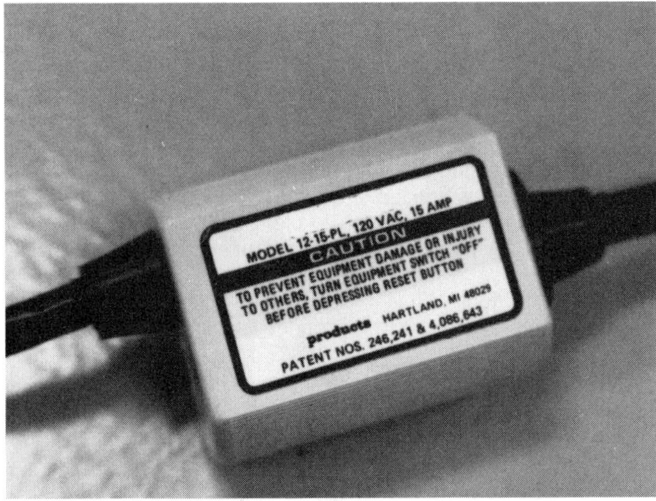

Figure 20–2. *Top*: When installed in the electrical cord of a machine, this device will prevent automatic restarting. *Bottom*: This magnetic switch has a ring guard around the START button to protect against accidental reactivation.

rendering the controls inoperative. If more than one person is involved in the maintenance or repair of the machine, each should install a separate padlock with a hasp. In addition to locking out the machine, the machine should be identified as inoperative. If the machine does not have a power disconnect to lock it in the OFF position, unplug the cord and insert a small padlock through the holes in the plug (Figure 20–3). (See Chapter 6, Safeguarding.)

- Install an electronic motor brake on machines that have excessive coasting time (Figure 20–4). This device can greatly reduce the exposure at the point of operation.

Guards

Enclose or guard all belts, shafts, gears, and other moving parts so no hazard is present for the operator. (See ANSI/ASME B15.1, *Safety Standard for Mechanical Power Transmission Apparatus.*) Because most woodworking operations involve cutting, it is necessary, although often difficult, to provide guards at the point of operation. On most machines, the point-of-operation guard must be (1) movable to accommodate the wood, (2) balanced so as not to impede the operations, and (3) strong enough to provide protection to the operator. Whenever possible, completely cover blades and cutting edges at the point of operation. Not all such areas can be fully covered while the tool is in the workpiece; for example, radial saws cannot be guarded in this manner. Management should use another method for protecting the worker and bystanders in these cases. (See Chapter 6, Safeguarding.)

Work Areas

Provide ample work space around each machine. Suggested typical heights and minimum work spaces are given in Table 20–A. The working surfaces of the machine should be at a height that will minimize fatigue. Make adjustments if the worker is taller or shorter than average (see Chapter 16, Ergonomics Programs, in the *Administration & Programs* volume). All accessory or feed tables should be the same height as the working surface of the machine.

Perform routine floor maintenance in the work area to prevent splintering and protruding nails. Keep floors level and free from holes and other irregularities. Install slip-resistant flooring in the work area near the machines. Mark aisleways with paint, railings, or other approved markings.

Maintain good housekeeping to prevent the accumulation of dust and chips. For instance, many exhaust vacuum systems are desirable and effective. A clean operation makes work easier and helps prevent fire and

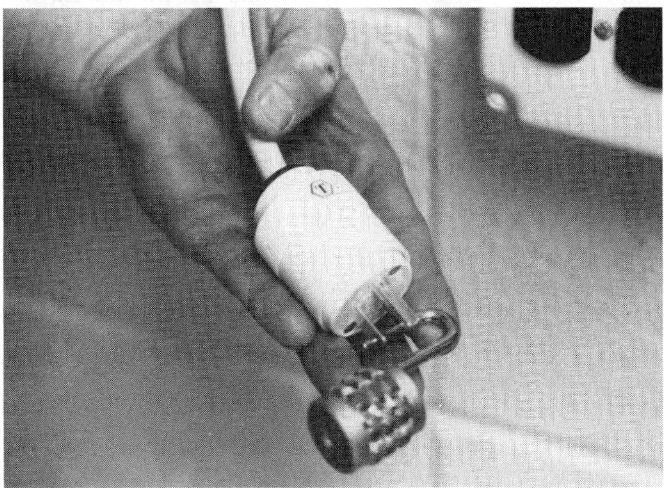

Figure 20–3. *Top*: Single-pole breaker lockout device. (Courtesy W. H. Brady Co., Signmark Division.) *Center*: Lockout with hasp, separate padlocks, and a tag. *Bottom*: Lockout for machines without a power disconnect.

Figure 20–4. On machines with excessive coasting time, this electronic motor brake greatly reduces exposure at the point of operation.

dust explosions. Because a number of fires originate and spread through ductwork, management should install automatic extinguishing systems in ducts, as well as in the collecting systems.

Adequately light the work area and the adjacent stock areas. Generally, 50 footcandles (538 lux) will be needed for work, but fine work may require 100 or more footcandles (1,076 lux). General illumination of 80 to 100 footcandles (861 to 1,076 lux) will pay dividends in both accident prevention and efficiency. There should be no shadows or reflected glare on the working surface.

Materials Handling

The facility's layout should encourage an even flow of materials and keep backtracking and crisscrossing to a minimum. Operators should not have to stand in or near aisles.

Arrange the machines so that the material handled by the operator and others requires a minimum of movement and changes of heights. This applies to both incoming supply and outgoing stock.

Inspection

Make safety checks by putting machines through trial runs before beginning a job and after each new setup. This usually is the responsibility of the setup or maintenance person.

The operator should inspect the machine at each new setup and at the start of each shift. The inspection process should follow the manufacturer's recommendations and the requirements and flow patterns of the workplace. This process would include inspecting the operating controls, safety controls, power drives, and sharpness of cutting edges and other parts. All cutting edges and tools must be kept sharp and be properly adjusted and properly secured.

Hearing Protection and Conservation

Most woodworking machinery creates high noise levels requiring that employers establish and maintain effective hearing conservation programs. Because some woodworking machines, especially saws, are noisy, management should have a qualified person or industrial hygienist take sound-level measurements. If the reading of the sound level dBA (slow response) exceeds 85 over an eight-hour period, that worker must be included in an effective hearing conservation program. If the sound level equals or exceeds 90 decibels (dBA) over an eight-hour period, then the worker's exposure to the sound level must be reduced. If the level is less than 85 dBA (legal in the United States), no action will be required. Ear protection may be desirable at 85 dBA and less. However, management should try to control the employee's exposure or somehow reduce the noise level. Some circular-saw blades are specifically designed to reduce noise levels. In other cases, large, sound-dampening washers can be used to keep noise at a safe level.

When a woodworking process creates fine dust, the safety professional should have the amount sampled. The threshold limit values (TLVs) and maximum permitted exposure (MPE) levels have been established for many materials and should be observed. Fine dust can be a health, fire, or explosion hazard. For workers' protection, respirators that reduce inhalation of various types of nuisance dust are available. (See National Safety Council), *Fundamentals of Industrial Hygiene,* 4th ed., in References at the end of this chapter.)

Personal Protective Equipment

All individuals in the work area should wear eye protection. Safety goggles complying with ANSI Z87.1, *Practice for Occupational and Educational Eye and Face Protection,* are excellent for operations that may generate flying objects. Face shields are not adequate if there are flying objects but do help if there is dust. On some operations, workers may need to use face shields in addition to goggles. Safety glasses with side shields may also be effective.

Workers should not wear loose clothing, gloves, and jewelry (especially rings, bracelets, and chains) that can become entangled in moving machinery. They should wear hair nets or caps to keep long hair away from moving parts and should keep their beards trimmed. Workers can protect their hands from splinters and rough lumber by wearing gloves. However, gloves should not be worn near rotating parts of the machine. Workers should wear approved protective footwear when handling heavy material or when there is a danger of injuring their feet. Where there is danger of a kickback—especially in ripping operations—workers should wear proper abdominal guards or antikickback aprons and always stand to one side of the saw.

Standards and Codes

There are a number of OSHA standards, such as 29 *CFR,* 1910.213, *Woodworking Machine Requirements,* that state required safety features for woodworking machines. Additionally, some states and other jurisdictions have codes that specify requirements. Management should consult all of these sources. The National Safety Council library has available a number of Occupational Safety & Health Data Sheets on woodworking machines. (See References at the end of this chapter.)

SAWS

All saws pose potential hazards for operators. Safety and health professionals can minimize these hazards by (1) providing training for operators, (2) ensuring that all machinery is properly guarded, and (3) making sure that all ANSI, NFPA, and government regulations are followed.

Circular Saws

Blade cuts or abrasions and kickbacks are among the most frequent incidents involving circular saws. These can be minimized by proper guarding and training and by enforcing safe working procedures.

Circular-saw operators are often injured when their hands slip off the stock while pushing it into the saw, or when holding their hands too close to the blade during the cutting operations. Other personnel can be injured by coming into contact with the blade when removing scrap or finished pieces from the table. Poor housekeeping practices and slippery floors are other sources of incidents involving circular saws.

Circular saws are designed to permit a wide range of cutting tasks. The problem with saws, as with most multiple-use equipment, is the difficulty in designing one guard that offers maximum protection for all types of tasks. The object is to prevent contact with the blade by using the proper type of hood guard, jigs, fixtures, combs, or other devices. Figures 20–5, 20–6, and 20–7 illustrate safety features and procedures.

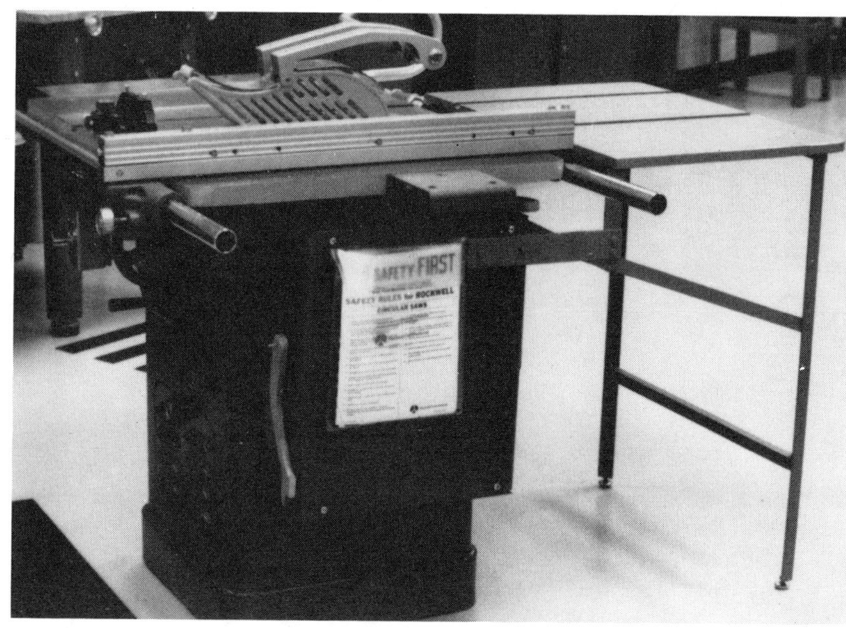

Figure 20–5. Features of this industrial model, tilting-arbor circular table saw include: (1) posted safety rules, (2) push stick, (3) tail-off table, (4) rip fence, (5) crosscut guide, (6) self-adjusting point-of-operation guard, and (7) enclosed power transmission.

Kickbacks and Ripping

A kickback occurs during a ripping operation when part or all of the workpiece is violently thrown back to the operator. Operators should keep their faces and bodies to one side of the blade, out of line with a possible kickback. To avoid kickbacks—and possible injury from them—operators should do the following:

- Maintain the rip fence parallel to the blade so the stock will not bind on the blade and be thrown.
- Keep the blade sharp. Replace or sharpen antikickback pawls when points become dull.
- Keep blades' guards, spreaders, and antikickback pawls in place and operating properly. The spreader must be in alignment with the blade, and the pawls must stop a kickback once it has started. Check their action before ripping.
- Cut only material that is seasoned, dry, and flat and that has a straight edge to guide it along the rip fence.
- Release work only when it has been pushed completely past the blade.
- Use a push stick for ripping widths of 2 to 6 in. (5 to 15 cm), and an auxiliary fence and push block for ripping widths narrower than 2 in. (5 cm).
- Allow the cut-off piece to be unconfined when ripping or crosscutting.

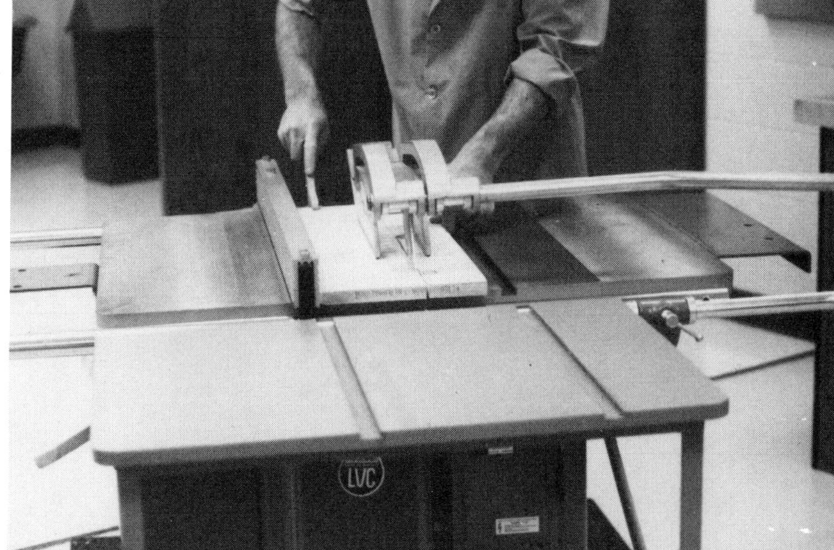

Figure 20–6. This operator is following safe operating procedures by: (1) standing to the side while ripping; (2) keeping sleeves rolled up; and (3) using the rip fence, blade guard, splitter and antikickback device, and tail-off table. When using a tail-off table, the operator has less tendency to reach over the blade to catch the stock before it falls to the floor.

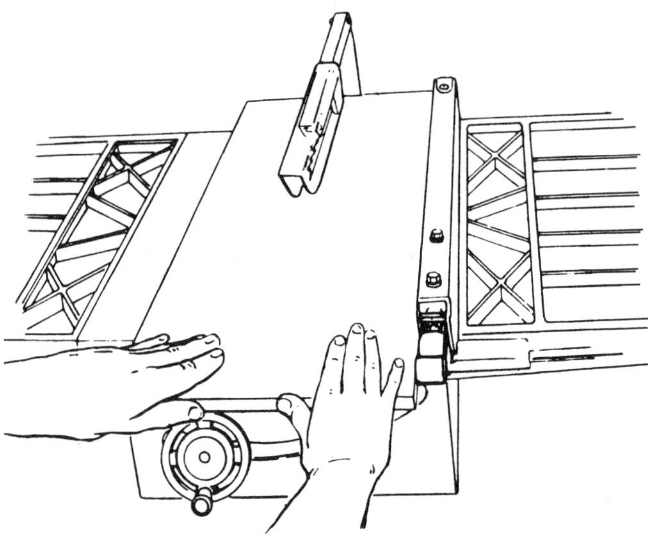

When the width of the rip is 6 in. or wider, use your right hand to feed the workpiece until it is clear of the table. Only ue the left hand to guide the workpiece—do not feed the workpiece with the left hand.

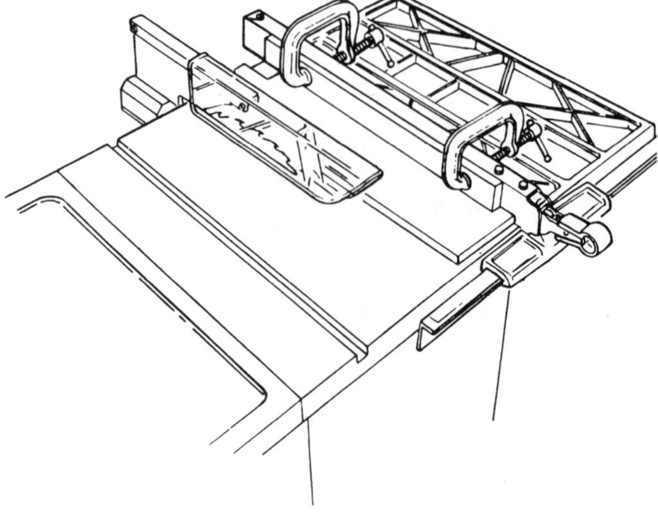

When the width of the rip is less than 2 in., the push stick cannot be used because the guard will interfere. Use the auxiliary fence-work support and push block. Use two C clamps to attach the auxiliary fence-work support to the rip fence.

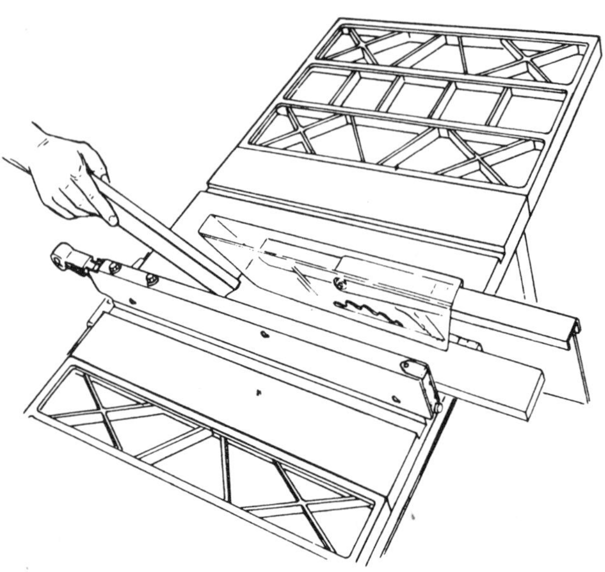

When the width of rip is 2 in. to 6 in., use the push stick to feed the work

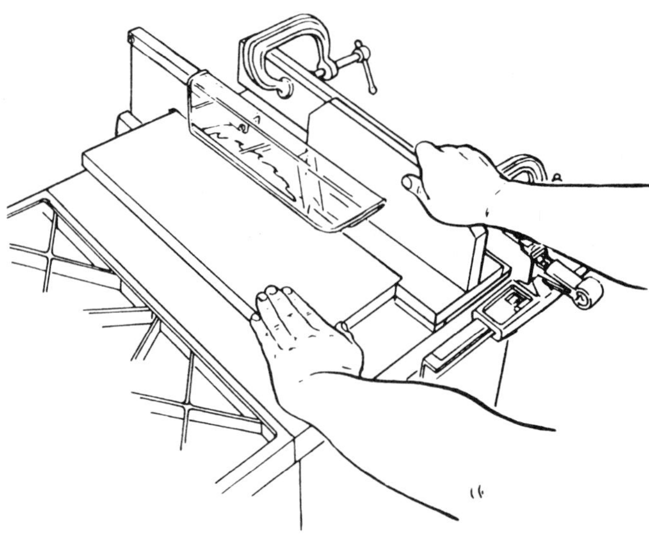

Feed the workpiece by hand along the auxiliary fence until the end is about 1 in. beyond the front edge of the table. Continue to feed using the push block. Hold the workpiece in position and install the push block by sliding it on top of the auxiliary fence-work support (this might raise the guard).

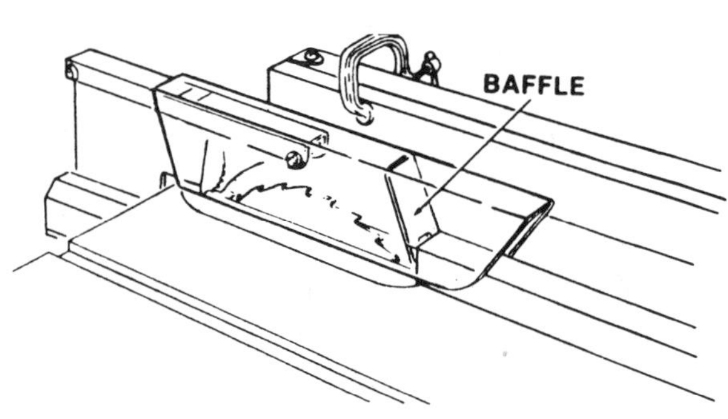

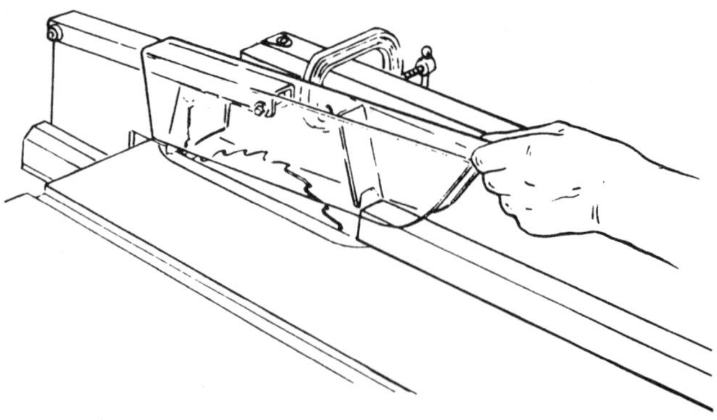

Figure 20–7. Safe ripping procedure.

- Apply the feed force to the section of the workpiece between the blade and the rip fence.

A safe ripping procedure is described in Figure 20–7.

Supplying Proper Equipment

Guards (Figures 20–8 and 20–9) greatly reduce the likelihood of injury and are now considered standard equipment. If they do not come with a saw, supply them when the saw is installed. The protection gained by using guards makes them essential. Management should be sure that the guards are practical and correct for the job, or else operators may be tempted to remove them.

Provide a circular tablesaw used for ripping with a spreader to prevent wood with internal stresses from clamping down or binding at the outfeed edge of the blade. In this way a spreader helps prevent kickbacks. It also keeps chips and slivers away from the back of the saw where they might be caught by the saw's teeth and be thrown.

Supervisors and operators should check that the spreader (1) is rigidly mounted, not more than ½ in. (13 mm) in back of the blade when the blade is fully elevated, and (2) is at least 2 in. (5 cm) wide at table level. It should conform to the radius of the saw as nearly as possible and be high enough above the table to penetrate the full thickness of the stock. The spreader should be attached so it will remain in true alignment with the blade, even when the table or arbor is tilted.

Guard a circular tablesaw, used for cutting, with a hood that completely covers the blade projecting above the table. Operators should let the guard ride the stock being cut, adjusting to the thickness of the stock (Figure 20–10). The hood should be strong enough to resist any blows it might sustain during reasonable operation, adjusting, and handling. It should be made of shatter-resistant material and should be no more flammable than wood. To be effective, the hood must remain in true alignment with the blade, even if the table or arbor is tilted.

Figure 20–8. Closeup of a properly functioning splitter and antikickback device during a ripping operation.

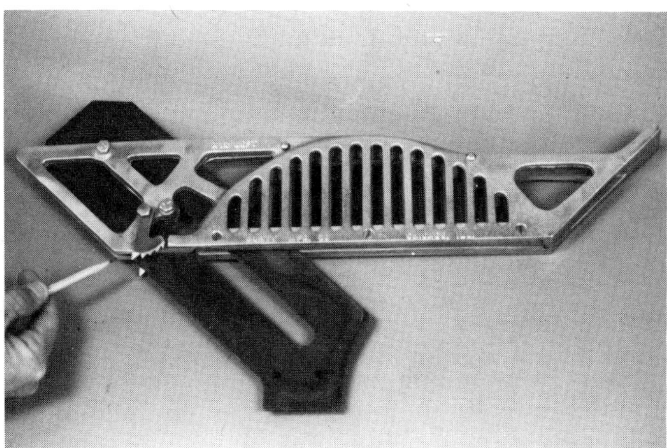

Figure 20–9. When sawing large pieces of stock, the support for the guard (see Figures 20–5 and 20–6) can be in the way and prevent sawing through the wood. The guard shown here permits sawing large pieces without interference. The splitter and antikickback dogs are built into the guard.

The hood may be suspended from a post attached to the side of the machine, or supported on the spreader. However, operators should secure and support the mounting so it will not wobble and strike against the blade. In strength and design, the hood must protect the operator against flying slivers or broken saw teeth. The mounting should resist reasonable side thrust or force.

Guard the part of the blade underneath the table so the operator cannot accidentally contact the blade. The enclosure, which may be part of the exhaust hood, should be constructed with a hinged cover so the blades can be easily changed.

Rabbeting and Dadoing

When rabbeting and dadoing, it is impossible to use a spreader and often impractical to use the standard hood guard. These operations can be effectively guarded by a jig that slides in the grooves of the transverse guide. In this way, the work is locked in the jig, and the operator's hands are kept well away from the saws or cutting head.

Because rabbeting and dadoing jobs vary, special jigs may be needed (Figure 20–11). The hazards of these jobs justify special guarding, especially when work is being done on small stock. If a shop does a lot of dadoing and rabbeting, supervisors should set aside one or more machines for this work. This will eliminate frequently removing standard guards from machines that are normally used for cutting and ripping.

Operators can use feather boards to hold the work to the table and against the fence as it is fed past the

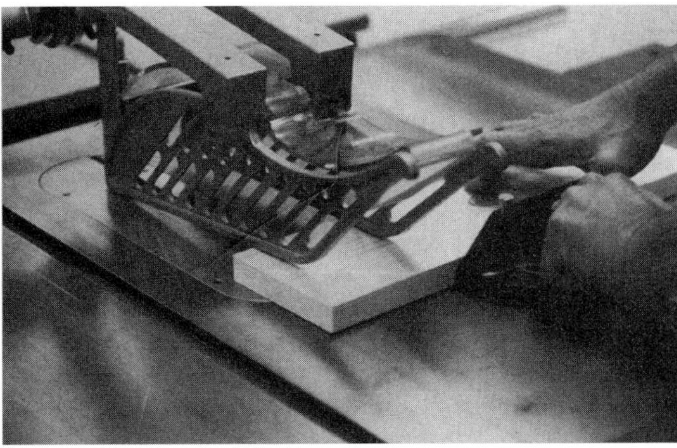

Figure 20–10. The overhead self-adjusting guard on this circular saw rides the stock as it is being cut and automatically adjusts to its thickness.

dado head. Feather boards are suitable for short runs because they can be quickly set up and are inexpensive. A feather board can be made from straight-grained stock, preferably hardwood. When using the board, the operator should make the parallel saw cuts (the comb) in the direction of the grain. The feather board should bear against the stock at an angle of 45 degrees to 60 degrees.

Proper Operating Methods

Only authorized persons should operate circular saws. A saw in good condition and running at the correct speed should cut easily. Operators should not have to cut freehand or crowd (push hard) the saw by forcing the stock faster than it can be cut. If the saw does not cut as fast as it should, or if it does not saw a clean, straight line, the saw blade or the running speed may be improperly set. Operators should check and correct these conditions—which are potential sources of incidents—before proceeding with the job.

Most hand- and power-feed saws run at about 3,450 rpm. This will give a 12 in. (0.3 m) blade a rim speed of 10,839 sfm (55 m/s); a 16 in. (0.4 m) blade, 14,451 sfm (74 m/s); and an 18 in. (0.46 m) blade, 16,258 sfm (83 m/s). Workers should always follow the manufacturer's instructions when operating these saws. Supervisors must find a way to prevent operators from placing a larger blade on the mandrel than is allowable for the mandrel's speed.

When the collar of a saw is tightly clamped in position, only the collar's outer edge should come in contact with the blade. If the inside of the collar has not been machined properly, it will force the rim of the saw out of line. When the blade comes in contact with the stock being cut, there will be a buckling effect on the saw. After the loose collar is securely fastened in place, the operator should test the saw with a straightedge. This test is important to conduct on circular saws as well as on edgers and trimmers.

When feeding a table saw, operators should keep their hands out of the line of the cut. Although the guard offers protection from the sides and from above, it does not protect workers from the front. When operators are ripping with the rip fence close to the saw, they should use a push stick between the blade and the fence to keep their fingers away from the blade. Operators should keep push sticks or blocks of various sizes and shapes near the machine. To make push sticks long enough to keep hands well away from the blade, add 6 in. (15.2 cm) to the blade's diameter. Because of kickbacks, operators should stand to the side of the stock they are ripping. A heavy leather or plastic apron or abdomen guard gives additional protection.

Operators should hold stock against a gauge; they should never saw freehand. Freehand sawing endangers the hands and may cause work to get out of line and bind on the saw. When ripping stock with narrow clearance on the fence side, the operator can gain more clearance by clamping a filler board flat to the table between the fence and the blade and by guiding the stock against it. Use of a filler makes the hazardous practice of removing the hood guard because of lack of clearance unnecessary.

The best height for the blade above the workpiece depends on the following considerations: (1) high-blade silhouette (the blade is as high as possible) and

Figure 20–11. When crosscutting several pieces to the same length, it is important to use a small block of wood clamped to the rip fence to allow room for clearance when the piece is cut off. If this is not done, the piece being cut off will bind between the fence and the blade and be thrown back toward the operator.

(2) low-blade silhouette (the saw blade extends just through the stock).

Advantages	
High-Blade Silhouette	*Low-Blade Silhouette*
Reduced kickback potential	Less exposure of the blade to the operator
Saw tooth cuts down nearly vertical to table	Smoother cut
Saw blade is closest to spreader	More table support needed for the workpiece in front of the saw blade
Less power needed	
Faster cutting	
Less saw blade wear	

Operators should use the right saw for the job. Using the wrong saw for the job makes the work harder and requires additional force when feeding the stock. Do not use a crosscut blade for ripping or a ripsaw for crosscutting. Using a general-purpose table saw for work that should be done on a special machine is also a poor practice. Work that can be done on special or power-feed machines should not be done on hand-feed, general-purpose machines. For example, a table saw is often used for hand-feed ripping operations. Instead, operators should perform this work on a power-feed ripsaw, which virtually eliminates the dangers of kickback and injury to the hands.

Long stock is sometimes crosscut on a table saw. Unless the stock is adequately supported, this, too, is a dangerous practice. The long stock extends beyond one or both ends of the table, interferes with other operations, and may be a hazard to other workers or trucks. Also, it is difficult to guide long pieces. Operators must exert considerable pressure while their hands are close to the saw. Such stock is more easily cut on a swing saw, pull saw, or radial saw.

Under no circumstances should operators adjust the fence while the saw is running. Parallel setting of the fence is particularly important. To enable the operator to set the fence accurately, mark the top of the saw table with a permanent, distinct line or other suitable device. Make the mark directly in front of and in line with the blade.

Operators should stop a circular table saw before leaving it. It is not sufficient to cut the switch and walk away. Workers have suffered amputations caused by saws still coasting with the power off. An electric brake attached to the motor's arbor offers fast, positive action.

Selecting and Maintaining Circular-Saw Blades

The characteristics and conditions of circular-saw blades are important safety factors for the operators who use them. Manufacturers of circular-saw blades have published valuable information on the selection, use, and care of these blades (Figures 20–12 and 20–13).

During designing, building, and tensioning, the maker gives a saw enough rigidity and tensile strength to cut without harmful distortion. When operators or others alter its original design, operate it at other than the rated speed, or change the balance or tension, they seriously affect the saw's efficiency and safety.

In addition, the following conditions of blades may cause unsafe, difficult, or unsatisfactory operation:

- Blade out of round. If some teeth are longer than others, the long teeth do most of the work. An unequal strain is imposed on the blade, which may cause it to run out of line, to heat up, and to warp.
- Blade not straight, out of plane. Lumps or warps can be checked with a straightedge across the length of the diameter of the unmounted blade. However, because of blade tensioning, it may not be flat, except under power.
- Blade out of balance. Too much of the blade was removed from one side during improper sharpening.
- Improper hook or pitch of teeth. The teeth of ripsaws and cutoff saws have different designs for different kinds of wood and for different purposes. Combination saws may be used for both crosscutting and ripping. There are other blades for certain kinds of woods, wood material, metals, and plastics.
- Improper or uneven set. A blade has to cut a kerf thicker than the blade to give adequate clearance for the saw to pass through the wood. The teeth can be given set or swage by bending alternate teeth right and left or by spreading the point of every tooth so each is slightly wider than the blade.
- Dull blades. Keep blades sharp so the saw works at top efficiency and the operator exerts minimum force when feeding the saw.
- Gummed blades. Also keep blades clean and free of pitch buildup so they will run at top efficiency and safety. A gummed blade can cause a kickback.
- Improper bushings. Bushings are provided to match the blade to the arbor's diameter. A bushing that is too large will cause a blade to be unstable at high-rotation speeds.
- Cracked blades. As soon as a crack is detected, remove the blade from service. Inspect blades for cracks every time the teeth are filed or set. Some cracks are so small that they may be invisible to the naked eye. For such cracks, use a nondestructive testing method, such as Magnaflux. If cracked blades are left in service, the crack frequently grows larger and eventually will cause partial fragmentation. Most cracks start in the gullets. Excessive heat and vibration cause saw blades to crack.

CIRCULAR SAW BLADES FOR CUTTING WOOD

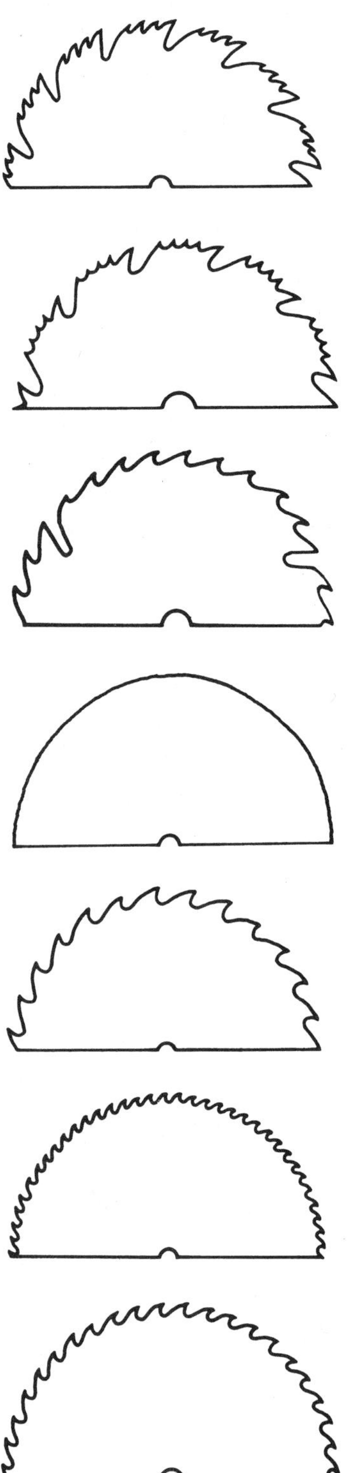

HOLLOW GROUND PLANER BLADES—The hollow ground planer blades are for precision cross cutting, mitering, and ripping on all woods, plywood, and laminates where the smoothest of cuts are desired.

MASTER COMBINATION BLADES—The master combination blades are for use on all woods, plywood, and wood base materials, such as fiberboard and chipboard. This type blade is better for cross cut and mitering than for ripping in solid woods. The teeth are set, and deep gullets are provided for cool and free sawing.

RIP BLADES—The rip blades are primarily intended for rip cuts in solid woods. The teeth are set and deep gullets are provided for cool and free cutting.

PLYWOOD BLADES—The plywood blade is a fine tooth cross cut type blade intended for cross cutting of all woods, plywood, veneers, and chipboard. It is especially recommended for cutting plywood where minimum of splintering is desired. The teeth are set and sharpened to give a smooth but free-cutting blade.

CHISEL TOOTH COMBINATION—The chisel tooth combination blade is an all-purpose blade for fast cutting of all wood where the best of finish is not required. Ideal for use in cutting of heavy rough timbers, in framing of buildings, etc. It cross cuts, rips, and miters equally well.

CABINET COMBINATION—The cabinet combination blade is for general cabinet and trim work in solid wood. It will cross cut, rip, and miter hard and soft wood to give good accurate cuts for moldings, trim, and cabinet work.

STANDARD COMBINATION—The standard combination blade is used for all hard and soft wood for cross cut, rip, or miter cut. It is especially recommended for use on power miter boxes and for accurate molding and framing work.

Figure 20–12. Choosing the correct saw blade for the job will increase operator safety. (Printed with permission from Sears, Roebuck and Company.)

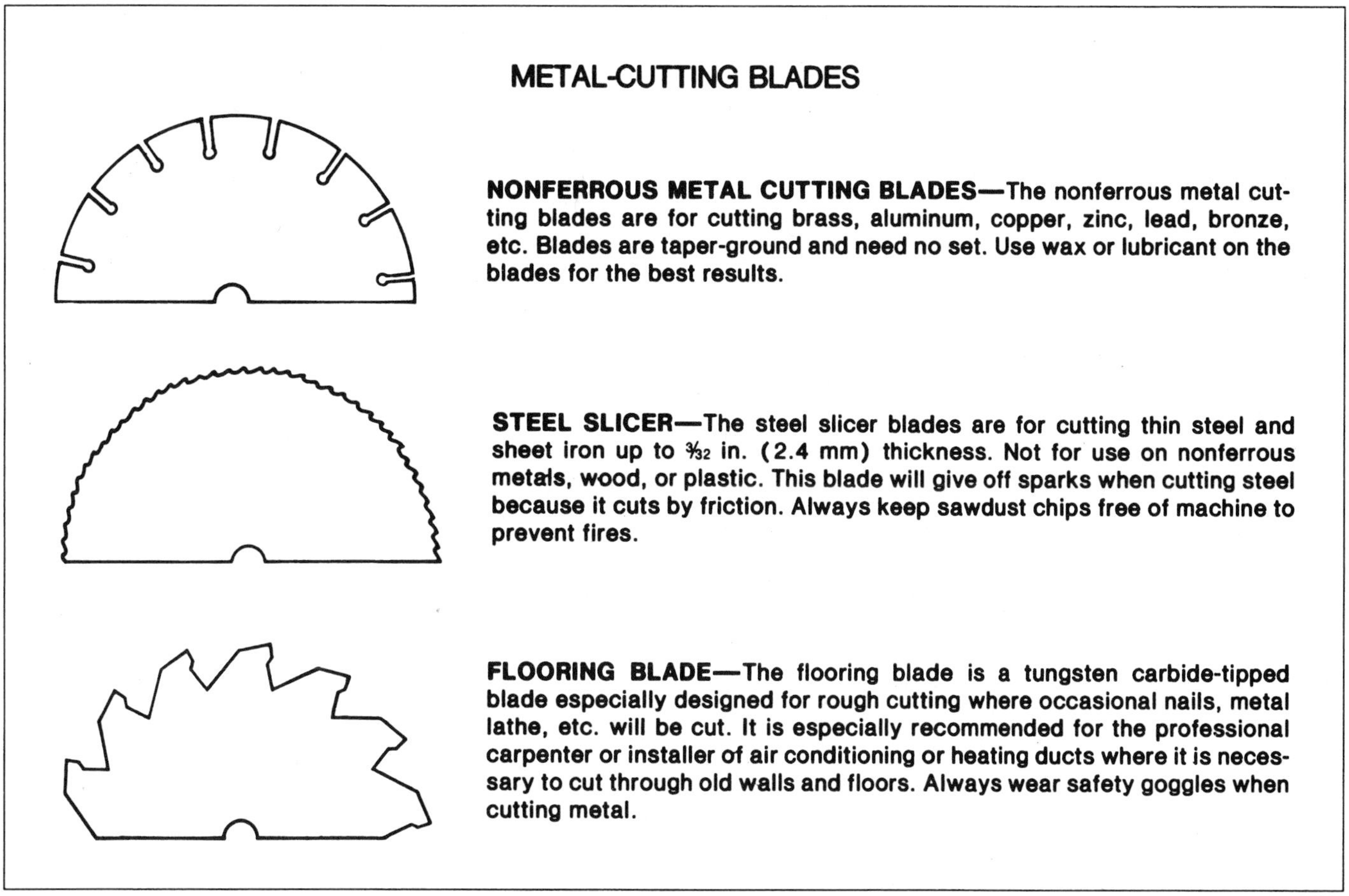

METAL-CUTTING BLADES

NONFERROUS METAL CUTTING BLADES—The nonferrous metal cutting blades are for cutting brass, aluminum, copper, zinc, lead, bronze, etc. Blades are taper-ground and need no set. Use wax or lubricant on the blades for the best results.

STEEL SLICER—The steel slicer blades are for cutting thin steel and sheet iron up to 3/32 in. (2.4 mm) thickness. Not for use on nonferrous metals, wood, or plastic. This blade will give off sparks when cutting steel because it cuts by friction. Always keep sawdust chips free of machine to prevent fires.

FLOORING BLADE—The flooring blade is a tungsten carbide-tipped blade especially designed for rough cutting where occasional nails, metal lathe, etc. will be cut. It is especially recommended for the professional carpenter or installer of air conditioning or heating ducts where it is necessary to cut through old walls and floors. Always wear safety goggles when cutting metal.

Figure 20–13. Metal-cutting saw blades.

To prevent cracking, operators should follow these precautions:

- Tighten the blade on the arbor for which it is designed.
- Operate the saw at speed specified by the manufacturer. If the saw is not tightened and operated according to the manufacturer's instructions, it may wobble, vibrate, heat, expand, and crack.
- Allow sufficient clearance (set or hollow grinding) for the teeth to prevent burning, and, thus, heating and cracking.
- Keep the blade in perfect round and balance.
- Keep the blade sharp at all times. A dull blade will not cut; rather, it will pound or burn itself through the wood, so that vibration, heating, and then cracking result.

After repairs have been made, the blade must be retensioned. This is a job for a sawsmith. Unless the company has the services of such a person, the blade should be repaired by the manufacturer.

Overhead Swing Saws and Straight-Line Pull Cutoff Saws

Overhead swing saws and straight-line pull cutoff saws cause hand injuries in several ways. Hands can be cut (1) while the blade coasts or idles, (2) when operators attempt to remove a sawed section of board or a piece of scrap, and (3) when operators measure boards or place them in position for the cut.

Operators' hands can be struck by a saw if it bounces forward from a retracted position or if it moves forward should the return device fail. Operators may pull the saw against their hands in the cutting path or may suffer body cuts from a saw that swings beyond its safe limits.

Guards

Cutoff saws must be guarded with a hood guard. The hood shall extend at least 2 in. (5 cm) in front of the saw blade when the saw is in the back position. Some guards cover the lower half of the saw when the saw is not cutting and ride on the top of the stock as the saw cuts.

Provide a counterweight or other device to automatically return the swing saw to the back of the table without rebounding when released. Secure the counterweight with a device designed to hold twice its weight. The counterweight shall be guarded, if within 7 ft (2.1 m) of the floor.

Install a limit chain or other device to prevent the saw from swinging beyond the back or front edges of the table. Another device should likewise keep the saw from rebounding from its idling position. A latch with a ratchet release on the handle is best, but in some instances a nonrecoil spring or bumper is adequate. A magnetic latch provides another way to prevent rebounding. Provide the saw table with a wood bumper to prevent bodily contact with the blade when it is extended the full length of the support arm.

Place STOP and START buttons for quick and easy access. STOP buttons should be easily contacted in an emergency. Mushroom-type buttons are recommended for this purpose. The STOP button should also have an easily identifiable color. Use protected START buttons, however, so accidental contact will not cause the saw to start. A collar around the button that extends ⅛ to ¼ in. (3 to 6 mm) above the top of the button is a recommended method (Figure 20–14). (See also Chapter 13, Ergonomics, in NSC's *Fundamentals of Industrial Hygiene,* 4th ed.)

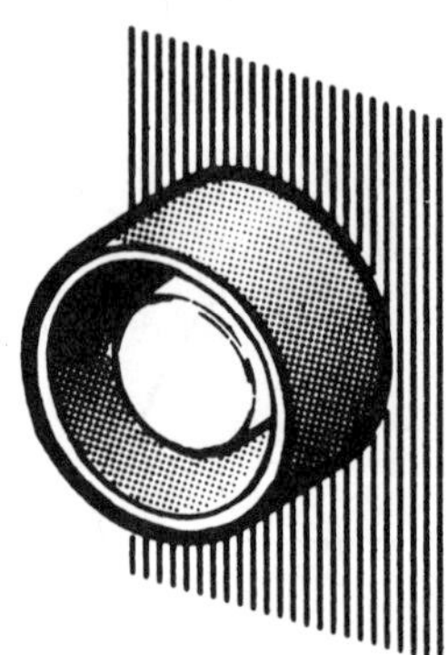

Figure 20–14. START button for a cutoff saw should be protected so accidental contact will not start the saw. (Printed with permission from *Machine Design Magazine,* June 23, 1977.)

Operating Methods

If the saw is pulled by a handle, the handle should be attached either to the right or left of the saw rather than in line with it. The operator should stand to the handle side and pull the saw with the hand nearer it.

Thus, if the handle is on the right side of the saw, boards should be pulled from the right with the right hand, and the saw should be pulled with the left hand. This method (1) makes it unnecessary for operators to bring their hands near the saw's path while the saw is cutting and (2) keeps the operators' bodies out of the line of the saw.

Saws may be ordered with either right or left handles. For a new saw, order it with the handle on the side from which the stock is to be pulled. If it is necessary to pull stock from the opposite side, place the handle on that side so the operator can stand in the correct position.

To measure boards, place their ends against a gauge stop. When it is necessary to measure the board with a scale while the board is on the table, move the board away from the blade.

At the completion of each cut, the operator should put the saw back to the idling position and make sure that all bounce has stopped before putting his or her hand on the table. Do not use automatic or constant-stroking saws unless the point of operation is guarded and there is no hazard to the operator.

Underslung Cutoff Saws

An underslung cutoff saw is usually operated by a treadle. Because its forward movement is fast, it should be completely enclosed in the noncutting position. For general work, it should also be covered by a movable hood guard that slides forward or drops to rest on the stock while the saw is cutting. A guard on the treadle ensures that the treadle will not be used accidentally.

Underslung cutoff saws are commonly used to cut knots out of narrow pieces such as flooring and molding. The stock is placed on the table by hand, and the hands are customarily held close to the line of the cut on either side. The movable guard gives little protection because the saw's action is so fast that the guard can ride over the top of the hands.

On either side of the line of travel, construct a barrier guard with enough clearance between the guard and the table top to admit the stock, but not the operator's hands or fingers. With practice, an operator can rapidly feed stock under this type of guard.

Radial Saws

When crosscutting, radial saws cut downward and pull the wood away from the operator and against a fence. These saws, like straight-line pull cutoff saws, require many adjustments to permit their full use. Adjustments should not be made when the saw is running. Lockout means should be provided.

The radial saw's head can be tilted to cut a bevel, or the supporting beam and track can be swung at an angle to make a miter cut. Both adjustments may be used to cut a compound bevel or miter. Likewise, the head may be turned parallel to the length of the table so the saw can be used for ripping. In this case, it is an overhead, stationary saw against which the stock is fed by hand.

Always guard the upper half of the saw, including the arbor end. The lower half of the saw should have an articulating guard for 90-degree crosscut operations. The lower guard should automatically adjust itself to the stock's thickness and should remain in contact with the stock being cut for the full working range. This prevents accidental contact with the sides of the blade (in an axial direction) when the cutting head is at rest (not in the cut) behind the fence and in the 90-degree crosscut mode. Under certain conditions, lower blade guards can cause additional hazards.

Provide some means so the cutting head will not roll or move out on the arm away from the column because of gravity or vibration (Figure 20–15). For repetitive crosscut operations, provide an adjustable stop to limit forward travel of the cutting head to that necessary to complete the cut.

The saw table should be large enough to cover the blade in any position (miter, bevel, or rip). Therefore, workers should never operate the saw with the blade in a position where it protrudes or extends beyond the table.

Obviously, only competent woodworkers should operate a saw with so many features. Operators should be well trained and aware of possible hazards. They need to know what to do when the machine is performing below standards.

The principal sources of injury connected with operating radial saws are those common to other power-driven saws. They include cuts to the arms and hands caused by the blade, by flying wood and chips, and by handling materials. As with other power saws, prevention of injuries requires proper use of the equipment.

Ripping

When ripping, rotate the radial saw's head 90 degrees, so the blade is parallel to the fence and is clamped in position. Then lower the blade until it will cut through the stock. Before ripping, position (1) the nose of the guard, or drop the guard down, (2) the spreader, and (3) the antikickback devices (Figures 20–16 and 20–17). Feed the stock against the direction of rotation of the revolving blade from the nose of the guard—the side at which the blade rotates upward toward the operator. For inrip, feed material from right to left; for outrip, feed from left to right (Figure 20–18).

Use a spreader in ripping to prevent the wood from immediately coming together after being cut. This action reduces the chances of the blade binding and causing a kickback. Mount the spreader in direct line with the blade. When properly adjusted, the spreader prevents wrong-way feed.

The antikickback device must be used in the ripping operation. This device is positioned so the antikickback fingers ride on the stock. Adjust the angle or height of the fingers so that if the stock is pulled out by hand, it will jam under the fingers and the stock cannot be moved. The antikickback device is adjustable for different thicknesses of stock. Check the antikickback fingers regularly for sharpness, making sure none of them are bent or in contact with the stock when sawing.

CAUTION: Operators should always follow the proper direction of blade rotation. The saw blade should always rotate downward as viewed from the operator position. Feeding from the wrong side (wrong-way feed) tends to grab the material away from the operator and throw it toward the infeed (nose) end of the guard. Wrong-way feed is prevented if a spreader is

Figure 20–15. A spring return is installed on this saw so the cutting head will not roll or move out on the arm—away from the column—because of gravity or vibration.

Figure 20–16. Before ripping, lower the guard on the infeed side until it almost touches the workpiece. Adjust the splitter-antikickback device on the outfeed side so the splitter rides within the saw kerf. Feed the workpiece against the blade rotation.

Figure 20–17. Warning sign on the outfeed side reads CAUTION—NEVER RIP FROM THIS END. Not to heed this warning would cause the workpiece to be pulled from the operator's grasp and sent flying across the room.

Figure 20–18. *Top*: In-ripping. *Bottom*: Out-ripping.

properly installed and correctly positioned.

Two possibilities of severe injury arise from feeding from the wrong side. The blade's direction of rotation makes it easy for the operator's hands to be drawn into the revolving saw. There is the additional danger to the helper, and to other people on the opposite side, or infeed side, of the saw. Flying stock can be thrown with enough force to drive the stock through a 1-in. (2.5-cm) board. There have been serious and fatal injuries because of failure to observe this precaution. Power-feed rolls are available for ripping operations. These not only greatly reduce kickback, but also speed up production.

When feeding the stock, hold it firmly against the table and fence. The blade should be sharp and parallel to the fence. Apply feed pressure between the blade and the fence. Use a push stick, longer than the blade's diameter, 6 in. (15.2 cm), when ripping narrow or short stock. Operators should never release the feed pressure until the cut is completed and the workpiece has fully cleared the blade.

Operators should exercise special care when ripping material with thin, lightweight, hard, or slippery surfaces because of the reduced efficiency of antikickback devices. When ripping, the operator should wear an antikickback apron.

Crosscutting

Radial saws used for crosscutting are pulled across the cutting area by means of a handle located to one side of the blade, rather than in line with the blade. Whenever possible, the operator should stand on the handle's side, pull the cutting head with the hand nearer the handle, and maneuver the lumber with the other hand. In this way, the operator's body is not in line with the blade. At the same time, the operator's hands are not near the blade's cutting area. Operators should never pull the blade beyond the point necessary to complete the cut because the back of the blade could lift the workpiece and throw it over the fence. Operators should always place the workpiece against either the fence or a special jig. They should never crosscut freehand.

Operators should never remove short pieces from the table until the saw has been returned to its position at the rear of the table. They should always use a stick or brush—never their hands—to remove scrap from the table.

Under normal circumstances, operators should measure by placing the boards to be cut against a stop gauge. However, in instances where it is necessary to measure with a rule, operators should turn the saw off until they have finished measuring.

At the conclusion of each cut, operators should always return the cutting head to the full-rearward position behind the fence. Operators should never remove their hands from the operating handle unless the cutting head is in this position.

Because the blade's direction of rotation and the feed's direction tend to cause the blade to feed itself through the work, operators should develop the habit of holding their right arms straight from the shoulder to the wrist. This will prevent the blade from grabbing and possibly stalling while in the workpiece.

Power-Feed Ripsaws

Because long stock is often ripped on power-feed ripsaws, the clearance at each working end of the saw table should be at least 3 ft (0.9 m) longer than the length of the longest material handled. Operators should adjust feed rolls to the thickness of the stock being ripped. Insufficient pressure on the stock can contribute to kickbacks (Figure 20–19).

Where multiple-cut, power-feed ripsaws are used, install a dado head alongside the last blade. This head disposes of the edging. The offbearer then does not have to handle any scrap pieces and can pay more attention to material coming from the saw.

A common hazard occurs with a power-feed, chain ripsaw that uses an overhead cutting saw with a solid chain having a rabbeted center trough. Unless care is taken to maintain the rabbet in the trough, very thin slivers may drop from a ripped edge into the center trough of the chain. These slivers can come flying out like bullets toward the operator. Regular inspection is necessary.

Band Saws

Although injuries from band saws are less frequent and less severe than those from circular saws, they still occur. The usual cause of band-saw injuries is operators' hands coming into contact with the blade. For instance, when hand-feeding stock, the operators' hands must come dangerously close to the blade. Operators should, therefore, use a push stick to control the workpiece when it is close to the blade. For this reason, it is especially important that the saw's table be well lit, yet free from glare.

The band saw's point of operation cannot be completely covered. However, install an adjustable guard—

Figure 20–19. Antikickback fingers should spread at least the full width of the feed rolls between the operator and the saw blade on a power-feed saw. This saw is equipped with a double set of antikickback fingers. (Printed with permission from Western Electric Company.)

designed to prevent operator contact with the front and right side of the blade above the upper blade guides—as close as possible to the workpiece. Be sure that the wheels and all nonworking parts of the blade are encased. Professional band saws have outside enclosures made of solid metal. The front and back of the enclosure should be made of solid material or sturdy mesh material. Some smaller lightweight band saws are made of metal and plastic materials.

A band saw should have a tension-control device to indicate proper blade tension. If it does not, the operator should test the blade for correct tension before beginning a job. An automatic tension-control device will help to prevent breakage of blades. Another device prevents the motor from starting if the tension on the blades is too tight or too loose.

A band saw, especially a large one, will run for a long time after the power is shut off. For this reason, it should have a brake that operates on one or both wheels to minimize the potential hazard of coasting when the machine is shut off and left unattended. Serious injuries occur when operators take hold of the running blade, not realizing it is in motion. Another reason for brakes is to stop the wheels in case the blade should break. The safety device shown in Figure 20–15 stops the saw if the blade breaks.

On band saws, be sure that a guard, formed to the curvature of the feed rolls, covers the nip point. Install this guard so the edge is ⅜ in. (9 mm) from the plane that is formed by the inside face of the feed roll, in contact with the stock.

When small pieces of stock are cut, use a special jig or fixture. A technique for safely sawing a sharp radius is shown in Figure 20–20.

Figure 20–20. When sawing a sharp radius, make several release cuts up to the cutting line. This prevents saw blade binding and possibly breaking the blade or causing it to jump off the guide wheels.

Jigsaws

Jigsaws are not normally considered as hazardous as other woodworking machinery. Occasionally, however, they also cause injuries, especially to the fingers and hands. Safe operating procedures for jigsaws require (1) the blade to be properly attached and secured; (2) the threshold rest, slotted foot, to be on the stock; (3) the guard to be in an effective position; and (4) the operators to keep their hands a safe distance from the blade. Also, all drive belts, pulleys, and other moving parts should be guarded.

In addition, operators should make turns slowly, with no sharp- or small-radius turns if working with a wide blade. Narrow blades should be used when small-radius curves are needed. Operators should plan clearance cuts to eliminate the need to back out of curves. Operators should clean the table with a long-handled brush after the blade has stopped.

WOODWORKING EQUIPMENT

Woodworking equipment such as jointer-planers, shapers, power-feed planers, sanders, and lathes all pose safety hazards for those operating them or working around or near the equipment. Workers must be trained in safe work practices and in emergency first aid and other procedures to prevent or minimize incidents and injuries.

Jointer-Planers

Second only to circular saws, hand-feed jointers or surface planers are the most dangerous woodworking machines (Figure 20–21). Most of the injuries are caused by the hands and fingers of operators in contact with these machines' knives. Many of these incidents occur when short lengths of stock are being jointed.

A jointer should be equipped with a horizontal cutter head, the knife projection of which extends beyond the body of the head not more than ⅛ in. (3.2 mm). The clearance between the path of the knife projection and the rear table should be not more than ⅛ in. (3.2 mm). The clearance is measured radially from the path of the knife's projection to the closest point on the table and with the rear table level with the path of the knife's projection. The clearance between the path of the knife projection and the front table shall not be more than 3/16 in. (4.8 mm). This clearance is measured the same as that for the rear of the table but with the tables in the same place (Figure 20–22).

Figure 20–21. This jointer is equipped with two guards: one on the working side of the fence, and the second on the back side of the blade.

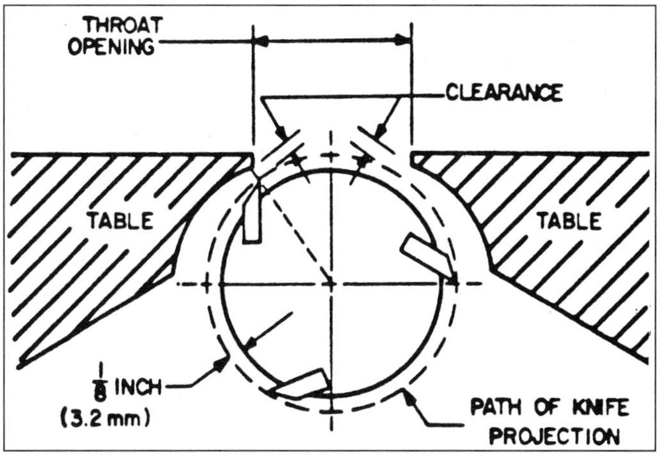

Figure 20–22. Table clearance for jointers. (Printed with permission from American National Standards Institute; Underwriters Laboratories Inc.)

The openings between the table and the head should be just large enough to clear the knife. In addition, the openings should be not more than 2 in. (5.1 cm) when the front and rear tables are set or aligned with each other for zero cut.

Cover the table opening on the working side of the fence with a spring-loaded, self-closing guard that adjusts itself to the moving stock. For good protection when edge jointing, install a swinging spring-loaded, self-closing guard or a guard that moves away from the fence along the axis of the other head. For surface planing, use only the swinging guard because it permits the use of hold-down push blocks, such as those pictured in Figure 20–23, to feed material smoothly over the other head. Use hold-down push blocks whenever the operator joints wood that is narrower than 3 in. (7.6 cm) (Figure 20–24).

Be sure that the unused end of the head, which is behind the fence, is enclosed at all times. A sheet-metal telescoping guard is acceptable for this purpose.

Jointer-planers are commonly used for planing off cupped or warped stock to make it flat. To do this job, power-feed attachments are available with resilient hold-down devices that simulate the pressure of the hands. If stock is properly conditioned, however, very little of it should have to be trued in this manner.

For doing surfacing work on the jointer, operators should have both hands on top of the stock, if it is thicker than 3 in. (7.6 cm). They should never place their hands over the front or back edges where they can easily come in contact with the head.

Shapers

The principal danger in using wood shapers is that operators' hands and fingers might strike against the revolving knives. Severe incidents can also result when broken knives are thrown by the machine. When one knife breaks or is thrown from the collar, the other knife is usually thrown as well. Thus, four or five pieces of heavy, sharp steel are thrown about the shop with sufficient speed to kill a person.

The greatest number of injury incidents occurs when shaping narrow stock, which if held in the hand brings it close to the knives. Use hold-down push blocks or jigs in these instances. Each shop should have a well-understood rule that stock narrower than a specified width must be held in a jig. Some shops put the limit at 6 in. (15 cm), others as high as 12 in. (30 cm). A cardinal rule is always feed against the direction of rotation of the cutter. Never back up the workpiece, or a kickback can occur.

Eliminate the danger from broken or thrown knives by using solid cutters that fit over the spindle. The initial cost is greater for cutters than for knives. However, on moderately long runs, cutters are less expensive. Carbide-tip or solid-carbide cutters are available. In all cases, cutters are safer than knives.

When knives are used, operators should take the following precautions to keep the blades from breaking or flying:

- Knives must meet rigid specifications for shaper steel.
- Knives must be sharpened and installed only by a fully qualified person.

Figure 20–23. Note the difference between a hold down *(left)* and a push block *(right)*. The push block has a piece of wood acting as a positive stop against the end of the workpiece; the hold down is flat on the bottom. Both are used to keep the operator's thumbs and fingers away from the cutter head.

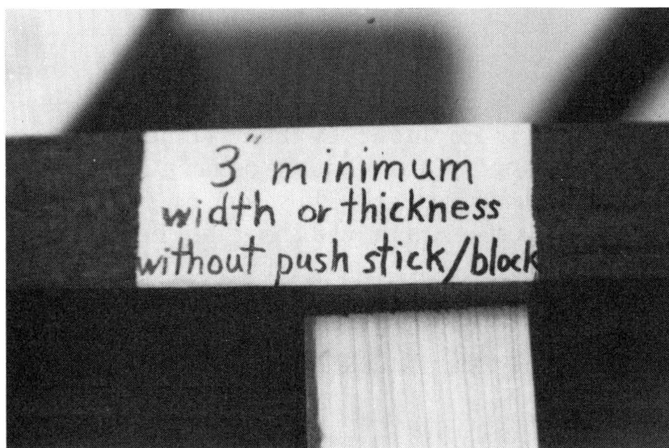

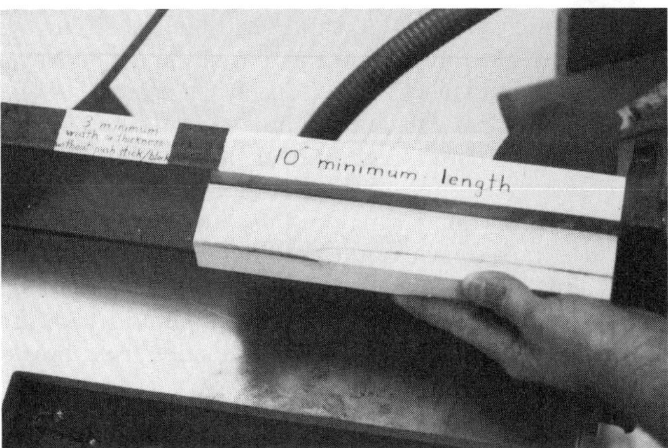

Figure 20–24. Painting a 3 in. (7.6 cm) strip *(top)* and a 10 in. (25 cm) strip *(bottom)* on top of the fence, and labeling them accordingly, serves as a quick reference for checking the width, thickness, or length of stock before jointing it.

- Knives and the grooves in the collars must fit perfectly and be free of dust.
- The two knives must balance perfectly. They must be weighted against each other each time they are set.
- A knife must not be used after it has become so short that the butt end does not extend beyond the middle point of the collar.
- Deep cuts should be avoided. It is safer and more efficient to take two light cuts than one heavy cut.
- During startup, operators should apply the power in a series of short starts and stops to slowly bring the spindle up to operating speed. They should listen carefully for chatter and should watch for other evidence that the knives are out of balance.

Various types of safety collars are in use that help prevent knives from flying. Although nothing should be done to discourage the use of such collars, do not consider them to be substitutes for perfectly balanced and fitted knives.

Another safety measure is to use some type of braking device to stop the spindle after the power is shut off. With double-spindle shapers, there should be starting and stopping devices for both spindles. The spindles should be started one at a time.

Shaper work must be held against guidepins for curved shaping or a fence for straight-line shaping. A feather board may be clamped to the fence (Figure 20–25). Keep the portion of the cutter, or knives, behind the fence covered. For curve shaping, adjust the overhead guard to just clear the stock. Use a starting pin when curve shaping. Operators should use a long-handled brush, or a vacuum system, to remove chips and scraps from the work table.

A number of guards are available to protect operators' fingers. For straight-line shaping, the fence frame or housing should contain the guard. A fence should have as small an opening as possible for the knives and should extend as far as possible, at least 18 in. (46 cm), on either side of the spindle. This assures good support at the start and finish of the cut. If the entire edge of the stock is to be shaped, adjust the portion of the fence beyond the cutter to receive the thinner stock and provide a stable bearing. This cannot be accomplished with a flat, continuous fence. It is best done with a split, adjustable fence.

Adjust the cutting-head guard for minimum head exposure. Provide a way to contain wood dust and chips. Jigs and hold-down clamps should be maintained to hold the stock securely.

Power-Feed (Thickness) Planers

Operators can reduce planer vibration by anchoring the planer on a solid foundation and by insulating it from

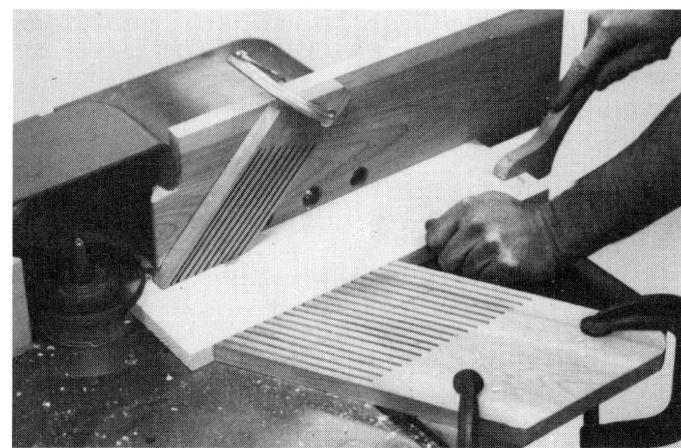

Figure 20–25. *Top*: A pair of feather boards clamped to the table top hold the workpiece down on the table and in against the fence. The operator is using a push stick. *Bottom*: This ring guard is installed over the cutter at the point of operation.

the foundation with cork, springs, or other vibration-absorbing material. The planer should have a three-point bearing and be bolted down without distortion. Distorting any woodworking machine will ultimately cause it to malfunction.

Because of the noise planers create, they should be isolated in a separate room or in a special soundproof enclosure built for this equipment. Helical cutter heads also will substantially reduce the noise levels. If neither soundproofing nor using helical cutter heads is practical, provide hearing protection for those working in the immediate area.

Completely enclose cutter heads in solid metal guards. These guards should be kept closed when the planer is running. Provide good local exhaust from the cutting heads.

Have operators stop, lock out, and tag out feed rolls, cutter heads, and cylinders before placing their hands in the bed plate to remove wood fragments, to make adjustments, or for any other reason. If planer parts are driven by belts running on the back side of the planer, completely enclose the belts and sheaves with sheet-metal or heavy-mesh guards, even though the planer is fenced at the back or is next to the wall.

Guard feed rolls with a wide metal strip or bar that will allow boards to pass but that will keep operators' fingers out of the rolls. Install antikickback fingers that operate on the infeed side across the entire throat (width of the machine).

Danger of kickbacks cannot be entirely overcome by mechanical means. Wood splinters and knots are frequently thrown out. Therefore, operators should not look into the back side to "watch" the operation. Operators also should always stand out of the line of the board's travel. Other people should not work or walk directly behind the feeding end of the planer. Use a barrier or guardrail when the machine is running. Operators must avoid feeding boards of different thicknesses at the same time. Thinner boards are not held by the feed rolls and can be kicked back from the heads. To keep workers from being struck by long, fast-moving boards, fence or mark off the space at the outrunning end. Operators should always wear safety goggles.

Sanders

Supervisors and others should make sure that drum, disk, or belt sanding machines are enclosed by dust exhaust hoods. The hood should enclose all portions of the machine except the portion designed to feed the stock. Personnel who operate sanders should wear goggles and dust respirators during sanding operations and cleanup.

On a belt sanding machine, place a guard at each inrunning nip point on both power transmission and feed roll parts. Guard feed rolls with a wide metal strip or bar that will allow boards to pass but that will keep the operators' fingers out. Guard the unused run of the abrasive belt on the operator's side of the machine to prevent contact with the operator.

All hand-feed sanders should have a work rest and be properly adjusted (1) to provide minimum clearance between the belt and the rest and (2) to secure support for the work. Small workpieces should be held in a jig or holding device.

Abrasive belts on sanders should be the same width as the pulley drum. When material is brought into contact with the moving abrasive belt, operators should adjust the drums to keep the abrasive belt taut enough to turn at the same speed as the pulley drum. Operators should inspect abrasive belts before using them and replace those found to be torn, frayed, or excessively worn.

Lathes and Shapers

Supervisors or operators should ensure that the rotating heads of lathes, whether running or not, are covered as completely as possible by hoods or shields. Hinge these hoods so they can be thrown back when adjustments are needed. The clearance between the workpiece and the tool rest should be about ⅛ in. (3 mm). Hold turning chisels firmly on the tool rest.

Operators should have lathes used for turning long pieces of stock that are held only between the two centers with long, curved guards. The guards should extend over the tops of the lathes to prevent the workpiece from being thrown out should it come loose.

Management should select and train lathe operators with care. Operators should carefully inspect all parts of the lathe for any defects and correct them before starting a job. Operators must not use stock that has checks, splits, cracks, or knots. They must give constant attention to stock being turned in order to discard any material likely to break. Sometimes stock can be trimmed out before turning. Operators must carefully place stock in the machine and feed the cutting tool slowly into the stock. They must wear safety goggles; a face shield is also required if the operation is dusty or if rough stock that may produce splinters is used. Operators should tie up long hair and not wear loose-fitting clothing to prevent becoming entangled in the machine. Figure 20–1 provides a summary of the safety rules for lathe use, as well as for other woodworking tools.

With shapers, operators should use jigs with handles to keep their hands at least 6 in. (15 cm) away from the cutting bit. Mount a ring guard cupguard around the cutting bit to reduce contact with it. Always feed stock into the rotation of the cutter. Always use a fence or a fixture with the stock.

SUMMARY

- All electrically driven machines should be adequately grounded, have a cutoff switch, and have some means of rendering the controls inoperative to prevent injuries. All belts, shafts, gears, and other moving parts must be guarded.
- Employees using woodworking machines should read the operator's manual prior to working with a new piece of equipment.
- Employees should maintain ample work areas around each machine, keep floors and work surfaces clear and free from hazards, provide adequate lighting and ventilation, and arrange machines to ensure a steady flow of materials. All machine controls, cutting edges, and power drives should be inspected at each new setup and at the start of each shift.
- Management should guard against hearing loss, airborne dust and contamination, explosion and fire hazards, and related problems. Workers should use personal protective equipment.
- Management can minimize the hazards posed by saws by (1) training operators, (2) ensuring all machines are guarded, and (3) making sure operators follow all safety procedures.
- To operate a circular saw safely, workers should keep their hands out of the cut line and use a holder to guide the stock. Hood guards are used for cutoff saws, while a counterweight will return swing saws to the proper position. Underslung cutoff saws can be guarded by constructing a barrier on either side of the line of travel. The upper half of radial saws should always be guarded, while the lower half should have an articulating guard for 90-degree crosscut operations.
- Because power-feed ripsaws are often used to cut long stock, operators should adjust feed rolls to the thickness of the stock being ripped to prevent kickbacks. Band saws must be adequately lighted and the point of operation completely covered.
- When operating wood shapers, workers should use push blocks or jigs to keep their hands away from the knives and use guards to prevent injuries from thrown knife blades.
- Power-feed planers should be secured to reduce vibration and isolated to help control noise hazards. Solid metal guards will enclose cutter heads to prevent injuries, and proper guards will also prevent kickbacks.
- To operate sanders safely, workers should use dust exhaust hoods; wear goggles and dust respirators during operation; inspect all belts; and be sure that hand-feed sanders have the proper distance between the sander and work rest.
- The rotating heads of lathes should be covered as completely as possible by hoods or shields. Lathe operators must be selected and trained with care to use, inspect, and maintain their equipment and to wear protective gear.
- Operators should clean their machines and work surfaces with a long-handled brush after the equipment has stopped and not use their hands or air nozzles.

REFERENCES

American National Standards Institute, 11 West 42nd Street, New York, NY 10036.

Safety Standard for Mechanical Power Transmission Apparatus, ANSI/ASME B.15.1–1992.

Safety Standard for Stationary and Fixed Electric Tools, ANSI/UL 987–1990.

National Fire Protection Association, 1 Batterymarch Park, Quincy, MA 02269.

Fire Protection Handbook, 17th ed., 1991.

Prevention of Fires and Explosions in Wood Processing and Woodworking Facilities, NFPA 664, 1993.

National Safety Council, 1121 Spring Lake Drive, Itasca, IL 60143-3201.

Fundamentals of Industrial Hygiene, 4th ed., 1996.
Safeguarding Concepts Illustrated, 6th ed., 1993.
Occupational Safety and Health Data Sheets: Tilting-Arbor and Tilting-Table Saws, 12304–0605, 1991.

Power Tool Institute, Inc., P.O. Box 818, Yachats, OR 97498.

U.S. Department of Health and Human Services, National Institute for Occupational Safety and Health, 4676 Columbia Parkway, Cincinnati, OH 45226.

Health and Safety Guide for Manufacturers of Woodworking Machinery, DHEW (NIOSH) Publication 79–131.
Health and Safety Guide for Millwork Shops, DHEW (NIOSH) Publication 76–111.
Health and Safety Guide for Plywood and Veneer Mills, DHEW (NIOSH) Publication 77–086.
Health and Safety Guide for Prefabricated Wooden Building Manufacturers, DHEW (NIOSH) Publication 76–159.
Health and Safety Guide for Sawmills and Planing Mills, DHEW (NIOSH) Publication 78–102.
Health and Safety Guide for Wooden Furniture Manufacturers, DHEW (NIOSH) Publication 75–167.

U.S. Department of Labor, Occupational Safety and Health Administration, 200 Constitution Avenue, NW, Washington, DC 20210.

Code of Federal Regulations, Title 29, Section 1910.213, *Woodworking Machinery Requirements.*
Code of Federal Regulations, Construction Industry Title 29, Section 1926.304, *Woodworking Machinery Requirements.*

REVIEW QUESTIONS

1. Companies should provide employees with equipment that meets the existing standards and regulations of what three organizations?
a.
b.
c.

2. On most machines, the point-of-operation guard must be:
a. Movable to accommodate the wood.
b. Balanced so as not to impede the operations.
c. Strong enough to provide protection to the operator.
d. All of the above.
e. Only b and c.

3. The working surfaces of the machine should be at a height that will minimize _______.

4. The machine inspection process should include inspecting what elements?
a.
b.
c.
d.

5. Name the two most frequent hazardous incidents involving circular saws.
a.
b.

6. When feeding a table saw, make push sticks long enough to keep hands well away from the blade by adding __________ to the blade's diameter.

7. What three actions can the operator of a saw take to seriously affect the saw's efficiency and safety?
a.
b.
c.

8. Name three ways overhead swing saws and straight-line pull cutoff saws can cause hand injuries.
a.
b.
c.

9. What two possibilities of severe injury arise from feeding from the wrong side of a saw?
a.
b.

10. Since long stock is often ripped on power-feed rip-saws, the clearance at each working end of the saw table should be at least __________ longer than the length of the longest material handled.

21

Welding and Cutting

Revised by

Philip E. Hagan, MPH, CIH, CHMM, CHCM

The purpose of welding is to join metal parts. All welding processes require heat and sometimes other substances to produce the weld. Because high heat is used to make the weld, a number of by-products result from the process, including fumes and gases that can be a serious health hazard to workers. Safety hazards are also associated with welding, such as the potential for fire or explosion and injuries from arc radiation, electrical shock, or materials handling.

Definitions used in this chapter are those of the American Welding Society. "Welder" and "welder operator" refer to the individual worker only. The machine used to perform the welding operation is referred to as the "welding machine." Equipment supplying current for electric welding is called either a "welding generator" or a "welding transformer." (See the References at the end of this chapter for the standards for welding and cutting operations. Also consult other regulations, where applicable.) The following topics are discussed in this chapter:

- Health hazards from airborne and liquid toxins
- Combustible materials and hazardous locations
- Exposure to fumes, noise, and radiation
- Protective equipment for workers
- Safety training
- Handling gases
- Resistance welding
- Arc welding

HEALTH HAZARDS

The most significant health hazard in the welding process is the generation of toxic metal fumes, vapors, and gases. The amount and type of fumes, vapors, and gases involved will depend on the welding process, the base material, the filler material, and the shielding gas used, if any. The toxicity of the contaminants depends primarily upon their hazardous properties, concentrations, and upon the physiological responses of the human body. Sampling by an industrial hygienist may be necessary to fully identify the toxic materials actually being given off in a specific operation.

Toxic Gases and Vapors

Exposure to various toxic gases and vapors generated during welding may produce one or more of the following effects:

- inflammation of the lungs (chemical pneumonitis)
- pulmonary edema (swelling and accumulation of fluids)
- emphysema (loss of elasticity of the lungs; only a small percentage of emphysema cases are caused by occupational exposure)
- chronic bronchitis
- asphyxiation

The major toxic gases associated with welding are classified as primary pulmonary and nonpulmonary.

Primary Pulmonary Gases

These gases can impair or injure the lungs and pulmonary system of workers who inhale these substances in hazardous amounts.

Ozone

Ozone is formed by electrical arcs and corona discharges in the air or by ultraviolet photochemical reactions. Welders who have had acute exposure at an estimated 9 parts per million (ppm) of ozone plus exposure to other air pollutants develop pulmonary edema. Ozone may become a problem when gas-shielded metal arc welding or arc gouging is conducted in enclosed or confined areas with poor ventilation.

Oxides of Nitrogen

Oxides of nitrogen are very irritating to the eyes and mucous membranes. Exposure to high concentrations may immediately produce coughing and chest pain. Death may occur within 24 hours of development of pulmonary edema.

Phosgene

Phosgene is produced when metals that have been cleaned with chlorinated hydrocarbons are heated to the temperatures used in welding. Inhalation of high concentrations of gas will produce pulmonary edema frequently preceded by a latent period of several hours' duration. Death may result from respiratory or cardiac arrest.

Phosphine

Phosphine or hydrogen phosphide is generated when steel that has been coated with a phosphate rustproofing is welded. High concentrations of the gas are irritating to the eyes, nose, and skin.

Nonpulmonary Gases

Although these gases do not directly injure the lungs or pulmonary systems of workers, they can threaten workers' lives by displacing or replacing oxygen in the human bloodstream. Management should routinely monitor the workplace to ensure these gases are not present in harmful amounts.

Carbon Monoxide

In some welding processes, carbon dioxide is reduced to carbon monoxide. In carbon-dioxide-shielded metal arc welding, carbon monoxide concentrations exceeding the recommended levels have been detected in the fumes near the arc; however, the concentration level rapidly decreases farther from the arc. With adequate ventilation, the carbon monoxide concentration in the welder's breathing zone can be maintained at acceptable levels.

Carbon Dioxide

Carbon dioxide is not usually considered a toxic gas. It is present in the atmosphere in a concentration of about 0.035% or 350 ppm and can be a higher concentration in occupied structures.

Particulate Matter

The deposition of particulate matter into the lungs is called "benign pneumoconiosis." Benign pneumoconioses associated with welding are aluminosis (aluminum), anthracosis (carbon), siderosis (iron oxide), and stannosis (tin oxide).

Aluminosis

The inhalation of aluminum, aluminum oxide, and aluminum hydrate does not appear to injure the pulmonary system. This conclusion is supported by both industrial experience and animal experimentation. Most of the complications associated with the inhalation of aluminum are probably produced by some co-inhalant such as silica. However, welding in inadequately vented confined spaces may produce ozone and oxides of nitrogen as well as aluminum oxides. The combined contaminants may produce effects on the lung.

Anthracosis

The term anthracosis refers to a blackish pigmentation of the lungs caused by the deposition of carbon particles. Some investigators still feel that this pneumoconiosis is due to impurities in coal such as silica. In most instances, inhalation of pure carbon does not produce a significant lung tissue reaction.

Iron Oxide

Siderosis is a benign pneumoconiosis resulting from the deposition of inert iron oxide dust in the lung. In general, there is neither fibrosis nor emphysema associated with this condition unless, as often occurs, there is a concomitant exposure to silica dust. Siderosis does not result in disability nor does it show any predisposition to pulmonary tuberculosis.

Tin

The inhalation of tin oxide dust over a long period of time will produce a benign pseudonodulation in the lungs known as stannosis. It is considered to be nonprogressive and nondisabling.

Pulmonary Irritants and Toxic Inhalants

The majority of the metal components contained in welding fumes do not produce radiographic changes in the lungs. Depending on the definition of pneumoconiosis, they can be classified in the separate categories of pulmonary irritant or toxic inhalants. These materials include cadmium, chromium, lead, magnesium, manganese, mercury, molybdenum, nickel, titanium, vanadium, zinc, and the fluorides.

Beryllium

The only nonfibrotic type of harmful pneumoconiosis is associated with the inhalation of beryllium. Inhalation of beryllium dust or fumes may result in an acute or chronic systemic disease. Chronic beryllium poisoning has been reported as resulting from exposure in facilities producing beryllium phosphors or beryllium compounds from the ore, in beryllium-copper founding, in ceramics laboratories, and in metallurgical shops.

This disease has been reported as occurring among individuals exposed to atmospheric pollution in the vicinity of facilities processing beryllium and in persons dwelling in the same household as beryllium workers. To date, there have been no known cases of acute or chronic beryllium poisoning caused by inhalation of beryl dust (the dust of beryllium ore).

There are specialized medical screening tests for beryllium exposure. This test is known as the lymphocyte transformation test (LTT) and is useful for screening beryllium-exposed workers. Beryllium is a suspected human carcinogen (ACGIH).

Cadmium

Operations that involve the heating of cadmium-plated or cadmium-containing parts, such as welding, brazing, and soldering, may produce a high concentration of cadmium oxide fumes. Inhalation of these fumes may cause respiratory irritation with a tender, sore, dry throat and a metallic taste, followed by cough, chest pain, and difficulty in breathing. A worker's liver or kidney and bone marrow may be injured by the presence of this metal.

A single exposure to cadmium fumes may cause a severe lung irritation that can be fatal. Most acute intoxications have been caused by the inhalation of cadmium fumes at concentrations that did not produce warning symptoms of irritation. Continued exposure to lower levels of cadmium has resulted in chronic poisoning characterized by irreversible lung injury and urinary excretion of multiple low molecular weight proteins that may be indicative of potentially progressive kidney impairment. Cadmium exposure via inhalation is associated with the development of cancer of the respiratory tract; however, direct oral ingestion of cadmium is not associated with carcinogenic toxins.

Chromium

The oxidation of chromium alloys can produce chromium trioxide fumes that are often referred to as chromic acid. These fumes react with water vapor to form chromic and dichromic acid. Contact with these fumes can produce small, painless cutaneous ulcers as well as dermatitis from primary irritation or allergic hypersensitivity. Inhalation of these fumes can produce bronchospasm, edema and hypersecretion, bronchitis, and hyperreaction of the trachea. Perforation of the nasal septa can also occur. Exposure to hexavalent chromium compounds seems to be related to an increased risk of lung cancer. Trivalent chromium compounds are suspected by some sources to be potentially carcinogenic.

Copper

Inhalation of copper fumes has been reported to produce symptoms of metal fume fever in welders. In chronic exposure, the liver, kidneys, and spleen may be injured and anemia may develop, although chronic poisoning like that from lead poisoning is unknown. Excessive exposure can cause nasal congestion and ulceration with perforation of the nasal septum.

Fluoride

The use of electrodes with fluoride-containing coatings creates a definite hazard. Fluoride compounds—fumes and gases—can burn eyes and skin on contact. When excessive amounts are inhaled, excretion lags behind the daily intake, resulting in a buildup of fluorides in the bones. If storage of fluorides continues over a sufficiently long period, the bones may show an increased radiographic density.

Lead

Lead poisoning in industry is almost always a result of inhalation of lead fumes from lead-containing materials or materials protected by lead-base paints. Inadvertent oral ingestion can also occur if poor workplace hygiene is present. The most common symptom of lead toxicity is abdominal pain. Lead levels can be easily measured in blood, and chelation therapy may be necessary for acute toxicity.

Magnesium

Oxide fumes from magnesium can produce metal fume fever, which can result in an irritation of mucous membranes. Experimental work with animals has failed to show any detrimental response in the lungs. Magnesium exposure is associated with metal fume fever.

Manganese

The fumes from manganese are highly toxic and can produce total disability after exposure as short as a few months. Such exposure is usually caused by inhalation of manganese dioxide dust. Chronic exposure will cause tiredness, uncontrollable laughter, and other nervous system effects. Manganese exposure is noted for the mask-like face reminiscent of parkinsonism and may produce subacute edema of the respiratory system, chronic bronchitis, or pneumonia.

Mercury

The welding of metals coated with protective materials containing mercury compounds produces toxic mercury fumes. Systemic mercury poisoning could result.

Molybdenum

Little is known concerning the exposure of humans to molybdenum or its compounds. Animal studies indicate molybdenum may produce nasal irritation, diarrhea, and weight loss at high concentrations.

Nickel

Nickel and its compounds are carcinogenic and toxic. A significant increase in cancer of the lungs and sinuses has occurred among employees in nickel smelting and refining facilities. Nickel fumes have been known to cause severe pneumonitis. Nickel carbonyl is highly toxic and can produce cyanosis, delirium, and death between 4 and 11 days after exposure. "Nickel itch" is a dermatitis that results from sensitivity to nickel.

Titanium

High concentration of titanium dioxide dust may produce irritations in the respiratory tract. Slight fibrosis, without disabling injury, has been observed in the lungs of industrial workers exposed to titanium dioxide dust.

Vanadium

Vanadium is present in some welding filler wires. It irritates the eyes and respiratory tract and may be responsible for asthmatic reactions.

Zinc

Zinc oxide fumes are formed during the welding, brazing, or cutting of galvanized metals. The inhalation of freshly formed fumes may produce a brief self-limiting illness known variously as zinc chills, metal fume fever, brass chills, and brass founders' fever. This condition is characterized by chills, fever, nausea, vomiting, muscular pain, dryness of mouth and throat, headache, fatigue, and weakness. These signs and symptoms usually abate in 12 to 24 hours with complete recovery following. Immunity from this condition is rapidly acquired if exposure occurs daily, but is quickly lost during holidays or over weekends. Because of this behavior, metal fume fever is sometimes known as "Monday morning sickness."

Cleaning Compounds

Because of their chemical properties, cleaning compounds can create health hazards if improperly mixed. They often require special ventilation precautions. Follow the manufacturer's instructions.

Chlorinated Hydrocarbons

Degreasing operations often employ chlorinated solvents that can decompose to toxic phosgene gas in the presence of the ultraviolet radiation emitted by the welding arc.

Asbestos

If welding or cutting involves asbestos, the regulations of the agency having authority must be consulted before beginning the job. Asbestos is associated with fibrotic lung disease, lung cancer, and a tumor known as a mesothelioma.

SAFETY HAZARDS

Management and workers should know the safety hazards involved in the workplace. They should also be trained to avoid, reduce, or eliminate them through the use of safe work practices, personal protective equipment, and safety equipment. The most common workplace safety hazards are discussed below.

Fire Protection

Because portable welding and cutting equipment creates special fire hazards, it should be used in a permanent welding and cutting location that can be designed to provide maximum safety and fire protection. Otherwise, the welding and cutting site should be inspected to determine what fire protection equipment is necessary (see ANSI Z49.1 and NFPA 51B).

It is advisable, particularly in hazardous locations, to require "hot work" permits issued by the welding supervisor, a member of the facility fire department, or some other qualified person before welding or cutting operations are started. Specifications for hot work permits are outlined in Chapter 11, Fire Protection, in this volume.

Floors and Combustible Materials

Where welding or cutting must be done near combustible materials, special precautions are necessary to prevent sparks or hot slag from reaching such material and starting fires. If the work itself cannot be moved, the exposed combustible material should, if possible, be moved a safe distance away. Otherwise, it should be covered with sheet metal. Spray booths and ducts should be cleaned to remove combustible deposits. Before welding or cutting is started, wood floors should be swept clean and, preferably, covered with metal or other noncombustible material where sparks or hot metal may fall.

If gas welding or oxygen cutting is done inside a booth provided for arc welding, the gas cylinders should be placed in an upright and secured position away from sparks to prevent contact with the flame or heat.

Hot metal or slag should not be allowed to fall through cracks in the floor or other openings, nor into machine tool pits. Cracks or holes in walls, open doorways, and open or broken windows should be covered with sheet metal guards. Because hot slag may roll along the floor, it is important that no openings exist between the curtain and the floor. Similar protection should be installed for wall openings through which hot metal or slag may enter when welding or cutting operations are conducted on the outside of the building.

If it is necessary to weld or cut close to wood construction or near combustible material that cannot be removed or protected, a fire hose, water pump tank extinguisher, or fire pails should be conveniently located. Portable extinguishers for specific protection against Class B and Class C fires should also be provided (see Chapter 11, Fire Protection). Pails of limestone dust or sand may be useful. It is good practice to provide a fire extinguisher, either dry chemical, multipurpose chemical, or carbon dioxide, for each welder.

A fire watcher equipped with a suitable fire extinguisher should be stationed at or near welding or cutting operations conducted in hazardous locations to see that sparks do not lodge in floor cracks or pass through floor or wall openings. The fire watch should be continued for at least 30 minutes after the job is completed to make sure that smoldering fires have not been started.

Hazardous Locations

Welding and cutting operations should not be permitted in or near rooms containing flammable or combustible vapors, liquids, or dusts. Nor should they be permitted on or inside closed tanks or other containers that have held such materials until all fire and explosion hazards have been eliminated. All of the surrounding premises should be thoroughly ventilated and frequent gas testing provided. Sufficient draft should be maintained to prevent accumulation of explosive concentrations. Local exhaust equipment should be provided for removal of hazardous gases, vapors, and fumes (present in the surroundings or generated by the welding or cutting operations) that ventilation fails to dispel.

Drums, Tanks, and Closed Containers

Workers should thoroughly clean closed containers that have held flammable liquids or other combustibles before welding or cutting materials. If the containers cannot be removed for standard cleaning procedures, two other practices are sometimes followed: (1) the containers are purged with an inert gas (Figure 21–1), or (2) they are filled with water to within an inch or two of the place where the work is to be done and a vent is left open. Either of these measures may also be used as an added precaution after standard cleaning. (See *Safeguarding of Tanks and Containers for Entry, Cleaning, or Repair,* NFPA 326 & 327, 1999. Also see Chapter 12, Flammable and Combustible Liquids, in this volume.)

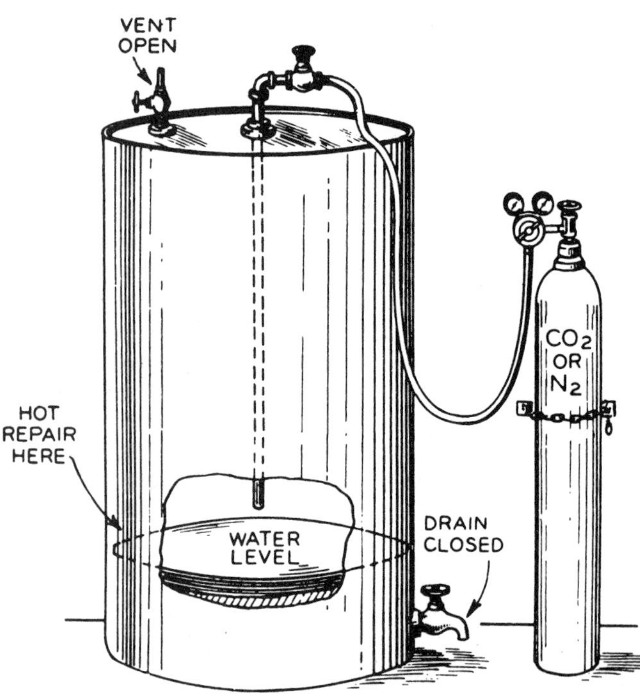

Figure 21–1. As an added precaution after cleaning, a container to be welded or cut may be filled with either carbon dioxide or nitrogen to dilute and render nonhazardous any remaining combustible gas or vapor. (Reprinted with permission from the American Welding Society.)

The accepted method for preparing tanks and drums for welding is:

1. Remove all sources of ignition (open flames, unguarded electric lights, and so on) from the vicinity of the drums to be cleaned.
2. Remove the bung with a special long-handled wrench.
3. Examine the inside for rags, waste, or other debris that might interfere with free draining. Use a portable electric hand lamp that is listed for hazardous locations, or an electric extension lamp protected by a guard of spark-resistant material.
4. Place the drums on a steam rack with the bung holes at the lowest possible point, and let the drums drain for 5 minutes.
5. Steam the drums for at least 30 minutes. Drums that have contained shellac, turpentine, or similar materials require longer steaming.
6. Remove the drums from the steaming rack, and fill them part way with caustic soda or soda ash solution. Rotate the drums for at least 5 minutes. Light hammering with a wood mallet will help to loosen scale.
7. Thoroughly flush the drums for at least 5 minutes with hot water. A water spray nozzle placed 6 or 8 in. (15 to 20 cm) inside the drum can be used. Drums should be placed so that water can drain out the bung openings during this operation.
8. Wash down the outside of the drum with a hose stream of hot water.
9. Dry the drum thoroughly by circulating warm air throughout the inside.
10. Thoroughly inspect the interior of the drum, using a light that is listed for hazardous locations, and a small mirror. If it is not clean, repeat the cleaning process.
11. Test the container for the presence of flammable vapors with a combustible gas indicator. Test for toxic contaminants and for oxygen sufficiency or enrichment if personnel are to enter.
12. Make similar tests just before welding repair operations are performed. If the operations extend over an appreciable period of time, make repeated tests.

Precautions for employee protection during container cleaning operations include the following suggestions:

- Wear head and eye protection, rubber gloves, boots, and aprons when handling steam, hot water, and caustic solutions. When handling dry caustic soda or soda ash, wear approved respiratory protective equipment, long sleeves, and gloves.
- To handle hot drums, wear suitable hand pads or gloves. Steam irons or other hot surfaces that may be touched should be insulated or otherwise guarded.
- Dispose of residue in a safe manner. In each instance, the method of disposal should be checked for hazards.
- If a vessel must be entered, wear respiratory protective equipment approved for the exposure and a safety harness with attached lifeline tended by a helper who is similarly equipped and stationed outside the vessel. Rescue procedures should be tested for adequacy before beginning work.

Many containers that have held toxic, flammable, combustible or explosive material present special problems. Detailed information can be secured from the manufacturer of such materials. For gasometers or gasholders for natural or manufactured gas, consult with the American Gas Association.

For cleaning and gas-freeing of tanks, bunkers, or compartments onboard ship, refer to NFPA Standard No. 306, *Control of Gas Hazards on Vessels,* and to the local fire marshal or Coast Guard marine inspection officer.

A word of caution: It may not be practical to purge and render gases inert in very large compartments due to the volumes of gas needed and the time required for such operations. Seek expert help from manufacturers or specialists for these cases.

CONTROLLING HAZARDOUS EXPOSURES

Certain materials, sometimes contained in the consumables, base metals, coatings, or atmospheres of welding or cutting operations, have low OSHA permissible exposure limits (PELs) and/or low American Conference of Governmental Industrial Hygienists (ACGIH) threshold limit values (TLVs). Among these materials are:

Antimony	Chromium	Mercury
Arsenic	Cobalt	Nickel
Barium	Copper	Selenium
Beryllium	Lead	Silver
Cadmium	Manganese	Vanadium

Refer to Material Safety Data Sheets provided by the manufacturer to identify any of the materials listed above that may be contained in the consumable. Whenever these materials are encountered as designated constituents in welding, brazing, or cutting operations, ventilation precautions must be taken to ensure that the level of contaminants in the atmosphere is below the limits allowed for human exposure. Unless atmospheric tests under the most adverse conditions have established that exposure is within acceptable concentrations, management and workers should observe the precautions discussed in the following section.

Ventilation

To keep fumes, vapors, and other toxic by-products of welding at safe levels, management must ensure that the workplace is properly ventilated. Natural and mechanical ventilation can be used to reduce the levels of airborne contaminants to acceptable levels.

Natural

Natural ventilation is acceptable for welding, cutting, and related processes where the necessary precautions are taken to keep the welder's breathing zone away from the plume and where sampling of the atmosphere shows that concentrations of contaminants are below mandated or recommended levels (see above). Taking air samples in workers' breathing zones is the only way to be sure that airborne contaminant levels are within allowable limits.

Natural ventilation often meets standards if the necessary precautions are taken to keep the welder's breathing zone away from the plume and all of the following specifications are met:

- Space of more than 10,000 ft^3 (284 m^3) per welder is provided.
- Ceiling height is more than 16 ft (5 m).
- Welding is not done in a confined space.
- Welding space refers to a building or an enclosed room in a building, not a welding booth or screened area that is used to provide protection from welding radiation; nor does the welding space contain partitions, balconies, or other structural barriers that obstruct cross-ventilation.
- Potentially hazardous materials, covered above, are not present as deliberate constituents.

Mechanical

Mechanical ventilation includes local exhaust, local forced, and general area mechanical air movement. Local exhaust ventilation is preferred. It means fixed or movable exhaust hoods placed as near as practical to the work and able to maintain a capture velocity sufficient to keep airborne contaminants below regulatory limits (Figure 21–2).

Figure 21–2. This welder wears personal protective equipment and is using a local exhaust system and fan. The protective curtain is pulled back for this photograph only. (Courtesy International Stamping Co./Midas International Corp.)

Local forced ventilation means a local air-moving system (such as a fan) placed so that it moves the air at right angles (90 degrees) to the welder (across the welder's face). It should produce an approximate velocity of 100 ft per min (30 m per min), and be maintained for a distance of approximately 2 ft (0.6 m) directly above the work area. Precautions must be taken to ensure that contaminants are not dispersed to other work areas.

General area mechanical ventilation includes roof exhaust fans, wall exhaust fans, and similar large area air movers. General mechanical ventilation is not usually satisfactory. It is often helpful, however, when used in addition to local ventilation. General mechanical ventilation may be necessary to maintain the general background level of airborne contaminants below the levels referred to or to prevent the accumulation of explosive gas mixtures.

Air Cleaners

Where permissible, air cleaners that have high efficiencies in the collection of submicron particles may be used to recirculate a portion of air that would otherwise be exhausted. Some filters do not remove gases, however. Therefore, adequate monitoring must be done to ensure that concentrations of harmful gases remain below allowable limits.

Other Factors

In addition to the factors listed above, there are a number of other items that must also be considered and evaluated to determine ventilation requirements:

- dimensions and layout of working areas
- number of welding stations or welders or both
- rates of welding

- tendency of air currents to dissipate or concentrate fumes in certain areas of the working space

The size and arrangement of the working facilities are major factors in ventilation requirements, especially when the working area is confined or divided into sections that limit air circulation. The number of welding stations and welders is also an important factor to be considered in designing ventilation systems. Welders who are paid by piece rate or incentive basis may weld a significantly higher number of pieces than those who are paid a fixed daily rate. As a rule of thumb, general ventilation should be provided at a minimum rate of 2,000 cubic feet per minute per welder except where local exhaust hoods are available.

The local exhaust hoods can be fixed or movable and should be provided with an air flow sufficient to maintain a velocity of 100 feet per minute in the breathing zone of the welder.

Special Equipment

In addition to conventional ventilating devices, special equipment has been developed to remove welding fumes at their source. For example, fume extraction nozzles using high-velocity, low-volume exhaust units are designed to fit over the end of the hand-held welding torch. The sleevelike fixture is usually connected to a small exhaust fan by a flexible hose.

Fume Avoidance

Welders and cutters must take precautions to avoid breathing the fume plume directly. This can be done by positioning of the work, the head, or by ventilation that directs the plume away from the face. Tests have shown that fume removal is more effective when the air flow is directed across the face of the welder, rather than from behind the person.

Nonionizing Radiation

Electric arcs and gas flames produce ultraviolet and infrared radiation that has a harmful effect on the eyes and skin upon continued or repeated exposure. The usual effect of ultraviolet is to "sunburn" the surface of the eye, which is painful and disabling but temporary in most instances. However, the effects of visible and near infrared radiation may cause permanent eye injury if the worker looks directly into a very powerful arc without eye protection. Ultraviolet may also produce the same effects on the skin as a severe sunburn.

Production of ultraviolet radiation is high in gas-shielded arc welding. For example, a shield of argon gas around the arc doubles the intensity of the ultraviolet radiation, and, with the greater current densities required (particularly with a consumable electrode), the intensity may be 5 to 30 times as great as with non-shielded welding such as covered-electrode or gas-shielded metal arc welding.

Infrared radiation has only the effect of heating the tissue with which it comes in contact. If the heat is not enough to cause an ordinary thermal burn, there is no harm. Exposure to certain intensities of infrared radiation is associated with the development of cataracts.

Whenever possible, arc-welding operations should be isolated so that other workers will not be exposed to either direct or reflected radiation. Arc-welding stations for regular production work can be enclosed in booths if the size of the work permits. The inside of the booth should be coated with a paint that is nonreflective to ultraviolet radiation. Portable flameproof screens similarly painted or flameproof curtains should also be provided. Booths should be designed to permit circulation of air at the floor level and adequate exhaust ventilation.

Chapter 24, Laboratory Safety, in the *Administration & Programs* volume, provides further coverage on ionizing and nonionizing radiation.

Noise

In welding, cutting, and the associated operations, noise levels can exceed the permissible limits. Hearing protective devices may be needed. (See Chapter 7, Personal Protective Equipment, in this volume.)

Chipping

Special slag-chipping hammers and chisels are part of a good welder's tool kit. Welders should never use an ordinary carpenter's hammer as a chipping hammer because the head of a carpenter's hammer can splinter and split.

Because slag can be sharp, welders should always wear safety glasses whenever they chip. Chipping can also be very noisy, especially when air-driven hammers are used. For this reason, welders should wear hearing protection as well as eye protection when chipping. (See ANSI/AWS F6.1–R1989, *Method for Sound Level Measurement of Manual Arc Welding and Cutting Processes*.)

PERSONAL PROTECTIVE EQUIPMENT

A baseline physical, including a chest x-ray and pulmonary function testing, is recommended for all persons engaged in welding. Periodic reexaminations should be made as recommended by the company or facility physician. In addition, workers should be trained in the type of personal protective equipment each job requires and in proper use and care of the equipment.

Respiratory Protection

If gases, dusts, and fumes cannot be kept below the applicable PEL or TLV, welders should wear respiratory protective equipment certified for the exposure by NIOSH. Where oxygen is also deficient, workers should wear a self-contained breathing apparatus.

Precautions for proper respiratory protection must be provided for workers doing inert-gas-shielded arc welding. Depending upon a number of factors, including the particular variety of gas-shielded arc welding to be done, the nature of the materials to be welded, and whether or not the work must be done in a confined space, workers will need positive ventilation, local exhaust removal, approved respirator equipment, or a combination of these precautions.

Eye Protection

Goggles, helmets, and shields that give maximum eye protection for each welding and cutting process should be worn by operators, welders, and their helpers. These items should conform to ANSI Z87.1–1989, *Practice for Occupational and Educational Eye and Face Protection,* and Z89.1–1986, *Protective Headwear for Industrial Workers.* Table 21–A is a guide for selecting the

Table 21–A. Guide for Selection of Lens Shade

Operation	*Electrode size 1/32 in. (mm)*	*Arc current (A)*	*Minimum protective shade*	*Suggested* shade no. (comfort)*
Shielded metal arc welding	Less than 3 (2.5)	Less than 60	7	—
	3-5 (2.5-4)	60-160	8	10
	5-8 (4-6.4)	160-250	10	12
	More than 8 (6.4)	250-550	11	14
Gas metal arc welding and flux cored arc welding		Less than 60	7	—
		60-160	10	11
		160-250	10	12
		250-500	10	14
Gas tungsten arc welding		Less than 50	8	10
		50-150	8	12
		150-500	10	14
Air carbon	(Light)	Less than 500	10	12
Arc cutting	(Heavy)	500-1000	11	14
Plasma arc welding		Less than 20	6	6 to 8
		20-100	8	10
		100-400	10	12
		400-800	11	14
Plasma arc cutting	(Light)**	Less than 300	8	9
	(Medium)**	300-400	9	12
	(Heavy)**	400-800	10	14
Torch brazing		—	—	3 or 4
Torch soldering		—	—	2
Carbon arc welding		—	—	14
	Plate thickness			
	in.	mm		
Gas welding				
Light	Under 1/8	Under 3.2		4 or 5
Medium	1/8 to 1/2	3.2 to 12.7		5 or 6
Heavy	Over 1/2	Over 12.7		6 or 8
Oxygen cutting				
Light	Under 1	Under 25		3 or 4
Medium	1 to 6	25 to 150		4 or 5
Heavy	Over 6	Over 150		5 or 6

* As a rule of thumb, start with a shade that is too dark to see the weld zone. Then go to a lighter shade which gives sufficient view of the weld zone without going below the minimum. In oxyfuel gas welding or cutting where the torch produces a high yellow light, it is desirable to use a filter lens that absorbs the yellow or sodium line in the visible light of the (spectrum) operation.

** These values apply where the actual arc is clearly seen. Experience has shown that lighter filters may be used when the arc is hidden by the workpiece. ANSI 249.1-1983.

correct filter lens for various welding and cutting operations. Goggles or spectacles should have side shields (see guidance for lens care given in this volume, Chapter 7, Personal Protective Equipment).

Protective Clothing

Some of the items of protective clothing needed by welders (Figure 21–3) include:

- flame-resistant gauntlet gloves—leather or other suitable material (may be insulated for heat)
- aprons made of leather or other flame-resistant material to withstand radiated heat and sparks

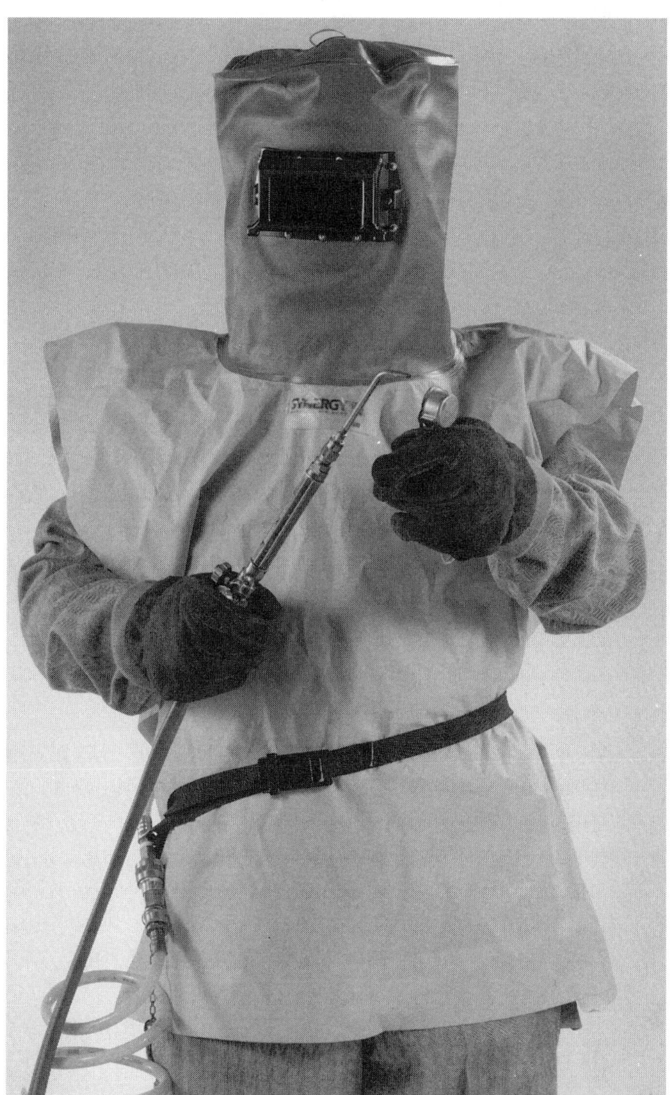

Figure 21–3. Proper personal protective equipment for a welder includes flash goggles worn under a helmet, chrome leather jacket, apron, gauntlet gloves, and leggings. (Courtesy Westinghouse Electric Corp.)

- for heavy work, fire-resistant leggings, high boots, or similar protection
- safety shoes, wherever heavy objects are handled (because of spark hazard, avoid using low-cut shoes with unprotected tops)
- for overhead work, capes or shoulder covers of leather or other suitable material. Skull caps of leather or flame-resistant fabric may be worn under helmets to prevent head burns. Also, for overhead welding, ear protection is sometimes desirable.
- safety hats or other head protection against sharp or heavy falling objects

Operators and other persons working with inert-gas-shielded arc-welding machines should keep all parts of the body that could be exposed to the ultraviolet and infrared radiation covered to protect against skin burns and other types of injuries. Dark clothing, particularly a dark shirt, is preferable to light-colored clothing in order to reduce reflection to the operator's face underneath the helmet. Woolen clothing is preferable to cotton because it is more resistant to deterioration and is not readily ignited.

CAUTION: Welders should never wear synthetic clothing and/or synthetic blends, including synthetic insulated underwear.

For gas-shielded arc welding, woolen clothing is also preferable to cotton. It is not readily ignited and protects the welder from changes in temperature. Cotton clothing, if used, should be chemically treated to reduce flammability. In either case, clothing should be thick enough to keep radiation from penetrating it.

Outer clothing should be reasonably free from oil and grease. Sleeves and collars should be kept buttoned. Aprons and overalls should have no front pockets, where sparks could be caught. For the same reason, trousers or overalls should not have turned-up cuffs.

Thermal-insulated underwear is designed only to be worn under other clothing and should not be exposed to open flames, sparks, or other sources of ignition.

TRAINING IN SAFE PRACTICES

Management should make sure that welders and cutters are well trained in the safe practices that apply to their work. The standards for training and qualification of welders set up by the American Welding Society are recommended. A training program should particularly emphasize that a welder or cutter can best provide for the safety of co-workers as well as the operator by observing safe practices that include:

- When possible, place work at an optimal height to avoid back strain or shoulder fatigue (Figure 21–4).

Figure 21–4. This welder uses a scissor lift table to place the work at an optimal height for improved posture that will reduce strain and fatigue. (Courtesy Presto Corp.)

- For work at more than 5 ft (1.5 m) above the floor or ground, use a platform with railings, or with fall protection equipment.
- Wear respiratory protection as needed and a safety harness with attached lifeline for work in confined spaces, such as tanks and pressure vessels. The lifeline should be tended by a similarly equipped helper whose duty is to observe the welder or cutter and effect rescue in an emergency.
- Take special precautions if welding or cutting in a confined space is stopped for some time. Disconnect the power of arc welding or cutting units, and remove the electrode from the holder. Turn off the torch valves of gas welding or cutting units. Shut off the gas supply at a point outside the confined area. If possible, remove the torch and hose from the area.
- After welding or cutting is completed, mark hot metal or post a warning sign to keep workers away from heated surfaces.
- Follow safe housekeeping principles. Don't throw electrode or rod stubs on the floor—discard them in the proper waste containers. Keep tools and other tripping hazards off the floor—put them in a safe storage area.
- Use equipment as directed by the manufacturer's instructions and practices.

Operators and management shall recognize their joint responsibilities for safety. Management shall ensure that welders and supervisors are trained and shall establish and enforce procedures to be used at off-plant locations and/or in hazardous locations. Only approved welding equipment shall be used, and the manufacturer's instructions must be followed. Supervisors shall be responsible for the safe operation of welding equipment and operators.

OXYFUEL WELDING AND CUTTING

An oxyfuel-welding process unites metals by heating; the heat source is a flame produced by the combustion of a fuel gas or gases. This process sometimes includes the use of pressure and a filler metal.

An oxygen-cutting process severs or removes metal by the chemical reaction of the base metal with oxygen at an elevated temperature. The temperature is maintained by heat from the combustion of fuel gases or from an arc. In the metal-powder-cutting process, a finely divided material, such as iron powder, is added to the cutting-oxygen stream. The powder bursts into flame in the oxygen stream and starts cutting without preheating the material to be cut. Metal-powder cutting is used on stainless and other steels, on many nonferrous metals, and on concrete in construction and demolition jobs. Plasma-arc cutting is now replacing metal-powder cutting.

Welding and Cutting Gases

Pure oxygen will not burn or explode. It supports combustion, causing other substances to burn when raised to the kindling temperature. Combustible materials burn much more rapidly in oxygen than in air. Oxygen forms explosive mixtures in certain proportions with acetylene, hydrogen, and other combustible gases.

Acetylene (C_2H_2) consists of 92.3% by weight of carbon and 7.7% by weight of hydrogen in chemical combination. It contains stored-up energy that is released as heat when it burns, as in the welding flame. This heat is in addition to that which would be obtained by combustion of equivalent amounts of elemental carbon and hydrogen.

Acetylene burned with oxygen can produce a higher flame temperature (approximately 6,000 F or 3,300 C) than any other gas used commercially. Acetylene, like other combustible gases, ignites readily, and in certain proportions forms a flammable mixture with air or oxygen. The range of flammable limits of acetylene (2.5% to 81.0% acetylene in air) is greater than that of other commonly used gases, with consequently greater hazard.

Other fuel gases are used with oxygen in torches, primarily for oxygen cutting. For example, propane, propylene, and their mixtures are supplied in cylinders in liquid form, generally under various trade names. These gases and LP-gas are discussed in Chapter 12, Flammable and Combustible Liquids, in this volume.

Compressed Gas Cylinders

Most of the oxygen and oxyfuels used for welding and cutting are supplied in pressured cylinders. These cylinders should be constructed and maintained in accordance with regulations of the U.S. Department of Transportation (DOT). The purchaser should make sure that all cylinders bear DOT, ICC (Interstate Commerce Commission), or CTC (Canadian Transport Commission) specification markings. The contents should be legibly marked on each cylinder in large letters.

Oxygen is supplied in steel cylinders; the usual size for welding contains 244 ft^3 (6.9 m^3) of oxygen under pressure of 2,200 psi (15.2 MPa) at 70 F (21 C). A cap should be provided to protect the outlet valve when the cylinder is not connected for use.

Hydrogen is furnished in cylinders under a pressure of about 2,000 psi (13.8 MPa) at 70 F (21 C). It may ignite in the presence of air or oxygen when in contact with a spark, open flame, or other source of ignition. Hydrogen-air mixtures ranging from 4.1% to 74.2% hydrogen are flammable.

Acetylene for welding and cutting is usually supplied in cylinders having a capacity up to about 300 ft^3 (8.5 m^3) of dissolved acetylene under pressure of 250 psi (1.7 MPa) at 70 F (21 C).

Acetylene cylinders should be completely filled with an approved porous material impregnated with acetone, the solvent for acetylene. The porous material should have no voids of appreciable size so that acetylene can be safely stored at the prescribed full cylinder pressure. Because acetylene is highly soluble in acetone at cylinder filling pressure, large quantities of acetylene can be stored in comparatively small cylinders at relatively low pressures.

Acetylene is either supplied in cylinders or generated as needed. It is a product of the reaction between water and calcium carbide, a gray crystalline substance made commercially by fusing lime and coke in an electric furnace. Calcium carbide is neither flammable nor explosive. It is stored and sold in air- and water-tight cans or drums. If the drums are damaged in handling and if water comes in contact with carbide, acetylene will be generated. There is a danger of ignition and explosion.

Handling Cylinders

Serious accidents may result from the misuse, abuse, or mishandling of compressed gas cylinders. Workers assigned to the handling of cylinders under pressure should be properly trained and should work only under competent supervision. Observance of the following rules will help control hazards in the handling of compressed gas cylinders.

- Accept only cylinders approved for use in interstate commerce for transportation of compressed gases.

- Do not remove or change the marks and numbers stamped on the cylinders.
- Cylinders that are heavy or difficult to carry by hand may be rolled on their bottom edge but never dragged.
- Transport cylinders weighing more than a total of 40 lb (18.2 kg) on a hand or motorized truck, securing them from falling.
- Keep the cylinders clean and protect them from cuts or abrasions.
- Do not lift compressed gas cylinders with an electromagnet. Where cylinders must be handled by a crane or derrick, as on construction jobs, carry them in a cradle or suitable platform and take extreme care that they are not dropped or bumped. Do not use slings.
- Do not drop cylinders or allow them to strike each other violently.
- Do not use cylinders for rollers, supports, or any purpose other than to contain gas.
- Do not tamper with safety devices in valves or on cylinders.
- Consult the supplier of the gas when in doubt about the proper handling of a compressed gas cylinder or its contents.
- Clearly write "EMPTY" or "MT" in chalk on empty cylinders that are to be returned to the vendor.
- Close cylinder valves and replace valve protection caps, if the cylinder is designed to accept a cap.
- Load cylinders to be transported to allow as little movement as possible. Secure them to prevent violent contact or upsetting.
- Always consider cylinders as being full and handle them with corresponding care. Accidents have resulted when containers under partial pressure were thought to be empty.

The fusible safety plugs on acetylene cylinders melt at about the boiling point of water. If an outlet valve becomes clogged with ice or frozen, it should be thawed with warm (not boiling) water, applied only to the valve. A flame should never be used.

Storing Cylinders

Cylinders need to be secured in the upright position in a safe, dry, well-ventilated place prepared and reserved for the purpose. Flammable substances, such as oil and volatile liquids, should not be stored in the same area. Cylinders should not be stored near elevators, gangways, stairwells, or other places where they can be knocked down or damaged.

Oxygen cylinders should not be stored within 20 ft (6 m) of highly combustible materials or cylinders containing flammable gases. If closer than 20 ft, cylinders should be separated by a fire-resistive partition at least 5 ft (1.5 m) high with a fire-resistance rating of at least 30 min.

Acetylene and liquefied fuel gas cylinders should be stored with the valve end up. If storage areas are within 100 ft (30.5 m) distance of each other and not protected by automatic sprinklers, the total capacity of acetylene cylinders stored and used inside the building should be limited to 2,500 ft^3 (70 m^3) of gas, exclusive of cylinders in use or connected for use. Quantities exceeding this total should be stored in a special room built in accordance with the specifications of NFPA 51, *Oxygen-Fuel Gas Systems for Welding, Cutting, and Allied Processes,* in a separate building, or outdoors. Acetylene storage rooms and buildings must be well ventilated, and open flames must be prohibited. Storage rooms should have no other occupancy.

Cylinders should be stored on a level, fireproof floor. One common type of storage house consists of a shed room with side walls extending approximately halfway down from the roof and a dividing wall between one kind of gas and another.

To prevent rusting, cylinders stored in the open should be protected from contact with the ground and against extremes of weather—accumulations of ice and snow in winter and continuous direct rays of the sun in summer.

Cylinders are not designed for temperatures in excess of 130 F (54 C). Accordingly, they should not be stored near sources of heat, such as radiators or furnaces, or near highly flammable substances like gasoline.

Cylinder storage should be planned so that cylinders will be used in the order that they are received from the supplier. Empty and full cylinders should be stored separately, with empty cylinders being plainly identified as such to avoid confusion. Group together empty cylinders that have held the same contents.

Storage rooms for cylinders containing flammable gases should be well ventilated to prevent the accumulation of explosive concentrations of gas. No source of ignition should be permitted. Smoking should be prohibited. Wiring should be in conduit. Electric lights should be in fixed position and enclosed in glass or other transparent material to prevent gas from contacting lighted sockets or lamps. Electric lights should also be equipped with guards to prevent breakage. Electric switches should be located outside the room.

Using Cylinders

Safe procedures for the use of compressed gas cylinders include the following:

- Use cylinders, particularly those containing liquefied gases and acetylene, in an upright position and secure them against being accidentally knocked over.

- Keep the metal cap in place to protect the valve when the cylinder is not connected for use unless the cylinder valve is protected by a recess in the head. A blow on an unprotected valve might cause gas under high pressure to escape.
- Make sure the threads on a regulator or union correspond to those on the cylinder valve outlet. Do not force connections that do not fit.
- Open cylinder valves slowly. A cylinder not provided with a handwheel valve should be opened with a spindle key or a special wrench or other tool provided or approved by the gas supplier.
- Do not use a cylinder of compressed gas without a pressure-reducing regulator attached to the cylinder valve, except where cylinders are attached to a manifold, in which case the regulator will be attached to the manifold header.
- Before making connection to a cylinder valve outlet, "crack" the valve for an instant to clear the opening of particles of dust or dirt. Always point the valve and opening away from the body and not toward anyone else. Never crack a fuel gas cylinder valve near other welding work or near sparks, open flames, or other possible sources of ignition.
- Small fires at the cylinder should be extinguished, if possible, by closing the cylinder valve. In case of a larger fire or if extinguishment is not possible, evacuate the area and immediately contact trained fire fighting personnel.
- Use regulators and pressure gauges only with gases for which they are designed and intended. Do not attempt to repair or alter cylinders, valves, or attachments. This work should be done only by the manufacturer.
- Unless the cylinder valve has first been closed tightly, do not attempt to stop a leak between the cylinder and the regulator by tightening the union nut.
- Leaking fuel gas cylinders should be taken out of use immediately and handled as follows:
    1. Close the valve, and take the cylinder outdoors well away from any source of ignition.
    2. Properly tag the cylinder, and notify the supplier. (A regulator attached to the valve may be used temporarily to stop a leak through the valve seat.)
    3. If the leak occurs at a fuse plug or other safety device, take the cylinder outdoors well away from any source of ignition, open the cylinder valve slightly, and permit the fuel gas to escape slowly. Tag the cylinder plainly.
    4. Post warnings against approaching with lighted cigarettes or other sources of ignition.
    5. Promptly notify the supplier, and follow instructions for returning the cylinder.
- Do not permit sparks, molten metal, electric currents, excessive heat, or flames to come in contact with the cylinder or attachments.
- Never use oil or grease as a lubricant on valves or attachments of oxygen cylinders. Keep oxygen cylinders and fittings away from oil and grease, and do not handle such cylinders or apparatus with oily hands, gloves, or clothing.
- Never use oxygen as a substitute for compressed air in pneumatic tools, in oil preheating burners, to start internal combustion engines, or to dust clothing. Use it only for the purpose that it is intended.
- Never bring cylinders into tanks or unventilated rooms or other closed quarters.
- Do not refill cylinders except with the consent of the owner and then only in accordance with DOT (or other applicable) regulations. Do not attempt to mix gases in a compressed gas cylinder or to use it for purposes other than those intended by the supplier.
- Before a regulator is removed from a cylinder valve, close the cylinder valve and release the gas from the regulator.
- Cylinder valves must be closed when work is finished.

Manifolds

Cylinders are manifolded to centralize the gas supply and to provide gas continuously and at a rate in excess of that which may be obtained from a single cylinder. Manifolds must be of substantial construction and of a design and material suitable for the particular gas and service for which they are to be used. Manifolds should be obtained from and installed under the supervision of a reliable manufacturer familiar with safe practices in construction and use of manifolds.

Portable manifolds connect a small number of cylinders (usually not more than five) for direct supply to a consuming device. The cylinders may be connected by individual leads to a single, common coupler block, or individual cylinders may be connected to a common line with coupler tees attached to the cylinder valves. A properly supported regulator serves the group of connected cylinders.

Stationary manifolds connect a larger number of cylinders for supply through piped distribution systems. This type of manifold consists of a substantially supported stationary pipe header that connects the cylinders by individual leads (Figure 21–5). One or more permanently mounted regulators serve to reduce and regulate the pressure of the gas flowing from the manifold.

Oxygen manifolds should be located away from highly flammable material and oil, grease, and the like. They should not be located in acetylene generator

Figure 21–5. This is a well-designed manifold system for acetylene cylinders. (Courtesy Linde Co., Division of Union Carbide Corp.)

rooms, or in close proximity to cylinders of combustible gases. There should be a 5 ft (1.5 m) high, 30 min. fire-resistant partition between an oxygen manifold and combustible gas cylinders, unless the manifold and such cylinders are separated at least 50 ft (15.2 m). Regulations of NFPA 51, *Oxygen-Fuel Gas Systems for Welding, Cutting, and Allied Processes,* should be followed.

Distribution Piping

All piping should be color-coded or clearly identified as to type of gas. Distribution piping carrying oxygen from a manifold or other centralized supply should be of steel, wrought iron, brass, or copper, as outlined in NFPA 51.

All pipe and fittings for oxygen service lines should be examined before use, and, if necessary, tapped with a hammer to free them from dirt and scale. They should, in every case, be washed out with a suitable nonflammable cleaner—hot water solutions of caustic soda and trisodium phosphate are effective.

Only steel or wrought iron piping should be used for acetylene distribution systems. Under no circumstances should acetylene gas be brought into contact with unalloyed copper except in a torch treated to prevent chemical reaction. Joints in steel or wrought iron pipe should be welded or made up with threaded or flanged fittings. Flanged connections in acetylene lines should be electrically bonded. Grey or white cast iron fittings should not be used.

Joints in brass or copper pipe may be welded, threaded, or flanged. A socket joint may be brazed with silver solder or similar high-melting-point material. Threaded connections in oxygen piping should be tinned or made up with litharge and glycerine or other joint compound approved for oxygen service.

In fuel gas distribution systems, a back-flow check valve or hydraulic seal should be used to prevent back flow at every point where gas is withdrawn from the piping system to supply a torch or machine. Such devices should be listed (or approved) by an agency such as Factory Mutual or Underwriters Laboratories Inc.

Portable Outlet Headers

Portable outlet headers are assemblies of valves and connections used for service outlet purposes and are connected to a permanent service piping system by means of hose or other nonrigid conductors. Devices of this nature are commonly used at piers and dry docks in shipyards where the service piping cannot be located close enough to the work to provide a direct supply. Their use should be restricted to outdoor locations and to temporary service where conditions preclude a direct supply, and they should be used in accordance with regulations in NFPA 51, previously referenced.

Regulators

Pressure regulators must be used on both oxygen and fuel gas cylinders to maintain a uniform gas supply to the torches at the correct pressure. The oxygen regulator should be equipped with a safety relief valve or be so designed that, should the diaphragm rupture, broken parts will not fly. Workers should stand to one side and away from regulator gauge faces when opening cylinder valves.

Only regulators listed by agencies such as Underwriters Laboratories Inc. or Factory Mutual should be used on cylinders of compressed gas. If unlisted regulators are used, they should be fully checked by a competent welding engineer. Each regulator (oxygen or fuel gas) should be equipped with both a high-pressure (contents) gauge and a low-pressure (working) gauge.

High-pressure oxygen dial gauges should have safety vent covers to protect the operator from flying parts in case of an internal explosion. Each oxygen dial gauge should be marked OXYGEN—USE NO OIL OR GREASE.

Serious, even fatal, accidents have resulted when oxygen regulators have been attached to cylinders containing fuel gas, or vice versa. To guard against this hazard, it has been customary to make connections for oxygen regulators with right-hand threads and those for acetylene with left-hand threads, to mark the gas service on the regulator case, and to paint the two types of regulators different colors. Cylinder valve outlet threads have been standardized for most industrial and medical gases (see *Compressed Gas Cylinder Valve Outlet and Inlet Connections,* ANSI/CGA V–1–1987). Different combinations of right-hand and left-hand threads, internal and external threads, and different diameters to guard against wrong connections are now standard.

The regulator is a delicate apparatus and should be handled carefully. It should not be dropped or pounded on. Regulators should be repaired only by qualified persons or sent to the manufacturer for repairs.

Leaky or "creeping" regulators are a source of danger and should be withdrawn from service at once for repairs. If a regulator shows a continuous creep, indicated on the low pressure (delivery) gauge by a steady buildup of pressure when the torch valves are closed, the cylinder valve should be closed and the regulator removed for repairs.

If the regulator pressure gauges have been strained so that the hands do not register properly, the regulator must be replaced or repaired before it is used again.

When regulators are connected but are not in use, the pressure-adjusting device should be released. Cylinder valves should never be opened until the regulator is drained of gas and the pressure-adjusting device on the regulator is fully released.

These procedures should be followed in detail when regulators or reducing valves are being attached to a gas cylinder:

1. To blow out dust or dirt that otherwise might enter the regulator, "crack" the discharge valve on the cylinder by opening it slightly for an instant and then closing it. On a fuel gas cylinder, first see that no open flame or other source of ignition is near; otherwise, the gas might ignite at the valve.
2. Connect the regulator to the outlet valve on the cylinder. Be sure the regulator inlet threads match the cylinder valve outlet threads. Never connect an oxygen regulator to a cylinder containing fuel gas, or vice versa. Don't force connections that do not fit easily. Be sure that the connections between the regulators and cylinder valves are gas-tight.
3. Release the pressure-adjusting screw on the regulator to its limit—turn it counter-clockwise until it is loose. Engage the adjusting screw and open the downstream line to the air to drain the regulator of gas.
4. Open the cylinder valve slightly to let the hand on the high-pressure gauge move up slowly. On an oxygen cylinder, gradually open the cylinder valve to its full limit, but on an acetylene cylinder, make no more than 1½ turns of the valve spindle.
5. Attach oxygen hose to outlet of oxygen regulator and to oxygen inlet valve on torch. Attach acetylene hose to outlet of acetylene regulator and to acetylene inlet on torch.
6. Test oxygen connections for leaks. Be sure torch oxygen valve is closed; then turn oxygen regulator pressure-adjusting screw clockwise to give about normal working pressure. Using soapy water or approved leak-test solution, check connections for leaks as in Figures 21–6a through 6c. At the same time, check regulator for creeping, indicated by an increase in the reading on the low-pressure (delivery) gauge. If the regulator creeps, have it replaced or repaired before it is used.
7. Test acetylene connections for leaks. Be sure torch acetylene valve is closed and proceed in manner similar to Step 6 above—except that acetylene regulator pressure-adjusting screw should be set to produce a pressure of about 10 psig (69 kPa) (Figures 21–6a through 6c).
8. If torch is to be used immediately, proceed to Step 9. If not, close cylinder valves, open torch valves to release pressure on regulator and gauges, close torch valves, and release pressure-adjusting screws on regulators.
9. To adjust pressures of oxygen and fuel gas prior to using torch, proceed as follows: with all torch valves closed, slowly open oxygen cylinder valve, open torch oxygen valve, turn in pressure-adjusting screw on oxygen regulator to desired pressure, then close torch oxygen valve. Open acetylene cylinder valve (1½ turns maximum), and with torch acetylene valve closed, turn in pressure-adjusting screw on acetylene regulator to desired pressure.
10. Purge each line individually. Open oxygen torch valve and release oxygen to the atmosphere for a few seconds before closing the valve; then open acetylene torch valve and release acetylene to the atmosphere for a few seconds and close the valve.
11. Open torch acetylene valve, light flame, and readjust regulator. Then close torch acetylene valve. (Acetylene pressure should first be adjusted with torch valve closed to prevent release of acetylene to air.)
12. Open torch valves and light torch according to procedure described in instructions provided with the equipment. The procedure for operating one torch is not necessarily best or even satisfactory for another.

Hoses and Hose Connections

Oxygen and acetylene hoses should be different colors or otherwise identified and distinguished from each other. Red is the generally recognized color for the fuel-gas hose and green for the oxygen hose (Figure 21–7). Black is used for inert gas and air hoses. The hose connections are usually marked STD-OXY for oxygen and STD-ACET for acetylene. The acetylene union nut has a groove cut around the center to indicate left-hand threads.

Connections for joining the hose to the hose nipple on the torches and regulators may be either the ferrule or clamp type. Gaskets should not be used on these connections. Special torch connectors with built-in shut-off valves are available.

Following are suggestions for the safe use of hose in welding and cutting operations:

- Do not use an unnecessarily long hose—it takes too long to purge. When a long hose must be used, see that it does not become kinked or tangled and that it is protected from being run over by trucks or

Figure 21–6a.

Figure 21–6b.

Figure 21–6c. How to test for leaks: With the pressure on and the torch valves closed, hold the hose (Figure 21–6a) and the torch tip under water. Bubbles indicate leaks. Use soapsuds to test for leaks in the torch values and hose-to-torch connections, as shown by the arrows. Separately test the oxygen cylinder (Figure 21–6b) and the acetylene cylinder and regulator connections (Figure 21–6c) for leaks at points marked by arrows. (Bottom photo courtesy J.I. Case.)

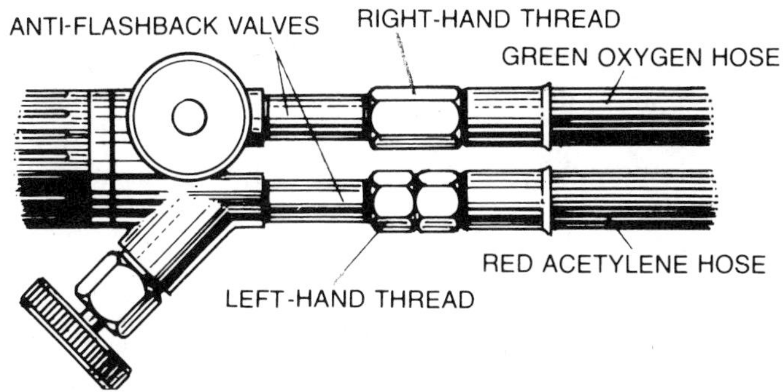

Figure 21–7. When attaching hoses to a welding or cutting torch, use the red hose for acetylene and the green hose for oxygen; then test connections for leaks.

otherwise damaged. Where a long hose must be used in areas exposed to vehicular or pedestrian traffic, suspend it high enough overhead to permit unobstructed passage.

- Repair leaks at once. Besides being a waste, escaping fuel gas may become ignited and start a serious fire; it may also set fire to the welder's clothing. Escaping oxygen is equally hazardous. Repair hose leaks by cutting the hose and inserting a splice. Don't try to repair leaky hose by taping.
- Examine hoses periodically and frequently for leaks and worn places, and check hose connections. Test for leaks by immersing the hose in water under normal working pressure. (Refer to Figure 21–6, top left.)
- Protect hoses from flying sparks, hot slag, other hot objects, and grease and oil. Store hose in a cool place.
- A single hose having more than one gas passage shall not be used. When oxygen and acetylene hoses are taped together for convenience and to prevent tangling, not more than 4 in. (10 cm) of each 12 in. (30.5 cm) of hose should be taped.
- The use of hoses with an external metallic covering is not recommended. In some machine processes and in certain types of operations, hoses with an inner metallic reinforcement that is exposed neither to the gas passage nor to the outside atmosphere are acceptable.
- Flashback devices (Figure 21–7) between the torch and hose can prevent burnback into hoses and regulators. If a flashback occurs and burns the hose, discard the burned section.
- A hose that has been subject to flashback, or that shows evidence of severe wear or damage, shall be tested to twice the normal pressure, but in no case less than 300 psi. A defective hose, or a hose in doubtful condition, shall not be used.

Torches

Torches are constructed of metal castings, forgings, and tubing. Usually, they are made of brass or bronze, but stainless steel may also be used. They should be of substantial design to withstand the rough handling they sometimes receive. It is best to use only those torches listed by an agency such as Underwriters Laboratories Inc. or Factory Mutual.

The gases enter the torch by separate inlets, go through valves to the mixing chamber, and then to the outlet orifice, located in the torch tip. Several interchangeable tips are provided with each torch and have orifices of various sizes according to the work to be done.

The cutting torch, unlike the welding torch, uses a separate jet of oxygen in addition to the jet or jets of mixed oxygen and fuel gas. The jets of mixed gases are for preheating the metal, and the pure oxygen jet is for cutting. The flow of oxygen to the cutting jet is controlled by a separate valve.

There are two types of torches in general use: the "injector" or low-pressure type, and the "pressure" or medium-pressure type. In the injector torch, the acetylene is drawn into the mixing chamber by the velocity of the oxygen. The acetylene may be supplied either from a low-pressure generator, a medium-pressure generator, or from cylinders. In the medium-pressure torch, both gases enter under pressure; therefore, the acetylene is supplied from cylinders or from a medium-pressure generator.

In the operation of torches, several precautions should be observed:

- Select the proper welding head or mixer and tip or cutting nozzle (according to charts supplied by the manufacturer), and screw it firmly into the torch.
- Before changing torches, shut off the gas at the pressure-reducing regulators and not by crimping the hose.
- To discontinue welding or cutting for a few minutes, closing only the torch valves is permissible. If the welding or cutting is to be stopped for a longer period (during lunch or overnight) proceed as follows:
    1. Close oxygen and acetylene cylinder valves.
    2. Open torch valves to relieve all gas pressure from hose and regulator.
    3. Close torch valves and release regulator pressure-adjusting screws.
- Do not use matches to light torches. Use a friction lighter, stationary pilot flame, or other suitable source of ignition. When lighting, point the torch tip so no one will be burned when the gas ignites.
- Never put down a torch until the gases have been completely shut off. Do not hang torches from a regulator or other equipment so that they come in contact with the sides of gas cylinders. If the flame has

not been completely extinguished or if a leaking torch ignites, it may heat the cylinder or even burn a hole through it.

- When extinguishing the flame, close the acetylene and oxygen valves in the order recommended by the torch manufacturer. If the oxygen valve is closed first, carbon soot will be deposited in the air. However, this ensures that the acetylene valve is closed tight when the flame is extinguished. If the acetylene valve is turned off first, no soot is formed, but there is no assurance that the fuel-gas valve is closed and that it is not leaking.

Powder Cutting

Powder-cutting processes for metal and concrete use similar equipment and gas supplies as do oxygen-cutting operations. The precautions previously discussed for safe handling and use of compressed gas equipment and cutting torches therefore apply. Manufacturers' recommendations for the operation and maintenance of the powder-dispensing apparatus—both pneumatic and vibratory—should be followed.

RESISTANCE WELDING

Because resistance welding equipment is normally permanently installed, the hazards are usually minimized if the equipment has been properly designed and safe operating practices have been established.

Certain hazards in the operation of this equipment—lack of point-of-operation guards, flying hot metallic particles, improper handling of materials, unauthorized adjustments and repairs—may cause eye injuries, burns, and electrical shock. Most of these hazards can be eliminated by safeguarding the equipment, wearing protective clothing, and strictly controlling operating practices.

Resistance welding is a metal-joining process whereby welding heat is generated at the joint by the resistance to the flow of electric current. The three fundamental parameters of resistance welding are current magnitude, current time, and tip pressure. Each of these must be accurately controlled (Figure 21–8).

Power Supply

Resistance welding usually employs a 60-hertz alternating current that is fed to the primary of the water-cooled welding transformer. The primary can vary from 150 to 10,000 amp, at 240, 440, or 550 volts. The output at the secondary of the transformer is a low-voltage (max 30 volts) and high-amp current (up to 200,000 amp) used for welding.

The welding current is sometimes furnished by the "stored energy" type of equipment. Energy is built up

Figure 21–8. This is an automatic resistance welding machine with a dial feed. The operator removes and places the work when the proper dial fixture comes to the front. (Courtesy General Electric Co.)

and stored either in capacitors or in a combination transformer-reactor during the nonwelding period, and then is discharged to form the weld. This process involves low primary currents and high voltages that must be guarded against.

To facilitate servicing the equipment, a safety-type disconnecting switch or circuit breaker of the correct rating for opening supply circuits should be installed near the welding machine. (See Control of All Energy Sources in Chapter 6, Safeguarding, in this volume, for lockout/tagout procedures.) Permanent injuries and several fatalities have resulted from neglecting to use the line-disconnecting switch before making adjustments inside enclosures. This precaution is imperative because the use of single-pole primary circuit breakers and electronic contactors leave one line to the welder "hot."

Cables

Abuse of the cables for resistance welding is severe. The production requirements demand the utmost of the cable materials used, and even the best cables need frequent replacement. In use, the cables are subjected to electrical pulsation, bending, and twisting that leads to fatigue and eventual breakdown. This condition is minimized by the use of concentric cables.

The secondary voltage presents little shock hazard, since the maximum voltage is about 30 volts; but the operator can be hurt by a cable blowout such as is caused by steam pressure due to overheating from faulty water cooling circulation or from electrical failure.

A periodic check for weak spots in the cable covering is good practice. The use of concentric welding cables is now common because they do not have the undesirable features of the pulsating cables. Portable welding machines, including the cables, should have proper weight balance to permit operation without undue strain to the operator.

Machine Installation

Installation of resistance welding equipment should conform to the NFPA *National Electrical Code,* Standard No. 70. Some items worthy of special attention are listed below.

- Control circuits should operate on low voltage, not exceeding 24 volts maximum for portable spot welders.
- Stored energy equipment (capacitor discharge or resistance welding) having control panels involving high voltage (more than 550 volts) should be completely enclosed. Doors should have locks and contacts wired in the control circuit to short circuit the capacitors when the door or panel is opened. A manually operated switch will serve as an additional safety measure, assuring complete discharge of the capacitors.
- Back doors of machines and panels should be kept locked or interlocked to prevent tampering.
- A fused safety switch or circuit breaker should be located conveniently near the welding machine so that power supply circuits may be opened before servicing the machine and its controls.
- The point-of-operation hazard should be eliminated by suitable guards. Enclosure guards, gate guards, two-hand controls, and similar standard guards as designed for punch press operations are applicable.
- A flash-welding machine should have a shield or hood to control flash and fumes, and a ventilating system to carry off the metallic dust and oil fumes.
- Where flying sparks are not confined, the operators and nearby persons should be protected by shields of safety glass or other fire-resistant material or by the use of personal eye protection.
- Foot switches, air or electrical, should be guarded to prevent accidental operation.

ARC WELDING AND CUTTING

For arc welding or cutting, two welding leads, the electrode lead and the work lead, are required from the source of current supply. Usually, one lead is connected to the work and the other to the electrode holder. The work lead (cable) is the most satisfactory means of providing the return (ground) circuit to the welding machine, but in some cases operating conditions may require the use of a grounded steel structure. The steel structure and connections should be capable of carrying the welding current.

Power Supply

Either AC or DC may be used for arc welding or cutting of any kind. With small-diameter electrodes used on thin sheets for manual arc welding, current values vary from 10 to 50 amp. For most manual welding, because the welder must withstand the heat, current values should not exceed 500 or 600 amp. Automatic machine arc welding may use current values up to 200 amp or even higher on special applications.

If a gasoline-powered welding generator is used inside a building or in a confined area, the engine exhaust should lead to the outside atmosphere. Otherwise, carbon monoxide and other toxic gases may accumulate.

Voltages

The voltage across the welding arc varies from 15 to 40 volts, depending on the type and size of electrode used. The welding circuit must supply somewhat higher voltage to strike the arc. This voltage is called the open-circuit or "no load" voltage. After the arc is established, the open-circuit voltage drops to a value about equal to the arc voltage plus the lead voltage drop. The open-circuit voltages on DC welding machines should be less than 100 volts. Constant voltage power supplies (welder or converter) are now also being widely used.

For AC transformer welding machines, ANSI Z49.1, *Safety in Welding and Cutting,* prescribes a maximum open-circuit voltage of 80 volts on manuals, 100 volts on automatics.

Heavy-duty AC welding machines (ratings usually over 500 amp for automatic machine welding) are also built with open-circuit voltages of 75 to 80 volts with a special tap to provide for 100 volts where necessary. The tap may be needed to obtain rated output from the machine if the line voltage is low or if the voltage of the secondary circuit drops. The tap should not be accessible to the welder for current adjustment but should be under the control of a responsible electrician or supervisor.

Open-circuit voltages should be as low as 50 volts on small AC welding machines used without expert supervision. Because these machines are often used with low arc voltage electrodes and with leads not more than 20 ft long (6 m), the probability of a large voltage drop in the welding circuit is small.

For other manual and automatic welding and cutting processes where work metal is connected electrically to one side of the circuit, an open-circuit voltage of 150 volts may be allowed if the following conditions are present:

1. All equipment and circuiting are fully insulated and the operator cannot make electrical contact other than through the arc itself, while the arc is maintained.
2. Disconnecting or voltage-reducing devices operate within a time limit not exceeding one second after breaking the arc.

Where neither side of the circuit is electrically connected to the work, open-circuit voltage of 300 volts is allowed if controls are present to prevent the operator from touching both sides of the circuit. Operators should use one hand to work the control devices. Also, the voltage should be disconnected automatically by a reliable switch instantly upon breaking the arc. For AC or DC welding under electrically hazardous conditions, a reliable automatic control device for reducing no-load voltage is recommended.

Cables

Several lengths of welding cable may be used in one circuit. To splice or connect cables, workers should use substantially insulated connectors of a capacity at least equivalent to that of the cable. Cable lugs used for ground and machine connections should be securely fastened to give good electric contact.

Welding cable is subjected to severe abuse if it is dragged over work under construction, across sharp corners, or run over by shop trucks. Workers should use special cable with high-quality insulation. The fact that welding circuit voltages are low may make workers lax about keeping the welding cable in good repair. Operators and maintenance personnel should make sure that defective cable is immediately replaced or repaired.

On large jobs, there is likely to be a great deal of loose cable lying around. Welders should keep this cable orderly and out of the way, preferably strung overhead to permit the passage of persons and vehicles. Welding cables should not lie in water or oil, in ditches or bottoms of tanks. Management should require that rooms where arc welding is to be done regularly should be permanently wired with enough outlets to prevent having extension cables strewn about.

Electrodes and Holders

Electrode holders for shielded-metal arc welding (SMAW) are used to connect the electrode to the welding cable supplying secondary current. Fully insulated holders (Figure 21–9) are preferred because there is less likelihood of shocking the welder or of accidentally striking an arc with such holders, particularly in close quarters.

Electrode holders will become hot during welding operations if holders designed for light work are used on heavy welding or if connections between the cable and the holder are loose. If workers cannot use a holder of the correct size for the electrode, they should have an extra holder so that one can cool while the other is in use. They should never dip hot electrode holders in water.

On light or medium-heavy work for which workers use light, extremely flexible cables, holders may be attached directly to the work lead running to the machine. On heavier work, welders generally prefer a short length attached to the holder, which is more flexible than the main work lead. Properly insulated cables of weight and flexibility that will not inconvenience the welder are available. Fully insulated connectors should be used to attach the short flexible cable to the main work lead.

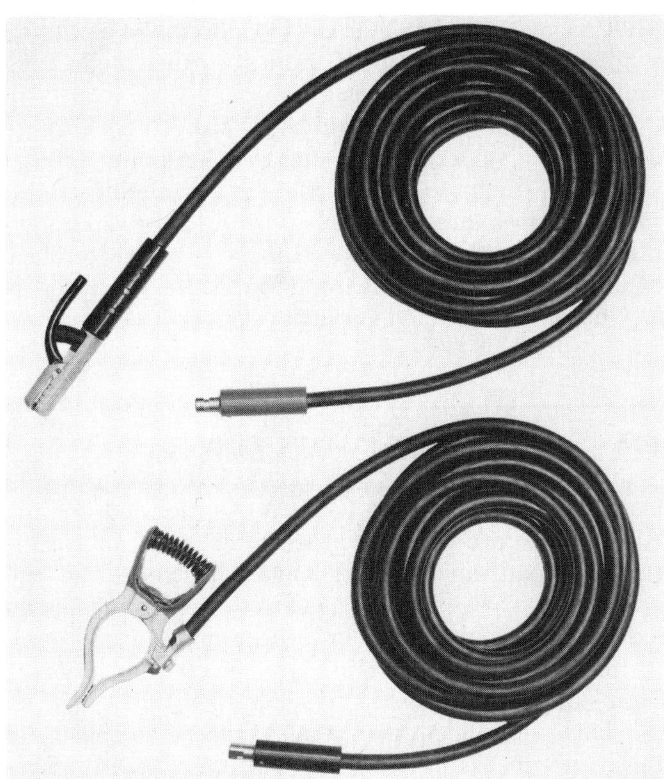

Figure 21–9. The fully insulated electrode holder on the electrode lead and the ground clamp on the work lead have insulated locking plugs for connection to receptacles on the welding machine. (Courtesy Westinghouse Electric Corp.)

Protection against Electric Shock

Although open-circuit voltages on standard arc-welding units are not high compared to those of other processes, they should be considered a potential hazard. Normally, the work setup is such that the work is grounded; unless care is exercised, the welder or operator can easily become grounded.

The voltage between the electrode holder and the ground, during the Off arc or No-Load period, is the open-circuit voltage.

The welder or welding operator shall be insulated from both the work and the metal electrode and holder. The bare metal part of an electrode or electrode holder should never be permitted to touch the operator's bare skin or wet clothing.

Consistent use of well-insulated electrode holders and cables, dry clothing on the hands and body, and insulation from ground will be helpful in preventing contact.

Pacemaker wearers should check with the manufacturer or medical person regarding welding operations.

Some specific precautions for prevention of electric shock are the following:

1. In confined places, cover or arrange cables to prevent contact with falling sparks.
2. Never change electrodes with bare hands or wet gloves, or when standing on wet floors or grounded surfaces.
3. Ground the frames of welding units, portable or stationary, in accordance with the *National Electrical Code*, NFPA 70. A primary cable containing an extra conductor, one end that is attached to the frame of the welding unit, can be used with a small welding unit. This ground connection can be carried back to the permanently grounded connection in the receptacle of the power supply by means of the proper polarized plug.
4. Arrange receptacles of power cables for portable welding units so that it is impossible to remove the plug without opening the power supply switch, or use plugs and receptacles that have been approved to break full load circuits of the unit.
5. If a cable (either work lead or electrode lead) becomes worn, exposing bare conductors, it may be repaired if the insulation repair on work-lead cables is equivalent in insulation to the original cable covering.
6. Keep welding cables dry and free of grease and oil to prevent premature breakdown of the insulation.
7. Suspend cables on substantial overhead supports if the cables must be run some distance from the welding unit. Protect cables that must be laid on the floor or ground so that they will not interfere with safe passage or become damaged or entangled.
8. Take special care to keep welding cables away from power supply cables or high-tension wires.
9. Never coil or loop welding cable around the body.

Gas-Tungsten Arc Welding, Plasma Arc Welding, and Cutting

In gas-tungsten arc welding (GTAW), the electrode does not melt and is not used for filler metal. The electrode is tungsten, which is highly resistant to heat and nonconsumable in the welding process. Filler metal may be added by using a cold (nonelectrical) welding rod that is introduced into the arc or molten weld puddle. The process can be used to weld nearly every weld.

Either AC or DC welding units can be used for gas-tungsten arc welding, depending on the following characteristics: (1) whether the weld is wide, deep, or narrow; (2) whether the job is to be performed on a permanent fixture or a portable machine; and (3) whether it is to be a manual or machine welding operation. For gas-metal arc, DC is used with a reverse polarity hookup, with current supplied by a generator or rectifier. The AC supply may be obtained through a transformer or high-frequency generator. The manufacturer of the welding equipment should be consulted as to the specific job before the equipment is installed, especially when high-frequency AC is used in order to avoid interference with radio transmission.

Argon, helium, and gas mixtures are supplied by manufacturers in cylinders similar to oxygen cylinders. Since cylinder pressures range from 2,200 to 2,640 psig (15.2 MPa to 18.2 MPa), argon and helium cylinders should be stored and handled like other high-pressure gas cylinders.

Carbon dioxide, although not strictly an inert gas, is sometimes used as a shield gas when steel is welded by the gas-metal arc-welding process. It is usually supplied in partially liquid and partially gaseous form in cylinders at approximately 835 psig (5.8 MPa) pressure. These cylinders should therefore be handled like other high-pressure gas cylinders.

To supply gas to the welding torch, a regulator must be used to lower the pressure to 25 psig (172 kPa) or less, and a flow-meter should measure the volume of gas being used. If more than one torch is used from the same gas line, a flow-meter should be installed at each torch connection.

Air is used to cool the torch and electric current cables. Water is also used for cooling, generally where the welding current is more than 250 amp. The water supply line, even if city water is used, should be equipped with a strainer to keep out impurities that might plug the water cooling passages.

In the gas-shielded tungsten arc torch (Figure 21–10), gas is conducted to the welding point through orifices in the torch around the electrode holder. Cooling water goes through passages through the torch handle and about the holder. In the smaller torches, ceramic cups are used. A torch for heavier work generally has a water-cooled gas cup.

GAS-METAL ARC WELDING

Gas-metal arc welding (GMAW) is defined by the AWS as "an arc-welding process that uses an arc between a continuous filler metal electrode and the weld pool. The process is used with shielding from an externally supplied gas and without the application of pressure." (See ANSI/AWS A3.0–89, *Welding Terms and Definitions.*)

This welding process is also known as the metal inert gas process (MIG). When GMAW was first introduced, it used inert gas to shield the arc. The gas was usually argon, or argon with a small amount of oxygen added. One company called the process the shielded-

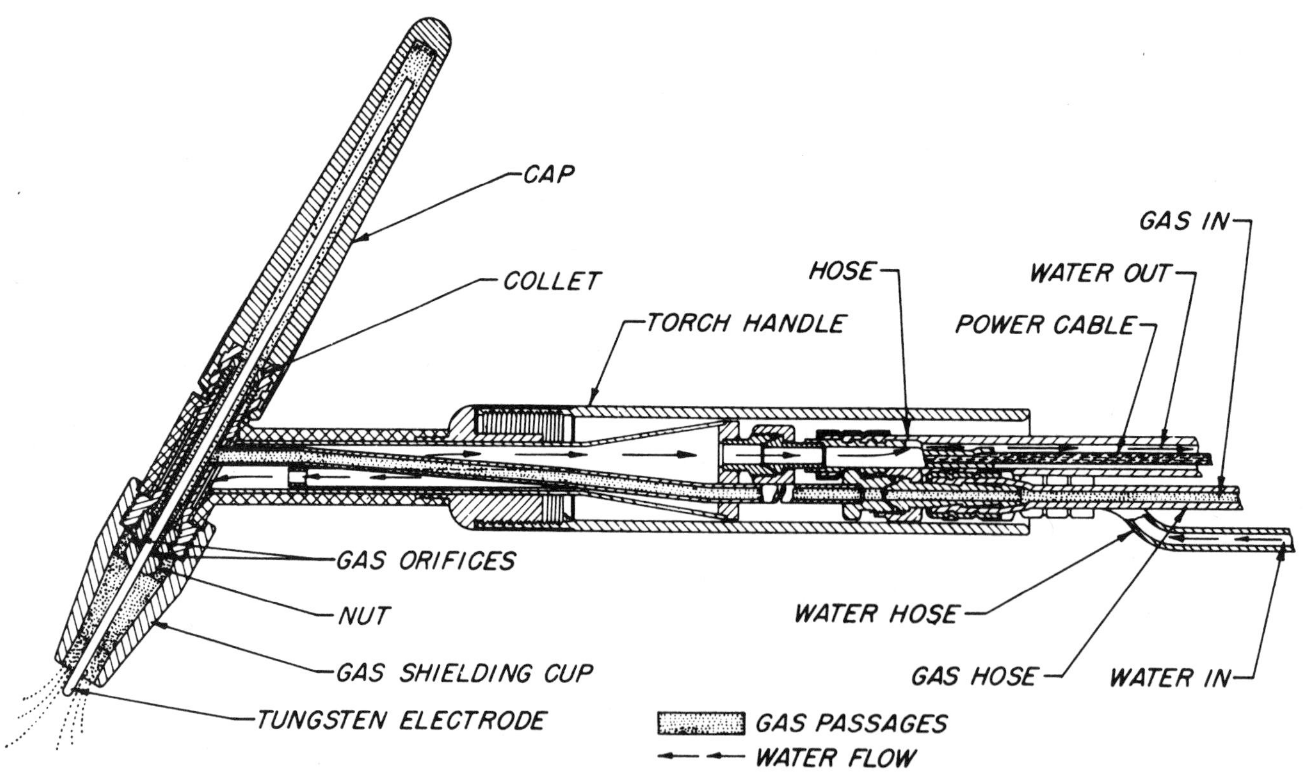

Figure 21–10. The electrode is nonconsumable in this gas-shielded tungsten arc welding torch. (Reprinted with permission from *Welding Handbook,* American Welding Society.)

inert-gas metal arc, or the SIGMA welding process. As the GMAW process was developed, other gases came into use, some of them not inert but chemically active. In Europe, the process using active gases came to be known as the metal active gas process (MAG).

One version of the MAG process uses CO_2 as the shielding gas. CO_2 is an active gas, but this variation of the process became known as the CO_2 welding process. All variations of the process use a consumable wire electrode that is fed through a welding gun. Most of the processes use an externally applied shielding gas of some type, or a mixture to protect the weld zone.

FLUX-CORED ARC WELDING

Flux-cored arc welding (FCAW) is defined by the AWS as "an arc-welding process that uses an arc between a continuous-filler metal electrode and the weld pool. The process is used with shielding gas from a flux contained within the tubular electrode, with or without additional shielding from an externally supplied gas, and without the application of pressure." (See ANSI/AWS A3.0.)

FCAW is sometimes called cored-wire welding by mistake. The term "cored wire" would cover any wire with a core. For example, it could be a piece of low-carbon steel wire covered by a jacket of high-alloy steel. Or it could be a steel wire with a core of aluminum. The term "flux-cored wire," however, is a wire with a core of flux. The wire is an inside-out, shielded metal arc electrode. Instead of the flux covering the outside of the electrode as it does in SMAW, the flux is on the inside. Because of its similarity to GMAW wire, FCAW wire is fed into the arc with similar equipment. Only slight variations are necessary because of the tubular and collapsible nature of the wire.

FCAW competes with SMAW as well as with GMAW as it is used mostly to weld steels. The flux acts to reduce porosity, control penetration, and stabilize and shield the arc. When used properly, the flux gives high-quality welds that have a clean, smooth appearance and that can be made in all positions.

Just as in SMAW, the flux produces a slag covering that must be removed. In addition, a great amount of fumes and gases are produced. Moreover, the process is dirtier than GTAW and GMAW. Because of its similarity to GMAW, it is easy to provide gas shielding to assist the flux. Some flux-cored electrodes do not provide enough gases and vapors to protect the weld metal from air. Then gas shielding, such as CO_2, is used to provide extra protection. Both shielded (with gas) and unshielded (without gas) FCAW are currently used in commercial processes. (See ANSI/AWS A3.0.)

OTHER WELDING AND CUTTING PROCESSES

There are several relatively new heat sources for welding and cutting, such as friction, ultrasonics, and lasers. Each of these special heat sources requires guarding and safe practices.

For example, the laser (Light Amplification by Stimulated Emission of Radiation) presents the hazard of eye damage from the optically amplified light beam. Because of its intensity, the beam can do damage even at great distances. All employees working with or near laser devices should be given preemployment and periodic follow-up eye examinations. Most companies require that employees work in pairs when using laser equipment.

To confine the laser beam, the company should develop and install suitable shields (Figure 21–11). Because a reflected laser beam is also hazardous, the work area should contain no glossy surfaces.

Power supplies for lasers and the electron beam are high voltage-equipment that should be operated with the precautions developed for this type of equipment. (See the discussion in this volume, in Chapter 7, Personal Protective Equipment; in the *Administration & Programs* volume, Chapter 24, Laboratory Safety, under Ionizing and Nonionizing Radiation; and in NSC's *Fundamentals of Industrial Hygiene,* 4th ed.)

SUMMARY

- In welding operations, the leading health risks are from toxic gases, particulate matter, pulmonary irritants and toxic inhalants, cleaning compounds, chlorinated hydrocarbons, and asbestos.
- Safety hazards common to all welding and cutting processes require measures such as fire protection, shielding, safe working areas, and safe handling and storage of all combustible or flammable materials in drums, tanks, and closed containers. Material Safety Data Sheets list the minimum exposure levels and operating standards for many hazardous welding materials.
- Personal protection for welding and cutting operations usually includes respiratory protection against dusts, gases, and fumes. Operators should be trained in the kind of protective gear to wear.
- The American Welding Society has established standards for training and qualification of welders. Supervisors are responsible for the safety of operators and the safe operation of welding equipment during all phases of work.
- In oxyfuel welding and oxygen cutting operations, workers must know how to handle and store cylinders safely to avoid explosion and fire. They should inspect their cylinders regularly to ensure that all parts are in good working order.

Figure 21–11. A self-standing safety shield provides close-quarter protection and rolls up for storage.

- Workers must strictly observe precautions for selecting and using welding equipment and for preventing fires and explosions.
- Risks associated with resistance welding can be eliminated by safeguarding equipment, using protective clothing, and strictly controlling operating practices.
- For arc welding or cutting operations, the power supply and welding structures should be grounded and insulated to prevent shock hazards, and all cables, connectors, and electrode holders should be well insulated.
- New heat sources for welding and cutting require guarding and safe practices to protect workers from risks posed by high-frequency sound waves and amplified light waves.

REFERENCES

American Conference of Governmental Industrial Hygienists, Committee on Industrial Ventilation, 1330 Kemper Meadow Dr., Cincinnati, OH 45240-1634.

Industrial Ventilation—A Manual of Recommended Practice.

American Gas Association, 1725 Jefferson Davis Highway, Suite 1004, Arlington, VA 22202-4102.

American Insurance Association, 85 John Street, New York, NY 10038.

Lasers and Masers, Special Hazards Bulletin Z–125.

American National Standards Institute, 11 West 42nd Street, New York, NY 10036.

Practices for Respiratory Protection, ANSI Z88.2-1992.
Safe Use of Lasers, ANSI Z136.1-1993.
Safety in Welding and Cutting, ANSI Z49.1–1994.
Safety Standard for Mechanical Power-Transmission Apparatus, ANSI/ASME B15.1–1992.
Safety Standard for Transformer-Type Arc-Welding Machines, ANSI/UL 551–1993.
Welding Terms and Definitions, ANSI/AWS A3.0–94.

American Petroleum Institute, 1220 L Street NW, Washington, DC 20005.

Cleaning Mobile Tanks Used for Transportation of Flammable Liquids, Accident Prevention Manual, No. 13, 1958.
Cleaning Petroleum Storage Tanks, RP2015.
Gas and Electric Cutting and Welding, RP2009.

American Society of Mechanical Engineers, 3 Park Ave., New York, NY 10016-5902.

ASME Boiler and Pressure Vessel Code, Section IX, "Qualification Standard for Welding and Brazing Procedures, Welders, Brazers, and Welding and Brazing Operators." 1965.

American Welding Society, 550 NW 42nd Ave., Miami, FL 33126.

Fire Prevention in Arc Welding and Cutting Processes, 249.2.
Recommended Practices for Gas Tungsten Arc Welding, C.5.5 (R1989).
Recommended Practices for Plasma Arc Cutting, C.5.2. (R1989).
Safety in Welding and Cutting, 249.1–1983, Z49.1–1983.
The Welding Environment, 1973.
Welding Fumes, 1–1981, 2–1981, 3–1983.
Welding Handbook, 7th ed. Section 1. 1976.
Welding Terms and Definitions, A3.0–1989.

Compressed Gas Association, Inc., 1725 Jefferson Davis Highway, Arlington, VA 22202-4100.

Oxygen-Deficient Atmospheres, SB–2.
Regulator Connection Standard.
Safe Handling of Compressed Gases in Cylinders.
Specification for Rubber Welding Hose.

Linde Division, Union Carbide Corp., 39 Old Ridgebury Road, Danbury, CT 06817.

"Plasma-Arc Process" Bulletins.

National Fire Protection Association, Batterymarch Park, Quincy, MA 02269.

Bulk Oxygen Systems at Consumer Sites, NFPA 50, 1996.
Safeguarding of Tanks and Containers for Entry, Cleaning, or Repair, NFPA 326 & 327, 1999.

Control of Gas Hazards on Vessels, NFPA 306, 1997.

Fire Prevention during Welding, Cutting, and Other Hotwork, NFPA 51B, 1999.

Gaseous Hydrogen Systems at Consumer Sites, NFPA 50A, 1999.

Liquefied Hydrogen Systems at Consumer Sites, NFPA 50B, 1999.

National Electrical Code,® NFPA 70, 1996.

Design and Installation of Oxygen-Fuel Gas Systems for Welding, Cutting, and Allied Processes, NFPA 51, 1997.

Solon LR. "Occupational Safety with Laser (Optical Maser) Beams." *Archives of Environmental Health,* 6:414–17 (March 1963).

U.S. Department of Health, Education, and Welfare, National Institute for Occupational Safety and Health, Division of Technical Services, Cincinnati, OH 45226.

Criteria for a Recommended Standard: Occupational Exposure to Ultraviolet Radiation, 73–11009.

Engineering Control of Welding Fumes, DHEW (NIOSH) Publication 75–115.

Safety and Health in Arc Welding and Gas Welding and Cutting, DHEW (NIOSH) Publication 78–138.

U.S. Navy, The Pentagon, Washington, DC 20350.

Bureau of Ships Manual, Chapter 92, "Welding and Allied Processes." NAVSHIPS 250–00–92.

Underwater Cutting and Welding Manual, NAVSHIPS 250–692–9.

REVIEW QUESTIONS

1. What is the most significant health hazard in the welding process?
2. List five health conditions a welder can experience as a result of exposure to various toxic gases generated during welding.
   a.
   b.
   c.
   d.
   e.
3. Ozone, one of the primary pulmonary gases that can injure the lungs, is formed by:
   a. Electrical arcs and corona discharges in the air
   b. Welding metals that have been cleaned with chlorinated hydrocarbons
   c. Ultraviolet photochemical reactions
   d. All of the above
   e. Only a and c
4. Which of the following inhalants is highly toxic and can produce total disability after exposure as short as a few months?
   a. Manganese
   b. Titanium
   c. Magnesium
   d. Chromium
5. If closed containers that have held flammable liquids cannot be removed for standard cleaning procedures prior to welding or cutting them, what two other practices can be followed?
   a.
   b.
6. Name the three general categories of personal protective equipment that should be used by welders.
   a.
   b.
   c.
7. Why is dark, woolen clothing preferred when welders work with inert-gas-shielded arc-welding machines?
8. The standards for training and qualification of welders was established by which organization?
9. What are the five steps in handling a leaking fuel gas cylinder?
   a.
   b.
   c.
   d.
   e.
10. Which of the following piping should be used for acetylene distribution systems?
    a. Unalloyed copper
    b. Steel
    c. Wrought iron
    d. All of the above
    e. Only b and c
11. The color that is generally recognized for a fuel-gas hose is:
    a. Green
    b. Red
    c. Black
    d. Blue
12. What device must be used on both oxygen and fuel gas cylinders to maintain a uniform gas supply to the torches at the correct pressure?
13. Define resistance welding.
14. Name three relatively new heat sources for welding and cutting.
    a.
    b.
    c.
15. Installation of resistance welding equipment should conform to what standard?

22

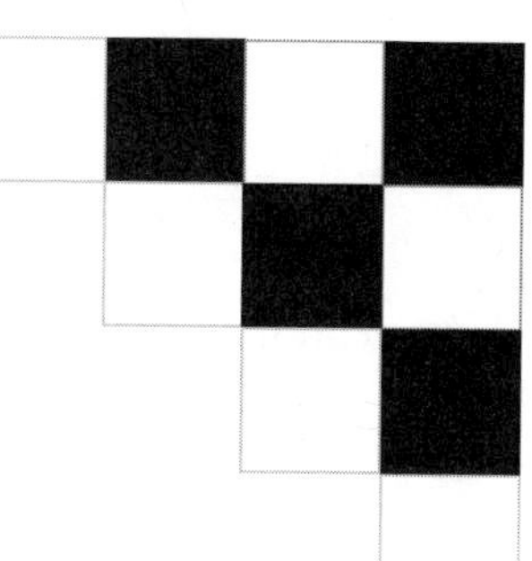

Metalworking Machinery

Revised by

Philip E. Hagan, MPH, CIH, CHMM, CHCM

Metalworking machinery includes all power-driven machines not movable by hand. They are used to shape or form metal by cutting, impact, pressure, electrical techniques, chemical techniques, or a combination of these processes. This chapter can help managers develop safety programs for metalworking machinery. The topics covered include:

- general safety rules for operating and maintaining machine tools
- specific hazards and safety practices for turning, boring, milling, planing, and grinding machines
- importance of proper safety training for operators of machine tools

The Association for Manufacturing Technology (AMT—formerly National Machine Tool Builders' Association) has classified some 200 types of machine tools into five basic groups: turning, boring, milling, planing, and grinding. Another classification is electrodischarge, electrochemical, laser, and machining tools. Some machines combine the functions of two or more groups.

Power presses, press brakes, and power squaring shears are also classified as metalworking tools. However, these tools are covered in Chapter 23, Cold Forming of Metals. Portable power tools, normally hand-held during operation, are not considered to be machine tools. They are covered in Chapter 19, Hand and Portable Power Tools.

Injuries with machine tools are most often caused by unsafe work practices or incorrect procedures. Insufficient training and inadequate supervision usually give rise to these problems. More rarely, injuries result when a machine fails mechanically or is operated after an unsafe condition develops. Only qualified and competent personnel should operate hazardous equipment with exposed moving parts and cutting edges. In addition, properly maintaining and operating equipment would reduce injuries from machine tools. Installing effective guarding devices that do not hamper operation or lower production would further reduce the number of injuries. Of course, certain guards are required by federal, state or provincial, and local regulations.

Good housekeeping in the work area can help establish good work habits when operating machine tools. These two factors—good housekeeping and good work habits—result in fewer accidents.

GENERAL SAFETY RULES

Emphasize safely operating metalworking machinery. Establish a written policy to eliminate unsafe practices by operators. Include the following provisions in the safety policy. Maintenance and repair personnel should also comply with this policy.

- Restrict operation, adjustment, and repair of any machine tool to authorized, experienced, and trained personnel.
- Ensure proper lockout/tagout procedures are followed whenever work is performed on equipment.
- Closely supervise all personnel during training.
- Establish and maintain safe work procedures; prohibit shortcuts and chance taking.
- When purchasing new equipment, make sure that specifications conform to all applicable standards, codes, and regulations concerning guarding, electrical safety, and other safeguards.
- Inspect new and modified equipment and make safety innovations before allowing operators to use the equipment.
- Devote full-time attention to the work in progress. If the operator must leave the machine, it should be shut down and locked out unless the machine tool has been designed to operate in this mode. For example, have interlocked guarding all around the machine and equip the machine with automatic shutdowns that work as soon as there is any deviation from normal operation.
- Make supervisors responsible for the strict enforcement of this policy.

Provide a tool rack for the convenience of operators and repair and maintenance personnel. Include all wrenches and tools needed for operation or adjustment as standard equipment. Provide necessary material handling equipment to avoid strains (Figure 22–1).

ANSI/NFPA 70, *National Electrical Code®*, by the American National Standards Institute and National Fire Protection Association, and ANSI C2, *National Electrical Safety Code,* should govern installation of electrical circuits and switches. Other metalworking standards are in NFPA 79, *Electrical Standard for Industrial Machinery.*

Electrical Controls on Machine Tools

In addition to the manufacturer-installed electrical controls on machine tools, each machine must have a disconnect switch that can be locked in the OFF position to isolate the machine from the power source. Do not permit maintenance or repair on any machine until the disconnect switch serving the equipment has been shut off, padlocked in the OFF position, and tagged. (See the discussion of lockout/tagout in Chapter 6, Safeguarding, in this volume.)

Figure 22–1. This machinist avoids back strain by sliding heavy materials onto a mobile lifter adjusted to the worktable height.

Rules for Safely Operating Machine Tools

The following rules apply to safely operating any machine tool. Be sure that operators know and follow these rules:

- Never leave machine tools running unattended, unless the machine has been designed to do so.
- Never wear jewelry or loose-fitting clothing, especially loose sleeves, loose shirt or jacket cuffs, and neckties.
- Keep long hair that could be caught by moving parts covered.
- Wear eye protection. This rule extends to others in the area, such as inspectors, stock handlers, and supervisors.
- Do not throw refuse into or spit into the machine tools' coolant. Such actions foul the coolant and might spread disease.
- Do not manually adjust and gauge (caliper) work while the machine is running.
- Use brushes, vacuum equipment, or special tools for removing chips. Do not use hands.
- Understand the differences in machining ferrous and nonferrous metals, and know the health or fire hazards of working with these metals.
- Use the proper hand tools for each job.

Safely Removing Chips, Shavings, and Cuttings

One of the major causes of accidents from machine tools, especially drilling equipment, is the careless use of high-pressure compressed air to blow chips, cuttings, or shavings from machines or workers' clothing. Brushes provide a less dangerous method, as do hand-held or air-operated vacuum units.

In cases where neither a brush nor a vacuum system is practical, it may be necessary to use air. Keep all unnecessary employees out of the area. Ensure employees wear applicable personal protective equipment (PPE): safety goggles, face shield, and full body protective clothing as determined by a PPE job evaluation. Many companies have found a nozzle pressure of 10 to 15 psig (70 to 100 kPa) is sufficient for most operations. The U.S. Occupational Safety and Health Administration (OSHA) specifies that the pressure be less than 30 psig (207 kPa). Nozzles meeting the OSHA requirements are available.

Isolate the machine tool operation so that nearby employees are not endangered. Place baffles or chip guards around the machine to shield the operator. Permit only reliable and trained employees to use compressed air. These employees should wear cup-type goggles and other personal protective equipment. Prohibit employees from using high-pressure compressed air to blow dust or dirt from their clothing or out of their hair. Such actions can cause damage to, or inject foreign materials into, their ears and eyes.

One company has almost eliminated injuries caused by removing chips, shavings, or cuttings. Employees wear leather palmed gloves and use a 3 ft (91 cm) long rod to pull shavings and cuttings from machines. The ¼ in. (6 mm) diameter round rod has a handle on one end for the operator to hold and a crook at the other to pull the shavings.

Personal Protection

The machine tool operator's safety depends largely upon following established safe working procedures and wearing proper protective clothing and equipment. Obviously, all machine tool operators should wear eye protection with side shields. Wearing close-fitting clothing is also vitally important to the operator's safety. Many serious injuries and fatalities have resulted when neckties, loose shirt sleeves, gloves, or other clothing have gotten caught in a belt and sheave, between gears, in a revolving shaft, or in the revolving workpiece held in the chuck (Figure 22–2). Operators also should not wear rings, necklaces, or other jewelry, as they can get caught in machinery. Because most machine operations involve handling heavy stock or heavy machine parts, such as faceplates and chucks, every operator should wear protective footwear.

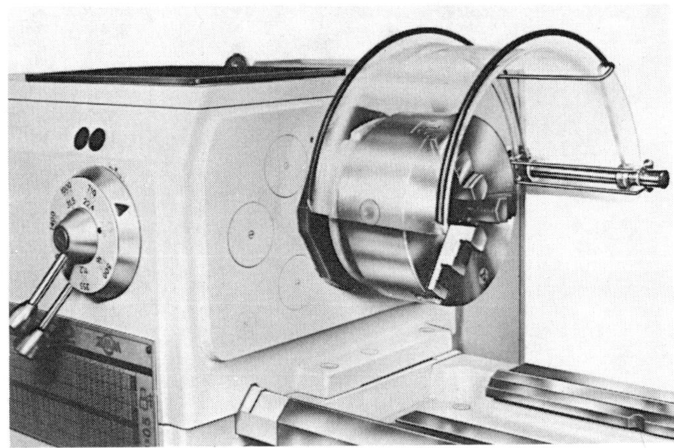

Figure 22–2. This lathe chuck shield has a semicircular construction and is made from high-impact-resistant, transparent acrylic. The shield is mounted to a chromium-plated extension tube fastened to the lathe headstock by mounting brackets. The shield can be lifted up and out of the way for quick and easy access to the piecepart.

Operators with long hair should wear caps, snoods, hairnets, or other protection that completely covers their hair. There have been many instances of people being partially, or entirely, scalped when their hair became entangled in the moving parts of a machine.

Use splash guards, shields, personal protective equipment, and other means to minimize exposure of workers to the irritating cutting oils and mineral spirits used to clean metal parts. Provide barrier creams, and encourage personal hygiene measures, such as thorough washings, to minimize skin irritations.

TURNING MACHINES

Shaping a rotating piece with a cutting tool, usually to give a circular cross section, is known as turning. This procedure is done on machines such as engine lathes, turret lathes, chuckers, semiautomatic lathes, and automatic screw machines.

Engine Lathes

To prevent accidents with engine lathes, allow only qualified personnel to operate them. Injuries are likely to result from incorrectly operating engine lathes in the following ways:

- having contact with projections on work or stock, faceplates, chucks, or lathe dogs, especially those with projecting setscrews
- being hit by flying metal chips

- hand braking the machine
- filing right-handed, using a file with an unprotected tang, or using the hand instead of a stick to hold an emery cloth against the work
- failing to keep both the center holes of taper work clean and true and the lathe's center true and sharp
- leaving the machines running unattended
- handling chips by hand instead of using a hooked rod
- calipering or gauging the job while the machine is running
- attempting to remove chips when the machine is running
- having contact with rotating stock projecting from turret lathes or screw machines
- leaving the chuck wrench in chuck
- catching rings, loose clothing, gloves, or rags for wiping on revolving parts

Several preventive measures will help operators run engine lathes safely. Use faceplates and chucks without projections whenever possible. Otherwise, install a simple shield formed to the contour of the chuck or plate and hinged at the back. This will prevent contact with the revolving plate or chuck.

Since safety-type lathe dogs are relatively inexpensive to install, substitute them for those with projecting setscrews. Chip shields, particularly on high-speed operations, help control flying chips. Use plastic or fine-mesh screen chip guards because they allow operators to see through them while confining the flying chips (Figure 22–2). These shields do not eliminate the need for protective eye equipment, however. Shields may need to be frequently replaced because of the abrasive action of the chips.

Provide mechanical means, such as an overhead hoist or a swinging, welded pipe fixture, to lift heavy faceplates, chucks, and stock on both lathes and screw machines.

Figure 22–3. This automatic screw machine has a chip shield to protect against cutting oil splashes and flying chips. (Printed with permission from Delco-Remy Division of General Motors.)

Turret Lathes and Screw Machines

Hazards associated with turret lathes and screw machines (Figure 22–3) are similar to those listed for other lathes. Additional hazards are caused by operators not moving the turret back as far as possible when changing or gauging work, or using machine power to start the faceplate or chuck onto the spindle. Other hazards result when operators fail to keep their hands clear of the turret's slide or permit a hand, arm, or elbow to strike the cutter while adjusting or setting up.

Install splash shields, especially on automatic machines, and keep them in good condition. Enclosure shields over the chuck confine hot metal chips and oil splashes. They also act as exhaust hoods to remove fumes.

When steel and some other materials are turned on lathes, the chip produced is in a continuous spiral that frequently causes injuries to hands and arms. Install chip breakers to protect against this.

Screw machines can create noise that exceeds noise limits. Try to reduce the noise level through engineering methods, such as commercially available sleeves for rotating bar stock. Use hearing protection whenever engineering and administrative controls are not sufficient to maintain noise exposure below regulated levels. (See the National Safety Council's (NSC) publication *Hearing Conservation in the Workplace: A Practical Guide.*)

Spinning Lathes

A spinning lathe is a forming tool rather than a cutting tool. It usually requires a specially skilled and qualified operator.

Unsafe practices that should be prohibited when operating a spinning lathe include:

- inserting blanks and removing the processed part without first stopping the machine
- failing to fully tighten the tailstock's handle and risking that the blank will work loose or ruin the stock or tool
- allowing the swarf (cuttings, turnings, particles) to build up into a long coil when trimming copper and certain grades of steel

This last practice has resulted in severed hands and severely cut arms. One fatality occurred when a coil became snarled around the operator's neck. Operators should remove the tool, when necessary, to allow the swarf to break off.

The spinning lathe's chuck is usually a form built up of hardwood, shaped exactly like the finished part. If the piece being worked must be "necked down" so the chuck cannot be taken out after the piece is formed, the chuck is made in sections held together by locking rings or locking grooves. A great hazard is possible if a chuck flies apart because the grooves have become worn or the rings break. Since spinning lathes reach speeds of from 500 to 2,000 rpm, the danger of flying chuck sections is obvious. Prevention of injuries from this cause lies in frequent inspection and maintenance of chucks. Similarly, inspect lathe tools frequently for cracks in the tools or handles.

The tailstock of spinning lathes can loosen during operation because of vibration or worn parts. The piece then will work loose or fly off. The solution to this problem also is frequent inspection and maintenance.

BORING MACHINES

Boring consists of cutting a round hole using a drill, boring cutter, or reamer. Drilling machines are equipped with rotating spindles, handles, and chucks that carry pointed or fluted cutting tools. Operations performed with drilling machines include countersinking, reaming, tapping, facing, spot facing, and routing. Boring mills use a cutter, either single- or multi-edged, that is mounted on a supporting spindle or shaft. The cutter trues up or enlarges a hole that has already been rough formed by drilling, casting, or forging.

Drills

Drill press accidents are more likely to occur during unusual jobs because special jibs or vises for holding the work are not usually provided. Radial drill accidents are frequently caused by incorrect manipulation. Properly clamp the drill's head and arm, as well as the workpiece, prior to cutting metal. The most common hazards in drilling operations are:

- contacting the rotating spindle or the tool
- being struck by a broken drill
- using dull drills
- being struck by insecurely clamped work
- catching hair, clothing, or gloves in the revolving parts
- sweeping chips, or trying to remove long, spiral chips, by hand
- leaving the key or the drift in the chuck
- being struck by flying metal chips
- failing to replace the guard over the speed-change pulley or gears

Observing safety precautions and using good operating habits can protect operators from these hazards. To guard an operator from contact with a spindle, use a plastic shield, a simple wire mesh guard, or other barrier (Figure 22–4). When necessary, guard the tool with a telescoping guard that covers the end of the tool. This leaves only enough of the tool exposed to allow easy placement into the piece being worked. The telescoping drill shield shown in Figure 22–4 has a stationary, ribbed cage. The outer sleeve rides up and down on the inner cage. This shield provides high visibility.

A frequent cause of breakage is using a dull drill. A thin drill, smaller than ⅛ in. (3 mm) in diameter will often break and cause injury. A larger drill may "fire up," freeze in the hole, and then break. Furthermore, a frozen tool may cause unclamped or insecurely clamped work to spin and injure the operator.

To avoid having a drill catch in thin material and spin it, clamp the work between two pieces of metal or

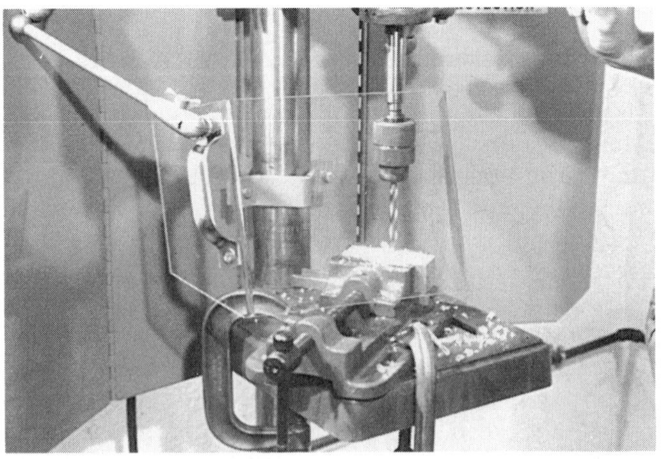

Figure 22–4. This transparent drill shield protects against flying chips and pieces of broken drills.

wood before drilling. When drilling thin ferrous stock, grind the drill point to an included angle of about 160 degrees, and thin the point of the drill by grinding the flutes. With nonferrous metals, a negative rake will further reduce chances of the drill grabbing or digging in.

When deep holes are being drilled, frequently remove the drill and clean out the chips. If chips are allowed to pile up, the tool may jam, with results similar to those of freezing. Maintain counterweight chains in good condition and install a shield around the counterweight.

Boring Mills

Some common causes of injury in boring mill operations are:

- being struck by insecurely clamped work or by tools left on or near a revolving table
- catching clothing or rags for wiping in revolving parts
- falling against revolving work
- calipering or checking work while the machine is in motion
- allowing turnings to build up on the table
- removing turnings by hand

Horizontal Boring Mills

The same accident prevention measures are effective on both table and floor types of horizontal boring mills. While the machine is in motion, the operator should never attempt to make measurements near the tool, reach across the table, or adjust the machine or the work.

Frequently inspect clamps and blocking to make certain the clamping is positive. Always avoid makeshift setups.

Before attempting to raise or lower a boring mill's head, the operator should make sure that the clamps on the column have been loosened. Otherwise, the boring bar can be bent or the clamps or bolts broken. This can cause damage to the machine and injury to the operator.

Before the boring bar is inserted into the spindle, the operator should make certain that the spindle's hole and the bar are clean and free from nicks. The operator should not attempt to drive the bar through the tailstock's bearing with a hammer or other heavy tool. Instead, the operator should use a soft metal hammer to drive the bar into the spindle. If a steel hammer or piece of steel must be used, the operator should hold a piece of soft copper or brass against the bar while driving it into the spindle.

Vertical Boring Mills

The same procedures apply to the safe operation of vertical boring mills. Each mill's table, particularly those tables 100 in. (2.5 m) or less in diameter, should have the rim enclosed in a metal band guard to protect the operator from being struck by the revolving table or by projecting work. Such guards should be hinged so they can be easily opened during setting up and adjustment (Figure 22–5).

If the table is flush with the floor, install a portable fence, usually of iron-pipe sections. Such fencing should conform to the state code or the specifications of the ANSI A1264.1—*Safety Requirements for Workplace Floor and Wall Openings, Stairs, and Railing Systems.*

While the machine is in operation, the operator should never attempt to tighten the work, the tool, or the caliper, nor measure the work, feel the edges of the cutting tool, nor oil the mill. The operator should never ride the table while it is in motion. There is one exception. On some large mills, like those used for boring turbine castings, the operator may have to ride the table in order to observe the work's progress. In such cases, he or she should always make sure that no portion of the body will come in contact with a stationary part of the mill.

In addition, steps or stairs that provide access to the machine or to the work should have a pitch of not more than 50 degrees and should have slip-resistant treads. Stairs with four or more risers must have a handrail.

MILLING MACHINES

Machining a piece of metal by bringing it into contact with a rotating multi-edged cutter is milling. This procedure is done by horizontal and vertical milling machines, by gear hobbers, profiling machines, circular and band saws, and a number of other types of related machines (Figure 22–6).

Many accidents with milling machines occur when operators unload or make adjustments. Other causes of injuries include:

- failure to draw the job back to a safe distance when loading or unloading
- using a jig or vise that prevents close adjustment of the guard
- placing the jig- or vise-locking arrangement in such a position that force must be exerted towards the cutter
- leaving the cutter exposed after the job has been withdrawn
- leaving hand tools on the worktable
- failing to securely clamp the work
- reaching around the cutter or hob to remove chips while the machine is in motion
- removing swarf cuttings, by hand instead of with a brush

Figure 22–5. This guard for a vertical boring mill is made in two sections of sheet metal. The sections are hinged to the machine. *Left:* the guards are closed. *Right:* the guards are opened to allow setup or adjustment.

Figure 22–6. The adjustable shield on this vertical milling machine protects against both flying metal chips and splashing coolant.

- adjusting the coolant's flow while the cutter is turning
- calipering or measuring the work while the machine is operating
- using a rag to clean excess oil off the table while the cutter is turning
- wearing gloves, rings, ties, or loose clothing; catching fingers, gloves, or clothing in power clamps
- using incorrectly dressed cutters
- incorrectly storing cutters
- attempting to remove a nut from the machine's arbor by applying power to the machine
- striking the cutter with a hand or an arm while setting up or adjusting the stopped machine
- misjudging clearances between the arbor or other parts
- cleaning the machine while it is in motion

Basic Milling Machines

Regardless of the classification, direction of movement, or special attachments that make varied operations possible on a milling machine, the safeguarding requirements are basically the same. To guard the cutter, employ one of several methods (Figure 22–7).

Mount hand-adjusting wheels, for quick or automatic traverse on some models, on the shaft by either clutches or ratchet devices. In this way, the wheels do not revolve when the automatic feed is used. As an alternative, provide compression-spring wheels with removable handles. These handles cannot remain in the wheels, unless held in place by the operator.

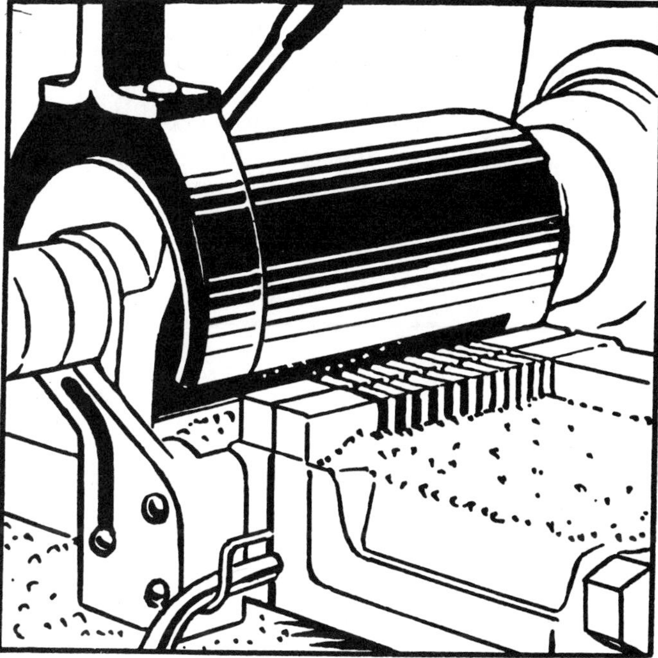

Figure 22–7. Self-closing guard for milling machine cutter. *Top:* the cutter is completely enclosed when the table is withdrawn. *Bottom:* as the table moves forward, the guard automatically opens.

The horizontal milling machine should have a splash guard and pans for catching thrown cutting lubricant and lubricant running from the tools. Direct the lubricant on the work so the distribution setup will not be drawn into the cut by the cutter's rotation. When possible, make all cuts into the travel of the table, rather than away from the direction of travel.

Metal Saws

Circular, swing, and band saws are metalworking, as well as woodworking, machinery. Provide suitable guarding for each type of saw. Chapter 20, Woodworking Machinery, in this volume, contains additional information about guarding circular, swing, and band saws.

Circular Saws

A circular saw for cutting cold metal stock should have a hood guard at least as deep as the roots of the teeth. The guard should automatically adjust itself to the thickness of the stock being cut.

Use a sliding stock guard when tube or bar stock is cut. Guard the portion of the saw under the table with a complete enclosure that provides for disposal of scrap metal. A plastic or metal guard placed in front of and over the saw provides protection against flying pieces of metal. Do not, however, consider a guard as a substitute for eye protection.

Swing Saws

For swing saws, adjust the length of the stroke so the blade will not pass the table at its forwardmost point. Locate the control so the saw can be operated with the left hand when fed from the left, or with the right hand if fed from the right. In this way, the operator is positioned to the side away from the moving blade.

Band Saws

Completely enclose the upper and lower wheels of metal-cutting band saws with sheet metal or a heavy, small-mesh screen mounted on angle-iron frames. To make the changing of blades convenient and safe, provide access doors equipped with latches. Except for the point at which the cut is made, completely enclose the portion of the blade between the upper wheel and the saw table with a sliding fixture attached to the slide.

The length of blade exposed should not be more than the thickness of the stock plus 3⁄16 in. (9 cm). Confine flying particles of metal with a metal or transparent plastic guard installed in front of the saw. On a hand-fed operation, take care at the end of a cut. Use a push block, not hands.

Gear Cutters

During operation of gear cutters and hobbers, both the tool and the workpiece move. Therefore, keep the point of operation guards simple and easily adjustable.

On operations where the workpiece (gear blank or rough-cut gear) is moved to the tool, a simple barrier guard, formed to cover the point of operation and sized to fit the workpiece, is satisfactory. Mount the guard on a spindle that carries the workpiece. This causes the guard to fit over the point of operation when the workpiece is brought into position.

When the tool is brought to the workpiece and when both the tool and workpiece are adjustable, attach an encircling type of guard to the tool's head. Such a guard can be an automatic drop-gate device. This device can be equipped with both a release latch to open the guard's enclosure and a spring release to return the guard to a position clearing the work. Each guard should have an automatic interlock so the machine will not operate except when the guard is in place.

On some makes of machines, the lever that controls the spindle's direction of operation is located so operators can catch their hands on the back gears driving the spindle. In such cases, install an auxiliary lever that can be operated at a point outside the danger zone created by these gears.

On large machines where the operator is not close to the regular control switch, install a pendant switch, mounted on an arm or sweep. This switch acts as a magnetic brake to stop the machine instantly.

When operators insert an arbor into the spindle, they should make sure that both the arbor and spindle holes are clean and free from nicks. Operators should draw the arbor firmly into place by a sleeve nut and securely tighten the nut. Before removing the arbor from the spindle, operators should make certain the machine is at a standstill.

Electrical Discharge Machining

The electrical discharge machine (EDM) process is designed to perform a variety of machining operations. This process makes simple or complex machining possible through hole boring or cavity sinking in any electrically conductive work material, including carbide, high-alloy steels, and many types of hardened metals.

The EDM Process

During the machining process, the workpiece is normally clamped to the table and an electrode is fastened to the vertical ram platen above the workpiece. The electrode is then brought near the workpiece so an accurately controlled electrical discharge takes place between the electrode and the workpiece. This discharge removes metal from the workpiece at the point where the gap is smallest.

As metal is removed, the electrode tool is fed into the workpiece and held in the correct cutting relationship by electrohydraulic servo control of the ram's workhead. The servo control automatically maintains a gap between the electrode and the workpiece.

During the process, a dielectric fluid should completely cover the workpiece in the work tank. During EDM, employ a flow of dielectric fluid through the machining gap whenever possible to increase machining efficiency (Figure 22–8).

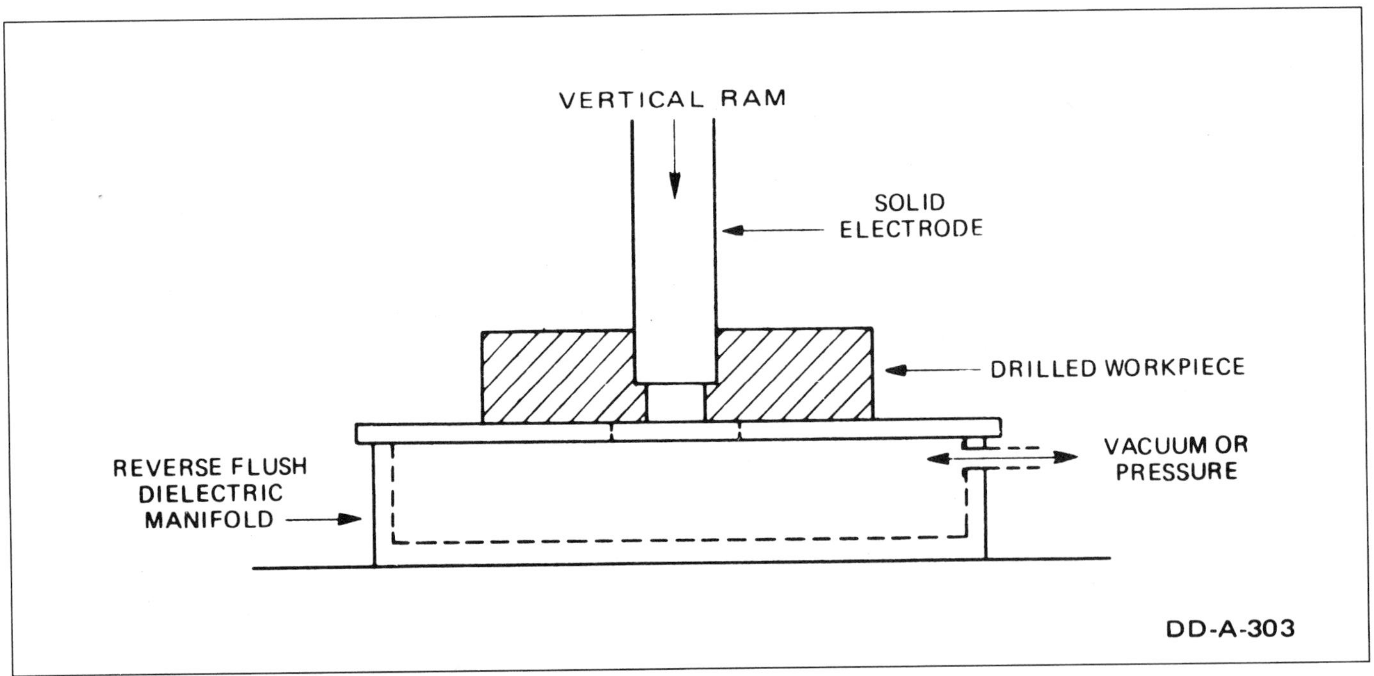

Figure 22–8. Electrical discharge machining (EDM) into manifolds (including a reverse flush dielectric manifold), tanks, domes, or any other structures capable of trapping discharge gases. (Printed with permission from Cincinnati Milacron Company.)

Hooking Up EDM Machines

Only a qualified electrician designated to work on machine tool circuits should maintain or hook up the electrical system of EDM machines. Before attempting any work, the electrician should read and completely understand the electrical schematics for the machine.

After the machine has been hooked up, operators should test all aspects of the electrical system to make sure it is functioning properly. Before considering the hookup job complete, operators should do the following:

- Be sure the machine is properly grounded, and check that all exposed electrical systems are properly covered.
- Place all selector switches in the OFF or neutral (disengaged) position.
- Be sure that the machine's push buttons, manual limit switches, or controls are set for a safe setup.
- Check that the doors of the main electrical cabinet are closed and that the main disconnect switch is in the OFF position.

Safety Precautions for EDM

To keep the operator from accidentally brushing against the live electrode or platen when the machine is operating at high voltages, install a clear plastic safety shield on the work tank. The shield must be in place before the power supply is turned on. Do not attempt to block out the wire around any electrical safety interlock. Before removing any cover, or before working on the machine or power supply, turn off and lock out the main electrical disconnect device. Tag the device and all START buttons with an OUT OF ORDER or DO NOT START tag. Always turn the electrical disconnect switches to the OFF position at the end of the working day.

Maintain the oil level above the highest portion of the electrode workpiece's working gap. Once the safe oil level above the part has been determined, adjust the safety float's switch to make certain that oil level is maintained. Maintain dielectric level at a minimum of 1 in. (2.5 cm) per 100 amperes of average current for flat geometry work. Since EDM is a heat-producing process, install an EDM machine only where there is adequate ventilation. The dielectric oil removes the concentrated heat from the machining gap and distributes it in the available oil. Air conditioners and electronic precipitators do not constitute adequate ventilation.

Discharge Gases Hazard

Operators and maintenance personnel should read and completely understand all the precautions before operating, setting up, running, or performing maintenance on EDM metal removal. Failure to comply with instructions can result in serious or fatal injury. The EDM machine operator must be aware of the possibility of discharge gases igniting. Turn off the electrical power to stop additional gas or hot metal particles from forming to extinguish the flame. Operators should have received training in how to use a carbon dioxide (CO_2) foam fire extinguisher. It is necessary that a CO_2 fire extinguisher be kept in the vicinity of the EDM machine.

Since all discharge gases are flammable, keep them away from sparks or flame. Allow discharge gases to escape without being trapped in a closed area. Avoid any setup that can result in trapped gases. Therefore, be sure that the EDM machine is adequately ventilated. When the dielectric level in a storage tank is suddenly raised, by dumping a work tank for example, a large volume of discharge gases will escape into the outside air. Ignition of such displaced gases can cause a fire that could backlash into the enclosed area and cause an explosion.

PLANING MACHINES

Planers machine metal surfaces. The cutting tool is held stationary while the workpiece is moved back and forth underneath it. With shapers, generally classified as planing machines, the process is reversed. The workpiece is held stationary while the cutting tool is moved back and forth. Other machine tools classified as planing machines are slotters and broaches.

Planers and Shapers

Accidents with planers frequently result from unsafe practices caused by inadequate training and supervision, such as the following:

- placing hand or fingers between the tool and the workpiece
- running the bare hand over sharp metal edges
- measuring the job while the machine is running
- failing to clamp the workpiece or tool securely before starting the cut
- riding the job
- having insufficient clearance for the workpiece
- coming in contact with reversing feed dogs
- failing, when magnetic chucks are used, to make sure the current is turned on before starting the machine
- unsafely adjusting the tool holder on the cross head.

To avoid these accidents, install guards on planers. Cover the reversing feed dogs on planers. If the planer bed, when fully extended, or any stock on the bed being processed, travels within 18 in. (46 cm) of a wall or fixed objects, close the space between the end of the travel and the obstruction with a guard on either side of

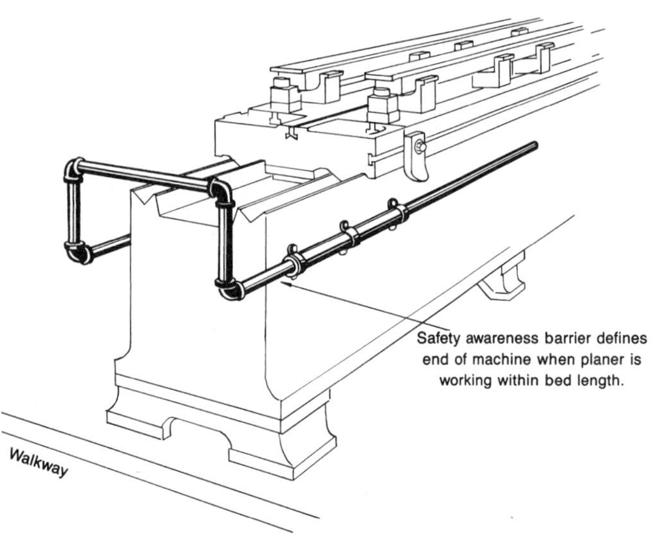

Figure 22–9. *Top:* a guardrail or similar barrier should close off any space 18 in. (46 cm) or less between a fixed object and parts of a fully extended planer or its stock. Any openings in the planer should be filled to eliminate shear hazards. *Bottom:* This self-adjusting planer table guard moves out with the table and is retained in position by friction sleeves.

the planer (Figure 22–9). Construct the guard so it will not cause an accident when the bed is extended.

Accidents with shapers have essentially the same causes as those with planers. In addition, injuries frequently result from contact with projections on the workpiece or with projecting bolts or brackets, especially when the table is being adjusted vertically. Leave the shaper's ram projecting over the table, to alert the operator that the table is high enough.

Failure to properly locate the stops or dogs can also injure shaper operators. Rigidly bolt the stops to the table, especially on heavy jobs.

The shaper operator should make sure the tool is set. That way, if it shifts away from the cut, it will rise away from the cut and not dig into the work. Remove the handle of the stroke-change screw before starting the shaper. To prevent injury to the operator and workers nearby from flying chips, install guards. Also, cover the reversing feed dogs on shapers.

Slotters

In slotter operations, the most serious accident is catching the fingers between the tool and the workpiece. Fingers can also be caught between the ram and the table when the ram is at the end of the downstroke. Since the ram works at a slow speed and the platen or machine table is small, operators may instinctively reach across the table and under the ram to pick up a tool or other object—thus catching their fingers. To avoid this kind of accident, enclose the ram's eccentric with a hinged guard made of sheet metal or cast iron.

Broaches

The broach's rated capacity should be equal to or greater than the force required for the job. The centerline of the work, the ram's head puller, and the follow rest, if used, should all line up. Check fixtures to make sure the workpiece is securely held. Do trial runs at slow speeds to make certain the chips do not pack tightly between the teeth.

During normal operations, safeguard broaches, like heavy production machine tools, through supplementary controls. The most widely used safeguard for electrically controlled broaches is a standard two-hand, constant pressure control. Install an EMERGENCY STOP button, preferably of the mushroom type, adjacent to one of the two-hand controls. Another type of safeguard for broaches is a foot-operated EMERGENCY STOP bar with a wide surface plate.

All pneumatically or hydraulically powered clamping equipment should be actuated by two-hand controls. Locate these controls so the operator's hand cannot reach the pinch area before the clamps close. Shield controls if they could be tripped by other parts of the body. If the hands are exposed in the clamping area, use tongs for loading and unloading.

GRINDING MACHINES

Grinding machines shape material by bringing it into contact with a rotating abrasive wheel or disk. Grinding includes surface, internal, external cylindrical, and centerless operations, as well as polishing, buffing, honing, and wire brushing. Portable machines that use small, high-speed grinding wheels are discussed in Chapter 19, Hand and Portable Power Tools.

The text and illustrations in this section have been adapted with permission from ANSI B7.1, *The Use, Care, and Protection of Abrasive Wheels.* Specifications for operation of grinding machines and construction of guards and safety devices are in this code.

Grinding Machine Hazards

Hazards associated with grinding machines include the following:

- failure to use eye protection in addition to the eye shield mounted on the grinder
- incorrectly holding the work
- incorrectly adjusting or not using the work rest
- using the wrong type, a poorly maintained, or imbalanced wheel or disk
- grinding on the side of a wheel not designed for side grinding
- taking too heavy a cut
- applying work too quickly to a cold wheel or disk
- grinding too high above the wheel's center
- failure to use wheel washers (blotters)
- vibration and excessive speed that lead to bursting a wheel or disk
- using bearing boxes with insufficient bearing surface
- using a spindle with incorrect diameter or with the threads cut so the nut loosens as the spindle revolves
- installing flanges of the wrong size, with unequal diameters, or with unrelieved centers
- incorrect wheel dressing
- contacting unguarded moving parts
- using controls that are out of the operator's normal reach
- using an abrasive blade instead of a grinder disk
- failure to run a wet wheel dry, without coolant, for a period of time before turning off the machine
- using an untested, broken, or cracked grinding wheel
- reaching across or near the rotating grinding wheel to load, unload, or adjust the machine during setup

Abrasive Disks and Wheels

An abrasive disk is made of bonded abrasive, with inserted nuts or washers, projecting studs, or tapped plate holes on one side of the disk. This side is mounted on the faceplate of a grinding machine. Only the exposed flat side of an abrasive disk is designed for grinding.

An abrasive wheel is made of bonded abrasive and is designed to be mounted, either directly or with adapters, on the spindle or arbor of a grinding machine. Only the periphery or circumference of many abrasive wheels are designed for grinding.

Inspecting Abrasive Disks and Wheels

When unpacking abrasive disks and wheels, inspect them for damage from shipment and have a qualified person give them the "ring" test. This test can be used for both light and heavy disks or wheels that are dry and free of foreign material. To conduct the ring test, suspend a light disk or wheel from its hole on a small pin or the finger, and place a heavy one vertically on a hard floor. Then gently tap the wheel or disk with a light tool, such as a wooden screwdriver's handle. A mallet may be used for heavy wheels or disks. Make the tap at a point 45 degrees from the vertical centerline and about 1 or 2 in. (2.5 or 5 cm) from the periphery (Figure 22–10). A wheel or disk in good condition will give a clear, metallic ring when tapped. Wheels and disks of various grades and sizes give different pitches.

Daily inspection of grinding machines should include those points shown in Figure 22–11. Thoroughly investigate grinding wheel and disk failures, preferably with the manufacturer's representative. This type of investigation, along with immediate corrective action, greatly reduces the possibility of recurrent failures.

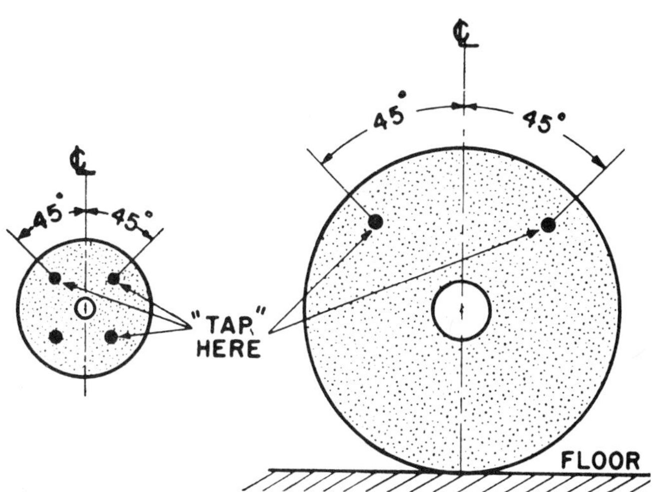

Figure 22–10. Tap points for ring test.

GRINDER CHECKLIST

TYPE ______________ RPM __________

SIZE ______ PERIPHERAL SPEED ______

Item	OK
WHEEL GUARD: securely fastened	☐
properly aligned	☐
GLASS SHIELD: clean	☐
unscored	☐
in place	☐
WORK REST: within ⅛ in. (3.2 mm) of wheel	☐
securely clamped	☐
FRAME: securely mounted	☐
no vibration	☐
WHEEL FACE: well lighted	☐
dressed evenly	☐
FLANGES: equal size	☐
correct diameter (½ wheel diam.)	☐
SPEED: correct for wheel mounted	☐
GUARD FOR POWER BELT OR DRIVE: in place	☐

DATE ________ DEPARTMENT ______

INSPECTED BY ________________

Figure 22–11. A summary of checkpoints for safe grinder operation.

Handling Abrasive Disks and Wheels

Abrasive disks and wheels require careful handling. Do not drop or bump them. Do not roll large disks and wheels on the floor. Transport disks and wheels too large or heavy to be carried by hand by truck or other means that provide the correct support.

Storing Abrasive Disks and Wheels

Store abrasive disks and wheels in a dry area not subject to extreme temperature changes, especially below-freezing temperatures. Wet wheels might break or crack if stored below 32 F (0 C). Breakage can occur if a wheel or disk is taken from a cold room and work is applied to it before it has warmed up.

Store abrasive disks and wheels in racks in a central storage area under the control of a specially trained person. The storage area should be as close as possible to the grinding operations to minimize handling and transportation.

The length of time abrasive disks and wheels may be stored and still be safe to use should be in accord with manufacturers' recommendations. Give the ring test to disks and wheels taken out of long storage. Follow this by a check for recommended speed, and a speed test on the machine that will be mounted. Check the speed of all grinding wheels against the spindle speed of the machine—some are designed only for low-speed use. The grinding wheel must be rated at the same RPM as the machine. If it is not, the wheel may explode and throw particles into the work area.

Mounting Wheels

Mount all abrasive wheels between flanges. Exceptions to this rule include mounted wheels, threaded wheels (plugs and cones), plate-mounted wheels, and cylinder, cup, or segmental wheels mounted in chucks.

Flanges should have a diameter not less than one-third of the wheel's diameter and preferably should be made in accord with ANSI B7.1, Sect. 5. Flanges for the same wheel should be of the same diameter and thickness, accurately turned to correct dimensions, and in balance. The requirement for balance does not apply to flanges made out of balance to counteract an unbalanced wheel.

Key, screw, shrink, or press the inner or driving flange onto the spindle. The bearing surface of the flange should run true with the spindle. The outer flange's bore should easily slide onto the spindle.

Schedule flange inspections frequently. Remove from the spindle a flange found to be sprung, not bearing evenly on the wheel, or defective in any other way. Replace it with a flange that is in good condition.

An incorrectly mounted abrasive wheel is the cause of much wheel breakage. Since rotational forces and grinding heat cause high stresses around the wheel's central hole, follow safety regulations concerning size and design of mounting flanges and mounting techniques.

Before a wheel is mounted, give it the same inspection and ring test as it was given when originally received and stored. In addition, check the bushings, particularly on wheels that have been rebushed by the user, for shifting or looseness.

Use compression washers to compensate for unevenness of the wheel or flanges. Blotting paper, not more than 0.025 in. (0.6 mm) thick, or rubber or leather compression washers, not more than 0.125 in. (3.2 mm) thick, may be used for this purpose.

Make allowance for the wheel mounting's fit in the wheel hole rather than in the arbor or wheel mount (Figure 22–12). Do not force the wheel onto the spindle. Forcing can loosen, or otherwise damage, the wheel's bushing; or it can crack the wheel. A wheel that is too loose on the spindle will run off-center, causing stress and vibration. Spindle end nuts should hold the wheel firmly but not too tightly. Too much pressure can spring

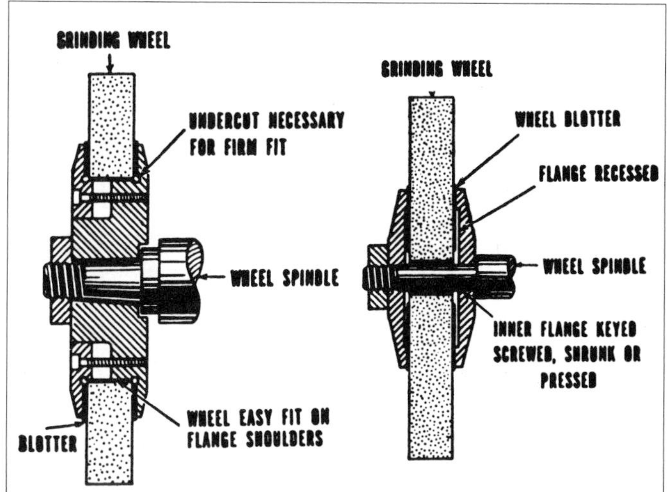

Figure 22–12. Correct methods of mounting abrasive wheels with large holes *(left)* and wheels with small holes *(right).*

or distort the flange or even break the wheel. If rebushing is necessary to make the wheel fit the spindle, have the manufacturer do it, or have an experienced employee with suitable equipment do it.

Immediately after mounting the wheel and before turning on the power, the operator should turn the wheel by hand for a few revolutions. At this time, check to make sure the wheel clears the hood guard and machine elements, such as work rests on work-holding equipment.

Operating a Grinding Machine

When starting a grinding machine, stand to one side away from the grinding wheel. Allow at least one minute of warm-up time before truing or grinding with the wheel. Always use coolant when truing the wheel or during normal grinding. Never allow coolant to flow on a stationary grinding wheel; coolant might collect on one portion of the wheel, causing an unbalanced condition. This unbalanced condition can cause the wheel to disintegrate upon restarting.

While the machine is running, never remove a guard fastener or guard. The guards are on the machine for the operator's safety; if they are removed, serious injuries to the operator or others can result.

Do not touch any moving part of the machine or the rotating grinding wheel to determine its smoothness or condition. Do not attempt to physically operate a machine that is in its automatic mode. Never alter or try to alter the machine, its wheel speed, or any of its safety equipment at any time.

Adjusting Safety Guards

The guard should enclose the wheel as completely as the nature of the work will permit. Adjust the peripheral guard to the constantly decreasing diameter of the wheel with an adjustable tongue or similar device. Doing this prevents the maximum distance between the wheel's periphery and the tongue or end of the peripheral band at the top of the opening from exceeding ¼ in. (6 mm) (Figure 22–13). Also, it maintains the angle of exposure specified for bench and floor-stand grinders throughout the life of the wheel. The maximum exposure angle varies with the type of grinding (Figure 22–14). Safety guards should also cover the exposed arbor ends.

On machines used for cutting, grooving, slotting, or coping stone or other materials, the safety guard or hood seldom offers adequate protection. On machines that permit a horizontal traverse between the wheel and the workpiece greater than 10 in. (25 cm) and on those that use solid cutting wheels 10 in. (25 cm) or more in diameter, provide an auxiliary enclosure in addition to the guard. This auxiliary enclosure can be a set of heavy screen panels, suspended approximately 8 ft (2.5 m) above the floor to or below the worktable. The panel screens should be ½-in. (13 mm) mesh or smaller, and the wire should be ⅛ in. (3.2 mm) or more in diameter. The framework of the panels should be made of 1 × 1¼ in. (2.5 × 3.1 cm), or heavier, structural steel angles or channels.

Safe Speeds

Do not operate abrasive wheels and disks at speeds exceeding those recommended by the manufacturer. In particular, unmarked wheels of unusual shape, such as deep cups with thin walls or backs with long drums, should be operated according to the manufacturer's recommendations.

As the wheel wears down, the spindle's speed (rpm) is sometimes increased to maintain the surface speed (sfpm). When the wheel is nearly worn down, the spindle is running at the highest rpm. When the worn wheel is replaced, adjust the spindle's speed. If the spindle's speed is not adjusted, the new wheel might break because the surface speed exceeds manufacturer's recommendations.

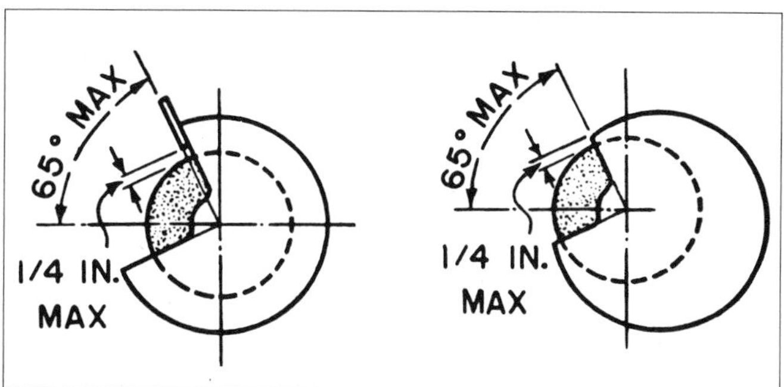

Figure 22–13. The correct wheel exposure can be maintained with an adjustable tongue *(left)* or a movable guard *(right).*

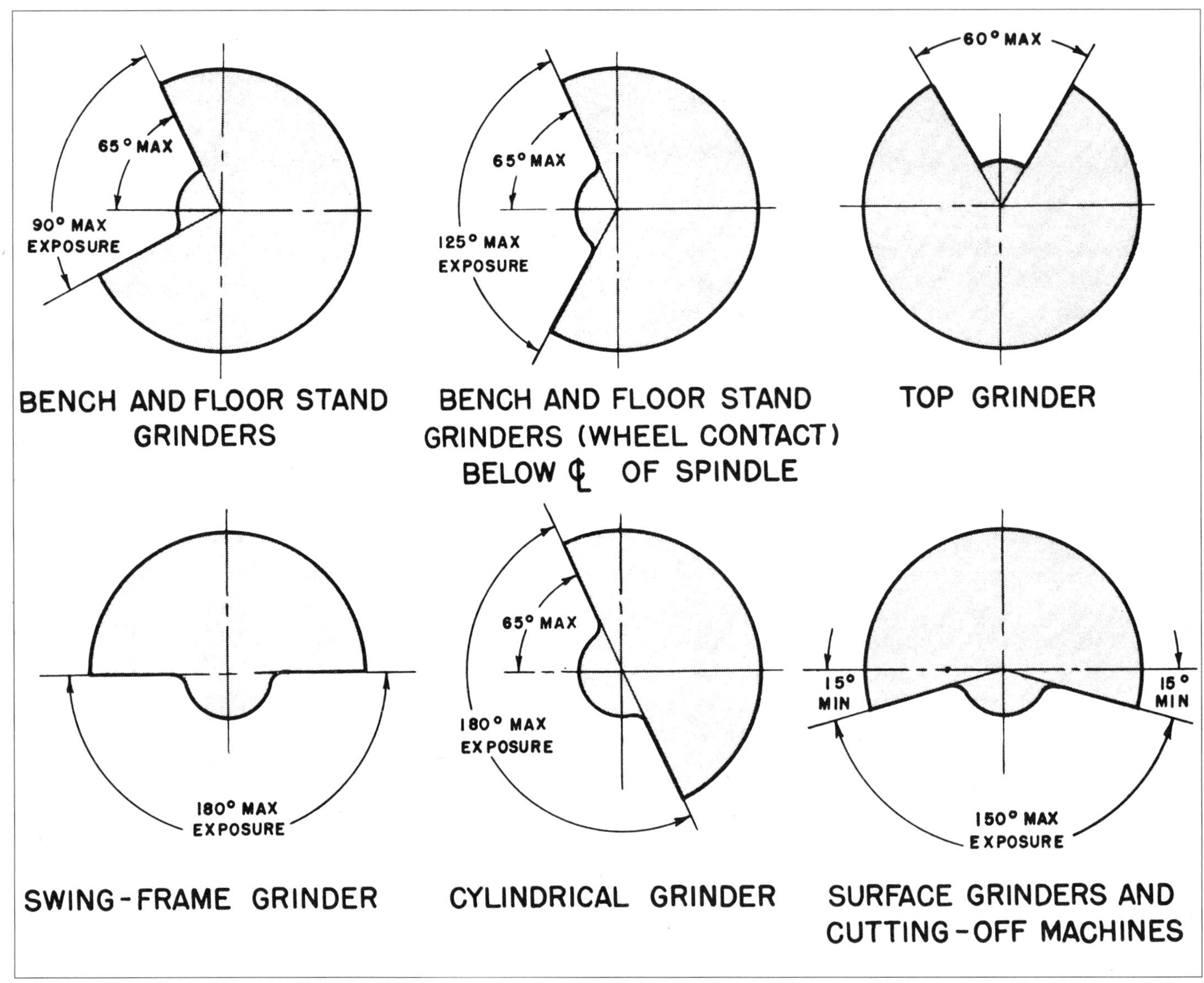

Figure 22–14. Maximum exposure angles for various grinding applications.

Grinding equipment for high-speed operation should be specially designed. Give special attention to spindle strength, guards, and flanges to eliminate mounting stresses. Such things as side-grinding pressure and the wheel's shape must also be considered. Proper maintenance and protective devices are also important for safe high-speed operations. Obtain the manufacturer's approval for all high-speed wheel and disk operations.

Work Rests

Because work has become wedged between the work rest and the wheel, many bench and floor-stand grinder wheels have broken, thus causing serious injury. The work rest should be substantially constructed and securely clamped not more than ⅛ in. (3.2 mm) from the wheel (Figure 22–15). Check the work rest's position frequently. The work rest's height must be on the horizontal centerline of the machine's spindle.

Never adjust the work rest while the wheel is in motion. The work rest might slip and strike and break the wheel, or the operator might catch a finger between the wheel and the work rest. To prevent work from adding twisting and bending stress to the wheel, operators should use guides to hold the work in position when slot grinding or performing similar operations.

Dressing Abrasive Wheels

Abrasive wheels that are not true or not in balance (Figure 22–16) will produce poor work. They can damage the machine and injure the operator. Keeping the wheels in good condition eliminates these possibilities, decreases wheel wastage, and lengthens the wheel's life.

Figure 22–15. *Top:* with a properly adjusted work rest, the operator can keep hands away from the wheel and still firmly hold the work in place. *Bottom:* there should be a safe space of no more than ⅛ in. (3.2 mm) between the tool rest and the wheel.

Figure 22–16. A badly rutted or out-of-balance grinding wheel should be taken out of service and dressed.

To recondition a rutted or excessively rough wheel, often it is necessary to dress it by removing a large area of the face. Equip wheel-dressing tools with hood guards over the tops of the cutters. They will protect the operator from particles flying from the wheel or pieces of broken cutters. The operator of a wheel dresser should use a rigid work rest set close to the wheel. The operator should move the wheel dresser back and forth across the wheel's face, while firmly holding the heel or lug—on the underside of the dresser's head—against the edge and not on top of the work rest (Figure 22–17).

Occasionally test wheels for balance, and rebalance them if necessary. Wheels that are too worn, or too out of balance, to be balanced by truing or dressing should be taken out of service.

Surface Grinders and Internal Grinders

Operating requirements for surface grinders (Figure 22–18) and internal grinders differ from those for other types of wheels. Insecurely clamped workpieces and unenergized magnetic chucks are common sources of injury

BENCH AND STAND GRINDERS

DRESSING

1. Wear a face shield over your safety glasses for protection against heavy particles.
2. Use a dressing tool approved for the job. Never use a lathe cutting tool.
3. Inspect star dressers for loose shaft and worn disks.
4. Round off the wheel edges with a hand stone before and after dressing to prevent the edges from chipping.
5. Use the work rest to support and guide the tool. Use a tool holder if one is available.
6. Apply moderate pressure slowly and evenly.
7. Always apply diamond dressers at the center or slightly below the center, never above.

Figure 22–17. Wheel dressing operations should adhere to these safety measures.

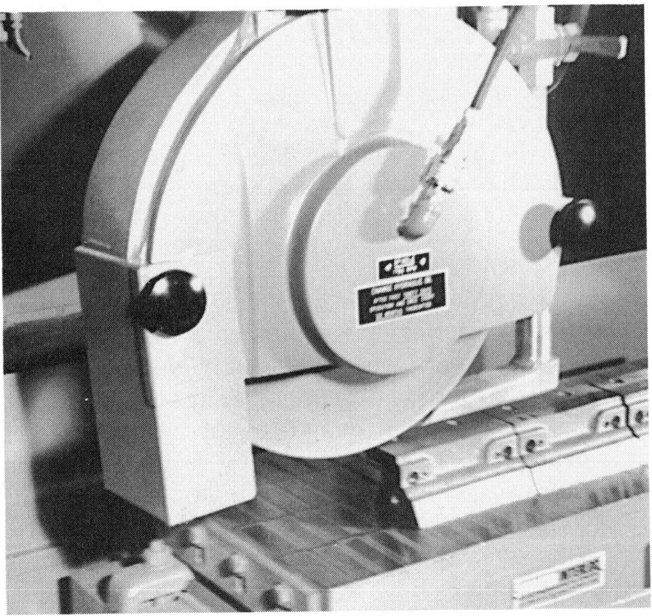

Figure 22–18. Guarded horizontal surface grinder.

to operators of surface grinders. Under these conditions, workpieces can be thrown with considerable force.

If the operator takes too deep a cut or too quickly traverses the table or wheel, the wheel can overheat at the rim and crack. Therefore, train operators, and supervise them to clamp work tightly. They must always properly adjust and turn on a magnetic chuck before applying the wheel. They must also control the work's speed and depth.

Baffle plates on each end of a surface grinder are usually standard equipment. They should also include some provision for exhausting the grinding dust.

Internal grinders can often be guarded with an automatic positioning hood. This kind of hood covers the grinding wheel when it is in the retracted or idling position.

Grindstones

When using grindstones, follow the manufacturer's suggested running speeds and operating procedures. Never run stones of unknown composition or manufacture at more than 2,500 sfpm (12.5 m/s), and ordinarily not more than 2,000 sfpm (10 m/s). The size and weight of grindstones require a stand that is rigidly constructed, heavy enough to hold the stone securely, and mounted on a solid foundation to withstand vibration.

Since grindstones are run wet, take all possible precautions to prevent slipping accidents near the stones. Use rough concrete or other slip-resistant floor material in grindstone-operating areas.

Carefully inspect grindstones for cracks and other defects, as soon as they arrive from the manufacturer. Store those not to be used immediately in a dry, uniformly heated room, where they will not be damaged.

Many grindstone failures result from faulty handling and incorrect mounting. Do not leave grindstones partially submerged in water. This practice causes an unbalanced stone that can break when rotated. Do not use wooden wedges on power-driven stones. Often, these wedges are too tightly driven or can become wet and swell. In either case, cracks start in the corners of a square center hole, radiate outward, and weaken the stone, causing ruptures to occur when operated at normal speeds.

After the stone has been centered, fill the central space about the arbor with lead or cement. Use double thicknesses of leather or rubber gaskets, rather than wood washers, wherever possible. If wood washers are used between the flanges and the stone, the washers should be ½ to 1 in. (1.3 to 2.5 cm) thick and the flanges should be clamped in place by heavy nuts.

To remove dust and wet spray or mist when dressing or operating power-driven grindstones (either wet or dry), provide an adequate exhaust system. Work rests should comply with the same requirements as those for grinding wheels.

Polishing Wheels and Buffing Wheels

Polishing wheels are either wood faced with leather or made of stitched-together disks of canvas or similar material. A coat of emery or other abrasive is glued to the periphery of the wheels.

Buffing wheels are made of disks of felt, linen, or canvas. The periphery is given a coat of rouge, tripoli, or other mild abrasive.

The softness of the wheel, built up of linen, canvas, felt, or leather, is determined by the size of the flanges used—the larger the flange, the harder the surface. When large flanges are used, it often is necessary to soften the working surface of the built-up wheels to conform to the contour of the object being polished. A safe procedure for softening the working surface is to place the wheel on the floor or other flat surface and pound the edges of the wheel with a hammer or mallet. Do not place the wheel on the spindle with a file or other object held against it. The file could catch in the wheel and be thrown with such force that the operator or nearby workers are injured.

Mounting

Mount polishing wheels and buffing wheels on rigid and substantially constructed stands that are heavy enough for the wheels used. Mounting procedures for polishing wheels and buffing wheels are the same as those for grinding wheels.

Speed

The polishing and buffing peripheral speed range is from 3,000 to 7,000 sfpm (15 to 35 m/s), with 4,000 sfpm (20 m/s) in general use for most purposes. If the motors that drive the polishing and buffing wheels are equipped with adjustable speed controls, install the controls in a locked case and only have an authorized person change the speed.

Safeguards

Hood guards should be designed to prevent the operator's hands or clothing from catching on protruding nuts or the ends of spindles. If working conditions require a hood that does not give the needed protection, then use a spanner wrench to install smooth nuts over the spindle's ends. Never substitute a prick punch and hammer for a spanner wrench. Exhaust hoods should be designed to catch particles thrown off by the wheels.

Operators of polishing wheels and buffing wheels should not wear gloves. A glove can catch and drag the operator's hand against the wheel. Operators should not attempt to hold a small piece against the wheel with the bare hands. Small pieces being polished or buffed can frequently be held in a simple jig or fixture. Some operators use a piece of an old linen or canvas wheel for holding small pieces.

When applying rouge or tripoli to a revolving wheel, hold the side of the cake lightly against the wheel's periphery. If a stick is used, apply the side of the stick to the off side so, if thrown, it will fly away from the wheel.

Wire Brush Wheels

Wire brush wheels or, more commonly, scratch wheels are used to remove burrs, scale, sand, and other materials. These wheels are made of various kinds of protruding wires, with different thicknesses.

The same machine setup and conditions that apply to polishing and buffing wheels apply to wire brush wheels. Use flanges or nuts to hold scratch wheels rigidly in place. Do not exceed the speed recommended by the manufacturer. The hood on scratch wheels should enclose the wheel as completely as the nature of the work allows. The hood should also be adjustable so protection will not lessen as the diameter of the wheel decreases. The hood should also cover the exposed arbor ends. If not, install a smooth nut on them. Adjust the work rest to about ⅛ in. (3 mm) from the wheel.

Personal protective equipment is especially important when operating scratch wheels because the wires tend to break off. Make it mandatory for operators to wear aprons made of leather, heavy canvas, or other heavy material; leather gloves; face shields; and goggles.

SUMMARY

- Injuries on machine tools are most often caused by unsafe work practices or incorrect procedures. Proper safeguarding on machines, good housekeeping in the work area, and good work habits help to reduce injuries and accidents.
- General safety rules should be founded on a policy to eliminate unsafe practices by workers and to train workers to operate machines safely.
- Turning machines should be properly shielded to prevent flying or spiral chips from striking or cutting operators. Screw machines should be shielded for excessive noise levels.
- Boring machine operators should be protected by shielding and guarding the drills and ensuring that drill bits are sharp and firmly attached to the drill head. Operators should never attempt to adjust work while the machine is running.
- Most accidents with milling machines occur when operators unload or make adjustments to the work while the machines are running. Machines in this class should be guarded to prevent injuries from contact or thrown chips and lubricant and have emergency shut-off switches.
- Only qualified electricians should maintain or hook up the electrical system of electrical discharge machines. Workers must maintain proper oil levels, make sure the area is adequately ventilated, prevent buildup of discharge gases, and be well trained in fire extinguishing procedures.
- Injuries from planers and shapers generally result from contact with projections on the workpiece or projecting bolts or brackets. To avoid accidents, the cutting edge should be guarded and operators should secure the tool or workpiece.
- Most accidents on slotters result from workers catching their fingers between the tool and workpiece. Broaches

should be safeguarded through supplementary controls, which should be shielded if they can be tripped by any part of the worker's body.

- Surface and internal grinders require different operating methods than do other grinding wheels. Injuries often occur when workpieces are thrown from the grinder. To prevent injuries, workpieces must be securely clamped and magnetic chucks energized.
- Grindstones, polishing and buffing wheels, and wire brush wheels must be operated at the manufacturer's suggested running speeds, be frequently inspected, and be mounted and handled correctly.

REFERENCES

American National Standards Institute, 11 West 42nd Street, New York, NY 10036.

Accuracy of Engine and Tool Room Lathes, B5.16–R1992.
Markings for Identifying Grinding Wheels and Other Bonded Adhesives, B74.13–1990 (R1995).
Milling Machine Arbor Assemblies, B5.47–1972 (R1991).
Milling Machines, B5.45–R1998.
National Electrical Code®, NFPA 70, 1996.
National Electrical Safety Code, ANSI C2–1997.
Rotary Table Surface Grinding Machines, B5.44–R1998.
Safety Requirements for the Construction, Care, and Use of Drilling, Milling, and Boring Machines, B11.8–1983 (R1994).
Safety Requirements for the Construction, Care, and Use of Gear Cutting Machines, B11.11–1994.
Safety Requirements for the Construction, Care, and Use of Lathes, B11.6–1994.
Safety Requirements for the Construction, Care, and Use of Machine Tools—Iron Workers, B11.5–1988 (R1994).
Safety Requirements for the Construction, Care, and Use of Metal Sewing Machines, B11.10–1990 (R1997).
Safety Requirements for Workplace Floor and Wall Openings, Stairs, and Railing Systems, ANSI A1264.1–1995.
Safety Requirements for the Construction, Care, and Use of Single- and Multiple-Spindle Automatic Screw/Bar and Chucking Machines, B11.13–1992 (R1998).
Specifications for Shapes and Sizes of Grinding Wheels, and Shapes, Sizes, and Identification of Mounted Wheels, B74.2–1992.
Spindle Noses and Tool Shanks for Horizontal Boring Machines, B5.40–1970 (R1991).

Grinding Wheel Institute, 30200 Detroit Road, Cleveland, OH 44145.

Safety Recommendations for Grinding Wheel Operations, 1991.

National Fire Protection Association, 1 Batterymarch Park, Quincy, MA 02269.

Electrical Standard for Industrial Machinery, NFPA 79, 1994.
National Electrical Code®, NFPA 70, 1996.

NMTBA—Association for Manufacturing Technology, 7901 Westpark Drive, McLean, VA 22102.

National Safety Council, 1121 Spring Lake Dr., Itasca, IL 60143-3201.

Hearing Conservation in the Workplace: A Practical Guide, 1992.
Safeguarding Concepts Illustrated, 6th ed., 1993.
Occupational Safety and Health Data Sheets:
Engine Lathes, 12304–0264, 1992.
Fire-Resistant, Water-in-Oil Emulsion Hydraulic Fluids, 12304–0543, 1990.
Gear-Hobbing Machines, 12304–0362, 1991.
Horizontal Metal Boring Mills, 12304–0269, 1991.
Manganese, 12304–0306, 1990.
Metal Planers, 12304–0383, 1991.
Metal Shapers, 12304–0216, 1991.
Metal-Working Drill Presses, 12304–0335, 1991.
Metal-Working Milling Machines, 12304–0364, 1991.
Portable Grinders, 12304–0583, 1990.
Vertical Boring Mills, 12304–0347, 1991.
Zinc and Zinc Oxide, 12304–0267, 1991.
Zirconium Powder, 12304–0729, 1991.

U.S. Department of Health and Human Services, National Institute for Occupational Safety and Health, Division of Technical Services, 4676 Columbia Parkway, Cincinnati, OH 45226.

Control of Exposure to Metalworking Fluids, DHEW (NIOSH) Publication 78–165.
Health and Safety Guide for the Screw Machine Products Industry, DHEW (NIOSH) Publication 76–165.

REVIEW QUESTIONS

1. What two factors result in fewer accidents when operating machine tools?
   a.
   b.
2. Identify one of the major causes of accidents from machine tools, especially drilling equipment.
3. List three preventive measures that will help operators run engine lathes safely.
   a.
   b.
   c.
4. What are six common causes of injury in boring mill operations?
   a.
   b.
   c.
   d.
   e.
   f.

5. Operators should do what four things after the electrical discharge machine (EDM) has been hooked up?
   a.
   b.
   c.
   d.

6. Injuries from shapers and planers frequently result from contact with projections on the workpiece or with projecting bolts or brackets, especially when the table is being adjusted ______________.

7. What type of machine shapes material by bringing it into contact with a rotating abrasive wheel or disk?
   a. Planing
   b. Boring
   c. Grinding
   d. Milling

8. Specifications for operation of grinding machines and construction of guards and safety devices are in what code?

9. What effect does cold and wetness have on grinding wheels?

10. Many grindstone failures result from faulty handling and incorrect ______________.

23

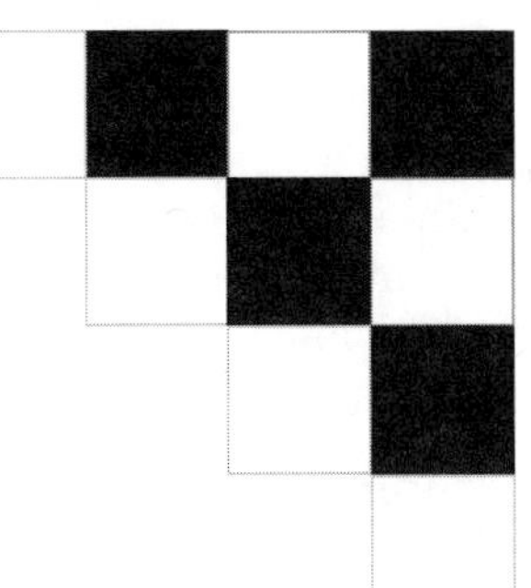

Cold Forming of Metals

Revised by

Philip E. Hagan, MPH, CIH, CHMM, CHCM

Fundamental to many products is the operation known as cold forming: the forming of blank metal into a shape. A variety of machines are used for this operation depending on the type of metal and the end product. The common machines, power presses, metal shears, and press brakes are discussed in this chapter, along with the following topics:

- primary/secondary operation of power presses
- safeguarding devices
- auxiliary tools and attachments
- feeding and ejecting mechanisms
- kick press hazards, safety precautions, and maintenance
- electrical control operations, construction and safeguarding
- safe procedures for die press setup and removal
- maintenance and troubleshooting
- metal shears
- guarding and safeguarding press brakes

POWER PRESSES

Power presses, with different dies attached, perform many types of metalworking operations that produce a variety of products. Since power press machines are so versatile, safeguarding the point of operation depends on (1) the die or tool component, (2) the type of press selected to power the die, and (3) the method selected to insert materials, to remove parts, and to dispose of scrap. (See Chapter 6, Safeguarding, in this volume.)

When designing the die or tool component, consider the following items:

- material used
- configuration of the finished part
- method of feeding the material
- method of removing parts
- method of scrap disposal
- ways to reduce noise

The method of feeding the material may be manual, semiautomatic, or automatic. Scrap disposal may be manual or automatic, and recycling should be considered.

The die component's design is very important since it determines the physical characteristics of the press, such as the tonnage required, dimensions of the die's space, and the speed and stroke of the press component.

Safety of power presses depends on (1) adequately safeguarding the point of operation, (2) properly training press operators, and (3) enforcing safe working practices. Setup and maintenance personnel must be trained to assure their safety while working in or around a press.

Primary Operations (Blanking)

Power press operations fall into two basic categories: primary operations and secondary operations. A primary operation is one in which stock material is processed to produce suitably sized and shaped flat blanks. Generally, these blanks require more forming and shaping, which takes place during a secondary operation.

Primary operations are easier to guard than secondary operations at the point of operation. The use of flat material permits construction or adjustment of the guards so the opening is only large enough for material to pass through the guard into the die. The trailing edge of strips can often be processed by pulling the stock through the die with the scrap skeleton, or by pushing the stock through the die with the leading edge of the next strip. If this is not possible, bend the guard inward to meet the die at the point of the stock's entry. In this way, the next strip is fully used to advance the processed strip manually. If the guard cannot be customized in this way, scrap the balance of the material unless other appropriate point-of-operation protection is used.

The remaining hazard of primary operations occurs during die setup or repair when guards, necessary for the operator's safety, have been removed. Use the following suggestions to avoid this hazard:

- On partial-revolution clutch presses, operate the machine in the inch or single-stroke mode, using two-hand buttons to stroke the press when the dies are being tested. Do not repair or modify work in the die when the machine is capable of being stroked. Provide protection by using interlocked safety blocks and energy isolation procedures (see Chapter 6, Safeguarding).
- On full-revolution and partial-revolution clutch machines, stop the flywheel. In addition, for partial-revolution clutch machines, shut off the control and the motor when work is to be done in the dies.
- If foot controls are used for production, disconnect and remove them from the area while work in the die is being performed. Before turning on the main motor to start a full-revolution clutch press after the work in the die has been performed, and at any other time, be sure that the slide is adjusted correctly and that everything has been removed from the die, including the operator's hands. This precaution is required in case the foot pedal is accidentally tripped while the machine is shut down. In such a case, the press would stroke immediately when the motor is turned on.

Stock Material for Primary Operations

The following materials are used in primary operations:

- **Coiled material.** Stock material initially in coiled form is frequently fed directly into a press which, with one or more dies, produces a finished part. Coiled material is sometimes fed into a cut-to-length line that produces strips. It can also be fed into a blanking line where shaped and sized blanks are produced for use in secondary operations.

- **Strip material.** Cut-to-length line is processed as stock material for subsequent primary operation or a combination primary and secondary operation.
- **Scrap and drop-off material.** Select scrap and drop-off material is a valuable source of primary material for producing some products. The material extending from the die varies in length and position with each stroke; this forces the operator to constantly reposition his/her hands. All die hazards, pinch points, and guide posts, as well as the point of operation at which blanking is performed, are potential hazards. To protect the operator, use a die with an attached guard that covers every hazard and allows only enough of an opening for material to pass through. This guard should also allow the operator to position the material close to the die for maximum yield. Such an operation requires constant ejection of parts and freedom of movement of the material within the die. Such a setup makes it absolutely unnecessary to remove guards except for sharpening dies.

Material-Handling Hazards

Hazards exist when handling the sharp edge or burr on strips, coiled stock, or scrap stock. A material-handling hazard is frequently encountered when unstrapping stock. The sudden release of a strap on a bundle of strips or on a restrained coil can injure any part of the body, including the eyes. When handling stock material, use gloves, eye protection, and arm gauntlets.

Feeding of Stock Material into Machines

Observe the following suggestions when feeding various stock material into machines:

Coil Feeds. When using coiled stock, place it into an uncoiler that will pay out, on demand, the required length of stock. Some uncoilers are powered to maintain a free loop of material from which the powered rolls on roll feeds or gripper jaws on gripper feeds pull a measured amount of stock. Other uncoilers are simple reels from which material is pulled directly. Heavy coils, fast feeding, and payout of long strips require powered uncoilers to keep the pulling effort to a minimum.

Strip Feeds. Although strip material is usually fed directly into the die by hand, it can be fed by machine. Although a roll feed can be used, gripper feeds are generally preferred. Gripper feeds can be automatically supplied with strips from an unstacker, or each strip can be manually loaded by an operator.

Blank Feeds. Blanks are usually inserted by hand into the die. However, blanks can be and frequently are loaded by robots.

Hand-Fed Material. Material is sometimes hand-fed into the die from a coil or from a stack of strips. In each case, material can be fed through a guard opening that is small enough so that the operator's hands cannot enter the point of operation. This type of operation is easy to safeguard by proper die-guard design with part ejection.

Feed-Machine Hazards. Pinch points, crush points on any type of feeder, and nip points at the inrunning side of rolls or gripper jaws are feed-machine hazards to be guarded. Keep openings for material large enough only for free movement of stock and too small to permit fingers to enter (Figure 6–7 in Chapter 6, Safeguarding). To guard inrunning points of rolls, it is sometimes necessary to use bell-mouthed guides up to the rolls for threading in the end of the stock.

Part and Scrap Ejection

Design the removal of parts and scrap into the operation so that there is never any accumulation within the die. Accumulation of parts and scrap makes guards difficult to use and discourages their use. Efficient, reliable ejection of parts and scrap is an important consideration in a fully guarded die since it eliminates access to the point of operation. At the first evidence of ejection failure, shut down and lock out the operation to resolve the problem. It is frequently necessary to provide inclined chutes and/or to incline the press, if possible, to assure ejection of parts and scraps.

The die's design should permit material to be removed or retracted from the side in which it enters, and the side at which it exits. Sometimes this requires beveled edges at both sides of component parts of the die so that edges in the material do not hang up on edges within the die. This would prevent the withdrawal or backup of the strip or scrap.

In the event that a die is used to recover usable drop-off material and scrap, the material should be fully mobile in the die without any hangup. Consequently, all edges in the die should be beveled. On some dies, it may be necessary to use supporting steel runners within the die.

Secondary Operations

Many secondary power press operations are adaptable for various feeding methods, including manual. With adequate safeguarding, the feeding operator will not allow the primary operator to reach into the danger area.

Gravity Feed

One such method uses gravity feed in which the part is placed on a chute and slides into the lower die (Figure 23–1). Provide pins, gauges, and stock guides to assure that parts nest in the proper position. Open-back, inclinable presses, because they can be inclined, can use gravity feed. The part can be placed into the die by gravity and ejected from the back of the press by gravity or air. Even if the opening in the chute is small enough to prevent a hand from entering the die area, install a full-barrier guard to prevent operators from reaching into the danger area. If the opening is large enough to allow the entry of a hand, use safeguarding devices such as type-A or type-B as defined in American National Standard Institute (ANSI) B11.1, movable barriers; two-hand controls; presence-sensing devices; or pullback devices. These devices are explained in Point-of-Operation Safeguarding Devices, later in this chapter.

Use various adaptations of gravity feeds to assure that only one part is placed in the die at a time, such as through the use of a single-piece feeder. When oil or other lubricant on the parts causes the parts to stick in the chute or slide, install wire or metal rods in the chute. They reduce friction and allow the parts to slide into the die without sticking.

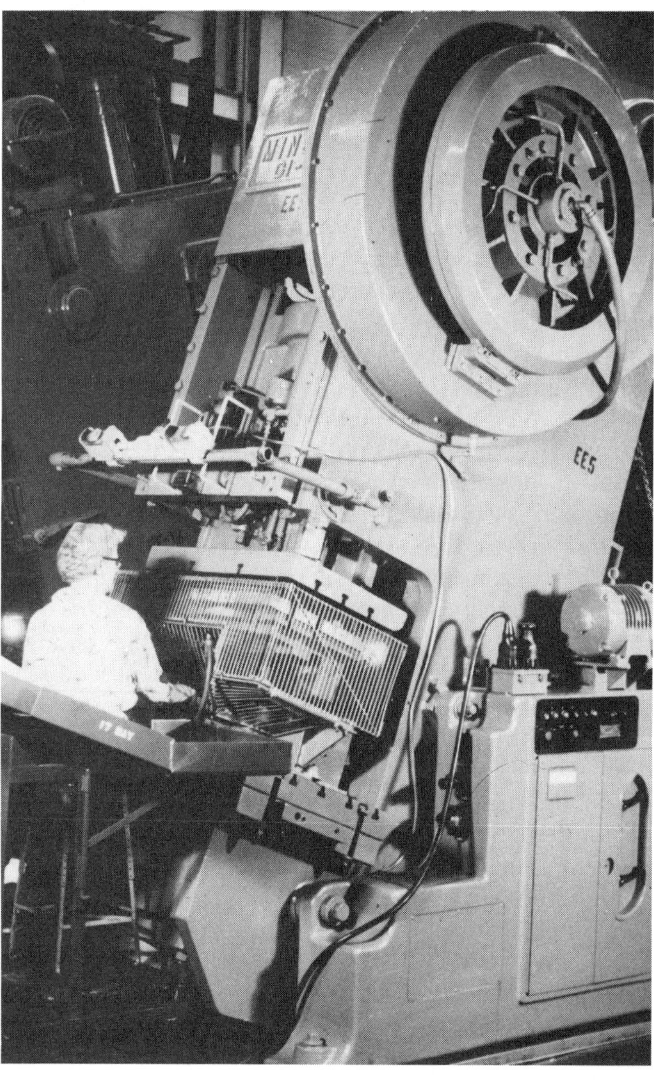

Figure 23–1. There are advantages to inclining a press. The fixed barrier guard is simple and economical to make. This operation allows either hand feed or automatic gravity feed direct from the blanking operation (via the chute).

Follow or Push Feed

A common type of semiautomatic feeding is the follow or push feed. This feed allows the operator to push parts that are on a tray into the die by pushing one part into the preceding part or using a stick to push the part into the die. This keeps the operator's hands well out of the danger area. This die is easily guarded with a die-enclosure guard. Even irregularly shaped parts can frequently be fed with a push feed to keep the operator's hands out of the danger area.

Magazine Feeds

Adding a magazine on the push feed can increase the rate of production as well as lessen the manual handling of blanks at the press operation. Blanks with various shapes and sizes are adaptable to magazine feeds. Magazine slide feeds can be made nearly automatic in operation by powering the slide with mechanical devices attached to the press ram or crankshaft. The operator then only needs to keep the magazine filled with blanks to feed the press. Other ways of providing power movement of slide feeds include air or hydraulic cylinder power that is controlled by solenoid valves timed with the press's cycle.

Hand Tools

Where parts must be placed and/or removed from the die manually, hand tools, such as soft metal pliers, tongs, tweezers, and suction cups, provide effective ways to keep the operator's hands out of the danger area (Figure 23–2). It is sometimes necessary to provide clearance holes or slots for hand tools to assist the operator when inserting or removing the workpieces. Providing clearance makes the job of grasping the parts safer and more efficient. Hand tools are not, by themselves, a means of safeguarding. For the operator's safety, install a barrier guard or some type of safeguarding device.

Cleaning Dies and Clearing Jammed Parts

Consider that these activities are going to happen during any press run and plan how to do them safely. Use cleaning and prying tools that keep hands out of the point of operation. If safeguards will need to be re-

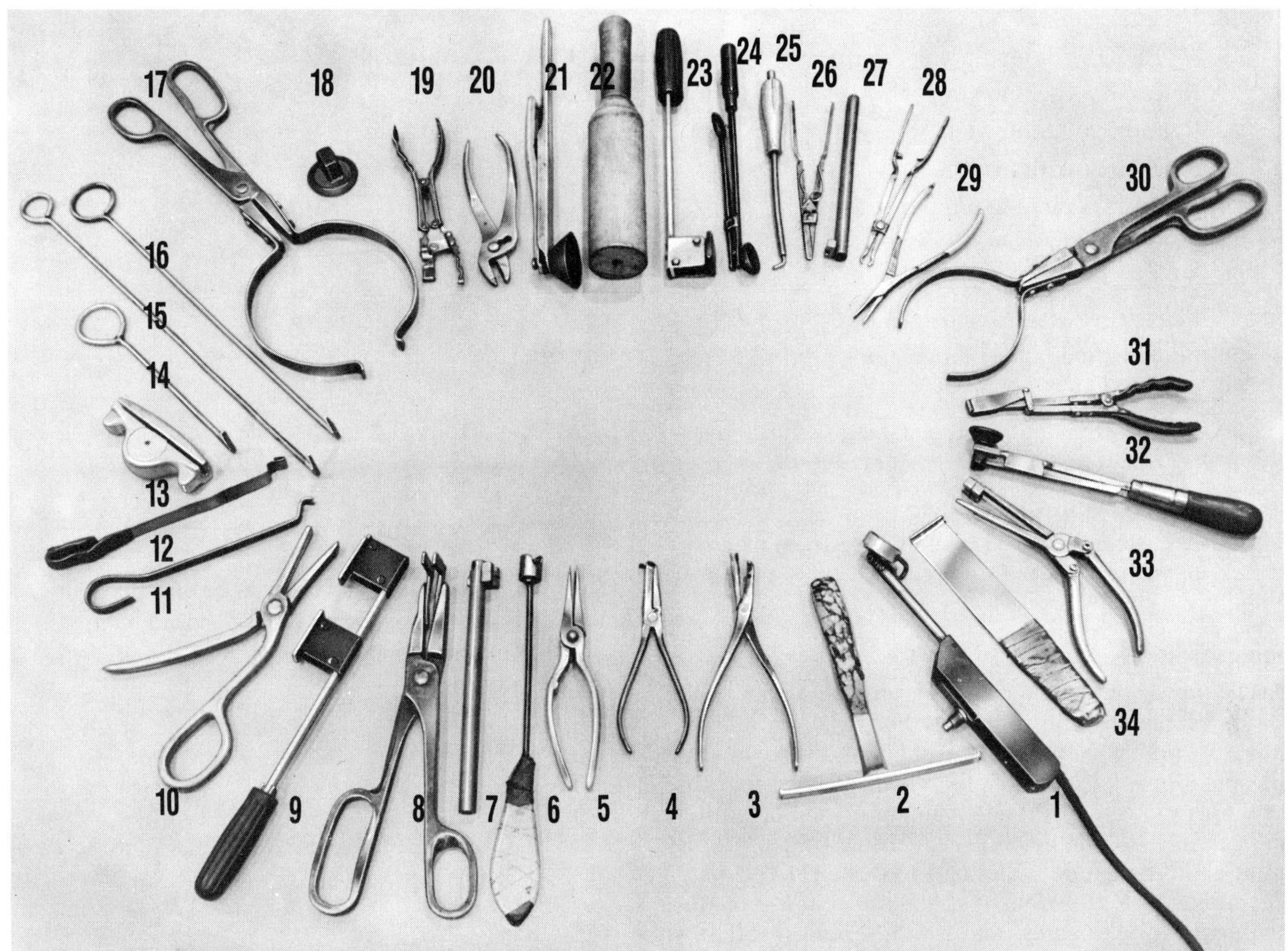

Figure 23–2. A sample of 34 "mechanical hands" that permit a NO HANDS IN DIE operation.

moved, have a lockout procedure or a well-planned alternative if lockout is not feasible.

Tool Identification

Figure 23–2 shows 34 simple, safe hand tools—all can be made in a shop, and all can save the hands and fingers of power press operators.

The tools shown include pliers up to 12 in. (30 cm) long that permit a hand to be kept out of the danger zone. Other pliers are designed to grasp a vertical flange with bent jaws, to pick up thin pieces, or to hold material in work. Tools also include vacuum cups for handling sheet metal at slitting and shearing machines, permanent magnets and electromagnets, pliers with magnets, steel hooks, and steel or brass pusher sticks.

Not only do tools save workers' fingers and hands but they also contribute to speed of operation. Studies show that it takes 1.4 seconds to load a press with 12-in (30 cm) pliers compared with 1.8 seconds by hand.

Keys to numbers on the accompanying photograph:

1. 110-V electric magnet for picking up sheet metal
2. Steel or brass pusher
3. Pliers with extra-long handles
4. Pliers with adapters for grasping vertical edges
5. Pliers with long handles and long grip
6. Alnico magnet on a stick
7. Fiber stick with Alnico magnet
8. Pliers with adapters for grasping vertical edges
9. Fiber stick with Alnico magnet
10. Pistol-grip pliers
11. Push stick
12. Push stick
13. Sheet-edge gripper

14–16. Hooks with 90-degree bend

17. Pliers with adapter for grasping vertical edges

18. Vacuum cup
19–20. Pliers with high-pressure grip
21. Releasable vacuum gripper
22. Cylindrical holding tool
23. Fiber stick with magnet
24. Releasable vacuum gripper
25. Hook with 90-degree bend
26. Normally closed pliers
27. Fiber stick with Alnico magnet
28. Pliers with adapters for grasping vertical edges
29. Long-nosed pliers
30. Pliers with adapters for grasping vertical edges
31. High-pressure pliers
32. Vacuum cup with handle
33. Adjustable-handled pliers with Alnico magnet
34. Push stick

Safeguarding

Base a power press's safeguarding program on an evaluation of the specific problems. Formulate a definite company policy to cover (1) use of guards or devices for all operations, (2) consideration of the safety factor in new operations, and (3) enforcement of safe operating standards. (See Chapter 6, Safeguarding.) Such a program should include adhering to ANSI B11.1, *Safety Requirements for the Construction, Care and Use of Mechanical Power Presses,* and to applicable regulatory standards.

Many press rooms are small, with only a foreman, a die setter, and a few press operators. In such a shop, a simple program consisting of personal supervision, the guarding of some dies, and the provision of proper safeguarding devices and hand tools will do.

Large power press shops that are divided into departments under various supervisors with several die setters usually have split responsibilities. These shops should develop companywide safeguarding standards. These standards should apply to and be followed by all groups concerned with power presses: supervisors, operators, maintenance personnel, and electricians.

The number and productivity of the dies are major factors in deciding the number of point-of-operation guards to be installed. First, guard high-production, long-run dies individually (Figure 23–3). Consider adjustable barriers for safeguarding only for dies used infrequently and for short-run jobs (Figure 23–4). If a shop has short runs, and thus many die changes each day, or uses dies owned by other manufacturers, install safeguarding devices for secondary operations. Install adjustable-barrier guards for primary operations.

Other factors that require attention for the safe operation of power presses include the following:

Figure 23–3. Fully enclosed dies can be operated with complete safety to personnel. The die shown in the top photo has plastic wrapped around three sides of die and is very effective.

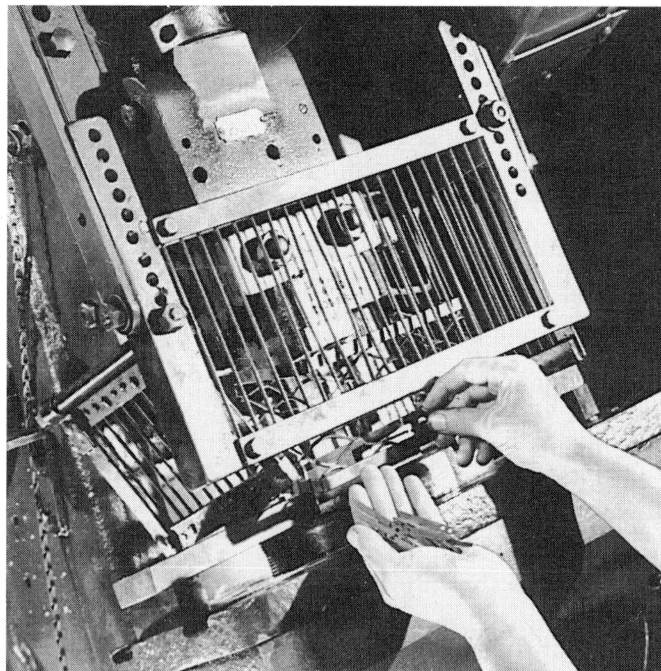

Figure 23–4. An adjustable barrier fits any size die used. This barrier is especially practical for short-run jobs.

- layout of machines
- machine and aisle space
- light and visibility

- containers for handling scrap and processed parts
- effective preventive maintenance program

Definitions

To properly understand this subject, it is important to know the definitions of various terms used in this chapter (see also Definitions in the Electrical Controls on Power Presses and the Press Brakes sections):

Antirepeat

The part of the clutch/brake control system designed to limit the press to a single stroke if the actuating means is held or stuck on "operate." Antirepeat requires release of all actuating mechanisms before another stroke can be initiated. Antirepeat is also called "single-stroke reset" or "reset circuit."

Bolster Plate

The plate attached to the top of the bed of the press component; it has drilled holes or T-slots for attaching the lower die or die shoes.

Brake

The mechanism used on a mechanical power press component to stop and/or hold the slides, either directly or through a gear train, when the clutch is disengaged.

Brake Monitor

A sensor that has been designed, constructed, and arranged to monitor the effectiveness of the press braking system.

Clutch

The coupling mechanism used on a mechanical power press component to couple the flywheel with the crankshaft to produce slide motion, either directly or through a gear train.

Full-Revolution Clutch. A type of clutch that, when tripped, cannot be disengaged until the drive mechanism (usually a crankshaft) has completed a full revolution and the slide, a full stroke.

Part-Revolution Clutch. A type of clutch that can be disengaged at any point before the drive mechanism (usually a crankshaft) has completed a full revolution and the press slide, a full stroke.

Direct Drive. The type of driving arrangement wherein no clutch is used; coupling and decoupling of the driving torque is accomplished by turning a motor on and off. Even though not employing a clutch, direct drives match the operational characteristics of part-revolution clutches because the driving power may be disengaged during the stroke of the slide.

Concurrent

Acting in conjunction, and used to describe a situation wherein two or more controls exist in an operated condition at the same time.

Continuous

Uninterrupted multiple strokes of the slide without intervening stops (or other clutch-control action) at the end of individual strokes.

Counterbalance

The mechanism that is used to balance or support the weight of the connecting rods, slide, and slide attachments.

Device

A press control or attachment that (1) restrains the operator from reaching into the point of operation, (2) prevents normal press operation if the operator's hands are within the point of operation, or (3) automatically withdraws the operator's hands, if they are within the point of operation as the dies close.

Presence-Sensing Device. A device designed, constructed, and arranged to create a sensing field or area and to deactivate the clutch control of the press component when an operator's hand or any other body part is detected within such field or area.

Type-A Movable Barrier Device. A self-powered movable barrier, which in normal operation is designed to: (1) close off access to the point of operation in response to operation of the press-tripping control; (2) prevent engagement of the clutch prior to closing of the barrier; (3) hold itself in the closed position; and (4) remain in the closed position until the slide has stopped at top of stroke.

Type-B Movable Barrier Device. A self-powered movable barrier, which in normal single-stroke operation is designed to: (1) close off access to the point of operation in response to operation of the press-tripping control; (2) prevent engagement of the clutch prior to closing of the barrier; (3) hold itself in the closed position during the downward portion of the stroke while the slide is in motion, but be permitted to open during the downward portion of the stroke if the slide is stopped due to clutch control action; and (4) in normal single-stroke operations, it may be able to open during the upstroke of the slide.

Holdout or Restraint Device. A mechanism, including attachments for the operator's hands, that when anchored and adjusted, prevents the operator's hands from entering the point of operation.

Pullback Device. A mechanism attached to the operator's hands and connected to the upper die or slide that is intended (when properly adjusted) to withdraw the operator's hands as the dies close.

Two-Hand Control Device. Actuating control that requires concurrent use of both hands of the operatur during a substantial part of the die-closing portion of the stroke.

Two-Hand Trip Device. Actuating control that requires the momentary concurrent use of both hands of the operator to trip the press.

Die

The complete (or portion of the) tooling component used for cutting, forming, or assembling material within its point of operation. This tooling component is used with the press and other components to comprise a complete machine capable of producing a part.

Die Builder

A person who builds dies for power press machines.

Die Set

A tool holder held in alignment by guide posts and bushings and consisting of a lower shoe, an upper shoe or punch holder, and guide posts and bushings.

Die Setter

An individual who places or removes tooling components in or from a mechanical power press machine and makes the necessary adjustments for safe and proper functioning of tooling and safeguarding.

Die Setting

The process of placing or removing tooling components in or from a mechanical power press machine, and the process of adjusting the dies and other tooling and safeguarding means for safe and proper functioning.

Die Shoe

A plate or block upon which a die holder is mounted. A die shoe functions primarily as a base for the complete die assembly and, when used, is bolted or clamped to the bolster plate or the face of slide.

Face of Slide

The bottom surface of the slide to which the punch or upper die is generally attached.

Feeding

The process of placing or removing material within or from the point of operation.

Automatic Feeding. Feeding wherein the material or part being processed is placed within or removed from the point of operation by a method or means not requiring action by an operator on each stroke.

Semiautomatic Feeding. Feeding wherein the material or part being processed is placed within or removed from the point of operation by an auxiliary means controlled by the operator on each stroke.

Manual Feeding. Feeding wherein the material or part being processed is handled by the operator on each stroke of the press.

Foot Control

The foot-operated control mechanism designed to be used with a clutch or clutch/brake control system.

Foot Pedal (Treadle)

The foot-operated lever designed to operate the mechanical linkage that trips a full-revolution clutch.

Guard

A barrier that prevents entry of the operator's hands, fingers, or feet into the point of operation.

Adjustable-Barrier Guard. A barrier requiring adjustment for each job or die setup.

Die-Enclosure Guard. An enclosure attached to the die shoe and/or stripper, in a fixed position.

Fixed-Barrier Guard. A die space barrier attached to the press component frame or bolster plate.

Interlocked-Barrier Guard. A barrier attached to the press frame and interlocked so the press stroke cannot be started normally unless the guard, or its hinged or movable sections, encloses the point of operation.

Guide Post

The pin attached to the upper or lower die shoe, operating within a bushing on the opposing die shoe, to maintain the alignment of the upper and lower dies.

Hand Feeding Tool

Any hand-held tool designed for placing within or removing from the point-of-operation material or parts to be processed.

Inch

An intermittent motion imparted to the slide (on machines using part-revolution clutches) by momentary operation of the inch-operating means. Operation of the inch-operating means engages the driving clutch so that a small portion of one stroke or indefinite stroking can occur, depending upon the length of time the inch-operating means is operated. Inch is a function used by the die setter for setup of dies and tooling. It is not intended for use during production by the operator.

Jog

An intermittent motion imparted to the slide by momentary operation of the drive motor, after the clutch is engaged with the flywheel at rest (positive operator pressure on control required).

Knockout

A mechanism for releasing material from either the upper or the lower die.

Liftout

The mechanism also known as knockout.

Manufacturer

Any person who constructs, reconstructs, or modifies the mechanical power press machine or any of its components.

Operator

Any individual performing production work on the mechanical power press and controlling the movement of the slide.

Operator's Control Station

The complement of controls used for stroking the press machine.

Pinch Point

Any point of the power press machine, except the point of operation, where a body part can be caught between the moving parts of a press component, or auxiliary equipment component(s), or between moving and stationary parts of a press component or auxiliary equipment component(s), or between the material and moving part(s) of the press component or auxiliary equipment component(s).

Point of Operation

The area of the die (tooling component) where material is positioned and work is performed during any process such as cutting, forming, or assembling.

Press or Mechanical Power Press

Press Component, or Mechanical Power Press Component. The basic press (component) of the mechanical power press machine; that portion devoid of the tooling component, the safeguarding component(s), and auxiliary feeding components.

Press Machine or Mechanical Power Press Machine. The combination of the press component, tooling component, safeguarding component(s), and feeding components; a complete machine capable of processing the specific job requirement for which it is outfitted by the user, i.e., the production system.

Primary Operation

Any preliminary press machine operation with material to be subsequently processed.

Repeat

An unintended or unexpected successive stroke of the slide resulting from a malfunction.

Run

Single stroke or continuous stroking of the slide.

Safeguarding

"Safeguarding the point of operation" is an umbrella term, under which all safeguarding methods fall and all types of die hazards exist. Two basic categories exist under this umbrella term: the guard and the device.

A *guard* is best understood as a physical barrier that absolutely prevents access to a point-of-operation die hazard when it is in place and while it remains in place during a production run of successful cycles. If a barrier allows any access to a point-of-operation die hazard during a production run, it is not a guard; it then serves only as an inadequate enclosure. An inadequate enclosure always requires a device in conjunction with its use to form an acceptable safeguarding system. (Some devices require the use of enclosures to form an acceptable system.)

A *device* is best understood as safeguarding means that control access to the point of operation. These means are (1) press-controlling devices, (2) operator-controlling devices, and (3) devices that control the operator and the power press.

Safety Block

A prop that, when inserted between the upper and lower dies or between the bolster plate and the face of the slide, prevents the slide from falling of its own dead weight.

Secondary Operation

Press machine operations in which a preworked part is further processed.

Single Stroke

One complete stroke of the slide, usually initiated from a full-open (or up) position, followed by closing (or down), and then returned to the full-open position.

Single-Stroke Capability

An arrangement wherein the operating means (lever, pedal, switch, or buttons), when held depressed, normally do not result in more than a single stroke of the slide. Release and reapplication of the operating means is required to obtain a successive stroke. Single-stroke capability is provided by antirepeat or by a single-stroke mechanism.

Single-Stroke Mechanism

A mechanical arrangement within the clutch assembly, or part of the trip mechanism on the full-revolution clutch, designed to provide single-stroke capability.

Slide

The main reciprocating press component member. A slide may be called a ram, plunger, head, or platen.

Stop Control

An operator control designed to immediately deactivate the clutch control and activate the brake to stop slide motion.

Top Stop Control

An operator control designed to delay its action after being operated to stop slide motion at a predetermined point in stroke. This predetermined point is independent of the instant at which button actuation was made.

Stripper

A mechanism or die part for removing the parts or material from the punch.

Stroking Selector

The part of the clutch/brake control that determines the type of stroking when the operating means is actuated. The stroking selector generally includes positions for OFF (clutch control), INCH, SINGLE STROKE, and CONTINUOUS (when CONTINUOUS is furnished).

Trip or Tripping

Activation of the drive mechanism to run the press machine.

Turnover Bar

A bar used in die setting to manually turn the crankshaft of the press machine.

Two-Hand Trip

An actuating means requiring the concurrent use of both hands of the operator to trip the press machine.

Unitized Tooling

A type of die in which the upper and lower members are incorporated into a self-contained unit so the die members are held in alignment.

POINT-OF-OPERATION SAFEGUARDING

Safeguarding the point of operation means protecting operating personnel, including helpers, after the dies have been installed, tested, and operated, and are ready for production. Protection of the die setter, who must have access to the point of operation while setting dies, is discussed later in this chapter, Power Press Setup and Die Removal. Companies must also protect passersby from power press hazards. Although engineering controls should be used to protect passersby, if not feasible, then administrative controls should be clearly outlined and enforced through company policy.

When determining point-of-operation safeguarding, consider all hazards in the die's space that may crush, cut, punch, sever, or otherwise injure personnel. There are two basic categories of safeguarding the point of operation. One category is the *guard;* the second category is the *device* (see Figure 6–13, in Chapter 6, Safeguarding).

A *guard,* or barrier guard, is a physical barrier that *prevents* access to a die's hazard during a production run of successive cycles. If a barrier allows access to a die's hazard during the production run, it is not a guard, it is an inadequate enclosure. An inadequate enclosure *always* requires using a *device* to form an acceptable safeguarding system. Recommended guard openings are shown in Figure 6–7, Chapter 6, Safeguarding.

A safeguarding device *controls* access to the point of operation. Devices can be divided into three types: (1) press-controlling devices, (2) operator-controlling devices, and (3) devices that control both the operator and the power press. Some devices require the use of enclosures to form an acceptable guarding system.

Guards

There are four main types of guards for safeguarding power presses at the point of operation: fixed die-enclosure guards, fixed-barrier guards, interlocked press-barrier guards, and adjustable-barrier guards.

Fixed Die-Enclosure Guards

Fixed die-enclosure guards provide the most complete protection for the operator since the die is completely enclosed and the guard is a permanent part of it. Die enclosures can be used on many types of press operations to prevent operators from putting their hands into the point of operation. Die enclosures are attached to the die's shoe, or stripper, in a fixed position. They are designed so that the operator's hands cannot reach over, under, through, or around the guard into the point of operation (Figures 23–5, 23–6, and 23–7).

Construct this type of guard to permit easy feeding of the stock, ejection of the part, and removal of scrap. Fixed die-enclosure guards should also afford the operator good visibility at all times.

Provide a minimum clearance of 1 in. (2.5 cm) between the top edge of the guard and the slide or any projection on the slide. In addition, the guard should extend at least 1 in. (2.5 cm) above the bottom of the punch holder to prevent a shearing hazard caused by travel of the slide.

Enclosure guards are made of various types of material, such as preslotted material. Build a metal frame, and weld or otherwise fasten rod stock to the frame. The openings should run vertically to lessen eye fatigue. Transparent polycarbonate also may be used. When properly maintained, it has the advantage of allowing operators to see through it. However, this material scratches easily and is damaged by oil and grease.

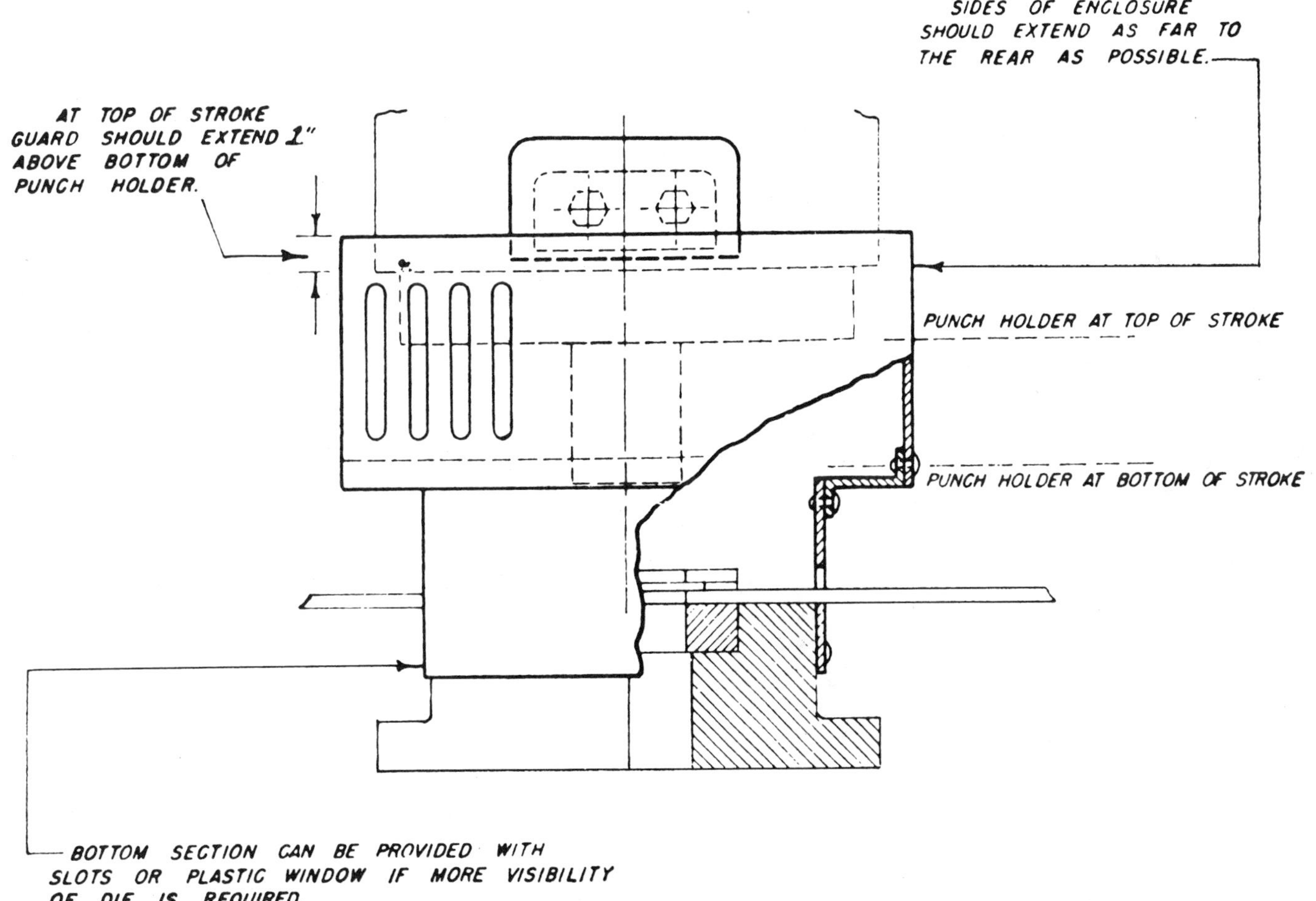

Figure 23–5. This die enclosure guard for strip or coil stock shows the vertical clearances required. (Courtesy Liberty Mutual Insurance Co.)

Figure 23–6. A die guard made of preslotted material can be easily fabricated in the shop.

Figure 23–7. The slide feed allows loading of the die outside the danger zone. The permanent plastic barrier guard permits full visibility of the operation. Note the separation of the guard at the top to permit die maintenance. Overlap of the guard at the separation eliminates shear hazard during travel of the slide. (Courtesy Allis-Chalmers Manufacturing Co.)

It should be at least ¼ in. (0.64 mm) thick and will last longer if mounted in a metal frame (Figure 23–7).

The expense of designing and installing fixed die enclosures is minimal and yields positive safety results. Many facilities that use this type of guarding install the die enclosure before the die leaves the toolroom.

Fixed-Barrier Guards

Fixed-barrier guards, when used, should be attached to the frame of the press or to the bolster plate.

Interlocked Press-Barrier Guards

Barrier guards can be designed with a pivoting, sliding, or removable section to allow ready access to the die. The pivoting or sliding section interlocks with the press's clutch control to prevent the machine from operating when the section is open.

This type of interlocked fixed-barrier guard is used successfully on automatic presses where the point of operation must occasionally be exposed to relieve jams. As a further safety measure, if the interlock fails to function, use hand tools or picks to relieve jams, or support the slide with safety blocks. Do not use the pivoting, sliding, or removable section for feeding on single-stroke applications since the simple interlocking is too easily bypassed.

Adjustable-Barrier Guards

When a die-enclosure guard or fixed-barrier guard would take too much time to complete or would be impractical, or both, provide an adjustable-barrier guard on each press. This type of guard prevents the operator's hands from entering the point of operation (Figure 23–4).

Adjustable-barrier guards are available commercially or may be made in the facility. They are attached to the frame of the press and have adjustable front and side sections for dies of almost any size (Figure 23–8). They are especially practical for short-run jobs.

This type of guard is usually constructed of rod stock or perforated metal. Interlock any pivoting or sliding sections, used for occasional access, with the press's control for maximum safety.

Unless feeding or ejection is automatic, it may be necessary to leave an opening in the barrier to insert a tool to remove a part from the die. The opening should not be wide enough to allow a hand or finger to extend into the point of operation.

In the case where the whole guard is removed to change dies or to adjust a job, the guard is usually not interlocked with the press's control. When adjusting sections of an adjustable-barrier guard, be sure that die setters are instructed to follow the dimensions for permissible openings given in Figure 6–7, Chapter 6, Safeguarding. Never allow the operator to make changes in the adjustments without the supervisor's approval.

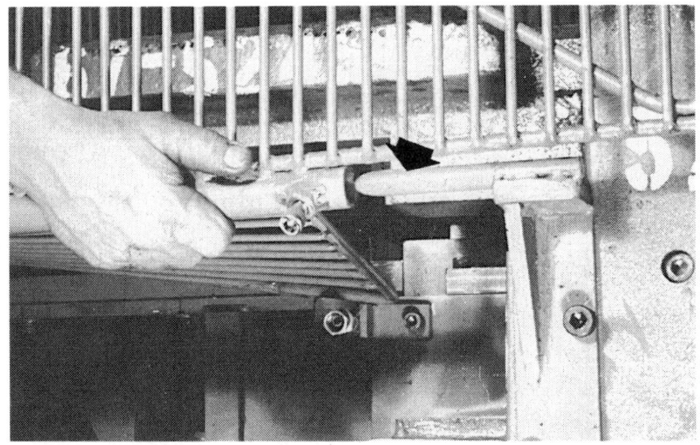

Figure 23–8a. Press guarding can be simplified by designing "universal" barrier guards. Two sections completely enclose the point of operation—the upper guard encloses any operation done on the press; the lower guard conforms to the upper guard, the feed device, and the lower die contour. A mounting pin is shown at the lower right and also at the black arrow in the top photo. Tack welding clips the bottom to the top and keeps employees from making their own adjustments. The guard shown here can be used for several different operations with similar feeding methods and die characteristics.

Figure 23–8b.

Point-of-Operation Safeguarding Devices

There are several types of point-of-operation safeguarding devices for power presses. All of the following devices control access to the point of operation.

Type-A, Movable-Barrier Devices

The type-A, movable-barrier device protects the operator by enclosing the point of operation before a press stroke begins. This barrier remains in the enclosed position until motion of the slide has ceased at the top of stroke (Figure 23–9).

Type-B, Movable-Barrier Devices

The type-B, movable-barrier device also protects the operator by enclosing the point of operation before a press stroke begins. This device makes it impossible to reach into the point of operation prior to the die's closing during the downstroke. The type-B device is permitted to open on the upstroke of the slide, or during the downstroke, if the slide is stopped on the downstroke. Do not use type-B devices on full-revolution machines.

Two-Hand Tripping Devices

Two-hand tripping devices are used on a full-revolution clutch machine. A two-hand tripping device should meet all of the following requirements:

Figure 23–9. A movable barrier device protects the operator by enclosing the point of operation before a press stroke can be initiated.

- protect the individual operator's hand controls against unintentional operation
- arrange the individual operator's hand controls—by design and construction, by separation, or both—to require the use of both hands to trip the press
- use a control arrangement requiring concurrent operation of the individual operator's hand controls
- include a single-stroke capacity
- be spaced far enough from the point of operation so the operator's hands cannot reach into the point of operation between the time the dies close and hands are removed from the controls. This distance may be specified by regulatory requirements and is prescribed to protect the operator from reaching into the point of operation before the downstroke is completed

Two-Hand Control Devices

Two-hand control devices must meet the requirements for protected buttons for two-hand tripping devices (Figure 23–10). They must also be arranged so concurrent pressure from both hands is required during a substantial part of the die-closing portion of the stroke. Where more than one operator is required on a press, two-hand controls must be provided for each operator. Locate the devices so only the operators' hands can operate them. Only use a two-hand control device on a partial-revolution clutch machine during the closing portion of the downstroke, so long as that machine can be stopped.

When used as a point-of-operation device, regulatory standards may specify a safe distance for mounting controls away from the point of operation. If hands-in-dies feeding is used, brake monitoring and additional controls are required.

Pullback Devices

Pullbacks, or pullouts, when properly used, adjusted, supervised, and maintained, always remove the operator's hands from the point of operation as the slide descends. This type of device is usually limited to secondary operations and jobs where the operator can remain at the feeding position. Attach pullback devices to the press's slide or upper die's shoe. Because of variations in the sizes of operators' hands and varying characteristics of different dies, adjust the pullback device to fit each operator after each change of die. When more than one operator is required on a press, provide pullbacks for each operator (Figure 23–11).

Restraints

Restraints, or holdouts, prevent the operator from reaching into the point of operation at any time. This is achieved by providing wrist straps and firmly anchored

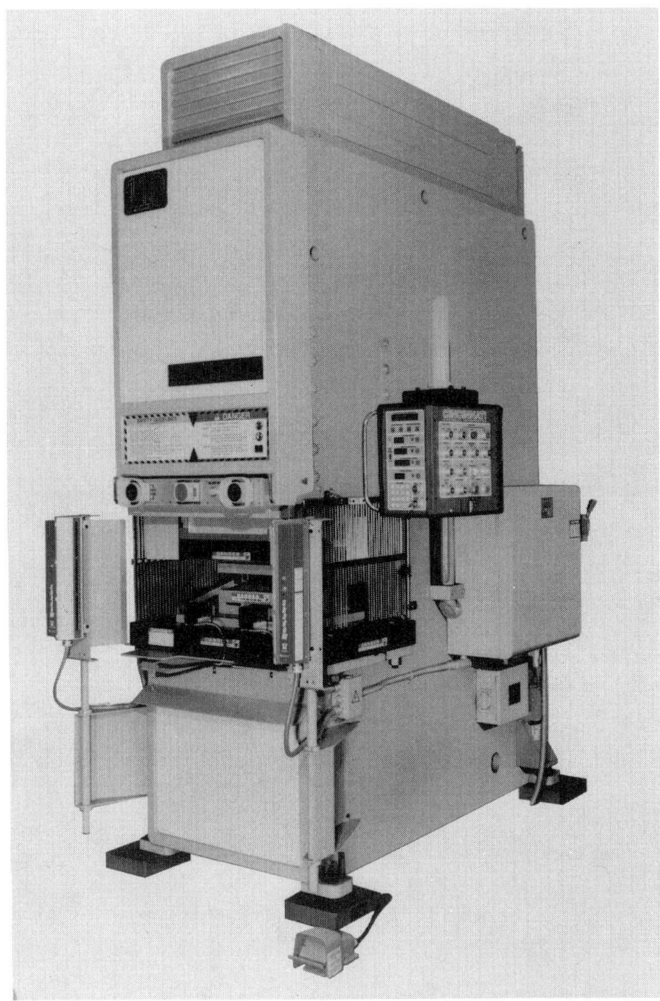

Figure 23–10. This hydraulic OBS press has protected buttons for two-hand control, hairpin side guards, and a photoelectric presence-sensing device at the front. Notice the interlocked slide blocks mounted at the rear of the side frame. (Courtesy Cincinnati Inc.)

restraint cords or cables. When more than one operator is required on a press, provide restraints for each operator (Figure 23–11).

Presence-Sensing Devices

Presence-sensing devices are designed, constructed, and arranged to create a sensing field or area. The clutch's control of the press deactivates when an operator's hand or any other body part is within such field or area (Figure 23–10).

Only use a presence-sensing device on a partial-revolution clutch machine. Regulatory standards may state that the sensing field be located at a safe distance from the nearest point-of-operation hazard depending on the slide's stopping time. Additionally, if hands-in-dies feeding is used, there must be brake monitoring and additional reliability for the control system.

Always supplement these devices with hand tools or feeding and ejection devices (Figure 23–2). In that way, operators need not place their hands in the point of operation. These devices, however, cannot protect against repeats.

Safeguarding Full-Revolution Clutch Power Presses

Power presses whose stroking cannot be interrupted—that is, controlled—during the closing or opening of the stroke cannot use a press-controlling device. However, a properly installed two-hand tripping device can serve as a means of both press activation and safeguarding. This is true of presses equipped with a full-revolution clutch that is simply tripped to start a stroke and must make a full revolution before it is forcibly disengaged through the cam's action. On such power presses, select from the following lists of safeguarding systems:

Guards

Remember, if hands can reach through, around, over, or under a guard to allow access to the point of operation, the guard is an inadequate enclosure and not acceptable by itself. Choose from the following guards for full-revolution clutch power presses:

- fixed die-enclosure guard
- fixed-barrier guard
- adjustable-barrier guard

Devices

Choose from the following operator-controlling devices for full-revolution clutch power presses:

- restraints, properly adjusted
- pullbacks, properly adjusted
- type-A, movable barrier, with an enclosure to prevent access through areas not protected by the movable barrier
- two-hand tripping located at a distance that exceeds the safe distance for the particular press

Safeguarding Partial-Revolution Clutch Power Presses

On a press whose stroke can be interrupted during the closing or opening of the stroke, use a press-controlling device, as well as an operator-controlling device. On such power presses, select from the following lists of safeguarding systems:

Figure 23–11. Properly set pullback devices will remove the operator(s) hands from the area near the dies to a safe distance by the time the ram has traveled through the first half of its down stroke. (Courtesy Cincinnati Inc.)

Guards

Again, remember that if hands can reach through, around, over, or under a guard to the point of operation, the guard is an inadequate enclosure and not acceptable by itself. Choose from the following guards for partial-revolution clutch power presses:

- fixed die-enclosure guard
- fixed-barrier guard
- adjustable-barrier guard
- interlocked press-barrier guard

Devices

Choose from the following operator-controlling devices for single-stroke operations only:

- restraints, properly adjusted
- pullbacks, properly adjusted

Use presence-sensing, machine-controlling devices for partial-revolution clutch power presses. Choose from the following operator- and machine-controlling devices for partial-revolution clutch power presses:

- two-hand control
- type-A, movable barrier
- type-B, movable barrier

AUXILIARY MECHANISMS

Many auxiliary mechanisms to protect operators of power presses are available. They can be purchased or designed and made in a plant's shop.

Feeding and Extracting Tools

A variety of special tools has been developed for feeding and extracting parts. These tools are made of soft metal, aluminum, or magnesium—some are magnetized. Pushers, pickers, pliers, tweezers of various types, forks, and suction disks are a few of these special tools (Figure 23–2).

Special tools for feeding and extracting provide protection only if operators always use them. Operators should never substitute them for proper safeguarding, however.

One way to make sure tools are used properly is to provide convenient storage for them in the workplace. Use a pegboard to mount various tools; paint a silhouette of each tool on the board to indicate the proper space for each tool. This encourages operators to keep tools in their correct place.

Foot Control and Shielding for Protection

Foot operation, by itself, is only an operating means and provides no protection for the operator unless hands and other body parts are kept a safe distance from the point of operation. When the operator is within reach of the point of operation, safeguard the point of operation. When barrier guards are used on foot-controlled presses, always ensure that they are replaced after die changes, cleaning, or jam clearing. Shield and position a foot-operated mechanism (1) to control its accessibility, and (2) to afford protection under circumstances, such as die setting, when safeguarding is necessarily removed from the point of operation.

Use shields that are large enough to allow room for operators to place a foot in the operating position without undue fatigue, and without striking or scraping their leg against sharp edges of the cover. A split length of ordinary garden hose around the opening of the pedal's guard will protect against such injury.

Ways to prevent accidental tripping of the press, in addition to placing a cover over the pedal, include the following:

- Install a safety spring on the trip rod of a full-revolution clutch press.
- Maintain aisles of ample width for trucking material to the presses.
- Use aisles adjacent to the presses only for necessary trucking of dies and stock.
- Allow enough working space between adjacent machines in the same line.
- Keep unauthorized persons away from the area.
- Protect operators from distractions.
- Keep the operator, and all other persons who are not concerned with it, away from the press when it is being set up.
- Prohibit operators from "riding" the pedal. Operators should remove their foot from the pedal immediately after the press is stroked, each time.
- Install flywheel brakes. Flywheels—especially on large presses—continue to coast for some time after the power source has been turned off. They have also been known to cycle a press when the starting controls were tripped. A flywheel brake will stop the flywheel within 10 to 20 seconds.

Foot Controls for Full-Revolution Clutch Presses

A mechanical foot pedal trips the clutch whenever it is depressed. It remains tripped until a stroke is made. If the press is not running, it will stroke immediately when started. Control access to the pedal, therefore, to prevent it from being turned on unintentionally.

An electric foot switch can be turned off to reduce the possibility of accidentally starting the press during off periods. However, during operating periods, the switch is more easily turned on than a pedal. Therefore, shield the switch so an operator's foot cannot accidentally turn the machine on.

A foot pedal should travel about 3 in. (7.6 cm). Connect the clutch rod to the pedal's lever so the distance between the clutch rod and the rear pivot on the pedal will be approximately ⅓ the length of the pedal. The travel of the connecting point will thus be about 1 in. (2.5 cm) when the pedal travel is 3 in. (7.6 cm).

The pedal's lever should not have side play. Maintain the tension of the pedal as recommended by the manufacturer. Do not add springs or counterweights to the pedal's shaft or to the pedal's shaft lever.

Foot Controls for Partial-Revolution Clutch Presses

When using foot switches with clutches able to be disengaged anywhere in the stroke, control the foot switches through an antirepeat circuit. This prevents a second stroke if the foot switch is depressed for the full cycle so that release of the foot switch on the downstroke will cause the ram to stop.

Single-Stroke Attachments

A press with a full-revolution clutch should have a single-stroke attachment that disconnects the pedal or operating lever after each stroke. A single-stroke spring device should depend on spring action only if it is a compression spring encased in a close-fitting tube or closely wound on a rod. The type wound on a rod is preferred because it permits easy detection of a broken spring. Be sure that the space between the spring's coils is less than the diameter of the wire to prevent parts from broken springs from interlocking.

Of necessity, a single-stroke device is made inoperative when the press is used for continuous operation. In this case, completely enclose the die, regardless of the method of feeding. When the press is set for other than continuous work, it is of the utmost importance to reconnect the single-stroke device.

FEEDING AND EJECTING MECHANISMS

A significant percentage of power press injuries, such as puncture wounds, lacerations, strains, and amputation or crushing of the hands or feet, occur when work is handled. Thus, any mechanism that eliminates handling of work should reduce exposure to those hazards, especially in operations that place the operator's hands in a danger zone.

Primary Operations

Primary operations are normally conducted with random lengths of strip stock or coiled stock. Strip stock is usually fed by hand. Coiled stock can be fed either by hand or by means of a roll or hitch feed.

Automatic roll feeds are often used on continuous operations of blanking from strip stock. Enclose gears on the feed rolls, and also guard the inrunning nip point of the roll feeds. These precautions are especially important when operators wear gloves or have long hair, even with hair nets or close-fitting caps.

Automatic push or pull feeds are similar to roll feeds and are used mostly in blanking larger pieces. When coiled stock is used with a reel and roll feed or hitch feed, a feed table is generally unnecessary. When coiled stock is used with a stock reel but not with a roll feed, a feed table helps the operator backgauge the stock and feed it to the die with minimum effort.

When feeding strip stock by hand, a feed table eliminates unnecessary motion and reduces operator fatigue. Adjust the feed table to the height at which the operator can work with minimum effort.

Use oiling rolls or pressure guns instead of a paintbrush system to lubricate strip or roll stock. Provide automatic or manual-controlled pressure guns to lubricate the punch and the die.

Secondary Operations

In secondary operations, selecting feeding and ejection methods that keep the operator's hands out of the point of operation is more difficult. When automatic feeding methods are used, safeguard the point of operation by limiting the slide stroke to ¼ in. (0.6 cm) or less. Also, provide a guard or device, according to ANSI B11.1, *Safety Requirements for the Construction, Care and Use of Mechanical Power Presses.*

Semiautomatic feeding mechanisms place the workpiece under the slide by a mechanical device that requires the attention of the operator at each stroke of the press. Such feeds have a distinct advantage because the operator is not required to reach into the point-of-operation area to feed the press. This feeding method also permits complete enclosure of the die. Semiautomatic feeding may not be adaptable for certain blanking operations or for nesting of odd-shaped pieces.

Semiautomatic Feeds

The six principal types of semiautomatic feeds are chute (both gravity and follow), slide or push, plunger, sliding die, dial, and revolving die. Each one will be discussed in turn.

Chute Feed

Of the six semiautomatic methods, the chute feed is probably the most widely used. It is a horizontal or inclined chute into which each workpiece is placed by hand. The workpieces then slide or are pushed, one at a time, into position in the lower die. The entire die may be enclosed, since it is unnecessary for operators to place their hands in the point-of-operation area if automatic ejection is also provided (Figure 23–12).

It is customary to use a soft metal pick or rod, or a wood stick, to remove pieces that jam in the die. Never use a steel rod. If caught in the die, it could shatter, causing injury to the operator and damage to the die.

Many hand-fed dies can be changed to chute feed by reversing the dies and inclining the press. A chute feed on an inclined press not only helps center the workpiece as it slides into the die but also simplifies the problem of ejection.

Slide or Push Feed

A variation of the chute feed, the slide or push feed is combined with magazines and plungers. The workpieces are stacked in the magazine. As each piece reaches the bottom, it is pushed into the die by means of a hand-operated plunger (Figure 23–13).

Plunger Feed

A variation of the push feed, the plunger feed may be semiautomatic or manual in operation. The semiautomatic plunger feed is a magazine or chute in which blanks or partly formed workpieces are placed. The blanks or pieces are fed, one at a time, by a mechanical plunger or other device that pushes them under the slide.

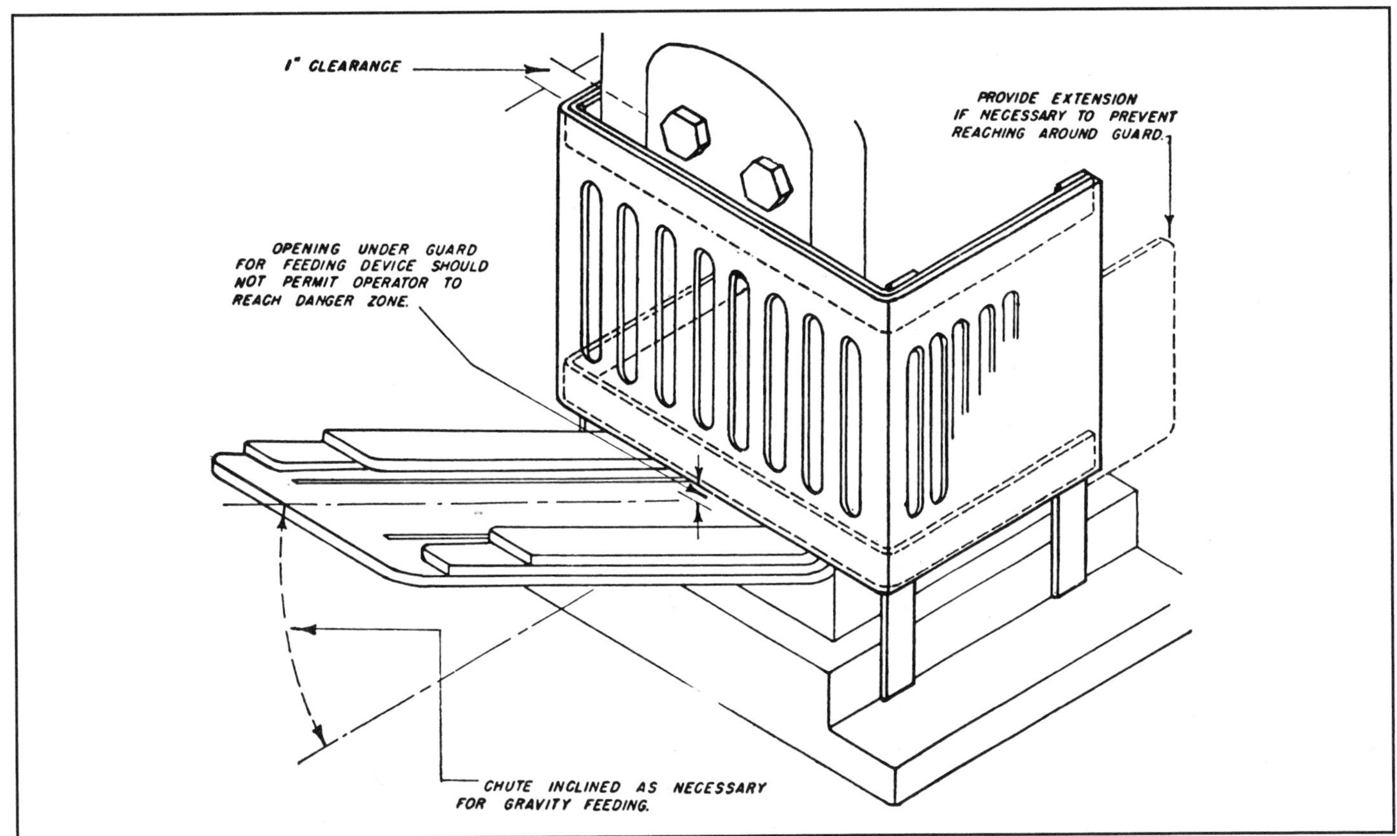

Figure 23–12. This die enclosure guard with an inclined chute for gravity feeding can also be used on an incline press with a straight chute. (Courtesy Liberty Mutual Insurance Co.)

Plunger feeds, operated by hand, are used for individual workpieces that, because of their irregular shape, will not stack in a magazine or will not easily slide down a gravity chute. Each workpiece is placed in the nest in the pusher and moved to the die by manually operating the pusher. To correctly locate the part in the die, an interlock is sometimes necessary. In that way, the press cannot be operated until the pusher has spotted the part accurately (Figure 23–14).

Sliding-Die Feed

With a sliding-die feed, the die is pulled toward the operator for safe feeding and then pushed into position under the slide, or ram, for the downstroke. The die may be moved in and out by hand or by means of a foot lever. Regardless of how the die is moved, interlock it with the press to prevent tripping when the die is out of alignment with the slide (Figure 23–15). Provide stops to prevent the die from being inadvertently pulled out of the slides.

Dial Feed

With a dial feed, two or more nests arranged in the form of a dial revolve with each stroke of the press so that the operator can safely feed the machine. The part to be processed is placed in a nest on the dial that is positioned in front of the die. The dial is indexed with each upstroke of the press to deliver the next nested part into the die.

When using a dial feed, enclose the point-of-operation area. Motion of the dial when operators' hands are free may create a hazard. An idle station may be required beyond the load station to prevent injury at the station that enters the guard.

The best method of ejection for a dial feed is usually by pickup fingers or compressed air. However, in many installations the operator both nests the part on the outside and removes it when it is returned on the dial. Two operators are sometimes used on a dial-feed press. One feeds the press and the other removes the processed parts.

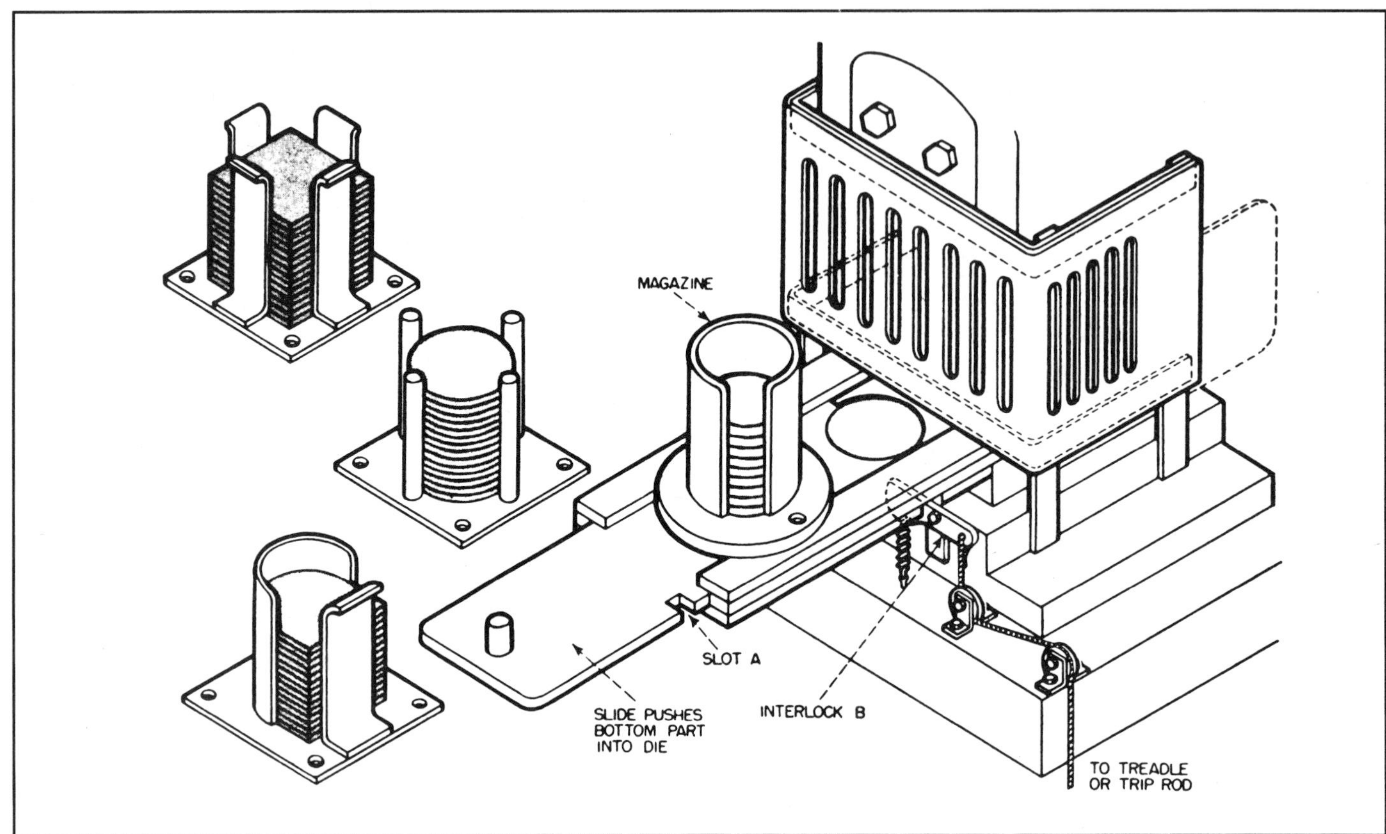

Figure 23–13. This magazine on a push feed enables the operator to catch every press stroke. Slot *A* in the pusher must be in alignment with interlock *B* before the press can be tripped. This feature assures proper positioning of the part in the die. (Courtesy Liberty Mutual Insurance Co.)

Revolving Die Feed

Operating on the same principle as the dial feed, the revolving die feed may consist of two dies or multiple dies.

Ejector Mechanisms

Properly designed and installed ejector mechanisms will eliminate many common hazards. Ejector mechanisms can automatically clear the press faster than humans can and with greater safety. In every facility, however, there will be times when parts will be moved manually. Not only is manual removal less efficient but it causes greater operator fatigue and exposes operators to more injuries than does mechanical removal. Follow the basic principle of keeping the operator out of danger zones even if parts are removed by hand. Properly guard the danger zones and have operators use suitable hand tools.

When designing ejection mechanisms, two problems must be solved: how to strip the piece from the punch or the die, and how to eject the piece from the die and the press to a container or conveyor. These problems can be solved in many ways. In some cases, a single mechanism performs both operations. In other cases, a separate mechanism is used for each operation.

Some of the more common mechanisms used, singly or in combination, to strip pieces from the die are:

- positive stripper plates
- spring-pressure pads or pins
- latch-type mechanical lift dogs
- compressed air jets
- pneumatic or hydraulic lift pins or pickup fingers

Air Jets

Single or multiple air jets can be used for effective removal of small pieces. Air ejection can be combined with other means of mechanical release or with gravity removal. Anchor all jets securely to direct the air stream effectively and to prevent jet tubes from shifting into the die's working area.

There are always hazards from flying particles in press operations, and the addition of air ejection increases these hazards. Using compressed air also increases the overall noise level. Be sure all operators wear eye and ear protection at all times.

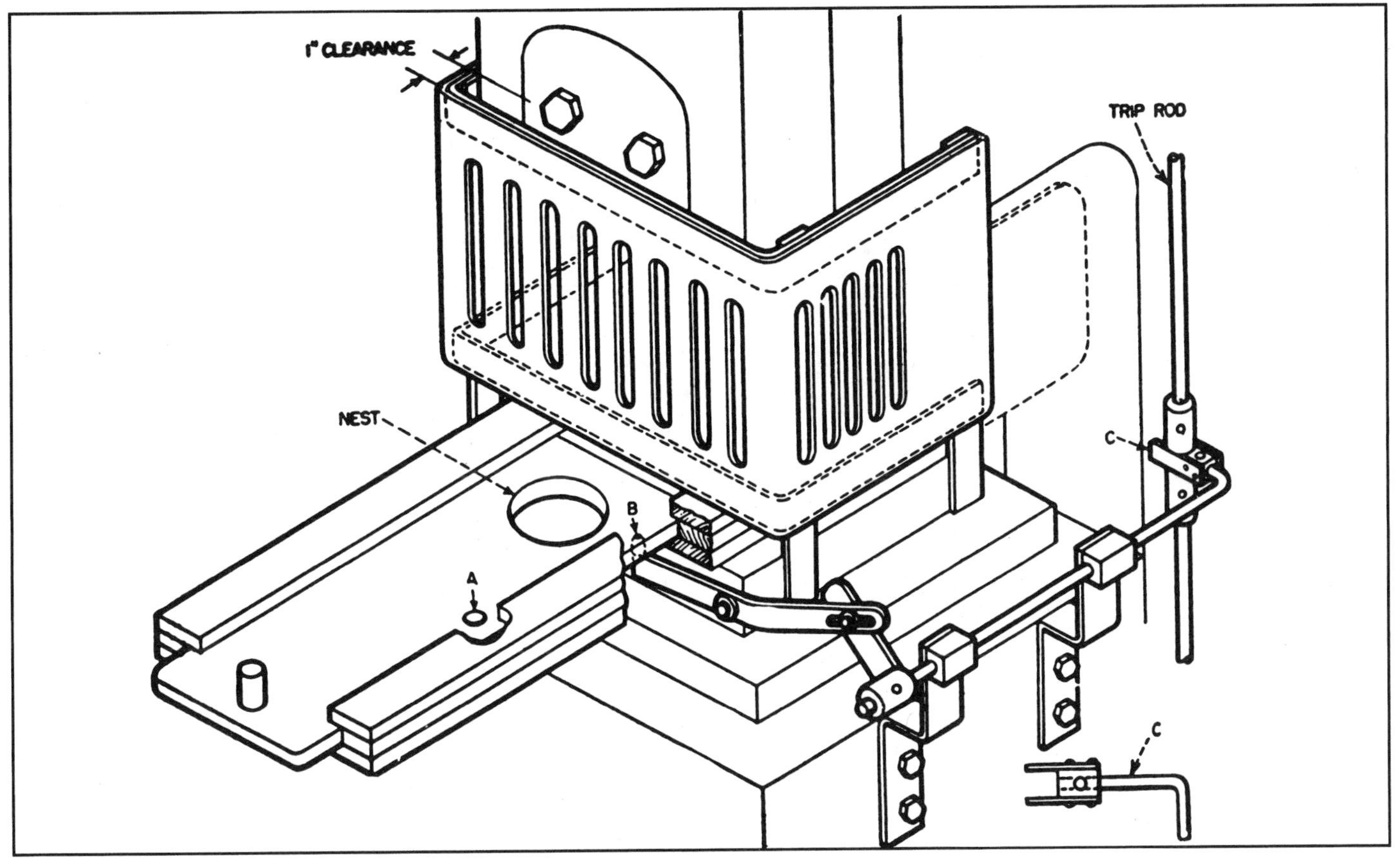

Figure 23–14. In this manually operated plunger feed, note that the press cannot be tripped until the part is pushed to the next location. At this point, hole *A* is directly over tapered pin *B*; it can rise and release yoke *C* so that the press can be tripped.

To reduce both air consumption and noise of air jets, incorporate one or more of the following suggestions:

- Locate the jet's discharge as close as possible to the workpiece to be removed.
- Limit the length of air discharge to the minimum time required to remove the piece.
- By means of flow valves or pressure regulators, reduce the discharge pressure to the minimum required. Reducing the discharge pressure also protects operators from flying particles. However, operators should also wear eye protection.
- Use an air-ejector nozzle with several openings.

Pneumatically Powered Cylinders

Pneumatically powered cylinders, operating sweeps or kickout pins timed with the upstroke of the press, are more effective than air jets for removing large and heavy pieces. For safety, when clearing jams or doing tryout operations, equip the pneumatic equipment with a valve to shut off the flow of air and to bleed any residual air pressure between the valve and the cylinders.

Clamp and Pan Shuttle Extractors

Clamp and pan shuttle extractors are used on presses of all sizes. Pivoting or straight-line clamps grip the part and remove it from the die's area to a pallet, bin, or conveyor for further processing. A pan shuttle (Figure 23–16) (1) catches the piece as it is stripped from the upper die by knockout pins or other means, and (2) removes it from the die's area.

Some extractors are mechanically powered by connection with the press's slide or shafting. However, independently powered extractors are used more frequently because they are more versatile and can be used with several presses. Interlock every independently powered extractor with the press's control circuit so the press cannot operate unless the extractor is in the home or out position.

Steel Chutes

Simple sheet steel chutes are generally provided along with air ejection to guide the parts into a container. Use chutes or slides for a controlled movement of pieces to

Figure 23–15. In this adjustable barrier with a sliding die, a locating mechanism is provided on the slide bolster to locate and lock the die slide in alignment with the punch holder. An interlock should be provided to prevent tripping of the press until the die slide is in the proper location.

subsequent operations. Eliminate sharp bends and inside projections, which may cause parts to pile up, from enclosed chutes.

Elevators and Conveyors

Use elevators and power and gravity conveyors to transport pieces of work to containers or to subsequent press operations.

Support all conveyors for stability on the sides and ends. A mobile conveyor, which can be moved from station to station, should have base supports of a width not less than one-half the height of the conveyor. Use adjustable conveyor sections, especially those of the gravity type, so that angles of slope can be set as the primary or auxiliary factor. This is important for controlling movement of parts at safe speeds and with orderly spacing.

Enclose the drive mechanisms of power conveyors and elevators as completely as possible. Eliminate pinch points formed by moving belts at pulleys or conveyors' structures. Provide electrically driven conveyors with equipment grounds and with On/Off controls located conveniently for emergency use.

KICK PRESSES

All power presses have counterparts in presses that are hand or foot powered. Many shops still use kick presses, foot shears, hand folders, and hand rollers. Since hazards do exist even though the operator is the source of power, guard such presses.

Kick presses are used principally for piercing, notching, forming, and shearing small parts, eyelet closings, and subassembly work. Although simple to operate, the kick press presents serious hazards. Since even the most experienced operator is subject to occasional failure of hand-foot coordination, the short job that is often performed without the use of guards is especially hazardous. Install guards on all kick presses, and thoroughly train all operators.

Safeguard foot-operated presses with an interlocking tripping mechanism. This mechanism requires the operator to use both hands simultaneously to release the slide's head before the pedal can be used. At all other times, the slide's head should come to rest in the top position.

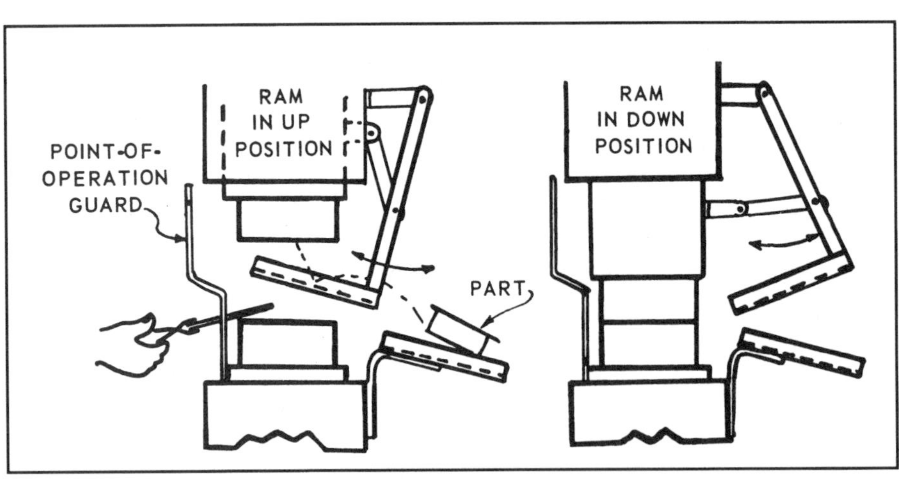

Figure 23–16. This pan shuttle mechanism moves under a finished part on the upside of the press slide. The shuttle then catches the part stripped from the slide by knockout pins and deflects the part into a chute. When the press slide moves down toward the next blank, the pan shuttle moves away from the die area.

Types of Injuries

The principal types of injuries resulting from unsafe operation of kick presses include the following:

- finger amputations and finger punctures from unguarded points of operation
- fatigue and abdominal strain resulting either from pressure required to perform the work or from improper posture
- strains from lifting materials
- eye injuries caused by small flying particles

General Precautions

When a new kick press is installed or an old one is moved to a new location, see that it is securely fastened to the floor or the bench. Otherwise, as the press is operated, it will shift. Also make sure that the gib is clean and lubricated and that the gib's screws are tightened.

Provide good general lighting. In some cases, local lighting may be required. Good visibility not only will increase production and reduce fatigue but also will lessen the possibility of injuries. To improve visibility, have the press painted in contrasting colors.

Guards

When a kick press is used for piercing and similar operations, guard the point of operation. Make a guard that fits around the punch from a small piece of perforated sheet metal, transparent plastic, or other material. The guard should allow enough room to insert the work into the press, but not enough to permit the fingers to come within the danger zone.

On some assembling jobs, when operators must place their hands under the punch, the point of operation cannot be guarded. In such cases, use a two-handed safety device, such as the ratchet mechanism shown in Figures 23–17 and 23–18. The ratchet mechanism was designed to be a thoroughly flexible protective device for kick presses. It operates in the following way:

1. A steel plate with ratchet teeth is securely fastened to the movable slide's head.
2. A pivot pin is anchored securely to each side of the head casting.
3. Two-hand levers with points shaped to fit the ratchet's teeth in the plate are designed so that from the pivot point, the top portion of the levers, when the levers are not held back, bears against the ratchet plate.
4. The slide's head is thus positively locked from downward motion.

Figure 23–17. In this ratchet mechanism for kick presses, the operator must use both hands to disengage the pawls from the ratchet. The ratchet teeth are cut into the plate attached to the slide head. Below the ratchet teeth, the plate tapers to the width of the slide and is secured to the slide by two hex-head bolts and washers. (Courtesy General Electric Co., Pittsfield, MA.)

To operate the kick press, the operator must use both hands to disengage the pawls from the ratchet and thus permit downward travel of the slide's head. This type of device not only prevents the operator from absentmindedly placing fingers beneath the punch, but also provides a brace when the operator applies pressure on the treadle. Should the operator release either hand, the pawls will engage the ratchet and stop the punch instantly. During the loading and unloading periods, the ram is locked in the top position.

Another simple, yet effective, type of two-hand device can be easily installed at small expense. Drill ½-in. (1.25-cm) holes through the sides of the frame and into the swinging arm of the ram. Then, when the ram is in the top position, automatic locking pins that can be withdrawn by the operator are inserted in the holes. Levers with springs attached to these pins are located conveniently for the operator to lock the ram when it is in the top position.

Eye Protection

Operators on kick-press operations should wear eye protection. When hard materials, such as brass and spring steel, are being shaved or notched, small particles thrown off with great force may cause serious injuries. Require goggles or spectacles with full side shields or face shields on such operations.

Fatigue and Strain

Since most kick presses are operated from a sitting position, provide a seat or chair that enables the operators to work with minimum fatigue. Operators should not have to make a long reach with their foot to perform the operation. In addition, arrange the seat so operators do not sit in a cramped position. The seat should have a comfortable back rest, and the base of the chair's legs should be wide enough so the chair will not tip over easily. (See Chapter 16, Ergonomics Programs, in the *Administration & Programs* volume of this *Manual.*

Proper adjustment of the counterweight and stop also lessens operator fatigue. Proper adjustment also reduces the travel distance of the pedal and results in a balanced operation.

If the pedal is worn smooth, the operator's foot can readily slip, thus causing a strained or sprained ankle. Placing flanges on both ends of the pedal and rubber or abrasive surfacing on the pedal will minimize this hazard.

There is also a possibility that the operator may suffer abdominal strain from operating a kick press. Occasionally, heavy operations are performed that cause unusual fatigue. In such cases, permit the operator to alternate kick-press work with light work. Allow only physically able employees to work a kick press. Do not permit any employee who is known to have weak abdominal walls, such as from a recent operation, to operate a kick press.

Figure 23–18. This side view of a ratchet mechanism for kick presses shows the simple ratchet principle of the guard. (Courtesy General Electric Co., Pittsfield, MA.)

Lifting heavy containers filled with parts may result in severe muscle strain, especially in continuous operations where fatigue may make lifting increasingly difficult. To eliminate this hazard, install conveyor belts to carry material along the bench to the press. If the volume of work does not warrant the use of a conveyor, fasten brackets to the press at truck height. This makes it possible to slide containers from trucks to the press and back again with little effort.

Maintenance

Check the kick press and its safeguard frequently. Lubricate and adjust all working parts. Replace immediately parts of the press or the guard that show wear or defects.

ELECTRICAL CONTROLS ON POWER PRESSES

Properly designed, applied, and installed electrical controls are an important element in press safety. This is particularly true when a two-hand control device or a two-hand tripping device is used for point-of-operation safeguarding.

Electrical controls may range from a simple power disconnect switch and a motor starter on a small full-revolution clutch press to an extensive system on a large partial-revolution clutch machine. The latter typically consists of:

- a main disconnect switch or circuit breaker for isolating the machine from electrical power
- starters for all motors including the main drive, slide adjustment, lubrication, and any auxiliary drives
- one or more control transformers with isolated secondary windings to step down line voltage to a level of not more than 120 volts AC for control circuits and solenoid coils
- overcurrent and ground-fault protective equipment or circuitry
- control relays or the electronic equivalent, for energizing magnetic valves and for interlocking, sequencing, and checking safety features
- operator controls, setup controls, indicating lights, limit switches, pressure switches, and other sequence, checking, and protective components
- solenoid-operated valves for clutch/brakes, flywheel brakes, and auxiliary equipment

Installation

The following information applies to all installations of two-hand controls and two-hand tripping devices, regardless of the means of point-of-operation safeguarding. Build controls for all machines in accord with National Fire Protection Association (NFPA) 79, Electrical Standard for Industrial Machinery, as well as recognized industry standards. Installation of machines should conform to NFPA 70, *National Electrical Code,* and any applicable local codes. The purpose of these codes and standards is to safeguard personnel and equipment. Consult ANSI B11.3 for specific application to press brakes.

Control Panel

The machine's control panel should contain the main disconnect switch, motor starters, clutch-control relays, and control transformer. Locate this panel on or immediately next to the machine it serves, but out of the operator's way during normal operation. Properly identify the control panel and have it easy to reach for maintenance (Figure 23–19).

Place the handle of the main disconnect switch on the exterior of the control's enclosure, and have the handle operable from floor level. The recommended maximum height for the centerline of the switch's handle is 6 ft (1.8 m) above the floor. Provide the switch with a means for complete lockout.

Control Components

Ground cases or frames of all control components to the press's frame. Mounting bolts are satisfactory for grounding, provided that all paint and dirt are removed from the joint's surfaces before assembly. Ground movable control components not secured to the press's frame by means of a grounding conductor in the connecting cable or conduit.

Rotary Limit Switch

The rotary limit switch is a vital element in the control system (Figure 23–20). Mechanical or electrical failure of this switch may cause the press to malfunction. The switch-driving mechanism—including all couplings, gears, sprockets, and associated parts—should be securely assembled and be of a rugged design. A constant, keyed connection is preferred to set screws since a single set screw has a tendency to vibrate loose. Use two separately driven switches or a drive-failure detector. Use cam switches in which the switch's contacts are forced open by the cam that starts the movement.

Flywheel Brake

When using a flywheel brake, electrically interlock it with the main drive's starter. This prevents simultaneous operation.

Safety Blocks

Provide safety blocks with interlock plugs. They disconnect the clutch- and motor-control circuits when blocks are removed from their storage pockets.

Magnetic Air Valves

For actuating the clutch, use three-way, normally closed, poppet-designed magnetic air valves. Never use spool valves for this purpose. Their pistons slide in close-fitting sleeves and can easily stick because of dirt, corrosion, or improper lubrication.

Figure 23–19. Presses must also be made safe for electricians, millwrights, pipe fitters, and others who must maintain them in a safe condition. Identification of components of this machine control panel are given in the chart on the inside of the door. Components are readily accessible for maintenance.

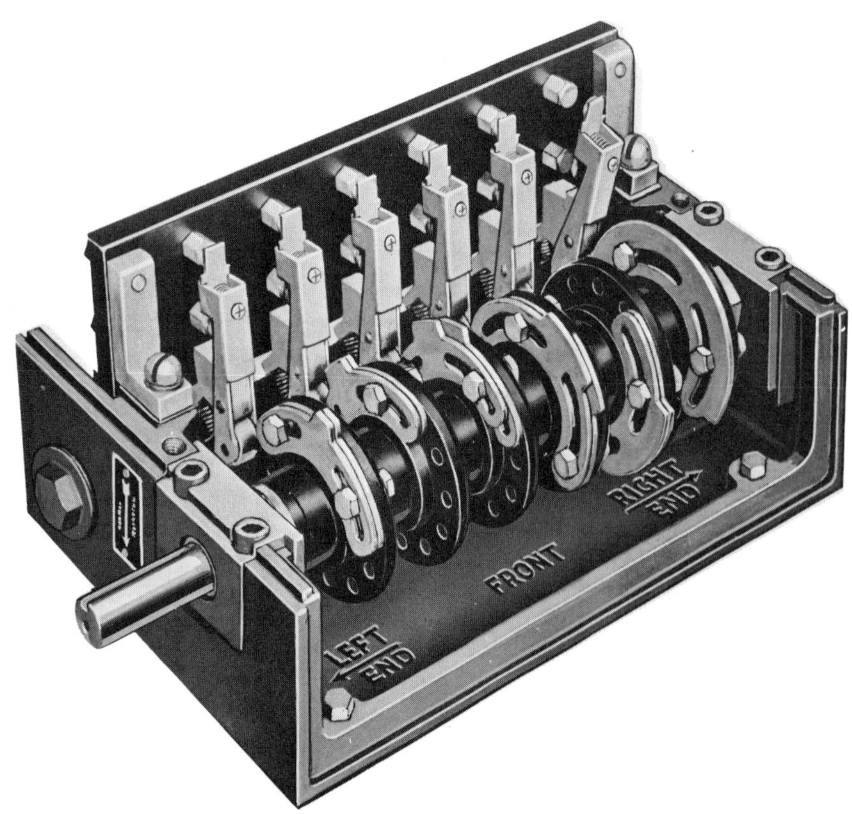

Figure 23–20. All rotary cam limit switches should be the type in which the contacts are forced open by the actuating cam.

Definitions

Two-Hand Trip

A clutch- or clutch/brake-actuating method requiring the momentary concurrent use of both hands of each operator to initiate a complete stroke of a mechanical press, generally of the full-revolution clutch type. Because only momentary pressure is necessary on hand buttons or other mechanisms, two-hand trip can be used as a point-of-operation safeguarding device. This method can be used for single-stroke operations and only when the location of the buttons or other hand mechanisms is of sufficient distance from the point of operation. This prevents moving either hand into a point-of-operation hazard prior to die closing. Two-hand trip is generally impractical on presses having speeds of less than 120 strokes per minute (spm). (See Construction Features for All Two-Hand Tripping Systems for design and construction requirements.)

Two-Hand Control

A clutch/brake-actuating and stroke-controlling method requiring the concurrent use of both hands of each operator on RUN buttons or other mechanisms during a substantial portion of the die-closing stroke of a part-revolution clutch press. Two-hand control can be used as a point-of-operation safeguarding device for the single-stroke mode of operation when properly designed, installed, and adjusted as outlined in Construction Features for Two-Hand Control Systems.

Safeguarding Exclusions for Two-Hand Systems

In the continuous mode of stroking, two-hand control and two-hand tripping do not qualify as point-of-operation safeguarding devices. Use a die-enclosure guard or a fixed-barrier guard to keep operators' hands out of the point of operation. In that way the operators can use the hand controls to initiate strokes.

If a foot control is connected in place of any hand-operated button or buttons, do not use the operator's foot control for point-of-operation protection—one or more hands are free to enter the die's area at any time. Instead, use a die-enclosure guard, a fixed-barrier guard, or a type-A, movable-barrier device for point-of-operation safeguarding on power presses.

Construction Features for All Two-Hand Tripping Systems

The individual operator's hand controls—RUN buttons or other two-hand tripping mechanism—for tripping the clutch must meet all of the following requirements:

- Protect each control against unintentional operation. Use protective rings around Run buttons or suitable barriers.
- Arrange each pair of controls so both hands are required. This means that buttons must be far enough apart and located so that a hand and elbow of the same arm, a knee, or any other part of the body cannot be used instead of both hands.
- When two-hand tripping is used on a multiple-operator press, provide each operator with a separate set of two buttons or other hand mechanism. Bypass buttons, not needed for particular operations, should be provided in complete sets of two—not by individual buttons. Supervisors should monitor bypassing.
- When two-hand tripping is used as a device for safeguarding the point of operation, fix buttons in position with a safe distance between the point of operation and each button. This distance should be great enough to prevent the operator from moving either hand from a button into a point of operation prior to the die's closure. Minimum safe distances are shown in Table 23–A and are based on the following formula:

D_m = 63 in. (160 cm) per second × T_m where:
D_m = minimum safety distance in inches 63 in. (160 cm) per second
= common hand movement speed (higher speeds have been measured)
T_m = maximum time in seconds for die closure after the press has been tripped.

For full-revolution clutch presses with only one engaging point, T_m is equal to the time necessary for 1½ revolutions of the crankshaft. For full-revolution clutch presses with more than one engaging point, T_m is calculated as follows:

$$T_m = \frac{1}{2} + \frac{1}{\left(\begin{array}{c}\text{number of}\\ \text{engaging points}\\ \text{per revolution}\end{array}\right)} \times \left(\begin{array}{c}\text{time in seconds}\\ \text{necessary to complete}\\ \text{one revolution of}\\ \text{the crankshaft}\end{array}\right)$$

The two-hand tripping system must include the following features:

- concurrent operation of all RUN buttons or other hand mechanisms
- an antirepeat feature for single-stroke operations
- electrical control circuit and valve-coil voltages not exceeding a nominal 120 volts AC or 240 volts DC isolated from higher voltages
- ground-fault detection to locate an accidental ground, and circuitry to prevent false operation due to an accidental ground
- features to minimize failures that would cause an unintended stroke

Table 23–A. Minimum Safety Distances for Two-Hand Trip Between Hand Controls and Nearest Point-of-Operation Hazard

Press speed in strokes per minute	Time in seconds (Tm) for one revolution of crankshaft	Minimum Safety Distance (Dm) in Inches: Full revolution clutch engaging points per revolution 1	2	3	4	14	Part revolution clutch Infinite
30	2.0	189	126	105	95	72	63
45	1.33	126	85	70	63	48	42
60	1.0	95	63	53	48	36	32
75	0.8	76	51	42	38	29	26
90	0.67	63	42	35	32	24	21
105	0.57	54	36	30	27	21	18
120	0.5	48	32	27	24	18	16
135	0.44	42	28	24	21	16	14
150	0.4	38	26	21	19	15	13
165	0.36	35	23	19	18	13	12
180	0.33	32	21	18	16	12	11
210	0.29	27	18	15	14	11	9
240	0.25	24	16	14	12	9	8

With adjustable-speed drive, use slowest speed. (Based on OSHA regulations 1910.217 (c) (3) (viii) of December 3, 1974, for presses with full- or part-revolution clutches.)

- a control system to prevent actuation of the clutch, on a multiple-operator press, if all operating stations are bypassed

Construction Features for Two-Hand Control Systems

The individual operator's hand controls—RUN buttons or other hand-control mechanisms—for engaging the clutch/brake must meet all of the following requirements:

- Protect each control against unintentional operation. Use protective rings around RUN buttons or suitable barriers.
- Arrange each pair of controls so both hands are required. This means that buttons must be far enough apart and located so that a hand and elbow of the same arm, a knee, or any other part of the body cannot be used instead of both hands.
- When a two-hand control is used on a multipleoperator press, provide each operator with a separate set of two buttons or other hand mechanisms. Bypass buttons, not needed for particular operations, should be provided in complete sets of two—not by individual buttons. Supervisors should monitor bypassing.
- When a two-hand control is used with a partial revolution clutch press as a device for safeguarding the point of operation, fix buttons in position with a safe distance between the point of operation and each button. This distance should be great enough to prevent moving either hand from a button into a point of operation while the slide (ram) is in motion during the die-closing portion of a stroke. Minimum safe distances shown in Table 23–B are based on the following formula:

D_S = 63 in. (160 cm) per second x T_S where:
D_S = minimum safety distance in inches 63 in. (160 cm) per second
= possible hand movement speed
T_S = longest stopping time of the slide (ram) in seconds, usually measured at approximately the 90-degree position of the crankshaft's rotation.

The stopping time includes the operating time of the press's control system, air exhaust from the clutch/brake, and braking time of the slide. Stopping-time measuring units are available, or a stopwatch can sometimes be used. If the minimum safe distance is not practical for production, use a guard or another safeguarding device.

Stop Control

Provide a red STOP button with the clutch/brake control system. Momentary operation of this button must immediately deactivate the clutch and apply the brake. This button must override all other controls. To put the clutch in motion again must require use of the RUN buttons or other operating mechanisms. Make a STOP button available to each operator. At least one STOP button must be connected and operative regardless of whether a hand or foot control is being used.

Table 23–B. Minimum Safety Distances for Two-Hand Control Between Hand Controls and Nearest Point-of-Operation Hazard

Stopping time of slide in seconds (Ts) at 90° point of stroke	*Minimum safety distance (D_s) in inches*
0.100	7
0.125	8
0.150	10
0.175	11
0.200	13
0.225	15
0.250	16
0.275	18
0.300	19
0.325	21
0.350	22
0.375	24
0.400	26
0.450	29
0.500	32
0.600	38
0.700	45
0.800	51
0.900	57
1.000	63

With adjustable-speed drive, use the stopping time for highest speed. (Based on OSHA regulations 1910.217 (c) (3) (vii) of December 3, 1974, for presses with part-revolution clutches only.)

Stroking Selector

Supply a means of selecting OFF, INCH, SINGLE STROKE, or CONTINUOUS (when continuous stroking is provided) with the clutch/brake control. Also include other positions when required for special features. Keep hand-foot selection separate, however, from the stroking selector.

Fixing the selection must be monitored by a supervisor. Key-operated selectors are commonly used to lock selectors in position. Additional precautions are needed for continuous stroking, as described under Continuous Control, later in this section.

To standardize the location of each mode of operation on rotary stroking selector switches, use the following sequence as the handle, knob, or key is rotated clockwise: OFF—INCH—SINGLE STROKE—CONTINUOUS (when continuous stroking is provided).

Inch Control

Never use the INCH mode of operation for production. There is generally no antirepeat circuitry, automatic top stop, or drive-motor interlock as there is in the single-stroke mode. Such features could interfere with the type of slide movement often needed during setup or tryout in the INCH mode. For example, it is sometimes necessary to INCH with the drive motor and flywheel coasting at reduced speed in order to get finer increments of the slide's motion.

To prevent exposure of personnel within the point of operation, require two hand controls to move the clutch. Use a single hand button only if it is protected from being started accidentally and is located so that the worker cannot reach into the point of operation while pressing the button. Never use a foot control for inching.

Single Stroke Control

In addition to the RUN button features listed in this section, provide RUN-button holding time and interrupted stroke protection for the control system. Always try to use dies and feeding methods that do not require placing hands in the point of operation at any time. However, there are instances in which it is not feasible to keep hands out of dies. Control reliability and brake monitoring are, therefore, also required with hands-in-dies operations.

Holding Time. If all RUN buttons are held down until the dies have closed sufficiently for safe release of the buttons, the slide must stop before a hand can enter the point of operation. The control should provide for an adjustment of the rotary limit switch's contacts to bypass the RUN buttons at a safe point in the stroke. This will vary according to speed, length of stroke, and the type of dies.

Interrupted Stroke Protection. Before an interrupted stroke can be resumed, again by pressing all RUN buttons, all such buttons should be released. The purpose of interrupted stroke protection is to minimize the possibility of accidentally restarting the slide after a button has been released during the holding time. This feature is accomplished through proper design of the circuitry.

Control Reliability. Construct the control system so that any control failure does not interfere with the normal stopping action when required. However, the initiation of successive strokes should not occur. Design the control system so a failure is easily detectable.

Brake System Monitoring. Incorporate a brake monitor into the control. It should prevent activation of a successive stroke if the stopping time or distance deteriorates to a point where the safe distance no longer complies with distances shown in Table 23–B. The monitoring action must take place during each cycle in single-stroke operations. The monitoring system should indicate any unsafe deterioration.

Automatic Single Stroke. Operating mode where the press control is given single-stroke actuating signals by an automatic feeding mechanism or other auxiliary equipment without action by an operator after initial start. Proper safeguarding is required for this type of automatic single-stroke operation because using only a two-hand control is not acceptable.

Continuous Control

The following three different types of continuous operations are used:

1. continuous—continuous stroking of the slide after initiation without the operator's controls being activated
2. maintained continuous—continuous stroking of the slide only as long as the operator's controls are activated
3. continuous on demand—continuous stroking of the slide after initiation without the operator's controls being activated, and with periodic stopping and automatic restarting controlled by auxiliary equipment

All of the continuous modes of operation have special control requirements. When a control system provides for both single-stroke and continuous functions, it must be designed for both of the following conditions:

- Supervision must be given when selecting the continuous mode. A key-operated lock selector is usually used for this.
- Initiation of a continuous run must require a prior action or decision by the operator different from that used for single stroke. A common method for this is using an ARMING or CONTINUOUS SETUP button that must be momentarily pressed a few seconds prior to activating the RUN buttons or other means of operation.
- These steps, or equivalent actions, are necessary to prevent inadvertent starting of the slide in the continuous mode. There is no RUN-button holding time and no automatic stop at the end of each stroke as in the single-stroke mode of operation. Another preventive measure is to install a die-enclosure guard or fixed-barrier guard to keep operators' hands out of the point of operation. Other guards or safeguarding devices may also be used.

Multiple-Operator Machine

On a multiple-operator machine, the control system must prevent activation of the clutch/brake if all operator stations are bypassed. This is frequently called dummy-plug protection because supervised dummy plugs or keyoperated selectors are generally used to bypass unneeded stations.

Stroking Control Systems' Component Failure Protection

In stroking control systems (clutch/brake-control circuits), incorporate design features to lessen the chance of failures that could cause an unintended stroke. The following features are critical:

- input elements, such as stroking selectors, RUN buttons, INCH buttons, foot controls, and STOP buttons
- sensing elements related to stroking, such as rotary limit switches or equivalent apparatus, including their drive mechanisms, motion or position detectors, valve failure detectors, and any associated relays or electronic components
- output elements to the slide-operating mechanism, such as valve relays and clutch/brake valves
- acceptable methods of controlling normal stopping of the slide, including interruption of the clutch's (and brake's, if separate) valve current through the cycle-control rotary limit switch, or series contacts of two or more independent relays or static circuits
- adjustable stopping controls, using counters or timers, used with suitable limit switch backup

Design the controls for air-clutch machines to prevent significant increases in the normal stopping time if an operating valve's mechanism fails. They should also be designed to inhibit further operation if such failure does occur. A self-checking assembly of two-valve elements in a common housing is usually the best way to meet this requirement.

If a machine has separate clutch and brake systems that require individual valves connected to a common manifold, both valves should be of the self-checking, two-valve type. Failure of a brake's valve is just as dangerous as a faulty clutch valve. Sometimes a failed brake valve is even more likely to cause a malfunction. The protective valve arrangement is not needed on machines intended only for continuous operation with automatic feeding.

Interlocks for Two-Hand Control Systems

The clutch/brake control must automatically deactivate in case electrical power or proper air pressure is lost, or if the clutch is air operated. If the machine has air counterbalance cylinders, loss of proper counterbalance air pressure must also deactivate the control. Reactivation of the clutch must require (1) restoration of normal electrical and air supply, and (2) use of the RUN buttons or other tripping means.

The control must include an automatic means to prevent initiation or continued activation of the single-stroke and continuous functions unless the drive's motor is energized and set forward. To meet this requirement connect an auxiliary contact on the main drive's forward contractor in the single-stroke and continuous control circuits.

Control-Circuit Voltage and Ground Protection for Two-Head Control Systems

The AC electrical control circuits and valve coils on all machines must be powered by not more than a nominal 120-volt supply obtained from a transformer with an isolated secondary winding. Isolate voltages above 120, which may be needed for particular mechanisms, from any control component handled by an operator. All DC control circuits must be powered by not more than a nominal 240 volt-DC supply isolated from any higher voltage.

Protect all clutch/brake control circuits against the possibility of an accidental ground in the control circuit. This could cause false operation of the machine (Figure 23–21a and b).

Foot Operation of Two-Hand Control Systems

If foot operation is an alternative to two-hand operation, provide point-of-operation protection and the required controls. A die-enclosure guard, fixed-barrier guard, or a type-A, movable-barrier device is recommended for safeguarding presses. Observe the following control requirements:

- Protect foot switches or foot valves and any attached mechanism. This prevents unintended operation caused by falling or moving objects, or by someone accidentally stepping on the foot control.
- Carefully supervise the selection of hand or foot operations. Use of a key-lock selector switch is the most common method of accomplishing this. A key that is always left in the lock is not supervised.
- When two-hand operation has been selected, deactivate the foot control and vice versa. Do not allow the use of RUN buttons and a foot control at the same time.
- Be sure that the antirepeat feature of the control system is in effect during foot operation.
- In the INCH mode, deactivate foot controls; do not use them in place of hand controls.

POWER PRESS SETUP AND DIE REMOVAL

Power press dies must remain rigidly accurate in spite of the pressure and stress they transmit during metal-stamping operations. As a result, they are usually heavy and difficult to handle. They vary in weight from a few pounds for small dies to several thousand pounds for large dies.

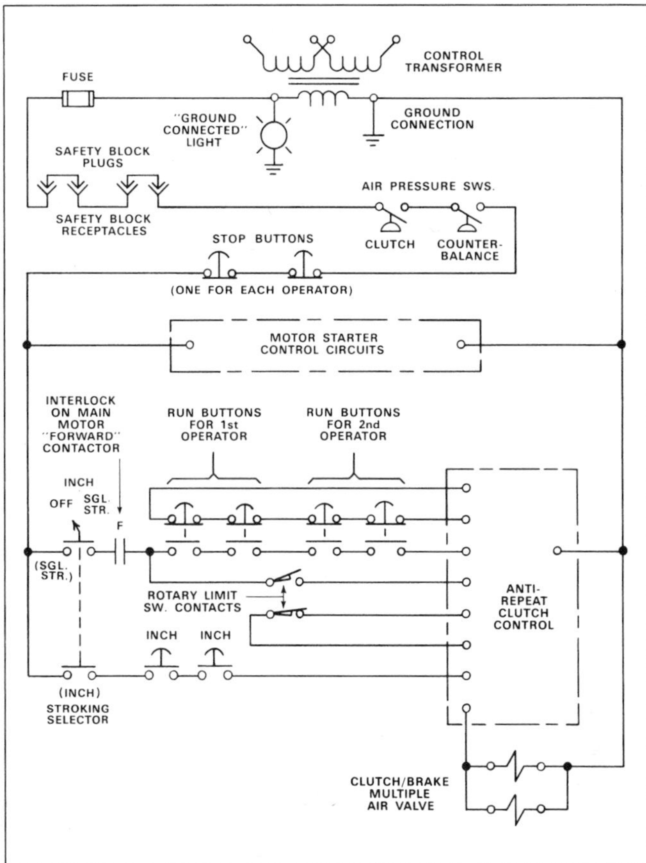

Figure 23–21a. This diagram shows a typical single break, grounded press-control connection.

Handling, setting up, and removing these dies are hazardous unless operators use proper equipment and methods. Entrust these operations only to experienced setup personnel whom the supervisor has instructed in detail about safe procedures.

Injuries likely to result when setting or removing power press dies are the following:

1. strains and hernias from inappropriate handling techniques
2. foot injuries from dies slipping off trucks, benches, bolster plates, or storage shelves
3. crushing injuries from body parts caught between the die and press, pinch points and other movements
4. hand injuries or amputations from sudden descent of the ram from brake failure, premature tripping during tryout, failure to lock out the switch, or failure to maintain proper pressure on pneumatic clutches
5. lacerations from wrenches slipping off worn nuts or from use of incorrect tools
6. eye injuries from pieces of shattered parts, dies, or material

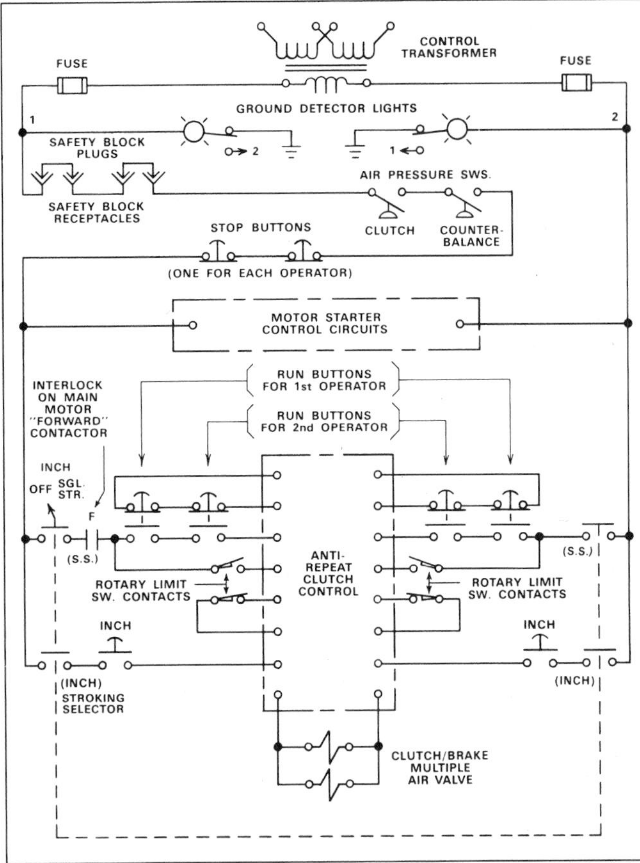

Figure 23–21b. This diagram shows a typical double break, ungrounded press control connection.

Transferring Dies Safely

Very light dies can be handled and carried manually. If proper die trucks are provided, set-up personnel can generally handle dies weighing up to about 100 lb (45 kg) without using lifting apparatus (Figure 23–22). Die trucks should have elevating tables adjustable to the height of storage shelves and press bolster plates. When transporting dies, personnel should carry them at the truck's lowest elevation. Use rollers, balls, or windows mounted in the top surface of the table to aid in sliding the die on or off the truck.

Heavy dies require more equipment for safe handling. Because they are often lifted and moved by hoists, these dies should have tapped holes and eye bolts (or lifting hooks), drilled holes and pins, chain slots, cast lugs in lower shoes, or clamping lugs to ease hookup and transfer. When tapped holes are used for securing lifting hooks or eye bolts, they should be either ¾ or 1 in. [1.9 or 2.5 cm] in diameter. (Do not use ⅞ in. [2.2 cm] holes since ¾ in. [1.9 cm] bolts might be used in them and appear to fit, only to pull out during the lift.) The thread's engagement between the hole and the bolt should be one and ½ times the bolt's diameter.

For moving dies that are about 1,000 lb (450 kg) or heavier, use special die-handling power trucks. These trucks have special equipment for pushing or pulling the dies, including power winches, roller tables, and hydraulic or pneumatic clamps. To minimize the effect of uneven floor surfaces, use trucks with large-diameter wheels.

When transferring dies, place the truck close to the storage shelf or press. Adjust the table to the same height as the storage shelf or bolster plate. Then, either chock the wheels, lock the brake, and chain the truck to the press, or use some other method to prevent movement (Figure 23–22). Next, push the die off the truck onto the storage shelf or bolster plate. To bring a die onto a truck, engage a hook in the die so it cannot slip, and exert a steady pull—do not jerk or tug on the hook. Where lifting is needed to transfer the die, use a hoist and never lift higher than is necessary for minimum clearance. At no time should an employee have hands, feet, or another body part underneath a suspended load. Only a person in charge should give signals for movement.

Procedure for Setting Dies

Safe procedures for setting dies vary slightly and depend on the press's size. Most of the difference occurs because the slide on light presses can be moved manually by turning the flywheel or crankshaft, while on heavy presses it must be power operated. Use the following safe method for setting and removing dies for all presses. Special procedures are given for light presses and heavy presses.

1. Dismantle or disconnect a point-of-operation safety device only if it is absolutely necessary. (See Chapter 6, Safeguarding, for lockout/tagout procedures.) For example, remove enclosure guards fastened to the press, gate or barrier guards, and other types of guards. However, do not generally remove pullback devices or presence-sensing devices. Store all parts of disconnected safety devices so they may be reinstalled in good condition as soon as the new dies have been installed.
2. Clean off the bolster plate, preferably with a vacuum system or brush. Keep all bolt holes clear of all obstructions. A pencil-shaped magnet is valuable for removing iron or steel particles. Remove burrs with a file.
3. Check the die to make sure that it contains no chips, tools, or parts, and that it is in good operating order. Then transfer the die from the truck to the press. (See Transferring Dies Safely, earlier in this section.)
4. Line up the die in the correct operating position, and remove the posts or blocks from under the slide. For heavy presses, reconnect the power to

Figure 23–22. A die truck with adjustable height and an adjustable angle die table can match the angle of inclined presses. The truck is secured to the press before moving the die. The holes in the upper die shoe indicate the drilled-hole-and-pin method of lifting.

operate the slide. For light presses, turn the flywheel by hand to move the slide, using a safety bar if necessary. Then, lower the slide until it fits firmly against the top die. It is extremely important not to put too much pressure on dies. Tighten all bolts and clamps to secure the top half of the die. Bolt the die to the ram with bolts through holes in the upper die's shoe (Figure 23–23). On heavy presses, if air-cushion pads are used, keep the ram close to the top of the die until the die is properly seated on the cushion's pins.

5. Shim or block up the lower half of the die to the proper level; bolt and clamp it to the bolster plate. Make sure that the bolts, bolt heads, and wrenches are the proper size and in good condition. Bolting the die's shoe to the bolster produces the most

Figure 23–23. Mounting bolts pass directly through the ram to fasten the upper die securely. The lower die is clamped to the bolster.

secure die setup. If clamps must be used, block up their outer ends slightly higher than the die's surface on which their inner ends will rest. Clamps should be of minimum length. Clamp-fastening bolts should be closer to the die than to the block end of the clamp.

6. Check all bolts and clamps to see that they are tight and that dies are securely fastened in the press. Remove all tools and equipment from the dies, bolster plates, or other areas on the press. For forming or drawing operations, adjust the ram down almost to its proper depth. For a pierce or trim die, raise the ram slightly so that the punch will not shear when the crank is again brought over and the die entered. Turn light presses several times by hand, checking to see that they are satisfactorily adjusted. On heavy presses, do not check adjustment at this point.
7. Raise the ram to its highest point and block it in this position. On heavy presses, disconnect the power before proceeding further. Wipe out the die and remove the safety blocks. Replace the safety device and check it for adjustment and operation. Properly adjust pullout devices, when used. In that way, an accident could not occur to an operator who has longer hands or arms than the previous operator.
8. Reconnect the power and try out several actual operations, using the proper stock. Make any necessary adjustments only after shutting off the power and blocking up the ram. After completing the adjustments, turn on the power and again try several actual operations on the press to assure safe operation.

Light Presses

Disconnect or shut off the electric (and/or other energy sources) power and lock out the switch. Lower the slide to its lowest position by turning the flywheel or crankshaft. If a safety bar is used as a lever to turn the crankshaft, it should have a spring-and-collar arrangement that will prevent the safety bar from being accidentally left in the crankshaft. Measure the clear height between the bottom of the slide and the bolster plate. This distance should be slightly greater than the height of the closed die. If not, raise the slide by means of the adjusting screw. When physical access is necessary, block the slide with timber, metal blocks, or posts provided for this purpose. Equip blocks with an electrical receptacle plug to hold the circuit open when blocks are in the die. Thus, they will not be left in the press when power is applied.

Heavy Presses

Jog or inch the press to bring the slide down to its lowest position. Measure the clear height between the slide and the bolster plate. Again, if this distance is not slightly greater than the height of the die in the closed position, adjust the slide until such clearance is assured. Raise the slide, block it in position, shut off the power, and lock out the switch.

Removing Dies

Follow safe die-removal methods at all times. Although modifications may be necessary in special cases, the safe procedure is as follows:

1. Make sure that the working space is cleared of all stock, containers, tools, and other items.
2. Disconnect or shut off the power and lock out the switch. Turn the flywheel by hand or by the safety bar until the ram is at the bottom of the stroke. If the press cannot be turned over by hand, jog it under power, then shut off the power, and lock out the switch.
3. Dismantle or disconnect the point-of-operation safety devices as required. Store the parts of the dismantled safety device so it can be reinstalled in good condition when the new die is in place.
4. Clean off the bolster plate, preferably with a vacuum system or brush.
5. If the die is to be operated with an air pad, shut off the air supply and open the release valve to permit the pins to go down. Also, shut off the air supply to the automatic blowout system used in the die.
6. Remove bolts and clamps holding the die in the press.
7. Make certain that the die is loose and that bolts, nuts, clamps, and other obstructions have been removed.
8. Raise the ram slowly—by hand on light presses, and by jogging or inching under power on large presses—and make sure that the die does not hang in the slide.
9. Block the ram in its highest position. If power was used, shut off the power and lock out the switch.
10. Place the die truck close to the press, adjust the table to the same height as the lower bolster plate, and chock the wheels or set the truck's brake to prevent movement. To pull the die onto the truck, use a device engaged so the die cannot slip.
11. Inspect, repair, and protect dies before storing them for the next run. Also, inspect the pins and bushings. Store hardened dies and punches in the closed position with a piece of soft wood between the edges to protect them. Injuries have occurred when hardened dies were handled when open or partially open. A sudden jolt can cause the dies to close and pinch a hand or finger.

INSPECTION AND MAINTENANCE

The best safety program for power presses cannot succeed, nor can maximum production be met, without good inspection and maintenance of presses and their safeguards. Proper inspection, adjustment, and repair of power presses and related equipment can only be done by competent, thoroughly trained employees. Be sure that these employees are completely familiar with the construction and operation of the equipment for which they are responsible. Also, provide them with proper tools and equipment. Clear the work area of all personnel not directly involved in maintenance. Erect flashing warning lights or other barriers or barricades used to mark the temporary maintenance area.

Troubleshooting with Power On

When it is necessary to locate and define problems with the power on, the employee can work on power presses with guards removed, or work within areas protected by barriers. However, such action should not place any body part in the path of any movable part of the press. A press may have to be stopped or locked out before removing a guard or barrier so that the press may subsequently be observed with power on. Ensure a written policy is in place to address these situations and ensure compliance with OSHA's Lockout/Tagout Standard, 29 *CFR* 1910.147.

Power presses, like all machinery, are subject to wear, breakage, and malfunction. Therefore, to prevent costly accidents and repairs and to promote maximum production, inspect the entire machine and its related equipment periodically. Make the required adjustments and repairs. The type of press, its related equipment, and its usage should determine the frequency of inspections.

Regulatory standards may require that each press be inspected weekly to determine the condition of the clutch/brake mechanism, antirepeat feature, and single-stroke mechanism. Necessary maintenance or repair or both shall be performed and completed before the press is operated. The weekly inspection is not required on those presses meeting the requirements for "hands-in-the-dies." Employers shall maintain records of these inspections and the maintenance performed.

Set up a checklist that details the frequency of inspection and maintenance for each press. Such a checklist gives immediate knowledge of the press's condition, and makes it easier to schedule production and to avoid downtime due to equipment failures. These checklists need not be complicated but should provide inspection frequencies for the following items: frames, bearings, drives, electrical controls, rams or slides, clutches and brakes, cushions and springs, die carriages, lubrication, and guards, and safeguarding devices.

Frames

Visually inspect the press's frame for cracks and broken parts. Check the fastenings of all brackets, guides, cylinders, covers, and other auxiliary parts. Check the tie-rods and nuts for fractures or stretching. Since the tie-rod's nuts on top of the press may fall if the tie-rod fractures, chain these nuts to the press's frame. A metal strap under each bottom tie-rod's nut will prevent the rod from dropping, should it fail. If a machine is bolted to a foundation, check hold-down fasteners for looseness and fractures.

Bearings

Check crankshaft, pinion-shaft, eccentric-gear, and toggle-link bearings for snug, nonrotating fit and for any loose caps or fastenings. Normally replace journal bearings, especially those supporting gearing, when worn to a shaft looseness of 0.0025 in. per inch (0.0635 mm per mm) of shaft diameter for bearings up to 12 in. (30.5 cm) in diameter. Larger bearings and some toggle-link bearings may require replacement when worn to a looseness of 0.0015 to 0.002 in. per inch (0.038 to 0.05 mm per mm) of diameter. Replace badly scored bearings. Check for proper lubrication.

Examine the slide or ram guide and gib surfaces for dirty or clogged lubricant grooves. Check for proper running clearances. These vary with the type of work but generally should total about 0.002 in. per foot (0.05 mm per meter) of ram's guide length, for any two opposing clearances such as front and back. This is the minimum clearance for cast iron and steel. These materials require more clearance than bronze and steel or brass and steel because they have a greater tendency to "pick up." Tighten all screws and lock nuts to hold the setting.

Check antifriction bearings for proper lubrication. Overlubrication can cause swelling of oil seals and overheating of the shaft. This can result in failure of bindings and shafts. Both this failure and the failure of flywheel bearings can cause unexpected descent of the ram. If lubrication of noisy flywheel bearings fails to silence bearings, replace bearings immediately.

Motor or Power Source and Drive

Properly adjust drive belts to prevent excess slippage or excess loads on the bearings and shaft, which could cause premature failure. Properly adjusted drive belts should slip slightly when the motor first starts but not when the press is operating. Check that all pins, slides, turnbuckles, jack screws, or other means of adjusting the motor are secure. Tighten all hold-down screws. Attach the motor to the press by a chain or wire rope for maximum safety. Inspect and lubricate the motor's shaft bearings.

Check gears for worn, pitted, and broken teeth, and for proper lubrication. Check bores and shafting for worn keys and keyways.

Whenever a crankshaft, or shaft carrying the flywheel or clutch and brake, is removed from a press, inspect it for fatigue cracks. Some companies inspect all drive shafts once a year for cracks, bending, or deformation. Fatigue cracks can be detected by numerous methods such as ultrasonic, radiographic, magnetic, or dye-penetrant techniques.

Have a supervisor monitor the selection of turnover bar operations. Use a separate push button to activate the clutch. Activate the clutch only if the drive motor is deenergized. Also use turnover bars that are spring loaded. They prevent the possibility of leaving the bar in the bar's hole.

Electrical Controls

Check all operating buttons for proper operation. All buttons must be depressed to start the press cycle and released at the end of the stroke. Holding time should be adequate for the operation and tooling involved.

Check for defective lamps in ground detector circuits where provided. Check ground connections on grounded controls or ground-detector connections on ungrounded controls.

Check the physical condition of wiring, relays, rotary-limit switch drives, pressure switches, valves, and other electrical and pneumatic devices. Follow manufacturer's recommendations for preventive maintenance.

Rams or Slides

Check the slide's structure visually for cracks, and check the die's mounting surface for evidence of overload or improper die-mounting damage. Inspect the slide's adjustment lock to be sure the die's setting can be maintained. On a machine with a motorized ram adjustment, check the motor for loose mounting bolts, loose drive chain or gears, excessive grease in the motor, and worn or frayed flexible electrical lead-in wires for motor control. Check that the slide's adjustment limit switches are operating properly.

Pin or chain knockout bars to prevent them from falling. If clearance between the slide's face and a right-to-left stationary support bar is less than 1 in. (2.5 cm) and the bar is within reach from the floor, install finger guards along the top of the bar.

If the ram is counterbalanced by springs, check them for breaks. If the ram is counterbalanced by air, check it for air leaks, air-line restrictions, correct operating pressure, loose piston rods, lubrication, and proper operation of pressure switches. Tighten all brackets. Note any rise in air pressure on the downstroke of the press. Any rise in excess of 20% probably indicates a surge tank filling up with condensate and lubricant. Periodically drain all surge tanks.

Visually check for fatigue cracks in the ram-adjusting screw and the connection. Check that the slide is securely fastened to the adjustment mechanism and that the connection cap is fastened to the connection. Look for evidence of rams being adjusted too high with subsequent interference with the frame. All mechanical power presses are capable of producing an overload force several times the press's tonnage at the top of the stroke as well as at the bottom. Sudden failure of any of the parts that attach the ram to the crank may cause an equally sudden and dangerous dropping of the slide.

Clutches and Brakes

There are several types of clutch and brake units. Each requires special inspection and maintenance.

Dry Friction. All dry-friction clutch units are air engaged and spring released. All brake units, except constant-drag types, are spring set and air released. Any other arrangement is not fail-safe. When the clutch and brake are combined into a single unit operated by a single air cylinder, only one unit can be engaged at a time due to the mechanical interlock. When the clutch is separated from the brake and each has its own air cylinder, both can be engaged at the same time. Prevent this from happening by either restricting the air flow into the clutch and out of the brake, or by limit-switch and/or pressure-switch timing of two air valves. Check to see that both the clutch and brake are not engaged at the same time.

Check the unit for loose fasteners, broken parts, lubrication leaks, air leaks, faulty or loose wiring, excessive accumulation of particles on the friction lining, and broken springs. Replace springs that have changed in free length more than 5%.

During inspection, check the action of the clutch and brake, both at rest and in motion. Travel on the friction disks will indicate the amount of wear. Adjust the disks according to the manufacturer's instructions. If not adjusted properly, they can cause a malfunction. It is also important to properly adjust the brake. There should be little or no coasting of the press's slide (ram) when the brake engages. The clutch and brake should operate smoothly and engage and disengage quickly. If the press is equipped with a brake-monitoring device, check it for proper operation according to the manufacturer's recommendations.

Check sliding surfaces, which keep parts in alignment, for excessive wear that might allow the parts to cock or wedge.

Inspection of the clutch and brake units will readily disclose leaks in the air-cylinder packings and in air glands. Air-line filters, lubricators, and moisture traps will increase the life of these packings and are necessary for safe operation. Check traps and strainers and clean them frequently. Do not clean plastic oil reservoirs on air-line lubricators with solvent. Some plastic reservoirs can be damaged by solvents. Refill lubricators regularly with the type of oil recommended by the press's manufacturer.

Magnetically operated air valves should operate smoothly without sticking or leaking. Valves may stick because of dirt or scale in the air line. Replace a leaky valve's packing. Inspect and clean valves according to the manufacturer's recommendations.

Electrical controls, although usually not part of the clutch and brake unit, affect the operation of the unit. Inspect push buttons, limit switches, relays, and contactors for excessive wear, broken springs, loose parts, loose or broken wires, bent magnetic-field surfaces, badly burned contacts, and dirt. Check circuit-grounding connections for continuity. Replace badly worn contacts. Specifically, inspect the rotary cam's limit-switch drive. Failure of this mechanism may result in dangerous repeat strokes.

A convenient method for checking a clutch/brake system is to measure the stopping time of a press. An increase in stopping time for a given press indicates the need for clutch/brake system maintenance.

Oil-Wet Clutches. Oil-wet clutches are a newer type of air-actuated friction clutch. They have many of the same maintenance characteristics as the dry-friction clutch with the following exceptions: (1) friction surfaces last much longer, (2) units are usually physically smaller because of higher rpm and may contain a set of planetary or other type of gears, and (3) the unit may require an oil cooler and pump. To prevent damage to the unit, lubricating oils should meet the manufacturer's requirements. These units may be serviced by qualified maintenance personnel.

Electric Clutch. The eddy-current electrical clutch has no friction surfaces to maintain. However, it does have slip rings and a special electrical control to maintain the torque and slip characteristics. Proper maintenance consists of lubricating bearings and taking care of electrical apparatus according to the manufacturer's recommendations.

Full-Revolution Clutch Presses. A full-revolution clutch is one which, when tripped, cannot be disengaged until it has completed its cycle. It is known by many names, such as pin, jaw, dog, positive, key, or spline, depending on the type and the manufacturer. Usually associated with this type of clutch is a drag brake on the crankshaft.

Typically, this clutch couples the flywheel to the press's crank by releasing the spring-loaded means of coupling, such as a pin, rolling key, or jaw arrangement. These are normally disengaged by a cam's action that extracts the engaging part, thus disconnecting the flywheel from driving the crank. The crank is held in its disengaged position by a braking system.

Examine the clutch for loose parts, worn pins, worn dogs, broken or weak springs, damaged lubrication seals, and excessive wear in the bearings. Replace worn or broken parts and adjust the clutch to throw out at the top or just before the top center stroke. This adjustment will affect the brake's setting.

On some presses, the clutch is tripped by foot or hand levers or by levers with a spring return. Other methods of tripping the clutch include electric, air, or a combination of air and electric. Examine all elements of the clutch's trip. Sources of trouble include broken or weak springs; worn pins and bushings; loose fasteners; leaking air packings, connections, or valves; loose or broken wires; poor electrical contacts; and defective relays and limit switches. After replacing defective parts, readjust the tripping mechanism and check it for smooth operation.

Drag or Band Brake. The continuous band brake is used on a large majority of full-revolution clutch presses. If the brake is not set or operating properly, the press may repeat and cause a serious accident. Therefore, replace worn, glazed, or oil-soaked brake linings. Rivets should not project above the linings. Remove, inspect, and replace anchor- and drag force-applying bolts, studs, springs, and other parts found defective.

Pneumatic Die Cushions and Springs

Examine pneumatic die cushions and springs for foreign or scrap material between the pressure pad and the bolster. Also check for faulty air packings, air leaks, improper lubrication, and fasteners on the supporting rods or plates.

Rolling-Bolster Die Carriages

Die carriages vary from small, two-position, manually moved die holders to large-capacity, eight-wheeled, four-directional holders. The latter are self-powered to change tracks.

The most common source of power for rolling-bolster die carriages is pneumatic; a less common power source is electricity. Inspect all hydraulic and air valves, hydraulic and air cylinders, and air or electric motors for normal operation. Inspect air-line filters and oilers. Inspect and lubricate all gears and bearings. Inspect all springs and latches for wear and breakage. Also inspect locating keys in the bed, retractable locating pins in the carriage, and locating pin holes in the floor, bed, and tracks.

Lubrication

Proper press and air-line lubrication is essential. Failures producing accidents and downtime are frequently directly traceable to either lack of lubrication or overlubrication. It is well known that lack of lubrication cannot be tolerated in machine parts. The results of overlubrication are less well known. Excess lubrication to flywheel and shaft bearings in the vicinity dry-friction clutch and brake linings is undesirable. It causes a loss of work capacity, and prevents stopping in emergencies.

Overlubrication of air cylinders can lead to sluggish clutch and brake action. Improperly mounted clutch and brake surge tanks will accumulate oil and water. This results in increased clutch slippage and wear. For specific lubrication information, refer to the press manufacturer's service manual.

Guards and Safeguarding Devices

Cover all gearing, belting, or other drive parts that can be accidentally contacted. Many safeguarding devices are synchronized with the action of the press. Because most of these devices will go out of adjustment through wear and vibration, they require periodic checking.

Wire ropes, leather straps, or steel springs used as parts of safeguarding devices will in time need replacing because of wear. In such cases, use nothing but the proper replacement parts. Follow the manufacturer's recommendations for maintenance and adjustment.

Keep all guards and covers on a power press in place. Properly adjust them after completion of each inspection and any necessary repairs. Check each press for all modes of operation before releasing for production.

METAL SHEARS

The following sections cover safeguarding methods and safe operating procedures for power squaring shears and alligator shears.

Power Squaring Shears

Equip the power squaring shear with safeguarding that will (1) prevent operators from placing their hands into the point of operation, (2) prevent or stop the operation of the shear if any part of the operator's body approaches the point of operation, or (3) provide awareness to operating personnel upon entry into a hazardous area. Follow the safeguarding requirements in ANSI B11.4, *Safety Requirements for the Construction, Care, and the Use of Shears.* According to ANSI B11.4, the point of operation includes the area between the upper and lower blades and the area between the hold-down, clamping mechanism, and the shear table.

Safeguarding should allow operators to see clearly into the point of operation to position material for shearing to a scribed line. Removal of the safeguarding is normally necessary for changing or adjusting the blades. Reinstall the safeguarding when this is completed. If a guard is used, provide sufficient clearance to allow the material to be fed. Normally, a recognized guideline for clearance is double the metal thickness of the material being sheared.

The design of the fixed guard should meet the requirements of ANSI B11.4. (See Figure 1 and Table 1 from ANSI B11.4. This table provides dimensional guidelines for a guard that will prevent operators from placing their hands into the point of operation.) When it becomes impractical to adhere to the guarding dimensions given in Table 1, ANSI B11.4 suggests using an awareness barrier. The dimensions for this barrier are in Figure 2 and Table 2 of ANSI B11.4. The design should assure that the barrier's movable sections are heavy enough so that operators would be aware of their hands entering the safeguarding. Operators should know, however, that this safeguarding may not prevent them from forcing their hands into the point of operation.

On shears with a throat, or end guard, provide a guard. It may be removed to provide for slitting material longer than the shear. However, it must be replaced when the slitting work is completed.

Provide a work chute or conveyor to discourage or eliminate the need for employees to be at the rear of the shear while it is being operated (Figure 23–24). Position no one at the rear of the shear; that is, within the area of moving machine parts.

New shears should be manufactured to comply with the construction and safeguarding requirements of ANSI B11.4. Users should update machines already in the field to meet these requirements (Figure 23–25).

Figure 23–24. Material transfer conveyors are used at the rear of shears to eliminate the need for personnel to enter this area to remove the product. (Courtesy Cincinnati Inc.)

Figure 23–25. This large-capacity plate shear is safeguarded with an awareness barrier at the point of operation. (Courtesy Cincinnati Inc.)

Some suggested guidelines for the safe operation of squaring shears include the following:

- Be familiar with the shear—its capacity, controls, operating modes, and safeguarding.
- Properly install safeguarding.
- Keep blades sharp and clearance correct.
- Keep clamping mechanism/hold-downs operating satisfactorily.
- Keep work area clean and free of obstructions in the machine area.
- Keep shear's table free of loose tools and materials.
- Keep hand tools and personal protective devices available, and use them. Wear safety glasses, gloves, protective footwear, and snug-fitting clothes.
- Keep hands out of the point of operation.
- Do not place hands between the material and shear table.
- Use the next piece to be sheared or a tool or pry bar to move small pieces that are on the shear's table and beyond the safeguarding.
- When leaving the shear, turn off the power and have the controls inoperative.
- Make certain all personnel are away from the shear's table before running the machine.
- Keep alert—pay attention to the job.
- Always turn the power off when servicing, maintaining, or removing jammed material.

At the start of each shift, check the shear for the following items:

- safeguarding at the point of operation properly adjusted
- pinch-point guarding properly installed
- operator station working properly
- operating modes functioning properly

- ram starting and stopping properly
- warning plates clean and easy to read
- electrical wiring in good condition
- caution color coding in good condition and clearly visible
- auxiliary equipment checked and working properly
- hand tools and personal protective equipment in good order and readily available
- safety manuals or operator manuals available
- normal maintenance work completed

Alligator Shears

Alligator shears perform a variety of cutoff operations. Their principal use is for cutting rods and bar stock to length.

Alligator shears can operate continuously. Therefore, under continuous operation, the operator must be trained to time movements with the opening and closing of the cutter. Because the machine is relatively simple and comparatively slow in its movement, management and operators often disregard this machine's hazards. Consequently, alligator shears are responsible for far more injuries than their inherent hazards or frequency of use warrants.

The principal hazards from operating an alligator shear include the following:

- finger or hand amputations at the shear point
- cut fingers
- hand or arm lacerations from handling material
- eye injuries caused from flying bits of metal
- foot and leg injuries from falling material
- injury to the operator or damage to the machine from failure to maintain the shear in good condition.

If possible, build a long bench to the right or left of the shear, depending upon the type of machine. The material should slide along the bench and through the cutter. Because the ragged edges are hazardous to handle, use care in piling the material on the bench.

Avoid finger and hand injuries by installing an adjustable guard. The wide variety of sizes and shapes of material to be cut makes it difficult at times to closely guard the point of operation. However, such a guard can often be used. When it is, set it far enough from the knife area, to prevent the fingers from entering this danger zone (Figure 23–26).

When stock size is such that the end held by operators may fly up and strike them, use hold-down guards or bar. They can be adjusted to fit any type of shear.

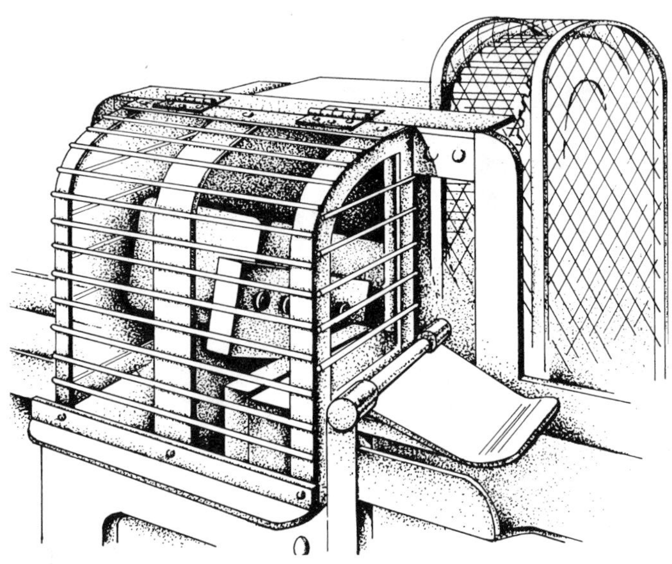

Figure 23–26. A well-designed guard for an alligator shear should permit easy maintenance and adjustment. Hinged section of the guard should be interlocked electrically to prevent shear operation if the hinged section is not in place. (Courtesy Jones & Laughlin Steel Corp.)

Keep material to be cut within the capacity of the machine. Do not attempt to cut hardened steel. Such action can result in damage to the machine and injury to the operator.

PRESS BRAKES

A power press brake is also sometimes called a bending brake or a brake press. Over the years, its design has evolved from the hand or folder brake because of the need for a power machine with enough capacity to bend thick sheet and heavy plate products. The primary function of the press brake is to cold form angles, channels, and curved shapes in plates, strip, or sheet metal stock. Press brakes can also be used for punching, trimming, embossing, corrugating, and notching, when manufactured and arranged to do so, even though these operations are considered power press operations.

Power press brake beds are typically long and narrow and are located in front of, and often extend beyond the machine side frames. The frames are gapped (cut out) to permit full-length use of the bed and ram. The piece-part component typically extends in front of the press brake and moves during the bending operation. Both the bed (or lower die holder) and the ram are equipped with a die-clamping arrangement along their full length to accept a standardized die tongue. Press brake beds are often equipped with an adjustable die holder that provides for aligning and adjusting the upper and lower dies. Backgauges or material-position gauges and stops in the front or rear are used with power press brakes to gauge the distance from the edge of the piece-part component's blank to the forming or bend line.

Follow ANSI B11.3, *Safety Requirements for the Construction, Care and Use of Power Press Brakes.* Consult appropriate regulatory agencies along with this standard when working with press brakes.

Power press brakes have been classified into two basic categories: general-purpose press brake and special-purpose press brake. General-purpose press brakes, both mechanical and hydraulic, are operated by one individual with a single operating control station (Figure 23–27). Special-purpose press brakes include all other types having mechanical, hydraulic, and other drive arrangements.

Planning the Production System

A power press brake is but one part of a production system. All parts of the system—power press brake, tools, feeding, and safeguarding components, operating personnel—must be brought together to perform any metalworking operation on a piece-part component. This can also be called a human factors-engineering approach to provide for the most efficient and safest method of performing a piece-part-bending operation.

The power press brake is the power component of such a system. Depending on the tooling component selected by the user, press brakes can bend, form, notch, punch, pierce, or perform other operations on the piece-part component. The piece-part component and the product being produced determine the feeding component of the production system. Feeding can be either mechanical or manual. Included in this element of the system is removal of parts and scrap.

The component that completes a functioning production system is the safeguarding component. Before selecting a suitable safeguarding component, the user should complete a thorough risk assessment, based on all the elements of the production system. Each new combination of production system elements requires that the user perform a new risk assessment to select a suitable safeguarding component.

A safe combination of components for one production system may not be a safe combination of components for another piece-part production system. Also, it may be necessary to change more than one component to provide a safe piece-part-bending production system, once it is determined that a change must be made.

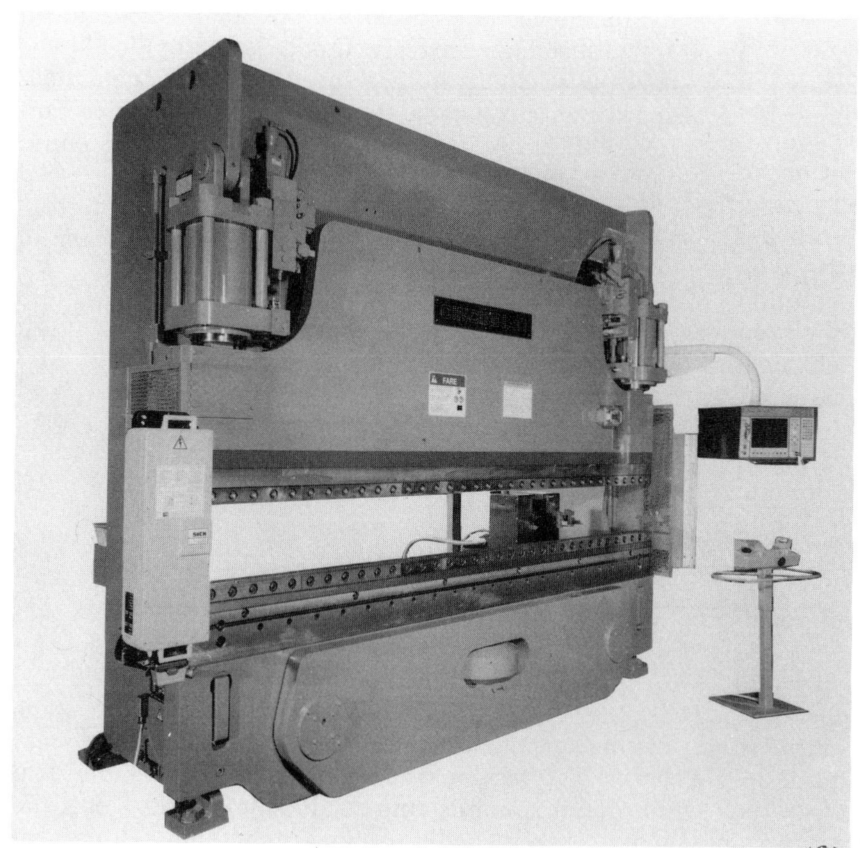

Figure 23–27. This hydraulic press brake has a pedestal-mounted palm button, operator control station, and photoelectric presence-sensing device for point-of-operation safeguarding. Note the computer control for programming the ram motion, backgauge, safeguarding device, and operator controls. (Courtesy Cincinnati Inc.)

Mechanical Press Brakes

A limited range of strokes is available in mechanical press brakes. Smaller units may have as little as a 1½ -in. (3.8-cm) stroke. Large press brakes can have as much as a 6-in. (15-cm) stroke or longer. Ram position is adjustable to accommodate the closed height of dies. This is accomplished by changing the length of the connections from the drive to the ram through the use of die-height adjustment screws.

Hydraulic Press Brakes

On a hydraulic press brake (Figure 23–27), stroke length is variable and can exceed 24 in. (61 cm). Speed changes, from high-speed advance to low-speed press, and upper and lower limits of ram travel, are generally established by limit switches. Operating strokes per minute can approach that of a mechanical press brake of equal capacity due to their variable stroke length. Rated tonnage can be exerted through the full downstroke on a hydraulic press brake.

General-Purpose Press Brakes

General-purpose press brakes are designed and built to be operated by one person, who controls the speed and movement of the ram by skillful use of the operator's control, usually a mechanical foot pedal. The ability and skill of the operator to control the speed of the ram permits slow bending of wide sheets, using general-purpose dies, without fast "whip up" of the extended edge of the sheet. Precise ram speed control is also required to permit the operator to control the ram to a partially closed position for line gauging, that is, bending to a previously scribed line. On single-speed mechanical press brakes, this is accomplished by slipping the mechanically actuated partial-revolution friction clutch to bring the ram to a partially closed position. Variable-speed and two-speed mechanical drive units and general-purpose hydraulic press brake units permit the same type of control. Operating a press brake at reduced speed (1) makes handling of the piece-part component by the operator easier and (2) can minimize the exposure of the operator to sheet or piece-part whip up.

Stroking control on a general-purpose mechanical press brake is managed by a foot pedal. A foot pedal-operated machine should be operated only from a safe distance. Determine the safe distance by the size and shape of the piece-part, unless point-of-operation safeguarding is provided. Position a foot control so the operator is not able to reach into the point of operation, unless safeguarding is provided. Locate the foot pedal above a "step-high" position to minimize the chance of operators accidentally stepping on it. Adjust the foot pedal (1) to require enough force to avoid accidentally running the machine and (2) to return the linkage to its normal Off position. Stroking control on a hydraulic press brake may be either by foot control or two-hand operator station.

Special-Purpose Press Brakes

Special-purpose mechanical or hydraulic press brakes can be constructed with many operational features or stroking options. The user must select a press brake component with the features that are suitable for its safe intended use in each and every piece-part operation.

Like the general-purpose press brake component, the special-purpose press brake can be used to bend, form, notch, punch, and pierce, if it is machined and constructed to do so. Special-purpose press brake components, however, can be operated by one or more operators. Each operator should have an operator control station appropriate to the piece-part production system in use. In this way, each operator and helper is able to exercise concurrent control of the press brake's ram cycle by activating an operator control station.

Stroking Controls

A variety of stroking controls and drive options are available for special-purpose press brakes, such as hydraulic-electric, air-electric, hydraulic-mechanical, single-speed, and two-speed brakes. Hydraulic-electric controls generally are two-speed brakes having a high-speed ram advance, a slow- speed work-forming portion of the stroke, followed by a high-speed ram return. Limit switches are used to control the speed changeover points in the stroke. Hydraulic-electric-controlled press brakes have an infinite number of stroke lengths within their range.

Air-electric clutch/brake controls are generally used on mechanical press brakes, both single speed and two speed. The mechanical two-speed drive is similar to the hydraulic stroke's control. However, the stroke's length is constant. The changeover point from high speed to slow speed is also adjustable.

Operating Modes

Special-purpose press brakes have various employer/supervisor-controlled modes of operation. They are designated in the ANSI B11.3, *Safety Requirements for the Construction, Care, and Use of Power Press Brakes*, as Off, Inch, Single Stroke, and Continuous. Off shuts off the operator's control station and stops the press brake. The person who sets up the die uses the Inch control; it is not used in production. In this mode, the ram may be inched down and up but only with a two-palm button, or a single control, firmly secured and located a safe distance from the point of operation.

The single-stroke mode is the standard operating mode during production. It can be initiated by the foot control or by a two-palm button control. The set-up person determines which of the controls to use after considering the various components of the production system in use. In this mode, the press brake is under the operator's control in the descent portion of the stroke. It automatically returns to the top position where it must stop. The operator's control station must be deactivated and then reactivated in order to initiate the next stroke.

In continuous operation control, used only with automatic feeds, the press brake does not stop after each stroke. It operates continuously until the STOP button is activated. There are several methods of initiating operation in this mode. Each method requires a positive, separate action on the part of the operator to minimize inadvertently placing the special-purpose press brake in the continuous-operating mode.

Tooling

Press brake tools, or dies, are generally divided into two categories: general-purpose tools and special-purpose tools. General-purpose tools are widely available, universal dies, used to perform bending and forming operations on a wide variety of piece-parts and products. Special-purpose tools are designed and built to perform specialized work on a specific piece-part of a product. Many special-purpose tools are designed to eliminate the need for an operator to hold the piece-part component's blank while one or more forming operations are performed on it during the ram's stroke of the press brake component.

When operating a general-purpose press brake, protect the operator's hands by locating them along the extended edge of the piece-part component's blank at a safe distance from the point of operation. If a safe distance, and point-of-operation safeguarding are used, the production system should involve the following three elements: (1) The operator must support the workpiece with both hands; (2) The operator must use material-position backgauges, and their stops must be large enough to keep the workpiece from slipping past them; and (3) Operators must be instructed to remove their foot from the foot pedal after each stroke of the ram.

The operator's controls should only be in place or operable when the operator intends to start the machine. At other times, the foot pedal should be removed and/or the operating linkage locked to prevent the machine from starting.

Do not use a general-purpose press brake to perform operations such as narrow pieces, angles, channels, Z-Bars, etc. If a general-purpose press brake is used because a special-purpose press brake with two-hand controls is unavailable, provide safeguarding such as pullbacks or restraints for feeding by hand. Tool setups for these types of parts require arrangements that support the part before and after it is formed. This prevents the part from falling behind the die and the operator from reaching between the dies for the fallen part. Material-position gauges that locate front and back edges are frequently required so operators do not place their hands near the point of operation during forming. If forming is done to a scribed line, use supports and a back stop to help operators maintain control over the part, and to prevent their hands from entering the die. Their hands could enter the die if they follow a part that could otherwise slip past the scribed line.

When the piece-part component is large and extends some distance in front of the die, the operator must hold the sheet so that hands or fingers are not exposed (1) to injury from impact with the moving piece-part, or (2) to pinch points or the point of operation while the bend is being made.

When operating a special-purpose press brake, protect the operator with either a two-hand operation control station or a safeguarded foot control (Figure 23–28). Depending upon the piece-part bending system, use a presence-sensing device, movable barrier device, or other means of safeguarding. A presence-sensing device may be useful for large or small parts for point-of-operation safeguarding. However, the piece-part's movement or requirements for holding during forming should not interfere with its function.

Special-purpose mechanical press brakes have air-electric, clutch-control mechanisms that provide a base for adapting many means of safeguarding. It is easier for the employer/user to provide and use many safeguarding means not available for a general-purpose press brake. Although some modifications are costly, they will (1) extend the use of the machine by permitting point-of-operation safeguarding devices, with piece-part bending systems, (2) provide protection for operators, (3) reduce costs from injuries, and (4) increase production.

In addition to the foregoing suggestions for safeguarding, observe the following precautions:

- Make periodic and thorough checks of the entire machine to assure that all parts are in good working condition.
- Retool or discard worn punches and dies; using worn tools causes excessive loads on press brakes.
- Provide barrier guards or devices wherever possible and provide hand-feeding tools for inserting or removing stock (Figure 23–28).
- Establish safe operating procedures and make sure that all operators follow the procedures.
- Do not allow personnel to bypass safety features.

Figure 23–28. *Top:* Hand tools are used for inserting and removing small piece parts. *Bottom:* The operator uses a two-hand control station as a point-of-operation safeguarding device. Note the use of partial barriers that guard the unused portion of the brake point of operation. (Courtesy Cincinnati Inc.)

- Provide adequate floor space for each machine so each worker works without interference from or to other workers.
- Provide adequate illumination for all machines and auxiliary lighting where necessary.
- Locate worktables or receptacles that may be required to store material being processed so they do not interfere with the operator's freedom of movement or access to the foot pedal or to the machine's controls. These tables or receptacles should be stationary and constructed so that material on them cannot be knocked to the floor and possibly activate the press brake's control.
- Assure adequate clearance for the changing shape of the stock being processed before ram motion is started.
- Assure that the material-position backgauge fingers are high enough to minimize the possibility of the material being gauged from slipping over them during operation.
- Select only personnel in good physical condition with excellent vision and hearing as operators.
- Never permit untrained personnel to operate press brakes.
- Require the operator to "test operate" the press brake a few times at the start of each work period and after each setup to be sure that it is operating properly.

Responsibility for Guarding and Safeguarding

The following hazards associated with power brakes require installing protective covers or other means of protecting operators and others in the vicinity:

- rotating components, such as flywheels, gears, sheaves, and shafts in close proximity to operating personnel
- inrunning pinch points associated with meshing gears, belts, and chains
- pinch points between the moving and stationary components of the power press brakes or auxiliary equipment

The manufacturer shall warn against the hazard, if the hazard cannot be eliminated or otherwise safeguarded.

ANSI B11.3 describes two major areas of hazards associated with power press brakes: (1) those related to the design and manufacture of the brake and (2) those associated with the point of operation. The manufacturing hazards are described in B11.3 Section 4.2.2.

Section 6.1.4, Safeguarding the Point of Operation, states:

> It shall be the responsibility of the employer, after selecting the tooling and the specific type of power press brake for producing a piece-part, to evaluate that operation before the piece-part is worked (bent, etc.) and to provide point-of-operation safeguarding according to provisions of section 6.1.4 (1).

Methods used to provide point-of-operation safeguarding for press brake-forming work include the following:

1. a point-of-operation guard, such as fixed barriers, die guards, and other means that do not allow access to the point of operation
2. a point-of-operation device, such as presence-sensing devices, gates or movable barriers, pullbacks, restraints, and two-hand operator controls
3. safe-distance methods when guards and/or devices cannot be used

Definitions

Due to differences between press brake machines and power presses, the following definitions are presented as applied specifically to press brakes (Source: B11.3, Power Press Brakes).

Brake

The mechanism used to stop the motion of the power press brake ram; when engaged, it holds the power press brake ram in a stopped position.

Clutch

An assembly that connects the flywheel to the crankshaft either directly or through a gear train; when engaged, it imparts motion to the power press brake ram.

Part-Revolution Clutch. A type of clutch that can be disengaged at any point before the drive mechanism (usually a crankshaft) has completed a full revolution and the press brake ram a full stroke.

Combined Stroking Control Systems

Two independent control systems on the same power press brake, only one of which is operable at a time.

Concurrent

Acting in conjunction, this term is used to describe a situation wherein two or more controls exist in an operated condition at the same time. The use of the term concurrent is not intended to imply that the individual two-hand controls must be actuated simultaneously.

Connection

The part of the power press brake that transmits motion and force from the revolving crank or eccentric to the power press brake ram. (Also see Pitman.)

Continuous

Uninterrupted multiple strokes of the ram without intervening stops (or other clutch control action) at the end of the individual stroke.

Counterbalance

The mechanism that is used to balance or support the weight of the connections, ram, ram attachments, and installed tooling components.

Cover Guard

An enclosure that covers moving machine parts (excluding point of operation).

Cycle

A full cycle is the complete movement of the ram from its open position through its closed position and back to its previous open position.

A half cycle is the movement of the ram from its open position to its closed position.

Device (Point of Operation)

A press brake control or attachment that does one of the following:

1. restrains the operator from reaching into the point of operation
2. inhibits normal press brake operation if the operator's hands are within the point of operation
3. automatically withdraws the operator's hands if they are within the point of operation as the dies close
4. maintains or restrains the operator or his or her hands, during the closing portion of the ram, a safe distance from the point of operation.

Hostage Control Device. A device designed, constructed, and arranged on a special-purpose mechanical and/or hydraulic power press brake to restrain and maintain the operator(s) at a control station located a safe distance from the point of operation or maintained by hand during the closing portion of the stroke.

Presence-Sensing Device. A device designed, constructed, and arranged to create a sensing field or area. When used with a special-purpose mechanical or hydraulic power press brake, it deactivates the ram motion control of the power press brake when an operator's hand or any other body part is within the sensing field or area.

Pullback Device. When used on general- or special-purpose mechanical and/or hydraulic power press brakes, a pullback device is a mechanism attached to the operator's hands and connected to the moving portion of the die or ram of the power press brake. When properly adjusted, it withdraws the operator's hands if they are inadvertently within the point of operation as the dies close.

Restraint Device. When used in conjunction with a general- or special-purpose mechanical and/or hydraulic power press brake, a restraint device is a mechanism, including attachments for the operator's hands, that when anchored and adjusted, inhibits the operator's hands from entering the point of operation.

Two-Hand Control Device. A stroke control-actuating means on a special-purpose mechanical and/or hydraulic power press brake requiring (1) the concurrent use of the operator's hands to start the ram movement and (2) concurrent pressure from the operator's hands during a substantial part of the die-closing portion of the power press brake stroke.

Die(s)

A term commonly used to describe the complete die, consisting of an upper and lower component, also known as tooling, which are the components used in a power press brake for bending or forming the piece-part material.

General-Purpose Dies. The universal dies used to perform bending and forming operations on a variety of piece-parts or products.

Special-Purpose Dies. Designed to perform work not normally done on general-purpose dies, or for performing a common bending or forming operation that eliminates piece-part whip up or the need for the power press brake operator to handhold the piece-part component.

Die Builder/Employer

Any person who builds dies for power press brakes. This is the person or entity totally responsible for the initial design of the die. This definition does not apply to a person who merely fabricates a die from detailed drawings furnished by another company or entity.

Die Holder

The heavy plate or rail to which the lower portion of the die is attached.

Die Set

A tool holder held in alignment by guide posts and bushings and consisting of a lower shoe, an upper shoe, guide posts, and bushings.

Die Setter

An individual who places or removes tooling components in or from a power press brake; and who, as a part of the job duties, makes the necessary adjustments for safe and proper functioning of the tooling in conjunction with the piece-part bending operation to be performed.

Die Setting

The process of placing or removing tooling components in or from a power press brake and the process of adjusting the dies, material position stops (backgauges), tooling, and safeguarding means, if required for bending the piece-part and for the operator to work properly and safely.

Die Space

The adjustable distance between the bed and the ram. The employer/user installs the tooling required for a specific piece-part bending operation into this space. Once the tooling has been installed and the proper adjustment for the material thickness made through the die space adjustment mechanism, it creates a point of operation, that is, the place where the actual piece-part bending will take place.

Eccentric

The offset portion of the crankshaft that governs the stroke or distance the ram moves on a mechanical power press brake.

Ejector

A mechanism for removing work or material from between the dies.

Feeding

Automatic Feeding. Feeding wherein the material or piece-part being processed is placed within, or removed from, the point of operation by methods or means not requiring action by an operator on each stroke of the power press brake.

Manual Feeding. Feeding wherein the material or piece-part being processed is located and placed in the point of operation by the operator after the power press brake has completed each stroke. Manual feeding, for inserting or removing piece-parts, shall only be used when the hands or other bodily members can be kept a safe distance from the point of operation.

Hand Feeding. A type of manual feeding wherein the material is placed within, and processed parts removed from, the point of operation by use of a hand-feeding tool.

Push or Slide Feeding (Hand Operated). A pusher or slide can be used to feed a blank under the upper die and withdraw it after the operation is performed. The pusher or slide may have a machined nest to fit the shape of the part. If the part neither drops through the die nor ejects by other means, it can be withdrawn by the pusher or slide.

Semiautomatic Feeding. Feeding wherein the material or piece-part being processed is placed within, or removed from, the point of operation by an auxiliary means controlled by the operator on each stroke of the power press brake.

Foot Control

The foot-operated control mechanism (other than mechanical foot pedal) designed to control the movement of the ram on mechanical, hydraulic, or special-purpose power press brakes.

Foot Pedal (Mechanical)

The foot-operated lever designed to operate the mechanical linkage that requires raising the foot on the mechanical foot pedal and applying a significant amount of pressure and travel to actuate and engage the clutch, and to disengage a mechanical power press brake while the mechanical foot pedal is depressed.

Foot Treadle Bar

A bar that is moved in a vertical direction when depressed by the foot of the operator at any point along its length. This bar is attached to two lever arms pivoted from the outside surface of the frame and is connected through linkage to the clutch and brake.

Gate or Movable Barrier Device

Constructed or arranged on a special-purpose mechanical and/or hydraulic power press brake, it encloses the point of operation before the press brake ram can be actuated. This device interlocks in the press brake control system to inhibit or stop activation of the ram movement whenever the device is prevented from being in a full-close position.

Gibs

The parts used for guiding the ram.

Guard

A barrier that prevents entry of the operator's hands, or any other body part, into the point of operation.

Holding Distance (Means of Operating)

The closing distance traveled by the ram during which time the operator(s) is compelled to hold the operating means depressed. Release of an operating means, before this holding distance is traveled, will stop the ram motion, whereas release of an operating means after this holding distance is reached will allow the ram motion to automatically continue as determined by the selected mode of operation.

Hostage Operator Control Station

A type of operator control station which physically maintains and restrains the operator at a sufficient distance from the point of operation. A sufficient distance is one where the operator cannot reach near the point of operation during the closing portion of the power press brake cycle. The use of the term "near the point of operation" means no closer than the distance referred to as "safe distance."

Housing

The stationary portion of the power press brake structure on which the ram is guided and to which the bed, crown, and drive are attached.

Human Factors Engineering

As it applies to this chapter, an analysis of the piece-part bending system to determine the most efficient and safest method of performing the operation. (See also Chapter 16, Ergonomics Programs in the *Administration & Programs* volume.) A piece-part bending system is an orderly arrangement of components that perform a specific task to a given piece-part. The components of a piece-part bending system are as follows:

1. the specific piece-part
2. the tooling designed or determined to perform the required bend or function
3. the power press brake utilized along with its operating control stations
4. the power press brake operator's function for loading, operating, and unloading the piece-part
5. the safeguarding means itself, which is the last but most important component necessary within this system.

Maintenance Personnel

Individuals who care for, inspect, and maintain mechanical power press machines or press brakes.

Material Position Gauge

A stop against which the material is placed to properly locate it at the point of operation.

Not Readily Removable

Refers to using fastening procedures requiring effort and time to remove rather than quick-release fasteners such as wing nuts, etc.

Operating Means

The mechanism depressed by the operator(s) that controls electric current or air flow (etc.), but excludes a mechanical foot pedal and other directly controlled mechanical linkage systems.

Operator

An individual who performs production work and who controls the movement of the ram.

Operator's Control Station

The control mechanism provided to each operator for the purpose of initiating the starting and stopping of the power press brake ram stroking. The control station may be fixed to the power press brake or it may be portable.

Pitman

That portion of the connection assembly that couples the eccentric to the ram.

Point of Operation

The area of the tooling or dies where material is positioned and work is performed.

Ram (Slide)

The powered movable portion of the power press brake structure, with die attachment surface, which imparts the pressing load through dies, piece-part, and against the stationary portion of the press brake bed.

Run

A single stroke or continuous stroking of a power press brake.

Safe Distance

A minimum distance between the press brake operator's hand (or hands) and the point of operation.

Shut Height

The distance between the bed and the ram when the ram is at the bottom of its stroke.

Shut Height Adjustment Screws

The screws used to set the shut height of the power press brake.

Single Stroke

One complete stroke of the ram, usually from a full-open position through a closed position back to a full-open position.

Stroke (Up or Down)

The vertical movement of the power press brake ram during half of the cycle, from full-open to full-closed position or vice versa.

Stroking Selector

That part of the control that determines the type of ram stroking when the operating means is actuated. Stroking selectors are normally furnished on hydraulic and special-purpose mechanical power press brakes.

SUMMARY

- Complete safety of power presses depends on adequately safeguarding the point of operation, training press operators and setup/maintenance personnel, and enforcing safe working practices.
- When determining point-of-operation safeguards, consider all hazards in the die's space that may crush, cut, punch, sever, or otherwise injure workers. A safeguarding device controls access to the point of operation and can be either press controlling, operator controlling, or a combination of the two.
- Many auxiliary mechanisms to protect power press operators are available either from manufacturers or from a company's own shop.
- A significant number of power press injuries occur when work is handled. Feeding and ejecting mechanisms can help to eliminate handling work and reduce exposure to those hazards.
- Kick presses present serious hazards to workers and must be carefully guarded and well lighted to prevent injuries. Kick press operators should also wear eye protection and should guard against fatigue and strain, which can affect their performance.
- Electrical controls on power presses should be properly designed, applied, and installed as an important part of press safety. This is particularly true when a two-hand control device or two-hand tripping device is used for point-of-operation guarding.
- Each power press should also have a stop control, stroking selector, inch control, and stroking control system. The clutch/brake control must automatically deactivate in case of loss of electrical power or proper air pressure.
- Workers often suffer injuries when handling, setting up, and removing power press dies. Only experienced personnel should perform these operations.
- Good inspection and maintenance of power presses are essential in achieving a company's safety goals and productivity levels. Workers must be completely familiar with the construction and operation of the equipment for which they are responsible.
- During maintenance work, the machine should be locked out to prevent accidental operation or electrical shock. When employees must work on a machine with the power on and guards removed, they should not place any part of their body in the path of any movable part of the press.
- Management should set up a checklist that outlines the frequency of inspection and maintenance for each press.
- Point-of-operation safeguards and backgauges for power press brakes should be specially designed for each press braking task. Operators must be able to control the speed and stroke of press brakes and have the correct hand tools to work the press brake safely.

REFERENCES

American National Standards Institute, 11 West 42nd Street, New York, NY 10036.

Safety Requirements for the Construction, Care, and Use of Hydraulic Power Presses, ANSI B11.2–1995 (R2000).

Safety Requirements for the Construction, Care, and Use of Mechanical Power Presses, ANSI B11.1–1988.(R1994)

Safety Requirements for the Construction, Care, and Use of Power Press Brakes, ANSI B11.3–1982 (R1994).

Safety Requirements for the Construction, Care, and Use of Shears, ANSI B11.4–1993.

National Fire Protection Association, 1 Batterymarch Park, Quincy, MA 02269.

Electrical Standard for Electrical Machinery, NFPA 79, 1997.

National Electrical Code,® NFPA 70, 1996.

National Safety Council, 1121 Spring Lake Dr., Itasca, IL 60143-3201.

Power Press Safety Manual, 4th ed., 1989.

Safeguarding Concepts Illustrated, 6th ed., 1993.

Occupational Safety and Health Data Sheets (available in the Council Library):

Alligator Shears, 12304–0213, 1990.

Electrical Controls for Mechanical Power Presses, 12304–0624, 1983.

Handling Steel Plates for Fabrication, 12304–0565, 1990.

Inspection and Maintenance of Mechanical Power Presses, 12304–0603, 1986.

Kick (Foot) Presses, 12304–0363, 1990.

Metal Cutting Shears, 12304–0328, 1993.

Concepts of Mechanical Power Press Point-of-Operation Safeguarding, 12304–0710, 1993.

Power Press Safeguarding: Presence-Sensing Devices, 12304–0711, 1993.

Power Press Safeguarding: Pullbacks and Restraint Devices, 12304–0713, 1986.

Power Press Safeguarding: Two-Hand Tripping Devices, 12304–0714, 1990.

Mechanical Power Press Safeguarding: Movable Barrier Devices, 12304–0712, 1993.
Press Brakes, 12304–0419, 1993.
Scrap Ballers, 12304–0611, 1989.
Setting Up and Removing Power Press Dies, 12304–0211, 1991.

United Auto Workers, 8000 East Jefferson Avenue, Detroit, MI 48214. (General.)

Wilson FW. (ed.). *Handbook of Fixture Design.* New York: McGraw-Hill Book Co., 1962.

REVIEW QUESTIONS

1. What does the safety of power presses depend on?

2. Briefly describe the antirepeat function of the clutch/brake control system.

3. Identify and describe the two types of clutches.

4. What is a presence-sensing device?

5. What is a pinchpoint?

6. What is the "point of operation"?

7. List the two basic categories of safeguarding the point of operation.
a.
b.

8. Name three of the four types of guards.
a.
b.
c.

9. Where is the two-hand tripping device used?

10. If more than one operator is used on a machine, how many pullback devices are necessary?

11. What is a kick press, and give two usages.
a.
b.

12. What is the recommended safeguard for a kick press?

13. Setting up, removal, and handling of power press dies can be very hazardous. Identify three of the six listed typical injuries that can be suffered by maintenance personnel.
a.
b.
c.

14. How should a power squaring shear be guarded?

15. What type of guard is commonly used on alligator shears?

16. What is the function of a power press brake?

17. When is the "safe distance" method used on power press brakes?

18. What is the "combined stroking control system"?

19. What is a cover guard?

24

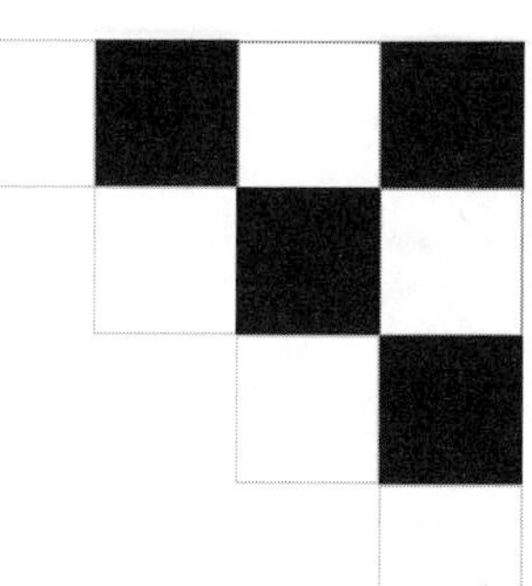

Hot Working of Metals

Revised by

Philip E. Hagan, MPH, CIH, CHMM, CHCM

This chapter discusses how to control materials handling hazards and environmental stresses (dust, fumes, gases, heat, and noise) that are present in foundries and permanent mold and die-casting facilities. Also covered are safeguarding methods and safe operating practices for forging and hot metal stamping operations. A survey of the use of nondestructive testing methods supplements these discussions.

An understanding of and knowledge about the environmental stresses are important, if the safety and health of employees are to meet regulatory standards. If the operation is safely implemented, very specific precautions and procedures must be followed and personal protective equipment must be worn. This chapter will cover the following topics:

- hazardous materials in foundries
- work environment, including housekeeping, ventilation, inspection, and maintenance
- substance handling and storage
- maintenance and repair of cupolas
- crucible storage and handling
- oven safety and inspection
- foundry production equipment
- cleaning and finishing foundry products
- hammer safety, inspection, and maintenance
- safe handling of dies and inspection and maintenance of forging upsetters
- basic precautions for forging presses
- nondestructive testing

HEALTH HAZARDS IN FOUNDRIES

In foundries and in permanent mold and die-casting facilities, metals are formed into finished castings. The overall foundry operation usually includes a pattern shop and sometimes a machine shop.

Safety and health professionals should refer to the National Safety Council's (NSC) *Fundamentals of Industrial Hygiene*, 4th edition, for principles and practices used to recognize, evaluate, and control health hazards in foundries. This book gives details on toxic and flammable hazards, general and local ventilation, and specific problems such as silicosis, dermatosis, and radioactivity. Before designing and installing equipment to control these hazards, consult with persons who are technically familiar with these hazards. Also consult these technical experts to test procedures and analyze new processes for the foundry's safety and health program. Besides NSC, other sources of help include the American Foundrymen's Society, the National Institute of Occupational Safety and Health, the American Industrial Hygiene Association, the American Conference of Governmental Industrial Hygienists (ACGIH), insurance carriers, safety and health consultants, and state or provincial and local governmental departments of industrial safety and health.

Hazardous Materials

Dust, solvents, and other materials present a health hazard in foundries. Their hazards and mode of production in foundry operations are discussed below.

Dust

Dust is generated in many foundry processes and presents a two-fold problem: (1) cleaning to remove deposits and (2) control at the point of origin to prevent further dispersion and accumulation. Vacuum cleaning is the best way to remove dust in foundries. The special equipment needed is well worth the investment. Once dust has been removed, prevent further accumulation by using local exhaust systems that remove it at the point of origin.

Solvents

Evaluate each solvent on the basis of its chemical ingredients. Proper labeling, substituting less hazardous for more hazardous chemicals, limiting the quantities in use, and using other methods of control can help minimize the toxic and flammable hazards involved in using solvents.

Other Materials

Many metals, resins, and other substances present safety and health hazards in foundries. Several of them are listed below:

- Acrolein occurs in foundry operations as a result of thermal decomposition of core oil, and is highly irritating.
- Aluminum, while usually not a toxic hazard in casting processes, presents a fire and explosion hazard in dust-collecting systems.
- Beryllium may produce pulmonary and skin disease, especially where plants cast beryllium-copper alloy.
- Carbon, as sea coal, is a common ingredient of molding sand used for facing. Carbon dust may cause anthracosis, a relatively harmless condition, but which produces characteristic lung shadows in an x-ray.
- Carbon monoxide gas is generated (1) during some cycles in the operation of a cupola, and (2) after pouring into green sand molds. Carbon monoxide is a toxic gas that preferentially binds to the hemoglobin molecule and significantly decreases the amount of oxygen to vital organs and tissue.

- Chromium is encountered in stainless-steel castings as the element or the oxide. Exposures occur during melting, gate and head burning, and grinding. Of the two common forms of chromium, trivalent and hexavalent, hexavalent is far more toxic and has a variety of health effects ranging from irritation to carcinogenicity.
- Fluorides, sometimes in cryolite form (sodium aluminum fluoride), are used in manufacturing ductile iron and magnesium castings. They are respiratory skin and eye irritants that can also cause diarrhea and abdominal pain. Further fluorides are associated with calcification—hardening of ligaments and bones.
- Iron-oxide fumes and dust are created during melting, burning, pouring, grinding, welding, and machining of ferrous castings. Exposure may be particularly high where manganese-steel castings or oxygen-lancing of the furnace is involved. Use a local exhaust to vent these fumes (Figure 24–1).
- Lead is a health hazard in nonferrous foundries. It forms the oxide in melting, pouring, and welding operations. Elemental lead dust is produced in cleaning and machining operations. Lead is a toxic material capable of adversely affecting a variety of organs and systems, including kidneys and nervous system.
- Magnesium dust or chips create hazards of fire and explosion. Physiological effects are confined to a form of metal-fume fever from inhaling finely divided magnesium. Magnesium-oxide fumes are generated when burned.
- Manganese is usually associated with steel castings and bronze alloys in foundry work. Constant exposure to high levels of manganese is associated with neurological disease.
- Phosphorus is used in the production of phosphor-copper. Acute cases of poisoning have not been reported and chronic cases are rare. The drying of phosphor-copper shot may produce phosphine gas, which is highly toxic.
- Resins—phenolformaldehyde and ureaformaldehyde—are used in shell molding. They create several hazards. Phenolformaldehyde resins contain hexamethylenetetramine (hex), a skin irritant that is highly explosive. When heated, this type of resin decomposes to produce a mixture of phenol and formaldehyde vapors. Ureaformaldehyde decomposes to give off ammonia and carbon dioxide. These materials are known to have good warning properties, but industrial hygiene evaluations should be conducted to evaluate potential exposure levels.
- Some types of resin dust, however, are highly explosive when suspended in air and require wet dust collectors. Alcohol, sometimes used for cold coating sand with resin, must be controlled to keep its concentration well below the lower flammable limit.
- Silica is usually encountered in the use of silica flour in molding sand or in core washes and sprays. Zircon, which is more dense and therefore settles more rapidly, is an effective substitute for silica flour in some applications. Sand-handling and -conditioning systems, shake-out operations, and sand-slinging constitute other sources of exposure to silica dust. Repeated exposure to high levels of silica dust is associated with the lung disease *silicosis* and exposures should be controlled to ensure compliance with regulatory exposure limits.
- Silicones are used as mold-release agents in shell molding. Hydrolyzing silicones are highly corrosive and hazardous if touched or inhaled. Care in handling them can eliminate the dangers of skin or eye contact. However, nonhydrolyzing silicones (methyl, mixed methyl, and phenylpolysiloxane) are better choices because they can be just as effective as mold-release agents, yet less toxic.
- Sulfur dioxide is the result of the oxidation of sulfur used in magnesium castings. In concentrations normally present in foundries, it is an irritant that can produce transient changes in pulmonary function.

Medical Program

Base the safety and health program for foundry workers on the recommendations and guidance of a safety and health professional, an industrial hygienist, and/or a physician. Such a program includes the following aspects:

- baseline physical examinations, including chest x-rays, audiometric tests, and pulmonary-function tests
- periodic physical examinations, including chest x-rays, audiometric tests, and pulmonary-function tests used to monitor employees' health, detect incipient disease, and help to reclassify workers as needed
- adequate first aid facilities, approved by the physician, and employee training in first aid
- regulatory requirements must be observed if respirators must be worn
- industrial hygiene monitoring where needed (Figure 24–2).

Make sure employees are aware of specific hazards to which they may be exposed and the proper control or emergency responses to those hazards. Make Material Safety Data Sheets (MSDS) available to all employees. (See, in the *Administration & Programs* volume, Chapter 5, Loss Control Programs; Chapter 14, Environmental Management; Chapter 18, Emergency Preparedness; and Chapter 8, Identifying Hazards. See also, in this volume, Chapter 11, Fire Protection; Chapter 14, Materials Handling and Storage; and Chapter 12, Flammable and Combustible Liquids.)

Figure 24–1. In these fume-diverting baffle furnaces, the hoods roll on a trolley so that correct positioning is made easy. (Courtesy American Brake Shoe Co.)

Personnel Facilities

Coreroom workers whose hands and arms may be exposed to sand and core oil mixtures are candidates for dermatitis. Prolonged contact with oil, grease, acids, alkalis, and dirt also can produce dermatitis. Encourage frequent washing with soap and water, and install adequate facilities.

Recommendations for toilets, washrooms, showers, locker rooms, and food service are given in Chapter 8, Industrial Sanitation and Personnel Facilities. Sanitary food preparation and service is especially important in nonferrous foundries. Prohibit eating in work areas where toxic materials are handled.

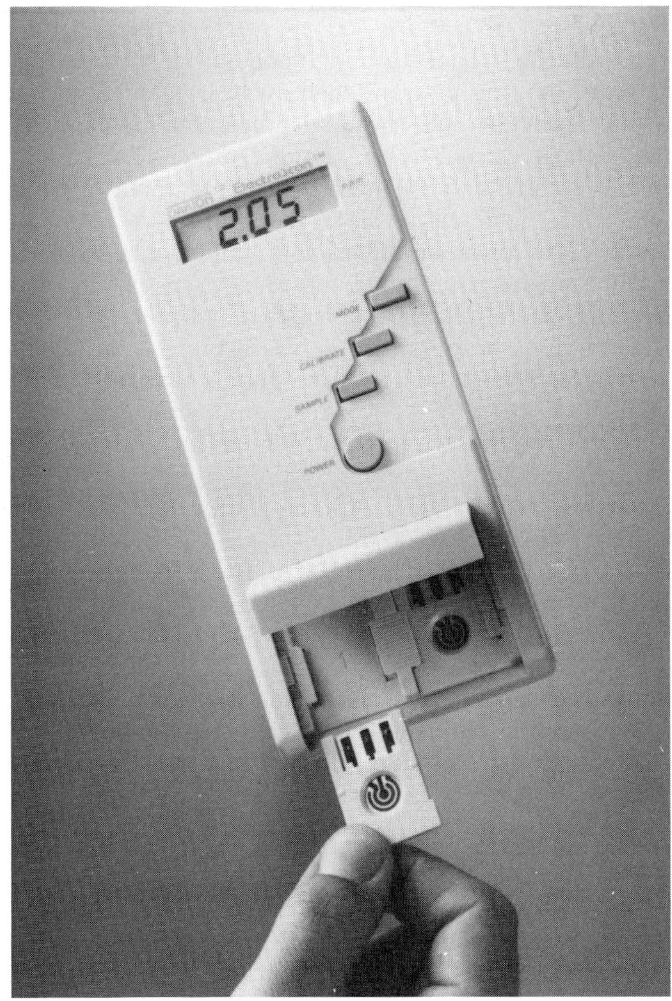

Figure 24–2. This hand-held meter detects trace metals in solution. It can be used for monitoring metal plating baths, industrial effluent, and water quality. It can measure a wide concentration range, meeting various international regulatory standards.

WORK ENVIRONMENT IN FOUNDRIES

Good housekeeping, ventilation, and lighting help maintain a safe and healthy work environment. Proper inspections, maintenance, and fire protection increase workers' safety in foundries.

Housekeeping

To achieve good housekeeping, hold each individual responsible for maintaining order in the work area. Set aside a specific time for housekeeping. Provide the necessary housekeeping equipment, and see that trash cans and special disposal bins are kept handy and emptied regularly.

Each worker should do the following chores:

- Clean machines and equipment after each shift, and keep them reasonably clean during the shift.
- Place all trash in the proper trash bins.
- Keep the floors and aisles in the work area unobstructed.
- Properly stack and store materials.

Floor Loading

Many buildings are used for purposes for which they were not designed. Mechanized movement of material introduces floor load problems caused by dead weight of platforms and lift trucks. Suspension of overhead cranes and hoists from wood ceiling joists severely taxes roof and floor members. Insurance engineers or local building inspectors can help determine safe floor load limits.

Ventilation

Control of air contaminants is the primary purpose of ventilation in foundries. The need for controlling ventilation may be determined by one or more of the following:

- applicable federal, state or provincial, and local regulations or standards, codes, and recommendations
- comparison with similar operations in a like environment
- collection and analysis of representative air samples taken by qualified personnel in the breathing zone of workers.

Noise Control

Controlling excessive levels of noise, over 85 dBA, may sometimes be difficult. Controlling noise through engineering is not always possible because of a lack of technology, or is impractical because of short-term, infrequent exposures or economic considerations. In such cases, develop a Hearing Conservation Program that provides approved hearing protection for each worker, and minimize exposure to identified high noise-level hazards. Chapter 7, Personal Protective Equipment, discusses elements of a successful hearing protection program as does NSC's *Fundamentals of Industrial Hygiene.*

Lighting

Good lighting is difficult to achieve in foundries because of the nature of the operations. Where craneways are used, light fixtures must be placed high and at considerable distances from the work areas. Nevertheless, provide good lighting for each work area. (See Appendix 1, Safety and Health Tables, for recommended levels of illumination.) Foundries having difficulty maintaining recommended levels of light can call on their local power companies or illumination consultants for expert information.

Inspection and Maintenance

Follow standard inspection and maintenance procedures in foundries. Carefully select maintenance personnel. Train them in safe practices, particularly in procedures for locking out controls and isolating other energy sources. (See Chapter 6, Safeguarding. See also *Safety Requirements for Sand Preparation, Molding, and Coremaking in the Sand Foundry Industry*, ANSI Z241.1.)

Fire Protection

Foundries make periodic fire inspections and perform emergency fire fighting drills. (Emergency Services, if available, should participate in drills.) A fire brigade, if present, will also aid the safety program by keeping its members, as well as other employees in the foundry, safety conscious.

Facility Structures

Entrances and exits, stairways, floors and pits, galleries, gangways, and aisles of foundries must meet special requirements. Observe the following suggestions and requirements for facility structures in foundries.

Entrances and Exits

To prevent drafts from reaching employees, provide entrances and exits to heated buildings with vestibules or enclosures. Make these enclosures large enough to permit the passage of trucks regularly used inside the facility. This provision does not apply to entrances used for railroad or industrial cars handled by locomotives, or for traveling cranes, trucks, and automobiles.

All doors, particularly double-acting, swinging doors, should have a window opening approximately 8 by 8 in. (20 × 20 cm). Locate the window at normal eye level to permit a view beyond the door.

Stairways

Provide substantial handrails, standard guardrails, and toeboards for all permanent and portable stairways having four or more risers. (See *Safety Requirements for Workplace Floor and Wall Openings, Stairs, and Railing Systems*, ANSI A1264.1-1995.)

Floors and Pits

Have the floor beneath and immediately surrounding foundry melting units pitched away from the melting units to provide drainage and to prevent incidents, especially those caused by spills and "run-outs" of molten metal. Clean floors frequently and keep them in good condition—firm and level. Workers should report worn spots, holes, or other defects and maintenance workers should repair them immediately. Install special types of flooring where fires or explosions may occur, or where other serious hazards may exist. To prevent an explosion hazard, keep the floor free of pools of water. Where water is needed to reduce dusty operations, use only enough to hold down the dust. Where tram or standard-gauge railroad tracks run into or through a foundry, keep the top of the rails flush with the foundry's floor, which should be maintained at this level.

Because of the danger of explosion, keep pits and other containers in which molten metal is handled or poured free from dampness. Protect pits connected with ovens, furnaces, and floor openings with either a cover or a standard guardrail when not in use.

Locate pig molds and receiving stations for excess molten metal from ladles clear of passageways and at least 1 ft (0.3 m) above floor level. Never allow pig holes in the floor near pouring areas; it is inherently unsafe.

Galleries

Where molten metal is poured into molds, provide the galleries with solid, leak-proof floors—concrete or sheet steel covered with sand—and with partitions of sheet steel. The partitions should be approximately 42 in. (1 m) high. Install them on the open sides of such galleries.

Where floor space is cramped, construct galleries to store ladles, flasks, flask boards, and other equipment. Equip these galleries with standard handrails and toeboards, and provide them with sturdy stairways—not ladders.

Keep concrete pavements around pouring floors coated with sand during pouring operations. This will reduce spalling of cement in case of a molten-metal spill.

Gangways and Aisles

Keep every gangway and aisle in good condition. They should be firm enough to withstand the daily traffic for which they are intended. Make sure they are uniformly smooth, without obstructions, and free from pools of water.

Compressed Air Hoses

The compressed air hose presents another foundry hazard. Do not use air hoses to clean clothes. Improper use of the hose and "horseplay" have caused severe injuries to internal organs and eardrums. To prevent injuries, reduce compressed air to less than 30 psig (210 kPa). Install whip checks at all joints, especially at quick-release couplings.

Prohibit unsafe practices such as blowing and brushing sand from new castings without regard for the cloud of dust produced, blowing dust off patterns, and removing parting compounds and other light materials. Substitute vacuum methods for cleaning molds with compressed air. Carefully instruct workers in the safe use of air hoses if the latter method is employed. Use of nonsilica partings eliminates the possibility of silicosis from this source.

MATERIALS HANDLING IN FOUNDRIES

Improper materials handling in foundries results in a wide variety of injuries. Injuries involving fingers, hands, toes, feet, legs, arms, and the back can occur during the manual handling of scrap metals, pig iron, and similar materials.

Many foundries have replaced manual handling of materials with mechanical means, reducing exposure to manual-handling hazards. However, mechanical materials handling usually brings with it hazards of its own. For instance, do not use magnets for lifting over areas where people are working. A break in the magnetic circuit would cause the load to be released without warning. In addition, pieces dangling from the magnet can be jarred loose.

Some of the precautions that can be taken to prevent materials-handling injuries include the following:

- Instruct workers in the safe methods of manual and mechanical materials handling.
- Provide personal protective equipment—eye protection; safety hats; face shields; leather mitts or gloves, preferably studded with steel, unless hot metal is to be handled; hand pads, aprons; protective footwear including metatarsal shoes; and other items such as flame-retardant clothing. Respiratory protection may also be needed.
- Plan the sequence and method of handling materials to eliminate unnecessary handling.
- Safeguard mechanical devices and set up inspection procedures to assure their proper maintenance.
- Keep good order at storage piles and bins, and pile materials properly.
- Keep ground and floor surfaces level so that workers handling materials will have good footing.
- Install side stakes or side boards on tramway or railroad cars to prevent materials from falling off.
- Chock railroad cars and flag tracks as required. Use dock plates with box cars when loading or unloading.

Handling Sand, Coal, and Coke

Avoid certain hazards in handling materials such as sand, coal, and coke, as follows:

- Prevent falls through hoppers while unloading bottom-dump railroad cars by requiring the use of fall protection equipment. Be sure that observers are on the scene; they should be prepared to perform rescues and/or summon help in emergencies.
- Use safety ratchet wrenches for hopper doors to keep the doors from swinging and striking workers.
- Prevent hand and foot injuries by using safety car movers instead of ordinary pinch bars to spot cars by hand. However, use a locomotive when available.
- To keep dump cars under repair from being moved, use locking switches and car chocks; at night use warning targets, derails, and red lanterns.
- To reduce the danger of cave-ins of loose material, prohibit the undermining of piles and avoid overhangs.
- Prevent electric shock by grounding portable belt-conveyor loaders.

Some foundries have eliminated double handling of materials by having raw materials taken directly from the cars, storage piles, or bins and placed into unit charging trays or boxes. The trays or boxes are then taken to the point of use and dumped mechanically. Properly trim trays or boxes to be carried overhead.

Ladles

Handshank ladles and mixing ladles are used for distributing molten metal or reservoir. Ladles are mounted on stationary supports or trucks, or are handled by overhead cranes or monorails. Such ladles have a capacity of not more than 2,000 lb (900 kg). Provide ladles with a manually operated safety lock (Figure 24–3). Construct the shanks of the ladles from solid material, and install shields. Provide suitable covers for portable ladles.

Use gear-operated ladles for ladles with over a 2,000 lb (900 kg) capacity. Equip such ladles and those that are mechanically or electrically operated with an automatic safety lock or brake to prevent overturning or uncontrolled sway.

On hand or bull ladles, build up the rim or lip above the top of the metal shell with fire clay no more than 2 in. (1.3 cm) if the refractory ladle's lining is less than 12 in. (3.8 cm) thick at the rim. In any case, the maximum height of the rim or lip should be not greater than 1 in. (2.5 cm).

Thoroughly dry out and heat ladles before use. Provide local exhaust to control vapors or fumes produced

Figure 24–3. This tilting ladle is equipped with a manually operated antitilt level. (Courtesy American Foundrymen's Society, Inc.)

during ladle drying. Some foundries perform all ladle-drying operations in a shed located outside the foundry. This segregates the exposures and typically makes control measures easier to implement.

Equip monorail ladles and trucks, used to transport molten metal ladles, with warning devices, such as bells or sirens. Sound the bell or siren whenever molten metal is being transported.

Construct trunnions and devices used to attach them to flasks, buckets, ladles, and other equipment with a safety factor of at least 10. The diameter of the head on the outside end of the trunnion shaft should be not less than one and one-half times the diameter of the trunnion shaft. Fillet the inside corners where the trunnion shaft joins the base and the head to prevent the sling or hook from riding the trunnion's base or head.

Inoculation, or treatment, of molten metal to desulfurize it or to change its composition or type, as in the making of an alloy or ductile iron, is done in the reservoir or in a pouring ladle. Install a hood to cover this operation so that the workers are effectively shielded from possible spatters of metal caused by the violence of the reaction. The hood should draw off the fumes that result, and should exhaust them through a baghouse-filtering system and then through a stack.

Hoists and Cranes

Hoists and cranes that handle molten metal require a preventive maintenance program, conducted by personnel trained on and thoroughly familiar with the equipment. This program is in addition to observations and inspections made by supervisors and operators. The degree to which the program is carried out depends on both the equipment being used and the tonnage moved. Gear the program to ensure that the operation is safe—much safer than simply to comply minimally with existing regulations. For example, an effective program for a 300-employee gray-iron foundry could require weekly visual inspections of crane and hoist structures, as well as an inspection of wire ropes and hooks before every shift.

Because of the severe stresses and demanding service in some high-tonnage operations, these operations may require more elaborate inspection programs. Conduct inspections on a weekly basis by trained specialists. Some programs regularly schedule nondestructive testing—ultrasonic testing of the crane's hoist shafts and parts, and dye penetrant inspection for surface cracks on bales, dumping chains, clevises, and pins. (See Nondestructive Testing later in this chapter.)

Conveyors

Conveyor systems are typically used in foundries. Sand mixed in the mixing room is carried by a belt conveyor to hoppers at molding stations where each hopper is filled by a movable plow. Surplus sand is carried to the end of the belt and returned by bucket to the storage bin.

An endless conveyor is used to handle molds. Empty flasks come from the shake-out machine to the mold operators, who remove them and make their molds on molding machines, taking sand from the overhead hoppers as required. Spilled sand goes through a grating onto a belt conveyor that returns the sand to the mixing room. Thus, all shoveling operations are eliminated.

The molds are placed on a conveyor and passed into the pouring area. The pourers get their metal from the cupola or other furnaces and then step onto a moving platform geared to an endless, single-rail conveyor that moves at the same rate of speed as the platform. Hand ladles can be supported on the conveyor. Pouring is done as the workers move along. Install an electric switch near the end of the conveyor, so that if workers ride that far, their feet will come into contact with the switch and the conveyor will stop.

The mold conveyors then pass into a cooling zone. Weights can be removed from the molds by a mechanical device and returned by another conveyor to the place where they were originally used. The molds then move to the shake-out machine, where sand is dumped from flasks onto a vibrating grating. The sand falls to a belt conveyor that returns it to the mixing room. Using a hook, a worker pulls the castings onto another conveyor, which takes them to the tumbling barrels. Empty flasks are brought back to the molding section by another conveyor.

This is a complete system for mass production where each worker performs one function rather than several. When installing a system, guard shear points, crush points, and moving parts. Where conveyor systems run over passageways and working areas, protect the employees beneath them with screens, grating, or guards. These protective devices should be strong enough to resist the impact of the heaviest piece handled by the conveyor.

Where chain conveyors operate at various levels other than in a fixed-horizontal plane, install a mechanism of safety dogs on both the upgrade and the downgrade, in accordance with applicable standards. In case the chain fails, the safety dogs will hold the chain and prevent the load from piling up at the bottom of the incline.

Scrap Breakers

Guard shears to protect operators and passersby from flying particles. Keep the working floor clear and level.

Prohibit the use of a drop to break castings or scrap inside foundry buildings during working hours, unless such operations are performed within a permanent enclosure made of planking or equivalent materials. The enclosure should be strong enough to withstand the most severe impacts from the drop or from flying scrap. Construct the enclosure high enough to protect workers in the vicinity from flying fragments of metal.

If a rope is used, extend it over pulleys to a point clear of the breaking area. This assures that operators will be at a safe distance, preventing their entanglement in the rope.

Storage

Store foundry materials and equipment not in regular use in a safe, orderly manner on level and firm foundations. When workers remove materials from bins located at floor level or from storage piles, they should not undermine the piles and thereby cause cave-ins.

Cover hopper bins containing material that is fed out at the bottom, either by hand or by mechanical means, with a grating that prevents workers from entering the bin. Allow no one to get on the rails of a bin or to enter a hopper to break down bridged material. A worker who must enter a bin should wear fall protection equipment with attached lifeline. A second worker may be equipped for the rescue, or to summon aid.

Buildings that store patterns should have racks and shelves strong enough to hold the loads placed upon them. Provide pattern keepers with a sound ladder so they can safely reach the patterns. The pattern keeper, who is likely to be alone in the pattern storage area, should report at regular intervals to the supervisor. Design the floors and stairways of pattern-storage areas well, and keep them in good condition. The storage area should also be well lit.

Store flammable liquids in accord with National Fire Protection Association's (NFPA) 30, *Flammable and Combustible Liquids Code*, and mark them according to 29 *CFR*, 1910.1200, the Hazard Communication Standard. (See Chapter 12, Flammable and Combustible Liquids, in this volume.)

Slag Disposal

Design furnaces and pits with removable receptacles into which slag and kish (separated graphite) may flow or be dumped. Unless slag is disposed of in the molten state, provide enough of these receptacles so slag can solidify before it is dumped.

To decrease the amount of slag that goes into the slag pits, use slag or cinder pots. The pots can be set aside and allowed to cool eliminating the danger of explosion when they are emptied.

Dump slag where there is absolutely no water or dampness. Slag coming in contact with water might cause an explosion if some of the slag is still molten. Before breaking up slag, allow it to stand for several hours to prevent encountering still molten slag in the center.

CUPOLAS

Cupolas are vertical cylindrical furnaces used to melt iron in a foundry. Charging and blasting that take place in cupolas and generate carbon monoxide (CO) present several hazards.

Charging

The dangers in the charging of cupolas are principally confined to handling material. Never unevenly load or overload barrows or buggies. "Tip-up" barrows, used for charging coke, are sometimes so poorly balanced that they will not stay in the tipped-up position after being emptied. Instead, they could fall back on the chargers' feet at the slightest touch. Lowering the center of gravity prevents this hazard.

To prevent an explosion in the cupola, break open scrap cylinders, tanks, and drums before charging. Be sure these containers are empty.

The use of mechanical devices for charging cupolas not only saves labor but reduces the number of materials handling injuries. Most foundry cupolas are now charged either by fully automatic charging machines equipped with crane- and cone-bottom buckets or lift trucks equipped with tilting boxes.

The charging opening on some cupolas is covered by a door or chain curtain, which should be kept closed except during charging. To prevent material from dropping onto workers during charging operations, install railings or other safeguards for the space underneath the cupola's charging elevators, machines, lift hoists, skip hoists, and cranes.

Occasionally during idle periods, workers might rest under the charging platforms close to the warm chambers and flues. Prohibit this practice because of the danger of objects falling from the platform and the possibility of carbon monoxide escaping from the flues. Also, do not place employees' lockers under these platforms.

Charging Floor

For charging floors, use steel floor plates that are heavy enough not to turn up; securely fasten these plates in place. Steel floor plates in the immediate vicinity of the furnace, however, become extremely hot. Therefore, install brick flooring laid on a solid steel framework in those areas.

Keep charging floors free from loose materials, and provide storage racks for equipment not in use.

Provide standard railings, 42 in. high, and 4-in. (10-cm) toeboards around all floor openings. (See 29 *CFR*, 1910.179(d)(3),(4)(ii).) Because railings on the charging floor receive much abuse, construct them of angle iron, which is more easily repaired than pipe railings. At the tapping platforms, provide hinged gates or chains that may be hooked in place.

Where cupolas are manually charged, place a guardrail across the charging opening. Where cupolas are charged with wheelbarrows or cars, provide a curb of a height equal to the radius of the wheel of the barrow or car. This prevents the barrow or car from pitching over and falling into the cupola.

Carbon Monoxide

Carbon monoxide is generated during some cycles in the operation of a cupola. CO is an explosion hazard if it gets into the wind boxes and blast pipes when the blowers are shut down. To eliminate this hazard, supply adequate natural or mechanical ventilation in back of the cupola, and open two or more tuyeres after the blowers are shut down.

The large amount of blast air in the cupola generally carries the CO out the stack. In some cases, it is burned in the stack before it can be discharged. Sometimes, however, CO may escape. Since CO gives no warning, locate around the cupola CO indicators that light and give a loud sound. Also, post signs that show the proper procedures to follow should the CO indicators' alarm sound.

See that approved breathing equipment is close by. Train workers in its use, and be sure that this equipment is in good condition. If the concentration of CO is more than 200 ppm (0.2%), an engineering assessment should be considered. In addition, positive-pressure, self-contained breathing equipment, or an air-line respirator with an emergency escape bottle should be provided. However, the OSHA Permissible Exposure Limit for an eight-hour TWA is 35 ppm and the ceiling limit is 200 ppm; therefore, all efforts should be directed toward engineering controls.

Blast Gates

Blast gates and explosion doors are successfully used to prevent damage from gas explosions. They are sometimes placed in front of the tuyeres, so they can be opened to admit fresh air when the blowers are shut down. Never close blast gates or explosion doors until the blast has entered the wind box and driven out all gas.

Provide blast gates in the blast pipe that supplies air to the melting equipment. Close the blast gates when the air supply fails or when the melting equipment is shut down. This prevents the accumulation of combustible gases in the air-supply system. In the cupola, omit the blast gate if alternate tuyeres are opened to permit circulation of air.

Locate blast gates, in relation to the cupola's wind box, to keep the duct's volume at a minimum. Install motorized dampers at centrifugal blowers so they will close automatically when the air supply fails.

Equip positive-pressure blowers with safety valves having liberal discharge areas. If these are not provided, clogging of the cupola with slag, or quick closing of the gate or damper, may produce sufficient pressure in the blast pipe to cause it to burst.

Every cupola should have at least one safety tuyere, with a small channel 1 or 2 in. (2.5 or 5 cm) below the normal level of a tuyere. This channel has a fusible plate that will melt through, should the slag and iron rise to an unsafe level.

Tapping Out

Tapping out with safety requires skill; have only experienced and dependable operators perform tapping out. In "botting-up" the hole, operators should not thrust the bott directly into the stream of molten metal, because that would cause spattering. To eliminate this hazard, have operators place the bott immediately over the stream of metal, close to the hole, and aim it down toward the hole at a sharp angle. Keep a supply of botts ready for use within convenient reach of the operator who does the tapping.

When the cupola is tapped, hold the back end of the tapping bar below the level of the hole. This prevents puncturing the sand bed and causing molten metal to run out through the bottom.

A tilting spout placed with one end directly beneath the stationary cupola's spout and mounted on trunnions on a stand increases safety and efficiency. Operators can tilt it back and forth with a foot lever. The rear end of the tilting spout is closed so that when that end is tilted down, it forms a reservoir to receive the molten metal from the cupola. When the supplementary spout is tilted forward, the metal runs from it into the waiting ladle. At the same time, more metal continues to run into the spout from the cupola. Thus, the stream of metal runs from the cupola continuously, and the tilting spout acts as a reservoir between loads from the ladle.

Equip the slag spout of the cupola with a shield or guard to protect workers from sprays of molten slag and to form a hood to collect slag wool. The slag wool is sometimes collected through a wet-slagging system in which slag is thrown off into a water-filled container, or trough, and flushed away.

Dropping the Cupola's Bottom Doors

When the cupola is in operation, support its bottom doors with one solid prop and two adjustable-screw props (of the required structural strength) on a metal prop base. The base should be set on a concrete footing, or other fabricated footing of equivalent strength (Figure 24–4).

Place temporary supports—timbers, blocking, etc.—under the cupola's bottom doors. This prevents the doors from falling on employees while the metal props are being adjusted to proper height. Provide mechanical means for raising the bottom doors of the cupola (Figure 24–5).

Dropping the bottom doors of a cupola requires extraordinary care. One of the best methods for doing this is to use a block and tackle with a wire rope and a chain leader attached to the props that support the doors. The props can then be pulled out by means of the block and tackle from a safe distance or from behind a suitable barrier. Special locking devices for bottom

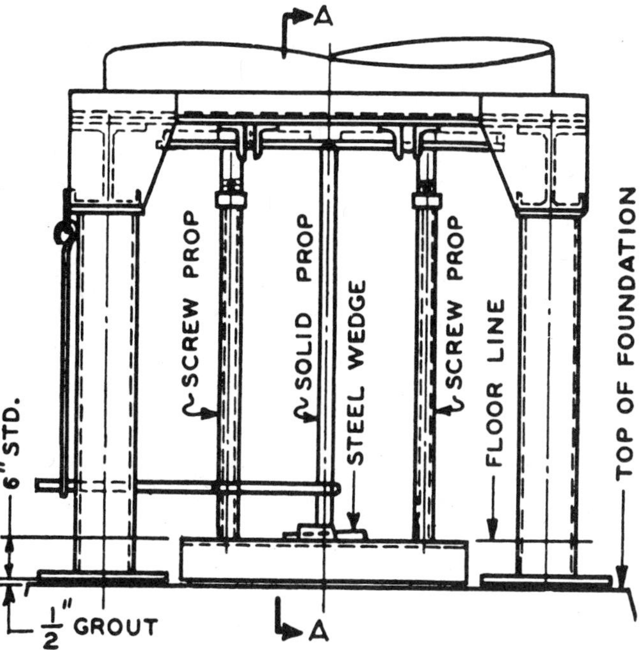

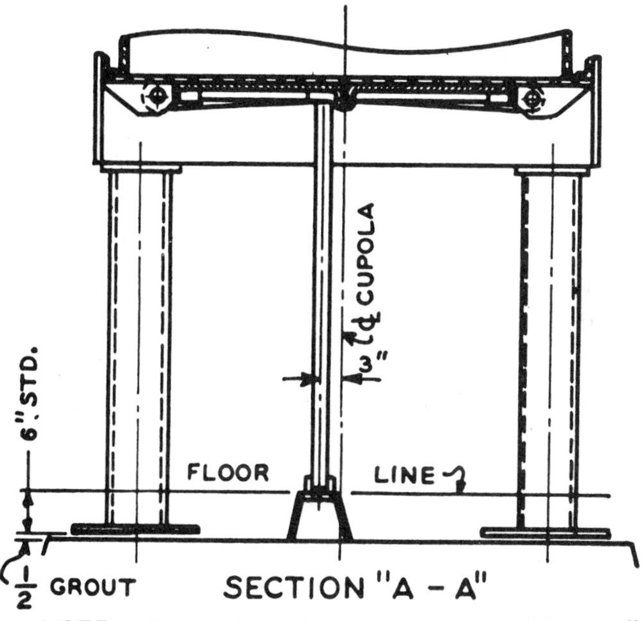

Figure 24–4. This is the proper method of supporting cupola bottom doors.

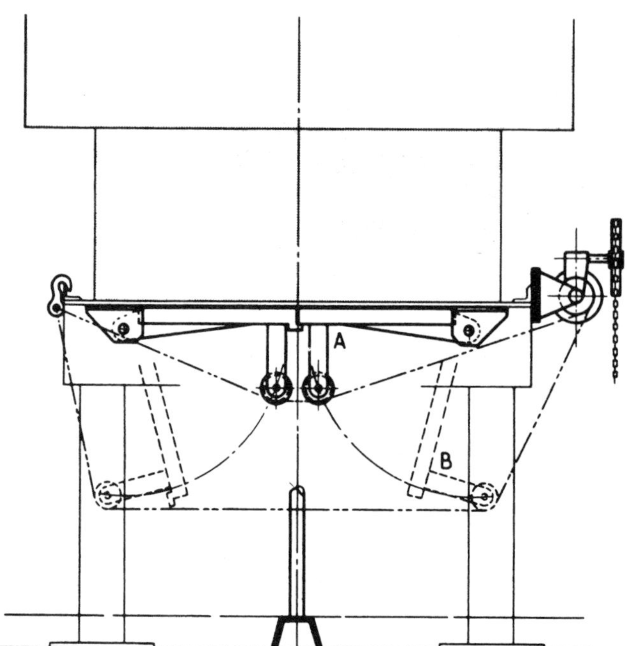

Figure 24–5. This is the suggested method of raising the bottom doors of the cupola by mechanical means.

doors may be used if the cupola's drop is to be caught in a container, car, or skid.

Before the bottom doors are dropped, carefully inspect the area underneath the cupola to see that no water has seeped under the sand. One worker should make sure that no one is in the danger zone and that workers stay away during the operation. Warn employees by means of a whistle or other signal before the bottom doors are dropped.

If the cupola's bottom doors fail to drop, or if the remaining charge inside the cupola bridges over, do not permit employees to enter the danger zone to force the doors or relieve the bridging. Relieve the bridging, instead, by turning on the blast fan. The vibration produced usually corrects the condition. A mechanical vibrator attached to the bottom doors is also effective. Another method of relieving the bridging is to drop a demolition ball from the charging door. If these methods fail, flame cut the doors with a lance, but only after the cupola has cooled to a safe temperature.

Repairing Linings

Allow only careful and experienced workers to repair a cupola's linings under special procedures and permits. Observe the following precautions:

1. Install a sturdy guard over the cupola's charging door to protect workers against falling objects. Construct the guard with heavy-gauge metal, and cover it with a screen or the equivalent. Securely support the guard by means of overhead slings or underpinnings to resist falling objects (Figure 24–6). An alternate method is to place a solid steel plate on the bucket ring in the cupola at the level of the charging door.
2. Implement a confined space program for workers entering cupolas—including personal protection equipment.
3. Provide approved respiratory protection for the workers relining the cupola, and place a blower fan in the bottom to keep the dust moving away from the workers. Exhaust the dust out the stack.
4. Place warning signs or crossbars at the charging door to indicate that workers are in a cupola.
5. Before relining of a cupola begins, have workers break all loose slag and bridges and allow these pieces to drop to the bottom of the cupola.

Figure 24–6. A screen placed over the charging door prevents falling objects from dropping on a worker who is repairing the cupola lining. (Courtesy Hamilton Foundry & Machinery Co.)

6. Check the condition of the shell while the cupola is down for relining. A weak shell is likely to increase the risk of a gas explosion.
7. Leave ample clearance, at least ¾ in. (2 cm), between the new brick lining and the shell to allow for expansion. Fill this space with dry sand to serve as a cushion to protect the shell against severe stresses.
8. Before the cupola is started up, make sure that all personnel have exited the cupola and the area beneath it, that the lining is thoroughly dry, and that all tools and other equipment have been removed.

CRUCIBLES

The principal danger in handling refractory clay crucibles is that one may break when full of molten metal. Therefore, have a trained inspector check all new crucibles for cracks, thin spots, and other flaws. Return to the manufacturer those showing signs of dampness. Examine the packages and the car in which they were shipped to find out whether or not they were exposed to moisture in transit.

Storing

Store crucibles in a warm (about 250 F or 120 C), dry place, and protect them from moist air as much as possible. It is generally best to place them in an oven built on top of a core oven, or at some other point where waste heat can be used. If all crucibles in stock cannot be kept in ovens, accurately date those stored elsewhere and use the oldest and best-seasoned crucibles first.

Annealing Process

In the annealing process, crucibles are brought up to red heat very slowly and uniformly, usually over a period of eight to ten hours. Do not allow crucibles to cool before they are charged because, as they cool, they may again absorb moisture. Moisture in the walls of crucibles that are heated quickly is converted into steam. The steam expands and causes cracks or ruptures and may also cause pinholes or "skelping." Do not use damp or high-sulfur coke or coal, or fuel oil containing excessive moisture, to heat crucibles.

Too high a percentage of sulfur in the fuel used in the drying or annealing process is also likely to cause fine cracks, sometimes called "alligator cracks." Too little oil, or too much air or steam, used at the burners of oil furnaces tends to oxidize a portion of the graphite in the crucible's wall. This leaves the binding material somewhat porous.

Charging

Proper care of crucibles is good economy as well as good safety. Since crucibles are costly, they should be made to last through as many "heats" as possible. To protect the crucible's lining or structure from damage, establish a process for cleaning crucibles. Improper cleaning may result in early failure of the crucible and thus injury or loss of the product.

Charge crucibles carefully. Do not throw in ingots with such force that they bend the bottom or walls of the crucible out of shape. Also do not force the ingots into the crucibles so they become wedged or jammed. Heat new crucibles very slowly for the first few runs, especially the first run. Because crucibles are soft at white heat and easily forced out of shape, handle them with great care.

Handling

To prevent damage to a crucible, select tongs of the proper size and shape for the particular crucible. Tongs should fit well around the bilge or belly of the crucible and should extend to within a few inches of the bottom. Provide at least two pair of tongs for each size of crucible so that if one pair becomes bent, the other will be available.

Before applying the tongs, check the sides of the crucible to see that no clinkers are adhering to them. Never drive tong rings down tight with a skimmer or other tool. This practice is almost certain to squeeze the crucible out of shape and produce cracks and fissures.

The blacksmith should have a complete set of cast-iron forms in the exact shapes of the crucibles used. Then the smith will have only to heat the tongs to red heat, clamp them onto the forms, and bring them into the exact shape with a heavy hammer.

Avoid ramming the fuel bed around a crucible. Should this become necessary, have it done cautiously and only by experienced workers. Support crucibles on foundations or pedestals of firebrick, graphite, or other infusible material.

The removal of heavy crucibles from furnaces calls not only for special skill but also for physical strength. If the workers are overstrained, serious injuries are likely to result. Where possible, therefore, use a mechanical device to remove heavy crucibles—those exceeding 100 lb (45 kg) in combined weight of crucible, tongs, and metal. When using air or electric hoists to move large crucibles, have one person at each sling and one operator controlling the hoist.

Crucible Furnaces

To make the operation of crucible furnaces relatively free from hazards, install suitable exhaust hoods on all furnaces used to melt metals that give off harmful fumes. Equip upright furnaces, having crown plates more than 12 in. (30 cm) above the surrounding floor, with metal platforms having standard rails. Such platforms should extend along the front and sides of the furnace, flush with the crown plates, and clear of all obstructions. From these platforms, mechanically lower crucibles containing molten metal.

Many crucible furnaces are oil fired. Unless the air supply for these furnaces and the motors driving the oil pump are connected to the same source of power, a considerable quantity of oil may flow onto the floor if the air line loses its power. One remedy for this is to put a gate valve in the oil supply line so that in case the air supply fails, oil can be shut off from the entire battery of crucible furnaces in one operation. If the pilot and burner are electrically controlled, arrange this valve so it is similar to an interlock.

Another preventive measure is to install a gate, lever-operated valve in the oil line. This valve is closed by the release of a weight. Because the weight is normally held up by an air cylinder, the oil supply stops at once when the air goes off. A small hole may be drilled in the cylinder so that air in it may be released promptly when the oil pressure goes off.

OVENS

The principal hazards in the construction and operation of core ovens and mold-drying ovens are excess smoke, gases, and fumes. Other unsafe conditions are unprotected firing pits; unguarded, vertical sliding doors or their counterweights, which may drop on workers; and flashbacks from fire boxes. To minimize these hazards, observe the following precautions:

- Guard firing pits, into which people may fall, with substantial railings or with grating covers. Grating covers should have trap doors to give access to the steps leading down into the pits.
- Install safe vertical sliding doors; wire ropes and chains, having a high safety factor; sturdy fastenings; large sheaves to prevent undue wear of ropes; and guarded counterweights. Thoroughly inspect all sliding doors at frequent intervals. Safety dogs may be used to hold the doors in raised position.
- On the wall near core ovens, place a caution sign that gives precautions, as well as the manufacturer's instructions for lighting the ovens.
- Many vertical core ovens are 60 to 70 ft (18 to 21 m) high and have a driving mechanism located at the top. To protect employees, some foundries have installed steel stairways with standard treads wherever possible. Where metal ladders must be used, maintain not less than 7 in. (18 cm) of clearance from the center of each ladder rung to the side of the oven.

Gas-Fired Ovens

Separate gas-fired ovens, whenever possible, from the molding floors and from the core-making room by a partition. This measure helps prevent equipment failures caused by sand in the controls.

Equip blast-tip pipe burners with baffles to keep sand out of the tip and also to spread the flame. Place tips horizontally to protect them from sand.

Install safety pilot valves on every gas-burning furnace or oven. They prevent the flow of unburned gas into the oven's combustion chamber, should the burner's pilot light go out, or should a cock or burner be opened unintentionally.

Install a bleeder valve in the line between two control valves close to the burner as an additional safety device. The operator can then allow gas to escape safely into the atmosphere instead of to the fire box, should there be leakage past the main control valve when the burner is not in use.

Ventilation

Where fumes, gases, and smoke are emitted from drying ovens, install near the oven's doors hoods and ducts, exhaust fans, or other means of removing these hazards. Such devices should be designed to keep the concentrations of fumes, gases, and smoke below toxic and irritant levels. Be sure that the composition of any emissions discharged outside the building complies with air pollution regulations.

To prevent flashbacks, install the proper size flues, and keep them free of soot. Then using oil burners, the type of equipment and the arrangement and control of drafts should ensure as perfect combustion as possible. In some installations, forced-draft equipment may be needed.

Equip core ovens with explosion vents. Lightweight panels may be installed on the top of the oven, or the oven may have hinged doors with explosion latches.

Natural-draft ventilation, however, is usually considered adequate for ovens under 500 cu ft (14,000 l) in volume. However, arrange the doors so they are wide open when the burners are being lit. Larger ovens, especially those with vertical sliding doors and other heavy construction, should have forced-draft ventilation. Interlock the ventilation system with the gas supply through a time relay. Arrange the relay to allow for at least three complete changes of air in the oven before the burners are lit.

Inspection

Before a foundry's core ovens are lit, thoroughly inspect the ovens and burners. Only trained and qualified personnel should do this work. Establish an inspection and preventive maintenance program for core ovens.

The following lists of questions represent a sample procedure of good inspection practices. First, shut off the main valve that controls the fuel supply. By doing this, the gas line to the burner's pilots are automatically turned off. Then check the following items:

- Do the safety cutoff valves for gas and oil close when the pilot burners fail? If not, check for stuck valves or unusually long flame electrodes.
- Do the red signal lights on the flame-detecting device light up? If not, check first for burned-out signal lamps and then for defective relays on the flame-detecting device.
- Does the oven-tender warning light on the back of the control panel light up? (This light is mounted so that it can readily be seen throughout the core room.)
- Does the warning horn blow? Can it be shut off by manipulating the safety cutoff valve's handle? It should not go off.
- Are spark plugs clean and ignition wires in good condition?
- Are OPEN and SHUT markings on gas and oil safety shutoff valves legible?
- Are there core oil resin deposits on safety shutoff valves or on the pilot's gas solenoid valve? Extensive deposits may cause valves to stick.
- Are all cover-retaining screws in place on flame-detecting devices and on flame electrode holders to prevent entry of dust or core oil resins?
- Are all motor controller, relay box, and wiring junction-box covers tightly in place and clearly marked as to what they control?
- Is there excess fuel oil or an oil-soaked accumulation of dirt or rags on the burner's deck?
- Are fuel oil atomizing heads removed from the burners, cleaned if necessary, and checked for burned tips?
- Is the atomizing air tube removed and are swirl vanes and the main burner's nozzle checked for burned parts or accumulations?
- If the oven is operated on oil, is the three-way cock between the gas control valve and the burners turned to permit outside air to be pulled into the normal gas inlet to the burners?
- Are there any broken refractory burner blocks?
- Are gas-, oil-, and air-control valves, valve control motor, and valve control linkage checked for loose valve-adjustment screws or broken valve cam springs?
- Is the combustion chamber checked for defective or burned-out arches, side walls, combustion walls, or metal tie bars that may span the combustion chamber above the arches?
- Are all electrical contacts in disconnect switches, motor controllers, and relays inspected? After inspection, replace all enclosure covers securely.
- Are temperature controller, controller relays, and control resistance lamp, if one is used, checked? Make sure that the burner control motor returns to low-burner position when the controller disconnect switch is in the OFF position.
- Are covers from all air-flow switches removed and gummy or worn parts, damaged wires, or damaged mercury tube elements removed?
- Has performance of flow switches been checked by starting the proper fans and observing the flow-switch operation?
- As fans are started, have they been checked for V-belt squeal or fan vibration?
- Is it certain that fans can be started only in the proper sequence: main circulation fan; power exhaust fan for drying zone; combustion air blower, similar fans for additional burner units, if used; and cooling-zone supply and exhaust fans?
- Have the operation of the purge duct's damper motor and linkage at the start and end of the purge cycle been checked? The purge cycle's time should be 10 minutes on all ovens, except horizontal drying oven-burner units, which have 5-minute cycles.
- Have red signal lights of the flame-detecting device come on at the end of the purge cycle?
- Do spark plugs and flame electrodes project approximately 2 in. (5 cm) beyond the burner-pilot casting? Center the flame electrode in the pilot's opening and set it to one side—about 3 in. (7.5 cm) at its tip.
- Have all explosion doors been checked for proper operation, using a light to check for evidence of damage to internal structures or ductwork? Make sure that the explosion panel in the roof is intact.

Next, turn on the main fuel valve, which opens the pilot's gas valve, press the ignition button, and check the following points:

- Do the pilot lights ignite readily?
- Are there fuel oil leaks in the piping around the burners?
- Is there an odor of leaking gas around the burners, the gas regulators, or the gas valves? Periodically test the various gas valves for leaks with a soap solution.

- Is there any indication of flame flashback to the outside from either the pilot's burner or the main burners?

Inspect all fans for these items:

- Using a tachometer, check the rpm against the proper speed shown on the metal tag on the fan drive's guard. If the fan's speed is too low, tighten the belts and recheck the speed.
- Check all V-belt drives for proper belt tension. A tag showing the proper V-belt size should be mounted on each belt guard.
- Be sure that all belt guards are in good condition and securely fastened in place.
- Check fans for vibration.
- Air leakage at fans on attached ductwork should be sealed.
- Be sure that fans' motors are clean and securely bolted in place.
- Check that damper motors, control valves, and control linkages are securely in place.

FOUNDRY PRODUCTION EQUIPMENT

On production-line equipment, fully guard moving parts, such as belts, pulleys, gears, chains, and sprockets, and other common machine hazards, such as projecting set screws in accord with standard practices (see Chapter 6, Safeguarding). Ground electrical equipment to eliminate shock hazards. Allow repairs only on equipment that is locked in the Off position and after all other sources of energy have been eliminated.

Some operations require mills, mixers, and cutters of such size that an employee can enter the machine to clean or repair it. In these cases, set up and enforce a lockout procedure. (See Lockouts in Chapter 10, Electrical Safety; and in Chapter 6, Safeguarding. Both chapters are in this volume.)

Sand Mills

The principal danger of sand mills, or mullers, exists when operators reach in for samples of sand or attempt to shovel out sand while the mill is running. In doing so, they may be caught and pulled into the mill. To protect against this hazard, one or more of the following measures may be used:

- Provide screen enclosures for charging and discharging the openings of mills.
- Install self-discharging mills, or equip mills with discharge gates or scoops.
- Provide sampling cones for taking samples of the sand during the mixing operation.
- Prohibit the shoveling of sand out of mills while they are running.
- Install an interlocking device so the mill cannot be operated until the doors are closed.

Dough Mixers

To prevent operators from reaching into a dough mixer while the blades are in motion, cover the top of the mixer with a sturdy grating made of 3/8-in. (1-cm) round bars, or an equivalent. Another method is to attach an interlocking device arranged so that neither the cover can be opened nor the bowl tilted until the blade's drive mechanism has been shut off. In such a setup, the blades cannot be set in motion again until the cover is in place.

If the dough mixer is driven by an individual motor, attach a small steel cable to the cover and extend it over a pulley to a counterweight. This cable is attached to a ring on the motor control switch's handle. In that way, when the cover of the mixer is lifted a predetermined distance, the switch is pulled open and cannot be closed again until the cover is back in place.

Sand Cutters

Sand cutters throw sand and pieces of tramp metal with bulletlike force, sometimes causing serious puncture wounds. If a guard that would not seriously impair the efficiency of the operation cannot be devised, then have operators wear suitable personal protective equipment.

It is often difficult to operate a power-driven cutter on a sand floor. Therefore, install parallel concrete strips to act as runways for the cutter's wheels.

Sifters

Guard rotary sand sifters with enclosures or with angle iron or pipe railings that are 42 in. (1.1 m) high and placed from 15 to 20 in. (38 to 50 cm) from the sifter. Place belt shifters and motor control switches within convenient reach of the operators. The control switches should be designed so that they cannot be unintentionally started.

Portable sand sifters equipped with pneumatic vibrators usually move slower than those equipped with electric vibrators. Oscillation of their heavy parts causes the entire machine to move around the floor in jerky fashion. Often, the machine's travel is limited only by the air hose. If the hose coupling breaks, the hose flails around and blows sand in every direction, presenting a hazard to workers' eyes. To prevent such incidents, anchor the sifter with a rope a little shorter than the hose.

Molds and Cores

The principal hazards in hand molding and core making include letting flasks down on feet, pinching fingers between flasks, dropping heavy core boxes on feet, cutting hands on nails and other sharp pieces of metal in the sand, and stepping on nails. Minimize hand and foot injuries by training workers to handle flasks and core boxes properly and to wear foot protection with stout soles. Screening or magnetic separation to remove nails and other sharp metal from the sand is also essential to safety.

In general molding and core making, gagger rods—pieces of iron used in a mold to keep the sand or core in place—and core wires are cut, straightened, and bent using hammers and cutting sets. This operation presents danger from flying pieces of metal and dirt. Machines are available for performing this work, but many of their hazards are similar to those found in the use of hand tools.

As the work progresses, carefully brace heavy cores in large molds to keep the cores from toppling over. Prohibit work underneath molds suspended from cranes. Sturdy tripod supports or wooden or steel horses will provide greater safety and efficiency.

Venting molds properly is essential to avoid explosions during pouring. However, when the sand in an undried mold is too wet, metal can boil and explosions may occur even though the molds are well vented.

In ramming a mold, do not place the peen of the ram too close to the pattern. Otherwise, a hard spot in the sand will be made and molten metal coming in contact with it will boil and tear the sand away to the depth of the hard spot. This will also occur if a gagger iron is rammed against a pattern and the sand between is pressed into a hard spot. When the molten metal reaches the wet gagger iron, an explosion usually results.

Molding Machines

Three types of molding machines are used in foundries: straight, semiautomatic, and automatic. Equip all molding machines with two-hand controls for each operator assigned to the machine. On automatic molding machines, install shields or apron-type metal guards to protect pinch points.

The carry-out person should stand clear of the squeeze at the back of the machine. Operators should never touch the frame while it is moving.

When patterns are changed, block the frame. This prevents the table from falling and trapping pattern-setup workers, should the dog fail or the stripping frame operate unintentionally.

If molding machines have sand delivered by elevator buckets, enclose the side on which the buckets return. Or install a railing with 42 in. (1 m) double railings placed 15 to 20 in. (38 to 50 cm) from the moving parts.

The jolt squeezer machine is used as a molding machine. The main hazard of this machine is that workers may keep their hands on the edge of the flask or place a hand between the head of the machine and the flask. Dual-squeeze controls and a knee-valve jolt control will eliminate this hazard.

Core-Blowing Machines

Straight, semiautomatic, and automatic core-blowing machines are used in foundries. On semiautomatic and automatic machines, guard core-box push cylinders, counterweight cable pulleys, wheel guides, and table-adjusting footpads. Install an automatic barrier guard between the operator and the machine. If the drier is lowered automatically from the rollover and then pushed and raised toward the operator, there can be a pinch point between the lowering table and the raised table. This hazard is eliminated by an automatic barrier guard.

Equip automatic and semiautomatic core machines with double-solenoid valves. Maintain the slide valve well, and lubricate it to prevent recycling or other malfunctions.

General Suggestions

To prevent sand blows, maintain parting lines in good condition. Also, guard the parting line of the core box with a dike seal (Figure 24–7).

Replace vents when ruptured. Install good feeding systems with vibrators to keep the magazine full. Keep core box and cover seals free of sand. When a core machine is to be repaired, shut off the air lines, lock and tag the controls in the Off position (being sure that the machine is energy isolated), and bleed the air from the machine. Make certain that no source of energy can be activated.

Safeguarding

Where practical, provide two-hand operating controls to prevent the operator from placing a hand or fingers between the top of the core box and the ram. Where two operators are employed on a core-blowing machine, provide two sets of two- hand control buttons. Equip all core boxes with handles so employees can move their boxes without placing their hands on top of them. If

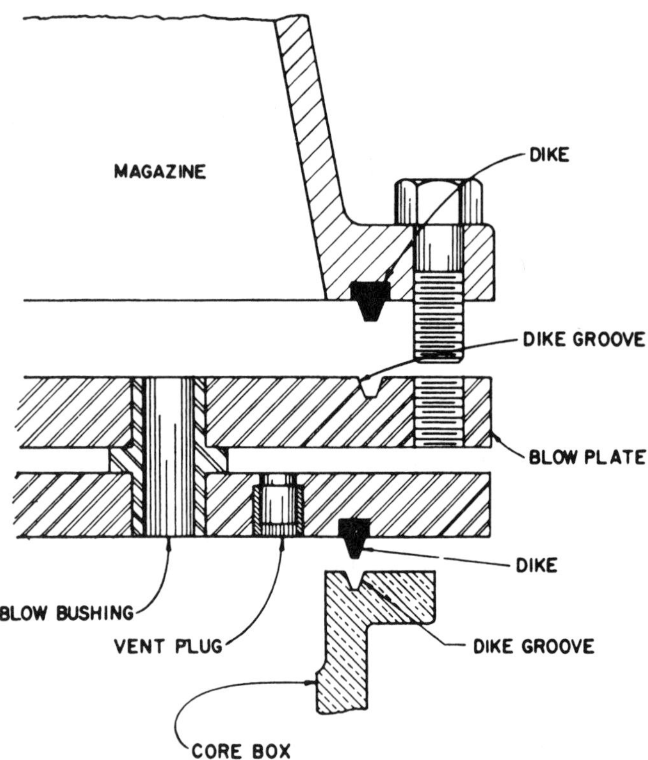

Figure 24–7. This section of a core box shows a rubber dike seal, which prevents sand blows and abrasion of the box. (Courtesy Dike-O-Seal Corp.)

driers are located above the rollover area for each core, place them high enough so they will not become entangled during the rollover process.

Cleaning

Some core dips contain substances capable of producing dermatitis on sensitive persons. Rubber gloves and plastic sleeves usually provide adequate protection. However, check employees engaged in core-dip operations at frequent intervals for sensitivity to the core-dip solution. Materials used in cleaning core boxes may also be toxic. Therefore, remove their emissions with a properly designed ventilation system.

Flasks

Iron or steel flasks are preferable to wood flasks. Wood flasks become worn, burned, or broken so that they do not fit together well and may let molten metal run out during pouring. Do not leave defective flasks in the foundry building or in outdoor storage piles because they may be put back into service without first having been repaired. Have competent inspectors carefully inspect flasks at frequent intervals. Inspectors should have the authority to have the defective ones destroyed or sent to the repair shop.

Flask trunnions should have end flanges at least twice the diameter of the trunnions to minimize the danger of hooks slipping or jumping. Trunnions should preferably be turned, or otherwise be smooth castings. It is sometimes best to cast the trunnions separately and bolt or weld them in place. This procedure speeds up machining operations and permits reuse of trunnions recovered from broken flasks. Trunnions cast separately should be of steel.

When trunnions are bolted or welded in place, the nuts should be inside the flask. If they project on the outside, slings are likely to catch on them and slip off with a jerk, which subjects both the sling and the trunnion to severe strain.

Large flasks should have loop handles made of wrought iron. On steel flasks, cast handles at frequent intervals to make chaining possible.

Design trunnions and handles for the loads they are to carry, and construct them with a safety factor of at least 10. Make sure that the bolts that fasten trunnions and handles to the flasks are of sturdy enough construction.

The diameter of the button should be equal to the diameter of the groove plus one and one-half times the diameter of the sling used to handle the flask. Inside corners shall be well filleted. To prevent the sling from sliding off and riding the button, (1) the radius of the corner between the groove and the button should be approximately equal to the radius of the sling used and (2) the remainder of the inside edge of the button should be straight.

Sandblast Rooms

Each foundry should have dust-tight sandblast rooms. Keep the doors to these sandblast rooms closed, and dust castings before they are removed from the rooms. Even small cracks in the walls or under doors will allow fine dust to escape and to contaminate air in the foundry. Equipment for workers in sandblast rooms should include air-supplied hoods and full body protection.

Tumbling Barrels

Tumbling barrels need frequent care to keep them dust tight. Enclose barrels that cannot be maintained dust tight in booths connected to an exhaust system. Barrels may be equipped with exhaust ducts through the trunnions. Safety precautions to observe with tumbling barrels include (1) placing a removable guardrail around the machine, and (2) locking barrels in a stationary position during loading and unloading.

Shake-Out Machines

Shaking out castings presents the danger of hands and feet being crushed or arms and legs being broken. For this reason, workers must wear steel-toed or metatarsal-guarded foot protection. If steel hooks or rakes are used to pull castings from the screen, instruct workers to stand with one foot behind the other. That way, they can keep their balance in case the hook slips from the casting while they are pulling.

Because this operation is also often a source of dust, install hoods on shake-out machines and provide local exhaust to draw the dust to a collector. In fact, many foundries perform shake-out operations at night so that as few people as possible are exposed to dust.

Design shake-out machines so that the flasks cannot fall off the plunger. Do not allow foundry workers to retrieve gagger irons while these machines are in operation.

The hazards of sand conveyors are also found at shake-out machines since the sand is collected under the machines on a conveyor belt that moves the sand to storage for reclaiming and reuse. Keep the area around the shake-out machine free of sand and scrap. Guard the conveyor belt's opening at the sides.

CLEANING AND FINISHING FOUNDRY PRODUCTS

Install and operate grinding, polishing, and buffing equipment for foundry use, as recommended in Chapter 22, Metalworking Machinery, in this volume. Have qualified personnel mount and change abrasive grinding wheels. Closely supervise the use of correct washers and wheel-mounting procedures. Keep required wheel guarding intact. Speed-test new wheels before allowing them to be used on the job.

Require operators to wear full personal protective equipment for eyes, face, hands, and feet. Excessive dust generated by dry abrasive wheels is a potential health hazard. Remove this dust with an exhaust system at the point of origin. Precleaning castings in a barrel, mill, or abrasive chamber also minimizes dust from grinding. Keep the space around the machines dry, clean, and as free as possible of castings and other obstructions. Note that silica dust is considered a suspect carcinogen.

Magnesium Grinding

The fundamental hazards of grinding magnesium are the possibilities of fire and explosion. To eliminate these hazards, use a proper dust-collection system (Figure 24–8).

Dust-Collecting System

In a dust-collecting system for magnesium, the dust should be wet down by a heavy spray of water and immediately washed into a sludge pit in which the dust is collected under water. Keep sludge pits or pans well ventilated because hydrogen evolves from the reaction of the collected dust with water. Frequently clean sludge pits or pans. Do not let wet magnesium dust stand and become partially dried since fire or explosion could result.

The dust-collecting system must not have filters or obstructions that allow dust to accumulate. Install pipes and ducts, and use the shortest possible route to eliminate bends or turns in which magnesium dust or fines could collect. As often as necessary, clean pipes and ducts connecting the grinder and the collecting device. Disconnect pipes and ducts while wheels are being dressed.

Also provide the following safeguards when grinding magnesium:

- a means for immediate quenching of sparks from grinding wheels, disks, or belts
- dust-proof motors to prevent the accumulation of static charges
- explosion doors on the collection system
- an automatic interlocking control on the collection system to assure its operation whenever grinding is started.

General Housekeeping

Good housekeeping is essential for safe handling of magnesium. Prevent accumulations of magnesium dust on benches, floors, window ledges, overhead beams and pipes, and other equipment. Do not use vacuum cleaners to collect the dust. Have it swept up and placed in covered, plainly labeled iron containers, and if it is not recycled into operations, dispose of in accordance with applicable federal, state, and local regulations. Do not allow magnesium dust to be mixed with regular floor sweepings.

Because sparks can be produced, it is dangerous to use equipment that grinds magnesium to grind other metals. Mark equipment for magnesium grinding FOR MAGNESIUM ONLY. Use benches made of wood grating for rough finishing operations.

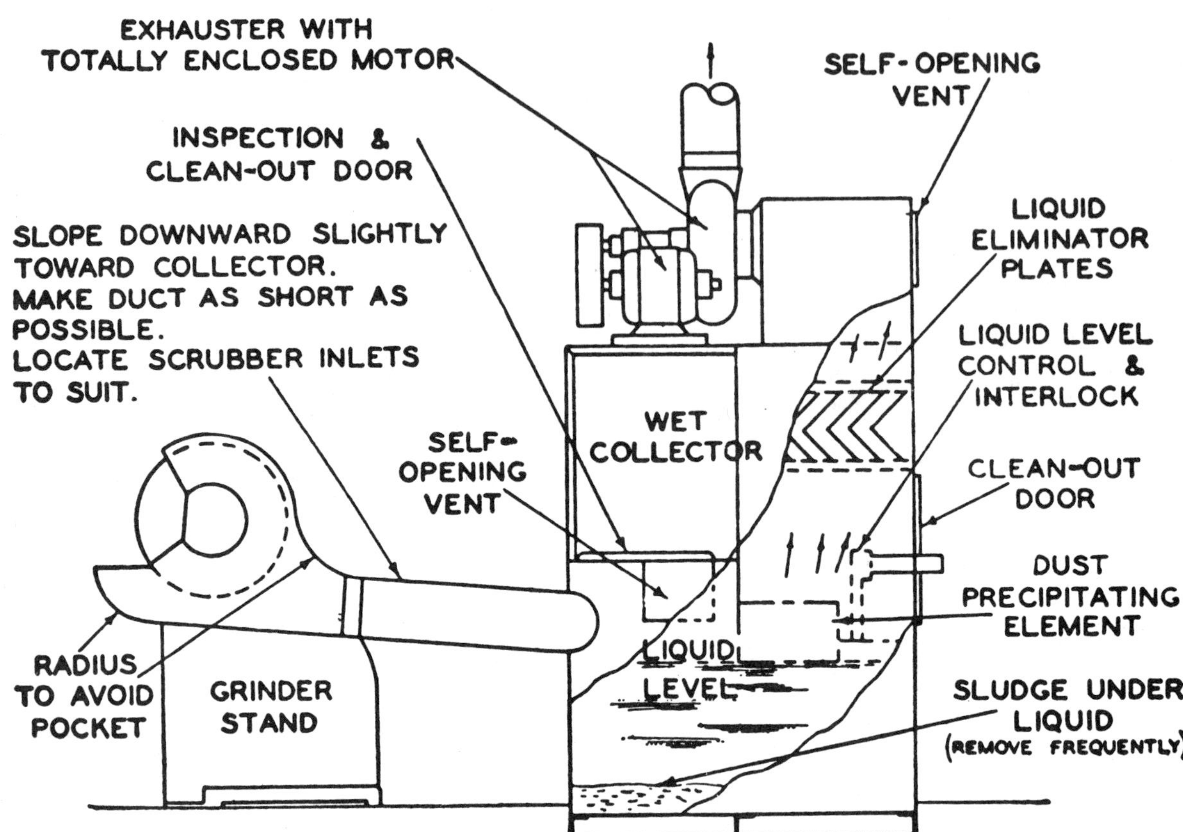

Figure 24–8. This magnesium dust-collection system converts dust into wet sludge for later removal. (Courtesy Dow Chemical Co.)

Prominently display warning signs inside and outside the grinding rooms or areas. Post signs that warn against smoking and against the use of water on magnesium fires. Signs should instruct how to use powdered graphite, limestone, or dolomite as an extinguishing agent.

Keep close to each grinding unit an ample supply of powdered graphite in plainly labeled and covered metal containers. Place a scoop inside each container. Keep the container's lid loose for easy access.

Prevention of Fires and Injuries

To prevent fires and injuries, observe the following safe work practices during the grinding of magnesium alloys:

- Start and run the grinder and exhaust system for a few minutes before beginning the grinding operation.
- Have operators of grinding equipment wear leather or smooth, fire-retardant clothing, not coarse-textured or fuzzy clothing. They should brush their clothing frequently.
- Also have operators wear goggles or a full-brim helmet with a face shield. They may wear a skull cap under the helmet. Recommend that they wear leather gloves with long gauntlets. Train workers in the use of self-contained breathing equipment.
- Keep machine tools sharp and properly ground for magnesium alloys, or friction sparks may cause fires.
- Use only neutral mineral oils and greases for cooling and lubrication. Animal or vegetable oils, acid-containing mineral oils, or oil-water emulsions are potentially hazardous.

Chipping

Where castings are cleaned or chipped, provide tables, benches, and jigs, or fixtures specially designed and shaped to hold the particular casting. Install screens or partitions to protect other employees from flying chips. Install hoods and exhaust systems in these areas to remove dust. Require workers to wear eye and face protection when cleaning or chipping castings.

Welding

Consider welding is done when cleaning or reclaiming castings. To help prevent fires in areas where welding operations are conducted, spread sand on the floor to a depth of 2 in. (5 cm). Sand, one of the best noncombustible materials, is plentiful in all foundries, but is also a health hazard.

Powder washing is a method of cleaning castings in which a stream of powdered iron oxide is introduced into a gas flame to intensify the heat produced. Perform powder washing according to the same safe practices as other carbon-steel or cast-iron welding and cutting. However, when this method is used to clean or cut sprues, gates, and risers from alloyed castings, use exhaust ventilation. (See Chapter 21, Welding and Cutting, in this volume for more safe welding practices.)

Power Presses

Power presses are used widely in finishing departments of foundries. For safety in power press operations, provide sufficient aisle space, good housekeeping, and effective lighting. Properly guard and maintain machines in good working order. Carefully select operators, and train them in the efficient and safe operation of power presses. Use mechanical feed and ejection equipment whenever practical. These topics are fully discussed in Chapter 23, Cold Forming of Metals.

FORGING HAMMERS

There are several types of forging hammers: open-frame, gravity-drop, and steam and air hammers. They have similar hazards in common and require special safeguarding and safe work practices.

Open-Frame Hammers

Open-frame or Smith forging hammers are constructed so the anvil's assembly is separate from the foundation of the frame and operating mechanism of the hammer. They may be single or double frames.

Flat dies are generally used in Smith hammers, and the work done allows for more machining of material. Smith hammers are used (1) when the quantity of forgings to be run is too small to warrant the expense of impression dies, or (2) when the forgings are too large or too irregular to be contained in the usual impression dies.

Gravity-Drop Hammers

Drop forgings in closed-impression dies are produced on gravity-drop hammers—both board-drop hammers and steam- or air-lift-drop hammers. Both types of gravity-drop hammers shape the hot metal in closed-impression dies. The impact of the hammer's blows shapes the forging through one or more states to the finished shape. On gravity-drop hammers, the ram and the upper die are raised to the top of the hammer's stroke. The impact blow comes from the free fall of the ram and the die.

Forgings on gravity-drop hammers may range in weight from less than an ounce (28 g) to 100 lb (45 kg) and may be made from any type of malleable metal, such as steel, brass, bronze, aluminum, or magnesium alloys. Gravity-drop hammers are designed so alignment between the dies can be maintained by using guides, die pins, die locks, or a combination of these devices.

The board-drop hammer, a more common type, uses hardwood boards that are secured to the ram and held in it by wedges. The boards pass between rotating rolls that grip the boards and raise the ram and the upper die for the successive blows. At the top of the stroke, the rolls release and the ram is held in this position by clamps until released by depression of the treadle. The impact of a board-drop hammer cannot be varied while the hammer is in operation.

A steam- or air-lift gravity-drop hammer is controlled by a valve that admits steam or air under the piston into the cylinder in the head of the hammer to raise the ram and the die. At the top point, the air or steam is exhausted and has no effect upon the hammer's blow, which is controlled entirely by the weight of the freely falling ram and die. Like the board-drop hammer, the steam- or air-lift hammer is operated by treadles or pedals, levers, or air valves.

The height to which the ram is raised, and consequently the impact, can be varied during the operation of an air- or steam-lift hammer. The falling weight of the ram assembly and upper die of gravity-drop hammers may range from 400 to 7,500 lb (181 to 3,402 kg).

Steam Hammers and Air Hammers

Steam hammers are also classified as drop hammers. Most steam hammers are double acting. They use steam pressure, or air pressure, that goes through a piston and cylinder to raise the ram and the die and to assist in striking the impact blow. Because steam or air power is used in addition to the weight of the falling ram and die, the steam hammer strikes a heavier blow than a gravity-drop hammer using an equivalent falling weight.

The falling weight of the ram assembly and upper die of double-acting steam hammers ranges from 1,000 to about 50,000 lb (454 to 22,680 kg). These hammers commonly produce forgings ranging from a few ounces to above 500 lb (227 kg), and sometimes up to 2,000 lb (907 kg).

Steam- and air-operated hammers are made with many built-in safety features. Some of the most outstanding are the following:

- Hammers are designed so the distance between the floor and the die seat is approximately 36 in. (90 cm).
- Safety latches prevent the ram from dropping when a job is being set up or when work is being done on the die. However, for complete protection, shut off the steam or air from the press and block up the ram.
- A safety cylinder head protects against piston overtravel.
- Many of the working parts are safely enclosed.
- Operating levers are placed in a safe position.

Hazards of Forging Hammers

For the most part, all types of forging hammers have identical hazards. The most frequent causes of injury include the following:

- being struck by flying drift and key fragments, or by flash or slugs
- using feeler gauges to check the guides, wear, or the matching of dies
- using material-handling equipment improperly, such as tong lifts
- having fingers, hands, or arms crushed between the dies
- having fingers crushed between tong reins
- receiving kickbacks from tongs
- using swabs or scale-blowing pipes with short handles
- being burned by hot scale
- dropping stock on the feet
- getting foreign objects, such as iron dust or scale, in the eye
- noise-induced hearing loss.

A hearing conservation program that includes proper hearing protection, annual audiometric examinations, as well as engineering controls, will greatly reduce or limit noise-induced hearing loss. (See *Fundamentals of Industrial Hygiene*, 4th ed.)

Injuries may also occur from a steam-drop hammer when the ram pulls off a new piston rod. Sometimes the rod must be set in the ram several times before it holds. If the piston rod breaks, the ram will fall. This hazard emphasizes the importance of operators using a safety prop to support the ram before reaching under it.

Operating a hammer with a worn cylinder sleeve is also hazardous. When the sleeve is so worn that the swing of the ram cannot be controlled at the throttle control, shut the hammer down and repair it.

Operating a hammer with broken piston rings is also dangerous. A piece of broken piston ring passing through the steam ports and lodging in the throttle's valve can cause the ram to drop out of control. When this happens, the operator's tongs or the transfer tool is often caught, thus causing serious injury.

Guarding

Maintenance personnel, in particular, are exposed to the potential danger of crushing injuries when they remove and install parts on the opt of the hammer and when they remove sow blocks, anvils, and columns. To avoid these injuries, provide and use means for locking out the power. To provide safe footing for personnel, install catwalks and guardrails on all hammers (Figure 24–9).

Figure 24–9. Permanent catwalks installed along the row of board-drop hammers make repair and servicing of hammers easy and safe.

Gravity-Drop Hammers

On gravity-drop steam- or air-lift hammers, use a hand lever rather than a treadle for cold restrike operations. Provide two-hand tripping controls (1) if the material being forged is not held by the hands or by hand tools, or (2) if a safety stop or tripping lever cannot be installed.

On board-drop hammers, provide a substantial guard around the boards above the rolls. This prevents the boards from falling should they break or come loose from the ram (Figure 24–10). Other standard protective features for a board-drop hammer include the ram stop and safety chain for the tie bolt and nut.

Steam and Air Hammers

Steam and air hammers should have a stop valve or quick-opening and -closing valve. Also, provide a safety head in the form of a steam or air cushion (if not already standard on the hammer) to prevent the piston from

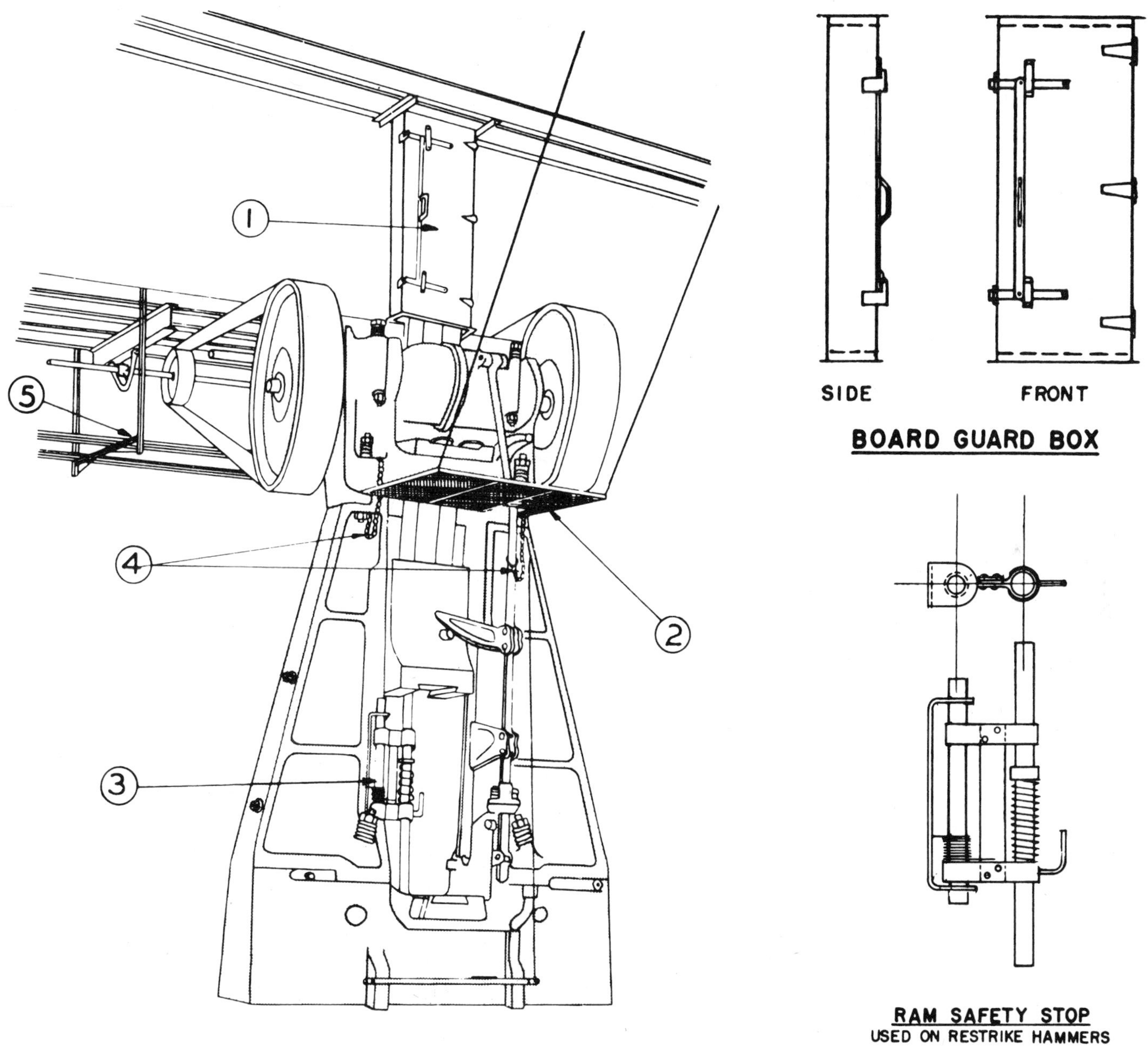

Figure 24–10. A well-guarded board-drop hammer features: (1) sheet steel board guard box, (2) screen platform made from No. 9 expanded metal, (3) steel ram safety stop that swivels on the left column, (4) safety chain to restrain tiebolt and nut, and (5) catwalk and belt catcher. Details of board guard box and ram safety stop are shown in drawings at the right. (Courtesy American Brake Shoe Co.)

striking the top of its cylinder. Connect the cylinder head's and safety head's bolts to an anchored wire rope.

If the hammer has no self-draining arrangement, install a drain cock, preferably the quick-acting type, in the lower part of the cylinder at the back of the hammer. Arrange this drain cock so that it can be opened without danger to anyone, or pipe it to discharge at a safe place. Ensure that steam lines are well trapped.

If air or steam is used to remove scale, provide a quick shutoff valve so the pressure can be regulated. The operator should adjust the scale guard to protect other employees from flying scale.

Key-Driving Rams

A pneumatic key-driving ram is superior to a manually operated one and offers a far greater margin of safety (Figure 24–11). Make the key-driving ram of properly hardened steel so that it will not chip on impact. A manually operated key-driving ram should be sturdy and well balanced. Do not allow the driving face to mushroom. Keep it in shape or replace it—do not burn it off with a cutting torch.

Scale Guards

Install a scale guard, to confine pieces of flying scale, as standard equipment on the back of every hammer. The guard should allow ample clearance for the ram and easy access to the dies. It may be installed in one of the following ways:

1. Hinged on one side to an upright post so that the guard can be swung closed or open, out of position, when access to the die area is required (Figure 24–12). This installation is considered the most efficient, and is widely used throughout the industry.
2. Supported on a floor standard.
3. Suspended from the ceiling or anchored to a rail.

Treadles and Pedals

Provide treadles and pedals with ample clearance. Guard them to prevent them from being unintentionally tripped by a falling object. Also guard any portion of a treadle or pedal at the rear of the hammer so that scrap or other material cannot interfere with the treadle's action.

Several methods for interlocking the treadle have been developed. A simple mechanical interlock is effective. When it is necessary to use a pry bar to remove stuck forgings from die cavities, lock the treadle. In order to lock the treadle, keep the pry bar in a sleeve-type holder with the bar's weight resting on the actuator of an air valve. If the pry bar is removed, the air valve opens and the air cylinder drives a wedge under the treadle's arm to prevent its movement.

Flywheels and Pulleys

Enclose flywheels or drive pulleys with a guard that is strong enough to prevent the pulley from falling to the floor should the shaft break. In this installation, the strength and location of the guard's bracket or the frame are more important safety factors than is the gauge of the sheet metal used for the enclosure. Bolt the brackets to the column of the hammer. In some instances, the guard enclosure is supported from the floor by an I-beam. Restrain all cylinder bolts, gland bolts, and guide bolts and liners, as well as the head assembly over the operator's working position, with wire ropes or chains (Figure 24–13).

Safety Props

Provide safety props equipped with handles at the middle. Require workers to use them when repairing, adjusting, or changing dies. The props should be held

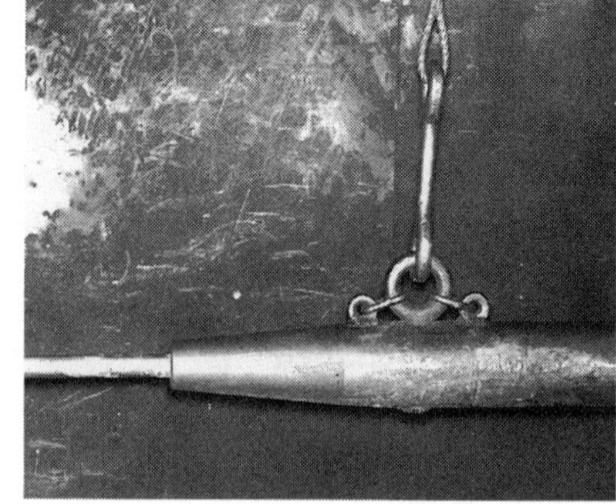

Figure 24–11. Mechanical key-driving rams, like the pneumatic model on the left, are preferred to manual ones. A manually operated key driver is shown on the right. (Right photo courtesy Tractor Works, International Harvester Co.)

Figure 24–12. This scale guard, closed in the left photo, is hinged to provide access to the drop hammer for housekeeping. The anchor post is set in a sleeve in the floor for easy removal. On the right, the hinged scale shield is open.

in place while power is released. This permits the weight of the upper die and the ram to rest on the props. Operators should never place their hands on top of a prop. The props can either be chained to the hammer, so they cannot slip out of position, or be hinged to the side of the hammer, so they are readily available and easily moved into and out of their blocking position (Figure 24–10). The material most commonly used for ram props is hardwood timber not less than 4 by 4 in. (10 cm by 10 cm) with a ferrule on each end. Ram props may also be made (1) of a section of magnesium or aluminum that is carefully squared at the ends, or (2) a section of steel tubing with not less than a 22 in. (5 cm) outside diameter with a wall thickness of 2 in. (1.3 cm). These metal materials must have the same load capacity and safety factor as the comparable hardwood prop.

Hand Tools

Use pliers, tongs, and other devices specially designed to feed the material so the operators need not place their hands under the hammer at any time. Tongs should be long enough that they can be held at the side of the body rather than in front. Tongs should fit the shape of the materials being held for forging. Oil swabs and scale brushes or pipes should also have handles long enough so operators do not have to place their hands or arms underneath the die.

Die Keys

Use die keys made of a suitable grade of medium, carbon-alloy steel that has been properly heat treated so it will not crack or splinter. Both ends of the key should be tapered for clearance in driving and removing the key. Never use mushroomed keys.

Die keys must be the correct length. They should not project more than 2 in. (5 cm) in front and 4 in. (10 cm) at the back of the hammer. Use shims if necessary. If keys project farther, they become a hazard to the operator working in front. They may also break off while the hammer is operating and fall between the dies in back.

Stock an adequate supply of die keys so drifts will be needed only when the end of a key becomes distorted and must be cut off before the key can be driven out. Block or securely hold the drift with a drift holder.

Design of Dies

Hammer dies are usually made of chrome, nickel, or molybdenum stellite—materials that have high resistance to heat, shock, and abrasion. Die blocks are commercially supplied in four different tempers. Selection of the proper die steel in the correct range of hardness is important in controlling checking and breakage of dies.

Size, amount of striking surface, and height are other pertinent factors in the safe design of dies. Allow the correct amount of striking surface in relation to the size of the die. Too little striking surface may cause breakage or an undersized forging when the dies pound down. Too much striking surface, especially if it is unbalanced, may cause a pull or misalignment.

Specify correct die height, especially for resinking forge dies. For the benefit of the hammer operator, maintain a normal work level. To prevent the bottom of the cylinder from being knocked out, never let the height be below a specific minimum. On each unit, label the maximum and minimum operating heights. This information is for the design engineer.

The dies must be made so they meet in precise alignment. Lay out the dies so the major portion of the heavy forge work is done in the center of the die under the center of the ram, where the maximum force of the hammer is transmitted. Each impression in the die must be backed up with enough die material to reduce the possibility of breakage, especially where multiple impressions or nesting methods are employed.

Figure 24–13. Safety ropes keep the cylinder head, tie plate bolts, and gland bolts from falling if they break. Note the master rope that circles the base of the steam chest. The tie plate bolts and the gland bolts are secured to the master rope. The gland bolt safety rope should be tight enough to prevent a broken gland bolt from swinging down and striking the ram.

Systematically arrange preliminary or breakdown operations so they do not create a hazard for the operator as the forging cycle is completed. Avoid radical bends or severe reductions in volume that might tend to jerk the tongs from the hands of the operator. Modify such operations or have them completed in additional operations.

Make the thickness and width of flash, gutter, and sprue ample enough so the flash or tong holds are not sheared off. The size of the gates is important—design in relation to the size of the stock and the tongs used. Gates should have enough width, depth, and clearance to allow safe handling.

Some dies, especially for smaller hammers, are designed with cutoffs that shear the completed forging from the end of the bar. If possible, place such cutoffs on one of the rear corners of the die for the operator's safety. If cutoffs are placed on the front of the die and are used by placing the stock across the knife's portion at an angle, provide enough clearance between the die and the hammer's frame or gib.

Because of the nature of forging work, and the abnormal abuse to which the dies are subject, maintenance of hammer dies is important. For example, fillet and corner radii in the impressions may be enlarged or sharpened by impact pressures, scale, abrasion, and wear. It is, therefore, necessary to inspect these radii and correct any alteration that may be hazardous.

Sometimes the face of a forging die may be welded to correct some defect or to maintain specifications. A hard, brittle, or thin-skinned weld can become a flying hazard under impact. Therefore, select the rod and the preheating and postheating methods with extreme caution.

Provisions for storing dies, such as racks and rails, are essential to safety, good housekeeping, and efficiency. Store dies in an area separate from the forge shop and away from vibration.

Setup and Removal of Dies

When forge dies are set up or removed, the hammer operator should act as leader of the group. The operator should see that all efforts are coordinated and that all safety rules are observed. That way, the work will be done efficiently and safely.

Pre-Setup Activities

Before setting up dies, clean the immediate area around the hammer and clear it of obstructions. Do not perform maintenance work on the equipment when setting up a die.

The hammer crew should make the following check of the equipment between setups:

- The hammer should be in good working order.
- The seats of both the sow block and the ram should be flat and clean.
- Dowels and die keys should be inspected for galls and burrs.
- Dies should be checked for burrs and other defects, such as cracks or sharp corners.
- The overall height of the dies should be greater than the shut height of the hammer.

Good lighting is essential for accurate setting of dies. It gives the operating crew a better view of potential hazards. Portable lights may be used. They should have heavy-duty cords, with bulbs protected by heavy-screen guards.

If lift trucks are used, be sure that the floor is level, in good condition, and free of obstructions. If cranes are used, check that the lift chains are in good condition and that the die pins have a snug, but free, fit.

Setting Up Dies

Dies are usually heavy and hazardous to handle without proper equipment. Drill uniform holes in both sides of each heavy die block, and insert pins to make lifting and moving them easier. The diameter of the pin and hole depends on the weight of the die. Standard practice in many companies is to have the pin 1/16 in. (1.6 mm) smaller in diameter than the hole. Keep the depth of the hole and the diameter of the hose and pin uniform in dies of a certain weight group. This assures there will be enough pressure to prevent the pin from falling out.

Do not use transfer boards to move dies between the work bench and the machine. Transfer trucks, preferably of the elevating type, are safer and more efficient. Cover the top or table of the truck with sheet metal at least 3 in. (7.5 cm) thick, and securely fasten it in place.

Use power lift trucks or die trucks for moving and installing dies. Block or secure lift trucks to the base of the hammer when dies are to be set or removed. Otherwise, the truck may slip from the hammer, causing the die to slip and fall. In addition, the operator should check the truck's safety controls prior to starting the die set. See that operators are trained to safely operate this equipment.

Many methods are used in setting dies in hammers. The type and size of dies and the type of hammer determine the method to be selected. To set dies:

1. Prop the ram securely and shut off and lock out the power—whether steam, electricity, or air.
2. Drive die dowels (made of a grade of steel that will not splinter or crack) into the dowel holes in the die shank. Dimensions should be accurate to ensure a tightly driven fit.
3. After the bottom die of a steam hammer has been set in place, drive the bottom key to help line up the die and partially tighten it.
4. Invert the top die and set it in position so the dies are face to face with the match lines aligned.

(Reverse this procedure for a gravity-drop hammer: set and key the top die first. Sometimes both dies can be set at once.)

5. Remove the safety prop between the ram and sow block.
6. Let the ram descend slowly until it engages the top die.

Using shims on the dowel in the top die creates an extra hazard. Normalized spring steel is used to shim dowels that must be set so they will fall into place when the ram engages the die. The hammer operator should record the number and location of shims (whether front or back) so succeeding shifts or different hammer crews can refer to the record of the setup for that specific set of dies.

If a die must be moved to match, use a prop after the ram is raised and before the operator reaches under the hammer to reset the shims. This prop must be strong enough to support the ram and long enough to extend from the top of the die to the ram.

If allowance is made for moving the dies, make the allowance on a steam hammer in the top die only—the bottom die should have a tight fit. On gravity-drop hammers, however, the general safe practice is to have the top die tight, allowing for movement in the bottom die.

At this stage of the setup, drive the die keys with a hand sledge only. Common practice is to drive the keys up tight with a sledge or light ram and then "bounce" the dies. The safe procedure for bouncing is as follows: bounce, shut off the power, ram the key, bounce, shut off the power, ram the key, etc. The impact helps align the dies but creates an additional hazard if lock dies are being set.

Take extra precautions and use special equipment for abnormally large or long dies. In setting such dies, the regular safety procedures for propping and handling may have to be changed. Get the approval of proper facility authorities for any changes.

After driving the die keys and before adjusting the gibs or column wedges, apply heaters to the dies if they have not been preheated. On deep impression jobs, it is a good practice to preheat dies in special low-temperature furnaces, in hot-water baths, or with hot scrap steel before setting them up. Driving die keys too tight when the dies are cold may crack the shanks, sow blocks, or rams.

After heating the dies to approximately 300 F (149 C), drive the die keys tight again by means of either a pneumatic ram or a light, suspended ram (Figure 24–11). If any further adjustment to the hammer is required, it can be done after a tryout forging has been made.

The hammer crew should use any waiting time to make the following final check before getting ready for production:

- Check the dies for proper alignment and for proper wing clearance—tight wings may cause breakage.
- Check that all tools have been put in their proper places.
- Move scale guards into position.
- Make final adjustments to the billet heating furnace.

Removing Dies

Before dies are removed, clear the immediate area around the hammer of overhead trolleys, suspended tongs, portable conveyors, tool and billet stands, and other equipment. Tie down overhead trolleys so they will not creep back into the work area. Move the scale guard back, and remove accumulated scale that would interfere with safe footing. Immediately move forgings away from the unit, and place them in the next work station.

If another set of forging equipment has been delivered, place it nearby but not directly in the area where the hammer crew will work. To eliminate unnecessary handling, make sure that service personnel (truckers, crane operators, and hookers) are familiar with the proper procedure.

Shut off and lock out the hammer's energy sources (electrical, air, steam, or hydraulic) before loosening the die keys. The top key is generally loosened first, usually with a mounted pneumatic ram (Figure 24–11). A light, well-balanced ram suspended from a cross beam or from an overhead crane or chain fall can also be used successfully.

Using a manually held drift pin or a knockout on a die key after it has been loosened and driven to a position even with the face of the ram or the sow block is a hazardous operation. Instead, use a special type of adjustable knockout (Figure 24–14) that is held in position mechanically rather than manually.

After the die keys have been driven out, raise the ram and prop it at once. The prop must be in good condition and must be placed on a clean surface. On a gravity-drop hammer, use a jack to raise the ram. A special prop may be required.

Do not attempt to raise the hammer to propping level if the top die has a tendency to "hang." Instead, first free the die within the shortest possible distance from the face of the bottom die. Prop the ram on an air-lift drop hammer with special care. After securely positioning the prop under the ram, shut off and lock out the power. Use special platform trucks with winches for this operation. They are practical and safe because the dies are horizontally winched or pulled out directly onto the table of the truck. Do not permit the dies to be dumped out of the hammer onto the floor.

After removing the dies from the hammer, extract the dowels. Two workers should drive out the dowels with the proper tools, usually a drift and a sledge. These tools should be in good condition and have sound handles. Because there is metal-to-metal contact, see that workers are careful and wear personal protective equipment.

Load the removed dies onto low, steel pallets and take them from the area as soon as possible. If dies need repair or modification, the hammer operator should notify the supervisor. The supervisor should then have the die-servicing department take care of any repairs before the next run.

Safe Operating Practices

Make the supervisor who directs the activities of workers in the hammer crew responsible for the following safe work practices:

- Keep all guards in place when the hammer is in operation. Make sure guards are in good repair.
- Move material and tools from the aisles and from the operator's work space, and store them in the proper place. Keep the floor area around hammers free of scale, oil, water, and other material, to ensure safe footing.
- Before starting work, have the hammer crew make its own inspection to see that the equipment and work area are in order. Have them frequently check for breakage at all critical points that are subject to severe strain. If an unsafe condition is found, the workers should inform the supervisor immediately.
- Never operate drop hammers when dies are cold. Dies should always be preheated by hot steel placed between them.
- Do not permit any adjustments, repairs, or service until (1) all energy sources (electric, air, steam, hydraulic and compression springs, etc.) have been isolated and locked out, (2) the treadle has been blocked to prevent unintentional tripping, and (3) the ram has been propped.
- When dies are being set on a board-drop, steam- or air-lift gravity-drop hammer, have operators fit the dowel in the upper die and the ram with as few shims as possible. First, move the bottom die, which should have enough shims so it can be lifted easily. This procedure is the opposite of that followed in setting dies in a double-acting steam-drop hammer.
- On steam-drop hammers, place a prop between the ram and the shank of the top die before the die is moved. When it is necessary to move the bottom die, place the prop between the sow block and the ram on the side containing the dowels, so the die can be moved.
- On steam-drop hammers, do not adjust the spool bolt until the main steam valve has been turned off and the treadle blocked. This prevents the ram from being unintentionally activated while the operator is adjusting the bolt.
- Laying liner stock between the dies to jar loose a stuck forging is dangerous. When a forging sticks, stop the hammer, remove the forging, relieve the die, and continue the operation. On some operations where this method is not possible, use a safety liner made of soft steel.
- Whenever operators leave the hammer, even if only for a few moments, they should leave the upper die resting on the lower die to prevent untentional tripping.
- Carefully observe flywheel speeds. As a rule, do not permit them to exceed the number of rpms given on specification sheets. This speed is the one upon which proper operation of the press is based.

Personal Protection

Operators of forging hammers and other employees in the vicinity of equipment should wear suitable personal protection equipment. This includes full eye protection, safety hats, protective footwear, leather leggings and aprons, and hearing protection in accordance with the plant's policy.

Operators should also wear cotton gloves. When the gloves get wet, they should be removed and allowed to dry. Operators should not wear leather gloves because perspiration may cause steam burns.

Maintenance and Inspection

A well-planned preventive maintenance program for forging hammers helps to reduce the number and severity of incidents by minimizing the breakage and wear of parts. Regular inspections disclose production units that are not properly operating so repairs or adjustments can be made.

The results of a good maintenance program can be measured in reduced operating costs that include:

- cost of machine downtime, breakage, and lost production
- cost of replacement parts and labor
- cost of incidents due to faulty equipment.

Maintenance checklists for hammers (Figures 24–15 and 24–16) can be the basis for formulating a definite, planned inspection program. A written checklist avoids the errors resulting from verbal reports that are often forgotten or misunderstood.

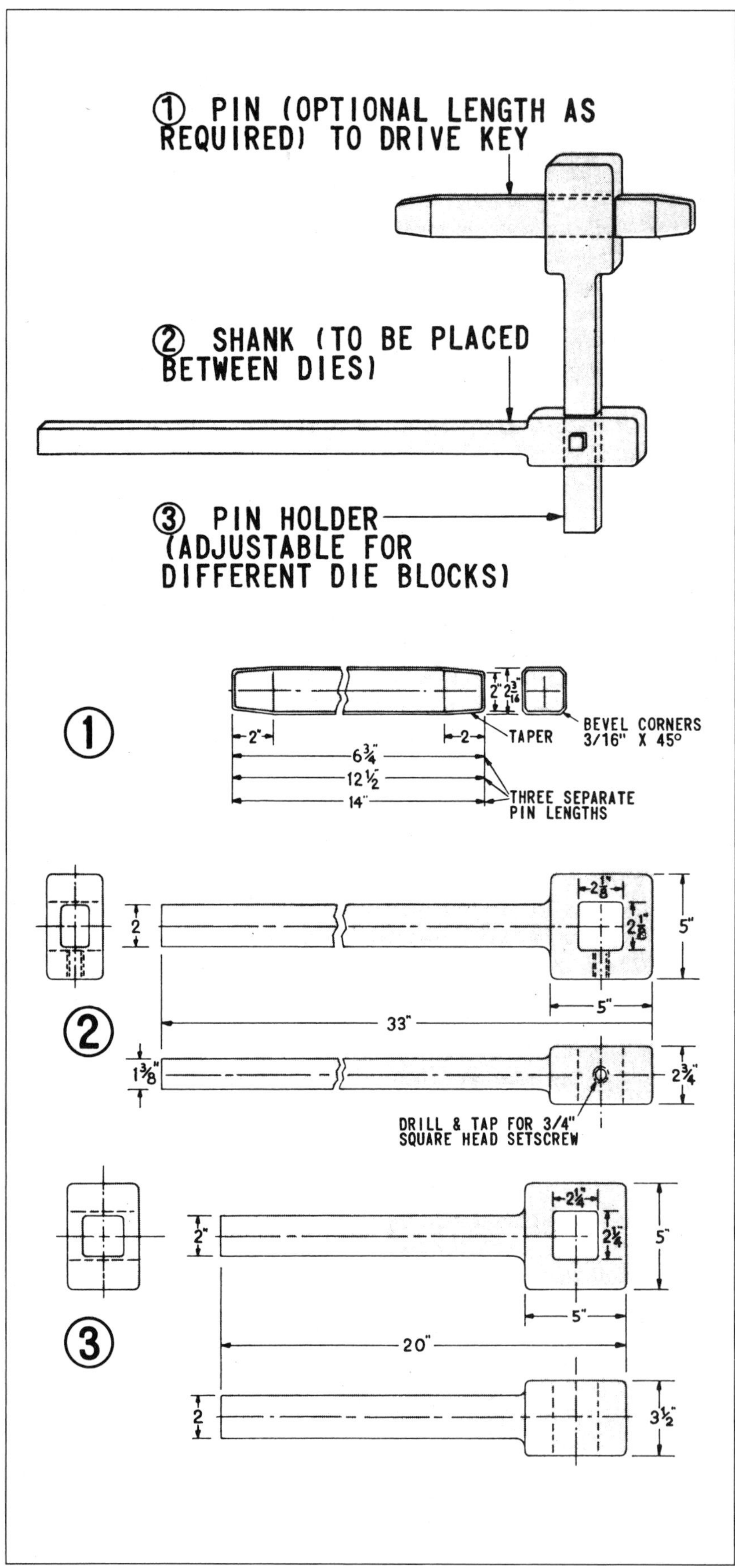

Figure 24–14. A key knockout, which can be easily made, is held in position mechanically. The only machining required is to drill and tap the ½ in. (2 cm) hole for the set screw.

Set up a work schedule for repairs based on data recorded on the checklists. Transfer data to the permanent records of the equipment and use it for future planning in the maintenance program, and to compare costs.

Since steam hammers constitute a considerable portion of the forging equipment, establish a definite maintenance program for them. Many steam hammers are not kept in as efficient condition as possible. Usually the cost of operation is not known, but upkeep costs are higher for units in poor condition. Waste of steam usually results from worn piston rings or sleeves, loose heads, blown head gaskets, and leaky glands. Replacing worn rings reduces costs. Worn piston sleeves and sloppy linkage make the hammer hard to control and create a hazard. Loose cylinder heads also are dangerous.

Periodic inspection of every forging hammer helps ensure the proper condition of bolts, screws, keys, valves, and other parts that may be loosened by vibration. Similarly, make thorough periodic inspections and adjustments of all parts of the treadle or pedal, clutch, and other operating mechanisms. Worn or loose treadle linkage, motion arm, crank arm, and treadle can cause the hammer to go out of control. Keep these parts in especially good repair.

BOARD HAMMER MAINTENANCE CHECK

Date________ Hammer No.______ Location______

ITEM CHECKED	CONDITION	TYPE OF REPAIR	EST. HRS. TO REPAIR
TIE BARS & SPRINGS			
FRAME STUDS & SPRINGS			
DIE KEYS & SHIMS			
SOW BLOCK KEY & SHIMS			
RAM CLEARANCE			
GUIDE BOLTS & ADJ.			
MOTOR MOUNTS			
MOTOR COUPLING			
ROLLSHAFT BEARINGS			
WIRING & CONTROLS			
FLYWHEEL & BEARINGS			
DRIVE GEARS			
ROLL ADJUSTMENT			
LUBRICATION			
STEAM LINES			
BOARDS & WEDGES			
AIR LINES			
AIR FOOT SWITCH			
AIR CYL. & LINKAGE			
TREADLE & LINKAGE			
BOARD CLAMPS & LINKAGE			
DOGS & STOPS			
KNOCKOUT ARM			
FRICTION RODS			
SAFETY RODS			
REMARKS:			

Figure 24–15. A maintenance checklist for board-drop hammers.

STEAM HAMMER MAINTENANCE CHECK

Date________ Hammer No.______ Location______

ITEM CHECKED	CONDITION	TYPE OF REPAIR	EST. HRS. TO REPAIR
CYL. HEAD BOLTS			
MOTION VALVE STEM			
MOTION VALVE CRANK			
MOTION VALVE CONNECT.			
WIPER BAR & CRANK			
THROTTLE CRANK			
THROTTLE LINKAGE			
RAM & SOW BLOCK			
DIE KEYS & SHIMS			
SOW BLOCK KEYS & SHIMS			
GUIDE BOLTS			
GUIDE WEAR			
GUIDE ADJUSTING BOLTS			
GUIDE WEDGE POSITION			
HOUSING BOLTS & SPRINGS			
TREADLE			
TIE PLATE LINER			
PISTON ROD GLAND PLATE			
GLAND BOLTS			
TREADLE PLATFORM			
STEAM CUSHION LINE			
SCALE HOSE			
SAFETY LINER			
SAFETY PROP			
STEAM LINES			
STEAM SHUTOFF VALVES			
REACH ROD			
COLUMNS FLAT ON BASE			
COLUMN WEDGE & BOLTS			
SPOOL BOLT & PIN			
SAFETY CABLES			
CRACKS IN BASE			
REMARKS:			

Figure 24–16. A maintenance checklist for steam hammers.

The clutch is also a vital part of the forging press. Keep it in good condition if the press is to operate efficiently. Replace a broken spring or part that shows wear at once.

Lead Casts

Lead casts are taken in practically every conceivable manner in the forging industry. If possible, take casts only in an isolated area, where there is no likelihood of interference from, or injury to, other workers. Make sure die impressions are dry because hot metal that contacts water produces flying particles of molten metal. Make sure that lead pots are properly ventilated. (See the National Safety Council's *Fundamentals of Industrial Hygiene*, 4th edition.)

FORGING UPSETTERS

The upsetter is a horizontal forging machine that forges hot bar stock, usually round, into a great many forms. The forms are made by squeezing action instead of impact blows, as in the case of forging hammers. Although numerous hazards are involved in the operation of an upsetter, the most serious problems are encountered in changing the dies.

Enclose the entire machine as much as possible, except for the feed area. Use heavy wire mesh, or expanded or sheet metal reinforced with structural steel. Cut doors into the enclosure to service the flywheel, brake, and other moving parts. Install a guard over the operating pedal.

For safe operating conditions, keep the area around the machine clean and clear of obstructions and litter. Especially keep the top of the machine clear of any objects, such as loose bolts, bars, nuts, or shims, that might fall into it or from it.

Note: Before attempting to adjust dies, heading tools, stock gauges, or backstops, the operator should shut off the power, lock the main power switch, and, after the flywheel has stopped completely, immobilize the flywheel.

Design of Dies

Dies and heading tools used in an upsetter or horizontal forging machine do not usually receive the severe abuse that hammer dies receive. For gripper dies, use a good grade of chrome, nickel, or molybdenum steel of the correct hardness.

When resinking dies for upsetters use inserts rather than cutting down the dies. The inserts most commonly used are made of a high-chrome, -nickel, or -molybdenum steel with silicon added.

Tool-quality steels with a vanadium and tungsten analysis are also used for inserts and headers. However, select them to satisfy the operating conditions because they have limited use where coolants are employed.

Design and locate front or back stock gauges or backstops to serve the specific purpose and to aid the operator. Take special care to eliminate all hazards, especially pinch points.

For abnormally heavy jobs (or jobs that would create an unbalanced condition when running) design and use balancing equipment that eases handling and reduces operator fatigue. Make sure that the grip sometimes provided on the upsetter's die impressions is strong enough to hold the stock securely. Check the grip after every run. This precaution is important for the operator's safety, especially on jobs where the heading tool could push the bar stock out of the impression toward the operator.

Setup and Removal of Dies

At the end of a run and before further work is done, move all skids of stock or forgings out of the area to allow as much room as possible for changing the dies. Die setters should inch the header slide forward to make a complete setup by measuring headers, strokes, and dies. See that this practice is especially followed on a worn machine that requires special shimming for proper alignment.

The proper sequence and procedure for removing and setting up dies are as follows:

1. Check to make sure that the correct tools and equipment are at hand.
2. Check dies and headers for defects that could develop into hazards.
3. Shut off all power and lock the master switch off.
4. Turn off the water or header lubricant and set the air brake. If the upsetter does not have an air brake, wait until the flywheel has stopped completely.
5. Loosen all set screws, lock nuts, and hold-down bolts on the die clamps. Do not remove hold-down bolts or die clamps until the die has been secured to the crane, hoist, or other device that will be used to remove the die. Even then, operators should watch out for falling or swinging dies.
6. Remove the dies by means of a swivel or arm crane. (Drill eyebolt holes 2 in. (5 cm) deep into all dies. Shouldered eyebolts or swivels should be stocked only in 3 in. (6 mm) graduations to a minimum of ¾ in. (2 cm) in diameter.)
7. Remove all die packing and thoroughly clean the die set's seat; inspect for burrs, especially along the die keys.
8. Remove headers and dummies and any shims that may have been used.
9. Measure the new dies to determine the amount of packing needed. To help protect the die seats, use at least an ⅛ in. (3 mm) liner.
10. Set in the new stationary die. Check to make sure that it is seated properly and packed correctly.
11. Tighten the hold-downs by hand before unhooking the die from the lifting device.
12. Turn on the power, open the safety, and release the brake and the flywheel. Set the machine on Inch and slowly inch the header slide forward to bring the tool holders into correct position for the headers.
13. Disconnect the power and lock out the master switch. Make certain that the flywheel has completely stopped.
14. Assemble the new headers, tool dummies, and tool holders. Be sure that the headers to the die cavity

are correctly matched and on correct center. To set up, use a die layout that shows all principal dimensions on the equipment.

15. Check to see if shims are needed behind the headers. Do not use horseshoe shims. Instead, use washer shims that fit around the shank of the header and cannot fall out.
16. Make a complete check of the assembly before finally tightening the header and dummy set screws.
17. Insert the moving die, and make sure that it is properly located. Then tighten the hold-down bolts, before unhooking the die from the lifting device.
18. Turn on the power. Set the machine on Inch, and slowly close the dies.
19. Check for match alignment and the proper amount of packing. If too much packing is used, the safety pin should open the dies.
20. Allow for expansion of the dies when they become hot. It may be necessary to remove a shim from in back of the die.
21. If everything checks out, shut off the power, lock out the master switch, and allow the flywheel to come to a complete stop.
22. Tighten hold-down bolts, lock nuts, and set screws.
23. If needed, attach a front gauge or backstop gauge.
24. Turn on the power and try out one forging. Check the dimensions. If dies or headers need further adjustment, again turn off the power, lock out the master switch, and wait until the flywheel comes to a complete stop.

Inspection and Maintenance

Because worn or defective upsetters can be dangerous to operate, keep these machines in top working order. Establish a definite program of inspection and maintenance.

At least once a week, the maintenance crew should check all working parts for wear and proper adjustment. Daily, the crew should inspect the air clutch and brake. Upsetters also require daily lubrication. If possible, install a means for automatic lubrication.

Operators should daily inspect air gauges, air lines, water lines, water valves, belts, pulleys, and tools. They should also check daily, and immediately report, any abnormal function. At each use, inspect all equipment for handling dies, such as chain, cables, and eyes or swivels.

Auxiliary Equipment

Design all equipment needed to safely operate upsetters, such as stock gauges, tongs, oil swabs, and scale removers, for the particular job.

Keep tools in good condition. Provide a complete set of wrenches to fit all sizes of nuts or bolts that are on the machine.

There are three basic types of stock gauges. The front gauge locates and swings away; the backstop gauge locates and helps hold the stock in place; and the special tong gauge or finger gauge locates and helps control the stock.

Use tongs made of tough, low-carbon steel so they will not harden from repeated quenching in water. The jaws of tongs should conform to the shape of the stock being handled.

Use oil swabs and scale removers with long handles. Thus, operators can reach the full length of the dies without having to put an arm or hand between the dies.

Air, electrical, water, and oil lines should have distinctly marked shutoffs. Locate safety valves or switches in a spot convenient for the operator.

FORGING PRESSES

Forging presses, because of their basic design, are similar to power presses discussed in Chapter 23, Cold Forming of Metals. Forging presses range from 500 tn to 6,000 tn (453,600 to 5,443,200 kg) capacity. However, since their work is quite different from the conventional cold-stamping operation, forging presses have their own operating technique, die setup, and maintenance problems.

Compared to the cold-stamping press, the forging press is designed with a faster-acting slide. The speed of the downward motion and of the pickup of the slide is one of the factors that determine the life of forging dies. Another factor is temperature control of the dies. Therefore, the action of the press should be fast enough to minimize the length of time that the dies are exposed to the billet's forging temperature.

Basic Precautions

The rapid action of the slide creates certain hazards that the operator must recognize and control. The size and shape of the forgings and the method of moving them into and out of the dies limit the use of point-of-operation guarding. The single most important factor for preventing injuries is the operator's control of the tools and methods used. Provide safety hats, hearing protection, eye protection, and protective footwear for operators at all forging press operations.

Tongs

Tongs, die swabs, and special handling tools are ordinarily the only tools used by an operator during the forging operation. Keep these tools in good condition at all times.

Because of the forging dies' temperature and the forging temperature of the billet forged, the operator must use tongs or special handling tools to move the billet into the dies. Operators should not hold the tongs' handles in front of their body nor place their fingers between the handles. Die swabs must have handles long enough to prevent hands and arms from entering between the dies.

Maintain proper clearance at the front of the press so operators have enough room for their hands in case of upward or downward motion of the tongs. This motion can be caused by improper spotting of the billets or tongs at the striking surfaces of the forging dies.

Scale

Properly locate steam lines, air lines, water headers, and splash aprons for scale and oil. If too near the working area of the operator's hands, they create pinch and shear points when tong handles are forced against them.

Unless properly confined, hot scale that is produced during the forging operation can cause serious burns and eye injuries. To prevent scale from coming out of the front of the press, locate air or steam curtains at the front of the die and direct them onto the die's facing. Install combination scale-and-smoke exhaust hoods at the back of the press to confine the scale and exhaust the smoke created by die lubricant.

Guarding

Equip all forging presses with pedal guards and nonrepeat devices. Never operate a forging press in the continuous-stroke mode. The operating controls should require depression of the pedal for every cycle of the slide. The press's controls should also permit inching of the slide for die setting or other press adjustments.

Die Setting

Because the setting of dies in forging presses is very different from the setting of dies in cold-stamping presses, certain extra precautions are required. Dies in forging presses are set in die holders designed in sets, each set consisting of an upper and a lower section. The upper section of the holder is secured to the press's slide by threaded bolts; the lower section, to the press's bed by threaded bolts. The forging dies are set or nested in die pockets that are recessed in both the upper and lower sections of the holder.

To solve the problem of setting the dies in the recessed pockets and of removing them, use an eyebolt attachment at the face of the die and a die truck with a boom attachment. Carefully use pry bars and blocking, since a pry bar could easily slip from the die and injure the operator's hands or fingers.

Secure the dies in the pockets of the holder with sectional flat clamps and a series of cap screws placed through the holder at all four sides of the die. These screws are primarily used to shift the die for matching and to prevent the die from floating after it has been properly aligned.

Before removing the clamps that secure the top die in the holder, install suitable blocking to hold the top die in the pocket. Do not rely on the adjustment cap screw to hold the die after removing the clamps. To prevent the motor from starting when the props are in use, equip the press with safety props that are interlocked with the press's motor circuit.

Keep the table wedges that are used for making vertical adjustment of the dies free of scale. In this way, the wedge can be raised and lowered with minimum pressure. Do not use trucks or driving rams for this operation. If wedges are kept free of scale, they can be raised or lowered by hand or by an air-powered motor wrench.

Maintenance

Maintenance of forging presses requires the same precautions used in maintenance of power presses used in cold stamping. First, lock out the energy source so the equipment cannot run while the maintenance is being done. Place safety props under the ram and surge tanks, and bleed the pressure lines so that the press's parts cannot be unintentionally activated. However, if adjustments must be made with the power on, see that the work is carried out under the direct supervision of the maintenance supervisor.

Install permanent work platforms for making brake adjustments and doing repair work at surge tanks and booster cylinders. To prevent falls, maintenance crews should not use portable straight ladders nor stand on parts of the press, such as the press's crown or the backshaft.

Major repair work on forging presses usually requires removal of bulky, heavy parts. Tearing down the rolling clutch, flywheel, slide, pitman, and crankshaft requires special heavy-duty rigging. Have only skilled workers, trained in tearing down, do this type of work. Observe the following maintenance procedures for forging presses:

- Properly adjust the brake, and keep it in good repair. To prevent the brake lining from becoming flooded with oil, place a sheet steel disk about 2 in. (5 cm) larger than the brake wheel on the eccentric shaft that is back of the brake wheel.
- Securely tighten pitman bolts. It may be necessary to design and make a socket, or box-end, wrench with a strong heavy handle for the specific press.
- Do not let flywheel hubs, clutch spline hubs, pinion gears, and brake wheels become loose on the pinion shaft. If they do become loose and run that way very long, keys and keyways will be ruined. Also, the keys will break out a portion of the pinion shaft on one side of the keyway. Periodically inspect the keys on each side of the pinion shaft, and tighten

them if they are loose. When a key will no longer stay tight, replace it with a carefully fitted new key.

- When the hub or gear on the shaft becomes too loose, it is impossible to hold the keys tight and the shaft will have to be turned down. Weld the inside of a steel hub, and bore it to a shrink-fit on the new shaft. Remachine the keyways, and make new keys. If the hub is of cast iron, make a tapered sleeve, or fit a new hub to the shaft. Experienced repair crews, however, warn against welding the flywheel end of the pinion shaft. If normalizing of the welded section is not complete, a crack may start and allow the shaft to break and drop the rotating flywheel.
- Tighten the friction slip on a press so it will not creep, even on heavy jobs. If the friction slip is allowed to move a small amount with every forging, this movement will polish or glaze the friction surfaces and soon the friction slip will move more with each forging. This movement will cause the friction surface to score the hub and clamp surfaces.
- The rotation of the flywheel on the friction hub will frequently wear or cut the inside diameter of the flywheel, causing it to run off center. To correct this condition, bore the inside diameter of the flywheel and fit it with a bronze bushing.
- Check the eccentric shaft for cracks in the fillet on each side of the eccentric. Cracks are usually noticed when the shaft is taken out to check the main bearings or to machine the shaft. Carefully examine any cracks to determine their length, depth, and direction of travel. If a crack is in an area where failure would permit the rolling clutch to drop off, replace the shaft.

OTHER FORGING EQUIPMENT

The following are examples of additional forging equipment:

- *Hot-trim dies and punches.* These dies and punches can be made from hardened chrome, nickel, carbon, and vanadium steels. They can also be made from medium-carbon steels with cutting edges that have been hard faced with a rod similar to a Haynes Stellite No. 6 rod. Use stellite on the trimmer only, and not on both the trimmer and the punch. A stellite punch should never work against a stellite punch die. One or the other, preferably the punch, should be softer.
- *Colt-trim dies and punches.* These dies and punches are usually made from a high-carbon, high-chrome, molybdenum-steel hardened and drawn to a 60 to 62 Rockwell C hardness. In both hot- and cold-trim dies, the designer should equalize the trim so that stresses are equally distributed. Provide proper working clearance for unloading cold-trim dies and punches. If possible, design cold-trims to work on guide pins for proper alignment.
- *Padding, bending, or straightening equipment.* This equipment is made from a good grade of wear-resistant carbon-nickel steel. Hot-pad dies are usually made with an opening between the die faces so that the hot forging acts as a cushion. Provide additional clearance in a hot- pad or restrike die, because the forging is constantly cooling and shrinking.
- *Cold-coin dies.* These dies are made from high-carbon, high-chrome steels hardened to a 62 to 64 Rockwell C hardness. The forging is coined cold. For best results, use either interlocking dies or dies with guide pins. Additional safety measures, such as magazine-type loaders, are sometimes made an integral part of the die for the operator's convenience.

  Grind the faces of cold-coin dies as smooth as possible. Do not use any lubricant—it could cause the forging to stick to one die face. The opening between the dies is sometimes limited mechanically so that no more than one forging can be loaded at a time.
- *Bulldozers.* The greatest danger in operating bulldozers is the possibility of a worker getting caught between the dies. To decrease this hazard, (1) attach a guard to the side of the moving head that travels with the moving head past the stationary head; (2) use telescoping rods or rails; (3) notch out the base plate to leave room for the operator's leg; (4) keep the clutch in good order so the machine will not repeat; and (5) guard the power-transmitting mechanisms.
- *Cold-heading machines.* Provide screen shields on cold-heading machines to protect workers from flying pieces. Guard relief springs to prevent the bolts and nuts from being thrown out, should they break.
- *Bolt headers and riveting machines.* Install treadle guards on these machines to prevent them from being unintentionally operated. Stop and block these machines before changing dies or making adjustments.
- *Hot saws.* Place tanks of water below the saws. Install 8 in. (20 cm) sheet-metal guards to stop flying sparks.

NONDESTRUCTIVE TESTING

Visual observation, even with magnification, cannot locate all small, below-the-surface defects in cast and forged metals, or in weldments, such as found in pressure vessels, boilers, and nuclear components. Proper nondestructive testing, however, reveals all such defects without damaging the parts being tested. Nondestructive testing methods locate the following defects:

- defects that are inherent in the metal, such as nonmetallic inclusions, shrinkage, and porosity
- defects that result from processing, such as high-residual stresses, cracks, and checks caused by handling, spruing, or grinding, or casting and forgings
- in-service defects, such as corrosion, erosion, and sharp changes in section.

The types of testing most commonly used for forged and cast metals are the following:

1. magnetic particle inspection
2. penetrant inspection
3. ultrasonic methods
4. riboelectric method
5. electromagnetic tests
6. radiography.

These methods, as well as others that apply to nonmetallic substances, are fully discussed in National Safety Council Occupational Safety and Health Data Sheet 12304-0662, *Ultrasonic Nondestructive Testing for Metals*. Recommendations for installation, inspection, and maintenance of the electrical equipment used in many of these testing procedures are given in Chapter 10, Electrical Safety, in this volume.

Magnetic Particle Inspection

Magnetic particle inspection is the most widely used testing method for forgings. It uses magnetism to attract and hold very fine magnetic particles right on the part itself. If a defect is present, it interrupts the magnetic field and is clearly shown by the pattern made by the particles. The part is magnetized in suitable directions by DC-line voltages transformed to low-voltage (4 to 18 volts), high-amperage AC, half-wave current, or three-phase full-wave current. Install and ground all electrical circuits according to NFPA 70, *National Electrical Code*.

Inspection materials are finely divided ferromagnetic particles that are selected, ground, and controlled to provide mobility and sensitivity. Materials are available in several forms and colors. The type of defect to be located and the condition of the surface to be inspected determine which form of material and which method—dry, wet, or fluorescent—is used. The color selected provides maximum contrast with the surface of the part.

Local exhaust is required to control the dust particles used for testing. If local exhaust is not feasible, operators should wear respiratory protection. They should also wear eye protection to guard against the irritating effects of the dust particles and of arcing. In addition, operators should wear personal protective equipment to prevent skin irritations from the dry powder and wet material used with this testing method.

Penetrant Inspection

Penetrant inspection is useful for revealing cracks, pores, leaks, and similar defects that are open to the surface in a metal or other solid material. The penetrant-inspection process is as follows:

1. Clean the part to be inspected.
2. Apply penetrant to the surface. Within a few minutes the penetrant is drawn into defects by capillary action.
3. Remove the penetrant from the surface. Depending upon the sensitivity of the material, the penetrant is removed by a water wash, a solvent cleaner, or an emulsifier followed by a water wash. The penetrant remains in the surface opening, however, until it is removed by the developer.

Use fluorescent penetrants to reveal defects under ultraviolet black light. Effectively shield ultraviolet equipment, or wear filter lenses of the correct shade. Defects may also be detected by a dye penetrant that contrasts with the surface color.

Since most penetrants are organic compounds that may cause dermatitis, avoid skin contact with penetrants. Wash exposed skin before smoking, eating, or drinking. Do not use or store smoking materials, food, or drink in the test area.

Ultrasonic Methods

Ultrasonic waves (above the audible range of 20,000 Hz) are created by an electronic generator that supplies high-frequency voltage to a piezoelectric-crystal transducer. Three basic ultrasonic methods have been developed for nondestructive testing of forged and cast metals: reflection method, through-transmission method, and resonant-frequency method.

Reflection Method

In the reflection method, that portion of the ultrasonic beam that strikes a flaw or discontinuity in the material is reflected; the rest of the beam goes on. The piezoelectric-crystal transducer radiates these waves through a coupling medium into the material. It also acts as a receiver to detect reflections, which are then picked up by an electronic amplifier and applied to a cathode-ray oscilloscope. The time intervals between the outgoing and the incoming waves are measured on the oscilloscope.

Through-Transmission Method

In the through-transmission method, a beam or wave is directed through a piece of material. If a flaw or discontinuity is found, the energy is absorbed, and the beam or wave does not get through. Fluids such as water, oil, and glycerine give better coupling than air. For that reason, they are generally used as the coupling medium between the transmitter, the material, and the receiver. In some applications, however, air or other gases can be used.

Resonant-Frequency Method

The resonant-frequency method is used primarily to measure the thickness of material. The equipment for resonant frequency consists of an electronic oscillator that supplies voltage of ultrasonic frequencies to a piezoelectric transducer. The transducer is pressed into contact with the part to be tested and includes lengthwise vibrations in the test piece under the area of contact. Another type of resonance instrument displays the thickness reading on a cathode-ray tube as a pip on a calibrated scale. Caution: Disconnect equipment from the power supply, and discharge the condensers whenever a cathode-ray tube must be adjusted or removed.

Triboelectric Method

The triboelectric method is used to detect minute quantities of current generated when two metallurgically or chemically unlike conductors are moved into frictional contact. If the conductors are alike, no current is generated. This method is designed to sort and identify metal parts of not more than four alloy types.

The equipment for the triboelectric method consists of a control unit and a portable sorting head that are connected by means of a cable. The sorting head contains the main controls for conducting the test and is designed as a one-hand operation.

Electromagnetic Tests

Two types of electromagnetic tests are currently being used in industry: magneto-inductive and eddy current. A third type, employing radar frequency, is also being used, but only to a limited extent.

Magneto-Inductive Tests

This method of electromagnetic testing uses variations in the porosity of magnetic materials to create variations in a pickup coil or probe.

Eddy-Current Tests

The second and most common type of electromagnetic testing uses alternating current in a coil or probe to induce eddy current into the part being tested. Defects and variations in properties or shape cause changes in the strength and distribution of the eddy current. The readout of eddy-current tests is presented on a cathode-ray tube, on a meter, by audible or visible alarm, or by a combination of these methods. Follow the manufacturer's recommendation to determine any leakage of electromagnetic radiation that may be present.

Radar-Frequency Tests

This third method uses high-frequency radar waves to measure the electromagnetic properties of thin coatings and surface layers of material. To make such tests, (1) a wave guide or cavity oscillator is coupled to the test object, and (2) high-frequency waves are then reflected from the object, thus indicating the surface's electrical resistance and the thickness of the nonconducting coatings.

In some radar-frequency testing installations, operators have been burned internally when they passed between the object being tested and the testing device. Formulate and enforce special regulations, and set up barriers to prevent operators and other workers from entering such areas. Explicitly follow the recommendations of the manufacturer of the equipment.

Radiography

Radiography uses x-rays and gamma rays. X-rays are unidirectional and their wavelengths can be varied, within certain limits, to suit the condition. Gamma radiography differs from x-ray radiography in that the gamma rays are multidirectional and their wavelengths, being characteristic of the source, cannot be regulated. Gamma rays for radiography usually are obtained from isotopes of cobalt-60 or iridium-192.

In some instances, gamma-ray exposures are inferior to x-ray exposures in sensitivity and contrast. Gamma radiography, however, has several advantages. Because of the nature of isotopes, a number of tests can be made at the same time, provided that specimens can be suitably located. Moreover, isotopes are independent of electrical power, their sources are portable, and the small size of the sources makes it possible to obtain radiographs in tight quarters.

Devices used to transform differences in intensity of the penetrating radiation into visible images are x-ray films, fluorescent screens, proportional-scintillation geiger counters, and ionization gauges. All sources of ionizing radiation are potentially dangerous. X-ray and gamma-ray sources may also produce hazardous secondary radiation. In addition, x-ray units involve both low- and high-potential electrical hazards. Appropriate radiation controls must be implemented.

SUMMARY

- Health hazards in foundries involve toxic and flammable materials, general and local ventilation, and specific problems such as silicosis, dermatosis, and radioactivity.

Dust, solvents, and other materials present serious health hazards in foundries.

- Companies must design and implement safety and health programs, safeguards, and safety devices to recognize, evaluate, and control or eliminate these hazards.
- Improper materials handling in foundries results in a wide variety of injuries. Supervisors and other managerial staff should instruct workers carefully in the use of proper safety practices and protective equipment.
- Because of the charging and blasting that takes place in cupolas and the carbon monoxide generated by them, these furnaces present several hazards. Only experienced workers should do tapping out operations or attempt to repair cupola linings.
- Refractory clay crucibles and ovens must be carefully inspected, operated, and maintained to prevent worker injuries. Good ventilation and guarding can protect workers from harmful fumes and hot materials.
- For safe operation of foundry production-line equipment; all grinding, polishing, and buffing equipment; and forging equipment, all exposed moving parts must be properly guarded and all electrical equipment grounded to eliminate shock hazards. Workers should wear personal protective equipment when using these machines.
- Nondestructive testing reveals all below-surface defects without damaging the parts being tested. This type of testing can locate defects inherent in metals and other solid materials or those that result from processing or inservice use.

REFERENCES

Foundries

American Conference of Government Industrial Hygienists, 6500 Glenway Avenue, Bldg. D7, Cincinnati, OH 45211.

American Foundrymen's Society, Golf and Wolf Roads, Des Plaines, IL 60016.

Engineering Manual for Control of In-Plant Environment in Foundries.

Health Protection in Foundry Practice.

Safety in Metal Casting.

Recommended Practices for Grinding, Polishing, and Buffing Equipment.

American Industrial Hygiene Association, 475 Wolf Ledges Parkway, Akron, OH 44311.

American National Standards Institute, 11 West 42nd Street, New York, NY 10036.

Safety Requirements for Workplace Floor and Wall Openings, Stairs, and Railing Systems, A1264.1–1995.

Safety Requirements for Melting and Pouring of Metals in the Metalcasting Industry, Z241.2–1999.

Safety Requirements for Sand Preparation, Molding, and Coremaking in the Sand Foundry Industry, Z241.1–1999.

Safety Requirements for the Cleaning and Finishing of Castings, Z241.3–1999.

National Fire Protection Association, 1 Batterymarch Park, Quincy, MA 02269.

National Electrical Code®, NFPA 70–1996.

Flammable and Combustible Liquids Code, NFPA 30, 1996.

National Safety Council, 1121 Spring Lake Drive, Itasca, IL 60143.

Fundamentals of Industrial Hygiene, 4th ed., 1996.

Occupational Safety Health Data Sheets (available in the Council Library):

Coated Abrasives, 12304–0452, 1991.

Ultrasonic Nondestructive Testing for Metals, 12304–0662, 1990.

U.S. Department of Health and Human Services, National Institute for Occupational Safety and Health, 4676 Columbia Parkway, Cincinnati, OH 45226.

U.S. Department of Labor, Occupational Safety and Health Administration, 200 Constitution Avenue NW, Washington, DC 20210.

Code of Federal Regulations, Title 29. "Hazard Communication Standard," and Section 1910.179(d)(3), (4)(ii).

Forging

National Safety Council, 1121 Spring Lake Drive, Itasca, IL 60143.

Occupational Safety and Health Data Sheets (available in the Council Library):

Handling Materials in the Forging Industry, 12304–0551, 1992.

Handling Steel Plates for Fabrication, 12304–0565, 1990.

Mechanical Forging Presses, 12304–0728, 1992.

Setup and Removal of Forging Hammer Dies, 12304–0716, 1993.

Ultrasonic Nondestructive Testing for Metals, 12304–0662, 1990.

Steam Drop Hammers, 12304–0720, 1993.

Upsetters, 12304–0721, 1993.

United Auto Workers, 8000 East Jefferson Avenue, Detroit, MI 48214.

Nondestructive Testing

McMaster R. *Nondestructive Testing Handbook*, 2nd ed. Columbus OH: American Society for Nondestructive Testing, 1982.

REVIEW QUESTIONS

1. Identify two ways of eliminating dust as a health hazard?
   a.
   b.

2. List three ways of minimizing the toxic and flammable hazards involved in using solvents.
   a.
   b.
   c.
3. Which of the following presents a fire and explosion hazard in dust-collecting systems?
   a. Aluminum
   b. Antimony
   c. Beryllium
   d. Phosphorus
   e. All of the above
4. What aspects should be included in the safety and health program for foundary workers?
   a.
   b.
   c.
   d.
   e.
5. Prolonged contact with oil, grease, acids, alkalis, and dirt can produce ______________.
6. The need for controlling ventilation in foundaries is determined by what three factors?
   a.
   b.
   c.
7. What hazard is sometimes generated when charging and blasting takes place in cupolas?
8. To eliminate a carbon monoxide explosion hazard, supply adequate natural or mechanical __________ in back of the cupola, and open two or more tuyeres after the blowers are shut down.
9. What is the principal danger in handling refractory clay crucibles?
10. List four hazards in the construction and operation of core ovens and mold-drying ovens.
    a.
    b.
    c.
    d.
11. Which of the following is usually considered adequate for ovens under 500 cu ft (14,000 L) in volume?
    a. Forced-draft ventilation
    b. No ventilation
    c. Natural-draft ventilation
12. Describe four safeguards that should be provided when grinding magnesium?
    a.
    b.
    c.
    d.
13. To help prevent fires in areas where welding operations are conducted, spread sand on the floor to a depth of:
    a. 1 in. (2.5 cm)
    b. 1.5 in. (3.75 cm)
    c. 2 in. (5 cm)
    d. 3 in. (7.5 cm)
14. Name the three types of forging hammers.
    a.
    b.
    c.
15. What are the differences between the conventional cold-stamping press and the forging press?
16. Identify the most widely used nondestructive testing method for forgings.
17. The resonant-frequency method falls under what general type of nondestructive testing for forged and cast metals?
    a. Ultrasonic
    b. Triboelectric
    c. Electromagnetic
    d. Radiography

25

Automated Lines, Systems, or Processes

Revised by

Philip E. Hagan, MPH, CIH, CHMM, CHCM

In a highly competitive manufacturing environment, more organizations are considering installing automated equipment and processes or expanding their use. A company can choose to set up islands of automation or to operate a dedicated, hard-automated assembly line with virtually no human intervention. Either way, the company's safety and health professional needs to address special safety concerns associated with the machines involved, the nature of automation, and the resulting change in manufacturing philosophies. Within the manufacturing community, any cultural shift or philosophical change has the potential to affect worker safety.

The following topics will be discussed in this chapter:

- safety planning and management
- identification of hazards
- hazard controls and maintenance and safety
- automated tasks and robotic equipment
- chemical processes

MANUFACTURING PHILOSOPHIES

The way that a company addresses its manufacturing processes has a direct bearing on safety in the workplace. If safety is regarded as a secondary concern, minimally factored into costs to comply with regulations, any pressure to increase production by redesigning a process or speeding up an assembly or production line is likely to make working conditions less safe.

Process Safety Management

One way to address the apparent conflict between maximizing capacity and keeping workers safe is to institute process safety management. This approach to safety involves everyone from operators and mechanics to the facility manager on a job site. As used within the chemical-processing industry, process safety management goes beyond traditional precautions. It is a means of managing process safety by recognizing and understanding the risks of production and by operating in a safe manner, so that hazards do not result in injuries and loss of life or property.

Elements of process safety management include:

- analyzing hazards and managing risks
- managing change in facility design or operation
- maintaining the integrity of equipment
- training and performance
- investigating incidents
- responding to and controlling emergencies
- auditing
- taking corrective actions

See Chapter 26, Process Safety Management, in the *Administration & Programs* volume for more detailed analysis of implementing process safety management.

Up-Front Planning for Safety

Up-front planning for safety represents another manufacturing philosophy with safety implications. One way of addressing safety concerns associated with automation is to categorize safety costs as part of the original installation. These costs are factored into management's buy/no-buy decision and capitalized as part of the original investment. At one company, worker teams (which are gradually replacing supervisors at some of the plants) get involved in determining how increased production goals can be met safely. They ask questions such as "Where is the line going to be placed?", "What equipment is needed?", or "How is the line going to be staffed?"

For example, if extra ventilation is needed for the new line, those costs are factored in at the beginning of the project. This approach assumes that it is easiest to get things approved up front, hardest to get them approved after installation. Training costs, including time required, also become part of the capital investment go/no-go decision. When a project is approved, it is approved in its entirety.

In addition, a company's safety and health professional may need to become a proactive safety advocate—arguing that concerns for worker safety during automated production should be included in the budget prioritization process. Such arguments can be more effective when they are backed up with lost time data, worker compensation costs, and legal expense figures from the facility's own history or from similar-industry estimates. Management must also factor the cost of compliance with current or forthcoming OSHA regulations and other industry standards into financial decisions on automated equipment.

Design in Safety

Another up-front approach to automation-related safety concerns is to ensure safety factors are primary considerations during the design process prior to the purchase or modification of automated equipment. Some companies encourage engineering and operations personnel to incorporate design-in safety into machine and tool procurement.

Although the comparisons in Table 25–A are not all-inclusive and are not necessarily appropriate to all

Table 25–A. Design-In Safety versus Traditional Philosophies

Traditional Approaches	Design-In Safety Approach
▪ common safeguards regardless of task	▪ safeguards designed to suit intended work
▪ zero risk	▪ recognize that zero risk does not exist—deal with task requirements, exposure, hazard and appropriate safeguards
▪ depend on employee to take corrective action	▪ where feasible, use automatic, passive controls such as a motion control timer, presence sensing, interlocked access, etc.
▪ local Health and Safety Committee attempts to oversee safeguards for each new machine	▪ determine feasible controls before awarding contract; assure that controls and cost are well defined
▪ heavy effort at runoff	▪ develop specifications—attend machine runoff as required
▪ periodic safety meetings for supervisors and employees	▪ periodic safety meetings for engineers, also
▪ compare injury statistics	▪ analyze injury statistics to determine specific problems
	Machine Safeguarding
▪ disconnects	▪ use control circuits for isolating energy, coupled with disconnects; provide for quick recovery to productive operation
▪ two-hand controls to initiate cycle	▪ use light screens and mats on many machines. Note: Mechanical power presses are special consideration.
▪ die blocks	▪ automatic slide locks
▪ guard pinch points	▪ eliminate pinch points in design
▪ use of perimeter guards or large guarded areas	▪ locate guards as close to hazard as possible; provide for observation of operation where required
▪ power is either ON or OFF	▪ design controls to isolate drive power and leave power on to limit switches for diagnostic work
▪ totally enclose robot and argue about power being locked out	▪ design multiple levels of safeguarding; i.e., walking through first interlocked door moves robot to home and allows ready access to change weld tip; going through second door into working envelope drops out drive
▪ guards located and interlocked such that necessary maintenance tasks cannot be performed	▪ guarding designed to facilitate maintenance in a safe manner
▪ no plan for access to elevations	▪ design in access and safeguarded platforms required for periodic maintenance
▪ service points for lubrication and adjustment inside safeguarded areas	▪ design in location of service points so safeguarding need not be interrupted

Source: General Motors

automated operations, they clarify the reasons why companies believe that engineering can take a leading role on safety issues in the concept and design phases of both product and processes. Design-in safety can contribute to safer working conditions because safety planning is targeted toward specific work procedures instead of being based on a one-size-fits-all philosophy.

Just-in-Time Method

Just-in-time (JIT) method, a manufacturing philosophy that reduces inventories and relies heavily on computerized scheduling, allegedly permits manufacturers to have more flexibility and to reduce costs. Yet this increasingly popular manufacturing philosophy has safety implications that safety and health professionals should consider.

Under traditional manufacturing practices, certain assumptions about production were widespread:

- large lots are efficient
- faster production is more efficient
- queues are necessary
- inventory "smooths" production and represents a "comfortable" position

However, many companies are questioning those assumptions, and are adapting JIT in varying degrees. Among claims made for JIT:

- inventories are reduced
- lead times are shorter
- quality control problems are uncovered as they occur
- manufacturers can react faster to demand changes

The downside of automation designed, implemented, or modified around JIT principles is that machines may be interdependent, with little or no provision for inventory to back up between any two stations. Consequently, if one automated element malfunctions—a robot, for example—and a line is shut down, there may be substantial pressure to fix the robot and move the line again. If safety and health professionals do not insist that the robot be carefully synchronized with the rest of the line before production resumes, worker safety may be compromised.

Computerized Maintenance Management Systems

With increasing automation, computerized maintenance management systems (CMMS) represents yet another manufacturing philosophy that can impact safety decisions. Under CMMS, maintenance management is automated; companies use microcomputers and sophisticated software to plan and control maintenance. Personnel, work orders, materials, purchasing, and scheduling are tracked through computer reports. Machine history can be traced through database sorting of work orders, so equipment failures can be analyzed. Should equipment be repaired or replaced? CMMS reporting can provide production downtime figures and cost data to support decisions.

As the use of automation increases, CMMS, which can be customized for industries and facilities, is becoming more popular because of its money-saving potential. As a single factory cell may handle from two to six or more processes, cell downtime is expensive. CMMS can shorten that downtime by providing easy-to-retrieve information from databases on replacement parts, vendor contact information, and repair instructions.

As part of CMMS, companies often develop modules on preventive maintenance (based on performing routine tasks at fixed intervals) and predictive maintenance (based on monitoring process conditions). Workers can use techniques such as vibration analysis, infrared scanning, thermography, and ultrasonic detection to identify potential trouble spots. They can remove equipment from service, examine it, and, if necessary, repair the part *before* it presents a serious safety hazard. Under older, traditional methods of maintenance management, work orders for machine repair were often filed chronologically, with multiple copies scattered throughout various offices. The work history for a specific machine was not always easy to trace, and potential safety problems were not always spotted in time to prevent incidents.

HAZARD IDENTIFICATION AND CONTROLS

Safety in automated manufacturing processes can be greatly increased in two ways: (1) by careful identification of hazards and (2) by the development of strategies to control the environment where the processes are taking place. For maximum effectiveness, such strategies also should include a training program addressing the specialized safety precautions required to operate and maintain various automated equipment, and any precautions workers need to know regarding the entire system.

Ensure strategies and training programs include requirements from applicable regulatory standards (for instance, the provisions of the Hazard Communication Standard, 29 *CFR* 1910.1200 for exposure to chemicals).

Types of Hazards

Safety precautions related to automation within a factory will vary for several reasons. These include the type of industry, the type and complexity of machines on the line, the way workers interact with the machinery, and the success of the automation used to integrate the various components of the production line. All of these factors influence the kinds of safety precautions a company must address.

Inhalation Hazards

Consider, for example, food processing and the complex safety precautions necessary to protect workers when production lines are automated. Spray dryers used in production of coffee, dairy substitutes, and sucrose can generate dust and powder. Workers who inhale the dust may experience irritation or be at risk for developing an occupational illness. In addition, if the concentration of dust achieves a critical level and if the dust has a high electrostatic charge, the danger of explosion can threaten all employees in the area.

Workers on a line also can be exposed to potentialy hazardous chemicals. Sodium hydroxide, phosphoric acid, and hydrochloric acid are used in food processing to adjust and change pH levels. Although facilities have in-line pH meters and in-line automatic injection mechanisms for chemical agents, workers who handle the chemicals for process preparation and feeding the line must wear appropriate personal protective equipment (PPE). Automatic processes for distillation of vegetable oil use extraction methods based on hexane or benzene, which give off vapors. Although the gases are part of a closed system, there may be a danger of leakage. Management should monitor vapor concentration carefully and provide good ventilation.

In aseptic packaging, product and package are sterilized separately and are brought together in an aseptic environment. Some sterilizing is done by heat, approximately 300–400 F (150–205 C), but plastic packaging materials for food products are frequently sterilized by hydrogen peroxide. Typically, an automated line has a roll sheet of the package material that enters an aseptic environment for forming. After forming, the container is sprayed with hydrogen peroxide. The hydrogen peroxide is later evaporated by hot air. Such a line is off limits to humans, so that the aseptic environment is not compromised. However, the hydrogen peroxide vapors exhausted from the aseptic unit could result in hazardous exposures if leaks or breakdowns of equipment occur. Personal protective equipment and good ventilation are important safety factors for those involved in feeding the line or handling the chemical in any way.

Burn Hazards

Products such as pet foods or certain cereals are often produced by automated extrusion lines that represent challenging technical safety problems. If pressure builds in the extruder to a level that is too high, if the extruder die is blocked, or if a plug builds up inside the barrel of the extruder, the material being extruded can escape at high temperature and pressure. However, the extrusion units usually have a safety switch that shuts off the main power when a sensor detects that torque exceeds predetermined conditions.

Evaporation, drying, cooking, sterilizing, and retorting—common operations in many food-processing facilities—often require steam as a heating agent. A leaking or broken steam line presents a serious safety hazard, and the supervisor should have it repaired immediately.

Cleaning in place (CIP) for food-processing equipment is frequently automated, and storage tanks with cleaning agents and pumps are connected directly to the automatic food-processing lines. Often pneumatic valves close the lines and switch them from the process mode to the cleaning mode. Because sulfuric acid and various alkalies are common cleaning solvents, the chemicals and the automatic switching process must be secured against incidents, a safety factor the engineer must take into account.

Utility regulation and safety also are a concern in safety planning. Electricity, steam, air, hot or cold water, compressed air, liquid carbon dioxide, and liquid ammonia—all used in various aspects of food-processing automation—carry their own potential problems, and each can become a hazard. For instance, heating a kettle of tomatoes with steam may require injecting steam into the jacket of the kettle. As a result of this heat exchange, the steam becomes condensate. The condensate is emitted by a steam trap. If a flow meter, a device that measures the amount of condensate going through the pipe per unit of time, malfunctions or fails to indicate a problem, the potential hazard can go unnoticed until an emergency develops.

Radiation Hazards

Radiation equipment presents several hazards in food processing automation if the equipment is not properly designed, installed, operated, and functioning. Microwaves often serve as the final stage for drying noodles and macaroni in automated lines. Workers should check the equipment periodically to ensure safety interlocks are functioning and that there is no microwave leakage. If a company uses gamma radiation to sterilize spices, employees involved with equipment should wear radiation badges and be checked regularly for radiation exposure. Management should take this precaution even if the company has a well-designed process and complies with regulations to keep employees a safe distance from the radiation chamber. (See also Chapter 24, Laboratory Safety, in the *Administration & Programs* volume of this *Manual.*)

Pinch Hazards

In classical primary packaging for food products, most production lines use heavily mechanized equipment, especially for conveying products, forming, and packaging. Because the combination of a machine's movement with its power can be hazardous for workers, the line should be enclosed wherever possible. Additionally, sensors that automatically stop the line whenever something gets caught or jams in the machinery should be installed.

Some of the recent American National Standards Institute (ANSI) standards for machine tools supply information on hazard identification and recommendations for a control strategy. Safety and health professionals should be sure all equipment complies with the latest standards. Be sure to check if subsequent revisions have been published (see References).

Boundaries Between Restricted/Nonrestricted Areas

It is important to identify the boundaries that separate restricted areas from areas where workers can move safely. One of the first steps is to learn the specific clearances required by each machine or component in the automated system. For instance, in laying out a system for automated guided vehicles (AGVs), all areas that involve pedestrian travel or interaction with workers require a minimum clearance of 18 in. (46 cm) between a vehicle with its transported load and any fixed object. That clearance allows a person to stand safely while a vehicle passes by.

However, AGVs often transfer loads automatically by powered roller belt or chain to and from fixed stations through a lift/lower fork truck design or through an automatic push or pull mechanism. At these load transfer stations, it is not always possible to maintain the 18 in. clearance. Consequently, load transfer stations need to be designated as restricted areas where no one can walk or stand. Areas used for battery-charging and battery-changing on AGVs also require special markings, designated by the following standards: *Internal-Combustion Engine-Powered Industrial Trucks,* ANSI/UL 558–1991; *Electric-Battery-Powered Industrial Trucks,* ANSI/UL 583–1991; and *Powered Industrial Trucks, Including Type Designations, Areas of Use, Maintenance, and Operation,* ANSI/NFPA 505–1992. Typically, restricted areas are identified by posted signs and striped floors.

Some ANSI standards specifically define the areas required for machine motion. For instance, *Safety Standard for Industrial Robots and Industrial Robot Systems,* ANSI/RIA R15.06–1992, defines the following:

- *Envelope (space), maximum:* The volume of space encompassing the maximum designed movements of all robot parts, including the end-effector, workpiece, and attachments.
- *Restricted envelope (space):* That portion of the maximum envelope to which a robot is restricted by limiting devices. The maximum distance that the robot can travel after the limiting device is actuated defines the boundaries of the restricted envelope (space) of the robot.
- *Operating envelope (space):* That portion of the restricted envelope (space) that is actually used by the robot while performing its programmed motions.

The National Safety Council's Occupational Safety and Health Data Sheet 12304-D717–1991, *Robots,* recommends that companies place warning signs around the robot at points of access to the operating envelope. These signs will warn those who enter about the normal hazards of the robot and/or any unusual hazards that may exist. Examples of unusual hazards include overlapping operating envelopes of two or more robots and other automated devices or machinery that can move into the work area.

Visual/Mechanical Warnings

Several systems of visual and mechanical warnings are commonly used to alert workers to present, or approaching, hazards. These warnings include signs, barriers, line markings on the floor, railings, or the equivalent. See Chapter 2, Buildings and Facility Layout, in this volume. Be sure to check the latest regulations.

Signs and In-Process Warnings

Make sure that employees whose native language is not English understand and follow the warnings. Wherever possible, use international signs. If colorblindness is a factor for some workers, management should consider installing flashing lights. Workers, supervisors, and maintenance personnel should be screened, trained, and tested to make sure they understand the meaning of signs, tags, and color codes. Remember that helpers and passersby may be exposed to hazards, so plan for additional safeguards.

Audible warning devices such as horns, bells, and electronic beepers must have a sound higher than the machine noise and greater than the noise level within the facility. Do not use these devices in conjunction with paging systems or as signals to indicate the start and stop of work.

Visual or audible warnings also can alert employees that work is in progress. Automatic turn signals on AGVs can show pedestrians which way the vehicle will turn at an intersection. An amber light, installed on a robot and easily seen from any angle, should be lighted any time the robot is energized. It signifies that the robot is "live," even when it is not moving.

Inspect warning signals and devices often to make sure they are operating properly. Workers may not always know that such devices have failed.

BARRIERS AND INTERLOCKED BARRIERS

Awareness barriers keep operators from reaching into hazardous areas unless they make a conscious effort to do so. Such awareness barriers should not only show the hazardous area but provide visual boundaries for the operator's movements. Inspect and test these parts regularly.

Perimeter guarding signs and barriers limit entrance to certain enclosed areas and protect personnel who work with and around robots. Management should post signs prominently at all entrances to prohibit unauthorized personnel from entering the area. In addition, locate the entrance so that authorized personnel can enter the area without inadvertently entering a restricted work envelope.

Interlocks are arrangements in which the operation of one control or mechanism automatically brings about or prevents the operation of another. Some of the most common forms of interlocks include contact switches activated by electrical, pneumatic, hydraulic, or magnetic mechanisms. Interlocked barrier guards ensure that a machine tool will not cycle, or continue to cycle, unless the guard itself, or its hinged or movable sections, enclose the hazardous area.

For maximum safety, if an interlocking device is disconnected, the machine and any automated process within the workstation should stop promptly. Further, connecting the device again should not start the automation cycle. Under certain conditions, a movable barrier device can function as an interlocked barrier guard when a machine tool has been set to operate continuously.

Hazard Controls

Once hazards have been identified in automated production, appropriate controls should be installed to facilitate worker safety. The ANSI B11 standards for machine tools, as well as B11.19–1990, *Performance Criteria for the Design, Construction, Care and Operation of Safeguarding When Referenced by the Other B11 Machine Tool Safety Standards,* provide recommendations for the placement and effectiveness of control devices (such as an EMERGENCY STOP button, a START button, or a selector switch) in various situations. Workers should follow these recommendations and manufacturers' instructions, and all personnel should be properly trained in the use of the controls.

If a machine or any other part of an automated production line has been modified, check the control devices to ensure that they still meet ANSI and other industry standards under the new setup. In particular, make sure that electromagnetic interference (EMI) or radio frequency interference (RFI) does not impede operation of the controls and that the power supply is constant and uninterrupted.

Controls Outside Boundaries

Some controls should be located outside boundaries. For instance, if a worker needs to reach a control for a robot while a machine is in automatic operation, the control should be placed outside the restricted envelope. This envelope represents the maximum space to which a robot is restricted by the devices that limit it. In that way, a person who needs to activate the control is also outside the restricted envelope. Other controls, located outside the restricted envelope, can ensure that if a robot is out of the automatic mode for maintenance, repair, or teaching purposes, personnel cannot reactivate the robot until they have left the restricted envelope, restored all safeguards required for automatic operation, and begun deliberate start-up procedures.

Controls Inside Boundaries

Other controls can be located inside boundaries. For instance, a hostage control device (such as a two-hand control, a single-hand control, a foot switch, or a safety mat) can hold the operator at a control station during the hazardous portion of the machine cycle. When a person is inside the robot's work envelope to teach, maintain, or repair the robot, and the power must be on, the robot system—as well as movement of other equipment in the restricted work envelope—should be under the sole control of the person in the envelope.

Controls that stop the system should override all other controls. They should interact so that if one component fails, the STOP control will still shut down the system immediately. For more information on guarding and controls, see Chapter 6, Safeguarding, in this volume.

Maintenance and Safety

Whether an automated manufacturing facility chooses to have maintenance and repair performed by its own personnel, by vendor personnel, or by an outside contract service, safety should be a major goal of all concerned. Be sure to follow the manufacturer's recommendations on maintenance and inspection of all components in the automated system—each part individually and as a component of an integrated assembly process. Each machine and automated materials-handling device must not only perform safely as a stand-alone unit, but must interface properly at the appropriate speed.

Safety training for maintenance and repair personnel should include, but not be limited to, a review of all applicable industry safety procedures and standards, manufacturers' recommendations, equipment-specific training, tasks, and responsibilities for each person, and identification of hazards associated with each task. Keep accurate records and documentation of all modifications and changes to components in the system, as well as a history of maintenance and repairs for each piece of equipment. Make sure this information is easily accessible to maintenance personnel.

Also, make sure maintenance and repair personnel working on the machines or system components have copies on site of the maintenance and operations manuals from vendors, as well as any updates. Keep a second copy of the manuals and updates at a supervisor's station or in the safety and health professional's office. Repair and maintenance workers should also know how to reach the appropriate person if problems develop that exceed workers' responsibilities or training.

Companies who move from crisis maintenance to planned maintenance greatly enhance safety in the workplace. Preventive maintenance can help supervisors anticipate and schedule production downtime. Consequently, there is less pressure to get a line or cell up and running again, possibly by cutting corners on safety.

Predictive maintenance, which gathers data automatically through monitoring conditions and integrates it with manufacturers' guidelines on tolerances and ranges, also can help boost safety procedures and practices. In this approach, computerized systems predict when a piece of equipment is likely to fail, allowing personnel to repair or replace it before the failure occurs.

Companies should train and retrain workers, as necessary, to be sure they are competent and knowledgeable about safety practices. If outside contract personnel or outside services are used for maintenance or repair, provide them with information on company safety procedures at the time the service is contracted. Make it clear that ALL personnel, not just the company's own employees, must follow the facility's safety practices at all times. If the firm uses an outside service for maintenance, consider writing contracts with outside services that require their personnel to receive both classroom and on-the-job training. Their training should be documented and meet industry standards and legal requirements.

Maintenance personnel also should be aware of regulations affecting the overall factory environment, as well as those dealing with particular equipment. For instance, the use and characteristics of protective coatings in manufacturing facilities now are subject to volatile organic compound (VOC) restrictions, which stipulate the solvent content percentage per liter of coating, and to regulatory right-to-know legislation. Although the federal government has set VOC limits for coatings, some state governments have enacted their own legislation. Consequently, companies with facilities in several states need to follow coating specifications that meet all restrictions established by the state governments for each of their facilities.

Energy Isolation—Lockout/Tagout

When workers are doing maintenance or repair work, make sure that they lock out and tag out all equipment in accordance with the OSHA Lockout/Tagout standard (29 *CFR*, 1910.147 and ANSI Z244.1–1982). For a detailed explanation and discussion of required procedures, see Chapter 6, Safeguarding, and Chapter 10, Electrical Safety, in this volume.

Do not allow workers to perform servicing, maintenance, or repair work on any equipment unless that equipment is isolated from all hazardous energy sources. Only designated authorized employees should complete the lockout/tagout procedures, carefully following the steps established by the company.

Machine-specific lockout steps are difficult to prescribe because they vary from machine to machine. In general, such data include the various types of hazardous energy and their magnitudes; the locations and identification numbers of switches and valves; whether locks, tags, or both are specified in each case; additional safety measures; a list of "designated authorized employees"; and a phone number to call if there are questions. Post the comprehensive lockout specifications at each machine. See the Energy Isolation section in Chapter 6, Safeguarding, of this volume.

Diagnostic Aids and Procedures

Various diagnostic aids and procedures will help solve maintenance problems. For example, vibration measurement and analysis monitor the condition of machinery and help diagnose machine problems. Measurement data, collected by portable, programmable, microprocessor-controlled vibration meters or permanently installed embedded sensors, are transmitted to a computer over an RS–232 serial interface. Workers can analyze the information later to determine specific component faults. Laser monitors are available that continuously monitor dust levels in the process industry. Software packages have been developed to track steam trap surveys and trap maintenance. These computer programs support ultrasonic, infrared, or standard pyrometry inspections of steam traps.

Companies also can use various techniques to help determine why bearings have failed. These techniques include measuring and analyzing the wear track, identifying and classifying surface damage, analyzing lubricant, and analyzing fretting corrosion patterns that form on bearing inner ring bores and outer ring surfaces.

Staff can derive equipment performance trends by evaluating equipment output. A standard statistical process control run chart enables a facility manager to collect data. If the demonstrated capability of a machine shifts significantly, adjustments may be necessary.

Employees who repair machines or components in the automated production line should document their actions. If problems recur, supervisors can evaluate those actions to determine where the trouble lies. Supervisors should maintain a history of problems and repairs on each machine to help identify performance trends and problem components. Some microcomputer-based management maintenance software, which runs on certain personal computers as well as on local area networks (LANs), allows users to "build" a problem list by entering appropriate keywords into a database. When a piece of equipment goes down, the user selects a particular problem from the list. A pop-up display on screen shows possible causes and suggested methods of correction.

AUTOMATED PRODUCTION

Automated manufacturing requires special safety precautions beyond those demanded by the use of single machines. Wherever workers and equipment interact—during operation, programming and maintenance, or through unintentional contact—the potential for injury exists. The best ways to reduce injuries are job and workstation design; careful compliance with existing industry standards, along with monitoring the development and implementation of new ones; and safety training, testing, and retraining (not only for the specific equipment but also for the automated system as a whole).

Effective safety training becomes even more important in automated production. Because automated manufacturing processes often require workers to interface with more than one system component, such as a program, terminal, machine, materials handling system, or robot, workers need to understand other parts of the system besides their own. Programmable automation, which lets a machine or group of machines make a wide range of parts or products, and automated manufacturing systems require that operators, maintenance personnel, and supervisors develop many skills to work safely and productively. Yet safety and health professionals cannot assume that all workers have the necessary skills required to read and understand technical manuals and specifications. The company may need to establish procedures to test and train workers in reading and comprehension or develop safety training programs that do not rely solely on written communications or English-language communications.

Some automated manufacturing involves continuous processes, such as those used in food or beverage processing, textiles, pulp and paper, metal refining, printing, pharmaceutical, and petrochemical industries. Another type of automated production, usually referred to as automated manufacturing systems/cells, generally includes machine tools and related machines and equipment brought together to form a new or modified manufacturing system. Such a system often is linked by a materials handling system that is interconnected with and operated by an electronic system. Management can program and reprogram the electronic system to control the manufacture of single parts or of assemblies.

Safety and health professionals involved with manufacturing systems or cells should use applicable standards, including *Electrical Safety Requirements for Employee Workplaces,* NFPA 70E–1988; *Electrical Standard for Industrial Machinery,* ANSI/NFPA 79–1994; and *Safety Requirements for Industrial Robots and Industrial Robot Systems,* ANSI/RIA R15.06–1992.

Computer Numerically Controlled Machines

A component of many automated manufacturing systems/cells is the numerically controlled (NC) machine, which staff can program and reprogram easily. Using older NC technology, a programmer writes instructions for the machine at a terminal that punches holes in a paper or special plastic tape. Next the tape is fed into the NC controller. The holes represent commands that are transmitted to the motors that guide the machine tool. The programmer can easily set the machine to make a different part when a different punched tape is fed to the machine.

The computerized numerically controlled (CNC) machines often have a screen and keyboard, so the machine operator can edit the program right at the machine. Off-line programming, which does not tie up expensive machines while programs are keyed in, is also used. In larger systems, mini- or mainframe computers program and run more than one NC tool. Although each machine tool has its own microcomputer, the systems are linked to a central computer that controls them.

Safety precautions for the machines are referenced in the appropriate ANSI B.11 machine tool standards, which include information on hazards, operation, and training.

Automated Materials Handling and Transport Systems

Companies can help make materials handling and transport systems safer by being familiar with current regulations and instituting safe work practices. Two areas that possess particular risks are conveyors and automated guided vehicles.

Conveyors

See Chapter 15, Hoisting and Conveying Equipment, in this volume for safety recommendations involving conveyors. Facility managers and supervisors should be familiar with *Safety Standards for Conveyors and Related Equipment,* ANSI/ASME B.20.1–1993, discussed in detail in that chapter. Management should safeguard all conveyors to protect employees and should train workers in safe work procedures, including the location and operation of all START/STOP devices. If the unit a conveyor feeds has been stopped or is blocked, make sure electrical or mechanical interlocking devices are provided to stop the conveyor automatically. Guard pinch points adequately, and ensure that workers use proper lockout/tagout procedures when performing maintenance and repairs.

Automated Guided Vehicles (AGVs)

Automated guided vehicles (AGVs), which provide a transport system for materials handling, are one of the key elements of automation in various industries (Figure 25–1). Since 1985, AGV usage has increased in each of the following industries (ranked in decreasing order of AGV users): automobile, electronics, paper, aircraft, textile, food, metal fabrication, farm machinery, printing, primary metals, rubber/plastics, and lumber/wood.

Figure 25–1. This automated guided vehicle uses a guidance system to track an invisible chemical path bonded to the floor surface. It is receiving material from a conveyor system. Note the EMERGENCY STOP at the top left of the AGV. (Courtesy Litton Corp.)

AGVs can tow pallets and integrate conveyors with other systems. They can also collect and dispose of refuse (Figure 25–2). In assembly line applications, AGVs often carry major subassemblies between assembly workstations. Because AGVs can be used in parallel operations, they are often cost effective in flexible manufacturing. In certain industries, such as the manufacture of printed circuit boards, AGVs offer advantages in product routing over conveyor systems because they can move material more easily between "islands" of automatic insertion equipment.

Traditionally, AGVs follow electromagnetic wires buried in the floor. However, "free-roaming" AGVs use competitive technology, such as optical guidance (using a chemical guidepath [Figure 25–1] or painted line), infrared guidance, inertial guidance (which uses a gyroscope), position-referencing beacons, and computer programming (Figure 25–3).

Safety and health professionals must be concerned with a variety of system conditions. Loads the AGVs are carrying should be appropriate for vehicle capabilities; check to be sure employees are following the manufacturers' specifications in loading AGVs. Workers should not stack loads any higher than specifications. In addition, if a load is stacked too loosely, it can tip over when on an incline. As a result, the vehicle will not be able to carry its rated level load capacity safely.

Figure 25–2. This self-guided lift truck can collect and dispose of chips from a machining cell. (Courtesy Caterpillar Inc.)

Every time the products, packaging, or loads that an AGV carries are changed, or the process is altered, workers should check the center of gravity for the load against the AGV system design parameters and make appropriate adjustments.

For the transport system to perform safely, the company needs to establish a program to monitor the floor conditions on the routes the AGVs travel. Floors should not sag or move. If the AGVs need to travel over or around expansion joints or rails in the floor, make sure the vehicles can do so before the AGV system is installed.

Floors should be as dry as possible, because AGVs can lose traction and behave unpredictably, especially if vehicles have to stop suddenly in an emergency. If floors are wet or slippery, apply slip-resistant floor coatings. Keep floors as clean as possible. Do not allow residues to build up, because they may prevent the AGVs from steering properly or stopping accurately.

Floor conditions become especially important for AGVs that operate with chemical guidepaths or painted lines. The lines can become obscured if dirt, oil, or grease accumulate. Keep floors clean, and repaint lines as often as needed, following manufacturers' recommendations. Keep all guidepath areas free from obstructions, such as boxes, stacked parts, or other vehicles.

Companies can design safety factors into the AGV system before installation. Establish provisions during the design phase not only to safeguard the transport of materials, but also to protect personnel in the immediate area of the vehicle system. Design-in safety has two objectives: to keep the AGVs from colliding with people and other vehicles, and, if an anticollision device fails to work, to prevent the AGV from operating until it can be repaired.

Safety devices for the AGV can include monitors for guidance and velocity, guards, deadman controls, turn signals, and sensors (infrared optics and ultrasonics) to anticipate and avoid collisions. Traffic control techniques include bumper blocking, in-floor zone blocking (that uses sensors in the floor at boundaries of zones and energizes a path to allow an AGV to enter a zone), and computer zone blocking.

Traditionally, AGVs are controlled by programmable controllers that "talk" to the vehicles through signals transmitted via a wire guidepath buried in the floor. Alternative technology for communication includes radio frequency and infrared guidance systems; however, the route must be clearly marked if guidewires are not used. The company should make some provision for communicating with the AGV when it is off the guidepath.

Refer to ANSI B56.5 for standards on bumper design, activation, and requirements. Bumpers should not need either software or hardware logic or signal conditioning in order to operate. In addition, if the AGV loses guidance in the automatic mode, ANSI B56.5 requires the vehicle to stop moving immediately.

Robotic Equipment

Organizations considering the installation or expansion of automated manufacturing are confronted with growing safety problems associated with robotics. In 1990, the United States robot population topped 40,000 units, second only to Japan. In terms of the number of robots per 10,000 production workers, however, the United States trails Japan, Sweden, Germany and Italy, according to Robotic Industries Association (RIA). RIA is a trade association of United States-based manufacturers, distributors, systems integrators, accessory equipment suppliers, users, research organizations, and consulting firms. RIA defines an industrial robot as "a reprogrammable multifunctional manipulator designed to move material, parts, tools, or specialized devices, through variable programmed motions for the performance of a variety of "tasks."

Three basic parts make up a robot: a manipulator, a power supply, and a system for controlling the robot. The robot arm, from the base of the robot through the

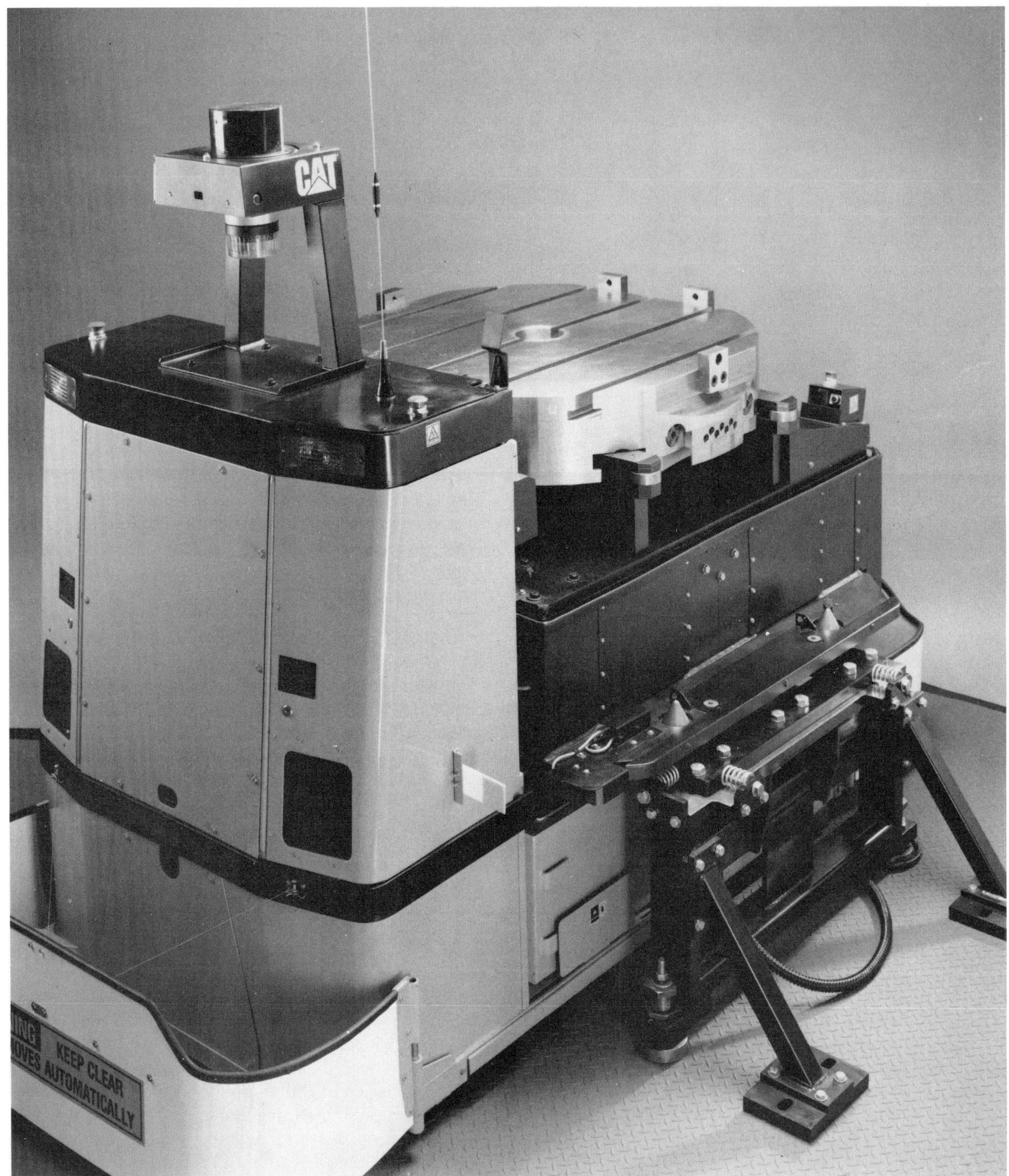

Figure 25–3. This free-roaming self-guided AGV requires no wires, tapes, or special flooring. It uses an on-board computer for guidance and can be reprogrammed for flexible manufacturing. (Courtesy Caterpillar Inc.)

wrist, is called the "manipulator." Actuators, drives, bearings, and feedback (all within that arm) make it possible for the arm to move in different directions. The extent that the robot hand, or working tool, can reach in all directions is called its "work envelope." The dimensions of the work envelope depend on the way the manufacturer has arranged the robot's axes of motion, or degrees of freedom.

Usually a robot has three major axes of motion: a vertical stroke that determines how high the robot can reach; a horizontal reach that lets the robot move in and out; and a swing, or rotation, around the robot's base. Some robots have additional axes of motion: pitch, yaw, and roll. A "wrist" at the end of a robot's arm contains components that allow the additional movement. Manufacturers who want to increase a robot's reach and ability to manipulate objects can provide additional axes of motion in end-of-arm tooling.

Industrial robots are powered pneumatically, hydraulically, or electrically. Three types of robot control systems exist: nonservo-controlled point-to-point robots (often used in pick-and-place applications); servo-controlled point-to-point robots (often used for loading and unloading); and servo-controlled continuous path robots (often used for spray painting and other finishing operations).

An industrial robot is almost never a stand-alone device. It is part of a system that includes industrial robots, end effectors, and any equipment, devices, and sensors required for the robot to perform its tasks, including communication interfaces that sequence or monitor the robot.

Companies use most industrial robots to perform regular, repetitive tasks. Often, robots have become the automation of choice in factory environments where working conditions are potentially hazardous or overly strenuous for human workers. Processes such as spray painting on an automotive assembly line or loading molten steel in extremely hot temperatures are good candidates for robot units.

Robots have many functions in industry. In the automotive sector, which, according to RIA, accounts for about half of all robots purchased each year in the United States, robots are used for arc and resistance welding, assembly, dispensing and applying sealants and adhesives, painting, inspection of vehicles, loading and unloading machines, transferring parts from one conveyor to another, packaging, and palletizing.

Robots also load and unload other machines in such fields as die casting, loading presses, forging and heat treating, and plastic molding. Robots that can handle tools or that use grippers to grasp special tools are especially useful in applications like grinding, drilling, and riveting in machining. The strong sales growth of assembly robots that are used primarily in nonautomotive markets indicates new users are emerging in a diverse range of industries. From aerospace facilities to appliance manufacturing to printed circuit board assembly, robots play an important part in automated manufacturing.

HAZARDS OF ROBOTICS

Those concerned with protecting workers in an environment where industrial robots are used have specific factors to consider because of the nature of robots and their installation and operation. Robots have a limited reach and mobility and are generally placed close to each other. Yet a wide area around a robot—in fact, anywhere within reach of its arm—can represent a potential danger zone (Figure 25–4). The danger zone can be even larger if a robot loses control of the object it is holding.

In 1982, the Labor Ministry of Japan conducted a survey of 190 factories with 4,341 robots installed. They reported 11 incidents, including two deaths; 37 cases of noninjury incidents (operators nearly contacting robot arms); and 300 cases of robot-related problems, including those unrelated to injuries. In 8 of the 11 incidents, operators reportedly entered the robot arm range while it was stationary and were hit by the arm when it moved unexpectedly. When 300 robot-related problems were analyzed to determine the relationship between robot trouble and the operation of the robot manipulator, the results showed that in 30.6% of the cases, the robot operated unnaturally; in 28.3%, the robot operation started unexpectedly; and in 8.9%, the piece the robot was working on was released or fell.

Another Japanese survey of 18 near-misses showed the following causes: erroneous action of the robot in normal operation (5.6%); erroneous action of peripheral equipment in normal operation (5.6%); careless approach to the robot by humans (11.2%); erroneous action of the robot in teaching and test operation (16.6%); erroneous action of peripheral equipment during teaching and test operation (16.6%); erroneous action during manual operation (16.6%); erroneous action during checking, regulation, and repair (16.6%); and other (11.2%).

In 1987, an analysis of cause and effect of 32 industrial robot incidents reported in Swedish, West German, Japanese, and American literature indicated that in 24 of the 32 cases, a definite cause associated with robot-human interaction could be reported. Seventeen incidents were caused by line workers; five by maintenance workers; and two by programmers.

The researchers grouped incident causes into four categories: human error, workplace design, robot design, and "other." Of the 32 incidents, 13 were attributed to human error. Workplace design accounted for

Figure 25–4. This robot is guarded by a chain-link fence with gates. Safe areas for maintenance and repair of the robot must be designed into the worksite.

20 incidents: 18 were related to guarding, and 2 were related to interfacing. The researchers attributed 7 of the 32 incidents to robot design and placed no incidents in the "other" category. The total of 40 exceeds the 32 incidents being analyzed because many of these incidents had more than one cause.

Of the 32 incidents recorded, 72% involved the operator or a nearby worker. None of the 32 incidents involved an unauthorized individual.

Safeguarding Robots

Robot safety can be divided into three major areas:

- safety in the process of manufacturing, remanufacturing, and rebuilding robots
- installation of robots
- safeguarding of workers exposed to hazards associated with the use of robots

Safety and health professionals need to be concerned with each of these three areas. Although one may assume that safety and health professionals would be interested primarily in installation and safeguarding, this is not necessarily true. Because robot users also rebuild robots, safety and health professionals must be sure that the rebuilt units conform to and perform within safety guidelines.

At the heart of any robot safety program is risk assessment. In this approach the end-user must determine the level and types of hazards presented by a specific robot application and develop procedures to minimize those hazards. Risk assessment is appropriate not only in production but also in the installation and testing of the robot before production begins (Table 25–B).

In addition, appropriate precautions for safeguarding personnel vary depending on the stage of development of the robot and the robot system. As RIA points out, elaborate safeguarding while the robot is being integrated at its manufacture may impede the development of the overall system. As a result it is generally impractical to debug the machine at this stage. However, when the robot is operating on a production line, full safeguarding and interlocks may be a requirement for the same system.

The industry recognizes four stages of robot and robot system development:

1. integration at the manufacturer or system developer
2. verification and buy-off testing
3. installation and testing on the site of the operation
4. operation in production

Table 25–B. Robot Sfety Program Risk Assessment

A risk assessment for a robot safety program should consider:

1. size, capability, and speed of the robot
2. application/process
3. anticipated tasks that will be required for continued operation
4. hazards associated with each task
5. anticipated failure modes
6. probability of occurrence and probable severity of injury
7. level of expertise of exposed personnel and the frequency of exposure.

Source: Robotic Industries Association

Protecting the Robot Teacher

Often, a robot is programmed by being physically guided by an operator through a desired sequence of tasks. Typically, a servo-controlled point-to-point robot is programmed through a TEACH pendant. This is a control box that resembles the remote control device for a television set. The operator uses it to "walk" the robot through the program steps slowly, making sure to record each step. When the operator is finished, he or she can switch the robot from TEACH to REPEAT.

Another way of teaching by guiding is often used for servo-controlled continuous-path robots. To teach such a robot, the operator typically takes the end of the robot arm and leads it through a pattern of motions. Meanwhile, the robot's control system is recording feedback data from the position sensors on the axes and storing it on a mass memory storage system.

In either case, the teacher must be within the operating envelope of the robot to program its movements within very close tolerance parameters. When a robot is in the TEACH mode, the highest degree of hazard exists. A Japanese survey in 1977 showed that the greatest risk of injuries involving robots occurs during programming, teaching, and maintenance. These are usually the times when a person is within the operating envelope of the robot.

ANSI/RIA R.15.06-1992 and the National Safety Council Data Sheet 717D1991 contain safety recommendations for the safe installation and operation of industrial robots in the TEACH mode. Consult these publications for a full description.

Before starting any operation, persons who are teaching the robot should check the machine and its envelope visually to be sure no hazardous conditions exist. Operators should check the TEACH pendant's EMERGENCY STOP and motion controls to be sure they are operating properly and should repair them if they are damaged or malfunctioning. The EMERGENCY STOP control should be hardwired into the drive-power stop circuit and should not be interfaced through a computer input/output register.

During teaching, preferably only the teacher should be in the restricted envelope. If more than one person is present, only the teacher should control robot motion. However, anyone else within the restricted envelope should use an enabling device that stops the robot motion when released. All personnel should leave the restricted envelope and restore all safeguards for automatic operation before starting the robot's AUTOMATIC mode.

Protecting the Operator

Special safety precautions may be needed when an operator interacts with the robot during each operating cycle. Safeguards should keep the worker from being in the restricted envelope when the robot is moving automatically, or should stop robot motion while any part of the operator's body is inside the restricted envelope. These safeguards can include devices that sense a person's presence, such as photoelectric cells, pressure-sensitive mats, laser and visual monitor sensors, and light or sound curtains. Additional safeguards can include detectors to indicate when robots or machines malfunction, barriers, awareness barriers, awareness signals, and safety training.

All persons involved with robots must remember two points:

- If the robot is motionless, do not assume it will remain so. Many programs have delays or waits, when the robot "sits" until told to do something.
- If the robot is repeating a pattern, do not assume it will continue the repetition. Computers can automatically modify the path a robot has been programmed to follow and thereby trap personnel within the work envelope when the robot moves in an unexpected direction.

A comprehensive cause-and-effect analysis of a 1984 pinch-point fatal injuries revealed both human error and workplace design problems. Although the worker had received one week of training three weeks before the incident, he disregarded safety procedures and entered the robot area when the robot was in normal operating mode. In addition, the fixed-pipe safety rail was not high enough and had two gaps. There were no interlocking gates or human sensing devices that would have shut down the robot. As a result, a fixed pole to limit the action in the robot work area created a pinch point. The worker, caught between the robot housing and the pole, was crushed to death.

Protecting the Maintenance/ Repair Personnel

Because robot design differs among manufacturers, specific instructions provided by the manufacturer for maintaining and repairing a particular robot should be followed. If the robot has been modified in any way, or if procedures or programming have changed, maintenance personnel must be informed of the modifications and their implications for safety. Rebuilt or previously owned robots should have all changes from their original design documented and available to maintenance personnel. Provide manuals, procedures, and instructions to employees who maintain or repair robots before such work is performed. Updated drawings and schematics should be available to train workers properly in safety procedures and applicable standards.

Workers should use proper lockout/tagout procedures for shutting off and locking out power sources and follow appropriate testing procedures before performing maintenance work or before repairing the robot. Refer to 29 *CFR*, 1910.147 and to a more detailed explanation in Chapter 6, Safeguarding, in this volume. Maintenance personnel should make sure any potentially hazardous stored energy is released or blocked before servicing the robot. This energy may be in the form of air and hydraulic accumulators, springs, counterweights, flywheels, and loads held by the robot.

When lockout/tagout is not possible, provide appropriate alternate safeguarding. These safeguards could include:

- EMERGENCY STOP and SLOW SPEED controls
- a second person monitoring the robot control panel to react to potential hazards immediately
- a control system that gives the maintenance employee who is within the work envelope control over robot movement, or blocking devices such as blocks and pins

If any safeguards must be bypassed while workers are maintaining or repairing the robot, provide alternative safeguarding. When the maintenance task is complete, return and check any bypassed safeguards to their active state before operations begin or before the robot is energized.

Users of robots and robot systems should establish and document an effective inspection and maintenance program. This program should include any preventive maintenance recommended by the robot's or system's manufacturer.

Computer-Integrated Controls

In automated production, machines and equipment work in a coordinated manner under computer control. Such computers facilitate information flow, coordinate factory operations, and increase efficiency and flexibility. The flexibility of programmable automation is becoming more attractive to firms seeking to compete effectively, because equipment for design, production, and management is linked together and can be reprogrammed.

However, computer-integrated manufacturing (CIM) has implications for safety that cannot be ignored. A 1984 report by the Office of Technology Assessment includes comments that the absence of standard programming languages, data formats, communications protocols, teaching methods, controls, and well-developed off-line programming capabilities are barriers to buying or using CIM. Although manufacturers are addressing these issues, this lack of standardization and the fact that individual machines use varying human-system interfaces make effective safety training even more imperative than for other manufacturing processes.

CHEMICAL PROCESSES

Chemical processes contain many hazards for workers that must be addressed through written safety policies, engineering and administrative controls, worker training, and personal protection equipment. The safety responsibilities are particularly acute in the chemical industry because of the many critical operations that must be handled by workers and because the nature and number of chemicals—and their byproducts—is constantly changing. Companies can help meet their safety responsibilities by informing workers of process safety information, conducting ongoing hazard/risk analyses and pre-start-up reviews, developing operating procedures manuals, managing change, and auditing processes. (See also Chapter 26, Process Safety Management, in the *Administration & Programs* volume of this *Manual.*)

Process Safety Information

Communicating the design basis for equipment and processes to all the affected personnel is not always as simple as it might seem. This fact is especially true in large, sophisticated chemical-processing facilities where pressures, temperatures, and other control parameters must be maintained within safe tolerances at thousands of critical junctures.

Process Document

One method that managers have used to explain design basis to their employees is to produce a "cookbook" for chemical processing, commonly known as the facility's Chemical Process Document. This basic manual, which must be understood by pertinent employees as a fundamental part of their job training, includes the parameters for safe operation of facility processes. Management must be aware of the consequences of operating outside those parameters in order to develop process controls to ensure safe operation.

Suggested Contents

A Process Document should contain the process safety information recommended by the American Petroleum Institute (API) in its *Management of Process Hazards, Recommended Practice* 750(1). This includes:

General Information

- an assessment of the hazards posed by materials used in chemical processing
- toxicity information
- permissible exposure limits
- physical data

- thermal and chemical stability data
- reactivity data
- corrosivity data
- data on hazardous effects of inadvertently mixing materials

Process Design Information

- block flow diagrams or simplified process flow diagrams
- process chemistry
- maximum needed inventory
- acceptable upper and lower limits, where applicable, for items such as temperatures, pressures, flows, and compositions

Mechanical Design Information

- piping and instrument diagrams, electrical area classification, design and basis of relief systems
- design of the ventilation system
- equipment and piping specifications, a description of the shutdown and interlock systems
- design codes employed
- required or mandatory inspections and maintenance activities

Where mechanical design deviates from applicable consensus codes and standards, the deviation and its design basis should be documented, API notes.

Information concerning additional data that could be included in a facility Process Document is provided in Chapter 8, Identifying Hazards, and Chapter 26, Process Safety Management, in the *Administration & Programs* volume of this *Manual*.

Hazards/Risk Analysis

Safe operation of any chemical process requires that the persons in charge ask the three "whats":

- What can go wrong?
- What is the probability that something will go wrong?
- What would be the consequences if something does go wrong?

In other words, identify the risks and resulting dangers of the system. By following the procedure recommended by Daniel A. Crowl and Joseph F. Louvar in *Chemical Process Safety: Fundamentals with Applications* (1990), managers can identify the hazards and risks and judge whether the danger or risk is acceptable (Figure 25–5).

If the danger is deemed acceptable, management can build and operate the process. When management deems the risks or hazards are unacceptable, design engineers must modify the design and repeat the hazard identification and risk assessment process.

Popular Methods

Hazard identification can be achieved using any one of several established methods. Four of the most widely used are:

1. hazard surveys
2. process checklists
3. hazards and operability studies (HAZOP)
4. safety reviews

Hazard Surveys

For small firms, a chemical hazard survey might consist of an inventory of hazardous materials. Managers can study this inventory to make certain, for example, that dangerous substances are stored safely and that plans are in place to avoid an explosion or escape of toxic substances in the event of a fire or other unintentional incident.

For larger concerns, however, many companies use Dow's rigorous *Fire and Explosion Index* (Figure 25–6). The Dow index constructs a relative ranking for a hazard and a mechanism for estimating dollar losses. A concise explanation (with examples) of how the index works is provided by Crowl and Louvar on pages 311–322 of *Chemical Process Safety*.

Surveys are useful for spotting hazards associated with material storage and design, but not for alerting management to hazards associated with human error. Still, most of these surveys can be conducted quickly and easily and may not require highly trained individuals.

Process Checklists

Good checklists tend to become longer and more specific with use, and the list can develop into an extensive document in a large chemical facility. One way to keep the hazard identification method from getting out of hand is to limit its scope, creating separate checklists as incident prevention requires. For example, one checklist might be geared toward starting up a new process, and another for modifying a process already under way. Checklists are good for only preliminary hazard identification, however, and should be followed up with a more complete hazard identification method.

Hazards and Operability (HAZOP) Studies

Commonly known as a HAZOP study, this is a formal procedure used to identify hazards in a chemical process facility. It often requires the attention of a committee made up of experienced chemical and facility process personnel; and a complete, thorough HAZOP review can require a significant investment in time and resources.

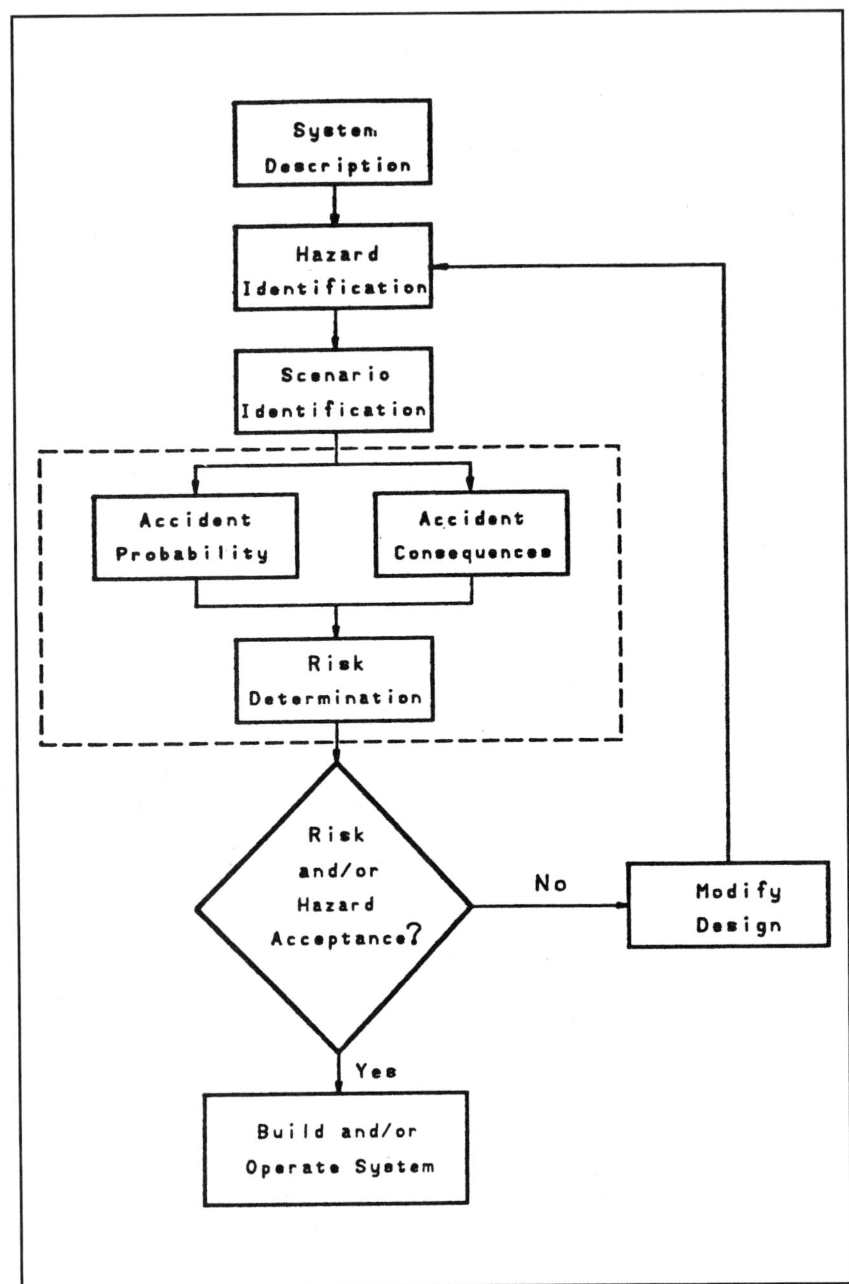

Figure 25–5. This procedural flowchart illustrates the process of hazard and/or risk assessment.

Briefly stated, the method encourages committee members to imagine all the ways process failures might occur and how they could be mitigated or minimized. Although a tedious process, it is nonetheless highly effective and is seeing wider use. For a full explanation of how a HAZOP review works, see the American Institute of Chemical Engineers' *Guidelines for Hazard Evaluation Procedures* or Crowl and Louvar, pages 323–327.

Safety Reviews

There are at least two types of safety reviews: An informal review involving a few people concerned about a small matter such as a change in procedure, and a formal review that usually results in a more detailed committee report. Of course, a report might also be written following an informal review. However, the chief goal of an informal review is to provide a forum where ideas can be aired and safety improvements developed.

Conversely, a company should establish a formal review committee and request an official written report before undertaking substantial updatings or changes in processes. The report should contain relevant technical data, equipment descriptions, an explanation of the process and attendant hazards, a safety checklist, material safety data sheets for each hazardous material used in the process under review, and a full accounting of all procedures. Crowl and Louvar provide a good example

FIRE AND EXPLOSION INDEX DOW

LOCATION	DATE

PLANT	PROCESS UNIT	EVALUATED BY	REVIEWED BY

MATERIALS AND PROCESS

MATERIALS IN PROCESS UNIT

STATE OF OPERATION
START UP SHUT DOWN NORMAL OPERATION

BASIC MATERIAL(S) FOR MATERIAL FACTOR

MATERIAL FACTOR (SEE TABLE I OR APPENDICES A OR B) Note requirements when unit temperature over 140 F)

1. GENERAL PROCESS HAZARDS	PENALTY	PENALTY USED
BASE FACTOR	1.00	1.00
A EXOTHERMIC CHEMICAL REACTIONS (FACTOR .30 to 1.25)		
B ENDOTHERMIC PROCESSES (FACTOR .20 to .40)		
C MATERIAL HANDLING & TRANSFER (FACTOR .25 to 1.05)		
D ENCLOSED OR INDOOR PROCESS UNITS (FACTOR .25 to .90)		
E ACCESS	.35	
F DRAINAGE AND SPILL CONTROL (FACTOR .25 to .50) ___ Gals		
GENERAL PROCESS HAZARDS FACTOR (F_1)		
2. SPECIAL PROCESS HAZARDS		
BASE FACTOR	1.00	1.00
A TOXIC MATERIAL(S) (FACTOR 0.20 to 0.80)		
B SUB ATMOSPHERIC PRESSURE (< 500 mm Hg)	.50	
C OPERATION IN OR NEAR FLAMMABLE RANGE ( ) INERTED ( ) NOT INERTED		
1 TANK FARMS STORAGE FLAMMABLE LIQUIDS	.50	
2 PROCESS UPSET OR PURGE FAILURE	.30	
3 ALWAYS IN FLAMMABLE RANGE	.80	
D DUST EXPLOSION (FACTOR .25 to 2.00) (SEE TABLE II)		
E PRESSURE (SEE FIGURE 2) OPERATING PRESSURE ____ psig RELIEF SETTING ____ psig		
F LOW TEMPERATURE (FACTOR .20 to .30)		
G QUANTITY OF FLAMMABLE/UNSTABLE MATERIAL QUANTITY ____ lbs. H_c = ____ BTU/lb		
1 LIQUIDS, GASES AND REACTIVE MATERIALS IN PROCESS (SEE FIG. 3)		
2 LIQUIDS OR GASES IN STORAGE (SEE FIG. 4)		
3 COMBUSTIBLE SOLIDS IN STORAGE, DUST IN PROCESS (SEE FIG. 5)		
H CORROSION AND EROSION (FACTOR .10 to .75)		
I. LEAKAGE – JOINTS AND PACKING (FACTOR .10 to 1.50)		
J. USE OF FIRED HEATERS (SEE FIG. 6)		
K. HOT OIL HEAT EXCHANGE SYSTEM (FACTOR .15 to 1.15) (SEE TABLE III)		
L ROTATING EQUIPMENT	.50	
SPECIAL PROCESS HAZARDS FACTOR (F_2)		
UNIT HAZARD FACTOR ($F_1 \times F_2 = F_3$)		
FIRE AND EXPLOSION INDEX ($F_3 \times MF = F\&EI$)		

Figure 25–6. This sample index can be used to assess hazards and estimate losses.

of a full report following a formal safety review in *Chemical Process Safety.*

A formal safety review committee can be convened in many cases with little difficulty. The formal review process often produces good results, although the process may take substantially more time than an informal review. If experienced safety reviewers are not available to form a committee for one reason or another, a company could use less experienced personnel and conduct a more structured HAZOP study and follow-up analysis.

Other Methods

Additional methods for pinpointing hazards include:

- *"What if?" analysis.* A hazard review team asks itself hypothetical questions about potential problems in the process, such as, "What if the pipe breaks?" The team then ponders the consequences and proposes solutions.
- *Failure mode, effects, and criticality analysis (FMECA).* A review team catalogs all the equipment used in a process under review and notes how each piece of equipment could fail. The team considers the consequences resulting from each type of failure and recommends safety measures.
- *Human error analysis.* The team identifies part of the process most likely to fall prey to human error. A review committee then makes safety recommendations or suggests preventive measures. This process is particularly useful in the layout and design of a control room's panels and instruments. The increased interactions of workers with processes in these locations greatly enhance the risk of injury.

Timing

Hazard identification and risk assessment should be an ongoing process. It should begin during the earliest conceptual phase of a new chemical process, when the firm is gathering basic data about the anticipated process. This is the time to ask basic questions, such as "Why are we contemplating the use of hazardous materials? Are less hazardous alternatives available?"

Another good time for asking questions comes during pre-authorization, when the design is less than half completed, and before the project has been officially authorized. Ask questions about hazards that could affect cost: "How are we going to handle steam venting?" for example.

Inserting Risk Assessment

In one major chemical company's process safety management plan, a four-part risk assessment is undertaken during the design stage. Reviewers ask:

- What potentially catastrophic incidents could possibly occur?
- What is the downwind dispersion likely to be in the event of a toxic gas release?
- What would the impact be on the workplace and community?
- Can we quantify or conduct a probability analysis for incidence occurrence?

Hazard identification also should occur during pre-start-up. Check preceding hazard reviews before chemical processing has begun, and use a pre-start-up checklist to make sure all elements of process safety management have been properly addressed.

Pre-Start-Up Reviews

All the elements of a process safety management program should be in place and running *before* beginning the warm-up phase of facility operation. A company should call in persons with no vested interest in facility start-up to assist in this effort.

A checklist of the various elements that make up a facility's process safety management plan should be created and should indicate who is responsible for review of each item, who is responsible for execution, and who is responsible for follow-through (Figure 25–7).

The American Petroleum Institute (API) recommends that pre-start-up safety reviews ensure that:

- construction conforms to specifications
- safety, operating, maintenance, and emergency procedures are in place and adequate
- process hazard analysis recommendations have been considered and implemented
- training of operating personnel is completed

Analysis for Existing Facilities

The American Petroleum Institute recommends that process hazard analyses be reviewed in order of priority and that they be updated every 3 to 10 years. In addition, the API says, the following factors should be considered when establishing priorities:

- high Substance Hazard Index value or large quantities of toxic, flammable, or explosive substances at that facility
- proximity to a populous area, or a facility location with a large workforce
- severe operating conditions, such as high temperatures or pressures, or conditions that cause severe corrosion or erosion

Pre-Start-Up Checklist

Is construction in accordance with design specifications?

Are key elements of process hazard management in place?

- safety information package
- completed process hazard reviews
- rules and procedures for operating, maintenance, safety, spill control, and emergencies
- training
- equipment tests and inspections

Are general subjects adequately addressed?

- fire protection
- means of exit
- availability and location of safety equipment
- equipment guards
- ventilations
- tripping hazards
- proper drainage

Does the checklist call for the inspection team to decide whether the facility is ready for start-up? Where recommendations are made, are responsibilities assigned to appropriate personnel for execution and follow-up?

Figure 25–7. This is a sample pre-start-up checklist.

The hazard analysis team should include persons knowledgeable in engineering, operations, design, and chemical processing.

Operating Procedures Manuals

Every chemical facility needs an operating manual that details the rules and guidelines for safe operation. It should contain the relevant safety, health, and environmental information needed to operate the facility without unintentional injuries.

Much has been written about what an operating manual should contain. The API, for example, recommends that operating procedures clarify the following points:

- position of the person(s) responsible for each of the facility's operating areas
- clear instructions for safe operation
- operating conditions and steps for initial start-up
- normal, temporary, and emergency operations
- normal shut-down
- start-up following a turnaround
- operating limits (where safety considerations are present)
- descriptions of the consequences of deviation
- steps to correct or avoid deviation
- safety systems and their functions
- occupational safety and health considerations

The operating procedures should explain to personnel the hazards involved in chemical operations and what the consequences could be if operations exceed safe limits. It is important that operating procedures be updated. The company should not authorize any change in technology, the facility, or procedural steps until it updates the operating procedures manual and trains appropriate personnel accordingly. A complete set of operating procedures, such as those given in Figure 25–8, is the basic document for operator training.

Management of Change

Any time someone proposes a change to the established technology, management should analyze the proposal thoroughly in light of its potential impact on the workplace and environment. Seemingly insignificant changes can have far-reaching effects.

For example, a worker repairs a piece of equipment using a different bolt. Later, different materials are used or process conditions are changed to improve efficiency. Perhaps raw materials begin to come from a different supplier. If there are no controls in place, change can build upon change until the equipment being operated is so far from the original design that the design review has become obsolete.

Changes must be consistent with established process technology. That means management must establish requirements for all field modifications before their implementation.

If management follows the suggestions of the American Petroleum Institute in this regard, it must:

- consider the process and mechanical design basis for change
- analyze the safety, health, and environmental considerations involved in the proposed change, along with the potential effects on facilities upstream and downstream
- consider the effect of modifications on operating procedures
- communicate the pending change and potential consequences to appropriate personnel
- consider the duration of the change
- acquire the necessary authorizations

Ten Rules for Safe Chemical Processing Operations

1. Never start adding a reactant at a lower or higher temperature than is called for in the process.
2. Do not add at a rate faster than specified.
3. During scale-up, make sure enough cooling capacity has been supplied.
4. Have specific, written plans on what to do if an emergency occurs. (See Chapter 18, Emergency Preparedness, in the *Administration & Programs* volume.)
5. If material is being added and the agitator has stopped, do not restart without first consulting with technical personnel.
6. Be extremely cautious with spill or low flash-point materials. Make sure the proper personnel understand and practice those plans.
7. In flammable and toxic liquid service, eliminate as much as practical the use of glass and other breakable materials.
8. Use nitrogen purges or other equivalent, recognized means on vapor spaces containing explosive materials.
9. Do not undercharge or overcharge reactants if a mistake has been made without consulting with technical personnel.
10. Watch for unusual incidents and for temperature, pressure, or other process parameter extremes that could make the reaction uncontrolled or dangerous.

Figure 25–8. These rules for safe chemical processing operations can provide basic topics for operator training.

Whenever a chemical process changes, corresponding equipment and personnel changes often occur, requiring additional personnel training. Make sure all the personnel involved understand the change, the reason for the change, how operations have been affected, and how personnel are to perform their new responsibilities. All these aspects are important to ensure safe facility operation.

Modifications of facilities or operating procedures should get prudent review, and the proper manager should document and authorize all changes.

Auditing

Auditing is management's way of making sure a process safety management program is operating as designed. One authoritative body, the API, recommends that audits be conducted at three- to five-year intervals. Regular audits should pinpoint deficiencies caused by such factors as changes in personnel or confusion over priorities. At the early stage, any corrective action needed is still minor.

Auditing of process safety management is primarily a line responsibility. It is easier to start corrective changes when workers view them as coming from an in-house source. Of course, outside audits are extremely useful, and often are used in conjunction with internal audits to assure an unbiased result. But internal auditing reinforces employees' understanding of priorities and the corporate mission or philosophy. Internal auditing also lays the groundwork for a successful outside audit. (For more information, see Process Safety Audits in Chapter 26, Process Safety Management in the *Administration & Programs* volume of this manual.

SUMMARY

- Safety in automated manufacturing processes can be enhanced by (1) careful identification of hazards and (2) development of appropriate strategies and worker training to control the workplace environment. Risks associated with automated processes include inhalation, burn, radiation, and pinch hazards.
- Companies should provide visual and mechanical warning signs, awareness barriers, interlocks, and controls to protect workers from hazards of automated equipment or systems.
- Preventive and predictive maintenance programs can enhance worker safety, particularly if maintenance procedures follow manufacturer and industry/government standards for the equipment and processes.
- Automated production requires special hazard precautions because of the nature of the machinery, the variety of human-machine interactions on the job, and the rapid development of automated technology.
- Components of many automated manufacturing systems or cells include computerized numerically controlled machines, materials handling and transport systems (automated guided vehicles), and robots. Each type of machine has its own hazards, and workers must be trained in proper safety procedures and housekeeping practices to operate this equipment safely.
- When safeguarding workers from robotic machinery, companies should focus on (1) manufacturing, remanufacturing, and rebuilding of robots; (2) robot installation; and (3) safeguards for workers exposed to hazards around robots.
- Automated chemical processes pose considerable health and safety risks for workers. Companies should conduct ongoing risk analyses to identify hazards in current, new, or redesigned processes and to design effective safety procedures and training programs.

- Proactive steps companies can take in hazard analysis and safety planning include pre-start-up reviews, pre-start-up checklists, preliminary analyses of existing facilities, updates of operating procedures manuals, and creation of strategies to manage change.
- Once a process safety management program is in place, the firm should conduct regular audits to ensure that the process is operating as designed.

REFERENCES

American Institute of Chemical Engineers. *Dow's Fire and Explosion Index Hazard Classification Guide,* 6th ed. New York: 1987.

American Institute of Chemical Engineers. *Guidelines for Hazard Evaluation Procedures,* 2nd ed. New York: 1992.

American National Standards Institute, 11 West 42nd Street, New York, NY 10036.

Coil-Slitting Machines/Systems, ANSI B11.14–1996.

Cold Headers and Cold Formers, ANSI B11.7–1995 (R2000).

Drilling, Milling, and Boring Machines, ANSI B11.8–1983 (R1994).

Electrical Safety Requirements for Employee Workplaces, ANSI/NFPA 70E–2000.

Electrical Standard for Industrial Machinery, ANSI/NFPA 79-1997.

Gear-Cutting Machines, ANSI B11.11–1985 (R1994).

Grinding Machines, ANSI B11.9–1975 (R1997).

Hydraulic Power Presses, ANSI B11.2–1995 (R2000).

Industrial Robots and Industrial Robot Systems, ANSI/RIA R15.06-1999.

Lathes, ANSI B11.6-1984 (R1994).

Machine Tools—Iron Workers, ANSI B11.5–1988 (R1994).

Machine Tools—Mechanical Power Presses, ANSI B11.1–1988 (R1994).

Machine and Machinery Systems for the Processing of Coiled Strip, Sheet, and Plate from Coiled Configurations, ANSI B11.18–1997.

Performance Criteria for the Design, Construction, Care, and Operation of Safeguarding When Referenced by the Other B11 Machine Tool Safety Standards, ANSI B11.19-1990 (R1997).

Pipe, Tube, and Shape Bending Machines, ANSI B11.15–R1994.

Powered Industrial Trucks, Including Type Designations, Areas of Use, Maintenance, and Operation, ANSI/NFPA 505–1999.

Power Press Brakes, ANSI B11.3-1982 (R1994).

Roll-Forming and Roll-Bending Machines, ANSI B11.12–R1996.

Shears, ANSI B11.4–1993.

Single- and Multiple-Spindle Automatic Bar and Chucking Machines, ANSI B11.13–1992 (R1998).

American Petroleum Institute. Management of Process Hazards, *Recommended Practice 750.* Washington, DC: 1990.

Bachelor BG, Waltz FM, eds. *Proceedings on Machine Vision Systems Integration in Industry.* Boston: SPIE—Society of Photo-Optical Instrumentation Engineers, 1990.

Chemical Manufacturers Association. *Process Safety Management (Control of Acute Hazards).* Washington, DC: 1985.

Chiantella NA, ed. *Management Guide for CIM (Computer-Integrated Manufacturing).* Dearborn, MI: The Computer and Automated Systems Association Society of Manufacturing Engineers, 1986.

Crowl DA and Louvar JF. *Chemical Process Safety: Fundamentals with Applications.* New York: Prentice-Hall, Inc., 1990.

Gainer CA, Jr. and Jiang BC. *A Cause-and-Effect Analysis of Industrial Robot Accidents from Four Countries.* Dearborn, MI: SME Technical Paper MS87–204, Society of Manufacturing Engineers, 1987.

Holland JR, ed. *Flexible Manufacturing Systems.* Dearborn, MI: Society of Manufacturing Engineers, 1984.

International Labour Office. *Major Hazard Control: A Practical Manual.* Geneva, Switzerland: 1988.

Kayes PJ, ed. *Manual of Industrial Hazard Assessment Techniques.* London, England: Technical Ltd., The World Bank, 1985.

Kletz TA. *Learning from Accidents in Industry.* London, England: Butterworth, 1988.

———. HAZOP and HAZAN: Identifying and Assessing Process Industry Hazards, 3rd Rev. and English Edition. Warwickshire, England: The Institution of Chemical Engineers, 1992.

———. What Went Wrong? Case Histories of Process Facility Disasters, 2nd ed. Houston: Gulf Publishing Co., 1988.

Lane JD, ed. *Automated Assembly,* 2nd ed. Dearborn, MI: Society of Manufacturing Engineers, 1986.

Lees FP. *Loss Prevention in the Process Industries.* London, England: Butterworth, 1980.

Miller RK. *Automated Guided Vehicles and Automated Manufacturing.* Dearborn, MI: Society of Manufacturing Engineers, 1987.

National Safety Council, 1121 Spring Lake Drive, Itasca, IL 60143. *Robots,* 12304–717, 1985.

Noro K, ed. *Occupational Health & Safety in Automation & Robotics.* London, England: Taylor & Francis, 1987.

Plog B, ed. *Fundamentals of Industrial Hygiene,* 4th ed. Itasca, IL: National Safety Council, 1996.

Roffel B and Rijnsdorp JE. *Process Dynamics, Control and Protection.* Ann Arbor, MI: Ann Arbor Science, 1982.

Strubhar PM, ed. *Working Safely with Industrial Robots,* 1st ed. Dearborn, MI: Robotics International Society of Manufacturing Engineers, 1986.

Zuech N, ed. *Machine Vision Capabilities for Industry.* Dearborn, MI: Society of Manufacturing Engineers, 1986.

REVIEW QUESTIONS

1. What is the disadvantage of automation being designed, implemented, or modified using the Just-in-Time (JIT) philosophy?
2. How can Computerized Maintenance Management Systems (CMMS) reduce company costs?
3. Safety in automated manufacturing processes can be greatly increased in what two ways?
   a.
   b.
4. List four factors that influence the kinds of safety precautions related to automation that companies must address.
   a.
   b.
   c.
   d.
5. A minimum clearance of ____________ is required between an automated guided vehicle (AGV), with its transported load, and any fixed object in all areas that involve pedestrian travel or interaction with workers.
6. Name the standard that specifically defines the areas, or envelopes, required for machine motion.
7. What two areas of automated materials handling and transport systems have particular risks?
   a.
   b.
8. Refer to ____________ for standards on bumper design, activation, and requirements for AGVs.
9. List the three basic parts that make up a robot.
   a.
   b.
   c.
10. According to Robotic Industries Association (RIA), the automotive industry accounts for about ____________ of all robots purchased each year in the United States.
    a. One-third
    b. Half
    c. Two-thirds
    d. Three-fourths
11. List the three major areas of robot safety.
    a.
    b.
    c.
12. Why does the highest degree of hazard exist when a robot is in the TEACH mode?
13. List the three "whats" that the persons in charge are required to ask to ensure the safe operation of any chemical process.
    a.
    b.
    c.
14. Name the formal procedure used to identify hazards in a chemical process facility.

APPENDIX 1

Safety and Health Tables

The safety and health professional requires information on many subjects. Specialized material must be sought out as required when the safety and health professional must deal with a specific subject in great detail and for a specific location. This latter information includes (1) applicable federal, state, or provincial, and local code requirements that are beyond the scope of this Manual and that must be checked locally, and (2) specific scientific and engineering information that generally is available in handbooks devoted to specific subject fields. For the reader's convenience in developing this additional information, a number of these handbooks are listed under References at the end of this appendix and in Appendix 1, Sources of Help, in the *Administration & Programs* volume.

Engineers and others have long recognized that equipment must be stronger than is really necessary. Actual work conditions are usually worse than testing, so a safety factor is often applied. By definition, this factor of safety is the ratio of the ultimate (breaking) strength of a member or piece of material to the actual working stress or to the maximum permissible (safe load) stress when in use. The magnitude of this factor depends on how great the cost of failure will be in terms of life or damage.

The selection of a safety factor is very important. The higher the number, the better, but a trade off is always required. Cost, weight, supporting structure, speed, or power requirement, size, and hazards involved are some of the common trade offs considered. The safety margin is that numerical value over 0 that results from strength divided by maximum stress. Typical factors of safety include:

	Factor of Safety
• stairs, landings	4
• standard railing	*
• scaffold, supporting member	4
• boiler - new	5
- used	5.5
- 10 years used	6
• unpressured vessel	5
• refrigerating system	5
• hydraulic pressure, piping, and hose	8
• cranes - hook	4–5
- gears	8
- structural steel	5
- hot work	10
• general hoisting equipment	5–8
• cast iron flywheel	10
• wood flywheel	20

*OSHA requires standard railings to withstand at least 200 lb (890 newtons) pressure applied in any direction at any point on the rail—§1910.23e(2)(v).

Table 1. Factors of Safety for Common Construction Materials

Material	*Steady Load*	*Load Varying from Zero to Maximum in one Direction*	*Load Varying from Zero to Maximum in both Directions*	*Suddenly Varying Loads and Shocks*
Cast iron	6	10	15	20
Wrought iron	4	6	8	12
Steel	5	6	8	12
Wood	8	10	15	20
Brick	15	20	25	30
Stone	15	20	25	30

Reprinted with permission from *Machinery's Handbook*, 16th ed., The Industrial Press.

Table 2. Safe Bearing Loads on Soils

Nature of Soil	*Safe Bearing Capacity (Tons per sq ft)*
Solid ledge of hard rock, such as granite, trap, etc.	25–100
Sound shale or other medium rock, requiring blasting for removal	10–15
Hardpan, cemented sand and gravel, difficult to remove by picking	8–10
Soft rock, disintegrated ledge; in natural ledge, difficult to remove by picking	5–10
Compact sand and gravel, requiring picking for removal	4–6
Hard clay, requiring picking for removal	4–5
Gravel, coarse sand, in natural thick beds	4–5
Loose, medium, and coarse sand; fine compact sand	1.5–4
Medium clay, stiff but capable of being spaded	2–4
Fine loose sand	1–2
Soft clay	1

Reprinted with permission from *Standard Handbook for Mechanical Engineers,* rev. 9th ed., Edited by T. Baumeister, III and E.A. Avallone. (Copyright 1987, McGraw-Hill Book Co.)
(Values approximate pressures allowed in major city building codes.)

Table 3. Specific Gravity of Gases and Liquids

Gas	*Sp Gr*	*Gas*	*Sp Gr*	*Gas*	*Sp Gr*
Air	1.000	Ether vapor	2.586	Methane	0.554
Acetylene	0.920	Ethylene	0.967	Nitrogen	0.971
Ethyl alcohol vapor	1.601	Helium	0.138	Nitric oxide	1.039
Ammonia	0.592	Hydrofluoric acid	2.370	Nitrous oxide	1.527
Carbon dioxide	1.520	Hydrochloric acid	1.261	Oxygen	1.106
Carbon monoxide	0.967	Hydrogen	0.069	Propane	1.554
Chlorine	2.423	Mercury vapor	6.940	Sulfur dioxide	2.250
Ethane	1.049	Marsh gas	0.555	Water vapor	0.623

1 cu ft of air at 32 F and atmospheric pressure weighs 0.0807 pounds.

Liquid	*Sp Gr*	*Liquid*	*Sp Gr*	*Liquid*	*Sp Gr*
Acetic acid	1.06	Fluoric acid	1.50	Palm oil	0.97
Alcohol, commercial	0.83	Gasoline	0.70	Petroleum oil	0.82
Ammonia	0.77	Glycerin	1.26	Phosphoric acid	1.78
Benzene	0.88	Kerosene	0.80	Rape oil	0.92
Bromine	2.97	Linseed oil	0.94	Vinegar	1.08
Carbolic acid	0.96	Mineral oil	0.92	Water	1.00
Carbon disulfide	1.26	Naphtha	0.76	Whale oil	0.92
Cotton-seed oil	0.93	Olive oil	0.92		

Reprinted with permission from *Machinery's Handbook,* 20th ed., The Industrial Press.
(Gases at 32 F. air = 1,000 liquids: water = 1,000)

1 cu ft water at 39 F weighs 62.43 pounds
1 ml water at 4 C weighs 1 g

Table 4. Approximate Specific Gravities and Densities

Substance	Specific gravity	Avg density, lb per cu ft	Substance	Specific gravity	Avg density, lb per cu ft
Metals, Alloys, Ores			Timber, air-dry		
Aluminum, cast-hammered	2.55–2.80	165	Apple	0.66–0.74	44
Aluminum, bronze	7.7	481	Ash, black	0.55	34
Brass, cast-rolled	8.4–8.7	534	Ash, white	0.64–0.71	42
Bronze, 7.9 to 14% Sn	7.4–8.9	509	Birch, sweet, yellow	0.71–0.72	44
Bronze, phosphor	8.88	554	Cedar, white, red	0.35	22
Copper, cast-rolled	8.8–8.95	556	Cherry, wild red	0.43	27
Copper ore, pyrites	4.1–4.3	262	Chestnut	0.48	30
German silver	8.58	536	Cypress	0.45–0.48	29
Gold, cast-hammered	19.25–19.35	1205	Fir, Douglas	0.48–0.55	32
Gold coin (U.S.)	17.18–17.2	1073	Fir, balsam	0.40	25
Iridium	21.78–22.42	1383	Elm, white	0.56	35
Iron, gray cast	7.03–7.13	442	Hemlock	0.45–0.50	29
Iron, cast, pig	7.2	450	Hickory	0.74–0.80	48
Iron, wrought	7.6–7.9	485	Locust	0.67–0.77	45
Iron, spiegel-eisen	7.5	468	Mahogany	0.56–0.85	44
Iron, ferrosilicon	6.7–7.3	437	Maple, sugar	0.68	43
Iron ore, hematite	5.2	325	Maple, white	0.53	33
Iron ore, limonite	3.6–4.0	237	Oak, chestnut	0.74	46
Iron ore, magnetite	4.9–5.2	315	Oak, live	0.87	54
Iron slag	2.5–3.0	172	Oak, red, black	0.64–0.71	42
Lead	11.34	710	Oak, white	0.77	48
Lead ore, galena	7.3–7.6	465	Pine, Oregon	0.51	32
Manganese	7.42	475	Pine, red	0.48	30
Manganese ore, pyrolusite	3.7–4.6	259	Pine, white	0.43	27
Mercury	13.546	847	Pine, Southern	0.61–0.67	38–42
Monel metal, rolled	8.97	555	Pine, Norway	0.55	34
Nickel	8.9	537	Poplar	0.43	27
Platinum, cast-hammered	21.5	1330	Redwood, California	0.42	26
Silver, cast-hammered	10.4–10.6	656	Spruce, white, red	0.45	28
Steel, cold-drawn	7.83	489	Teak, African	0.99	62
Steel, machine	7.80	487	Teak, Indian	0.66–0.88	48
Steel, tool	7.70–7.73	481	Walnut, black	0.59	37
Tin, cast-hammered	7.2–7.5	459	Willow	0.42–0.50	28
Tin ore, cassiterite	6.4–7.0	418			
Tungsten	19.22	1200	Various Liquids		
Zinc, cast-rolled	6.9–7.2	440	Alcohol, ethyl (100%)	0.789	49
Zinc, ore, blende	3.9–4.2	253	Alcohol, methyl (100%)	0.796	50
			Acid, muriatic (HCl), 40%	1.20	75
Various Solids			Acid, nitric, 91%	1.50	94
Cereals, oats, bulk	0.41	26	Acid, sulfuric, 87%	1.80	112
Cereals, barley, bulk	0.62	39	Chloroform	1.500	95
Cereals, corn, rye, bulk	0.73	45	Ether	0.736	46
Cereals, wheat, bulk	0.77	48	Lye, soda, 66%	1.70	106
Cork	0.22–0.26	15	Oils, vegetable	0.91–0.94	58
Cotton, flax, hemp	1.47–1.50	93	Oils, mineral, lubricants	0.88–0.94	57
Fats	0.90–0.97	58	Turpentine	0.861–0.867	54
Flour, loose	0.40–0.50	28	Water, 4 C, max. density	1.0	62.428
Flour, pressed	0.70–0.80	47	Water, 100 C	0.9584	59.830
Glass, common	2.40–2.80	162	Water, ice	0.88–0.92	56
Glass, plate or crown	2.45–2.72	161	Water, snow, fresh fallen	0.125	8
Glass, crystal	2.90–3.00	184	Water, sea water	1.02–1.03	64
Glass, flint	3.2–4.7	247			
Hay and straw, bales	0.32	20	Ashlar Masonry		
Leather	0.86–1.02	59	Granite, syenite, gneiss	2.4–2.7	159
Paper	0.70–1.15	58	Limestone	2.1–2.8	153
Potatoes, piled	0.67	44			
Rubber, Caoutchouc	0.92–0.96	59	Marble	2.4–2.8	162
Rubber goods	1.0–2.0	94	Sandstone	2.0–2.6	143
Salt, granulated, piled	0.77	48	Bluestone	2.3–2.6	153
Saltpeter	2.11	132			
Starch	1.53	96	Rubble Masonry		
Sulfur	1.93–2.07	125	Granite, syenite, gneiss	2.3–2.6	153
Wool	1.32	82			

(At room temperature with reference to water at 39 F)

Table 4. *(Continued.)*

Substance	*Specific gravity*	*Avg density, lb per cu ft*
Rubble Masonry		
Limestone	2.0–2.7	147
Sandstone	1.9–2.5	137
Bluestone	2.2–2.5	147
Marble	2.3–2.7	156
Dry Rubble Masonry		
Granite, syenite, gneiss	1.9–2.3	130
Limestone, marble	1.9–2.1	125
Sandstone, bluestone	1.8–1.9	110
Brick Masonry		
Hard brick	1.8–2.3	128
Medium brick	1.6–2.0	112
Soft brick	1.4–1.9	103
Sand-lime brick	1.4–2.2	112
Concrete Masonry		
Cement, stone, sand	2.2–2.4	144
Cement, slag, etc.	1.9–2.3	130
Cement, cinder, etc.	1.5–1.7	100
Various Building Mat'ls		
Ashes, cinders	0.64–0.72	40–45
Cement, portland, loose	1.5	94
Portland cement	3.1–3.2	196
Lime, gypsum, loose	0.85–1.00	53–64
Mortar, lime, set	1.4–1.9	103
		94
Mortar, portland cement	2.08–2.25	135
Slags, bank slag	1.1–1.2	67–72
Slags, bank screenings	1.5–1.9	98–117
Slags, machine slag	1.5	96
Slags, slag sand	0.8–0.9	49–55
Earth, etc., Excavated		
Clay, dry	1.0	63
Clay, damp, plastic	1.76	110
Clay and gravel, dry	1.6	100
Earth, dry, loose	1.2	76
Earth, dry, packed	1.5	95
Earth, moist, loose	1.3	78
Earth, moist, packed	1.6	96
Earth, mud, flowing	1.7	108
Earth, mud, packed	1.8	115
Riprap, limestone	1.3–1.4	80–85
Riprap, sandstone	1.4	90
Riprap, shale	1.7	105
Sand, gravel, dry, loose	1.4–1.7	90–105
Sand, gravel, dry, packed	1.6–1.9	100–120
Sand, gravel, wet	1.89–2.16	126
Excavations in Water		
Sand or gravel	0.96	60
Sand or gravel and clay	1.00	65
Clay	1.28	80
River mud	1.44	90
Soil	1.12	70
Stone riprap	1.00	65

Substance	*Specific gravity*	*Avg density, lb per cu ft*
Minerals		
Asbestos	2.1–2.8	153
Barytes	4.50	281
Basalt	2.7–3.2	184
Bauxite	2.55	159
Bluestone	2.5–2.6	159
Borax	1.7–1.8	109
Chalk	1.8–2.8	143
Clay, marl	1.8–2.6	137
Dolomite	2.9	181
Feldspar, orthoclase	2.5–2.7	162
Gneiss	2.7–2.9	175
Granite	2.6–2.7	165
Greenstone, trap	2.8–3.2	187
Gypsum, alabaster	2.3–2.8	159
Hornblende	3.0	187
Limestone	2.1–2.86	155
Marble	2.6–2.86	170
Magnesite	3.0	187
Phosphate rock, apatite	3.2	200
Porphyry	2.6–2.9	172
Pumice, natural	0.37–0.90	40
Quartz, flint	2.5–2.8	165
Sandstone	2.0–2.6	143
Serpentine	2.7–2.8	171
Shale, slate	2.6–2.9	172
Soapstone, talc	2.6–2.8	169
Syenite	2.6–2.7	165
Stone, Quarried, Piled		
Basalt, granite, gneiss	1.5	96
Limestone, marble quartz	1.5	95
Sandstone	1.3	82
Shale	1.5	92
Greenstone, hornblende	1.7	107
Bituminous Substances		
Asphaltum	1.1–1.5	81
Coal, anthracite	1.4–1.8	97
Coal, bituminous	1.2–1.5	84
Coal, lignite	1.1–1.4	78
Coal, peat, turf, dry	0.65–0.85	47
Coal, charcoal, pine	0.28–0.44	23
Coal, charcoal, oak	0.47–0.57	33
Coal, coke	1.0–1.4	75
Graphite	1.64–2.7	135
Paraffin	0.87–0.91	56
Petroleum	0.87	54
Petroleum, refined (kerosene)	0.78–0.82	50
Petroleum, benzene	0.73–0.75	46
Petroleum, gasoline	0.70–0.75	45
Pitch	1.07–1.15	69
Tar, bituminous	1.20	75
Coal and Coke, Piled		
Coal, anthracite	0.75–0.93	47–58
Coal, bituminous, lignite	0.64–0.87	40–54
Coal, peat, turf	0.32–0.42	20–26
Coal, charcoal	0.16–0.23	10–14
Coal, coke	0.37–0.51	23–32

1 lb per cu ft = 16.02 kg/m³

Reprinted with permission from *Standard Handbook for Mechanical Engineers,* rev. 9th ed. Edited by T. Baumeister, III and E.A. Avallone. (Copyright 1987. McGraw-Hill Book Co.)

Table 5. Levels of Illumination

This material is adapted from ANSI/IES standard RP1, RP7–1983. The values represent guides, and other factors, such as light source, colors in the work area, employee abilities, etc. must also be considered. They are not regulatory minimum nor maximum standards. The reader is advised to consult RP7–1983, *Illumination Levels for Industry.* Some currently recommended illuminances for industrial facility exteriors include:

Area/Activity	Lux	Footcandles
• Building (construction)		
General construction	100	10
Excavation work	20	2
• Building exteriors		
Entrances		
Active (pedestrian and/or conveyance)	50	5
• Dredging	20	2
• Loading and unloading platforms	200	20
• Freight car interiors	100	10
• Lumber yards	10	1
• Parking areas		
Main plant parking	20	2
Secondary parking	10	1
• Ship yards		
General	50	5
Ways	100	10
Fabrication areas	300	30
• Storage yards		
Active	200	20
Inactive	10	1
• Warehousing	100-500	10-50
• Aircraft Maintenance		
Close up		
Install plates, panels, fairings,	750	75
Seal plates	750	
Paint (exterior or interior of aircraft) where plates, panels, fairings, cowls, etc., must be in place before accomplishing	750-1000	75
• Docking		
Position doors and control surfaces for docking	300	30
Move aircraft into position in dock	500	50
• Maintenance, modification and repairs to airframe structures	750-1000	
• Specialty shops		
Instruments, radio	1500-1000	150
Electrical	1500	150
Hydraulic and pneumatic	1000	100
Components	1000	100
• Aircraft Manufacturing		
Fabrication (preparation for assembly)		
Rough bench work and sheet metal operations such as shears, presses, punches, countersinking, spinning	500	50
Drilling, riveting, screw fastening	750	75
Medium bench work and machining such as ordinary automatic machines, rough grinding, medium buffing and polishing	1000	100
Fine bench work and machining such as ordinary automatic machines, rough grinding, medium buffing and polishing	5000	500
Extra fine bench and machine work	10000	1000
• Final assembly such as placing of motors, propellers, wing sections, landing gear	1000	100
• General		
Rough easy seeing	300	30
Rough difficult seeing	500	50
Medium	1000	100
Fine	5000	500
Extra fine	10000	1000
• First manufacturing operations (first cut)		
Marking, sheering, sawing	500	50
• Flight test and delivery area		
On the horizontal plane	50	5
On the vertical plane	20	2
• Automotive Industry Facilities		
Elevators, steel furnace areas, locker rooms, exterior active storage areas	200	20
Waste treatment facilities (interior), clay mold and kiln rooms	300	30
Frame assembly, powerhouse, forgings, quick service dining, casting pouring and sorting.	500	50
Control and dispatch rooms, kitchens, large casting core and molding areas (engines), machining operations (engine and parts)	750	75
Chassis, body and component assembly	1000	100
Parts inspection stations	1500	150
Final assembly, body finishing and assembly, difficult inspection, paint color comparison	2000	200
Fine difficult inspection (casting cracks)	5000	500
Iron and Steel Industry		
Open hearth		
Stock yard	100	10
Charging floor	200	20
Mold yard	50	5
Hot top	300	30
Hot top storage	100	10
Rolling mills		
Blooming, slabbing, hot strip hot sheet	300	30
Motor room, machine room	300	30
Inspection		
Black plate, bloom and billet chipping	1000	100
Tin plate and other bright surfaces	2000	200
• Rubber Tires and Mechanical Rubber Goods		
Rubber tire manufacturing		
Banbury	300	30
Calendering		
General	300	30
Letoff and windup	500	50
Stock cutting		
General	300	30
Cutters and splicers	1000	100
Rubber goods — mechanical		
Stock preparation		
Plasticating, milling, Banbury	300	30
Calendering	500	50

Table 6. Hot and Cold Work Environments

Air Temperature and Relative Humidity versus Apparent Temperature

Apparent Temperature

Air Temperature (°F)	Relative Humidity (%) 0	5	10	15	20	25	30	35	40	45	50	55	60	65	70	75	80	85	90	95	100
140	125																				
135	120	128																			
130	117	122	131																		
125	111	116	123	131	141																
120	107	111	116	123	130	139	148														
115	103	107	111	115	120	127	135	143	151												
110	99	102	105	108	112	117	123	130	137	143	150										
105	95	97	100	102	105	109	113	118	123	129	135	142	149								
100	91	93	95	97	99	101	104	107	110	115	120	126	132	138	144						
95	87	88	90	91	93	94	96	98	101	104	107	110	114	119	124	130	136				
90	83	84	85	86	87	88	90	91	93	95	96	98	100	102	106	109	113	117	122		
85	78	79	80	81	82	83	84	85	86	87	88	89	90	91	93	95	97	99	102	105	108
80	73	74	75	76	77	77	78	79	79	80	81	81	82	83	85	86	86	87	88	89	91
75	69	69	70	71	72	72	73	73	74	74	75	75	76	76	77	77	78	78	79	79	80
70	64	64	65	65	66	66	67	67	68	68	69	69	70	70	70	70	71	71	71	71	72

Wind-Chill Chart

Estimated Wind Speed MPH	Actual Thermometer Reading °F 50	40	30	20	10	0	-10	-20	-30	-40	-50	-60
	Equivalent Temperature °F.											
Calm	50	40	30	20	10	0	-10	-20	-30	-40	-50	-60
5	48	37	27	16	6	-5	-15	-26	-36	-47	-57	-68
10	40	28	16	4	-9	-21	-33	-46	-58	-70	-83	-95
15	36	22	9	-5	-18	-36	-45	-58	-72	-85	-99	-112
20	32	18	4	-10	-25	-39	-53	-67	-82	-96	-110	-124
25	30	16	0	-15	-29	-44	-59	-74	-88	-104	-118	-133
30	28	13	-2	-18	-33	-48	-63	-79	-94	-109	-125	-140
35	27	11	-4	-20	-35	-49	-67	-82	-98	-113	-129	-145
40	26	10	-6	-21	-37	-53	-69	-85	-100	-116	-132	-148
Wind speeds greater than 40 MPH have little additional effect	Little Danger for Properly Clothed Person				Increasing Danger			Great Danger (Exposed flesh may freeze within 30 seconds)				
					Danger from Freezing of Exposed Flesh							

The wind-chill chart indicates the importance of wearing proper attire to combat injury to exposed skin, even though the temperature may be comparatively mild. To convert Fahrenheit degrees to Celsius, subtract 32 from the Fahrenheit reading; then take five-ninths of that figure (Courtesy—American Petroleum Institute.)

Table 7. Coefficients of Friction of Floors and Shoes

Coefficient of Friction	*Floors*	*Floor Clean*	*Floor Soiled*	*Shoes: Soles*
1.0	Soft rubber pad	.8	.6	Rubber-cork
.8	End grain wood	.75	.55	U.S.Army-U.S. Air Force std.
.7	Concrete, rough finish	.7	.5	Rubber-crepe
.65	Working decorative, dry	.6	.4	Neoprene
.5	Working decorative, soiled	.5	.3	Leather
.4	Steel			
				Shoes: Heels
		.7	.5	Neoprene
		.65	.45	Nylon

(Developed from Kroemer and Robinson, 1971; Kroemer, 1974.)

Table 8. Selected Common Abbreviations

SELECTED COMMON ABBREVIATIONS

Å	Angstrom unit of length
abs	absolute
amb	ambient
amp	ampere
app mol wt	apparent molecular weight
atm	atmospheric
at wt	atomic weight
Bé	degrees Baumé
bp	boiling point
bbl	barrel
Btu	British thermal unit
Btuh	Btu per hour
c	cycles per second (see Hz)
cal	calorie
cfh	cubic feet per hour
cfm	cubic feet per minute
cfs	cubic feet per second
cg	centigram
cm	centimeter
cgs system	centimeter-gram-second system
conc	concentrated, concentration
cc, cm^3	cubic centimeter
cu ft, ft^3	cubic foot
cu in.	cubic inch
° or deg	degree
C	degree Centigrade, degree Celsius
F	degree Fahrenheit
K	degree Kelvin
R	degree Reaumur, degree Rankine
dB	decibel
ET	effective temperature
ft	foot
ft-c	foot-candle
ft lb	foot pound
fpm	feet per minute
fps	feet per second
fps system	foot-pound-second system
fp	freezing point
gal	gallon
gr	grain
g	gram
gpm	gallons per minute
Hz	hertz (cycles per second)
hp	horsepower
hr	hour
ID	inside diameter
in.	inch
kcal	kilocalorie
kg	kilogram
km	kilometer
liq	liquid
L	liter and lambert(s)
LP-gas	liquefied petroleum gas
log	logarithm (common)
ln	logarithm (natural)
m, M	meter
ma	milliampere
MAC	maximum allowable concentration
max	maximum
mp	melting point
μ	micron
mks system	meter-kilogram-second system
mph	miles per hour
mg	milligram
ml	milliliter
mm	millimeter
mm (Hg)	mm of mercury
mμ	millimicron
mppcf	million particles per cu ft
mr	millirem
mR	1/1000 Roentgen
min	minute or minimum
mol wt, MW	molecular weight
N	newton
OD	outside diameter
oz	ounce
ppb	parts per billion
pphm	parts per hundred million
ppm	parts per million
lb	pound
psf	pounds per square foot
psi	pounds per square inch
psia	pounds per square inch absolute
psig	pounds per square inch gage
Rem	Roentgen equivalent man
rpm	revolution per minute
sec	second
sp gr	specific gravity
sp ht	specific heat
sp wt	specific weight
sq	square
scf	standard cubic foot
STP	standard temperature and pressure
temp	temperature
TLV	threshold limit valve
v	volt
W	watt
wt	weight

Note: Symbols are always written in singular form. Unabbreviated units form plurals in the usual manner.

Table 9. Table of Unit Prefixes

Multiples and submultiples	Prefixes	Symbols
1,000,000,000,000 = 10^{12}	tera-	T
1,000,000,000 = 10^{9}	giga-	G
1,000,000 = 10^{6}	mega-	M
1,000 = 10^{3}	kilo-	k
100 = 10^{2}	hecto-	h
10 = 10	deka-	D
0.1 = 10^{-1}	deci-	d
.01 = 10^{-2}	centi-	c
.001 = 10^{-3}	milli-	m
.000001 = 10^{-6}	micro-	μ
.000000001 = 10^{-9}	nano-	n
.000000000001 = 10^{-12}	pico	p

Data from National Bureau of Standards.

Table 10. Signs and Symbols

$+$	plus, addition, positive
$-$	minus, subtraction, negative
$\pm$	plus or minus, positive or negative
$\mp$	minus or plus, negative or positive
$\div$, /, ——	division
$\times$, $\cdot$, ( ) ( )	multiplication
( ) [ ]	collection
$=$	is equal to
$\neq$	is not equal to
$\equiv$	is identical to
$\cong$	equals approximately, congruent
$>$	greater than
$\ngtr$	not greater than
$\geqq$	greater than or equal to
$<$	less than
$\nless$	not less than
$\leqq$	less than or equal to
::	proportional to
:	ratio
$\sim$	similar to
$\propto$	varies as, proportional to
$\rightarrow$	approaches
∞	infinity
$\therefore$	therefore
$\sqrt{}$	square root
$\sqrt[n]{}$	nth root
a^n	nth power of a
log, $\log_{10}$	common logarithm
ln, $\log_e$	natural logarithm
e or ϵ	base of natural logs, 2.718
π	pi, 3.1416
$\angle$	angle
$\perp$	perpendicular to
$\parallel$	parallel to
n	any number
$\lvert n \rvert$	absolute value of n
$\bar{n}$	average value of n
a^{-n}	reciprocal of nth power of a, of a, or $\left\{\frac{1}{a^n}\right\}$
$n°$	n degrees (angle)
n'	n minutes, n feet
n''	n seconds, n inches
f(x)	function of x
Δx	increment of x
dx	differential of x
Σ	summation of
sin	sine
cos	cosine
tan	tangent

REFERENCES TO USEFUL HANDBOOKS OF ENGINEERING TABLES AND FORMULAS

NOTE: Because engineering and scientific handbooks are revised often, as frequently as once a year in some cases, no attempt has been made to list the latest edition in the following bibliography. In ordering, however, the most recent edition should be requested.

Alexander, JM, and RC Brewer, *Manufacturing Properties of Materials.* Princeton, NJ 08540. VanNostrand Co., Inc.

Allegheny Ludlum Steel Corp, Pittsburgh, PA 15222. *Tool Steel Handbook.*

American National Metric Council, 1625 Massachusetts Avenue, NW, Washington, DC 20036. "Metrication for the Manager."

American Society for Testing and Materials, 1916 Race Street, Philadelphia, PA 19103. "Standard Metric Practice Guide," E 380 (ANSI Z2101).

American Society of Heating, Refrigerating and Air Conditioning Engineers, 1791 Tullie Circle, NE, Atlanta, GA, 30329. *Guide and Data Books: Applications, Handbook of Fundamentals, Systems and Equipment.*

Baumeister, T III and EA Avallone, eds. *Standard Handbook for Mechanical Engineers.* New York, NY 10036, McGraw-Hill Book Co.

Bennett, H, ed. *Concise Chemical and Technical Dictionary.* New York, NY 10003, Chemical Publishing Co., Inc.

AM Best Company, Oldwick NJ 08858. *Best's Safety Directory,* 2 vols.

Compressed Air Magazine Co, Phillipsburg, NJ 08865. *Compressed Air Data.*

Eshbach, OW, *Handbook of Engineering Fundamentals.* New York NY 10016. John Wiley & Sons.

Gardner, W, EI Cooke, and RWI Cooke. *Handbook of Chemical Synonyms and Trade Names.* Cleveland, OH 44128, CRC Press.

Hudson, RG. *The Engineer's Manual.* New York, NY 10016, John Wiley & Sons, Inc.

Illuminating Engineering Society, 345 East 47th Street, New York, NY 10017. *IES Lighting Handbook (The Standard Lighting Guide).*

Kidder, FE, and H Parker. *Kidder-Parker Architects' and Builders' Handbook.* New York, NY 10016, John Wiley & Sons, Inc.

Knowlton, AE, ed. *Standard Handbook for Electrical Engineers.* New York, NY 10036, McGraw-Hill Book Co.

Kurtz, EB and TM Shoemaker. *The Linemans' and Cablemans' Handbook.* New York, NY 10036, McGraw-Hill Book Co.

LaLonde, WS Jr, and MF Janes. *Concrete Engineers Handbook.* New York, NY 10036, McGraw-Hill Book Co.

LeGrand, R. ed. *The New American Machinists' Handbook.* New York, NY 10036, McGraw-Hill Book Co.

Liebers, A. *The Engineer's Handbook Illustrated.* Los Angeles, CA 90047, Key Publishing Co.

Lindsey, FR. *Pipefitters Handbook.* New York, NY 10016, The Industrial Press.

Mantell, CL, ed. *Engineering Materials Handbook.* New York, NY 10036, McGraw-Hill Book Co.

Maynard, HB. *Industrial Engineering Handbook.* New York, NY 10036, McGraw-Hill Book Co.

Morris, IE. *Handbook of Structural Design.* New York, NY 10022, Reinhold Publishing Corp.

Morrow, LC, ed. *Maintenance Engineering Handbook.* New York, NY 10036, McGraw-Hill Book Co.

National Association of Home Builders. *Construction Dictionary.*

National Fire Protection Association, Batterymatch Park, Quincy, MA 02269. *Fire Protection Handbook.*

Oberg, E, and FD Jones. *Machinery's Handbook.* New York, NY 10016, The Industrial Press.

Pender, H, *et al. Electrical Engineers' Handbook.* New York, NY 10016, John Wiley & Sons, Inc.

Perry, JH, and RH Perry. *Engineering Manual.* New York, NY 10036, McGraw-Hill Book Co.

Perry, RH, *et al.*, eds. *Chemical Engineers' Handbook.* New York, NY 10036, McGraw-Hill Book Co.

Stanier, W. *Mechanical Power Transmission Handbook.* New York, NY 10036, McGraw-Hill Book Co.

———. *Plant Engineering Handbook.* New York, NY 10036, McGraw-Hill Book Co.

Urquhart, LC, ed. *Civil Engineering Handbook.* New York, NY 10036, McGraw-Hill Book Co.

Wilson, FW and PD Harvey, eds. *Tool Engineers Handbook.* New York, NY 10036, McGraw-Hill Book Co.

APPENDIX 2

Conversion of Units

All physical units of measurement can be reduced to three basic dimensions—mass, length, and time. Not only does reducing units to these basic dimensions simplify the solution of problems, but standardization of units makes comparison between operations (and between operations and standards) easier.

For example, air flows are usually measured in liters per minute, cubic meters per second, or cubic feet per minute. The total volume of air sampled can be easily converted to cubic meters or cubic feet. In another situation, the results of atmospheric pollution studies and stack sampling surveys are often reported as grains per cubic foot, grams per cubic foot, or pounds per cubic foot. The degree of contamination is usually reported as parts of contaminant per million parts of air.

If physical measurements are made or reported in different units, they must be converted to the standard units if any comparisons are to be meaningful.

To save time and space in reporting data, many units have standard abbreviations. Because the metric system (SI) is becoming more frequently used, conversion factors are given for the standard units of measurement.

FUNDAMENTAL UNITS

Conversion factors for various measurement units are listed in the tables in this section. To use a table to find the numerical value of the quantity desired, locate the unit to be converted in the first column. Then multiply this value by the number appearing at the intersection of the row and the column containing the desired unit. The answer will be the numerical value in the desired unit.

Various English system and metric system units are given for the reader's convenience. The new system of measurement, however, is the International System of Units (SI). The official conversion factors and an explanation of the system are given to 6- or 7-place accuracy in ASTM Standard E 380-76 (ANSI Z210.0-1976).

Briefly, the SI System being adopted throughout the world is a modern version of the MKSA (meter, kilogram, second, ampere) system. Its details are published and controlled by an international treaty organization, the International Bureau of Weights and Measures (BIPM), set up by the Metre Convention signed in Paris, France, on May 20, 1875. The United States and Canada are member states of this Convention, as implemented by the Metric Conversion Act of 1975 (Public Law 94-168).

HELPFUL ORGANIZATIONS

The following four groups in the U.S. and Canada are deeply involved in planning and implementing metric conversion:

American National Metric Council
5410 Grosvenor Lane
Bethesda, MD. 20814

Metric Commission Canada
240 Sparks Street
Ottawa, Ontario, Canada K1A 0H5

U.S. Metric Association, Inc.
Boulder, CO 80302

Office of Metric Programs
U.S. Dept. of Commerce
Washington, DC 20230

CONVERSION OF UNITS

Fahrenheit-Celsius Conversion Table

Fahrenheit-Celsius Conversion.—A simple way to convert a Fahrenheit temperature reading into a Celsius temperature reading or vice versa is to enter the accompanying table in the center or boldface column of figures. These figures refer to the temperature in either Fahrenheit or Celsius degrees. If it is desired to convert from Fahrenheit to Celsius degrees, consider the center column as a table of Fahrenheit temperatures and read the corresponding Celsius temperature in the column at the left. If it is desired to convert from Celsius to Fahrenheit degrees, consider the center column as a table of Celsius values, and read the corresponding Fahrenheit temperature on the right.

To convert from "degrees Fahrenheit" to "degrees Celsius" (formerly called "degrees centigrade"), use the formula:

$$t_c = \frac{(t_f - 32)}{1.8} \text{ or } \frac{5}{9}(t_f - 32)$$

$$\text{Conversely, } t_f = 1.8t_c + 32 \text{ or } \frac{5}{9}t_c + 32$$

Example, convert the boiling point of water in F to C:

$$212\text{ F} - 32 = 180$$

$$\frac{5}{9}(180) = 100\text{ C}$$

Fahrenheit–Celsius Conversion Table

Deg C		*Deg F*	*Deg C*		*Deg F*	*Deg C*		*Deg F*	*Deg C*		*Deg F*
−273	**−459.4**	...	−129	**−200**	−328	−13.9	**7**	44.6	1.1	**34**	93.2
−268	**−450**	...	−123	**−190**	−310	−13.3	**8**	46.4	1.7	**35**	95.0
−262	**−440**	...	−118	**−180**	−292	−12.8	**9**	48.2	2.2	**36**	96.8
−257	**−430**	...	−112	**−170**	−274	−12.2	**10**	50.0	2.7	**37**	98.6
−251	**−420**	...	−107	**−160**	−256	−11.7	**11**	51.8	3.3	**38**	100.4
−246	**−410**	...	−101	**−150**	−238	−11.1	**12**	53.6	3.9	**39**	102.2
−240	**−400**	...	−96	**−140**	−220	−10.6	**13**	55.4	4.4	**40**	104.0
−234	**−390**	...	−90	**−130**	−202	−10.0	**14**	57.2	5.0	**41**	105.8
−229	**−380**	...	−84	**−120**	−184	−9.4	**15**	59.0	5.6	**42**	107.6
−223	**−370**	...	−79	**−110**	−166	−8.9	**16**	60.8	6.1	**43**	109.4
−218	**−360**	...	−73	**−100**	−148	−8.3	**17**	62.6	6.7	**44**	111.2
−212	**−350**	...	−68	**−90**	−130	−7.8	**18**	64.4	7.2	**45**	113.0
−207	**−340**	...	−62	**−80**	−112	−7.2	**19**	66.2	7.8	**46**	114.8
−201	**−330**	...	−57	**−70**	−94	−6.7	**20**	68.0	8.3	**47**	116.6
−196	**−320**	...	−51	**−60**	−76	−6.1	**21**	69.8	8.9	**48**	118.4
−190	**−310**	...	−46	**−50**	−58	−5.6	**22**	71.6	9.4	**49**	120.2
−184	**−300**	...	−40	**−40**	−40	−5.0	**23**	73.4	10.0	**50**	122.0
−179	**−290**	...	−34	**−30**	−22	−4.4	**24**	75.2	10.6	**51**	123.8
−173	**−280**	...	−29	**−20**	−4	−3.9	**25**	77.0	11.1	**52**	125.6
−169	**−273**	−459.4	−23	**−10**	14	−3.3	**26**	78.8	11.7	**53**	127.4
−168	**−270**	−454	−17.8	**0**	32−	−2.8	**27**	80.6	12.2	**54**	129.2
−162	**−260**	−436	−17.2	**1**	33.8	−2.2	**28**	82.4	12.8	**55**	131.0
−157	**−250**	−418	−16.7	**2**	35.6	−1.7	**29**	84.2	13.3	**56**	132.8
−151	**−240**	−400	−16.1	**3**	37.4	−1.1	**30**	86.0	13.9	**57**	134.6
−146	**−230**	−382	−15.6	**4**	39.2	−0.6	**31**	87.8	14.4	**58**	136.4
−140	**−220**	−364	−15.0	**5**	41.0	0−	**32**	89.6	15.0	**59**	138.2
−134	**−210**	−346	−14.4	**6**	42.8	0.6	**33**	91.4	15.6	**60**	140.0

Deg C		*Deg F*	*Deg C*		*Deg F*	*Deg C*		*Deg F*	*Deg C*		*Deg F*
16.1	**61**	141.8	50.0	**122**	251.6	83.9	**183**	361.4	276.7	**530**	986
16.7	**62**	143.6	50.6	**123**	253.4	84.4	**184**	363.2	282.2	**540**	1004
17.2	**63**	145.4	51.1	**124**	255.2	85.0	**185**	365.0	287.8	**550**	1022
17.8	**64**	147.2	51.7	**125**	257.0	85.6	**186**	366.8	293.3	**560**	1040
18.3	**65**	149.0	52.2	**126**	258.8	86.1	**187**	368.6	298.9	**570**	1058
18.9	**66**	150.8	52.8	**127**	260.6	86.7	**188**	370.4	304.4	**580**	1076
19.4	**67**	152.6	53.3	**128**	262.4	87.2	**189**	372.2	310.0	**590**	1094
20.0	**68**	154.4	53.9	**129**	264.2	87.8	**190**	374.0	315.6	**600**	1112
20.6	**69**	156.2	54.4	**130**	266.0	88.3	**191**	375.8	321.1	**610**	1130
21.1	**70**	158.0	55.0	**131**	267.8	88.9	**192**	377.6	326.7	**620**	1148
21.7	**71**	159.8	55.6	**132**	269.6	89.4	**193**	379.4	332.2	**630**	1166
22.2	**72**	161.6	56.1	**133**	271.4	90.0	**194**	381.2	337.8	**640**	1184
22.8	**73**	163.4	56.7	**134**	273.2	90.6	**195**	383.0	343.3	**650**	1202
23.3	**74**	165.2	57.2	**135**	275.0	91.1	**196**	384.8	348.9	**660**	1220
23.9	**75**	167.0	57.8	**136**	276.8	91.7	**197**	386.6	354.4	**670**	1238
24.4	**76**	168.8	58.3	**137**	278.6	92.2	**198**	388.4	360.0	**680**	1256
25.0	**77**	170.6	58.9	**138**	280.4	92.8	**199**	390.2	365.6	**690**	1274
25.6	**78**	172.4	59.4	**139**	282.2	93.3	**200**	392.0	371.1	**700**	1292
26.1	**79**	174.2	60.0	**140**	284.0	93.9	**201**	393.8	376.7	**710**	1310
26.7	**80**	176.0	60.6	**141**	285.8	94.4	**202**	395.6	382.2	**720**	1328
27.2	**81**	177.8	61.1	**142**	287.6	95.0	**203**	397.4	387.8	**730**	1346
27.8	**82**	179.6	61.7	**143**	289.4	95.6	**204**	399.2	393.3	**740**	1364
28.3	**83**	181.4	62.2	**144**	291.2	96.1	**205**	401.0	398.9	**750**	1382
28.9	**84**	183.2	62.8	**145**	293.0	96.7	**206**	402.8	404.4	**760**	1400
29.4	**85**	185.0	63.3	**146**	294.8	97.2	**207**	404.6	410.0	**770**	1418
30.0	**86**	186.8	63.9	**147**	296.6	97.8	**208**	406.4	415.6	**780**	1436
30.6	**87**	188.6	64.4	**148**	298.4	98.3	**209**	408.2	421.1	**790**	1454
31.1	**88**	190.4	65.0	**149**	300.2	98.9	**210**	410.0	426.7	**800**	1472
31.7	**89**	192.2	65.6	**150**	302.0	99.4	**211**	411.8	432.2	**810**	1490
32.2	**90**	194.0	66.1	**151**	303.8	100.0	**212**	413.6	437.8	**820**	1508
32.8	**91**	195.8	66.7	**152**	305.6	104.4	**220**	428.0	443.3	**830**	1526
33.3	**92**	197.6	67.2	**153**	307.4	110.0	**230**	446.0	448.9	**840**	1544
33.9	**93**	199.4	67.8	**154**	309.2	115.6	**240**	464.0	454.4	**850**	1562
34.4	**94**	201.2	68.3	**155**	311.0	121.1	**250**	482.0	460.0	**860**	1580
35.0	**95**	203.0	68.9	**156**	312.8	126.7	**260**	500.0	465.6	**870**	1598
35.6	**96**	204.8	69.4	**157**	314.6	132.2	**270**	518.0	471.1	**880**	1616
36.1	**97**	206.6	70.0	**158**	316.4	137.8	**280**	536.0	476.7	**890**	1634
36.7	**98**	208.4	70.6	**159**	318.2	143.3	**290**	554.0	482.2	**900**	1652
37.2	**99**	210.2	71.1	**160**	320.0	148.9	**300**	572.0	487.8	**910**	1670
37.8	**100**	212.0	71.7	**161**	321.8	154.4	**310**	590.0	493.3	**920**	1688
38.3	**101**	213.8	72.2	**162**	323.6	160.0	**320**	608.0	498.9	**930**	1706
38.9	**102**	215.6	72.8	**163**	325.4	165.6	**330**	626.0	504.4	**940**	1724
39.4	**103**	217.4	73.3	**164**	327.2	171.1	**340**	644.0	510.0	**950**	1742
40.0	**104**	219.2	73.9	**165**	329.0	176.7	**350**	662.0	515.6	**960**	1760
40.6	**105**	221.0	74.4	**166**	330.8	182.2	**360**	680.0	521.1	**970**	1778
41.1	**106**	222.8	75.0	**167**	332.6	187.8	**370**	698.0	526.7	**980**	1796
41.7	**107**	224.6	75.6	**168**	334.4	193.3	**380**	716.0	532.2	**990**	1814
42.2	**108**	226.4	76.1	**169**	336.2	198.9	**390**	734.0	537.8	**1000**	1832
42.8	**109**	228.2	76.7	**170**	338.0	204.4	**400**	752.0	565.6	**1050**	1922
43.3	**110**	230.0	77.2	**171**	339.8	210	**410**	770.0	593.3	**1100**	2012
43.9	**111**	231.8	71.8	**172**	341.6	215.6	**420**	788	621.1	**1150**	2102
44.4	**112**	233.6	78.3	**173**	343.4	221.1	**430**	806	648.9	**1200**	2192
45.0	**113**	235.4	78.9	**174**	345.2	226.7	**440**	824	676.7	**1250**	2282
45.6	**114**	237.2	79.4	**175**	347.0	232.2	**450**	842	704.4	**1300**	2372
46.1	**115**	239.0	80.0	**176**	348.8	237.8	**460**	860	732.2	**1350**	2462
46.7	**116**	240.8	80.6	**177**	350.6	243.3	**470**	878	760.0	**1400**	2552
47.2	**117**	242.6	81.1	**178**	352.4	248.9	**480**	896	787.8	**1450**	2642
47.8	**118**	244.4	81.7	**179**	354.2	254.4	**490**	914	815.6	**1500**	2732
48.3	**119**	246.2	82.2	**180**	356.0	260.0	**500**	932	1093.9	**2000**	3632
48.9	**120**	248.0	82.8	**181**	357.8	265.6	**510**	950	1648.9	**3000**	5432
49.4	**121**	249.8	83.3	**182**	359.6	271.1	**520**	968	2760.0	**5000**	9032

Above 1000 in the center column, the table increases in increments of 50. To convert 1,462 degrees F to Celsius, for instance, add to the Celsius equivalent of 1,400 degrees F 5/9 of 62 or 34 degrees, which equals 794 C.

LENGTH

To Obtain → / Multiply Number of ↓ by ↘	*meter (m)*	*centimeter (cm)*	*millimeter (mm)*	*micron (μ) or micrometer (μm)*	*angstrom unit, (A)*	*inch (in.)*	*foot (ft)*
meter	1	100	1000	10^6	10^{10}	39.37	3.28
centimeter	0.01	1	10	10^4	10^8	0.394	0.0328
millimeter	0.001	0.1	1	10^3	10^7	0.0394	0.00328
micron	10^{-6}	10^{-4}	10^{-3}	1	10^4	3.94×10^{-5}	3.28×10^{-6}
angstrom	10^{-10}	10^{-8}	10^{-7}	10^{-4}	1	3.94×10^{-9}	3.28×10^{-10}
inch	0.0254	2.540	25.40	2.54×10^4	2.54×10^8	1	0.0833
foot	0.305	30.48	304.8	304,800	3.048×10^9	12	1

AREA

To Obtain → / Multiply Number of ↓ By ↘	*square meter (m^2)*	*square inch (sq in.)*	*square foot (sq ft)*	*square centimeter (cm^2)*	*square millimeter (mm^2)*
square meter	1	1,550	10.76	10,000	10^6
square inch	6.452×10^{-3}	1	6.94×10^{-3}	6.452	645.2
square foot	0.0929	144	1	929.0	92,903
square centimeter	0.0001	0.155	0.001	1	100
square millimeter	10^{-6}	0.00155	0.00001	0.01	1

DENSITY

To Obtain → / Multiply Number of ↓ By ↘	*gm/cm³*	*lb/cu ft*	*lb/gal*
gram/cubic centimeter	1	62.43	8.345
pound/cubic foot	0.01602	1	0.1337
pound/gallon (U.S.)	0.1198	7.481	1

1 grain/cu ft = 2.28 mg/m³

FORCE

To Obtain → Multiply Number of ↓ By ↘	*dyne*	*newton (N)*	*kilogram-force*	*pound-force (lbf)*
dyne	1	1.0×10^{-5}	1.02×10^{-6}	2.248×10^{-6}
newton	1.0×10^{5}	1	0.1020	0.2248
kilogram-force	9.807×10^{5}	9.807	1	2.205
pound-force	4.448×10^{5}	4.448	0.4536	1

MASS

To Obtain → Multiply Number of ↓ By ↘	*gram (gm)*	*kilogram (kg)*	*grains (gr)*	*ounce (avoir) (oz)*	*pound (avoir) (lb)*
gram	1	0.001	15.432	0.03527	0.00220
kilogram	1,000	1	15,432	35.27	2.205
grain	0.0648	6.480×10^{-5}	1	2.286×10^{-3}	1.429×10^{-4}
ounce	28.35	0.02835	437.5	1	0.0625
pound	453.59	0.4536	7,000	16	1

VOLUME

To Obtain → Multiply Number of ↓ By ↘	*cu ft*	*gallon (U.S. liquid)*	*liters*	*cm^3*	*m^3*
cubic foot	1	7.481	28.32	28,320	0.0283
gallon (U.S. liquid)	0.1337	1	3.785	3,785	3.79×10^{-3}
liter	0.03531	0.2642	1	1,000	1×10^{-3}
cubic centimeters	3.531×10^{-5}	2.64×10^{-4}	0.001	1	10^{-6}
cubic meters	35.31	264.2	1,000	10^{6}	1

VELOCITY

To Obtain → / Multiply Number of ↓ By ↘	cm/s	m/s	km/hr	ft/s	ft/min	mph
centimeter/second	1	0.01	0.036	0.0328	1.968	0.02237
meter/second	100	1	3.6	3.281	196.85	2.237
kilometer/hour	27.78	0.2778	1	0.9113	54.68	0.6214
foot/second	30.48	0.3048	18.29	1	60	0.6818
foot/minute	0.5080	0.00508	0.0183	0.0166	1	0.01136
mile per hour	44.70	0.4470	1.609	1.467	88	1

PRESSURE

To Obtain → / Multiply Number of ↓ By ↘	lb/sq in. (psi)	atm	in. (Hg) 32 F 0 C	mm (Hg) 32 F 0 C	kPa (kN/m²)	ft (H_2O) 60 F 15 C	in. (H_2O)	lb/sq ft
pound/square inch	1	0.068	2.036	51.71	6.895	2.309	27.71	144
atmospheres	14.696	1	29.92	760.0	101.32	33.93	407.2	2,116
inch (Hg)	0.4912	0.033	1	25.40	3.386	1.134	13.61	70.73
millimeter (Hg)	0.01934	0.0013	0.039	1	0.1333	0.04464	0.5357	2.785
kilopascals	0.1450	9.87×10^{-3}	0.2953	7.502	1	0.3460*	4.019	20.89
foot (H_20)(15 C)	0.4332	0.0294	0.8819	22.40	2.989*	1	12.00	62.37
inch (H_20)	0.03609	0.0024	0.073	1.867	0.2488	0.0833	1	5.197
pound/square foot	0.0069	4.72×10^{-4}	0.014	0.359	0.04788	0.016	0.193	1

*at 4 C

APPENDIX 3

Glossary

Abrasive blasting. A process for cleaning surfaces by means of high-pressure air or water with sand, alumina, nonfree silica, or organic materials.

Absorbent. A substance that takes in (absorbs) other material.

Accident. That occurrence in a sequence of events that produces unintended injury, death, or property damage. *Accident* refers to the event, not the result of the event (See Unintentional injury).

Accident causes. Hazards and those factors that, individually or in combination, directly cause accidents.

Accident prevention. The application of countermeasures designed to reduce accidents.

Accident rate. Accident experience in relation to a base unit of measure. For example:

- number of accidents per worker hours of exposure
- number of accidents per worker days worked
- number of accidents per miles traveled
- number of accidents per 100 employees.

Acclimation. The process of becoming adjusted to new climatic conditions (e.g., heat, cold, humidity, altitude, etc.).

Accuracy (instrument). Accuracy refers to the agreement of a reading or observation obtained from an instrument or a technique with the true value. Quite often used incorrectly as precision. (See Precision.)

ACGIH. American Conference of Governmental Industrial Hygienists. ACGIH develops and publishes recommended occupational exposure limits for chemical substances and physical agents. (See TLV.)

Acoustic, Acoustical. Containing, producing, arising from, actuated by, related to, or associated with sound.

Action level. Term used by U.S. OSHA and NIOSH to express the level of toxicant that requires medical surveillance, usually one-half the Permissible Exposure Limit. (See PEL.)

Acute effect. An adverse effect on a human or animal body, with several symptoms developing rapidly and coming quickly to a crisis.

Acute toxicity. The acute adverse effects resulting from a single dose of or exposure to a substance.

ADA. Americans with Disabilities Act. Legal standard requiring protections and accommodations be extended to the disabled, including employees and patrons of businesses.

Administrative controls. Methods of controlling employee exposures by job rotation, varying tasks, work assignment, operational procedures, or time periods away from the hazard(s).

Adsorption. The condensation of gases, liquids, or dissolved substances on the surfaces of solids.

Affirmative action. Positive action taken to assure nondiscriminatory treatment of all groups in employment regardless of sex, religion, age, handicap (disabilities), or national origin.

Agency or agent. The principal object, such as a tool, machine, or material, involved in an accident that inflicts injury, illness, or property damage.

Air. The mixture of gases that surrounds the earth; its major components are as follows: 78.08% nitrogen, 20.95% oxygen, 0.03% carbon dioxide, 0.93% argon, and varying amounts of water vapor. The mixture changes with altitude. (See Standard air.)

Air cleaner. A device designed to remove atmospheric airborne impurities, such as dusts, gases, vapors, fumes, and smokes.

Air-line respirator. A respirator that is connected to a compressed breathing air source by a hose.

Air monitoring. The sampling for and measuring of contaminants in a free or captive atmosphere.

Air-powered tools. Tools that use air under pressure to drive various rotating or percussion attachments.

Air-purifying respirator. Respirators that use filters or sorbents to remove harmful substances from the air.

Air-regulating valve. An adjustable valve used to regulate air pressure and flow rate, such as to the facepiece, helmet, or hood of an air-line respirator.

Air-supplied respirator. Respirator that provides a supply of breathable air from a clean air source.

Alloy. A mixture of metals.

Alpha particle (alpha ray, alpha radiation). A small electrically charged atomic particle of very high velocity thrown off by many radioactive materials, including uranium and radium. It is made up of two neutrons and two protons. Its electric charge is positive.

Aluminosis. A form of pneumoconiosis due to the presence of aluminum-bearing dust in the lungs, especially that of alum, bauxite, or clay.

Ambient noise. The all-encompassing noise associated with a given environment, being usually a composite of sounds from many sources.

Ampere. The standard unit for measuring the strength of an electrical current.

Anemometer. A device to measure air velocity.

Anneal. To treat by heat with subsequent cooling for drawing the temper of metals, that is, to soften and render them less brittle.

ANSI. The American National Standards Institute is a nonprofit voluntary membership organization that coordinates the U.S. Voluntary Consensus Standards System and approves American National Standards.

Anthropometric evaluation. A study of human body sizes and modes of action to better design tools and machines to human capabilities.

Anthropometry. The science of measuring the human body for differences in various characteristics.

Approved. Tested and/or listed as satisfactory; meeting predetermined requirements of some qualifying organization.

Arc welding. One form of electrical resistance welding using either uncoated or coated rods.

Arc-welding electrode. A component of the welding circuit through which current is conducted between the electrode holder and the arc.

Asbestos. A fibrous hydrated magnesium silicate.

Asbestosis. A disease of the lungs caused by the inhalation of fine airborne fibers of asbestos.

Asphyxia. Suffocation from lack of oxygen.

Asphyxiant. A vapor or gas that can cause unconsciousness or death by suffocation (lack of oxygen).

ASTM. American Society for Testing and Materials; voluntary membership organization with members from broad spectrum of individuals, agencies, and industries concerned with materials.

Atmosphere-supply respirator. A respirator that provides breathing air from a source independent of the surrounding atmosphere.

Atmospheric pressure. The pressure exerted in all directions by the atmosphere. At sea level, mean atmospheric pressure is 29.92 in. Hg, 14.7 psi, or 407 in. w.g.

Atomic energy. Energy released in nuclear reactions. The energy is released when a neutron splits an atom nucleus into smaller pieces (fission) or when two nuclei are joined together under millions of degrees of heat (fusion).

Audible range. The normal frequency range for human hearing is approximately 20 Hz through 20,000 Hz. Above the range of 20,000 Hz, the term ultrasonic is used. Below 20 Hz, the term subsonic is used.

Audiogram. A record of hearing loss (i.e., hearing level measured at several different frequencies—usually 500 to 6,000 Hz). The audiogram may be presented graphically or numerically. Hearing level is shown as a function of frequency.

Audiometer. An instrument that measures a person's ability to hear a pure tone at various frequencies. (See Frequency.)

Aural insert. Usually called earplugs or inserts. The pliable material is inserted into the ear canal, to reduce the amount of noise reaching the inner ear.

Autoignition temperature. The lowest temperature that a flammable gas-air or vapor-air mixture will ignite from its own heat source or contact with a hot surface, without spark or flame.

Background noise. Noise coming from sources other than the particular noise source being monitored.

Bag house. Term commonly used for the housing containing bag filters for recovery of airborne particulates from the exhausts of industrial operations.

Barrier guard. Physical protection for operators and other individuals from hazard points on machinery and equipment.

- *Fixed barrier guard*—A nonmovable physical enclosure attached to the machine or equipment.
- *Interlocked barrier guard*—An enclosure attached to the machinery or equipment frame and interlocked with the power switch so that the operating cycle cannot be started unless the guard is in its proper position.
- *Adjustable barrier guard*—An enclosure attached to the frame of the machinery or equipment with front and side sections that can be adjusted.
- *Gate or movable barrier guard*—A device designed to enclose the point of operation to exclude entry prior to equipment operation.

Base. A compound that reacts with an acid to form a salt. It is another term for alkali. It turns litmus paper blue.

Benign. Not malignant. A benign tumor is one that does not metastasize or invade tissue. Benign tumors may still be lethal, due to pressure on vital organs.

Biodegradable. Capable of being broken down into innocuous products by the action of living things.

Bioengineering. Designing equipment, machines, and other structures to fit the characteristics of people. (See Ergonomics.)

Biohazard. A biological hazard. Organisms or products of organisms that present a hazard to humans.

Biohazard area. Any area (a complete operating complex, a single facility, a room within a facility, etc.) in which work has been or is being performed with biohazardous agents or materials.

Biohazard control. Any set of equipment and procedures utilized to prevent or minimize the exposure of humans and their environment to biohazardous agents or materials.

Biomechanics. The study of the human body as a system operating under two sets of law: the laws of Newtonian mechanics and the biological laws of life.

Black light. Ultraviolet (UV) light radiation between 3,000 and 4,000 angstroms (0.3 to 0.4 micrometers).

Boiler codes. Standards prescribing requirements for the design, construction, testing, and installation of boilers and unfired pressure vessels. (e.g., *American Society of Mechanical Engineers Boiler and Pressure Vessel Code.*)

Boiling point. The temperature at which a liquid changes to a vapor state, expressed in degree.

Bonding. The interconnecting of two objects by means of an electrical conductor. Its purpose is to equalize the electrical potential between objects. (See Grounding.)

Braze. To solder with any alloy that is relatively infusible.

Breathing tube. A tube through which air or oxygen flows to the facepiece, helmet, or hood.

Breathing zone. The area encompassed by an imaginary globe of two-foot radius surrounding the head.

Btu. British thermal unit.

Bubble tube. A device used to calibrate air-sampling pumps.

Buffer. Any substance in a fluid that tends to resist the change in pH when acid or alkali is added.

Building code. An assembly of regulations that set forth the standards to which buildings must be constructed.

Bulk density. Mass of powdered or granulated solid material per unit of volume.

Bulk plant. That portion of a property where flammable or combustible liquids are received by tank vessel, pipelines, tank car, or tank vehicle, and are sorted or blended in bulk for the purpose of distributing such liquids by tank vessel, pipeline, tank car, tank vehicle, or container.

Bump cap. A hard shell cap, without an interior suspension system, designed to protect the wearer's head in situations where the employee might bump into something.

Burns. See Chemical burns, Thermal burns.

cc. Cubic centimeter; a volume measurement in the metric system, equal in capacity to one milliliter (ml)—approximately twenty (20) drops. There are 16.4 cc in one cubic inch.

Calender. An assembly of rollers for producing a desired finish on paper, rubber, artificial leather, plastics, or other sheet material.

Capture velocity. Air velocity at any point outside of an exhaust opening necessary to overcome opposing air currents and to capture the contaminated air by causing it to flow into the exhaust opening.

Carbon monoxide. A colorless, odorless toxic gas produced by any process that involves the incomplete combustion of carbon-containing substances.

Carcinogen. A substance or agent that can cause a growth of abnormal tissue or tumors in humans or animals.

Carcinogenic. Cancer producing.

Carpal tunnel. A passage in the wrist through which the median nerve and many tendons pass between the hand and the forearm.

Carpal tunnel syndrome. An affliction caused by compression of the median nerve in the carpal tunnel.

CAS number. Identifies a particular chemical by the Chemical Abstract Service, a service of the American Chemical Society that indexes and compiles abstracts of worldwide chemical literature called "Chemical Abstracts."

Casting. The pouring of a liquid material into a mold and permitting it to solidify to the desired shape.

Catalyst. A substance that changes the speed of a chemical reaction but undergoes no permanent change itself.

Catwalk. A narrow footway constructed usually for inspection or maintenance purposes.

Causal factor (of an accident). One or a combination of simultaneous or sequential circumstances directly or indirectly contributing to an accident. Modified to identify several kinds of causes such as direct, early, mediate, proximate, distal, etc.

Caustic. Something that strongly irritates, burns, corrodes, or destroys living tissue.

Ceiling limit (C). In ACGIH terminology, an airborne concentration of a toxic substance in the work environment that should never be exceeded. (See TLV.)

Celsius. The Celsius temperature scale is a designation of the scale previously known as the centigrade scale.

Centrifuge. An apparatus that uses centrifugal force to separate or remove particulate matter suspended in a liquid.

Ceramic. A term applied to pottery, brick, and the tile products molded from clay and subsequently calcined.

CERCLA. Comprehensive Environmental Response, Compensation and Liability Act.

CEU. Continuing education unit, needed by individuals for some educational programs.

CFR. *Code of Federal Regulations.* A collection of the regulations that have been promulgated under U.S. law.

CHCM. Certified Hazard Control Manager, issued by the Board of Certified Hazard Control Management.

Chemical burns. Generally similar to those caused by heat. After emergency first aid, their treatment is the same as that for thermal burns.

Chemical cartridge. A changeable container filled with various chemical substances for removal of low concentrations of specific vapors, mists, gases and fumes from the air passing through it.

Chemical cartridge respirator. A respirator that uses changeable cartridges containing various chemical substances to purify inhaled air of certain gases, vapors, mists and fumes.

Chemical engineering. That branch of engineering concerned with the development and application of manufacturing processes in which chemical or certain physical changes of materials are involved.

Chemical reaction. A change in the arrangement of atoms or molecules to yield substances of different composition and properties. Common types of reaction are combination, decomposition, double decomposition, replacement, and double replacement.

CHEMTREC. Chemical Transportation Emergency Center. (See Appendix, Sources of Help.)

Circuit. A complete path over which electrical current may flow.

Circuit breaker. A device that automatically interrupts the flow of an electrical current when the current exceeds a specified level.

Citation. A written change issued by regulatory representatives alleging specific conditions or actions that violate maritime, construction, environmental, mining, or general industry laws and standards.

Clean Air Act. U.S. law enacted to regulate/reduce air pollution. Administered by EPA.

Clean Water Act. U.S. law enacted to regulate/reduce water pollution. Administered by EPA.

Coated electrode. A composite filler metal electrode consisting of a core of bare electrode or metal-cored electrode to which a covering (sufficient to provide a slag layer on the weld metal) has been applied—the covering may contain materials providing such functions as shielding from the atmosphere, deoxidation, and arc stabilization and can serve as a source of metallic additions to the weld.

Coated welding rods. Welding rods coated with various materials such as manganese, titanium, and a silicate, for the purpose of facilitating a solid welding bond on various kinds of iron and steel. (See Coated electrode.)

Codes. Rules and standards that have been adopted by a government agency as mandatory regulations having the force and effect of law. Also used to describe a body of standards.

Code of Federal Regulations. The rules promulgated under U.S. law and published in the *Federal Register* and actually in force at the end of a calendar year are incorporated in this code (*CFR*).

Combustible. Able to catch fire and burn.

Combustible liquids. Combustible liquids are those having a flash point at or above 37.8 C (100 F), and below 93.3 C (200 F).

Common name. Any designation or identification such as code name, code number, trade name, brand name, or generic name used to identify something other than by its proper name.

Communicable. Refers to a disease whose causative agent is readily transferred from one person to another.

Competent person. One who is capable of identifying existing and predictable hazards in the surroundings or working conditions that are unsanitary, hazardous, or dangerous to employees, and who has the authorization to take prompt corrective measures to eliminate them. (29 *CFR* 1926.32).

Compound. A substance composed of two or more elements joined according to the laws of chemical combination. Each compound has its own characteristic properties different from those of its constituent elements.

Compressed gas cylinder. A cylinder containing vapor or gas under higher than atmospheric pressure, sometimes to the point where it is liquified.

Conductive hearing loss. Type of hearing loss; not caused by noise exposure, but due to any disorder in the middle or external ear that prevents sound from reaching the inner ear.

Confined space. Any area that has limited openings for entry and exit that would make escape difficult in an emergency, has a lack of ventilation, contains known and potential hazards, and is not intended nor designated for continuous human occupancy.

Consensus standard. A standard developed through a consensus process or general opinion among representatives of various interested or affected organizations and individuals.

Contact dermatitis. Dermatitis caused by skin contact with a substance—gaseous, liquid, or solid. May be due to primary irritation or an allergy.

Corrective lens. A lens ground to the wearer's individual prescription to improve vision.

Counter. A device for counting.

CPR. Cardiopulmonary resuscitation.

Cps. Cycles per second (frequency), in electricity called hertz.

CPSC. Consumer Products Safety Commission; U.S. agency with responsibility for regulating hazardous materials when they appear in consumer goods.

Critical pressure. The pressure under which a substance may exist as a gas in equilibrium with the liquid at the critical temperature.

Critical temperature. The temperature above which a gas cannot be liquefied by pressure alone.

Crucible. A heat-resistant barrel-shaped pot used to hold metal during melting in a furnace.

Cry-, cryo- (prefix). Very cold.

Cryogenics. The field of science dealing with the behavior of matter at very low temperatures.

CSP. Certified Safety Professional, a designation from the Board of Certified Safety Professionals.

Cubic centimeter (cc). Cubic centimeter, a volumetric measurement that is also equal to one milliliter (ml).

Cubic meter (m^3). A measure of volume in the metric system.

Cumulative trauma disorder (CTD). A disorder caused by one or more of the following: repetitive excessive motion of a body part, excessive force, or awkward body posture.

Current. Flow of electrons in an electrical circuit measured in amperes (amps). (See Ampere.)

Dampers. Adjustable sources of airflow resistance used to regulate airflow in a ventilation intake or exhaust system.

Dangerous to life or health, immediately (IDLH). Used to describe very hazardous atmospheres where employee exposure can cause serious injury or death within a short time or serious delayed effects.

dBA. Sound level in decibels read on the A-scale of a sound level meter. The A-scale discriminates against very low frequencies (as does the human ear) and is therefore better for measuring general sound levels. (See also Decibel.)

Decibel (dB). A unit used to express sound power level (L_w). Sound power is the total acoustic output of a sound source in watts (W). By definition, sound power level, in decibels, is: $L_w = 10 \log W/W_o$ where W is the sound power of the source and W_o is the reference sound power.

Decontaminate. To make safe by eliminating poisonous or otherwise harmful substances, such as noxious chemicals or radioactive material.

Density. The mass (weight) per unit volume of a substance.

Dermatitis. Inflammation of the skin.

Dermatosis. A broader term than dermatitis; it includes any cutaneous abnormality, thus encompassing folliculitis, acne, pigmentary changes, and nodules and tumors.

Die. A (hard metal or plastic) form used to shape material to a particular contour or section.

Differential pressure. The difference in static pressure between two locations.

Diffusion rate. A measure of the tendency of one gas or vapor to disperse into or mix with another gas or vapor.

Dike. A barrier constructed to control or confine solid or liquid substances and prevent their movement.

Direct costs (Insured costs). Those costs that are paid by the organization for accidents. Usually include compensation insurance, medical, damage, etc. (See Indirect costs.)

Direct-reading instrumentation. Those instruments that give an immediate indication of the concentration of aerosols, gases, or vapors or magnitude of physical hazard by some means such as a dial or meter.

DOL. U.S. Department of Labor; includes the Occupational Safety and Health Administration (OSHA) and Mine Safety and Health Administration (MSHA).

Dose. A term used (1) to express the amount of a chemical or of ionizing radiation energy absorbed in a unit volume or an organ or individual. Dose rate is the dose delivered per unit of time. (2) Used to express amount of exposure to a chemical substance.

Dose equivalent, maximum permissible (MPD). The largest equivalent received within a specified period that is permitted by a regulatory agency or other authoritative group on the assumption that receipt of such dose equivalent creates no appreciable somatic or genetic injury. Different levels of MPD may be set for different groups within a population. (By popular usage, "dose, maximum permissible," is an accepted synonym.)

DOT. U.S. Department of Transportation.

DOT hazard class. DOT requires that hazardous materials offered for shipment be labeled with the proper DOT hazard class. These classes include corrosive, flammable liquid, organic peroxide, ORM-E, poison B, etc. The DOT hazard class may not adequately describe all the hazard properties of the material.

Double insulated. A method of encasing electric components of tools so that the operator cannot touch parts that could become energized during normal operation or in the event of tool failure.

Drop forge. To forge between dies by a drop hammer or drop press.

Dry chemical. A powdered fire extinguishing agent usually composed of sodium bicarbonate, monoammonium phosphate, potassium bicarbonate, etc.

Duct. A conduit used for conveying air at low pressures.

Dust collector. An air-cleaning device to remove heavy particulate loadings from exhaust systems before discharge to outdoors; usual range is loadings of 0.003 grains per cubic foot (gr/ft^3) (0.007 mg/m^3) and higher.

Dusts. Solid particles generated by handling, crushing, grinding, rapid impact, detonation, and decrepitation of organic or inorganic materials, such as rock, ore, metal, coal, wood, and grain. Dusts do not tend to flocculate, except under electrostatic forces; they do not diffuse in air but settle under the influence of gravity.

EAP. Employee assistance program.

Ear. The entire human hearing apparatus, consisting of three parts: external ear; middle ear or tympanic cavity, membrane and eustachian tube; and the inner ear or labyrinth.

Effective temperature. An arbitrary index that combines into a single value the effect of temperature, humidity, and air movement on the human body's sensation of warmth and cold.

Electrical current. The flow of electricity measured in amperes.

Electrical precipitator. A device that removes particles from an airstream by applying a positive charge to the particles and collecting the charged particles on a negatively charged surface.

Element. Solid, liquid, or gaseous matter that cannot be further decomposed into simpler substances by chemical means.

Emergency Plan. A plan of action for an anticipated, unwanted occurrence/disaster.

- *Shower.* A water shower for an employee when the employee has had chemical contamination that needs to be washed off quickly.
- *Stop (switch).* A switch or other device that, when activated, disengages the power source of and quickly stops the controlled mechanisms.

Emery. Aluminum oxide; natural and synthetic abrasive.

Emission. The release of some by-product or product from an operation or process.

Emission standards. The maximum amount of pollutant emissions permitted to be discharged into the water or air from a single polluting source.

Encapsulate. To cover or coat over with another substance.

Energy-control program. A program consisting of an energy-control procedure and employee training to ensure that a machine or equipment is isolated and inoperative before servicing or maintenance, thus protecting the employee from unexpected machine start-up or energizing.

Energy isolation. See Energy control program and Energy isolation device.

Energy-isolating device. A mechanical device that physically prevents the release or transmission of energy. Some examples of energy-isolating devices include: a manually operated circuit breaker, a disconnect switch, a line valve, a block, and other similar devices. The following are *not* energy-isolating devices: push buttons, selector switches, and other circuit control devices.

Engineer. A person who can apply scientific principles creatively to design, operate, and maintain structures, machines, and apparatus.

Engineering controls. Methods of controlling employee exposures by modifying the source, the means of exposure or reducing the quantity of hazards.

Environmental toxicity. Information obtained as a result of conducting environmental testing designed to study the effects on aquatic and plant life.

EOE. Equal opportunity employer.

EPA. U.S. Environmental Protection Agency.

EPA number. The number assigned to chemicals regulated by the U.S. Environmental Protection Agency.

Ergonomics. The study of human characteristics for the appropriate design of living and work environments.

Exhalation valve. A device that allows exhaled air to leave a respirator and prevents outside air from entering through the valve.

Exhaust ventilation. The removal of air (usually by mechanical means) from any space. The flow of air between two points is due to the occurrence of a pressure difference between the two points. This pressure difference will cause air to flow from the high pressure to the low pressure zone.

Explosion. A reaction that causes a sudden, almost instantaneous release of pressure, gas, and heat.

Explosive limit. See Lower explosive limit and Upper explosive limit.

Exposure. Contact with a chemical, biological, or radiological hazard. Also, the near proximity to an unprotected physical hazard.

Extinguishing media. The firefighting substance used to stop combustion. It is usually referred to by its generic name, such as CO_2, foam, water, dry chemical, etc.

Extrusion. The forcing of raw material through a die or a form in either a heated or cold state, in a solid state, or in partial fluid.

Eyepiece. Gas-tight, transparent window(s) in a full facepiece through which the wearer may see.

Eye protection. "Safety" glasses, goggles, face shields, etc., used to protect against physical, chemical and nonionizing radiation hazards.

Face velocity. Average air velocity into the exhaust system measured at the opening into the hood or booth.

Facepiece. That portion of a respirator that covers the wearer's nose and mouth (in a half-mask facepiece) or, nose, mouth, and eyes in a full facepiece respirator.

Facilitator. A person who makes learning easier, assists interactions and the execution of tasks, and clarifies goals and processes.

Factor of safety. The ratio of ultimate strength of a material or structure to the specified stress allowable.

FDA. The U.S. Food and Drug Administration. The FDA establishes requirements for the labeling of foods and drugs to protect consumers from misbranded, unwholesome, ineffective, and hazardous products. FDA also regulates materials for food contact service and the conditions under which such materials are approved.

Federal Register (FR). Official publication of U.S. government documents and other communications promulgated under the law, documents whose validity depends upon the publication. (See *Code of Federal Regulations.*)

Filter. (1) A device for separating components of a signal on the basis of its frequency. It allows components in one or more frequency bands to pass relatively unattenuated, and it attenuates greatly components in other frequency bands. (2) A fibrous medium used in respirators to remove solid or liquid particles from the airstream entering the respirator. (3) A sheet of material that is interposed between patient and the source of x-rays to absorb a selective part of the x-rays. (4) A fibrous or membranous medium used to collect dust, fume, or mist air samples.

Filter, HEPA. High-efficiency particulate air filter that is at least 99.97% efficient in removing thermally generated monodisperse dioctylphthalate smoke particles with a diameter of 0.3 μm.

Fire brigade. An organized group trained in firefighting operations.

Fire doors. Doors tested and rated for resistance to various degrees of fire exposure and utilized to prevent the spread of fire through horizontal and vertical openings.

Fire resistant. See Flame proof.

First aid. The immediate care given to the injured or suddenly ill person.

Fission. The splitting of an atomic nucleus into two parts accompanied by the release of a large amount of radioactivity and heat.

Flame proof. Material incapable of burning. The term "fire-proof" is false. No material is immune to the effects of fire possessing sufficient intensity and duration. The term is commonly, although erroneously, used synonymously with "fire resistive."

Flame propagation. See Propagation of flame.

Flammable. Any substance that is easily ignited, burns intensely, or has a rapid rate of flame spread. Flammable and inflammable are identical in meaning; however, the prefix "in" indicates negative in many words and can cause confusion. Flammable, therefore, is the preferred term.

Flammable liquid. Any liquid having a flash point below 37.8 C (100 F).

Flammable range. The difference between the lower and upper flammable limits, expressed in terms of percentage of vapor or gas in air by volume, and is also often referred to as the "explosive range." (See Lower explosive limit and Upper explosive limit.)

Flashback. Occurs when flame from a torch burns back into the tip, the torch, or the hose.

Flash blindness. Temporary visual disturbance resulting from viewing an intense light source.

Flash ignition. See Flash point.

Flash point. The lowest temperature at which a liquid gives off enough vapor to form an ignitable mixture with air and produce a flame when a source of ignition is present.

Floor load. (1) The weight that may be safely placed on a floor without danger of structural collapse. (2) The actual load (weight) placed on a floor.

Flow meter. An instrument for measuring the rate of flow of a fluid or gas.

Fluid. A substance tending to flow or conform to the outline of its container. It may be liquid, vapor, gas, or semisolid (like raw rubber).

Fluorescent screen. A screen coated with a fluorescent substance that emits light when irradiated with x-rays or electromagnetic radiation.

Fly ash. Finely divided particles of ash entrained in flue gases arising from the combustion of solid fuel.

Fog. The visible presence of small water droplets suspended in air.

Foot candle. A unit of illumination.

Foot-pounds. A unit of work equal to the energy required to raise 1 lb, 1 ft.

Force. That which changes the state of rest or motion in matter.

Frequency (in hertz or Hz). Rate at which oscillations are produced. One hertz is equivalent to one cycle per second.

Friable. Readily crumbled or crumbling state.

Fume. Airborne particulate formed by the evaporation of solid materials, e.g., metal fume emitted during welding. Usually less than one micron in diameter.

Fuse. A wire or strip of metal with known electrical resistance, usually set in a plug, placed in an electrical circuit as a safeguard. As the electrical current increases, the metal's resistance to flow causes it to heat until it reaches the point where the metal melts, breaking the current at the rated amperage.

Fusion. The joining of atomic nuclei to form a heavier nucleus, accomplished under conditions of extreme heat (millions of degrees). If two nuclei of light atoms fuse, the fusion is accompanied by the release of a great deal of energy. The energy of the sun is believed to be derived from the fusion of hydrogen atoms to form helium. In welding, the melting together of filler metal and base metal (substrate), or of base metal only.

Galvanizing. An old but still used method of providing corrosion protection for metals by dipping them in a bath of molten zinc.

Gas. A state of matter in which the material has very low density and viscosity; can expand and contract greatly in response to changes in temperature and pressure; easily diffuses; and is neither a solid nor a liquid.

Gas metal arc-welding (GMAW). An arc-welding process that produces coalescence of metals by heating them with an arc between a continuous filler metal (consumable) electrode and the work. Shielding of this process from surrounding air is required and is obtained from an externally supplied gas or gas mixture. Some variants of this process are called MIG or CO_2 welding.

Gas tungsten arc-welding (GTAW). An arc-welding process that produces coalescence of metals by heating them with an arc between a tungsten (nonconsumable) electrode and the work. Shielding is obtained from a gas or gas mixture. Pressure may or may not be used and filler metal may or may not be used. (This process has sometimes been called TIG welding.)

Gauge pressure. Pressure measured with respect to atmospheric pressure.

General exhaust. A system for exhausting air from a general work area, accomplished mechanically by air handling units that drain air from the space.

General ventilation. System of exchanging air in a general work area by either natural or mechanically induced fresh air movements to mix with the existing room air and escape by natural means.

Generic name. A nonproprietary name for a material or product.

GFCI. See Ground fault circuit interrupter.

Glove box. A sealed enclosure in which all handling of items inside the box is carried out through long impervious gloves sealed to ports in the walls of the enclosure.

GMAW. See Gas metal arc-welding.

Grab sample. A sample that is taken within a very short time period during which atmospheric concentration is assumed to be constant throughout the sample.

Gram (g). A metric unit of weight equal to 0.035 ounce (avoir).

Gravity, specific. The ratio of the mass of a unit volume of a substance to the mass of the same volume of a standard substance at a standard temperature. Water at 4 C (39.2 F) is the standard substance usually referred to. For gases, dry air, at the same temperature and pressure as the gas, is often taken as the standard substance.

Gravity, standard. A gravitational force that will produce an acceleration equal to (9.8 m/sec^2) or (32.17 ft/sec^2). The actual force of gravity varies slightly with altitude and latitude. The standard was arbitrarily established as that at sea level and 45 degrees latitude.

Ground. A contact with the ground that becomes part of the electrical circuit.

Ground fault circuit interrupter (GFCI). A device that measures the amount of current flowing to and from an electrical source. When a difference between the two is sensed, indicating a leakage of current, the device very quickly breaks the circuit.

Grounding. The procedure used to carry an electrical charge to ground through a conductive path. (See Bonding.)

GTAW. See Gas tungsten arc-welding.

Guard. A generic term applied to physical barriers, extraction and presence-sensing devices designed to prevent contact with hazards.

Halogenated hydrocarbon. A chemical substance that has carbon plus one or more of these elements: chlorine, fluorine, bromine, or iodine.

Hammer mill. A machine for reducing the size of stone or other bulk material by means of hammers usually placed on a rotating axle inside a steel cylinder.

Hand protection. Coverings worn over the hands to protect against physical, chemical, biological, thermal, and electrical hazards.

Hard hat. A helmet so constructed as to help prevent head injuries from falling objects of limited size.

Hazard. An unsafe condition or activity that, if left uncontrolled, can contribute to an accident.

Hazard analysis. An analysis performed to identify and evaluate hazards for the purpose of their elimination or control.

Hazard control. A program to recognize, evaluate, eliminate or control the existence of and exposure to hazards.

Hazardous material. Any substance or compound that has the capability of producing adverse effects on the health and safety of humans.

Health. Personal freedom from physical or mental defect, pain, injury, or disease.

Health hazard. A chemical, biological or radiological material for which there is statistically significant scientific evidence that acute or chronic health effects may occur in exposed employees.

Hearing conservation. The prevention or minimizing of noise-induced hearing loss through the use of hearing protection devices; the control of noise through engineering and administrative methods, audiometric tests, and employee training.

Hearing level. The deviation in decibels of an individual's threshold from the zero reference of the audiometer.

Heat stress. Relative amount of thermal strain from the environment.

Heat stress index. Index that combines the environmental heat and metabolic heat into an expression of stress in terms of requirement for evaporation of sweat.

Helmet. A device that shields the eyes, face, neck, and other parts of the head.

HEPA filter. See Filter, HEPA.

Hertz. Frequency of oscillation measured in cycles per second. 1 cps = 1 Hz.

High frequency loss. Refers to a hearing deficit starting with frequencies of 2,000 Hz and higher.

Hold harmless. A written agreement in which a party absolves or is absolved by another for liability arising from a specified cause.

Hood. (1) Enclosure, part of a local exhaust system; (2) a device that completely covers the head, neck, and portions of the shoulders.

Horsepower. A unit of power, equivalent to 33,000 foot-pounds per minute (746 w).

Hot. In addition to meaning having a relatively high temperature, this is a colloquial term meaning highly radioactive.

Human-equipment interface. Areas of physical or perceptual contact between man and equipment. The design characteristics of the human-equipment interface determine the quality of information. Poorly designed interfaces may lead to excessive fatigue or localized trauma.

Human factors. See Ergonomics.

Human factors engineering. See Ergonomics and the expressed application of engineering to human factors.

Humidify. To add water vapor to the atmosphere; to add water vapor or moisture to any material.

Humidity. (1) Absolute humidity is the weight of water vapor per unit volume, pounds per cubic foot or grams per cubic centimeter; (2) relative humidity is the ratio of the actual partial vapor pressure of the water vapor to the saturation pressure of pure water at the same temperature.

Hydrocarbons. Organic compounds, composed solely of carbon and hydrogen.

ICC. U.S. Interstate Commerce Commission.

IDLH. See Immediately dangerous to life or health.

Ignitable. Capable of being set afire.

Imminent danger. An impending or threatening hazard that could be expected to cause death or serious injury to persons in the immediate future unless corrective measures are taken.

Impervious. A material that does not allow another substance to pass through or penetrate it.

Impingement. In air-sampling, impingement refers to a process for the collection of particulate or gaseous matter in which the gas containing the contaminant is directed into the collecting solution and the particles or gas are retained by the liquid.

Inches of mercury column. A unit used in measuring pressures. One inch of mercury column equals a pressure of 1.66 kPa (0.491 lb per sq in.).

Inches of water column. A unit used in measuring pressures. One inch of water column equals a pressure of 0.25 kPa (0.036 lb per sq in.).

Incidence rate (as defined by U.S. OSHA). The number of injuries and/or illnesses or lost workdays per 100 full-time employees per year or 200,000 hours of exposure.

Incident. An unintentional event that may cause personal harm or other damage. In the United States, OSHA specifies that incidents of a certain severity be recorded. (See also Near-miss incident and Accident.)

Indirect costs. Losses ultimately measurable in a monetary sense resulting from an accident other than those costs that are insurable. (See Direct costs.)

Industrial hygiene. The science (or art) devoted to the recognition, evaluation, and control of those environmental factors or stresses (i.e., chemical, physical, biological, and ergonomic) that may cause sickness, impaired health, or significant discomfort to employees or residents of the community.

Inert gas. A gas that does not normally combine chemically with other substances.

Inert gas welding. An electric welding operation utilizing an inert gas such as helium to shield the metal being welded from exposure to air, preventing oxidation.

Infrared radiation. Electromagnetic energy with wavelengths from 770 nm to 12,000 nm.

Ingestion. (1) The process of taking substances into the stomach, as food, drink, medicine, etc. (2) With regard to certain cells, the act of engulfing or taking up bacteria and other foreign matter.

Inhalation. The breathing in of a substance in the form of a gas, vapor, fume, mist, or dust.

Inhalation valve. A device that allows respirable air to enter the facepiece and prevents exhaled air from leaving the facepiece through the intake opening.

Injury. Physical harm or damage to the body resulting from an exchange of mechanical, chemical, thermal, or other environmental energy that exceeds the body's tolerance.

Inrunning nip (point). A rotating mechanism that can seize loose clothing, belts, hair, body parts, etc. It exists when two or more shafts or rolls rotate parallel to one another in opposite directions. (See Nip point and Pinch point.)

Insoluble. Incapable of being dissolved.

Inspection. Monitoring function conducted in an organization to locate and report existing and potential hazards having the capacity to cause accidents in the workplace.

Interlock. A device that interacts with another device or mechanism to govern succeeding operations. For example, an interlocked machine guard will prevent the machine from operating unless the guard is in its proper place. An interlock on an elevator door will prevent the car from moving unless the door is properly closed.

Ionizing radiation. Refers to (1) electrically charged or neutral particles, or (2) electromagnetic radiation that will interact with gases, liquids, or solids to produce ions. There are five major types: alpha, beta, x (or x-ray), gamma, and neutrons.

Irradiation. The exposure of something to radiation.

Irritant. A substance that produces an irritating effect when it contacts skin, eyes, nose, or respiratory system.

Jigs and fixtures. Often used interchangeably; precisely, a jig holds work in position and guides the tools acting on the work, while a fixture holds but does not guide.

Job safety analysis. A method for studying a job in order to (a) identify hazards or potential accidents associated with each step or task and (b) develop solutions that will eliminate, nullify, or prevent such hazards or accidents. Sometimes called Job Hazard Analysis.

Kilogram (kg). A unit of weight in the metric system equal to 2.2 lb.

L. See Liter.

Laser. The acronym for Light Amplification by Stimulated Emission of Radiation.

Laser light region. A portion of the electromagnetic spectrum, which includes ultraviolet, visible, and infrared light.

Lathe. A machine tool used to perform cutting operations on wood or metal by the rotation of the work piece against a blade.

LC. Lethal concentration; a concentration of a substance being tested that will kill a test animal.

LD. Lethal dose; an amount of a substance being tested that will kill a test animal.

Lead poisoning. Lead compounds can produce poisoning when they are swallowed or inhaled. Inorganic lead compounds commonly cause symptoms of lead colic and lead anemia. Organic lead compounds can attack the nervous system.

LEL. Lower explosive limit. (See Lower explosive limit and Upper explosive limit).

Lethal. Capable of causing death.

LFL. Lower flammable limit. (See Lower explosive limit.)

Liability. The state of being bound or obliged in law to do, pay, or make good on something. As to the law of torts, usually based on the law of negligence.

Liability, strict. The imposition of liability for damages resulting from any and all defective and hazardous products without requiring proof of negligence. Disclaimers are not valid; traditional warranty concepts, privity, and notice of injury are eliminated.

Liquefied petroleum gas. A compressed or liquefied gas usually composed of propane, some butane, and lesser quantities of other light hydrocarbons and impurities; obtained as a byproduct in petroleum refining. Used chiefly as a fuel and in chemical synthesis.

Liquid. A state of matter in which the substance is a formless fluid that flows in accord with a law of gravity.

Liter. A metric measure of capacity—one quart equals 0.908 l (dry measure); one liter equals 1.057 quart (liquid).

Load limit. The upper weight limit capable of safe support by a vehicle, floor, or roof structure.

Local exhaust. A system for capturing and exhausting contaminants from the air at the point where the contaminants are produced.

Local exhaust ventilation. A ventilation system that captures and removes contaminants at the point they are being produced before they escape into the workroom air.

Lockout/tagout. A program or procedure that prevents injury by eliminating unintentional operation or release of energy within machinery or processes during set-up, start-up, or maintenance. (See Energy-control program.)

Long-term sample. Sample taken over sufficiently long period of time that the variations in exposure cycles are averaged.

Loss control. A program designed to minimize accident-based financial losses. The concept of total loss control is based on detailed analysis of both indirect and direct accident costs. Property damage as well as injurious and potentially injurious accidents are included in the analysis.

Loss prevention. A before-the-loss program designed to identify and correct hazards before they result in incidents that produce actual financial loss or injury.

Lost workday. The number of workdays (consecutive or not), beyond the day of injury or onset of illness, that an employee was away from work or limited to restricted work activity because of an occupational injury or illness.

Loudness. The intensity of an auditory sensation, in terms of which sounds may be ordered on a scale extending from soft to loud. Loudness depends primarily upon the sound pressure of the stimulus, but it also depends upon the frequency and wave form of the stimulus.

Lower explosive limit (LEL). The lower limit of flammability of a gas or vapor at ordinary ambient temperatures expressed in percent of the gas or vapor in air by volume. (See Upper explosive limit and Flammable range.)

LP-gas. See Liquefied petroleum gas.

Lumen. The luminous flux on one square foot of a sphere, one foot in radius, with a light source of one candela at the center that radiates uniformly in all directions.

m^3. Cubic meter; a metric measure of volume, about 35.3 cubic feet or 1.3 cubic yards.

Makeup air. Clean, tempered outdoor air supplied to a work space to replace air removed by exhaust ventilation or some industrial process.

Manometer. Instrument for measuring pressure; essentially a U-tube partially filled with a liquid (usually water, mercury, or a light oil), so constructed that the amount of displacement of the liquid indicates the pressure being exerted on the instrument.

Maser. Microwave Amplification by Stimulated Emission of Radiation.

Maximum permissible concentration (MPC). These concentrations are set by the National Committee on Radiation Protection (NCRP). They are recommended maximum average concentrations of radionuclides to which a worker may be exposed, assuming that he or she works 8 hours a day, 5 days a week, and 50 weeks a year.

Maximum permissible dose (MPD). Currently, a permissible dose is defined as the dose of ionizing radiation that, in the light of present knowledge, is not expected to cause appreciable bodily injury to a person at any time during their lifetime. NRC and OSHA have established a maximum permissible dose of 5 rem per year for persons over age 18 and a lifetime dose of 5(N-18) where N is a person's present age.

Maximum use concentration (MUC). The product of the protection factor of the respiratory protection equipment and the permissible exposure limit (PEL).

Mechanical filter respirator. A respirator used to protect against airborne particulate matter like dusts, mists, metal fume, and smoke. Mechanical filter respirators do not provide protection against gases, vapors, or oxygen-deficient atmospheres.

Mechanical ventilation. A powered device, such as a motor-driven fan or vacuum hose attachment, for exhausting contaminants from a workplace, vessel, or enclosure.

Mega. One million—for example, megacurie = 1 million curies.

Melting point. The temperature at which a solid substance changes to a liquid state.

Meter (m). A unit of length in the metric system. One meter is about 39.37 in.

Mev. Million electron volts.

mg. Milligram; a metric unit of weight. There are 1,000 milligrams in one gram (g) of a substance. One gram is equivalent to almost 4/100 of an ounce.

mg/kg. Milligrams per kilogram.

mg/m^3. Milligrams per cubic meter.

Mica. A large group of silicates of varying composition, but similar in physical properties. All have cleavage characteristics that allow them to be split into very thin sheets. Used in electrical insulation.

Microphone. An electroacoustic transducer that responds to sound waves and delivers essentially equivalent electric waves.

Milliampere. 1/1000 of an ampere.

Milligram (mg). A unit of weight in the metric system. One thousand milligrams equal one gram (see mg).

Milligrams per cubic meter (mg/m^3). Unit used in the measurement of concentrations of dusts, gases, mists, and fumes in air.

Milliliter (ml). A metric unit used to measure volume. One milliliter equals one cubic centimeter or about $\frac{1}{16}$ cubic inch.

Millimeter of mercury (mmHg). The unit of pressure equal to the pressure exerted by a column of liquid mercury one millimeter high at a standard temperature.

Mist. Suspended liquid droplets generated by condensation from the gaseous to the liquid state or by breaking up a liquid into a dispersed state, such as by spraying or atomizing.

Mixture. A combination of two or more substances that may be separated by mechanical means. The components may not be uniformly dispersed. (See also Solution.)

ml. See Milliliter.

mmHg. Millimeters (mm) of mercury (Hg).

Monaural hearing. Refers to hearing with one ear only.

Monitoring. Testing to determine if the parameters being measured are within acceptable limits. This includes environmental and medical (biological) monitoring in the workplace.

MORT. Management Oversight and Risk Tree.

MPC. See Maximum permissible concentration.

MPD. See Dose equivalent, maximum permissible.

MPE. Maximum permissible exposure.

MPL. May be either maximum permissible level or limit, or dose. It refers to the tolerable dose rate of humans exposed to nuclear radiation.

MSDS. Material Safety Data Sheet. A document prepared by a chemical manufacturer, describing the composition, properties, and hazards of a chemical along with recommended safeguards to handling, storage, and use.

MSHA. The Mine Safety and Health Administration of the U.S. Department of Labor; federal agency with safety and health regulatory and enforcement authority for the mining industry established by the Mine Safety and Health Act.

MUC. See Maximum use concentration.

Muff. A covering over the outside ear to reduce noise exposure.

Nature of injury. The type of injury inflicted, such as: sprain, burn, contusion, laceration, etc.

Near-miss incident. For purposes of internal reporting, some employers choose to classify as "incidents" the near-miss incident; an injury requiring first aid; the newly discovered unsafe condition; fires of any size; or nontrivial incidents of damage to equipment, building, property, or product.

Negligence. The lack of required, expected, or reasonable conduct or care that a prudent person would ordinarily exhibit. There need not be a legal duty.

NEISS. National Electronic Injury Surveillance System collects data from 119 representative hospital emergency rooms on product-related injuries receiving emergency room treatment. A part of the U.S. Consumer Product Safety Commission.

Neutral wire. Wire carrying electrical current back to its source, thus completing a circuit.

NFPA. The National Fire Protection Association is a voluntary organization whose aim is to promote and improve fire protection and prevention.

NIOSH. The National Institute for Occupational Safety and Health is a branch of the U.S. Department of Labor. It conducts research on health and safety concerns, tests and certifies respirators, and trains occupational health and safety professionals.

Nip point. The point of intersection or contact between two or more surfaces when one or more are moving.

Noise. Any unwanted sound.

Noise-induced hearing loss. The slowly progressive inner ear hearing loss that results from exposure to continuous noise over a long period of time as contrasted to acoustic trauma or physical injury to the ear.

Nonflammable. Not easily ignited, or if ignited, not burning with a flame (smolders).

Nonionizing radiation. Electromagnetic radiation that does not cause ionization. Includes ultraviolet, laser, infrared, microwave, and radiofrequency radiation.

Nonsparking tools. Tools made from beryllium-copper or aluminum-bronze that produce no sparks, or low energy sparks, when used to strike other objects.

Nonvolatile matter. The portion of a material that does not evaporate at ordinary temperatures.

NRC. (1) U.S. National Response Center; a notification center in the Coast Guard Building in Washington, DC, (2) Nuclear Regulatory Commission.

NRR—Noise Reduction Rating. As applied to ear protection, the amount of sound intensity reduction afforded by the device, measured in dB.

Nuisance dust. Has a long history of little adverse effect on the lungs and does not produce significant organic disease or toxic effect when exposures are kept under reasonable control.

Ohm. The unit of electrical resistance.

Ohm's Law. The current (I) through an electrical circuit is directly proportional to the applied electromotive force (voltage) (E) and the resistance (R) of the conductor. $I = E/R$

Orifice. The opening that serves as an entrance and/or outlet. May apply to a body cavity, organ, or some types of equipment, especially the opening of a canal or a passage.

Orifice meter. A flowmeter, employing as the measure of flow rate the difference between the pressures measured on the upstream and downstream sides of a restriction within a pipe or duct.

OSHA. The U.S. Occupational Safety and Health Administration of the Department of Labor; federal agency with safety and health regulatory and enforcement authorities for general U.S. industry and business.

Oxidation. Process of combining oxygen with some other substance; technically, a chemical change in which an atom loses one or more electrons whether or not oxygen is involved. Opposite of reduction.

Oxygen deficiency. An atmosphere containing a lower percentage of oxygen by volume than is contained in free air at sea level.

Particulate matter. A suspension of fine solid or liquid particles in air, such as dust, fog, fume, mist, smoke, or sprays. Particulate matter suspended in air is commonly known as an aerosol.

PAW. See Plasma arc welding.

PEL. See Permissible exposure limit.

Permanent disability or permanent impairment. The partial or complete loss or impairment of any part or function of the body.

Permissible exposure limit (PEL). The legally enforced exposure limit for a substance established by the U.S. OSHA. The PEL indicates the permissible concentration of air contaminants to which nearly all workers may be repeatedly exposed eight (8) hours a day, forty (40) hours a week, over a working lifetime (30 years) without adverse health effects.

Personal protective equipment (PPE). Devices worn by the worker to protect against hazards in the environment.

Pesticides. General term for that group of chemicals used to control or kill such pests as rats, insects, fungi, bacteria, weeds, etc., that prey on man or agricultural products. Pesticides include insecticides, herbicides, fungicides, rodenticides, miticides, fumigants, and repellents.

PF. See Protection factor.

Physical hazards of chemicals. A chemical for which there is scientifically valid evidence that it is a combustible liquid, a compressed gas, explosive, flammable, an organic peroxide, an oxidizer, pyrochloric, unstable (reactive), or water-reactive.

Pinch point. Any point at which it is possible to be caught between the moving parts, stationary parts, or the material being processed. (See Nip point and Inrunning nip [point].)

Plasma arc welding (PAW). An arc welding process that produces coalescence of metals by heating them with a constricted arc between an electrode and the workpiece (transferred arc) or the electrode and the constricting nozzle (nontransferred arc). Shielding is obtained by the hot, ionized gas issuing from the orifice, which may be supplemented by an auxiliary source of shielding gas. Shielding gas can be an inert gas or a mixture of gases. Pressure may or may not be used, and filler metal may or may not be supplied.

Poison, Class A. A U.S. DOT hazard class for extremely dangerous poisons, that is, poisonous gases or liquids of such nature that a very small amount of the gas, or vapor of the liquid, mixed with air is dangerous to life. Some examples: phosgene, cyanogen, hydrocyanic acid, nitrogen peroxide.

Poison, Class B. A U.S. DOT hazard class for liquid, solid, paste, or semi-solid substances—other than Class A poisons or irritating materials—that are known (or presumed on the basis of animal tests) to be so toxic to man as to afford a hazard to health during transportation. Some examples: arsenic, beryllium chloride, cyanide, mercuric oxide.

Pollution. Contamination of soil, water, or atmosphere beyond that which is natural.

Potential energy. Energy due to position of one body with respect to another or to the relative parts of the same body.

Power. Time rate at which work is done; units are the watt (one joule per second) and the horsepower (33,000 foot-pounds per minute). One horsepower equals 746 watts.

PPE. See Personal protective equipment.

ppm. Parts per million part of air by volume of vapor or gas or other contaminant.

Precision. The degree of agreement of repeated measurements of the same property, expressed in terms of dispersion of test results about the mean result obtained by repetitive testing of a homogeneous sample under specified conditions.

Presence-sensing device. A device designed to detect an intrusion into a defined danger zone and to cause the potentially harmful action to cease.

Pressure. Force applied to, or distributed over a surface; measured as force per unit area. (See Atmospheric pressure, Gauge pressure, Standard air, and Static pressure.)

Pressure vessel. A storage tank or vessel designed to operate at pressures greater than 15 psig (103 kPa).

Preventive maintenance. The systematic actions performed to maintain equipment in normal working condition and prevent failure.

Probe. A tube used for sampling or for measuring pressures at a distance from the actual collection or measuring apparatus. It is commonly used for reaching inside stacks or ducts.

Product liability. The liability a merchant or a manufacturer may incur as the result of some defect in the product sold or manufactured, or the liability a contractor might incur after job completion from improperly performed work.

Propagation of flame. The spread of flame through the entire volume of the flammable vapor-air mixture from a single source of ignition.

Protection factor (PF). With respiratory protective equipment—the ratio of the ambient airborne concentration of the contaminant to the concentration inside the facepiece.

Protective atmosphere. A gas envelope surrounding the part to be brazed, welded, or thermal sprayed, with the gas composition controlled with respect to chemical composition, dew point, pressure, flow rate, etc.

Protective coating. A thin layer of metal or organic material, as paint applied to a surface primarily to protect it from oxidation, weathering, and corrosion.

psi. Pounds per square inch. For technical accuracy, pressure must be expressed as psig (pounds per square inch gauge) or psia (pounds per square inch absolute; that is, gauge pressure plus sea level atmospheric pressure, of psig plus about 14.7 pounds per square inch). (Also mmHg).

psig. Pounds per square inch gauge.

Quality assurance (Quality control). A management function to assure that the products or goods are produced as intended.

Radar (Radio Detection and Ranging). A radio detecting instrument, able to measure distance to an object, among other characteristics.

Radiation (nuclear). The emission of atomic particles or electromagnetic radiation from the nucleus of an atom.

Radiation (thermal). The transmission of energy by means of electromagnetic waves longer than visible light. Radiant energy of any wavelength may, when absorbed, become thermal energy and result in the increase in the temperature of the absorbing body.

Radiation protection guide (RPG). The radiation dose that should not be exceeded without careful consideration of the reasons for doing so; every effort should be made to encourage the maintenance of radiation doses as far below this guide as practicable.

Radiation source. An apparatus or a material emitting or capable of emitting ionizing radiation.

Radiator. That which is capable of emitting energy in wave form.

Radioactive. The property of an isotope or element that is characterized by spontaneous decay and emission of radiation.

Rated line voltage. The range of potentials in volts, of the supply line.

Reaction. A chemical transformation or change; the interaction of two or more substances to form new substances.

Relative humidity. See Humidity.

Reliability. The degree to which an instrument, component, or system retains its performance characteristics over a period of time.

Resistance. (1) In electricity, any condition that retards current (or the flow of electrons); it is measured in ohms. (2) Opposition to the flow of air, as through a canister, cartridge, particulate filter, or orifice. (3) A property of conductors, depending on their dimensions, material, and temperature, that determines the current produced by a given difference in electrical potential.

Respirable size particulates. Particles in the size range that permits them to penetrate deep into the lungs upon inhalation.

Respirator. A device to protect the wearer from inhalation of harmful contaminants.

Respiratory protection. Devices that will protect the wearer's respiratory system from overexposure by inhalation to airborne contaminants.

Respiratory system. Consists of (in descending order) —the nose, mouth, nasal passages, nasal pharynx, pharynx, larynx, trachea, bronchi, bronchiole, air sacs (alveoli) of the lungs, and muscles of respiration.

Risk. (1) An insurance term for insured value and, another name for the insured or prospective insured. (2) A term applied to the individual or combined assessments of "probability of loss" and potential amount of loss.

Route of entry. The path by which chemicals can enter the body, primarily inhalation, ingestion, skin absorption, and injection.

RPG. See Radiation protection guide.

Safe. A condition of relative freedom from danger.

Safeguarding. The term used to cover all methods of protection against injury or illness.

Safety. The control of recognized hazards to attain an acceptable level of risk.

Safety belt. A life belt worn by linesmen, window washers, etc., attached to a secure object (window sill, etc.) to prevent falling. A seat or torso belt securing a passenger in an automobile or airplane to provide body protection during a collision, sudden stop, air turbulence, etc.

Safety can. An approved container, of not more than 19 l (5 gal) capacity, having a spring-closing lid and spout cover, and so designed that it will safely relieve internal pressure when subjected to fire exposure.

Safety factor. See Factor of safety.

Safety program. Activities designed to assist employees in the recognition, understanding, and control of hazards in the workplace.

Safety shoes. A term commonly used to describe protective footwear meeting ANSI Z41 requirements.

Salamander. A small furnace usually cylindrical in shape, without grates, used for heating. Also a term used to refer to open barrels on a construction site used to provide a heat source.

Salt. A product of the reaction between an acid and a base.

Sampling. A process consisting of the withdrawal or isolation of a fractional part of a whole.

Sanitize. To reduce the microbial flora in or on articles, such as eating utensils, to levels judged safe by public health authorities.

SARA. Superfund Amendments and Reauthorization Act.

SCBA. See Self-contained breathing apparatus.

Self-contained breathing apparatus. A respiratory protection device that consists of a supply or a means of respirable air, oxygen, or oxygen-generating material, carried by the wearer.

Self-ignition. See Autoignition temperature.

Self-insurance. A term used to describe the assumption of one's own financial risk.

Sensible. Capable of being perceived by the sense organs.

Serious violation. Any violation in which there is a substantial probability that death or serious physical harm could result from the violative condition (OSHAct).

Shakeout. In the foundry industry, the separation of the solid, but still not cold, casting from its molding sand.

Shielded metal arc welding (SMAW). An arc-welding process that produces coalescence of metals by heating them with an arc between a covered metal electrode and the work. Shielding is obtained from decomposition of the electrode covering. Pressure is not used and filler metal is obtained from the electrode.

Shock. The physical effects of trauma to the body.

Short-term exposure limit (STEL). See TLV.

SIC. See Standard Industrial Classification.

Silicon. A nonmetallic element being, next to oxygen, the chief elementary constituent of the earth's crust.

Skin toxicity. See Dermal toxicity.

Sludge. In general, any muddy or slushy mass.

Slurry. A thick, creamy liquid resulting from the mixing and grinding of limestone, clay, and other raw materials with water.

SMAW. See Shielded metal arc welding.

Smelting. One step in the procurement of metals from ore—hence to reduce, to refine, to flux, or to scorify.

Smog. Irritating hazard resulting from the sun's effect on certain pollutants in the air.

Smoke. An air suspension (aerosol) of particles, originating from combustion or sublimation.

Solder. A material used for joining metal surfaces together by filling a joint or covering a junction.

Solution. Mixture in which the components lose their identities and are uniformly dispersed. All solutions are composed of a solvent (water or other fluid) and the substance dissolved, called the "solute." A true solution is homogeneous, as salt is in water.

Solvent. A substance that dissolves another substance.

Soot. Agglomerations of particles of carbon impregnated with tar, formed in the incomplete combustion of carbonaceous material.

Sorbent(s). (1) A material that removes toxic gases and vapors from air inhaled through a canister or cartridge. (2) Material used to collect gases and vapors during air-sampling. (3) Nonreactive materials used to clean up chemical spills. Examples: clay and vermiculite.

Sound. An oscillation in pressure, stress, particle displacement, particle velocity, etc., that is propagated in an elastic material, in a medium with internal forces (e.g., elastic, viscous), or the superposition of such propagated oscillations.

Sound level. A weighted sound pressure level, obtained by the use of metering instruments using weighting scales specified in ANSI S1.4.

Sound-level meter and octave-band analyzer. Instruments for measuring sound pressure levels in decibels referenced to 0.0002 microbars.

Sound pressure level (SPL). The level, in decibels, of a sound is 20 times the logarithm to the base 10 of the ratio of the pressure of this sound to the reference pressure. The reference pressure must be explicitly stated.

Specific gravity. The weight of a material compared to the weight of an equal volume of water; an expression of the density (or heaviness) of the material.

Specific weight. The weight per unit volume of a substance; same as density.

SPL. See Sound pressure level.

Spontaneously combustible. A material that ignites as a result of retained heat from processing, or which will oxidize to generate heat and ignite, or which absorbs moisture to generate heat and ignite.

Spot welding. One form of electrical-resistance welding in which the current and pressure are restricted to the spots of metal surfaces directly in contact.

Spray coating painting. The result of the application of a spray in painting as a substitute for brush painting or dipping.

Stamping. Many different usages in industry, but a common one is the cutting or forming of sheet metals with a power press.

Standard. A written guide that may or may not be a legal requirement.

Standard air. Air at standard temperature and pressure. The most common values are 21.1 C (70 F) and 101.3 kPa (29.92 in. Hg).

Standard conditions. In industrial ventilation, 21.1 C (70 F), 50% relative humidity, and 101.3 kPa (29.92 in. of mercury) atmosphere pressure.

Standard Industrial Classification (SIC). A U.S. Government classification system for places of employment according to business activity.

Standard man. A theoretical physically fit man of standard (average) height, weight dimensions, and other parameters (blood composition, percentage of water, mass of salivary glands, to name a few).

Standard temperature and pressure. See Standard air.

Static pressure. The potential pressure exerted in all directions by a fluid at rest.

STEL. See Short-term exposure limit and TLV.

Sterilization. The process of making sterile; the killing of all forms of life.

Stress. (1) A physical, chemical, or emotional factor that causes bodily or mental tension and may be a factor in disease causation or fatigue. (2) An applied force or system of forces that tends to strain or deform a body.

Stressor. Any agent or thing causing a condition of stress.

Strict liability. See Liability, strict.

Superfund. See CERCLA.

Supplied-air respirators. Air line respirators or self-contained breathing apparatus.

Supplied-air suit. A one- or two-piece suit that is impermeable to most particulate and gaseous contaminants and is provided with an adequate supply of respirable air.

Suspect carcinogen. A material that is believed to be capable of causing cancer but for which there is limited scientific evidence.

Sweating. (1) Visible perspiration; (2) the process of uniting metal parts by heating solder so that it runs between the parts.

Synthetic. (From Greek word *synthetikos*—that which is put together.) "Man-made 'synthetic' should not be thought of as a substitute for the natural," states *Encyclopedia of the Chemical Process Industries;* it adds, "Synthetic chemicals are frequently more pure and uniform than those obtained naturally."

Systemic toxicity. Adverse effects caused by a substance that affects the body in a general rather than local manner.

Temporary total disability. An injury that does not result in death or permanent disability, but which renders the injured person unable to perform regular duties or activities on one or more calendar days after the day of injury. (This is a definition established by U.S. OSHA.)

Tenosynovitis. Inflammation of the connective tissue sheath of a tendon.

Teratogen. A substance or agent to which exposure of a pregnant female can result in malformations in the fetus. An example is thalidomide.

Thermal burns. Result of the application of too much heat to the skin. First degree burns show redness of the unbroken skin; second degree, skin blisters and some breaking of the skin; third degree, skin blisters and destruction of the skin and underlying tissues, which can include charring and blackening.

Thermal pollution. Discharge of heat into bodies of water to the point that increased warmth activates all sewage, depletes the oxygen the water needs to cleanse itself, and eventually destroys some of the fish and other organisms in the water.

Threshold. The level where the first effects occur; also the point at which a person just begins to notice the tone is becoming audible.

Time-weighted average concentration (TWA). Refers to concentrations of airborne toxic materials that have been weighted for a certain time duration, usually 8 hours (ACGIH).

TLV. Threshold limit value; a term used by ACGIH to express the airborne concentration of a material to which *nearly* all persons can be exposed day after day, without adverse effects. ACGIH expresses TLVs in three ways:

- *TLV-C.* The Ceiling limit—the concentration that should not be exceeded even instantaneously.
- *TLV-STEL.* The Short-Term Exposure Limit, or maximum concentration for a continuous 15-minute exposure period (maximum of four such periods per day, with at least 60 minutes between exposure periods, and provided that the daily TLV-TWA is not exceeded).
- *TLV-TWA.* The allowable Time-Weighted Average concentration for a normal 8-hour work day or 40-hour work week.

Tort. A civil wrong, other than breach of contract, for which the law allows compensation by payment of money damages.

Toxicity. The sum of adverse effects resulting from exposure to a material, generally by the mouth, skin, or respiratory tract.

Toxic substance. Any substance that can cause acute or chronic injury to the human body, or which is suspected of being able to cause diseases or injury under some conditions.

Toxin. A poisonous substance that is derived from an organism.

Trade name. The commercial name or trademark by which a chemical is known.

Trade secret. Any confidential formula, pattern, process, device, information or compilation of information (including chemical name or other unique chemical identifier) that is used in an employer's business, and that gives the employer an opportunity to obtain an advantage over competitors who do not know or use it.

Trauma. An injury or wound brought about by an outside force.

TSCA. Toxic Substances Control Act; U.S. environmental legislation, administered by EPA, for regulating the manufacture, handling, and use of materials classified as "toxic substances."

TWA. Time-weighted average exposure. (See TLV.)

UEL. See Upper explosive limit and Lower explosive limit.

Ultraviolet. Those wavelengths of the electromagnetic spectrum that are shorter than those of visible light and longer than x-rays, 10^{-5} cm to 10^{-6} cm wavelength.

Unintentional injury. The preferred term for accidental injury in the public health community. It refers to the result of an accident.

Upper explosive limit (UEL). The highest concentration (expressed in percent vapor or gas in the air by volume) of a substance that will burn or explode when an ignition source is present. (See Lower explosive limit and Flammable range.)

USC. United States Code is the official compilation of federal statutes.

USDA. U.S. Department of Agriculture.

Vapors. The gaseous form of substances that are normally in the solid or liquid state (at room temperature and pressure).

Ventilation. Circulating fresh air to replace contaminated air.

- *Dilution.* Airflow designed to dilute contaminants to acceptable levels.
- *Mechanical.* Air movement caused by a fan or other air moving device.
- *Natural.* Air movement caused by wind, temperature difference, or other nonmechanical factors.

Vibration. An oscillating motion about an equilibrium position produced by a distributing force.

Volatile. Percent volatile by volume; the percentage of a liquid or solid (by volume) that will evaporate at an ambient temperature of 70 F (unless some other temperature is stated). Examples: butane, gasoline, and paint thinner (mineral spirits) are 100% volatile; their individual evaporation rates vary, but over a period of time each will evaporate completely.

Volt. The practical unit of electromotive force or difference in potential between two points in an electrical field.

Warranty. A promise that a proposition of fact is true, and if not true, a consideration is available.

- *Expressed warranty.* A written warranty.
- *Implied warranty.* A generally nonwritten warranty but expressed to the other party in the action.

Watt (w). A unit of electrical power, equal to one joule per second.

Weight. The force with which a body is attracted toward the earth.

Weld (welding). A localized coalescence of metals or nonmetals produced either by heating the materials to suitable temperatures, with or without the application of pressure, or by the application of pressure alone, and with or without the use of filler material.

Welding. The several types of welding are electric arc-welding, oxyacetylene welding, spot welding, and inert or shielded gas welding utilizing helium or argon. The hazards involved in welding stem from (1) the fumes from the weld metal such as lead or cadmium metal, or (2) the gases created by the process, or (3) the fumes or gases arising from the flux.

Welding rod. A rod or heavy wire that is melted and fused into metals in arc welding.

Wellness. The practice of a healthy lifestyle.

Wind load (force or pressure). The pressure exerted on a building or structure from moving air.

Work. When a force acts against resistance to produce motion in a body, the force is said to work. Work is measured by the product of the force acting and the distance moved through against resistance. The units of measurement are the erg (the joule is 1×10^7 ergs) and the foot-pound.

Workers' Compensation. An insurance system under law, financed by employers, that provides payment to injured and diseased employees or relatives for job-related injuries and illnesses.

Work hours. The total number of hours worked by all employees.

Work injuries. Injuries (including occupational illnesses) that arise out of or in the course of gainful employment regardless of where the accident occurs. Excluded are work injuries to private household workers and injuries occurring in connection with farm chores, which are classified as home injuries.

Work stress. Biomechanically, any external force acting on the body during the performance of a task. Application of work stress to the human body is the inevitable consequence of performance of any task, and is, therefore, only synonymous with "stressful work conditions" when excessive. Work stress analysis is an integral part of task design.

Zero energy state. See Zero mechanical energy

Zero mechanical energy (ZME). An old term, now called energy isolation, indicates a piece of equipment without any source of power, that could harm someone.

APPENDIX 4

Answers to Review Questions

1—DESIGNING FOR SAFETY

1. e
2. Safety through design is defined as the integration of hazard analysis and risk assessment methods early in the design and engineering stages and the taking of the actions necessary so that the risks of injury or damage are at an acceptable level. This concept encompasses facilities, hardware, equipment, tooling, materials, layout and configuration, energy controls, environmental concerns, and products.
3. a. Design for minimum risk.
   b. Incorporate safety devices.
   c. Provide warning devices.
   d. Develop and implement operating procedures and employee training programs.
   e. Use personal protective equipment.
4. a. Preoperational design stage
   b. Operational stage
   c. Postincident stage
5. A proactive approach involves asking safety-related questions while a building, work system, or equipment is being designed. This is the most effective and economic way for a company to deal with hazards. In a reactive response, a company looks at design only after an incident has occurred.
6. Alan D. Swain's approach to improving job safety suggests that it is management's responsibility to provide the worker with a safety-prone work situation, and to avoid the temptation to place the burden of accident prevention on the individual worker.
7. The central point in Dr. Alphonse Chapanis' work is that the greatest impact on improving system performance comes from the redesigning of equipment rather than the selection and training of personnel.
8. a. Establishing objectives
   b. Assessing hazard probability and severity
   c. Conducting hazard analysis/risk assessment
   d. Using proper hazard analysis techniques
   e. Implementing design review practices
   f. Using a project review checklist
9. A good risk assessment model is essential because it enables decision makers to understand and categorize the risks and to determine the methods and costs to reduce risks to an acceptable level.

2—BUILDINGS AND FACILITY LAYOUT

1. Any five of the following:
   a. Illumination
   b. Noise and vibration control
   c. Product flow
   d. Ventilation (particularly of dust, vapors, and fumes)
   e. Control of temperature and humidity
   f. Work positions and movements of employees

 g. Supervision and communication
 h. Support requirements for such things as vehicles, portable ladders, material handling devices, monitoring and controlling systems, and cleaning and maintenance equipment

2. d
3. The specific safety code for electric wiring and electrical installations is NFPA 70, *National Electrical Code,* issued by the National Fire Protection Association (NFPA).
4. a. Construction and procedures
 b. Visual displays, signs, and labels
 c. Protective features and guards
 d. Controls and handles
 e. Maintenance and service needs
 f. Safety signs
5. To protect pedestrians when their entrances are located near railroad tracks or busy thoroughfares, fence part of the right-of-way, install traffic signals, and build subways or pedestrian bridges.
6. b
7. a. Identify all confined spaces.
 b. Identify all potential hazards for each confined space and the methods to eliminate them.
 c. Develop a confined space work permit form.
 d. Train personnel on the dangers and proper observance of confined spaces.
 e. Ensure that a trained and equipped rescue team is available to respond to an emergency.
8. a
9. a. Keep the number of openings to a minimum.
 b. Secure all windows.
 c. Use protective lighting.
 d. Have entrances and service doors lead to a reception area.
 e. Install alarm systems that detect fire, fumes, vapors, and intruders.
 f. Limit access to docks and other receiving areas.
10. Safety engineers use warm colors to call attention to dangerous machine parts or hazards because warm colors (reds, yellows, and oranges) have longer wavelengths than cooler colors. Due to the human eye's reflective response to color, warm colors with longer wavelengths seem to move toward the observer.
11. Neutral colors of low-light reflectance values should be used in laboratories where reflected color might prevent accurate observation of materials being tested and analyzed.
12. d
13. a. Load
 b. Durability
 c. Maintenance
 d. Noise
 e. Dustiness
 f. Drainage
 g. Heat conductivity
 h. Resilience
 i. Electrical conductivity
 j. Appearance
 k. Chemical composition
 l. Slip-resistance

3—CONSTRUCTION OF FACILITIES

1. The primary reference manual used for most construction operations is the federal OSHA Construction Safety Standard 29 *CFR 1926.*
2. a. Building codes
 b. Safety standards
 c. Communication
 d. Training
3. a. Proper placement and servicing
 b. Safe clearance from combustible material
 c. Close surveillance
 d. Safe fuel storage and refueling
 e. Proper maintenance
 f. Ventilation and determination of gaseous contamination or oxygen deficiency
4. d
5. Any five of the following:
 a. Require employees to use personal protective equipment.
 b. Do not permit employees to ride loads, hooks, or "headache" balls.
 c. Do not permit employees to work near electric wires unless the wires are fully insulated.
 d. Remove from the job any worker who is under the influence of alcohol or drugs or who is too sick (in a doctor's opinion) to work.
 e. Do not allow employees to work on wet, freshly painted, or slippery steel construction.
 f. Have workers wear eye protection while cutting out rivets, chipping, or doing similar work.
 g. Where it is impractical to provide temporary floors, suspend safety nets below points where employees are working, or have them use fall protection equipment.
 h. Ensure guy cables or braces, used to hold steel while it is being erected, are guarded.
6. a
7. U.S. OSHA 1926.503 requires scaffold users to receive training for fall protection systems.
8. a. Depth of cut
 b. Possible variation in water content of the material while the excavation is open
 c. Anticipated changes in materials from exposure to air, sun, water, or freezing
 d. Loading imposed by structures, equipment, overlaying material, or stored material
 e. Vibration from equipment, blasting, or traffic
9. c
10. a
11. a. Hold on with both hands when going up or down. If material must be handled, raise or lower it with a rope either before going down or after climbing to the desired level.

   b. Always face the ladder when ascending or descending.
   c. Never slide down a ladder.
   d. Be sure shoes are not greasy, muddy, or slippery before climbing.
   e. Do not climb higher than the third rung from the top on straight or extension ladders, or the second tread from the top on stepladders.
   f. Carry tools in a tool belt, not in the hand.

12. Warning lines protect employees by alerting them that they are working within six feet of a roof edge. These lines also protect employees who are working within the six-foot area between the warning line and the roof edge, creating a zone for higher awareness of the fall hazard. Warning lines are restricted to low slope or flat roof work. They can be used alone or in conjunction with rails or nets.

4—MAINTENANCE OF FACILITIES

1. d
2. a. Footings
   b. Column bases
   c. Foundation walls
   d. Pits
3. c
4. b
5. a. Clear a path from the center of the roof to the drains to avoid ice and snow buildup near drain areas.
   b. Clear a path leading to the roof's edge to allow drainage on a pitched roof with no drains.
   c. Never use blowtorches or similar devices to melt ice from drains or roof surfaces.
   d. Use care when removing ice and snow to avoid puncturing the roof.
6. a. Exit signs should be appropriately placed.
   b. Exits should not serve as storage areas.
   c. Exits should be well-lit with smooth floors.
   d. Exit doors should move freely with no obstructions.
   e. Exit signs and emergency lighting, designed to operate in the dark in case the lighting system fails, should be tested.
7.
   a. Inspect and thoroughly clean the heating systems.
   b. Annually inspect chimneys and vent pipes for cracks, missing mortar, and rusted holes.
   c. Keep the inside of buildings at a minimum temperature of 40 F.
   d. Do not leave buildings unattended for long periods of time.
   e. Each day check that the heating equipment is operating properly.
8. d
9. Any five of the following:
   a. Choose a heating, ventilating, and air conditioning (HVAC) system that fits the building size and anticipated uses.
   b. Allow for a generous number of intake and exhaust vents.
   c. Locate intake vents where they will receive the largest supply of fresh air away from cars, buildings, and process exhausts, and as close as possible to trees and bushes.
   d. Fit the HVAC system with regulating generators that are flexible enough to adjust to the varying air pressures of intake and exhaust vents.
   e. Use only steam humidifiers.
   f. Use prefilters to clean the air before it passes over higher-efficiency filters.
   g. Institute a preventive-maintenance program. Provide for regular inspections of drain pans, filters, and any area of the HVAC that is accessible and that might fall prey to germs.
10. a. Accidents and injuries from landscaping tools and machines
    b. Poisonous vines, shrubs, fruits, insects, and reptiles
    c. Pesticide poisoning
    d. Snow-shoveling injuries
11. a. Changing hand position while the trimmer is running
    b. Holding branches away from the cutting bar
    c. Removing debris from the trimmer
    d. Holding the trimmer with only one hand
    e. Failing to wait for the blades to stop after turning the trimmer off
12. A CPM program is a preventive measure that monitors equipment before malfunctioning or failure begins. In addition to alerting the proper personnel to potentially hazardous conditions, a CPM program also provides the record keeping required by state and federal safety regulations.
13. a. Their experience
    b. Their alertness
    c. Their mechanical ability
14. b
15. When employees work underneath pipelines that carry chemicals, isolate or cover the overhead pipelines so they will not drip on workers or materials. Issue special protective equipment to workers, such as chemical-protective goggles, protective suits, rubber gloves, or respiratory-protective equipment. Provide emergency showers with plainly marked locations.

5—BOILERS AND UNFIRED PRESSURE VESSELS

1. The ASME Code provides guidelines for inspection during the construction of boilers and pressure vessels. The NB Code provides guidelines for inspection after the installation, repair, or alteration of boilers and pressure vessels.
2. a. Errors in design, construction, and installation
   b. Improper operation, human failure, and improper operator training
   c. Corrosion or erosion
   d. Mechanical breakdown, failure, or blocking of safety devices

   e. Failure to inspect thoroughly and frequently
   f. Improper application of equipment
   g. Lack of planned preventive maintenance
3. b
4. A boiler is a closed vessel in which water is heated by combustion of fuel or other heat source.
5. d
6. When it is suspected that there is a fire in the gas passages
7. To relieve excess pressure or vacuum
8. As soon as the boiler has cooled
9. Direct the water to the outside of the pile and move toward the center. Water aimed directly at the center of an ash pile can cause an explosion.
10. d
11. a. Proper ventilation
    b. Proper equipment
    c. Proper protection
12. Water in a closed system under high pressure that remains liquid instead of turning to steam when the temperature exceeds 212 F (100 C)
13. A vessel designed to withstand pressure or vacuum, but not external heat sources, such as burning fuel or electric heaters
14. a. Blueprints
    b. Manufacturer's data reports and instructions
    c. Design data
    d. Installation information
    e. Records of process changes
    f. Historical profile, including repair records and inspection reports
15. If the vessel cannot be inspected internally
16. When the vessel's contents are dangerous (e.g., toxic, flammable)
17. It is a U-pipe filled with water, with one end connected to the pressure side of the pressure vessel and the other vented to the atmosphere. It is used on pressure vessels that operate on low pressure or slight vacuum.
18. Closure devices, and failure or blocking of automatic control

6—SAFEGUARDING

1. Safeguarding is any means of preventing personnel from coming in contact with the moving parts of machinery or equipment that would potentially cause physical harm.
2. a. Integrated as part of the machine
   b. Well-constructed, durable, and strong
   c. Able to accommodate workpiece feeding and ejection
   d. Protective
   e. Easy to inspect and maintain
   f. Relatively tamper-proof or foolproof
3. b
4. a. Built-in safeguards
   b. Barrier guards
   c. Interlocking barrier guards
   d. Automatic safeguarding devices
5. a. Built-in safeguards conform more closely to the contours of the machine, making them superior in appearance and placement.
   b. Built-in safeguards eliminate hazards completely and permanently while withstanding daily wear and handling.
   c. Built-in safeguards tend to cost less because the cost is usually spread over a large number of machines.
6. c
7. a. It must guard the hazardous area before the machine can be operated.
   b. It must stay closed until the rotating equipment is at rest.
   c. It must prevent operation of the machine if the interlocking device fails.
8. b
9. Nonmetal barriers are less expensive and resist the effects of splashes, vapors, and fumes from corrosive substances that would react with metal.
10. a. The workplace should require a minimum amount of strenuous lifting and traveling.
    b. The work height of workstations should be the optimal height for stand-up or sit-down operations.
    c. Controls should be standardized and readily accessible.
    d. Materials handling aids should be provided to minimize manual handling of materials.
    e. Factors contributing to operator fatigue should be minimized, including excessive speed-up, boredom from monotonous operations, and awkward work motion or operator position.
11. a. Check the machine or equipment and the surrounding area to ensure that nonessential items have been removed and that the machine or equipment components are operationally intact.
    b. Check the work area to ensure that all employees have been safely positioned or removed from the area.
    c. Verify that the controls are in neutral.
    d. Remove the lockout devices and reenergize the machine or equipment.
    e. Notify affected employees that the maintenance is completed and the machine or equipment is ready to use.
12. d
13. a. Photoelectric cells
    b. Pressure-sensitive mats
    c. Light or sound curtains

7—PERSONAL PROTECTIVE EQUIPMENT

1. a. Engineering controls
   b. Administrative controls

c. Personal protective equipment (PPE)

2. Personal protective equipment is referred to as the use of respirators, special clothing, safety glasses, hard hats, or similar devices whose proper use reduces the risk of personal injury or illness resulting from occupational hazards.
3. e
4. a
5. The standard established for eye and face protection is ANSI Z87.1 1989, *Practice for Occupational and Educational Eye and Face Protection.*
6. b
7. a. Enclosure—completely surrounds the head, such as an astronaut's helmet
   b. Aural insert—acts as a plug, commonly called earplug
   c. Superaural—cap seals the external edge of the ear canal, also called canal cap
   d. Circumaural—cup covers the external ear, also called earmuff
8. a. Passive
   b. Active
9. a. General all-purpose nets
   b. Personnel nets
   c. Debris nets
10. b
11. a. Safety belts
    b. Fall arresters and shock absorbers
    c. Harnesses
    d. Lifelines
    e. Fall-arresting systems
12. a. Identify the hazard.
    b. Evaluate the hazard.
    c. Select the appropriate, approved respiratory equipment.
13. a. Air-supplying respirators
    b. Air-purifying respirators
14. c
15. a. Compression resistance
    b. Impact resistance

8—INDUSTRIAL SANITATION AND PERSONNEL FACILITIES

1. e
2. d
3. One of the most common ways a water supply becomes contaminated is by accidental entry of sewage or septic water into a drinking water supply.
4. a
5. A worker can easily disinfect a drinking water system by filling it with water containing not less than 100 mg/l of available chlorine. This solution should remain for 24 hours in either a new system or one that has not previously carried treated water.
6. a. Filtration
   b. Chemical disinfection
7. a. Glass
   b. Metals
   c. Paper
   d. Batteries
   e. Chemicals
   f. Cardboard
8. a. Drinking fountains
   b. Washrooms
   c. Locker rooms
   d. Showers
   e. Toilets
9. a
10. a. Nonabsorbency
    b. Durability
    c. Sanitation
    d. Slipping and falling hazards
11. c
12. a. All repairs should be made by manufacturer authorized repair personnel.
    b. Persons with cardiac pacemaker units should be warned against coming too close to microwave ovens.
13. a. Cafeterias
    b. Canteens or lunchrooms
    c. Mobile canteens
    d. Box-lunch services
    e. Vending machines
14. c
15. Food-service equipment should receive the approval from or meet established standards of the National Sanitation Foundation (NSF).

9—OCCUPATIONAL MEDICAL SURVEILLANCE

1. The information obtained from the surveillance can then lead to a successful public health intervention, such as a routine radio allergo sorbent test (RAST) for healthy workers and removal of workers who test positive from exposure.
2. a. Individuals' exposures and doses
   b. Toxicity of the hazard
   c. Effectiveness of available screening tests
   d. How clinical outcomes will be affected by early detection
3. e
4. a. OSHA standards and surveillance requirements related to the hazardous substance
   b. The worker's job description
   c. The worker's exposure history and industrial hygiene monitoring data
   d. Use of personal protective equipment data

e. Data from prior medical examinations

5. Biomarkers are indicators of events—such as a symptom of disease or physical change—taking place in biological systems or samples.

6. a. Difficulty correlating a health risk with exposure once the exposure information is known
   b. Short biological half-lives of some substances, which prevent accurate exposure assessment except within a limited time of exposure
   c. Ineffective monitoring for surface active agents
   d. Interference of tobacco, alcohol, and other agents with some test results
   e. Measurement may reflect multiple exposure sources (air, food, water, soil, and skin contact), preventing accurate determination of occupational exposure

7. a. Chemicals
   b. Certain working populations (such as hazardous waste workers and fire brigade members)

8. When a worker is exposed to more than one chemical or substance, the total effect becomes more than the sum of the parts. As chemicals or substances combine in the body, they greatly increase the potential for an individual to develop an injury or disease. For example, when workers who smoke are exposed to asbestos, their chances of developing lung cancer are greatly increased.

9. Some of the ethical and legal concerns that have been raised in the use of biological markers include the confidentiality of workers' medical information and how that information will be used. The potential for discrimination exists against workers on the basis of racial or cultural characteristics and acquired or inherited genetic susceptibility.

10. The International Agency for Research in Cancer (IARC) has classified chemicals and industrial processes with respect to their potential to cause cancer in humans.

10—ELECTRICAL SAFETY

1. Skin surface
2. Prevents the victim from breaking contact with the circuit
3. Resistance
4. a. Current flow: the amount of current that flows through the victim
   b. Time: the length of time that the body receives the current
5. Current-carrying parts are exposed and an arc forms when the switch is opened.
6. Size, material, insulation, and the way they are installed
7. To interrupt the current flow when it exceeds the conductor capacity
8. A strip of fusible metal that links two terminals of a fuse block
9. Thermal and magnetic
10. Thermal element is connected in the power circuit to the motor
11. Prevent the occurrence of excessive voltages from sources such as lightning, line surges, or accidental contact with higher voltage lines
12. Bonding the identified conductor to a grounding electrode by means of an unbroken wire called a grounding electrode conductor

11—FIRE PROTECTION

1. a. Immediately detecting the fire and promptly transmitting an alarm
   b. Initiating evacuation of the building
   c. Confining the fire
   d. Extinguishing the fire
2. The National Fire Protection Association (NFPA) and the Society of Fire Protection Engineers can provide the safety and health professional with the names of experts in the area of fire protection.
3. e
4. a. Close every window that faces the burning building.
   b. Station fire brigade workers with fire extinguishers or fire hoses at each window nearest the fire.
   c. Station fire fighters on the exposed building's roof with hose lines to keep the roof wetted down and with extinguishers to put out any burning embers.
5. a. Heat can be taken away by cooling.
   b. Oxygen can be taken away by excluding the air.
   c. Fuel can be moved to an area where there is not enough heat for ignition.
   d. The chemical reaction of the flame fire can be interrupted by inhibiting the rapid oxidation of the fuel and the concomitant production of free radicals.
6. a. If the fire is clearly spreading beyond its point of origin
   b. If the fire could block the exit
   c. If the employee is unsure of how to use the extinguisher
7. c
8. a. Life safety
   b. Continuity of operations
   c. Property protection
9. a. Stair enclosures
   b. Fire walls
   c. Fire doors
   d. Dividing a building into smaller units
10. Fire doors should be installed in accordance with NFPA 80, *Fire Doors and Windows.*
11. a. Electrical equipment
    b. Smoking
    c. Friction
    d. Foreign objects or tramp metal
    e. Open flames
    f. Spontaneous ignition
    g. Housekeeping
    h. Explosive atmospheres
12. a. Sensitivity—the device is able to detect what it is meant to detect

b. Reliability—the device is able to perform an intended function when needed
c. Maintainability—the device is easy to service and keep at operating efficiency
d. Stability—the device can sustain its sensitivity over time

13. The dry-chemical extinguishers work by interrupting the chemical flame's chain reaction.
14. c
15. b

12—FLAMMABLE AND COMBUSTIBLE LIQUIDS

1. A flammable liquid, as defined by NFPA 30, *Flammable and Combustible Liquids Code* is any liquid having a closed-cup flash point below 100 F (37.8 C) and having a vapor pressure not exceeding 40 psia (1,276 kPa) at 100 F.
2. a. The flash point of the liquid
   b. The concentration of vapors in the air (whether the vapor-air mixture is in the flammable range or not)
   c. The possibility of a source of ignition at or above a temperature or energy level high enough to cause the mixture to burst into flame
   d. The amount of vapors present
3. Autoignition temperature is the minimum temperature at which a flammable gas-air or vapor-air mixture will ignite from its own heat source or a contacted heated surface without the presence of an open spark or flame.
4. Flash point is the minimum temperature at which a liquid gives off vapor concentrated enough to form an ignitable mixture with air near the surface of the liquid within a vessel specified by the appropriate testing procedure and apparatus.
5. a
6. Static electricity is generated by the contact and separation of dissimilar materials.
7. Bonding eliminates a difference in the static-electrical-charge potential between two or more objects. Grounding eliminates a potential difference between an object and the ground (earth). Bonding and grounding are effective only when the bonded objects are conductive materials.
8. d
9. f
10. a. Set the brakes.
    b. Stop the engine (unless power takeoff is required for unloading).
    c. Turn off the lights.
    d. Make the bonding connection before the dome cover is opened for inspection or gauging.
11. a. If a person is killed
    b. If a person is injured and requires hospitalization
    c. If estimated carrier or property damage exceeds $50,000
    d. If fire, breakage, spillage, or contamination involving a radioactive material occurs
12. Use dry chemicals or carbon dioxide on a flammable liquid fire.
13. a. Cover with at least 3 ft (0.9 cm) of earth.
    b. Cover with 18 inches (0.46 m) of tamped earth plus 6 inches (15 cm) of reinforced concrete
    c. Cover with 18 inches (0.46 m) of tamped earth plus 8 inches (20 cm) of asphalt concrete.
14. Aluminum, pastel, and white paint reflect the heat and help reduce the internal vapor pressure of tanks that are exposed to the sun.
15. a. An unblanked line or connection
    b. A break in the bottom of the tank
    c. Sludge, sediment, or sidewall scale
    d. Wood structures soaked with the liquid

13—WORKERS WITH DISABILITIES

1. a. Title I—Employment Provisions
   b. Title II—State and Local Government Provisions
   c. Title III—Public Accommodations and Services Operated by Private Entities
   d. Title IV—Telecommunications
   e. Title V—Miscellaneous Provisions
2. a. The job would put the individual in a hazardous situation.
   b. Other employees would be placed in a hazardous situation if the person were on the job.
   c. The job requirements cannot be met by an individual with certain physical or mental limitations.
   d. Accommodation of the job cannot reasonably be accomplished.
3. a. Many individuals with disabilities were hired to help fill job vacancies left by employees who joined the military.
   b. Companies established affirmative action programs to help each returning disabled veteran to become an employable person.
   c. The U.S. Department of Labor published a study that showed disabled workers were as productive as other workers, had lower frequency and severity of injury rates, and were absent from work only one day more per year than other workers.
4. a. The disabled individual
   b. The disabled veteran
   c. The qualified disabled individual
5. f
6. a. Maintaining close liaison with the Equal Employment Opportunity manager-coordinator and with medical and personnel departments when they are placing disabled employees
   b. Making job safety hazard analysis of existing work based on the job responsibilities and the abilities and limitations of the disabled employee or applicant when employing, promoting, transferring, and selecting workers with disabilities

c. Making recommendations for safety modifications of machine tools, established processes and procedures, and existing facilities and workplace environment when the company must make reasonable accommodations for disabled employees
d. As required, cooperating with the plant or building engineer or mechanical engineer and the planning, production, and maintenance departments when disabled employee accommodations are being evaluated

7. The safety evaluation form for disabled employees should be kept for at least one year after the employee leaves the company.
8. a. Physical requirements
   b. Working conditions
   c. Health hazards
   d. Accident hazards
9. a. Cafeteria, washroom, and restroom facilities
   b. Width of doors
   c. Height of plumbing fixtures
   d. Electrical controls
   e. Phones
   f. Drinking fountains
10. b
11. a
12. One means of safely evacuating wheelchair users and permanently or temporarily disabled persons is through the use of an evacuation chair.

14—MATERIALS HANDLING AND STORAGE

1. b
2. a. Task repetition
   b. Load location
   c. Load weight
3. It lessens the stress on the lower back.
4. a
5. a. Hook—the hook glancing off a hard object and injuring the worker
   b. Crowbar—slippage
   c. Rollers—fingers or toes may be pinched or crushed between the roller and the floor
6. Metal-to-metal contact between the jack head and the load can be avoided by using a hardwood shim between the jack and the load.
7. Such extenders should not be used
8. Strap or chain the items to the hand truck
9. It causes less stress to the lower back and protects the worker's heel from being caught under the truck back
10. a. Running wheels off bridge plates or platforms
    b. Colliding with other trucks or obstructions
    c. Hands may be jammed between the truck and other objects
11. a. Stabilize the lumber pile
    b. Allow for air circulation though the lumber pile
12. a. The material being piped
    b. The hazards involved
    c. Directions for safe use
13. c
14. c
15. Moisten the material
16. b
17. No smoking or open flame
18. b
19. An open-mouthed, nonpressurized, vacuum-jacketed vessel used to hold liquid oxygen, nitrogen, or helium
20. Four times the heaviest expected load

15—HOISTING AND CONVEYING EQUIPMENT

1. A load should be lifted only when it is directly under a hoist because stresses for which the hoist was not designed could be imposed upon it. If the load is not properly centered, it can swing, and injury could result.
2. The three general types of chain hoists include spur geared, differential, and screw geared. The spur geared is the most efficient because it can pick up a load with the least effort on the part of the operator.
3. A spring return guarantees that if the operator releases a lever it will automatically move into the OFF position.
4. The Stop movement control hand signal should be obeyed even if it is being given by someone other than the signaler in charge.
5. c
6. a. The crane must be moved to a location where there will be minimum interference with other cranes and operations.
   b. All controllers should be placed in the OFF position.
   c. The main power source should be disconnected/deenergized and locked, tagged, or flagged in the deenergized position. All power sources should be neutralized so that they are in a state of energy isolation.
7. a. Hand-operated monorail—material is raised with a hand-powered hoist, and the trolley is propelled by hand.
   b. Semi-hand operated monorail—has a power hoist and is moved horizontally by hand.
   c. Power-operated monorail—electrically actuated for both vertical and horizontal movements.
8. Hoists and cranes should not be used to lift, support, or transport people because they do not provide an alternate means of supporting the load if the suspension element fails.
9. b
10. Light service is operation with loads that are usually half or less of the rated load.

11. Frequent inspections are usually completed by the operator with no record-keeping requirement. Periodic inspections require records and are done by an appointed person.
12. a. Operator error
    b. Support failure
    c. Failure to use outrigger
    d. Crane failure
    e. Rigging
13. c
14. a. Boom length
    b. Boom angle
    c. Capacity
15. d
16. a. To ensure that the boom configuration and load lines are adequate
    b. To ensure that no interference of any kind exists and protection devices are working
17. a. Warn others
    b. See that switches are properly set
    c. See that the track is free of obstructions
18. c
19. a. It allows for errors in load weight estimation
    b. It allows for vibration or shock during load handling
    c. It allows for loss of strength at knots or bends
    d. It allows for deterioration of the rope due to wear
20. d
21. a. Tow conveyors
    b. Trolley conveyors
22. Roller conveyors are similar to chute conveyors, except that the angle of slope is 2% to 4% less than chute conveyors.
23. The American National Standards Institute/American Society of Mechanical Engineers (ANSI/ASME) A17.1, *Elevators, Escalators, and Moving Walks*, also referred to as the Elevator Code, governs the use and design of elevators.
24. Car safeties are not required for electrohydraulic elevators because they can come down no faster than the fluid can be forced out of the cylinder by the descending plunger.

16—ROPES, CHAINS, AND SLINGS

1. a
2. The properties of Manila fiber make it the best-suited natural fiber for cordage and is often recommended for capstan work.
3. a. Nylon
   b. Polyester
   c. Polyolefin
4. b
5. To make a good estimate of the strength of fibers in a rope, and to test for chemical damage, scratch the fibers with a fingernail—fibers of poor strength will readily part.
6. b
7. d
8. a
9. a. The type of service required
   b. The federal, state or provincial, or local codes covering the particular hoisting operation
10. a. Corrosion
    b. Wear
    c. Kinks
    d. Fatigue
    e. Drying out of lubrication
    f. Overloading
    g. Overwinding
    h. Mechanical abuse
11. In the United States, OSHA regulations and other industrial and construction codes prohibit the use of knots in wire rope.
12. In the United States, OSHA requires wire rope or cable to be inspected when installed and weekly thereafter.
13. a. Material used (fiber rope or wire rope)
    b. Fittings of suitable strength for the load
    c. Method of fastening the rope to the fittings
    d. Type of sling
    e. Type of hitch
    f. Regular inspection and maintenance
14. Alloy steel has become the standard material for chain slings because it has high resistance to abrasion and is practically immune to failure since the metal is cold worked.
15. d
16. Synthetic web slings are useful for lifting loads that need their surfaces protected. Metal mesh slings can safely handle sharp-edged materials, concrete, and high-temperature materials.

17—POWERED INDUSTRIAL TRUCKS

1. a. Lift trucks
   b. Straddle trucks
   c. Crane trucks
   d. Tractors and trailers
   e. Motorized hand trucks
   f. Automated Guided Vehicles (AGVs)
2. a. Backup alarm lights
   b. Headlights
   c. Turn signals
   d. Enhanced front and rear vision
   e. Noise-reducing insulation
   f. Fail-safe brakes
   g. Comfortable, wrap-around seats that provide protection
3. e
4. Since operators in saddle trucks sit so high off the ground, their angle of sight is reduced immediately to the front and rear, posing a hazard to pedestrians.
5. a. The operator being pinned between the truck and a fixed object

b. The truck running up on the operator's heels

6. Operators of powered industrial trucks can prevent traffic accidents by using the same safe practices that apply to highway traffic.

7. a. Bumping skids
   b. Pushing piles of material out of the way
   c. Using makeshift connections to move heavy objects
   d. Using the forks as a hoist
   e. Moving other trucks

8. a. Lift trucks are generally steered by the rear wheels.
   b. Lift trucks steer more easily loaded than empty.
   c. Lift trucks are driven in the reverse direction as often as in the forward.
   d. Lift trucks are often steered with one hand—the other hand being used to operate the controls.

9. a. Company policies
   b. Operating conditions
   c. Types of trucks used

18—HAULAGE AND OFF-ROAD EQUIPMENT

1. b

2. a. So the driver can see the signals easily
   b. So the driver will have the helper in sight
   c. So the helper will be clear of the backing vehicle and falling material

3. Twice the weight of the machinery

4. a. No one should go between the vehicles while either one is in motion.
   b. Parked vehicles should have their brakes set, their wheels blocked, or both.
   c. Before moving, the driver should receive an all-clear signal, to indicate that no one is between the vehicle and a solid and immovable object.
   d. Two bars are safer than tow ropes.
   e. Equipment towed on trailers should be secured to the trailer.

5. c

6. Boom stops limit the travel of the boom beyond the angle of 80 degrees above the horizontal plane and prevent the boom being pulled backward over the top of the machine.

7. a. The hook should be centered over the load to keep the load from swinging.
   b. Hands should be kept out of the pinch point when holding the hook or slings in place while the slack is taken up.
   c. The hooker, rigger, and all other personnel must be in the clear before a load is lifted.
   d. Tag lines should be used for guiding loads.
   e. Hookers, riggers, and others working around cranes must keep clear of the swing of the boom and cab.
   f. No load should be lifted without a signal.

8. a. When driving off the road and beyond the shoulder, on steep grades, or at rough places
   b. In congested areas
   c. Under icy or slippery conditions

9. Look down to make sure footing is secure and no vehicles are approaching

10. d

19—HAND AND PORTABLE POWER TOOLS

1. b

2. a. Provide proper protective equipment and have employees wear it.
   b. Select the right tool for the job.
   c. Know if a tool is in good condition and keep it in good condition.
   d. Use tools correctly.
   e. Keep tools in a safe place.

3. Central tool control ensures uniform inspection and maintenance by a trained employee.

4. b

5. a

6. When striking another tool, the striking face of the hammer should have a diameter approximately ⅜ in. (0.9 cm) larger than the struck face of the tool.

7. The knife is more frequently the source of disabling injuries than any other hand tool.

8. a. Electrical
   b. Pneumatic
   c. Gasoline
   d. Hydraulic
   e. Powder actuated

9. a. They can easily come in contact with the operator's body.
   b. It is difficult to guard such equipment.
   c. There is the possibility of breakage because the tool may be dropped or roughly handled.
   d. The source of power is brought closer to the operator, creating additional potential hazards.

10. a. Always disconnect the tool from the source of power before changing accessories.
    b. Never leave a tool in an overhead place where there is a chance that the cord or hose can be pulled, causing the tool to fall.
    c. Use proper hearing protection when using power-loaded equipment for driving anchors into concrete, or when using air-driven hammers or jacks.

11. a. Set up an inspection schedule and a system for keeping records for each tool.
    b. Tag defective tools and withdraw them from service until they are repaired.
    c. Provide a visual or external inspection at the tool-room each time a tool is returned. Inspect as specified by the OSHA electrical safety-related work standard.
    d. Use colored tags to tell when the tool was last inspected.

e. Instruct and train employees to inspect tools and to recognize and report defects.
f. Clean power tools with a recommended nonflammable and nontoxic solvent.
g. Use air drying in place of blowing with compressed air.

12. Grounding portable electrical tools and using a ground-fault circuit interrupter (GFCI) provide the most convenient way of safeguarding the operator.
13. Any of the following:
    a. Be sure the trigger switch works properly.
    b. Check carefully for loose power-cord connections and frays or damage to the cord.
    c. Be sure the chuck is tightly secured to the spindle.
    d. Tighten the drill bit securely.
    e. Check auxiliary handles to ensure they are securely installed.
    f. Always wear safety goggles, or safety glasses with side shields, that comply with current national standards and a full face shield when needed.
    g. Always hold or brace the tool securely.
    h. If the drill binds in the work, release the trigger immediately, unplug the drill from the power source, and then remove the bit from the workpiece.
    i. Never attempt to free a jammed bit by starting and stopping the drill.
    j. As the hole is about to be broken through, grip or brace the drill firmly, reduce pressure, and allow the bit to pass easily through the hole.
    k. Unplug the drill before changing bits, accessories, or attachments.
    l. Do not raise or lower a drill by its power cord.

20—WOODWORKING MACHINERY

1. a. U.S. Occupational Safety and Health Administration (OSHA)
   b. American National Standards Institute (ANSI)
   c. National Fire Protection Association's (NFPA) National Electrical Code
2. d
3. The working surfaces of the machine should be at a height that will minimize fatigue.
4. a. Operating controls
   b. Safety controls
   c. Power drives
   d. Sharpness of cutting edges and other parts
5. a. Blade cuts or abrasions
   b. Kickbacks
6. When feeding a table saw, make push sticks long enough to keep hands well away from the blade by adding 6 in. (15.2 cm) to the blade's diameter.
7. a. Alter its original design.
   b. Operate it at other than the rated speed.
   c. Change the balance or tension.
8. a. While the blade coasts or idles
   b. When operators attempt to remove a sawed section of board or a piece of scrap
   c. When operators measure boards or place them in position for the cut
9. a. The blade's direction of rotation makes it easy for the operator's hands to be drawn into the revolving saw.
   b. Flying stock can be thrown with enough force to drive the stock through a 1 in. (2.5 cm) board.
10. Since long stock is often ripped on power-feed ripsaws, the clearance at each working end of the saw table should be at least 3 ft (0.9 m) longer than the length of the longest material handled.

21—WELDING AND CUTTING

1. The most significant health hazard in the welding process is the generation of toxic fumes and gases.
2. a. Inflammation of the lungs (chemical pneumonitis)
   b. Pulmonary edema (swelling and accumulation of fluids)
   c. Emphysema (loss of elasticity of the lungs)
   d. Chronic bronchitis
   e. Asphyxiation
3. e
4. a
5. a. The containers can be purged with an inert gas.
   b. The containers can be filled with water to within an inch or two of the place where the work is to be done with a vent left open.
6. a. Respiratory protection equipment
   b. Eye protection equipment
   c. Protective clothing
7. Dark, woolen clothing is preferred when welders work with inert-gas-shielded, arc-welding machines because it reduces any reflection to the operator's face underneath the helmet. Woolen clothing is more resistant to deterioration and is not readily ignited.
8. The standards for training and qualifications of welders was established by the American Welding Society (AWS).
9. a. Close the valve, and take the cylinder outdoors well away from any source of ignition. A regulator attached to the valve may be used temporarily to stop a leak through the valve seat.
   b. Properly tag the cylinder, and notify the supplier.
   c. If the leak occurs at a fuse plug or other safety device, take the cylinder outdoors well away from any source of ignition, open the cylinder valve slightly, and permit the fuel gas to escape slowly. Tag the cylinder plainly.
   d. Post warnings against approaching with lighted cigarettes or other sources of ignition.
   e. Promptly notify the supplier, and follow instructions for returning the cylinder.
10. e
11. b

12. Pressure regulators must be used on both oxygen and fuel gas cylinders to maintain a uniform gas supply to torches at the correct pressure.
13. Resistance welding is a metal-joining process where welding heat is generated at the joint by the resistance to the flow of electric current.
14. a. Friction
    b. Ultrasonics
    c. Lasers
15. The installation of resistance-welding equipment should conform to NFPA *National Electrical Code*, Standard No. 70.

22—METALWORKING MACHINERY

1. a. Good housekeeping
   b. Good work habits
2. One of the major causes of accidents from machine tools, especially drilling equipment, is the careless use of high-pressure compressed air to blow chips, cuttings, or shavings from machines or workers' clothing.
3. a. Use faceplates and chucks without projections whenever possible.
   b. Install plastic or fine-mesh screen chip shields, particularly on high-speed operations, because they allow operators to see through them while confining the flying chips.
   c. Provide an overhead hoist or a swinging, welded pipe fixture to lift heavy faceplates, chucks, and stock on both lathes and screw machines.
4. a. Being struck by insecurely clamped work or by tools left on or near a revolving table
   b. Catching clothing or rags for wiping in revolving parts
   c. Falling against revolving work
   d. Calipering or checking work while the machine is in motion
   e. Allowing turnings to build up on the table
   f. Removing turnings by hand
5. a. Be sure the machine is properly grounded, and check that all exposed systems are properly covered.
   b. Place all selector switches in the OFF or neutral (disengaged) position.
   c. Be sure that the machine's push buttons, manual limit switches, or controls are set for a safe setup.
   d. Check that the doors of the main electrical cabinet are closed and that the main disconnect switch is in the OFF position.
6. Injuries from shapers and planers frequently result from contact with projections on the workpiece or with projecting bolts or brackets, especially when the table is being adjusted vertically.
7. c
8. Specifications for operation of grinding machines and construction of guards and safety devices are in ANSI B7.1, *The Use, Care, and Protection of Abrasive Wheels.*
9. Cold and wetness can cause grinding wheels to break or crack.
10. Many grindstone failures result from faulty handling and incorrect mounting.

23—COLD FORMING OF METALS

1. a. Adequate safeguarding of the point of operation
   b. Proper training of operators
   c. Enforcing safe working practices
2. Antirepeat requires release of all actuating mechanisms before another stroke can be initiated.
3. a. Full-revolution clutch: A type of clutch that, when tripped, cannot be disengaged until the drive mechanism has completed a full revolution and the slide, a full stroke.
   b. Part-revolution clutch: A type of clutch that can be disengaged at any point before the crankshaft has completed a full revolution and the press slide, a full stroke.
4. A device designed, constructed and arranged to create a sensing field or area and to deactivate the clutch control when an operator's hand or any other body part is detected in the area
5. Any point of the power press machine, except the point of operation, where a body part can be caught between the moving parts of a press component or between material and moving parts of the press
6. The area of the die or tooling where material is positioned and work is performed during any process such as cutting, forming or assembly
7. a. Guard
   b. Devise
8. a. Fixed die-enclosure guards
   b. Fixed barrier guards
   c. Interlocked press-barrier guards
   d. Adjustable-barrier guards
9. On a full-revolution clutch machine
10. Each operator must have their own set.
11. A kickpress is a press that is hand or foot powered, and is used for piercing, notching, forming, or shearing.
12. Interlocking tripping mechanism
13. Strains and hernias, foot injuries, crushing injuries, hand injuries or amputations, lacerations from wrench slippage and eye injuries
14. a. Prevent operators from placing their hands into the point-of-operation.
    b. Prevent or stop the operation if any part of the body approaches the point-of-operation.
15. Adjustable guard
16. Cold bend angles, channels, and curved shapes in plate, strip, or sheet stock
17. The safe distance method is used when guards and devices cannot be used.

18. Two independent control systems on the same power press brake, only one of which is operable at a time
19. An enclosure that covers moving machine parts (excluding point-of-operation)
20. The three main criteria are safeguarding the point-of-operation, training press operators, and enforcing safe work practices.

24—HOT WORKING OF METALS

1. a. Vacuum cleaning to remove dust deposits
   b. Using local exhaust systems that remove dust at the point of origin
2. a. Proper labeling
   b. Substituting less hazardous for more hazardous chemicals
   c. Limiting the quantities in use
3. a
4. a. Baseline physical examinations, including chest x-rays, audiometric tests, and pulmonary-function tests
   b. Periodic physical examinations, keeping track of employees' health, detecting incipient disease, and helping to reclassify workers as needed
   c. Adequate first aid facilities, approved by the physician, and employee training in first aid
   d. Observing regulatory requirements if respirators must be worn
   e. Monitoring industrial hygiene where needed
5. Prolonged contact with oil, grease, acids, alkalis, and dirt can produce dermatitis.
6. a. Applicable federal, state/provincial, and local regulations or standards, codes, and recommendations
   b. Comparison with similar operations in a like environment
   c. Collection and analysis of representative air samples taken by qualified personnel in the breathing zone of workers
7. Carbon monoxide (CO) is sometimes generated when charging and blasting take place in cupolas.
8. To eliminate a carbon monoxide explosion hazard, supply adequate natural or mechanical ventilation in back of the cupola, and open two or more tuyeres after the blowers are shut down.
9. The principal danger in handling refractory clay crucibles is that one may break when full of molten metal.
10. a. Excess smoke, gases, and fumes
    b. Unprotected firing pits
    c. Unguarded, vertical sliding doors or their counterweights
    d. Flashbacks from fire boxes
11. c
12. a. A means for immediate quenching of sparks from grinding wheels, disks, or belts
    b. Dust-proof motors to prevent the accumulation of static charges
    c. Explosion doors on the collection system
    d. An automatic interlocking control on the collection system to assure its operation whenever grinding is started
13. c
14. a. Open-frame
    b. Gravity-drop
    c. Steam and air hammers
15. The differences between the conventional coldstamping press and the forging press are that the forging press has its own operating technique, die setup, and maintenance problems, and is designed with a faster-acting slide.
16. The most widely used nondestructive testing method for forgings is the magnetic particle inspection.
17. a

25—AUTOMATED LINES, SYSTEMS, OR PROCESSES

1. The disadvantage of automation being designed, implemented, or modified using the JIT philosophy is that machines may be interdependent, with little or no provision for inventory to back up operations between any two stations.
2. CMMS can shorten factory cell downtime by providing easy-to-retrieve information from databases on replacement parts, vendor contact information, and repair instructions.
3. a. By careful identification of hazards
   b. By the development of strategies to control the environment where the processes are taking place
4. a. The type of industry
   b. The type and complexity of machines on the line
   c. The way workers interact with the machinery
   d. The success of the automation used to integrate the various components of the production line
5. A minimum clearance of 18 in. (46 cm) is required between an automated guided vehicle (AGV), with its transported load, and any fixed object in all areas that involve pedestrian travel or interaction with workers.
6. The standard that specifically defines the areas, or envelopes, required for machine motion is *Safety Standard for Industrial Robots and Industrial Robot Systems*, ANSI/RIA R15.06-1992.
7. a. Conveyors
   b. Automated guided vehicles
8. Refer to ANSI B56.5 for standards on bumper design, activation, and requirements for AGVs.
9.
   a. A manipulator
   b. A power supply
   c. A system for controlling the robot
10. b
11. a. Safety in the process of manufacturing, remanufacturing, and rebuilding robots

b. Installation of robots
c. Safeguarding of workers exposed to hazards associated with the use of robots

12. The highest degree of hazard exists when a robot is in the Teach mode because the teacher must be within the operating envelope of the robot to program its movements within very close tolerance parameters.

13. a. What can go wrong?
    b. What is the probability that something will go wrong?
    c. What would be the consequences if something does go wrong?

14. The formal procedure used to identify hazards in a chemical process facility is commonly known as a hazards and operability (HAZOP) study.

Index